The Palgrave Handbook of the History of Human Sciences

David McCallum
Editor

The Palgrave Handbook of the History of Human Sciences

Volume 2

With 45 Figures and 7 Tables

Editor
David McCallum
Victoria University
Melbourne, VIC, Australia

ISBN 978-981-16-7254-5   ISBN 978-981-16-7255-2 (eBook)
https://doi.org/10.1007/978-981-16-7255-2

© Springer Nature Singapore Pte Ltd. 2022
This work is subject to copyright. All rights are reserved by the Publisher, whether the whole or part of the material is concerned, specifically the rights of translation, reprinting, reuse of illustrations, recitation, broadcasting, reproduction on microfilms or in any other physical way, and transmission or information storage and retrieval, electronic adaptation, computer software, or by similar or dissimilar methodology now known or hereafter developed.
The use of general descriptive names, registered names, trademarks, service marks, etc. in this publication does not imply, even in the absence of a specific statement, that such names are exempt from the relevant protective laws and regulations and therefore free for general use.
The publisher, the authors and the editors are safe to assume that the advice and information in this book are believed to be true and accurate at the date of publication. Neither the publisher nor the authors or the editors give a warranty, expressed or implied, with respect to the material contained herein or for any errors or omissions that may have been made. The publisher remains neutral with regard to jurisdictional claims in published maps and institutional affiliations.

This Palgrave Macmillan imprint is published by the registered company Springer Nature Singapore Pte Ltd.
The registered company address is: 152 Beach Road, #21-01/04 Gateway East, Singapore 189721, Singapore

Contents

Volume 1

Part I What Is the History of Human Sciences 1

1 **What Is the History of the Human Sciences?** 3
   Roger Smith

2 **Kant After Kant: Towards a History of the Human Sciences from a Cosmopolitan Standpoint** 29
   Steve Fuller

3 **History of the Human Sciences in France: From Science de l'homme to Sciences Humaines et sociales** 57
   Nathalie Richard

4 **Narrative and the Human Sciences** 79
   Kim M. Hajek

5 **Diltheyan Understanding and Contextual Orientation in the Human Sciences** 109
   Rudolf Makkreel

6 **Durkheimian Revolution in Understanding Morality: Socially Created, Scientifically Grasped** 133
   Raquel Weiss

7 **Problematizing Societal Practice: Histories of the Present and Their Genealogies** 159
   Ulrich Koch

8 **Human Sciences and Theories of Religion** 181
   Robert A. Segal

9 **Historical Studies in Nineteenth-Century Germany: The Case of Hartwig Floto** 207
   Herman Paul

Part II Visualizations ... **227**

10 Anatomy: Representations of the Body in Two and Three Dimensions ... 229
Anna Maerker

11 History of Embryology: Visualizations Through Series and Animation ... 259
Janina Wellmann

12 Visualizations in the Sciences of Human Origins and Evolution ... 291
Marianne Sommer

Part III Self and Personhood ... **321**

13 Made-Up People: Conceptualizing Histories of the Self and the Human Sciences ... 323
Elwin Hofman

14 Inner Lives and the Human Sciences from the Eighteenth Century to the Present ... 349
Kirsi Tuohela

15 Michel Foucault and the Practices of "Spirituality": Self-Transformation in the History of the Human Sciences ... 375
Nima Bassiri

16 Human Sciences and Technologies of the Self Since the Nineteenth Century ... 401
Fenneke Sysling

17 The Sex of the Self and Its Ambiguities, 1899–1964 ... 423
Geertje Mak

Part IV Anthropology ... **455**

18 Economic Anthropology in View of the Global Financial Crisis ... 457
Timothy Heffernan

19 Anthropology, the Environment, and Environmental Crisis ... 483
María A. Guzmán-Gallegos and Esben Leifsen

20 On the Commonness of Skin: An Anthropology of Being in a More Than Human World ... 505
Simone Dennis

21	Indigeneity: An Historical Reflection on a Very European Idea Judith Friedlander	533

Part V Historical Sociology 559

22	The Past and the Future of Historical Sociology: An Introduction Marta Bucholc and Stephen Mennell	561
23	Power and Politics: State Formation in Historical Sociology Helmut Kuzmics	583
24	Organized Violence and Historical Sociology Christian Olsson and Siniša Malešević	625
25	Historical Sociology of Law Marta Bucholc	651
26	Sport and Leisure: A Historical Sociological Study Dominic Malcolm	677
27	Norbert Elias and Psychoanalysis: The Historical Sociology of Emotions Robert van Krieken	699
28	Identity, Identification, Habitus: A Process Sociology Approach Florence Delmotte	725
29	Hidden Gender Orders: Socio-historical Dynamics of Power and Inequality Between the Sexes Stefanie Ernst	749
30	Collective Memory and Historical Sociology Joanna Wawrzyniak	775

Part VI History of Sociology 805

31	Historiography and National Histories of Sociology: Methods and Methodologies Fran Collyer	807
32	The History of Sociology as Disciplinary Self-Reflexivity George Steinmetz	833
33	Locating the History of Sociology: Inequality, Exclusion, and Diversity Wiebke Keim	865

| 34 | Colonialism and Its Knowledges | 893 |

Sujata Patel

| 35 | Knowledge Boundaries and the History of Sociology | 917 |

Per Wisselgren

| 36 | Social Theory and the History of Sociology | 935 |

Hon-Fai Chen

| 37 | José Carlos Mariátegui and the Origins of the Latin American Sociology | 961 |

Fernanda Beigel

Volume 2

Part VII   History of Psychology **975**

| 38 | Psychologies: Their Diverse Histories | 977 |

Roger Smith

| 39 | Social Psychology: Exemplary Interdiscipline or Subdiscipline | 1005 |

James M. M. Good

| 40 | Of Power and Problems: Gender in Psychology's Past | 1035 |

Elissa N. Rodkey and Krista L. Rodkey

| 41 | Indigenous Psychologies: Resources for Future Histories | 1065 |

Wade E. Pickren and Gülşah Taşçı

| 42 | Children as Psychological Objects: A History of Psychological Research of Child Development in Hungary | 1087 |

Zsuzsanna Vajda

| 43 | A History of Self-Esteem: From a Just Honoring to a Social Vaccine | 1117 |

Alan F. Collins, George Turner, and Susan Condor

| 44 | Vygotsky, Luria, and Cross-Cultural Research in the Soviet Union | 1145 |

René van der Veer

| 45 | Values and Persons: The Persistent Problem of Values in Science and Psychology | 1167 |

Lisa M. Osbeck

| 46 | Politics and Ideology in the History of Psychology: Stratification Theory in Germany | 1195 |

Martin Wieser

Part VIII History of Psychiatry 1221

47 The Mental Patient in History 1223
Peter Barham

48 Asylums and Alienists: The Institutional Foundations of
Psychiatry, 1760–1914 1253
David Wright

49 Forensic Psychiatry: Human Science in the Borderlands
Between Crime and Madness 1273
Eric J. Engstrom

50 Psychiatry and Society 1305
Petteri Pietikäinen

51 Early Child Psychiatry in Britain 1331
Nicola Sugden

52 Human Experimentation and Clinical Trials in Psychiatry 1357
Erika Dyck and Emmanuel Delille

53 Colonial and Transcultural Psychiatries: What We Learn From
History 1379
Sloan Mahone

54 Geriatric Psychiatry and Its Development in History 1403
Jesse F. Ballenger

55 Antipsychiatry: The Mid-Twentieth Century Era
(1960–1980) 1419
Allan Beveridge

Part IX History of Economics 1451

56 History of Thought of Economics as a Guide for the Future 1453
Dieter Bögenhold

57 Classical Political Economy 1473
Heinz D. Kurz

58 Neoclassical Economics: Origins, Evolution, and Critique 1501
Reinhard Neck

59 The Austrian School and the Theory of Markets 1541
David Emanuel Andersson and Marek Hudik

60 Joseph Schumpeter and the Origins of His Thought 1563
Panayotis G. Michaelides and Theofanis Papageorgiou

61 Learning from Intellectual History: Reflection on Sen's
    Capabilities Approach and Human Development 1585
    Farah Naz

Part X History of Ethnography and Ethnology **1611**

62 History of Ethnography and Ethnology: Section Introduction ... 1613
    Bican Polat

63 Before Fieldwork: Textual and Visual Stereotypes of Indigenous
    Peoples and the Emergence of World Ethnography in Hungary
    in the Seventeenth to Nineteenth Centuries 1621
    Ildikó S. Kristóf

64 Scientists and Specimens: Early Anthropology Networks in
    and Between Nations and the Natural and Human Sciences 1651
    Brooke Penaloza-Patzak

65 Center and Periphery: Anthropological Theory and National
    Identity in Portuguese Ethnography 1679
    João Leal

66 Making and Unmaking of Ethnos Theory in Twentieth-Century
    Russia 1699
    Sergei Alymov

67 Assessing Ethnographic Representations of Micronesia Under the
    Japanese Administration 1729
    Shingo Iitaka

68 Boasian Cultural Anthropologists, Interdisciplinary Initiatives,
    and the Making of Personality and Culture during the
    Interwar Years 1757
    Dennis Bryson

Part XI Gender and Health in the Social Sciences **1783**

69 Gender and Health in the Social Sciences: Section
    Introduction 1785
    Meagan Tyler and Natalie Jovanovski

70 Our Bodies, Ourselves: 50 Years of Education and Activism 1791
    Amy Agigian and Wendy Sanford

71 Reclaiming Indigenous Health Research and Knowledges
    As Self-Determination in Canada 1805
    Carrie Bourassa, Danette Starblanket, Mikayla Hagel, Marlin Legare,
    Miranda Keewatin, Nathan Oakes, Sebastien Lefebvre,
    Betty McKenna, Margaret Kîsikâw Piyêsîs, and Gail Boehme

| 72 | The Limits of Public Health Approaches and Discourses of Masculinities in Violence Against Women Prevention 1831
Bob Pease |
| --- | --- |
| 73 | The Madness of Women: Myth and Experience 1853
Jane M. Ussher |
| 74 | Feminine Hunger: A Brief History of Women's Food Restriction Practices in the West 1877
Natalie Jovanovski |
| 75 | Systems of Prostitution and Pornography: Harm, Health, and Gendered Inequalities 1897
Meagan Tyler and Maddy Coy |
| **Index** .... 1921 | |

About the Editor

David McCallum is Emeritus Professor of Sociology at Victoria University, Melbourne, Australia.

He publishes work in the fields of historical sociology, law and society, history of psychiatry, Indigenous studies, and sociology of education, and has most recently published a book titled *Criminalizing Children: Welfare and the State in Australia*, published in the Cambridge Studies in Law and Society series (Cambridge University Press, 2017).

Victoria University (VU) acknowledges, honors, recognizes, and respects the Ancestors, Elders, and families of the Boonwurrung (Bunurong), Woiwurrung (Wurundjeri), and Wadawurrung (Wathaurung) people of the Kulin Nation on the Melbourne campuses, and the Gadigal and Guring-gai people of the Eora Nation on our Sydney Campus. These groups are the custodians of university land and have been for many centuries.

Section Editors

Elwin Hofman
Cultural History Research Group
KU Leuven
Leuven, Belgium

Fran Collyer
Department of Social and Psychological Studies
University of Karlstad
Karlstad, Sweden

Dieter Bögenhold
Department of Sociology
Klagenfurt University
Klagenfurt, Austria

Stephen Mennell
School of Sociology
University College Dublin
Dublin, Ireland

Roger Smith
Reader Emeritus, History of Science
Lancaster University
Lancaster, UK

Bican Polat
Department of Philosophy
Boğaziçi University
Bebek, İstanbul, Turkey

Section Editors

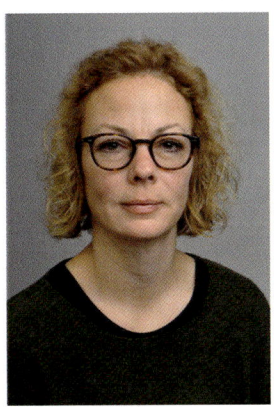

Marianne Sommer
Department for Cultural and Science Studies
University of Lucerne
Luzern, Switzerland

Eric J. Engstrom
Department of History
Humboldt University
Berlin, Germany

Natalie Jovanovski
Centre for Health Equity | Melbourne School of
Population and Global Health
University of Melbourne
Melbourne, VIC, Australia

Marta Bucholc
Faculty of Sociology
University of Warsaw
Warsaw, Poland

Centre de recherche en science politique
Université Saint-Louis Bruxelles
Brussels, Belgium

Meagan Tyler
Centre for People, Organisation and Work (CPOW)
RMIT University
Melbourne, VIC, Australia

Muller-Wille Department of History and Philosophy of Science, University of Cambridge, Cambridge, UK

Contributors

Amy Agigian Department of Sociology and Criminal Justice, Suffolk University, Boston, MA, USA

Sergei Alymov Institute of Ethnology and Anthropology, Russian Academy of Sciences, Moscow, Russia

David Emanuel Andersson IBMBA Program, College of Management, National Sun Yat-sen University, Kaohsiung City, Taiwan

Jesse F. Ballenger College of Nursing and Health Professions, Drexel University, Philadelphia, PA, USA

Peter Barham London, UK

Nima Bassiri Duke University, Durham, NC, USA

Fernanda Beigel Consejo Nacional de Investigaciones Científicas y Técnicas (CONICET), Mendoza, Argentina

CECIC- Universidad Nacional de Cuyo, Mendoza, Argentina

Allan Beveridge Royal College of Physicians of Edinburgh, Edinburgh, UK

Gail Boehme File Hills Qu'Appelle Tribal Council, Fort Qu'Appelle, SK, Canada

Dieter Bögenhold Department of Sociology, Universitat Klagenfurt, Klagenfurt am Wörthersee, Austria

Carrie Bourassa University of Saskatchewan, Regina, SK, Canada

Dennis Bryson Department of American Culture and Literature, Faculty of Humanities and Letters, İhsan Doğramacı Bilkent University, Ankara, Turkey

Marta Bucholc University of Warsaw, Warsaw, Poland

Hon-Fai Chen Lingnan University, New Territories, Hong Kong

Alan F. Collins Lancaster, UK

Fran Collyer Sociology and Social Policy, University of Sydney, Sydney, NSW, Australia

Susan Condor Emeritus Professor of Social Psychology, Loughborough University, Loughborough, UK

Maddy Coy Center for Gender, Sexualities and Women's Studies Research, University of Florida, Gainesville, FL, USA

Emmanuel Delille Department of Contemporary History, University Johannes Gutenberg, Mainz, Germany

Centre Marc Bloch, Berlin, Germany

Florence Delmotte Fund for Scientific Research (FNRS) / Université Saint-Louis – Bruxelles, Bruxelles, Belgium

Simone Dennis School of Archaeology and Anthropology, College of Arts and Social Sciences, Australian National University, Canberra, ACT, Australia

Erika Dyck Department of History, University of Saskatchewan, Saskatoon, Saskatchewan, Canada

Eric J. Engstrom Department of History, Humboldt University, Berlin, Germany

Stefanie Ernst Institut für Soziologie, University of Münster, Münster, Germany

Judith Friedlander Hunter College of the City University of New York, New York, NY, USA

Steve Fuller Department of Sociology, University of Warwick, Coventry, UK

James M. M. Good Department of Psychology, Durham University, Durham, UK

María A. Guzmán-Gallegos VID Specialized University, Oslo, Norway

Mikayla Hagel University of Saskatchewan, Regina, SK, Canada

Kim M. Hajek Institute for History, Leiden University, Leiden, The Netherlands

Timothy Heffernan School of Built Environment, Faculty of Arts, Design and Architecture, University of New South Wales, Sydney, NSW, Australia

Elwin Hofman Cultural History Research Group, KU Leuven, Leuven, Belgium

Marek Hudik Faculty of Business Administration, Prague University of Economics and Business, Prague, Czechia

Center for Theoretical Study, Charles University, Prague, Czechia

Shingo Iitaka Faculty of Cultural Studies, University of Kochi, Kochi, Japan

Natalie Jovanovski University of Melbourne, Melbourne, VIC, Australia

Miranda Keewatin University of Saskatchewan, Regina, SK, Canada

All Nations Hope Network, Regina, SK, Canada

Wiebke Keim CNRS/SAGE (University of Strasbourg), Strasbourg, France

Margaret Kîsikâw Piyêsîs All Nations Hope Network, Regina, SK, Canada

Ulrich Koch George Washington University, Washington, DC, USA

Ildikó S. Kristóf Institute of Ethnology, Hungarian Academy of Sciences, Budapest, Hungary

Heinz D. Kurz Graz Schumpeter Centre, University of Graz, Graz, Austria

Helmut Kuzmics University of Graz, Graz, Österreich

João Leal CRIA, Universidade Nova de Lisboa, Lisboa, Portugal

Sebastien Lefebvre Laurentian University, Sudbury, ON, Canada

Marlin Legare University of Saskatchewan, Regina, SK, Canada

Esben Leifsen Department of International Environment and Development Studies, Norwegian University of Life Sciences (NMBU), Aas, Norway

Anna Maerker History, King's College London, London, UK

Sloan Mahone University of Oxford, Oxford, UK

Geertje Mak NL-lab, KNAW Humanities Cluster, Amsterdam, The Netherlands

Amsterdam School of Historical Studies, University of Amsterdam, Amsterdam, The Netherlands

Rudolf Makkreel Atlanta, GA, USA

Dominic Malcolm School of Sport, Exercise and Health Sciences, Loughborough University, Loughborough, UK

Siniša Malešević School of Sociology, University College, Dublin, Ireland

Betty McKenna University of Regina, Regina, SK, Canada

Stephen Mennell School of Sociology, University College Dublin, Dublin, Ireland

Panayotis G. Michaelides Laboratory of Theoretical and Applied Economics and Law, National Technical University of Athens, Athens, Greece

Farah Naz Department of Sociology and Criminology, University of Sargodha, Sargodha, Pakistan

Reinhard Neck Department of Economics, Alpen-Adria-Universität Klagenfurt; Karl Popper Foundation Klagenfurt; Kärntner Institut für Höhere Studien; and CESifo, Klagenfurt, Austria

Nathan Oakes University of Saskatchewan, Regina, SK, Canada

Christian Olsson Université libre de Bruxelles, Bruxelles, Belgium

Lisa M. Osbeck Department of Anthropology, Psychology, and Sociology, University of West Georgia, Carrollton, GA, USA

Theofanis Papageorgiou University of Patras, Patras, Greece

Sujata Patel Umea University, Umea, Sweden

Herman Paul Institute for History, Leiden University, Leiden, The Netherlands

Bob Pease Institute for Social Change, University of Tasmania, Hobart, TAS, Australia

School of Humanities and Social Sciences, Deakin University, Geelong, VIC, Australia

Brooke Penaloza-Patzak Department of History and Sociology of Science, University of Pennsylvania, Philadelphia, PA, USA

Wade E. Pickren Dryden, NY, USA

Petteri Pietikäinen History of Sciences and Ideas, University of Oulu, Oulu, Finland

Bican Polat Tsinghua University, Beijing, China

Nathalie Richard Department of History, Le Mans Université, TEMOS CNRS UMR 9016, Le Mans, France

Elissa N. Rodkey Crandall University, Moncton, NB, Canada

Krista L. Rodkey Santa Barbara, CA, USA

Wendy Sanford Our Bodies, Ourselves, Newton, MA, USA

Robert A. Segal School of Divinity, History and Philosophy, King's College, University of Aberdeen, Aberdeen, UK

Roger Smith Institute of Philosophy, Russian Academy of Sciences, Moscow, Russian Federation

Marianne Sommer Department for Cultural and Sciences Studies, University of Lucerne, Luzern, Switzerland

Danette Starblanket University of Saskatchewan, Regina, SK, Canada

George Steinmetz Department of Sociology, University of Michigan, Ann Arbor, MI, USA

Nicola Sugden Centre for the History of Science, Technology and Medicine, University of Manchester, Manchester, UK

Fenneke Sysling Leiden University, Leiden, The Netherlands

Gülşah Taşçı Istanbul 29 Mayıs University, Istanbul, Turkey

Kirsi Tuohela University of Turku, Turku, Finland

George Turner Preston, UK

Meagan Tyler RMIT, Melbourne, VIC, Australia

Jane M. Ussher Translational Health Research Institute, Western Sydney University, Penrith, NSW, Australia

Zsuzsanna Vajda Department of Psychology, Károli Gáspár University of the Reformed Church in Hungary, Budapest, Hungary

Robert van Krieken University of Sydney, Sydney, NSW, Australia

School of Sociology, University College Dublin, Dublin, Ireland

René van der Veer University of Leiden, Leiden, The Netherlands

Joanna Wawrzyniak University of Warsaw, Warsaw, Poland

Raquel Weiss Department of Sociology, Federal University of Rio Grande do Sul (UFRGS), Porto Alegre, Brazil

Janina Wellmann Institute for Advanced Study on Media Cultures of Computer Simulation (MECS), Leuphana Universität Lüneburg, Lüneburg, Germany

Martin Wieser Faculty of Psychology, Sigmund Freud Private University Berlin, Berlin, Germany

Per Wisselgren Department of History of Science and Ideas, Uppsala University, Uppsala, Sweden

David Wright Department of History and Classical Studies, McGill University, Montreal, QC, Canada

Part VII
History of Psychology

Psychologies: Their Diverse Histories

38

Roger Smith

Contents

Introduction: The Diversity of Psychologies 978
The Diverse Roots of Psychologies 984
The Critical, "Reflexive" Implications of Diversity 992
Conclusion 997
References 999

Abstract

"Psychology" is a family name for many different kinds of knowledge and activity, as well as a collective name for states people have. Because of this breadth and multi-sidedness, it is not a speciality of interest only to psychologists but appropriately discussed as part of the history of the human sciences. This chapter relates the diversity of psychologies to the diversity of historical conditions in which psychological knowledge, practice, and states have become defining features of the modern age. It does this not by attempting to condense "the history of psychology" into a few pages but by examining the intellectual significance the history has both for contemporary psychologists and for a wider comprehension of the social function and structural position of psychologies in the human sciences. An introductory section clarifies why the chapter focuses on diversity rather than on hopes for theoretical unity or on any one particular characterization of what psychology "really is" in social and institutional terms. It begins with the question of the definition of psychology, which inevitably leads to the inquiry into whether psychology is or should be a unified natural science. The second section moves from recognition of psychology's modern diversity to argue that the roots of psychologies are also diverse. There can be no history of psychology; rather, there are histories. Wide-ranging

R. Smith (✉)
Institute of Philosophy, Russian Academy of Sciences, Moscow, Russian Federation
e-mail: rogersmith1945@gmail.com

© The Author(s), under exclusive licence to Springer Nature Singapore Pte Ltd. 2022
D. McCallum (ed.), *The Palgrave Handbook of the History of Human Sciences*,
https://doi.org/10.1007/978-981-16-7255-2_77

comments on recent scholarship illustrate the argument. The third section turns to the value of the history of psychology as a source of perspective for psychologists and as a source of critique in forms of human self-understanding in their relations with practice. This requires comment on the notion of critique and on notions of reflection and reflexivity closely associated with it. The conclusion restates the argument, founded in the theory of historical knowledge, that all knowledge is for a purpose. The intellectual purposes of the history of psychology, inquiry into the conditions in which the many varieties of psychological knowledge and practice have come about, are central to the human sciences. This is not an optional extra to science but intrinsic to the achievement of rational knowledge about "the human."

Keywords

Psychology · History · Definitions of psychology · Reflection · Reflexivity · Critical psychology

Introduction: The Diversity of Psychologies

References to psychology are so pervasive in contemporary life in so many parts of the world that it is tempting to identify psychology as an essential feature of modernity. Modernity appears to be a condition in which people characterize themselves in psychological terms; and where people characterize themselves psychologically, they are modern. Such statements, however, will not take us very far, for psychology and modernity are categories so ill-defined and of such a high level of generality that the characterization of one by the other says little. What indeed *is* psychology? What is it that *all* psychologies, as fields of knowledge, or as social practices, or as states of being human (or animal), share in common?

Introducing an attempt to clarify "psychology's territories," Mitchell Ash wrote: "Psychology has become a protean discipline that occupies a peculiar place among the sciences, suspended between methodological orientations derived from the physical and biological sciences and a subject matter extending into the social and human sciences. At the same time, modern societies and cultures have become permeated with psychological thinking and practices, much of which relates tenuously at best to what goes on in the discipline" (Ash and Sturm 2007: 1). Psychology, understood as a discipline, appears to be both a natural science and a social science. Understood as something more than a discipline, it is a host of beliefs and practices found in everyday life; in common language, psychology is also a set of states someone has – a person's psychology. With such complex usages, it is not even clear it is possible to recognize "family resemblance" (to use Wittgenstein's phrase) among everything denoted by "psychology." Ash therefore did not define "psychology's territories" but instead introduced chapters discussing a number of manageable topics. This is the common practice: discussions of psychology, with the noteworthy exception of textbooks, do not attempt comprehensiveness. It is a practice commonly accompanied by hopes for interdisciplinarity, and it is just such hopes that references to psychology in the context of the human sciences often represent. Considering psychology in the framework of the human sciences is surely a constructive response to the seeming impossibility of defining psychology.

Many people have of course proposed definitions over the years. It would be a project in its own right to analyze them, a worthwhile project, not "mere" play with a word, and it would inescapably enter into history. For instance, William James, in 1890, opened his *The Principles of Psychology* (1950, vol. 1:1) with the definition, "Psychology is the Science of Mental Life, both of its phenomena and their conditions." He evidently emphasized the word "mental." In contrast, Svend Brinkmann (2020: 3) emphasized the word "life": "my contention is that psychology as a science of life must be a biographical science – a science of what it is to live a life." Both definitions in turn markedly contrast with current enthusiasm for psychology understood as the science of brain functions. The point need not be pressed further. What concerns this chapter, an introduction to psychology viewed within the frame of the history of the human sciences, is the implication of the existence of strikingly different definitions: there can no more be a comprehensive study of the history of psychology than there can be a comprehensive study of psychology (whatever it is). This chapter, therefore, explores the implications of the diversity and sheer scale of activity in psychology for the study of its history, taking a broad view encouraged by working with the notion of history of the human sciences.

Even if we turn away from definition and say, in nominalist fashion, that, well, psychology is what people who are members of the world's largest psychological institution, the American Psychological Association (APA), and similar institutions, do, we would be no better off, since such organizations have many diverse sections. Moreover (as Wade Pickren's and Gülsah Taşçi's chapter in this section of the *Handbook* outlines; also Baker 2012; Brock 2006), there are movements around the world advancing the claims of different "indigenous psychologies" and seeking to break from "coloniality," the mind-set assuming western, or specifically the United States, dominance in science and activities in its name. It is a substantial, and also divisive, question whether this break could be achieved while determinedly claiming to do "psychology." The arguments are ongoing. They run in parallel with arguments about the relations between professional and everyday psychological activity. Since many ordinary people think and act in self-expressed psychological terms, who controls and has authority over what in psychology? What is the relation of "scientific" psychology and "popular" psychology?

If psychology is diverse both as a matter of social fact and because, for conceptual reasons, there are different meanings of the word "psychology," then *history of psychology is necessarily diverse*. This deserves emphasis. History of psychology is not a narrowly focused preoccupation, perhaps of interest only to psychologists, but raises general questions about the diversity of human self-understanding and governance. For this reason, it belongs with the history of the human sciences, a heading that promotes study of the particular historical and social forms of how and why and in what form one group or another conceives of its culture and activity in psychological terms.

The chapters in this section illuminate critical understanding of psychology's projects, and they do so for the most part through specific historical example. This is the way historians work. (Michael Billig (2008, 2019), a social scientist interested in social psychology, has specifically recommended the value of historical examples over generalities generated by "theory.") The chapters also deal with questions and methods related to historical practice. There is no expectation that the authors who contribute to this section will agree with all the arguments of the introduction. But their chapters go some way to explore what in intellectual practice diversity implies. Other sections of the *Handbook*, particularly those on medicine and psychiatry, and on the self and personhood, substantially add to the larger picture.

As Nadine Weidman (2016, writing in her capacity at the time as editor of the journal, *History of Psychology*) observed, historians of psychology broadly fall into two groups. Much the larger is composed of psychologist-historians, academics working in psychology departments and owing allegiance to psychology as a discipline for their teaching, research, and careers. It is therefore inevitable that the great majority of publications concern activity clearly identifiable as part of specialist psychology fields in the modern period. The smaller group consists of historians of science, trained in or at least identifying either with history as a discipline or with science, medicine, and technology studies (studies with significant input from the social sciences). It is a valid description, though it leaves out of the account significant, if problematic, relations of history in general and history of psychology in particular with philosophy. In the past, English-language teaching in psychology subsumed consideration of these relations under "theory and systems" in psychology. In the present, especially in North America, both history of science and history of psychology have become rather divided from philosophy, though the journal *Theory & Psychology* has fostered overlap. The Brazilian historian and philosopher of psychology Saulo de Freitas Araujo (2016, 2017) has argued persuasively for the necessity of a philosophical understanding on the part of historians; indeed, there are settings, as in intellectual history, or in the section of the British Psychological Society (BPS) for History and Philosophy of Psychology, where the separation is not so evident. The notion of the history of the human sciences, to make a point argued for in the introductory section of the *Handbook*, also suggests a framework in which history and philosophy work together.

A glance at publications in the history of psychology indicates that many authors are concerned with psychology as social practice, rather than with psychology as scientific knowledge. To be sure, knowledge forming is social practice too, and any practice presupposes some kind of knowledge. The fact is, however, that writers on the history of psychology are often concerned first and foremost with what psychologists do in social context, rather than with what psychologists have done to advance science. This, indeed, is the source of some discord within the psychology profession, as those who profess science often do not see the point of other activity, especially history.

The present section of the *Handbook* might seem to confirm such scientists' assessment of history, as it contains no discussion on psychology as a natural science. Three lines of argument may counter such a negative reaction. The first is simply pragmatic: large and wide-ranging reference works exist for the established branches of scientific psychology, and it would be pointless to duplicate them. By contrast, the social relations of psychology are much less well served, and this section provides significant discussion and working examples of historically informed approaches to these social relations. It thereby complements other sections of the *Handbook* concerned with the human rather than natural sciences. The second line of argument relates to the interest in critique, that is, to the virtues of history and philosophy as disciplined means for reflection, by individuals or groups (nations, branches of science, gender groupings, or other classes of people), on the shaping of psychological activity. Numerous writers appear to be attracted to the history of psychology because of the interest and ethical or political appeal of acquiring perspective. (For one list of possible reasons for turning to history of psychology, see Smith 2016.) Many such authors, it is important if prosaic to note, have also found themselves required to teach history of psychology; this, for example, has played an important part in the notable flourishing of the field in Brazil. The third line of argument is to suggest that there is little profit in returning to argument about whether psychology is or is not a science, meaning in English a natural science, or whether it belongs to *die Naturwissenschaften* (natural sciences) or to *die Geisteswissenschaften* (in different contexts, human sciences or cultural sciences, or humanities). Whatever the arguments, there are *psychologies*, in the plural, and discussion of the nature of scientific psychology belongs in discussion of that diversity.

It may appear self-evident to many psychologists that *the* task is to establish psychology as a science, a body of reliable knowledge unified by a coherent, rational-theoretical understanding. It may appear that history, philosophy, and social studies of psychology have value if, and even only if, they help with this scientific project. (The argument is stated firmly and thoughtfully, e.g., in Toomela 2016.) This is not the place to assess the claims of cognitive or neuro-fields to provide this unification, but only to note that many psychologists would think, rather, that social psychology should have a central position in any integrative project, since, whatever else psychology might be, it is about social relations. Jaan Valsiner (2012) and co-workers, for instance, have placed hopes in developing the legacy of L. S. Vygotsky in cultural psychology into such a unifying theory. For yet other psychologists, perhaps following Gregory Bateson (2000), discussion of social or cultural psychology is not enough, and they think it necessary to conceive of mind as immanent in the brain-person-environment system as a whole, to which only some kind of systems thinking could do justice.

The search for unifying theory is compatible with recognition of psychology's past and present diversity, if one understands diversity as a marker of psychology's scientific immaturity. It is a historical fact, though, that at various times in the modern history of psychology, psychologists have reacted to the patent absence of unified theory by declaring a "crisis" (Mülberger and Sturm 2012; Proietto and Lombardo 2015). "Crisis" thinking is revealing historically, as it has promoted attempts to clarify and agree principles and standards about what constitutes a science, and hence about what methods and content would constitute psychology as a science. On precisely this there has been, and is, marked disagreement. Proposing, in 1934, to give a course of public lectures on "general psychology," C. G. Jung stated that "psychology should be taught in its biological, ethnic, medical, philosophical, cultural-historical, and religious aspects ... general psychology should communicate to the student the possibility of a culture of the psyche" (application for funding support, quoted in introduction to Jung 2019: iii). Considered without reference to context, the proposal appears so broad as to be empty. In context, however, the proposal was full of meaning, as Jung did indeed claim to offer a unifying science. (For the formation of Jung's idea of psychology as science, see Shamdasani 2003.) Jung intended to lecture on the collective unconscious, to spread knowledge of the science that, he trusted, was necessary for individual well-being and the harmonious integration of psychic forces in collective culture. Yet, of course, the great majority of his contemporaries committed to the creation of a science of psychology would not have defined psychology as "culture of the psyche." An understanding of these differences, and of the influence different stances had, requires knowledge of biography and institutional, national, and intellectual context, *and* knowledge of the nature of different claims about what it is to be human and to construct a science of the human. The open-ended width of the knowledge required for such understanding might suggest reasons for concluding that unification of something called "the theory of psychology" is not viable. Moreover, social studies of science have seriously questioned the possibility of finding universal criteria for separating science and nonscience. If, as Isabelle Stengers stated, "it is pointless to search for a noncontextual, general definition of the difference between science and non-science" (Stengers, quoted in Shamdasani 2004: 28), it seems even more pointless to search for a general, noncontextual way to separate psychological science (or sense) and psychological nonscience (or nonsense).

Is psychology like or unlike other natural or social sciences in its actual diversity? While psychologists have commonly lamented the divisions among psychologists, in comparison, say, with among physicists, some observers of science argue that sharp divisions exist in all fields. In physics, for example, there is marked disagreement over the possibility of a general field theory (Galison and Stump 1996), while among social scientists there is wide acceptance of a diversity of sociologies. Agreement is characteristic of particular social communities, not of fields abstractly conceived.

The chronic issue of the scientific standing of psychology is bound up with the question of the relation of academic psychology to everyday psychologies, that is, nonspecialist, or public, or popular ways of representing self and others in psychological terms. American-English sometimes calls these nonspecialist forms "folk psychology," though the German word *Alltagspsychologie* ("everyday psychology"; Ash and Sturm 2007: 9–10) provides a better characterization. Fictional writing, with reason, is often seen as an influential resource of nonprofessional psychology. Moreover, it is a social reality that very large parts of what goes on under the heading of psychology are concerned with practice and not directly with research or theory. A majority of psychologists find employment in clinical and educational or other occupations, rather than in scientific research, and they therefore constantly interface with everyday psychology. Some psychologists, notably, at present, those neuropsychologists who advocate a strong program to translate psychological statements into statements about evolved brain states, look forward to the elimination of "folk psychology" by science (the argument stated formally in Churchland 1989). But there are strong arguments against this program (Kusch 1997, 1999: part 2), backed by knowledge of the pragmatic rationality and emotional richness of ordinary people's psychological understanding in the context of particular ways of life (with their social rules, interpretive languages, semiotic practices, and so forth). If the relation of everyday to scientific psychology continues to be a pressing one for professional psychologists, it is largely because professional authority and the legal status of claimed expertise have been tied rhetorically to dependence on the standing of psychology as science. (For an attempt to rethink the relationship between academic psychology and its publics, see Petitt and Young 2017.)

Rather than trying to separate either theoretical and applied psychology or scientific and popular psychology, it may be constructive to think about the issues at stake in terms of critical versus noncritical thinking. (On the psychology of critical thinking, see Lamont 2013, 2020.) This would have a number of benefits. It would focus on a key question: the relation between the understanding of rationality and the achievement of social authority, integrating studies of knowledge formation and normative social processes. (There is considerable precedent for such an integration in science and technology studies.) Divisions within psychology, such as pure-applied, scientific-popular, normal-paranormal phenomena, culture-biology, and so on, would themselves become objects of study. A concern with critical thinking also inevitably fosters reflection – which historians naturally have as much need of as anyone else.

Earlier projects to write the history of psychology as a single narrative, with a beginning (conventionally Aristotle) and an end (the present activities of psychologists of a favored school), depended on belief that psychology had become, or was becoming, or should become, a science. That belief appeared to give psychology a core identity, ideal if not actual, and it thereby provided its history with a plot. It provided a criterion with which to judge the worth of a historical narrative from the point of view of psychologists: did it illuminate the core identity? It has, however, been the effect of history, as of social studies of psychology, to render this notion of a core identity problematic.

These kinds of issues do not matter most of the time to most professional psychologists since their activity and identity comes in practice from the speciality in which they work, and they have more than enough to do to keep pace with the speciality. For many, the history of psychology is itself a specialist field, about which it is not necessary to know much. The issue of diversity, however, does not go away, and it resurfaces, for instance, when one speciality claims to colonize or replace another, when it is necessary to introduce students to "general psychology," or when psychologists seek to sustain the legal or intellectual authority of "scientific" approaches in the face of "popular" practices or beliefs.

This chapter sets out a case, drawing on historical studies, for arguing that there neither is nor could be one definition of psychology. While it is just about possible to imagine circumstances in which one institutionalized authority, commanding enough power, imposes one body of psychological knowledge and associated practice throughout all human arrangements, these are not, nor ever have been, the circumstances of actually existing psychology. That such power does exist in local settings, where it imposes and sustains one version of psychology, is another matter – and this is central to what historians study.

The Diverse Roots of Psychologies

A straightforward but inadequately appreciated implication for the practice of the history of psychology follows from psychology's diversity. If psychology is diverse now, it also has *diverse roots, diverse histories*. The different forms taken by psychology require different histories: it is not a field with one origin or founding moment, let alone a founding father. It is a matter for historical work to determine the different roots of different psychologies; and it is wrong to judge the character of these roots in the light of one modern understanding of scientific psychology. As historians are indeed showing, it is possible to trace contemporary psychologies to many roots – religious, philosophical, medical, jurisprudential, economic, scientific, and perhaps above all the practical means of organizing daily life (in child care, in enforcing moral or economic practice, in sustaining personal relations, in education, and so on). There is a striking range of topics in academic articles on the history of psychology, and this in itself demonstrates the parallel between diversity in psychology and diversity in history of psychology.

Of course, it is eminently possible to study the particular roots of a particular psychology. The purpose, however, must be clear. Talk about "*the* origin" or "*the* founders" *in general* is misguided. This calls into intellectual question a whole class of textbook histories, whose purpose comes with teaching requirements rather than with settling questions about the proper subject matter of histories. Given the large size of student audiences for psychology, there are many textbook histories on the market. The more defensible intellectually avow a distinctive purpose: Jeroen Janz and Peter van Drunen (2004) judged that psychology students first needed a broad introduction to modern social history; Wade Pickren and Alexandra Rutherford (2010) gave much emphasis to the social organization of the field and the context of sociopolitical issues in the United States; Graham Richards (2010) took subjects designed to expose students to reflective, critical thinking; Bruce K. Alexander and Curtis P. Shelton (2014) to a highly selective history of western civilization in order to subsume psychology in a moral project. Though not a textbook but a general history, Roger Smith (2013a) took diversity as its theme. (There are also large-scale reference works, including Baker (2012), Pickren (2021); and Sternberg and Pickren (2019), which present the history of topics or specialities that professional psychologists will at once recognize as central to the contemporary field.)

The matter of psychologies' diverse roots overlaps with the matter of "indigenous psychologies," just as historical questions about the life of past societies overlap with questions originating with anthropological studies of peoples living today. While everyone concedes that the ancient Greeks did not have a word for "psychology," nor name such a division of knowledge, there is considerable disagreement about the propriety of referring to "Greek psychology." Both textbooks and common language find it natural to do so, as ancient writers discussed topics like memory, moral reasoning, dreams, and sensation, topics that appear as the subject matter of modern psychology. Speaking of Aristotle, while leaving the nature of psychology unexamined, John D. Greenwood (2015: 24) wrote, "he was also the first Greek theorist to devote a whole work to psychology (*De Anima*)." Philosophers, not just psychologists, use such language (e.g., Sorabji 2006). Simon Kemp and other scholars writing on ancient or medieval "psychology" made no fuss about using the word (Kemp 1990, 2018). Yet, is it accurate to describe ancient or medieval discussions as "psychology" (or, e.g., to refer to "Tibetan psychology"), especially when by doing so it makes earlier writing appear continuous with modern understandings?

There are many dimensions of argument here (discussion continues in the following section on critique). Philosophically, the argument is about the presence and relation of what the Canadian philosopher of science Ian Hacking (1995a) has distinguished as natural kinds and human kinds, descriptive categories held to correspond to naturally existing entities, denoting universals, or descriptive categories held to have a contingent, constructed existence in human culture, denoting particulars. This contributes to a very large discussion. Indeed, the historian and philosopher of science Nick Jardine (2000: 263) concluded: "The problem of human universals is surely the most intractable of all [... the issues] raised. ... Indeed, it is entirely unclear what kind of 'science of human nature' or 'philosophical anthropology' could settle such issues." Hacking himself preferred to think about the questions through specific historical, empirical studies, rather than restating long undecided philosophical debates. That is an encouragement for the historian of psychology and intellectual historian, because it shows that it is viable to get on with analysis, linking past and present, while yet (of course!) not settling what appear fundamental questions of philosophy. Historians may contribute empirical material and clarify concepts in studying psychological categories, just as a speciality in empirical psychology, inter-cultural psychology (Bender and Beller 2016; Kitayama and Cohen 2007; Lloyd 2007) hopes to do. But it remains the case that though comparative inter-cultural psychology has made sophisticated efforts to settle the empirical facts, it has proved hard to separate questions about facts from conceptual questions. The same is true in historical research.

The relationship between categories and what is said to be "there" in nature has been subject to much philosophical debate. Categories, in some sense, are the outcome of human historical activity and not given "in nature." Faced by the intricate complexities of "realism" and "anti-realism" as theories of knowledge, however, historians of psychology, like many historians of science, will sympathize with the decision to get on with what Hacking (2002) called "historical epistemology" and what Lorraine Daston (2000) called "biographies of scientific objects." This was also Kurt Danziger's decision, studying the history of "the scientific objects" of personality and of memory (Danziger 1990, 2008), though he used this material to support a strong argument for belief in the social construction of categories. One consequence for practice is the need for inter-disciplinary work to assimilate the range of argument (Ash and Sturm 2007). The work of Hacking has also been of interest to psychologists, as he provided sophisticated studies of the historical constitution of self-identity (Hacking 1995b, 1999, 2007).

It is helpful for historians to link judgment about the universality or temporality of categories to more accessible judgment about anachronism. Substituting the word "psychological" for "biological," some historians would therefore endorse Jardine's conclusion (2000: 261): "because there were no institutional arrangements for the systematic conveying from generation to generation of a body of practices of empirical enquiry ... it is viciously anachronistic to write of Aristotle's 'biological [or psychological] investigations'." It is all too easy to equate ancient and modern concepts and descriptions – and, in a parallel way, to equate psychological states and concepts from around the modern world. One widely understood example involves Plato's view of memory, which he characterized as the recovery of the soul's knowledge in its ideal condition (Danziger 2008: 92–94; Draaisma 2000). He did not present a psychological understanding independent of highly elaborated Socratic philosophizing, with its idealist conception of reason. He understood memory as the intuition of the realm of ideas, the true realm, which it is the business of wisdom to restore to human kind. Similarly, through the subsequent centuries, memory meant different things.

With comparable historical analysis, it may be shown that the notion of a free will came with the late Stoics and especially Augustine (Frede 2011), and the concept of child development with the first centuries of Christian thought (Goodey 2021). We might, for example, seek to make the same point about aggression. What *is* aggression? Statements about aggression as a universal factor in human nature do little in any precise way to explain the particular and very varied forms of expression the supposed universal characteristic actually has had. The philosophical response, consistent with our knowledge of the variety of psychologies of the past few centuries, is that, as Sonu Shamdasani (2004) has expressed the point, psychologies are "ontology-making practices," that is, historically located ways of bringing into existence the states of being that they themselves study. An instrumental, working principle for historians sympathetic to this line of argument would be to presume that traditional ways of tracing back the history of psychology, seeking out a line leading towards modern science, have taken far too much for granted. Instead, there should be particular histories for particular things – beginning with the presumption of diversity rather than unity.

The foremost subject for the application of this principle is the category of psychology itself. The vexed issue of whether it is valid to refer to ancient Greek psychology has been mentioned. Highly valuable contributions to discussion of this have been made in books by Paul Mengal (2005) and Fernando Vidal (2011) that have detailed the way in which a science of the soul, associated with *physica*, or the science of natural bodies rather than the higher science of theology, developed in the sixteenth century. For Mengal, this amounted to the establishment of the science of psychology. Vidal discounted this and instead emphasized the diversity of language used in studies of the soul, elements of which some authors called psychology. Then, during the eighteenth century, the shaping of a field of "anthropology" and the differentiation of physical and moral dimensions in "the science of Man" consolidated studies of the physical and everyday relations of the soul into a domain more and more often addressed as "psychology." The US philosopher and historian of science Gary Hatfield (1992, 1995, 1997) also turned to the early modern and eighteenth-century literature to describe a field of knowledge of psychology, even though there was no socially delineated discipline. The historian of psychology Horst Gundlach (2006, 2012) pointed to the gradual separation of different lines of inquiry in the German-speaking world. Following such work, and seeking to link it with large-scale historical change creating modern individualism, Sven Hroar Klempe (2020) argued for the specific influence of Christian Wolff in shaping a new field.

This kind of research has concentrated on intellectual sources. The challenge for historians is then to relate the sources to the historical conditions establishing modernity. This in turn leads to the history of the individual states that we would now simply call "psychological" states, like emotion, motivation, cognition, unconscious activity, and so on. The sense in which such states have a social and cultural history, in addition to or as opposed to an evolved biological nature, is central to the history of social psychology (Jahoda 2007). Writing this makes it necessary to recognize the difference between the practice of "historical psychology," a field which studies "change over time in human psychological capacities and seeks to understand differences in mentality and emotionality of people in the past, as a function of their time, place, and historical context" (Weidman 2016: 250), and the practice of "the history of psychology." (For psychologists discovering the difference within history of psychology, see Henley 2020, and following discussion papers.) "Historical psychology" has roots as a branch of historical study in the anthropology of culture of the late eighteenth century (Carhart 2007; Zammito 2002), in the *Völkerpsychologie* of the second half of the nineteenth century (Diriwächter 2004), and in the sociological and historical work creating French scholarship on the history of *mentalité*. (For Ignace Meyerson and the link between psychology and the French historians, see Pizarroso 2013, 2018.) René van der Veer's chapter in this volume well illustrates the relevance of history for the cross-cultural studies with which psychologists attempt to study the issues (Chap. 44, "Vygotsky, Luria, and Cross-Cultural Research in the Soviet Union"). And if we turn to the work of historians, we find a very substantial contemporary interest in the history of emotions and of the senses, reflecting the current preoccupation of disciplines in the humanities with the embodied nature of the human subject. (For argument to combine studies of emotions and senses, see Boddice and Smith 2020.) There is indeed an audience in the humanities (including the history of science and medicine) for a historical understanding of psychology.

The more thoroughly research engages with the wider historical setting, however, the more professional psychologists may cease to attend. As busy specialists, their identity and interest is bound, seemingly naturally, to a narrow range of contemporary questions. This leaves the defense of historical work to those who sustain the general defense of history as part of the humanities (Guldi and Armitage 2014).

Danziger, who did want psychologists to pay attention to the history, therefore focused on themes intended to speak to and to shift thinking about contemporary psychology (especially social psychology, which is discussed in Jim Good's chapter in this volume), as in his books on the category of personality, on psychological terms, and on memory (Danziger1990, 1997a, b, 2008). He accompanied these studies with provoking reflections on the past and future of the history of psychology (e.g., Danziger 1994, 2004) informed by work in the social studies of science.

The interest in the humanities for history of psychology is illustrated by work on Victorian psychology, including studies of the psychological conceptions of poets and novelists (Anger 2018; Shuttleworth 1984, 1996). Jacqueline Carroy (1993) discussed the close relationship of psychological and literary studies in France, and there are also studies of individual writers, like Kierkegaard (Pind 2016; Klempe 2014). Editing an anthology of texts to introduce Victorian psychology, Jenny Bourne Taylor and Sally Shuttleworth (1998) paid special attention to the medical, self-help, and moralistic physiological literature so prominent at the time, much of it written for the general public, and this led to work in the historical psychology and medicine of anxiety (Bonea et al. 2019). In the nineteenth century, there was a huge literature relating psychology, morals, and medicine in much of Europe and North America in journals read by the general public; there was, for example, substantial debate about the philosophy, moral implications, spiritual dimension, social importance, and psychology of free will (Smith 2004, 2013b). The English-speaking public continued in the twentieth century to think of psychology as centrally concerned with self-help and self-improvement (Thomson 2006), and even though many psychologists have insisted on a difference between "professional" or scientific and "popular" forms of psychology, the boundaries have continued to be permeable and cross-fertilization has continued to be common.

It is impossible to discuss this nineteenth-century formation of a large public interest in psychology, by this time clearly so named, without taking into account the interest in hypnotism and spiritualism. The significance of these practices has been recognized in both studies of the background of psychoanalysis (especially Ellenberger 1970, brought up to date in detailed studies, e.g., Crabtree 2019; and in histories of psychoanalysis, Makari 2010) and in histories of public representation of social questions in psychological terms (e.g., Hayward 2007). Public fascination with "the paranormal" was very visible in early conferences on psychology, including the first international meeting in Paris in 1889 (which eventually led to the founding of the International Union of Psychological Science, IUPsyS), presided over by the medical neurologist J.-M. Charcot. Modern professional psychological organizations, indeed, developed in significant part as the institutional means for self-styled scientific psychologists to differentiate themselves from this public form of psychology (Baker and Benjamin 2014; Carroy et al. 2006; Parot 1994; Plas 2000). Annette Mülberger (2012: 441, quoting Rudolf Willy) described the seeming chaos that psychology presented at the Third International Congress in 1897: "even at a glance the conference program gave us the impression of an undirected army recruiting people by chance from the street."

Historical studies have indeed shown just how large a place the forms of psychology that nineteenth- and twentieth-century rationalists dismissed as "pseudoscience" or "superstition" had in developing psychology. Practices that look like, or have become, psychological practices, such as physiognomy, phrenology, mesmerism, spiritualism, confession, prayer, memory training, rhetoric, romantic writing, teaching techniques, self-help guides, diaries, spiritual healing, and much else besides, have a place in the history. Such practices helped create a public sensible of psychological states and ways of framing the human condition, and they helped form and demarcate new areas of science. Phrenology, for instance, substantially contributed to acceptance of knowledge that the brain is the organ of mind, to search for the localization of mental functions, and to practices of assessment of character in naturalistic terms (Young 1970; van Wyhe 2004). There are no sharp lines between the history of such practices and the history of psychology understood as a science – a circumstance James's career and interests famously illustrates (Bordogna 2008; Taylor 1996). The extent and influence of psychological enthusiasms that many modern scientists would hold to be "unscientific" is such that Mathew Thomson (2001, 2006), a social historian, attributed the rise of the twentieth-century public interest in psychological practices to social changes taking place almost entirely outside the sphere of academic psychology.

Another area of boundless public interest in psychology has concerned the upbringing and education of children. This has a long history in many parts of the world, and it may be that significant elements of the roots of modern psychology lie in this area of human activity. Chris Goodey (2011) argued that the category of intelligence, understood as a psychological category, originated in early modern debates in Calvinism concerned with the capacity of individual people to discern divine grace. He then extended the historical scope of the argument to encompass the conception of child development itself, going back to the ancient Greek and early Christian legacy (Goodey 2021). At least from the late eighteenth century, letters and diaries, including diaries kept about children, became significant elements of psychology forming practices. In a study of late Victorian child psychology, Sally Shuttleworth (2010) described the range of the interest and importance of concerns transcending the emerging professional and public divide. Nineteenth-century national policies to enact compulsory education were of the greatest importance, creating conditions in which a competence to measure and rank-order psychological capacities, initially intelligence but later many other things, legitimated claims to establish a psychology profession offering social services. (Zsuzsanna Vajda's chapter in the present volume studies this history as it shaped psychology in Hungary; other studies have shown the connections in, e.g., Russia and Spain – Byford 2014, 2020; Mülberger et al. 2014, 2019.) Many historians have taken up this theme, not least because of its connection with the development of statistics, the specialist background for making sense of intelligence tests, and the subsequent importance of statistics for policing scientific methods in psychology. Other modern concerns, part and parcel of the modern psychological culture of social relations, also invite historical interpretation, for example, adaptation, self-esteem (discussed in this volume by Alan Collins, George Turner, and Susan Condor), empathy (Lanzoni 2018), equanimity (Mckay 2019), and inhibition (Smith 1992, 2020a).

Changes in the practice of the history of science generally, in the years around 1970, led in academic writing to a decisive turn towards the study of science in social and historical context. While this generated some adverse reaction from natural scientists, some of whom implied that historians began to leave out "the real science," the consequence was a mass of studies of the past and present conditions of scientific activity, breaking down distinctions between knowledge and practice, pure and applied science (and technology), and formal distinctions between science and nonscience. The result was an enlarged and vibrant professional speciality, to which, institutionally, the history of psychology both does and does not belong (reflecting the employment of historians of psychology in psychology departments, as referred to earlier). This inevitably raised the question, whether it was desirable and possible to show how such diverse studies, whether of science in general or of psychology in particular, contributed to an inclusive story or to an integrated body of knowledge. It was also necessary to ask about the interest such studies might or might not have for nonspecialist audiences, whether scientists, other historians, or "the public." What was the relation between particular scholarly studies, in all their diversity, and ideals of reason, truth, objectivity, and human well-being that have been said to inspire science in general and psychology in particular?

Might serve to prompt the jibe that history describes one damn fact after another. In fact, though, every study is written for a purpose (or for a number of purposes), and that purpose (or purposes) is at work in every step, from the selection of a topic to ordering detail, to the framing of concepts, to the choice of rhetorical voice. When, half a century ago, historians of science took the step of adopting from mainstream historians the standards of archival research and contextual understanding, they often argued against "presentism." This in practice meant distancing the writing of history as the interpretation of past events from the shaping of the historical record in a way solely relevant to some element of present knowledge or practice. The historians of science wrote against presentism in order to achieve a historical discipline. With more philosophical reflection, however, it became clear that all historical knowledge is knowledge constructed by a historian working in a particular time or place and that the historian therefore brings to bear presumptions and values characteristic of that time and place. Contemporary historians of psychology apply the standards of their speciality – veracity to all the available empirical historical evidence, coherence, contextual considerations, sensitivity to the forms of reasoning of their subjects, and so forth. At the same time, they have purposes originating in the situations in which they work. (And for this reason, the concept of presentism is analytically not helpful.) The purpose may be the mundane one of writing a piece meeting certain standards for career reasons. It may be an intellectual ambition to show the social rather than biological sources of a taken-for-granted concept or theory in psychology. It may be part of an attempt to understand the meaning and applicability of categories like autism, dyslexia, and depression which appear rather suddenly to have acquired social prominence. It may be a critical contribution to recognizing the place that prejudice or bias about race or gender has had in psychology. (For the concept of race in psychology, see Richards 1997; for gender, see Rodkey's chapter in the present volume.) The purposes are myriad – but they are there, and they all connect some present understanding with knowledge about the past. There can be no understanding of the nature of historical knowledge without reference to these purposes.

The issue is the more confused because there has been a very influential practice called "the history of the present." This originated with the purpose to do seriously researched history, but to do so in order to intervene in current debates in the social and psychological sciences about social problems, rather than primarily to contribute to professional history. The purpose, in effect, was politically engaged history. Its intellectual inspiration was the "genealogical" approach to knowledge promoted by Michel Foucault, which turned into the microanalysis of power relations, especially as they worked themselves out in embodied subjects (as "bio-power"). The work of the London sociologist Nikolas Rose (1985, 1998, 1999), work influential in the history of psychology, gave it form as a seemingly concrete program of research. Given that many people have been attracted to the history of the human sciences because of disquiet about the direction of the social or psychological sciences in which they were trained, the notion of the history of the present, combining history with political relevance, was bound to appeal. Other sections of the *Handbook*, for example, on gender and health, demonstrate this.

Rose and others linked modern psychological practices to ways of organizing social life since the early nineteenth century, ways greatly expanded in the twentieth century and embedded in the institutions of education, business, health, and welfare of liberal democracies generally. There were also contributions from influential cultural critics of the commercialization of the self and loss of moral social relations in contemporary forms of "psychological society," society dominated by individualistic excess and an overly, or even exclusively, psychological discourse about the self and its values. The shaping of psychology began to appear part and parcel of the shaping of late modernity. Though Foucault himself had said little directly about psychology in his reanalysis of power as local, dispersed, and embodied, his studies of madness, discipline, governance, and sexuality proved richly suggestive in building an understanding of power into the work of historians of testing procedures, therapy, norms of individual development, gender, the construction of categories like autism and PTSD, and more.

At the same time, many other purposes continue to inform historical work. One large group raises philosophical questions about the nature of psychological knowledge and the coherence of psychological descriptions. There continues to be an interest in the intellectual history of science, including psychology, alongside the shift to contextualizing historical explanations. This is home for meta-level historical questions relating past and present, since intellectual historians necessarily face questions about the relation between the historians' reasoning processes and the reasoning encountered in past writing. These matters have, for instance, informed this author's work (Smith 2019a, 2020b) about the history of the sense of movement (kinesthesia), giving contemporary embodied knowledge and movement practices like dance a long history in opposition to mind-body dualism and mechanistic science, an opposition that contemporary performers and therapists widely celebrate as "new."

The Critical, "Reflexive" Implications of Diversity

The British psychologist and historian of psychology L. S. Hearnshaw (1987) once hoped that history would serve to hold together the different parts of the field he thought in danger of falling apart through overspecialization. In fact, it would seem, dashing Hearnshaw's hopes, historical work has itself contributed to the disparateness of psychology's activities: it has created yet another speciality rather than linked existing ones. The bewildering range of contents of leading journals in the area points to this conclusion.

All the same, for the student encountering demanding courses on statistics, neurophysiology, and computer-related cognitive science, a historical perspective may provide a sense of the human motives and relations that lie behind it all. Further, historical work, in its nature, involves some degree of reflection, for researcher or for student, on activity in the present that is otherwise taken for granted. In this sense, history is *critical* work. This indeed is visible politically in the efforts of institutions, in this respect like nations, to control historical knowledge, and equally visible in the efforts of movements for change to rewrite history. "Critique," however, has a number of meanings.

The critical character of perspective comes with the acquisition of some level of awareness – "raised consciousness" – beyond unexamined acceptance of statements of knowledge or unquestioned execution of practice. History of psychology is critical in this sense: by seeking to understand the reasons and causes for why people and institutions and societies do what they do, it makes visible values at work. (For a systematic overview of the relations of science, psychology, and values, see ▶ Chap. 45, "Values and Persons: The Persistent Problem of Values in Science and Psychology," by Lisa Osbeck's in this volume; on perspective, see Osbeck 2019: 83–107.) These values include the values held characteristic of scientific practices in general – the pursuit of objectivity, translated into specific norms such as replicability of experiments, right use of statistics, coherence of statements, and so forth. This has been much discussed in the extensive literature in social studies of science. But references to the role of history of psychology in critique, and specifically references to "critical psychology," implicate values in further ways. Explaining the phrase "critical psychology," Thomas Teo (2005: 2) wrote: "in its generic meaning, it would suggest a perspective that does not identify itself with the field but looks from a historical and theoretical distance at the development of psychology." That is, it not only examines the values at work in scientific practices as science but uses history to question the political, social, ethical, religious, or aesthetic values at work in the field. Put another way, history exposes to view the plot of the social stories in which psychologists participate. Studies of the use of normative concepts in psychology (such as self-esteem, discussed in this volume, or personality, or depression) are very characteristic of this kind of critical work. Choice about which "hi-stories" to write in its turn becomes an evaluative practice, an important dimension of activity in the history of psychology. To celebrate the anniversary of a laboratory is one kind of politics; to discuss the use of racial typologies in psychology is quite another.

At once stage, a Berlin-based group of psychologists went beyond these "liberal" conceptions of critique and embraced "critical psychology" as a Marxist praxis. This group explored the interpretation of psychological knowledge and theory as the scientific-cultural outcome of a certain stage of history, and in the light of this set out to separate a true science based on knowledge of its conditions of production from science constructed as if reason engaged nature without mediation (van Ijzendoorn, and van der Veer 1984; Schönplug 2021; Teo 2013). Understanding this kind of argument is also necessary for incorporating the history of Soviet psychology into histories of the field (Proctor 2020; Sirotkina and Smith 2012; Todes 2014; Yasnitsky 2020). The challenge of the Berlin group to "liberal" conceptions of objectivity in psychology was perpetuated in the longer term in the practice of "historical psychology," the attempt to relate psychological states to historical conditions, and in the turn to recognize "indigenous psychologies" as genuine alternatives to western formations (Staeuble 1991, 2006). This linked up with the aforementioned anthropological, psychosociological, and historical tradition of examining mental states (and, more recently, embodied mental states) as historical processes.

Gaining perspective, or critically examining the conditions of knowledge production and use, requires some appreciation of notions of reflection and/or reflexivity. Humans are at one and the same time biological organisms and reflective subjects creating knowledge and practices. For this reason psychologies belong in both natural science and human science, and the question of the relation between the two kinds of knowledge is fundamental in discussion of the identities of psychologies. The field encompasses the person as biological phenomenon; at the same time the field encompasses the person as a socially and historical individual representation of the phenomenon. Emotion, for instance, is most certainly embodied and requires a brain, but it is embodied in the posture and gait and habits of whole people acting within a framework of social rules, customs, and language. Knowledge requires both biology and history. Moreover (as discussed above), psychological categories themselves have a history. Fernando Vidal (2015) made the incisive observation that the scientists who advocate the development of fields like neuroethics and neurohistory – and we might also add neuropsychology – which advocate explanation in terms of neuronal relations, actually take their categories, the descriptive language for their subject matter (ethics, history, or psychology) from fields developed in the humanities or their predecessors ("the science of the soul," "the moral sciences," and "the science of man" in the case of psychology). Human culture established discourse about ethics, history, and psychology, and then created communities of people who argued that these categories should be analyzed in neuro-terms. This is critical knowledge in contemporary circumstances where, to put it bluntly, the neuro-disciplines command the cash.

The notions of reflection and reflexivity require further comment. (Graham Richards 2010, introduced terminology, sometimes adopted, to assist this, distinguishing between the psychological states people have and the subject field they have created. Psychology, the institutionalized subject field is signaled by upper case "P"; the subject matter is signaled by lower case "p.") Descartes, to take the classic philosophical instance, undertook with his reason to reflect on the reasoning self, somewhat by analogy to the way a mirror reflects a ray of light back on itself (a form of physical reflection in which Descartes was closely interested). Kant denied that such reflection could reveal the true being of the self, and he therefore turned to transcendental reason and moral intuition to do this work. The philosophical idealists in turn reacted against Kant and reasserted the capacity of reflective reason to determine its ground in being (Tauber 2005). All this, along with other strands of culture, especially characteristic of Romanticism, embedded the notion of reflection in thought about the nature of the self (Seigel 2005; Hofman 2016; Smith 2019b). At several steps removed, this was the tradition reworked (he would have said overthrown) by Heidegger in the form of an irreducible claim about *Dasein*, or the Being or Presence that has being as its subject. A number of twentieth-century humanist philosophers, such as R. G. Collingwood and Ernst Cassirer, distinguished reflective philosophical awareness as the phenomenon differentiating the human from the nonhuman world. This made what was said about reflection central to the very wide debate about the possibility of maintaining "the human" as a distinct category. For those scholars who did uphold the distinctness of the category, human reflexive activity explained the difference between the study of the human world in the humanities disciplines and the study of the physical world in the natural science disciplines. For Collingwood (1961: 83–84; a passage explored in Smith 2007: 114–121): "If that which we come to understand better is something other than ourselves, for example, the chemical properties of matter, our improved understanding of it in no way improves the thing itself. If, on the other hand, that which we understand better is our own understanding, an improvement in that science is an improvement not only in its subject but its object also ... Hence the historical development to the science of human nature entails an historical development in human nature itself." The argument underwrote the claim that history was the key to human self-understanding, since it posited history as the process of human self-formation. As Hacking (2002: 39) commented, "there is a time-honored opinion that history matters to the very content of the human sciences, while it does not matter much to the natural sciences." History "matters" because "the human" is a historical formation.

While it may be debated whether it is possible to demarcate the human and the natural sciences in this way, the point for the present purpose is clear: there are reasons in the theory of knowledge to engage psychologies critically as historical formations. In part, this is easy to comprehend. All psychologists, like everyone else, build what they say and do on existing literature and practices; that is, they are historical actors. Making a discipline out of study of this historical action is therefore an essential part of science; it is epistemic "reflection."

Words confuse, since "reflection" also denotes introspection understood as a psychological topic. But the wider connotations of the word explain why the long debate about the viability of introspection as a method of inquiry into the subject matter of psychology, the method so repeatedly attacked by psychologists seeking natural science methods, has been at the heart of the field's self-understanding. This debate, for good reasons, has been the source of considerable diversity of view, as a number of studies (e.g., Wann 1964) comparing phenomenological and natural science approaches clearly illustrate.

The word "reflexivity" is often used interchangeably with "reflection." But it is a term with a number of different, if related, meanings, and has proved not especially helpful for analysis (in spite of an earlier attempt to make it central to the history of the human sciences: Smith 2007). "Reflexivity," in the sense relevant to psychology in which it does denote something more than "reflection," refers to the phenomenon in which the beliefs people hold about themselves are understood to be psychological states, with the consequence that if beliefs change, psychological states also change, and vice versa, so that changes in psychological experience change psychological knowledge. Hacking (1995a), considering concrete exemplifications of this phenomenon, called it "looping." His construction of the point drew on the distinction between natural kinds and human kinds mentioned earlier: a natural kind stays what it is whatever is thought about it, while a human kind changes with what is thought about it. Discussing this, Hacking made a claim central for the history of psychology: "In social phenomena we may generate kinds of people and kinds of action as we devise new classifications and categories. My point is that we 'make up people' in a stronger sense than we 'make up' the world" (Hacking 2002: 40). If indeed, in the sense that Hacking delineated, we "make up people," then the diversity of psychologies directly relates to the diverse historical conditions of their formation.

"Looping" effects are familiar and significant in daily life in which psychological knowledge changes a person's psychological activity. In psychotherapy, it is precisely the purpose of verbalized thought to change the nature of a subjectively held presumption embodied in activity. Something of the logic of "looping" can also be seen in E. G. Boring's much quoted aphorism (1963: 18), that intelligence is "as the tests test it." Though Boring's theme was a discussion of tests (he wrote in 1923) as stages on the way to scientific and objective definition of intelligence, awareness of reflexivity might lead to reading Boring's statement as saying that the identification of intelligence requires understanding of the socially and historically specific means psychologists use to describe intelligence. While psychologists went on to consolidate their references to intelligence as if it were a natural kind, Boring, if inadvertently, pointed out that it might be a human kind.

Social processes of "looping" tie the diverse psychologies logically, as well as factually, to the diverse conditions of their production. There are reciprocal relations between the content of psychological approaches and institutional, methodological, material, and ideological settings in which they develop. In the postwar years, the use of statistical methods in experimental practices encouraged the theory that humans are statistical reasoners (Gigerenzer 1992). On the national stage, scientists in the United States held up their understanding of their own reasoning with "an open mind," free, as they claimed, of ideology, as a model of rational cognition in general (Cohen-Cole 2005, 2014: for Cold War rationality, Review Symposium 2018). Other historical studies have discussed the material "reflexivity" of instrumentation, the reciprocal relations of the technical means employed in research and the content of the resulting knowledge claims. (For an introduction to the role of instruments, see Sturm and Ash 2005.) The history, in recent decades, of representations of mental functions in terms of brain scans illustrates the process. The demand to use brain scanners, once institutions had acquired such expensive equipment for research purposes, had large consequences for belief about the nature of mental functions. It reinforced belief in the propriety of thinking that the proper subject matter of psychology was the activity of mental functions and the identification of self and brain (Vidal and Ortega 2017). Other psychological technology, such as B. F. Skinner's once notorious "baby-box," teaching machines, and related behavior modification techniques have a history implicating circles of material and conceptual practice (Rutherford 2009, 2017).

Critical psychologists have certainly advocated reflexivity, in the sense that they have challenged their fellow professionals, individually and collectively, to be self-consciously reflective about what they themselves do, to use history and social science to look at their own activity, and thus to gain perspective on what they do in relation to political, ethical, epistemological, or other values (Capshew 2007; Morawski 1992, 2005, 2007, 2020). The development of professional ethical standards has been one long dialogue in this kind of reflection. In this connection, "reflexivity" denotes individual and collective self-inquiry, with the intention of regulating or changing something that psychologists do.

Many scientific psychologists, one imagines, would be sympathetic with some degree of reflection understood in this way, for instance, when used to reassess "classical" experiments in the light of new historical data (e.g., Brannigan 2021; Rodkey 2015). For a smaller number of psychologists, such reflection leads directly to questions about political context (e.g., Pickren 2018). Critical reflection, however, in the argument put forward here, implicates the argument that different forms of knowledge about human "nature" or about the "being" of being human change, change reflexively, with change in states we would now commonly call psychological. The attempt to intervene in these relations is indeed, like intervention in psychotherapy, the point of critique. Historical understanding is intrinsic to this activity.

Here we should also recognize the very great value of biography, the most concrete and empirical, and often enough the most literate, of all forms of historical writing with psychological relevance. Biography's capacity to make an integrated whole out of the life, social context, *and* thought of individuals in certain ways sets a standard for reflective practice (e.g., for John Dewey, Dalton 2002; for Melanie Klein, Grosskurth 1986; for Ivan Pavlov, Todes 2014).

Reference to biography leads on to yet another way to think about reflexive effects as they relate to psychology's diversity. This concerns the proposition that psychology's diversity is a necessary consequence of the diversity of the psychological character of psychologists themselves. Jung (1923: 627), for one, thought this: "The necessity of a plurality of explanations, however, in the case of a psychological theory is definitely granted, since, unlike any other natural science theory, the object of psychological process has to explain another." Jung's interest was the psychological explanation of diversity of belief through understanding the different individual and collective character types of people. Since he claimed knowledge of an empirically specifiable range of character types, types expressing the structure of the shared or collective psychic unconscious, he thought he could specify the psychological basis of different forms of knowledge. Countering Jung, one may envisage the argument that different forms of life, based on different ways of organizing production, different technologies, and different principles of belief, create different psychological states, or even new human "types," as well as different forms of psychological knowledge. Some futuristically oriented observers argue that this is currently taking place, with the creation of "transhuman" activity mediated by new digital and biomedical technologies. In the past, psychologists held that biological and social conditions presented objective circumstances, a kind of baseline, to which scientists could refer in order to discover the real, nonrelative causes of human "nature." Certainly, such belief supported, and still supports, psychologists in getting on with the empirical inquiries with which they were, and are, engaged. Yet, at the same time, everyone is well aware that new technologies are reconstructing the possibilities for "human nature." In this case, technology is adding further weight to the argument that psychologies are necessarily diverse: their subject matters change with time.

Conclusion

If the word "reflexivity" is overdetermined and has multiple meanings, this does not invalidate argument linking history of psychology, being human, and the diversity of psychologies. The form of knowledge thought appropriate for psychology depends on the way of life and the purposes of the way of life for which different people seek knowledge. Knowledge and ways of life do not interact but are constitutive parts of each other. The sciences of psychology, along with psychological practices, are dependent on the ways of life and purposes of the ways of life of which they are part, and sustaining ways of life is dependent on what is held to be true in knowledge about them. All knowledge is *for* something, something dependent on values and ways of life.

The history of psychology has become a speciality within psychology. The last two or three decades have seen the publication of many studies of particular theories, experiments, and practices in psychology viewed in contextual detail, aligning work with the research current in the history of science generally. Such work tends to create its own in-turned community, publishing its own version of "normal science" (to use T. S. Kuhn's term) in its own journals and holding its own meetings, though this has been counterbalanced by the need to teach psychology students and by the association of history of psychology with the values of a critical perspective. Because such research often speaks about the history of local communities, it may have considerable local impact. Locating the history of psychology in the history of the human sciences, rather than treating it solely as a speciality in psychology, it has been argued here, has the value of drawing out the larger questions, philosophical and political, which often remain implicit and unnoticed in narrowly focused studies. It also drags psychologists into transdisciplinary work.

The past and present and "western" and "indigenous" diversity of psychologies confronts psychologists and those who work in the human sciences with a challenge that they may sidestep only by staying firmly within the confines of a speciality. Understanding the diversity is not possible without history, and a turn to history rather than to natural science or to practice in the field of psychology is in its nature a critical step, even where the historian-author concentrates to the apparent exclusion of everything else on the empirical content of the story. Doing historical work recognizes, if only implicitly, the claims of the humanities disciplines to be a source of knowledge of "the human" along with the claims of the natural and social sciences.

It follows from the previous discussion that it depends on the purpose in view whether it is legitimate to refer to ancient Greek psychology, to Tibetan psychology, or to the psychology of Islamic civilization, though blanket and unthinking reference to such psychologies is not acceptable in scholarly work. Greek peripatetics, Tibetan monks, and Islamic scholars of earlier centuries, in this like many other people, did not shape their world with the category, psychology. This is surely relevant to current discussion around the world as to whether a person, to be a psychologist, needs to have adopted the practices of an international community dominated by European and even more North-American forms of knowledge, training, and professional organization. Can an untrained person be a psychologist? The answer obviously depends – and that dependency is most certainly central to what historians study. If there were one uniquely scientific psychology, integrated by theory, or even if there were a delimited set of scientific psychologies, professional psychologists would think it necessary to fall into line. But if it can be said that psychological states of being, along with knowledge about them, vary with time and place, with cultural tradition and context, then it can be argued that local knowledge and local practices may, or should, have a place in psychology. This will also permit recognition of the "psychological" knowledge of nonspecialists. The periphery may have a say. However, it is also an option, and a more radical one, to refuse to call local ways of thought or practices "psychological," and instead to promote them as *alternatives to* psychology. That, it scarcely needs to be said, involves strong political and cultural resistance to the ideological import of western psychological science and the natural scientific worldview of which it is part. The discussion of diversity goes some way to open up these issues.

This chapter has tried to set out a case for viewing actually existing diversity in psychologies as a positive condition. The case does not necessarily oppose projects to establish more integrated and comprehensive theory than now exists. The purpose of discussing psychology in the context of the broad field of the history of the human sciences is to provide analytic resources and perspective on the diverse activities, including the claims for united theory, which have existed and continue to exist. Historical studies of the diverse roots of the diverse practices categorized in diverse ways as psychology are central to this. History writing in general, and of course not only the history of psychology, is a major resource as a critical voice. There are disciplined reasons for this, since the examination of where we are now and why, or where earlier people were and why, cannot but construct knowledge with purposes other than those that lie behind the preferred narratives of specialist or dominant groups of people.

References

Alexander BK, Shelton CP (2014) A history of psychology in western civilization. Cambridge University Press, Cambridge

Anger S (2018) The Victorian mental sciences. Vic Lit Cult 46:275–287

Araujo S d F (2016) Wundt and the philosophical foundations of psychology: a reappraisal. Springer, Cham

Araujo S d F (2017) Toward a philosophical history of psychology: an alternative path for the future. Theory Psychol 27(1):87–107

Ash M, Sturm T (eds) (2007) Psychology's territories: historical and contemporary perspectives from different disciplines. Lawrence Erlbaum Associates, Mahwah

Baker DB (ed) (2012) The Oxford handbook of the history of psychology: global perspectives. Oxford University Press, Oxford

Baker DB, Benjamin L (2014) From séance to science: a history of the profession of psychology in America, 2nd edn. Akron University Press, Akron

Bateson G (2000) Steps to an ecology of mind. University of Chicago Press, Chicago

Bender A, Beller S (2016) Current perspectives on cognitive diversity. Front Psychol. https://doi.org/10.3389/fp-syg.206.00509

Billig M (2008) The hidden roots of critical psychology: understanding the impact of Locke, Shaftesbury and Reid. Sage, London

Billig M (2019) More examples, less theory: historical studies of writing psychology. Cambridge University Press, Cambridge

Boddice R, Smith M (2020) Emotion, sense, experience. Cambridge University Press, Cambridge, e-print

Bonea A, Dickson M, Shuttleworth S, Wallis J (eds) (2019) Anxious times: medicine and modernity in nineteenth-century Britain. Pittsburgh University Press, Pittsburgh

Bordogna F (2008) William James at the boundaries: philosophy, science, and the geography of knowledge. University of Chicago Press, Chicago

Boring EG (1963) Intelligence as the tests test it. In: Watson RI, Campbell DT (eds) History, psychology, and science: selected papers. Wiley, New York, pp 187–189

Brannigan A (2021) The archival turn in classical social psychology: some recent reports. Theory Psychol 31(1):138–146

Brinkmann S (2020) Psychology as a science of life. Theory Psychol 30(1):3–17

Brock AC (ed) (2006) Internationalizing the history of psychology. New York University Press, New York

Byford A (2014) The mental test as a boundary object in early-20th-century Russian child science. Hist Hum Sci 27(4):22–58

Byford A (2020) Science of the child in late Imperial and early Soviet Russia. Oxford University Press, Oxford

Capshew JH (2007) Reflexivity revisited. In: Ash MG, Sturm T (eds) Psychologies territories: historical and contemporary perspectives from different disciplines. Lawrence Erlbaum Associates, Mahwah, pp 343–356

Carhart MC (2007) The science of culture in enlightenment Germany. Harvard University Press, Cambridge, MA

Carroy J (1993) Double and multiple personality: between science and fiction. Presses Universitaires de France, Paris

Carroy J, Ohayon A, Plas R (2006) History of psychology in France: 19th to 20th centuries. La Découverte, Paris

Churchland PM (1989) Eliminative materialism and the propositional attitudes. In: A neurocomputational perspective: the nature of mind and the structure of science. MIT Press, Cambridge, MA, pp 1–22

Cohen-Cole J (2005) The reflexivity of cognitive science: the scientist as model of human nature. Hist Hum Sci 18(4):107–139

Cohen-Cole J (2014) The open mind: cold war politics and the science of human nature. University of Chicago Press, Chicago

Collingwood RG (1961) The idea of history. Oxford University Press, Oxford

Crabtree A (2019) 1784: the marquis de Puységur and the psychological turn in the West. J Hist Behav Sci 55(3):199–215

Dalton TC (2002) Becoming John Dewey: dilemmas of a philosopher and naturalist. Indiana University Press, Bloomington

Danziger K (1990) Constructing the subject: historical origins of psychological research. Cambridge University Press, Cambridge

Danziger K (1994) Does the history of psychology have a future? Theory Psychol 4:467–484

Danziger K (1997a) The historical formation of selves. In: Ashmore RD, Jussim L (eds) Self and identity: fundamental issues. Oxford University Press, New York, pp 137–159

Danziger K (1997b) Naming the mind: how psychology found its language. Sage, London

Danziger K (2004) Concluding comments. In: Brock AC, Louw J, van Hoorn W (eds) Rediscovering the history of psychology: essays inspired by the work of Kurt Danziger. Kluwer Academic, Plenum, New York, pp 207–231

Danziger K (2008) Marking the mind: a history of memory. Cambridge University Press, Cambridge

Daston L (ed) (2000) Biographies of scientific objects. University of Chicago Press, Chicago

Diriwächter R (2004) Völkerpsychologie: the synthesis that never was. Cult Psychol 10(1):85–109

Draaisma D (2000) Metaphors of memory: a history of ideas about the mind (trans: Vincent P). Cambridge University Press, Cambridge

Ellenberger H (1970) The discovery of the unconscious: the history and evolution of dynamic psychiatry. Allen Lane, Penguin Press, London

Frede M (2011) A free will: origins of the notion in ancient thought (ed: Long AA). University of California Press, Berkeley

Galison P, Stump DJ (eds) (1996) The disunity of science: boundaries, contexts and power. Stanford University Press, Stanford

Gigerenzer G (1992) Discovery in cognitive psychology: new tools inspire new theories. Sci Context 5:329–350

Goodey CF (2011) A history of intelligence and "intellectual disability": the shaping of psychology in early modern Europe. Ashgate, Farnham

Goodey GF (2021) Development: the history of a psychological concept. Cambridge University Press, Cambridge

Greenwood JD (2015) A conceptual history of psychology: exploring the tangled web, 2nd edn. Cambridge University Press, Cambridge

Grosskurth P (1986) Melanie Klein: her world and her work. Harvard University Press, Cambridge, MA

Guldi J, Armitage D (2014) The history manifesto. Cambridge University Press, Cambridge

Gundlach H (2006) Psychology as science and as discipline. Physis 43:61–90

Gundlach HUK (2012) Germany. In: Baker DB (ed) The Oxford handbook of the history of psychology: global perspectives. Oxford University Press, Oxford, pp 255–288

Hacking I (1995a) The looping effects of human kinds. In: Sperber D, Premack D, Premack AJ (eds) Causal cognition: a multidisciplinary debate. Clarendon Press, Oxford, pp 351–394

Hacking I (1995b) Rewriting the soul: multiple personality and the sciences of memory. Princeton University Press, Princeton

Hacking I (1999) Mind travellers: reflections on the reality of transient mental illness. Free Association Books, London

Hacking I (2002) Historical ontology. Harvard University Press, Cambridge, MA

Hacking I (2007) Kinds of people: moving targets. Proceedings of the British Academy 151:285–318

Hatfield G (1992) Descartes' physiology and its relation to his psychology. In: Cottingham J (ed) The Cambridge companion to Descartes. Cambridge University Press, Cambridge, pp 335–370

Hatfield G (1995) Remaking the science of mind: psychology as natural science. In: Fox C, Porter R, Wokler R (eds) Inventing human science: eighteenth-century domains. University of California Press, Berkeley, pp 184–231

Hatfield G (1997) Wundt and psychology as science: disciplinary transformations. Perspect Sci 5: 349–382

Hayward R (2007) Resisting history: religious transcendence and the invention of the unconscious. Manchester University Press, Manchester

Hearnshaw LS (1987) The shaping of modern psychology. Routledge & Kegan Paul, London

Henley TB (2020) On prehistoric psychology: reflections at the invitation of Göbekli Tepe. Hist Psychol 23(3):211–219

Hofman E (2016) How to do the history of the self. Hist Hum Sci 29(3):8–24

van Ijzendoorn MH, van der Veer R (1984) Main currents of critical psychology: Vygotskij, Holzkamp, Riegel (trans: Schoen M). Irvington, New York

Jahoda G (2007) A history of social psychology: from the eighteenth-century enlightenment to the second world war. Cambridge University Press, Cambridge

James W (1950) The principles of psychology. Dover reprint, New York

Janz J, van Drunen P (2004) A social history of psychology. Blackwell, Malden

Jardine N (2000) Uses and abuses of anachronism in the history of the sciences. Hist Sci 38:251–270

Jung CG (1923) Psychological types or the psychology of individuation (trans: Baynes HG). Routledge & Kegan Paul, London

Jung CG (2019) History of modern psychology: lectures delivered at ETH Zurich, volume 1 1933–1934 (ed: Falzeder E; trans: Kyburz E, Peck J, Falzeder E). Princeton University Press, Princeton

Kemp S (1990) Medieval psychology. Greenwood Press, New York

Kemp S (2018) Quantification of virtue in late medieval Europe. Hist Psychol 21(1):33–46

Kitayama S, Cohen D (eds) (2007) Handbook of cultural psychology. Guilford Press, New York

Klempe SH (2014) Kierkegaard and the rise of modern psychology. Transaction, New Brunswick

Klempe SH (2020) Tracing the emergence of psychology 1520–1750: a sophisticated intruder to philosophy. Springer, Cham

Kusch M (1997) The sociophilosophy of folk psychology. Stud Hist Phil Sci 28(1):1–25

Kusch M (1999) Psychological knowledge: a social history and philosophy. Routledge, London

Lanzoni S (2018) Empathy: a history. Yale University Press, New Haven

Lloyd GER (2007) Cognitive variations: reflections on the unity and diversity of the human mind. Oxford University Press, Oxford

Makari G (2010) Revolution in mind: the creation of psychoanalysis. Duckworth Overlook, London

Mckay F (2019) Equanimity: the somatization of a moral sentiment from the 18th to the late 20th century. J Hist Behav Sci 55(4):281–298

Mengal P (2005) The birth of psychology. L'Harmattan, Paris

Morawski JG (1992) Self-regard and other-regard: reflexive practices in American psychology, 1890–1940. Sci Context 5:281–308

Morawski JG (2005) Reflexivity and the psychologist. Hist Hum Sci 18(4):77–105

Morawski JG (2007) Scientific selves: discerning the subject and the experimenter, as they effect the experimental situation. In: Ash MG, Sturm T (eds) Psychologies territories: historical and contemporary perspectives from different disciplines. Lawrence Erlbaum Associates, Mahwah, pp 129–148

Morawski JG (2020) Psychologists' psychologies of psychologists in a time of crisis. Hist Psychol 23(2):176–198

Mülberger A (2012) Wundt contested: the first crisis declaration in psychology. Stud Hist Philos Sci Part C 43(2):434–444

Mülberger A, Sturm T (eds) (2012) Psychology: a science in crisis? A century of reflection and debate. Special section. Stud Hist Philos Sci Part C 43(2)

Mülberger A, Balltondre M, Graus A (2014) Aims of teachers' psychometry: intelligence testing in Barcelona (1920). In: Mental testing after 1905: uses in different local contexts (ed: Mülberger A). Hist Psychol 17(3):206–222

Mülberger A, Gomez C, Cervantes M, Cañas AM, Anglada L (2019) Testing the intelligence of Barcelona's schoolchildren in 1908. Revista de Historia de la Psicología 40(1):2–11

Osbeck LM (2019) Values in psychological science: re-imagining epistemic priorities at a new frontier. Cambridge University Press, Cambridge

Parot F (1994) The banishment of spirits: the birth of an institutional border between spiritualism. In: Les territoires de la psychologie (ed: Parot F). Revue de synthèse 4th series 115(3–4):417–443

Petitt M, Young JL (eds) (2017) Psychology and its publics. Special issue. Hist Hum Sci 30(4)

Pickren WE (ed) (2018) Psychology in the social imaginary of neoliberalism. Special issue. Theory Psychol 28(5)

Pickren WE (ed) (2021) The Oxford research encyclopedia of the history of psychology. Oxford University Press, New York

Pickren WE, Rutherford A (2010) A history of modern psychology in context. John Wiley, Holden

Pind JL (2016) The psychologist as a poet: Kierkegaard and psychology in nineteenth-century Copenhagen. Hist Psychol 19(4):352–370

Pizarroso N (2013) Mind's historicity: its hidden history. Hist Psychol 16(1):72–90

Pizarroso López N (2018) Ignace Meyerson. Les Belles lettres, Paris

Plas R (2000) Birth of a human science: psychology, the psychologists and "psychic marvels". Presses Universitaires de Rennes, Rennes

Proctor H (2020) Psychologies in revolution: Alexander Luria's 'romantic science' and Soviet social history. Palgrave Macmillan, Cham

Proietto M, Lombardo GP (2015) The "crisis" of psychology between fragmentation and integration: the Italian case. Theory Psychol 25(3):313–327

Review Symposium (2018) Cold war rationality. Hist Hum Sci 31(3)

Richards G (1997) 'Race', racism and psychology: towards a reflexive history. Routledge, London

Richards G (2010) Putting psychology in its place: critical historical perspectives, 3rd edn. Routledge, London

Rodkey EN (2015) The visual cliff's forgotten menagerie: rats, goats, babies, and myth-making in the history of psychology. J Hist Behav Sci 51(2):113–140

Rose N (1985) The psychological complex: social regulation and the psychology of the individual. Routledge & Kegan Paul, London

Rose N (1998) Inventing our selves: psychology, power, and personhood. Cambridge University Press, Cambridge

Rose N (1999) Governing the soul: the shaping of the private self, 2nd edn. Free Association Books, London

Rutherford A (2009) Beyond the box: B. F. Skinner's technology of behavior from laboratory to life, 1950–1970s. Toronto University Press, Toronto

Rutherford A (2017) B. F. Skinner: technology's nation' Hist Psychol 20(3):290–312

Schönplug W (2021) Beyond narratives: German critical psychology revisited. Hist Psychol 24. https://doi.org/10.1037/hop0000183

Seigel J (2005) The idea of the self: thought and experience in Western Europe since the seventeenth century. Cambridge University Press, Cambridge

Shamdasani S (2003) Jung and the making of modern psychology: the dream of a science. Cambridge University Press, Cambridge

Shamdasani S (2004) Psychologies as ontology-making practices: William James and the pluralities of psychological experience. In: Carrette J (ed) William James and 'the varieties of religious experience'. Routledge, London, pp 27–44

Shuttleworth S (1984) George Eliot and nineteenth-century science: the make-believe of a beginning. Cambridge University Press, Cambridge

Shuttleworth S (1996) Charlotte Brontë and Victorian psychology. Cambridge University Press, Cambridge

Shuttleworth S (2010) The mind of the child: child development in literature, science, and medicine, 1840–1900. Oxford University Press, Oxford

Sirotkina I, Smith R (2012) Russian Federation. In: Baker DB (ed) The Oxford handbook of the history of psychology: global perspectives. Oxford University Press, Oxford, pp 412–441

Smith R (1992) Inhibition: history and meaning in the sciences of mind and brain. Free Association Books, London

Smith R (2004) The physiology of the will: mind, body, and psychology in the periodical literature, 1855–1875. In: Cantor G, Shuttleworth S (eds) Science serialized: representations of the sciences in nineteenth-century periodicals. MIT Press, Cambridge, MA, pp 81–110

Smith R (2007) Being human: historical knowledge and the creation of human nature. Manchester University Press, Manchester

Smith R (2013a) Between mind and nature: a history of psychology. Reaktion Books, London

Smith R (2013b) Free will and the human sciences in Britain, 1870–1910. Pickering & Chatto, London

Smith R (2016) History of psychology: what for? In: Klempe SH, Smith R (eds) Centrality of history for theory construction in psychology, Annals of theoretical psychology, vol 14. Springer, Cham, pp 47–73

Smith R (2019a) The sense of movement: an intellectual history. Process Press, London

Smith R (2019b) Individuality, the self and concepts of mind. In: Claeys G (ed) The Cambridge companion to nineteenth-century thought. Cambridge University Press, Cambridge, pp 141–162

Smith R (2020a) Inhibition and metaphor of top-down organization. Stud Hist Philos Sci Part C 83. https://doi.org/10.1016/j.shpsc.2020.101253

Smith R (2020b) Kinaesthesia and a feeling for relations. Rev Gen Psychol 24(4):355–368. https://doi.org/10.1177/1089268020930193

Sorabji R (2006) Self: ancient and modern insights about individuality, life, and death. Clarendon Press, Oxford

Staeuble I (1991) "Psychological man" and human subjectivity in historical perspective. Hist Hum Sci 4:417–432

Staeuble I (2006) Decentering western perspectives: psychology and the disciplinary order in the first and third world. In: Brock AC, Louw J, van Hoorn W (eds) Rediscovering the history of psychology: essays inspired by the work of Kurt Danziger. Kluwer, New York, pp 183–205

Sternberg RJ, Pickren WE (eds) (2019) The Cambridge handbook of the intellectual history of psychology. Cambridge University Press, Cambridge

Sturm T, Ash MG (2005) Roles of instruments in psychological research. In: The roles of instruments in psychological research (ed: Sturm T, Ash MG), special issue. Hist Psychol 8(1):3–34

Tauber AI (2005) The reflexive project: reconstructing the moral agent. Hist Hum Sci 18(4):49–75

Taylor E (1996) William James on consciousness beyond the margin. Princeton University Press, Princeton

Taylor JB, Shuttleworth S (eds) (1998) Embodied selves: an anthology of psychological texts 1830–1890. Clarendon Press, Oxford

Teo T (2005) The critique of psychology: from Kant to postcolonial theory. Springer, New York

Teo T (2013) Backlash against American psychology: an indigenous reconstruction of the history of German critical psychology. Hist Psychol 16(1):1–18

Thomson M (2001) The popular, the practical and the professional: psychological identities in Britain, 1901–1950. In: Bunn GC, Lovie AD, Richards GD (eds) Psychology in Britain: historical essays and personal reflections. British Psychological Society, Leicester, pp 115–132

Thomson M (2006) Psychological subjects: identity, culture, and health in twentieth-century Britain. Oxford University Press, Oxford

Todes DP (2014) Ivan Pavlov: a Russian life in science. Oxford University Press, Oxford

Toomela A (2016) Six meanings of the history of science: the case of psychology. In: Klempe SH, Smith R (eds) Centrality of history for theory construction in psychology, Annals of theoretical psychology, vol 14. Springer, Cham, pp 47–73

Valsiner J (2012) A guided science: history of psychology in the mirror of its making. Transaction, New Brunswick

Vidal F (2011) The sciences of the soul: the early modern origins of psychology (trans: Brown S). University of Chicago Press, Chicago

Vidal F (2015) Are we heading towards a brain-centered culture? An interview to Fernando Vidal, conducted by Csíri P, Mantilla J. Culturas Psi/Psy Cultures 5:5–12

Vidal F, Ortega F (2017) Being brains: making the cerebral subject. Fordham University Press, New York

Wann TW (ed) (1964) Behaviorism and phenomenology: contrasting roots for modern psychology. University of Chicago Press for William Marsh Rice University, Chicago

Weidman N (2016) Overcoming our mutual isolation: how historians and psychologists can work together. Hist Psychol 19(3):248–253

van Wyhe J (2004) Phrenology and the origins of Victorian scientific naturalism. Ashgate, Aldershot

Yasnitsky A (ed) (2020) A history of Marxist psychology: the golden age of Soviet science. Routledge, Abingdon

Young RM (1970) Mind, brain, and adaptation in the nineteenth century: cerebral localization and its biological context from Gall to Ferrier. Clarendon Press, Oxford

Zammito JH (2002) Kant, Herder, and the birth of anthropology. University of Chicago Press, Chicago

Social Psychology: Exemplary Interdiscipline or Subdiscipline

39

James M. M. Good

Contents

Introduction: The Disciplinary Status of Social Psychology	1006
Social Psychology, Its Objects, and Its Parent Disciplines	1008
Social Psychology's Dual Disciplinary Heritage	1011
A Dialectical Tension Between the Social Psychologies and Their Parent Disciplines	1017
The Quest for an Interdisciplinary Social Psychology	1019
The Post–Second World War Development of European Social Psychology	1020
Conclusion: Whether the Disciplines of Social Psychology?	1025
References	1027

Abstract

This chapter explores the history and disciplinary status of twentieth-century social psychology. In keeping with the theoretical commitments of this Handbook, the analysis is framed within a history of the human sciences perspective grounded in the mutual development of the human science disciplines. It focuses on whether the history of social psychology should be seen as that of a subdiscipline of psychology or of sociology, or, that of an exemplary interdiscipline. Following a preliminary consideration of the disciplinary status of social psychology, social psychology's objects of study, its relations with its parent disciplines, and its units of analysis are discussed. The central section of the chapter explores social psychology's dual disciplinary heritage. A dialectical tension between the disparate traditions of social psychology and their parent disciplines is noted. The intermittent quest for an interdisciplinary psychology is then reviewed. Attention is given to the post–Second World War development of European social psychology. A final section reflects on the significance of the diversity of social psychological traditions, concluding that over the twentieth century, the boundaries of these disparate traditions can be seen as fluid,

J. M. M. Good (✉)
Department of Psychology, Durham University, Durham, UK
e-mail: j.m.m.good@durham.ac.uk

contingent, local, and contestable, reflecting the thematic preoccupations, disciplinary origins, and meta-theoretical commitments of social psychologists, of the parent disciplines, and of those who represent disciplinary practices.

Keywords

History of social psychology · Disciplining of social psychology · Interdisciplinary social psychology · Interdisciplinarity · Social psychology's dual heritage history

Introduction: The Disciplinary Status of Social Psychology

While the academic field of social psychology is clearly present in the division of labor that currently is to be found in the human sciences, what exactly is the disciplinary domain of social psychology? Is it a branch or subdiscipline of psychology or of sociology? Or should it, perhaps, be seen as an exemplary interdiscipline, drawing theoretical and methodological inspiration from both its parent disciplines, psychology, and sociology. If seen as a branch of psychology, what is its relationship to sociology and to other human science disciplines such as anthropology? As will be discussed below, a credible case can also be made for social psychology as a historically legitimate subdiscipline of sociology. In light of this potentially contestable and dual heritage, social psychology should perhaps be seen as having two major candidate parent disciplines – psychology and sociology.

For any aspiring historian of social psychology, the above passage poses a procedural dilemma. Indeed, over 20 years ago, Rob Farr had observed that an adequate history of social psychology must also, in part, be a history of sociology and of psychology (Farr 1996). This has proved to be a prescient remark. Recent historiographical writing in the social sciences has emphasized the need for cross-disciplinary histories or at least histories of social sciences that take into account the histories of related fields (Backhouse and Fontaine 2010 2014; Fontaine 2015), a view that had already been enshrined as a central notion in the history of the human sciences project (Smith 1997).

The aim of this chapter is to examine historically one aspect of such a project, the disciplinary status of social psychology as a subdiscipline of psychology or sociology or as an exemplary interdiscipline. The analysis begins with a discussion of what has been called the disciplining of social psychology (Good 2000). After a review of the changing and diverse representations within social psychology of the "psychological," the "social," and their interfacing, there follows a dual-heritage history of the parallel development of sociological and psychological social psychology. The chapter will take note, especially, of the impact of both the Second World War and of subsequent Cold War issues on the development of social psychology. As has been widely noted, the Second World War provided many opportunities for interdisciplinary projects involving sociologists and psychologists. Indeed, the American sociologist William Sewell had described the period from roughly 1939 to the mid-1960s as the "golden age of interdisciplinary social psychology" (Sewell 1989). Such attempts to foster a more interdisciplinary social psychology are reviewed. Sewell also noted, however, a decline of interest in the possibility of developing such an interdisciplinary field in the last third of the twentieth century, citing social psychology's weak position in the university structure and inadequate funding from university and federal sources as possible causes. The subsequent advent of a more cognitively-oriented social psychology led to what Carl Graumann described as an "individualization of the social" and a "desocialization of the individual" (Graumann 1986) – a further impoverishment of the nature of the "social" in social psychology (Greenwood 2004b, 2014). One of the ambitions of post-war European social psychology in the late 1960s and 1970s had been to reinstate a more "social" social psychology. It is noted that one of the outcomes of the Cold War was the fact that many of these post-war initiatives promoted experimental forms of social psychology to the detriment of observational and more qualitative modes of enquiry. The involvement of eminent American social psychologists in these activities has often been referred to as the "Americanization" of European social psychology. For some such as Pieter van Strien (van Strien 1997) this engendered aspects of "colonization," or at least "neo-colonization"; for others this represented a genuine attempt to help reconstruct and develop a distinctively European social psychology (Markova 2012).

A definition of social psychology influential in the United Kingdom suggested that "social psychology is...a discipline which aims at an *integration* of the *psychological functioning of individuals* with the *social settings, small and large,* in which this functioning takes place" (Tajfel and Fraser 1978, p. 17, emphasis added). Thus, insofar as the disciplines of psychology and sociology provided repositories of knowledge about and methodologies for the investigation of the psychological functioning of the individual and social settings, respectively, then the task of social psychologists would seem to require at least two sets of disciplinary resources. Indeed, this might suggest that social psychology was an interdisciplinary field, a view endorsed by Hans Gerth and C. Wright Mills when they wrote that "one of the special obligations of the social psychologist...[is] to bridge the *departmentalized gap* which unfortunately separates the sociological and psychological approaches" (Gerth and Mills 1954, p. xviii, emphasis added).

The view that an adequate social psychology needed to bring together expertise about psychological functioning and social settings did not go unchallenged. From rather different theoretical perspectives – the social behaviorism of George Herbert Mead and social constructionism respectively – Rob Farr (1978) and Kenneth Gergen (2008) argued that what was needed was a more robust conception of the social which embraced the idea that social factors were in fact constitutive of mind, i.e., incorporating mind within a monistic social ontology.

Such views serve as a reminder that social psychology as a human science had also embodied one of the central features of the legacy of Cartesian philosophy – an individualistic representational theory of mind. It had thus wrestled with a variety of dualisms that have dominated the intellectual life of the human sciences – mind/body, individual/society, subject/object, organism/environment, knowledge/action, fact/value. The impact of Cartesian dualism and methodological individualism on the development of the social sciences has been discussed by Manicas (1987, Chap. 14). Brian Fay has also noted the perils of "pernicious" dualistic thinking in the social sciences (Fay 1996, Chap. 11).

The author has previously explored the links between the disciplines of sociology and psychology and the changing status of social psychology since the early 1930s (Good 1981; Good and Still 1992). In a contribution to a special issue of the *Journal of the History of the Behavioral Sciences* published in 2000, he concluded:

> From such reflections emerges a variety of social psychological disciplines – not just psychological or sociological traditions. Thus both modernist and postmodernist tendencies in social psychology embodied different practices and reflected different styles of social psychological reasoning, each carrying distinctive evidential requirements and criteria of objectivity. The tasks of social psychology were clearly diverse and there were thus many "messages of social psychology." (Good 2000, p. 398)
>
> Accordingly, in this chapter the author uses the term "social psychologies" to refer the variety of traditions of social psychology differentiated in terms of disciplinary affiliation (sociological or psychological) or metatheoretical commitments.

This chapter is, in part, an exercise in what has been called "participant history" where dual historiographic and ethnographic perspectives are required (Meier 1999). This is so because the author is attempting to provide account of events, some of which he has been a part of and to which he has contributed. Thus, there is an "ethnographic" dimension to the research reported here insofar as the author has been both "participant" and "observer."

The chapter also draws upon a mutualist approach to historiography. Mutualism is a perspective which has its source in Hegel and in the anti-mechanistic philosophy of Romanticism (Still and Good 1992, 1998). From this perspective mind, body and environment cannot be understood in isolation but are constructions from the flow of purposive activity in the world. In mutualist terms, there can be no certain foundations, no privileged points of observation. Histories are always positioned and positioning, written from some standpoint (Good and Roberts 1993; Rouse 1987; Jenkins 1991).

Social Psychology, Its Objects, and Its Parent Disciplines

Although the early links between the embryonic disciplines of sociology and psychology were very close, with many scholars contributing to the development of both, e.g., Wilhelm Wundt, Georg Simmel, William James, John Dewey, and George Herbert Mead, over the twentieth century the two disciplines had grown apart (Good 1981). Despite this there remained a clear "psychological" dimension in modern sociological theory, not least in terms of a continuing preoccupation with issues concerning human agency and the links between "micro" and "macro" analysis. As Alan Dawe observed, "Here then, is the problematic around which the entire history of sociological analysis could be written: the problematic of human agency. And it is this which gives social action its status as the single most central concept in sociology" (Dawe 1978). The Borgatta and Borgatta (1992) *Encyclopedia of Sociology* listed some 28 entries on psychological topics.

It is not just that within sociology we can find social psychological work but rather that American sociology, certainly during the first three decades of the twentieth century, embodied what has been called a "psychic" approach focusing on the "interests, purposes, needs and activities of human beings... [and] the modes of interaction and group process which they develop" (Fine 1979, pp. 108–109). This approach is to be found in the writings of such major figures as Lester Ward, Albion Small, Franklin Giddings, and John Dewey as well as in the more obviously social psychological writings of Charles Ellwood, Charles Horton Cooley, Edward Ross, William Thomas, and Florian Znaniecki. The "psychic" perspective drew upon sources as diverse as American religious thought, elements of idealism and romanticism, European social psychology (including folk psychology), early functional psychology, and pragmatism.

The manifestations of the "social" in psychology are much less clear. The clearest links are indeed to be found within social psychology (to be discussed in a later section), despite the reservations about the adequacy of the social expressed above. Nonetheless there were a number of twentieth-century attempts to promote an approach to psychology as a discipline that was thoroughly social (e.g., Brown 1936; Edwards 1997; Farr 1996, Harré 1992a, b; Harré and Secord (1972); Himmelweit and Gaskell 1990; Margolis et al. 1986; Sarason 1981; Gergen 2008).

Furthermore, there were a number of topics of common interest to both sociologists and psychologists: self and identity, emotion, social interaction, sex and gender, interpersonal relationships, groups and crowds are all phenomena which could be seen as functioning as "boundary objects." Boundary objects are objects that are "plastic enough to adapt to local needs and the constraints of the several parties employing them, yet robust enough to maintain a common identity" (Star and Griesemer 1989, p. 393). A recent symposium on the bridging of the two social psychologies was organized in terms of a sociological social psychologist and a psychological social psychologist reflecting on the approach of their respective domains to a series of bridging objects such as identity, cognition, emotion, gender, inequality, and culture (Eagly and Fine 2010).

Many of the disciplinary debates over the past century have concerned social psychology's "objects of study" as revealed, for example, in changing conceptions of the "psychological," of the "social," and of the "interfacing" of the "social" and the "psychological" (Good 2000).

Since the end of the Second World War, psychological approaches to social psychology have had an increasingly cognitive flavor. A distinctive feature of post-war social psychology was the work on cognitive consistency (Festinger 1957; Heider 1958; Newcomb 1953), followed closely by the development of the attribution theories of Kelley (1967) and Jones and Davis (1965). The 1970s and 1980s saw a "marriage" of social and cognitive psychology in which the concepts and procedures of cognitive psychology were applied to the domain of the social. Indeed it was observed that "Social psychology and cognitive social psychology are today nearly synonymous" (Markus and Zajonc 1985, p. 137). A cognitive appropriation of the psychological took place, not just in social psychology, but in the parent disciplines as well (Gardner 1985). Although motivation, emotion, and context were for some time seriously neglected, Schwarz (1998) discerned a "warmer and more social view" within cognitive social psychology and Fiske (1992) belatedly acknowledged that "thinking might be for doing." The claim that "the cognitive approach is now clearly the dominant approach among social psychologists, having virtually no competitors" (Markus and Zajonc 1985 p. 137) had been vitiated, however, by the development of ecological, socio-cultural, discursive, and Q-methodological approaches to cognition, all of which rejected an individualistic, mental representational theory of mind.

John Greenwood (2000) identified and lamented the loss of the social conception of cognition, emotion, and behavior held by early American social psychologists. In the 1970s and 1980s, the psychological and sociological traditions of the social psychologies were viewed by some commentators as insufficiently "social." For others, too little attention had been paid to the "emergent" (social) products of interaction – relationships, groups, and crowds and too much attention paid to the (individual) attributes of attitudes, attributions, schemas, conversation analysis, and the minutiae of social interaction.

Yet as Greenwood has also pointed out (1997, 2000, 2004a, 2019), all too often the "mark" of the social had been none too clearly specified. The range of meanings of the term social was broad, encompassing the effects of other people on thought, feeling, and behavior, as in Allport's (1954) Handbook definition, discussed in Lubek and Apfelbaum (2000), mutual influence or interaction (Newcomb 1950), "emergent" products such as roles and norms (Sherif 1936), shared identification (for two contrasted views see Brown (1988) on the definition of a group, and Gilbert (1989, 1997) on "plural subjects" as the mark of the social), and shared knowledge – reflected especially in the study of social representations (Farr and Moscovici 1984), discourse (Potter and Wetherell 1987), and Q-methodology (Stainton Rogers 1995; Stephenson 1953).

Earlier in this chapter, the need for social psychology to harness the resources of psychology and sociology in the articulation of appropriate conceptions of the psychological and the social was noted. A further key task in the history of social psychology has been the linking of the "individual" and the "social." At issue here is the fundamental question of the unit of analysis in social psychology. Throughout much of the history of social psychology the balance (in psychological traditions of social psychology at least) had tilted in the direction of the individual as the unit of analysis, as in the G. Allport's (1954) definition cited earlier. Nonetheless, there had been several attempts to formulate a unit of analysis which more effectively linked the "individual" to a variety of properties of the "social" as identified above. Coutu's (1949) notion of "tinsit" (tendency in the situation); Lewin's (1951) field theory; Kantor's (1959) interbehavioral field; Newcomb's (1950) focus on interaction; Krech, Crutchfield, and Ballachy's (1962) interpersonal behavior event, and Jackson's "social act" (Jackson 1988) provide just a few examples.

Vygotskian-inspired socio-cultural theory had drawn attention to the role of culture in the constitution of mind. In 1981, Mike Cole noted that

> [W]e cannot avoid the centrality of culturally organized experience in the operation not to say the definition of human psychological processes. If culture and individual functioning are inextricably woven....new kinds of observations and new theories that are neither psychological nor anthropological in currently understood senses of those terms must be created. (Cole 1981 p. 334)

Subsequent socio-cultural research focused on "activity" as the unit of analysis, thus uniting persons, situations, and action (Lave 1988; Wertsch et al. 1995). Jan Valsiner, a key figure at the center of this socio-cultural tradition, had shown how culture both reflects and shapes psychological processes (Valsiner 2014). The importance of the notion of culture for the development of social psychology had been much earlier promoted by Gustav Jahoda (Jahoda 1984, 2007, 2016; Valsiner 2018).

The ecological approach to cognition utilized the Gibsonian notion of "affordance," a relational construct linking a suitably equipped (i.e., attuned) perceiver to the environment. This had been a key notion in the attempt to develop an ecological approach to social psychology (Charles 2011; Good 2007; Heft 2020; McArthur and Baron 1983; Valenti and Good 1991).

Social Psychology's Dual Disciplinary Heritage

This section begins with an examination of what has been called the "two social psychologies" thesis (Britt 1937a): the idea that there were two broad traditions of social psychology, differentiated by origins, institutional base, topics, theories and concepts, and disciplinary practices. Some of the boundary work discourse generated in the quest for an interdisciplinary social psychology will then be reviewed. Attempts to differentiate the social psychologies in terms of metatheoretical commitments will be discussed. This section concludes with an examination of some of the ways in which boundary work in social psychology has been influenced by developments in the parent disciplines of sociology and psychology.

The disciplines of social psychology and the historical representation of their practices can be seen as reflecting the varied agendas for social psychology that were beginning to crystallize in the 1930s (Apfelbaum 1986). And as noted above, any adequate history of social psychology must also, in part, be a history of sociology and of psychology (Farr 1996). Yet various official or legitimating histories of social psychology had been written by either sociologists (Karpf 1932) or psychologists (G. Allport 1954; Jones 1985) in which they had reflected their own unitary disciplinary preoccupations. In these disciplinary-bound histories, the boundaries of social psychology were often drawn in such a way as to exclude or marginalize work in the other disciplinary field of social psychology. As a partial corrective, a number of social psychology texts had gone some way to redress this imbalance (Cherry 1995; Collier et al. 1991; Farr 1996; Jackson 1988).

At this juncture it is important to give some consideration to the origins of social psychology's dual disciplinary heritage. Social psychology's disciplinary "origins" had often been traced to the year 1908 in which two books were published in English whose titles included the name social psychology (William McDougall's *Introduction to Social Psychology* and E. A. Ross's *Social Psychology*). This particular "origin myth" failed to take account of earlier "social psychological" developments such as the *Völkerpsychologie* of Moritz Lazarus, Hermann Steinthal, and Wilhelm Wundt and the crowd psychology of late nineteenth-century Italian and French writers such as Gabriel Tarde and Gustav LeBon. Nonetheless, as Charles Camic had pointed out, the highlighting of these two early classics has served the purpose of enabling the reconstruction of social psychology's dual heritage history (Camic 2008).

What is clear is that by the mid-1920s when both psychology and sociology were well-established as academic disciplines in the USA, sociological and psychological traditions of social psychology were already developing independently from one another. Lundberg 1931-1932 noted that by 1930 social psychology was the most frequently listed field of special interest of the members of the American Sociological Society. Indeed by the mid-1930s, concern was being expressed about the relative isolation of the two traditions from one another and proposals were being made to facilitate more communication and closer cooperation between the "two social psychologies." In 1937 a paper was published, unusually, in both the *Journal of Abnormal and Social Psychology* and the *American Sociological Review* (Britt, 1937a, b). This paper represented one of the earliest expressions of concern about the negative consequences of the existence of "two social psychologies."

Steuart Britt presented three lines of evidence for social psychologists being divided into two separate camps: the division of social psychologists as to membership in the American Sociological Society and the American Psychological Association; the separate meetings of the two groups and lack of formal contact; and their divergent reading and research habits. Britt drew attention to "the lack of understanding, and sometimes even jealousy, between the two groups." "Surely," he wrote, "the time has come for the two groups of social psychologists to form a united front to attack their mutual problems. Differences in points of view should not prevent coordination of interests and activities" (Britt 1937a, p. 318). In his paper Britt clearly anticipated three elements in subsequent analyses of the "two social psychologies" literature: descriptive, evaluative, and prescriptive; he described the nature of the two social psychologies and their differentiation, he expressed the view that this state of affairs was undesirable and that it needed to be changed, and he offered a proposal to remedy the situation. Britt returned to this theme in a further short note in the *American Sociological Review*, reporting the results of a survey of social psychologists about the matter. He also wrote a successful textbook of social psychology which attempted to provide an integrated approach (Britt 1941). The first edition ran to 11 printings and a second edition appeared in 1949. Sadly, for the project of a united social psychology, Britt left the academic world for that of commerce in the late 1940s.

Britt, perhaps innocent of the realities of academic politics, conjectured as to whether the field had been dubbed "social psychology" because "social" was a more convenient adjective than "psychological" and wondered whether it really would have made any difference if the word order had been reversed and the field were called "psychological sociology." Britt concluded by proposing a "united group" – a "Society of Social Psychologists," a proposal which was to be made on more than one occasion as a solution to concerns similar to Britt's (e.g., echoed in Lundberg 1931-1932, and in Karpf 1952-1953),

Following the publication of Britt's (1937a, b) papers, a recurring theme of disciplinary debates in social psychology was the notion that there were indeed "two social psychologies," differentiated by origins, institutional base, topics, theories and concepts, and disciplinary practices: a psychological social psychology (hereafter PSP) – focusing on the psychological properties and processes of individuals and studied predominantly in the laboratory; and a sociological psychology (SSP) – focusing on the reciprocal relations of the individual and society, the mutuality of social interaction and on the "emergent" products of such interaction. Samples of such debates could be found in Boutilier et al. (1980); House (1977); McMahon (1984); Saxton and Hall (1987); Stryker (1977).

In both the USA and Europe, in the last decade of the twentieth century, there were far more social psychologists whose professional disciplinary affiliation was in psychology than sociology. Backman (1991) estimated on the basis of the relative size of the American Psychological Association (APA) Division of Personality and Social Psychology (Division 8) and the Social Psychology Section of the American Sociological Society (ASA) that the ratio of psychologically trained social psychologists to their counterparts was about eight to one. The figure was around four to one on the basis of 1999 figures. However, the Division Eight numbers included personality psychologists and the ASA figures omitted social psychologically inclined members who might be members of other sections such as Collective Behavior and Social Movements. In the APA, an almost equal number of social psychologists were members of Division 9, the Society for the Psychological Study of Social Issues (SPSSI). Whatever the ratio, sociological social psychologists could be seen to constitute a small minority.

Nonetheless, it was still possible to separately identify SSP from PSP in terms of patterns of disciplinary socialization, institutional affiliations (e.g., membership of the Social Psychology Section of the ASA versus Division 8 of the APA), textbooks, reference books, journals, and awards.

According to its website, the purpose of the ASA Section on Social Psychology "[was] to foster the development of this branch of sociology through stimulating research and communicating knowledge. Social psychology [was] interpreted according to its inclusive and traditional use in sociology to encompass such broad fields as socialisation, interpersonal relations and social interaction, attitudes and public opinion, and collective behavior." Membership in 1999 was just under 700. It was the seventh largest section (out of 40) and was about the same size as that of the Society for the Study of Experimental Social Psychology (SESP).

Influential SSP textbooks included McCall (1992), Hewitt (1999, 1st ed. 1976); Lindesmith et al. (1999, 1st ed. 1949), and deLamater and Ward (2013, 1st ed. 2006), Stephan and Stephan (1990) was a rare example of a textbook providing an introduction to social psychology that attempted to integrate psychological and sociological perspectives.

Two ASA-sponsored volumes functioned as unofficial handbooks (Rosenburg and Turner 1990 1st ed. 1981; Cook et al. 1995).

Social Psychology Quarterly (formerly *Sociometry*), published by the ASA, provided theoretical and empirical papers on the link between individual and society. This included the study of the relations of individuals to one another, to groups, collectivities, and institutions. It also included the study of intraindividual processes insofar as they substantially influenced, or were influenced by, social structure and process. *Symbolic Interaction,* the journal of the Society for the Study of Symbolic Interaction was for scholars interested in qualitative, especially interactionist research. The Society held an annual Couch-Stone Symposium and also sponsored the research annual *Studies in Symbolic Interaction.*

The Social Psychology Section of the ASA established in 1978 the Cooley-Mead Award "to recognize persons who have made substantial and lasting contributions to social psychology, particularly from a sociological perspective." The first recipient of the award, appropriately, was Muzafer Sherif. Subsequent recipients included: in 1979, Erving Goffman; in the 1980s: George Homans, Theodore M. Newcomb, Alex Inkeles, Robert F. Bales, Herbert Blumer, Howard Becker, Sheldon Stryker, Ralph Turner, William Sewell, and Morris Rosenberg; in the 1990s: John Clausen, Thomas Berger, Melvin L. Kohn, Glen H. Elder, Anselm Strauss, Harold Garfinkel, Melvin Seeman, Robert K. Merton, David R. Heise, and Harold H. Kelley; in the 2000s: Linda Molm, James House, Lynn Smith-Lovin, Karen Cook, and Peter Burke; and in the 2010s: William Corsaro, Carmi Schooler, Thomas Pettigrew, and Gary Alan Fine.

There was little agreement as to how these traditions of PSP and SSP should be described in more substantive content-oriented terms. Some sensitivity to differentiation within the psychological and sociological traditions was to be found, with PSP tending to be presented in a more monolithic way. But there always has been much greater diversity within each of these allegedly monolithic social psychologies than such a view would suppose (Good 1981; Pettigrew 1991). It was mistaken on the one hand to identify PSP with experimental social psychology, and on the other to identify SSP with symbolic interactionism. In the case of PSP, in doing so one excluded a significant nonexperimental tradition, and in the case of SSP one excluded work on social structure and personality (also known as "psychological sociology" or contextual social psychology), behavioral sociology, exchange theory, ethnomethodology and conversational analysis, and more than one significant experimental tradition in sociology. On social structure and personality, House (1981, 1995), Pettigrew (1997); on behavioral sociology, Michaels and Green (1978); on ethnomethodology, Boden (1990) and Garfinkel (1996); on experimental approaches, Bonacich and Light (1978), Couch (1987), Katovich (1984), Meeker and Leik (1995); on symbolic interaction, Blumer (1969) and Fine (1990).

The most detailed discussion of the "two social psychologies" thesis can be found in *The Future of Social Psychology* (Stephan et al. 1991), In this even-handed volume with contributions from seven prominent American social psychologists – four sociological social psychologists (Carl Backman, James House, Cookie W. Stephan, and Sheldon Stryker) and three psychological social psychologists (Thomas Pettigrew, Carmi Schooler, and Walter G. Stephan) – there was much agreement that a major schism still existed between the sociological and psychological traditions of social psychology and that its continuation had "negative consequences" for both traditions, including duplication of effort, loss of valuable theoretical and methodological resources, intellectual isolation, etc.

In an authoritative, but appropriately ecumenical final chapter titled "Is unity possible?", Pettigrew (1991) identified various problems and barriers to greater communication. These included the now familiar factors of disparate histories, intellectual traditions and epistemologies, one-way communication (especially evident in journal citations – articles in sociological social psychology journals such as *Social Psychology Quarterly* being much more likely to cite psychological sources than were PSP articles likely to cite sociological sources) and above all, the way these differences were embedded in institutional arrangements – separate departments, journals, professional organizations and meetings, research funding councils, etc. (Pettigrew 1991).

Pettigrew rejected the idea of separate "interdisciplinary" departments of social psychology and, bringing together a number of proposals that have been made over the years, suggested instead that attention be focused on the "gate-keeping nodes of the discipline" – graduate training, research institutes, the journals, the research funding agencies, and professional organizations and meetings. Each of these had, at one time or another, already been singled out as having a contribution to make to the solution of social psychology's problems. For a sample of such proposals: Britt (1937a, b); Jaspars and Ackermans (1966); Lundstedt (1968); Newcomb (1951).

Subsequently, Pettigrew focused his attention on the development of a contextual approach to social psychology that would enable social psychologists to place psychological phenomena in their wider, and proper social contexts (Pettigrew 2018, 2020, 2021),

Some sociological social psychologists, although keen proposals to pursue greater communication (e.g., Stryker), were opposed to integration – especially in the form of single departments of social psychology – fearing the loss of both their own distinctive disciplinary perspective and of contacts with the parent discipline. Stryker (1991, p. 96) suggested that "thorough institutional integration would be disastrous" because sociological social psychology's "core message" (that social psychological processes are crucially affected by social structural settings) would be lost in the process of integration. Stryker beautifully captured the tension between disciplinary and interdisciplinary forms of knowing:

> the two social psychologies need one another, and they need one another in something approximating their current form rather than in a blend of the two ... the dialectic that I claim is healthy requires that communication between the two takes place, almost certainly on a

> level more extensive than is currently accomplished, and that means more forums bringing together representatives of the two social psychologies. Their coming together, however, must be against a context of support from their parent disciplines, and the continued separate disciplinary identification that implies. What we should not be looking towards is a disappearance of the two social psychologies. (Stryker 1991, pp. 96–97)

However, he did propose that dialogue between the two social psychologies (indeed between various social psychological "traditions") should be seen as an essential part of the discipline of social psychology.

Revisiting these issues in 2008, James House noted the decline of social psychology as an area of specialization for departments and graduate faculty in the top 30 American Graduate Programs between 1970 and 2007. He also noted that the percentage of the total ASA Section membership for the ASA Social Psychology Section had declined from 18.5% to 2.7% (House 2008, pp. 235–236).

House suggested three strategic goals for the revival of sociological social psychology: the need to formulate and articulate why sociology needs to have as one of its core elements a sociological wing of an interdisciplinary social psychology; the need to reconnect with social psychological colleagues in psychology and to jointly recognize and articulate how a sociologically and psychologically informed and balanced social psychology is foundational to the more integrated social science needed to understand and explain and deal with major social phenomena and problems of the twenty-first century; and the need to significantly articulate with and be able to effectively incorporate and utilize developments in other disciplines within and outside of social psychology, most notably economics but also the biomedical and even physical sciences (House 2008, p. 252).

Because some PSPs (working on discourse, for example) might have more in common with some sociological colleagues than with some of their own departmental colleagues, it was also possible to attempt to identify styles of social psychological practice in terms of metatheories or methodology. If the labels PSP and SSP no longer were useful in identifying different styles of work, was there any way of classifying the different perspectives which were to be found within each of the "social psychologies" (defined in terms of institutional affiliation)? Some 40 years ago a tentative answer to this question, focusing on metatheoretical or philosophical commitments, was proposed drawing on the work of the social theorist and philosopher Richard Bernstein (Good 1981). In his influential analysis of social and political theory (Bernstein 1976), Bernstein had identified three major traditions informing social theory – *empirical, interpretive, and critical*. At the time these labels fitted work in SSP more closely than that in PSP, but there was then work within PSP which reflected the metatheoretical commitments of each of these intellectual traditions. It was important to note that these labels were not meant to represent exclusive choices. In the final analysis we were not confronted with exclusive choices: either empirical theory or interpretive theory or critical theory – "an adequate social theory must be *empirical, interpretive* and *critical*" (Bernstein 1976, p. 235, emphasis in original).

In the half century since Bernstein described social theory in such terms, a wider range of choices had become available for the student of social life, partly as a consequence of the impact of theories from the humanities – literary theory and philosophy, for example – and elsewhere, e.g., Clifford Geertz's "blurred genres" (Geertz 1980). Henderikus Stam (2000), commenting on sources of change in recent theorizing in psychology, noted the impact of ideas from James Gibson's ecological psychology and also influences from the philosophy of Ludwig Wittgenstein.

Andrew Weigert and Viktor Gecas (1995) made a similar proposal in their attempt to move beyond the "two social psychologies" debate. They acknowledged that "the sources of different traditions of social psychology are only partly disciplinary" (Weigert and Gecas 1995, p. 141) and they formulated a metatheoretical semantic space for locating social psychologies within the semantic realm. This generated four broad styles of work: psychological, biographical, structural, and cultural. The "psychological" related to "internal dynamics and components of individuals" such as attitudes, cognitions, perceptions, and emotions with a view to developing general principles of psychological functioning. The "biographical" studied persons in their socio-historical settings over time and with an emphasis on qualitative and situational aspects of interpretation and understanding. The "structural" focused on the impact of institutional and formal structures on individual functioning. And the "cultural" emphasized the mutual impact of historical forces and cultural contexts on selves and identities, e.g., ideologies and religions, to formulate general characterizations of selves in the context of larger units such as ethnic groups or nation states. Such a framework, Weigert and Gecas believed, could enable social psychologists "to interpret [their] perspectives with explicit epistemological criteria rather than the accidental contingencies of departmental grouping or historical location" (1995 p. 148). Although Weigert and Gecas made a promising start, much further work would be required to develop an adequate taxonomy of the metatheoretical commitments of the diverse social psychologies.

A Dialectical Tension Between the Social Psychologies and Their Parent Disciplines

Thomas Gieryn's (1983) notion of boundary work has been employed to demarcate different disciplines of social psychology, illustrating some of the ways in which their cognitive authority has been restricted, protected, expanded, and enforced (Good 2000).

In such boundary work major factors were the disciplinary affiliation of the social psychologist and related developments within the parent disciplines. Within sociology in the 1960s and the early 1970s, we can see sociological traditions of social psychology (such as symbolic interactionism and ethnomethodology) providing a means of overcoming the "oversocialised conception" of human nature in social theory (Wrong 1961). Later in the 1980s, Viktor Gecas drew attention to the need to "rekindle the sociological imagination" (Gecas 1989), a term coined by C. Wright Mills, to capture what he saw as the essence of sociology – the linking of biography and history, the private and the public (Wright Mills 1959). Sheldon Stryker commented on the urgent need for a re-vitalization of symbolic interactionism following a period of decline (Stryker 1987).

At this time, other SSPs were seeking to ensure that SSP's "key message" was not lost: that social psychological processes were crucially affected by social structural settings (Stryker 1991, 1997). In Melvin Kohn's words, "SSP was insufficiently sociological, too little concerned with larger social structures." He added that "the *sins of commission* of his fellow SSP's (too strong a focus on micro-social factors) pale by comparison with the *sins of omission* of other sociologists" (Kohn 1989 p. 27, emphasis in original). SSPs "at least recognise the existence of other people; other sociologists sometimes seem to act as if they thought that social institutions function without the benefit of human participants" (Kohn 1989).

In the early 1960s, PSPs on the other hand were inclined to note that one of social psychology's central lessons concerned the impact of *situational* factors on human behavior, the classic studies of Philip Zimbardo (Haney et al. 1973) and especially Stanley Milgram (1965) being noteworthy here. Not more than a decade later, however, especially in Europe, social psychologists such as Henri Tajfel, Serge Moscovici, and Joachim Israel, among many others, were calling for a more "social" social psychology, for a stronger social and cultural dimension in the human sciences (Israel and Tajfel 1972; Markova 2015; Tajfel 1972b).

In disciplinary debates about the nature of social psychology we sometimes find instances of boundary work in persuasive discourse that employed "rhetorics of exclusion, incorporation and integration" (Klein, 1990a, b) to which could be added rhetorics of "rehabilitation" (Moscovici 1993), of "colonisation" (van Strien 1997), and "relevance" (Parker and Spears 1996).

Rhetorics of Exclusion. Edward E. Jones in his article on major developments in social psychology written for the 1985 edition of the *Handbook of Social Psychology* (and posthumously reprinted in 1998) focused mainly on PSP and mainly on the period after the Second World War (Lubek and Apfelbaum 2000). He characterized social psychology as primarily a *subdiscipline* of psychology, and excluded SSP from serious consideration on a variety of grounds, including the proportionate volume of social psychological literature in psychological (as opposed to SSP) journals, a history of failed interdisciplinary enterprises, and because "the ultimate interests of the two approaches are distinctive" (Jones 1985, p. 50) – i.e., sociological social psychologists were not real social *psychologists.*

Serge Moscovici provided a European example. In a reply to a review of his 1976 book on *Social Influence and Social Change*, he suggested that the reviewer (Peter Kelvin) had a "wrong notion of what social psychology is by including in or grafting on it a *sociological* trend [symbolic interactionism] which has had no influence on it and does not belong to it" (Moscovici 1979, p. 449). Kelvin had suggested in his review that Moscovici had overlooked the SSP literature on deviance and labelling theory (Kelvin 1979).

Rhetorics of Incorporation. Rob Farr argued for the incorporation of the sociological in the psychological in terms of the development of an essentially "social" theory of mind (the symbolic interactionist model of the individual) rather than dealing with the social and the psychological in terms of different *levels of analysis*.

Farr observed that "the attempt initially to distinguish social psychology from psychology, and subsequently also sociology from psychology in terms of a difference in levels, is the *cause* rather than the antidote of the problems" (Farr 1978, p. 513).

Indeed many of the so-called "new social psychologies" which caught the imagination of PSPs in the 1970s could be seen as reflecting a discovery by psychological social psychologists of the social behaviorism of G. H. Mead and of such SSP perspectives as symbolic interactionism, ethnomethodology, critical theory, etc., i.e., PSPs had "incorporated" these perspectives into their work.

Rhetorics of Rehabilitation. Although Serge Moscovici had been at the heart of the 1970s European quest for a more social and cultural psychology (Moscovici 1972), in his book *The Invention of Society* (Moscovici 1993), we find a concern to rehabilitate *psychological* causes of human affairs, to restore a proper balance between psychology and sociology (in analyses of religion, innovation and money), a recognition that if psychology did not provide the human sciences with an adequate theory of mind, then these sciences would invent their own (as they often had done in the past). Of course, there had been a longstanding tradition that when looking for a theory of human nature, sociologists and anthropologists were much more likely to turn to psychoanalysis than academic psychology; in postmodern times, Freud was sometimes pushed aside by Lacan and others (Giddens 1984).

The Quest for an Interdisciplinary Social Psychology

Britt's (1937a, b) reflections on the disciplinary status of social psychology discussed above provided one of the earliest examples of the attempt to promote bridge-building and synthesis. Since then there had been intermittent calls for the development of a properly interdisciplinary social psychology. Backman (1983), Cook and Pike (1988), Jaspars and Ackerman (1966) Moir (2015) and Stephan, Stephan and Pettigrew (1991) provide a few examples.

The Second World War brought a rapid expansion of social psychology and fostered a period of cross-disciplinary activity. The roots of this "golden age of interdisciplinary social psychology" lay in the experiences of social psychologists engaged in interdisciplinary research on the adjustments of the American soldier and on the US strategic bombing surveys in Germany and Japan (Sewell 1989). After the war, in the USA, interdisciplinary graduate training programs in social psychology were established at Michigan, Harvard, Yale, Cornell, Berkeley, Columbia, Minnesota, Wisconsin, and other universities. In addition, a variety of interdisciplinary research centers were set up (e.g., the *Institute for Social Research* at the University of Michigan and the *Bureau of Applied Social Research* at Columbia University). As Sewell noted, however, by the late 1960s, these interdisciplinary programs in social psychology had all but vanished, leaving "a fractionated set of social psychological traditions returning to their original disciplinary moorings once the interdisciplinary arrangements had faltered" (Sewell 1989, p. 3). Several chapters in Lundstedt's (1968) volume on graduate social psychology showed the fragility of and need to modify social psychology graduate programs such as the joint interdisciplinary Michigan program. Despite the continued expansion of social psychology, a period of disciplinary soul-searching began – the so-called "crisis of social psychology."

There were clearly powerful disciplinary (institutional) constraints which stood in the way of synthesis and integration. At times of financial or intellectual uncertainty, "disciplinary" loyalties tended to prevail. But there was a more personal dimension as well. Reflecting on the factors which led to the dissolving of the University of Michigan's highly successful interdepartmental doctoral program in social psychology (which he had co-directed for over 20 years), Theodore Newcomb noted that though the "structural" fact that far more psychologists than sociologists were available to run the program had, from the beginning, been a source of potential threat to some members of the sociology department, "the crucial irreversible steps were not explainable primarily in such terms... Individuals of strong determination, on both sides, moved toward irreconcilable positions on issues that had previously been negotiable... a budgetless organization, serving to promote the overlapping interests of two strong parent organizations, cannot long survive beyond the time when those organization's representatives, who must work together, have ceased to trust one another" (Newcomb 1974, p. 384).

In their historical and rhetorical analysis of the idea of an interdisciplinary social psychology, James Good and Arthur Still reluctantly rejected the feasibility and desirability of an interdisciplinary social psychology but believed that it was necessary to foster a strategic "undisciplining" of the human sciences. By this they meant a strategic and selective focus on limited topic areas, often but not always of a social problem or social policy nature. They suggested that such a strategy was consistent with Serge Moscovici's (1990) proposals for dealing with the "crisis of representation of the social" by means of a "flexible union of sociology, economics, psychology, history." In the social sciences and the humanities there have been recent signs of a revival of interest in such "flexible unions" of disciplines, reinforced by the policy of national Research Councils and funding agencies in increasingly prioritizing such problem-focused research (Callard and Fitzgerald 2015).

Good and Still concluded, nonetheless, that the intermittent quest for an interdisciplinary social psychology had served a useful purpose in the development of social psychology in drawing attention to the boundary work within and between the sociological and psychological traditions and to the false sense of self-sufficiency of "disciplined" social psychology, sociological, or psychological (Good and Still 1992).

The Post-Second World War Development of European Social Psychology

While some of the extant literature on the history of psychology/social psychology made occasional reference to the impact of Cold War issues, e.g., Ash (2010), Dafermos (2015), Faye (2012), Herman (1995), House (2008), Jahoda (2007), Markova (2012), Oishi, Kesebir, and Snyder (2009), Schruijer (2010, 2012), and van Strien (1997), only the Moscovici and Markova (2006) monograph directly addressed the changing disciplinary status of social psychology in the context of the Cold War. Likewise, although the literature on the impact of the Cold War on the development of the social sciences did make reference to some aspects of psychology and sociology, there had been no systematic discussion of the impact of the Cold War on the disciplinary status of social psychology (Cohen-Cole 2014; Solovey and Cravens 2012; Isaac 2007, 2011; Solovey 2014).

This section focuses attention on the post-war development of European social psychology. As has become abundantly clear from some of the studies of the impact of the Cold War on the development of the social sciences noted above, there was substantial post-war aid to promote the recovery and development of European social sciences. There was a recognition of the need for international involvement, leading to the creation of the International Social Science Research Council (ISSRC) in 1952.

In 1954 a UNESCO study of the nature of social science teaching at higher education institutions in various parts of the world was published (de Bie et al. 1954). In this, social psychology was grouped with sociology and cultural anthropology. In his contribution to this report, the Belgian social psychologist Jozef Nuttin recommended that at both undergraduate and graduate level social psychology should be closely associated with psychology on the one hand and with cultural anthropology and sociology on the other. Nuttin specifically voiced his fear of the undesirable possibility that social psychology might be split between its psychological and sociological variants (Nuttin 1954).

In the early 1950s, social psychology was still relatively underdeveloped in the USA. Nonetheless, European social psychologists were attracted by the activity of their North-American counterparts and quite a number of those who were to play a prominent role in the subsequent development of European social psychology traveled to the USA to work in American laboratories, including Henri Tajfel, Claude Faucheux, and Ragnar Rommetveit. American social psychologists were also to play an important role supported by funding from the Office of Naval Research, NATO, the Ford Foundation, *Volkswagen Stiftung*, and other bodies.

The Committee on Transnational Social Psychology (the Transnational Committee) was set up in 1964 by the New York-based Social Science Research Council (SSRC). Over the next 10 years this committee was to play a pivotal role in the future development of social psychology. Moscovici and Markova (2006) provide a detailed account of its workings. For the first 5 years it was chaired by Leon Festinger, perhaps the most eminent and influential American social psychologist of his generation. The Transnational Committee, comprised of a mixture of American and European social psychologists, was tasked with "the stimulation of international cooperation and developments in experimental social psychology." Although Festinger had often been described as one of the key architects of American experimental social psychology, he also felt it important that social psychology employ other (nonexperimental) methods. Indeed one of his most celebrated works was *When Prophecy Fails*, a participant observation study of a prophecy cult (Festinger et al. 1956).

An important step forward in the mission to stimulate international cooperation and developments in experimental social psychology occurred in 1963, when two American social psychologists, John Lanzetta and Luigi Petrullo, established a Planning Committee for a "European Conference on Social Psychology" which was to be held in Sorrento in 1963. John Lanzetta was an experimental social psychologist, Professor of Social Psychology at the University of Delaware and stationed in London as a representative of the Office of Naval Research (ONR). Luigi Petrullo was the head of the ONR in Washington, DC.

Of the 28 participants, 21 came from eight European countries, two from Israel, and five from America. Lanzetta used the Sorrento Planning Committee to help prepare a more ambitious "Proposal for Contributions to the Development of Experimental Social Psychology in Europe." This was submitted to SSRC for financial support. It included provision for a further European conference, a summer school, the possibility of an exchange program, and a scheme for specialized seminars. Key elements of this proposal were approved by the Transnational Committee, and a second conference was held in Frascati in December 1964. At this conference a European Planning Committee was approved with Gustav Jahoda, Max Mulder, Jozef Nuttin, Serge Moscovici, and Henri Tajfel as members. It was chaired by Serge Moscovici. This committee was invited "(a) to plan some form of organizational structure for the continuing activities; (b) to plan the detail of these activities; (c) to explore the possibility of finding funds in Europe; (d) to report to the next conference" (Nuttin 1990, p. 366).

Under Moscovici's direction, the Planning Committee proposed the formation of an Association, devised a name and organizational structure for it, and set up the third European Conference on Experimental Social Psychology which was held at Abbaye de Royaumont, near Paris, 27 March–1 April 1966. There the foundation of the European Association for the Advancement of Experimental Social Psychology was formally approved (Jahoda and Moscovici 1967). The European Planning Committee became the first Executive Committee and Serge Moscovici the first president of the Association.

Prior to the setting up of the Transnational Committee there were two bodies keen to establish some form of training for young European social psychologists. A European Foundation for Summer Schools had been set up in 1964 by Marie Jahoda, Fred Emery, and Henri Tajfel. The European Board was also planning to set up summer training institutes in 1965 and 1966. This was comprised only of European social psychologists – Novak (Poland), Jezernik (Yugoslavia), Iacono (Italy), Ardoino (France), Tajfel (England), Thorsud (Norway), Hutte and Koekebakker (Holland), and Irle (Germany). There was also involvement from the Provisional committee set up for the Frascati conference. In the end, to avoid potential conflict some members of the Provisional Committee were added to the European Foundation for Summer Schools.

The first summer school held at "Het Oude Hof," a former Royal Palace in The Hague, 15 July–11 August, 1965. The theme was the "Social-Psychological Aspects of Organisations." The Director was Jaap Koekebakker (Leiden) and the Faculty were Morton Deutsch (Columbia), Fred Emery (Tavistock), Claude Faucheux

(Paris), Claude Flament (Aix-en-Provence), Philip Herbst (Trondheim), and Robert Zajonc (Michigan). There were morning sessions of lectures and discussions; afternoons were devoted to work on group projects; and there were also evening seminars. The participant list certainly reflected the organizers dual aims of broad geographical representation and some bridging of eastern and western Europe. Participants were drawn from Belgium (4), Ireland (2), Czechoslovakia (2), Italy (1), France (4), Netherlands (5), Germany (3), Poland (3), Great Britain (7), Yugoslavia (6), and Greece (1).

Although this inaugural summer school was deemed to have been a success, some important lessons were learned which were incorporated in the planning for the second summer school held in Louvain in 1967 (van Gils and Koekkebakker 1965). In general, the participants felt that there was a mismatch between the content of the morning lecture sessions – predominantly focused on experimental work on social facilitation (e.g., Zajonc) or interpersonal conflict (e.g., Deutsch) and the more applied afternoon group projects focused on aspects of organizations. The east European participants in particular remarked on this discrepancy and it quickly became clear that there was quite a gulf between the eastern and western Europeans in terms of both their theoretical commitments and methodological sophistication. One of the UK participants also drew attention to some variations among the western European participants, quipping in his report on the conference that the Dutch were more American than the Americans in their rigor and experimental zeal (Pugh 1966). The syllabus for the second summer school was revised substantially to take account of these and some other issues – most notably to reduce the amount of formal lecturing and to provide more continuity of the permanent staff. It was also felt that it would be desirable to select more advanced research students with less emphasis on geographical representation.

Nonetheless, despite these initial setbacks, the Association flourished, establishing its journal the *European Journal of Social Psychology* and the European Monographs in Social Psychology in 1971. The *European Review of Social Psychology* followed in 2003 as an outlet for theoretical work and review articles. It continued to promote summer schools and introduced a series of small group meetings. A series of East-West meetings was initiated – the first two being held in Vienna (1965) and Prague (1968).

At this point it is, perhaps, appropriate to ask two questions. Just how European was the European Association? And to what extent did it foster transdisciplinary work? Sandra Schruijer had some doubts on both counts. In an article in *History of the Human Sciences* published in 2012, she provided some results from a study of the ambitions involved in the founding of the EAESP. Drawing on a variety of sources, including findings from interviews with eight key actors (four of whom who were on the original Executive Committee), EAESP archives, and data regarding EAESP membership and its house journal, the *European Journal of Social Psychology,* Schruijer acknowledged that although the institutionalization of social psychology in Europe had been successfully achieved, the original intellectual and ideological ambitions of fostering and promoting a distinctively European social psychology were not.

In the interviews, concern was expressed about the limited conception of the "social" in American social psychology and about how that was beginning to permeate some European social psychological work on cognition. Although interviewees generally endorsed the Association's attempts to develop a scientific social psychology, it was felt that experimental work was beginning to eclipse other styles of work despite the fact that Tajfel had himself written that the term "experimental" was "meant to represent our preoccupation with developing work of a fundamental kind....the term 'experimental' did not really express what we meant; many of us felt ... that 'experimental' social psychology is not necessarily the only, or even the best way to pursue knowledge in this field" (Tajfel 1972a, p. 310). Some concern was also expressed about the success of the attempts to foster links between social psychologists from western and eastern Europe.

In summing up, Schruijer surmised that her findings might be seen as a reflection of the "Americanization" of European social psychology and of "neo-colonization" (van Strien 1997; Markova 2012). At the same time, she did acknowledge some impact of European ideas on American social psychology – especially in the areas of social identity theory and minority influence. Nonetheless, she concluded that such European influences had not made group research any more social – it was *individual* social cognition and not *shared* social cognition that was foregrounded in American research. As to why this might have come about, Schruijer pointed to the hold of "impact factors" in the contemporary academy and the fostering of publication in English-language journals of short-term, experimental work (mainly with students). Schruijer did acknowledge that work representing a more "social" approach to social psychology did exist (including social representations research) but that it was to be found in different journals, such as *Social Psychology Quarterly*, the *Journal of Community and Applied Psychology*, or the *British Journal of Social Psychology*. Miles Hewstone and Peter Smith provided a more positive assessment of the impact of the EASP (Hewstone et al. 2012; Smith 2005).

In their indispensable account of the making of modern social psychology in Europe, Moscovici and Markova had ended their narrative by referring to "an unfinished task" for social psychology – the development of an international social psychology:

> [social psychology] should ... find a form that addresses itself more easily and naturally to people all over the planet. We speak here about the experience, ideas and achievements that were not completed by the Transnational Committee and which await to be rediscovered and adopted by a younger generation with passion and vitality. It is possible that another generation will pursue the vast problems that social psychology must try to solve: our destiny in society and sense of life in community. (Moscovici and Markova 2006, p. 265)

As Carl Graumann noted in his history of the EAESP, the later development of the Association took place in a period of relative affluence in western Europe leading to the foundation of many new universities (Graumann 1999). This impacted on social psychology with the setting up chairs and programs of its own. With an established place in the curriculum, social psychology attracted more students, and this led to the development of textbooks and handbooks. At the same time, the Vietnam war generated strong reactions against the politics of the US administration among students, one of the factors responsible for much of the unrest that was to be manifest among social psychology students in a questioning and criticizing of established ways of teaching and research, which for some came to be associated with the alleged "crisis" of social psychology. Here again, diversity was in evidence as there were multiple putative crises – relating to confidence, experimentation, methodology, relevance, and ethics. Contrasting views about the crises in social psychology can be found in Armistead (1974), Elms (1975), Faye (2012), Dowd (1991), and Parker (1989). Sturm and Mülberger (2011) drew attention to the need to identify the variety of different crises encountered in psychology over the course of the twentieth century. At the turn of the century, social psychologists had to address new crises involving the falsification of data (Callaway 2011; Levelt et al. 2012; Strobe 2013) and of replicability (Baumeister 2016; Earp and Trafimow 2015).

Martin Wieser has recently suggested that the crisis notion can be seen as a fruitful historical category that has not just negative connotations in terms of the risk of an end that was nearing, but more positively of a chance that was opening up (Wieser 2020).

Conclusion: Whether the Disciplines of Social Psychology?

The aim of his chapter was to provide a historical narrative that did justice to social psychology's dual disciplinary heritage. Diverse representations of the "psychological," the "social," and their interfacing have been discussed. Attention was given to the intermittent attempts at synthesis and bridge-building, especially the changing fortunes of the quest to foster an interdisciplinary social psychology. Note was taken of the impact of the Second World War and subsequent Cold War influences on the development of the social sciences.

The pivotal role of the USA in the postwar reconstruction of European social psychology was outlined. In Europe, the importance of the Transnational Committee in the setting up of the European Association for Experimental Social Psychology was noted. Through its training institutes, conferences, journals, and monographs, the Association was to have a significant influence on subsequent generations of European social psychologists. It was clear that at the outset it was the intention of the Association's first President, Serge Moscovici, that a truly European social psychology should be interdisciplinary and address issues that were of particular concern to European social psychologists (Markova 2015). While it is clear that the latter objective was at least partially realized, the dominant form of social psychology continued to be experimental and the geographical center tended to be northwestern Europe. And the attempts to bridge the East-West divide seemed to be only partially successful. Although in the early years there was, perhaps, too much of an emphasis on experimentation, as evidenced from the summer schools and initial conferences, with time there was some broadening of scope.

Debate has continued about these issues (Doise 2012), but there is reason to believe that members of the European Association were still committed to Moscovici's original vision. An article published in 2013 in the *European Bulletin of Social Psychology*, arising out of an EASP Small Group Meeting held in Lausanne in June 2013, explored whether the EASP had lived up to its mission statement to promote and develop "empirical and theoretical social psychology within Europe and the interchange of information relating to this subject between the members and other associations throughout the world." The article addressed three related concerns: underrepresentation; social irrelevance; and intellectual narrowness. A variety of proposals were put forward to tackle these concerns (Ben Alaya et al. 2013).

Although UK social psychologists were prominent members of the EASP, the position of British social psychology was, as with so many other matters, rather out of line with continental European developments. Although not describing itself as explicitly interdisciplinary, UK social psychology was much more pluralistic and eclectic, open to quantitative and qualitative research, and also evidenced an openness to input from other disciplines (Wetherell 2011). In this regard, Wetherell pointed to the intellectual openness of the *British Journal of Social Psychology* (BJSP) in helping to foster such developments, a view endorsed by Stephen Reicher (2011), who saw the BJSP as having promoted a culture of innovation in British social psychology.

The final three decades of the twentieth century saw the development and steady growth in the UK and continental Europe of social identity and intergroup theory (Tajfel 1974; Tajfel and Turner 1979; Brown 2020a, b).

Michael Billig documented the subsequent origins of discursive psychology as a distinct tradition in social psychology, originally associated with the Discourse and Rhetoric Group (DARG) at Loughborough University (Billig 2012). He identified three key texts, all published in 1987, *Discourse and Social Psychology* (Potter and Wetherell), *Common Knowledge* (Edwards and Mercer), and *Arguing and Thinking* (Billig) which could be seen as marking the emergence of a distinctive new body of work on social psychology and language, the "Loughborough school." Augoustinos and Tileaga (2012) also provided an authoritative review of the first twenty five years of discursive psychology and of its relationship to social psychology.

The conference on *The Vision of Social Psychology* held at the LSE (London School of Economics) in 2015 to mark the 50th anniversary of its Social Psychology Department provided a timely illustration of both the diversity of the subject and also its ambitions for a transdisciplinary social psychology (Obradovic and Nicholson 2015).

In the conclusion of his 2008 article, an expanded version of his 2007 American Association Cooley-Mead Award address, James House provided a reminder of the importance of, and indeed necessity for, a deeper understanding of the history of social psychologies:

> Social psychology, in both its interdisciplinary and intradisciplinary forms, has had better times in the past. It can and should have them in the future, because it is integral to the future of sociology, psychology, economics and the broader social sciences, and to all of these

realizing their potential for understanding and improving social policy and social life. But we can only get to that goal by recognizing and confronting where we are now and where we have been, for better or worse. (House 2008, p. 253)

This chapter began with an acknowledgment of a central theme of this Handbook – the diversity of psychological practices. As Roger Smith noted in his chapter in this Handbook on psychologies and their diverse histories, "If psychology is diverse now, it also has *diverse roots, diverse histories*. The different forms taken by psychology require different histories. . . . it is a matter for historical work to determine the different roots of different psychologies" (Smith 2022). Accordingly, this chapter has attempted to illustrate the particular relevance of this consequence for an understanding of the diverse social psychologies that have waxed and waned over the twentieth century.

Additionally, like all human science disciplines, these social psychologies have been seen to be domains with contested concepts. As a consequence, this chapter has documented an *endemic disunity of social psychology*. Further, it is inevitable that the histories of these diverse social psychologies should reflect such disagreements and reveal a broad array of social psychological traditions, distinguished not just in terms of disciplinary affiliations but in terms of more meta-theoretical commitments that cross disciplinary borders. Over the twentieth century, the boundaries of these disparate traditions can be seen as fluid, contingent, local, and contestable, reflecting the thematic preoccupations, disciplinary origins, and meta-theoretical commitments of social psychologists, of the parent disciplines, and of those who represent disciplinary practices.

Moreover, as Joseph Rouse has observed in discussing the relation between history and scientific practice:

To do science is also to construe its history, while to offer an account of its history (or its supposedly ahistorical nature) is to project particular ways to do science . . . the futural dimension of the narrative reconstruction of science [in historical work] is crucial here. Actions and explicit reinterpretations of them do not merely construe the past but project a future. (Rouse 1999, p. 452)

Historical narratives, then, direct us to possible futures for social psychology and, as Smith has also reminded us, insofar as the histories of the human sciences are themselves a *form* of human science, they also provide a way of exploring the nature of human possibility (Smith 1997).

References

Allport GW (1954) The historical background of modern social psychology. In: Lindzey G (ed) Handbook of social psychology, vol 1. Addison-Wesley, Reading
Apfelbaum E (1986) Prolegomena for a history of social psychology: some hypotheses concerning its emergence in the 20th century and its raison d'être. In: Larsen KS (ed) Dialectics and ideology in psychology. Ablex, Norwood, pp 3–13
Armistead N (1974) Reconstructing social psychology. Penguin, Harmondsworth

Ash M (2010) Psychology. In: Backhouse RE, Fontaine P (eds) The history of the social sciences since 1945. Cambridge University Press, Cambridge, pp 16–37

Augoustinos M, Tilieaga C (2012) Editorial. Twenty five years of discursive psychology. Br J Soc Psychol 51:405–412

Backhouse RE, Fontaine P (eds) (2010) The history of the social sciences since 1945. Cambridge University Press, Cambridge

Backhouse RE, Fontaine P (2014) Introduction. In: Backhouse RE, Fontaine P (eds) A historiography of the modern social sciences. Cambridge University Press, Cambridge, pp 1–28

Backman CW (1983) Toward an interdisciplinary social psychology. Adv Exp Soc Psychol 16:219–261

Backman CW (1991) Interdisciplinary social psychology: prospects and problems. In: Stephan CW, Stephan WG, Pettigrew TF (eds) The future of social psychology: defining the relationship between sociology and psychology. Springer-Verlag, New York, pp 61–70

Baumeister RE (2016) Charting the future of social psychology on stormy seas: winners, losers, and recommendations. J Exp Soc Psychol 66:153–158

Ben Alaya et al (2013) Opinions and perspectives. Developing diversity as a means to achieve a vibrant and relevant social psychology. Eur Bull Soc Psychol 25(2):5–9

Bernstein RJ (1976) The restructuring of social and political theory. Blackwell, Oxford

Billig M (2012) Undisciplined beginnings, academic success, and discursive psychology. Br J Soc Psychol 51:413–424

Blumer H (1969) Symbolic interactionism: perspective and method. Prentice-Hall, Englewood Cliffs

Boden D (1990) The world as it happens: ethnomethodology and conversation analysis. In: Ritzer G (ed) Frontiers of social theory: the new syntheses. Columbia University Press, New York, pp 185–213

Bonacich P, Light J (1978) Laboratory experimentation in sociology. Annu Rev Sociol 4:145–170

Borgatta EF, Borgatta ML (eds) (1992) Encyclopedia of sociology. Macmillan, New York

Boutilier RG, Roed JC, Svendsen AC (1980) Crises in the two social psychologies: a critical comparison. Soc Psychol Q 43:5–17

Britt SH (1937a) Social psychologists or psychological sociologists – which? J Abnorm Soc Psychol 32:314–318

Britt SH (1937b) Social psychologists or psychological sociologists – which? Am Sociol Rev 2:898–902

Britt SH (1941/1949) Social psychology of modern life. New York: Rinehart & Co.

Brown JF (1936) Psychology and the social order: an introduction to the dynamic study of social fields. McGraw-Hill, New York

Brown R (1988) Group processes: dynamics within and between groups. Blackwell, Oxford

Brown R (2020a) The social identity approach: appraising the Tajfellian legacy. Br J Soc Psychol 59:5–29

Brown R (2020b) Henri Tajfel: explorer of identity and difference. Routledge, London

Callard F, Fitzgerald D (2015) Rethinking interdisciplinarity across the social sciences and neurosciences. Palgrave Macmillan, London

Callaway E (2011) Report finds massive fraud at Dutch universities. Nature 479(7371):15

Camic C (2008) Classics in what sense? Soc Psychol Q 71(4):324–330

Charles E (2011) Ecological psychology and social psychology: it is Holt, or nothing! Integr Psychol Behav Sci 45(1):132–153

Cherry F (1995) The "stubborn particulars" of social psychology. Routledge, London

Cohen-Cole J (2014) The open mind: cold war politics and the sciences of human nature. The University of Chicago Press, Chicago

Cole M (1981) Cross-cultural psychology: a combined review. Am Psychol 26:334

Collier G, Minton HL, Reynolds G (1991) Currents of thought in American social psychology. Oxford University Press, New York

Cook KS, Fine GA, House JS (eds) (1995) Sociological perspectives on social psychology. Allyn and Bacon, Boston

Cook KS, Pike KC (1988) Social psychology: models of action, reaction and interaction. In: Borgatta EF, Cook KS (eds) The future of sociology. Sage, Newbury Park, pp 236–254

Couch CJ (1987) Researching social processes in the laboratory. JAI Press, Greenwich
Coutu W (1949) Emergent human nature: a symbolic field interpretation. Knopf, New York
Dafermos M (2015) Rethinking the crisis in social psychology: a dialectical perspective. Soc Person Compass 9(8):394–405
Dawe A (1978) Theories of social action. In: Bottomore T, Nisbet R (eds) A history of sociological analysis. Heinemann, London
de Bie P, Lévi-Strauss C, Nuttin J, Jacobsen E (1954) Les Science Sociales dans l'enseignement supérieur: Sociologie, Psychologie, et Anthropologie Culturelle. UNESCO, Paris, pp 134–159
DeLamater J, Ward A (eds) (2013) Handbook of social psychology. Springer, Dordrecht
Doise W (2012) On a persistent shortcoming in the recent policy of the EASP. Eur Bull Soc Psychol 24(1):8–10
Dowd JT (1991) Social psychology in a postmodern age: a discipline without a subject. Am Sociol 22(3/4):188–209
Eagly A, Fine GA (2010) Bridging social psychologies: an introduction. Soc Psychol Q 73(4):313–315
Earp BD, Trafimow D (2015) Replication, falsification, and the crisis of confidence in social psychology. Front Psychol 6:621
Edwards D (1997) Discourse and cognition. Sage, London
Elms AC (1975) The crisis of confidence in social psychology. Am Psychol 30(10):967–976
Farr RM (1978) On the varieties of social psychology: an essay on the relationships between social psychology and other social sciences. Soc Sci Inf 17:503–525
Farr RM (1996) The roots of modern social psychology. Blackwell, Oxford
Farr RM, Moscovici S (1984) Social representations. Cambridge University Press/Maison des Sciences de l'Homme, Cambridge/Paris
Fay B (1996) Contemporary philosophy of social science: a multicultural approach. Blackwell, Oxford
Faye C (2012) American social psychology: examining the contours of the 1970s crisis. Stud Hist Phil Biol Biomed Sci 43:514–521
Festinger L (1957) A theory of cognitive dissonance. Row Peterson, Evanston
Festinger L, Riecken HW, Schachter S (1956) When prophecy fails. University of Minnesota Press, Minneapolis
Fine (1979) Progressive evolutionism and American sociology, 1890–1920. UMI Research Press, Ann Arbor
Fine GA (1990) Symbolic interactionism in the post-Blumerian age. In G. Ritzer (Ed.), Frontiers of social theory: The new syntheses (pp. 117–157). New York: Columbia University Press
Fiske S (1992) Thinking is for doing: portraits of social cognition from daguerreotype to laserphoto. J Pers Soc Psychol 63:877–889
Fontaine P (2015) Introduction: the social sciences in a cross-disciplinary age. J Hist Behav Sci 51(1):1–9
Gardner H (1985) The mind's new science: a history of the cognitive revolution. Basic Books, New York
Garfinkel H (1996) Ethnomethodology's program. Soc Psychol Q 59:5–21
Gecas V (1989) Rekindling the sociological imagination in social psychology. J Theory Soc Behav 19:97–115
Geertz C (1980) Blurred genres. The refiguration of social thought. Am Sch 49(2):165–179
Gergen KJ (2008) On the very idea of social psychology. Soc Psychol Q 71(4):331–337
Gerth H, Mills CW (1954) Character and social structure: the psychology of social institutions. Routledge & Kegan Paul, London
Giddens A (1984) The constitution of society: outline of the theory of structuration. Polity Press, Cambridge
Gieryn TF (1983) Boundary work and the demarcation of science from non-science: strains and interests in the professional ideologies of scientists. Am Sociol Rev 48:781–795
Gilbert M (1989) On social facts. Routledge, London
Gilbert M (1997) Concerning sociality: the plural subject as paradigm. In: Greenwood JD (ed) The mark of the social: discovery or invention? Rowman & Littlefield, Lanham

Good JMM (1981) Sociology and psychology – promise unfulfilled? In: Abrams P, Lethwaite P (eds) Development and diversity: British sociology. British Sociological Association, London, pp 1950–1980

Good JMM (2000) Disciplining social psychology: a case study of boundary relations in the history of the human sciences. J Hist Behav Sci 36(4):383–403

Good JMM (2007) The affordances for social psychology of the ecological approach to social knowing. Theory Psychol 17(2):265–295

Good JMM, Roberts RH (1993) Persuasive discourse in and between disciplines. In: Roberts RH, Good JMM (eds) The recovery of rhetoric: persuasive discourse and disciplinarity in the human sciences. The Bristol Press/University of Virginia Press, Bristol/Charlotte

Good JMM, Still AW (1992) The idea of an interdisciplinary social psychology: an historical and rhetorical analysis. Can Psychol 33:563–568

Graumann CF (1986) The individualization of the social and the desocialization of the individual: Floyd H. Allport's contribution to social psychology. In: Graumann CF, Moscovici S (eds) Changing conceptions of crowd mind and behavior. Springer-Verlag, New York

Graumann CF (1999) History of the European Association of Experimental Social Psychology. Retrieved from https://www.easp.eu/getmedia.php/_media/easp/201707/401v0-orig.pdf

Greenwood JD (ed) (1997) The mark of the social. Discovery or invention? Rowman and Littlefield, Lanham

Greenwood JD (2000) Individualism and the social in early American social psychology. J Hist Behav Sci 36:443–455

Greenwood JD (2004a) The disappearance of the social in American social psychology. Cambridge University Press, New York

Greenwood JD (2004b) What happened to the "social" in social psychology? J Theory Soc Behav 34(1):19–34

Greenwood JD (2014) The social in social psychology. Soc Personal Psychol Compass 8(7):303–313

Greenwood JD (2019) Social cognition, social neuroscience, and evolutionary social psychology: What's missing? J Theory Soc Behav 49:161–178

Haney C, Banks C, Zimbardo P (1973) Interpersonal dynamics in a simulated prison. Int Criminol Penol 1:69–97

Harré R (1992a) The discursive creation of human psychology. Symb Interact 15:515–527

Harré R (1992b) The second cognitive revolution. Am Behav Sci 36:5–7

Harré R, Secord PF (1972) The explanation of social behaviour. Blackwell, Oxford

Heft H (2020) Ecological psychology as social psychology. Theory Psychol 30(6):813–826

Heider F (1958) The psychology of interpersonal relations. Wiley, New York

Herman E (1995) The romance of American psychology. Political culture in the age of experts. University of California Press, Berkeley

Hewitt JP (1999) Self and society: a symbolic interactionist social psychology, 8th edn. Prentice Hall, New York

Hewstone M et al (2012) An evaluation of the impact of the European Association of Social Psychology: a response to Schruijer (2012). Hist Hum Sci 25(3):117–126

Himmelweit H, Gaskell G (eds) (1990) Societal psychology. Sage, Newbury Park

House J (1977) The three faces of social psychology. Sociometry 40:161–177

House JS (1981) Social structure and personality. In: Rosenberg M, Turner RS (eds) Social psychology: sociological perspectives. Basic Books, 2nd edn 1990. New York, pp 525–561

House JS (1995) Introduction: social structure and personality: past, present, and future. In: Cook KS, Fine GA, House JS (eds) Sociological perspectives on sociological psychology. Allyn & Bacon, Boston, pp 387–395

House JS (2008) Social psychology, social science, and economics: twentieth century progress and problems, twenty-first century prospects. Soc Psychol Q 71(3):232–256

Isaac J (2007) The human sciences and cold war America. Hist J 50:725–746

Isaac J (2011) Introduction: the human sciences and cold war America. J Hist Behav Sci 47(3):225–231

Israel, Tajfel (1972) The context of social psychology. Academic Press, London

Jackson JM (1988) Social psychology, past and present: an integrative orientation, Hillsdale
Jahoda G (1984) Do we need a concept of culture? Cross Cult Psychol 15:139–151
Jahoda G (2007) A history of social psychology. Cambridge University Press, Cambridge
Jahoda G (2016) Seventy years of social psychology: a cultural and personal critique. J Soc Polit Psychol 4:364–380
Jahoda G, Moscovici S (1967) European association of experimental social psychology. Soc Sci Inf 6:297–305
Jaspars JMF, Ackermans E (1966) The interdisciplinary character of social psychology: an illustration. Sociol Neerl 4:62–79
Jenkins R (1991) Rethinking history. Routledge, London
Jones EE (1985) Major developments in social psychology during the past five decades. In: Lindzey G, Aronson E (eds) Handbook of social psychology, vol 1, 3rd edn. Random House, New York, pp 47–108
Jones EE, Davis KE (1965) From acts to dispositions: the attribution process in person perception. In: Berkowitz L (ed) Advances in experimental social psychology, vol 2. Academic Press, New York
Kantor JR (1959) Inter-behavioral psychology: a sample of scientific system construction. Principia Press, Granville
Karpf FB (1932) American social psychology: its origins, development, and European background. Macmillan, New York
Karpf FB (1952-1953) American social psychology – 1951. Am J Sociol 58:187–193
Katovich MA (1984) Symbolic interactionism and experimentation: the laboratory as a provocative stage. Stud Symb Interact 5:49–67
Kelley HH (1967) Attribution theory in social psychology. In: Levine D (ed) Nebraska symposium on motivation, vol 15. Lincoln, University of Nebraska Press
Kelvin P (1979) Review of S. Moscovici, social influence and social change. Eur J Soc Psyochol 9: 441–454
Klein JT (1990a) Across the boundaries. Soc Epistemol 4:267–280
Klein JT (1990b) Interdisciplinarity: history, theory and practice. Wayne State University Press, Detroit
Kohn ML (1989) Social structure and personality: a quintessentially sociological approach to social psychology. Soc Forces 68:26–33
Krech D, Crutchfield RS, Ballachey EL (1962) Individual in society: a textbook of social psychology. McGraw-Hill, New York
Lave J (1988) Cognition in practice: mind, mathematics and culture in everyday life. Cambridge University Press, Cambridge
Levelt WJM, Drenth P, Noort E (Eds.) (2012) Flawed science: The fraudulent research practices of social psychologist Diederik Stapel. Tilburg: Commissioned by the Tilburg University, University of Amsterdam and the University of Groningen. https://pure.mpg.de/rest/items/item_1569964/component/file_1569966/content
Lewin K (1951) Field theory in social science: selected papers. Harper & Row, New York
Lindesmith AR, Strauss AL, Denzin NK (1999) Social psychology, 8th edn. Holt, Rinehart and Winston, New York
Lubek I, Apfelbaum E (2000) A critical gaze and wistful glance at handbook histories of social psychology: did the successive accounts of Gordon Allport and successors historiographically succeed? J Hist Behav Sci 36(4):405–428
Lundberg G (1931-1932) The interests of the members of the American Sociological Society, 1930. Am J Sociol 37:458–460
Lundstedt S (ed) (1968) Higher education in social psychology. Case-Western Reserve Press, Cleveland
Manicas PT (1987) A history and philosophy of the social sciences. Blackwell, Oxford
Margolis J, Manicas PT, Harré R, Secord PF (1986) Psychology: designing the discipline. Blackwell, Oxford
Markova I (2012) 'Americanization' of European social psychology. Hist Hum Sci 25(3):108–116
Markova I (2015) Serge Moscovici's vision of social psychology. Eur Bull Soc Psychol 27(1):28–30
Markus, Zajonc (1985) The cognitive perspective in social psychology. In: Lindzey G, Aronson E (eds) Handbook of social psychology, vol 1, 3rd edn. Erlbaum, Hillsdale, pp 137–230

McArthur LZ, Baron RM (1983) Toward an ecological theory of social perception. Pychol Rev 90: 215–238
McCall GJ (1992) Social psychology. Free Press, New York
McDougall W (1908) Introduction to social psychology. Methuen, London
Meeker BF, Leik RK (1995) Experimentation in sociological social psychology. In: Cook KS, Fine GA, House JS (eds) Sociological perspectives on social psychology. Allyn & Bacon, Boston, pp 630–649
Meier CS (1999) Preface. In: Commemorative Practices in Science: Historical Perspectives on the Politics of Collective Memory, Osiris 14(1):ix–xii.
McMahon AM (1984) The two social psychologies: post-crises directions. Annu Rev Sociol 10: 121–140
Michaels JW, Green DS (1978) Behavioral sociology: emergent forms and issues. Am Sociol 13: 23–29
Milgram S (1965) Some conditions of obedience and disobedience to authority. Hum Relat 18: 57–76
Mills CW (1959) The sociological imagination. Oxford University Press, New York
Moir J (2015) Social psychology: discipline, interdiscipline or transdiscipline? In: Mohan B (ed) Construction of social psychology. inScience Press, Lisbon, pp 15–24
Moscovici S (1972) Society and theory in social psychology – what is "social" in social psychology? In: Israel J, Tajfel H (eds) The context of social psychology. Academic Press, London
Moscovici S (1979) Rejoinder. Eur J Soc Psychol 9:441–454
Moscovici S (1990) Questions for the twenty-first century. Theory Cult Soc 7:1–19
Moscovici S (1993) The invention of society: psychological explanations for social phenomena. Polity Press, Cambridge
Moscovici S, Markova I (2006) The making of modern social psychology: the hidden story of how an international social science was created. Polity Press, Cambridge
Newcomb TM (1950) Social psychology. Dryden Press
Newcomb TM (1951) Social psychological theory: integrating individual and social approaches. In: Rohrer J, Sherif M (eds) Social psychology at the crossroads. Harper and Brothers, New York, pp 31–49
Newcomb TM (1953) An approach to the study of communicative acts. Psychol Rev 60:393–404
Newcomb TM (1974) Theodore M. Newcomb. In: Lindzey G (ed) A history of psychology in autobiography, vol 6. Prentice-Hall, New York, pp 365–392
Nuttin (1954) Rapport spécial sur l'enseignement de la psycholgie social dans les sciences sociales. In: de Bie P, Lévi-Strauss C, Nuttin J, Jacobsen E (eds) Les Science Sociales dans l'enseignement supérieur: Sociologie, Psychologie, et Anthropologie Culturelle. UNESCO, Paris, pp 134–159
Nuttin (1990) In memoriam: John T. Lanzetta. Eur J Soc Psychol 20:363–367
Obradovic S, Nicholson C (2015) Reflections on the vision of social psychology. Retrieved from http://www.easp.eu/news/itm/reflections_on_the_vision_of_social_psychology-12.html
Oishi S, Kesebir S, Snyder BH (2009) Sociology: a lost connection in social psychology. Personal Soc Psychol Rev 13(4):334–353
Parker I (1989) The crisis in modern social psychology – and how to end it. Routledge, London
Parker I, Spears R (eds) (1996) Psychology and society: radical theory and practice. Pluto Press, London
Pettigrew TF (1991) Is unity possible? A summary. In: Stephan CW, Stephan WG, Pettigrew TF (eds) The future of social psychology: defining the relationship between sociology and psychology. Springer-Verlag, New York, pp 99–121
Pettigrew TF (1997) Personality and social structure: social psychological considerations. In: Hogan R, Johnson J, Briggs S (eds) Handbook of personality psychology. Academic Press, San Diego, pp 417–438
Pettigrew TF (2018) The emergence of contextual social psychology. Personal Soc Psychol Bull 44(7):963–971

Pettigrew TF (2020) History of social psychology at mid-20th century. Oxford research encyclopedia of psychology. Retrieved 22 March 2022, from https://doi.org/10.1093/acrefore/9780190236557.013.509

Pettigrew TF (2021) Contextual social psychology: reanalyzing prejudice, voting and intergroup contact. American Psychological Association, Washington

Potter J, Wetherell M (1987) Discourse and social psychology: beyond attitudes and behaviour. Sage, London

Pugh D (1966) The first European summer school in social psychology, 1965. Bull Br Psychol Soc 19(64):35–37

Reicher S (2011) Promoting a culture of innovation: BJSP and the emergence of new paradigms in social psychology. Br J Soc Psychol 50:391–398

Rosenberg M, Turner RH (eds) (1981) Social psychology: sociological perspectives, 2nd edn. Basic Books, New York

Ross EA (1908) Social psychology: an outline and source book. Macmillan, New York

Rouse J (1987) Knowledge and power: toward a political philosophy of science. Cornell University Press, Ithaca

Rouse J (1999) Understanding scientific practices: cultural studies of science as a philosophical program. In: Biagioli M (ed) Science studies reader. Routledge, New York

Sarason SB (1981) Psychology misdirected. New York: Free Press

Saxton SL, Hall PM (1987) Two social psychologies: new grounds for discussion. Stud Symb Interact 8:43–67

Schruijer SGL (2010) Is the EAESP a cold war baby? An investigation into the political context of its formation. Unpublished manuscript

Schruijer SGL (2012) Whatever happened to the "European" in European social psychology? A study of the ambitions in founding the European Association of Experimental social psychology. Hist Hum Sci 25(3):88–107

Schwarz N (1998) Warmer and more social: recent developments in cognitive social psychology. Annu Rev Sociol 24:239–264

Sewell WH (1989) Some reflections on the golden age of interdisciplinary social psychology. Annu Rev Sociol 15:1–17

Sherif M (1936) The psychology of social norms. Harper & Row, New York

Smith P (2005) Is there an indigenous European social psychology? Int J Psychol 40(4):252–262

Smith R (1997) The Fontana history of the human sciences. Fontana, London

Smith R (2022) Psychologies: Their Diverse Histories. In: McCallum D. (eds) The Palgrave Handbook of the History of Human Sciences. Palgrave Macmillan, Singapore. https://doi.org/10.1007/978-981-15-4106-3_77-2

Solovey M (2014) Project Camelot. In: Teo T (ed) Encyclopedia of critical psychology. Springer-Verlag, Berlin, pp 1515–1520

Solovey M, Cravens H (eds) (2012) Cold war social science: knowledge production, liberal democracy, and human nature. Palgrave Macmillan

Stainton Rogers R (1995) Q methodology. In: Smith J, Harré R, Van Langenhove L (eds) Rethinking methods in psychology. Sage, London, pp 178–192

Stam H (2000) Ten years after, decade to come: the contributions of theory to psychology. Theory Psychol 10(1):5–21

Stam HJ, Radke HL, Lubek I (2000) Strains in experimental social psychology: a textual analysis of the development of experimentation in social psychology. J Hist Behav Sci 36(4):365–382

Star SL, Griesemer JR (1989) Institutional ecology, 'translations' and boundary objects: amateurs and professionals in Berkeley's Museum of Vertebrate Zoology, 1907–39. Soc Stud Sci 19:387–420

Stephan CW, Stephan WG (1990) Two social psychologies: an integrated approach, 2nd edn. Dorsey Press, Homewood

Stephan CW, Stephan WG, Pettigrew TF (eds) (1991) The future of social psychology: defining the relationship between sociology and psychology. Springer-Verlag, New York

Stephenson W (1953) The study of behavior: Q-technique and its methodology. University of Chicago Pres, Chicago

Still AW, Good JMM (1992) Mutualism in the human sciences: towards the implementation of a theory. J Theory Soc Behav 22:105–138

Still and Good (1998) The Ontology of Mutualism, Ecological Psychology 10(1):39–63

Strobe (2013) Scientific fraud: Lessons to be learnt. European Bulletin of Social Psychology 25 (1):5–12

Stryker S (1977) Developments in "two social psychologies": toward an appreciation of mutual relevance. Sociometry 40:145–160

Stryker S (1987) The vitalization of symbolic interactionism. Soc Psychol Q 50:83–94

Stryker S (1991) Consequences of the gap between the "two social psychologies". In: Stephan CW, Stephan WG, Pettigrew TF (eds) The future of social psychology: defining the relationship between sociology and psychology. Springer-Verlag, New York, pp 83–98

Stryker S (1997) "In the beginning there is society": lessons from a sociological social psychology. In: McGarty C, Haslam SA (eds) The message of social psychology: perspectives on mind in society. Blackwell, Oxford, pp 315–327

Sturm T, Mülberger A (2011) Crisis discussions in psychology – new historical and philosophical perspectives. Stud Hist Phil Biol Biomed Sci 43(2):425–33

Tajfel H (1972a) Some developments in European social psychology. Eur J Soc Psychol 2:307–322

Tajfel H (1972b) Experiments in a vacuum. In: Israel J, Tajfel H (eds) The context of social psychology. Academic Press, London

Tajfel H (1974) Social identity and intergroup behaviour. Social Science Information 13(2):65–93

Tajfel H, Fraser C (1978) Introducing social psychology. Penguin, London

Tajfel H, Turner J (1979) An integrative theory of intergroup conflict. In Austin WG, Worchel S (Eds.), The social psychology of intergroup relations (pp. 33–47). Monterey, CA: Brooks/Cole

Valsiner J (2014) Invitation to cultural psychology. Sage, London

Valsiner (2018). Culture, mind, and history: Building on the contribution of Gustav Jahoda, Culture & Psychology 24(3):398–400.

Valenti SS, Good JMM (1991) Social affordances and interaction 1: introduction. Ecol Psychol 3: 77–98

van Gils MR, Koekebakker J (1965) The first European summer school on social psychology, The Hague, July 15-August 11, 1965. Items 19(4):50–54

van Strien P (1997) The American "colonization" of northwest European social psychology after World War II. J Hist Behav Sci 33(4):349–363

Wallerstein et al (1996) Open the social sciences. Report of the Gulbenkian commission on the restructureing of the social sciences. Stanford University Press, Stanford

Weigert AJ, Gecas V (1995) Multiplicity and dialogue in social psychology. J Theory Soc Behav 25:141–174

Wertsch JV, del Rio P, Alvarez A (1995) Sociocultural studies: history, action and mediation. In: Wertsch JV, del Rio P, Alvarez A (eds) Sociocultural studies of mind. Cambridge University Press, Cambridge, pp 1–34

Wetherell M (2011) The winds of change: some challenges in reconfiguring social psychology for the future. Br J Soc Psychol 50:399–404

Wieser M (2020) The concept of crisis in the history of Western psychology. In: Oxford research encyclopedia of psychology. Retrieved 25 Mar 2022, from https://oxfordre.com/psychology/view/10.1093/acrefore/9780190236557.001.0001/acrefore-9780190236557-e-470

Wrong DH (1961) The oversocialized conception of man in modern sociology. Am Sociol Rev 26: 183–193

Of Power and Problems: Gender in Psychology's Past

Elissa N. Rodkey and Krista L. Rodkey

Contents

Introduction: Of Power and Problems: Gender in Psychology's Past	1036
Problems: The Research	1037
Womanless Psychology: Gender in Early Academic Psychology	1038
Woman as Problem: Gender in Clinical Contexts	1042
Historiographic Approaches to Psychology's Women	1043
Power: The Material Conditions of Science	1046
Laboratories and Other Necessary Resources	1048
Reproductive Labor	1051
Intellectual Community and Gendered Spaces	1053
Conclusion	1058
References	1059

Abstract

This chapter traces the history of gender in American psychology by analyzing the gendered nature of both psychology's problems (the research questions considered urgent) and its power structures. This approach goes beyond the more traditional compensatory approach to women in psychology to examine the ways in which psychology is itself gendered, i.e., how its foundational concepts are distorted by androcentric bias and gender stereotypes and its scientific culture is primarily unmarked masculine. An important part of this approach is a focus on the material resources necessary for doing science, which counters a celebratory approach to scientists and reveals the ways in which women and non-white men were systematically denied such resources. Often these resources are social, such as the requirement for intellectual community. The allocation of

E. N. Rodkey (✉)
Crandall University, Moncton, NB, Canada
e-mail: elissa.rodkey@crandallu.ca

K. L. Rodkey
Santa Barbara, CA, USA

© The Author(s), under exclusive licence to Springer Nature Singapore Pte Ltd. 2022
D. McCallum (ed.), *The Palgrave Handbook of the History of Human Sciences*,
https://doi.org/10.1007/978-981-16-7255-2_79

resources tends to happen according to existing power structures and creates hierarchies and exclusive groups who use their "cognitive authority" to shape the discipline. For the sake of specificity, this chapter explores these dynamics through the life of Eleanor Gibson, and two exclusive male-only societies – The Experimentalists and The Psychological Round Table.

> **Keywords**
>
> Sexism · Power · Gender · Hierarchies · Authority · Women · Feminism · Masculinity · Intellectual Community · Material resources · Reproductive labor · Academic sociality · Cognitive authority · Objectivity · Gender essentialist · Celebratory

Introduction: Of Power and Problems: Gender in Psychology's Past

A sense of déjà vu is an occupational hazard to those who devote any significant time to studying women in the history of psychology. The first exciting observation of repetitions, of common patterns in these women's lives, gives way to a dull sense of inevitability – of course she wasn't allowed in the lab, of course anti-nepotism rules limited her career, of course she was double-crossed by a male colleague. There is a curious sameness in these stories that persists despite diverse locations, theoretical schools, and historical periods. In short, there is a unity to these stories.

This seems ironic in the face of psychology's much-lamented lack of unity. One can make sense of these repetitions by understanding them as evidence of shared ideology or cultural norms invisibly at work. To state it clearly, one of the few ways in which the sprawling field of psychology has historically been unified was its unmarked masculine culture. Even as psychology grew and diversified, it remained constant in one thing – it uncritically adopted and reinforced common cultural understandings of gender. This phenomenon is best understood with reference to the characteristics of psychology that Roger Smith explores in the introduction to this volume. The history of gender in psychology provides powerful evidence for the importance of "looping," "reflexivity," and the idea that psychologies are "historically located ways of bringing into existence the states of being that they themselves study" (▶ Chap. 1, "What Is the History of the Human Sciences?," by Smith, this volume).

Thus, the task of the historian of gender in psychology is broader than mere recovery. Rescuing forgotten female psychologists from oblivion is not enough; even this noble task must be done with the aim of discovering the ways in which psychology is itself gendered. Those engaged in such gendering projects have examined such things as psychology's gendered academic culture, theories, and practices (Furumoto 1988; Rutherford et al. 2015; Rutherford 2020), sexual harassment as part of academic culture and as experimental technique (Young and Hegarty 2019), gendered views of eminence and appropriate work/life balance (Rutherford 2015; Johnston and Johnson 2017), psychology's construction of sex/gender (Morawski 1985; Lewin 1984; Shields 2007), the masculinity of particular subfields and in the construction of scientific identity (Nicholson 2001), and the performance of masculinity in famous psychology experiments (Nicholson 2011).

The perspective of such work is articulated most boldly by Naomi Weisstein (1968) in "Kinder, Kirche, Kuche as Scientific Law: Psychology Constructs the Female." Weisstein's statement, "Psychology has nothing to say about what women are really like...because psychology does not know" (1968: 197) is one way of making a point that could be made with any number of examples – that the androcentric bias in psychology has profoundly shaped its research program and findings. The view that psychologists' blindness around the issue of gender has spoiled its research, or at least curtailed its usefulness, is increasingly understood, if not fully accepted.

In addition to understanding the gendered nature of research programs and concepts, a gendering psychology project must also analyze the role of power in psychology. This includes such things as the spaces in which science is done, its cultural norms and social practices, and its necessary material resources. In addition to analyzing the obvious operations of power within psychology (who held prestigious positions, received funding, had graduate students, etc.) attention must be paid to the subtle operation of power to establish the spaces and roles in psychology available to women.

Thus, in addition to attending to "problems," i.e., the research questions that psychology considered urgent, gendering psychology projects also look at power, the unequal distribution of the material conditions necessary for science. To put it concretely, in addition to analyzing the questionable scientific status of Freud's "penis envy," such a project also means considering the brass penis that served as a gavel in the secretive Psychological Round Table meetings. Of this grotesque object, more later. But for the moment let it – a casual masculine assertion of power in a completely male space – serve as a symbol for the invisible but nonetheless potent operation of power in psychology.

First, a couple technical notes. This chapter focuses almost exclusively on gender in American psychology. This is to a certain degree necessitated by the available scholarship – research on gender in the history of psychology is very American-centric; taking a more international view requires reliance on more general history of science scholarship. An American focus has been chosen here to allow the analysis be concrete and specific, and because American psychology has been so influential in shaping and defining psychology internationally. While the details of psychology's gender dynamics differ across national boundaries, it is likely that the larger patterns of power and inequality will resonate beyond an American context. Another important note: the concept "gender" itself has a history (Rutherford 2019) and for much of psychology's history the modern distinction between sex and gender did not exist. Nevertheless, for the sake of clarity, this chapter uses the modern distinction throughout.

Problems: The Research

There are two routes into the intellectual history of gender in psychology. One starts with experimental psychologists, tracing how they largely ignored the existence of women, blithely conceptualizing the humanity they studied as male. The other route is through the preoccupations of the physicians and therapists for whom women were often the subject of intense interest, usually as a problem to be solved. Although these two routes would seem to be contradictory approaches, both have their origins in the gender ideologies of the Victorian era. After Crawford and Marecek (1989) we can describe the former as "womanless psychology" and the later as the "woman as problem" framework. The irony is that there were women who participated in "womanless psychology" and that sex differences was a popular research topic in this type of psychology. To clarify, Crawford and Marecek use this term to describe pre-feminist psychology, characterized by the assumption that women's experiences are "too unimportant to be a focus of inquiry" (1989: 149). There are modern examples of research operating with this assumption, but with the advent of psychology of women as a recognized disciplinary focus, womanless psychology officially came to an end. In contrast, "woman as problem" is a widely applicable term, describing not only early pre-feminist psychology, but modern research, including even ostensibly feminist work.

Womanless Psychology: Gender in Early Academic Psychology

Womanless psychology arose from a culture of strong gender stereotypes, clearly defined gender roles, and Darwinian evolutionary ideas. These combined in early psychology theorizing in ways are challenging to untangle. For example, take the statement of G. Stanley Hall, the first president of the American Psychological Association (APA): "To be a true woman means to be yet more mother than wife. The madonna conception expresses man's highest comprehension of woman's real nature" (Hall 1918: 297). This belief was the product of cultural ideas about gender and Hall's genetic psychology perspective that emphasized high-quality children and childrearing. Darwin is a significant influence here, both as the source of evolutionary theory that was spun off into a number of variations (with varying degrees of scientific accuracy) and as the ur-scientist, the model of leisurely gentlemanly observation.

Hall exemplifies how psychologists took enthusiastically to integrating evolutionary theory into psychology. Hall's embrace of the idea of recapitulation – the idea that children relived the various stages of human evolutionary development – profoundly shaped his psychological thinking, including his most famous contribution, the concept of adolescence. Recapitulation reinforced Victorian gender dynamics. Adolescence was a critical moment of transition (religious conversion was best at this age [Vande Kemp 1992]) and it was vital for the adolescent girl to be educated toward motherhood and not coeducationally (Diehl 1986; Russett 1989). In contrast, boys were considered to be reliving the primitive age, and needed the freedom to express their violent and sexual tendencies in order to progress to civilized manliness (Bederman 1995). As Bederman argues, this theory arose out of Hall's personal experience – the convergence of repressive Victorian Protestant views of sexuality and Hall's struggle for self-mastery against the temptations of masturbation during his youth. Such "self-abuse" was illicit, not only for moral reasons, but because it was thought to waste nerve force, a limited resource, and might result in illness (Bederman 1995).

While Hall's religious and evolutionary theories have been abandoned, adolescence as a time of "storm and stress" remains, invisibly shaped by Hall's beliefs about gender as well as his adolescent struggles with his male body. Although Hall is perhaps unusual in his intense interest in masculinity, he is typical in letting his gendered assumptions influence his research. The observation that psychologists often used exclusively male experimental subjects to draw conclusions about human nature has been made by many; less frequently acknowledged is that even when subjects included women, the experimenter's imagined universal human was likely male. Culturally influenced gender stereotypes were not limited to human subjects; heterosexuality was read onto rats, with male heterosexuality seen as "a robust entity capable of surviving all manner of adversity" (Pettit 2012: 219). Similarly, the "harem" analogy was used to describe animal colonies (Bharj and Hegarty 2015). As this term suggests, psychology embraced a colonial view of the world: the imaged universal human was more specifically a white male.

The "doctrine of separate spheres" – the idea that women belonged in private, domestic spaces, and men in public, intellectual spaces – was another Victorian idea which influenced early psychology. It was undergirded by gender essentialist beliefs that men and women possessed essentially different human natures, with women inclined more toward nurturance and moral excellence; men toward intellectual and political achievements. Such assumptions were incorporated into early individual differences testing, when psychologists began to measure everything from grip strength to wingspan (Daston 1992).

The rise of mental testing led to the acceptance of the variability hypothesis, the belief that men were more variable than women. Accordingly, early psychologists argued that there were more male "geniuses" and male "imbeciles" and more women of average intelligence, with the greater number of men in asylums mustered as supporting evidence (Shields 1975, 1982). This view conveniently justified the status quo – men occupied positions of cultural power because they were more fit to rule. Such findings ignored gendered patterns of childrearing, treating any differences found as natural. This faulty reasoning was also apparent in another pastime of early psychologists: discovering persons of eminence in the past. The role of nurture – child rearing, education, financial resources, etc. – was ignored in favor of heredity when considering why the Darwin family, for example, had produced several "geniuses."

It was left to the first generation of female psychologists to point out the logical and methodological flaws of such scholarship. Although many women faced barriers to higher education, those allowed to study academic psychology often brought a critical eye to sex differences research. Both Mary Whiton Calkins and Leta Stetter Hollingworth debunked the variability hypothesis (Nevers and Calkins 1895; Hollingworth 1914) and Helen Thompson Woolley took up sex differences in her dissertation *The Mental Traits of Sex* (Thompson 1903). Woolley summed up the field of psychology of sex in the following terms:

> There is perhaps no other field aspiring to be scientific where flagrant personal bias, logic martyred in the cause of supporting a prejudice, unfounded assertions, and even sentimental rot and drivel, have run riot to such an extent as here. (Woolley 1910: 340–341)

Aware of the intense scrutiny their work would face, Woolley and other female psychologists avoided confirming evidence in existing data sets; they subjected such claims to experimental test. Their research generally found small, if any, sex differences. And these researchers pointed out that even large differences would have been impossible to interpret since social influences had not been controlled for. In 1896 Amy Tanner expressed the problem with looking to society for evidence of women's psychological qualities, "The real tendencies of women cannot be known until they are free to choose, any more than those of a tied-up dog can be" (Pettit 2008: 150).

Yet many male psychologists resisted criticisms and dismissed these findings, so convinced were they of the existence of large sex differences. Ten years after completing her sex differences research, Woolley found little had changed in the field: "the truest thing to be said at present is that scientific evidence plays very little part" (Woolley 1910). As late as 1951, E. G. Boring was drawing on traditional gender stereotypes to account for "the Woman Problem" (1951). "The Woman Problem" was that there were fewer women leaders and researchers in psychology. Boring thought this could be explained by the preference of women to deal with the personal and specific and to eschew the theoretical questions that brought scientific recognition. This even led Boring to exaggerate gender differences in the survey that his co-author Alice Bryan conducted on the interests of male and female psychologists (Rutherford 2015).

Yet throughout his work on this topic, Boring conceptualized his own approach as objective and data-driven and saw Bryan and others who suggested the differences might be the result of discrimination as subjective and personally motivated (Rutherford 2020). Boring's perspective on this topic highlights the importance of scientific objectivity as a key concept for gendering psychology projects. Boring, influenced by E. B. Titchener, saw objectivity as a core scientific virtue, but also a specifically masculine characteristic. To be female was to be inherently subjective and partial. Boring's vision of experimental psychology is particularly relevant, since the man popularly known as "Mr. Psychology" had an outsized effect on shaping the identity of American psychology through his influential *A History of Experimental Psychology* (1929). Given early psychology's insecurity about its scientific status, perceived objectivity would play an outsized role in conferring legitimacy upon particular specialities and methods (Green 2010; Daston and Galison 2007).

Still gender essentialist views persisted and were written into psychology's authoritative texts, for example, personality tests, as in the case of Lewis Terman and Catharine Cox Miles's Attitude–Interest Analysis Test, which they designed to reliably distinguish between male and female participants (Terman and Miles 1936). Their psychometric approach was a bi-polar measure that discarded any items that men and women did not usually answer differently, thus reifying common gender stereotypes: "masculine minds were adventurous, mechanically and object-oriented, aggressive, self-assertive, fearless, and rough; feminine minds were aesthetically and domestically oriented, sedentary, compassionate, timid, emotional, and fastidious" (Rutherford 2020: 350). This type of research not only erased sexual minorities, but reinforced heterosexuality as the norm. Intelligence testers like Terman were suspicious of bright children whose personalities did not "match" their sex, a possible symptom of homosexuality (Hegarty 2007, 2013). The assumption that male and female psyches differed significantly meant that anyone who did not conform to gender stereotypes was in danger of being pathologized (Hubbard 2019).

Essentialist assumptions can be seen in research from psychology subfields not explicitly gender-related, and this affected popularization efforts. The public generally shared psychologists' beliefs about gender, and the prospect of an enthusiastic audience no doubt influenced psychologists' decisions about what avenues to pursue. For example, Nicholson (2011) demonstrates how the Milgram experiments resonated with Cold War anxieties about masculinity; their "spectacle of masculinity" (257) serving as a morality tale, summoning modern men to resist the emasculating effects of post-war life and heroically reject authoritarian orders. Similar landmark studies, such as Solomon Asch's conformity experiment, the Stanford Prison Experiment, and the Robber's Cave Experiment, also used exclusively male subjects and thus presented an unmarked masculinity in the guise of "human nature." The stark moral choices the subjects faced – between conformity or truth-telling, cruelty or compassion, competition or cooperation – thus played into cultural expectations or worries about masculinity.

Research findings that did not match cultural gender roles were often modified in the retelling. Vicedo (2009) shows how Harry Harlow's multiple experiments with rhesus monkeys and complex theories about multiple affectional systems were simplified into a single experiment that demonstrated the instinctual need for mother love. Even the choice of images reproduced for articles and textbooks was selective, resulting in historical amnesia, as in the visual cliff experiment. Photographs depicting the diverse animal experiments on the cliff were abandoned to obscurity in favor of images imbued with middle class domesticity: a mother calling to a baby over a checked cloth surface (Rodkey 2015). Sometimes it was not straightforward rejection or acceptance, as Rutherford (2003) shows with B. F. Skinner's Air Crib and teaching machines. These inventions were compelling because of mid-century cultural trends around domesticity and consumption but also repellant because they threatened key feminine nurturing roles (mother and teacher).

Ultimately the end of womanless psychology was foreshadowed in the critiques of early sex differences research. The methodological sloppiness of this research opened up a chink in the armor: psychology could be critiqued for failing to live up to its own objective standards, and those standards could be further be revealed to be inappropriate. The Association for Women in Psychology (founded 1969), the Report of the Task Force on the Status of Women in Psychology (published 1970) and APA's Division 35 (founded 1973) helped define psychology of women and feminist psychology (two overlapping but distinct subfields), as important parts of psychology. Ironically, even as the field's practitioners have become increasingly female, the variability hypothesis lives on in evolutionary psychology, and psychological theories built on gender essentialist assumptions remain popular (Greene 2004). In contrast to the field of sociobiology, where feminist critiques were incorporated into evolutionary theory, in evolutionary psychology the two have remained at odds (Liesen 2007). The imagined evolutionary past is frequently reconstructed using modern gender stereotypes. For example, Weissman (2020) shows how the concept of aggression as an innate and inescapable human characteristic (e.g., killer-instinct, the alpha male, etc.) arose in the 1960s more out of good storytelling than strong evidence, yet remains seen as a fact of human nature (see also Schmidt 2020 on the origins of the midlife crisis).

Woman as Problem: Gender in Clinical Contexts

To return to psychology's origins, another piece of the "sentimental rot and drivel" early female psychologists had to contend with were assumptions that women's bodies and minds were more fragile and ill. This idea had deep historic roots, as can be seen in the ancient diagnosis of hysteria, and often drew on medical theories that were no longer regarded as scientifically valid. For example, Victorian worries that higher education would threaten women's reproductive capacity seem influenced by outdated humoral theories (blood would be drawn away from the uterus to the brain) (Diehl 1986). Nineteenth-century precursors to psychology, such as physiognomy, traded in age-old gender stereotypes, locating women's unique propensity to illness and criminality in their sexed bodies (e.g., Lombroso and Ferrero 1895).

Even someone like Jean-Martin Charcot, who argued that hysteria was not a peculiarly female illness, still embodied in his dramatic demonstrations with hysterics the role of the male expert manipulating the distressed female body. As Danziger (1990) notes, the psychology that emerged in the French context was influenced by the medical field, pairing an authoritative medical-experimenter and a subordinate subject/patient, in contrast to the more egalitarian German psychological model of students taking turns experimenting upon each other. This French model was influential in American psychology's clinical domains – a dynamic one can spot in everything from the typical gendered dynamics of mesmerism (Winter 1998) to Freud's female patients. Freud contributed to the gendering of clinical psychology, not only in his dependence upon female patients, but in his sexual theories. He introduced concepts (penis envy, Oedipus complex, seduction theory) which would become mainstays in the discipline, influencing how the public and therapists alike thought about gender (Chodorow 1991). Even Karen Horney's addition of womb envy (Horney 1967) did not change the basic dynamic of locating psychological distress in the sexed body and mind.

The long-lived dynamic of the male scientific authority and the female subject can be understood in the context of another long history: the practice of personifying science as male and nature as female. Carolyn Merchant's *The Death of Nature* (1980) traces the tradition of the association between women and nature, and the sexually-freighted metaphors of early modern scientists, especially Francis Bacon. Merchant connects the tradition of depicting nature as a veiled woman with Bacon's descriptions of the domination and even torture of nature in pursuit of knowledge. Although Bacon sometimes uses marital images to describe the relationship between the masculine inquirer and feminine nature, the model for the scientific method is clearly the subjugation of an unruly wife, even by violent sexual means.

In this model of science woman is a problem to be solved; the exercise of power is necessary for science. Nature must be pinned down and stripped for inspection, her veil or skirts raised. In this view, feminine disease is no surprise – it is part of the disorder and mystery of nature. This tradition can be seen even in the most basic biological conceptualizations, such as the depiction of eggs as passive and sperm as active and competitive (Martin 1991). Standards of wellness or normality were generally constructed using male subjects, with women prone to being pathologized once their data was added. Whether it was their messy hormones or deficient moral reasoning, women were a problem.

"Woman as problem" had many twentieth century iterations: mother-blaming for their childrens' problems, the construction of Postpartum Depression and PMS to account for feminine distress in hormonal terms, and naturalistic explanations for women's greater needs for anti-depressants (Rutherford et al. 2012; Held and Rutherford 2012). Often clinicians were complicit in the "adjustment" of their female clients to their sexist social reality, the goal middle-class respectability.

In 1963, Betty Friedan's *The Feminine Mystique* (Friedan 1963) turned such reasoning on its head, locating the cause of suburban feminine malaise – the "problem that has no name" – outside women's bodies, and in society and in women's disempowerment. Female psychologists resonated with this analysis and contributed their own academic equivalents, such as Phyllis Chesler's *Women and Madness* (1972), which argued that women often only seemed mad because of sexist norms. The feminist movement of the 1970s allowed a reframing of women's bodies, minds, and even the goals of counseling. Such popular works as *Our Bodies, Ourselves* (Boston 1973) were accompanied by important academic contributions on sex differences (Maccoby and Jacklin 1974), androgyny (Bem 1974), and the distinction between "sex" and "gender" in research (Unger 1979). Feminist psychologists critiqued counseling psychology (Broverman et al. 1970; Marecek and Hare-Mustin 1991), including advocating for the creation of the first APA ethics rule banning therapist-patient sex (Kim and Rutherford 2015), and working to popularize terms like date and marital rape (Rutherford 2011). More recent efforts have targeted the DSM's pathologization of women's sexual problems (Kaschak and Tiefer 2001), though the "women as problem" perspective remains deeply entrenched. Yet feminist psychology challenged the default association between women and problem and popularized the idea that women might be sick thanks to sexism. Psychological studies of masculinity have further supported this perspective, radically identifying men as the problematic sex, i.e., subject to numerous unique pathologies.

Historiographic Approaches to Psychology's Women

As a result both of the end of "womanless psychology" in the 1970s and the development of history of psychology as a distinct discipline, historians worked to rectify the lack of female representation in psychology's histories. Initially scholars highlighted forgotten exemplary women, featuring Anna as well as Sigmund Freud, Mary Ainsworth alongside John Bowlby. These types of histories were necessarily very biographic-centric, often "Great Woman" narratives written using the model of "Great Man" histories. This celebratory approach had a natural appeal, since the challenges women faced in pursuing a career in psychology provided compelling stories of triumph over adversity. This "Up with our foremothers" perspective (Bernstein and Russo 1974) has been described variously as "recovery," "compensatory," and "re-placing women in psychology."

Scholars working on "recovery" did the difficult basic research to identify many women who had been forgotten or neglected, despite their achievements (see Russo and O'Connell 1980; O'Connell and Russo 1980; Russo and Denmark 1987; Furumoto and Scarborough 1986; Scarborough and Furumoto 1987; Stevens and Gardner 1982). As Stevens and Gardner's subtitle, "A bio-bibliographic sourcebook" makes clear, these works were frequently census-like, offering biographies that ranged from the comprehensive to mere brief preliminary sketches, with the intention of spurring further scholarship. Sometimes scholars went beyond this and offered analysis of the gendered cultural contexts that led to women's struggles. For example, Scarborough and Furumoto's (1987) *Untold Lives* used the biographies of five early American women in psychology to identify common challenges, or "the difference being a woman made."

Untold Lives focused on women in the first generation of American women in psychology (those who had PhDs by 1906) and this led to a generational analysis of women. Some continued to define the first generation (Milar 2000), who were likely to have had difficulty receiving graduate training in psychology and to struggle to balance familial obligations with a career. Johnston and Johnson (2008) used this generational framing to explore the second generation (PhDs between 1906 and 1945). These women, although less likely to be excluded from graduate studies, faced new, more subtle challenges such as anti-nepotism rules and "old-boy" network hiring practices. Regardless of preference, female psychologists were often tracked into less prestigious fields, such as developmental and applied psychology (Johnson and Johnston 2015). Others scholars have extended this generational analysis to other countries (for Europe see Gundlach et al. 2010; for Japan see Takasuna 2020; for England see Valentine 2020; for Canada see Gul et al. 2013) or have analyzed the common career struggles of women in particular subdisciplines (for child development see Cameron and Hagen 2005, for educational psychology see Aldridge et al. 2014).

The "recovery" of women in psychology largely succeeded: women were re-placed in psychology textbooks and the *Portraits of Pioneers in Psychology* series. Yet this approach's aims proved too modest, as seen by the manner in which women were introduced into textbooks. Female psychologists are usually located inside boxes kept apart from the text's narrative flow, limited to a decorative function, not requiring a radical rewriting of psychology's history. This lack of transformation is in part the result of the conservative nature of celebratory history. Rather than questioning the "Great Man" account that describes psychology as moving from triumph to triumph, the "Great Woman" approach reinforced this narrative.

While the impulse to recover forgotten women was necessary for the sake of justice and laid the groundwork for more complex analyses, continuing to pursue this historiographic approach has several downsides. For one thing, although recovery expands psychology's "Who's Who" to include eminent women, this approach requires women to fit male models of scientific success. This can lead to a misreading of historical evidence, as in the case of Milicent Shinn, who in *Untold Lives* was used an example of a daughter's duties keeping her from pursuing a career in psychology. In fact, following the completion of her PhD, Shinn continued her work in child development from home, recording observations of her nieces and nephews and mentoring a group of mothers observing their babies (Rodkey 2016; von Oertzen 2013), evidence that Scarborough and Furumoto seemed to have missed because they expected a masculine academic paradigm, not research conducted via informal networks and in domestic spaces. Even well intentioned celebratory feminist readings of the past can result in historians missing "unfamiliar feminisms" (Johnson and Johnston 2010) or in distorting a woman's life to fit modern standards of female empowerment. Evaluating women using a "Great Man" standard means curtailing narrative possibilities and missing more interesting stories (cf. Hegarty 2012b; Rutherford 2012).

While it is tempting to focus on eminent women, especially since the overwhelming career barriers meant that those women who did succeed in psychology were often extremely talented, such a focus reinforces the myth of meritocracy. This perspective ignores intersectional realities such as the impact of race and class operating with gender to influence access to academic and career opportunities. It was no accident that the few women who did succeed in the first generation of psychology were almost exclusively white and reasonably well off. This framing of forgotten women as geniuses not given their due also reinforces an inaccurately individualistic perspective on science, and it is limited to traditional sites of academic power and prestige. A truly inclusive psychological history would rescue from oblivion not only the successful scientists, but would be influenced by a history from below perspective (Porter 1985), including patients and experimental subjects (Hoffman et al. 2015; Morawski 2015) as well as the lowly secretaries, wives, graduate students, social workers, and research assistants that made psychology possible. This approach means adopting a critical distance from psychology's disciplinary norms and embracing Lerner's call to conduct history as "seen through the eyes of women and ordered by the values they define" (1979: 178).

Although scholarship doing simple rediscovery of forgotten women continues to be popular, a small number of scholars have used psychology's women less as biographical subjects and more to gain a critical perspective on the field. This often means adopting a contextual history (Bohan 2016) rather than a presentist orientation to trace modern intellectual categories. For example, Johnson (2015) explores how Florence Goodenough championed observation and rich contextual description in an era of intense pressure toward behaviorist objectivity; Pettit (2008) explores Amy Tanner's embodied methodology and eclectic research interests, which were both side effects of her exile to margins of psychology. Thus, rather than simply tracing the patterns of barriers women faced, this approach respects the agency of the women and highlights the countervailing attitudes and approaches that women brought to the field. Such scholarship usually comes from an explicitly critical feminist perspective, with the aim of challenging positivistic psychology. This approach, which one might call critical psychology through a historical lens, often emphasizes the operation of power and structural forces at work in the experiences of historical women in psychology.

Power: The Material Conditions of Science

The Great Man approach to the history of science promotes the myth of the individual genius who through brilliance heroically overcomes all obstacles to scientific discovery. This perspective renders invisible the material conditions required for conducting science. To use the example of the ur-scientist, it was Darwin's wealth, supportive wife, servants, and children that allowed him the devoted physical space, materials (instruments and specimens), leisure, and mental focus needed to develop his revolutionary theory. Yet a celebratory or Great Man approach downplays such advantages. In contrast, the common struggles of women in science reveal the operation of power. All too often women lacked the power to order their lives to pursue science or to join the spaces where science was taking place. What the repeating patterns in women's biographies indicate is that this disempowerment was not accidental: it was the result of structural features, not individual failures.

Hegarty (2007) notes that the analysis of power has largely been neglected in psychology, an ironic state of affairs since psychology today is "a site where power and knowledge are transformed into each other in particularly dense ways" (77). Hegerty calls for histories that attend to the connection between psychology's function in carving up people "into their properly natural and enculturated parts" to keep people in their place (2007: 75). Examples of this abound; perhaps most obvious is the behaviorist interest in the "prediction and control" of behavior. This is also visible in many less explicit ways, in almost any sort of social intervention, or in the discipline's obsession with expertise. This theme comes through clearly in psychologists' uncensored dreams: of the four utopias of early psychologists that Morawski (1982) highlights, three identify a scientific expert as the organizer of ideal societies. Although these three psychologists describe this masculine figure differently – "superman" researcher, and "high priest of souls" (G. S. Hall), "Seer" (William McDougall), or "behaviorist physician" (J. B. Watson) – all see him as wise, meriting deference. This expert is someone who ought to be entrusted with the power to organize society. These utopian visions reveal that, despite protests to the contrary or gestures toward an objective neutrality, psychologists have always aspired to change society, which necessarily means the exercise of power.

Importantly, psychology's disciplining works on people within psychology as well as without. This has meant that certain types of people have been excluded from working within psychology or limited to particular roles, kept in what is considered their proper place. Such a dynamic is the result of psychology having, from its beginnings, been imbedded within systems of patriarchy and white supremacy. Both of these systems create hierarchies, and social hierarchies are by nature unstable, requiring constant maintenance to keep from toppling. This results in the creation of particular structural or systemic mechanisms which make maintaining the hierarchy more sustainable.

The hierarchical nature of psychology reveals itself in a variety of ways, not least in the proliferation of binaries that function as the basis for hierarchies: experimental vs. applied, hard vs. soft, objective vs. subjective. These hierarchical organizations create less desirable spaces to which lower status members can be exiled, allowing the discipline to benefit from their work while still withholding legitimacy or status.

Psychology's hierarchical nature is also evident in the actions of eminent psychologists. Henry Tajfel created devoted "acolytes" via severe initiation (Young and Hegarty 2019: 465); Boring's extensive job recommendations gave him the power to act as strict gatekeeper for the discipline (Rutherford 2015; Winston 1998); and Titchener served as a "feudal lord" of his domain (Rutherford 2015: 253), summoning students to his home for meetings and requiring them to use his own idiosyncratic system of introspection for research. Indeed, when Titchener condescended to visit campus, for occasional, highly anticipated lectures, the hierarchical organization of the department was made explicit. Wearing his Oxford academic regalia, he would enter the classroom "in a very magnificent fashion" followed by a retinue of instructors and assistants who sat down in the front row "eager to find out if Titchener had discovered anything new" and "drank in every single word" (Friedline 1960). These power dynamics were not lost on Titchener's student Cora Fredline, who recalled that in a meeting she realized his cigar had set his beard on fire but waited to tell him because "I couldn't interrupt such a very important man in the middle of a sentence" (Bazar 2010).

When historians have considered these disciplinary dynamics they have often attributed them to psychology's insecurity as a young science (i.e., physics envy). As Carolyn Wood Sherif recalled of early twentieth century psychology, since there was a preexisting hierarchy of scientific prestige, each specialty in psychology "sought to improve its status by adopting ... the perspectives, theories, and methodologies as high on the hierarchy as possible" for an "appearance of rigor" (Sherif 1979/1998: 98). This insecurity can also be seen in early psychologists' hostility to psychoanalysis: they feared that "psychoanalysis attempts to creep in wearing the uniform of science and to strangle it from the inside" (Hornstein 1992: 256). In other words, what was at stake was scientific status. Anything that might drag psychology down, whether it was psychoanalysis or "soft" methods, was to be renounced.

Psychology was hardly unique in its insecurity. Shteir (1997) notes how in Victorian botany, a field with many female participants and perceived as feminized, John Lindley worked to differentiate polite botany ("an amusement for ladies") from botanical sciences (a serious and professional endeavor for men). Thanks to Lindley's work "botanist" eventually became masculine, and female contributors were pushed out. These exclusionary practices can be seen as a typical stage in the process of the professionalization of science. There are numerous examples of this phenomenon in the Victorian era, in which sciences in the process of professionalizing were busy "demarcating disciplinary practices, hiving off professional from amateur pursuits, distinguishing hierarchies of practice and articulating appropriate levels of discourse" (Shteir 1997: 33). Given existing patriarchal cultural norms, it is not surprising that this process was gendered.

Of course, often women were not completely excluded but their activity curtailed to particular lower-status domains. With the professional enclosure of medicine, for example, women were allowed to remain in nursing. Rather than ejecting women from professions, "the encirclement of women within a related but distinct sphere of competence in an occupational division of labor" (Witz 1990: 47) allowed for a convenient source of cheap labor. The hierarchical organization of science allows for a "division of labor that allocates scientific creativity and decision making to scientists and laboratory work to those assigned the role of technician or student" (Reskin 2007: 20). Within psychology, while the first generation of women were regularly denied access to universities or laboratories, following WWI women were generally accepted in the lower status roles of student and technician, with the elite professional roles reserved for men (Johnston and Johnson 2008; Furumoto 1987).

This social organization of psychology resulted in an uneven distribution of the material conditions required to do science. Access to well-equipped laboratories, to human or animal subjects, to government or academic funding and to elite colleagues, to permanent university positions, light teaching loads, and even the promise of public recognition in the event of success, were skewed toward male scientists. Women were tracked into what Rossiter (1980) has called "women's work" in science, jobs that were "so low-paying or low-ranking that competent men would not take them, which often required great docility or painstaking attention to detail, and those that involved social service, as working in the home or with women or children" (383; see also Furumoto 1987). This had a dual advantage for male scientists: they gained cheap labor while at the same time reinforcing their superior status.

The biographies of women who had ambitions to experimental psychology record the barriers they faced and the resources denied them. Yet reference to resources and barriers is euphemistic: what these women lacked was power. It was the greater power given to elite male psychologists that allowed them agency – to do scientific work of their choosing, to pick students and protégées, and to give or deny the material conditions required for science to others. It allowed their theoretical questions to be taken seriously, and their metaphysical commitments ignored. This agency, so unequally distributed, had long-term consequences for the form and direction of the discipline.

Laboratories and Other Necessary Resources

To illustrate more concretely the expansive implications of a lack of power we turn to an extended look at the life of Eleanor J. Gibson (1910–2002). Her life exemplifies challenges in accessing laboratory resources, reproductive labor, and intellectual community, three major types of material resources necessary for science. Gibson began her career in the egalitarian environs of Smith College, where she met and married James J. Gibson. Eager to do animal research and wanting "a superscientific, strongly experimental atmosphere" (Gibson 1979a), Gibson went to

Yale for her Ph.D., hoping to do primate research with Robert Yerkes. However, upon meeting to request his supervision, Gibson was shown the door, Yerkes proclaiming, "I have no women in my laboratory" (Gibson 2002: 21). Yerkes' comment about his laboratory may seem a non sequitur, but the gendered nature of science made the laboratory a masculine domain. When Ethel Puffer Howes first called on her professor Hugo Münsterburg, his maid was disapproving: "the young lady probably does not wish to go into the laboratory" (Scarborough and Furumoto 1987: 79). The notion that the laboratory environment was incompatible with femininity was not limited to the Victorian period. In 1961, as a graduate student at Harvard, Naomi Weisstein was told she couldn't use the school's equipment for her research because the men in the department needed it more, and because she might break it. Such behavior was the result of gender essentialist beliefs, as can be seen by Boring's 1967 comments that though women had "thoroughly infiltrated psychology," they "avoid the laboratory, so many of them, only because nature has given them the genes of the sociotrope" – i.e., they were too social (321).

In Gibson's case, Yerkes' prejudice not only barred her from his lab, but, practically speaking, from animal research altogether. As Gibson tellingly put it, at Yale, "Yerkes had the most impressive empire" (1980: 246) with his chimpanzees loudly proclaiming his dominance from the Institute of Human Relations rooftop. Yerkes was strategically located in comparative research networks (Pettit et al. 2015), with a virtual monopoly on chimpanzee research within psychology, making his refusal to supervise a definitive closed door. Instead Gibson did her PhD under Clark Hull, even though she was not a behaviorist and had to disguise her functionalist views (Gibson and Levin 1975).

Gibson returned to Smith after her PhD, but WWII was shortly to take her away from its women-friendly environment. Like so many male psychologists, J. J. Gibson was recruited to do his research for the military, and Eleanor went along for the ride as "camp follower." It was in the real-world context of training pilots that J. J. Gibson began to develop his ground-breaking ecological theory of perception. Eleanor had to content herself with weekend gatherings with the other researchers that offered "some psychological shop talk after my rather tiresome week of coping with wartime domestic problems and intellectual aridity" (Gibson 1991: 201).

Eleanor Gibson was hardly unique in her experience of WWII as a "latent period" in her career (Gibson 1980: 252). Despite the fact that the war offered women unprecedented opportunities for work outside the home, employment rates for US female psychologists declined between 1940 and 1944; at one point during the war 14% of women with doctorates in psychology were unemployed (Napoli 1981). This pattern of male psychologists unequally benefitting from the war was no accident. The Emergency Committee in Psychology (ECP), a group formed in 1940 to promote psychologists' contribution to the war effort, was an exclusively male group, and actively resisted efforts to include female psychologists. When a group of women approached the ECP protesting the lack of roles for women, they were told to be "good girls," and to wait patiently for an appropriate need to arise (Schwesinger 1943: 299). As Gladys Schwesinger recalled, they were told "our job was to keep the home fires burning" (1943: 298) and to do volunteer rather than paid war work

(Capshew and Laszlo 1986). Wearying of this passive role, in 1941 a group of women reluctantly established the National Council of Women Psychologists (NCWP), as a female equivalent to the ECP. Despite the NCWP's lack of feminist aims, the group received a hostile reception. Karl Dallenbach, chair of the ECP, responded, "If the women want it, let them have it, God bless them! But I think it is too bad, because I feel that there is no restriction on a woman anywhere in the profession" (Capshew and Laszlo 1986: 168). Dallenbach's feelings were contradicted by the data. In contrast to the over 1000 male psychologists who served in the US military during WWII, fewer than 40 women psychologists so served, and only half of these were employed as military psychologists (Capshew 1999). As a result, female psychologists tended to miss out on training and research opportunities in organizational and clinical psychology. Or even worse, opportunities tailored to their specific research interests, as in James Gibson's assignment as director of a research unit on perception. This gendered exclusion no doubt had cascading effects into the coming decades, making it more difficult for female psychologists to benefit from the lavish military and government research funding during the Cold War era. Russo and Denmark (1987) note that WWII also facilitated opportunities for male psychologists in business and industry, resulting in the creation of the male dominated industrial/organizational psychology. The powerful Veterans Association and the psychological needs of returning soldiers also contributed to a male-oriented knowledge base in counseling psychology (Marecek and Hare-Mustin 1991).

Following the war, the Gibsons moved to Cornell University, attracted by the fact that James could supervise graduate students. It was there that Eleanor would encounter her most persistent career challenges, exemplifying the problems of second generation women. The Cornell offer was to James alone and a spousal hire was out of the question because Cornell had an anti-nepotism rule. Nepotism rules, which were common at mid-century, were in theory gender neutral, simply keeping a married couple from serving on the same faculty. In practice, however, anti-nepotism rules disproportionately impacted women, who were less likely than their husbands to be offered full time faculty positions (Rossiter 1995). Gibson's case provides a clear cut example of this phenomenon. For the next 16 years, her only status at Cornell would be as an unpaid "research associate" – affiliated with the school, but with no laboratory access or funding. For the first 4 years she was not even able to apply for grants in her own name; this only changed when James threatened to leave Cornell. Gibson's blunt retrospective highlights her lack of agency in this period: "Couldn't I just set out to work on the research of my choice? No, I couldn't. One needs a lab, and I didn't have one" (Gibson 1977).

This lack of power is also clear in Gibson's first Cornell research project, working as an assistant at Howard Liddell's Behavior Farm. Once again, Gibson was working with a behaviorist without being in sympathy with his intellectual project, this time repeatedly shocking goats to induce experimental neurosis. However, Gibson made the best of the situation and used the experimental set up to design two studies of her own. The fate of these projects highlights a further risk of a lack of power within academia: even if one creatively adapted to a marginalized position, ownership of one's intellectual project remained tenuous. In the case of her first independent research project, Gibson was floored when Liddell attempted to commandeer her conference presentation, saying "You know, if you don't mind I think I'll just give the paper for you" even though "he had nothing to do with it, absolutely nothing" (Gibson 1998: 6). Gibson's second project was a study of development, using twin goats as experimental and control animals. This study came to an abrupt end when Gibson returned to the farm to discover that her control kids had been given away as Easter presents, an event that made her "absolutely furious" and prompted her to quit the Behavior Farm for good (Gibson 1998: 7).

What followed was a series of research projects that creatively used the resources available to Gibson but which resulted in research which one biographer described as being "on the barest fringes of the issues that concerned her" (Caudle 1990: 107). Gibson herself acknowledged this reality: "I did have a theme, a sort of direction, and opportunities, even very unlikely ones, can sometimes be bent to one's theme" (Gibson 1977). Gibson's actions in this period exemplify the "adapt and evolve" strategy common in the second generation (Johnson and Johnston 2010). The hostile academic environment and women's generally lower status roles required them to become "creative opportunists" (Johnston and Johnson 2008: 63) to survive. In Gibson's case, she frequently employed a strategy of "accepting a lesser short-term goal so that she could continue, uninterrupted, towards long-range objectives" (Caudle 1990: 105).

Gibson's creative opportunism sometimes involved collaborating with Cornell professors who had access to lab space. Her most famous research, the visual cliff experiment, was the result of her initiating research with a junior male colleague with laboratory access (Rodkey 2015). In these collaborations Gibson seems to have strategically exercised a gendered social style. Her husband noted that her success took:

> a lot of stubborn persistence. But she had to collaborate with me, and with many others, especially at first, so she had to be tactfully stubborn – or mildly, or sweetly, perhaps – but not plain stubborn. She was bound to be a threat to her male collaborators, of course, including me, so she had to cope with it by never seeming to threaten. (Gibson 1979b: p. xii)

This "charming stubbornness" meant disguising or downplaying her ambitions and opinions while waiting patiently for the opportune moment (Rodkey 2010b). Although Gibson was uncomfortable with a feminist reading of her life (Gibson 1976), she clearly was aware of the threats to her professional standing – for example, she refused to learn to type, lest she be asked to take meeting minutes. Similarly, "charming stubbornness" can be understood as a strategy of resistance. Thanks to the structural barrier of the nepotism rule, Gibson could not help being marginalized, but her charming stubbornness allowed her to operate smoothly within the margins of an elite university.

Reproductive Labor

In addition to her lack of access to a laboratory Gibson also lacked other, less obvious, material resources. Inequities in these areas have traditionally been less visible, arising as they do from socially accepted gendered patterns of work. Reproductive labor, a Marxist concept, describes the labor required to produce and sustain the next generation of laborers (Fraiser 2013). This caregiving work, which extends to care for the old or infirm, is usually financially uncompensated, despite its necessity for a functioning society. Women shoulder the majority of this type of work, not only physically bearing children but caring for their many needs. Reproductive labor is relevant even for unpartnered people and non-parents, since it extends to all the mundane chores necessary to sustain a worker between shifts, from meal preparation to home maintenance. In contrast to the heavy expectations of caregiving and household work for academic women, academic men have often been allowed to escape such duties, usually by means of female surrogates, who help free their time to focus on their academic work.

Even in the absence of other systemic barriers, this dynamic results in profound inequalities in academic life. Tenure expectations were built around the male academic, who it was assumed, had both a secretary and a wife, one to deal with professional mundanities, one to look after domestic matters. As the "thanks for typing" Twitter hashtag documents, often the personal/professional boundary was permeable, with wives serving their husbands as unpaid typists, editors, data analysts, and even uncredited co-authors. Although such helpful wives might be thanked in the preface to a book, the labor they (and paid help) performed remained largely invisible, rather than an explicit part of a male academic's job requirements.

Reflecting the traditional masculine academic norm, Boring once claimed that eminence in psychology required working 168 hours a week (Boring 1951), best done by members of a "scientific unmarried priesthood" (Rutherford 2015: 266, see also Noble 1992 on the clerical tradition in science). As Johnston and Johnson (2017) note, this level of concentration on work meant outsourcing all housekeeping and parenting duties to one's spouse or paid help. In contrast, women rarely had such help from partners. In the first generation, many academic women avoided such challenges by choosing not to marry, often substituting female housemates and social groups. Women's colleges provided a particularly supportive environment for such single women (Palmieri 1983).

If women did marry, they devoted much thought to how to they could order their domestic situation to allow them time for science. Ethel Puffer Howes (1872–1950) provides a striking example of this phenomenon. Howes struggled against the "intolerable choice" (Howes 1929: 16) between a career and family life. In the 1920s she wrote a series of articles which championed reorganizing the gendered domestic work load so as to demolish this dilemma: "The man demands of life that he have love, home, fatherhood, and the special work which his particular brain combination fits. Shall the woman demand less?" (Howes 1929: 19). At Smith College, Howes took practical action, designing innovative programs, such as childcare cooperatives and an inexpensive meal service, to lessen women's domestic drudgery (Johnston and Johnson 2017). These programs' wild popularity points to how universal this issue was. Gibson, thanks to some well-timed births and cheap domestic labor, seems to have gotten off relatively easy – her WWII camp follower years conveniently happened when her children were young and needed most care;

for the rest of her career, the Gibsons could afford a housekeeper or nanny to meet their children after school. Yet even so, domestic concerns remained something for Eleanor to manage. One of the striking similarities in biographies of women in psychology is their explanations of how they solved the challenges of work-life balance. These passages are noticeably missing from male psychologists' biographies.

Intellectual Community and Gendered Spaces

Rodkey and Rodkey (2020) recently introduced the idea of intellectual community as a resource necessary for producing science, one often denied marginalized scientists. They differentiate between official and unofficial intellectual communities – the official being places like universities or academic societies, the unofficial being more informal groups with less well defined boundaries and aims. These more open groups often are formed to compensate for the lack of access to more closed, official audiences. Ideas do not develop in a vacuum; every academic needs the feedback and encouragement of an engaged audience of peers (no matter how small) to produce fully formed ideas and disseminate them.

Considering the Gibsons through this lens provides another contrast between Eleanor and James. From this perspective, Eleanor, although technically a part of the Cornell community, remained marginal to it for her 16 years as research associate. Paradoxically, she both benefited from and was harmed by her marriage. Although she had access to Cornell circles thanks to James, due to nepotism rules her marriage meant she also was kept from accessing such sources of intellectual community as having her own graduate students. In contrast, James held weekly meetings with his graduate students to discuss his "Purple Perils" – short position papers that he used to develop his pioneering perception theory (Gibson 2002).

This divergence in the Gibsons' experience can be traced back to their graduate school training. Whereas Eleanor was frozen out by Yerkes, James, on the strength of his Princeton connections, was allowed to attend meetings of The Experimentalists (later the Society for Experimenting Psychologists or SEP), an exclusive all-male society, even while a non-member and graduate student. These professional contacts smoothed the way for James to be elected a member of The Experimentalists in 1939; Eleanor was not inducted until 1958, ending a 20 year streak of exclusively male members. Elite scientists use their influence on their students' behalf to gain entrance to selective intellectual communities, among other career benefits. As Addelson (1983) points out, this allows the ideas favored by such scientists to retain "cognitive authority," the stamp of legitimacy and rationality. Interestingly, the example Addelson gives of this is how Yerkes helped Clarence Ray Carpenter obtain funding and appointments that allowed him to pursue his primatology research. In stark contrast, for women like Eleanor Gibson, shut out at various points in her life from graduate mentorship, elite societies, and permanent university positions, unofficial intellectual community had to substitute. For Gibson, her main source of intellectual community appears to have been her husband, though she often downplayed this dynamic to avoid evoking the scientist's wife trope (Gibson 2002). Other marginalized scientists found intellectual community in friends, religious groups, and family members (Rodkey and Rodkey 2020). Yet while such unofficial groups might supply a stimulating intellectual environment, their lack of power meant that ideas honed in such groups did not carry the same cognitive authority as ideas generated in more official spaces.

The gendered nature of access to powerful spaces has fundamentally shaped the contours of the discipline, gendering particular spaces and specialities. The most obvious feminized space is developmental psychology, which since its beginnings in child study has had a strongly maternal association (Noon 2004). The feminine-marked nature of this sub-field meant that sometimes women who were not particularly interested in development nevertheless found themselves tracked into it. Canadian Lila Braine, who received her PhD in neuropsychology under Donald Hebb, encountered these pressures. When Braine asked for an assignment teaching neuropsychology, she was instead offered a developmental psychology class, despite the fact that she had never even taken a developmental class. When she objected, she was told "you'll learn, you're a woman. Just keep a chapter ahead" (Braine 2009: 13). Despite her initial ambivalence, Braine soon found there were advantages to being in a female-coded domain. Braine recalled that she "found it so much more hospitable. There were people to talk to in a different way. I didn't feel I had to be careful about what I said. I felt more welcomed" (2009: 13). This hospitable environment ultimately meant that Braine moved into developmental psychology, a decision that, according to Braine, was due to sexist pressures rather than her own preferences (Gul et al. 2013). Although ending up in the more feminized regions of psychology was rarely a first choice, nonetheless such career trajectories were often read as a sign of natural feminine interest or inclination rather than attributed to a lack of agency.

In contrast, the male-only regions of psychology were without exception elite spaces that men fought to enter. Of course, nearly all of psychology was invisibly masculine, but the strength of the masculinity varied – the field's "soft" to "hard" science spectrum corresponded roughly with a feminine/masculine binary, but with default masculine spaces predominating. The "hard science" regions were masculine both in terms of how the subject matter was perceived and because the regions were reserved for male scientists. Women who wanted to enter these spaces had to fight for admission, and were expected to accommodate themselves to an overtly masculine culture, sometimes to the extent of tolerating sexual harassment from peers or mentors (Young and Hegarty 2019).

The masculine nature of psychology in the early twentieth century can be seen in perhaps its purest form in its all-male societies, where women were deliberately excluded. Titchener specified his required conditions for The Experimentalists as "no women, smoking allowed, plenty of perfectly frank criticism and discussions" (Rutherford 2015: 256). The idea that rigorous male discourse would be impossible with women present was made explicit by one member: "they would undoubtedly interfere with the smoking and to a certain extent with the general freedom of a purely masculine assembly" (Furumoto 1988: 104). This all-male elite social space operated according to a "masculine code of professional sociability" (Rutherford 2015: 256; Nye 1997) that required women's absence. Boring depicts this dynamic in his recollection of his first time attending an Experimentalists meeting. "Dodge and Holt attacked Titchener on introspection. My wife-to-be and Mabel Goudge secreted themselves in a next room with the door just ajar to hear what unexpurgated male psychology was like" (Boring 1967: 322). Notable in this account are both the group's aggressive masculine intellectual style and the exclusion of interested women (both listening women were at the time completing PhDs in psychology) that added to the society's mystique and prestige. Another incident involving Boring and The Experimentalists illustrates how sex-based boundaries were strictly applied, even after the organization elected two women in 1928, after Titchener's death. One of these women, Margaret Floy Washburn, made the faux pas of joining her fellow members of the society for a meal at the Harvard Faculty Club by using the men's-only entrance. Although she wrote a letter of apology to Boring ("I hadn't the smallest intention of claiming a right in the Faculty Club's penetralia. I have never done the slightest thing to advance the cause of feminism" [M. F. Washburn to E. G. Boring, 10 April 1934]), as Rutherford points out, 15 years later Boring was still complaining that Washburn "would insist on going in at some man's entrance to a men's club or something of that sort, and just made a nuisance of things" (2015: 259). Washburn was to him an example of an aggressive femininity ("Just setting your teeth and barging ahead" Boring to Helen Peak, 31 January 1949) that did not respect proper gendered boundaries or decorum.

The Experimentalists' ethos was influenced by Titchener's British educational background. He was inspired by memories of Oxford "Smokers" (Rossiter 1982), informal single-gender discussion groups that merged the culture of the all-male education system with culture of the men-only club, which allowed men an enclave for smoking, drinking, and frank talk. This atmosphere worked in conjunction with paternal sexism, with the exclusion of women justified by chivalrous codes. Titchener's Oxbridge formality was seen as somewhat old fashioned during his lifetime, but even as Titchner's style of masculinity declined, there remained an appetite for exclusive masculine spaces.

The Psychological Round Table (PRT) illustrates how such exclusive masculine groups persisted, yet morphed to fit the surrounding culture. This group was formed in 1936, in reaction to The Experimentalists – "a youth-fired rebellion" (Benjamin 1977: 542) by early career psychologists who felt their seniors had gotten too far away from their experimental days. The PRT was to be limited to active experimenters, "those who still labored with their hands and were not too old to be sh[it] upon by monkeys" (Hardcastle 2000: 365), and to those under age 40. There were other differences from The Experimentalists. It was a secret society, and the membership, known as "The Brethren," were governed by an "autocratic minority" – "the Secret Six" who issued the invitations for the meetings and changed the rules of the society as desired, often to be more exclusive (Benjamin 1977: 543). No rule explicitly barred women, but no women were invited, even though, as Hardcastle notes, there was an obviously qualified candidate whom members would have been familiar with: Eleanor Gibson. James Gibson was one of the Secret Six between

1938–1941 and hosted several meetings, yet Eleanor recalled: "Needless to say, *I* was not [at the 1941 meeting of the PRT] – no women – it was a very sexist group. . . . I got to pay the bills and make arrangements often, but that was it" (Hardcastle 2000: 369).

No feminist, Eleanor Gibson's bar for calling something sexist was high. In this case, she was likely not simply referring to the exclusion of women but to what has euphemistically been called the PRT's "informal and irreverent" (Hardcastle 2000: 347), "raucous and scatological" atmosphere (Benjamin 1977: 544). Former PRT member Charles Gross' recollections of the PRT meetings in the 1960s helps cut through these euphemisms:

> Benjamin refers to the Hunt Memorial Lectures as "scatological," "humorous," and "satirical," and as an "instrument for fun and relaxation." In my time, although these descriptions might still apply, the visual content was almost entirely slides and films of women undressed or undressing. Breasts were a dominant theme, although if I recall correctly, various sexual activities became more common in later years. To the extent that one can distinguish erotica from pornography, the material was usually on the latter end of the dimension. (Gross 1977: 1121)

The Hunt Memorial Lectures at the PRT banquet were a reoccurring feature of the meetings, following William Hunt's speech "The Spontaneous Burrowing Habits of Phallus Domesticus" which skewered Carl Murchison's work on the common hen, *Callus domesticus*, at the inaugural meeting (Benjamin 1977). Gross reports that when he attended a PRT meeting again in 1974 (in company with Naomi Weisstein, to raise the issues of the group's sexist, elitist, and anti-democratic norms), he discovered little had changed. While there were now women members, "The Hunt Lectures continued, but male genitalia were added" (Gross 1977: 1121). Gross ends his damning account of the PRT with an extremely on the nose detail: The Psychological Round Table's "gavel was a brass penis" (1977: 1121).

While Hardcastle frames the driving motivation of the group as "the cult of the experiment" and casts the exclusion of women as due to the desire to "discuss experiments in an informal, "no-holds-barred" atmosphere" (Hardcastle 2000: 357), it is initially puzzling why pornography would be necessary to such an endeavor. One important clue comes from the group members' perception of themselves as disempowered relative to more established psychologists: Hunt recalled PRT members' belief that "we young bucks were doing the work while our elders in the SEP. . . reaped the glory for our efforts" (Hardcastle 2000: 364). Given this context, the power dynamics of the situation become clear – PRT was engaged in a power struggle with the establishment – caricatured by Hunt as "some gray beard with a reputation" but no ideas (Hardcastle 2000: 364). Rather than destabilizing the existing hierarchy through solidarity with groups also excluded from elite circles (e.g., women; non-white men; applied psychologists) the PRT members instead created an elite society with a form of sociality hostile and toxic to such groups, thus excluding them and elevating the PRT by contrast. Another clue is to be found in scholarship on the inherent fragility of patriarchal masculinity and the centrality of sexualized male bonding in cementing masculine identity. The crude humor of the meetings bound the group together as brothers, ready for the fight against the elite, and paradoxically, emphasized their existing power. The PRT allowed them to shape a safe space for themselves, a space where they could indulge their least professional impulses in perfect freedom. The penis gavel proclaimed their dominance and ability to keep order over at least some in psychology.

Despite their differences, The Experimentalists and PRT both used female exclusion and masculine informality to emphasize their dominance. Both groups exemplify the masculine honor culture that Nye (1997) says regulated professional sociability in science in the nineteenth and twentieth centuries. A model of masculine sociality that emphasized "rationality, calculation, and orderliness" (Connell 1990: 522) came to replace an earlier male honor culture that involved violence (i.e., dueling or warfare). The result was that "the masculine codes that regulated professional sociability provided additional obstacles to both men and women ignorant of their inner workings" (Nye 1997: 72). To be welcome in elite spaces one needed to be both "a gentleman and a scholar" as Hunt described J. J. Gibson, when recommending him for PRT membership (Hardcastle 2000: 360).

Some historians have pushed back on considering the men who ran such exclusive groups sexist and pointed to their record of mentoring women in science (e.g., Proctor and Evans 2014). However the question of personal misogyny is missing the point. Whatever the personal views of the men involved, the effect of such groups was sexist, creating a permanent class of "ineligibles" (Witz 1992: 42) for full professional acceptance. The women excluded from such groups understood their significance. When in 1912 Titchener denied her request to read a paper at The Experimentalists, Christine Ladd-Franklin wrote him multiple protests of the policy she described as "so unconscientious, so immoral,—worse than that—so unscientific!" (Furumoto 1994: 97).

Proctor and Evans (2014) argue that excluded women had other venues open for presenting papers (e.g., the APA), and Hardcastle attempts to downplay the impact of networking among PRT members "as they sought jobs, fellowships, publications and awards" (2000: 362). Yet this is at odds with the description of these groups' virtues, e.g., Titchener's own aspirations that the group facilitate mentorship: "the youngsters taken in on an equality with the men who have arrived" (Furumoto 1994: 97). Or take Hardcastle's description of the PRT as "*a distinct social space*—suited to an intricate and communal tending of scientific experiment" (2000: 351). From an intellectual community perspective, the exclusion of women from these spaces was significant, denying them both access to the movers and shakers of psychology and the exploration of ideas in an informal, collegial environment that Hardcastle rightly notes especially benefits solitary experimenters. Despite the PRT's conception of themselves as scrappy underdogs, the group in fact contained "individuals who were to become the prominent figures of American psychology in the 1950s" and the group served as "a major feeder" for the Experimentalists (Benjamin 1977: 545). As Gross notes, analysis of the PRT membership's "subsequent role in grant-awarding committees, editorial boards, and open organizations" (1977: 1121) would be enlightening.

The dynamics of the PRT and the Experimentalists reveal that it was no accident that the elite spaces of psychology were masculine, they were deliberately kept masculine. Even when given the opportunity to create a social space from scratch, with no existing rules (e.g., no nepotism rules), elite men chose exclusionary rules and practices. Historians of psychology would do well do attend to the way a concept like "gentleman" operated in different ways for male and female psychologists. In the same years that the gentleman scholars of the PRT had their first few meetings and were enjoying complete freedom from social constraint, Helen Thompson Woolley, a gifted psychologist with a career characterized by top notch applied work, was experiencing a very different reality of the term "gentlemen." In 1930 she had learned that the "gentlemen's agreement" guaranteeing her continued employment at Columbia University if her Institute of Child Welfare Research funding was ever cut, would not be honored, and she would be fired. She later wrote that she had learned "when one party in a gentleman's agreement is a woman, with no written evidence of the agreement, it counts for little" (Morse 2002: 137). Left jobless during the Great Depression, Woolley spent the 30s desperately writing to her male psychologist friends for work, but to no avail. She never worked in psychology again, and lived dependent on her daughter until her death.

Conclusion

The traditional approach to gender in the history of psychology assumes that focusing on gender is necessary primarily to do justice to forgotten women. What a gendering psychology approach reveals is the extent to which study of this topic is necessary for any adequate historical understanding of psychology. The lives of women in the history of psychology – their struggles, disappointments, and triumphs – illuminate the social mechanisms that lie at the heart of science.

Science involves the allocation of the material resources necessary for research, and this happens according to existing power structures. The repeating patterns in the biographies of women psychologists help make visible the operation of power and privilege within the discipline.

Going forward, it is critical that historians of psychology also reevaluate the biographies of prominent men through the lens of gender. Doing so will reveal the extent to which their academic success was not a purely meritocratic rewarding of brilliant ideas, but a function of how their power positioned them to take advantage of material and social resources. Supplemental academic and domestic labor, access to exclusive societies, epistemic authority among their peers, the ability to mentor students – all these resources facilitated their ability to conduct research and to secure a lasting reputation in psychology. In addition to analyzing how women in psychology navigated career barriers, attention must be paid to men's use of their agency. For example, how did they navigate hierarchies? Were they complicit in creating or maintaining hierarchies or did they act in solidarity with the less powerful? It might be appropriate to apply the generational analysis to men, but with a focus on how hierarchies and the operation of power changed across time. The differences between PRT and the Experimentalists suggest that sexism, while perennial, changes forms according to its historic context. In summary, an emphasis on the material conditions of science can help demythologize scientists and allow for a properly contextualized history of the discipline.

Finally, historians and psychologists alike must interrogate the unmarked masculine nature of much of psychology. Today shifting gender dynamics in psychology can obscure this historical reality. For example, the fact that clinical psychology is now a majority female profession tends to mask the reality that the major theories and concepts of the field were androcentric in origin. Since gender stereotypes were baked into core psychology concepts and shaped its research questions and methods an ahistorical approach to psychology risks perpetuating distorted views of human nature. Instead, doing psychology critically requires recognizing the social mechanisms of science and how they interact with existing power structures, both historically and in the present day.

References

Addelson KP (1983) The man of professional wisdom. In: Harding S, Hintikka MB (eds) Discovering reality, Synthese Library, vol 161. Springer, Dordrecht, pp 165–186

Aldridge J, Kilgo J, Jepkemboi G, Bruton A (2014) Forgotten women of educational psychology: redeeming their contributions in the 21st century. Int J Case Stud 3:55–61

Bazar JL (2010). "Your whiskers are on fire": a history of Titchener's lab from below [Conference session]. APA

Bederman G (1995) Manliness and civilization: a cultural history of gender and race in the United States, 1880–1917. University of Chicago, Chicago

Bem SL (1974) The measurement of psychological androgyny. J Consult Clin Psychol 42:155–162

Benjamin L (1977) The psychological round table: revolution of 1936. Am Psychol 32:542–549. https://doi.org/10.1037/0003-066X.32.7.542

Bernstein M, Russo N (1974) The history of psychology revisited: or, up with our foremothers. Am Psychol 29:130–134. https://doi.org/10.1037/h0035837

Bharj N, Hegarty P (2015) A postcolonial feminist critique of harem analogies in psychological science. J Soc Political Psychol 3:257–275

Bohan J (2016) Contextual history: a framework for re-placing women in the history of psychology. Psychol Women Q 14:213–227. https://doi.org/10.1111/j.1471-6402.1990.tb00015.x

Boring EG (1929) A history of experimental psychology. Century, New York

Boring EG (1951) The woman problem. Am Psychol 6:679–682

Boring EG (1967) Titchener's Experimentalists. J Hist Behav Sci 3:315–325

Boston Women's Health Book Collective (1973) Our bodies, ourselves: a book by and for women. Touchstone/Simon & Schuster, New York

Braine LG (2009). Interview by L Ball, A Rutherford, & A Karera [Video Recording]. Psychology's Feminist Voices Oral History and Online Archive Project. Newport

Broverman IK, Broverman DM, Clarkson FE, Rosenkrantz PS, Vogel SR (1970) Sex-role stereotypes and clinical judgments of mental health. J Consult Clin Psychol 34:1–7. https://doi.org/10.1037/h0028797

Cameron C, Hagen J (2005) Women in child development: themes from the SRCD oral history project. Hist Psychol 8:289–316. https://doi.org/10.1037/1093-4510.8.3.289

Capshew JH (1999) Psychologists on the march: science, practice and professional identity in America, 1929–1969. Cambridge University, Cambridge

Capshew JH, Laszlo AC (1986) We would not take no for an answer': women psychologists and gender politics during World War II. J Soc Issues 42:157–180

Caudle FM (1990) Eleanor Jack Gibson (1910–). In: O'Connell AN, Russo NF (eds) Women in psychology: a bio-bibliographic sourcebook. Greenwood, New York, pp 104–116

Chesler P (1972) Women and madness. Doubleday, Garden City

Chodorow N (1991) Freud on women. In: Neu J (ed) The Cambridge companion to Freud. Cambridge University, Cambridge, pp 224–248

Connell RW (1990) The state, gender, and sexual politics: theory and appraisal. Theory Soc 19: 507–544

Crawford M, Marecek J (1989) Psychology reconstructs the female 1968–1988. Psychol Women Q 13:147–165. https://doi.org/10.1111/j.1471-6402.1989.tb00993.x

Danziger K (1990) Constructing the subject: historical origins of psychological research. Cambridge University, Cambridge

Daston L (1992) The naturalized female intellect. Sci Context 5:209–235. https://doi.org/10.1017/S0269889700001162

Daston L, Galison P (2007) Objectivity. Zone Books, New York

Diehl LA (1986) The paradox of G. Stanley Hall: foe of coeducation and educator of women. Am Psychol 41:868–878

Fraiser N (2013) Fortunes of feminism: from state-managed capitalism to neoliberal crisis. Verso, Brooklyn

Friedan B (1963) The feminine mystique. Dell, New York

Friedline C (1960) Lecture to Frederick Rowe's class at Randolph Macon Women's College. [Audio recording]. Cora Friedline papers (Box OH5, folder 2). Cummings Center for the History of Psychology

Furumoto L (1987) On the margins: women and the professionalization of psychology in the United States 1890–1940. In: Ash MG, Woodward WR (eds) Psychology in 20th century thought and society. Cambridge University, Cambridge, pp 93–113

Furumoto L (1988) Shared knowledge: the experimentalists, 1904–1929. In: Morawski JG (ed) The rise of experimentation in American psychology. Yale University, New Haven, pp 94–113

Furumoto L (1994) Christine Ladd-Franklin's color theory: strategy for claiming scientific authority? In: Adler HE, Rieber RW (eds) Aspects of the history of psychology in America: 1892–1992. The New York Academy of Sciences, New York, pp 91–100

Furumoto L, Scarborough E (1986) Placing women in the history of psychology: the first American women psychologists. Am Psychol 41:35–42. https://doi.org/10.1037/0003-066X.41.1.35

Gibson EJ (1976) Oral history interview conducted by Margaret Savino, Cornell University. Rare and Manuscript Collections, Carl A. Kroch Library, Cornell University Library

Gibson EJ (1977) History of developmental psychology seminar at Cornell. Eleanor Gibson Papers (#14-23-2658, Box 11, Folder 3). Cornell University Library.2

Gibson EJ (1979a) EPA Symposium: women in academia. Eleanor Gibson Papers (#14-23-2658, Box 1, Folder 11). Cornell University Library

Gibson JJ (1979b) Forward: a note on E. J. G. by J. J. G. In: Pick AD (ed) Perception and its development: a tribute to Eleanor J. Gibson. Lawrence Earlbaum Associates, Hillsdale, pp ix–xii

Gibson EJ (1980) Autobiography. In: Lindzey G (ed) History of psychology in autobiography, vol 7. W.H. Freeman, San Francisco, pp 239–272

Gibson EJ (1991) An odyssey in learning and perception. MIT Press, Cambridge

Gibson EJ (1998) Oral history interview conducted by Marion Eppler, Society for Research in Childhood Development Oral History Project. http://www.srcd.org/sites/default/files/documents/gibson_eleanor_interview.pdf. Accessed 25 May 2021

Gibson EJ (2002) Perceiving the affordances: a portrait of two psychologists. Lawrence Erlbaum Associates, Mahwah

Gibson EJ, Levin H (1975) Psychology of reading. MIT, Cambridge

Green C (2010) Scientific objectivity and E. B Titchener's experimental psychology. Isis 101:697–721. https://doi.org/10.1086/657473

Greene S (2004) Biological determinism: persisting problems for the psychology of women. Fem & Psychol 14:431–435. https://doi.org/10.1177/0959353504044648

Gross C (1977) Psychological round table in the 1960s. Am Psychol 32:1120–1121. https://doi.org/10.1037/0003-066X.32.12.1120.b

Gul P, Korosteliov A, Caplan L, Ball LC, Bazar JL, Rodkey EN, Sheese K, Young J, Rutherford A (2013) Reconstructing the experiences of first generation women in Canadian psychology. Can Psychol 54:94–104. https://doi.org/10.1037/a0032669

Gundlach H, Roe R, Sinatra M, Tanucci G (eds) (2010) European pioneer women in psychology. FrancoAngeli Psicologia, Milano

Hall GS (1918) Youth, its education, regimen and hygiene. Appleton, New York

Hardcastle G (2000) The cult of experiment: the psychological round table, 1936–1941. Hist Psychol 3:344–370. https://doi.org/10.1037/1093-4510.3.4.344

Hegarty P (2003) Homosexual signs and heterosexual silences: Rorschach research on male homosexuality from 1921 to 1969. J Hist Sex 12:400–423

Hegarty P (2007) Getting dirty: psychology's history of power. Hist Psychol 10:75–91. https://doi.org/10.1037/1093-4510.10.2.75

Hegarty P (2012a) Beyond Kinsey: the committee for research on problems of sex and American psychology. Hist Psychol 15:197–200. https://doi.org/10.1037/a0027270

Hegarty P (2012b) Getting Miles away from Terman: did the CRPS fund Catharine Cox Miles's unsilenced psychology of sex? Hist Psychol 15:201–208. https://doi.org/10.1037/a0025725

Hegarty P (2013) Gentleman's disagreement: Alfred Kinsey, Lewis Terman, and the sexual politics of smart men. University of Chicago, Chicago. https://doi.org/10.7208/chicago/9780226024615.001.0001

Held L, Rutherford A (2012) Can't a mother sing the blues? Postpartum depression and the construction of motherhood in late 20th-century America. Hist Psychol 15:107–123

Hoffman E, Myerberg N, Morawski J (2015) Acting otherwise: resistance, agency, and subjectivities in Milgram's studies of obedience. Theory & Psychol 25:670–689. https://doi.org/10.1177/0959354315608705

Hollingworth LS (1914) Variability as related to sex differences in achievement. Am J Sociol 19:510–530. https://doi.org/10.1086/212287

Horney K, Association for the Advancement of Psychoanalysis (1967) Feminine psychology: [papers]. W.W. Norton, New York

Hornstein G (1992) The return of the repressed: psychology's problematic relations with psychoanalysis, 1909–1960. Am Psychol 47:254–263. https://doi.org/10.1037/0003-066X.47.2.254

Howes EP (1929) The meaning of progress in the women movement. Ann Am Acad Pol Soc Sci 143:14–20

Hubbard K (2019) Queer ink: a blotted history towards liberation. Routledge, London

Johnson A (2015) Florence Goodenough and child study: the question of mothers as researchers. Hist Psychol 18:183–195. https://doi.org/10.1037/a0038865

Johnson A, Johnston E (2010) Unfamiliar feminisms: revisiting the national council of women psychologists. Psychol Women Q 34:311–327. https://doi.org/10.1111/j.1471-6402.2010.01577.x

Johnson A, Johnston E (2015) Up the years with the Bettersons: gender and parent education in interwar America. Hist Psychol 18:252–269. https://doi.org/10.1037/a0039521

Johnston E, Johnson A (2008) Searching for the second generation of American women psychologists. Hist Psychol 11:40–69. https://doi.org/10.1037/1093-4510.11.1.40

Johnston E, Johnson A (2017) Balancing life and work by unbending gender: early American women psychologists' struggles and contributions. J Hist Behav Sci 53:246–264. https://doi.org/10.1002/jhbs.2186

Kaschak E, Tiefer L (2001) A new view of women's sexual problems. Women Ther 24:1–8

vande Kemp H (1992). G. Stanley Hall and the Clark school of religious psychology. Am Psychol 47:290–298 https://doi.org/10.1037/0003-066X.47.2.290

Kim S, Rutherford A (2015) From seduction to sexism: feminists challenge the ethics of therapist-client sexual relations in 1970s America. Hist Psychol 18:283–296. https://doi.org/10.1037/a0039524

Lerner G (1979) The majority finds its past: placing women in history. Oxford University, Oxford

Lewin M (ed) (1984) In the shadow of the past: psychology portrays the sexes. Columbia University, New York

Liesen L (2007) Women, behavior, and evolution: understanding the debate between feminist evolutionists and evolutionary psychologists. Politics Life Sci 26:51–70. https://doi.org/10.2990/21_1_51

Lombroso C, Ferrero W (1895) The female offender. The criminology series, Morrison, WD (ed). Am J Psychiatry 52:119–120. https://doi.org/10.1176/ajp.52.1.119

Maccoby EE, Jacklin CN (1974) The psychology of sex differences. Stanford University, Stanford

Marecek J, Hare-Mustin RT (1991) A short history of the future: feminism and clinical psychology. Psychol Women Q 15:521–536

Martin E (1991) The egg and the sperm: how science has constructed a romance based on stereotypical male-female roles. Signs 16:485–501. https://doi.org/10.1086/494680

Merchant C (1980) The death of nature: women, ecology, and the scientific revolution. Harper Collins, San Francisco

Milar K (2000) The first generation of women psychologists and the psychology of women. Am Psychol 55:616–619. https://doi.org/10.1037/0003-066X.55.6.616

Morawski JG (1982) Assessing psychology's moral heritage through our neglected utopias. Am Psychol 37:1082–1095. https://doi.org/10.1037/0003-066X.37.10.1082rt5

Morawski JG (1985) The measurement of masculinity and femininity: engendering categorical realities. J Pers 53:196–223. https://doi.org/10.1111/j.1467-6494.1985.tb00364.x

Morawski JG (2015) Epistemological dizziness in the psychology laboratory: lively subjects, anxious experimenters, and experimental relations, 1950–1970. Isis 106:567–597. https://doi.org/10.1086/683411

Morse JF (2002) Ignored but not forgotten: the work of Helen Bradford Thompson Woolley. NWSA J 14:121–147. https://doi.org/10.2979/NWS.2002.14.2.121

Napoli DS (1981) Architects of adjustment: the history of the psychological profession in the United States. Kennikat Press, Port Washington

Nevers CC, Calkins MW (1895) Wellesley College psychological studies: Dr. Jastrow on community of ideas of men and women. Psychol Rev 2:363–367. https://doi.org/10.1037/h0074975

Nicholson I (2001) Giving up maleness: Abraham Maslow, masculinity, and the boundaries of psychology. Hist Psychol 4:79–91. https://doi.org/10.1037/1093-4510.4.1.79

Nicholson I (2011) "Shocking" masculinity: Stanley Milgram, "obedience to authority," and the "crisis of manhood" in Cold War America. Isis 102:238–268. https://doi.org/10.1086/660129

Noble DF (1992) A world without women: the Christian clerical culture of western science. Alfred A. Knopf, New York

Noon D (2004) Situating gender and professional identity in American child study, 1880–1910. Hist Psychol 7:107–129. https://doi.org/10.1037/1093-4510.7.2.107

Nye R (1997) Medicine and science as masculine "fields of honor". Osiris 12:60–79

O'Connell AN, Russo NF (Eds) (1980) Eminent women in psychology: models of achievement. Human Sciences Press, New York

O'Connell AN, Russo NF (Eds) (1990) Women in psychology: a bio-bibliographic sourcebook. Greenwood Press, Westport

Palmieri P (1983) Here was fellowship: a social portrait of academic women at Wellesley College, 1895–1920. Hist Educ Q 23:195–214

Pettit M (2008) The new woman as "tied-up dog": Amy E. Tanner's situated knowledges. Hist Psychol 11:145–163

Pettit M (2012) The queer life of a lab rat. Hist Psychol 15:217–227. https://doi.org/10.1037/a0027269

Pettit M, Serykh D, Green CD (2015) Multispecies networks: visualizing the psychological research of the Committee for Research in Problems of Sex. Isis 106:121–149. https://doi.org/10.1086/681039

Porter R (1985) The patient's view: doing medical history from below. Theory Soc 14:175–198

Proctor RW, Evans REB (2014) Titchener, women psychologists, and the experimentalists. Am J Psychol 127:501–526. https://doi.org/10.5406/amerjpsyc.127.4.0501

Reskin B (2007) Sex differentiation and the social organization of science. Sociol Inq 48:6–37. https://doi.org/10.1111/j.1475-682X.1978.tb00815.x

Rodkey EN (2010a) Eleanor Gibson and the visual cliff myth: the biography of a scientific object. Unpublished master's thesis. York University, Toronto

Rodkey EN (2010b) Charming stubbornness: Eleanor Gibson's second generation leadership strategy. Paper presented at the Eastern Psychological Association

Rodkey EN (2015) The visual cliff's forgotten menagerie: rats, goats, babies, and myth-making in the history of psychology. J Hist Behav Sci 51:113–140. https://doi.org/10.1002/jhbs.21712

Rodkey EN (2016) Far more than dutiful daughter: Milicent Shinn's child study and education advocacy after 1898. Genet Psychol 177:209–230. https://doi.org/10.1080/00221325.2016.1237235

Rodkey KL, Rodkey EN (2020) Family, friends, and faith-communities: intellectual community and the benefits of unofficial networks for marginalized scientists. Hist Psychol 23:289–311. https://doi.org/10.1037/hop0000172

Rossiter MW (1980) "Women's work" in science, 1880–1910. Isis 71:381–398

Rossiter MW (1982) Women scientists in America: struggles and strategies to 1940. John Hopkins, Baltimore

Rossiter MW (1995) Women scientists in America: before affirmative action, 1940–1972. Johns Hopkins University Press, Baltimore

Russett CE (1989) Sexual science: the Victorian construction of womanhood. Harvard University, Cambridge

Russo N, Denmark F (1987) Contributions of women to psychology. Annu Rev Psychol 38:279–298. https://doi.org/10.1146/annurev.ps.38.020187.001431

Russo N, O'Connell A (1980) Models from our past: psychology's foremothers. Psychol Women Q 5:11–54. https://doi.org/10.1111/j.1471-6402.1980.tb01032.x

Rutherford A (2003) B. F. Skinner's technology of behavior in American life: from consumer culture to counterculture. J Hist Behav Sci 39:1–23. https://doi.org/10.1002/jhbs.10090

Rutherford A (2011) Sexual violence against women: putting rape research in context. Psychol Women Q 35:342–347

Rutherford A (2012) Problems of sex and the problem with nature: a commentary on "Beyond Kinsey". Hist Psychol 15:228–232. https://doi.org/10.1037/a0027668

Rutherford A (2015) Maintaining masculinity in mid-twentieth-century American psychology: Edwin Boring, scientific eminence, and the "woman problem". Osiris 30:250–271. https://doi.org/10.1086/683022

Rutherford A (2019) Psychological perspectives on gender: an intellectual history. In: Sternberg R, Pickren WE (eds) The Cambridge handbook of the intellectual history of psychology. Cambridge University, Cambridge, pp 345–370

Rutherford A (2020) Doing science, doing gender: using history in the present. J Theor Philos Psychol 40:21–31. https://doi.org/10.1037/teo0000134

Rutherford A, Marecek J, Sheese K (2012) Psychology of women and gender. In: Freedheim DK, Weiner IB (eds) Handbook of psychology, vol 1 2E. Wiley, New York, pp 279–301

Rutherford A, Vaughn-Johnson K, Rodkey EN (2015) Does psychology have a gender? The Psychol 26:508–510

Scarborough E, Furumoto L (1987) Untold lives: the first generation of American women psychologists. Columbia University, New York

Schmidt S (2020) Midlife crisis: the feminist origins of a chauvinist cliché. University of Chicago Press, Chicago

Schwesinger G (1943) Wartime organizational activities of women psychologists. II The National Council of Women Psychologists. J Consult Psychol 7:298–301. https://doi.org/10.1037/h0053771

Sherif CW (1979/1998) Bias in psychology. Fem Psychol 8:58–75

Shields SA (1975) Functionalism, Darwinism, and the psychology of women. Am Psychol 30:739–754. https://doi.org/10.1037/h0076948

Shields SA (1982) The variability hypothesis: the history of a biological model of sex differences in intelligence. Signs 7:769–797

Shields SA (2007) Passionate men, emotional women: psychology constructs gender difference in the late 19th century. Hist Psychol 10:92–110

Shteir A (1997) Gender and "modern" botany in Victorian England. Osiris 12:29–38

Stevens G, Gardner S (1982) The women of psychology. Schenkman Publishing, Cambridge

Takasuna M (2020) The first generation of Japanese women psychologists. Geneal 4:61. https://doi.org/10.3390/genealogy4020061

Terman LM, Miles CC (1936) Sex and personality. Yale University Press

Thompson HB (1903) The mental traits of sex: an empirical investigation of the normal mind in men and women. University of Chicago, Chicago

Unger RK (1979) Toward a redefinition of sex and gender. Am Psychol 34:1085–1094

Valentine E (2020) Philosophy and history of psychology: selected works of Elizabeth Valentine. Psychology, London

Vicedo M (2009) Mothers, machines, and morals: Harry Harlow's work on primate love from lab to legend. J Hist Behav Sci 45:193–218. https://doi.org/10.1002/jhbs.20378

von Oertzen C (2013) Science in the cradle: Milicent Shinn and her home-based network of baby observers, 1890–1910. Centaurus 55:175–195. https://doi.org/10.1111/1600-0498.12016

Weismann N (2020) Do humans really have a killer instinct or is that just manly fancy? Psyche. https://psyche.co/ideas/do-humans-really-have-a-killer-instinct-or-is-that-just-manly-fancy. Accessed 25 May 2021

Weisstein N (1968) Kinder, kuche, kirche as scientific law: psychology constructs the female. New England Free, Boston

Winston A (1998) The defects of his race: E. G. Boring and antisemitism in American psychology, 1923–1953. Hist Psychol 1:27–51. https://doi.org/10.1037/1093-4510.1.1.27

Winter A (1998) Mesmerized: powers of mind in Victorian Britain. University of Chicago, Chicago

Witz A (1990) Patriarchy and professions: the gendered politics of occupational closure. Sociol 24:675–690. https://doi.org/10.1177/0038038590024004007

Witz A (1992) Professions and patriarchy. Routledge, London

Woolley HT (1910) A review of the recent literature on the psychology of sex. Psychol Bullet VII 9:335–342

Young J, Hegarty P (2019) Reasonable men: sexual harassment and norms of conduct in social psychology. Fem Psychol 29:453–474. https://doi.org/10.1177/0959353519855746

Indigenous Psychologies: Resources for Future Histories

41

Wade E. Pickren and Gülşah Taşçı

Contents

Introduction: Complicating the Histories of Indigenous Psychologies	1066
Ancient to Modern: The Ground of the Psychological	1069
Modernity/Coloniality and Psychological Knowledge	1072
Conclusion	1077
Cross-References	1078
References	1079

Abstract

The number and variety of Indigenous psychologies has grown immensely in the 75 years since the end of World War II. To date, there has been a dearth of histories of these psychologies, though brief descriptive historical introductions in articles are common. The approach of this chapter is to provide a critical rationale for understanding the challenges that face Indigenous psychologies, as well as the challenge of writing critical histories. Just as there are a multiplicity of Indigenous psychologies and many possible histories, so there are multiple rationales for writing histories of Indigenous psychologies. The chapter provides one intellectual and critical rationale based in decolonization approaches and decolonial theory that will be of use for future historians. The chapter also provides a substantial list of published resources to aid the development of future histories.

Keywords

Indigenous psychology (IP) · Decolonizing psychology · Decolonial theory in psychology · History resources for IP

W. E. Pickren (✉)
Dryden, NY, USA
e-mail: rgpwade@gmail.com

G. Taşçı
Istanbul 29 Mayıs University, Istanbul, Turkey

© The Author(s), under exclusive licence to Springer Nature Singapore Pte Ltd. 2022
D. McCallum (ed.), *The Palgrave Handbook of the History of Human Sciences*,
https://doi.org/10.1007/978-981-16-7255-2_80

Indigenous psychology's voice, status, and ability to speak and engage has mostly occurred within the terms and rules set by the colonial powers (Bhatia 2002; Macleod and Bhatia 2008; Yeh and Sundararajan 2015). *A decolonizing perspective would begin by charting the entire history of indigenous psychology in the modern and ancient era to show its various manifestations from precolonial to colonial times, and from the time of enlightenment to the present-day conditions of neoliberal globalization.* (Bhatia 2019, p. 111, emphasis added)

Introduction: Complicating the Histories of Indigenous Psychologies

This chapter seeks to provide resources for others to write histories of Indigenous psychologies. The bulk of the chapter provides an intellectual framework for such histories, as well as caveats and cautions about writing histories. After the conclusion, the reader will find a list of published articles and books about Indigenous psychologies in various settings in the section "Further Reading."

Indigenous psychology is one of the disciplines that try to understand psychological phenomena in a cultural context (Yang and Lu 2007). The Indigenous psychology perspective continues to be debated as to whether Indigenous psychology represents a more universalist or a more relative approach (Chakkarath 2012).

Indigenous or Indigenization when used in reference to psychology does not have a fixed meaning or denote a fixed process. One of the authors (Pickren) described it elsewhere as, "The process whereby a local culture or region develops its own form of psychology, either by developing it from within that culture, or by importing aspects of psychologies developed elsewhere and combining them with local concepts" (Pickren and Rutherford 2010, p. xxii). In the latter sense, psychology in the USA is historically an Indigenized psychology. In the late nineteenth century, a number of US citizens traveled to various sites in Europe, primarily to Germany, to learn about the new experimental psychology and to gain credentials for a professional career in the new science. The history of this has been told by many able historians and psychologist-historians who have shown that whatever the actual interest in the new science of psychology, such training was also linked inextricably with larger social processes and population growth, with the consequence in an increasingly urbanized society that the emerging division of intellectual and expert labor required credentials in order to work, have status, and forge a career (Abbott 1988; Haber 1964; Larson 1984). While the initial theory and experimental praxes were borrowed from elements of various European cultures, the new discipline and its experts in the USA were successful as they were able to modify the imported theory and praxes to better suit American cultural, intellectual, and social norms, such as individuality and autonomy, a belief in progress and self-improvement, and a faith in science and technology to solve human problems (Danziger 1990; O'Donnell 1985; Smith 1997). It is critically important, however, to understand that the other reason for the rapid growth of psychology in the USA was that there was already an everyday psychology present when European scientific psychology was introduced (Pickren 2000; Pickren and Rutherford 2010; Schmit 2005, 2010). In one sense, then, there are histories of Indigenous/Indigenized psychologies, as psychology in the USA has been well documented, but the historical literature on the topic is scant for other regions. Because of its growth into the dominant force in disciplinary and professional psychology, US psychology has become a source for an Indigenization process in many other sites around the world since WWII. The history of that process is problematic.

Sunil Bhatia's description of the challenges facing Indigenous psychology (IP) in the quote above is apt and insightful. The growth of Indigenous psychologies since World War II has been marked by fits and starts, with many IPs taking the path, as Bhatia states, of "countering and aligning with the dominant frameworks of psychological science" (2019, p. 111). His call for charting the entire history of Indigenous psychology both contemporary and ancient is welcome. This chapter addresses some of his concerns and broadens them.

First, it is important to note that a singular history of Indigenous psychology is problematic. There is no Indigenous psychology in the singular, only in the plural – so, no history, but histories. And to cover them all is well beyond the scope of this chapter and the ability/knowledge of its authors. Perhaps the best way to approach this subject is via histories otherwise of psychologies, which is the path chosen.

But there are complicated matters to address before delving into the substance of the chapter. Is the main interest in histories of disciplinary/professional psychologies? Only? Or should the psychologies of everyday life also be included? What are the relations between the two (or three) psychologies? Are disciplinary/professional psychologies possible without the presence of everyday psychologies? Do they reflexively cocreate each other, as suggested by Smith in the introduction to this section (Chap. 1, "What Is the History of the Human Sciences?," by R. Smith, this volume)? If they do cocreate each other, what are the common foundations? At this point, it is provident to offer partial answers for partial histories, as definitive answers to such questions remain outside the possible.

Further complications are present. What is indicated by the term history or histories? History (capitalized) is a modernist project that increasingly is suspected as an instrument of coloniality of being and knowledge intended to retain the right of intellectual authority over what counts as truth regarding events, people, ideas of the past, and who its legitimate practitioners are (Maldonado-Torres 2007). This extends even to judgment about what practices and which countries/societies are modern, with Europe serving as the measure of modernity (Goody 2006; Mignolo 2011). History's practitioners have been predominately white, male, and of a certain social class and of a certain training. History in the European Enlightenment tradition's practice is to reserve the right to name lands, make official maps, and determine what is history and what is tradition or myth (see also, Chakrabarty 2000, and Mignolo 2011). As Maori scholar Linda Tuhiwai Smith noted in regard to histories of Indigenous peoples, modernist histories have largely determined that Indigenous peoples are not the "final arbiters of what counts as the truth" about their own histories (Smith 1999, p. 34). Kagitcibasi (2010) also emphasized that, as psychology is an imported discipline, it is alien to important social events in non-Western societies, so instead of producing knowledge in these societies, it focuses on transferring Western ways of understanding theory. However, Turkish studies in the field of psychology date back to very old times. In Islamic psychology, the subjects (*ilm-i ruh / ilm-i nefs*) examined under various headings such as the soul, and the relationship between the soul and body, are remarkable in terms of the history of psychology. Moreover, when psychology (*ilm-i ruh / ilm-i nefs*) is mentioned, it is necessary to deal with the concept of soul. The first prominent people to do so in this respect are Farabi, İbni Sina, and El-Muhâsibî (Kırklaroglu 2018).

Yet, histories are important, as Indigenous scholars recognize (e.g., Couture 1987; Dunbar-Ortiz 2014; Morales 1998; Smith 1999; Trask 1999). As Linda T. Smith argues, "To hold alternative histories is to hold alternative knowledges. The pedagogical implication of this access to alternative knowledges is that they can form the basis of alternative ways of doing things" (Smith 1999, p. 34). Indigenous histories provide empowering counterstories that can undermine the histories written by settler colonialists and thus decolonize received or canonical knowledges (Solórzano and Yosso 2002). Such histories of Indigenous psychologies may be particularly important, given the WEIRDness (White, Educated, Industrialized, Rich, Democratic – Henrich et al. 2010) of psychology and its substantial role in supporting and maintaining patriarchal neoliberalism (Bhatia and Priya 2018; Rutherford 2018; Sugarman 2015).

An additional complication lies in the use and understanding of key terms: Indigenous, aboriginal, Indian, and First Nations. Although only the first is referenced in the title of this chapter, the others are implied. The point is that each of the terms, while not identical, is problematic, and the problems vary within and between the referenced peoples (Peters and Mika 2017). Even the use of "tribe(s)" is often problematic, invoking as it does racist anthropology and the sorting of "primitive" and "modern" peoples into stratified groups (Béteille 1998). In settler colonialist societies, such as the USA, Canada, and Australia (not an exhaustive list), the peoples who were already there when Europeans arrived were targets of attempted genocide. Thus, even the terms used can raise troubling memories for settlers and for the peoples being referenced by those terms. As several scholars and activists have pointed out, all of these terms are ones assigned by colonizers and their descendants. Typically, the terms have been used for policing, control, and legal purposes that perpetuate oppression. In Canada, where First Nations and Aborigine have been used as legal descriptors, Chiefs of the Anishinabek Nation in 2008 issued a statement that

> government agencies, NGOS, educators, and media organizations should discontinue using inappropriate terminology when they are referring to the Anishinabek. We respect the cultures and traditions of our Métis and Inuit brothers and sisters, but their issues are different from ours. . . . .Chief Patrick Madahbee of Aundeck Omni Kaning said: '*Referring to ourselves as Anishinabek is the natural thing to do because that is who we are. We are not Indians, natives, or aboriginal. We are, always have been and always will be Anishinabek*'. (Cited in Peters and Mika 2017, p. 1230–1231, emphasis added, see https://intercontinentalcry.org/anishinabek-outlaw-term-aboriginal/)

Yet, it is also true that "First Nations" and "Indigenous" are embraced and used by members of groups who were on the land prior to colonization. This is often true in academia, as evidenced by journals such as *Decolonization: Indigeneity, Education & Society, AlterNative,* and many scholarly books (e.g., Battiste 2000; Byrd 2011; Minthorn and Shotton 2018; Smith et al. 2019), and some scholars prefer Native or Indian (e.g., Mann 2016). The term, Indigenous Psychology, also is used in multiple ways to serve diverse ends, including as part of what has been called the "indigenous strategy" (Gagne 2015).

The final complication is the authors' identities. They are not Indigenous, aborigine, First Nations, or Indian. Writing this without the active collaboration of Indigenous scholars means that our biases and limitations will be on display. The authors are mindful of the phrase, "Nothing about us, without us," by the disability rights activist and author, James Charlton (1998).

Ancient to Modern: The Ground of the Psychological

As Bhatia noted, histories of Indigenous psychology(ies) are few. One of the earliest was *Indigenous Psychologies: The Anthropology of the Self* by Paul Heelas and Andrew Lock (1981). The editors chose authors who offered both anthropological and psychological perspectives, which included acknowledgments of the role of religion and myth in creating psychologies. *Psychology moving East: The status of Western psychology in Asia and Oceania,* a volume edited by Geoffrey Blowers and Alison Turtle, appeared in 1987, with a new publication of the original contents in 2109. After a sparkling introduction by Alison Turtle, authors from 19 different countries that cover an immense geographical area with very different cultural traditions and political histories relate the development and influence of what is generally termed Euro-American psychology. Some of the countries were former colonies of European countries and in the case of the Philippines, of Spain, and then of the USA. Some of the countries had only recently become countries, such as Malaysia (1963). In general, the question of finding ways to make psychology locally relevant, while also conforming to the psychology exported by the West or by the USSR, was present in most of the chapters. Other volumes on Indigenous psychology provide historical context for aspects of their development, while the main focus is on the internal dynamics of the creation of psychology, typically in the image of US or European psychology (e.g., Kim and Berry 1993; Kim et al. 2006). A similar approach is offered in most scholarly journal accounts of IP. For the most part, these sources indicate how IPs were formed in reaction or resistance to hegemonic psychological science/practice and serve as examples of what it means to develop an IP through the processes of "countering and aligning with the dominant frameworks of psychological science" (Bhatia 2019, p. 111).

Historical analyses of IPs written in a more critical vein of scholarship are even fewer. The most notable is that by Irmingard Staeuble, "De-centering Western perspectives: Psychology and the disciplinary order in the First and Third World" (Staeuble 2006). There are several chapters in Brock's volume, *Internationalizing the History of Psychology* (2006), that offer a critical perspective on the development of psychology outside the Euro-American axis. Particularly compelling are the chapters by Danziger, Moghaddam & Lee, Paranjpe, and Staeuble. Given that theoretical lenses available for analysis change frequently, perhaps what is needed is an updated critical analysis. Here, the authors offer one through the use of a decolonization/decolonial lens to help readers understand how hegemonic psychological science creates barriers to the acceptance of nonhegemonic psychologies wherever they arise, Global South or Global North. Such barriers have had the effect, as Bhatia noted, of constricting IPs in such a way that their creation has occurred in settings primarily determined by "the terms and rules set by the colonial powers" (Bhatia 2019, p. 111).

The chapter now turns to a brief examination of the historical background of ideas and practices that were important in creating the possibility of the psychological, or, to use Smith's phrase from the introduction to this section, alternatives to psychology. Bhatia's language in the opening epigraph bears repeating here:

> A decolonizing perspective would begin by charting the entire history of indigenous psychology in the modern and ancient era to show its various manifestations from pre-colonial to colonial times, and from the time of enlightenment to the present-day conditions of neoliberal globalization.
> ...
> One reason why we do not have detailed intellectual and social histories of indigenous psychology is because it has often been considered as *deeply rooted in local practices and relegated to the realm of the mythological, collective, religious, traditional, philosophical, irrational, primitive, imaginative, and cultural.* (Bhatia 2019, p. 111, emphasis added)

In the view of dominant or hegemonic psychological science, Indigenous psychologies are not *modern*, but "other" in their genesis and foundation. The use of primitive is an othering term to indicate that what is primitive is outside modernity, so disallowed. This is the dark side of Western modernity, to borrow Mignolo's phrase (Mignolo 2011). Countering the othering of IPs is especially troublesome in regard to the "database" of what can be used to create histories of Indigenous psychologies, such as oral traditions, myths, or findings from archeology, to name a few of many possible sources. For many psychologists, such sources are unacceptable as they are supposedly other than the sources of true psychological science/practice. To paraphrase Hegarty and Pratto, Indigenous psychologies are "that which needs to be explained" because they do not fit within the standard narrative of contemporary psychology (2001), or hegemonic psychology, to use Glenn Adams' label (Adams et al. 2019). The use of the terms hegemonic psychology or hegemonic psychological science rather than Western psychology is for two reasons. First, because of the coloniality of knowledge practices of "Western" psychology since WWII, Western psychology is now practiced everywhere and is the dominant mode of practice pretty much everywhere. Second, Western psychology is not uniform. There are many strands of "Western" psychology, some of which are constructed on theoretical bases and engage in disciplinary practices that are markedly not the same as hegemonic psychological science.

The stance of hegemonic psychological science of being free from such contaminating origins obscures at least two critical facts. First, psychological science has been constructed on what Danziger called a "narrow social basis. That entailed a very considerable narrowing of epistemic access to the variety of psychological realities" (Danziger 1990, 197). Second, hegemonic psychological science also has a foundation built on spiritual ideas, myths, religion, mind science, custom, commerce, and, in the case of the USA, intense individualism, as numerous authors have shown and which was alluded to earlier in regard to the everyday psychology that existed in the USA at the time of the importation of European scientific psychology (e.g., Albanese 2007; Cooter 1984; Fuller 1982; Harrington 2008; Hoopes 1989; Schmit 2005, 2010; Taves 1999; Taylor 1999; Thomson 2006; Vidal 2011; White 2009; Winter 1998). As Adams, Bhatia, and others have pointed out, those who practice scientific psychology prefer to chart the history of psychological science as moving from the local cultural base of knowledge to the claim of the universal, with the end point being positivist research data that show their originary point in neural processes (e.g., Vidal and Ortega 2017). Hegemonic psychological science claims that its findings are true of all people (and even animals), everywhere, and at all times (Bhatia 2019), despite its evidentiary basis of data being primarily drawn from studies of undergraduates in privileged situations who represent less than 5% of the world's population (Arnett 2008; Henrich et al. 2010). Perhaps the occlusion of hegemonic psychology's mythic/religious/commercial past is an example of what Santos calls a sociology of absences (2014). One of the likely unintended consequences is that many of psychology's knowledge claims are divorced from the daily experiences of most human beings (Escobar 2020; Santos 2014).

The foundations, if you will, of all psychologies are the stories told in every culture, every society, and for as long as humans have had language. Hunter-gatherer groups, noted for their emphasis on cooperation for survival, tell stories that place cooperative behavior within their cosmologies, which in turn supports their behavior (Bird-David 1990; Ingold 1999; Mann 2016; Smith et al. 2017; Suzman 2017). Myths are key stories. One of the reasons that some discount myths as relevant to modern psychologies and other sciences is that they think of myths as fairy tales. Serious scholars of mythology understand it differently. Estonian/Canadian linguist and comparative mythologist, Jaan Puhvel, has argued, "Myth in the technical sense is a serious object of study, because true myth is by definition deadly serious to its originating environment. In myth are expressed the thought patterns by which a group formulates self-cognition and self-realization, attains self-knowledge and self-confidence, explains its own sources and being and that of its surroundings..." (Puhvel 1989, p. 128). Other scholars of myth and religion also see myths as key to cultural self-understanding and allow us to explore ideas that are shared between cultures (e.g., Doniger 1998). There is a long history in psychology (broadly defined) of the psychological import of myths, from Freud and Jung to our contemporary world (Downing 1975; Shamdasani 2003). Our point is not that Freud, Jung, and their followers are correct that myths are universal in theme or meaning. Rather, what is important is that enduring myths are part of the scaffolding for the eventual construction of the kind of psychological society that now seems pervasive (e.g., Glavenau 2005; MacLeod 2018). Other foundational resources for the emergence of the psychological, such as oral traditions (Datta 2018; Hale 1998), are also key to the transmission of knowledge and values. The archeological findings are the material evidence of these resources, although there is typically no one-to-one correspondence with them and the myths and oral traditions of a group (see Rautman 2000). It is these and related sources of the human stories that make the psychological possible. The term, "the psychological," is used to indicate that the references are not to disciplinary or professional psychology anywhere. Rather, what is meant is the long train of events over millennia that created the possibility of psychology or, if the reader prefers, alternatives to psychology. Thus, ultimately, even hegemonic psychology is also deeply rooted in the local, to borrow Bhatia's phrase. The difference, and it is an important one, is that since the end of WWII, hegemonic psychology has been able to project its local roots into a worldwide dominance. The content, legitimacy, and authority of this psychology is not simply underwritten by its claims of scientific veracity or articulation of some singular truth about human/nonhuman animal psychology; rather, it came to dominance as part of the package of modernity embedded in the projection of economic, military, and political power of the US-led postwar world, which was/is the most recent demonstration of Euro-American control of what it means to be modern.

Modernity/Coloniality and Psychological Knowledge

The construction of modern disciplinary/professional psychology began in the nineteenth century, as noted above. Before this period, there was no general delineation between philosophy and research/knowledge production. But over the course of the century and into the twentieth century, barriers were erected between the specialized knowledge holders, demarcated over time as disciplines, and the public, including the educated public (Markus 1987). The result was that the members of the emergent discipline of psychology held the final word on what counted as psychology. In most places in the European world, this meant that religion, spirituality, phrenology, New Thought, and multiple other bodies of knowledge and practice common in everyday experiences of the psychological could be excluded. That which could be appropriated and subsumed into disciplinary knowledge was kept but made the preserve of elites. This is a kind of intellectual or psychic enclosure. Enclosure classically refers to the conversion of what is traditionally held in common and accessible by all into personal or private property. The classic example is the conversion of the commons [this was conversion, but the social process is usually known as "enclosure"] in England from the thirteenth century onward, but with rapidity from the sixteenth century onward and with particular state-sanctioned violence during the Industrial Revolution. Millions of acres traditionally held in common and in small landholdings became commodified – converted to private property that could be bought and sold at profit. Likewise, that which was psychologically in common that every person held, perhaps even common sense, now was uncertain unless validated by men of science. This was part of the intellectual division of labor that arose on the model of the division in the manufacture and distribution of goods. Philosopher Lewis Gordon refers to this process as the colonization of knowledge (Gordon 2011).

What has this meant for the discipline of psychology and its professional client services domain that arose soon after? It implies that psychology, like other disciplines, became a science and profession whose main function is preserving its place through intellectual and professional enclosure. To preserve its standing, its members take on the advancement of careers and earning a livelihood as central to their lives, acknowledged or not – thus the sharp defenses and the guarding of boundaries and what counts as real psychology and allowable praxes. Much of the early public discourse of psychology in the USA was devoted to defending its intellectual and practical territory from those cognate fields with which it could be and was often confused, such as the Emmanuel Movement, psychical research, mesmerism, phrenology, spiritualism, New Thought, and certain expressions of religious faith and practice (Caplan 2001; Pickren 2000). The success of psychology was predicated, as noted in the Danziger quote above, on "a very considerable narrowing of epistemic access to the variety of psychological realities" (Danziger 1990, 197). As Danziger later pointed out, "The Western cultural roots of Cartesian psychology are glaringly obvious, yet its preconceptions are so deeply enmeshed with the procedural norms of traditional scientific psychology that alternative approaches have long been subject to disciplinary marginalization" (2006, p. 272).

This leads to the other apt term for how the institutionalized form of psychology has operated in regard to psychologies that become or threaten to become competitors: colonization. When Indigenous psychologies began to emerge in the post-WWII period of decolonization, it was at the same time as the rise of US psychology, that is, as a major practice of one of the two dominant world powers. The dominance was so great that the adjective, American, did not have to be used. Real psychology was American psychology, particularly psychology committed to scientific reductionism.

The hegemony of American psychology in the post-WWII era was not due only to its power of explanation of all things psychological. Rather, the sociopolitical influence of the USA, supported by American military might and the economic development programs of the World Bank and International Monetary fund, played key roles, still continuing, in the emergence and continued dominance of US scientists in many fields. This was a deliberate strategy on the part of the USA to use its strength – political, military, economic, and intellectual – to increase its influence in many parts of the world. Programs meant to win friends among new countries emerging from former colonies covering the spectrum of state power, including economic influence, educational programs, funds for recruiting and supporting foreign students to American graduate programs, and covert use of intelligence agencies to undermine and overthrow governments perceived as hostile, among others, were all deployed.

The influence strategies were not only aimed at decolonized states but were also used in Europe. Well-known examples include the Marshall Plan and the US Agency for International Development. After the war, many European countries were in desperate shape, many had lost a significant part of their populations, the physical infrastructure was badly damaged, and at many universities there was little money to repair damaged laboratories or support faculty and student research. Funds from the Marshall Plan and other US sources were used to help rebuild much of the scientific infrastructure. This brought with it American influence on science, including in psychology.

Within Europe, the structures of science were rebuilt, often with massive amounts of American aid and with the concomitant American influence. Pieter van Strien (1997) has written about the impact of US social psychology in the Netherlands after the war, where American influence was critical in redirecting much of social psychology into a strongly empirical model of research, with little reference to theoretical underpinnings. The author referred to the result of this influence as "American Colonization of Northwest European Social Psychology."

Outside Europe and especially among the new nation-states, many countries developed extensive programs, often with additional support from one of the two Cold War powers, the USA and the USSR, to send students abroad for training in the sciences and other academic disciplines. Many national leaders hoped science would help their countries make up the deficits in intellectual infrastructure that was one of the legacies of colonization. Social sciences were among the disciplines that attracted large numbers of students. What many students found when they returned home was that the psychology they had learned in the laboratories of the West simply did not fit their home situation. This was the nexus for the emergence of Indigenous psychologies.

Accounts of the development of IPs by their originators are, of course, very useful. What they often do not include is a consideration of larger historical and sociopolitical contexts. Reading many of the IP creation accounts, it is obvious that some failed to consider that creating a psychology that fits local conditions could be an obstacle to being accepted as a legitimate psychology in the eyes of the dominant powers. As Sumie Okazaki and her colleagues argued, one of the main problems of many extant accounts of the creation of Indigenous psychologies is that they are ahistorical. That is, they often fail to account for the role of political and economic power, especially in colonial and postcolonial situations, as factors in the spread of dominant modes of psychology (Okazaki et al. 2008). To explore this further, it may be helpful to invoke analyses based on coloniality of being and knowledge and their implications for decolonization and decoloniality.

Nelson Maldonado-Torres has clearly articulated the concept of coloniality as it relates to knowledge and self-understanding:

> Coloniality...refers to long-standing patterns of power that emerged as a result of colonialism, but that define culture, labor, intersubjective relations, and knowledge production well beyond the strict limits of colonial administrations. Thus, coloniality survives colonialism. It is maintained alive in books, in the criteria for academic performance, in cultural patterns, in common sense, in the self-image of peoples, in aspirations of self, and so many other aspects of our modern experience. In a way, as modern subjects we breathe coloniality all the time and everyday. (2007, p. 242)

Coloniality, then, survives colonization and decolonization. It informs the way persons think and act, both personally, and in social, educational, and political settings. Coloniality of being and knowledge may well be the basis for what has often been observed about Indigenous psychologies that whatever the arguments made for the cultural veracity and need for a psychology that fits local conditions, most end up positing the hope that their IP can or does contribute to a supposed universal psychology. Kurt Danziger described this best in his invited commentary on a journal issue devoted to Indigenous psychologies:

> The notion of a universal psychology appears to hold an abiding attraction for most contributors [to the journal]. Possibly this attraction derives from a faith in the psychic unity of humankind, which is certainly preferable to the sorts of prejudice that flourished in the period of colonialism. What is questionable is any implication that a *particular historically constituted version* of the norms of scientific practice represents the royal road to the discovery of humankind's psychic unity. (2006, p. 272, emphasis added)

There have been multiple responses to the challenges of hegemonic psychological science. For example, critical social theory has offered critiques that are relevant to Indigenous psychology. Domination and obedience in society arose from the need to understand the issues of equality and inequality and how these issues are shaped by historical factors (Kincheloe and McLaren 2002). In this context, critical social theory and indigenous postcolonial theory are important in understanding the dominant or Western ways of conceptualization.

Over the last several years, it has become common in psychology to call for decolonization of psychological science. Social psychologist Glenn Adams, along with colleagues from Costa Rica, put together an exemplary issue on the topic in the *Journal of Social and Political Psychology* (2015, v. 3(1), https://doi.org/10.5964/jspp.v3i1.564). Hegemonic psychology, Adams et al. argue, has served the interests of the privileged of the world. Psychologists,

> typically proceed with academic business as usual with few opportunities to reflect on our participation (as both intellectuals and citizens) in ongoing processes of domination that promote exploration, expansion, and growth for a privileged few but undermine sustainable well-being for the vast majority of humanity. (Adams et al. 2015, p. 214)

Among the many IPs developed over the last 60 plus years, there is a recognition and resistance to this hegemony, but most frequently, the efforts to resist the hegemon end up using intellectual and institutional resources, as Bhatia wrote, "countering and aligning with the dominant frameworks of psychological science"... and finding their status and their work still being defined "within the terms and rules set by the colonial powers" (Bhatia 2019, p. 111). This is similar to the dilemma experienced by researchers who study topics such as diversity and inclusion, often in the pursuit of social justice. To be heard, they must address those in privileged social positions, such as white heteronormative, cisgender individuals. Doing so may well "overempower" the already privileged rather than empower those who are historically oppressed (e.g., Becker and Aiello 2013; Sohl 2018). So it is with those involved in developing IPs; in order to be recognized, they feel they have to address the normative status of hegemonic psychology. Doing so can subvert their claims to legitimacy by overempowering the dominant discourse. As a result, hegemonic psychology continues to be considered the "real" psychology.

Decolonizing psychology as both phrase and action has become a favored term to change these power relations. Bhatia's book of that title (2018) is one of the best examples to date of thoughtful analyses of how hegemonic psychology, especially of a variety that reflects US corporate interests, has changed the working lives of millions of Indians and how it can be challenged. Here, we wish to unpack decolonization a bit in order to clarify what it seeks to do. This approach may vary from other common understandings, but it is especially relevant to Indigenous psychologies. Decolonization is aimed at states, institutions of the state, land, place, or what is instantiated or material – thus Tuck and Yang's argument that decolonization is not a metaphor (2012), but is always already tied to land and materiality. For academics, materiality means universities, journals, disciplines, societies, etc. To decolonize is to seek to replace or supplant the governance of those institutions or at least gain recognition from them. Indigenous psychologies are among the most common examples of efforts at decolonization to be found in psychology. Decolonization work in psychology typically is predicated upon working within established institutional and academic frameworks to make psychology more inclusive as a discipline (see Bhatia 2018). In the case of Indigenous psychologies, the demand is that local knowledge, such as that regarding unique personality styles or local research methodologies, be included in what counts as psychological knowledge or methods.

Decoloniality, or decolonial theory, offers a different approach. Readers, who are more familiar with postcolonial stances on these issues or historical work on Europe that challenges the received narrative of Europe's rise to dominance, may note that the argument of the modernity/coloniality group has its corollaries in the scholarship of Jack Goody (2006), J. M. Blaut (1993), and John M. Hobson (2004). Postcolonial historiography also is complementary. See the special issue of *Postcolonial Studies, vol. 11(2), 2008*. For links and divergences of postcolonialism and decoloniality, see G. K. Bhambra 2014, *Postcolonial and decolonial dialogues*.

Decolonial theory is a body of theoretical and scholarly work developed over the last 20 plus years under the rubric of modernity/coloniality. It has grown out of the work of a loose affiliation of scholars based in or with close ties to Latin America. The reader is referred to the issue of *Cultural Studies* 2007, Volume 21, nos. 2–3, for an excellent introduction to decolonial theory. A key element of their intellectual argument is the need to delink from the ontological and epistemological claims that stem from the Enlightenment. It is important to note that the work is not a rejection of rationality but a rejection of European Enlightenment claims that it is the only rationality and its concomitant stance that Europe and its Eurocentric derivatives, such as the USA, are the center of even the possibility of true knowledge. This, they argue, can be seen in the Eurocentric world claim of the right to determine what is modern and what is primitive, what is intellectually valid and what is not (Mignolo 2011; Mignolo and Walsh 2018), even to the point of determining compass directions – East, South, West, and North

(Gordon 2011; Mignolo 2014). The claim of privilege to name modernity and what counts as knowledge is based on coloniality. Without the European colonization of much of the world, there would be no modernity. The European colonization endured for approximately 500 years beginning around 1500 CE. The outcome has been a European definition of modernity on the basis of their own provinciality (Chakrabarty 2000; Quijano 2007), rather than the alternative of a pluricentric, noncapitalist world of multiple civilizations and cultures (Escobar 2007, 2017, 2020).

A way to counter this One-World World (Law 2011) with its insistence on intellectual and practical privilege is to de-link from it via a decolonial turn. Decoloniality "means to change the terms of the (assumptions, rules, principles) conversation," in ways that counter the division of knowledge and practice, in the academy and in social and political life (Mignolo and Walsh 2018, p. 113). It requires an ontological and epistemic disobedience that makes it possible to think, feel, and live otherwise. Elsewhere, Pickren (in press) has written about psychologies otherwise and what that might look like. For Indigenous psychologies, a decolonial strategy would make it possible to de-link from the frame of hegemonic psychology dominated by Eurocentric notions of rationality, ontology, and epistemology that enforce the coloniality of knowledge in ways both subtle and violent (epistemological, geophysical, interpersonal, intercultural violence). Decoloniality would make it possible to develop Indigenous psychology without being constrained by Global North sensibilities of modernity. To echo Bhatia again, it would mean being able to fully embrace what the One-World World calls the sign of the primitive, that which is "deeply rooted in local practices and relegated to the realm of the mythological, collective, religious, traditional, philosophical, irrational, primitive, imaginative, and cultural" (Bhatia 2019, p. 111). Taking the decolonial option would bring the reward of other possibilities, other ontologies, epistemologies, and methodologies, and thus other realities, worlds otherwise, that is, the pluriverse (Escobar 2020; Mignolo and Walsh 2018). The real-world practical benefits would include a psychology that is actually true to its place, its people, and as importantly, a way to live that does not depend on domination and exploitation of the earth and of other beings, both human and other than human.

Conclusion

Just as there are a multiplicity of Indigenous psychologies and so many possible histories, so there are multiple rationales for writing histories of Indigenous psychologies. This chapter has provided one intellectual and critical rationale that will be of use for future historians. Before turning to the very practical matter of where to find the materials to write such histories, some perspectives are offered that will stand in for a short conclusion.

A difficulty in writing histories of Indigenous psychologies will lie in the overempowering problem that was noted earlier. If future historians are not careful, their work will end up serving the current status of hegemonic psychology as the real psychology. After all, in a neoliberal era already marked by gross inequities of wealth, income, and health, to name only a few (Wilkerson and Pickett 2010), it is likely that humans face a future where these crises are amplified by the crises of ecological devastation, climate change, biodiversity loss, and the many sequelae that are already making life ever more difficult, such as the coronavirus pandemic that began in late 2019 (Pickren in press). It can be anticipated that current elites, those who hold resources out of all proportion to the rest of humanity, and the professional classes that serve their interests will move to ensure their own survival. Given that hegemonic psychology is in the service of elites and neoliberal ends, one can imagine that those entities which challenge its preeminence, such as Indigenous psychologies, will be maligned and threatened. As Pickren wrote elsewhere,

> The social imaginary that has emerged in this regime [of neoliberalism] defines personhood as self- reliant individualism which achieves success through an entrepreneurialism of the self...The psychology that arose in Western societies generally, but especially in the United States, since the late 19th century accords with this view and holds a reflexive relationship with neoliberalism. One could make the case that the ascendance of American psychology since World War II facilitated neoliberal thinking, but even more that it prepared individuals to think and act neoliberally, if I may coin a term (see Pickren and Rutherford 2010, Chapter 10). Regardless, it is now the case that psychology is embedded in and reflective of—perhaps even a driver of—the neoliberal imaginary. (Pickren 2018, 576)

If this relationship of psychology to the elite power structure remains central, then any challenge is likely to be a target of delegitimization efforts, at the least.

This critique and the published resources that are cited, especially those by Glenn Adams and his colleagues, the modernity/coloniality group, especially the work of Arturo Escobar, the critical research methodologies of Indigenous peoples themselves, along with allies such as Michelle Fine and Maria Torre, and the small number of critical psychologists found throughout the world, may be resources for developing Indigenous psychologies and writing their histories that help create psychologies otherwise and bring psychology into the pluriverse (Escobar 2017; Pickren in press). The work presented here can help those who are creating Indigenous psychologies and those who wish to write their histories to make the decolonial turn and de-link from the dominant psychology of our time. Making such a turn will require that scholars and activists develop a habit of ontological and epistemological disobedience in order to create a psychology that is linked to the reality of place, which is the core of any Indigenous psychology. Can we make the decolonial move to de-link from a rather narrow embrace of what psychology is as defined in the canon we were educated in? Can we engage in ontological disobedience, Or perhaps even onto-epistemic disobedience? As Anand Paranjpe wrote some years ago, "what is particularly needed is a psychology guided by emancipatory interests" (Paranjpe 2002, p. 29). May this chapter serve such a psychology.

Cross-References

▶ Psychologies: Their Diverse Histories

References

Abbott A (1988) The system of professions: an essay on the division of expert labor. University of Chicago Press, Chicago

Adams G, Dobles I, Gómez LH, Kurtiş T, Molina LE (2015) Decolonizing psychological science: introduction to the special thematic section. J Soc Polit Psychol 3:213–238

Adams G, Estrada-Villalta S, Sullivan D, Markus HR (2019) The psychology of neoliberalism and the neoliberalism of psychology. J Soc Issues 75:189–216

Albanese CL (2007) A republic of mind and spirit: a cultural history of American metaphysical religion. Yale University Press, New Haven

Allwood CM (2011) On the foundation of the indigenous psychologies. Soc Epistemol 25:3–14

Allwood CM (2018) The nature and challenges of indigenous psychologies. Cambridge, New York

Allwood CM (2019) Future prospects for indigenous psychologies. J Theor Philos Psychol 39:90–97

Allwood CM, Berry JW (2006) Origins and development of indigenous psychologies: an international analysis. Int J Psychol 41:243–268

Arnett JJ (2008) The neglected 95%: why American psychology needs to be less American. Am Psychol 63:602–614

Battiste M (ed) (2000) Reclaiming indigenous voice and vision. University of British Columbia Press, Vancouver

Becker S, Aiello B (2013) The continuum of complicity: "studying up"/studying power as a feminist, anti-racist, or social justice venture. Women's Stud Int Forum 38:63–74

Béteille A (1998) The idea of indigenous people. Curr Anthropol 39:187–192

Bhambra GK (2014) Postcolonial and decolonial dialogues. Postcolonial Stud 17:115–121

Bhatia S (2018) Decolonizing psychology: globalization, social justice, and Indian youth identities. Oxford University Press, New York

Bhatia S (2019) Searching for justice in an unequal world: reframing indigenous psychology as a cultural and political project. J Theor Philos Psychol 39:107–114

Bhatia S, Priya KR (2018) Decolonizing culture: euro-American psychology and the shaping of neoliberal selves in India. Theory Psychol 28:645–668

Bird-David N (1990) The giving environment: another perspective on the economic system of gatherer-hunters. Curr Anthropol 31:189–196

Blaut JM (1993) The colonizer's model of the world: geographical diffusionism and Eurocentric history. Guilford, New York

Blowers GH, Turtle AM (eds) (1987) A Westview special study. Psychology moving east: the status of Western psychology in Asia and Oceania. Westview Press; Sydney University Press

Blowers GH, Turtle AM (eds) (2019) Psychology moving east: the status of Western psychology in Asia and Oceania. Routledge, New York

Brock AC (ed) (2006) Internationalizing the history of psychology. New York University Press, New York

Byrd JA (2011) The transit of empire: indigenous critiques of colonialism. University of Minnesota Press, Minneapolis

Caplan E (2001) Mind games: American culture and the birth of psychotherapy. University of California Press, Berkeley

Chakkarath P (2012) The role of indigenous psychologies in the building of basic cultural psychology. In: Valsiner J (ed) The Oxford handbook of culture and psychology. Oxford University Press, New York, pp 71–95

Chakrabarty D (2000) Provincializing Europe: postcolonial thought and historical difference. Princeton University Press, Princeton

Charlton J (1998) Nothing about us without us: disability, oppression, and empowerment. University of California Press, Berkeley

Ciofalo N (ed) (2019) Indigenous psychologies in an era of decolonization. Springer, New York

Cooter R (1984) The cultural meaning of popular science: phrenology and the organization of consent in nineteenth-century Britain. New York, Cambridge

Couture JE (1987) What is fundamental to native education? Some thoughts on the relationship between thinking, feeling, and learning. In: Stewin LL, McCann SJH (eds) Contemporary educational issues: the Canadian mosaic. Copp Clark Pitman, Toronto, pp 178–191

Danziger K (1990) Constructing the subject: historical origins of psychological research. New York, Cambridge

Danziger K (2006) Comment. Int J Psychol 41:269–275

Datta R (2018) Traditional storytelling: An effective Indigenous research methodology and its implications for environmental research. AlterNative: An International Journal of Indigenous Peoples 14(1):35–44

Doniger W (1998) The implied spider: politics and theology in myth. Columbia University Press, New York

Downing C (1975) Sigmund Freud and the Greek mythological tradition. J Am Acad Relig 43:3–14

Dunbar-Ortiz R (2014) An indigenous peoples' history of the United States. Beacon Press, Boston

Escobar A (2007) Worlds and knowledges otherwise: the Latin American modernity/coloniality research program. Cult Stud 21:179–210

Escobar A (2017) Designs for the pluriverse: radical interdependence, autonomy, and the making of worlds. Duke University Press, Durham

Escobar A (2020) Pluriversal politics: the real and the possible. Duke University Press, Durham

Fuller RC (1982) Mesmerism and the American cure of souls. University of Pennsylvania Press, Philadelphia

Gagne N (2015) Brave new words: the complexities and possibilities of an "indigenous" identity in French Polynesia and New Caledonia. Contemp Pac 27:371–402

Glavenau V (2005) From mythology to psychology-an essay on the archaic psychology in Greek myths. EJOP:1. https://doi.org/10.5964/ejop.v1i1.351

Goody J (2006) The theft of history. Cambridge University Press, New York

Gordon L (2011) Shifting the geography of reason in an age of disciplinary decadence. Transmodern J Peripher Cult Prod Luso-Hispanic World 1:95–103

Haber S (1964) Efficiency and uplift: scientific management in the progressive era, 1890–1920. University of Chicago Press, Chicago

Hale TA (1998) Griots and griottes: masters of words and music. Indiana University Press, Bloomington

Harrington A (2008) The cure within: a history of mind-body medicine. Norton, New York

Heelas P, Lock A (eds) (1981) Indigenous psychologies: the anthropology of the self. Academic Press, London

Hegarty P, Pratto F (2001) The effects of social category norms and stereotypes on explanations for intergroup differences. J Pers Soc Psychol 80:723–735

Henrich J, Heine SJ, Norenzayan A (2010) The weirdest people in the world? Behav Brain Sci 33: 61–83

Hobson JM (2004) The eastern origins of Western civilization. New York, Cambridge

Hoopes J (1989) Consciousness in New England: from puritanism and ideas to psychoanalysis and semiotic. Johns Hopkins University Press, Baltimore

Ingold T (1999) On the social relations of the hunter-gatherer band. In: Lee RB, Daly R (eds) The Cambridge encyclopedia of hunters and gatherers. Cambridge, New York, pp 399–410

Kim U, Berry JW (1993) Indigenous psychologies: experience and research in cultural context. Sage, Newbury Park

Kim U, Yang K-S, Hwang K-K (eds) (2006) Indigenous & cultural psychology: understanding people in context. Springer SBM, New York

Kincheloe JL, McLaren P (2002) Rethinking critical theory and qualitative research. Ethnogr Schools Qual Approach Study of Educ:87–138

Kırklaroglu H (2018) Bilim tarihi açisindan psikoloji ve bilimselliği üzerine tartişma. Gazi Üniversitesi Sosyal Bilimler Dergisi 5:194–210

Larson MS (1984) The production of expertise and the constitution of expert power. In: Haskell TL (ed) The authority of experts. Indiana University Press, Bloomington, pp 25–64

Law J (2011, September 25) What's wrong with a one-world world? Heterogeneities. http://www.heterogeneities/publications/Law2011WhatsWrongWithAOneWorldWorld.pdf

MacLeod SP (2018) Celtic cosmology and the otherworld: mythic origins, sovereignty, and liminality. McFarland, Jefferson

Maldonado-Torres N (2006) Cesaire's gift and the decolonial turn. Radic Philos Rev 9:111–138

Maldonado-Torres N (2007) On the coloniality of being. Cult Stud 21:240–270

Mann BA (2016) Spirits of blood, spirits of breath: the twinned cosmos of indigenous America. New York, Oxford

Markus G (1987) Why is there no hermeneutics of natural sciences? Some preliminary theses. Sci Context 1:5–51

Mignolo WD (2011) The darker side of Western modernity: global futures, decolonial options. Duke University Press, Durham

Mignolo WE (2014) The north of the south and the west of the east: a provocation to the question. IBRAAZ. https://www.ibraaz.org/essays/108/ (downloaded February 10, 2021)

Mignolo WD, Walsh CE (2018) On decoloniality: Concepts, analytics, praxis. Durham, NC: Duke University Press

Minthorn RS, Shotton HJ (eds) (2018) Reclaiming indigenous research in higher education. Rutgers University Press, New Brunswick

Misra G, Sanyal N, De S (eds) (in press) Psychology in modern India: historical, methodological, and future perspectives. Springer SBM, New York

Morales AL (1998) The historian as curandera, Working paper #40. Institute for Cultural Activism, Berkeley

O'Donnell JM (1985) The origins of behaviorism: American psychology, 1870–1920. New York University Press, New York

Okazaki S, David EJR, Abelmann N (2008) Colonialism and psychology of culture. Soc Personal Psychol Compass 2:90–106

Peters MA, Mika CT (2017) Aborigine, Indian, indigenous or first nations? Educ Philos Theory 49:1229–1234

Pickren WE (2000) A whisper of salvation: psychology and religion at the turn of the twentieth century. Am Psychol 55:1022–1024

Pickren WE (2018) Psychology in the social imaginary of neoliberalism: critique and beyond. Theory Psychol 28:575–580

Pickren WE (in press) Psychologies otherwise/earthwise: pluriversal approaches to crises of climate, equity, and health. In: Dege M (ed) The psychology of global crises and crisis politics: intervention, resistance, decolonization. Palgrave, New York

Pickren WE, Rutherford A (2010) A history of psychology in modern context. Wiley, Hoboken

Puhvel J (1989) Comparative mythology. Johns Hopkins University Press, Baltimore

Quijano A (2007) Coloniality and modernity/rationality. Cult Stud 2(3):168–178

Rautman AE (2000) Reading the body: representation and remains in the archeological record. University of Pennsylvania Press, Philadelphia

Rutherford A (2018) Feminism, psychology, and the gendering of neoliberal subjectivity: from critique to disruption. Theory Psychol 28:619–644

Santos B d S (2014) Epistemologies of the south: justice against epistemicide. Routledge, New York

Schmit D (2005) Re-visioning American antebellum psychology: the dissemination of mesmerism, 1836–1854. Hist Psychol 8:403–434

Schmit D (2010) The Mesmerists inquire about "oriental mind powers:" west meets east in the search for the universal trance. J Hist Behav Sci 46:1–26

Shamdasani S (2003) Jung and the making of modern psychology: the dream of a science. New York, Cambridge

Smith D, Schlaepfer P, Major K, Dyble M, Page AE, Thompson J et al (2017) Cooperation and the evolution of hunter-gatherer storytelling. Nat Commun 8:1–9

Smith LT (1999) Decolonizing methodologies: research and indigenous peoples. Zed Book/University of Otago Press, New York/Dunedin

Smith LT (2012) Decolonizing methodologies: research and indigenous peoples, 2nd edn. Zed Books, Ltd., London

Smith LT, Tuck E, Yang KW (eds) (2019) Indigenous and decolonizing studies in education. Routledge, New York

Smith R (1997) The Norton history of the human sciences. Norton, New York

Smith R (xxxx) Introductory chapter to section on psychology. In: Palgrave handbook of the history of the human sciences

Sohl L (2018) Feel-bad moments: Unpacking the complexity of class, gender and whiteness when studying 'up'. Eur J Women's Stud. https://doi.org/10.1177/1350506818762232

Solórzano DG, Yosso TJ (2002) Critical race methodology: counter-storytelling as an analytical framework for education research. Qual Inq 8:23–44

Staeuble I (2006) Decentering Western perspectives: psychology and the disciplinary order in the first and third world. In: Brock AC, Louw J, van Hoorn W (eds) Rediscovering the history of psychology: essays inspired by the work of Kurt Danziger. Kluwer, New York, pp 183–205

Sugarman J (2015) Neoliberalism and psychological ethics. J Theor Philos Psychol 35:103–116

Suzman J (2017) Affluence without abundance: the disappearing world of the bushmen. Bloomsbury, London

Taves A (1999) Fits, trances, and visions: experiencing religion and explaining experience from Wesley to James. Princeton University Press, Princeton

Taylor E (1999) Shadow culture: psychology and spirituality in America. Counterpoint, Washington, DC

Thomson M (2006) Psychological subjects: identity, culture, and health in twentieth-century Britain, New York/Oxford

Trask H-K (1999) From a native daughter: colonialism and sovereignty in Hawai'i. University of Hawaii Press and Kamakakūokalani Center for Hawaiian Studies, Honolulu

Tuck E, Yang KW (2012) Decolonization is not a metaphor. Decolon Indig Educ Soc 1:1–40

van Strien PJ (1997) The American colonization of northwest European social psychology after world war II. J Hist Behav Sci 33:349–363

Vidal F (2011) The sciences of the soul: the early modern origins of psychology (Trans. S Brown). University of Chicago Press, Chicago. (First published in French in 2006)

Vidal F, Ortega F (2017) Being brains: making the cerebral subject. Fordham University Press, New York

White CG (2009) Unsettled minds: psychology and the American search for spiritual assurance, 1830–1940. University of California Press, Berkeley

Wilkerson R, Pickett K (2010) The spirit level: why greater equality makes societies stronger. Bloomsbury Press, New York

Winter A (1998) Mesmerized: powers of mind in Victorian Britain. University of Chicago Press, Chicago

Yang CP, Lu FG (2007) Indigenous and cultural psychology: understanding people in context a book review from the transpersonal psychology perspective. Pastor Psychol 56:105–113

Further Reading

Published Resources of Contemporary Accounts of Indigenous Psychology(ies)

The authors take quite seriously the caution of James Charlton quoted above, "Nothing about us, without us" (2018). So, rather than write a history or histories of the otherwise of Indigenous psychologies, the chapter offers a set of published accounts that may serve as places to begin such histories. The sets of articles noted below are indicative rather than comprehensive, so the serious reader will need to explore further

The list of readings does not include articles that specifically theorize Indigenous psychology, though many of the authors do include some measure of theoretical justification for why they are developing an Indigenous psychology. The Allwood volume (2018) includes references to a number of such articles, and they are worth perusing. Theory is also at the forefront of the special issue of the Journal of Theoretical and Philosophical Psychology edited by Louise Sundarajan (2019). Some of the most powerful arguments, theoretical and otherwise, can be found in the writings of Enriquez listed below, along with work by Martin-Baro. Girishwar Misra and many of his colleagues in India have been at the forefront of developing Indigenous psychologies in India. There is a volume forthcoming that features historical and theoretical work on Indigenous psychologies in India (Misra, Sanyal, & De, in press)

One of the best places to begin in a literature search on Indigenous psychology is Carl Martin Allwood's recent *The Nature and Challenges of Indigenous Psychologies* (2018). Allwood is a Swedish cognitive psychologist who has published several key pieces on Indigenous psychologies (e.g., Allwood, 2011, 2018, 2019; Allwood & Berry, 2006). Allwood employs an anthropology of knowledge and science studies framework in his analyses of IPs. He is sensitive to situate IPs in historical and cultural context, but his accounts are not histories, per se. Allwood uses a geographical division so that IPs are grouped by region and/or country. It is worth keeping in mind that the demarcation of IPs by country can be problematic. Many countries have multiple racial and ethnic groups, so it would be difficult to imagine one Indigenous psychology for an entire country would cover the waterfront, so to speak

Indigenous Psychologies by Book Volumes (authored and edited; some already noted in text)

Starting Point:

Smith LT (1999) Decolonizing methodologies: research and indigenous peoples. Zed Books/ University of Otago Press, New York/Dunedin

Smith LT (2012). Decolonizing methodologies: research and indigenous peoples, 2nd ed. Zed Books, Ltd., London

Bhawuk DPS (2011) Spirituality and Indian psychology: lessons from the Bhagavad-Gita. Springer SBM, New York

Blowers GH, Turtle AM (eds) (1987) A Westview special study. Psychology moving east: the status of Western psychology in Asia and Oceania. Westview Press/Sydney University Press. [Reprint of original: Blowers GH, Turtle AM (eds) (2019) Psychology moving east: the status of Western psychology in Asia and Oceania. Routledge, New York

Bond MH (ed) (2010) Oxford handbook of Chinese psychology. Oxford University Press, New York

Bond MH (1997) Working at the interface of culture: eighteen lives in social science. Routledge, London. [Accounts of individual scientists working in a culture other than their own. Several of the chapters are by psychologists who have been important in developing indigenous psychologies.]

Enriquez VG (1989) Indigenous psychology and national consciousness. Institute for the Study of Languages and Cultures of Asia and Africa, Tokyo

Enriquez VG (1992) From colonial to liberation psychology. University of the Philippines Press, Quezon City

Heelas P, Lock A (eds) (1981) Indigenous psychologies: the anthropology of the self. Academic Press, London:

Hwang K-K (2012) Foundations of Chinese psychology: confucian social relations. Springer SBM, New York

Kagitcibasi C (2010) Changing life styles – changing competencies: Turkish migrant youth in Europe. Hist Soc Res 35:151–168

Misra G, Mohanty AK (eds) (2002) Perspectives on indigenous psychology. Concept Publishing Company, New Delhi

Montero M, Sonn CC (eds) (2009) Psychology of liberation: theory and applications. Springer SBM, New York

Rao R (1962) Development of psychological thought in India. Kavyalaya, Mysore

Rao KR, Marwah SB (eds) (2005) Towards a spiritual psychology: essays in Indian psychology. Samvad, New Delhi

Rao KR, Paranjpe AC, Dalal AK (eds) (2008) Handbook of Indian psychology. Cambridge University Press, New Delhi

Sinha D (1986) Psychology in a third world country. Sage, New Delhi

Indigenous Psychologies in Journal Special Issues

International Journal of Psychology: Special Issue on Indigenous Psychologies

Guest Editors: C. M. Allwood & John W. Berry

Comment by Kurt Danziger; Articles by Hwang, Kao, Kim & Park, Nsamenang, Yang, plus brief accounts by Pawel Boski; Fanny M. Cheung; Kwang-Kuo Hwang; Henry Kao; Uichol Kim & Young-Shin Park; Leo Marai; Fathali M. Moghaddam; Linda Waimarie Nikora, Michelle Levy, Bridgette Masters, & Moana Waitoki; A. Bame Nsamenang; Elizabeth Protacio-De Castro (formerly Marcelino), Melecio C. Fabros, & Reginald Kapunan; T. S. Saraswathi; Jai B. P. Sinha; Kuo-Shu Yang

Journal of Theoretical and Philosophical Psychology: Special Issue on Indigenous Psychologies: What's the Next Step

Guest Editor: Louise Sundarajan

Articles and Comments by L. Sundarajan, S. Bhatia, K-K. Hwang, A. Dueck, W. Long, C. Allwood

Indigenous Psychologies in Journal Articles

Africa

Abdi YO (1975) The problem and prospects of psychology in Africa. Int J Psychol 10:227–234

Asante KO, Oppong S (2012) Psychology in Ghana. J Psychol Afr 22:473–476

Mate-Kole CC (2013) Psychology in Ghana revisited. J Black Psychol 39:316–320

Mayer S (2002) Psychology in Nigeria: a view from the outside. Ife PsychologIA 10:1–8

Nsamenang AB (2006) Human ontogenesis: an indigenous African view on development and intelligence. Int J Psychol 41:293–297

Nsamenang AB (2013) Cameroon black psychologists. J Black Psychol 39:307–310

Nwoye A (2015) What is African psychology the psychology of? Theory Psychol 25:96–116

Oppong S (2016) The journey towards Africanising psychology in Ghana. Psychol Thought 9:1–14

Indigenous Psychologies in Journal Articles (Asia)

China and Taiwan

Gabrenya WK Jr, Kung M-C, Chen L-Y (2006) Understanding the Taiwan indigenous psychology movement. A sociology of science approach. J Cross-Cult Psychol 37:597–622

Ho DYF (1998) Indigenous psychologies: Asian perspectives. J Cross-Cult Psychol 29:88–103

Hwang K-K (2005) From anti-colonialism to postcolonialism: the emergence of Chinese indigenous psychology in Taiwan. Int J Psychol 40:228–238

Jing Q, Fu X (2001) Modern Chinese psychology: its indigenous roots and international influences. Int J Psychol 36:408–418

Kim U, Park Y-S, Park D (1999) The Korean indigenous psychology approach: theoretical considerations and empirical applications. Appl Psychol Int Rev 48:451–464

Yang K-S (2012) Indigenous psychology, westernized psychology, and indigenized psychology: a non-Western psychologist's view. Chang Gung J Human Soc Sci 5:1–32

India

Adair JG, Pandey J, Begum HA, Puhan BN, Vohra N (1995) Indigenization and development of the discipline: perceptions and opinions of Indian and Bangladeshi psychologists. J Cross-Cult Psychol 26:392–407

Bhatia S (2002) Orientalism in euro-American and Indian psychology: historical representations of "native" in colonial and post -colonial contexts. Hist Psychol 5:376–398

Chaudhary N, Sriram S (2020) Psychology in the "backyards of the world": experiences from India. J Cross-Cult Psychol 51:113–133

Dalal AK, Misra G (2010) The core and context of Indian psychology. Psychol Dev Psychol 22:121–155

Padalia D (2017) Why indigenous psychology? A review article. Int J Indian Psychol 4:78–82

Paranjpe AC (2002) Indigenous psychology in the post-colonial context: a historical perspective. Psychol Dev Soc 14:27–43

Sinha C (2016) Decolonizing social psychology in India: exploring its role as emancipatory social science. Psychology & Society 8:57–74

Sinha D (1965) Integration of modern psychology with Indian thought. J Humanist Psychol 5:6–17

Sinha D (1981) Non-western perspectives in psychology: why, what and whither? J Indian Psychol 3:1–9

Sinha D (1994) Origins and development of psychology in India: outgrowing the alien framework. Int J Psychol 29:695–705

Sinha D (1998) Changing perspectives in social psychology in India: a journey towards indigenization. Asian J Soc Psychol 1:17–31

Philippines

Church AT, Katigbak MS (2002) Indigenization of psychology in the Philippines. Int J Psychol 37:129–148

Enriquez VG (1977) Filipino psychology in the third world. Philipp J Psychol 10:3–18

Gastardo-Conaco MC (2005) The development of a Filipino indigenous psychology. Philipp J Psychol 38:1–17

Pe-Pua R, Protacio-Marcelino E (2000) *Sikolohiyang Pilipino* (Filipino psychology): a legacy of Virgilio G. Enriquez. Asian J Soc Psychol 3:49–71

San Juan E Jr (2006) Toward a decolonizing indigenous psychology in the Philippines: introducing Sikolohiyang Pilipino. J Cult Res 10:47–67

Australia and New Zealand

Dudgeon P (2015) Decolonising Australian psychology: discourses, strategies, and practice. J Soc Polit Psychol 3. https://doi.org/10.5964/jspp.v3i1.126

Dudgeon P (2017) Editorial Australian indigenous psychology. Aust Psychol 52:251–254

Dudgeon P, Bray A, D'Costa B, Walker R (2017) Decolonizing psychology: validating social and emotional wellbeing. Aust Psychol 52:316–325

Nikora LW (2007) Māori and psychology: indigenous psychology in New Zealand. In: Weatherall A, Wilson M, Harper D, McDowall J (eds) Psychology in Aotearoa/New Zealand. Pearson Education New Zealand, Auckland, pp 80–85

Latin America/Mexico

Burton M, Kagan C (2005) Liberation social psychology: learning from Latin America. J Community Appl Soc Psychol 15:63–78

Diaz-Loving R (2005) Emergence and contributions of a Latin American indigenous social psychology. Int J Psychol 40:213–227

Diaz-Guerrero R (1977) A Mexican psychology. Am Psychol 32:934–944

Martín-Baró, I. (1996) Writings for a liberation psychology. Harvard University Press, Cambridge, MA

Montero M (2002) On the construction of reality and truth. Towards an epistemology of community social psychology. Am J Commun Psychol 30:571–584

Nahuelpan H (2016) The place of the "Indio" I social research: considerations from Mapuche history. AlterNative Int J Indig Peoples 12:3–17

Sanzez Sosa JJ, Valderrama-Iturbe P (2001) Psychology in Latin America: historical reflections and perspectives. Int J Psychol 36:384–394

Children as Psychological Objects: A History of Psychological Research of Child Development in Hungary

42

Zsuzsanna Vajda

Contents

Introduction: The History of Psychology and Psychology in History	1088
The Birth of Child Psychology in Hungary	1089
Introduction of Pedology and Psychometry into Child Psychology	1092
Observing Children	1094
Child Psychology Between the World Wars	1095
The Budapest School of Psychoanalysis	1098
Child Psychology After World War II	1105
Restart After the Revolution of 1956	1107
Children and Psychology After the Regime Change	1110
Conclusion	1112
References	1113

Abstract

The Hungarian experience exemplifies the significance of local context and social concerns in the history of psychology, specifically in the field of child development. Child psychology, which gained great importance with the introduction of compulsory schooling, has been perhaps the single most important practical area of psychology. In Hungary, moreover, it was the only area of psychology with sustained continuity through social and political changes.

The first researchers of child psychology were schoolteachers who shared the view that children's characteristics have to be studied for the sake of a common and effective education. By the end of the nineteenth century, with the growing popularity of pedology, children's feelings also became of significance. Psychoanalysis increased the interest, and Budapest psychoanalysts – two or three decades Freud's junior – challenged some important views of the Master. Alice Bálint, Imre Hermann, Géza Róheim, and others drew important conclusions about child

---

Z. Vajda (✉)
Department of Psychology, Károli Gáspár University of the Reformed Church in Hungary, Budapest, Hungary
e-mail: vajdazsuzsanna@gmail.com

© The Author(s), under exclusive licence to Springer Nature Singapore Pte Ltd. 2022
D. McCallum (ed.), *The Palgrave Handbook of the History of Human Sciences*,
https://doi.org/10.1007/978-981-16-7255-2_84

development on the basis of their own observations and research. Unfortunately, their work was interrupted by World War II and emigration. The remaining professionals could not continue their work in the era of dictatorship from 1949 to 1956, when psychology – and in particular psychoanalysis – was considered as ideologically harmful. From the 1960s there was a revival of child psychology, especially in the field of practice, but after the regime change in 1989 it confronted new risks: the devaluation of children's perspective and commercialism.

Keywords

Child psychology in Hungary · Schooling, pedagogy and psychology · Budapest psychoanalysts' views on child development · Psychology and education in the one-party system and after

Introduction: The History of Psychology and Psychology in History

Be their reasoning well-founded or groundless, residents in small countries often tend to highly estimate native scientists' and artists' achievements, which may be less known and acknowledged internationally. Hungarian historians of psychology share the view that certain achievements of their compatriots have failed to receive the attention they deserve (Szokolszky 2016; Pléh 2004). Professionals and the interested audience alike may well be familiar with the activity of psychoanalysts like Sándor Ferenczi and Mihály Bálint, though Hungarian psychology can boast other important contributions. Hungarian psychologists' achievement met international standards in several fields and kept abreast of international trends till World War II. In the first half of the twentieth century, a number of Hungarian psychologists previously working in their home country also played an important role in higher education, research, and practical work in other countries.

This chapter describes psychology emerging in Hungary as a practice that regarded childhood as a research object. This social process took place in many parts of the world, and certainly elsewhere in central Europe, not necessarily in centers most associated with academic psychology. While the details concern local events and personalities, these particulars actually exemplify a very important dimension of the shaping of modern psychology.

The focus of this chapter, for several reasons, is the research conducted by Hungarian experts in the field of child development. The first is the special interest of Hungarian intellectuals regarding child development and education. Even professionals primarily researching other areas, such as the psychology of labor and experimental or medical fields, were intensely interested in the specific psychological characteristics of child development. An overview of the history of Hungarian child psychology is all the more reasonable as this is the only area of psychology with sustained continuity. Even in the 1950s, when psychology generally was regarded as imperialistic pseudoscience by the reigning Stalinist regime, child psychology was granted legitimate status, if to a limited degree.

Furthermore, child psychology is a very special field of the discipline. First of all it is one of the important, if not the most important, practical areas of psychology. It is worth recalling that in the public mind psychology primarily appears in practice, yet theoretical and historical works mainly focus on research, treating practice as a secondary factor. The chapter will therefore illustrate the importance of activity outside what have conventionally been thought to be the centers of academic psychology, both in terms of geography and in terms of practical commitment.

Contradictions that prevent psychology from becoming a coherent science exert a particular strain in the study of child development. Although images of the child in western (and nonwestern) cultures are different (Richards 2002), there is a huge amount of data supporting the fact that children's physical and mental development have universal characteristics. Stages of motion and cognitive development of children are – within certain limits – similar all over the world. Although less quantifiable, vulnerability, susceptibility, and limited skills of foresight of young children are also universal features. In fact, there is no society that can completely ignore these facts. However, there are very significant differences in the interpretation of them. Do we consider a child to be innocent or warped; is childhood ignorance a virtue or a handicap? This study will show through the example of Hungarian child psychology that the changes in childhood interpretations that professionals attribute to new scientific findings are due to the changing social conditions.

Owing to its geographical and historical circumstances, scientists in Hungary may not in fact have had the serenity associated with research. This can be illustrated well with the history of child development in a country where the past 150 years have seen several changes regarding its official language, borders, and forms of state from kingdom through monarchy to republic and people's republic. The Carpathian basin, home to Hungarians for more than a thousand years, belonging to a nation with a strange and unique language and rather eccentric behavior (in both the positive and the negative sense of the word) has never been a quiet corner of the world. In addition to Hungarians, several ethnic groups inhabited the territory of historic Hungary, which was part of the Austro-Hungarian Monarchy up to the Treaty of Trianon putting an end to World War I. Constant mobility was characteristic of most Hungarian science professionals, meaning not only regular study tours abroad but also migration from country to country, from region to region. The latter was especially typical of the Jewish professionals proportionally overrepresented in the fields of medicine and psychology. In the course of the nineteenth century, they moved in great numbers from Galicia in the east of the Monarchy to the politically more liberal Hungary, Austria, and Bohemia, hoping for better chances of economic and social advancement, benefitting from modernization. The second and third generation of the new arrivals successfully assimilated and played a significant role in local culture and science.

The Birth of Child Psychology in Hungary

At the beginning of the nineteenth century Hungary had the largest population of all states of the Habsburg Empire, with growing national consciousness. The independence movements of 1848–1849 led to revolutionary struggles, then to an unsuccessful war of independence lasting for almost 2 years, followed by despotic rule. After the Compromise of 1867 with the Habsburg dynasty, a dualistic form of state was created, the Austro-Hungarian Monarchy. This resulted in significantly stronger independence affecting, among other things, education and language use. Bilingualism, typical of the middle class, was a great advantage in the field of science and culture. Hungarian intellectuals maintained close relationships with their German-speaking colleagues. Owing to the common language, cultural diversity had a fruitful, creative influence on arts and science. It was also a progressive period in the formation of the nation state, in which intellectuals laid the foundations of Hungarian specialist language and literature.

If, as Roger Smith wrote, "something changed in the second half of the nineteenth century, enabling psychology to flourish in the twentieth century on the scale and with the many identities it then acquired" (1997: 493), one of these changes was that school education became widespread. This significantly contributed to the interest in child development. Although most works on the history of psychology trace this interest back to the first experiments in the psychology of sensation, it is a fact that as early as half a century before the famous Weber-Fechner experiments, various works and diaries recorded observations related to child development and rearing. The formation of European school systems was greatly furthered by nation-states aiming to encourage their citizens to identify themselves with their home country and cooperate with its institutions. The purpose of public education was not merely to raise the population to a higher cultural level, since schools became the most important channels of mass communication, propaganda, and moral education (Pukánszky et al. 2005).

Attending school became a daily routine for children, and it was accompanied by a broadening of job opportunities and increasing social mobility. All this brought about a significant turn in the structure of society and in individuals' mental frame of mind. It is impossible to overemphasize the significance of the fact that, by the first decades of the twentieth century, a decisive proportion of the population had acquired the skills of writing, reading, and counting. The new public education professionals and schoolteachers form the most populous group of intellectuals up to the present day. The discovery of the characteristics of childhood is mainly due to their efforts, and they may be considered the first representatives of developmental psychology. Moreover, schools were an ideal ground for psychological experiments and observations. Organizing children into age groups made it possible to identify features which were characteristic of the given age and were suitable for generalization and measurement.

Psychological works published at the end of the eighteenth century aimed to define the subject of the newly born field of science and outline its professional boundaries. Péter Bárány (1763–1829), a physician, the author of the first work on psychology written in the Hungarian language, aimed to create an experimental psychology which does not only originate from "speculation" (Bárány 1790). The way to the discovery of the mind leads through self-awareness, the knowledge of others, and the observation of history and animals. The experiences ("true stories") must be recorded and analyzed considering physiological experiences. Bárány was interested in child development; he translated the German teacher-reformist, Heinrich Campe's book for children about the "science of the psyche" into Hungarian, adding his own ideas. Since the Hungarian language still lacked special terminology, he complemented his translation with a list of vocabulary entailing the Hungarian equivalents of Latin terms. Another early author, Ádám Pálóczi Horváth (1760–1820), compiled his work entitled "Psychology" from mainly German and French authors' works, and focused on a more precise definition of the subject of psychology. As opposed to Bárány, he emphasized that psychology only involves the study of human psyche and does not deal with the inner phenomena of "ignorant animals" (1790: 2–3).

During the following decades, a surprisingly high number of works were published in Hungarian on the characteristics of mental processes, the majority of which discussed questions connected with development and child-rearing. The mass attendance at school meant that children spent an increasing proportion of their time with activities organized for them and among their peers. Thus school made a significant contribution to the separate treatment of childhood. Medicine also accumulated more and more knowledge about the different characteristics of the children's body. All this seemed to prove the claim attributed to Rousseau (who otherwise is rarely mentioned in contemporary writings about children) that children are not little "grown-ups." Teachers and physicians of the era attempted to prove and illustrate the difference. They even suggested that children have an alternative nature that is closer to the "real" one, which is obscured by the norms of civilization. In this early period of child psychology the main aim was to assist moral education. Psychology was to be instrumental in making individuals work for the benefit of mankind and encouraging them to learn that it is not material goods that make a person esteemed but a good heart and professional skills. At the same time, children's abilities and their motivation were becoming increasingly important.

Early works continued to attempt to clarify the subject and boundaries of psychology, especially its relationship to pedagogy. Intellectuals and public officials shared the view that children need a common and effective education for the sake of goals that they also shared – literacy which leads to social mobility, patriotism, and good morals. István Varga (1776–1831) was against formalism in didactics, reasoning that abilities should be developed taking into consideration the natural development of the individual. The necessity of educating socially disadvantaged children was also often mentioned. In 1839, Mihály D. Mocsi emphasized that it is not just the privileged few but members of all social classes who should be endowed with fundamental knowledge and skills. This he wanted to facilitate by discussing the role of mimicry in learning as well as how mental processes stem from the nervous system. Early, "self-made" psychologists of the nineteenth century were not at all familiar with the role of the central nervous system in producing psychological phenomena. According to Mocsi, "the substance of the psyche is not connected to the brain," but in spite of this, the brain is the organ of the psyche, although the heart, the abdominal cavity, and the thoracic cavity significantly affect the "workings of the psyche." János Warga (1804–1875), headmaster of a country secondary school with a good reputation, expressed his idea that education and child rearing should be adjusted to pupils' individual abilities as well as a syllabus designed to their individual pace. It is a particular contradiction – which would be discovered only by psychoanalysts a hundred years later – that teachers of the nineteenth century wanted to apprehend the real nature of children to make acquisition of social norms more effective (Deák 2000). The special terminology of child psychology had been born by the beginning of the twentieth century.

The first comprehensive work on child psychology was published in 1868. The author, Márton Nagy (1804–1873), took a stand on the segmented character and the spontaneity of child development. According to his view, only abilities and skills that were already present in the child have the chance to develop. He drew a parallel between the processes of physical and mental development and paid special attention to the physiological basis of speech acquisition. Another author, Lajos Felméri (1840–1895), emphasized the superiority of humans over animals and regarded children as "atoms," the basic units of society (Felméri 1890). The first woman author, Irén Quirtsfeld (1884), studied the imagination of children using her own observations. Lajos Donner (1849–1923) like other experts gave an account of his experiences with his own child in a book (Donner 1898). According to his view, children have especially lively emotions, as opposed to those of animals, which are controlled by instincts. Relying on his observations, he stated that children are born with "dispositions," but they are not fully determined by them; infants are characterized by curiosity, unusual liveliness, tool use, and instability. He also mentioned that "in western nations" the available findings are inconsistent and contain too many particulars, making them difficult to systematize. János Pethes (1857–1928), author of *Child Psychology* (1901), identified his own approach as physiological child psychology. He gave a detailed description of the results obtained by experts from other countries, underlining that America, and especially Stanley Hall, was playing the leading role. However, he thought the results were difficult to systematize, in spite of the increasing number of scientific works and growing practical demand.

Introduction of Pedology and Psychometry into Child Psychology

The period from 1867 to World War I, the period of the Austro-Hungarian Monarchy, was an age of economic and social prosperity. In the decades around 1900, in addition to thriving industry, commerce, and urban development, compulsory schooling was in effect realized. A higher and higher number of children started attending primary schools, in which the mother tongue of the local population had by then become the language of tuition. By the 1930s, illiteracy had fallen below 10%. A network of high-quality secondary schools was established, built upon primary schools of rather uneven level.

At the end of the nineteenth century, a new approach appeared in Hungary: pedology, created by Stanley Hall. It was the subject chosen for a national congress of education organized in 1896 on the occasion of millenary celebrations of the State. The supporters of pedology advocated the methods of the natural sciences, referring to Hugo Münsterberg and William James as well, later, to Pavlov's reflex theories. In pedology, children were no longer subjects to be educated for the sake of certain moral values; their feelings were also of significance. László Nagy (1857–1931), the first author whose activity is sometimes referred to even nowadays, introduced pedology into Hungary. He was an active participant in scientific life both in Hungary and abroad and the founder of the Paedological Society (1906), which soon became a highly esteemed professional body consisting of several sections, playing a role in decision making concerning education. According to the opinion of János Waldapfel, however, a representative of the "Herbartians" (a group of teachers for whom psychology could only be an instrument of moral education), the study of children is only reasonable in the case of mentally handicapped ones or those with abnormal behavior. In normal circumstances, children may be disturbed if they "are experimented with." In the course of the discussions, evergreen questions arose: due to individual uniqueness, it is difficult to classify people into categories; the majority of psychological phenomena cannot be measured by physiological means; measurement results are affected by the personality and the expectations of those taking the measurements (Deák 2000).

In the debate, "child-centered" pedology seems to have prevailed over pedagogy. László Nagy became a central representative of child psychology. He founded the journal on developmental psychology *Child ("Gyermek")* and established the museum and library of pedagogy. He also instituted a laboratory within the teacher training college, which contemporary researchers regarded as indispensable for scientific psychology. Laboratory research aimed at the examination of psychological questions arising in connection with pedagogy. Nagy's comprehensive research activity involved several fields of developmental psychology, such as children's fields of interest, the analysis of children's drawings and the measurement of intelligence. Among other things, he examined the effect of the war on children's minds by using questionnaires (Nagy 1905, 1908, 1915).

In 1906, at a meeting of the Hungarian Philosophical Society, Nagy presented a detailed report on the state of pedology in which he declared that although pedology had all the charms and magical powers of a new field of science, it bore the features of immaturity. He mentioned Münsterberg's opinion that the existence of child psychology in itself is not justified: it is only necessary to study childhood development in order to better understand the psychological characteristics of adults. According to Nagy, the prime effect of Munsterberg's view was to cause indignation among German researchers, and Nagy shared this view. Contrary to the opinion of American pedologists, the Hungarians and Germans left the anatomical, physiological, and anthropological examination of childhood to the "true investigators" of their given fields and restricted themselves to research on psychological development. The study of child development is necessary in order to understand children themselves, and the research must be conducted using the methods of natural sciences. At the same time, Nagy shared the widely accepted view that children can be regarded as creatures of nature rather than adults; for this reason, the investigation of children takes researchers closer to the "true" cognition of the adult psyche. Besides, studying children throws light upon two questions: what are the boundaries of natural development, and how far is "artificial" interference in social factors justified? Although Hungarian psychology is often described as "Prussian-like," the approach now called "child centered" was built on notable traditions. Nagy himself was a committed supporter of reform pedagogy, and his students were the founders of the first experimental schools working with the principles of reform-pedagogy (Domokos 1925).

Long before Binet, in Hungary as well as elsewhere, attempts were made to measure intellectual skills, though there was no basis for comparison or for the interpretation of data. After the introduction of compulsory schooling, the measurement of skills gained practical importance: it was judged that methods were needed to diagnose mental disability and to test the efficiency of school marks in reflecting children's skills. At the beginning of the twentieth century various experiments were carried out to test children's intellectual skills, but the results showed significant differences even in peer groups. Urban children proved to be better informed than pupils from villages in several respects, but there were also considerable differences between the children of poor day-laborers and well-educated parents (Eperjessy 1906).

Binet's test, a test based on a statistical sample size with a rising difficulty level based on age, offered a solution in this situation. As early as 1911, soon after it was publicized, it gained practical application in several countries, including Hungary. First of all, however, it had to be adapted, since the results showed significant deviation among countries. In Hungary, testing was restricted to diagnostic purposes and used mainly for vetting mental disorders. However, Hungarian experts had doubts about the validity of standardized measurements. As a result of research conducted in the 1930s, Pal Harkai Schiller (1908–1949) – who died young in emigration – came to the conclusion that unschooled people are characterized by analogical thinking rather than by the use of logic. For this reason, different kinds of measurement tools should be used. Another expert, Erzsébet Baranyai, emphasized that statistical data were hardly appropriate for describing the specific individual and qualitative characteristics of intelligence (Baranyai 1939; Harkai Schiller 1935).

Interestingly, from the beginning Hungarian experts held the view that the results reached by researchers from "more advanced countries" surpassed those of their compatriots in the field of psychology. They followed scientific progress, participated in international programs, and maintained relationships with colleagues, mainly in German-speaking regions. From the second half of the nineteenth century on, an overwhelming number of books and periodicals on psychological and pedagogical subjects were published. Moreover, articles dealing with child development appeared from time to time in journals of other fields (*Hungarian Philosophical Review*, *Hungarian Review of Social Sciences*). In 1906, the pioneer of Hungarian pedology, László Nagy, listed 52 Hungarian publications as well as 220 studies and presentations. There was a strong focus on the observation, recording, and systematization of concrete aspects of child development, and the debates still continue.

Observing Children

An author who relied upon actual observations when drawing conclusions in connection with child development was, interestingly, not a physician or educator but a researcher of ancient languages and linguistics, Emil Ponory Thewrewk

(1838–1917). In 1870, decades ahead of William Stern and Noam Chomsky, he states that language acquisition is not a matter of mimicry but partly a "natural development" – children themselves create their own language. The first syllables uttered by the infant are produced by physiological capabilities. Ponory Thewrewk stated that it is the adults caring for the infant who learn motherese from the child and not the other way round, as was generally thought. He adds that motherese plays an intermediate role between infant and later speech. He observed that the first words used by the infant consist of only one syllable, which has more than one meaning, and he noted the phenomenon Chomsky later called "overcontrol" (Ponory 1871).

The other "founding father" of Hungarian psychology was Pál Ranschburg (1870–1945). Like many other Jewish intellectuals, in spite of being the son of a rabbi, he graduated from an excellent high school of the Benedictine Order. Ranschburg started his career as a psychiatrist at the medical university of Budapest but from the beginning showed an intense interest in questions of psychology, and he even visited Wundt's laboratory in Leipzig. Though Ranschburg is primarily known as an experimental psychologist, he made a considerable contribution to the study of infant cognitive development (1905). Several of his articles and studies about infant memory, the development of counting and reading skills, and mental disorders in infants appeared in German and French. He founded the Experimental Laboratory of Pathology and Therapy at the special education teacher's training college then being created, where several excellent representatives of the field were later employed. In 1928 he established the Hungarian Psychological Society, in which he later served for some time as president.

In addition to these two significant representatives of psychology, a great number of educators and medical doctors made contributions. Ábrahám Léderer, for example, argued that the psychological examination of children should also involve the study of their character and personality (Léderer 1899). Partly following Claparède, Károly Lázár compiled a list of methods used in child observation. Károly Ballai also dealt with methods of cognition and observation, stressing that the speculative methods applied in the case of adults are not suitable with children, whose investigation requires empirical methods derived from anthropology and physiology (Ballai 1929). Lipót Nemes (1886–1960) was one of the first to conduct research into the relationship between social circumstances and psychological development. Writing about children living in the slums of suburbs and industrial plants, he underlined that all kinds of psychological and pedagogical efforts fail if children are brought up in such circumstances (Nemes 1913).

Child Psychology Between the World Wars

The peaceful development of the country preceding World War I was followed by the serious damage caused by the war and then the collapse of the Monarchy. After four centuries, Hungary again became an independent nation state – but on an area of only one third of the historical territory. Several million citizens with Hungarian as their mother tongue, important roads of communication, and economic centers as well as cities with Hungarian cultural traditions were annexed to neighboring countries. The use of German and the proportion of bilingual speakers decreased significantly. After some months in power of the Council Republic, created following the Soviet pattern, a right wing conservative Miklós Horthy became regent. In its efforts made to regain possession of the annexed territories, Hungary was supported by Germany, and with this the country again joined the losers' side in World War II. The result was defeat and destruction of the country and the death of masses of people, including four-fifths of Hungarian Jewry.

Nevertheless, interwar Hungary was a country with lively intellectual and scientific life. The loss of territories increased a sense of national identity and, thus, the necessity for public education. Village elementary schools were built, one after the other, and universities from the annexed regions moved back "home" to major towns. There was an unbroken interest in child study, and the period was also a significant one for psychoanalysis. The birth rate decreased generation by generation, but so did infant mortality, and most children who were born reached adulthood. The rising requirements for professional child rearing were manifold. Many children lived in poverty and worked from a tender age, but children's rooms appeared in wealthy bourgeois families' homes in towns and children's literature flourished. There was consciousness that children are the guarantee of a better future for individuals and society alike and must therefore be reared with care and expertise. At the same time, a great number of children were in need of state support and education due to the postwar damage and economic difficulties. In 1918, the Paedological Society proposed that courses in child psychology be introduced into the curriculum of art, law, and medical students. Nevertheless, government educational bodies did not at that time consider this a field of science suitable for practical application, and internationally education in psychology was not sufficiently spread to offer an example worth following. In 1919, however, during the short-lived Council Republic, left-wing politicians held psychology in high esteem. The University of Budapest established a department of psychoanalysis, with Sándor Ferenczi as its Head. László Nagy and Lipót Nemes were trusted with organizing institutes of education, while Ranschburg, Géza Révész, an experimental psychologist, and Géza Róheim, a psychoanalyst and anthropologist, were given roles in planning new higher education and science policy (Kovai 2016). Afterwards, they faced discrimination, and finding employment was very difficult or impossible for them. Further, discrimination against Jews grew stronger (Hárs 2009; Erős 2019). The elderly Nagy was forced to apply for a supplementary pension, Révész emigrated in 1920, while Ferenczi and Roheim lived by private psychoanalytic consultation.

However, child psychology survived, and the pedagogue reformist, Sándor Imre, an enthusiastic supporter of the Paedological Society, was minister of education in the 1920s. The general secretary of the society was Elemér Kenyeres (1891–1933), who spent a year studying the subject in Geneva in 1924. Thanks to his efforts, Claparède's and Piaget's achievements became known in Hungary. In the late 1920s, László Nógrády (1871–1939), an associate of Nagy, organized a nationwide collection of data to examine children's health status and intellectual development. He also wrote a comprehensive book about the actual issues of child-rearing, including sexual education, which was of great concern to the public. He warned parents not to wish to raise their children as subordinate but also not to become dependent on them. Nógrády criticized parents who had a single child because of their "love of comfort," claiming that such parents spoil "the ancient clarity of the Hungarian family" (Nógrády 1913) Károly Ballai, in the work mentioned earlier (Ballai 1929), assured his readers that Hungarian children are extremely smart and have a natural interest for the history of and a commitment to the nation. Since strengthening nationalism showed up among researchers of childhood, conservative nationalists also sympathized with movements for reform–pedagogy, hoping for more effective education for patriotism (Sáska 2001).

A significant representative of Hungarian developmental psychology was a Benedictine monk, Hildebrand Várkonyi (1888–1971). He took his degree in Budapest, and then won a scholarship to the Sorbonne. In 1929, he was appointed head of the Department of Pedagogy-Psychology, the first of its kind in the country, at the University of Szeged. The department had a school of its own operating on the principles of reform pedagogy. Várkonyi was well acquainted with outstanding representatives of psychology and psychoanalysis. He took a critical stand toward the spread of quantitative approaches in cognition and paid special attention to the study of activity, publishing several books and in *The School of Activity*, a specialist journal of reform pedagogy published in Szeged from 1933 to 1944. His fields of interest covered ability measurement, mental disorders, and the psychological aspects of development. After World War II, till 1954, he worked as a tutor at ELTE – Eötvös Loránd University, still the main university of Hungary, but he was neglected as an expert (Pukánszky 1999).

In the 1920s, the laboratory founded by Ranschburg – which can be regarded as the cradle of Hungarian psychology – was split in two. One half became the Institute of Child Psychology, which was later transformed into a research institute of the Hungarian Academy of Sciences (HAS). This institute provided assistance to those who earlier had not received medical care since they were not handicapped or characterized by deviant behavior. During the 1920s, due to job shortages or to their Jewish descent, several professionals with various qualifications in this way found unpaid employment. They were pleased to have the chance to work in their field even without being paid. One of them was Ferenc Mérei (1909–1986), a substantial figure of postwar psychology in Hungary. His life-story is more than romantic. He came from of a Jewish petty-bourgeois family and in the 1930s conducted studies in the Sorbonne, Paris, as a student of Henry Wallon, though he sided with Piaget who disputed Wallon's views. In France he became a member of the French Communist Party. In 1934 Mérei came back to Budapest and worked as a researcher of child psychology. In 1938 he started to work (unpaid as well) in the laboratory of special education under the leadership of Lipót Szondi. Children and young adults with various disabilities were examined, with the special aim of identifying the role of inheritance. Several scientists working in the laboratory later played important roles in Hungarian psychology. Szondi organized weekly seminars in his home, possibly following Freud's example. Among participants of these seminars were psychoanalysts and other leading intellectuals. It was here that Hungarian experts became acquainted with August Aichorn, whose institute served as pattern for the home for destitute children that was established a little later by the psychiatrist, Júlia György. Anna Freud was also among the invited presenters (Kovai 2016).

The followers of the pedological movement, as well as several left-wing intellectuals who were also committed social reformers, had connections with Adler's individual psychology. The followers of individual psychology were primarily interested in its practical applications, and they provided educational counseling services and ran homes for destitute children.

The spreading of applied psychology in the field of education highlighted unsolved theoretical problems. In 1939, János Schnell (1893–1973), head of the Institute of Child Psychology, expressed his fears that references to psychology often masked a lack of necessary expertise. His fears were all the more justified since psychologists were not yet being educated in Hungarian universities. Well-meaning amateurs, as well as greedy chiromancers and clairvoyants, he said, were ruining the authenticity of psychology and arousing public skepticism (Kovai 2016). Some experts, including Ranschburg, expressed their concern whether the "huge building of education" could be laid on foundations as unstable and as unstructured as psychological knowledge. Nevertheless, theoretical considerations were pushed into the background by growing practical demand. According to the distinguished conservative educator, Ödön Weszely, all fields of science had become psychologized, and this, "so to speak, has threatened pedagogy with the danger of being swallowed up." Weszely hoped that the traditional intuitive, and the positivist approach of pedagogy would meet. In 1928, in one of his writings he made the following surprising statement in reference to psychology: "This transformation of mentality is not some kind of external change. It is not merely a change of terminology, but a change in the essence of things. *Things have occurred deep in people's minds*" (1935: 7).

The Budapest School of Psychoanalysis

The school of psychoanalysis in Budapest evolved in the period following World War I, operated for approximately one and a half decades, and became the most widely known achievement of Hungarian psychology. The school's members maintained relationships with the cultural intelligentsia. Due to these influences, psychoanalysis significantly affected almost every area of Hungarian culture, including the culture of child development. Involvement of psychoanalysts meant a radical change in emphasis of child-rearing and education: now the child, the individual, came into focus. New concepts like neurosis, trauma and mental health were introduced into public discourse. Significant effort was made to discover the purely biological basis for these phenomena, but to no avail.

The special social conditions characteristic of Central Europe played an important part in the formation of the circle of psychoanalysts. Medical science and later psychoanalysis created new chances of social advancement for women and people of Jewish descent. The offspring of Jews, who had immigrated in the course of the nineteenth century and gradually entered the middle class, graduated from universities and became intellectuals in numbers disproportionate to the population as a whole. This contributed to the increase of anti-Semitism in Hungary (Borgos 2018).

Jewish families who moved to Hungary from the East of the Monarchy and became wealthy due to their business enterprises also supported their daughters' education. In the first decades of the twentieth century, an independent career was rather unusual for educated women, in spite of the fact that more and more girls attended universities. As Anna Borgos points out, among psychoanalysts women and Jews were overrepresented partly owing to the fact that within this area no professional hierarchies, which presented an obstacle to newcomers or forced them to the periphery in other fields, had yet built up (Borgos 2018). At first, female psychoanalysts participated in public life mainly in their husbands' company, sometimes even failing to use their own name. Mihály Bálint, for example, wrote after the death of his wife, Alice, that his writings were, in fact, born as a result of their common thinking and mental effort. The Hermanns' case is similar, although Alice Hermann carried out individual research work as early as the 1930s. From this era on, more and more women took part in adult and child therapy, discussions, and writing case studies and popular articles related to psychoanalysis. It was partly due to the presence of women that child development had priority within psychoanalysts' activity in Budapest. Most representatives of the pedological movement and experimental psychology, the experts in charge of institutional education, came from different social groups, that is, teaching dynasties or the lower nobility. Women were less represented here. Although the proportion of female school teachers and educators rose, only a few publicized their professional experiences. It is important to emphasize that aversion and distancing by origin were not characteristic of professionals working in psychology and neighboring disciplines.

Psychoanalysts' idea of humanity and their views about mental processes were differed radically from contemporary views of psychology. Psychoanalysts also failed to meet the requirements set for contemporary scientists, including measurements and experiments. Freud, the Master, who kept a tight hold over his followers, did not even accept the latter as a method of investigation. It is a curious paradox that psychoanalysts, who regarded their own research work as strictly based on natural science, did not pay any attention to the neurological background of the conscious mind, the ego, or the unconscious. Ferenczi dealt with the special standpoint of psychoanalysis, arguing that psychoanalysis "stirred up the waters of psychology stagnant for decades and becoming ever shallower." (Ferenczi 1990:11.) He added that philosophers overestimated the role of consciousness. He thought it Freud's merit to have recognized that mental diseases and neuroses are "caricatures of normal life" (Ferenczi 1914: 158). Further on, he said that before this time psychiatrists had exclusively dealt with anatomy or attempted to apply the results of experimental psychology to mental diseases, with poor results. He mentioned Charcot, Janet, and Breuer among those who had not left research on mental phenomena, which cannot be studied by means of clocks or scales, to "belletrists."

By adopting a psychological approach to hysteria, they had opened the way for Freud's psychoanalysis.

However, the Budapest psychoanalysts – two or three decades Freud's junior – were concerned about the methodological and theoretical isolation of psychoanalysis. Led by their personal and professional experiences, they were convinced that it is impossible to understand mental processes exclusively by the results gained in the course of psychoanalytic therapy. From the 1930s on, their criticism became more and more explicit – while repeatedly claiming that the Master himself had the same opinion. In 1908, the most senior, Ferenczi, who was at this time otherwise loyal to Freud, wrote an often-quoted study about the application of psychoanalysis in pedagogy. The question of child development "was in the air," though study of the development of children's intellectual abilities, memory, imagination, and drawing skills did not provide much information about inner drives. Although pedologists repeatedly expressed the need for the consideration of children's individual characteristics in the course of child rearing and education, it remained unclear which characteristics to consider and how to help children morally and socially.

Freud's original ideas handled this problem in a distinctive way. In his view, meeting social expectations is the result of constraints imposed by civilization. It is needed, since childrens' innate characteristics result in their inclination not to adapt. Ferenczi initially shared the Master's views: infants are creatures controlled by instincts, and characterized by "unlimited self-assertion," "wild, often ferocious violence and cruelty," which "can alternate with subservience" (Ferenczi 1908). Yet, in Ferenczi's opinion, society, which believes in the repression of these bad qualities, is neurotic as a whole (Ferenczi 1914). The essence of the problem is repression. He suggested that teachers should mold children's character by using psychoanalytic knowledge and purposeful work so that they gratify their instincts for the good of society. The "egoistic instincts freed from their fetters" will not destroy human civilization because, due to the use of psychoanalytic methods, the individual's conscious consideration will replace repression that was earlier produced by constraint. In Ferenczi's view, sooner or later, repressionless education will lead to a change of society and transformation of hierarchies of power (Ferenczi 1914). The relationship between power and repression resulting in neurosis appears in his later writings as well, without any reference to politics. After a time, however, his views on children's nature changed, and his estrangement from Freud after World War I had a role in this.

By this time, in light of the new natural-scientific facts, the assumption that development is spontaneously determined in biologically controlled ways could not be maintained any longer. It became impossible to ignore the important role that the human environment, primarily mothers, plays in infant development and in performing the essential tasks of care. At the end of the 1920s, in a number of his writings Ferenczi mentioned that motherly rejection and indifference lead to neurosis and a self-destructive disposition. In "The family's adaptation to the child" (Ferenczi 1929), he called attention to the role of the family. He was also concerned about another inadequacy of psychoanalytical theory, namely its account of the child's "sense of reality" – while studying infant development, the role of the environment cannot be ignored. Ferenczi held that the sense of reality does not develop spontaneously but as the consequence of the constraints imposed by reality. Finally, by the 1930s, his views related to infant nature underwent a complete change, introducing the aspect of power into analysis of the parent-child relationship. While expounding his view that the long-term dependence of the child on parents is harmful, it is obvious that he drew on his experience of his conflicts with Freud and during analysis: "Fathers need to descend from their unstable throne of alleged perfection, say goodbye to their almost divine almightiness and authoritarianism towards their children; they do not need to conceal their weak human nature" (Ferenczi 1924:172). As he states in his famous study leading to his estrangement from Freud, the conflict between the individual and the environment is not the process of becoming antisocial though repressed instincts but evolves if the child does not experience loving treatment in the environment.

Ferenczi's young followers basically criticized Freud in relation to infant development. They emphasized the specific role of the mother in early development. Imre Hermann (1889–1984) and his wife, Alice Hermann (1895–1975), started their career as researchers in Révész's experimental laboratory. The couple survived World War II and continued their career in Hungary, where Imre Hermann played a key role in the survival of psychoanalysis, while Alice Hermann, an outstanding expert on early childhood education, worked for the modernization of nursery and kindergarten education. Her influence, discussed later, can still be felt.

Imre Hermann explained babies' innate grasping and embracing reflexes by the fact that primate offspring spend the first days and weeks of their life clinging to their mothers' hair. He assumed that human reflexes originated with this primate behavior, to which he gave the name "clinging instinct." These innate clinging reflexes clearly prove the primates' and human babies' biological drive to create physical contact with the mother. This relationship was later verified by Harlow's experiments. According to Hermann, babies are aware of themselves as members of a dual unity with their mother. This led to some debate with the Bálints, who accepted that the mother's role is decisive, but attributed more independence to the child. Neither view was consistent with Freud's development theory, according to which the sources for babies' gratification are narcissistic and autoerotic stimuli.

Although he was a physician, Hermann made a considerable one-off effort to integrate psychoanalysis and experimental psychology, and to integrate conscious and logical thinking into the system of psychoanalysis. As shown by one of his early experiments, when mentally healthy persons were asked to select elements from a pile, they tended to pick those in the middle, from within the pile, while children and persons with mental disorders preferred elements near the edges (Hermann, 1923/2011). Another of his experiments, with deviant adolescents, led to the conclusion that the inner, moral disorder, the inclination to break the rules, is demonstrated by the subjects' inability to draw an exact copy of a regular pattern. In his book *Psychoanalysis and Logic* (Hermann 1928), he pointed out that logic also has unconscious, intuitive elements, and he assumed that logic could be detected in unconscious processes. It was probably due to the unfavorable historic conditions that these promising experiments passed almost unnoticed. Later, Hermann built up a complex but not very plausible theory that human babies are traumatized by the lack of motherly hair, since they have nothing to cling to. He gave a summary of this theory in *The Ancient Instincts of Man*, but he did not live to see its publication in 1984 (Hermann 1984).

In the 1920s, the world famous Polish anthropologist, Bronislaw Malinowski, relying on his findings from fieldwork, expressed doubts that the Oedipus complex could be identified in the matriarchal society he had studied. At the end of the 1920s, motivated by Malinowski's work, Géza Róheim travelled to Africa, Australia, and an island in Oceania, where peoples then considered to be "natural" were still living (Vajda 2020). He started his career by studying Hungarian folklore, and later he established contact with the members of the Budapest school and became an enthusiastic follower of psychoanalysis. His fieldwork aimed to underpin belief in the uniformity of the stages of development, and the presence in each culture of the Oedipal conflict and castration anxiety. Nevertheless, while studying the life of faraway peoples he found that the oral and anal stages of psychosexual development were not present in different cultural conditions. Australian Aborigines did not retreat to relieve nature or forbid their children's sexual games, and hence there was no sexual latency. Although Róheim claimed to detect traces of the Oedipus complex and castration anxiety in almost every case in children's rites and games, his explanations are often embarrassingly unnatural. In a grotesque way, as though against his intention, his observations make it obvious that infant caring habits (and the nature of development) are dependent on the changing circumstances of civilization. The children who were always fed when they were hungry did not think of creating reserves as grown-ups even in miserable conditions; those who neglected toilet training as infants did not find being dirty embarrassing as adults. It became obvious that the basic elements of future adaptation are formulated by the conditions that the adult members of the given society create for the child, not a biological program or natural environment (Róheim 1929: 101).

Róheim, who spoke several foreign languages and published in English and German as well as Hungarian, made serious efforts to prove the universality of psychological development, but the conclusions he drew were difficult to integrate into the system of psychoanalysis. The most important component of human development is the period of childhood: "human culture as a whole is the consequence of our elongated childhood," he stated. Yet, prolonged childhood increases the effect of environmental influences. This is supported by Róheim's suggestion according to which the difference between cultures results from the differences in the way childhood is prolonged ("retardation").

Róheim and Imre Hermann's activity had a significant influence on two members of the Budapest school, Mihály Bálint and his wife, Alice Bálint, who openly criticized the Freudian theory of infant development. Alice Bálint was born in 1898, the daughter of Vilma Kovács, a psychoanalyst trained by Ferenczi. In secondary school, the radiantly intelligent girl's classmate was Margit Schönberger, later Margaret Mahler, the world famous child analyst. Alice Bálint was widely educated, but from the very beginning she took an interest in psychoanalysis. Owing also to their common interest, Alice married Mihály Bálint in 1921. They received psychoanalytic training in Berlin and then returned to Hungary, where they belonged to the most active circle of psychoanalysts until their emigration in 1938 (Borgos 2018).

Alice Bálint named the special mother-infant relationship "archaic love," which is not identical with libido controlled by erotic instincts, but an individual drive. She did not share Freud's views on the superego and the discomforts of civilization but thought the superego, the representation of society, became locked in the ego in the early years. Repression in its own right does not bear harmful consequences, unless it is taken to extremes. The renunciation of pleasures, she wrote, is not exclusively characteristic of humans; animals are often forced to choose, for example, between mating and the desire to escape. But she thought it important that repression should not prevent people, more than necessary, from gratifying their desires.

Furthermore, Alice Bálint challenged the view that family life necessarily leads to neurosis, or, as stated by left-wing critics, trained people to accept subject positions. She believed that the family represents reality; unusual, sterile circumstances bear greater risks. "Parents and children are natural life partners and are primarily those," she writes. Tensions present in each community and negative influence are like "necessary bacilli," by means of which a child can acquire immunity, in order to become a "whole and true person" (Bálint 1936: 23). There is no recipe for education, in which "a bit more or a bit less would lead to fatal consequences." Referring to Róheim's experience, she inferred that human infants' "natural world" is an artificial one; they acquire their basic knowledge and the fundamental models of behavior when every element of this reality is still determined by their human carers and not by object reality (123).

Alice Bálint's most radical conclusion fully contradicted Freud's theory, according to which the stages of child development are of universal character. There are no abstract rules or biological laws either to provide guidance on assessing the value of a method of education or to guide human life, she wrote. All this can only be determined by the scale of values existing in a given society and which controls the conditions of a child's primary adaptation from the beginning. There is not and there cannot be an absolute recipe for education; there is no mental hygiene valid for all periods of history. Children have been treated in every possible way, from extreme lenience to the most severe tyranny, and they have borne all this. It is not human nature that cannot tolerate certain kinds of educational methods but individual cultures and societies (Bálint 1990a).

The Bálint couple's experience as parents and therapists and the Hermanns' and Róheim's results were finally presented by Mihály Bálint at two international conferences organized in Hungary in 1935 and 1937, with the participation of professionals from neighboring countries.

In two of his presentations, prepared in collaboration with his wife, he challenged Freud's assumption that the oral, anal, and phallic stages of libido development are universally present and biologically determined. We do not know anything about the infants' thinking, he stated. The assumptions drawn from the experience of psychoanalyses cannot be verified, and they do not meet the conditions of a "biologically orientated theory": only the careful observation of infants can provide guidance concerning the mental processes during the early years. Bálint stated that psychoanalysis needs to reexamine the assumption of the primary nature of narcissism and attach greater importance to early object relations, which means the particular human environment (Bálint 1936).

Several other experts, who laid great emphasis on introducing certain elements of theory into practice, joined the circle of psychoanalysts. Their regularly held lectures and seminars were open to the public. Several of them published in the journal, *On Future Paths* (*A jövő útjain*), issued by the international League of New Education. Interested parents were offered free consultation. During the period 1936 to 1939, psychoanalysts even edited their own journal, *Child Rearing*. The articles in these journals aimed to provide answers to questions concerning infant aggression, sexual education, coping with death and tragic events in childhood, and the influence of contemporary mass communication.

Alice Bálint's articles were the most original and of the highest standard. She stuck to her view that repression does not necessarily cause harm. Renouncing momentary pleasure cannot always be avoided, but similar experiences strengthen the ego and increase the individual's stress tolerance. The same result can be achieved by observing some seemingly useless formal rules, like greeting, expressing thanks, following elements of the daily routine, politeness. In fact, she thought these to be "mental gymnastic exercises"; through them individuals learn that in certain situations they have to control their instincts in order to be able to cleverly exploit the pleasures available to them (Bálint 1990b). Although the authors writing in the spirit of psychoanalysis took a firm stand on the importance of understanding the child, Alice Bálint's approach was more subtle. She thought that permitting everything deprives children of the opportunity to rebel, just in the same way as extreme strictness (Bálint 1990c). The problem of childhood dependence cannot be solved by creating parent-child equality, since children expect to be protected and supported. To cease dependence is the final goal of the long-lasting rearing process, resulting in a new, friendly relationship between parent and child. Alice Bálint's opinion on sex education, a problem evidently provoking great interest, was similarly specific. While experts almost exclusively advised parents to limit themselves to answering the child's questions and to avoid going into details, Alice Bálint noted that children often fail to inquire about problems they are most interested in, and in this way their questions may remain unanswered (Bálint 1990e).

Several representatives of institutional pedagogy, including Weszely quoted above and the monk-teacher, Várkonyi, were of the opinion that teachers should be familiar with the theory of psychoanalysis. Criticism of psychoanalysis came primarily from the left in the first-rate journal, *Korunk* ("Our Age"), issued in Transylvania. The left-wing authors of the journal held that it is not civilization in general but oppressing capitalist society that educates by repression resulting in neurosis. Erik Jeszenszky, Minister of Culture in the communist government after World War I, believed that psychoanalysis considers people to be irrational creatures. He did not accept "infantile fatalism," referring to the theory according to which the first years of people's lives have a decisive influence on later development (Jeszenszky 1934).

The productive co-thinking of the Budapest psychoanalysts came to an end after hardly two decades. During the 1930s, most Jewish professionals left Hungary as a consequence of increasing anti-Semitism and neglect. Alice Bálint died in England in 1939, Mihály Bálint never returned to Hungary; Lipót Szondi pursued his career in Switzerland. Others tried to start a new life and make a living in Sweden, the United States, and Brazil, with uneven success. Regrettably, the survivors and the ones who came back home did not have the chance to enjoy the pleasures of peaceful activity for long.

Child Psychology After World War II

In Hungary, during the years 1945–1949, the shock of the war was followed by a relatively peaceful period with a thriving cultural life. In spite of the enormous losses, the survivors started the reconstruction of the country with optimistic fervor. Child psychology was at the center of attention. Owing to the war and the poverty, tens of thousands of children were in need of care. Although the child centered approach had both right and left-wing supporters in Hungary, the psychological approach to education was primarily attributed to leftists. Thanks to this fact, the approach to child psychology created by the Budapest psychoanalysts was integrated into state education in nurseries and kindergartens. The year 1946 saw two books published in this spirit: Lilian Kertész-Rotter, *The Child's Inner Life*, and Alice Hermann, *Raising Humans* (Kertész Rotter 1946. Hermann 1946).

This period can be regarded as a golden age in the history of the Institute of Child Psychology, which grew out of Ranschburg's laboratory in 1934. In 1947, it won scientific institute status and was entrusted by the Ministry of Education with building up a network of child psychological centers throughout the country. By 1948, branches of the institute were operating in 15 towns. They were mainly involved in collecting psychological and sociological data, with a view to supporting public education, but they also provided educational and career counseling and organized programs for workers' children (Kovai 2016).

Further, the National Institute for Education was established for the purpose of research into childhood and youth under the leadership of Ferenc Mérei. From 1942, Mérei, like his Jewish compatriots, performed labor service at the Ukrainian front. However, he escaped under unlikely circumstances and returned to Budapest as an officer in the Soviet army. Then he threw himself with great enthusiasm into the organization of Hungarian psychological life (Pléh, 2020). His famous work on social psychology, *The Collective Experience*, which is also referred to in English language textbooks, was written at this time. Mérei's experiments with children aged 3–5 are especially worth mentioning, since the conclusions basically contradicted those of Kurt Lewin related to leadership models (as well as the later findings of Milgram's and Zimbardo's experiments), which claimed that superiors and leaders exert a decisive influence on subordinates' behavior. Mérei's experiment seemed to prove the opposite: the leader cannot enforce his will indefinitely; the individuals together, organized into a group, are stronger than the authoritative leader striving to oppress them. But his results failed to provoke any response in the atmosphere of the escalating Cold War (Mérei 1948). During the years when the communist dictatorship was being built, his institute was characterized by a dual approach: while the traditions of pedology were kept alive and the regularities of child groups were studied, the achievements of Stalinist pedagogy were referred to more and more frequently. However, this attempt at juggling did not work: although Mérei was even awarded the State Prize in 1949, he soon became unemployed and his institute was closed down in the same year. In 1949, the Paedological Society, which had started operating again in 1946, faced legal procedures, which ended in condemnation by the Party, following the Soviet pattern of 1936 (Kovai 2016).

The Stalinist dictatorship in the period from 1949 to 1956 penetrated every area of life. Psychology, and in particular psychoanalysis, was considered as ideologically harmful. Several experts surrendered: some yielded to pressure, while others were motivated by their own former convictions. Nevertheless, in their later lists of publications, they failed to mention their works written during this period of the "dictated recipe." The majority of professional periodicals ceased publication, and the *Hungarian Psychological Review* did not appear for more than a decade. The operation of the Psychoanalytical Association was terminated under dubious circumstances. Psychoanalysts maintained their relationships by holding secret meetings in private apartments, but the atmosphere was poisoned by suspicion and fear. Those who kept psychoanalytical traditions alive took refuge in the children's clinic of the university of medicine, led by Professor Pál Gegesi Kiss, a pediatrician of Transylvanian descent, who took an interest in child psychology. The educational counseling unit within the children's clinic was the only institute of this kind to have been continuously operating since 1937. The unit was run by Lucy Lieberman, a psychoanalyst, who maintained relations with Mihály Bálint, living in London at the time. Through him, she was able to inform a select circle of her colleagues about the events of psychoanalytic life. Her informal seminars, which operated without interruption even during the 1950s, played a significant role in the rebirth of psychology. Another refuge for psychologists was the laboratory once run by Szondi, which belonged to the College of Special Education.

There was no independent training in psychology, apart from psychological courses available for future teachers. In these courses, a large part of the teaching material was occupied by the "socialist" Pavlov's work, which was introduced as a psychological theory covering the totality of mental functions. The former university departments of psychology were abandoned by the experts; for years, the only scientist of international renown at the University of Budapest was Lajos Kardos. Oddly, the experts who participated in the resuscitation of psychology after 1956 had pursued their studies in the Soviet Union – naturally not in psychology but in pedagogy or philosophy. Another group of Hungarian psychologists of the years 1960–1980 attended the Stalinist, fundamentalist Lenin Institute of Budapest, which operated from 1952 to 1956.

Left-wing ideologists could not openly take a stand in defense of autocratic education, in spite of the dictatorial feature of their ideology. Unconditional political assimilation was a requirement, but corporal punishment was prohibited at schools in 1949. Official educational ideology rejected strict discipline, the enforcement of unquestioning obedience, punishment, and humiliation. Those who remained loyal to their professional integrity, at least to a certain degree, strove to ease the tension caused by the menacing atmosphere. Their most important representative was Alice Hermann, author of several educational articles for the general public. In her contemporary articles, apart from the unavoidable propaganda, she emphasized the importance of understanding children and considering the characteristics of the age group. Nevertheless, the almighty leaders of the one-party system sensed the ideological adversary in her: in the 1950s she lost her job of school inspector, in spite of the fact that she had been a committed leftist already before the war. Finally, in 1955, together with Mérei and other psychologists, she had an important role in the movement of intellectuals, the Petőfi Circle, which laid the ideological foundations for the revolution of 1956.

Restart After the Revolution of 1956

Csaba Pléh, the historian of psychology in Hungary, divides the 30 years of the development of Hungarian psychology between 1956 and the change of regime, into three phases (Pléh 2018):

- 1960–1970 Psychology appears again
- 1970–1980 Spreading and differentiation
- 1980–1990 Social acceptance of psychology

In fact, the first signs of the reappearance of psychology can be noticed in the late 1950s, and the final date, 1990, is more fluid. The first years following 1956 were those of reprisal against the participants in the revolution: fear remained, as well as awkward acceptance of the official dictums of the Party. However, the atmosphere became less tense, and living standards continuously rose in the decades between 1960 and 1990. After total isolation, foreign travel became liberalized, and finally in 1988 – that is before the change of regime – Hungarians were granted foreign passports. In spite of all the contradictions, like the network of informers and the arbitrariness of the privileged of the party-state, this was a period of boom in culture and science. Hungarian photography and film-making won an international reputation, new literary journals as well as periodicals in various topics of public life appeared, and science received a new impetus. In spite of the improving living conditions and the lasting social peace, the ideology of the party-state failed to affect the population. Toward the end, the state officials themselves hardly believed in it. The strong dislike of the party-state manifested itself in the idealization of "the West," "the world over the Leitha," which could also be experienced in science.

It was not possible to return to pre-war professional life. The former social network disappeared without trace; several important persons had emigrated or fallen victims to the war. Furthermore, psychology was initially still regarded as a suspicious field of science, to be tolerated but not supported. Although university training in psychology began at the university of Budapest in 1963, enrolments were very limited, in spite of the high number of applicants. No books on psychoanalysis came out before the 1970s, and formerly published books could only be borrowed by professionals from the library of the Parliament. References to Soviet authors dominated university courses and the articles published in specialist journals.

Nevertheless, fear gradually eased and psychology occupied an increasing role in scientific and public life. Influenced by growing demand, various forms of training (evening classes and correspondence classes) were organized to compensate for the low full-time enrolment numbers at universities. In 1975, a new training center was established. By the end of the 1970s, both the research infrastructure and the applied fields, including child psychology, developed remarkably. Very few people had faith in Marxist science, the majority tried to get rid of dogmatism. The acceptance of a psychological approach was a kind of implicit political critique to the one-party system. All this contributed to "western science" being received without any criticism, "western" meaning mainly American by this time. As early as the mid-1960s, the number of references to English language publications in Hungarian journals surpassed those written in Russian, and from the early 1970s on, English and American references became dominant, though few people had a command of English (Pléh, 2018). Up to the change of regime or even after it, "the West" was taken to be homogeneous, resulting from the fact that all films, books, and scientific works from abroad underwent preliminary selection, and the works which were deemed conservative or retrograde never reached a Hungarian public. This qualitative selection worked spontaneously in connection with professional events and discussions: if these failed to fit into the prescribed categories of "advanced West, oppressed East," they were left out of consideration. Thus, the IQ debate generated by Arthur Jensen's famous study hardly provoked any response from Hungarian experts: the views concerning the decisive effect of inheritance did not fit the image of advanced American science. Skinner's book, *Beyond Freedom and Dignity*, as well as the debate following its publication, passed unnoticed: the society designed by means of engineering methods bore too strong a resemblance to the social conceptions of National Socialism or dogmatic Stalinism. For some reason, Piaget was disliked by party-state ideology, and up to the end of the 1960s, students could only become acquainted with his views in "concealed" seminars. Thus, his apotheosis in American psychology and the criticism he attracted afterwards – which would not necessarily have been shared by local researchers – remained "invisible" for Hungarian experts.

After the long years of disruption it was not easy to find university staff when training did commence. The first psychology professor at Budapest university was Lajos Kardos (1899–1985), who was equally familiar with German and American psychology (Pléh, 2000). In his book written in 1957, he discussed the relationship between Pavlov's activity and contemporary behaviorism. Obviously, the well-informed scientist knew what uneducated party functionaries did not: Pavlov's activity had a great influence on American behaviorism and the two bore certain similarities. Nevertheless, in the 1960s, Pavlov's activity – often in a grotesque and arbitrary way – was referred to in every field of the training of psychologists. Kardos, who was otherwise acknowledged as an excellent teacher and researcher of animal memory, in his *General Psychology* ("Általános pszichológia," 1965) clearly attempted to meet ideological expectations (he even mentioned Lenin!). In spite of being extremely boring owing to these references, the book had 11 editions, illustrating the remarkable interest in psychology at the time – though it must be borne in mind that this was the only book on psychology available in the 1960s. On the other hand, there was a chapter of special interest and worth mentioning even today, discussing will and intentional action. The will is a non-category according to the idea of science currently dominating psychology, and in the textbooks published since then the word "will" does not even turn up as a subject heading.

Despite his particularly tumultuous life story Mérei became a decisive figure in Hungarian psychological life. After being rehabilitated in 1956, he continued his group investigations. His unparalleled experiments with the role of pairs in a group should have been part of the international canon, but the author was imprisoned for seditious activity in 1957. At the age of 54, in 1963, he was released, thanks to a UN amnesty. Afterwards, owing to the help of Miklós Kun, an enthusiastic supporter of psychology and professor of psychiatry, who had good relations in the Party, he was offered premises for the purpose of research in the National Institute of Psychology and Neurology, in fact the largest asylum in the country. In the spirit of former traditions, he called this place a laboratory. Although he was prohibited from teaching for at least half a decade, his laboratory, where several of his disciples were later employed, became a place of pilgrimage. Mérei was active in numerous fields of psychology, and psychologists of the 1960s and 1970s all studied from his books about group life, sociometry, child psychology, and clinical psychology, and participated in group therapies led by him or his followers (Erős 1989).

In this period, less attention was paid to child development: the Hungarian Academy of Science (which meant official recognition) gave recognition to educational not to developmental psychology. According to Ferenc Lénárd's report, it was obviously the task of pedagogy to control education and to define the aims, while psychology was responsible for providing the means. Left-wing ideologies were suspicious of individuality and it seemed that in the debate concerning the competence of pedagogy and psychology, the former took priority. But his was only on the surface. True, in the 1970s and 1980s psychological experts did not have a large role in specific professional decisions, but they were active and influential in public debate and gained ground in applied fields concerning children. In addition to consultation rooms created in pediatric clinics, which were also involved in educational tasks, a counseling center was founded in the capital with a view to providing help with problems related to child psychology. The legendary "Faludi Clinic" became a refuge for psychoanalysts, where they finally had the chance to make use of their knowledge within legal though limited circumstances. In parallel with this, thanks to some committed and enthusiastic experts, a network of child guidance centers was again instituted throughout the country. These centers provided a free service in the field of psychological diagnostics and psychotherapy. They also offered developmental sessions, gave advice and support in problems of divorce, and performed psychological activity in schools (Horányi 2019).

The development was also spectacular in the field of clinical psychology. Psychotherapy still existed in informal circumstances, but, from the 1960s on, state owned centers of psychotherapy were created and offered individual and group therapy. The creation of laboratories for purposes of occupational therapy becomes trendy in state owned factories and companies. Due to lack of legal ways for workers to advance their interests, these laboratories became channels through which they could express dissatisfaction. The Institute for Career Counselling was officially involved in occupational psychology, but in fact also provided therapeutic consultation for adolescents.

Children and Psychology After the Regime Change

In 1990 in Hungary, like in other states previously belonging to the Soviet zone of power, the one-party system ceased to exist; the institutions of parliamentary democracy were established, and the command economy was replaced by a market economy based on private ownership. These favorable conditions brought about a new upsurge in social and intellectual life, including psychology, finally resulting in the revival of its glorious past, after about half a century. However, the changes are open to different views.

Hungary has failed to catch up either economically or socially with the most developed countries. Although it has approached the global trend in many respects, new circumstances (e.g., a sharp increase in social inequality) have not necessarily improved the conditions of those who live in the country. The situation of children has also changed, partly due to local conditions and partly following global changes in the concept of the child.

As a consequence of the economic crisis following the change of regime, the GDP per capita of 1989 was only reached again in 1999. Millions lost their jobs, while part of the middle class became richer and enjoyed bigger freedom for travel and speech. Children's age specificities became less important: their treatment began to depend more on social class. The school system was transformed in a way that was advantageous for the middle class, even at state schools, and a significant proportion of educational institutions was privatized and/or handed over to the churches, while the education of poor children remained a state task.

From the 1970s and 1980s the extremely liberal educational conceptions primarily represented by Carl Rogers – who visited Hungary several times – exerted an ever growing influence on Hungarian pedagogy. A high number of reforms were introduced with a view to making school education more flexible. After the change of regime, however, owing to the neoliberal approach – which could not at first be identified – every kind of state intervention was interpreted in a negative way. As a result, the "Prussian" spirit, in spite of the reforms supposedly characteristic of Hungarian state schools, was contrasted with the "Anglo-Saxon" way of education supposedly focusing on democracy. Teachers and parents were informed how to educate children to become true democrats in the course of training sessions, which were organized using multilevel marketing methods. The most successful sessions were Thomas Gordon's, which were financed by state organs and local government. In Gordon's view, parents used two thousand-year-old methods, which could not go on. Like advocates of other, seemingly new training methods, Gordon promoted his product by stating that they would be a means to stop international conflicts (Gordon 1991).

Undoubtedly, good processes were initiated too. In the 1980s, the role of the party-state ideology was reduced to a minimum, and psychology was granted full emancipation. During the 1990s, training in psychology was launched at four further universities. There seemed to be no more obstacles in the way of candidates interested in psychology – except for the lack of financial means (since nowadays more than half of university students pay very high tuition fees). Indeed, in the past decades the number of students taking their degree in psychology has multiplied, and psychoanalysis has been fully rehabilitated. Back in 1988, right before the change of regime, Budapest hosted the annual conference of the European Society of the History of the Human Sciences (ESHHS), then Cheiron – Europe. The conference was the largest event in the history of the society, with over 300 participants. Specialists from Western Europe as well as from the Soviet successor states met one another in euphoric mood in the hope of the forthcoming changes.

However, the decreasing role of the state in science resulted in unpleasant consequences. As György Csepeli points out, the financial resources once provided by state socialism became limited. As a result, "borders having been opened, the scientific relationships with the West intensified, which in several cases did not happen in the spirit of equality but followed the logic of colonisation" (Csepeli 1998). It was very fashionable to write about socialism and the change of regime, but local experts were not given priority as authors. They had the chance only to help with the collection of data, doing interviews; they were familiar with the local circumstances, but they had no say in defining the basic conception of the research. The authentic analysis of the human and social psychological effects of the change of regime is still missing, and this situation probably cannot be remedied.

The institutional system of science appeared to have been transformed, but in fact the former structure survived, with some modifications. The persons holding significant positions in scientific life did not change. Young researchers, therefore, were ready to accept employment in American and Western European universities, motivated not just by higher salaries and better conditions but also by the rigid professional structure in Hungary.

Important professional knowledge was lost because the institutions involved in child rearing and education – nurseries, kindergartens, schools – stopped running research institutes of their own due to the lack of state financing. The valuable experience gathered there, for example concerning the rearing of infants aged under three, has not subsequently been studied. The gap between theoretical research and practice widened. Theoretical research was connected to the dynamically developing field of cognitive psychology, far away from children's everyday life, while practice was soon penetrated by fast spreading commercial enterprises. Educational counseling centers lost their significance, to be replaced by thousands of private service providers practically without any professional control.

Hungarian professionals, unaware of the evolving concept of medicalization, handled the peculiarities of individual behavior as if they were the symptoms of illness (hyperactivity, autism, etc.). When in the mid-1990s a Hungarian specialist wrote an article about the long lasting scandal abroad concerning the use of antidepressants, Hungarian journals refused to publish it (Szendi 2004). Further, there was public support for the process of medicalization. Middle-class and wealthy parents do not tolerate critique of their manner of child-rearing and, rather, are eager to pay huge sums for the treatment of "innate mental disorders" of their children. Politicians, in public discourse, also prefer to see mental disorders as the main causes of the disabilities of the poor children who's retardedness is the consequence of bad material conditions. Unfortunately, the disorders caused by deprivation cannot be treated by psychological means alone – as it was stated by Lipót Nemes 100 years ago.

In the past decades, the concept of the child in "mainstream" developmental psychology has also changed. There was a significant delay before Hungarian experts became familiar with the paradigm shift which took place in international specialist literature. The debate initiated by William Kessen (1991) challenged the normative approach to child development. Kessen and his followers were convinced that the changes in childhood cannot be regarded as development, since the latter means a kind of change with direction that involves reference to cultural values and ideology which are not objects of a scientific approach. Kessen rejected Piaget's theory of cognitive development and the theory of stages too. According to him and his students, it is only the amount of experience and not a different logic that makes a difference between children's and adults' cognitive skills.

The problem with this is that it leaves the question, why "emancipated" children need special circumstances and protection? Indeed, the rejection of the concept of childhood as a process of development is more than just a change in scientific approach. Children and adults became more alike in everyday life, not only in the cognitive chapters of textbooks. In the past century, childhood was a distinct period of preparation for adulthood – and this was why children became important as research objects. Nevertheless, in the present these conditions tend to disappear. As a result, the process of development in childhood seems to be less and less predictable and interpretable for scientific observation. The same is true in relation to the influence of adults' manner of treatment of children's future behavior. It is to be feared that the psychology of child development is gradually losing its objective, unless it restricts itself to just supporting the growing generation in accepting – possibly unconditionally – consumerism and conformity.

Conclusion

Those studying the history of Hungarian psychology may find it remarkable that Hungarian psychology relies so little on its own rich traditions. As Szokolszky (2016) has argued, it would be important to identify the reasons for this. It would be equally important for Hungarian psychology to find the position it deserves in international science, though at the present moment there seems to be little chance of this happening. The results achieved by Hungarian psychology are rooted in the European tradition of modernity. The focus of their approach was "subjects, individual people of value of their own right" (Smith 2013: 265). In the case of developmental psychology, subjects were children as a human group with special characteristics. It is more and more difficult to adapt this approach to both of the present child-concept and the perceived contemporary role of psychology.

References

Bálint A (1936) A szeméremről [About shyness]. Gyermeknevelés 3:33–35
Bálint A (1990a) Nevelési rendszerünk alapjai [The basics of our child-rearing principles]. Anya és gyermek:122–123. Budapest, Animula
Bálint A (1990b) Tiltás és megengedés a nevelésben [Permissiveness and discipline in child-rearing]. Anya és gyermek:91–102. Budapest, Animula
Bálint A (1990c) Szabad – nem szabad [Should be done – should not be done]. Anya és gyermek:103–108. Budapest, Animula
Bálint A (1990e) A gyermeket csak akkor világosítsuk fel, ha kérdez (Children's sexual education is recommended only when they ask questions) Anya és gyermek:135–141. Budapest, Animula
Ballai K (1929) A magyar gyermek. Eredeti mérések és lélektani adatok alapján [The Hungarian child. On the base of original measurements and psychological data]. MGYT, Budapest
Bárány P (1790) Jelenséges lélek-mény [About the mind] Manuscript, Magyar Tudományos Akadémia kézirattára, Ir. lev. 8 r. 10
Baranyai E (1939) Értelmesség és egyéniség [Intelligence and individuality]. Magyar Pszichológiai Szemle, január–december 22–30
Borgos A (2018) Holnaplányok. Nők a pszichoanalízis budapesti iskolájában [Women of tomorrow. Women members of Budapest school of psychoanalysis]. Noran Libro, Budapest
Csepeli Gy (1998) Szociálpszichológia [Social psychology]. Osiris, Budapest
Deák G (2000) A magyar gyermektanulmányi társaság története [The history of Hungarian Association of Paedology]. AKG Junior, Budapest
Domokos L (1925) Tájékoztató a gyermektanulmányi alapon álló új iskoláról [Information about the new school based on the principles of paedology]. MGYT, Budapest
Donner L (1898) A gyermek értelmi fejlődése gyermekpszichológiai szempontból [Cognitive development of children from the point of view of child psychology]. Corvina, Békéscsaba
Eperjessy K (1906) A népiskolába lépő gyerekek ismeretvilága [Literacy of children starting elementary school]. Népmívelés, 125–135, Vol. II. booklet 7–12
Erős F (1989) Mérei Ferenc fényében és árnyékában [In the light and shadow of Ferenc Mérei]. Interart, Budapest
Erős F (2019) Sándor Ferenczi, Géza Róheim and the University of Budapest, 1918–19. In: Borgos A, Erős F, Gyimesi J (eds) Psychology and politics. Intersections of science and ideology in the history of psy-sciences. CEU Press, Budapest/New York, pp 81–94
Felméri L (1890) A neveléstudomány kézikönyve [Textbook of education]. Ajtai, Kolozsvár
Ferenczi S (1908 (1982)) Pszichoanalízis és pedagógia [Psychoanalysis and pedagogy]. Lelki problémák a pszichoanalízis tükrében, pp 41–49. Magvető, Budapest
Ferenczi S (1914 (1982)) A pszichoanalízisről és annak jogi és társadalomtudományi következményeiről [About psychoanalysis and its relationship with the law and social sciences]. Lelki problémák a pszichoanalízis tükrében, pp 158–175. Magvető, Budapest
Ferenczi S (1924 (1982)) A valóságérzés fejlődésfokai és patologikus visszatérésük [Development of the perception of reality and their pathological return]. Lelki problémák a pszichoanalízis tükrében, pp 124–146. Budapest, Magvető

Ferenczi S (1929) A család alkalmazkodása a gyermekhez [Adaptation of the family to the child]. Korunk 9:593–599
Ferenczi S (1990) Foreword to the book of Alice Bálint: A gyermekszoba pszichológiája (Psychology of the children's room) Kossuth, Budapest
Gordon T (1991) P.E.T. A szülői eredményesség tanulása [How to be effective parent?]. Gondolat, Budapest
Harkai Schiller P (1935) Érettségizettek értelemvizsgálata [Mental testing of graduated high school students]. Magyar Pszichológiai Szemle, január – június, 425–435
Hárs Gy P (2009) A vörös Róheim [The Red Róheim]. Thalassa 20(4):45–74
Hermann A (1946) Emberré nevelés [Education in humanistic way]. Tankönyvkiadó, Budapest
Hermann I (1923 (2011)) A széli preferencia, mint primér folyamat (Preference of edges as a primary process) In: Gondolkodáslélektani tanulmányok. Animula, Budapest, pp 123–128
Hermann I (1928) Psychoanalízis és logika [Psychoanalysis and logic]. Pantheon, Budapest
Hermann I (1984) Az ember ősi ösztönei [The ancient instincts of man]. Magvető, Budapest
Horányi A (2019) A nevelési tanácsadók fél évszázada [50 years of the child guidance centres]. In: Pléh C, Mészáros J, Csépe V (eds) A pszichológiatörténetírás módszerei és a magyar pszichológiatörténet. Gondolat, Budapest, pp 403–420
Jeszenszky E (1934) Gyermeklélektan, jellem és irracionalizmus. Korunk, 11. 814–820
Kardos L (1965) Általános pszichológia [General psychology]. Akadémiai, Budapest
Kertész Rotter L (1946) A gyermek lelki fejlődése [Mental development of children]. Cserépfalvi, Budapest
Kessen W (1991) Nearing the end: a lifetime of being 17. In: Kessel FS, Bornstein MH, Sameroff AJ (eds) Contemporary constructions of the child. Essays in honor of William Kessen. Lawrence Erlbaum Associates
Kovai M (2016) Lélektan és politika. Pszichotudományok az államszocializmusban [Psychology and politics. Psy- sciences in the era of state socialism]. L'Harmattan, Budapest
Léderer Á (1899) A kisdedek lelki képességeinek megismerési módjai [Methods of investigating mental ableness of the little children]. Egyetemi Nyomda, Budapest
Mérei F (1948) Az együttes élmény [Collective experience]. Budapest, Officina
Mocsi MD (1839) Elmélkedések a physiologia és a psychologia körében, különös tekintettel a polgári és erkölcsi nevelésre [Meditations about physiology and psychology, in particular about the civil and moral education]. Egyetemi nyomda, Budán
Nagy L (1905) Fejezetek a gyermekrajz lélektanából [Chapters from psychology of children's drawings]. Singer és Wolfner, Budapest
Nagy L (1908) A gyermek érdeklődésének lélektana [Psychology of the interest of children]. Franklin, Budapest
Nagy L (1915) A háború és a gyermek lelke [The war and children's psyche]. Gyermektanulmányi Könyvtár, Budapest
Nemes L (1913) A kültelki gyermek lelki világa [Psychology of the Inner city child]. A Gyermek, 1912:176–180
Nógrády L (1913) Az egyke gyermek [The only child]. A Gyermek, 1913:81–96
Pálóczi Horváth Á (1790) Psychológiy az az A lélekről való tudomány, íratott az 1789. esztendőben (Psychology or the science of the mind written in the year of 1789) Trattner, Budapest
Pethes J (1901) Gyermekpszichológia [Child psychology]. Lampel, Budapest
Pléh C (2000) Egy magyar tudós az alaklélektani gondolkodás peremvidékén: Kardos Lajos. (A Hungarian scholar at the borderland of Gestalt psychology: Lajos Kardos). In: A lélektan története. Osiris, Budapest, p 477
Pléh C (2004) Magyar hozzájárulások a modern pszichológiához. (Hungarian contributions to modern psychology) In Pléh Cs.-Boross O.: Bevezetés a pszichológiába. Osiris, Budapest
Pléh C (2018) Intézmények és gondolkodásmódok fél évszázad magyar pszichológiájában [Institutions and mindset in the half century of Hungarian psychology]. In: Pléh C, Mészáros J, Csépe V (eds) A pszichológiatörténetírás módszerei és a magyar pszichológiatörténet, pp 133–160

Pléh C (2020) Mérei Ferenc a polgári és a szocialista embereszmény közepette. (Ferenc Mérei amid the bourgeois and socialist ideal of man) Educatio, 29. pp 545–566

Ponory Thewrewk E (1871) A gyermeknyelvről [About the children's language]. Aigner Lajos bizománya, Pest

Pukánszky B (1999) Pedagógia és Pszichológia [Pedagogy and Psychology]. l. In: A Szegedi Tudományegyetem múltja és jelene: 1921–1998 [Past and present of Szeged University]. JATE, Szeged, Officina, pp 219–221

Pukánszky B, Mészáros I, Németh A (2005) Neveléstörténet: bevezetés a pedagógia és az iskoláztatás történetébe [History of education: introduction into the history of pedagogy and schooling]. Osiris, Budapest

Quirtsfeld I (1884) A képzelő tehetség hatása az akaratra s ezáltal a jellemre [Impact of imagination on will and character]. Máramarossziget

Ranschburg P (1905) A gyermeki elme fejlődése és működése [Development and functioning of children's mind]. Atheneum, Budapest

Richards G (2002) Putting psychology in its place. A critical historical overview. Psychology Press, Sussex

Róheim G (1929) Australian totemism. A psychoanalytic study in anthropology. George Allen & Unwin, London

Sáska G (2001) Az alternatív pedagógia posztszocialista győzelme [Victory of the alternative pedagogy after socialism]. Beszélő, 2004/12. (Lábjegyzet). http://beszelo.c3.hu/04/12/05saska.htm

Smith R (1997) The Fontana history of the human sciences. Fontana Press Original, London

Smith R (2013) Between mind and nature. A history of psychology. Reaktion Books, London

Szendi G (2004) A depresszióipar [Depression industry]. Mozgó Világ 10:9–31

Szokolszky Á (2016) Hungarian psychology in context. Reclaiming the past. Hung Stud 30(1): 17–56

Vajda Z (2020) Róheim Géza a gyermekkori fejlődésről a terepmunkák tükrében [Géza Róheim about child development in the light of fieldwork]. Imago 9:18–37

Weszely Ö (1935) A tudatalatti lelki élet és a nevelés [Education and the unconscious]. Magyar Gyermektanulmányi Társaság, Budapest. (My italics)

A History of Self-Esteem: From a Just Honoring to a Social Vaccine

43

Alan F. Collins, George Turner, and Susan Condor

Contents

Introduction	1118
The Emergence of Self-Esteem	1120
Early Uses of Self-Esteem	1120
Self-Esteem, Agreeableness, and Social Order	1124
Mental Philosophers and Self-Esteem	1127
A Formulaic Approach to Self-Esteem	1129
Inferiority, Damage, and Discrimination	1131
Therapy and Self-Realization	1135
Tests of Self-Esteem: Quantity and Measurement	1137
Conclusion	1138
References	1140

Abstract

In the twenty-first century, self-esteem has become an almost routine way of talking about, thinking about, and acting upon human relations and on ourselves. Now firmly established in the lexicon of academic psychology, self-esteem was first used in largely theological and moral debates around pride and self-regard. However, from its inception, there were concerns about deficit and more particularly excess. The elusive nature of the "appropriate amount" of self-esteem has made it a constant resource for anxious debates about the state of the self, the action of psychological functions, and relations between the self

---

A. F. Collins (✉)
Lancaster, UK
e-mail: a.collins8@lancaster.ac.uk

G. Turner
Preston, UK

S. Condor
Emeritus Professor of Social Psychology, Loughborough University, Loughborough, UK
e-mail: s.condor@lboro.ac.uk

© The Author(s), under exclusive licence to Springer Nature Singapore Pte Ltd. 2022
D. McCallum (ed.), *The Palgrave Handbook of the History of Human Sciences*,
https://doi.org/10.1007/978-981-16-7255-2_85

and social structures. While debate over amount may have been a constant, the chapter shows how there have been significant shifts in beliefs about whether self-esteem was beneficial or not. A second constant in the history self-esteem has been its perceived role as some form of prompt to action, though that too has undergone change, and this facet of self-esteem has led to debates over how best to manage it for the benefits of individuals and groups. As a normative concept that has been considered as an impetus to act, as a source of inner pleasure, and as deriving from and contributing to social relations, self-esteem may be a concept over which psychologists now claim particular expertise, but it has been used in wider spiritual, moral, and sociopolitical debates. Developed as a scientific term within psychology, the history presented in this chapter also examines the enmeshment of the psychology of self-esteem with moral and social concerns.

Keywords

Self-esteem · Psychology · History · Normative · Moral order · Motive · Needs · Rights

Introduction

In 1990 in the report of *The California Task Force to Promote Self-Esteem and Personal and Social Responsibility,* Senator John Vasconcellos was nothing if not optimistic: *"*We have initiated an historic and hopeful search for a 'social vaccine'" (California Task Force 1990, p. ix). In 2020, the word "vaccine" acquired a renewed salience that surely accentuates the force of Vasconcellos's words and captures vividly the enormity of what he felt was being discovered. He believed that modern psychology had unearthed a "vaccine" that would protect against a range of social ills embracing crime, violence, substance misuse, premature pregnancy, child abuse, chronic dependence on welfare, and educational failure. The vaccine that would protect us and ensure a healthier society was self-esteem. Or, more specifically, high self-esteem. Ensuring the healthy psychological development of Californians, Vasconcellos reasoned, would result in a healthier, happier, wealthier California and, after California, the World. The report was equally clear about where responsibility lay: it was "our individual and personal responsibility to be part of the solution that we also realize higher 'self-esteem'" (1990, p. viii). When the report was written, self-esteem was firmly embedded in the vocabulary of academic psychology: it was the subject of a vast number of psychological publications and a central topic in popular books on personal well-being. By the 1990s, for some academic psychologists, enthusiasm for self-esteem was waning, and it was dismissed by a number of prominent critics who urged policymakers, parents, teachers, and therapists to forget about self-esteem, and Scheff and Fearon concluded that others who described the research field was a "whaling expedition that has, at best, caught minnows" (Baumeister et al. 2005; Scheff and Fearon 2004).

Despite some serious criticism, self-esteem has proved resilient and mutable. In the twenty-first century, the critiques of the likes of Baumeister are already themselves objects of criticism, with researchers offering defenses of self-esteem both in itself and as a predictor of a variety of outcomes (e.g., Swann et al. 2007; Orth et al. 2018). And there are conceptual and methodological debates such as that attempting to distinguish self-esteem from other forms of self-regard, notably narcissism (Brummelman et al. 2016). The research literature on self-esteem is now so extensive that it defies one overarching review, at least within the format of a journal paper. Instead, researchers review self-esteem literatures in relation to topics as diverse as employee well-being, obesity, parenting style, crime, self-injury, exercise, various forms of identity, and many aspects of mental health. How is it that self-esteem has become such a major concept and measure in modern psychological research and writing? Pronouncing on the self has never been the sole preserve of psychology and psychologists, but in the twenty-first century, this is not true of self-esteem, and psychology is the discipline firmly associated with its academic study. This chapter examines how self-esteem became a psychological concept and one judged pertinent to such a wide range of social debates.

While the chapter examines the continuities and discontinuities in the concept of self-esteem and the conditions that allowed it to become a major idea in modern psychological thinking, three things should be noted about the approach taken: the analysis takes a long view beginning in the seventeenth century, it focuses largely on self-esteem in academic writings and some religious writings rather than in practices or other popular works, and it is limited to the literature in English. As regards the last of these limitations, there is no presumption that self-esteem has the same connotations in other languages or that other languages and cultures necessarily have an equivalent term. Self-esteem may have been assumed to be a universal by many of the writers discussed but it is not an assumption made in this chapter. Many histories of self-esteem by psychologists begin with William James in 1890 and trace developments from there. While understandable, for reasons outlined below, this approach neglects significant developments in the concept before James that remained relevant beyond James, and ignores aspects of its meaning that were radically different from current connotations of the term. By ignoring how the term was used in poetry, fiction, self-help literatures, magazines, and other sources, the current chapter does not examine its meanings in such popular outputs. This is largely a matter of space, and it should be recognized that its uses in those sources are crucial in establishing common understandings and usage of the term. By considering self-esteem as a concept or idea, the chapter only touches upon what Foucault termed technologies of self, that loose assemblage of knowledge, instruments, places, frames of reference and of judgement that shape human conduct in particular directions and which help constitute our subjectivity or consciousness of self (Cruikshank 1993; Foucault 1988). Nevertheless, in common with a number of historians of the human sciences, the approach taken here emphasizes the role of language, including the language of the human sciences, in constituting shared understandings of how we comprehend and act on our selves and others (Danziger 1997a; Smith 1992). This chapter argues that psychologists have changed how we think of self-esteem, and its status as a part of the self and of our selves. Equally, however, self-esteem was and is a term with both descriptive and prescriptive aspects to its meanings, and despite its use as a scientific term, psychology has not been stripped it of its moral connotations.

The Emergence of Self-Esteem

Early Uses of Self-Esteem

One of the earliest uses of self-esteem as an English compound noun was an insult. In 1619, when Coffin passed judgement on Joseph Hall, later to become the Bishop of Norwich, he concluded:

> such was the mans misfortune, his wit being so shallow, and selfe esteeme of his owne worth and works so great that as before he neuer more bragged ...where he had least cause
> (Coffin 1619, p. 353)

Coffin's suspicion of inflated self-esteem was typical of many of its early uses. The Reverend Richard Baxter (1615–1691) was a Protestant minister who was one of the principal early users of the term and he was clear on the danger of excess self-esteem, which he saw as a sin akin to pride:

> The next faculty that *self* hath corrupted, is the *Understanding*; and here we meet with the sin of *self-esteem*, which is the second part of *selfishness* to be mortified. It is not more natural for man to be sinful, vile, and miserable, than to think himself vertuous, worthy and honourable. All men naturally over-value themselves and would have others over-value them. This is the sin of Pride.
> (Baxter 1675, p. 93)

Baxter argued that self-esteem involved a misunderstanding and evaluation of the self that contributed to the sin of pride. These were not marginal views and Baxter was no obscure rural minister. In his *The Protestant Ethic*, Max Weber described him as "one of the most successful ministers known in history," he debated with Oliver Cromwell and ministered to Cromwell's troops, was prolific in his writing and was widely read by contemporaries (Cooper 2016). When John Evans delivered his sermon on the "character of Solomon's vertuous woman," in Croydon in 1695, he too regarded self-esteem as a problem stemming from a human nature "prone to swell upon a conceit of its own Perfections," (Evans 1695, p. 17). Bishop Joseph Butler (1691–1752), a prominent eighteenth-century English minister, was concerned by the moral threats that had their roots in self-esteem, or as he preferred self-love, arguing that it could foster a partiality to ourselves that was insufficiently examined and undeserved (Butler 1726). This was not a condemnation of those who never doubted themselves or their actions, for they were "too far gone to have any thing said to them," but of the more mundane failing, as he saw it, to assess one's moral character accurately.

The dangers of excessive and under scrutinized self-esteem in Baxter, Evans, and Butler were presented as uncontentious largely because of longstanding Christian fears around the sin of pride, which as, as Baxter summarized, was the sin to be most feared and "the surest and nearest way to hell" (Baxter 1675). The Christian arguments against pride were well rehearsed and had been since Augustine. Its sinfulness lay, among other things, in taking credit for qualities or achievements that were due to God, an attribution of merit to ourselves that denied our place in a divinely ordained order, and in seeking to want to be God or even above God. Falsely inflating one's self-esteem risked a corrupt conscience that excused further sin and behaving badly towards others. For Baxter, the very concept of "self" was problematic, because, for him and other writers, it implied a self-centeredness, and selfhood was "something to be overcome not realized" and something that should sound in Christian ears "as very terrible, wakening words, that are next to the names of sin and satan" (Baxter 1675).

The early association of self-esteem with pride, sin, and the self made it a human quality to be treated with suspicion, and its threat was made sharper by the Protestant emphasis on self-scrutiny. In the Lutheran tradition, the dominant Protestant belief was that salvation could not be gained through deeds but by God's grace alone and that good works should be undertaken as gratitude to God and as living faith (Brauer 1996). The articles of the Church of England agreed in 1571 asserted that justification (broadly, God's moving a person from a state of sin to righteousness) by faith alone was a "most Wholesome doctrine." Baxter sought to modify this view of justification informed in part by his experience with Cromwell's troops, some of whom inferred from "by faith alone" a certain liberty of action that threatened obedience to any form of moral law. Although interpretations of Baxter's views are contested, as a response to the danger of libertine behavior, he appears to have become more nonconformist and argued that meritorious works *were* important to salvation (Cooper 2016). Within this, what led to good deeds was critical and performing good works solely for the sake of one's pride or self-esteem was sinful. Nevertheless, it seeded the idea that self-esteem could be harnessed to good deeds and be beneficial both to individual salvation and to ensuring moral order in a society, seventeenth-century England, that was riven by divisions. The role of self-esteem in a synergy between works and grace prefigured later attempts to give it a role aligning individual improvement and social harmony, though here the basis was faith and doctrine.

An early defender of the potential merits of self-esteem and a contemporary of Baxter's was the poet John Milton (1608–1674). In his *The Reason of Church Government Urg'd against Prelacy* of 1641, Milton recommended esteeming the self as a pious and virtuous matter, second only to the love of God:

> an esteem, whereby men bear an inward reverence toward their own
> persons. And if the love of God as a fire sent from heaven to be ever kept
> alive upon the altar of our hearts, be the first principle of all godly and

vertuous actions in men, this pious and just honouring of our selves is the second,
(Milton 1641, p. 53)

A year later, in his *An Apology Against a Pamphlet*, a text littered with "I" and containing several uses of esteem, Milton wrote of how:

a certaine nicenesse of nature, an honest haughtinesse, and self-esteem either of what I was, or what I might be, (which let envie call pride) and lastly that modesty, whereof though not in the Title page yet here I may be excus'd to make some beseeming profession, all these uniting the supply of their naturall aide together, kept me still above those low descents of minde, beneath which he must deject and plunge himself, that can agree to salable and unlawfull prostitutions (Milton 1642, p. 17)

For Milton, the proper management of self-esteem together with other feelings provided a means of ensuring one acted virtuously and made virtue an individual undertaking that spoke to humans as free and autonomous beings (Fallon 2007). Milton believed self-esteem, together with other qualities, had kept him above low descents and ensured a self-restraint that had its roots in self-knowledge and self-respect. By preferring the comparatively unfamiliar term self-esteem, Milton distanced his ideas from those of self-love and the long problematic associations of that feeling for many Christians. Self-esteem here was not an all-embracing assessment of one's self; it was limited in scope to one's moral worth or virtue (Darwall 1977).

Milton continued to make use of self-esteem and in *Paradise Lost*, he again argued for its merits:

Oft times nothing profits more,
Then self-esteem grounded on just and right
Well manag'd; of that skill the more thou know'st,
The more she will acknowledge thee her Head.
(Milton 1667, Book 8, lines 571/4)

When matters of self-respect and self-esteem were "well manag'd" not only was acceptance of God's rule enhanced and individual virtuous action made possible but repressive earthly social control was rendered unnecessary – both topics of great salience in seventeenth-century England and to Milton personally. Milton derided mindless conformity to tradition and superstition arguing that each man's "proper ethical regard for himself" was the guarantee that he conducted himself properly (Jordan 2016). Duty, and more widely moral and social order, were guaranteed through a willingly undertaken inspection of oneself, not through unthinking obedience.

Though scholars of Milton disagree on the exact role of gender in his writings, most are agreed that there was a strongly gendered aspect to his work and this extended to his views on self-respect: good citizens exhibited manly virtues, including the possession of self-respect (Froula 1983). By contrast, he wrote of a femininity that was domesticated, sexualized, and subordinate to masculine liberty, and by implication one less imbued with self-respect. Femininity, he and many of his male contemporaries argued, had much to answer for: he blamed man's "First Disobedience [that] brought

Death into the World and all our woe" (Book 1, lines 1–3) not only on Eve but on "Mans effeminate slackness" (Book 11, line 634), a slackness that arose because Adam did not possess the self-esteem proper to manliness. Milton's self-esteem was a masculine virtue.

Mary Astell (1668–1731) is now most famous for her writings on marriage, but she too saw merit in self-esteem (Hill 1986). Astell was taught at an early age by her uncle, Ralph Astell, an ex-clergyman who was associated with the Cambridge Platonists who believed in the role of free will in virtuous action, the compatibility of faith and reason, and, contra thinkers such as Hobbes, the eternal nature of moral principles. In 1693, the year before she published her first work, Astell began a lifelong correspondence with John Norris, another who saw self-esteem as having the capacity to promote virtue. Astell portrayed self-esteem as a positive quality:

The Humblest Person that lives has some Self-Esteem, nor is it Fit or Possible that any one should be without it.
(Astell 1697, p. 288)

She argued fiercely that women were deserving of respect and, as part of God's creation, this entailed respecting oneself for "God's sake," not one's own, which was in accord with the common view on pride.

Scrutiny of the self, of one's actions, and the responses of others were all integral to self-esteem. As Kurt Danziger has remarked, new possibilities around the self were opened up towards the end of that century by the philosopher John Locke's argument that it could be known empirically in the same way as other objects and in ways that the immortal soul could not (Danziger 1997b). In Locke's analysis, the right and wrong of one's actions could be judged by divine law, civil law, or by the law of opinion. The law of opinion provided a basis for deciding between virtue or vice and that decision could be based either on inherent right or wrong or on the custom and practice of the society. Consistent with his more general theory, these outcomes gained force from people learning that the good opinion of others was frequently advantageous. Commendation and disgrace, or praise and blame, became powerful social forces, so much so that he argued that when men broke divine law, they could console themselves with the belief of some reconciliation with God yet they could not bear the "constant dislike and condemnation" of their contemporaries. Locke's idea held out the promise that estimation of the self was possible in a new way. He may not have used the term self-esteem but these ideas prepared the way for self-estimation as a knowable aspect of self.

From the outset, self-esteem was something that required recognition and management by the individual: one's natural endowment may have included a taste for self-esteem, but that was to be as much struggled against as indulged. In seventeenth-century England, those suspicious of self-esteem saw the natural propensity to find pleasure in esteeming oneself a thin justification for action and likely to lead to sin; if self-esteem was allowed to exist, it should at the very least be restrained. While fear of excess self-esteem was the dominant view, these fears were balanced by the idea that there was a need to esteem oneself in order to act and to act in a moral, devout manner consistent with religious doctrine and with the standards of virtue common in one's society. This was not a project of self-perfection but it was consistent with the more modest ambitions of personal advancement and of advancing the good of mankind. Finally, many writers were ambivalent about self-esteem and their views were complex: Baxter wrote against it frequently, and at the same time he saw the possibility of merit in it; Norris recognized that despite it being a force for good, it could also be damaging and lead us into sin (Norris 1687, p. 355). This ambivalence reinforced the need for self-esteem to be treated with care or well-managed, as Milton advocated. Towards the end of the century, Locke proposed that the self could be known in the manner of natural objects and that in turn made possible the idea that the estimation of the self could be known in the same way.

Self-Esteem, Agreeableness, and Social Order

In Milton, Baxter and other seventeenth-century writers, self-esteem was almost invariably discussed in relation to religious matters. While through the eighteenth-century these strong associations between self-esteem and religion remained, new dimensions to the concept also emerged that gave it a role in proposals for living a good life, maintaining individual integrity and sustaining social order that did not rely so explicitly on religious faith and doctrine. To illustrate this, this section examines two contrasting sources: the philosophy of David Hume and phrenology.

The Scot David Hume (1711–1776) has become recognized as the most influential English language philosopher of the eighteenth century and a central figure in the Enlightenment (Harris 2015). Famously, he attempted to develop a moral philosophy not explicitly motivated by religion or religious doctrine. He understood the implications of his efforts, and reflecting on his life, he concluded that "I have no enemies; except indeed all the Whigs, all the Tories, and all the Christians." His *Treatise of Human Nature* (1739–1740) and his *Enquiries Concerning the Principles of Morals* (1751) formed part of a project to develop a general account of the relations between human nature and society, and in that project, Hume made occasional but telling use of self-esteem.

In the *Treatise*, Hume's first use of self-esteem was conventional enough: "the very same qualities and circumstances, which are the causes of pride or self-esteem, are also the causes of vanity or the desire of reputation" (Hume 1739, p. 332). But Hume was quick to point out that self-esteem and pride were of great significance in ensuring virtuous and honorable conduct: "a genuine and hearty pride, or self-esteem, if well conceal'd and well founded, is essential to the character of a man of honour" (p. 598). Such was its potential as an impetus to action that it could inspire heroic virtue, which was "nothing but a steady and well-establish'd pride and self-esteem, or partakes largely of that passion. Courage, intrepidity, ambition, love of glory, magnanimity, and all the other shining virtues of that kind, have plainly a strong mixture of self-esteem in them" (pp. 599–600). In a famous contrast, Hume affirmed that the value of virtues usually thought of as counterweights to self-esteem, such as humility, was somewhat overstated and though he deemed it necessary that self-esteem be concealed, he listed humility among what he called the monkish virtues, that is, virtues that are neither pleasant nor useful.

Self-esteem sat alongside pride in Hume's moral philosophy, and although he used pride more frequently, the similarities in his uses of the two are striking (whereas there are marked contrasts between, for example, his uses of pride and vanity, Galvagni 2020). For Hume, reason alone was not sufficient to lead to acts of will, which required the motivating force of the direct and indirect passions, with pride being an instance of the latter. Pride itself rested on the notion of sympathy, an ability that allowed for the recognition of resemblance, the communication of sentiment, and an understanding of others' opinions of us. Certain qualities – Hume cites wealth, power, and certain mental and physical attributes – were reliably associated with pride, whereas other qualities associated with pride were much more susceptible to local variation. Pride was fundamentally an externally driven, socially oriented feeling where the self was not the cause of pride but its object, a quality it shared with its near synonym, self-esteem (Wahrman 2006). Acts became virtuous when they met with approval or esteem and became vices when they met with disapproval. The agreeableness of approval, pride, and self-esteem encouraged us to act virtuously though this did not preclude the possibility of an "ill-founded" pride becoming a vice. Hume was well aware of the long acknowledged temptations to act insincerely to attract flattery or to deceive oneself simply to make oneself feel good (Hume 1751). Nevertheless, in Hume's thought, esteeming oneself was a means of achieving virtuous behavior and wider social order that did not depend on an underlying religious belief or doctrine; rather than being threats, pride and self-esteem could be guarantors of good character, a good society, and something that "capacitates us for business" (Hume 1739, p. 628). Nor was Hume alone; his near contemporary and friend, Adam Smith, also made powerful economic arguments that drew heavily on notions of esteem (Brennan and Pettit 2004).

Hume's philosophy seems far removed from religious anxieties over self-esteem and equally far removed from phrenology. But just as Hume was a dominant figure in the history of moral and mental philosophy, so phrenology had a prominent place in the development of psychological science (Cooter 1984). As first developed by the Viennese physician Franz Joseph Gall (1758–1828) in the late eighteenth and early nineteenth-century, phrenology made a number of claims that allowed the assessment of individual differences in character. It claimed that the mind consisted of a set of distinct psychological abilities or functions, that those functions had discrete locations in the brain, that the greater the volume of the brain region associated with a particular function the more powerful the influence of that faculty on the character of the person, and finally that the surface of the skull was shaped by the volume of the underlying brain areas allowing a phrenologist to measure the strength of a psychological ability by measurements of skull areas. Although it did allow for the cultivation of character by exercising some faculties and seeking to counteract the effects of others, the structure of character was something endowed: in phrenology, an individual's character was some way from the tabula rasa of Locke. The list of functions used by phrenologists was more a reflection of early nineteenth-century values than anything else, and given this, it is not surprising that self-esteem was listed among those functions (Young 1970). Gall claimed to have identified the faculty and its location following an encounter with a beggar who had said he was too proud to work. In early translations, the faculty was labelled as "pride" or "arrogance" or "self-love" (Spurzheim 1815). While self-love continued to be used in some phrenological texts, self-esteem came to be preferred by most, and its use distanced phrenological thinking from the longstanding and fevered religious debates over self-love.

Phrenology was extremely popular in Britain, where George Combe's *The Constitution of Man*, which was inspired by Gall and Spurzheim's writings, sold an impressive 350,000 copies between the first publication in 1828 and 1900. One of the great attractions of phrenology was its accessibility and the familiarity of the concepts, if not always the given names, described by the faculties, and Combe included self-esteem as one of the mental faculties. He discussed it briefly in *The Constitution*, mainly in relation to the faculties of Adhesiveness and Love of Approbation, but in his later *A System of Phrenology*, he presented a more extended discussion and argued that "due endowment" of the faculty led to "excellent effects." With echoes of Milton and Hume, for Combe, the links between self-esteem and other moral sentiments, such as dignity, respect, and restraint, made it a positive facet of character. Inevitably, it could be overdone and, like almost everyone else who wrote on the topic, Combe conceded that self-esteem could be damaging when unfounded or overblown or not sufficiently counteracted by other moral faculties. Importantly, deficiency too could be a problem, and he wrote of how a small faculty of self-esteem meant that a person lacked "a due sense of his own importance" (Combe 1853, p. 344). Combe's general position was that people were not inherently morally flawed, instead "each [faculty] has a legitimate sphere of action and, when properly gratified, is a fountain of pleasure" (Combe 1841, p. 104). While some may have baulked at this particular elevation of pleasure, like most of the early writers on self-esteem, Combe discussed self-esteem using qualifiers such as "due" or "just" or "properly": self-esteem was healthy only if justified. That justification, of course, relied on the particular values of the society, and Combe was well aware that what he was promoting was a moral science where moral values were made part of the natural world: for many phrenologists, theirs was a form of natural theology, a revealing of the divine in the natural, that allowed for a harmony between its precepts and the moral and religious nature of people (Fowler 1848).

Phrenology became popular for a variety of reasons that included its accessibility, its promise as a vehicle of social reform, and the insight and scientific certainty it appeared to provide in a rapidly changing society (Cooter 1984). Despite its apparently rigid structure, the self-knowledge it claimed to provide was compatible with the hopes expressed in the growing self-improvement literatures which aligned self-improvement with social improvement. Consistent with this, self-esteem was endowed but could be managed as part of attempts to improve one's character, and character itself was an enduring and pervasive theme in nineteenth-century Britain where it was treated as consequential in all aspects of life from economic success to political and social order (Collini 1985). The debates around character frequently conflated the descriptive and the prescriptive in ways mirrored in phrenology. Having described a person's self-esteem as "large," that could soon become a recommendation from the phrenologist to cultivate competing faculties and to develop modesty through a combination of habit and will in order to avoid self-esteem developing into arrogance or undeserved pride. Self-esteem could be nurtured through various actions according to social status: the mercantile, manufacturing, and agricultural classes could gain it through "accumulating wealth," while the more "intelligent gentlemen" could obtain it through "political, literary, or philosophical eminence" (Combe cited in van Wyhe 2004). The laboring classes, apparently, were too busy gratifying "the inferior propensities." While there were wider debates over whether character was made by us or for us, there was a growing literature on the idea one had a duty to cultivate facets of character and to achieve an appropriate balance between potentially conflicting aspects of character (Collini 1985). Self-improvement was a form of governing the self and extended to moral and intellectual improvement, education, professional or technical training, social status, and reputation. In these various fields that were so valued in the rising industrial capitalism, self-esteem could act as both motivator and reward.

Both Hume and the phrenologists believed they were proposing schemes that had sufficient scope to explain the basis for a morally ordered society. For both, maintenance of social and moral order and the means of personal and social change required an understanding of individual psychology. Reason alone was insufficient to maintain order: what was also required were qualities that, rightly managed, could impel us to act and to act appropriately. Self-esteem was a quality that appeared to satisfy such requirements: it appeared to produce agreeable sensations, it was a prompt to action, it was satisfied by adhering to what they believed to be universal values, but it could also be shaped by social mores and opinions, it varied in amount and could be managed so as to improve one's self and society.

Mental Philosophers and Self-Esteem

The entry of self-esteem into psychological writings and into the lexicon of psychologists has been most associated with William James's formulation in 1890, of which more shortly. However, its position as a psychological topic was recognized and promoted earlier in the nineteenth century. The Scottish philosopher Alexander Bain (1818–1893) was an important advocate of a science of mind in works such *The Senses and the Intellect* and *The Emotions and the Will*. William James acknowledged him as a considerable influence on his thinking, and in his *Principles of Psychology*, James recommended Bain's remarks on self-esteem or self-feeling (James 1890). Bain's own thinking on the emotional faculties had, in turn, been influenced by the phrenologists, and he emphasized the importance of their grounding of the mental faculties in the brain (Young 1970). When it came to self-esteem, Bain's views encapsulated many of the tensions that arose when thinking about self-esteem. He characterized it as a settled opinion of our merits that was also susceptible to fluctuation, something that might be warranted or not, and something that could arise from a specific ability or from a more general assessment of a person's character and conduct. It was "exceedingly likely to over-indulgence," but equally a sense of worthlessness or humiliation might arise when a person surrendered a "previous estimate of self" (Bain 1865, p. 112, p. 106).

Repeating claims already often made, Bain accepted that whatever the challenges posed by self-esteem, it was both pleasurable and an imperative to action: there was the "warm eye" of self-esteem and a prompt to "self-cultivation and active usefulness" (Bain 1865, p. 100, p. 112). What those actions were and the standards by which they were judged was, of course, a matter for debate and change: Bain's reference to the proper pride of the workman struck a very different chord from references to diligence in prayer. Bain did not consider self-esteem one of the primitive emotions but one derived from what he described as the "tender emotion," and when the tender emotion was turned inwards to the self, it was a source of pleasure. Bain claimed that in applying the tender emotion to oneself, other psychological abilities or achievements were required such as the recognition of self, making comparisons of one's own actions with those of others, reflecting upon those actions, formulating expectations, and making judgements of merit. Bain's emphasis on the psychological abilities that self-esteem entailed highlighted how complex understanding self-esteem might prove. As Bain also recognized, many of these qualities, such as the kind of person one might strive to be, how one was expected to behave, and recognizing the actions that attracted the commendation of one's fellows, made self-esteem a psychological quality saturated with social values.

Emerging evolutionary thought, the increasing grip of industrial capitalism, and faith in progress contributed significantly to shifts in the meaning and status of self-esteem. Herbert Spencer (1820–1903) attempted to produce a unified account embracing individuals and societies as subject to the same natural laws, an approach that facilitated analogies across domains. Spencer's ideas about competition, the adjustment of the internal to the demands of the external environment, success, failure, and the survival of fittest combined to create a worldview that saw evolution as essential and essentially progressive (though it could be hampered by human interference). In this grandest of grand schemes, self-esteem played a small part, although at a personal level, a phrenological reading concluded that Spencer's own self-esteem was "very large," something that prompted a friend to remark "he might have arrived at the same conclusion without feeling your head" (Spencer 1904, p. 230). Self-esteem in Spencer became closely linked to success, failure, and a progressive view of evolution and society (Spencer 1855). The pleasure that accompanied esteeming oneself was to be welcomed and not viewed with suspicion. Here was a new and perhaps more acceptable reason for why self-esteem made us feel good: it had evolved that way. It prompted us to act in ways that promoted our chances of survival and survival was necessary for reproductive success. In Spencer's thinking, the adjustment of conduct to which self-esteem led may have had moral and social consequences but it was not driven by them; it was tied to conformity to values only to the extent these benefited the individual. This was a long way from self-esteem as a result of faith or as a route to salvation.

In accord with Spencer's general view of laws, self-esteem was not limited to individual organisms. He wrote that self-esteem, or egoism, and its survival value extended to groups, where success in war and in mercantile trade could boost the esteem, and something akin to self-esteem, of nations. This was not quite the jingoistic position it might seem as he maintained that the "corporate conscience" of nations was not yet as developed as that of individuals. Consequently, while the self-scrutiny that made it possible for an individual to recognize excess of egoism was a good quality, such criticism was not permitted when it came to one's nation as it amounted to something like treason. For Spencer, this was a fault and, presumably, one that the further evolution of nations would eliminate: an evolved and superior nation should be able to scrutinize itself critically while still sustaining its self-esteem. What Spencer did, however, was to highlight how self-esteem could be something associated with groups and group membership.

By the second half of the nineteenth-century, self-esteem was grounded in the need for survival while its excesses were kept in check by the "rules of polite intercourse" (Sully 1876, p. 159). For the English psychologist James Sully (1842–1923), the impulse to feel good about oneself was ancient enough to be written into physiology: "below the surface of distinct consciousness, in the intricate formation of ganglion cells and nerve fibre, the connections between the idea of self and this emotion of esteem have been slowly woven" (Sully 1876, p. 164). With his interests in child development, Sully argued for the simultaneous development of the idea of self and feeling about the self: the two were entwined physiologically and psychologically. In his 1876 article on self-esteem, his repeated references to personal attractiveness (for women), moral worth, and critical powers as sources of self-esteem exemplified a shift in register from the depths of the nervous system to the salons of polite society that certain strands of evolutionary thought made possible and that self-esteem seemed able to bear. Sully doubted people's ability to make accurate judgements of their own worth and argued that there was a natural tendency to think too highly of one's self. The natural tendency resulted in a belief about oneself that was overblown, unquestioned, intense, and absolute (Sully 1876). This tendency could be advantageous, and, unlike Spencer, Sully believed that inaccurate assessment could have survival value and hence became the usual state of affairs (Sully 1881). What was less natural were those rare people "who are disposed to make a very low estimate of themselves" and who were "quick and sensitive in finding holes in their [own] moral garments" (Sully 1876, p. 161). Although Sully allowed that such self-deprecation could be an affectation, he also implied that underestimation could be genuinely felt and lead to "mental distress." Many earlier writers, including the phrenologists, had linked self-esteem to insanity, though in most cases the link was with an overblown self-esteem. However, Sully's brief aside on those sensitive to their own moral weaknesses pointed to a line of thinking that was to dominate twentieth-century psychological thinking on self-esteem: the dangers of low self-esteem for one's psychological well-being.

A Formulaic Approach to Self-Esteem

If Bain, Spencer, and Sully are examples of writers who continued to treat self-esteem as an intrinsic part of human nature, when self-esteem became a more widely research topic in the 1970s and 1980s, it was William James (1842–1910) who was more commonly cited as the person who brought self-esteem firmly into the psychological literature. While the discussion so far has made it clear that self-esteem was understood as an attribute of persons long before James, his position in American academic and cultural life, the range and depth of his writings on psychology, and his formulation of the nature of self-esteem have all contributed to him being positioned as some kind of founder of the psychology of self-esteem. In his *Principles of Psychology* of 1890, James discussed it briefly in his chapter on "Consciousness of Self" where, like Sully and others, he treated self and awareness of self as topics too entwined to be easily separated. Self-feelings, for James, were "elementary endowments of our nature" that divided into two broad classes, self-satisfaction and self-dissatisfaction, which were as primitive and as fundamental to human nature as feelings of anger or pain. He regarded self-esteem and self-satisfaction as broadly equivalent, and in his most direct and celebrated summary of the former, he reduced it to an equation in which Self-esteem = Success/Pretensions, going on to remark that this formulation meant that self-esteem could be as easily raised by decreasing the denominator as by increasing the numerator. In highlighting pretensions, "what we back ourselves to see and do," James was well aware that while there was cultural pressure to aspire, there was also the religious tradition that it was not the place of individuals to pretend to too much. As he himself acknowledged, James's formulation owed much to earlier writers, especially the Scottish historian and moralist Thomas Carlyle (1795–1881), who James quoted as urging people to: "Make thy claim of wages a zero, then thou hast thou the world under thy feet" (James 1890, p. 311). Carlyle's preceding words from *Sartor Resartus*, which James did not quote, were just as revealing about the value of pretension:

> ...the Fraction of Life can be increased in value not so much by increasing your Numerator as by lessening your Denominator. Nay, unless my algebra deceive me Unity itself divided by Zero will give Infinity. Make thy claim of wages...
> (Carlyle 1831, p. 136)

Carlyle's and James's formulations brought together achievement and aspiration as qualities necessary for self-satisfaction and as fundamental to societies that prized individual striving and success. By making it an equation, James held out the seductive promise of quantification and, on the surface at least, the equation stripped self-esteem of the notion of justification or deservedness with which it had been so closely linked historically. While James's thinking was much more subtle than this would imply, the equation provided a point of reference that connected self-esteem to two valued aspects of nineteenth century life yet distanced it somewhat from other complex questions of moral value and how one should behave. Just as concepts such as intelligence and motivation had been reformulated in ways that made them less closely linked to moral judgements and more amenable to a science, so too with self-esteem (Danziger 1997a).

The considerable wider social changes in Britain and the USA in the nineteenth-century also made aspects of Carlyle and James's characterization sharper. The increasing urbanization of the population, for example, has been associated with greater material aspiration and exposure to a far wider range of models to aspire to. As the sociologist Émile Durkheim (1858–1917) remarked of his own lifetime, it was a period that through the "malady of infinite aspiration" led to a "Pessimism that always accompanies unlimited aspiration." (Durkheim 1925/2012, p. 40). However, aspiration was integral to industrialized capitalism. Despite James's injunctions, such rapid social developments made the task of lessening the denominator that much more challenging and any rise in the denominator, of course, made success even more necessary to sustaining self-esteem.

Remaining, as it did, relevant to discussions of character, conduct, and religious doctrine, the emerging psychologists of the nineteenth century could not claim self-esteem as "their" concept or as something about which they had particular expertise, and the concept retained its relevance to discussions of the basis for value-laden actions and social order. However, in the writings of Bain, Spencer, Sully, and James, it acquired new connotations; self-esteem was less about moral good or bad and more about survival, personal achievement, and aspiration, qualities that chimed, not coincidentally, with the demands of industrial capitalism. Although the preceding names are a familiar rollcall of historical figures from the nineteenth-century, self-esteem was used much more widely than this would suggest. There is not room to explore that in detail here, but self-esteem was not a specialist term and it appeared in literatures beyond philosophy and academic psychology. It was, in other words, an accessible term that could mediate between psychological accounts of human nature that sought to develop scientific terminology that shed moral baggage and wider social debates about what kinds of acts should be esteemed.

Inferiority, Damage, and Discrimination

Some historians of self-esteem have noted that despite James being feted as the founder of the psychological study of self-esteem, his influence on the field was limited, and after 1890, self-esteem largely disappeared as a subject of psychological research until the 1960s (Mruk 2006). Others have acknowledged that there were significant developments in psychological thinking about self-esteem in the period and cite G.H. Mead, Cooley, a series of psychoanalytic writers such as Adler, Horney, Sullivan, Fromm, as well as Allport, Maslow, Rogers, and Rosenberg as significant contributors (Wells and Marwell 1976). Both characterizations have merit. Self-esteem was not a major concept in psychological research or writing in the period 1890–1960, yet there were social developments and changes in psychological thinking well before the 1960s that were critical to its subsequent rapid rise. As with so much psychological research in the Anglophone tradition, these developments were concentrated in the USA. Two crucial changes that had begun in the nineteenth century gathered momentum through the twentieth century: a change from an emphasis on the dangers of high self-esteem to the dangers associated with a sense of inferiority and low self-esteem, and an increasing examination of the links between self-esteem and groups.

The growing interest in understanding the psychological role of inferiority is most famously associated with the work of Alfred Adler (1870–1937). Adler was born into a Jewish family who lived near Vienna, and after early medical training in that city, he became prominent in psychoanalytic circles. With persecution of Jews on the rise in Germany and Austria, like many other Jewish academics, he left for a post in the USA in the 1930s. In Adler's characterization, a sense of inferiority was a fundamental aspect of human nature, one that was rooted in the physical vulnerability of humans, particularly children, in the face of nature and in social contexts (Adler 1917). The implications of inferiority were not all negative as it produced a desire to overcome inferiority, vulnerability, and adversity in ways that benefited the individual and simultaneously, in Adler's view, society as well (Adler 1930). While Nietzsche was a clear influence, many of Adler's supporters and interpreters have seen his ideas around striving as something more akin to a will to achieve than a will to power (Ansbacher and Ansbacher 1964). In *Understanding Human Nature*, Adler wrote of self-esteem as something to be saved, as something that could be dilapidated, deficient, or injured, and as something that was malleable and susceptible to elevation (Adler 1927). In *Demoralized Children* he commented that during education, "under the weight of accumulated difficulties [the child] is afflicted with there emerges a painful feeling of being unable to perform the task others achieve, and finally to be a witness of wounded and offended self-esteem!" (Adler 1925, p. 346). A deficiency of self-esteem was a danger to the individual, to social groups, and to society (although he also wrote of the "pathological desire to succeed"). In guarding against an inferiority complex, self-esteem was something to be protected and nurtured; it was vulnerable, precious, and susceptible to injury. Overcoming the sense of inferiority was a task for the individual. In line with his arguments about the individual as fundamentally interdependent, he thought the means of doing so lay in achieving a feeling of value through what he described as the "common welfare" (Ansbacher and Ansbacher 1964). In Adler's work, the preoccupation with human superiority and strength was replaced with a view of vulnerability and overcoming as the great drivers of human action.

Adler's theoretical work treated people as social products and the interdependence of people was integral to what he called his individual psychology (Ansbacher and Ansbacher 1964). Inferiority and group identity were more explicitly brought together by another Jewish émigré to the USA in the 1930s, Kurt Lewin (1890–1947). Lewin was not the first to use the problematic phrase "Jewish self-hatred" but he was its most high profile user within psychology, and it stirred some controversy (Glenn 2006; Lewin 1941). Jewish self-hatred was self-evidently related to the idea of inferiority and group membership, and equally self-evidently it was something to be overcome. Lewin's solution required forms of education and child rearing that instilled in Jewish children a more positive view of their Jewishness. Self-esteem was not a problem of the individual but a product of wider social relations whereby "only by raising their self-esteem as group members to the normal level can a remedy be produced." (Lewin and Lewin 1948, p. 214). Lewin did not limit his analysis to Jews and extended it to other minority groups, particularly black Americans, concluding that "raising the self-esteem of the minority groups is one of the most strategic means for the improvement of intergroup relations" (Lewin 1946, p. 46). By making self-esteem something produced in individuals by group relations and not simply interpersonal relations, and something that could be a means to improve social harmony, Lewin's work presented it as a powerful psychological tool for addressing pressing social issues.

Kurt Lewin's interest in groups, and minority groups in particular, were given considerable impetus and salience by the large-scale immigration the USA had experienced in the early twentieth-century, by the atrocities leading up to and during World War II, and by discrimination against black Americans. It is also difficult to ignore the role of Lewin's personal experiences as a Jew in early twentieth-century Europe in the development of his interests in minority groups and discrimination. In 1933, with the Nazi Party seizing power in Germany and shortly before his departure to the USA, Lewin drafted a letter to his colleague and fellow psychologist Wolfgang Köhler. In that letter, which was never sent, Lewin expressed many of his concerns and fears; he wrote of how in the Prussian village in which he grew up "100% anti-Semitism of the coarsest type was taken for granted," of the singling out that young Jewish children experienced, and of "how the fate of an individual Jew has never only been a personal fate" (Lewin 1933/1986). The antisemitism in Europe early in the century and in Nazi Germany were immediate and personal confirmations to Lewin that membership of groups had a profound impact on individual psychology. Lewin's interests in self-esteem were part of his attempt to make the group an entity worthy of psychological study in its own right and to develop an embracing theory of group dynamics (Lewin and Lewin 1948). They wedded self-esteem to groups, group membership, and the position of groups, very general ideas that later were developed further, particularly in areas such as social identity and the social psychology of intergroup relations (Tajfel 1974).

In the 1930s, the black psychologist Howard Long commented that concentrating on the material effects of discrimination had obscured its psychological effects and that the latter were deserving of attention (Long 1935). Postwar, racial segregation became a particular issue of debate, and although it had its supporters in the academic community, there was a growing number of psychologists arguing that it was psychologically damaging. The general notion of a "damaged black psyche" gained traction in psychological work from the late 1930s onwards (Scott 1997, though there was also a more positive assertion of "race pride" that gathered momentum during and after the Civil Rights movements, Rhea 1997). Such damage represented the kind of neglected psychological effects to which Long had been referring, and self-esteem accordingly became a major concept in specifying what was meant by damage and what kinds of repair were required. The link between psychological damage and segregation has been extensively discussed in relation to the case of *Brown* vs. *Board of Education*, where the court concluded that segregation in schools deprived children in minority groups of equal education opportunities and should, therefore, be abandoned. There is a considerable literature on the case and there remain differing views on whether it was a success or failure for race relations and, more specifically, on how influential the cited social scientific work was in the case (Jackson 2004). It is moot whether the court really saw the psychological evidence and opinion as decisive. Nevertheless, in the court judgement Justice Warren chose to refer to the detrimental effects of segregation on the mental development of black American children and supported the claim by pointing to the weight of "modern authority," which referred to the work and opinion of a range of social scientists.

When the case against the Board of Education was being constructed, the psychoanalyst Otto Klineberg recommended that those challenging segregation contact Kenneth Clark, a psychologist who was regarded as an authority on racial identification, prejudice and the effects of segregation. Together with his wife, Mamie Clark, Clark had begun research in these areas in the 1930s and he made frequent use of the term self-esteem, though it would be an overstatement to claim that the concept was front and center of the psychological work relating to the Brown case. Instead, vaguer more embracing terms or phrases such as adjustment, inferiority, and damage to personality featured prominently. However, by the time Clark wrote his *Prejudice and Your Child* in the 1950s, he relied much more heavily on the concept of self-esteem and it became central to his understanding of the psychological effects of prejudice and discrimination (Clark 1955). Postwar, prejudice was increasingly a topic of social psychological investigation in the USA and the association of self-esteem with that issue was fundamental to establishing it as both potential cause and significant consequence of social problems (Samelson 1978).

One of the best known and most controversial works on the negative psychological effects of discrimination was Abram Kardiner and Lionel Ovesey's *Mark of Oppression* (Kardiner and Ovesey 1951). During the postwar unrest about racial discrimination, they developed an account of psychological damage done to black Americans that made use of the concept of self-esteem. Typical of their approach was their case study of W.S., a young black American man whose life "begins and ends with low self-esteem" and "whose compensatory efforts ['pretentiousness', 'isolation' etc.] fail and terminate in a reinforcement of the originally low self-esteem with which he started" (p. 216; 178). They concluded that WS's "low self-esteem is not due to his persistent incompetence, but to a situation for which he is blameless – the caste system" (p. 179). Self-esteem was a facet of personality damaged by circumstances; black Americans were being required to adapt psychologically to a situation that was disadvantageous, morally reprehensible, politically controversial, and which produced mental pathology. While Kardiner and Ovesey have been criticized for reducing the problem of the damaged psyche to one of matriarchal family relations and for focusing on a negative view of self in an oppressed group, the conclusion of *Mark of Oppression* was unequivocal about what was required and why: segregation should be abandoned and the justification for that lay in the psychological concept of self-esteem.

> Negro self-esteem cannot be retrieved, nor Negro self-hatred destroyed, as long as the status is quo. What is needed by the Negro is not education, but *re-integration*.
> (Kardiner and Ovesey 1951, p. 387, original emphasis)

The reasons for preferring self-esteem over other terms in relation to prejudice and discrimination are speculative. One prominent competitor, "inferiority," allowed for easy slippage between potentially racist discourses of "real" inferiority. Another, "adjustment," was wider in scope and invited critiques of what were people being asked to adjust to, which might extend to an unjust society. Politically, self-esteem may have seemed a less loaded term. Whatever the role of social relations, the unambiguous interiority of the feeling of esteeming oneself made it unequivocally a psychological concept. Simultaneously, the phenomenological implications it carried conveyed a sense of tackling the experience of the marginalized. Finally, self-esteem had the merit of remaining an accessible term that lacked the obscurity of some more specialist terms, and this allowed it to enter comparatively easily into political and legal debates and into psychological works intended for a wide audience, such as Clark's *Prejudice and Your Child*.

As a means of expressing the negative consequences of circumstance, self-esteem was now implicated in debates about social relations extending from the home to school to questions of social organization and justice. It remained descriptive and prescriptive and now had relevance not just to the individual citizen and her self-governance but to some of the most debated topics in twentieth-century USA. Far removed from the early fears over excess, psychologists now advocated that we should be more concerned about deficiency. That deficiency was a part of an individual's self, though theorists such as Adler, Lewin, and Kardiner all stressed that it was never only an individual quality. Deficiency was, however, treated as unequivocally bad, and discussions of the possible merits of low self-esteem were rare indeed, as were discussions of the dangers of high self-esteem.

Therapy and Self-Realization

Alongside the growing debate over race relations, in the postwar period, there was a raised awareness of mental health issues being faced by US war veterans, a group who created what Engel described as a crying need for help (Engel 2008). Despite the experiences of World War I, in 1941, the US military had been ill-prepared for supporting the number of troops experiencing mental problems, while the dire postwar situation was revealed through a series of press exposés of the conditions in mental hospitals (e.g., Maisel 1946). Prompted by these revelations, the needs of the returning veterans, and growing acceptance of psychotherapy, in the late 1940s provision of psychiatric care became a prominent social and political issue. Psychiatrists writing about the needs of war veterans could draw upon an eclectic, almost bewildering, range of psychological therapies when considering treatment. Nevertheless, much of the writing was infused by the psychoanalytic ideas that had transformed much American psychiatric thinking in the interwar years (Burnham 1988). In his wartime article, "Psychological Adjustments of Discharged Service Personnel," Carl Rogers identified six major problems that discharged service personnel were likely to face and which therapists needed to address (Rogers

1944). The third of these problems was disturbances of self-esteem. Once discharged, soldiers lost their social purpose, their daily routine, and any rank they held, and, psychologists argued, they suffered a trauma to their sense of themselves: "With some ...the blow to self-esteem is very great, and the psychological reaction correspondingly deep" (p. 691). Reflecting on the challenges of psychotherapy, the American psychoanalyst and psychiatrist Lewis Wolberg (1905–1988) argued that dealing with "inordinately low self-esteem" was one of the major obstacles in psychotherapy (Wolberg 1943, p. 1415). Often set within the larger discourse of adjustment, the restoration of self-esteem became an acknowledged therapeutic aim (e.g., Ackerman 1945). The long-claimed connection between self-esteem and purposive action was now relevant to restoring the veteran as a productive citizen, to repairing their sense of well-being and to their rights to dignity and respect.

Picking up on the historical thread of self-esteem as a spur to act, the contemporary theorist of self-esteem, Mark Leary, has characterized the more recent history of the concept as development from instinct to need to motive (Leary 2005). Post-World War II, Abraham Maslow (1908–1970) was perhaps the most cited source positioning self-esteem as a need. Maslow's postgraduate research in the 1930s largely revolved around dominance in primates, work in which he treated the pursuit of dominance as a motivating force (e.g., Maslow 1937). His focus on dominance was complementary to the earlier work on inferiority and he acknowledged the importance of Adler's ideas. Following a series of lectures by Erich Fromm, Maslow wrote on the authoritarian character in which he claimed dominance was a key feature of the authoritarianism associated with Nazis and Fascists (Maslow 1943a). However, partly because of this analysis, he began to move away from using high dominance, which he saw as too readily aligned with authoritarianism and superiority, and instead began to prefer high self-esteem, which allowed for a greater range of leadership styles. By the time he wrote his best known papers proposing a set of hierarchically organized needs, he used the term "esteem needs" (Maslow 1943b). Maslow's work represented a strand of thinking that made the monitoring and cultivation of self-esteem a matter for all and not only for oppressed minorities. Maslow's proposals implied that anyone whose esteem needs were not fully met might be faced with psychological problems, and therefore that uncovering low self-esteem was a step in preventing psychological difficulties. Rather than repeat what he saw as the mainstream preoccupation with negatives, Maslow attempted to develop what he described as the healthy half of psychology, with satisfied self-esteem being a good for all, an approach that contributed to the wider change from self-sacrifice as a defining feature of character to self-affirmation as part of healthy personality development (Burnham 1988). Self-esteem was no longer only something that was the result of aspiration and achievement, it was also an aspiration and achievement in its own right that need not lead to arrogance or a sense of superiority (a distinction also made in the recent literature on narcissism).

The emerging humanist psychology of the 1960s was closely associated with the work of both Rogers and Maslow. Maslow famously made self-actualization the rarely achieved pinnacle of human development in his needs hierarchy. In order to achieve this newly recognized form of human perfection, Maslow believed that certain social conditions had to be in place, and the same, presumably, was true of the immediately preceding need for self-esteem (Maslow 1954, p. 99). This inescapably linked his theory to political debates and Rogers too saw the connection between his ideas and politics. There were substantial differences in their personal politics, with Rogers embracing the liberal left while Maslow saw his theory as more consistent with a form of elitism, but for both the lines between psychological needs, political debate and human rights were blurred. The development of Rogers' and Maslow's ideas and similar approaches overlapped with movements such as the Civil Rights and Women's Rights movements. As was highlighted by discussions of the Brown vs. Board case, many people thought that arguments for rights need not, and many thought should not, rest principally on psychological needs or effects. Historically, there had been a move away from treating rights around children as stemming from the child as property or future source of income and towards rights based on treating the child as person and all that that entailed (Hart 1991). However, there was a long history of rights not only being grounded in principles, such as freedom and equality, but also in material interests, social cohesion, and status (Eekelaar 1986). Understood as a universal psychological need, preventing damage to self-esteem became something akin to a right and social institutions might be judged by how they affected the self-respect of individuals (Bay 1982). Psychologists such as Stuart Hart, Maria Brassard, and David Hardy made the case for the idea of "psychological rights" for children, and there was momentum behind the idea that protection from maltreatment included being protected from psychological harm (e.g., Brassard et al. 1987). In 1989, the United Nations issued the Convention on the Rights of the Child, and Article 19 stated that "States Parties shall take all appropriate legislative, administrative, social and educational measures to protect the child from all forms of physical or mental violence." On some interpretations, people now had the right to be protected from low self-esteem and to opportunities to realize their full psychological potential, including satisfying their need for self-esteem. Despite the plea for protection and opportunity to be matters of state, the plea often appeared to place responsibility for meeting those needs on the individual, something fueled by publications in the popular self-discovery, self-help, and self-improvement literatures. Self-esteem became a matter of both psychological and political concern, a connection that was furthered by psychological theories stressing self-esteem as an individual need that societies and individuals should satisfy.

Tests of Self-Esteem: Quantity and Measurement

Throughout its history, self-esteem was associated with notion of "amount," though in the early literature this did not equate to the kind of quantification that became integral to psychology as a science. As with so many other concepts, the expansion of interest in it as a psychological attribute depended on the development of tests accepted as measuring that attribute. As has been well rehearsed, tests made individual differences measurable and visible and helped establish psychology as a discipline offering practical expertise (Rose 1990). Add this to the ongoing struggles to get psychology accepted as a science and the growth of interest in personality, it is unsurprising that as the concept of self-esteem gained traction in the 1930s and 1940s, psychologists also began to develop tests to make it measurable. Carl Rogers and colleagues were involved in developing tests to help establish the efficacy of therapy, particularly when it came under attack in the early 1950s. Many of the items for those tests were ultimately derived from statements in therapeutic encounters that were judged to relate to an evaluation of self-worth (Butler & Haigh in Rogers and Dymond 1954). Several of these items then made their way into one of the first self-esteem tests to be widely used, Stanley Coopersmith's Self-Esteem Inventory (CSEI, Coopersmith 1957, 1959). The CSEI also included items drawn from the Minnesota Multiphasic Personality Inventory, which was a test developed in the late 1930s and early 1940s with the main aim of detecting forms of mental illness, and which subsequently became one of the most widely used personality tests. The method of development of the CSEI ensured that the relationship between low self-esteem and mental illness was built into the test at the outset, though as with other tests, use of the CSEI developed from something aimed at detecting extremes to a test on which all people could be graded and sorted (Rose 1990). The method of testing itself also changed. Earlier techniques were based on William Stephenson's Q-sort method where the response to an item such as "I understand myself" might be judged as "like me or not like me" and the focus was on the discrepancy between the ideal self and actual self (the importance of the ideal self-actual self distinction and the discrepancy between them having been developed and emphasized by a number of theorists, including Carl Rogers). By contrast, the CSEI developed a simpler, universal marking scheme where rejecting statements such as "I understand myself" simply contributed to a low self-esteem score. Rosenberg simplified testing still further and reduced it to the convenience of 10 items (Rosenberg 1962). Coopersmith's inventory and Rosenberg's Self-Esteem Scale became the two most widely used tests as the study of self-esteem gathered pace during the 1960s. The construction of these and of other tests allowed self-esteem to become a variable and part of the increasingly entrenched methodological language of psychology. Often used as predictor variable, self-esteem became available as a means to highlight profound and widespread relations between it and a range of social phenomena. Self-esteem tests allowed the identification of deficit and excess, the grading of all, and the kinds of investigation that could establish its value as John Vasconcello's vaunted "social vaccine."

Conclusion

Most psychological theory would seem to require some concept of what Gordon Allport referred to as a determining tendency, be it goal or instinct or need or motive or reward or something else. What impels us to act and to act in socially valued ways is not a modern concern, but self-esteem has become a favored candidate in explaining our will to act according to social demands and expectations. The pleasure experienced from the esteem of others and the pleasure gained from thinking well of our selves are also issues of longstanding interest. While they are viewed in largely positive ways in the twenty-first century, in early writings on self-esteem, this agreeableness was viewed with deep suspicion. For seventeenth-century commentators, allowing that self-esteem was "natural" was no protection against it as a source of sin. Writers frequently complained of the dangers of excess. Nevertheless, some early writers, such as Milton and Astell, saw it as a welcome prompt to act in virtuous ways and as a link between socially valued action and self-scrutiny. With the emergence of more secular accounts of moral rectitude, the reservations about excessive self-esteem remained but its agreeableness was viewed with less suspicion, and as evolutionary theory gained ground, so too did the idea of self-esteem as something that enhanced the survival prospects of the individual and the cohesion of the group. The need for self-esteem was now regarded as something fundamental to survival and the pleasure secondary. It was with the rise of such formulations that deficit in self-esteem gradually became viewed as at least as dangerous as excess.

Implicitly or explicitly, many authors implied that possessing the appropriate amount of self-esteem was beneficial. However, discussion more often focused on the threats posed by possessing inappropriate amounts, and characterizations of it took a familiar historical path of transformation from sin to vice to mental illness to mental well-being. In early writings, the "right amount" was about whether the esteem was merited or not, but gradually discussions became more concerned with whether it was beneficial to individual survival and success and eventually to mental well-being. The notion of avoiding excess and deficit connected self-esteem to familiar and valued narratives around cultivation of character and the restraint of natural urges. In these discussions of excess and deficit, the implied point at which self-esteem was at its optimum was simultaneously vague and fine, and terribly easy to miss. The result was that deficit or excess were the norm and were easily linked, at least in theory, to a vast array of mental health and social problems. Working back from a host of problems – be it mental illness, poverty, crime, prejudice, or educational failure – it was never difficult to construct them in such a way that the wrong amount of self-esteem both contributed to and resulted from them.

By the early twentieth-century, self-esteem was a recognized if minor concept in the emerging empirical psychology in the USA and Great Britain. The much more prominent concept of the self was firmly established, and apart from marginalization during the behaviorist period, it was an accepted psychological topic. It was, however, a desperately complex concept that was not easily reduced to a number. Self-esteem by contrast was a simpler concept and aided by developments such as James's formula and the growth of psychological testing more generally, it offered the possibility of measurement, an attractive quality in a discipline increasingly wedded to quantification. The pressures to measure self-esteem also arose from concerns to which it seemed relevant, extending from the efficacy of therapy to the desire to assess any psychological damage inflicted by social inequalities. The development of tests in relation to mental health soon expanded from tests intended to detect deficiency to tests that graded all, something encouraged by the growth of the idea of self-esteem as integral to personal development and burgeoning ideas around satisfying psychological needs, maximizing self-esteem, and achieving self-realization. In psychological research, self-esteem became a measurable entity that was frequently taken as a proxy for the self and personal well-being.

The threats and opportunities offered by self-esteem were never seen as wholly individual. In the twentieth century, in the context of social concerns linked to groups, and with psychology as a strengthening academic discipline and source of expertise, self-esteem became strongly associated with matters of group membership and relations. Understood as a mediator between individuals and society and between the language of values and the language of science, self-esteem was a flexible concept readily recruited to different debates and contexts, an entity that was both universal and responsive to historical changes. Its developing association with the values of aspiration, achievement, adjustment, and self-improvement made it a psychological concept that chimed with the demands and values of liberal democracies. While it is tempting to treat self-esteem and its management as solely the responsibility of the individual, and another example of how in liberal democracies governance of the self has been made largely the responsibility of the individual, the cultivation of self-esteem was never seen as a wholly individual affair. Instead, psychologists argued that particular social arrangements beyond the power of the individual to control were responsible for levels of self-esteem, and, although contentious, this formulation brought self-esteem once again into contact with discussions of values, rights, and entitlements.

References

Ackerman NW (1945) Some theoretical aspects of group psychotherapy. Sociometry 8(3/4):117–124
Adler A (1917) The neurotic constitution: outlines of comparative individualistic psychology and psychotherapy (trans: Gluck B, Lind J). Moffat, Yard & Co., New York
Adler A (1925) The practice and theory of individual psychology. Kegan, Paul, Tench, Trubner & Co, London
Adler A (1927) Understanding human nature: a key to self-knowledge. W. Beran Wolfe trans, Greenberg, New York
Adler A (1930) The science of living. Allen & Unwin, London
Ansbacher HL, Ansbacher RR (1964) The individual psychology of Alfred Adler. Harper & Row, New York
Astell M (1697) A serious proposal to the ladies for the advancement of their true and greatest interest by a lover of her sex, Part 2. Printed for R. Wilkin, London
Bain A (1865) The emotions and the will, 2nd edn. Parker and Son, London
Baumeister RF, Campbell JD, Krueger JI, Vohs KD (2005) Exploding the self-esteem myth. Sci Am 292(1):84–91
Baxter R (1675) A treatise on self-denial. Nevil Simmons, London
Bay C (1982) Self-respect as a human right. Hum Rights Q 4:53–75
Brassard MR, Germain R, Hart SN (1987) Psychological maltreatment of children and youth. Pergamon, New York
Brauer S (1996) Protestantism. In: Hillebrand HJ (ed) Oxford encyclopedia of the reformation. Oxford University Press, Oxford, pp 357–359

Brennan G, Pettit P (2004) The economy of esteem: an essay on civil and political society. Oxford University Press, Oxford

Brummelman E, Thomaes S, Sedikides C (2016) Separating narcissism from self-esteem. Curr Dir Psychol Sci 25:8–13

Burnham JC (1988) Paths into American culture: psychology, medicine and morals. Temple University Press, Philadelphia

Butler J (1726) Fifteen sermons preached at the Rolls Chapel. J and J Knapton, London

California Task Force to Promote Self-esteem and Personal and Social Responsibility (1990) Towards a state of esteem. California State Department of Education, Sacramento

Carlyle T (1831) Sartor Resartus: the life and opinions of Herr Teufelsdrockh. Chapman and Hall, London

Clark KB (1955) Prejudice and your child. Beacon Press, Boston

Coffin E (1619) A refutation of M. Joseph Hall his apologeticall discourse, for the marriage of ecclesiasticall persons directed unto M. John Whiting &c. English College Press, Saint Omer

Collini S (1985) The idea of 'Character' in Victorian political thought. Trans R Hist Soc 35:29–50

Combe G (1841) The constitution of man in relation to external objects, 5th edn. MacLachlan Stewart & Co Ltd, Edinburgh

Combe G (1853) A system of phrenology, 5th edn. Maclachlan & Stewart, Edinburgh

Cooper T (2016) John Owen, Richard Baxter and the formation of nonconformity. Routledge, London

Coopersmith S (1957) Self-esteem as a determinant of selective recall and repetition. PhD thesis, Cornell University

Coopersmith S (1959) A method for determining types of self-esteem. J Abnorm Soc Psychol 59(1): 87–97

Cooter R (1984) The cultural meaning of popular science: phrenology and the organization of consent in nineteenth century Britain. Cambridge University Press, Cambridge, UK

Cruikshank B (1993) Revolutions within: self-government and self-esteem. Econ Soc 22:327–344

Danziger K (1997a) Naming the mind: how psychology found its language. SAGE, London

Danziger K (1997b) The historical formation of selves. In: Ashmore RD, Jussim L (eds) Self and identity: fundamental issues. Oxford University Press, Oxford, pp 137–159

Darwall SL (1977) Two kinds of respect. Ethics 88(1):36–49

Durkheim E (1925/2012) Moral education: a study in the theory and application of the sociology of education. Dover Publications, New York

Eekelaar J (1986) The emergence of children's rights. Oxf J Leg Stud 6:161–182

Engel J (2008) American therapy: the rise of psychotherapy in the United States. Gotham Books, New York

Evans J (1695) Some thoughts on the character of Solomon's Vertuous woman. Sam Crouch, London

Fallon SM (2007) Milton among the philosophers: poetry and materialism in seventeenth-century England. Cornell University Press, Ithaca

Foucault M (1988) Technologies of the self. In: Martin L, Gutman H, Hutton P (eds) Technologies of the self: a seminar with Michel Foucault. University of Massachusetts, Amherst, pp 16–48

Fowler OS (1848) Religion, natural and revealed: or, the natural theology and moral bearings of phrenology and physiology. Fowler, New York

Froula C (1983) When eve reads Milton: undoing the canonical economy. Crit Inq 10(2):321–347

Galvagni E (2020) Hume on pride, vanity and society. J Scott Philos 18(2):157–173

Glenn SA (2006) The vogue of Jewish self-hatred in post-World War II America. Jew Soc Stud 12: 95–136

Harris JA (2015) Hume: an intellectual biography. Cambridge University Press, Cambridge, UK

Hart SN (1991) From property to person status: historical perspective on children's rights. Am Psychol 46(1):53–59

Hill B (ed) (1986) The first English feminist: reflections upon marriage and other writings by Mary Astell. St. Martins Press, New York

Hume D (1739) A treatise of human nature. John Noon, London
Hume D (1751) Enquiries concerning the principles of morals. A. Millar, London
Jackson JP Jr (2004) The scientific attack on Brown v. Board of Education, 1954–1964. Am Psychol 59(6):530–537
James W (1890) Principles of psychology. Henry Holt & Co, New York
Jordan M (2016) Milton and modernity: politics, masculinity and paradise lost. Springer, New York
Kardiner K, Ovesey L (1951) Mark of oppression: a psychosocial study of the American negro. Norton, New York
Leary MR (2005) Sociometer theory and the pursuit of relational value: getting to the root of self-esteem. Eur Rev Soc Psychol 16:75–111
Lewin K (1933/1986) "Everything within me rebels": a letter from Kurt Lewin to Wolfgang Kohler. J Soc Issues 42:39–47
Lewin K (1941) Jewish self-hatred. Contemp Jewish Rec 4(1):941–950
Lewin K (1946) Action research and minority problems. J Soc Issues 2(4):34–46
Lewin K, Lewin GW (1948) Resolving social conflicts; selected papers on group dynamics. Harper & Row, New York
Long HH (1935) Some psychogenic hazards of segregated education of Negroes. J Negro Educ 1:336–350
Maisel AQ (1946) Bedlam 1946: most U.S. Mental Hospitals are a shame and a disgrace. Life Mag, 6th May, pp 102–118
Maslow AH (1937) Dominance-feeling, behavior, and status. Psychol Rev 44(5):404–429
Maslow AH (1943a) The authoritarian character structure. J Soc Psychol 18:401–411
Maslow AH (1943b) A theory of human motivation. Psychol Rev 50(4):370–396
Maslow AH (1954) Motivation and personality. Longman, New York
Milton J (1641) The reason of church-government urg'd against prelaty. John Rothwell, London
Milton J (1642) An apology against a pamphlet call'd A modest confutation of the animadversions upon the remonstrant against Smectymnuus. John Rothwell, London
Milton J (1667) Paradise lost: Aorth poem written in ten books. Peter Parker, London
Mruk CJ (2006) Self-esteem research, theory, and practice: toward a positive psychology of self-esteem. Springer, New York
Norris J (1687) A collection of miscellanies consisting of poems, essays, discourse and letters. John Crossley, Oxford
Orth U, Erol RY, Luciano EC (2018) Development of self-esteem from age 4 to 94 years: a meta-analysis of longitudinal studies. Psychol Bull 144(10):1045–1080
Rhea JT (1997) Race pride and the American identity. Harvard University Press, Harvard
Rogers CR (1944) Psychological adjustments of discharged service personnel. Psychol Bull 41(10):689–696
Rogers CR, Dymond RF (eds) (1954) Psychotherapy and personality change. University of Chicago Press, Chicago
Rose NS (1990) Governing the soul: the shaping of the private self. Routledge, London
Rosenberg M (1962) The association between self-esteem and anxiety. J Psychiatr Res 1:135–152
Samelson F (1978) From "race psychology" to "studies in prejudice": some observations on the thematic reversal in social psychology. J Hist Behav Sci 14(3):265–278
Scheff TJ, Fearon DS Jr (2004) Cognition and emotion? The dead end in self-esteem research. J Theory Soc Behav 34(1):73–90
Scott DM (1997) Contempt and pity: social policy and the image of the damaged black psyche 1880–1996. University of North Carolina Press, Chapel Hill
Smith R (1992) Inhibition: history and meaning in the sciences of mind and brain. University of California Press, Berkeley
Spencer H (1855) Principles of psychology. Longman, Brown, Green and Longmans, London
Spencer H (1904) An autobiography. Appleton & Co, New York
Spurzheim JG (1815) The physiognomical system of Drs. Gall and Spurzheim: founded on an anatomical and physiological examination of the nervous system in general, and of the brain in

particular; and indicating the dispositions and manifestations of the mind. Baldwin, Cradock and Joy, London

Sully J (1876) Self-esteem and self-estimation. Cornhill Mag 33:159–171

Sully J (1881) Illusions: a psychological study. Kegan Paul and co, London

Swann WB Jr, Chang-Schneider C, Larsen McClarty K (2007) Do people's self-views matter? Self-concept and self-esteem in everyday life. Am Psychol 62(2):84–94

Tajfel H (1974) Social identity and intergroup behaviour. Soc Sci Inf 13(2):65–93

Van Wyhe J (2004) Was phrenology a reform science? Towards a new generalization for phrenology. Hist Sci 42(3):313–331

Wahrman D (2006) The making of the modern self: identity and culture in eighteenth-century England. Yale University Press, Yale

Wells LE, Marwell G (1976) Self-esteem: its conceptualization and measurement. Sage, London

Wolberg LR (1943) The problem of self-esteem in psychotherapy. N Y State J Med 43:1415–1419

Young RM (1970) Mind, brain and adaptation in the nineteenth century: cerebral localization and its biological context from Gall to Ferrier. Clarendon Press, Oxford

Vygotsky, Luria, and Cross-Cultural Research in the Soviet Union

44

René van der Veer

Contents

Introduction	1146
The Announcement of a New Study	1147
Pedological Research and Mental Testing	1148
Cross-Cultural Research Within the Soviet Union	1151
Climate Change	1156
Conclusions	1159
References	1161

Abstract

This chapter presents the first overview and analysis of the Russian scientific studies and debates that preceded Vygotsky and Luria's famous study of the thinking of Uzbek people in the early 1930s. It is shown that these earlier studies used similar methods, found similar results, and reached similar conclusions. Moreover, it is shown that such cross-cultural studies of Soviet minorities became increasingly suspect in the eyes of leading scientists and ideologists. The use of mental tests for cross-cultural comparison, in particular, was heavily criticized but gradually any difference found in the thinking of minority groups became politically unacceptable because it seemed to imply Russian chauvinism. In addition, Vygotsky and Luria's theory of cultural development, which they used to interpret their findings, was claimed to be anti-Marxist. This historical background explains why the results and conclusions of Vygotsky and Luria's study, which essentially found that some Uzbek subjects were incapable of abstract thought, were met with harsh criticism and could not be published at the time.

R. van der Veer (✉)
University of Leiden, Leiden, The Netherlands
e-mail: veer5821@gmail.com

© The Author(s), under exclusive licence to Springer Nature Singapore Pte Ltd. 2022
D. McCallum (ed.), *The Palgrave Handbook of the History of Human Sciences*,
https://doi.org/10.1007/978-981-16-7255-2_82

Keywords

Soviet Union · Cross-cultural psychology · Minorities · Vygotsky · Luria · Mental testing

Introduction

In the summers of 1931 and 1932 Vygotsky and Luria organized two psychological studies of the thinking of Uzbek people who were living in a rural area during the period of collectivization. Part of the results of these studies were published with a delay of more than 40 years (Luria 1974, 1976) and have been widely discussed by modern researchers (e.g., Tulviste 1988; Yasnitsky and Van der Veer 2016; Van der Veer and Valsiner 1991). Basically, Luria found that higher cognitive processes change under the impact of social and cultural changes, whereas lower psychological processes (e.g., elementary visual perception) remain unaffected, which confirmed Vygotsky and Luria's theory of cultural development advanced from 1928 onwards (Vygotsky 1928; Vygotsky and Luria 1930). Thus, Uzbek subjects with little schooling seemed unable to assign objects to general categories on the basis of their essential features and focused on their use in practical contexts. Also, these subjects seemed incapable of making logical inferences on the basis of language propositions when the content of the premise did not form part or even contradicted their practical experience. Subjects with more schooling and the rudiments of literacy did better on such tasks and, thus, it seemed that the acquisition of literacy and the general human experience as established in language changes subjects' discursive thinking. In sum, Luria and Vygotsky claimed to have found that little educated, illiterate subjects engage in concrete, practical thinking, whereas schooled, literate subjects become capable of abstract categorical thought.

Cross-cultural psychology seeks to find a path between the unthinking assumption, often embedded in psychological language that "all people think that" or "that's human nature," and an extreme relativism, that no psychological categories have universal human application. It has proved extremely challenging to conduct cross-cultural investigations and to arrive at results commanding authority. While trained psychologists have attempted to devise empirical studies, as they have of color perception and of intelligence, the underlying theoretical and methodological issues are shared across the human sciences. For this reason, historical studies of projects to study the distribution of a psychological category in individuals and groups have interest both for their own sake and for the lessons they may have to offer in an area of research where questions, methods, and results continue to diverge. This chapter, on fieldwork to compare intelligence in different ethnic groups in the early years of the markedly multiethnic Soviet Union, speaks to these issues, and it also adds to the cross-cultural history of testing. It shows that an understanding of what went on in this area of the human science requires considerable knowledge of specific social and political circumstances. In addition, the chapter contributes to knowledge of the history of the celebrated thought of Lev Vygotsky, through his research together with Aleksandr Luria, to a psychology ("cultural psychology") integrating knowledge of individual development and social realities. Vygotsky's biography and research has been much discussed, but the historical detail presented in this chapter, detail about comparative studies preceding Luria's fieldwork in Uzbekistan, is substantially new. The chapter therefore exemplifies historical research at work in psychology.

As said above, Luria published only part of his results and we still have no good idea why he chose to publish some results and not others. It is known from other sources (Zavershnev and Van der Veer 2018), for example, that he also subjected Uzbek adults and children to various intelligence and personality tests (e.g., Kohs block design test and the Rorschach inkblot test) and found results that diverged from the average findings for Russian subjects. What we do know, however, is that in the early 1930s, when Luria's initial findings became known, the reactions were not entirely favorable. A committee of the Workers-Farmers Inspection was installed to study Luria's data and interview the participants of the expeditions and reached devastating conclusions (Razmyslov 1934; Van der Veer and Valsiner 1991). What stuck the members of the committee was that active and politically conscious Uzbek kolkhoz members with little or no education, 15 years after the October Revolution with all its societal changes, were still designated as subjects who were incapable of hypothetical reasoning and categorical thought. Apparently, by 1932 the interpretation of cross-cultural findings had become such a sensitive issue in the Soviet Union that Vygotsky and Luria, when letters of defense to Aleksey Stetskiy, the party official responsible for culture and propaganda (*Kultprop*), and Andrey Bubnov, minister of education (*Narkompros*), failed to elicit a favorable response (Luria 1992, p. 36), deemed it wise not to publish their findings after the committee had reached its negative conclusion.

Luria's findings were indeed highly interesting and controversial and still merit careful analysis. However, this chapter is focused on the scientific and sociopolitical context of Luria's studies. To interpret Luria's investigations, we need to know somewhat more about previous cross-cultural studies of Russian minorities, about their findings, and about the scientific and political debates that ensued. We also need to know whether the methods he and his collaborators used in their comparative research (e.g., mental tests and personality tests) were generally accepted or, on the other hand, hotly debated. Finally, it is instructive to know somewhat more about the status of Vygotsky and Luria's theory of cultural development, which they used to interpret their findings. In sum, this chapter provides historical data that allow us to appreciate various aspects of Luria's investigations. After all, we can only know whether a study was novel when we know something about its predecessors and we can only know whether its results are politically sensitive when we have an idea of the contemporary debates. Finally, we better understand why the results of a study are rejected when we have an idea about the appreciation of the theory that lies behind it.

The Announcement of a New Study

On October 16, 1931, *Science* published a short note by "professor A.R. Luria," which stated that in July of that year the Uzbek Research Institute of Samarkand and the Moscow Institute of Experimental Psychology had organized "the first expedition of the Soviet Union" for the study of psychological characteristics of people in "various stages of cultural development" (Luria 1931a). The same note was published in the *Zeitschrift für angewandte Psychologie* (Luria 1931b) and the *Pedagogical Seminary and Journal of Genetic Psychology* (Luria 1932). Luria explained that the aim of the expedition, which had been prepared by a 2-month seminar conducted by him in Samarkand in May–June 1931, was to investigate the mental processes of people living under "very primitive" circumstances in "very backward" communities with a corresponding low cultural level. Luria also mentioned that the group wished to develop new methods to determine the intellectual level of these people, because the usual methods of determining intelligence were "inapplicable."

Luria's note raises several interconnected questions. Dictionaries agree that the word "expedition" is used to describe a carefully organized journey made to explore an unfamiliar (and possibly dangerous) place. But in the case of an anthropological or psychological expedition the word certainly also suggests a difference in social status between the visitors and the group visited: Few people would say that a group of aborigines visiting London is engaging in an expedition. Luria's whole terminology suggests that he indeed thought along these lines: Members of a superior culture wished to investigate an inferior, primitive group. So, the question becomes what Luria exactly meant by levels of cultural development and also why he believed that the usual methods to determine intelligence could not be applied in these specific groups. To answer these questions, we have to go back in time and to familiarize ourselves with the theory of cultural development elaborated by Vygotsky and Luria, to take a look at the methods of determining intelligence used in the Soviet Union of that time, and to examine previous attempts to study "primitive" people in the Soviet Union. In doing so, we will also answer the question whether Luria was right in claiming that his was indeed the first psychological expedition to take place in the Soviet Union.

Pedological Research and Mental Testing

The discipline of pedology or child studies became popular in Russia around 1900 (Van der Veer 2020) and after the October revolution it was enthusiastically accepted by the new regime. The Soviet government faced the formidable task of educating millions of adults and children who previously had no access to schools and pedology seemed to have the tools to enable or facilitate this process. Pedologists studied the physical and mental development of children in relation to their environment and soon became an integral part of the whole educational system. After all, to see what kind of education was needed one first had to know something about the children's level of development and knowledge. Pedologists rapidly spread over the vast territory of the Soviet Union, investigated and tested the children and adults of various ethnic groups, and found immense differences in their living circumstances, habits, physical condition, body build, and test scores (Byford 2020). The primary instrument of the pedologists, the mental test, became very popular in the mid-1920s and was widely adopted to allocate children to various school types and to select people for jobs. In this respect, the Soviet Union did not differ from many other countries and it was also no novelty that the differences found between various groups within the population (e.g., urban vs rural children; cf. the first issue of *Pedologiya* in 1929) became the subject of scientific and, eventually, sociopolitical discussion. For a number of years, however, these discussions remained relatively mild and researchers were not forced to accept a specific worldview.

In 1924 and 1925, for example, the vocational psychologist Isaak Shpil'reyn and his colleagues studied the language comprehension, vocabulary, etc. of Red Army soldiers (who mostly had enjoyed just a few years of schooling) to check whether they could understand the abstract language of their political instructors (Shpil'reyn et al. 1928). They concluded that "purely formal logic is not suitable for this audience. No syllogisms will be understood and in any case they will not seem innerly convincing when the correctness of the main premise is not obvious to the listeners... abstract examples will always be met by arguments that hold in daily life." For example, to the syllogism "All club members subscribe to the library – Ivanov is a club member – Hence, Ivanov subscribes to the library' the Red Army soldiers objected that Ivanov perhaps hadn't yet found the time to subscribe" (ibid., p. 34). And in a talk for the Academy of Communist Education on December 2, 1929, Shpil'reyn added that Red Army soldiers had difficulty in remembering abstract forms: "A circle is more or less OK. The Red Army soldiers call it a wheel, and a wheel the Red Army soldier remembers. [But] a triangle, a hexagon he does not remember. Such an abstract test form is not suitable for the farmers' youth" (Kurek 1999, cf. Shpil'reyn 1930). Shpil'reyn's results, i.e., the unwillingness of subjects to think in abstract syllogisms and their tendency to refer to abstract geometrical figures with the names of concrete objects resembling them, anticipated Luria's findings by some 8 years. But even in 1929, when Shpil'reyn finally published his results, they were not yet seen as politically damaging, despite the unflattering portrait of the Red Army soldier. This was also true for the cross-cultural studies by Shtilerman (1927, 1928) and Petrov (1928), which raised much more criticism in the early 1930s than at the time they were published.

Shtilerman was a medical doctor working at the school clinic of old-town Tashkent who subjected Uzbek children to the standard pedological investigation, i.e., he measured their body proportions, determined their health condition, living conditions, and schooling environment, and tested their mental capacities with the Binet-Simon and short Rossolimo (Byford 2016; Kurek 2004). It was the measurement of the children's mental development that would draw huge criticism sometime later. Although the tests were adapted to the Uzbek culture in terms of text and pictures, the 164 Uzbek schoolchildren from 8- to 15-years-old tested by Shtilerman obtained rather low scores: Just 17% scored normal, 63% were found to be slightly retarded, and 20% seemed deeply retarded. Moreover, with the short Rossolimo, Shtilerman found that what he called lower mental processes (e.g., attention, observation, and visual memory) differed less from the norm than what he called higher processes (thinking and combination). To explain his findings, Shtilerman turned to environmental factors. He suggested, for example, that the Uzbek infants' prolonged stay (up to 2 years) on the cradleboard inhibited their mental development. That Uzbek girls scored on average higher than boys he explained by pointing to the fact that girls attended boarding schools where their mental faculties were constantly being trained, while boys entered school relatively late, if at all. Shtilerman concluded that it was essential to improve the socioeconomic conditions of the Uzbek subjects and proposed to lower the age norm for the mental tests for this group. That is, he did not so much analyze why children found it difficult to solve specific problems but suggested that in the case of Uzbek children lower sum scores should be sufficient for the average IQ score of 100.

A somewhat similar study was done by Petrov (1928), who investigated children from the Chuvash, a Turkic ethnic minority, in 1926 and 1927. Petrov physically examined 1398 children in the age from 3 to 13 years old and submitted them to the Binet-Simon. Like Shtilerman, he found the children to score suboptimally, i.e., below the European norms. The mental age of some 15% of the younger Chuvash children lagged 2 years behind what was considered normal, while for older children this percentage even increased to 50%. Petrov's explanation of the deviating results and his solution also resembled Shtilerman's proposals: The "backwardness" of Chuvash children was caused by socioeconomic conditions, and schooling, in Petrov's case sending the most backward children to special schools, would solve the problem. Meanwhile, the norms of mental tests should be lowered for the children who scored suboptimally.

The results of Shtilerman, Petrov, and others (e.g., Shirokova-Divaeva 1927; Shishov 1927, 1928; Solov'ev 1929) highlighted a general problem in the interpretation of interethnic differences (cf. Byford 2016): Whereas it seemed totally acceptable to adjust the norms for different national or ethnic groups in the case of anthropometric measures (e.g., chest or skull circumference), and growth and weight curves, this seemed more problematic for the norms for mental development. Accepting lower IQ scores on mental tests seemed equivalent to accepting that various groups differed in intelligence, whereas using different norms for various groups seemed to violate the principle that the same property should always be measured using the same units (e.g., length in meters). In itself, this was not a problem specific for the Soviet Union; in the 1920s, throughout the world researchers were testing various groups and many did not shy away from attributing differences in average mental test scores to genetic between-group differences. In their view, some groups were on average simply less tall and intelligent than others and it seemed self-evident that Western Europeans score highest in all positive respects. Others rejected this view and argued that something was fundamentally wrong with the cross-cultural use of mental tests (cf. Gould 1981; Van der Veer 2007, pp. 75–78; Van der Veer 2020).

One of the first who tried to formulate a provisional answer to these specific questions and who addressed the problem of the pedological study of national minorities at large was Vygotsky (1929). In a talk that sketched a plan for the pedological study of minorities in the next 5 years, Vygotsky stated that national minorities are different from Russians, stand on an economical and cultural lower level, and will have to make "a grandiose leap on the ladder of their cultural development and skip a whole series of historical levels" to reach the Russian level. The fact, however, that minority children are culturally backward does not imply they are mentally backward, and mental tests that suggest otherwise are wrong. Here Vygotsky, without mentioning his name, referred to Shtilerman's study which found that high percentages of Uzbek children were slightly or deeply retarded. In his view, Shtilerman rightly attributed the results to pedagogical neglect and inhibiting influences in the child's environment and was also right in adjusting the content of the test to make it more understandable. But this was not enough, because the problem went deeper. It is not sufficient to establish what minority children lack, we must study the particular expression of general laws in their specific cultural environment and in order to do that the local pedological centers must gather massive data and pedological expeditions should become as normal as field work in ethnology (Vygotsky 1929, p. 374). Using nonadapted tests in such research leads to absurd results: Whole nations are classified as 5- to 7-year-olds. All European countries have adapted the Binet scale and this is a good first step. But what we should do, Vygotsky argued, is to leave the tests temporarily aside, because the test method rests upon a number of assumptions that do not necessarily hold in other cultures (ibid., p. 375). First, we must study the structure, dynamics, and content of the child's social environment, which determines their means of thinking and behavior: Muslim children cannot be expected to be able to draw as European children, and people who never saw a pencil will show a delay in writing. Cultures develop and suppress different genetic capacities and create unique sociopsychological types of children. Second, we must study the cultural or historical development of mind and behavior. In this respect, Vygotsky mentioned that studies such as Shtilerman's confused cultural primitivism with mental deficiency. Third, we must clarify the role of racial biological properties, "which no doubt exist and exert their influence on development" (ibid., p. 377).

All in all, it seems that Vygotsky's talk contained sound scientific advice: Study the minority culture in detail, realize that cultures promote and inhibit different capacities in children, understand that "bad answers" do not always reflect low intelligence, and see that Western tests are based on hidden assumptions (e.g., that children will try their best; that it makes sense to them to respond to the questions of a person who evidently already knows the right answer; and that it is not impolite to give answers). But, as we will see, Vygotsky's proposal to study ethnic minorities proved not radical enough for the later critics of cross-cultural research.

Cross-Cultural Research Within the Soviet Union

When Luria (1931a) internationally announced his psychological expedition to Uzbekistan, he suggested this was the first Soviet expedition of its kind and also in his later accounts (Luria 1974, 1976) he never referred to previous Russian attempts to study the mode of thinking or intelligence of Soviet minorities. Luria's expedition was indeed unprecedented because of its scale (the large group of psychologists who participated, the great number of aspects of psychological functioning being investigated) and because of its explicit goal to compare different groups within a minority. However, psychological expeditions to distant regions of the Soviet Union, including Uzbekistan, had been undertaken before and much empirical work had already been done which anticipated both Luria's methods and results. The leading journal *Pedologiya*, with Vygotsky as a member of the editorial board, and other journals regularly published the findings of pedologists working in various parts of the Soviet Union (e.g., Kapusto 1928; Rybnikov 1928; Efimov 1931b; Leventuev 1932) and cross-cultural papers (e.g., Ostrovskiy 1929; Solov'ev 1929). Moreover, in 1930, *Pedologiya* published a special issue about the "pedology of the national minorities" with 11 contributions. Most of these contributions dealt with the study of the cognitive development and way of life of non-Russian ethnic groups in remote areas of the Soviet Union and thus were directly relevant for Luria.

Luria was, of course, well aware of this research and in many cases knew the participants personally. Take, for example, the two expeditions that were undertaken in the Summer of 1929 under the supervision of the psychologist Anna Mikhaylovna Shubert (1881–1963) of the Institute of Educational Method and Practice in Moscow (Khronika 1930). Shubert was an expert in mental tests for children, who had published brochures and books on mental tests and had adapted them for the Russian population (Shubert 1913, 1922, 1923, 1924, 1926a, b, 1928; Bukhgol'ts and Shubert 1926). She also published on the drawing abilities of minority (Oyrotsky or Altay, Tungus or Evenki) children, for example, in a paper in *The Pedagogical Seminary and Journal of Genetic Psychology* (Schubert 1930), which was accepted for publication in April 1929 by Luria himself. The accounts of the two expeditions were published in five papers in the special issue of *Pedologiya*, preceded by an introductory paper by Shubert (1930a). In that paper, Shubert explained that one expedition (with Shepalova, Usova, and Bulanov) went to the Northern Baykal region to study Tungus children, while the other went to the Altay region to investigate Oyrotsky children. Enormous difficulties had to be overcome: No foreign literature, except for a brief guide by Thurnwald (1912), was available, no Soviet anthropological data was at hand, enormous distances had to be covered (in part on horseback), just July and August were available, the budget was very tight, and the languages not known. In view of these difficulties, it comes as no surprise that experienced pedologists could not be found and in the end just five students of pedology left Moscow for the adventure of their life. Shubert explained that the group decided to be flexible and modest and that the results should be taken with a grain of salt given the inadequate preparation and the local peculiarities (e.g., Shubert mentioned, for example, that for the Oyrotsky it is good taste to reply "I don't know," when asked a question).

The expedition to the Northern Baykal region landed more than 5.500 km from Moscow on the northeastern shore of Lake Baykal, in Nizhneangarsk, which the students described as a humid, unhealthy village. The local population, the Tungus, consisted of three groups, of which two were nomadic. The mountain Tungus could not be reached as it required another trip of 180 km on horseback and swimming across five rivers (Shepovalova 1930, p. 173). Investigating the children of the other two groups was fraught with difficulties: The students did not speak the language and had to learn their questions and some standard answers by heart, the children were quickly bored by the test questions (ibid., p. 183) and regularly ran away, the interviews were conducted in the yurt with other family members present, etc. Shepovalova (1930) described the living conditions of the Tungus in some detail. Infant mortality rate was (IMR) estimated at 50%, infants were breastfed for 2 years and sometimes switched to tobacco immediately afterwards (ibid., p. 178). Food consisted largely of meat, dairy products, and large amounts of tea, and the hygienic circumstances and habits were poor. Remarkably for the time, Shepalova openly wrote that, in particular, the mountain Tungus disliked the Russians (ibid., p. 185). All in all, she described the Tungus as a peaceful group who treated their children lovingly.

Shepalova's fellow student Usova (1930) focused on schooling and literacy in the region. There were just two schools in old and dirty buildings without enough facilities. Just 7% of the school children were Tungus: Many parents resisted school as something Russian (as Usova put it, they did not yet understand that "the interests of all workers are the same," p. 192) and their children often played truant. However, when Tungus children attended school, they rapidly caught up with the Russian children and proved perfectly capable of learning, arithmetic being among their favorite subjects. Usova recommended teaching the Tungus in small groups and mixed with Russian children. This way they would widen their horizon and lose their antagonistic feelings.

Finally, Bulanov (1930) focused on the cognitive skills and knowledge of the Tungus children. He first probed their general knowledge, finding, for example, that they knew the origin of butter and bread, had an exquisite knowledge of the local rivers, but were less familiar with technological inventions, such as radio and train, and did not know the names of exotic animals shown to them on pictures. Their knowledge of ideology also left something to be desired: About half of the children knew that the USSR was a "good power," many knew that "the bourgeois do not work," and 89% identified the portrait of Lenin. However, they had no idea about the Revolution and the Communist Party (Bulanov 1930, p. 199). Bulanov also administered three mental tests (Binet, Rossolimo, and Pintner) to several children. On the Binet test they scored between 65 and 80, and Bulanov suggested that this test depends too much on school culture. Probably, he hypothesized, they would do much better on a test adapted to their culture. In addition, the children were not used to the testing situation and became quickly bored. More in general, Bulanov warned against international comparisons of mental test results, also because we have no idea how children arrive at their answers, as Zaporozhets (1930) also remarked (ibid., p. 204). In conclusion, Bulanov attached very little value to the low scores obtained by Tungus children, because to him they seemed lively, skilled, social, and perfectly normal.

The second expedition went to Biysk in the Oyrotsky Autonomous Oblast in Southern Siberia, northeast of Kazakhstan. This was a journey of more than 5000 km by train and 100 km on horseback. Golubeva (1930) focused on the daily life and mentality of the local people, which consisted of various groups with a Turkish background. It is obvious that she was appalled by the living conditions of the Oyrotsky people: All yearlong the families lived in incredibly dirty yurts, which in wintertime they shared with animals; they never washed themselves and wore the same dirty clothes until these fell apart. IMR was very high and children of 3 or 5 years old sometimes already smoked. Boys and girls were hardly distinguishable: They wore the same clothes and had their head shaven. Schools were of low quality and the local children attended school very irregularly. Like in the Northern Baykal region, the Oyrotsky population had no knowledge of Western ideas about proper food intake (e.g., the advantage of eating fruit and vegetables) and largely survived on meat, tea, and dairy products. Overall, Golubeva described the Oyrotsky as culturally backward (e.g., many were illiterate) but very friendly and social.

It was Zaporozhets (1930) who focused on the application of mental tests to children living on this "economically and culturally low level." Fifty-two children were investigated with Shubert's version of the Binet-Simon and the results were as follows: Four children were excluded as being mentally retarded and the others scored an average of 67 (SD = 8.5). Zaporozhets noted that this was significantly below the Western average but that we should be very careful in interpreting these numbers. The testing took place under difficult circumstances (in the yurt with all the family), with children who refused to answer or tried to run away, and took up to 4 h, leaving both experimenter and child exhausted. The children failed to comprehend abstract verbal questions, did not understand why a stranger wanted to know these things, and the instruction had to be repeated frequently. Asked to define objects they referred to their use (e.g., a knife is "a thing to cut with") and abstract questions were constantly turned into concrete ones (e.g., "What would you do if you broke another person's thing?" – "Which thing?" "Whose thing?"; "What should you do before you do something important?" – "Which important thing?"). Children hardly noticed contradictions in the verbal-logical plane. Confronted with a contradictory statement (e.g., "A flood reached the village but nothing serious happened. Just 60 people were killed.") they refused to accept it ("People would run away") or came with arguments ("The water cannot reach the village"). The Pintner-Paterson test, a nonverbal intelligence test for children largely consisting of wooden jigsaw puzzles (Pintner and Paterson 1917), produced somewhat better results (Zaporozhets 1930, p. 226) and the short Rossolimo, a profile test for 11 cognitive capacities, yielded mixed results: The subtest measuring voluntary attention caused considerable difficulty (just 3 of 50 children obtained a sufficient score) and the description of pictures produced very strange results, because the children did not recognize the circumstances (e.g., a picture of a man in a prison cell was described as "Lenin in his cabin"). Zaporozhets concluded that we face a methodological problem: The same children who seem totally normal in daily life score low on Western mental tests. To interpret this finding, he turned to Vygotsky's theory of cultural development: The problem of the Oyrotsky children is that their cultural development is somewhat backward. They have a phenomenal memory for local things but have difficulty memorizing things at will; they have lots of energy but it is difficult for them to engage in sustained work without external stimulation; the mnemonic tool of writing is hardly developed because of the socioeconomic circumstances and hence their cultural memory is less developed (ibid., p. 229); they rely on primitive counting methods (e.g., notches in wood), tend to think concretely, and fear to pronounce "dangerous" words like "bear," which points to the mixing up of sign and meaning (ibid., p. 231). In sum, the children do not just lack academic knowledge, but, rather, their mind seems to function differently. The whole picture led Zaporozhets to doubt Blonskiy's (1930[1925] p. 113) claim that the Binet could be applied everywhere: It is not just a matter of replacing items by more familiar ones; we cannot just lower the standards because of the lack of school knowledge. The whole point is that these children think in a qualitatively different way and although under the guidance of the Soviet national policy the children now made rapid progress which they otherwise would not have made in centuries, we should take their particular way of thinking into account.

Thus, the results of Shubert's two expeditions to two remote regions of the Soviet Union seemed to confirm the picture that the local inhabitants of these regions obtain low scores on the standard intelligence tests. However, the students were inclined to blame the tests and to explain the scores by referring to local cultural peculiarities and inadequate schooling. Zaporozhets, in particular, submitted the possibility of an inadequate cultural development as advanced by Vygotsky. In no way did the students believe that the members of the primitive societies they visited were innately less intelligent or gifted than Russians.

Other accounts published in the special issue of *Pedologiya* in 1930 largely confirmed these findings. Granat and Zagorzhel'skaya (1930) visited the Buryats, another ethnic group living near lake Baikal. They found the same living conditions as Shepovalova (1930): People lived in yurts with cattle and large numbers of insects, personal hygiene did not exist (the authors claim they met adolescents who had never washed since birth), and diet was very one-sided. Nevertheless, the Buryats proved surprisingly healthy and obtained a normal score on the short Rossolimo (Granat and Zagorzhel'skaya 1930, p. 254). Lavrova-Bikchentay (1930) and Bikchentay and Karimova (1930) studied Tatar children living in Moscow in 1927 and 1928. They measured their body proportions in various ways, using Blonskiy's indices (cf. Van der Veer 2020), and tested the children using various adapted Binet tests. In a sample of 380 children from five schools the IQ varied from 82–99 and was believed to depend on the profession of the parents, school attendance, and other social circumstances. Baranova (1930a, b) pointed out that mental test results reflect both innate abilities and cultural background and argued that great care is therefore needed in applying such tests in other cultures. She mentioned, among several other things, that Uzbek or Kazak children may not be familiar with the use of perspective in drawings or obtain low scores on color tests because they have one word for both yellow and blue. To express their performance on mental tests in one number is misleading, Baranova argued; we always need careful qualitative analysis to understand the results.

In her concluding overview of the international literature, Shubert (1930b) raised the question whether various ethnic groups might differ in mentality. With Thurnwald, she suggested that some groups might be less able to organize their behavior for longer periods of time (hence their distractibility and spontaneity) and with Thurnwald, Boas, and Wertheimer, she posited that the difference between various ethnic groups might reside in the "instruments, tools, cultural auxiliary means of thinking, such as language, writing, the counting system, etc., i.e., in the ways they utilize their natural thinking mechanisms." Here she referred to the "excellent" explanation of this view in Vygotsky and Luria's *Studies on the history of behavior: Ape, primitive, and child* (1930). That did not yet explain the cultural differences and Shubert went on to discuss a great number of ethnographic studies (e.g., Lévy-Bruhl, Mead, Porteus, and Thurnwald), the use of verbal and performal tests in studying intelligence, the need to study not just intellect but temperament and character as well, etc., without reaching definitive conclusions.

Climate Change

The many minority groups, subcultures, and nations constituted a problem for the government of the Soviet Union from its inception. The problem was to find a middle way between monolithic unification and extreme diversity, or, as Stalin put it at Twelfth Party Congress in 1923, between "great power chauvinism" and "nationalism." Stalin was initially inclined to see the first option as the greater evil and for years no strong pressure was exerted on minorities to adopt a common (read: Russian) culture as long as one adopted the communist worldview. Indeed, the policy of *korenizatsiya* (nativization) encouraged or even prescribed the promotion of local languages, with the ultimate goal to integrate non-Russian nationalities in the government of their Soviet republics. Of course, this was a lengthy process with scores of problems and it is fascinating to see how the Russian officials wrestled with the question as to what was acceptable and what was inacceptable in the minority cultures. Language was one thing, but religion quite another, and remnants of a bourgeois mentality (e.g., exploitation and unequal income distribution) had, of course, to be eradicated (e.g., Dimanshteyn 1929). Gradually, however, the Party tightened its grip on the scientific debates and the right answers to the sensitive dilemmas became increasingly prescribed. Whereas the pedologists and vocational psychologists involved in measuring and testing children were first relatively free to advance their hypotheses to explain interethnic differences and could, for example, refer to differences in genetic makeup, this now became anathema. Explanations based on children's socioeconomic background became increasingly mandatory and the existence of differences in mental makeup was called into doubt.

As often, it was the leading pedologist Aron Zalkind who set the tone. At the First All-Union Congress for the Study of the Behavior of Man, held from 25 January to 1 February 1930, he boldly stated that if Western tests show children in Turkestan to be idiots, it is because these tests were idiotic themselves (Zalkind 1930). Western children would presumably obtain low scores on test developed in Turkestan, Zalkind stated, and he added some phrases about the suppression that took place in bourgeois colonies and the necessity to use a social class approach. When Stalin – during a visit at the Institute of Red Professors in 1931 – called for the intensified struggle with the distortions of Marxism in the social sciences, the debate became increasingly vicious (e.g., Gur-Gurevich 1931). Pedology and psychology were threatened by, on the one hand, "idealism" and "mechanicism" on the other. Soon researchers engaged in a process of criticism and self-criticism where it was often difficult to separate scientific from ideological arguments and both past and contemporary research were subjected to scrutiny (Rezolutsiya 1931).

Efimov (1931a) now attacked Petrov's (1928) study mentioned above and claimed that Petrov was still infected by the tsarist spirit. Literacy in Chuvash subjects had grown from 20% to 85% in the period from 1917 to 1931 and Chuvashiya now prided itself with two higher Institutes. Hence, to call Chuvash children "backward" just betrayed Petrov's prejudice. Moreover, Petrov's attempt to explain the differences by referring to the different social-economic conditions was mistaken: These were created by the Soviet regime under the leadership of the Communist Party and, thus, Petrov objectively accepted the viewpoint of the counterrevolution.

Leventuev (1931), a participant in Luria's first expedition, fulminated against an Uzbek schoolmaster who had used a survey from the Leningrad Herzen Pedagogical State Institute in his school. The survey was meant to reveal the children's worldview and asked the children, among other things, who were the "best" people (minority) and why, and whether they wanted to be rich or believed in God. Of course, the children turned out to have several "prejudices" and a commission was installed to investigate the issue. In Leventuev's view such surveys did immense harm to the highly suggestible children, and he pleaded for much more political control on pedological research to prevent such harmful practices.

One year later, Leventuev et al. (1932) retrospectively attacked several of the cross-cultural studies mentioned above, arguing that the spread of socialist culture meets with vicious resistance by chauvinism and local nationalism. Shirokova-Divaeva (1927) was accused of "great power chauvinism," because measurements with the Pignet index in her view showed that the Uzbeks constituted a "weak" group. Shtilerman (1928) was attacked because he made the Uzbek children seem like "a mass of idiots" (Leventuev et al. 1932, p. 48), and Baranova (1930a, b) was criticized for her suggestion that Uzbek children have problems with the perspective used in the images in mental test. In the view of Leventuev and his colleagues, this showed she believed the Uzbek children had some organic defect, which again suggested chauvinism and racism. After some equally dishonest remarks about the work of Bikchentay, Leventuev concluded with the remark that all this was terrible given that the Party had decided to liquidate the classes and cultural inequality.

In sum, by 1932 the scientific journals were filled with highly political accusations and self-accusations, and monographs and textbooks (e.g., Gur'yanov et al. 1930) that went unnoticed several years before now became the subject of heavy and often unfair criticism. The interpretation of mental test results in pedology and vocational psychology was hotly debated and drawing inferences from cross-cultural or interethnic differences had become a political issue. In this climate it took some courage to organize a cross-cultural study, all the more so as Vygotsky and Luria's theory of cultural development, which lay at the basis of Luria's expeditions, did not escape criticism either (Van der Veer and Valsiner 1991).

Talankin (1931a, b) warned that the "group of Vygotsky and Luria" uncritically accepted foreign theories and claimed that their concepts of "instrument" and "culture" were decidedly un-Marxist. After having criticized his own Freudian views, Zalkind (1931, p. 13), the main editor of *Pedologiya*, noted that "the serious critical assessment of the works of L.S. Vygotsky and A.R. Luria has begun. These comrades should not wait for 'attacks' and are invited to reconsider their very grave mistakes as a form of proactive self-criticism on the pages of our journal." To leave no room for doubt, Zalkind added in a footnote that "comrade Luria's psychological specialization does not exempt him from his responsibility before pedology." In the next issue, Mukovnin (1931, p. 80) mentioned that Vygotsky and Luria were slow to admit their mistakes. A committee had visited the Psychological Institute and found their "so-called theory of cultural development" wanting. It was a theory that did not take social class into account and its authors did very little to fight behaviorism, *Ganzheitspsychologie*, Freud, and Adler. Also, Vygotsky and Luria's ideas suffered from academism and had little relevance for praxis. Several issues later, Bolotnikov et al. (1931), in a longer article about the "situation at the pedagogical front," argued that Vygotsky's combination of "behaviorism, reactology, and the basically idealist Gestalt theory" was a clear example of "eclecticism." In that same issue, Bikchentay (1931) and Kostin (1931) followed suit. Kostin just affirmed that Vygotsky neglected the leading role of the collective in personality growth. Bikchentay, who with his wife himself was involved in the study of Tatar schoolchildren (Bikchentay and Karimova 1930; Lavrova-Bikchentay 1930), voiced more explicit criticism, although he did not mention Vygotsky by name. Bikchentay noted that 80 languages were spoken in the Soviet Union and that pedological knowledge was largely based on the study of Russian children. Meanwhile, he said, we did not know enough about the ways to develop children into the new builders of a communist society. We must not study "memory," said Bikchentay (1931, p. 32), but children's readiness to join in the building of socialism. "We are not so much interested in the level of the primitive on the biological or even the historical ladder (all bourgeois researchers write about this). We are interested in the level of the former hunter in socialist production… we do not need tests for Tatar or Chuvash children, we need yardsticks to establish the productive level of children" (ibid., p. 33). This was no easy thing, however, and different pernicious foreign theories had to be avoided. Some Soviet theorists, for example, adhered to evolutionary theories, placing "savages" between ape and "cultural man." Such ideas, inspired by Spencer and Tylor, are racist and reactionary. Others, said Bikchentay, advanced what he called "historico-culturological ideas," in which primitive people are only capable of imitation and prelogical thinking in the spirit of Lévy-Bruhl. As Bikchentay noted, "several ethnologists and psychologists who wrote '*Studies*' and '*Outlines*' worked in this direction" (ibid., p. 35). Still others belonged to the "culturological" current based on the ideas of Spranger. This current "strongly influenced our psychologists who study cultural memory, cultural attention, etc." Finally, we may discern the historico-labor approach, which does not study the psychology of the kulak or the kolkhoz member but the influence of the environment on backward people. This was equally wrong, Bikchentay argued, because "we must remember that all regions of the USSR have already entered the period of socialism. There is no feudalism, no capitalism in the USSR" (ibid., p. 35). It was easy, of course, to recognize aspects of Vygotsky and Luria's theory (e.g., as described in their *Studies on the history of behavior: Ape, primitive, and child*; Vygotsky and Luria 1930, 1993) in the various approaches that Bikchentay condemned.

In a particularly vicious paper in the journal of the Communist Academy (*Vestnik Kommunisticheskoy Akademii*) by Sapir (1931, p. 43), the author observed that "idealist and vitalist ideas from the rotten bourgeois world penetrate also here in the form of uncritical flirts with Freud, Adler, Gestalt psychology, Thurnwald's ethnopsychology, etc.; they separate psychological laws from sociohistorical conditions and neurology or abstractly reduce the social determination of behavior to the influences of cultural-historical development, taken without its organic link with the economic base and the processes of the class struggle (advancing this theory together with positivism is the anti-Marxist mistake of Vygotsky and Luria, in particular)."

Feofanov (1932) published a lengthy and somewhat incoherent criticism of Vygotsky's *Pedology of the school age* (cf. Van der Veer 2020) in *Pedologiya*. In his view Vygotsky defended a biological view of development and did not take its social class environment into account (Feofanov 1932, p. 22). Feofanov particularly objected to Vygotsky's distinction between "natural" and "cultural" behavior, arguing that there is no such thing as natural behavior in humans, because they are born into a specific social class environment (ibid., p. 27). Hence, we cannot talk about a natural stage that precedes cultural development; the developmental stages all form part of the same social dialectical process, where the distinction between natural and cultural has no meaning (ibid., p. 29). The same holds for cross-cultural comparisons: The memory of a savage is just as "cultural" as ours in the sense that it is a product of his sociocultural milieu (ibid., p. 32). Feofanov concluded that Vygotsky and Luria's theory of cultural development was based on the idealist ideas of Spranger, Stern, Bühler, and Adler and should be considered very harmful. The editorial board added a note that Feofanov's paper was not fully correct and that further criticism of Vygotsky and Luria's theory was needed.

Finally, Abelskaya and Neopikhonova (1932) repeated the criticism that Vygotsky and Luria's theory of cultural development neglected the concrete sociohistorical conditions. Their view of auxiliary means – instruments, tools, and signs – Abelskaya and Neopikhonova (1932, pp. 33–34) considered to be isolated from the "relationships of production," the "concrete labor activity," and the specific "social-class environment." Hence, their theory of child development suffered from formalism and biologism. Again, the editorial board added a note saying that the theory also had its serious methodological defects.

In sum, from 1931 onwards (see also the posthumous attacks by Georgiev 1936; Kozyryev and Turko 1936; Rudneva 1937) Vygotsky and Luria's theory of cultural development was heavily criticized and deemed anti-Marxist and harmful. The fact that, to the best of our knowledge, Vygotsky and Luria never recanted or publicly defended their points of view did not make things much better. In view of this circumstance, Razmyslov's (1934, p. 83) judgment that their study in Uzbekistan was based on a "pseudoscientific, reactionary anti-Marxist theory" cannot have come as a big surprise.

Conclusions

When Luria (1931a) announced his cross-cultural study as the "first expedition in the Soviet Union," this was somewhat misleading. Several pedological expeditions to remote areas of the Soviet Union had already been undertaken, including one in which his own student Aleksandr Zaporozhets participated. The results of these studies were published in the main journal, *Pedologiya*, where Vygotsky was one of the editors and which Luria knew very well. Moreover, from the 1920s onwards pedologists increasingly used foreign and national intelligence tests (e.g., Binet-Simon, Rossolimo) to compare the mental capacities of the various ethnic groups living within the Soviet Union. For some reason, data from Uzbekistan were especially well known. From about 1931 the interpretation of mental tests became the topic of increasingly vehement debates. The use of mental tests for vocational selection or the allocation of children to various school types became more and more contested. Differences in average scores could no longer be attributed to putative differences in genetic makeup and had to be attributed to differences in the subjects' socioeconomical background. But even this gradually became a risky card to play as negative socioeconomical circumstances by definition could no longer exist in the Soviet Union. Hence, low average group scores on mental tests had to be attributed to the tests' inadequacy and any other interpretation was liable to be seen as the manifestation of "great power chauvinism." Thus, differential psychology and cross-cultural psychology became increasingly sensitive research areas. In this respect, it is quite telling that even in 1974, that is, 40 years after a committee deemed his results "anti-Marxist," Luria still did not mention the IQ scores he found in the Uzbek population.

The results that Luria obtained had in part already been anticipated by other Soviet researchers. That subjects with little or no schooling tended to interpret images as the pictures of concrete objects ("a plate") and not as a specimen of abstract geometrical categories ("a circle") had been found before. That such subjects had difficulty dealing with abstract syllogisms (e.g., of the form "All A are B, x is A, hence x is B") had also been shown before (e.g., the study by Shpil'reyn mentioned above). That these persons tend to define objects in terms of their practical use and turn abstract questions into concrete ones had been demonstrated by Luria's student Zaporozhets. That it can be very difficult to test children or adults from foreign cultures (e.g., individual testing is often impossible, children tend to run away, the researcher needs to speak the native language) had been demonstrated multiple times in the existing Soviet cross-cultural studies.

Vygotsky and Luria explained their findings by arguing that the Uzbek subjects were culturally disadvantaged. When cultures do not provide the training in certain cultural instruments (e.g., categorical classification; logical reasoning) subjects will find it difficult or impossible to attain the heights of abstract verbal reasoning. This does not mean that these persons are inherently less intelligent ("mentally backward") but implies that education should offer them the opportunity to appropriate new cultural tools (e.g., literacy) and allow them to make the next step in their "cultural development." In this sense, Vygotsky and Luria spoke of "superior" and "inferior" cultures, which was an unfortunate term since it suggested that cultures can be compared on a global scale and that European culture is best. The distinction they made between "mentally backward" and "culturally backward" or "primitive" children or adults was somewhat subtle, but in the sociopolitical climate of the 1930s Vygotsky and Luria's theory of cultural development became the subject of heavy criticism. As we have seen, to point out any differences between ethnic groups living in the Soviet Union became highly suspect, in particular when the persons belonging to these groups were politically active communists, and in the end not even explanations that referred to different living circumstances were acceptable, because it was claimed that negative living circumstances no longer existed in the socialist state. In this sense the 1930s differed dramatically from the 1920s when the popular authors Ilf and Petrov could still portray extreme poverty, poor housing conditions, and begging street children during the New Economic Policy (Ilf and Petrov 1980[1928]).

In conclusion, with the benefit of hindsight, the expeditions to Uzbekistan organized by Vygotsky and Luria seem ill-timed and unlikely to be acclaimed by the most vocal ideologists of the time. Their methods, results, and interpretations were similar to those of previous Soviet studies and expeditions, which had become the subject of heated debate. Because the theory behind their cross-cultural study had been deemed anti-Marxist by the leading ideologists, it was unlikely that their study would be published. Interviewing kolkhoz members in the period of dekulakization was sensitive enough, but questioning their abilities in the field of abstract reasoning was asking for trouble. In the present chapter the background of the expeditions to Uzbekistan (cf. Van der Veer and Valsiner 1991) has been more fully provided in order to better appreciate how contemporaries may have received their findings and why it became well-nigh impossible to publish their results. Now, almost 90 years later, it seems Luria was in the wrong place with the wrong study at the wrong time.

References

Abelskaya RS, Neopikhonova OV (1932) Problema razvitiya v nemetskoy psikhologii i ee vliyanie na sovetskuyu pedologiyu i psikhologiyu [The problem of development in German psychology and its influence on Soviet pedology and psychology]. Pedologiya [Pedology] 5(3):27–36. [In Russian]

Baranova T (1930a) Prisposoblenie testovoy metodiki i izmereniya umstvennogo razvitiya k usloviyam Sredney Azii [Adaptation of the test method and the measurement of mental development to the conditions in Central Asia]. Pedologiya [Pedology] 3(2):255–262. [In Russian]

Baranova T (1930b) Printsipy prisposobleniya testovoy metodiki i izmereniya umstvennogo razvitiya k usloviyam Sredney Azii [Principles of the adaptation of the test method and the measurement of mental development to the conditions in Central Asia]. In: Zalkind AB (ed) Psikhonevrologicheskie nauki v SSSR (Materialy 1 Vsesoyuznogo s'ezda po izucheniyu povedeniya cheloveka) [Neuropsychological sciences in the USSR (Proceedings of the 1st All-Union congress on the study of human behavior)]. Gosudarstvennoe Meditsinskoe Izdatel'stvo, Moscow/Leningrad, pp 229–231. [In Russian]

Bikchentay IN (1931) Ocherednye zadachi natspedologii [The next tasks of the pedology of nationalities]. Pedologiya [Pedology] 4(7–8):31–36. [In Russian]

Bikchentay IN, Karimova Z (1930) Intellektual'nyy uroven' tatarskikh shkol'nikov Moskvy po kollektivnomu metodu Bine [The intellectual level of Tatar students in Moscow according to the Binet group test]. Pedologiya [Pedology] 3(2):271–278. [In Russian]

Blonskiy PP (1930 [1925]) Pedologiya v massovoy shkole pervoy stupeni [Pedology in the mass school of the first level], 7th edn. Izdatel'stvo 'Rabotnik Prosveshcheniya', Moscow. [In Russian]

Bolotnikov AA, Zalkind AB, Levina MA, Poberezhskaya MS, Tatulov G, Vilenkina RG, Levin L, Mukovnin AI, Simonova AV, Fedosenko AF (1931) O polozhenii na pedagogicheskom fronte [About the situation at the pedagogical front]. Pedologiya [Pedology] 4(7–8):8–11. [In Russian]

Bukhgol'ts (Buchholz) NA, Shubert (Schubert) AM (1926) Ispytaniya umstvennoy odarennosti i shkol'noy uspeshnosti: Massovye Amerikanskie testy [Intelligence tests and school performance: American mass tests]. Novaya Moskva, Moscow. [In Russian]

Bulanov I (1930) Materialy po izucheniya povedeniya rebenka-tungusa [Data from the study of the Tungus child]. Pedologiya [Pedology] 3(2):194–207. [In Russian]

Byford A (2016) Imperial normativities and the sciences of the child: the politics of development in the USSR, 1920s–1930s. Ab Imperio 2:71–124

Byford A (2020) Science of the child in late Imperial and early Soviet Russia. Oxford University Press, Oxford

Dimanshteyn S (1929) Problemy natsional'noy kul'tury i kul'turnogo stroitel'stva v natsional'nykh respublikakh [Problems of national culture and cultural development in the national republics]. Vestnik Kommunisticheskoy Akademii [Herald Commun Acad] 31:113–143. [In Russian]

Efimov M (1931a) Retsenziya na F.P. Petrov-Opyt issledovaniya intellektual'nogo razvitiya chuvashkikh detey po metodu Bine-Simon, 1928 [Review of F.P. Petrov, studying the intellectual development of Chuvash children with the method of Binet-Simon]. Pedologiya [Pedology] 4(7–8):127–128. [In Russian]

Efimov M (1931b) Rabota pedologo-pedagogicheskogo kabineta Chuvashkogo nauchno-issledovatel'skogo institute [The work of the pedological-pedagogical lab of the Chuvash scientific research institute]. Pedologiya 4(7–8):147–149. [In Russian]

Feofanov MP (1932) Teoriya kul'turnogo razvitiya v pedologii kak elektricheskaya [sic] kontseptsiya imeyyushaya v osnovnom idealicheskie korni [The theory of cultural development in pedology as an electric [sic] conception with mainly idealist roots]. Pedologiya [Pedology] 5(1–2):21–34

Georgiev FI (1936) O sostayanii i zadachakh psikhologicheskoy nauki v SSSR: Otchet o soveshchanii psikhologov pri redaktsii zhurnala 'Pod Znamenem Marksizma' [About the situation and tasks of the psychological science in the USSR: Account of a meeting of psychologists at the editorial office of the journal 'Under the Banner of Marxism']. Pod Znamenem Marksizma [Under Banner Marxism] 9:87–99. [In Russian]

Golubeva AP (1930) Izucheniye oyrotskogo rebenka na Altae [The study of the Oyrot child in the Altai region]. Pedologiya [Pedology] 3(2):208–221. [In Russian]

Gould SJ (1981) The mismeasure of man. Penguin, Harmondsworth

Granat EE, Zagorzhel'skaya EI (1930) Mediko-pedologicheskaya ekspeditsiya v Buryat-Mongoliyu [A medical-pedological expedition to Buryat-Mongolia]. Pedologiya [Pedology] 3(2):235–254. [In Russian]

Gur'yanov EV, Smirnov AA, Sokolov MV, Shevarev PA (1930) Shkala Bine Termen dlya izmereniya umstvennogo razvitiya detey [The Binet-Terman scale for the measurement of children's mental development]. Izdatel'stvo 'Rabotnik Prosveshcheniya', Moscow. [In Russian]

Gur-Gurevich VM (1931) Levatskie izvrashcheniya v pedologii [Leftist perversions in pedology]. Pedologiya [Pedology] 4(7–8):46–48. [In Russian]

Ilf I, Petrov E (1980 [1928]) Dvenadtsat' stul'yev [Twelve chairs]. Khudozhestvennaya Literatura, Moscow. [In Russian]

Khronika (1930) V sektsii pedagogiki narodnostey nauchno-pedagogicheskogo instituta metodov shkolnoy raboty [In the section of the pedagogics of nationalities at the pedagogical institute of the methods of schoolwork]. Pedologiya [Pedology] 3(2):287. [In Russian]

Kostin NN (1931) Ob'ekt izucheniya pedologii [Pedology's object of study]. Pedologiya [Pedology] 4(7–8):23–30. [In Russian]

Kozyryev AV, Turko AP (1936) Pedologicheskaya shkola prof. L.S. Vygotskogo [The pedological school of prof. L.S. Vygotsky]. Vysshaya Shkola [Higher School] 2:44–57. [In Russian]

Kurek NS (1999) Razrushenie psikhotekhniki [The destruction of psychotechnics]. Novyy Mir [New World] 2:46–50. [In Russian] See https://magazines.gorky.media/novyi_mi/1999/2/razrushenie-psihotehniki.html

Kurek NS (2004) Istoriya likvidatsiya pedologii i psikhotekhniki [The history of the liquidation of pedology and psychotechnics]. Aleteyya, St. Petersburg. [In Russian]

Kapusto EV (1928) Opyty raboty antropometricheskogo kabineta Tashkentskoy detskoy ambulatorii [Results of the anthropometric lab of the Tashkent child ambulatory]. Meditsinskaya Mysl' Uzbekistana [Med Thought Uzbekistan] 4:52–60. [In Russian]

Lavrova-Bikchentay ZG (1930) Fizicheskoe razvitie moskovskikh shol'nikov-tatar [The physical development of Moscow Tatar school children]. Pedologiya [Pedology] 3(2):263–270. [In Russian]

Leventuev PI (1931) Politicheskie isvrashcheniya v pedologii: O nekotorykh pedologicheskikh issledovaniyakh [Political perversions in pedology: about several pedological studies]. Pedologiya [Pedology] 4(3):63–66. [In Russian]

Leventuev PI (1932) K pervomu vypusku pedologo-pedagogicheskogo otdela Uzbekistanskoy Pedagogicheskoy Akademii [About the first graduates of the pedological-pedagogical section of the Uzbek Pedagogical Academy]. Pedologiya [Pedology] 5(3):7–9. [In Russian]

Leventuev PI, Bagautdinov A, Musael'yants AR, Mangusheva Z, Nugmanov S, Tillya Khodzaev S, Usmanov AA, Khalilov V (1932) Protiv velikoderzhavnogo shovinizma v pedologii [Against great power chauvinism in pedology]. Pedologiya [Pedology] 5(1–2):46–49. [In Russian]

Luria AR (1931a) Psychological expedition to Central Asia. Science 74:383–384

Luria AR (1931b) Psychologische Expedition nach Mittelasien [Psychological expedition to Central Asia]. Zeitschrift für angewandte Psychologie [J Appl Psychol] 40:551–552. [In German]

Luria AR (1932) Psychological expedition to Central Asia. J Genet Psychol 40:241–242

Luria AR (1974) Ob istoricheskom razvitie poznavatel'nykh protsessov [About the historical development of cognitive processes]. Nauka, Moscow. [In Russian]

Luria AR (1976) Cognitive development: its cultural and social foundations. Harvard University Press, Cambridge, MA

Luria EA (1992) Fergana, milaya Fergana [Fergana, lovely Fergana]. Vestnik MGU. Seriya 14. Psikhologiya [Herald of Moscow State University. Series 14. Psychology] 2:27–37. [In Russian]

Mukovnin A (1931) K itogam smotra kafedr pedologii i psikhologii v akademii komvospitaniya im. Krupskoy [On the results of the inspection of the departments of pedology and psychology in the Krupskaya academy of communist education]. Pedologiya [Pedology] 4(4):79–83. [In Russian]

Ostrovskiy AD (1929) Izucheniya byta i sredy v nats-oblastyakh severnogo Kavkaza [A study of the life and environment of the national territories in the Northen Caucasus]. Pedologiya [Pedology] 2(1–2):214–220. [In Russian]

Petrov FP (1928) Opyt issledovaniya intellektual'nogo razvitiya chuvashkikh detey po metodu Bine-Simona [Results of a study of the intellectual development of Chuvash children with the Binet-Simon method]. Narkompros ChASSR, Cheboksary. [In Russian]

Razmyslov P (1934) O "kul'turno-istoricheskoy teorii psikhologii" Vygotskogo i Luriya [About Vygotsky and Luria's "cultural-historical theory of psychology"]. Kniga i Proletarskaya Revolyutsiya [Book Proletarian Revolut] 4:78–86. [In Russian]

Rezolutsiya (1931) Rezolyutsiya po dokladu obshchestva pedagogov-marksistov v prezidiume komakademii [A resolution about the account of the society of Marxist pedagogues in the presidium of the communist academy]. Vestnik Kommunisticheskoy Akademii [Herald Commun Acad] 47–48:21–23. [In Russian]

Rudneva EI (1937) Pedologicheskie isvrashcheniya Vygotskogo [Vygotsky's pedological perversions]. Gosudarstvennoe Uchebno-pedagogicheskoe Izdatel'stvo, Moscow. [In Russian]

Rybnikov N (1928) Pedologicheskie uchrezhdeniya respubliki [The pedological institutes of the republic]. Pedologiya [Pedology] 1(1):181–191. [In Russian]

Sapir D (1931) Institut vysshey nervnoy deyatel'nosti na novom etape [The institute of higher nervous activity in a new stage]. Vestnik Kommunisticheskoy Akademii [Herald Commun Acad] 46:41–48. [In Russian]
Schubert AM (1930) Drawings of Orotchen children and young people. Pedagogical Seminary J Genet Psychol 37(2):232–244. [Received for publication by ARL, April 23, 1929]
Shepovalova A (1930) Sotsial'no-bytovaya sreda tungusskikh detey na severnom Baykale [The everyday social environment of Tungush children in the northen Baikal]. Pedologiya [Pedology] 3(2):172–186. [In Russian]
Shirokova-Divaeva VP (1927) Opyt primeneniya indeksa Pin'e v otnoshenii uchenikov korennogo naseleniya UzbSSR [Results of applying the Pignet index with schoolchildren of the indigenous population of the Uzbek SSR]. Meditsinskaya Mysl' Uzbekistana [Med Thought Uzbekistan] 2: 73–77. [In Russian]
Shishov A (1927) Pokazatel' Pignet v pedometricheskoy praktike [The Pignet index in pedometrical practice]. Meditsinskaya Mysl' Uzbekistana [Med Thought Uzbekistan] 5:42–45. [In Russian]
Shishov A (1928) Mal'chiki-uzbeki. Antropometricheskie issledovaniya [Uzbek youngsters. Anthropometric studies]. Meditsinskaya Mysl' Uzbekistana [Med Thought Uzbekistan] 4: 16–27. [In Russian]
Shpil'reyn IN (1930) Psikhotekhnika v rekonstruktivnyy period [Psychotechnics in a period of reconstruction]. Vestnik Kommunisticheskoy Akademii [Herald Commun Acad] 39:166–198. [In Russian]
Shpil'reyn IN, Reytynbarg DI, Netskiy GO (1928) Yazyk krasnoarmeytsa: Opyt issledovaniya slovarya krasnoarmeytsa Moskovskogo garnizona [The language of the Red Army soldier: results of a study of the vocabulary of the Red Army soldier in the Moscow garrison]. Gosudarstvennoe Izdatel'stvo, Moscow/Leningrad. [In Russian]
Shtilerman A (1927) Byt i zdorov'e uzbekskogo shkol'nika starogo goroda Tashkenta [Life and health of the Uzbek schoolchild in the old town of Tashkent]. Meditsinskaya Mysl' Uzbekistana [Med Thought Uzbekistan] 2:115–129. [In Russian]
Shtilerman A (1928) Materialy psikhologicheskogo issledovaniya uzbekskikh shkol'nikov starogo goroda Tashkenta po pereredaktirovannomu kratkomu Rossolimo [Data of the psychological study of Uzbek schoolchildren in the old town of Tashkent with the revised short Rossolimo]. Meditsinskaya Mysl' Uzbekistana [Med Thought Uzbekistan] 4:42–51. [In Russian]
Shubert AM (1913) Kratkoe opisanie i kharakteristika metodov opredeleniya umstvennoy otstalosti u detey: Bine i Simona, Veiganota [sic], Norsvorsi-Goddara, Pitstsoli, Rossolimo, Sante-de-Sanktisa i Tsiegena [A concise description and characterization of methods to determine mental backwardness in children: Binet and Simon, Weiganot [sic], Norsworthy-Goddard, Pizzoli, Rossolimo, Sante-de-Sanctis, and Ziehen]. K.I. Tikhomirov, Moscow. [In Russian]
Shubert AM (1922) Kratkoe opisanie i kharakteristika metodov issledovaniya umstvennoy odarennosti detey: Bine i Simona, Bernshteyna i Veigandta, Nechaeva, Pitstsoli, Rossolimo, Sante-de-Sanktisa, Tsiegena i dr [A concise description and characterization of methods to study children's mental capacity: Binet and Simon, Bernstein and Weigandt, Nechaev, Pizzoli, Rossolimo, Sante-de-Sanctis, Ziehen, etc.]. K.I. Tikhomirov, Moscow. [In Russian]
Shubert AM (ed) (1923) Metricheskaya skala Bine i Simona: Posobie dlya issledovaniya umstvennoy odarennosti [The metric scale of Binet and Simon: A manual for the study of mental capacity]. Zadruga, Moscow. [In Russian]
Shubert AM (1924) Kak izuchat' shkol'nika: Lichnaya karta shkol'nika [How to study the schoolchild: the schoolchild's personal card]. Tsentral'nyy Pedologicheskiy Institut, Moscow. [In Russian]
Shubert AM (1926a) Metricheskaya skala Bine i Simona: Posobie dlya ispytaniya umstvennoy odarennosti. S pril. testov Kul'mana dlya mladencheskikh vozrastov [The metric scale of Binet and Simon: a manual for the testing of mental capacity. Including Kuhlman's tests for the younger age groups]. Novaya Moskva, Moscow. [In Russian]
Shubert AM (ed) (1926b) Shkol'nye testy: Dlya pervykh chetyrekh grupp trudovoy shkoly. Rukovodstvo dlya uchitel'ya [School tests: for the first four groups of the labor school. A guide for the teacher]. Novaya Moskva, Moscow. [In Russian]

Shubert AM (1928) Besslovesnye testy ispytaniya umstvennogo razvitiya Pintnera i Patersona: Metodicheskoe rukovodstvo [The nonverbal tests of mental development by Pintner and Paterson]. Int, Moscow. [In Russian]

Shubert AM (1930a) Opyt pedologo-pedagogicheskikh ekspeditsii po izucheniyu narodov dalekikh okrain [Results of a pedological-pedagogical expedition to study the people in remote regions]. Pedologiya [Pedology] 3(2):167–171. [In Russian]

Shubert AM (1930b) Problemy pedologii po dannym literatury: Kritiko-bibliograficheskiy obzor [Pedological problems in the literature: a critical bibliographical overview]. Pedologiya [Pedology] 3(2):279–286. [In Russian]

Solov'ev VK (1929) Godichnyy opyt ispytaniya obshchey odarennosti uzbekov i metodicheskiy analiz serii VSU RKKA [Annual results of the testing of the general intelligence of Uzbeks and the methodical analysis of the Red Army series]. Psikhotekhnika i Psikhofiziologiya Truda [Psychotech Psychophysiol Work] 2–3:151–167. [In Russian]

Talankin AA (1931a) O povorote na psikhologicheskom fronte [About the turnabout at the psychological front]. Sovetskaya Psikhonevrologiya [Soviet Psychoneurol] 2–3:8–23. [In Russian]

Talankin AA (1931b) O 'marksistkoy psikhologii' prof. Kornilova [About prof. Kornilov's 'Marxist psychology']. Psikhologiya [Psychology] 4(1):24–43. [In Russian]

Thurnwald R (1912) Vorschläge zur psychologischen Untersuchung primitiver Menschen [Proposals for the psychological study of primitive man]. Barth, Leipzig. [In German]

Tulviste P (1988) Kul'turno-istoricheskoe razvitie verbal'nogo myshleniya [The cultural-historical development of verbal thinking]. Valgus, Tallinn. [In Russian]

Usova KI (1930) Rebenok-Tungus v shkole [The Tungus child at school]. Pedologiya [Pedology] 3 (2):187–193. [In Russian]

Van der Veer R (2007) Lev Vygotsky. Bloomsbury, London/New York

Van der Veer R (ed) (2020) Vygotsky's Pedology of the school age. Information Age Publishing, Charlotte

Van der Veer R, Valsiner J (1991) Understanding Vygotsky: a quest for synthesis. Blackwell, Oxford

Vygotsky LS (1928) Pedologiya shkol'nogo vozrasta [Pedology of the school age]. Byuro Zaochnogo Obucheniya pri Pedfake 2 MGU, Moscow. [In Russian]

Vygotsky LS (1929) K voprosu o plane nauchnoy-issledovatel'skoy raboty po pedologii natsional'nykh menshinstv [About the question of the plan for scientific research in the pedology of national minorities]. Pedologiya [Pedology] 2(1–2):367–377. [In Russian]

Vygotsky LS, Luria AR (1930) Etyudy po istorii povedeniya: Obez'yana. Primitiv. Rebenok [Studies in the history of behavior: ape. Primitive man. Child]. Gosizdat, Moscow/Leningrad. [In Russian]

Vygotsky LS, Luria AR (1993) Studies on the history of behavior: ape, primitive, and child. Lawrence Erlbaum Associates, Hillsdale

Yasnitsky A, Van der Veer R (eds) (2016) Revisionist revolution in Vygotsky studies. Routledge, London

Zalkind AB (1930) O psikhonevrologicheskom izuchenii natsional'nykh menshinstv [About the psychoneurological study of national minorities]. Pedologiya [Pedology] 3(2):165–166. [In Russian]

Zalkind AB (1931) Differentsirovka na pedologicheskom fronte [Differentiation at the pedological front]. Pedologiya [Pedology] 4(3):7–14. [In Russian]

Zaporozhets AV (1930) Umstvennoe razvitie i psikhicheskie osobennosti oyrotskikh detey [The mental development and psychic characteristics of Oyrot children]. Pedologiya [Pedology] 3(2): 222–235. [In Russian]

Zavershnev E, Van der Veer R (2018) Vygotsky's notebooks: a selection. Springer, Singapore/New York

Values and Persons: The Persistent Problem of Values in Science and Psychology

45

Lisa M. Osbeck

Contents

Introduction: The Complex Landscape of Science-Values Relations	1168
Aim and Overview	1170
The Broader Context of Questions Concerning Values and Science	1171
Definitions and Distinctions	1172
Values Underlying the Valuing of Value-Free Science	1174
Values Underlying Denial of "Value-Free Science"	1177
What Kinds or Categories of Value Are Intrinsic to Science?	1178
Where (or How) Values Impact Science	1180
Values and Psychological Science	1181
Methods as Value Systems	1184
Problems for Interdisciplinary and Intradisciplinary Collaboration	1186
Towards Compromise and Negotiation	1187
Conclusion	1189
References	1190

Abstract

That values are inescapable in science is now widely acknowledged, yet the admission raises many questions. Among the most important is what establishes the basis for following recommendations of scientists, for example, regarding climate and public health concerns. The chapter foregrounds the importance of values to science *and* psychology but emphasizes the complex nature of the questions and the historical situation of values. Hence the aim is to circumvent binary positioning. The chapter is organized in two parts. The first part broadly concerns science-values relations, covering difficulties in clearly defining both

---

I have no known conflicts of interest to disclose.

L. M. Osbeck (✉)
Department of Anthropology, Psychology, and Sociology, University of West Georgia, Carrollton, GA, USA
e-mail: losbeck@westga.edu

© The Author(s), under exclusive licence to Springer Nature Singapore Pte Ltd. 2022
D. McCallum (ed.), *The Palgrave Handbook of the History of Human Sciences*,
https://doi.org/10.1007/978-981-16-7255-2_87

science and values and the importance of distinguishing different values that underlie aspiration to value-neutrality as well as its rejection. It examines different kinds of values and considers problems with some existing categorization strategies. Finally, it considers what values are "good for science" and proposes that interdisciplinary collaboration is important to good science in the current context. The second part of the chapter analyzes implications of topics discussed for psychological science, within which very different methodological traditions pose special concerns and illustrate difficulties imposed by competing values. Methods are described as internally coherent value systems with distinctive sets of constraints. However, questions persist as to how persons within divergent value systems can productively communicate and collaborate. Suggesting some steps forward, the chapter concludes with emphasis that problems of values and science underscore both the social and the personal dimensions of science, and thus the responsibility imbuing the production and use of knowledge claims in any science, including psychology.

Keywords

Values · Science · Persons · Interdisciplinary collaboration

"The problem of value, or more generally of requiredness, is gradually becoming the outstanding difficulty or the eminent task of human thought" (Köhler 1938, p. 35).
"What in the name of Heaven and Psychology can we do about it?" (Tolman 1948, p. 207)

Introduction: The Complex Landscape of Science-Values Relations

Given the impact of science in every aspect of human life, the study of the values that enable and surround science is an enduringly important topic for human science, including psychology. The topic of values and science gains additional urgency in a contemporary context marked by polarized and complex attitudes toward science, both within the academy and outside of it. For example, politization of the Covid-19 pandemic forged a deep cultural cleavage, especially in the USA, where assaults on expertise and scientific authority included threats of violence, and stoked resistance to recommended behavioral changes needed to mitigate viral spread. Manipulation of knowledge for oppressive political agendas and suppression of research findings that oppose corporate and political interests are hardly limited to recent history, but the current public health and climate crises have magnified transgressive patterns. As a corrective to the blatant abuse of power, the notion that "data do not lie," continues to entice scientists and citizens looking for a reliable guide in the morass. Trust in science to illuminate in darkness has long been rooted in conviction of its essential neutrality, freedom from the biases that muddle more pedestrian human affairs. We may consider the quest for authoritative neutrality misguided or naïve yet continue to appreciate the hope and fear behind it.

Such hope Gerald Holton describes as an "implicit contract" forged in the aftermath of World War II "between science and society," and a view of science remains "the dominant image among the majority of scientists," (Holton 1996, p. 4):

> For a few decades the pursuit of scientific knowledge was widely thought – above all by scientists themselves - to embody the classical values of Western civilization, starting with the three primary virtues of truth, goodness, and beauty. That is, science tended to be praised as a central truth-seeking and enlightening process in modern culture, what one might call the Newtonian search for omniscience. Science and scientists were also thought to embody the ethos of practical goodness in an imperfect world, both through the largely self-correcting practice of honor in scientific research, and through applications that might improve the human condition and ward off the enemies of society, a sort of Baconian search for a benign omnipotence. Finally, science as also thought of as a Keplerian enchantment; the discovery of beauty in the structure, coherence, simplicity, and rationality of the world was the highest reward for the exhausting labor the discipline required. (Holton 1996, p. 4)

Whatever the self-representation and beliefs about the integrity of science itself, scientists like all persons inevitably bring agendas and priorities to even their activities, including the most high-level and complex of them. To the extent that science consists of human practices, it cannot be stripped of values and valuing, with more obvious examples including the valuing of logic, systematicity, and prediction. Values impact the selection of a research question and the identification of data gathered to address it; they play a vital role in the selection of analytic procedures and justificatory practices, as well as conventions for disseminating knowledge once produced. A view that remains influential is that all scientific theories are "value statements," such that "putative 'facts' are viewed not only through a theory window but through a value window as well" (Guba and Lincoln 1994:107).

The idea of theories as value statements is entirely consistent with acknowledgment that science consists in human activity and is therefore value laden. Yet questions then arise as to whether science consists *merely* in statements of value; that is, whether *value-laden* is equivalent to *reducible to* values. If reducible to values, we have reason to question the epistemic standing of theories, models, and findings established through even the most scrupulous and systematic of procedures. Even if justified from several intellectual vantage points, the genuine pragmatic consequences of the reduction must give one pause. On what basis can we distinguish scientific claims from the bare assertion of will and desire? What counts as evidence in favor of any point of view? What separates the results of any investigation from those of "wishful thinking" or political maneuvering? (Elliott 2017, p. 13)

Clearly, too, we must distinguish in some way the generation and production of new scientific knowledge from its application by practitioners (e.g., physicians, technicians, therapists), as well as from the dissemination of scientific knowledge on a broader scale (e.g., public health recommendations to mitigate the spread of disease). Values play an important role at each of these levels. But if we were to declare science reducible to values, what would be the grounds for trusting scientific evidence and the recommendations of the experts who offer it, whether to follow medical advice, observe public health guidelines, or heed the warnings of climate scientists? On the other hand, if science is *not* reducible to values, of what else does it consist, and where are the lines to be drawn around its distinctive aspects? That is, what is the basis for maintaining "epistemic trust" in science as is necessary both for science and humankind (Wilholt 2013)? For our purposes, can human science, especially psychology, shed light on these seemingly conflicting if not incompatible positions: recognition of the value-saturation of science with a requisite faith in science in the service of human survival?

Aim and Overview

There are many excellent resources available to help navigate the complex landscape of value-science relations, including the vast critical and historical scholarship aimed specifically at psychological science. Rather than reproduce their arguments, for the present purposes the intent is to foreground the importance of values to science and psychology and emphasize the complex nature of the questions that surround these relations, with the additional aim of circumventing a too common binary positioning around the topic.

In that spirit, the first section of the chapter will first address the broader context of questions relating to values and science, consider the meaning of values, including variations in the meanings attributable to them, and acknowledge the various kinds or categories of values that have been described as "entering" science. As others have noted, the idea that science can and should be "value-free" itself reflects values is important to acknowledge (e.g., Douglas 2013; Slife 2009). Thus, the chapter will then address some of the values that underlie this broader value, as well as some of the values that contribute to the rejection of the value-free ideal. In turn, it is fitting to reflect on ways values enhance rather than detract from science, that is, how values contribute to what is conceived as "good science." Finally, "good science" in contemporary context may mean science that can participate in broad scale interdisciplinary and collaborative problem-solving, so discussion will turn to the question of how values might impact the potential for collaboration of this kind.

The second part of the chapter examines implications of topics discussed for the discipline of psychology more specifically. Psychological science partakes in the problems that beset science in general but also invites additional questions given a lack of clarity concerning its subject matter and deep disagreements about the most appropriate way to investigate it. In turn, the specific problems psychology faces given its own vastly different methodological traditions relate directly to the problem of value differences that may complicate interdisciplinary collaboration. The conception of methods as value systems for understanding the relation between these traditions is helpful to a point. However, the current demand for interdisciplinary and *intra*disciplinary collaboration requires some new analysis of the values-science relationship if differing values compromise the potential for collaborative exchange. An increase in emphasis on interdisciplinary scholarship over the past several decades accompanies recognition that the complexity of global challenges such as climate change, nuclear proliferation, and extreme poverty requires ensembles of concepts, models, methods, and instruments from different disciplinary vantage points to generate novel and comprehensive solutions.

The Broader Context of Questions Concerning Values and Science

Questions concerning science and values are embedded in a much wider dialogue connected to questions about objectivity, perspective, knowledge, authority, truth, and "the real" (Douglas 2009, 2013; Elliott 2017; Lauden 1984; Lacey 1999; Machamer and Wolters 2004). They form a network of intricate and interconnected problems, each with its own historical dimensions. Although the basic dilemma of how we can trust scientific conclusions if drawn by persons who cannot escape their values endures across generations of scholars, it does so with different concepts, foci, and emphases relevant to the academic, cultural, economic, technological, religious, and political contexts at hand. The complexity and embeddedness of value questions steer one away from a choice between some version of "data don't lie" and "science offers only one possibility on par with many others."

At a most basic level, we must come to some agreement on the very subject matter we are considering, which requires not only understanding what values are but also what they are not. That is, what is the precise nature of the contribution to science of that which is something other than values? Here disagreements are many, and the answer we give depends on whether our emphasis is ontological, epistemic, or ethical. The qualifications of the psychologist to address the ontological dimensions of science are minimal and should be limited to the ontology of persons and practices. Those who dedicate their lives to studying observed regularities in natural phenomena are in the best position to address the composition and causes of these phenomena, whatever the science in question. But the process by which we arrive at knowledge, and the application and use of that knowledge invoke epistemology and ethics. They concern human practices and as such cut across personal (e.g., cognitive) and social realms of analysis.

A central question is whether science includes activities that are in some way "special," that demarcate science from all the ordinary kinds of sense-making in everyday life. In what specific ways does scientific reasoning differ from that of other reasoning? The position we take in relation to this question will inform the question of values, because those who believe that science is a truly special activity are more likely to aspire toward value-neutrality in science, while those who acknowledge no dividing line between science and other contexts of reasoning are more likely to deny the possibility of any such neutral zone. Moreover, a view that science is a unique form of human practice does not commit one to an ontological realism. It is possible to focus entirely on cultural dimensions of science yet to maintain that the normative frameworks within scientific cultures have a special character that distinguishes science from other forms of human practice. The idea that science may be "culturally unique and demarcated sharply from other intellectual pursuits" (Lauden 1984, p. 2) may highlight unique forms of communication and more binding justificatory standards, and distinct values that are shared within that culture.

If we consider scientific reasoning to be prototypical of human thought, as it has been characterized (Feist 2008), the importance of a distinction between two divergent meanings associated with "prototype" and "prototypical" becomes apparent. One sense is prototype as an ideal, the other meaning conveys a typical representative of a class. Thus, in one case something we reference something rarefied, but in the other we can acknowledge the continuity between scientific and everyday forms of reasoning (Osbeck et al. 2010). A related and hugely important question concerns whether "prototypical" in either sense applies principally to Western thought or extends to all human thought, thereby calling into question the relation of such "prototypical" thought processes to indigenous psychologies around the world, contemporary and historical.

Definitions and Distinctions

Further complicating the discussion at hand is that there is no single understanding of science or scientific method, despite popular opinion and mystique. One finds little consensus on what it means to think scientifically, even among those upholding the position that science represents an especially rarefied human practice. There are various models of the logical structure of science and the principal form of reasoning that enables it: inductive, hypothetico-deductive, abductive, Bayesian, and model based (e.g., see Haig 2014). Each of these forms in turn can be delineated further; also there is little consensus on their relation to one another or to different epistemic activities (e.g., creativity and explanation). In addition, historical, autobiographical, and ethnographic analyses of science reveal discrepancy between science as actually practiced and the a priori accounts forwarded over the centuries to elevate science above ordinary human affairs. Acting in their natural habitats (laboratories, field sites, observatories), scientists exhibit wildly divergent practices, despite the widely held (and taught) opinion that scientists follow a recipe and code. Henry Cowles provides an excellent historical analysis of "the scientific method," underscoring its symbolic, even mythical standing in the public realm. He distinguishes these "stories we tell about science" from the everyday world of science as practiced: "Scientists will tell you that there is no single method that characterizes all that they do, much less a simple set of steps that binds everything together. Scientific labor is complex and diverse, brutally difficult and impossible to encapsulate" (Cowles 2020, p. 1). He notes that in "the real world, we make mistakes and get bogged down; it is only in hindsight that thinking seems clean and rational" (Cowles 2020, p. 9). This point is also central to Feyerabend's *Against Method* (1970), and Lauden (1984) locates the origin of this position (one which emphasizes discontinuity across science as practiced) to transformative intellectual developments situated in the 1960s and 1970s based on the collapse of the positivist theory of knowledge as a viable program. Gaps between the "storybook" (Mitroff 1974, p. 8) or "legend" (Kitcher 1993, p. 3) conception of science and the actual practices of scientists "in the wild" bode poorly for our ability to draw lines around science and scientific reasoning with any degree of precision. If we cannot say distinctly what science is, how can we analyze how values contribute to it?

The meaning of "values" is equally elusive. George Howard, in one of the first important statements on the role of values in psychological science, describes value as "a catchall category for an enormous array of very different judgments, decisions, preferences, and orientations" (Howard 1985, p. 255). Tjeltveit offers detailed historical review of the concept of values, tracing the path through which discussion of values entered nineteenth century ethical theory principally through German thought, and noting the importance of Lotze's primary distinction between the "reals" of fact and of value (see also Rescher 1969). He calls the term "remarkably elastic" in contemporary context, "put to use by psychotherapists, sociologists, philosophers, and the general public in a wide assortment of ways" (Tjeltveit 1999, p. 82).

Contemporary definition of values typically includes some acknowledgement of (1) emotion or "desire" (along with preferences, pleasure, or displeasure) and (2) appraisal, judgment, or evaluation. For example, "something that is desirable or worthy of pursuit" (Elliott 2017, p. 11), in which desire connects with evaluation of worth. Therein lies much of the problem in definition, for we can easily see that one can desire something but not deem it *worthy of pursuit* (a day in bed; a binge of junk food) or can deem something worthy of pursuit but not *desire* it or find it pleasant (a biopsy). Values are also sometimes conceptualized as related to action or motivation, or at least to judging the worth of action. For example: "Put simply, values tend to concern what matters or has merit. To value something is to judge that it has merit for some *enterprise*" (Slife 2009, p. 10, emphasis added). Vernon and Allport's pioneering psychological study of values similarly emphasized both motivational and evaluative aspects (Vernon and Allport 1931; see Campbell et al. 2015). Allport himself later stated that he found most useful a conception of values as tied to interest, indeed that objects have value if they are of interest (Allport 1955). If connected to interest, values overlap with the aims and goals of inquiry. In turn, aims and goals implicate emotional investment. This web of associated concepts and meanings imposes barriers to envisioning a value-free science, even as an intellectual exercise. On purely logical grounds, how could any form of scientific inquiry occur without interest, aims, and goals?

Values are also sometimes defined in a way that overlaps with the idea of one's comportment, style, and general outlook, for example, as "preferred ways of understanding and being in the world," (Duffy and Chenail 2009, p. 24). This very general way of viewing values aligns them with the similarly broad concepts of "perspective," "identity," and "worldview," all of which occupy different conceptual and scholarly territory and all of which are notoriously murky. Further complicating the meaning of values is that they are often taken to be implicit – hidden from view but profoundly directive (Slife 2009).

Given the ambiguity entailed in conceptualizing *both* science and values, unproblematic understanding of their relation remains elusive. We may, in fact, identify five distinct possibilities concerning the relation of science and values: (1) Science is reducible to values, from which we may conclude that there is nothing beyond values to confer special epistemic standing on science; (2) Values do substantively impact scientific practice and therefore the construction of scientific knowledge, but science does not reduce to values. Note the implication here that there is something else occurring besides values, which in turn requires us to address what that is and appraise its epistemic status in relation to claims of other kinds; (3) Values *can* influence science but *need not* do so; (4) Values can impact science in constructive ways; and (5) Science qua science is value-free. A variation on (3) and (4) is that *some* kinds of values enhance science while others detract from it. Other positions may be discernable, but they are likely to be modifications of one of these possibilities, differing in degree or adding qualifications.

Each of these possibilities in turn is organized around various themes which receive greater emphasis at one time or another. That is, each reflects a specific idea or concern, something held to be important and thus given emphasis – a value upholding the value in a broad sense. Although different themes are prominent at different times, locating the themes historically is not a straightforward task. Gerald Holton's insightful analysis links themes underlying the veneration of science to different episodes in Western thought and culture, though he is careful to remind us that any era is characterized by multiple competing narratives and thus resists categorization into simple refrains, even within the cultural context in question (Holton 1996). Nevertheless, in the service of demonstrating the complexity of questions concerning the relation of values to science, it is useful to distinguish possible themes and consider the concepts that connect to them.

Values Underlying the Valuing of Value-Free Science

Epistemic Authority

Clearly the prospect of reliance on a trustworthy authority is a value underpinning the cultural and professional faith in science characteristic of many historical periods. For many laypersons and professionals in the contemporary era, epistemic authority is crystallized in the sentiment that "data don't lie," that evidence speaks for itself, or that "listening to science" will provide guidance in daily life. In addition to providing the bedrock condition of advancement within each science, the idea that science is authoritative structures many aspects of existence for countless persons within what we may count as the dominant globalized culture, including dietary and activity choices, approaches to childcare, and response to unexpected problems, including crises (e.g., pandemics).

Predictive Power

Such authority tends to come with a proviso of universality, a unified view of the world that can be trusted, and regularity, such that the phenomenon of interest may be expected to behave in a similar way under highly similar conditions (e.g., Proctor 1991; Proctor and Capaldi 2006). The underlying value is less in claiming universality for its own sake (whatever that would mean) but is rooted in the predictive power that accompanies claims of regularity. In turn, we may distinguish two important aspects of prediction, both of which are valued for their service in human purposes. The first is that prediction enables us to control our actions in anticipation of events. For example, if we can accurately predict threatening weather conditions, we are able to adjust our own activity accordingly (take shelter). The other aspect is the use of prediction to manipulate or control circumstances, to intervene with some process such that the natural course of action is changed through human effort. For example, predicting the natural pattern of growth in cancer cells is required to develop effective agents to hinder that growth. Much scientific and technological innovation relies upon predictive accuracy, a point we may glean in Bacon's third aphorism (Book I): "Knowledge and human power are synonymous...for nature is only subdued by submission" (1902/1620, p. 11). Bacon's explicit mention of power and submission strikes us as sinister in the contemporary context of devastation wrought through the subjection of nature. Yet of course the same predictive power is necessary for the mitigation of suffering and disease and to combat the destructive impact of human carelessness and disregard through pro-environmental action.

Problem-Solving

Closely related to predictive accuracy is the value of solving problems. The value of problem-solving in scientific achievement may be understood in two senses. There is first the personal value of the challenge or intrigue – the puzzle – so aptly characterized by Dewey in *How We Think:*

> The best, indeed the only preparation is arousal to a perception of something that needs explanation, something unexpected, puzzling, peculiar. When the feeling of a genuine perplexity lays hold of any mind (no matter how the feeling arises), that mind is alert and inquiring, because stimulated from within...It is the sense of a problem that forces the mind to a survey and recall of the past to discover what the question means and how it may be dealt with. (Dewey 1910, p. 207)

With scientific reasoning, the idea that one is solving difficult problems can be a powerful incentive for sustained effort. Interdisciplinary problem-solving increasingly dominates contemporary science; a multi-perspectival effort to tackle complex problems is the overarching purpose of many laboratories, including systems biology and bioengineering (MacLeod and Nersessian 2016).

The assumption that some forms of innovation are made possible through interdisciplinary cooperation bears on the second sense in which the value of problem-solving is important in science. Problem-solving on the collective level, that is, aimed at improving human or other forms of life, is a hugely important value undergirding much scientific effort. Recognition of this value cuts across several schools of thought in both cognitive psychology and philosophy of science, and the emphasis on problem-solving may be seen in such otherwise diverse scholars as Karl Popper, Herbert Simon, and Jean Piaget.

Progress

The societal value of problem-solving instantiated in much scientific achievement overlaps with the closely related idea that science enables human progress.

This assumption and the associated concept of progress has been the target of much critique. Central questions include whether scientific advance and the technological achievement that accompanies it increases or diminishes human flourishing, given the increasing alienation from the natural world and altered human relationships that accompany it (e.g., Turkle 2012). Other targets include the destructive assaults on human welfare and environmental damage wrought by scientific ends, for which the hydrogen bomb serves as a ghastly exemplar. Yet for decades the idea of progress through science went generally unchallenged. For many scientists, the idea of societal progress (or the progress of humankind) is coextensive with the idea of scientific progress – the accumulation and increasing systematization of scientific knowledge over time. Holton points out that scientific progress occurs both through "analysis and accretion" as well as through synthesis, with the former understood as a breaking down through disagreement between scientists, or through the accumulation of disconfirming evidence, requiring modification or abandonment of theories. Synthesis, in turn, "is more transformative." With synthetic activity, "progress is equivalent to an increase in inclusiveness (a wider range of phenomena is accounted for by the new theory) or an increase in parsimony or restrictiveness (fewer separate fundamental terms and assumptions are needed)," for which the allure of synthesis to Copernicus, Galileo, and Newton are exemplary (Holton 1996, p. 52).

Important to note, too, is the way in which an idea of moral progress made possible through scientific advance is tied to the idea of bettering society through scientific discovery and application – an idea promoted explicitly through nineteenth century positivist philosophy (Comte 1848; see also Mill 1865) but evident even in the twentieth century's Vienna Circle, a group that included both scientists and philosophers and purposefully aimed to avoid contaminating science with metaphysics (e.g., Neurath 1973).

Understanding

Although a hard distinction between "pure" and applied science no longer seems tenable, we can identify many cases, indeed whole branches of science for which understanding (in a broad sense) is more centrally valued than predictive power, though accuracy of knowledge remains an essential aim. Paleontology, some applications of geology and astronomy, or any science with an obvious historical dimension could serve as a ready example.

Truth and Beauty

Tied to the goal of understanding for its own sake (i.e., without apparent connection to human problem-solving) is appreciation for truth, the aim of arriving at a true picture of the universe independent of the agenda of any person, group, or culture – to be in possession of true ideas of it. There is no need to review the themes of the Enlightenment that are so frequently tied to naïve views of scientific objectivity and the value-free ideal. Washburn (2020), for example, offers an excellent overview and analysis in relation to the concepts of "truth" and "post truth." Expression of the search for truth undergoes complex variations in different times and contexts, including the variation evident in any scholar's own evolving thought, even within the scholarly era most typically associated with it. One thing less frequently emphasized by those who position against "Enlightenment rationality" is the connection of science to truth and truth to beauty, linking science to important aesthetic considerations, including symmetry, balance, harmony, and elegance. To align subjectivity and subjective expression with aesthetic values instantiated only in the humanities is to miss the deeply rooted ties between science and the aesthetic dimension and as it has been theorized philosophically (Hadamard 1945; McAllister 1999, 2002; Polanyi 1974/1958) and as it reveals itself in the private notes and musings of practicing scientists (Holton 1996, 1998).

Mystery and Imagination

Valuing the aesthetic dimension of science also overlaps with valuing mystery and imagination in connection with the idea of neutrality and a value-free stance. The idea that through correct application of scientific principles and methods one can connect with underlying structures, capture a dimension unknowable to the naked eye, or unlock the key to secret worlds of esoteric insight attracts some scientists. Imagination is required to adequately conceptualize many phenomena, for example, through analogy and thought experiment, and it is certainly valued in discovery. It is important to recognize that this concern with the role of imagination in science does not equate to an interest in "making things up"; rather, it reflects recognition of the human capacity to connect with dimensions of reality that are in some sense hidden from view. Polanyi, for example, expresses this idea most clearly: "Such knowing is indeed *objective* in the sense of establishing contact with a hidden reality; a contact that is defined as the condition for anticipating an indeterminate range of yet unknown (and perhaps yet inconceivable) true implications" (1974/1958: viii). If the word is in some way helpful in expressing a value, we might consider this a spiritual dimension tied to the project of knowing the world.

Values Underlying Denial of "Value-Free Science"

Having examined some of the values "beneath the value" of value-free science, it is only fair to give representation to the other side. Associated literature is too vast to cover adequately, but important to acknowledge is that the denial of value-free science is rarely merely descriptive, rooted in empirical analysis of scientific practice, or even based on the conceptual incoherence of such a position. Instead, there are clear values, often with clear political connections, informing the view that value-free science is *not* ideal. To avoid veering too far from topic, they bear mentioning here to attempt some rough categorization for the sake of illustration. Representative values associated with the denial of value-free science can be moral or ethical, in which case they may focus on the oppression associated with elevation of a singular way of viewing the world (one truth), assumed in connection with the value-free ideal, or the value of giving voice to those traditionally underrepresented or oppressed. The values may also be epistemic, focusing on the enhanced observation attainable through multiple perspectives on a phenomenon of interest (with perspectives understood as implicating different values), or on the creative affordances of divergent (perhaps multi or interdisciplinary) values as these are manifest in inquiry. They may also emphasize the aesthetic dimensions of value expression, or even the practical, associated with the difficulty of keeping values out of science.

The point is that in addition to being *upheld* for different reasons (or different values), the value-free conception of science is *rejected* for different reasons, in different contexts, at different times. If the idea that science cannot and indeed should not be value-free seems like an obvious conclusion at present, it is worth remembering that issues raised by the controversy surrounding values are rooted in the intellectual and social history of the twentieth century, including but not limited to the devastation of two world wars. The cynicism wrought by these events was preceded by powerful philosophical critique aimed at dismantling the idea of science as an enterprise capable of transcending the assertion of will and desire or escaping the "residue" of religious values (e.g., Nietzsche 2000/1909). Yet prior to that century, the standard of science as value-free, that is, the aim of science that avoids the imposition of human bias (equated with value) was largely taken for granted, as was the idea of science as devoid of passion or personal investment ("dispassionate") (e.g., see Holton 1996; Köhler 1938; Polanyi 1974/1958), despite some exceptions to this position (Gaukroger 2020).

What Kinds or Categories of Value Are Intrinsic to Science?

The brief overview of some of the important considerations, sentiments, or values underlying and bolstering perpetuation of the myth of value-free science (and against it) helps to underscore that different value kinds – different categories of value –are intrinsic to science, and these at different times and in a multitude of ways. Such complexity in identifying values is consistent with patterns evident in late twentieth century and twenty-first century philosophy of science. Although controversies concerning the proper relation of values to science continue, there is little contemporary philosophical interest in defending the position that science is value free (see Douglas 2009, 2013). Instead, attention has turned to developing appropriate schemes to demarcate value kinds or to provide case-based analysis in the interests of demonstrating precisely how, when, and where science becomes infused with values. In turn, this categorization is related to the project of determining what might be acceptable or appropriate scientific values (Lacey 1999).

As a first cut, distinguishing epistemic from "non-epistemic" values is a fundamental, though controversial, classificatory strategy (e.g., Dorato 2004; Douglas 2009; Howard 1985; Lacey 1999; Lauden 1984). Concern with truth claims, adequate evidence, and appropriate inferential strategies (logical foundations) are the province of the former. Therefore, few philosophers or scientists would argue that epistemic values are not important to science. The latter category (non-epistemic) attracts more controversy and debate. At a conceptual level alone is the problem of its utter vagueness. As a negative (non), it is a container for every other category of value – moral, pragmatic, cultural, political, psychological, aesthetic – making it difficult to coherently analyze. As value kinds these relate to different concerns and may require different practices, even if they overlap at some level. For example, practical values might include the ease of administration (e.g., of a vaccine); aesthetic values may relate to theoretical elegance. Values also influence the very categorization scheme used to distinguish and catalog values, and in turn, these values reveal their historical situation. For example, Douglas locates the epistemic/non-epistemic distinction in debates taking place in the 1950s (Douglas 2009, 2016).

Intended as a corrective is a basic separation of "cognitive" from "social" values (e.g., Douglas 2013; Lacey 1999). However, this is also problematic. More fine-grained value categories defy categorization as "either" cognitive or social. For example, aesthetic values in science are both cognitive in reflecting a concern for simplicity or symmetry, and social, in reflecting standards of simplicity negotiated within communities. In countless ways cognitive and social practices operate in an integrated fashion (Rouse 2002), including in problem-solving laboratories (Danziger 1994; Machamer and Douglas 1999; Nersessian 2008).

The idea that values are shared within scientific communities accords with analyses of science that foreground social processes in knowledge construction rather than the intellectual achievement of individuals reasoning in isolation. Drawing upon the earlier insights of microbiologist Ludwig Fleck (see Fleck and Kuhn 1981), Kuhn (1962) emphasized the shared (social) nature not only of theories but of scientific values. Values are instantiated in what are upheld as exemplars of good practice (exemplary methods and achievements) within a community of scientists. Values, that is, are part of the "paradigm" that might undergo a seismic "shift" following a revolutionary discovery, after which the scientific community begins to perceive the problem and the very subject matter of investigation differently. The most controversial idea entailed in this formulation is that paradigms are incommensurable, enabling no metric for comparison within systems: "If the intellectual action in science is in the paradigm shift, and if paradigms are incommensurable, then our traditional notions of scientific progress are clearly unsupportable" (Oreskes 2019, p. 43).

After Kuhn, the tradition of science studies, especially through contributions of the Edinburgh school (e.g., Bloor 1976; Shapin 1996), emphasized to an even greater extent the social and political undergirding of Western science, including the epistemic values evident in any historical period. In strongest form, it is through irreducibly social processes that scientific knowledge is achieved and established, calling into question its very authority. Justificatory processes and the underlying logic of science itself are construed in terms of socially sanctioned habits and negotiated rules. These rules in turn develop from agendas that are fundamentally economic and that uphold social hierarchies, functioning to keep the powerful in power. Yet even a modest acknowledgement of social influences in science must recognize that values specific to a practice community (e.g., a lab) and the larger disciplinary groupings of which it is a part (e.g., biochemistry) are negotiated through discipline-specific communicative practices. Communal decisions determine the form of data collected, the methods of analysis, and the wider and immediate research goals, blurring any distinct boundary between social and epistemic values (e.g., see Danziger 1994; Machamer and Osbeck 2004). Therefore, according to one version of this view, abstract reasoning about the essence of scientific foundations should be replaced by methods drawn from social science, including anthropology. Study of science as communal practice in situ (e.g., a working laboratory) offers a basis for description of the construction of facts and the values upheld in scientific communities (e.g., Latour and Woolgar 1986/1979). With ethnographic analysis of in vivo scientific practices, demarcating values by kind becomes exceedingly challenging.

But whatever their classification or by whatever means they are studied, the question of how much "non-epistemic" or "social" values *should* influence science remains a target of dispute. One direction of thought entails reflection on ways values can benefit science, including which values are most important for scientific integrity (a broader value). Honesty and integrity, (Lauden 1984), diversity (Harding 2015), passionate engagement (Polanyi 1974/1958), and democratic participation (Longino 1990) have all been suggested in connection with values facilitative of good science. Of course, to identify values that are good for science requires a particular notion of the good and summons reflection on the end and aim of science (e.g., see Kitcher 1993, 2001). Could one give a general answer to this question or must it always be answered as relative to a given science and agenda? Even within a particular science, how much agreement can be expected within scientific communities? We might expect more agreement in physics than in psychology, wherein even the subject matter of the discipline has long been in dispute, though that there are those who argue that a plurality of theory is endemic in science in general, whereby psychology may be a model and not the exception (e.g., see Gaukroger 2020). In any case, the increasing interdisciplinary character of contemporary science complicates the question of what the science in question should be aiming toward; what "good" science means for the interdisciplinary collaborators. For example, ethnographic analysis of bioengineers revealed important differences between computational modelers (engineers) and biologists in relation to their conception of good science and what they valued, with differences centering around on the one hand, the value of system level analysis and concern with possibilities inherent in matter and on the other hand, the value of experimental rigor and fidelity to the living organism (Osbeck and Nersessian 2017).

Where (or How) Values Impact Science

A question related to the kinds of values that are present in science is where such values may be found. We can easily illustrate different value kinds at different levels or with different types of scientific activity by considering a recent episode impacting most of the world: the development of a vaccine to combat Covid-19. Clearly, interest in developing a vaccine is tied to the value of preserving human lives, especially vulnerable populations, for which reason one could acknowledge a moral dimension in the very question of whether a vaccine might be developed, and later in the effort to determine its efficacy and safety, not to mention decisions concerning its fair distribution. However, economic and political values just as obviously and quickly appear: the value of returning workers to their jobs and packaging warp-speed vaccine development as a victory in an election year. Practical values are obvious, and if it is difficult to find aesthetic value, we can certainly acknowledge narrative value in the race to discover and effectively distribute a life-saving miracle during a period of global chaos. That most episodes of scientific discovery and innovation are more mundane does not diminish the clarity of this example, especially in revealing the way categories of value overlap in practice.

The example also points to the difficulty of determining which values are good for science, and which operate to its detriment. Because political and economic values impact its development, is a vaccine less effective or useful for this reason? Are the epistemic values compromised by these other categories of value? That is, does concern for utility compromise or enhance the epistemic value of the achievement and the activities that enable it?

Might we draw from this example any principles to inform the question originally posed, namely how we could trust science as an epistemic authority if it achieved only through the activity of valuing beings? Clearly the value of the scientific accomplishment in question will be vetted on the street, tested by its ability to mitigate the spread of the virus, and this with fairness and equity. To receive a vaccine is a demonstration of at least some degree of trust in the integrity of science, yet this does not deny the value-laden nature of both the product and the act. How we make decisions about epistemic trust, how we place faith in scientific achievement is not easily addressed without affirming the integrity of the very methods critiqued as fallible. To conclude this brief overview of questions relating to science and values in general, the main message is that more nuanced understanding of the values-science relation does not resolve essential tensions or provide definitive answers to how we should understand values in science, but it does offer some guidelines for analyzing values in science in a more fine-grained way, in the service of opening conversations and seeking common ground for collaboration in the face of seemingly incompatible values. This topic is addressed in the following section, with reference to the discipline of psychology.

Values and Psychological Science

Having examined some of the questions relevant to the relation of values and science in general, the focus will now turn more particular, though will circle back to the general topic periodically. Here we will turn briefly to implications for the special case of psychology: psychology as a science or "psychological science," not only to stress that values infuse the very conception and practice of psychological science, but to consider some ways of moving beyond unhelpful polarities and impasses that may have relevance for similar impasses in interdisciplinary endeavors more generally. The topic here is distinct from the measurement of values in psychology or the role of values in the practice of psychotherapy, that is, the ethical considerations associated with clinical practice decisions. Also avoided is "research ethics" as such.

Instead, the focus is the role of values in the broader context of psychological inquiry, a practice that includes the empirical study of values and informs clinical practice but extends beyond these to all psychological research. Of course, this focus begs the question of the meaning of psychology and the psychological. That topic takes us too far afield, so for the purposes at hand the subject is psychological science as self-identified and practiced in contemporary context.

If it is a relatively straightforward thing to acknowledge diverse values implicated in vaccine development, it is even easier to identify values all over and in psychological science, making psychology an easy target for critique. The point has been made for decades (e.g., Burr 2015; Gergen 1985; Giorgi 2000; Howard 1985; Slife 2009; Slife and Williams 1995; Teo 2015), but there is now little space for denial even among those most sanguine about the scientific integrity of the discipline. The enlistment of psychologists to aid in administering torture in black sites, the impassioned rebuke of that involvement on the part of other psychologists, a crippling pandemic the origin and spread of which are tied directly to human behavior, and a gradual institutional awakening to ongoing, systemic racism should finally abolish any idea that psychology can or should understand itself in terms of a value-neutral stance. Moreover, it is obvious that psychologists display views about "good psychological science" that range from emphasis on the unification of theory or method to elimination of personal suffering to demonstrable intervention at the community level to exploration of human potential.

More fundamentally, the idea that psychology, as a human science, must be treated as distinct because of the investigator's dual role as subject and object of investigation is an idea found throughout the discipline's history in various calls to denounce natural science methods for the study of its own elusive subject matter. It appears in early analyses of what holds psychology together as a discipline – what we might mean by general psychology (Ward 1904), in mid-century critiques of psychological method, especially phenomenological (e.g., Giorgi 1970), critical psychology (e.g., Danziger 1994; Fox et al. 2009; Hook 2004; Parker 1999; Richards 2009; Teo 2015), and in Hacking's depiction of human science as a "looping effect," whereby the act of describing and classifying human behavior is transformative because of the human response to these classifications (Hacking 2007, p. 286). As Howard puts it, psychological scientists (and practitioners) are "agents in the formation of human beings" (1985, p. 262). In short, analysis of the values inherent in psychological science has tended to be subsumed in discussions that center on reflexivity, a word that Smith (2005) acknowledges has multiple meanings, and which is analyzed historically and philosophically by Smith and others, including Morawski (2005). It has also assumed a central place in pioneering feminist critiques of science (e.g., Keller 1982) and of psychology specifically (e.g., Hare-Mustin and Marecek 1988, 1990). Smith (2005) acknowledges as a descriptive reality the dual position of the human scientist and therefore underscores moral responsibility in every act of psychological theorizing, a point also made in the Ward's early analysis (1904).

Moreover, acknowledging the ubiquity of values in psychology does not inform us on how we are to deal with them as researchers or theorists. As a descriptive fact, we must recognize that many if not most psychologists, especially in North America, position their activity as well within the bounds of scientific practice, typically without undue reflection on what that entails. From Cronbach's alignment of experimental methods and actuarial assessment measures as definitive of the psychologist's identity, and by means of which they qualify the psychologist as a scientist rather than an artist, to the Strategic Plan of the American Psychological Association (APA) that establishes a scientific foundation as one of the discipline's core values, there is little evidence that the self-representation of most psychologists has incorporated the central messages of decades of critique. Psychology as a discipline has not done much to confront what seems to be an incompatibility between its mainstream and critical trajectories.

Disciplinary tradition advances the assumption that measurement precision and robust research design can minimize the influence of values, enabling the science itself to speak authoritatively through the authoritative use of measurement, statistics, and probability estimates (see Ezzamel and Willmott 2014; Grice et al. 2020; Hohn 2020; Lamiell 2020; Michell 2003). It follows logically that if values infiltrate all scientific practice, rigorous procedures do not successfully isolate us from the influence of values in our own science or any other, as is obvious even in the allegiance to a scientific foundation *as* a disciplinary value. Duffy and Chenail articulate values undergirding reliance on experimental methods in psychology as follows:

> (a) adherence to the rigorous procedures required in an experimental design; (b) researcher detachment and impartiality, as much as possible, to avoid undue influence over the research participants or subjects; (c) careful analysis of the data; (d) presentation of the findings without overstating their significance so that policy changes are not made lightly; and (e) disclosure of methodological and analytic procedures so that the study can be replicated by other experimental researchers. (Duffy and Chenail 2009, p. 25)

Similarly, especially fuzzy borders distinguish basic from applied science in the case of psychology. Several authors make explicit the central position of values and ethics inherent in the provision of service to the suffering (Miller 2004; Slife 2009; Tjeltveit 1999; Smith 2020). Yet as scientists, practitioners and clinical researchers must square their ethical entanglement with the images of science bequeathed by ideological tradition, with "scientism" as it has been named (Williams and Robinson 2014). As for any applied science, guidance for practice and justification for decision-making is sought from an evidence base, and with tacit agreement that evidence convincingly speaks for itself to inform treatment. Of course, the very concept of evidence-based practice begs the question of what counts as evidence, in what proportions, and toward what specific ends. The APA policy document acknowledges this very problem, that experts in psychological science disagree about what is to count as evidence and how different forms of evidence should be weighted and compared (APA Presidential Task Force on Evidence-Based Practice 2006).

It will not suffice to draw a contrast between quantitative and qualitative methods on the matter of values. Qualitative traditions, especially in more their more recent instantiations, certainly recognize a more intimate and co-constructive relation of researcher and participant, foregrounding the centrality of values. Yet although an *explicit concern* with values may legitimately distinguish qualitative and quantitative traditions, values are no less implicated in the analysis of variance, null hypothesis significance testing, or structural equation modeling than they are in phenomenological research or discourse analysis.

The past several decades bear witness to an extraordinary proliferation of methodological approaches to psychology, including qualitative and mixed-method approaches. More explicitly, quantitative procedures continue apace, some continuing in problematic practices (Grice et al. 2020; Lamiell 2020), but also offering innovative new approaches to dynamic and integrative modeling, blurring the boundary between quantitative and qualitative reasoning on some dimensions (Osbeck 2014). Yet as communities, the divide between qualitative and quantitative researchers in psychological science is stark; they operate almost as independent disciplines with different criteria and aims. Calls for epistemic or methodological pluralism in psychology are an appropriate response to recognition of methods as value systems, but even pluralism does not remove responsibility. There remain important questions about the affordances of various methods, about their relation to one another, and finally about the value of the interpretations and conclusions reached through their use. All methods cannot be equal to all purposes, thus the need to hierarchically order their merits in response to specific tasks arises inevitably, as does the need to evaluate the knowledge claims produced.

Methods as Value Systems

As noted, many resources from philosophy of science and theoretical psychology are available to inform any discussion of values and science, including the problem of how to conceptualize the intradisciplinary differences in value we see most readily in differences between quantitative and qualitative psychological research. Wolfgang Köhler's *The Place of Value in a World of Facts* (1938) is one such source, though underappreciated for its contemporary relevance. Written following the first world war, in a context of exceptional social and political upheaval and human tragedy, Köhler's text confronts the increasing public doubt about the capability of science to address the questions that matter most profoundly. Köhler acknowledges that the concept of value is vague, as are the questions we have about it, yet he describes science as "utterly imbued" with value; values are "at the bottom" of all human activities (Köhler 1938, p. 25). He depicts values as "vectors" (Köhler 1938, p. 74) to convey both that they are bear a directive relation to our activity and that they are irreducible: there is nothing more fundamental from which values are derived. *The Place of Values* also has relevance for the science of psychology as currently practiced given the profound methodological diversity already discussed. An idea we might draw from Köhler is useful for addressing the problem of how best to respect and preserve the integrity of different methodological traditions, that of construing methods as distinct value systems.

For any problem-solving context, including those of any science, values bear a close relation to what Köhler terms "requiredness," similar in meaning to conditions of possibility of the purposeful activity undertaken. Köhler notes that no scientific procedure can be accomplished "without at least the requiredness of logic, the distinction between essential and unessential facts, and so forth. A science, therefore, which would seriously admit nothing but indifferent facts even in its own procedure could not fail to destroy itself" (Köhler 1938, p. 35). Requiredness is tied to the specific aims that structure any problem situation or inquiry and it sets up conditions that are not arbitrary. There may be a range of possible ways to solve a problem, but these solutions are not limitless in number. They are bound and constrained by material conditions, social conditions, cognitive conditions, and conditions of value. In addition to the situation of any problem-solving context (for example, the problem addressed within a research lab), we can think of methodological traditions and frameworks as imposing forms of requiredness, constituting distinguishable value systems, within which norms specific to that system are binding, but without an accompanying assumption that they are binding outside of that system. In science studies, Longino (2002) offers a contemporary version of this idea, emphasizing the historical dimension of the normative framework that distinguishes systems: standards established within a scientific community are enforced within that system, even if the standards show variation across time and with the particularities of the scientific community in question.

For psychology, characterization of methodological traditions as value systems is compatible with the broader challenges to positivism Polkinghorne characterizes as "postpositivist" inasmuch as the conception of distinct sets of procedural requirements represents a departure from the project of a unified psychological science. Polkinghorne notes that "[i]n postpositivist science various systems of inquiry, each providing internal coherence and meaning to a research project, can be useful in developing knowledge. Methods, then, take their validity and reliability from their participation in a particular system of inquiry" (Polkinghorne 1983, p. 5). He calls his approach pragmatic, and credits earlier methodological analysis of Kaplan (1964) as the source of inspiration for the idea. An important difference in Polkinghorne's account is that methods are themselves not systems; "the use of a method changes only as a researcher uses it in different systems of inquiry." So cast, methods are merely instruments or tools to be used in various employments toward particular ends, various forms of problem-solving, though "the method is shaped by its implicit or explicit reference to a particular system of inquiry (Polkinghorne 1983, p. 6). It is the overarching value of usefulness, then, that determines what methods are to be used to investigate any question. Usefulness, however, must be determined by persons – in keeping with their values.

It is not clear where a system of inquiry ends and a method begins, nor are methods easily exchanged between frameworks, as is quickly witnessed when qualitative researchers are asked by reviewers to tabulate coding instances or provide reports on significance measures. It seems clearest to merely call methods themselves value systems, each tied to be sure to a set of specific assumptions about the nature of the subject matter and the goal of inquiry. This is in keeping with the assumptions forwarded by many contemporary qualitive researchers and with efforts to develop coherent standards by which the products of qualitative inquiry should be evaluated, especially when the evaluators hold different ideas about the aims of psychological science (e.g., Levitt et al. 2017, 2018). The relation between qualitative and quantitative traditions in psychology, revealing *intra*disciplinary differences, is comparable in important ways to that of disciplines that struggle to understand each other, and that grapple with the problem of values.

Viewing methods as value systems with distinct normative frameworks and dissimilar underlying assumptions, reflective of different historically sanctioned traditions of practice, enables us to at least aim to reconcile the goals and standards that may vary between methods. For example, broad scale "generalization" as traditionally conceptualized may not be the goal of narrative inquiry (though this is itself debatable). Importantly, however, identifying different methods within the parameters of a value system (whether the method is the system or merely participates in it) can give rise to the idea that these systems are incommensurable. The idea that different methods and their accompanying value sets reflect distinct and even incompatible world views is sometimes offered, often as a basis for defending the legitimacy of qualitative research (e.g., Guba and Lincoln 1994; Slife et al. 2017) or the overt foregrounding of religious values, for example, in counseling practice (e.g., Slife and Whoolery 2006).

Problems for Interdisciplinary and Intradisciplinary Collaboration

One problem is that appeals to incommensurability to defend methodological difference pose roadblocks to collaboration – whether between disciplines or within a discipline such as psychology that features different methodological traditions. There is a risk of missing the potential for productive engagement with excuses focused on difference, armed with an excuse akin to "you wouldn't understand," or worse, "I don't need to care if you don't understand." There are debilitating human problems of many varieties confronting us, and new surprises around every corner (e.g., pandemics) for which it would be helpful to foster flexible problem solving on the part of psychologists and participation with other disciplines toward innovative ends. Arguably the magnitude of the problems makes the project of intradisciplinary and intradisciplinary collaboration a moral imperative in our contemporary context.

In psychology, despite increasing interest in mixed method designs on the part of many journals and funding agencies, the conceptual and procedural grounds upon which productive intradisciplinary collaboration have not been adequately theorized (Wertz 2020). Yet intellectual polarization in the interest of value differences is hardly limited to psychology, and it is not always organized around completely different agendas or goals. In an ethnographic study of bioengineering laboratories, researchers, all of whom demonstrated commitment to the problem-solving agenda of the laboratory, and all of whom identify as natural scientists, demonstrated differences in their approach to problem-solving and method. Although scientists with different disciplinary backgrounds assumed different perspectives (levels of analysis) and used different instruments and methods and to address the problems undertaken, there was not an attitude akin to letting a hundred flowers bloom. Rather, researchers' descriptions of their collaborators were not neutral in affective or evaluative tone. They were sometimes marked by implicit critical appraisal of the colleagues' research goals, even the form of science practiced by other scientific disciplines. As described in greater detail elsewhere, this was especially evident in computational modelers' descriptions of experimental work as following recipes, and in turn biologists (using experimental methods) characterizations of high-level computational practices as out of touch with the concrete living cell. Similar examples emerged across four different laboratory analyses (Osbeck and Nersessian 2010, 2017). More encouragingly, other instances in the coding suggested that researchers were able to adopt a different perspective (e.g., that of a modeler when one has been trained as an experimentalist) when offered a targeted educational experience that put them in the position of the other discipline.

There are important implications for psychology and for the broader topic of values at hand. These align with the emphasis of scholars who argue for gains in understanding, even objectivity, with plurality of perspective (e.g., Harding 2015). As an activity, perspective-taking is important to both interdisciplinary and intradisciplinary collaboration, facilitating not only flexibility in thinking but a more comprehensive view (understanding) of the subject matter or problem under consideration. In turn, such flexibility and comprehensiveness of view establishes the condition for innovation: new concepts, methods, tools, and research agendas forged through comingling of disciplinary traditions, and a corresponding increase in problem-solving capability (Chandrasekharan and Nersessian 2015; Nersessian 2008). The rise of interdisciplinary science in recent decades reflects recognition that any hope of addressing problems of great complexity (e.g., climate change, global poverty, cancer) requires new channels for and forms of collaboration and cross-fertilization to stimulate innovation.

Towards Compromise and Negotiation

If effective collaboration is a goal, the question of what can be done in the face of deep differences between systems of value now presents itself. An assumption that open communication about values will facilitate mutual understanding and respect is core to the conclusions reached by several contemporary philosophers of science who study values, including Heather Douglas (2009, 2013) and Michael O'Rourke and colleagues (e.g., O'Rourke and Crowley 2013). They acknowledge that unhelpful impasses may be managed best by explicit attention to and open exploration of the values inherent in any deliberate epistemic activity, including the many complex activities constitutive of science. O'Rourke and colleagues have developed interventions (workshops) that make use of philosophical tools aimed at enhancing communication between collaborators with different epistemic emphases. Surely psychologists similarly have disciplinary resources available to aid in the project of encouraging clearer identification and articulation of values.

Regarding intradisciplinary collaboration specifically, an open discussion may be of limited use if many psychologists who self-identify as scientists persist in minimizing, even denying, the centrality of values to psychological science. Moreover, it is important to recognize that the position one takes on the question of the relation of values to science is not itself a matter of knowledge but of values, and thus it may be optimistic to imagine that communication alone can establish the grounds of effective collaboration when epistemic differences are pronounced.

Open-communication and understanding are clearly important, but so too is more fine-grained analysis and comparison of values across differing disciplinary systems or across methodological traditions in psychology. The choices are not limited to ignoring values or wallowing in the ubiquity of values to the detriment of further discrimination. Although we must recognize important ways they overlap, specification of kinds of value, that is, the value categories implicit in different methodological frameworks is important for interrogating the precise nature of the boundaries between systems. Here we might find important commonalities that establish conditions for collaborative potential, and perhaps challenge an assumption of incommensurability or discordant purposes. For example, differences in openness to psychoanalysis displayed by two psychological researchers might be properly understood as a conflict in aesthetic values in some cases – a preference for complexity of explanation and linguistic flourish versus a valuing of directness of expression and conceptual simplicity. These aesthetic values may masquerade as epistemic values, igniting conflict, when greater appreciation of Freud's own epistemic values might alleviate some concern on either side. Moreover, it is likely that psychologists who hold differing epistemic values share an overarching set of moral or ethical values, and that their activity is directed toward the shared purpose of alleviating human suffering. George Kelly offers a touching account of his own deepened appreciation for Freud's compassion despite the deep epistemic (and aesthetic) gulf between them (Kelly, in Maher 1969).

Notwithstanding the merits of conceptualizing methods as value systems, even as the idea is defended here, there is rarely a one-to-one correspondence between a given value and a given method. The Kelly example underscores that even when there are epistemic differences, there may be common social, ethical, or practical values in common between different traditions, including but not limited to those underlying therapeutic practice. For example, although social justice is traditionally more openly embraced as a value within critical psychology, which advocates the use of qualitative methods to "give voice" to underrepresented communities, many psychologists hold social justice as a primary value while also valuing the idea of objectivity or of science as epistemically authoritative (see Held 2020a). Moreover, epistemic lines of division between traditions may become more blurred upon closer analysis. Barbara Held also offers an excellent recent case study analysis of forms of evidence upheld as epistemic standards in the humanities (Held 2020b), calling into question the assumption that science (or social science) can be distinguished from the humanities based on its orientation to evidence gathering and appraisal. From this vantage point, psychological science and "psychological humanities" (e.g., Teo 2020) may have underlying correspondences in value that await yet more specific analysis.

Conclusion

Questions remain, with much hinging on the answers. How are we to think about and what are we to do about values in science? How can we harness the predictive power of science in an era of tremendous challenge and possibility while retaining sensitivity to its blatant abuses? In short, what non-dichotomous options present themselves, options that veer from the unhelpful extremes of embracing the value-free ideal or disparaging science as archaic and oppressive, leaving us no platform for informed decisions?

One answer lies in the very social foundation of science, acknowledging the *social contracts* that make it possible, and the social uses to which it is directed. For Longino, it is the democratic society of science, one that relies on the deliberation and scrutiny of the community: hypotheses become knowledge, and their proper use is determined only when there is sufficient agreement within a collective of evaluators who have negotiated a set of criteria (Longino 1990, 2002). Similarly, Oreskes (2019) emphasizes that the very openness to criticism established by the broader scientific community is the basis for our ability to trust the recommendations of its experts. In the face of decades of scrutiny and critique, what remains is a view of "science as a communal activity of experts, who use diverse methods to gather empirical evidence, and critically vet claims deriving from it" (Oreskes 2019, p. 246). Oreskes notes that this was indeed the most important implication of the social turn in science studies.

But experts are persons, and attention must also be paid to the personal dimension of science practice, including the kind of thinking that generates good and fruitful ideas able to pass the tests established within scientific communities. In the interests of enhanced understanding and increased capacity for flexible and adaptive reasoning, it is important to analyze common epistemic activities across disciplinary traditions and associated value systems, such as the importance of observation, imaginative sense-making, and perspective-taking across the humanities and sciences as well as self-corrective strategies (Osbeck 2018). A variation on this point is evident in the conclusion Cowles reaches after a lengthy and nuanced historical study of the scientific method. His coverage of Dewey's analysis of scientific method focuses on the continuity Dewey found between scientific reasoning and the forms of reasoning required for any other-directed activity: "It is not a peculiar development of thinking for highly specialized ends; it is thinking so far as thought has become conscious of its proper ends and of the equipment indispensable for success in their pursuit" (Dewey 1910, quoted in Cowles 2020, p. 260). Noting that "champions and critics" of science are alike currently preoccupied with questions about the potential for scientific method to escape bias, political or otherwise, Cowles underscores the historical and contingent situation of this very debate yet ties it to an array of contemporary crises wrought by the deeply divided attitudes toward science. He likewise hints at the affordances of a return to the earlier view of science demonstrated in Dewey. "It is possible," Cowles emphasizes, "to think of science as the flawed, fallible activity of some imperfect, evolving creatures *and* as a worthy, even noble pursuit" (Cowles 2020, p. 279, emphasis mine). Although this conclusion has merit, it does not fully address the question of the grounds upon which our trust in science should be placed. Moreover, to bring in consideration of what worthiness of pursuit consists in, whether for science or any other activity, we are opened to a much larger conversation. It is one that recalls the virtues of epistemic humility or modesty (Oreskes 2019; Teo 2019) and requires greater attention to the ways the personhood of the researcher impacts the production and use of scientific knowledge (including psychology) for good or ill. These matters in turn call for reflection on the meaning and cultivation of wisdom, something we must put off for another occasion but can never avoid entirely.

References

Allport GW (1955) Review of Realms of value: a critique of human civilization [Review of the book Realms of value: a critique of human civilization, by R. B. Perry]. J Abnorm Soc Psychol 50(1):154–156. https://doi.org/10.1037/h0038319

APA Presidential Task Force on Evidence-Based Practice (2006) Evidence-based practice in psychology. Am Psychol 61(4):272–285

Bacon F (1902) Novum organum. J Devey (Ed) PF Collier. Original work published 1620

Bloor D (1976) Knowledge and social imagery. Routledge/Kegan Paul

Burr V (2015) Social constructionism. Routledge

Campbell JB, Jayawickreme E, Hanson EJ (2015) Measures of values and moral personality. In: Measures of personality and social psychological constructs. Academic, pp 505–529

Chandrasekharan S, Nersessian NJ (2015) Building cognition: the construction of computational representations for scientific discovery. Cogn Sci 39(8):1727–1763

Comte A (1848) Republic of the West order and progress: a general view of positivism or, summary explosion of the system of thought and life. Academic Reprints, Stanford

Cowles HM (2020) The scientific method: an evolution of thinking from Darwin to Dewey. Harvard University Press

Danziger K (1994) Constructing the subject: historical origins of psychological research. Cambridge University Press

Dewey J (1910) How we think. Heath

Dorato M (2004) Epistemic and nonepistemic values in science. In: Machamer P, Wolters G (eds) Science, values, and objectivity. University of Pittsburgh Press, pp 52–76

Douglas H (2009) Science, policy, and the value-free ideal. University of Pittsburgh Press

Douglas H (2013) The value of cognitive values. Philos Sci 80(5):796–806

Douglas H (2016) Values in science. In: Humphries P (ed) The Oxford handbook of philosophy of science. Oxford University Press, pp 609–632

Duffy M, Chenail RJ (2009) Values in qualitative and quantitative research. Couns Values 53(1):22–38

Elliott KC (2017) A tapestry of values. An introduction to values in science. Oxford University Press

Ezzamel M, Willmott H (2014) Registering 'the ethical' in organization theory formation: towards the disclosure of an 'invisible force'. Organ Stud 35(7):1013–1039

Feist GJ (2008) The psychology of science and the origins of the scientific mind. Yale University Press

Feyerabend P (1970) Against method. Verso

Fleck L, Kuhn T (1981) Genesis and development of a scientific fact (eds: Trenn T, Merton R; trans: Bradley F). University of Chicago Press

Fox D, Prilleltensky I, Austin, S. (Eds.). (2009) Critical psychology: an introduction. SAGE

Gaukroger S (2020) Civilization and the culture of science. Oxford University Press

Gergen K (1985) The social constructionist movement in modern psychology. Am Psychol 40(3): 266–275

Giorgi A (1970) Psychology as a human science: a phenomenologically based approach. Harper & Row

Giorgi A (2000) Psychology as a human science revisited. J Humanist Psychol 40(3):56–73. https://doi.org/10.1177/0022167800403005

Grice JW, Huntjens R, Johnson H (2020) Persistent disregard for the inadequacies of null hypothesis significance testing and the viable alternative of observation-oriented modeling. In: Problematic research practices and inertia in scientific psychology. Routledge, pp 55–69

Guba EG, Lincoln YS (1994) Competing paradigms in qualitative research. In: Handbook of qualitative research, vol 2(163–194). SAGE, pp 105–117

Hacking I (2007) Kinds of people: moving targets. Proc Br Acad 151(p):285–318

Hadamard J (1945) The mathematician's mind: the psychology of invention in the mathematical field. Princeton University Press

Haig BD (2014) Investigating the psychological world: scientific method in the behavioral sciences. MIT Press

Harding SG (2015) Objectivity and diversity: another logic of scientific research. University of Chicago Press

Hare-Mustin RT, Marecek J (1988) The meaning of difference: gender theory, postmodernism, and psychology. Am Psychol 43(6):455

Hare-Mustin RT, Marecek J (eds) (1990) Making a difference: psychology and the construction of gender. Yale University Press

Held BS (2020a) Epistemic violence in psychological science: can knowledge of, from, and for the (othered) people solve the problem? Theory Psychol. https://doi.org/10.1177/0959354319883943

Held BS (2020b) Taking the humanities seriously. Rev Gen Psychol. https://doi.org/10.1177/1089268020975024

Hohn RE (2020) Intransigence in mainstream thinking about psychological measurement. In: Lamiell J, Slaney K (eds) Problematic research practices and inertia in scientific psychology. Routledge, pp 39–54

Holton G (1996) Einstein, history, and other passions. Addison-Wesley Publishing

Holton GJ (1998) The advancement of science, and its burdens: with a new introduction. Harvard University Press

Hook D (2004) Frantz fanon, Steve Biko, 'psychopolitics' and critical psychology. In: Hook D, Collins A, Burman E, Parker I, Kiguwa P, Mkhize N (eds) Critical psychology. UCT Press, pp 84–114

Howard GS (1985) The role of values in the science of psychology. Am Psychol 40(3):255–265. https://doi.org/10.1037/0003-066X.40.3.255

Kaplan A (1964) The conduct of inquiry: methodology for behavioral science. Chandler Publishing Company, San Francisco

Keller EF (1982) Feminism and science. J Women Cult Soc 7(3):589–602

Kelly G (1969) Autobiography of a theory. In: Maher B (ed) Clinical psychology and personality: the selected papers of George Kelly. Wiley, pp 46–92

Kitcher P (1993) The advancement of science: science without legend, objectivity without illusions. Oxford University Press, Oxford

Kitcher P (2001) Science, truth, and democracy. Oxford University Press, Oxford

Köhler W (1938) The place of value in a world of facts. Liveright

Kuhn T (1962) The structure of scientific revolutions. University of Chicago Press

Kvale S (1992) From the architecture of the psyche to the architecture of cultural landscapes. In: Kvale S (ed) Psychology and postmodernism. SAGE, pp 1–16

Lacey H (1999) Is science value-free? Values and scientific understanding. Routledge

Lamiell JT (2020) On the systemic misuse of statistical methods within mainstream psychology. In: Lamiell J, Slaney K (eds) Problematic research practices and inertia in scientific psychology: history, sources, and recommended solutions. Routledge, pp 8–22

Latour B, Woolgar S (1986) Laboratory life: the construction of scientific facts. Princeton University Press. (Original work published 1979)

Lauden L (1984) Science and values. University of California Press

Levitt HM, Motulsky SL, Wertz FJ, Morrow SL, Ponterotto JG (2017) Recommendations for designing and reviewing qualitative research in psychology: promoting methodological integrity. Qual Psychol 4(1):2–22. https://doi.org/10.1037/qup0000082

Levitt HM, Bamberg M, Creswell JW, Frost DM, Josselson R, Suárez-Orozco C (2018) Journal article reporting standards for qualitative primary, qualitative meta-analytic, and mixed methods research in psychology: the APA publications and communications board task force report. Am Psychol 73(1):26–46. https://doi.org/10.1037/amp0000151

Longino H (1990) Science as social knowledge: value and objectivity in scientific inquiry. Princeton University Press

Longino H (2002) The fate of knowledge. Princeton University Press

Machamer P, Douglas H (1999) Cognitive and social values. Sci Educ 8(1):45–54

Machamer P, Osbeck L (2004) The social in the epistemic. In: Machamer P, Wolters G (eds) Science, values, and objectivity. University of Pittsburgh Press, pp 78–89

Machamer P, Wolters G (2004) Introduction. In: Machamer P, Wolters G (eds) Science, values, and objectivity. University of Pittsburgh Press, pp 1–13

MacLeod MAJ, Nersessian NJ (2016) Interdisciplinary problem-solving: emerging modes in integrative systems biology. Eur J Philos Sci 6(3):401–418. https://doi.org/10.1007/s13194-016-0157-x

McAllister JW (1999) Beauty and revolution in science. Cornell University Press

McAllister JW (2002) Recent work on aesthetics of science. Int Stud Philos Sci 16(1):7–11

Michell J (2003) The quantitative imperative. Theory Psychol 13(1):5–31. https://doi.org/10.1177/0959354303013001758

Mill JS (1865) Auguste Comte and positivism. Trubner

Miller R (2004) Facing human suffering: psychology and psychotherapy as moral engagement. American Psychological Association

Mitroff I (1974) The subjective side of science: philosophical inquiry into the psychology of the Apollo Moon Scientists. Elsevier

Morawski JG (2005) Reflexivity and the psychologist. Hist Hum Sci 18(4):77–105

Nersessian N (2008) Creating scientific concepts. MIT Press

Neurath O (1973) Empiricism and sociology. Reidel, Dordrecht. (Originally published as Carnap, Hahn, and Neurath (1973). Wissenschaftliche Weltauffassung: Der Wiener Kris, 1929, Verlag)

Nietzsche F (2000) The will to power: an attempted transvaluation of all values (trans: Ludovici A, vol 1, books 1–2). Macmillan, New York. (Originally published 1909)

O'Rourke M, Crowley SJ (2013) Philosophical intervention and cross-disciplinary science: the story of the toolbox project. Synthese 190(11):1937–1954

Oreskes N (2019) Why trust science? Princeton University Press

Osbeck L (2018) Values in psychological science: re-imagining epistemic priorities at a new frontier. Cambridge University Press

Osbeck L, Nersessian N (2010) Forms of positioning in interdisciplinary science practice and their epistemic effects. J Theory Soc Behav 40(2):136–161

Osbeck LM (2014) Scientific reasoning as sense-making: implications for qualitative inquiry. Qual Psychol 1(1):34–46

Osbeck LM, Nersessian NJ (2017) Epistemic identities in interdisciplinary science. Perspect Sci 25(2):226–260

Osbeck LM, Nersessian NJ, Malone KR, Newstetter WC (2010) Science as psychology: sense-making and identity in science practice. Cambridge University Press

Parker I (1999) Critical reflexive humanism and critical constructionist psychology. In: Nightingale DJ, Cromby J (eds) Social constructionist psychology: a critical analysis of theory and practice. Open University Press, Buckingham, pp 23–36

Polanyi M (1974) Personal knowledge. Toward a post-critical philosophy. University of Chicago Press. Originally published 1958

Polkinghorne D (1983) Methodology of the human sciences: systems of inquiry. State University of New York Press

Proctor R (1991) Value-free science? Purity and power in modern knowledge. Harvard University Press, Cambridge

Proctor R, Capaldi E (2006) Why science matters: understanding the methods of psychological research. Blackwell

Rescher N (1969) Introduction to value theory. Prentice-Hall

Richards G (2009) Putting psychology in its place: critical historical perspectives, 3rd edn. Routledge

Rouse J (2002) How scientific practices matter: reclaiming philosophical naturalism. University of Chicago Press

Shapin S (1996) A social history of truth: civility and science in seventeenth-century England. University of Chicago Press

Slife BD (2009) A primer of the values implicit in counseling research methods. Couns Values 53(1):8–21

Slife BD, Whoolery M (2006) Are psychology's main methods biased against the worldview of many religious people? J Psychol Theol 34(3):217–231

Slife BD, Williams RN (1995) What's behind the research? Discovering hidden assumptions in the behavioral sciences. SAGE, London

Slife BD, O'Grady KA, Kosits, R. D. (Eds.). (2017) The hidden worldviews of psychology's theory, research, and practice. Taylor & Francis

Smith R (2005) Does reflexivity separate the human sciences from the natural sciences? Hist Hum Sci 18(4):1–25

Smith KR (2020) Therapeutic ethics in context and in dialogue. Routledge

Teo T (2015) Critical psychology: a geography of intellectual engagement and resistance. Am Psychol 70(3):243–254. https://doi.org/10.1037/a0038727

Teo T (2019) Academic subjectivity, idols, and the vicissitudes of virtues in science: epistemic modesty versus epistemic grandiosity. In: O'Doherty K, Osbeck L, Schraube E, Yen (eds) Psychological studies of science and technology. Palgrave, pp 31–48

Teo T (2020) The primacy of critical theory and the relevance of the psychological humanities. In: Cultural-historical and critical psychology. Springer, Singapore, pp 63–76

Tjeltveit A (1999) Ethics and values in psychotherapy. Routledge

Tolman EC (1948) Cognitive maps in rats and men. Psychol Rev 55(4):189–208. https://doi.org/10.1037/h0061626

Turkle S (2012) Alone together. Basic Books

Vernon PE, Allport GW (1931) A test for personal values. J Abnorm Soc Psychol 26(3):231–248. https://doi.org/10.1037/h0073233

Ward J (1904) The present problems of general psychology. Philos Rev 13(6):603–621

Washburn P (2020) Is truth relative? J Constr Psychol. https://doi.org/10.1080/10720537.2020.1727389

Wertz FJ (2020) Objectivity and eidetic generality in psychology: the value of explicating fundamental methods. Qual Psychol. https://doi.org/10.1037/qup0000190

Wilholt T (2013) Epistemic trust in science. Br J Philos Sci 64(2):233–253

Williams RN, Robinson DN (eds) (2014) Scientism: the new orthodoxy. Bloomsbury Publishing

Politics and Ideology in the History of Psychology: Stratification Theory in Germany

46

Martin Wieser

Contents

Introduction: The Ambivalent Relationship Between Psychology and History	1196
Showdown in Montreal: The Beginning of the Demise of Stratification Theory	1199
"The Great Chain of Being": Ancient Roots of Personality Psychology	1201
A Forgotten Universe of Layers in Philosophy, the Sciences, and Psychology	1203
The Uncovering of the Layers of Personality: German Psychology on the Rise	1207
Stratification Theory and the Search for Holism in the Weimar Republic	1208
Stratification Theory in Wartime	1211
After the Fall: Continuity and Breaks at the Dawn of the Cold War	1212
Conclusion	1214
References	1215

Abstract

This chapter is concerned with the interaction between Western academic psychology, politics, and ideology in the twentieth century. Stratification theory is used as a case example to illustrate how political and ideological factors influenced the development of psychological thought and practice through different historical contexts. Based on metaphysical ideas from Greek and Christian cosmology, stratification theory emerged during the Enlightenment as successor to the "great chain of being." Stratification theory reached its greatest influence in the interwar period, when its concepts became popular across many different disciplines in the natural sciences and the humanities. In German psychology, it served as a theoretical fundament for the analysis and description of personality. After the rise of National Socialism, some proponents of stratification theory emphasized the compatibility of their theories with racist doctrines and aimed to prove the practical usefulness of their knowledge for the regime. Despite the involvement of its

---

M. Wieser (✉)
Faculty of Psychology, Sigmund Freud Private University Berlin, Berlin, Germany
e-mail: martin.wieser@sfu-berlin.de

© The Author(s), under exclusive licence to Springer Nature Singapore Pte Ltd. 2022
D. McCallum (ed.), *The Palgrave Handbook of the History of Human Sciences*,
https://doi.org/10.1007/978-981-16-7255-2_48

proponents with National Socialism, stratification theory prevailed for another decade in the German academic landscape after end of World War II, until it was finally superseded by cybernetics at the dawn of the Cold War.

Keywords

Stratification theory · History of psychology · Psychology of personality · World War II · Cold War · Cybernetics

Introduction: The Ambivalent Relationship Between Psychology and History

From the outside, it might appear quite puzzling that academic psychology, one of the most influential and authoritative human sciences today, has little to say about the historical dimension of its subject matter. Experimental methods and practically oriented bodies of knowledge (e.g., testing, counselling, and clinical and therapeutic areas of practice) came to dominate the psychological curricula in most universities over the course of the twentieth century (Baker and Benjamin 2014). This blind spot in psychological thinking has not gone completely unnoticed. The complex and ambivalent relationship of Western academic psychology with history – whether it is directed towards psychology's subject matter or the past of the discipline – has been the subject of regular and controversial discussion over the last 50 years. Criticisms that highlight the lack of historical perspectives in academic psychology were not just voiced by historians (Smith 2007), but were also frequently published by psychologists as well (Gergen 1973; Danziger 1997; Holzkamp 2013; Teo 2015). Nevertheless, the majority of the profession appears not too impressed by these debates (Danziger 1994). One may have become accustomed to the fact that historical analyses are not usually part of the methodological canon of psychology today, but the question remains why this is the case. Taking into consideration that there exists a long and rich tradition of historical thought in neighboring sciences like biology, sociology, and philosophy, it remains surprising how little recognition is given to historical perspectives in Western academic psychology today.

If we take a step back from the current situation and look at the state of the field during the second half of the nineteenth century, when psychology was yet to become a distinct, institutionalized discipline in Europe and North America, we can easily see that historical and psychological thinking were much closer together at that time. In Germany, the philosopher Georg Wilhelm Friedrich Hegel promoted the idea of an "objective spirit" (*"objektiver Geist"*), which did not just represent a supra-individual mental structure but also a *process* that unfolds throughout history. Hegel's concept of "objective spirit" and its reinterpretation as *"Volksgeist"* ("mind of the nation") was fundamental for a field of study named *"Völkerpsychologie"* (which can be roughly translated as "folk psychology" or "psychology of the nation."). It was first popularized by Moritz Lazarus and Heymann Steinthal from the 1850s onwards (Klautke 2013). Wilhelm Wundt adopted many of Lazarus' and

Steinthal's ideas and carried them forward in his version of "*Völkerpsychologie*" (Wundt 1916). The fact that Wundt is often cited today as a founding figure of modern academic psychology is somewhat ironic, since from his point of view, only one half of his inheritance was carried on. While the experimental side of the discipline (which he named "physiological psychology") soon had many followers, his extensive historical treatises found very little support among his colleagues (Danziger 1990).

While most of academic psychology began to move away from philosophy, historical thinking gained widespread popularity in the emerging field of "*Geisteswissenschaften*." Wilhelm Dilthey, one of its most influential spokespersons, promoted the use of "hermeneutic" methods, which build on "understanding" meaningful "wholes" instead of "explaining" and analyzing isolated processes of cause and effect. Dilthey advocated implementing these perspectives also in psychology and moving away from causal explanation and the search for universal laws (Dilthey 1894), but representatives of experimental psychology such as Hermann Ebbinghaus (1896) strongly rejected Dilthey's suggestion. In the German-speaking world, debates concerning "Geisteswissenschaftliche Psychologie" continued for several decades (Bühler 1927). It remained a somewhat visible part of the discipline until the 1950s. Since then, experimental methods and applied branches of the discipline have begun to dominate the field, for reasons that will be explained further below.

A critical reader who is somewhat familiar with academic psychology today might object that there are quite a few firmly established and areas where temporality plays an important theoretical role. One example can be found in areas where the influence of biological theories of evolution was particularly strong, such as in the English-speaking world from the middle of the nineteenth century onwards (Spencer 1855; Wright 1877; Morgan 1891; Baldwin 1894). In the US, Darwin's theory of evolution served as the theoretical foundation of for representatives of functionalism (James, 1892; Angell 1909) and early behaviorism (cf. Wozniak 1993). Those who recognized "no dividing line between man and brute" (Watson 1913) shared the fundamental proposition that mental functions and behavior are shaped through the course of evolution by the principles of adaption and selection (cf. Green 2009). This line of thought is very much still alive, especially in the context of modern evolutionary psychology (Buss 2019). Another example is the subdiscipline of developmental psychology, which is concerned with the growth and development of mental functions and personality over the course of the lifetime. The most prominent models used in this field, such as those of Jean Piaget, Sigmund Freud, and Erik Erikson, present mental development (whether it is concerned with modes of thinking, psychosexual development, or the capability for moral judgment) as a universal series of "stages" or "phases" that appear at a certain age and in a predefined temporal order.

Evolutionary psychology and developmental psychology provide theoretical frameworks for psychologists who aim to conceptualize the *becoming* of their subject matter. However, as these perspectives promote a temporal perspective on the human being as a *species* (its phylogenesis) or as an *individual* (its ontogenesis), a significant part of history remains left out: the historical becoming of human subjectivity in connection with its societal context. Changing political, economic, and cultural conditions that facilitate and shape, foster and limit human thinking, feeling, and acting have no real place either in evolutionary psychology or developmental psychology. For this reason, critics such as Kurt Danziger observed that, "as for history in the usual sense, it is not regarded as having any significance for current psychological investigation or its products" (Danziger 1997, p. 9).

Why are proposals to integrate historical thinking so vehemently rejected in academic psychology? Several hypotheses can be taken into consideration. Rose (1988) and Holzkamp (2013) argued that the supposed lack of historical thinking and critical reflection in psychology is intrinsically connected to its function as a body of "governmental" knowledge in modern capitalism. Others emphasized that the historical roots of psychologists' orientation towards the natural sciences are connected with the need to demarcate the discipline from philosophy (Danziger 1990; Kusch 1999). Furthermore, some critics argue that when the past of the discipline is taken into account, it is often in a "jubilatory" manner, celebrating the "great men" and their "achievements" while paying little attention to those schools or currents that were discarded or forgotten (Young 1966; Smith 1988; Danziger 1994).

The complex question why psychology has become an ahistoric discipline surely extends beyond the scope of this chapter. Instead, a historical and context-focused perspective on the intertwining of psychological theorizing and changing political and ideological contexts is taken up. In doing so, we follow Roger Smith's suggestion that "historical case studies about the establishment of authoritative knowledge, or about the failure of this goal, may have great contemporary relevance" (Smith 2022). Our analysis of the historical connections between history, politics, ideology, and academic psychology serves two major functions. Firstly, it is intended to deepen our understanding of the specific political and ideological dynamics that facilitated the transformation of academic psychology from a field of research to a profession in the Western world. Secondly, by contextualizing theories and concepts in use today, we go beyond generating knowledge about the social foundation of scientific theorizing *and* contributes to a historical understanding of psychology's subject matter. In contrast to "natural kinds," such as molecules and chemical substances, "human kinds" can read, interpret, acknowledge, or deny scientific theories about themselves. In the human sciences, an interaction between theories and their subject matter is constantly taking place – a phenomenon which the philosopher Ian Hacking named "looping" (Hacking 1995). As psychology gradually turned into an independent academic discipline during the early twentieth century, many of its representatives also began to explore new territories for psychological practices (Napoli 1981). The more successful they were, the more psychological and psychoanalytic categories (such as "motivation," "intelligence," "repression," or the "inferiority complex") became part of everyday language (Smith 2005). Practices such as testing mental capabilities, the assessment of personality, or the interpretation of dreams were quickly integrated into many areas of modern living, from the work place to spare time activities (Pickren and Rutherford 2010; Furedi 2004). Therefore, the story that will be told in the following is not only concerned with the history of academic knowledge. It also tells us what kind of knowledge was offered by psychologists and how it was subsequently acquired, reinterpreted, adapted, and criticized by others. As this knowledge was transformed and put into practice, those who adopted psychological knowledge became different "human kinds."

The case study that follows focuses on the transformation of Western academic psychological thinking and practice between the 1930s and the 1960s. Even though the time span may appear short, there are good reasons to describe it as an "axial age" (Jaspers 1953) of substantial change and development in the history of Western academic psychology. In Germany, the structure of the discipline changed profoundly during these three decades on the level of theory, concepts, and methods, as well as in regards to its practical function in society. The period from the 1930s to the 1950s saw the rise and fall of two major intellectual currents, stratification theory and cybernetics, while the rise of fascism led into the catastrophe of World War II. The events that will be recounted exemplify how the fate of psychological currents can be determined by political and ideological dynamics that are beyond their control. This, however, does not say that individual psychologists had no other choices. While the focus of the following narrative is on Germany, the moral of the story goes far beyond this context. Even if the situation depicted was historically unique, the dynamics behind it are still visible today. Their analysis deepens our understanding of how our concept of "being human" is intrinsically linked to (and dependent on) specific political, economic, technological, and societal conditions.

Showdown in Montreal: The Beginning of the Demise of Stratification Theory

In June 1954, the 14th International Congress of Psychology was held in Montreal, Canada. The conference was organized by the International Union of Scientific Psychology and brought together renowned representatives from Europe and North America (David and Bracken 1957). The list of participants included eminent names such as Gordon Allport, Henri Ellenberger, Hans-Jürgen Eysenck, Else Frenkel-Brunswik, Philipp Lersch, and Albert Wellek, all of whom had contributed to the field of personality psychology or psychopathology in their academic work. Since 1889, this series of conferences had been held regularly in various European cities. As Gordon Allport noted in the introduction to the conference proceedings, the relationship between European and American psychologists had significantly changed since that time:

> a generation ago scarcely any American psychologist considered himself adequately trained until he had spent a predoctoral or postdoctoral year in Europe. Now the tables seem turned. European psychologists are visiting American shores in greater numbers than ever before, while fewer American students go abroad for psychological training (Allport 1957, p. 4)

The Montreal meeting was the third gathering after the end of World War II and the first one that was set on North American soil. Psychological communities that had been separated during the war come together again, and even though the conference had started with the aim to foster the intellectual exchange between national schools and currents, it quickly became clear that there was a considerable amount of disagreement on the conceptual and methodological foundations of the discipline between the German and US-American representatives.

The German participants such as Helmut von Bracken, Philipp Lersch, Hans Thomae, and Albert Wellek seemed to be particularly fond of the concept of "layers" or "strata" [*Schichten*] which most of their colleagues from North America seemed to be unaware of. Their psychological theory of stratification suggested that the structure and dynamics of the adult personality is the result of the unfolding and interaction of cognitive, volitional, emotional, and psychophysiological "layers" over the course of lifetime. Stratification theory and characterology (in combination with elements of holistic psychology [*Ganzheitspsychologie*] and psychoanalysis) represented the theoretical foundations of the German approach as it was presented in Montreal. On a practical level, they promoted the use of expression psychology [*Ausdruckspsychologie*] to decipher the "expression" of the personality by analyzing the mimicry, gesture, speech, writing, and other forms of overt behavior. During the war, personality psychology and its applications had come to dominate the discipline in Germany. Lersch and many of his colleagues took over leading roles during the expansion of psychology as a profession, especially in the field of military aptitude testing (Geuter 1992). When the conference was held in Montreal, the involvement of German psychologists in the Wehrmacht was still considered a success story by its representatives (Wieser 2020a). However, most of their US-American colleagues seemed quite skeptical in regard to the methodological soundness as well as the empirical foundation of what their German colleagues presented. Hans-Jürgen Eysenck, an avid proponent of statistical and experimental methods, stood out as one of the harshest critics of the German approach. He argued that "the discussion of stratification theory is extremely obscure, fails to come to a sharp focus, and leaves the reader without any clear-cut definition of the meaning of the terms used" (Eysenck 1957, p. 323). Operationalism and behaviorism had come to dominate psychology in the English-speaking world since the 1930s. As an outspoken advocate of this tradition, Eysenck saw German psychologists lapsing into metaphysical aberrations. To him, their position

> is essentially anti-scientific; a late growth of the idiographic, geisteswissenschaftliche, "understanding" approach so favored by German philosophers. As such it will not be easily acceptable to psychologists of the Anglo-Saxon countries who wish to treat the study of behavior and personality as a branch of science. (Eysenck 1957, p. 324)

From Eysenck's perspective, the only valid methodological basis for psychology is the observation, measurement, prediction, and control of behavior. To get there, psychology needs to rely on the systematic construction and careful testing of hypotheses. However, his German colleagues based their approaches on unreliable methods (such as empathic "understanding" of the "expression" of personality) to promote unfalsifiable propositions about the "layers of personality." As these assumptions could not be directly observed, tested, or measured, he saw little use in dealing with stratification theory, characterology, or psychoanalysis until their proponents agreed to operationalize their terms and standardize their methods. Wellek countered that mathematical analysis was not suitable to fully grasp the richness, dynamics and ambivalences of the human personality. Instead, he promoted the use of trained intuition and phenomenological perspectives to analyze the unique "wholeness" of the personality (Wellek 1957). Some participants from both sides, such as Gordon Allport, Albin Gilbert, and Helmut Bracken, looked for common ground and aimed to build bridges between these traditions. Nonetheless, it became obvious that this was not an easy task, as differences in methodology, epistemology, and metaphysics were quite profound.

Without knowing what came later, the conference report from Montreal might suggest that what happened in Montreal was not much different from any other critical exchange between two competing intellectual currents within one discipline. From today's perspective, the confrontation appears much more like the last stand of stratification theory in the psychological field: Lersch, Bracken, Thomae, and Wellek were to become the last generation of a school of thought that rapidly lost its influence after the 1950s. As debates at the Montreal conference were centered mostly on questions of methodology, the historical roots, political contexts, and societal function of the English-speaking and German traditions was not seriously up for debate (a pattern that is still not uncommon in psychological fields today). Yet, even though these aspects were not openly discussed, they still played an important role. In retrospect, some readers might assume that stratification theory was simply falsified on scientific grounds. In fact, the situation was much more complex, as many different factors contributed to its demise. To understand the reasons and consequences of this turnaround that started in Montreal, it is necessary to first discuss the roots and emergence of stratification theory in the psychological field.

"The Great Chain of Being": Ancient Roots of Personality Psychology

Many centuries before German psychologists clashed with their US-American colleagues in Montreal, their intellectual precursor emerged in ancient Greece. Since the middle ages, it was known as the "great chain of being," though the basic principles of this idea can already be found in the writings of Plato and Aristotle. In his dialogue *Timaios* Plato suggested that the divine "absolute," through its omnipotence, emanates "being" into the world. From "above," the absolute "fills" the universe with existence. In this metaphysical framework, everything that *can* exist *does* exist: there are no "holes" or gaps in nature, as they would prove that god missed out on something that could have been created. While Plato only differentiated between the (lower) realm of ephemeral physical objects and the (higher) realm of immortal ideas, Aristotle's *History of Animals* expanded this vertical ontology to a linear order that reaches from the lowest, most simple and inanimate elements up to the divine. There is an eternal order in our world, Aristotle (1902) argued, which reaches from minerals, plants, animals, and human beings up to the stars, as there is a continuous increase of perfection in nature.

In his classic study, "The great chain of being," the philosopher and historian Arthur Lovejoy named three basic axioms of this metaphysical system that deeply influenced Western thinking for many centuries to come: the principle of "plenitude" (everything that can exist does exist), the principle of "continuity" (there are no holes or gaps in nature), and the principle of "participation" (the higher a creature is located on the ladder, the closer it is to god and the more it participates in his perfection). Neoplatonist and scholastic authors such as Plotinus, Augustine, Ramon Llull, and Thomas Aquinas connected the "chain" with Christian thought and teachings of the bible (Lovejoy 2001). These writers fused Christian lore and Greek philosophy to promote the idea of a hierarchical and rational cosmos that was created by the will of the almighty god. According to this ontology, everything that exists depends on the highest, most perfect of all beings and has its proper place in nature. This metaphysical system had an epistemological as well as a political and ethical dimension: As human kind is located in the middle of the "chain," below the spiritual world and above animals, plants, and matter, it reflects the rational order of the cosmos and contains the highest and the lowest parts in itself. Reason, the rational part of the soul, enables human beings to recognize the eternal laws of the universe and god as well. But as the human soul is trapped in a body made out of flesh and blood, it is also bound to the lower realm of matter. Therefore, it is assumed that human's ability to gain true knowledge is also limited by their mortality. The "chain" includes a moral imperative as well, as it suggests searching for the way "up" towards wisdom. By increasing their knowledge about all the elements of the "chain," humans come to know about and to come closer to their creator. Humans also recognize and learn to accept their position in the universe – as well as in society. Everything that exists shows the same hierarchical, vertical structure: from the pope to the peasant, the order of the feudal society is legitimized by the hierarchical order of the universe presented in the "chain."

The cosmology of the "great chain" conveyed the image of a static cosmos in which everything that exists has its predetermined place since the beginning of time. However, this hierarchical ontology represented more than just a syncretic mixture of metaphysical dogmas from Greek philosophy and Christianity. Empirical research also had its proper place within this worldview, since many elements of the "chain" waited to be discovered: beyond all known kinds of birds and bees, fish and trees, apes and the human, there was an unknown number of species that still were still unknown. After the middle ages, the "great chain" was not abandoned, but it was gradually transformed and gained a new role during the eighteenth and nineteenth centuries. The *scala naturae* was taken up in the writings of Gottfried Wilhelm Leibniz, Carl Linnaeus, Charles Bonnet, and the Comte de Buffon. As they chased the enlightened dream to complete the encyclopedic knowledge of the world, philosophers and naturalists alike built on the "principle of continuity" to fill all unknown "gaps" that waited to be filled.

In the late eighteenth century, the idea of the "chain" was transformed in two ways. First, instead of representing an eternal order of being that remained unchanged since the moment of creation, the idea of a continuous emergence of its elements was introduced. Second, the direction of the "chain" and the source of being were relocated from the top to the bottom: in natural history, being does not spontaneously "emanate" from the divine above, but gradually unfolds from matter "below." Jean-Baptiste Lamarck, Herbert Spencer, and Charles Darwin, despite their different understandings of the evolution of species, all acknowledged the idea that nature does not make any jumps or leaps, but gradually transforms from one state to the other. Instead of a spontaneous act of creation, they promoted the idea of natural history as a continuous process that develops and unfolds over time.

The transformation of the "chain" also corresponded to a rupture in the political sphere in Europe. As the power of the church was called into question in the aftermath of the French Revolution, the emergence of the national states fostered a new perspective on the political order, which for many centuries was presented as god-ordained, eternal, and unchangeable. In the wake of these turnovers, the static image of the "chain" lost its adherents – but the search for knowledge of the "gaps" between plants, animals, and mankind lived on, as it served as a guiding principle for natural philosophy.

A Forgotten Universe of Layers in Philosophy, the Sciences, and Psychology

At the end of the nineteenth century, theories of stratification began to spread out across the academic landscape. The French philosopher Émile Boutroux was the first author who worked out a comprehensive ontological theory of stratification in his dissertation thesis, "The Contingency of the Laws of Nature" from 1874. Overall, Boutroux acknowledged the successes of modern science and technology in the Western world. As a Catholic, however, Boutroux was also an ardent opponent of materialism, positivism, and determinism. Boutroux's theory of stratification was designed to reconcile science and religion. He argued that the laws of nature, as they are presented in physics and biology, do not necessarily stand in opposition to the ideas of spontaneity and free will. From his perspective, the universe is not just made out of one singular material substance. Instead, he describes several "worlds, forming, as it were, stages superposed on one another" (Boutroux 1916, p. 151). Boutroux distinguished between seven "layers" that are characterized by an ascending degree of "contingency": the world of pure necessity, the world of causes, the world of notions, the mathematical world, the physical world, the living world, and the world of thinking. Altogether, these layers form a harmonious, structured, coherent, and hierarchical "scale of beings." According to Boutroux, the relationship between these layers follows certain regularities: each layer "rests" and builds upon on those beneath it. Higher layers cannot exist without the lower ones, but they cannot be reduced to them either. Their connection is not one of necessity, because each layer has its own, distinct qualities: "Each given world, then, possesses a certain degree of independence as regards the lower worlds" (Boutroux 1916, p. 154). Based on these considerations, Boutroux opposed any attempt to reduce higher layers to lower ones (e.g., reducing biological processes to physical laws, or mental functions to physiological layers). Science, faith, and religion can be reconciled, he argued, because determinism, responsibility, and moral freedom do not necessarily contradict each other. From Boutroux's point of view, each one of them must be assigned to different "layers" of the world (Nye 1976).

Boutroux's philosophy of the "contingency" of the laws of nature served as an inspiration for many renowned philosophers and researchers such as William James, Henri Bergson, and Émile Durkheim. In Germany, it was Nicolai Hartmann, a German philosopher who held chairs in Marburg, Cologne, Berlin, and Göttingen from 1920 until his death in 1950. Hartmann positioned himself as a "critical realist" aiming to renew metaphysics (rather than abandon it, as was demanded by Kant) by drawing on Boutroux's works as well as Plato and Aristotle (Hartmann 1943). Hartmann took up Boutroux's ideas and developed them further in his major work "The structure of the real world," where Hartmann distinguished four "stages" of being: the material, the biological, the soul, and the mind (Hartmann 1940). Like Boutroux, Hartmann aimed to overcome materialism, idealism, and Cartesian dualism. Hartmann's theory of stratification does not set one layer as absolute, nor does it put two of them in an irreconcilable dichotomy. Instead, he argues that there exists a recurring pattern in the relationship between all layers: The biological "builds" upon matter, without which it could not exist, but it also "overforms" and reorganizes matter according to its own laws. Physical and mechanical laws of cause and effect cannot explain processes of growth and reproduction in the biological realm, and the same goes for the relationship between mental processes and biological laws. Every layer has distinct and unique characteristics (e.g., only matter is completely determined by physical effects of cause and effect; "mind" is the only layer which does not exist in space). The higher the layers, the more autonomy and control over the layers below can be observed. No higher layer can exist without those beneath it – there is no life without matter, Hartmann argues, and no thinking without a feeling and perceiving organism. In the following, the term "stratification theory" will only be used for theories that fulfil the criteria mentioned by Hartmann and which use the term "layers" or "strata" [*Schichten*]. There exist countless theories which have a hierarchical structure or shed light on the development of their subject matter, but only if they meet these requirements are they considered relevant for our investigation.

Hartmann's philosophy was not only designed to serve as an alternative to materialism or idealism. His theory was also of relevance to the theory of science, as Hartmann assigned each layer to a specific scientific discipline. Physics and inorganic chemistry are concerned with matter, physiology and biology explore the "layer" of the living, while psychology measures and describes the functions of the "soul" (thinking, perceiving, and feeling). Philosophy, the arts, and humanities are concerned with the "mind" and its objectifications in literature, art, or architecture. Due to Hartmann's "law of the autonomy of the layers," no single discipline can claim to generalize its findings onto other "layers" without taking into account their specific characteristics (Hartmann 1940). As specialization increased across all scientific disciplines and the relationship between them was becoming less and less clear, Hartmann aimed to revive metaphysics in order to reveal the connection of all disciplines by uncovering the layered structure of the "real world."

Hartmann laid the foundation of a "layered" perspective on metaphysics and epistemology, but the popularity of theories of stratification during the first half of the twentieth century extended much further than that. A popular strain of stratification theories that aimed to connect perspectives from science and philosophy was philosophical anthropology, as it was presented in the works of Max Scheler (1928), Helmut Plessner (1928), and Erich Rothacker (1941). All three authors presented stratified perspectives of man's place in nature by embedding results of the behavioral sciences, evolution theory, and biology in a larger philosophical framework. Plessner held that there is a hierarchical order of mental functions that runs through all that lives, from plants up to humankind. The oldest and most primitive layer is defined by the capability to strive towards an object (e.g., food or sunlight) or move away from it (e.g., from an enemy). Going up the evolutionary "ladder," we can find "instinctual behaviour" that is set in motion when certain stimuli appear, and after that "associative memory," which can be observed when organisms are capable of adapting to new environmental circumstances. "Practical intelligence" can be found wherever organisms anticipate circumstances and modify the environment based on their inner representation of it. While human beings share all these functions with the animal kingdom, only they possess "mind," the ability to analyze objects and situations independently from current needs and desires.

Since the late nineteenth century, theories of stratification became increasing popularity in the medical field as well. The British neurologist John Hughlings Jackson (1884) interpreted the "disintegration" of "higher" nervous structures as a regression to "earlier" stages of cortical development (as occurs in epileptic seizures or dementia). When "lower" layers take over control, the conscious control of behavior is severely limited, and behavior becomes more automatic and reflex-driven. Theodor Meynert (1892), Ludwig Edinger (1912), and Constantin Monakow (1914) also built on the concept of cortical "layers" to distinguish between different species and describe the ontogenetic unfolding of the nervous system. They presented phylogenetic development of species as well as the development of the individual as an unfolding of layers with increasing complexity (e.g., from the limbic system to the neocortex). Just as in Hartmann's system, higher layers can control and inhibit those beneath them (e.g., adult humans can consciously control their breathing) but they are also dependent on them (without breathing, higher mental functions soon stop working; cf. Smith 1992). The whole system can only work if all layers interact in harmony with each other, and disturbances such as the effects of alcohol or cerebral diseases are pictured as a successive dismantling of layers from the "top" to the "bottom."

Meanwhile, in sociology, German-Danish Theodor Geiger (1932) worked out his theory of social stratification as an attempt to overcome the class dichotomy of Marxism. Based on his own empirical findings and statistical analysis, Geiger argued that Western industrialized societies do not just consist of a capitalist and a proletarian class, but that there exist also an old and new "middle class" as well as "proletaroids." Each social layer has a "typical" mentality that is represented in its norms, beliefs, language, and habits. In modern societies, Geiger argued, the picture has become much more complex than was assumed in classical Marxism, because the mentality of the members of each "layer," according to Geiger, cannot be deduced directly from their socioeconomic position. Just like Hartmann, Geiger argued that each layer needs to be understood in terms of its own structure and development, and any reduction to the underlying economic layer should be rejected. Another strain of stratification theories can be found in the works of Roman Ingarden (1931) and Erwin Panofsky (1932), who both argued that works of art such as literature and paintings can be understood as compositions of different layers of meaning. From basic geometric shapes or plain symbols up to the complex and significative structure of myths or paintings, they recognized a multileveled system of layers of symbolization which increase in complexity and richness.

At the end of the nineteenth century, the interaction between the natural sciences and psychiatry became stronger than ever in Western medicine. This reorientation of psychiatry towards biology also brought stratification theories into this field. Although the suggested number of layers and their localization varied considerably between different neurological theories of stratification, their overall influence on psychiatric thinking became quite apparent at the turn of the century. Karl Kleist, a former student of the famous neurologist Carl Wernicke and head of the Frankfurt Neuropsychiatric Clinic, proposed a system of six "psychobiological" layers that emerge through the course of development of the individual (Kleist 1934). He suggested that the sequence of appearance as well as the function of each layer represents a repetition of the evolutionary past of the species. Based on examinations of World War I veterans who had suffered from brain injuries, he argued that lesions of "higher" parts lead inevitably to a "falling back" to earlier stages of development, resulting in less differentiated and coordinated forms of behavior. A very similar argument was made by Herrmann Hoffmann, Professor of Psychiatry in Gießen, in his book titled "The Theory of Stratification. A view of nature and life" (Hoffmann 1935). The early career of Sigmund Freud in physiology and neurology, when he worked in Ernst Brücke's laboratory and thereafter under Theodor Meynert's supervision in Vienna in the 1880s, also left its mark in his later theories about the mental apparatus. Just as contemporary neurology promoted the image of a stratified system of neuronal cells, Freud imaged the mind as a set of layers which unfold over time. The primitive layer of the "id" prevails in adulthood, but in the sane adult, it is – at least to some degree – under control by the "ego" through various mechanism of "defence" which are acquired through growing up as a "civilised" human being (Wieser 2013).

Many other examples from various branches of theories of stratification could be named in this line (cf. Wieser 2018) but for our context, it suffices to point out that in the interwar era, German intellectuals across many disciplines argued that theories of stratification provides a theoretical framework connecting philosophy, science, arts, and humanities. In times when the natural sciences grew rapidly and developed many different specialized terminologies and methodologies that made it seemingly impossible to find a unified system of knowledge, stratification theory was supposed to provide a general answer to what "being human" means. Like representatives of the "great chain," stratification theorists shared the idea that there exists a rational order in the universe which can be recognized by human reason. This order was imagined as a continuous unfolding from simple structures to "higher" ones, which become increasingly complex. Each layer has its distinct laws, qualities, and characteristics, and any attempts to reduce higher layers to lower ones (or vice versa) are rejected as reductionist. From a "stratological" perspective, the sciences, philosophy, and the humanities are not supposed to compete with each other, as every discipline has its own realm that needs to be described in its own terminology. In the following section, we will come to one of the largest branches of stratification theory in the psychology of personality, and take a closer look at the political and ideological dimension of this theory.

The Uncovering of the Layers of Personality: German Psychology on the Rise

Psychology played a special role in the sprawling and fast-growing landscape of theories of stratification in the early 20th century. In German academia, experimental psychology grew rapidly during the interwar era. But "personality" was a topic that was difficult to fit into Wundt's experimental program of the measurement of "lower" mental functions (such as sensory discrimination and reaction times). It was mostly through representatives of *"Geisteswissenschaftliche Psychologie,"* such as Ludwig Klages, Eduard Spranger, and proponents of holistic psychology [*Ganzheitspsychologie*], such as Philip Lersch and Felix Krueger, that stratified perspectives took hold in German psychology – even though theories of stratification had such a strong connection to other natural sciences such as biology, physiology, and neurology. The list of psychological works that built upon or referred to theories of stratification published between the 1930s and the 1950s is quite extensive (cf. Wieser 2018, pp. 20–21). Two of the most most influential works at this tine were Erich Rothacker's "The layers of the personality" (Rothacker 1941, first published in 1938) and Philip Lersch's "The structure of the character" (Lersch 1942) which were both first published in 1938. Rothacker and Lersch were influential figures in German academic psychology in the post-Wundt era, with Rothacker holding his chair in Bonn from 1928 to 1956, and Lersch working as a Professor in Breslau, Leipzig, and Munich from 1937 until his retirement in 1966.

Lersch's and Rothacker's theories had very similar outlines. Both suggested that older, primitive layers are gradually overgrown by higher, more complex layers on the way to adulthood. Lersch only differentiated between three layers, while Rothacker named six, but the vertical order and the relationship between the layers of the personality were identical. Structures that are closely tied to vital functions are dominated by layers that control instinctual and drive-based behavior. Towards the top, emotions, the will, the intellect, and finally reason reside above one another. Older layers retain their function, but they are supposed to be kept under control in the sane and adult person. Thus, stratification theory was the first major psychological theory that combined three main features:

As stratification theory included a temporal perspective on the emergence of mental functions and structures over time, it established a connection between the psychology of personality and developmental psychology. Heinz Werner (1926) and Oswald Kroh (1936) were among the authors who put special emphasis on this connection, but it can also be found in the works of Freud (1961) and Jung (1989).

Secondly, proponents of psychological stratification theory promised to fill a gap that was left open by experimental psychology: instead of measuring isolated mental functions, the structure of the character was supposed to be grasped as a whole by exploring the development, relationship, and dynamics of the layers of personality. On this level, stratification theory was in line with the Leipzig school of "*Ganzheitspsychologie*" [holistic psychology] which rose to prominence in Germany after Wundt's death through the works of Friedrich Sander and Felix Krueger, as well as through Ludwig Klages' system of "Characterology" (Klages 1926) which was very popular among intellectual circles in the interwar era (Alksnis 2015).

Thirdly, representatives of stratification theory pursued more than a purely academic or theoretical claim. During the 1930s, the practical side of their knowledge became increasingly important, as methods for personality testing had become increasingly in demand. At that time, quantitative methods of psychological testing had already arrived in industry as well as in education and the welfare system (Jäger and Stäuble 1981). As was the case in most other countries that were involved in World War I, the first steps of applied psychology were taken in the context of the military, where quantitative methods of aptitude testing were used for the examination of motorists, artillerists, and other specialists (Moede 1926). Psychologists successfully demonstrated that they could – and would – prove useful in modern warfare. With few exceptions, the apparatuses and methods used for military aptitude testing were taken from the laboratory. But later a different kind of knowledge was in demand: it was not one which allowed the assessment of "lower" mental functions, but one which would enable the examiner to evaluate the overall suitability of candidates to be trained as leadership personnel.

As the size of the German military was officially under restriction after the Treaty of Versailles, only the most suitable candidates were admitted to officer training. German psychologists took a different direction than their US-American colleagues, where mass-testing was widely used in World War I (Kevles 1968). Lersch was among a group of only five psychologists began working for the military in 1925. In the following eight years, he developed and tested a new system of personality assessment for the military by connecting stratification theory with the practical requirements of personality testing in the military. In the years to follow, Lersch's habilitation thesis, "face and soul" [*Gesicht und Seele*] (Lersch 1932), came to play a very influential role.

Stratification Theory and the Search for Holism in the Weimar Republic

The vertical structure of the layers of personality that was presented by Lersch, Rothacker, and their colleagues represented more than just a conceptual framework to determine the suitability of a person for a particular task, training or occupation.

It also implicated a normative dimension about the evaluation of personality. Although its proponents repeatedly claimed to remain purely descriptive, their system in fact reflected a well-known hierarchy in Western thinking: the new is located above the old, the "civilised" above the "primitive," the sane above the insane. Reason and intellect tower above emotions, drives and instincts, and men above women, children, and the animal kingdom. Lersch argued that "insofar as a man is capable of feeling, he seeks to justify the decision of feeling [*Gefühlsentscheidung*] before the judgement seat of reason [...] For him, feeling is not, as for the woman, an instance that is justified by itself and does not require justification by reason" (Lersch 1942, p. 285). Rothacker wrote that "it can generally be said that children, women, [...] and artists in general live more out of the id, the 'unconscious,' the 'archaic man' [...] than the distinctly modern-rational-technical type of man" (Rothacker 1941, p. 84). Heinz Werner proposed that the workings of the older layers can still be openly observed in the "magical thinking" of so-called "primitives" – but from time to time, their remnants become visible in Western adult as well: "the individual experiences and thinks, as it were, in different layers of his soul, sometimes behaving and behaving in a more 'primitive,' sometimes in a more 'cultivated' and civilised manner" (Werner 1926, p. 3). In "Totem and Taboo," Freud compared the "savage" with the "neurotic" and the child, who all supposedly fail in differentiating between reality and imagination because "wishes and impulses have the full value of facts" to them (Freud 1955, p. 160). Sexist and racist stereotypes like these have existed before, after, and outside psychological theories of stratification (Gould 1996; Teo 2008). Nevertheless, their proponents have supported and spread an ideology of the adult, sane, civilized, and intellectual man as the supposed "top layer" of humankind in the academic sphere. Even if not all of their representatives were in favor of this, this line of thinking was also used as a scientific legitimization for repression and violence against individuals who supposedly belonged to "lower" layers of civilization.

Like many of their colleagues who made a career in academia in the Weimar Republic, Lersch, Rothacker, Krueger, and Kroh were politically conservative-minded. The loss of World War I had shattered their hopes and beliefs in the renewal of the German nation, and many of them were haunted by the question why Germany had lost its emperor, parts of its territory, and a war they thought they would win for sure. Many of them saw the Treaty of Versailles as a disgrace to their country, and the economic crisis that shattered Europe at the end of the 1920s was seen as sign that the "Decline of the West," as Oswald Spengler's (1926) best seller from 1918 was titled, was inevitable. In this context, many German academics – both in the humanities and the natural sciences – joined a quest for knowledge that would provide answers and guidance in times of radical societal change. Science and technology go astray, German intellectuals argued, if they are unleashed on society without any epistemological principles and moral norms that guide human reason and action. Rapid urbanization and industrialization took off in the second half of the nineteenth century and catapulted a nation which was not prepared into the modern age. This process was anything but harmonious: the rise of the worker's movement and the loss of public influence of the church and academic scholars, the emergence of the entertainment industry, cinemas, and mass media contributed to a rapid change and decay of traditional values and ways of living. This process of a rapid, contradictory, and often brutal modernization of the society was depicted in such iconic novels as Alfred Döblin's *Alexanderplatz* (Döblin 1929) and Erich Kästner's *Fabian* (Kästner 1932), and intellectuals who lived in this cultural climate inevitably responded to it.

In the US, the emergence of corporate capitalism and the rise of pragmatism fueled the emergence of applied psychology, predominantly in business and education, since the late 19th century (Baker and Benjamin 2014). In Russia, after the Communist Party took power, historical and dialectical materialism were considered to represent the theoretical foundation of academic psychology (Joravsky 1989). Neither in the US nor the USSR did stratification theory ever really take hold. But the German context during the interwar era was very different. Besides the criticism by public intellectuals of the downside of modernity, concerns were also raised within psychology. An increasing number of psychologists saw their discipline in a state of fragmentation and crisis at the end of the 1920s (Driesch 1926; Bühler 1927; cf. Wieser 2020b). To overcome this threatening situation, the academic elite of Germany in the interwar era who were called the "German mandarins" by the historian Fritz Ringer (1969), found their common enemies in Marxism, materialism, mechanicism, and psychologism. In opposition to these currents, which they saw as a sign of moral disintegration and decay, they joined in the "search for holism" (Harrington 1996) – a search for unity, for meaning and completeness, in a world that appeared to be falling apart in disconnected mechanical elements, political fractions, and incompatible worldviews. Besides holistic psychology proper, Gestalt Psychology, as it was developed in Frankfurt and Berlin by Max Wertheimer, Wolfgang Köhler, Kurt Koffka, can also be understood as part of this movement (Ash 1998). The works of the biologists Johann von Uexküll and Hans Driesch and the physician Viktor von Weizsäcker (one of the founders of psychosomatic medicine in Germany), the pedagogist and philosopher Eduard Spranger show that this search for the "whole" extended far beyond the disciplinary boundaries of psychology (cf. Harrington 1996).

Against this background, stratification theory turns out to be more than merely another theory or model to describe and analyze the human personality. Stratification theory responded to a disciplinary and a cultural-political crisis at the same time. Just like proponents of the "great chain," stratification theorists argued that there is a rational order in the universe which can be recognized by human reason. The development of the individual and of the human species, they argued, is not random or without purpose. Both developments follow a direction from the simple to the complex, from the instinctual to the rational, from emotion to reason. Against mechanicism, materialism, and determinism, they argued that the laws of physics, chemistry, and biology are well-placed in their "layers" of the universe – but they are not to be mistaken for the whole. Against Darwinism, they argued that man rises above the animal kingdom – we may share some "layers" with our ancestors, but what really makes humankind special and puts it at the "top" is its ability to reason. Teleology is not necessarily implied here – in fact, Hartmann (1940) vehemently rejected the assumption that there was some kind of final determination from the lower to the higher realm that is inherent in nature. A divine creator is not necessarily implied in this system, as being is supposed to rise from "below" and is not "gifted from above." Nevertheless, the order of nature and the realm of the cultural still show a similarity in structure, which supposedly provides orientation and guidance to the human being.

While stratification theory blossomed in the interwar era, the rise of National Socialism brought a sudden end to the peaceful exploration of the layers of nature, society, and soul. The violence that became omnipresent the Nazi party soon also became visible at the universities and psychological institutes. About one third of the psychological academic community was expelled or forced into migration (Ash 1984). For all of those who remained, the pressure to demonstrate their political and ideological support increased. While some branches of stratification theory – especially psychoanalysis – suffered severely from political, racist, and anti-Semitic persecution, others managed to survive – and even grow under the Swastika.

Stratification Theory in Wartime

The Nazi leaders showed little regard for the contractual arrangements of the Treaty of Versailles and began – first in secret, and from 1935 in the open – to rearm and enlarge the military. Psychologists had proven their usefulness in World War I, and now they profited from the rapid increase in resources that were channeled into the military. In the mid-1920s, there were only five psychologists working in the military. Until 1933, their number slowly grew to 33. After the attack of Nazi Germany on Poland in September 1939, the numbers of psychologists in the military grew rapidly. It is estimated that a maximum number of 450 posts in the military was reached in 1941 (Geuter 1992, p. 161). In less than a decade, the military became the largest employer for (male) psychologists, who were hired as experts in the field of personality assessment. While traditional experimental methods of quantitative performance diagnostics were still in use in the military, the most advanced psychological expertise was to be found in the examination of the personality of leading personnel. The assessment of officer candidates included a 2- to 3-day examination that combined extensive medical, physical, biographical, and psychological tests. Besides military psychologists, military officers and medical personnel were also part of the selection committees. These were located in the inspection offices of the military distributed all over the territory of the Third Reich. To come to a recommendation whether a candidate was fit for officer training, German military psychologists relied on their system of "expression psychology" [*Ausdruckspsychologie*]. Instead of measuring mental function and performance, they observed, analyzed, and interpreted gesture, mimicry, voice, handwriting, social interaction, and behavior in various situations. As Lersch and his colleagues argued, overt conduct must always be interpreted as the result of the interaction of all layers of the personality, from the basic vital functions up to reason. In every single act, all layers are involved, but not to the same extent, depending on the situation as well as the permanent structure of the personality. One person's personality is dominated by the emotional layer, another's by the intellect, and a third one's impulses and drives might come out uninhibited when they are under pressure. The ideal soldier, in the eyes of the military psychologists, is not merely controlled by rational thinking. What they were looking for was a unity of will, emotion, and thinking, not a heady intellectual. The ideal candidate would be strong-willed and masculine, self-sacrificing and captivating, even in the face of immediate danger. Even though military psychologists argued that their assessment is "directed towards what is and not its value" (Simoneit 1938, p. 7), the standard of their evaluation was clearly taken from the traditional Prussian stereotype of the soldier whose fate in battle is supposed to be dependent on the strength of his will and courage.

In Germany, stratification theory served as the theoretical basis of aptitude testing for officer candidates, and as long as the war seemed to go in Germany's favor, psychologists profited from their usefulness in the military. In 1941, the first diploma curriculum for psychology was established in Germany. The military played an important role in its construction, as subjects such as characterology, expression psychology, applied psychology, and psychological diagnostics now became an official part of the curricula for psychology students (Lück 2020). Personality psychology and methods of diagnostics reached their climax in Germany during the war – until the tides turned against them. As the replenishment of military personnel could no longer make up for the losses on the Eastern Front, methods of personnel selection soon became unnecessary and too expensive. In July 1942, almost the entire organization of military psychology was almost completely disbanded. Psychology's attachment to the military made it dependent on the course of the war, and psychologists had to search for new occupations quickly if they wanted to avoid being sent to the front. Many found employment in industry, psychiatric wards, youth detention centers, psychotherapeutic practice, and other areas (Wieser 2020c).

After the Fall: Continuity and Breaks at the Dawn of the Cold War

After the surrender of the German Wehrmacht and the occupation of Nazi Germany, the territory of the former Third Reich was divided by the occupying powers. Notwithstanding the political and ideological collaboration of many psychologists with the fascist regime, at first it looked as if stratification theory might live on after the war. In East Germany, proponents of stratification theory, like the Gestalt Psychologist Kurt Gottschaldt, managed to maintain academic positions – in his case, it was the renowned chair of psychology at Humboldt University. Rebuilding the institute from the ground up, Gottschaldt became one of the founding fathers of psychology in the GDR (Ash 1995). Even though he increasingly came into trouble for his "bourgeois" and "idealistic" perspectives at the end of the 1950s, he was still allowed to teach Gestalt Psychology and stratification theory in Berlin until he fled to West Germany in 1962 (Gottschaldt 1946).

On the other side of the iron curtain, things were not too different. In West Germany, the diploma curriculum was kept in place, with only a few minor adaptations and change of titles that were implemented in 1955 (Lück 2020). Only few careers were permanently affected by the short and superficial measures of "denazification." Lersch remained one of the leading figures of academic psychology in post-war Germany. The advancement of military psychology and the establishment of the diploma curriculum were remembered as big successes among psychologists, who blamed politicians and arrogant military officers who would not accept being overruled by scientific judgement in connection with the dissolution of military psychology in 1941. Even the US Army showed interest in psychological methods in the German military (Ansbacher 1951). In the post-war period, the Wehrmacht was officially perceived in the German and Austrian public as "clean" participant of the war, in contrast to the SS and the leaders of the Nazi party. Passages from psychological textbooks that praised racial ideology, euthanasia, the Führer, or warfare were removed and titles of books were changed, but their basic structure and content remained largely unchanged. Lersch's "Structure of the Character" was retitled "Structure of the Person" in 1952, and it saw several new editions down till 1970. Erich Rothacker also remained professor until his retirement in 1956.

On both sides of the Iron Curtain, it was not the political and ideological entanglements of representatives of stratification theory with National Socialism that brought an end to their careers. It was the rise of science and technology in the Cold War and the competition between two economic systems that heralded the final days of stratification theory. While the "Marshal Plan" helped to bring the West German economy back to its feet, East Germany was hit severely by war damages and struggled hard to keep pace with its capitalist neighbour. At the beginning of the Cold War, the two economic systems entered into fierce competition. The question of how to increase production and human efficiency now became crucial, and any psychological theory that would contribute to this had good prospects.

In the meantime, a new transdisciplinary field of knowledge emerged in the US. Through the pioneering works of Norbert Wiener (1948), Claude Shannon (1948) and the participants of the "Macy Conferences" that were held from 1946 to 1953, cybernetics and information technology became popular and was adopted soon in Europe as well (Pias 2004). Psychology was part of this movement from the beginning, and cybernetic terminology and imagery began to spread in German psychology from the 1950s. Cybernetics had a very different perspective on the mind and personality compared to stratification theory. Donald Broadbent, pioneer of cognitive psychology in the UK, described what "being human" means in the age of cybernetics as follows: "Continuously the skilled man must select the correct issues from the environment, take decisions upon them which may possibly involve prediction of the future, and initiate sequences of responses whose progress is controlled by feed-back" (Broadbent 1958, pp. 295–296). The storage, selection, filtering, processing, and executing of digital information became the key concern of the cybernetic age, and stratification theory seemed hopelessly pre-modern compared to that.

In West Germany, the "Americanization" (Métraux 1985) of psychology that started in the late 1950s heralded the demise of stratification theory. Many representatives of the next generation of aspiring young researchers were introduced to the new and exciting methods of American experimental psychology. Stratification theory and methods of expression psychology seemed rather old-fashioned and outdated by comparison. The fact that its representatives remained silent about their involvement with the Nazi regime did not help either. After a short period of hesitation, cybernetics was also received with great enthusiasm during the 1960s in the Soviet Union and East Germany. Some East German philosophers and scientists even interpreted cybernetics as a renewed and modernized form of Marxism-Leninism (Klaus 1964), while psychologists hoped to put cybernetics concepts and models to use to fulfil the political mandate to improve the modern working environment (Skala and Hacker 1963) and advance psychological experimentation (Klix 1968).

After the 1950s, stratification theory did not suddenly vanish from the scenery. But it lost its former function as a theoretical framework that connected disciplines across the academic landscape and reconnected human kind with nature. In some disciplines such as neurology, psychoanalysis, and sociology, reference to concepts of "layers" remained common, but few were aware that these "layers" were once elements of a picture that extended far beyond theories of one specific discipline.

Conclusion

This chapter began by adressing the ambivalent relationship between contemporary Western psychology and historical thought. As was pointed out, historical perspectives played an integral part in the founding period of psychology during the second half of the nineteenth century. After the Second World War, however, the discipline took a different direction. Although there are still currents and subdisciplines today that include temporality, such as evolutionary psychology or developmental psychology, psychological research is mostly limited to very specific dimension of the past: it is either concerned with the development of the individual or of human kind as a species. The case example of stratification theory was presented at length to show how a specific line of historical thinking, which had roots in the ancient "great chain of being," was revived in academic psychology during the interwar period in Germany. It has been argued that the reasons for its resurgence are to be found in the specific historical situation that the discipline faced during this period: Wundt's experimental program of physiological psychology was perceived as too narrow by many of his colleagues, and it was also considered inadequate to contribute to the study of human personality. Stratification theory was supposed to address both issues: it was supposed to facilitate the analysis and description of personality as a "whole" and also provide the conceptual foundation for practical application. Psychological stratification theory promised to connect psychological knowledge with other disciplines that were also concerned with the development and unfolding of different "layers" of reality, be it the universe as a whole, the stratification of society, man's place in nature, or the layers of the human brain. As long as the search for "holism" and the need for a unified and harmonious worldview was widespread in German academia, stratification theory flourished across the academic landscape.

After the rise of National Socialism, the conditions for psychological research, teaching, and practice changed drastically. While some psychological schools and currents suffered heavily from political and racist persecution, many proponents of stratification theory actually managed to hold their academic positions and even made a career in Nazi Germany. The historical roots of their thinking went back to the origins of Western thought: one key legacy of Greek and Christian metaphysics was the notion of a hierarchically ordered cosmos, at the top of which was to be found the highest of all beings. As the "great chain" was replaced by stratification theory in the course of the Enlightenment, the origin of all beings was relocated from the top to the bottom, while human kind was now to be found where once there had been the divine. Racial ideology enhanced and perverted this idea by placing one particular human race on top, in a position of dominance of everything else. Not all proponents of stratification theory openly supported this hierarchy of races, but the hierarchical structure of their theories made them susceptible to the devaluation of women, the "uncivilised," and people who suffer from mental illnesses or impairments.

After the war, most key figures of German psychology managed to keep their positions. At first, it looked like stratification theory would live on as well. But the rise of cybernetics and operationalism set an end to the era of stratification theory in the 1960s. From then on, cybernetics and cognitive psychology served as means for psychologists to identify, measure, and solve technical problems at the intersection of humans and information-processing devices. Operationalism was introduced in the 1930s to standardize methods of psychological research and leave as little space to subjectivity and intuition as possible. Proponents of psychological stratification theory did not aim to fulfill these requirements, nor could they, since stratification theory came out of a intellectual climate that rebelled precisely against this mechanization and standardization. Stratification theory had its heyday during a time when sentiments against modernity, against industrialization and materialism were strong in German academia. Against this background, it becomes clear that the implementation of a historical dimension in science and philosophy served a specific purpose. It was supposed to reconcile man, nature, and society by revealing a similar process of stratification in the universe and in nature, in society, in the life of the individual, and in the human psyche. After World War II, politics in West and East Germany was directed towards rebuilding the economy and society, and stratification theory was not designed to keep up with to the newest technological and economic innovations. It is no wonder then that the English-speaking and the German fraction of psychologists could not find agreement in Montreal: what was discussed on the surface of the debate – methodological issues and questions of operationalization of the assessment of personality – was only a small element of what was really at stake. What was much more decisive for the development of the discipline was the fact the political and ideological contexts had changed so drastically after the war that the question what "being human" means had to be answered anew.

References

Alksnis G (2015) Chthonic gnosis. Ludwig Klages and his quest for the pandaemonic all. Theion, München

Allport G (1957) European and American theory of personality. In: David H, Bracken H (eds) Perspectives in personality theory. Basic Books, New York, pp 3–24

Angell J (1909) The influence of Darwin on psychology. Psychol Rev 16:152–169

Ansbacher HL (1951) The history of the leaderless group discussion technique. Psychol Bull 48(5): 383–391

Aristotle (1902) Aristotle's history of animals (trans: Cresswell R). George Bell, London

Ash M (1984) Disziplinentwicklung und Wissenschaftstransfer – Deutschsprachige Psychologen in der Emigration [Discipline development and the transfer of science – German-speaking psychologists in the emigration]. Berichte zur Wissenschaftsgeschichte [History of Science and Humanities] 7:207–226

Ash M (1995) Übertragungsschwierigkeiten: Kurt Gottschaldt und die Psychologie in der Sowjetischen Besatzungszone und in der Deutschen Demokratischen Republik [Transfer difficulties: Kurt Gottschaldt and psychology in the Soviet occupation zone and the German Democratic Republic]. In: Jaeger S, Staeuble I, Sprung L, Brauns HP (eds) Psychologie im soziokulturellen Wandel – Kontinuitäten und Diskontinuitäten [Psychology in socio-cultural transformation – continuities and discontinuities]. Peter Lang, Frankfurt am Main, pp 286–294

Ash M (1998) Gestalt psychology in German culture, 1890–1967. Holism and the quest for objectivity. Cambridge University Press, Cambridge

Baker DB, Benjamin LT (2014) From séance to science. A history of the profession of psychology in America, 2nd edn. University of Akron Press, Ohio

Baldwin JM (1894) Imitation: a chapter in the natural history of consciousness. Mind 3:26–55

Boutroux E (1916) The contingency of the laws of nature. Open Court, Chicago

Broadbent D (1958) Perception and communication. Pergamon, Oxford

Bühler K (1927) Die Krise der Psychologie [The crisis of psychology]. Fischer, Jena

Buss S (2019) Evolutionary psychology: the new science of the mind. Routledge, New York

Danziger K (1990) Constructing the subject. Historical origins of psychological research. Cambridge University Press, Cambridge

Danziger K (1994) Does the history of psychology have a future? Theory Psychol 4(4):467–484

Danziger K (1997) Naming the mind. How psychology found its language. Sage, London

David H, Bracken H (eds) (1957) Perspectives in personality theory. Basic books, New York

Dilthey W (1894) Ideen über eine beschreibende und zergliedernde Psychologie [Ideas Concerning a Descriptive and Analytic Psychology]. Sitzungsberichte der königlich preußischen Akademie der Wissenschaften zu Berlin 26–28:1309–1407

Döblin A (1929) Berlin Alexanderplatz. Fischer, Berlin

Driesch H (1926) Grundprobleme der Psychologie. Ihre Krisis in der Gegenwart [Fundamental problems of psychology. Its crisis in the present]. Reinicke, Leipzig

Ebbinghaus H (1896) Über erklärende und beschreibende Psychologie [On analytic and descriptive psychology]. Zeitschrift für Psychologie und Physiologie der Sinnesorgane 9:161–205

Edinger L (1912) Einführung in die Lehre vom Bau und den Verrichtungen des Nervensystems [Introduction to the study of the structure and function of the nervous system]. Vogel, Leipzig

Eysenck H (1957) Characterology, stratification theory and psychoanalysis: an evaluation. In: David H, Bracken H (eds) Perspectives in personality theory. Basic books, New York, pp 323–335

Freud S (1955) Totem and Taboo. In: Strachey J (ed) The standard edition of the complete psychological works of Sigmund Freud, vol 13. Hogarth, London, pp 1–162

Freud S (1961) The standard edition of the complete psychological works of Sigmund Freud. In: Strachey J (ed) The ego and the id, vol 19. Hogarth, London, pp 3–66

Furedi F (2004) Therapy culture. Cultivating vulnerability in an uncertain age. Routledge, New York

Geiger T (1932) Die soziale Schichtung des deutschen Volkes [The social stratification of the German nation]. Enke, Stuttgart

Gergen K (1973) Social psychology as history. J Pers Soc Psychol 26(2):309–320

Geuter U (1992) The professionalization of psychology in Nazi Germany. Cabridge University Press, Cambridge
Gottschaldt K (1946) Die Pädagogische Psychologie im Universitätsstudium der Lehrer [Educational psychology in the university study of teachers]. Pädagogik 1(5):1–23
Gould SJ (1996) The mismeasure of man. Norton, New York
Green C (2009) Darwinian theory, functionalism, and the first American psychological revolution. Am Psychol 64(2):75–83
Harrington A (1996) Reenchanted science. Holism in German culture. From Wilhelm II to Hitler. Princeton University Press, Princeton
Hacking I (1995) The looping effects of human kinds. In: Sperber D, Premack D, Premack J (eds) Causal cognition. A multidisciplinary debate, Clarendon, Oxford, pp 351–383
Hartmann N (1940) Der Aufbau der realen Welt [The structure of the real world]. Hain, Weisenheim
Hartmann N (1943) Die Anfänge des Schichtungsgedankens in der Alten Philosophie [The beginnings of the concept of stratification in old philosophy]. Akademie der Wissenschaften, Berlin
Hoffmann H (1935) Die Schichttheorie [A theory of layers]. Enke, Stuttgart
Holzkamp K (2013) In: Schraube E, Osterkamp U (eds) Psychology from the standpoint of the subject. Selected writings of Klaus Holzkamp. Palgrave Macmillan, New York
Hughlings Jackson J (1884) The Croonian lectures on evolution and dissolution of the nervous system. Br Med J 1:591–593, 660–663, 703–707
Ingarden R (1931) Das literarische Kunstwerk [The literary work of art]. Niemeyer, Halle
Jäger I, Stäuble S (1981) Die Psychotechnik und ihre gesellschaftlichen Entwicklungsbedingungen. [Psychotechnics and its social development conditions]. In: Stoll F (ed) Kindlers Psychologie des 20. Jahrhunderts [Kindler's psychology of the 20th century], vol 1. Beltz, Weinheim, pp 49–91
James W (1892) A plea for psychology as a natural science. Philos Rev 1(2):146–153
Jaspers K (1953) The origin and goal of history (trans: Bullock M). Yale University Press, New Haven
Joravsky D (1989) Russian psychology. A critical history. Blackwell, Oxford
Jung CG (1989) Analytical psychology: notes on the seminar given in 1925 (ed: W McGuire). Princeton University Press, Princeton
Kästner E (1932) Fabian. Die Geschichte eines Moralisten [Fabian. Story of a moralist]. Deutsche Verlags-Anstalt, Stuttgart
Kevles D (1968) Testing the army's intelligence: psychologists and the military in World War I. J Am Hist 55(3):565–581
Klages L (1926) Zur Ausdruckslehre und Charakterkunde [On the study of expression and character]. Kampmann, Heidelberg
Klaus G (1964) Kybernetik und Gesellschaft [Cybernetics and society]. Deutscher Verlag der Wissenschaften, Berlin
Klautke E (2013) The mind of the nation. Völkerpsychologie in Germany, 1851–1955. Berghahn, New York
Kleist K (1934) Gehirnpathologie [Pathology of the brain]. Barth, Leipzig
Klix F (1968) Kybernetische Analysen Geistiger Prozesse [Cybernetic analyses of mental processes]. Deutscher Verlag der Wissenschaften, Berlin
Kroh O (1936) Die Gesetzhaftigkeit geistiger Entwicklung [The principles of mental development]. Zeitschrift für pädagogische Psychologie und Jugendkunde 37:1–108
Kusch M (1999) Psychological knowledge. A social history and philosophy. Routledge, London
Lersch P (1932) Gesicht und Seele [Face and soul]. Reinhardt, Munich
Lersch P (1942) Der Aufbau des Charakters [The structure of the character], 2nd edn. Barth, Leipzig
Lovejoy A (2001) The great chain of being. A study of the history of an idea. Harvard University Press, Cambridge
Lück H (2020) Die Diplomprüfungsordnung für Studierende der Psychologie – eine nationalsozialistische Prüfungsordnung? [The diploma examination regulations for students of psychology –

a National Socialist Examination Regulation?]. In: Wieser M (ed) Psychologie im Nationalsozialismus [Psychology in National Socialism]. Peter Lang, Frankfurt am Main, pp 47–72

Métraux A (1985) Der Methodenstreit und die Amerikanisierung der Psychologie in der Bundesrepublik 1950–1970 [The methodological controversy and the Americanisation of psychology in west Germany 1950–1970]. In: Ash M, Geuter U (eds) Geschichte der deutschen Psychologie im 20. Jahrhundert [History of German psychology in 20th century]. Westdeutscher Verlag, Opladen, pp 225–251

Meynert T (1892) Sammlung von populär-wissenschaftlichen Vorträgen über den Bau und die Leistungen des Gehirns [Collection of popular scientific lectures on the structure and activities of the brain]. Braumüller, Vienna

Moede W (1926) Kraftfahrer-Eignungsprüfungen beim Deutschen Heer 1915–1918 [Driver aptitude tests in the German military 1915–1918]. Industrielle Psychotechnik 3(1):23–34

Monakow C (1914) Die Lokalisation im Großhirn und der Abbau der Funktion durch kortikale Herde [The localization in the cerebrum and the degradation of function through cortical lesions]. Bergmann, Wiesbaden

Morgan CL (1891) Animal life and intelligence. Edward Arnold, London

Napoli DS (1981) Architects of adjustment. Kennikat, London

Nye MJ (1976) The moral freedom of man and the determinism of nature: the catholic synthesis of science and history in the *Revue des Questions Scientifiques*. Br J Hist Sci 9(3):274–292

Panofsky E (1932) Zum Problem der Beschreibung und Inhaltsdeutung von Werken der bildenden Kunst [On the problem of describing and interpreting works of visual art]. Logos 21(1):103–119

Pias C (ed) (2004) Cybernetics/Kybernetik. The Macy-conferences 1946–1953. diaphanes, Zürich

Pickren W, Rutherford A (2010) A history of modern psychology in context. Wiley, Hoboken

Plessner H (1928) Die Stufen des Organischen und der Mensch [The levels of the organic and man]. De Gruyter, Berlin

Revill J (2009) Emile Boutroux, redefining science and faith in the Third Republic. Mod Intellect Hist 6:485–512

Ringer F (1969) The decline of the German mandarins. The German academic community 1890–1933. Harvard University Press, Cambridge

Rose N (1988) Calculable minds and manageable individuals. 'History' and the psychological sciences. Hist Hum Sci 1(2):179–200

Rothacker E (1941) Die Schichten der Persönlichkeit [The layers of personality], 2nd edn. Barth, Leipzig

Scheler M (1928) Die Stellung des Menschen im Kosmos [The human and his place in the cosmos]. Reichl, Darmstadt

Shannon C (1948) A mathematical theory of communication. Bell Syst Tech J 27:379–423, 623–656

Simoneit M (1938) Leitgedanken über die psychologische Untersuchung des Offiziers-Nachwuchses in der Wehrmacht [Guiding principles for the psychological examination of officer candidates in the Wehrmacht]. Bernard & Graefe, Berlin

Skala H, Hacker W (1963) Die Arbeitspsychologie und ihre Aufgaben in der DDR [Industrial psychology and its mission in the GDR]. Einheit. Zeitschrift für Theorie und Praxis des wissenschaftlichen Sozialismus 7:48–58

Smith R (1988) Does the history of psychology have a subject? Hist Human Sci 1(2):147–177

Smith R (1992) Inhibition. History and meaning in the sciences of the mind and brain. University of California Press, Berkeley

Smith R (2005) The history of psychological categories. Stud Hist Phil Biol Biomed Sci 36(1):55–94

Smith R (2007) Being human. Historical knowledge and the creation of human nature. Columbia University Press, Columbia

Smith R (2022) What is the history of the human sciences? The Palgrave handbook of the history of the human sciences, vol I. Springer, Cham

Spencer H (1855) The principles of psychology. Longman, London

Spengler O (1926) The decline of the west (trans: Atkinson CF). George Allen & Unwin, London

Teo T (2008) From speculation to epistemological violence in psychology: a critical-hermeneutic reconstruction. Theory Psychol 18(1):47–67

Teo T (2015) Historical thinking as a tool for theoretical psychology: on objectivity. In: Martin J, Sugarman J, Slaney K (eds) The Wiley handbook of theoretical and philosophical psychology: methods, approaches and new directions for social sciences. Wiley, New York

Watson J (1913) Psychology as the behaviourist views it. Psychol Rev 20(2):158–177

Wellek A (1957) The phenomenological and the experimental approaches to psychology and characterology. In: David H, Bracken H (eds) Perspectives in personality theory. Basic books, New York, pp 278–299

Werner H (1926) Einführung in die Entwicklungspsychologie [Introduction to developmental psychology]. Barth, Leipzig

Wiener N (1948) Cybernetics or control and communication in the animal and the machine. MIT Press, Cambridge

Wieser M (2013) From the eel to the ego. Psychoanalysis and the remnants of Freud's early scientific practice. J Hist Behav Sci 49(3):259–280

Wieser M (2018) Buried layers. On the origins, rise and fall of stratification theories. Hist Psychol 21(1):1–32

Wieser M (2020a) Einleitung [Introduction]. In: Wieser M (ed) Psychologie im Nationalsozialismus. Peter Lang, Frankfurt am Main, pp 7–22

Wieser M (2020b) The concept of crisis in the history of western psychology. In: Pickren W (ed) Oxford research encyclopedia of psychology. Oxford University Press, Oxford

Wieser M (2020c) Zur Geschichte der angewandten Psychologie in der "Ostmark" [On the history of applied psychology in the "Ostmark"]. In: Wieser M (ed) Psychologie im Nationalsozialismus [Psychology in National Socialism]. Peter Lang, Frankfurt am Main, pp 167–193

Wozniak RH (ed) (1993) Theoretical roots of early behaviourism. Functionalism, the critique of introspection, and the nature and evolution of consciousness. Routledge/Thoemmes, London

Wright C (1877) Evolution of self-consciousness. In: Norton CE (ed) Philosophical discussions. Burt Franklin, New York

Wundt W (1916) Elements of folk psychology. Outlines of a psychological history of the development of mankind (trans: Schaub EL). George Allen & Unwin, London

Young R (1966) Scholarship and the history of the behavioural sciences. Hist Sci 5(1):1–51

Part VIII
History of Psychiatry

The Mental Patient in History

47

Peter Barham

Contents

Introduction	1224
More Than a "Pathological Attitude to Life"	1228
"What Kind of Object or Life Is This?"	1230
Psychopathology, Art, and Politics	1231
"Africa in Their Natures": Black and Insane as Double Jeopardy	1233
Mental Suffering, Psychosis, and Witnessing	1234
Precarity, Precariousness, and Mental Patient Lives	1240
Social Dangers and the Human Sciences	1241
The Human Sciences and Coercive Rehabilitation	1242
Institutionalized Lives	1243
Extramural Lives	1245
Conclusion	1248
References	1248

Abstract

This chapter aims to throw light on the multifaceted and contested terrain of the mental patient in history and to bring into focus the strengths and weaknesses of the diverse kinds of knowledge-making that have gathered around the trajectories of distressed psyches. Though the histories of mental patients may sometimes occupy an adversarial, or antagonistic, relationship to psychiatric orthodoxies, equally they may provide the stimulus for new epistemologies within psychiatry. The discussion ranges widely across a number of disciplines, notably anthropology, psychiatry, psychology, and history, and engages with creative artists, writers, and activists, such as Vincent Van Gogh, Daniel Paul Schreber, Bessie Head, Friedrich Krauss, Dimitri Tsafendas, Rodrigo Souza Leao, and Lemn Sissay. A critical concept in this discussion is that of precarity which names a politically induced condition under which certain groups may become differentially exposed to injury and suffering.

P. Barham (✉)
London, UK
e-mail: j.peterbarham@icloud.com

Topics considered include psychopathology, art, and politics, notably the contribution and significance of art historian Hans Prinzhorn; the double jeopardy of race and insanity; and the exploration of how reflections on action and suffering may become a window on the times in which protagonists live, revealing how "things of the soul" are inextricably inwoven with "questions of power." The precariousness of the psyche in the power fields of history is also illustrated on a wider canvas, at the level of the changes and conditions that affect a whole society in a given period, notably China, India, Japan, and the Soviet Union/Russian Federation, so as to highlight dimensions of what has been described as the politics of social suffering.

Keywords

Anthropology · Human · Mental patient · Mad person · Power · Precarity · Psyche · Psychiatry · Psychosis · Social suffering · Trauma

Introduction

The terrain that is the concern of this chapter can equally go by other names such as "the mad person in history" or "mad lives in history." These are, almost invariably, precarious lives, possessing only a tenuous foothold in history, "shifting" subjects who are "hard to pin down" (Coleborne 2020), and "elusive," where "what it meant to be a patient was forever slipping out of view" (Pinto 2014), recounted in stories that are mostly "fractured, incomplete, riddled with gaps and inconsistencies" (Kim Hopper in Luhrmann and Marrow 2016: xii). At the same time, there are enormous variations in what it means to be a mad person, in the opportunities and career paths that are available, and in how things humanly stand for mad people, at different times and places. However, these designations share a recognition that, regardless of their infirmities and misfortunes, mental patients or mad persons are fellow human beings or agents, embedded in, and influenced by, a diversity of cultural, social, and political circumstances and conditions.

The vicissitudes of the psyche invariably involve power, the life courses, and trajectories of psyches in the power fields of history. The "things of the soul," proposes the South African writer Bessie Head (1973), are really "a question of power." Under some social conditions, the disturbed psyche may come to possess enormous cultural and political resonance, giving rise to entanglements between psychiatric and political forms of action and thought that historian David Freis (2019) terms "psycho-politics." Moreover, inflamed cultural and political currents and complexes have sometimes resulted in an intermingling of concepts such as "modern," "mentally ill," "racially inferior," and "primitive" (Peters 2014: 43—45). The physician and anthropologist Didier Fassin argues that precarious lives must be defined, not in the absolute of a condition (for instance, in an earlier era of psychiatry, "constitutional inadequacy") but in relation to those who have power over them (Fassin 2012). In considering the lives of mad persons, we must, therefore, move beyond the patient-illness nexus towards a two-way relationship between mental illness and its context. As the anthropologist and psychiatrist Arthur Kleinman has argued, mental illness can frequently be considered as a form of social suffering that is brought about "under the impress of large scale transformations that shape an era or a place" (Kleinman 1999). As cultural historian Sander Gilman (1985) stresses, historically an ineradicable sense of innate difference attends cultural reactions to, and representations of, the mad person. Though radical artistic currents may sometimes have embraced the mad, it would be rash to conclude that they have done much to mitigate, or assuage, this sense of difference and isolation. Typically, the existential and ontological concerns that arise for the mad person may hover around questions such as What is at stake for us? What is our status? Are we human?

In a growing body of scholarship preoccupied with social suffering and marginalization, the concept of precarity names a politically induced condition in which certain groups or populations become differentially exposed to injury, violence, and death (Butler 2009). All human lives are ontologically precarious but the examination of how precarity is created and lived requires attention to the political force fields in which the precariousness of life is exploited (Han 2018; Kasmir 2018). Precarity may, for instance, refer to the effects of an epistemic power that keeps the lid on vulnerable subjectivities by annulling their significance and inhibiting them from exerting a claim on the present. With regard to the mentally ill, historically, public psychiatry has tended to deprecate the human value of its patients, functioning as a form of pauper burial that assimilates a plethora of unfortunate souls into categories and frameworks in which their individualities are to a large extent dismembered, relegated to the nonhistorical, and discarded as meaningless, in official asylum ledgers and case histories, as surplus people of no significance (Barham 2004; Speed 2017). "When is life grievable?" asks the political philosopher Judith Butler. Certain lives do not really qualify as lives, so they are never really lost or grievable in the full sense. "We might think of war," Butler proposes, "as dividing populations into those who are grievable and those who are not" (Butler 2009). The emotional and psychological traumas of the First World War brought into focus struggles for recognition, and poignant questions about human worth, for vulnerable or mentally afflicted individuals, that were to resonate across the next decades through into the next war (Barham 2004).

It is appropriate to emphasize that this topic does not in its entirety properly constitute an established field of knowledge and what is described here draws, at points, on various instances of counter-knowledge, or counter-discourse, producing "alternative truths" that in due course may be assimilated into established knowledge in psychiatry or other branches of the human sciences (Jacob et al. 2014: 8). At the same time, though, these various examples of counter-knowledge unfold within cultural, social, and epistemological currents that have been the subject of close study and one of the aims of this chapter will be to identify, and summarize, such currents and complexes where apposite, so as to provide depth and anchorage to this discussion.

It will be helpful, first, to set this discussion in the context of new approaches to people with disabilities that have emerged in recent decades and also to new forms of knowledge-making, more responsive than heretofore to the voices and viewpoints of the mentally ill themselves, that have emerged from the human sciences, notably anthropology. As critics have underscored, the traditional paradigm of knowledge-making about mad persons, and about the history of madness, has mostly forefronted the viewpoint of the powerful, tending to compromise or negate the humanity of mad persons. To this day, people with mental disabilities may be stigmatized because of their documented psychiatric pasts. By the late nineteenth century already, the English lunacy certificate had become the standard for regulating admissions into lunatic asylums in numerous jurisdictions around the globe. Creating a record that was difficult to erase from bureaucratic and social memory, certification, and the broad category of the "certified insane" restricted the rights of people with disabilities in determining their lives and, in line with the ideal of preventing the reproduction of the "unfit," also served as a spur to scientific inquiry and eugenic speculation about racial degeneration, permanent segregation, and, in some cases, sterilization (Sposini 2020).

During the past four decades, disabilities rights movements have advocated for a fundamental shift in the way society looks at people with disabilities, from "objects of welfare and science" to active citizens. The United Nations Convention on the Rights of Persons with Disabilities (CRPD), adopted in 2006, with its emphasis on "nondiscrimination," has set the conditions for denouncing stigmatizing attitudes and behaviors. The CRPD adopts a human rights model of disability and calls for the abrogation of detention, substitute decision-making, and compulsory treatment. It claims to "take to a new height the movement from viewing persons with disabilities as 'objects' of charity, medical treatment and social protection towards viewing persons with disabilities as 'subjects' with rights, who are capable of claiming those rights and making decisions for their lives based on their free and informed consent as well as being active members of society" (United Nations 2006). The Committee charged with implementing the CRPD stipulates that the existence of an often scientifically quantifiable impairment (to include a physical, mental, sensory, or psychosocial impairment) must never be grounds for denying legal capacity and the imposition of a "substitute decision-making." Though for some people with disabilities, the exercise of that capacity will require support – sometimes a great deal of support – the Committee takes the view that all persons retain legal capacity, rejecting impaired "mental capacity" as a basis for denial of legal capacity, and believing that with the right level of support, people with disabilities will be able to express their "will and preferences." To ensure that persons with disabilities enjoy the right to legal capacity on an equal basis with others, the Committee also insists that the "will and preference" paradigm must replace the paternalistic "best interests" paradigm.

The late historian of psychiatry Roy Porter (1985) famously called for the reconstruction of the people's history of suffering and historian Catherine Coleborne (2020) asks that we engage with the field of mad histories and with the global history of what it has meant to be mad at different times and places. Roy Porter's call to reclaim the voice of the patient has proved controversial, with some casting doubt on the viability of a history of psychiatry "from below," but in a critical re-appraisal of the field Bacopoulos-Viau and Fauvel (2016) seek to revitalize the writing of psychiatric history "from below" through approaches that lend credence to histories of collective "mad" cultures and comparable cultures of "patienthood." "The very identity of the 'patient' has changed over time," Coleborne remarks. (For a bibliography of recent first-person narratives of madness, see Jasyasree Kalathil (ed.) 2013). Despite "a two-centuries-long alliance with medicine," lament psychiatrists and historians Ivana Markova and German Berrios (2012), psychiatry, "its structure, objects, language, and praxis, remains as opaque as ever," and they urge a reconsideration of the epistemology of psychiatry to take account of the hybrid nature of psychiatric knowledge, combining questions and methods pertaining to both the human and the natural sciences.

As the anthropologist Robert Desjarlais (1997) observes, the type of people considered mentally ill have mostly been studied as solitary clinical entities, not as persons immersed in the stream of social life. By contrast, following in the path of thinkers such as Mikhail Bakhtin, Lev Vygotsky, Gregory Bateson, Jacques Lacan, and Jean Laplanche, human lives are necessarily dialogic, for a person is, by the very nature of being a person, enmeshed in social, communicative, and moral relations with others. Thus, Desjarlais proposes moving away from identifying people primarily through the all-defining characterizations of psychiatric diagnoses, and, instead, to "heed the subtle, multiplex, pragmatically attuned and occasionally contradictory dimensions of people's lives, minds and actions." In the shelter for the homeless mentally ill which he studied, though the residents were "often tormented and opaque," nonetheless for the most part they were "socially astute and ethically motivated eccentrics," whose "identities never fused completely with diagnoses, for ailments were only one strand in very complicated lives." Similarly, psychiatrist and historian Jonathan Metzl (2009) argues that wholesale acceptance of psychiatric terms and frameworks can result in submersion into a potentially racially subjugating symbolic order in which persons with diagnoses such as schizophrenia may become doubly stigmatized, both by a diagnosis that carries poor prognosis and by a medical system that enforces acceptance of hegemonic descriptors of well-being at the expense of autobiographies of protest or survival. South African clinical psychologist and historian Sally Swartz (2014) has also highlighted the epistemic violence of the case history in which lives are wholly absorbed into the discourse of psychiatry and the voice of the patient may be entirely lost. It may, however, be a mistake to assume that the discourse of psychiatry was inevitably dominant, for Sarah Chaney (2016) and others have shown that in the nineteenth century patients were not always victims of "psychiatric power," sometimes influencing the formulation and circulation of medical constructs by serving as intermediaries between medical and lay perceptions of madness.

In distancing from the "objective" categories of clinical psychiatry, and reaching out towards the idea of a person embedded in a wider, and mutable, cultural, social, and political context, Desjarlais, Metzl, and Swartz exemplify a trend in anthropology and in the social sciences more widely, in which the study of the primitive, the savage, and the radically "other" is displaced by a focus on what anthropologist Joel Robbins (2013) calls the "suffering subject," "the subject living in pain, in poverty or under conditions of violence or oppression." A notable example of this approach is Joao Biehl's ethnographic study *Vita* (2005) in which Biehl recounts the life story of Catarina, a Brazilian woman suffering from a neurological disorder, who is rejected as mentally ill by her family and left to wither by the state in an "ex-human" condition in an institution for those whom Biehl terms the "socially abandoned." Another example is Peter Wilson's classic study of Oscar Bryan on the Caribbean island of Providencia in the 1950s (Wilson 1976). More recently, a classic historical example of a combative mentally distressed "suffering subject" (though he did not see himself as being "mentally ill") is Friedrich Krauss (1791–1868), a travelling salesman from Southern Germany, the author of a voluminous self-narrative of madness, *Notschrei eines Magnetisch-Vergifteten [Cry of Distress by a Victim of Magnetic Poisoning]* that he published in 1852 accusing a Flemish family of trying to assassinate him and that told of how he had been the victim of mesmerist forces since 1816, of his experiences (harsh but only temporary) in private madhouses in Antwerp, of his efforts (largely in vain) to enlist doctors, lawyers, and others to assist him, and of the orthodox and nonorthodox forms of medical knowledge that he drew upon to try to make sense of his experiences (Bruckner 2016). Anthropologists Tanya Luhrmann and Jocelyn Marrow (2016) have assembled a formidable body of evidence to argue that people in the West with serious psychotic disorders are vulnerable in many different ways to "the sense that they have been defeated at the hands of others." The "most troubling madness" of schizophrenia is, above all, shaped by its social context and what the course and outcome of schizophrenia principally reveals is a social rather than a natural history. Schizophrenia is the "story of the way that poverty, violence, and being on the wrong side of power drives us mad." When life beats people up, they are more at risk of developing psychosis and even those who have been "successfully" treated may feel a sense of loss and betrayal and take the label of their condition as signs of their lesser humanity. At the same time, the historicization of mad subjectivities in these and related accounts should not be taken to imply that mental disorders are essentially a matter of social construction, free of biological constraints. Rather, mad subjectivity may be understood as located at the juncture between the subject, their biology, and local regimes of normalcy and power (Biehl et al. 2007).

More Than a "Pathological Attitude to Life"

In considering the mental patient in history in the modern period, the figure of Daniel Paul Schreber, author of *Memoirs of My Nervous Illness* (*Denwürdigkeiten eines Nervenkranken*) (2000), first published in 1903, but known to posterity mainly through Sigmund Freud's notorious case study, "*Psychoanalytic Notes on an Autobiographical Account of a Case of Paranoia (Dementia Paranoides)*" (1911), is of enormous significance. Born in Leipzig in 1842, and reared in a rarified somatic environment reflecting the orthopedic pedagogy and mechanical inventions of his influential physician father, Schreber became a highly successful and acclaimed jurist who in June 1893 was nominated to the position of presiding judge in the

Saxon Supreme Court in Wilhelmine Germany. Not long afterwards he succumbed to a profound psychotic crisis and collapse, for which he was interned in an asylum for almost 9 years. In the teeth of concerted opposition from the physicians who presided over his case, he sought to redeem his status and reputation through the publication of his personal memoirs in which he describes and analyzes, in intimate and unsettling detail, not merely what has befallen him, but also what it may signify. Schreber's psychotic breakdown brought him intense, and to a considerable extent unrelieved, suffering – he was to be subject to the "bellowing" that had been afflicting him for the rest of his life – but at the same time it possessed a profound resonance that transcended his individual case. As Schreber himself recognized, his exposition of his illness was to be read, not just as an expression of his own psychopathology, as his physicians were all too eager to maintain, but also as an expression of an alternative stance within life, and an opening onto truths about the society of his period that were generally hidden or denied.

His psychic decomposition commenced with an experience of female *jouissance* that led inexorably to the delusion that he was turning into a woman and merging with the figure of the unmanned Wandering Jew. Moreover, this was accompanied by a profound and anguished sense that he was entangled in a malevolent divine conspiracy directed against him, largely controlled by his chief physician Paul Emil Flechsig, who was able to maintain nerve-contact, and speak directly to his nerves, without being physically present, inducing a terror in Schreber that he was no longer in possession of his own mind and that "*inner voices*" moving "like long threads in his head" were directed from without (Santner 1996: 76).

But what exactly was it that provoked Schreber's crisis? According to Eric L. Santner (1996), in an influential study, Schreber was experiencing a crisis of investiture in which the discovery was borne upon him that the symbolic order in which he participated both as a judge and as a German man of his period was sustained through the repetition of meaningless rituals that possessed no ultimate grounding and was to all intents and purposes a sham. From quite an early stage in his crisis, Schreber started to compile his memoirs, largely completing them by 1900, guided by a compelling belief that his investiture crisis, and other dimensions of his experience, held some wider relevance and that by sharing his crisis, he would, in Eric Santner's words, be "disseminating its significance as a general state of emergency of symbolic authority that touches everyone."

In desiring so strongly to publish his *Memoirs,* Schreber was wagering that, indiscrete or shameful though they might appear, his experiences of psychosis, and his *Memoirs* as a record of them, bespoke an intimate and searing account of what life really amounted to below the surface façade of the society of his time that possessed vital significance well beyond the confines of an individual case record. His *Memoirs* were not just a pathography or psychological narrative, they must also be permitted to resonate in a wider cultural sphere. Schreber possessed a strong proprietorial sense of the integrity and significance of his narrative and would not permit it to be defined by the likes of Flechsig. His legacy now rested on the publication of his *Memoirs*, since for lack of them his aborted legal career, terminating in an asylum case register, would have warranted no more than an historical footnote. He would, to a large extent, have become a forgotten person. Only a decade or so later, during the First World War, a British surgeon who had latterly turned to psychosomatic medicine and to psychology was to write about psychosis as "action in which the psyche takes part" which describes admirably the outlook that had been distilling in Schreber's imagination and that he was anxious to convey in his *Memoirs* (Barham 2004).

Fortuitously, however, Schreber was experiencing his crisis at an historical moment in Wilhelmine Germany when the authorities were inclined to look sympathetically upon grievances over maltreatment in the psychiatric system and very likely this favorable climate of opinion assisted him in brokering his release and securing a publisher for his memoirs. For some years already, a lunatics' rights movement had been gathering pace in Germany, subsequently to be dubbed Europe's first antipsychiatry movement, led mainly by middle class psychiatric patients, who had been bringing before a receptive public, in numerous books and pamphlets, the details of the abusive practices to which they had been subjected (Goldberg 2003). Though a receptive audience may already have been created for his *Memoirs,* his struggle to escape the confines of the casebook, and to secure a standing as a mental patient in history on his own terms, was not done just yet. He may have succeeded early on in eluding the grasp of Paul Flechsig, but the legacy of his *Memoirs* would for the best part of a century have to reckon with the stamp imposed on it by Sigmund Freud in his essay on Schreber's case that was published in 1911. In Freud's account, Schreber remained captive to a discourse that pivoted about his personal psychology, and it was only towards the end of the last century, in the work of scholars such as Eric Santner and notably also Zvi Lothane (in his book published in 1992 and in numerous articles published between 1989 and 2014) that it became possible to restore or transfigure the psychopathology in which he was immersed into an understanding of the historical dynamics from which it derived and so to embrace an understanding of Daniel Paul Schreber as a compelling and instructive historical agent in his own right.

"What Kind of Object or Life Is This?"

A recent biography of Vincent Van Gogh concentrates the ambiguity and uncertainty around a life like that of Schreber over which there hangs a psychotic intensity. "What *kind* of life is this?" we may ask of the nine hundred and fifty pages of *Van Gogh: The Life* by Steven Naifeh and Gregory White Smith (2012). Only with hindsight can we declare it to be the life of a great artist, for as we are reading and living it, we have no idea how to define what it is we are reading, where it is leading, and how it is going to resolve, for it could very plausibly conclude as a pathography or a case history in a lunatic asylum. Indeed, for some of the immediate protagonists, it remained unresolved, Van Gogh's mother comparing her son's errant life to a death in the family, systematically discarding any paintings and drawings he had left at home, as if disposing of rubbish and, even after his death, when fame belatedly found him, never amending or regretting her verdict that his art was "ridiculous." Is he a real artist or a make-believe artist? Not until Van Gogh reaches Arles does the uncertainty around this question start to abate, and even there it is vulnerable and in tension, for the idea that Van Gogh might now be a "celebrity" seems more like a flimsy, and competing, delusional vision than a reality. In Arles, in March 1889, he found himself in a dark place in every sense, unable to paint, holding his head in his hands. A few months later, after he has started to paint again, he writes to his favorite sister Wilhelmina: "I am beginning to consider madness as a disease like any other and to accept the thing as such. Almost everyone we know among our friends has something the matter with him." Shortly afterwards, he asks to be admitted to an asylum: "I am unable to look after myself and control myself. . . . I have been "in a hole" all my life." Soon he will go to St Remy Asylum, where he discovers in the inmate community a fellowship of mutual consolation and a serenity that he had never found in the world outside. A month after he arrives there, he declares: "I think my place is here!" "The more my health comes back to normal," he writes in October 1889, "the more foolish it seems to me, and a thing against all reason, to be doing this painting which costs us so much and brings in nothing." He asks to stay at least another year in the asylum, blaming himself for his illness and for not seeking treatment sooner. Around then, he is for the first time extolled in a series of articles by the young critic Albert Aurier as "the mad artist," an "epileptoid genius," with the "nerves of a hysterical woman and the soul of a mystic. . . .always at the brink of the pathological." Van Gogh will have none of this, however, believing himself to be a fraud and failure, tolling his sins in a letter to his mother in a desperate jumble of words, and brooding on the precariousness of sanity and life. Though other accounts of his death have been proposed, Van Gogh is alleged to have shot himself in the chest with a revolver on July 27, 1890. In the final months of his life, he had become increasingly worried about his future. His mental instability had forced him to reconsider his ambition and he believed that he failed as an artist. "My life is attacked at the very root," he wrote shortly before his death. Dr. Felix Rey, one of Van Gogh's more sympathetic doctors, believed that he suffered from a form of epilepsy caused by his nervous nature and self-neglect. Along the way, numerous other scientific diagnoses have been advanced including: epileptic attacks with acute mania, schizophrenia, episodic twilight states, neuro-syphilis, psychopathy, bipolar disorder, cycloid psychosis, and borderline personality disorder (Marije Vellekoop et al. 2016). From the privileged vantage point of the present day, it may seem natural to deride the reactionary judgments of family members and others, notably Van Gogh's mother, who were seemingly oblivious to, or even decried, his creative genius. But this is, surely, a deeply anachronistic and ahistorical assessment, for in the lived moment of the late 1880s, the answer to the question "What kind of person or life is this?" was far from obvious, and even today, it must be admitted, it still attracts controversy.

Psychopathology, Art, and Politics

The mad have frequently been denied their own voice and forced to speak through institutions such as psychiatry which have historically often denied them any insight and in which even their humanity has been subject to dispute. At points, the artistic productions of the insane may have played a role in propagating an alternative view of their humanity. Thus, Pliny Earle, an American alienist, one of the founders of what became the American Psychiatric Association, published an essay in 1845 on the artistic productions of the insane in which he fiercely defended their innate humanity and ability to convey truths that are repressed in normal people. However, though the Romantics extolled the verbal creations of the mentally ill, by the late nineteenth century the art of the insane had become a symptom of degeneration, rather than a model of inspiration. For the Italian forensic psychiatrist Cesare Lombroso, the artistic productions of the insane constituted an atavistic or "primitive" form of representation, providing proof of his thesis that sociopathic and psychopathic conditions were a throwback to a more primitive stage of human development. In the early twentieth century, however, avant-garde currents such as the Dadaist movement appropriated the identity of the mental patient, and of specific diagnostic types such as "the schizophrenic," to articulate a critique of bourgeois society. If the state found it necessary to isolate the insane, the avant garde taunted, they would integrate them, or at least the myth of insanity, into their ideal world (Gilman 1985).

German psychiatrist and art historian Hans Prinzhorn played a notable role in the revaluation of the art of the mentally ill in the early part of the twentieth century after he became curator of an art collection at the psychiatric hospital of the University of Heidelberg, where Emil Kraepelin had once worked. In 1922, he published the book *Bildnerei der Geisteskranken* (*Artistry of the Mentally Ill*) for which he is chiefly known and which became hugely influential among the avant garde in opening a window on the creative potential to be found among the so-called mad. At the same time, the title of Prinzhorn's work, with its stress on "artistry" rather than "art," gestures at his diplomatically equivocal position, amidst the volatile and rancorous cultural politics of the 1920s, over the status of the art of the mentally ill, maintaining that though the mentally ill might not be true artists themselves, they did point the way to a new creative outlook. Prinzhorn was successful, it has been suggested, in bringing about a shift in values and rescuing previously despised works from the psychopathological clutches of his colleagues, for they could now be discussed without the contemptuous tone that doctors commonly assumed in referring to "this rubbish" in their clinical notes (Brand-Claussen et al. 1996).

The productions of the insane in this period existed in a curious limbo somewhere between the pathological exhibit and the work of art. Like the life histories of their makers, creative objects made by people with experience of mental illness are frequently ontologically complex, proving resistant to any single answer to the question: "What kind of thing or life is this?" (Jones et al. 2010, p. 12). In the interwar years, however, such equivocation or uncertainty was prone to attract a disregard and opprobrium that could be life threatening. The art critic Hal Foster hints at the tense political undercurrents here, when he remarks that "a reversibility haunts the modernist re-evaluation of the art of the mentally ill, for if this art could be revalued as somehow modernist in affinity, the art of the modernists could also be branded as somehow pathological in tendency" (Foster 2001).

The failed Austrian water-colorist, Adolf Hitler, perceived the entire interest in the art of the insane as proof of the "crazy" direction which the avant garde had taken. The Nazi answer to the question of the creativity of the insane was to repudiate it, and so to reduce the insane to a subhuman level, to deny them the status of membership of a cultural entity, and eventually to murder them. Psychiatrist Carl Schneider served as one of the most important experts in the action against "degenerate art," denying that the insane were capable of any true esthetic sensibility and playing a major role in the sterilization and murder of the mentally ill. In common with Jews and blacks, the insane were perceived as unable to communicate at any level. Such a deprivation of meaning facilitated the "euthanasia" of asylum inmates and since, argues Sander Gilman, the mad and the Jews had for a good while been embraced in the same paradigm in Germany, it was but a short step from killing the insane to killing Jews (Friedlander 1995).

"Africa in Their Natures": Black and Insane as Double Jeopardy

"My name is Ann Pratt. I entered the asylum on the 14th of January last. It was on a Saturday, about eleven o clock..." Thus commence the revelations, part oral and part written, of Ann Pratt, a young Mulatto woman from a slave background, who was an inmate of Kingston Lunatic Asylum in Jamaica in 1860 and the reputed authoress of a pamphlet entitled *"Seven Months In The Kingston Lunatic Asylum, And What I Saw There."* In her testimony, Pratt challenged both the nature of the treatment that was given her and also the very basis for her confinement. *"I was there for treatment,"* she says. *"I was sent out as cured –so they said—if I was ever mad at all. I don't think there is any curing effected there –the treatment is more likely to drive you mad than to cure you."* Unsurprisingly, there were many who bridled at Ann's polemics and sought to discredit her. Dr. Scott, the Principal Medical Officer, insisted that in the asylum she was in "a highly insane condition" which was "painfully apparent." Fortunately for Pratt, she found an advocate and ally in Dr. Lewis Bowerbank, a Jamaican born and Edinburgh-trained physician, who was a stern critic of medical abuses on the island and, thanks largely to his advocacy, the goings on at the asylum were made public. A commission of enquiry held in 1861 found that Pratt's accusations (subsequently corroborated by the testimonies of other inmates) were largely true and that random beatings from the matron and attendants *"with fists, feet, sticks, broomsticks, straps, umbrellas and any other weapon which happened to be in the way"* were commonplace. Though the asylum authorities made every effort to discredit her testimony, the unintended but cumulative effect of the inquiry was to subvert the credibility of benevolent, or humanitarian claims by the colonial authorities (Fryar 2018).

It has been argued that liberalism has predominantly been a *racial* liberalism in which an egalitarian humanism has co-existed with an hierarchical racism. At its most benevolent, imperial ideology viewed the "native" and "lunatic" as childlike innocents, in need of humane care and of civilizing influence. However, influenced by Social Darwinism and the scientific theories that supported it, the lunatic was viewed as a regression to a more primitive stage of human development, or as a waste product of civilization. A double jeopardy could attach to being both black and insane. "It was Africa, hitherto dormant, that had broken out in their natures," intoned the London *Times* following the Morant Bay uprising by black Jamaicans in 1865. Historians have examined how "blackness" became medicalized and how assumptions about the existence of distinctive black and white psyches have shaped diagnostic and therapeutic regimes, leaving a legacy of poor treatment of African American patients, even after psychiatrists had begun to reject racialist conceptions of the psyche. Enormously influential was the evolutionary framework adopted by the early twentieth-century psychiatrist William Alonson White who identified psychological well-being primarily with the white, male social role, presaging representations of Black Americans as primitive and culturally atavistic.

Forging alliances that challenged these perils required courage. One former mental patient who was determined to advocate for such alliances, and paid for it with his life, was William "Bill" Lewis Moore (1927–1963), a white postal worker who in April 1963 set out on a one-man civil rights march to protest against the pervasive presence of white supremacy in American society, pulling a two-wheeled postal trolley and wearing two sandwich-board placards, one of which read "End Segregation in America." While a graduate student at John Hopkins University, Bill suffered a breakdown and between 1953 and 1955, he was institutionalized at Binghampton State Mental Hospital with a diagnosis of schizophrenia. He later became an activist for the mentally ill and a civil rights activist for African Americans, in the early 1960s undertaking three civil rights protests in which he marched to the capital to deliver hand-written letters denouncing racial segregation. On April 23, 1963, Bill Moore was shot to death at close range alongside a highway in northern Alabama. Black studies scholar George Lipsitz was 15 years old at the time and he can still remember the impression that Moore's murder made on him. A man who found white supremacy an abomination, even though he was white himself, Bill Moore was "a rare individual" for "few white people are willing to risk their lives in the fight against white supremacy" (Lipsitz 1998).

Mental Suffering, Psychosis, and Witnessing

All four of the individuals discussed in this section, three of them writers, Rodrigo Souza Leao, Bessie Head, and Lemn Sissay, and also Dimitri Tsafendas, who on September 6, 1966, became the assassin of Dr Hendrik Verwoerd, the founding prime minister of the Republic of South Africa, known to posterity as the architect of the infamous apartheid policy, turned their reflections on their actions and sufferings into a window on the times in which they lived. "I forced myself to talk about the system that surrounds me," opens the acclaimed autobiographical novella, *All Dogs are Blue*, by the Brazilian Souza Leao, a ribald and kaleidoscopic portrayal of life in a Rio de Janeiro lunatic asylum. Bessie Head, Lemn Sissay, and Dimitri Tsafendas also had long-term, intimate, and distressing relationships to the "systems" that surrounded them (Sissay refers to it in his memoir as "The Authority"), that for all them held a meaning that exceeded their own personal sufferings.

These narratives all place the sufferers center stage, and in so doing they exhibit graphically how they are all constrained to manoeuver, and to negotiate their cultural/social/institutional identities, in circumstances where the odds are frequently stacked against them and, above all, where control over their own biographies, the narratives that have been told them about their own lives, over what they know about themselves and their origins, and what they are authorized to say about themselves, may have been wrested from them, or never granted to them in the first place. So, in large part, this is about a nexus of knowledge-making and knowledge-keeping, in a very basic, intimate, and emotive sense, that must contend with forces of deception, disavowal, and denial, and where it becomes palpable that the "things of the soul" are, indeed, inextricably inwoven with "questions of power," and that all along the line there are questions here that the history of the human sciences must be willing to engage and refuse to take for granted or to foreclose.

The system does more than surround the narrator in *All Dogs are Blue*, it also intrudes upon, and interferes with, his feeling, his being, and the way in which he thinks about himself: "this was Brazil, a total mess." It is much greater than the asylum, or even psychiatry, it is the controlling forces in society that humiliate and reject him, stripping him of who he is, leaving him reduced to just another nutcase. "In the old days," the narrator says, "anyone who was different or who appeared to be a threat was crucified. Nowadays you wind up in places like asylums, which is the best way to not get better.....I had my first attack at fifteen, at thirty-six I've still got problems....I'm a walking problem." Yet, still, he answers back: "I'll never be nothing. I can't want to be nothing...I'll always be the one who waited for a door to open up for him in a wall without a door." Later in the narrative, outside the asylum now, he founds, and becomes the leader, of a religious cult called Todog that promises to unite all beings but all too quickly, amidst dissent among its members, splinters and turns to violence, attracting the attention of the authorities, with the consequence that the narrator finds himself back in the asylum.

Journalist, musician, poet, writer, and painter, the life of Rodrigo Souza Leao was marked by his prodigious creative activity and output, but equally by his unrelenting mental suffering, occasioning frequent admissions to psychiatric clinics in Rio de Janeiro. Though he was all the while prolific in his writing and creative activity, maintaining a foothold and influence in the social scene via email and social media through publishing, co-founding and editing one of Brazil's most important poetry magazines, due to his psychological vulnerability, mostly manifest in paranoid and psychotic experiences, he was rarely able to leave his house. Turning his own personal suffering into a window on the suffering of his own society, Souza Leao's intense and haunting narrative never tidies away the messiness and raggedness of a psychotic experience of life, as to try to recompose it from a detached standpoint, but instead lives it relentlessly from within, even while it discloses a longing for recognition and acceptance, and a quest for an alternative ethos, that may breach an opening in a seemingly impenetrable wall. Alas, the door remained closed to him, for Rodrigo de Souza Leao died on July 2, 2009, at the age of 43 in a psychiatric clinic in Rio, a few months after *All Dogs Are Blue* was published.

The Southern African writer Bessie Head and Rodrigo de Souza Leao both share something in common with Daniel Paul Schreber: communicating as they both do, to a considerable extent, not out of choice, but perforce, from within an experience of psychosis, each of them seeks to appropriate their experience of psychosis as an outlook on life and as a mode of observation and critique. As a result, they turn it into something that quite transcends their personal psychopathologies, providing instead the basis for their creative and historical legacies. They may exemplify, in this respect, the idea advanced by the Swiss psychiatrist Manfred Bleuler that, in some cases, recovery from a psychotic condition such as schizophrenia may not be desirable since a form of psychotic life has become an inner necessity for them (Barham 1995, pp. 82—84).

Much of Bessie Head's writing is visited by a psychotic intensity that directly reflects the emotional, political, and racial currents and fissures that she was having to navigate in her life. All the while she felt herself to be traversing, and frequently becoming ensnared in, the wires and cross currents of fields of power, to the point of inducing in her a crisis of such proportions that she could only communicate in a prose of psychotic intensity. "We, as black people," she wrote, "could make no appraisal of our own worth; we did not know who or what we were, apart from objects of abuse or exploitation" (Head 1990). "The whites," she later observed, "have imposed a whole range of jargon to define their humanity as opposed to the nonhumanity of black people." The ordeal of her crisis stimulated her to embark on what became *A Question of Power* and to muse on what might be at stake "if the things of the soul are really a question of power" (Eilersen, p. 142). Though Head had already achieved some recognition as a writer, the composition and publication of *A Question of Power* proved a great struggle for her. The novel closely tracks Head's own experience through the crisis of the central protagonist Elizabeth, a young colored South African woman with a family background closely resembling Head's own, her nightmares in a "shared journey into hell" in the company of Dan and Sello, two hallucinated figures, and finally her gradual, but decidedly partial, recovery in the village of Motabeng in Botswana.

Written in a heightened form of prose poetry, the psychotic intensity of the novel proved a stumbling block for agents and publishers: one rebuked it as bordering "on the meaningless," another rejected it on account of her "misuse of the English language." As a crucial determiner of the meaning and value of a work, the publisher inevitably played a critical role as gate-keeper here in deciding whether Head could be permitted to transcend her status as a mental patient, or ex-mental patient, and achieve recognition for her creation as a literary, rather than exclusively a pathological, product. And in setting about this, she was very much alone, lacking the "social capital," or "relational power," on which metropolitan writers frequently depend in making their way in the world and securing credibility for their creations. Fortunately, she at last found in James Currey an editor who responded positively to the work, even though he was baffled by how to categorize the work and found himself fumbling for a consensual framework in which to locate it: "*A Question of Power* numbs me. I go back to it and back to it." Head had broken with the convention of African social realism, and what she had produced instead was "more closely related to the mainstream of Anglo-American internal writing, though the whole race thing gets across.... It is big. You throw the lot on us and I really can feel, feel, feel though I cannot always understand. I know you have laid the inside of your head out on the paper" (letter to Bessie Head 24 August 1972, cited in Davis 2018). Like Souza Leao, Bessie Head is striving to find an opening in a seemingly impenetrable wall (Pearse 1983).

The predicament of Dimitri Tsafendas is not entirely dissimilar to that of Bessie Head, for like her his utterances and actions are at risk of being, or already have been, construed and discredited as those of a mad person except that, in the first instance, it is not a literary editor or publisher who plays the critical role in receiving and validating alternative meanings and understandings for him but Greek Orthodox priests, who later vouchsafed their testimonies to Tsafendas's biographer. Born in 1918 in Lourenco Marques, Mozambique, the unintended offspring of a liaison between a Greek marine engineer from Crete and his mixed-race housemaid, her mother an African and her father a German, supposedly an aristocrat, who vanished when he learnt of the pregnancy, for many years Dimitri Tsafendas, imbued from an early age with the communist leanings of his father who was a passionate anarchist, lived a varied and adventurous life.

Returning to South Africa in 1965, shortly before the assassination, he applied unsuccessfully for reclassification from "white" to "colored" and in July 1966, at the age of 48, he obtained a temporary position as parliamentary messenger in the House of Assembly in Cape Town. After the assassination, Tsafendas was found unfit to stand trial in the Cape Supreme Court on the grounds that he suffered from schizophrenia, and so was spared the death penalty. Psychiatrist Harold Cooper testified that, as early as 1935 or 1936, Tsafendas began to believe he had a tapeworm inside him that controlled his life. According to Cooper, Tsafendas "felt that the reasons underlying his killing of Dr Verwoerd were far too complex. He couldn't explain to me why. And then he started talking about frustration, frustration, and the tapeworm, and not holding jobs, and having nowhere to live, and the whole thing became jumbled in his mind." After the trial, headlines across the world announced that Verwoerd's assassin believed he had a tapeworm that was responsible for the killing. The official, and heavily promoted, position was that Verwoerd had been murdered by a madman. The Court ordered that he be incarcerated "at the pleasure of the State President" and, until his death in 1999, he was detained, first in prison, and subsequently in a mental hospital. For several decades, Tsafendas was considered a simple schizophrenic and a crazy loser, completely lacking in political motivation, who scarcely elicited any sympathy from the public or from history books which typically wrote off Verwoerd's murder in a single line such as "stabbed to death by a deranged parliamentary messenger."

Recently, however, a biography of Dimitiri Tsafendas by Harris Dousemetzis (2018), based on thorough research over a number of years, has appeared that challenges the official portrait. An examination of Tsafendas's life history discloses a man who was articulate and knowledgeable – he spoke several languages – and politically engaged over many years. As early as 1939, he entered South Africa illegally and joined the South African Communist Party, and during the Greek Civil War, he fought against the Royalists with the military wing of the Greek communist

Party. Labeled a "Communist and a half-caste," for almost 12 years, he was banned by the authorities from entering Mozambique or South Africa because of his supposed political allegiances, and during this period he roamed across Europe and the Middle East, sometimes working as a teacher and translator. During this time already, so it is alleged, when hard up, he would sometimes claim to be mentally ill so as to secure food and shelter, displaying a skill for dissembling that he would draw upon later on. The two statements that Tsafendas made to the police, giving lucid political reasons for killing Verwoerd and making no reference to a tapeworm, were concealed at the trial. Dousemetzis contends that the diagnosis of mental illness has since been dismissed by several forensic psychiatrists and that it had suited the powers of the day to portray Tsafendas as a crazy man rather than as a political assassin. This claim is convincing up to a point, and there can now be no doubt that, considered as a historical figure, Dimitri Tsafendas is a far richer, and more interesting, subject than had previously been supposed, but at the same time the choice between "insane killer" and "political assassin" is not entirely clear cut, and arguably is, in any case, a misleading and simplifying fiction, in as much that the domain of the "political" is itself riven by contradictions, ambiguities, paradoxes, and irrationalities.

Bessie Head, Lemn Sissay, and Dimitris Tsafendas have something else in common which is that they were all deceived, and told untruths, about their origins and they were all separated at a very early age from their birth mothers. Dimitri Tsafendas only discovered the identity of his birth mother when in his late teens he had to request a copy of his birth certificate in order to apply for a new job. In Bessie Head's own words: "I was born on the 6th July 1937 in the Pietermaritzburg Mental Hospital in South Africa. The reason for my peculiar birthplace was that my mother was white and she had acquired me from a black man. She was judged insane and committed to the Mental Hospital while pregnant." Soon after her birth, Bessie was given out for adoption to a white family who quickly returned her because she looked "strange." She was then passed to a couple classified as colored living in the poorest part of the city, shortly after which contact with her mother's white family terminated. Her mother died in the mental hospital a few years later bearing a diagnosis of "dementia praecox." Only when she was 12, after she had been placed in a boarding school for orphans, did the authorities abruptly tell her that she was the daughter not of her colored foster mother, Nelly Heathcote, as she had supposed, but of a white woman. One day an English missionary produced a huge file: "Most of what she said was that my mother had been insane and my father was 'a native'. That was the big horror, but then there were so many of us at that orphanage who were horrors." Head felt that she had an obligation to share the stigma of insanity that her mother bore, she could not permit her to carry it alone. "A birth such as I had," she wrote, "links me to her in a very deep way and makes her belong to that unending wail of the human heart" (Eilersen 1995: 103). All along the way, she had to contend with the meaning and traumatic circumstances of her birth and upbringing, and the questions about being "human" that they raised, not just for her personally, but amidst the struggles of her time. "I am extremely prone," Bessie Head later said, referring to this period, "to having emotional storms and a very turbulent destiny." Existence in South Africa, she remarked, was like living "with permanent nervous tension."

In a different place and time, Lemn Sissay had to confront rather similar questions about the circumstances of his birth and upbringing. Sissay was born in 1967 in Wigan, Lancashire, to an Ethiopian unmarried mother who had come to England to study for a short period of time at a college in the south, but when the authorities became aware of her condition was hurriedly dispatched to an institution for unmarried mothers in the north, and soon after the birth was compelled to return to Ethiopia to take care of her dying father, leaving Lemn to be cared for by what he subsequently names "The Authority." The traumatic history of his origins and upbringing produced episodes of acute emotional upheaval and distress in which he came under the care of a psychiatrist, though he was never given a formal psychiatric diagnosis or made to assume an identity as a psychiatric patient. Throughout his formative years, until the age of 18, he was absorbed into a care system that took away his name and identity (Lemn was quickly expunged by "Norman," and he became either Norman Sissay or Norman Greenwood, the family name of his foster parents), making him over into someone divested of any connection to his origins, shaping his life through an institutional action driven by a combination of prejudice, incompetence, and bureaucratic inertia, which left him disoriented, confused, and questioning his worth.

In 1983, at the age of 15, his real name and origins, and the letter from his mother, were finally revealed to him. Soon afterwards, he decided that he would now call himself Lemn and ask others to do so. However, so much of what he was now learning about his identity diverged so starkly with what had been told him over so many years that the consternation and confusion that now built up in him is scarcely surprising: "December 1983. Christmas was close, too close. My mental health deteriorated rapidly. I started to walk via back streets. I found main roads too intense. I didn't want to be looked at.... My head was falling apart and the people in the home didn't seem to know. They couldn't see what any parent would have seen in their own child –a mental breakdown" (127). When he was 14, he had acquired Chalky, or Chalky White, as a nickname but now: "Chalky became a haunting, not a name. I needed it to stop.... Behind the veil of the race joke was hatred. There was no one to guide me on these matters. I had my instinct. I am a black man, I said to myself.... I am not Chalky White. I am not a nigger, a coon, a wog, I am a black man. I changed, seemingly overnight, from the cheeky chappy, the happy-go-luck joker, into a threat" (120). Norman Mills, his long suffering social worker, was desperate to find a new situation for him but the prospects were bleak. "I became more and more insular. More and more broken" (146). A meeting within "The Authority" reported: "Norman is confused about his identity and no longer wants to be 'Chalky White' –everyone's favourite coloured comedien. However, he appears to have a very low self-image, sees himself as being stupid and can't talk to others freely." Brought before Leigh Juvenile Court for kicking a garage door, Lemn explained that he committed the offence in retaliation for being described previously as a 'wog' by the owner of the garage door in question (153). On June 21, 1984, "The Authority" decided to imprison him in Wood End Assessment Centre. "The officers were mainly men. White men, men who wanted to be somewhere else.... Men with anger issues. Broken men. Hurt men. Dangerous, white men." Wood End was "a violent and toxic place where the short, sharp shock treatment was an alternative term for abuse.... It was institutionalised violence for voyeurs. I made myself a witness. When a fight broke out between the boys I watched the men, the staff, ejaculate" (172).

Lemn Sissay, Bessie Head, Souza Leao, and Dimitri Tsafendas all made it an obligation to become witnesses to the abuses that were happening around them. Though in her novel *A Question of Power,* which Bessie Head considered "almost autobiographical," the protagonist Elizabeth comes to recognize that it is the fact of her being mixed race that provides her with a unique purpose in life, for Head herself that awakening only came at a considerable cost. Her mental health was always volatile, and in 1969 she suffered a major psychotic episode, resulting in a period of hospitalization in Lobatse Mental Hospital. In Lemn Sissay's case, it took him 30 years of persistent pestering to retrieve his files from "The Authority" (more lies along the way), but having secured them he decided to compose a memoir reproducing items from the case files that followed the trajectory of his movements through the care system, the decisions that were made about him, the altercations and conflicts they provoked, and not least the massive toll that his experience in the system took on his mental health, interspersed with his own wry comments and elaborations. "Secret meetings were held. The folders were taken out and placed on tables surrounded by men and women from The Authority. Decisions were made: *Put him here, move him there. Shall we try drugs? Try this, try that."* "Once I received my files from The Authority their subterfuge became apparent. Everyone had been lying. The Greenwoods had lied from the start." Alongside a letter from Lemn's mother, written in 1968, requesting that he be returned to her in Ethiopia, it is suggested: "perhaps Norman should be made aware of this?" followed by a reply in ink: "Not yet I think."

*My Name is Why (*the title is precisely what the name Lemn signifies in the Amharic tongue of his Ethiopian birth mother) is a record of Sissay's relationship with "The Authority," which wrote reports about him for the first 18 years of his life, placing him with incapable foster parents, moving him from institution to institution, imprisoning and lying to him, and to his birth mother, all along the line. Written very much in the tradition of the hybrid genre, combining personal opinion and commentary with primary documents (medical opinions, legal briefs, judicial rulings, etc.), to provide additional authenticity and gravitas, that was greatly favored by ex-mental patients struggling to secure a voice in bringing their accounts of the injustices they had suffered before a receptive public in the lunatic rights' movement, this is a compelling and instructive memoir. Becoming re-connected with his origins, enabled Lemn Sissay to forge a sustainable creative path, and he is now an acclaimed writer, and poet, and the Chancellor of Manchester University.

Precarity, Precariousness, and Mental Patient Lives

The precariousness of the psyche in the power fields of history, and of precarity as a socially and politically induced condition in which a population group such as the mentally ill becomes differentially exposed to deprivation, injury, and trauma, may also be explored on a wider canvas, at the level of the structural changes and conditions that affect a whole society in a given period and of the diverse influences – cultural, social, and political – both on the status of mental suffering and on the diverse, and sometimes competing, forms of knowledge that converge on the situation of the mentally ill in society. It is therefore worthwhile comparing, with special reference to mental patients with severe psychoses, the precarious life circumstances over the last century of the mentally ill in different societies. Brief consideration of China, India, Japan, and the Soviet Union/Russian Federation may be thought of as illustrating varying dimensions of what has been described as the politics of social suffering (Marques 2018).

Social Dangers and the Human Sciences

As the Soviet Union was nearing its demise, the political scientist Peter Reddaway (1991) observed that for decades the Soviet system had regarded the mentally ill as an "unredeemable, essentially incurable, unproductive, and expensive nuisance" to be kept out of sight and "handled at lowest possible cost to the state's resources." In distinctive, but closely connected, ways, this observation holds true not only for the Soviet Union but also for China, Japan, and India where the hallmarks of the psychiatric system over the best part of the last century have been economic frugality, apprehensions over the social dangers posed by the mentally ill, a disregard for human rights, and a knowledge base anchored in notions of deficit and of psychopathology. The concept of "social danger" had become increasingly prominent in Europe in the first decades of the twentieth century in legal and psychiatric circles, giving rise to a range of legal policies known as "social defense." Criminological and psychiatric theories helped to sanction modernization projects, to articulate cultural or ideological anxieties in a language of medical science, and to highlight the threats posed by social constituencies perceived to have failed in their adaptation to the social order.

Ideas stimulated by the mental hygiene movement which had developed in the United States under the leadership of the psychiatrist Adolf Meyer were also adopted in China and in the Soviet Union but in both countries the movement was given a more political slant. Though in the United States and in Europe there were two distinct vectors to the mental hygiene project, a humanitarian concern over the welfare of the mentally ill and an authoritarian concern over order and social control, in China and the Soviet Union the latter prevailed. Some physicians may have been receptive to discourses stressing the humane treatment of the mentally ill, but human scientists were disposed to interpret mental illness as primarily a socio-political problem, and a transgression against social norms, rather than as a problem of individual health. In Japan also, social defense thought held sway for a considerable period, based on eugenics and the Eugenic Protection Law which was effective between 1948 and 1996. During this time, the hospitalization of psychiatric patients was promoted to rid society of those who were considered socially harmful. Economic pressures were also in play here, for by hospitalizing mentally ill family members, caregivers could re-join the work force.

In the Soviet Union, perhaps more than in China, mental hygiene became the focus of political struggles between rival factions over the right to treat, direct, regulate, and interfere in all the complex relationships of life (Sirotkina 2002: 176). By the 1930s, in China and in the Soviet Union, mental illness no longer referred exclusively to the obviously insane, and those who displayed transgressive behavior, or ideological nonconformity, might also be classified as mentally ill. Mental illness did not sit comfortably in the socialist landscape for it was seen as a characteristic of capitalist societies and not something that should exist in communist ones. In the Soviet Union, people who showed signs of mental illness were marked as underdeveloped, or incomplete, and faced much discrimination. Generally, these people were institutionalized and removed from their normal lives and their families were urged to abandon them. Mental health institutions were known to abuse patients and use ineffective and cruel treatment methods.

The hallmark of mental health care in contemporary China is a perpetual struggle between the provision of adequate care for an expanding population and the maintenance of social order, in which the latter has mostly prevailed. It is estimated that around 92% of severely mentally ill Chinese do not receive any treatment, and their families are frequently shunned and isolated (Jie Yang 2018). Such psychiatric care as exists mostly relies exclusively on medication. Patients and families in search of complementary resources, or a progressive social justice approach to care for mentally distressed people on the margins of Chinese society, must look beyond the mainstream to find them. The new mental health law in China of 2012 is actually a mental hygiene law and Chinese psychiatry remains committed to the political ideal of mental hygiene. The lasting appeal of mental hygiene in China derives from the satisfaction it provides to the enduring political desire for conformity and control. Though the new law promises to protect the rights and interests of psychiatric patients, it mostly highlights the need for interventionist state policies to prevent mental disorders and to mobilize social forces to target the various social dangers believed to be associated with psychiatric disorders (Emily Baum 2018; Chiang 2014; Zhang et al. 2010).

The Human Sciences and Coercive Rehabilitation

Thanks to the transforming powers of mental hygiene in a socialist state, it was envisaged that a new, and better, type of human being was about to be born. However, such utopian longings soon gave way to a sterner appraisal of social realities in which the biomedical sciences played a central role in investigating the problems, and potentials, of human beings and populations. Degeneration theory, and the associated concepts of social pathology, moral contagion, and psychopathology, helped to conceptualize a mutually constitutive relationship between the individual and social world. The consolidation of a biomedical discourse of individual deviance from the disciplines of criminology, psychiatry, and related fields helped to structure Bolsheviks' attempts to reclassify society and to provide substance to their fears about the residual effects of capitalism on the mentality and morality of Russian citizens after 1917, with the result that entire constituencies were invested with the dangerous moral defects of the prerevolutionary class of habitual criminals and psychopaths (Beer 2008: 170).

According to the progressive psychiatrist Vladimir Mikhailovich Bekhterev (1857–1927), individuals were all the time subject to the effect of mental microbes in the society that surrounded them and constantly in danger of being mentally infected. In periods of heightened social passion, especially, members of society "infect each other with their thoughts every minute." Failures to adjust to the new order now became evidence of an inability to adapt to the revolution itself and of an antipathy towards its values. Within the human sciences in the 1920s, coercive rehabilitation was one of the principle measures of progress toward the harmonious socialist society of the future. "People who manifest anti-social reactions, and are therefore enemies of society, should be isolated with the aim of coercively healing or re-educating them," it was recommended in 1927. The Bolsheviks' transformative program came increasingly to be understood in terms of a coercive rehabilitation of those amenable to treatment and the isolation (and even elimination) of those who were not. In 1929, Ivan Vvedenskii argued that psychiatry was the branch of medicine in which coercion was deployed specifically and systematically, and was the rule rather than the exception, since mental illness distorted the individual's capacity to understand his situation and his interests, rights, and obligations. The human sciences provided powerful theoretical legitimation for state repression and coercive rehabilitation became part of the Stalinist "civilizing process" (Beer 2008: 197).

Institutionalized Lives

Historically mental institutions have frequently provided, and in many places they still do, the fulcrum for the distribution and management of mental lives, for the legal frameworks that have arisen around them, and for the types of knowledge that have been created about the mentally ill. Thus, decades after independence, mental health policies and laws in India still reflect the normative framework developed by the British in which the indigenous mentally ill were interned in lunatic asylums, mostly built between the 1850s and 1900s. Strongly penal in character, this regime reflected colonial principles of control by segregation and a legal framework providing for the involuntary civil commitment of persons with "mental illness" founded on notions of "incapacity," whereby the sufferer became a nonperson in law and was denied human rights (Bhargavi Davar 2014).

The general picture of the Indian mental hospitals in the early twentieth century was grim and disturbing. There were notable reports by Edward Mapother in 1938 and by Moore Taylor in 1946. Mapother referred especially to overcrowding, indifference, desolation, ugliness, and corruption. An evaluation of the mental hospitals by the National Human Rights Commission in 1999 found that most admissions were involuntary, psychosocial interventions were almost nonexistent, and apart from medical management little happened to patients, and overall it was almost a throwback to Mapother's report of 1938. The report of an inspection committee in 2016 also resonated strongly with the recommendations made by Mapother and Taylor in the last century. Little progress has been made in developing community-based services, there is a major problem with long-stay destitute patients, and the rehabilitation of the chronic mentally ill continues to be an area of serious concern. Though the Mental Healthcare Act of 2017 purports to recognize the rights of mentally ill persons to determine their course of treatment, in actuality the right to informed consent is undercut by Section 4 of the Act which permits a mental health professional to determine whether a person has the capacity to make mental healthcare and treatment decisions. Despite shifts in terminology in line with "modernization," persons with psychosocial disabilities in India continue to be nonpersons in law. States Dr Bhargavi Davar, a noted activist and commentator, and director of the Pune-based Bapu Trust: "Informed consent is a distant dream, people are just grabbed and brought."

At the same time, the mental hospital may sometimes provide an essential refuge. In India there remains a gap between the rhetoric of new policy proclamations and their actual implementation. Many states are alleged still to follow the obsolete India Lunacy Act of 1912 and have yet to issue a gazette notification for the promulgation of the Mental Health Act of 1987. Overall, the mental health sector is driven by the archetype of the "mental asylum," which continues to possess relevance as a place of treatment and refuge for many persons with serious mental illness who might otherwise have been abandoned by helpless, or uncaring, families or died due to neglect. For instance, the Government Mental Hospital in Thiruvananthapuram in Kerala, one of the more progressive states in mental health care, has a bed strength of 507 but there are 830 patients. Of these around 170 patients consider the hospital to be home and have lived there for decades in many cases, even though many of them have at one time been certified fit for discharge. Though conditions may be described as subhuman, patients do not complain as they are desperate for a place to live and someone to look after them. "People think of the stigma attached to having a mental patient at home," remarks the hospital superintendent. "We regularly get letters from VIPs requesting us to retain such patients." Krishna, a former Carnatic music teacher, has been in the hospital for 37 years. Though he was discharged in 1985 and returned home to his brother, "he sent me back the same day, I have not seen him since." Conditions elsewhere are likely to be similar or worse.

Mental institutions are not exclusively associated with the public sector, however, and in some societies, they have become big business. In Japan in the 1950s, for instance, there was a "mental hospital boom" in which private institutions became lucrative businesses by absorbing the contradictions caused by postwar economic growth, and psychiatrists were able to expand their power through a network of private mental hospitals in which involuntary long-term hospitalization was frequently justified on the grounds of economic hardship, with little imperative to provide therapeutic treatment or try to return patients to the community. As late as 1987, over 90% of the inpatients in these hospitals were involuntary. It transpired subsequently that the Mental Hygiene Law of 1950 which had provided incentives for the development of private mental hospitals had been drafted by members of the Japanese Association of Psychiatric Hospitals which represented the interests of the private mental hospitals. A revised Mental Hygiene Law was enacted in 1965 which purported to promote community-based care, but in actuality mainly sought to safeguard the public from people with "dangerous mental disorders" and to assist the police in maintaining peace and order in society. Overall, people suffering from mental illness in Japan have traditionally been considered weak and outside the norm, and their families have been spurned (Suzuki 2003). The ignominy of the mentally ill in Japan, and the attendant problem of long-term psychiatric patients, has been constructed socially and politically, it has been argued, through the exploitation of the private mental hospital sector to obstruct, or prevent, the discharge of mental patients, in concert with the ideological use of "social defense thought" (stressing the dangers of the mentally ill), and the political priority given to economic growth above welfare and therapeutic measures.

Extramural Lives

Historically there is enormous divergence in what it has meant to live a life as a mentally ill person outside the confines of a formal mental institution and in the opportunities for extramural living. In Japan, for instance, for a considerable period, the confinement of the mentally ill in the domestic home was a state policy. In the Meiji period (1868–1912), Japan moved from the age of feudalism to a period of accelerated modernization. At this time, the psychiatric profession was preeminently an arm of the state, providing expertise mainly to classify and exclude those who were deemed a threat to the establishment of the new social order. Reluctant to increase public expenditure on asylums, the state instead drew upon the prevailing family ideology to make the head of the household responsible for confining mentally ill family members at home. The Law for the Care and Custody of the Mentally Ill, enacted in 1900, sanctioned the private confinement of mentally ill people in their family homes under the supervision of the police. Based on a regime of fear, shame, and secrecy, mental illness was treated in the same way as infectious diseases. In the Edo period, those considered harmful to the public were confined in a so-called "zashiki-rou," a cell for confining lunatics floored with tatami mats. The number of privately imprisoned mentally ill continued to increase in the interwar period at a rate that exceeded that of the growth of the population as a whole (Yumi Kim 2018).

In the early 1900s, psychiatry sought to distance itself from the state and to acquire an independent voice. Shuzo Kure, who became known as "the father of Japanese psychiatry," promoted the value of the mental hygiene movement and of moral therapy and in 1918 he produced a report drawing on the "state of private confinement in Japan" in which he famously stated: "Hundred of thousands of mental patients in our country not only have to suffer the misfortune that brought this illness, but they also have to suffer the misfortune of being born in this country." However, this humanistic vision failed to bear fruit and little changed during this period in the conditions of care for the mentally ill. In the tradition of university-based academic psychiatry, there was a sharp split between medical science and humanistic care in which psychiatrists were encouraged to see themselves primarily as scientists, helping the mentally ill by pursuing the organic cause of, and cures for, mental illness (Kitanaka 2012).

Moreover, the extramural prospects for the mentally ill were further hindered in the interwar period by the increasing salience of racial concerns emphasizing the ideology of the nation-as-family and the need for medicine to contribute to strengthening the "body of the Japanese race" so as to withstand the "struggle of the fittest" between nations. Pressure was placed on psychiatrists to champion the racial improvement movement which emphasized heredity over environment and biology over social conditions. In 1937, policies to support the war effort were introduced by the eugenics section of the newly created Ministry of Health and Welfare. The resulting wartime psychiatric discourse of the mentally ill person as a reject and aberration, linking family, biology and heredity, proved a lasting legacy until it was later challenged by the antipsychiatry movement. However, during the war some psychiatrists rejected the eugenics discourse of racial hygiene and, due to their opposition, the sterilization of the mentally ill was not carried out to the extent that the Eugenics Law of 1940 required. Though some studies have shown that, as late as the 1970s, fear of mental illness as a hereditary, genetic disease still permeated people's minds in rural and urban areas alike, there is also evidence to suggest that psychiatric discourse has not penetrated evenly across Japan, or been received uniformly.

The term "domestic citizenship" has been used to describe settings where disability is managed, disciplined, contained, and treated within domestic space rather than within civil society (Luhrmann and Marrow 2016: 210). In India there are enormous variations in the way in which families interact with, and manage, mentally disturbed members. In North India, it has been suggested, doctors may play down the diagnosis and try not interfere with families own interpretations of mental illness (Amy June Sousa, 'Diagnostic neutrality in psychiatric treatment in North India' in: Luhrmann and Marrow 2016: 43). A condition like schizophrenia may be defined through its impact on other household members rather than as an intrapsychic phenomenon like delusion or hallucination. A range of outcomes are possible here, not all of them benign. People are not always safe in families in India and families are not always caring. For example, it has been proposed that if in the United States the disturbed person is liked to be expelled from the family to the streets, in India the disturbed person may be marginalized, and forcefully "hidden" and abandoned, within the family (Marrow and Luhrmann 2012).

In Japan in the 1980s, mental illness induced by stress at the workplace became a significant national problem through the discovery of the stressed-out salesman (the term "karoshi" was coined to identify cases where people have all but worked themselves to death). Where in the West, the depression victim has typically been represented as a melancholic housewife, in Japan it is the burnt-out salesman who occupies the limelight. Responding to these circumstances, psychiatry in Japan started to gain influence by becoming more critical, siding with the depressed patient by questioning the social order in which the depressed person must live. Until quite recently, claims Junko Kitanaka (2012), the history of psychiatry in Japan had been characterized by a radical disconnection from subjective pain but the debate about depression has reversed the meaning of mental illness by turning it from a mark of inherent individual defect to a sign of an existential crisis that calls for urgent social action. From around 1969 until 2000, the antipsychiatry movement was particularly long-lasting and vehement in Japan and, whether they agreed with it or not, psychiatrists had to confront the argument debated daily that all mental illnesses are in some ways socially produced.

In the early 1950s, the prominent Soviet psychiatrist Andrei Snezhnevskii began developing a classification of mental illness that diverged from Western models, greatly facilitating the pathologization of healthy citizens by psychiatrists, and trapping individuals inside a vicious circle of diagnosis. Only by redefining diagnosis from within could this circle be broken. Dissidents such as Vladimir Bukovsky contributed to a widely shared perception that the coercive potential of psychiatry was greatly assisted by the ambiguity of psychiatric discourse itself. Bukovsky and others endeavored to turn psychiatric discourse back on the state and to redefine madness as "thinking differently," using the ambiguities of psychiatric discourse to creative advantage so as to pathologize Soviet society and to provide some legitimacy and breathing space for the extramural lives of "thinking differently" people. By pathologizing the state in conversation with psychiatrists and in samizdat, dissidents depathologized themselves and asserted their authority over psychiatric discourse. Bukovsky and Semen Gluzman produced a *"Manual on Psychiatry for Differently Thinking People"* that exposed the social norms on which psychiatrists based their diagnoses. The "normal" citizen did not think differently, indeed, he barely thought at all. The less he understood of modern art, the saner he appeared to be.

In China, the moral conflicts and emotional struggles produced in the wake of rapid and extreme modernization are reflected in the phenomenon of *bei jingshenbing*, which refers to those individuals who were subjected to unnecessary psychiatric treatment during the first decade of the twenty-first century. Ironically, however, the sometimes insurmountable difficulties in proving that some of those who were considered mentally ill may have been "differently thinking" but still sane have sometimes resulted in a shift, at the grassroots, from defending the rights of self-identified victims of psychiatry to promoting the rights of actual mental patients (Harry Yi-Jui Wu 2016).

Juliane Fürst (2017) and Rebecca Reich (2018) both explore the subjectivity of dissidents and hippies within the Soviet psychiatric system. For both hippies and dissidents, albeit in different ways, madness became a weapon as much as a sentence, and abnormality could be used as a defense. Soviet hippies in the late 1960s and 1970s embraced madness and did not necessarily fight their diagnoses of schizophrenia. Even so, their experience of the psychiatric hospital instilled a wariness in them and hippy declarations of "difference" were invariably double edged. Psychiatric hospitals revealed to them corners of the Soviet system that were hidden from ordinary citizens, a knowledge that "ensured that those who had seen Soviet psychiatry would never be 'normal' again" (Fürst 2017; Brintlinger and Vinitsky 2007).

There are plentiful examples of local knowledge in different places, mainly deriving from community-based NGOs, that testify to alternative life prospects for people with severe mental health disabilities. Anthropologist Sarah Pinto (2014) alludes to a deeply embedded disposition in Indian culture to think about what it means to suffer, and to seek to heal, in a plurality of different ways and to a willingness to embrace spiritual and religious frameworks and modalities alongside the imports of Western medicine. For example, a 40-year-old woman with a schizophrenia diagnosis, known as Lokkhi, was found eating garbage, and talking to herself, in a locality of Kolkata in India by the field staff of an NGO a year after she was turned out of her in-laws' home, but once relocated to a shelter for homeless women with psychosocial disabilities, she recovered quite quickly and was soon able to resume an occupation and to save money to support her sons. Though there are undoubted risks in the absence of proper safe-guards, available evidence endorses the positive aspects of healing and recovery through holistic methods in shrines that offer safe spaces for experiencing life at moments of vulnerability and crisis (Davar and Lohokare 2009). Overall, people with serious psychoses may do better in India than in Western societies and there may be some positive over-lap between psychotic hallucinations and standard religious practices (Luhrmann and Marrow 2016). Anthropologists have described how people with severe psychiatric histories may occupy healing roles at temples in which symptoms that may initially have been tormenting now possess a more positive valence (Anubha Sood, "Madness experienced as faith: Temple healing in North India 2016," in: Luhrmann and Marrow 2016).

Conclusion

This chapter has endeavored to throw some light on the multifaceted and contested terrain of the mental patient in history and to bring into focus some of the strengths and weaknesses of the different kinds of knowledge-making that have gathered around the trajectories of distressed psyches in different times and places. The vicissitudes of the psyche are inevitably imbricated in political force fields that not infrequently bring about a condition of differential vulnerability to illness and suffering that may be identified as precarity. Under some circumstances, however, the conditions of structural precarity imposed and maintained by major power groups may to a modest, but still significant, degree be allayed, and counter-balanced by the re-discovery and renewal of local or indigenous cultural resources that assign value to mad persons and make it possible to assimilate, and sustain, them within a more generic cultural understanding of human precariousness. Necessarily, the conclusions of an inquiry of this kind are strictly provisional, and subject to constant revision, as new historical circumstances may stimulate and permit.

References

Bacopoulos-Viau A, Fauvel A (2016) The patient's turn: Roy Porter and psychiatry's tales, thirty years on. Med Hist 60(1):1–18

Barham P (1995) Schizophrenia and Human Value, 2nd edn. Free Association Books, London
Barham P (2004) Forgotten Lunatics of the Great War. Yale University Press, London & New Haven
Baum E (2018) The Invention of Madness: State, Society and the Insane in Modern China. University of Chicago Press, Chicago
Beer D (2008) Renovating Russia: the human sciences and the fate of liberal modernity, 1880–1930. Cornell University Press, Ithaca
Biehl J (2005) Vita: life in a zone of social abandonment. University of California Press, Oakland
Biehl J, Good B, Kleinman A (eds) (2007) Subjectivity: ethnographic investigations. University of California Press, Oakland
Brand-Claussen B et al (eds) (1996) Beyond Reason: Art and Psychosis –Works from the Prinzhorn Collection. Hayward Gallery, London
Brintlinger A, Vinitsky I (eds) (2007) Madness and the Mad in Russian Culture. University of Toronto Press, Toronto
Bruckner B (2016) Animal magnetism, psychiatry and subjective experience in nineteenth century German: Friedrich Krauss and his Nothschrei. Med Hist 60(1):19–36
Butler J (2009) Frames of War: when is life Grievable? Verso, London
Chaney S (2016) 'No "sane" person would have any idea': patients' involvement in late nineteenth-century British asylum psychiatry. Med Hist 60(1):37–53
Chiang H (ed) (2014) Psychiatry and Chinese history. Pickering & Chatto, London
Coleborne C (2020) Why talk about madness? Bringing history into the conversation. Palgrave Macmillan, London
Davar B (2014) Globalizing psychiatry and the case of "vanishing" alternatives in a neo-colonial state. Disabil Glob South 1(2):266–284
Davar B, Lohokare M (2009) Recovering from psychosocial traumas: the place of Dargahs in Maharashtra. Econ Polit Wkly 44(16):60–67
Davis C (2018) A question of power: Bessie Head and her publishers. J South Afr Stud 44(3):491–506
De Souza Leao, Rodrigo (2013) All Dogs Are Blue. And Other Stories, High Wycombe, Bucks
Desjarlais R (1997) Shelter Blues: Sanity and Selfhood Among the Homeless. University of Pennsylvania Press, Philadelphia
Dousemetzis H (2018) The Man Who Killed Apartheid: The Life of Dimitri Tsafendas. Jacana Media, Auckland Park, South Africa
Eilersen GS (1995) Bessie Head–Thunder Behind Her Ears: Her Life and Writing. James Currey, London
Fassin D (2012) Humanitarian Reason: A Moral History of the Present. University of California Press, Oakland
Foster H (2001) Blinded insights: on the modernist reception of the art of the mentally ill. October 97:3–30
Freis D (2019) Psycho-politics Between the World Wars: Psychiatry and Society in Germany, Austria, and Switzerland, Palgrave Macmillan, London
Friedlander H (1995) The Origins of Nazi Genocide: From Euthanasia to the Final Solution. University of North Carolina Press, Chapel Hill
Fryar C (2018) The narrative of Ann Pratt: life-writing, genre & bureaucracy in a post-emancipation scandal. Hist Work J 85:265–279
Furst J (2017) Liberating madness –punishing insanity: soviet hippies and the politics of craziness. J Contemp Hist 53(4):832–860
Gilman S (1985) The mad man as artist: medicine, history & degenerate art. J Contemp Hist 20(4):575–597
Goldberg A (2003) A re-invented public: "Lunatics' rights" and bourgeois populism in the Kaiserreich. Ger Hist 21(2):159–182
Han C (2018) Precarity, precariousness and vulnerability. Annu Rev Anthropol 47:331–343
Head B (1973) A Question of Power. Davis-Poynter, London
Head B (1990) Social and political pressures that shape literature in Southern Africa. In: Abrahams C (ed) The Tragic Life: Bessie Head and Literature in Southern Africa, African World Press, London

Jacob JD, Perron A, Holmes D (2014) Power and the Psychiatric Apparatus Routledge, London
Jones K et al. (2010) Framing Marginalised Art, Cunningham Dax Collection. University of Melbourne, Melbourne
Kalathil J (2013) Personal narratives of madness. In: Fulford KWM, Davies M et al (eds) The Oxford Handbook of Philosophy and Psychiatry. Oxford University Press, Oxford. http://global.oup.com/booksites/content/9780199579563/narratives/
Kasmir S (2018) Precarity. In: Stein F, Lazar S, Candea M, Diemberger H, Robbins J, Sanchez A, Stasch R (eds), The Cambridge Encyclopedia of Anthropology. Cambridge University Press, Cambridge
Kim Y (2018) Seeing cages: home confinement in early twentieth century. Jpn J Asian Stud 77(3): 635–658
Kitanaka J (2012) Depression in Japan: Psychiatric Cures for a Society in Distress. Princeton University Press, Princeton, New Jersey,
Kleinman A (1999) Experience and its moral codes: culture, human conditions and disorder. In: Peterson G (ed) The Tanner Lectures on Human Values. University of Utah Press, Salt Lake City
Lipsitz G (1998) The Possessive Investment in Whiteness,Temple University Press,Philadelphia
Lothane Z (1992) In Defense of Schreber: Soul murder and Psychiatry. The Analytic Press, Hillsdale & London
Luhrmann T, Marrow J (2016) Our Most Troubling madness: Case Studies in Schizophrenia Across Cultures   University of California Press, Oakland.
Markova IS, Berrios GE (2012) The epistemology of psychiatry. Psychopathology 45:220–227
Marques TP (2018) Illness and the politics of social suffering: towards a critical research agenda in health & science studies. Revista Critica de Ciencias Sociais, numero especial, Novembro 2018: 141–164
Marrow J, Luhrmann TM (2012) The zone of social abandonment in cultural geography: on the streets in the United States, inside the family in India. Cult Med Psychiatry 36:493–513
Metzl JM (2009) The Protest Psychosis: How Schizophrenia Became a Black Disease. Beacon Press, Boston
Naifeh S, White Smith G (2012) Van Gogh: The life. Profile Books, London
Pearse A (1983) Apartheid and madness: Bessie Head's a question of power. Kunapipi 5(2)
Peters O (ed) (2014) Degenerate Art: The Attack on Modern Art in Nazi Germany, 1937. Prestel, Munich, London, & NewYork
Pinto S (2014) Daughters of Parvati: Women and Madness in Contemporary India. University of Pennsylvania Press, Philadephia
Porter R (1985) The patient's view: doing medical history from below. Theory Soc 14(2):175–198
Prinzhorn H (1922) Bildnerei der Geisteskranken (Artistry of the Mentally Ill) Galerie Rothe, Heidelberg
Reddaway P (1991) Civil society & Soviet psychiatry. Probl Communism 40(4):41–48
Reich R (2018) State of Madness: Psychiatry, Literature and Dissent after Stalin. Northern Illinois University Press, Illinois.
Robbins J (2013) Beyond the suffering subject: toward an anthropology of the good. J R Anthropol Inst 19(3):447–462
Santner EL (1996) My Own Private Germany: Daniel Paul Schreber's Secret History of Germany. Princeton University Press Princeton, New Jersey
Sirotkina I (2002) Diagnosing Literary genius: A Cultural History of Psychiatry in Russia, 1880–1930 John Hopkins University Press, Baltimore
Speed C (2017) As little regard in life as in death: a critical analysis of subjugation and accountability following deaths in psychiatric detention. Illn Crisis Loss 25(1):27–42
Sposini FM (2020) Confinement and certificates: consensus, stigma and disability rights. Can Med Assoc J 192(48):E1642–E1643
Suzuki A (2003) The state, family and the insane in Japan, 1900–1945. In: Porter R, Wright D (eds) The Confinement of the Insane: International Perspectives. Cambridge University Press, Cambridge

Swartz S (2014) Multiple voices and plausible claims: historiography and colonial lunatic asylum archives. In: Bala P (ed) Medicine and Colonialism. Pickering & Chatto, London

United Nations (2006). www.un.org/development/desa/disabilities/convention-on-the-rights-of-persons-with-disabilities.html

Vellekoop M, Luijten H, Jansen L (eds) (2016) On the Verge of Insanity: Van Gogh and his Illness Yale University Press, London & New Haven

Wilson, Peter (1976) Oscar:An inquiry into the nature of sanity. Waveland Press, LongGrove, Illinois

Wu HY-J (2016) The moral career of "outmates": towards a history of manufactured mental disorders in post-socialist China. Med Hist 60(1):87–104

Yang J (2018) Mental Health in China: Change, Tradition and Therapeutic Governance. Polity Press, Cambridge

Zhang E, Kleinman A, Weiming T (eds) (2010) Governance of Life in Chinese Moral Experience: The Quest for an Adequate Life Routledge, London

Asylums and Alienists: The Institutional Foundations of Psychiatry, 1760–1914

48

David Wright

Contents

Introduction	1254
The Rise of the Lunatic Asylum	1254
Lunacy Reform	1256
Professionalization	1259
Models of Madness	1261
In Search of Biological Causes	1264
Conclusions: The Legacy of Asylum Psychiatry	1266
References	1268

Abstract

In the 150 years leading up to World War I, lunatic asylums became the largest and most controversial medical institutions in the Western World. It is within these facilities that the discursive formulations and clinical practices that eventually became known as "psychiatry" took shape. Over the course of the 1800s, asylum medical officers underwent a decades-long process of professionalization – creating national associations, founding journals devoted to the study of mental diseases, and finally integrating "psychological medicine" into university medical training. The legal requirements of medical certification and the institutional bureaucracies of record keeping created the foundations upon which new taxonomies of mental illness were first constructed. This chapter outlines the institutional origins of psychiatry as a human science, focusing on what historians refer to as the "long" nineteenth century – that is, the period from 1760 to 1914.

Keywords

Lunatic asylums · Psychiatry · Madness · Institutional confinement · Alienists

D. Wright (✉)
Department of History and Classical Studies, McGill University, Montreal, QC, Canada
e-mail: david.j.wright@mcgill.ca

© The Author(s), under exclusive licence to Springer Nature Singapore Pte Ltd. 2022
D. McCallum (ed.), *The Palgrave Handbook of the History of Human Sciences*,
https://doi.org/10.1007/978-981-16-7255-2_100

Introduction

This chapter outlines the institutional origins of psychiatry as a human science, examining the rise of the lunatic asylum as the primary space in which medical practitioners were given legal monopoly over the treatment of the insane and consolidated their professional expertise. The chapter will address several interrelated themes that emerged in the nineteenth century, including: the establishment and expansion of public lunatic asylums; the articulation and development of the social movement known as "lunacy reform"; the evolution of professional associations and their journals devoted to diseases of the mind; the general classification of mental disorders; and, finally, the emergence of turn-of-the-century concerns over heredity and degeneration. The chapter will examine, in particular, English-language scholarship and will not address, for purposes of length, certain related topics, such as the advent of forensic psychiatry or the development of institutions for the intellectually disabled.

The Rise of the Lunatic Asylum

Although there were a handful of welfare institutions in Europe dating back to the Medieval period that included "lunatics" among their diverse clientele (Andrews et al. 1997), specialist hospitals constructed solely for the treatment of lunatics overwhelmingly date from the 1700s (Winston 1994). The rise of private and public lunatic asylums dedicated to the control and treatment of insanity reflected a confluence of forces unleashed by a nascent industrial society. These included the rapid accumulation of wealth, the rising authority of doctors, the emergence of a professional middle class, and the embrace of Enlightenment ideas about the improvability of humankind (Porter 2004). Institutional solutions to accommodate and control mad behavior manifested themselves in different forms and in diverse locations, in part a function of the social, religious, and geographical context of the household and communities affected (Philo 2004).

Specialist institutional care became pronounced during the eighteenth century, when, amidst the thriving marketplace for medical services, a growing number of prosperous families began to purchase extramural nursing and medical care for mentally troubled kin (Porter 1987). In response to this demand, a network of private homes for the insane dotted the countryside of Britain, the United States, and Continental Europe. These so-called private madhouses were run both by medical men as well as lay members of the community (Parry-Jones 1972; Andrews and Scull 2001). One historian estimates that, by 1815, no fewer than 70% of the institutionalized insane in England were in private, for-profit institutions (Smith 2020). In the United States, exclusive institutions, such as the Hartford Retreat, were also popular among those of privilege and means (Goodheart 2003). However, private homes for the mad also courted negative press. Critics accused proprietors of profiting from the misery of others or, worse still, being complicit in the intentional misuse of private homes for nefarious legal purposes (Jones 1993). Over time, these institutions would become overshadowed by the larger state facilities discussed below; however, private homes, despite ongoing public misgivings, never really disappeared (Scull et al. 1996).

With the rise of general hospitals in major cities and provincial towns during the 1700s, there was a growing demand for medical institutions designed specifically for lunatics, particularly to serve respectable families who could not afford private care but who were not otherwise paupers. Within the British system of voluntary hospitals, the insane – considered to be long-term patients – were regularly prohibited from being admitted (Woodward 1974). In response to this lack of provision for institutional care of insane persons of the middling ranks of society, a small number of charitable lunatic asylums emerged in some of the larger cities by mid-century – St. Luke's Hospital in London, for example, or St Patrick's in Dublin (Malcolm 1989). In the American colonies, the Philadelphia Hospital, under the leadership of Benjamin Rush, accommodated a significant number of lunatics in separate wards in the 1790s (Grob 1973). A parallel arrangement of lunatic wards existed from 1806 in Berlin's Charité Hospital (Shorter 1998). Religious denominations also occasionally administered their own institutions, including the Quaker York Retreat, in the north of England (Digby 1985). In jurisdictions where the Catholic church was dominant, religious orders staffed and administered many lunatic asylums. In France, for example, the *Frères de la Charité* (Brothers of Charity) were best known for managing, among others, the famous asylum of Charenton (Goldstein 1987).

For centuries, the poorest of families were forced to rely on a myriad of poor laws and local customs that obliged local authorities to provide rudimentary shelter and care for the destitute and disabled of their communities. There were also, more coercively, vagrancy acts, that targeted "lunatics at large" and sequestered them in local jails or houses of industry. Poor law workhouses, or local almshouses, were, by definition, places of last resort and included a diverse array of residents suffering from privation and dependence due to old age, physical disability, abandonment in childhood, and from debilitating mental disorders. Those researching the "Old Poor Law" in England (also known as the Elizabethan poor law) have identified a variety of local responses, including both outdoor and indoor relief to lunatics and their families (Pelling 1985). Somewhat surprisingly, some poor law parishes contracted with private madhouses and even charitable institutions to shelter and control their more troublesome inmates. By contrast, in rural Wales and Scotland, there appears to have been a longstanding practice of "boarding out" the insane, whereby poor families would be paid a modest stipend to shelter and feed "harmless" lunatics (Sturdy and Parry-Jones 1999).

As the general population grew rapidly during the industrial revolution, more and more purpose-built institutions were constructed and supported directly by taxpayers. In England and Wales, the Asylums Act of 1808 provided a framework permitting local magistrates to raise funds from parishes in order to construct public lunatic asylums at the county level. The first generation of pauper lunatic asylums included rural counties, including Cornwall, as well as more urbanized communities, such as the first Middlesex Asylum at Hanwell (Smith 1999). In Ireland, following the Act of Union with Britain (1801), legislation dating from 1817 divided the island into districts (groups of counties) with a similar responsibility for raising funds for the construction and maintenance of new public institutions, such as the establishment of asylums in Armagh and county Derry in the 1820s (Finnane 1981). In Scotland, asylum building was no less pronounced, with royal and district asylums being constructed in diverse rural and urban locations, including Montrose, Edinburgh, and Dundee (Andrews 2004). In America, medical and welfare institutions were the prerogative of individual states, thus inhibiting any centrally administered asylum system. Nevertheless, between the 1770s and the 1830s, a diverse cluster of states – including Virginia, Kentucky, South Carolina, and New York – all opened their own lunatic asylums for "public" (that is, nonpaying) patients, pursuant to their own respective state legislation (Rothman 1971). In continental Europe, asylum construction was also underway, if in a less pronounced manner, as witnessed by some early nineteenth-century German states during the *Vormärz* (pre-1848) period, including the Eberbach asylum, in the western German territory of Nassau (Goldberg 1999), Winnenthal, near Stuttgart, and Illenau in Baden (Kramer 1998).

Lunacy Reform

The growing network of institutional solutions was informed by a revolution in thinking about the nature of madness. Prior to the 1700s, there was a certain amount of fatalism about its potential curability, as well as competing explanations as to the origins and natural history. The attributed causes of madness might include any one, or combination, of a constellation of factors, including the climate, diet, physical injury, economic reversal, sudden shock, inebriety, personal loss, infectious disease fevers, and religious impulses. Although some medical men might claim a certain experience in treating madness, the belief that lunacy was somehow narrowly "medical" was far from accepted or universal (MacDonald 1981). What is clear is that communities had a long knowledge of, and experience responding to, mad behavior. Some households and communities observed individuals who appeared to remit and return to their former states of being; others were observed to repeatedly relapse into unsoundness of mind and household dependency. Lunacy was a social reality involving diverse members of local communities. We know, from local court records of lunacy inquiries, for example, that local juries and judges were asked to adjudicate mental competence in cases concerning the management of family property (Houston 2000; Moran 2019).

By the end of the 1700s, the ideas of the French and Scottish Enlightenments animated discussions among the educated classes about a cluster of intersecting topics, eliciting passionate debates about responsible government, the trans-Atlantic slave trade, and the effectiveness of penal servitude. All human experience and behavior, including the diverse forms of madness, came be understood, to educated men and women of the Enlightenment, as phenomena susceptible to scientific study (Alexander and Selesnick 1966). The French Revolution, in particular, unleashed radical new ideas about citizenship and the institutional care of citizens suffering from a variety of medical and mental diseases. Ideas about a new and "enlightened" manner of treating lunatics found a champion in Philippe Pinel, a French medical doctor who came to occupy the post of superintendent of two large Parisian welfare institutions of the 1790s – the Bicêtre and the Salpêtrière (Castel 1988). Critics at the time lamented that lunatics were too often reflexively considered incurable, relegating them to a social and moral status similar to animals, as witnessed by the cruel treatment in local jails and the common recourse to physical restraints in both domestic and institutional settings (Foucault 1965). Pinel famously advanced an interpersonal approach to the amelioration of insanity which he called *traitement moral* (moral treatment, sometimes also referred to in English as moral therapy) (Pinel 1800). Moral treatment – a type of psychological therapy – advocated that *all* mental illness was curable, given systematic scientific investigation and humane attendance. Proponents of moral treatment vilified the use of mechanical restraints, recommending instead sympathetic, interpersonal therapeutic encounters.

The principal ideas of moral treatment fanned out across continental Europe, into Britain and North America, finding a receptive audience in progressive politicians, religious leaders, and doctors. Across the Atlantic, moral treatment was advocated by Benjamin Rush, who, in addition to being a signature to the American Declaration of Independence, also became the chief medical officer of the Philadelphia Hospital, where he initiated and supervised a large, separate wing for the treatment of lunatics. A physician dedicated to the older tradition of "heroic medicine," Rush nevertheless championed the abolition of restraints and encouraged what he considered to be a more enlightened management of the insane (McGovern 1986). Perhaps the most famous iteration of moral treatment involved the Quaker religious leader William Tuke who, in the 1790s, embraced this new approach in his founding of an institution near York (England) restricted to his coreligionists in the Society of Friends. The published account of the founding and principles of what became known as the Quaker York Retreat by Samuel Tuke, his grandson, acted as an inspiration within the English-speaking world in the later 1800s. Indeed, the York Retreat took on mythic status during the nineteenth century, as an idealized version of small therapeutic communities located in the bucolic countryside (Digby 1985).

Notwithstanding the notable support of certain progressive Protestant dominations and Catholic orders, the movement to reform the treatment of lunatics and establish public hospitals for their treatment was slowly appropriated by members of the medical profession during the first decades of the 1800s. Doctors increasingly employed the language of "illness" and "disease" to describe the ambiguous and contested state of lunacy (Scull 1975). Pinel himself began to self-identify as an "alienist" – that is, a doctor who specialized in the institutional management and treatment of lunatics (*les aliénés*, in French). Over time, the term alienist became popular in some English-speaking jurisdictions as well, where it took on an air of respectability and even sophistication. Medical treatises of the time repeatedly advocated for medical institutions for the insane, funded by the state. The lunatic asylum was deemed central, for these proponents, to a modern, scientific, and humane system of care, surveillance, empirical study, and treatment.

Several key works would appear to legitimize the asylum as the locus for treatment and scientific inquiry. The intellectual origins of the therapeutic asylum have been traced to William Battie, the superintendent of St. Luke's Hospital in London, who emphasized the fact that the asylum itself could be an effective device for treating insanity (Battie 1758). His ideas gathered momentum by the early 1800s. Pinel's protegé, Jean-Étienne Esquirol, would author his own *Question médico-légale sur l'isolement des aliénés* in 1832 (Goldstein 1987), promoting the asylum as a social good for the lunatic and for society at large. In Scotland, William Browne, formulated his own ideas of the asylum "as utopia" shortly after taking over one of the first Scottish institutions, in Montrose, in the 1830s (Browne 1837). Ten years later, John Conolly, the most famous English asylum doctor of mid-century and then superintendent of the Middlesex Asylum, published his own manual of best practices, outlining the administration of an ideal therapeutic asylum in 1847 (Conolly 1847). His counterpart across the Atlantic, Thomas Kirkbride, did much the same in the United States several years later (Kirkbride 1854). These successive publications acted as blueprints for separate institutional treatment under medical control, providing details as to the ideal ratio of staffing, diets that were best suited to specific mental disorders, the importance of age-, class- and gender-specific occupations, as well as such topics as leisure activities that were most conducive to ease the mental states of directionless frenzy, confused thinking, and emotional distress. So influential were these works, that asylums in America that followed these prescriptions were often designated approvingly as "Kirkbride" institutions (Tomes 1985).

The optimism of Enlightenment thought is evident in the expressions of curability circulating in the first half of the 1800s. Alienists, and their growing legion of supporters and well-wishers, confidently predicted that madness could be empirically studied and understood, that the laws of the human mind and behavior would eventually be unlocked, and that learned men could ultimately devise methods to treat the brain in a manner analogous to the advances already occurring in other areas of medicine. The lunatic asylum, according to its proponents, protected society from possibly violent lunatics, relieved poor families from the burden of caring for violent or dependent members, and rescued sufferers from environments that might have caused mental breakdown in the first place (Conolly 1830). Unsurprisingly, the medical men who authored these treatises insisted that these institutions must be placed under the supervision and authority of members of the orthodox medical profession. In this respect, the drive for institutional specialization paralleled that of other spheres of medicine in the nineteenth century, such as medical institutions for children, fever hospitals, lying in hospitals, and separate institutions for the blind and deaf (Weisz 2005).

The results of lunacy reform could be seen in the remarkable century of asylum construction throughout Western Europe and North America, one that accelerated by the middle decades of the 1800s (McCandless 1979). Part of the impetus came directly from compulsory national legislation, where decades of advocacy for lunacy reform resulted in tangible political victories. France, for example, passed a comprehensive lunacy law in 1838, obliging *départements* to establish asylums for their respective insane poor (Castel 1988). In 1845, England made it obligatory for counties and boroughs to erect, or combine to erect, pauper lunatic asylums across the country (Bartlett 1999). Scotland would have similar compulsory legislation passed in 1857, a delay that Andrews attributes to the Scots not wanting to appear to be simply copying models from south of the border (Andrews 1998). From 1817, the British administration in Dublin was empowered to construct an asylum system at the regional or "district" level, encompassing groups of Irish counties (Cox 2012). And asylums were hardly a British or French cultural invention. Lunatic asylums began to appear across Western Europe, in states of the Germany Confederation, to Imperial Russia, to the pre- and post-unification Italian states, the cantons of Switzerland, throughout Scandinavia and the Iberian peninsula (Porter and Wright 2003). Colonial asylums, based on European models, also appeared in India, French Indo-China, the British Caribbean, Dutch Indonesia, and throughout Australasia (Mahone and Vaughan 2007; Coleborne 2009; Smith 2014; Edington 2019: Leckie 2019).

The staggering pace of asylum construction reflected an unprecedented number of individuals resident in these institutions, both in absolute numbers as well as a proportion of the general population. By the dawn of the twentieth century, England, France, and the United States all had in excess of 70,000 in-patients residing in licensed psychiatric institutions (Rothman 1980; Scull 1993; Quétel 2009). Although the pace and timing varied, it is noteworthy that many Western countries, with the exception of Ireland (which was even more elevated), plateaued at about 1 resident per every 300 persons in the general population by the end of the period covered by this chapter (Fuller Torrey and Miller 2002). Unsurprisingly, the "success" of the lunatic asylum led some prominent doctors, like the English alienist Henry Maudsley, to muse publicly about the future of society due to what appeared to be the relentless increase in insanity (Maudsley 1872). Others recognized that some, perhaps most, of the numeric explosion was a statistical artifact, as lunatics in workhouses and jails were transferred to asylums once they had been created, thus appearing as newly reported patients when they had, in fact, been merely transferred from one carceral authority to another. Nevertheless, asylums were clearly expanding at a remarkable rate. As a consequence, some historians have concluded that asylums may well have lowered the threshold for what families were willing to tolerate, thus widening the boundaries of what was considered pathological behavior (Scull 1979). A small number of historians have even speculated that the real rate of insanity was on the increase in the century leading up to the age of antibiotics (Shorter 1998; Fuller Torrey and Miller 2002). Quite apart from these historical debates over the reason for the acceleration in the number of asylum inmates during the 1800s, the relentless increase in the number and size of asylums would occasion a certain disquiet in medical circles throughout the second half of the nineteenth century.

Professionalization

Lunatic asylums required a cadre of medical practitioners employed in what were some of the largest state institutions in the Western world. Over time, the sheer number of asylum doctors gave rise to professional associations that appeared concurrently across different jurisdictions. For example, the Medico-Psychological Association of Great-Britain and Ireland was founded in 1841; 2 years later, the *Société médico-psychologique* (first of Paris, later extended to all of France) and *Verein Deutscher Irrenärzte* in Germany followed suit. By 1844, the United States inaugurated its own professional group – the Association of Medical Superintendents of American Institutions for the Insane. These professional associations gave birth to scholarly journals from the 1850s, including the British *Asylum Journal of Mental Science,* the French *Annales médico-psychologiques,* as well as the *American Journal of Insanity* and the German *Allgemeine Zeitschrift für Psychiatrie* (Paradis 1994). Medical periodicals performed various organizational and scientific functions, such as reporting on annual congresses, presenting original scientific papers, aggregating statistics derived from asylums, as well as circulating noteworthy articles in translation. It is important to remember that these national-based societies did not prevent a degree of cross-fertilization of ideas between doctors in different countries. Indeed, it was not uncommon for asylum superintendents to tour leading institutions in other countries and form friendships and alliances with colleagues in different national contexts. In the era before postgraduate qualifications (which were not formalized until after World War I), medical practitioners sought an experience of specialization through visiting fellowships and research collaborations (Schmiedebach 2010). Indeed, sometimes prominent alienists, like Daniel Hack Tuke, took it upon themselves to tour and pronounce upon the state of asylums in other national jurisdictions (Tuke 1885).

These proto-psychiatric communities were anchored in legislation that formalized the status and privilege of orthodox medical practitioners in institutional management and legal protocols, a noteworthy point considering that these organizations, and their journals, were often established before the medical profession itself had achieved full professional closure and unification. For example, Britain's foundational asylum legislation occurred in 1845, 13 years before the British Medical Registration Act (1858). Nor was there any clear evidence that doctors were any more successful in curing madness than lay individuals (Scull 1979). Nevertheless, legislation across the Western World required that public institutions for the insane, described above, be under the supervision of a medically trained individual, as defined and understood at the time. The asylum superintendents were overwhelmingly resident superintendents, living on the grounds of these sprawling institutions and responsible for all matters administrative and medical. In the community, licensed medical practitioners were given the responsibility over certification and involuntary hospitalization. It is not by chance, then, that some scholars have characterized this transitional period as representing the "medicalization of madness" or, to use a more anti-psychiatry turn of phrase, the "manufacture of madness" (Szasz 1997).

The legal requirements governing asylum management thus obliged local authorities to hire medical men as resident superintendents, most often on salary. In the volatile medical marketplace of mid-century, when orthodox medicine was still attempting to consolidate its power and close the profession to threats from homeopathy and other medical modalities, graduates of medical schools were only too happy to take on these unusual institutional positions. The bucolic asylum grounds, retinue of servants, and superintendent's house, gave alienists the trappings of country gentlemen. Meanwhile, many of these asylum doctors had very limited experience in the practical aspects of psychological medicine, since "diseases of the mind" were not part of the standard medical curriculum of the time. The expertise of alienists was very much one that was gained "on the job," supplemented by some public lectures on the subject as well as doctors' occasional participation in the aforementioned nascent professional associations. In some instances, the sheer size of asylums necessitated an enlarged medical staff, with assistant medical men (and a very small number of medical women) performing a lot of the daily administrative work. It was common for junior asylum doctors to work for a number of years in their first institution, gaining experience and seniority, until they were hired as resident superintendents elsewhere, in the ever-expanding asylum system of the second half of the nineteenth century (Hide 2014). Frustrated by a lack of opportunity at home, some would venture to distant colonial institutions where there was less competition for senior positions (Parle 2007).

The sheer size of lunatic asylums by the last third of the nineteenth century had important ramifications for the growing number of asylum medical doctors who found employment therein. Although early lunacy reformers envisaged small institutions with close contact between doctor and patient, the reality on the ground was that a small group of asylum doctors – often two, three, or four in number – found themselves responsible for institutions serving hundreds of residents. Institutions originally built for 200 or 300 patients were quickly bursting at the seams, adding new wings and additional buildings over time. The arrival of dozens of new patients every year added to the rapid expansion in the in-patient populations, distancing the alienists from individual patient care. In this environment, daily attendance and surveillance devolved to the asylum's female nurses and male attendants, the training of which varied considerably between institutions. Complicating matters, many patients arrived at the asylum with a diverse range of nonpsychiatric problems comorbid with their mental troubles. As a consequence, a great deal of the daily activity of asylum doctors was occupied by attending to nonpsychiatric ailments, as exemplified by the entries in asylum medical case books. Meanwhile, the not inconsiderable paperwork required by lunacy legislation (medical inspections of new admissions, death reports, discharge reports, entries in medical case books on the status of individual patients that were required to be completed at regular intervals) also occupied much of the attention of asylum medical officers.

Models of Madness

The legal obligations of performing certification, compiling medical case books, updating asylum admission registers, and submitting consolidated annual reports were central to the generation of data and new ideas about the etiology and classification of insanity. Lunacy laws required the medical certification of persons confined in licensed medical institutions or, privately, in a home other than their domicile. Certification involved the direct interview and observation of the person in question by a medically qualified individual, most often a local doctor in the community. This involved taking testimony from family and friends, statements that would justify the claim that the person was insane and in need of institutional treatment. The doctor would then add his own observations, derived from a brief interview with the prospective patient. Certification varied between jurisdictions but most often involved two separate certificates carried out independently (Sposini 2020). Medical certification, it should be remembered, was a privilege given to all licensed medical practitioners and not to a subset of individuals who happened to work in asylums and claim specialist expertise therein. Indeed, legislation in some jurisdictions even prohibited asylum medical superintendents from engaging in the process, lest it give the impression of collusion and possible wrongful confinement. Information from these lay and medical testimonials embedded in the certificates of insanity was then transcribed into medical case books, forming the first section of an ongoing, if often perfunctory, account of the patient's progress in the institution (Wright 1998).

These certificates of insanity constituted a critical medical encounter, where the patient's "history" was constructed from community members and encoded in a medico-legal document. The short narrative descriptions retold instances of delusions, strange manners of talking, violent or unusual behavior, emotional withdrawal, threats of self-harm, and sudden refusals to observe expected social norms and household duties. Doctors reiterated these comments from kin and community members, but also added more "physical" comments, such as the affect, gait, or general physical ailments of the alleged lunatic at the time of interview. A second document that was common to this process (sometimes referred to as a "warrant" or "reception order") detailed the general socio-demographic information related to the lunatics, often including their name, age, marital status, previous place of abode, religious persuasion, information on previous "attacks" of insanity, as well as specific questions asking whether the future patient was an epileptic, delusional, or suicidal. These data were transcribed into the admission register of the lunatic asylum, along with the date of admission. It is noteworthy that the asylum doctor, who was required to conduct a physical on all newly admitted patients, would then be charged with attributing a diagnosis to the patient (Andrews 1998a). It is worth remembering, then, that the categorization of different types of madness, as well as hypothesizing about their underlying causes, did not arise out of 'thin air', but were rather the product of this complicated process of knowledge exchange between medical doctors and the lay public.

Prominent asylum doctors, some of whom were founding members of the aforementioned periodicals used these institutional protocols, and the data generated therefrom, to establish their own credentials by publishing medical treatises that included the first attempts at systemic classification. In 1801, Pinel published his *Traité medico-philosophique sur l'aliénation mentale*. This treatise attempted to map out the "natural science of mental disease" by identifying possible causes, which for Pinel were multitudinous and might include heredity, social environment, physical factors, uncontrolled passions, and irregular ways of life (Ackernecht 1959). Benjamin Rush published his own *Medical Inquiries and Observations upon the Diseases of the Mind* in 1812 (Rush 1812), an effort that earned him the later title as being the "father" of American psychiatry. Jean-Étienne Esquirol, who had studied and worked under Pinel, authored his own *Mental Maladies: A Treatises on Insanity* in 1838, which appeared in English translation in 1845 (Esquirol 1845). John Charles Bucknill, the first editor of the *Asylum Journal of Mental Science,* published, with his colleague Daniel Hack Tuke, *A Manual of Psychological Medicine* in 1858 (Bucknill and Tuke 1858) which became the standard textbook for Victorian alienists in Britain and the British colonies. In Germany, the standard textbook was Wilhelm Griesinger's *Die Pathologie und Therapie der psychischen Krankheiten* (1845). These works represented learned opinions on interrelated topics, such as proposed systems of classification, suggested modalities of treatment, well as justifications for increased financial support for institutions devoted solely for the study and treatment of insanity.

Asylum doctors saw a diverse array of patients over the course of their clinical careers. But the administrative and legal structures encouraged them to develop broad categories of mental diseases that are worth summarizing here. The first sharp distinction that arose was the conceptual division between "lunatics" and "idiots," one that actually dated back several centuries in European legal deliberations. **Lunatics** were those who had been of sound mind and then suffered some mental derangement or decline in adult life that might prove temporary or permanent. **Idiots**, by contrast, were individuals who, from birth or a very early age, had mental impairment that was considered permanent and beyond treatment (though subject to some improvement through specialized education). Although those labelled as idiots (or another common term, imbeciles) were resident, in small numbers, in many institutions for the insane, it was widely argued that public lunatic asylums were not appropriate places for "harmless idiots," and many asylum superintendents encouraged authorities to build separate institutions suited to their needs (Wright 2001).

Among those deemed to be lunatics, there were several defining features that appear repeatedly in the medical case books and treatises of the nineteenth century. The first, and perhaps the most straightforward, was the medical diagnosis of **melancholia**. Melancholic patients were "depressed in spirits" and who had undergone a sudden and often inexplicable change in their mood, appetite, and regular household or employment activities. Individuals who had previously been active and functional were described as losing interest in everyday things, suffering from a lack of energy and motivation, staying in bed all day, or crying for no apparent reason. Family members, friends, neighbors, and doctors associated some acute cases with individuals threatening, or attempting, to take their own lives. Indeed, prominent alienists warned about the association between acute melancholia and suicidal behavior and recommended institutional precautions against self-murder occurring in the asylums (Shepherd and Wright 2003). Asylum superintendents also recognized the occurrence of a subset of women who suffered from melancholia immediately after childbirth, labelling them as suffering from **puerperal melancholia**. These cases, no doubt distressing to contemporaries who were not accustomed to responding to women who threatened to harm their own newborn children, were also conditions that were considered very amenable to improvement (Marland 2004).

The other principal category was **mania**, an elastic term that, in English, most often referred to individuals who had some sort of active and unusual disinhibition, often including certain recognizable types of delusion or hallucination (delusions of grandeur, for example). But mania might also be frustratingly vague, a catch-all term that included a diverse range of behaviors. Mania was also subject to sub-categorization. Those who exhibited alternating manic and depressive episodes might be labelled **mania with melancholia** or what the French alienists Jean-Pierre Falret and Jules Baillerger described in the 1850s as *folie circulaire* to describe the alternation between euphoria and depression (Berrios 1996). Another common term used among Victorian alienists, and one that gained currency over the course of the century, was the diagnosis of **dementia**. Unlike its twenty-first century connotation, dementia in the 1800s referred to a state of progressive cognitive decline, that was either related to senility and old age (**senile dementia**) or, in unusual cases, with premature or precocious dementia, a concept that would later be reformulated in the early twentieth century, as schizophrenia.

A final cluster of categories that came into popular and institutional usage by mid-century was the idea of **moral insanity** and **hysteria**. Moral insanity was actually coined by Pinel, but further articulated by the English alienist, James Prichard, in 1829 and again in 1835 in his *Treatise on Insanity and other disorders affecting the mind*. It gained in popularity in the second half of the nineteenth century, describing abnormal emotions or behaviors in the absence of delusions hallucinations and other known impairments (thus distinguishing it from **mania**, as described above). The controversy associated with moral insanity lay in the fact that sufferers could be fully functional in some attributes, but deficient in one or two key areas. **Hysteria**, about which much as been written, was a little-used designation in public lunatic asylums, despite its claim to fame through the writings of the Parisian alienist Charcot (Micale 1995). Like the common, if vague, employment of the terms **nerves** and **neurasthenia** (Oppenheim 1991), hysteria was a diagnosis most often used with a sensitive, middle-class clientele and did not figure prominently in lunatic asylum diagnoses prior to the First World War.

In Search of Biological Causes

Throughout the 1800s, asylum doctors remained convinced that different types of insanity had biological causes, even if they disputed what they were or simply admitted that the underlying and precipitating causes were unknown. Early in the century, some alienists were drawn to phrenological inquiry, where the allure of cerebral localization gave the potential of finding direct pathological causality (Cooter 1985). Over time, the mass production of the microscope, and the growing centrality of postmortem dissections in medical schools, provided the scientific justification and practical means for more sustained asylum-based bodily investigations. As Elizabeth Herren has demonstrated, the sheer size of the patient populations made asylums an ideal place for a new generation of practitioners to conduct postmortems in search of pathological lesions (Hurren 2012). As the largest welfare institutions by mid-century, lunatic asylums elicited a passionate debate within the medical profession as to whether or not lunatic cadavers should be included, or excluded, from Anatomy laws that sought to regulate the supply in dead bodies and help suppress the trade in illegal corpses for pedagogical purposes.

In general, research into mental diseases reflected broad changes going on in medical education. Central to this was the rise of the laboratory medicine, inspired by the advances in bacteriology associated with Louis Pasteur and Robert Koch in the 1860s and 1870s, as well as the advent of antiseptic techniques made famous by Joseph Lister (Bynum 1994). This new "scientific medicine," underpinned by a preoccupation with benchside research and quantification, had implications for psychiatric research and clinical practice. These could be seen in the work of Wilhelm Griesinger, who founded the journal *Archiv für Psychiatrie und Nervenkrankheiten* (*Archive for Psychiatry and Nervous Diseases*) in 1868. Griesinger attempted to systematize the pluralistic environment of psychiatric taxonomy and focus clinical research on the study of neuropathology (Engstrom 2003). It was at this time that a specific condition – **general paralysis of the insane** (g.p.i. also often shorted to **general paresis** or even **paresis**) became central to the widely held view that all acute mental disorders had a biological basis, since paresis represented the psychiatric manifestation of the final (tertiary, or "neuro-") stage of syphilis (Braslow 2007; Davis 2008). So common was general paresis, especially among male asylum admissions, that some historians of psychiatry have even suggested that there was a real increase in its prevalence in the decades leading up to the era of antibiotics (Fuller Torrey and Miller 2002).

Griesinger's work reflected an important divide that had emerged between continental European approaches to scientific enquiry and those typically found in Anglo-America. In Germany, psychiatric research became institutionalized in university-affiliated clinics, where academic-oriented psychiatrists sought to wrest control and authority from asylum superintendents (Engstrom 2003). The German approach, in Edward Shorter's words, "truly dominated the field" of research-based biological psychiatry in the last third of the nineteenth century (Shorter 1998). The biological model found its champion in Emil Kraepelin, whose *Compendium der Psychiatrie: Zum Gebrauche für Studirende und Aerzte* (*Compendium of Psychiatry: For the Use of Students and Physicians*), published in 1883, sought to reaffirm psychiatry's rightful role as a medical discipline, and the chart a consolidated classification of psychiatric disorders that would prove deeply influential in the two decades leading up to World War I. It was the fifth edition of the *Compendium* that famously enunciated the distinction between manic-depressive psychosis and dementia praecox based on Kraepelin's interest in the longitudinal course of severe mental disorders (Engstrom 1995).

The attraction of the German model of university clinics arose, in part, from the fact that the traditional lunatic asylum had become, in many countries, the backwater of modern medicine. Institutions had grown dramatically in size; many exceeded 1000 patients by the dawn of the twentieth century. Far from the intentions of those who had articulated moral treatment in small, intimate rural retreats a century earlier, most public lunatic asylums at the dawn of the twentieth century suffered from severe overcrowding and chronic underfunding, with unmotivated staff and self-defeating subcultures. While major advances were occurring in other areas of medicine, asylum psychiatry appeared to be lagging behind. Moreover, the ever-rising cost of these public institutions led to the inevitable speculation as to what accounted for the apparent increase in insanity. Some practitioners began to embrace hereditarian explanations, ones that, in the context of colonial settler societies, also focused on the perceived problem of unfettered immigration (Bashford 2013). It was easy, if somewhat self-serving, for commentators to suggest that the mentally troubled were being "dumped" on them by European countries. Psychiatrists, such as C.K. Clarke, the medical superintendent of the Toronto Asylum and future President of the American Association of Officers for Asylums for the Insane, obsessed about the disproportionate numbers of new immigrants among asylum patients. He and others played a key role in the implementing measures to screen new arrivals in North America for insanity (Dowbiggin 1997).

The idea of a degenerative "hereditary taint" was hardly a turn-of-the-century intellectual invention. Many medical treatises throughout the 1800s had suggested a certain "constitutional" tendency in some individuals that, with the appropriate trigger, might result in full-blown psychosis or melancholia. But the idea that heredity was an overpowering force that, despite the best therapeutic intentions of medical men, would lead inevitably to chronic insanity, was clearly a stance that the first generation of alienists were reluctant to embrace. Nevertheless, with the ever-growing numbers of long-stay patients in public institutions by the end of the century, hereditarianism began to become discussed and validated more widely (Dowbiggin 1991). The celebrated English alienist Henry Maudsley, among others, introduced the idea an "inherited insane diathesis" in his *Physiology and Pathology of the Mind* (1867), hypothesizing a cascade of increasingly worse mental states through successive generations. Such theories of trans-generational degeneration began to gain ground by the turn of the century, when they received support from prominent members of national eugenics societies (Pick 1989). Over time, some asylum psychiatrists embraced mental testing and became important boosters of eugenic-inspired legislation that also promoted, by the interwar period, sexual sterilization laws as one solution to asylum overcrowding.

Conclusions: The Legacy of Asylum Psychiatry

The First World War represented a turning point in the history of institutional psychiatry. It was at this time that the asylum system had reached its peak, if measured by the percentage of the general population who found themselves resident in psychiatric facilities. By the outbreak of the hostilities in 1914, state-run "mental hospitals" (as they were by then increasingly known) were widely denigrated as a professional dead end, where the early optimism of moral therapy had given way to disappointment and frustration. Once heralded as progressive therapeutic institutions for the socially marginalized, long-stay psychiatric facilities were characterized by disillusionment and custodialism. By the interwar years, the first comprehensive programs of discharge and extra-institutional care were being trialed, if imperfectly, in Britain and North America. Treatments for psychiatric disorders, throughout the 1800s, had been most often limited to tinkering with diets, employing strategies to prevent suicide, and with the liberal use of sedatives for troublesome and violent patients. Many medical superintendents simply hoped that the ordered environment of the asylum, through its architecture, daily routine, leisure, and occupational therapies, would create situations amenable to remission, or at least amelioration. Many patients did indeed leave the asylum, either experiencing a full or partial remission of their symptoms, or because families and communities demanded their discharge. Indeed, one of the great misunderstandings of the lunatic asylum during this period is a belief that, once admitted, all residents were there for life. In reality, over 40% of all admissions were eventually discharged, a figure that does not suggest an unrecognized success of psychiatric treatments, but rather a more complicated dynamic involving multiple social and medical uses of these institutions in the past.

World War I also coincided with a stark duality in psychiatric treatment that would last for most of the twentieth century. A group of modalities would emerge which were self-consciously positioned *outside* of the world of asylum psychiatry and raised broader questions as to whether they were "medical" or not. During the war, talk therapy arose in response to the war trauma (commonly referred to in English as "shell shock") that had psychologically damaged so many soldiers who had spent months suffering in the trenches (Micale and Lerner 2009). These new psychotherapeutic approaches were continued after the war in the civilian population, and were supplemented by other "talk therapies," most famously by the psychodynamic approaches associated with Sigmund Freud and his followers. Psychotherapy, located in private offices in the community, and overwhelmingly directed towards an urban (paying) middle-class clientele, provided an alternative to the lunatic asylum as a locus of psychiatric practice and knowledge production. With its origins in central Europe, psychoanalysis was exported to Britain and the Americas by European psychiatrists in the interwar period and by refugee doctors during and immediately after World War II.

Despite, or perhaps because of, the perceived failures of therapeutics prior to World War I, the first four decades of the twentieth century also saw an upswing in institutional clinical experiments, of what became known as the "somatic" therapies. Spurred on by the apparent success of bacteriology, and lured by the growing caché of scientific awards and funding (as exemplified by the Nobel Prize, beginning in 1901), institutional-based psychiatry desperately looked for the "magic bullet" that would eradicate certain common mental illnesses. The apparent success of other medical advances emboldened individual clinicians to tinker with a variety of somatic interventions, including hydrotherapy (cold and hot water baths) popular in the 1910s, prophylactic dental extractions (Scull 2007) and fever therapies (in the 1920s), the experimental use of insulin (first used in 1927, but common by the 1940s) (Braslow 2007), unmodified electroconvulsive therapy (ECT) from the late 1930s to the 1960s (Shorter and Healy 2007; Sadowsky 2017), followed shortly thereafter by leucotomy and transorbital lobotomy (Pressman 1998). These various modalities of somatic psychiatric interventions would dominate psychiatric experimentation and practice in asylums for the first half of the twentieth century until most were slowly superseded and displaced by new psychopharmacological agents (Healy 1999).

This divide between biological and psychotherapeutic psychiatry would haunt the profession during the first two-thirds of the twentieth century, despite the attempt to unify and standardize psychiatric practice and research through an agreed upon universal system of classification. It had become apparent that the use of such labels as "schizophrenia" and "depression" varied widely between different institutional and national contexts. As a consequence, the American Psychiatric Association established a Diagnostic and Statistical Manual (DSM), the first edition of which appeared in 1952. Meanwhile the World Health Organization included in the sixth edition (1949) of its International Classification of Disorders (ICD), a section on mental disorders for the first time. The DSM and ICD would provide two competing, transnational standardized approaches to psychiatric classification. They, of course, reflected "Western" biomedical categories and approaches to mental disorders that fit awkwardly when applied to non-Western communities (Wu 2021).

Finally, with the growing specialization of medicine and heightened importance of research for those attached to university training programs, psychiatrists were also successful in establishing formal, multiyear postgraduate training programs (also known as residencies) in psychiatry, paralleling the advent of postgraduate programs in other medical specialties, such as pediatrics or surgery. Areas of subspecialization – such as child psychiatry, liaison psychiatry, geriatric psychiatry – would permit even further clinical specialization and research. By the time of their establishment (in the 1950s), psychiatric practice would be slowly transformed by the advent of the first generation of neuroleptics which would, combined with other nonclinical factors, further decenter the asylum in clinical practice and social policy. Western society would slowly turn away from traditional psychiatric facilities as a part of a long process that is often referred to as deinstitutionalization. Although far from the only reason, the advent of psychopharmacology marked the beginning of the end of the lunatic asylum.

References

Ackerknecht E (1959) A short history of psychiatry. Hafner, New York

Alexander F, Selesnick S (1966) The history of psychiatry: an evaluation of psychiatric thought and practice from prehistoric times to the present. Harper & Row, New York

Andrews J (1998) "They are in the trade…of lunacy, they 'cannot' interfere" – they say: The Scottish Lunacy Commissioners and Lunacy Reform in Nineteenth-Century Scotland. The Wellcome Trust, London

Andrews J (1998a) Case notes, case histories, and the patient's experience of insanity at Gartnavel Royal Asylum, Glasgow, in the nineteenth century. Soc Hist Med 11(2):255–282

Andrews J (2004) The rise of the asylum in Britain. In: Brunton D (ed) Medicine transformed: health, disease and society in Europe, 1800–1930. Manchester University Press, Manchester, pp 298–330

Andrews J, Scull A (2001) Undertaker of the mind: Dr John Monro and the mad-doctoring trade in eighteenth-century England. University of California Press, Berkeley

Andrews J, Briggs A, Porter R, Tucker P, Waddington K (1997) A history of Bethlem hospital. Routledge, London

Bartlett P (1999) The poor law of lunacy: the administration of Pauper Lunatics in mid-nineteenth-century England. Leicester University Press, London
Bashford A (2013) Insanity and immigration restriction. In: Cox C, Marland H (eds) Migration, health and ethnicity in the modern world. Palgrave Macmillan, London, pp 14–35
Battie W (1758) A Treatise on Madness. Winston & White, London
Berrios G (1996) The history of mental symptoms: descriptive psychopathology in the nineteenth century. Cambridge University Press, Cambridge
Braslow J (2007) Mental ills and bodily cures: psychiatric treatment in the first half of the twentieth century. University of California Press, Berkeley
Browne WAF (1837) What asylums were, are and ought to be. AC Black, Edinburgh
Bucknill JC, Tuke DH (1858) A manual of psychological medicine. John Churchill, London
Bynum W (1994) Science and the practice of medicine in the nineteenth century. Cambridge University Press, Cambridge
Castel R (1988) The regulation of madness: the origins of incarceration in France. University of California Press, Berkeley
Coleborne C (2009) Madness in the family: insanity and institutions in the Australasian colonial world, 1860–1914. Palgrave Macmillan, New York
Conolly J (1830) Inquiry concerning the indications of insanity, with suggestions for the better protection and care of the insane. John Taylor, London
Conolly J (1847) The construction and government of lunatic asylums and hospitals for the insane. John Churchill, London
Cooter R (1985) The cultural meaning of popular science: phrenology and the organization of consent in nineteenth-century Britain. Cambridge University Press, Cambridge
Cox C (2012) Negotiating insanity in the southeast of Ireland, 1820–1901. Manchester University Press, Manchester
Davis G (2008) 'The cruel madness of love': sex, syphilis and psychiatry in Scotland, 1880–1930. Rodopi, Amsterdam
Digby A (1985) Madness, morality and medicine: a study of the York Retreat, 1796–1914. Cambridge University Press, Cambridge
Dowbiggin I (1991) Inheriting madness: professionalization and psychiatric knowledge in nineteenth-century France. University of California Press, Berkeley
Dowbiggin I (1997) Keeping America sane: psychiatry and eugenics in the United States and Canada, 1880-1940. Cornell University Press, Ithaca
Edington C (2019) Beyond the asylum: mental illness in French Colonial Vietnam. Cornell University Press, Ithaca
Engstrom E (1995) Kraepelin: social section. In: Berrios G, Porter R (eds) A history of clinical psychiatry: the origin and history of psychiatric disorders. Athlone, London, pp 292–301
Engstrom E (2003) Clinical psychiatry in Imperial Germany: a history of psychiatric practice. Cornell University Press, Ithaca
Esquirol J-É (1845) Mental maladies: a treatises on insanity. Lea & Blanchford, Philadelphia
Finnane M (1981) Insanity and the insane in post-famine Ireland. Croom Helm, London
Foucault M (1965) Madness and civilization: a history of insanity in the age of reason. Random House, New York
Fuller Torrey E, Miller J (2002) The invisible plague: the rise of mental illness from 1750 to the present. Rutgers University Press, New Brunswick
Goldberg A (1999) Sex, religion, and the making of modern madness: the Eberbach asylum and German Society, 1815–49. Oxford University Press, New York
Goldstein J (1987) Console and classify: the French psychiatric profession in the nineteenth century. Cambridge University Press, Cambridge
Goodheart L (2003) Mad Yankees: the Hartford retreat for the insane and nineteenth-century psychiatry. University of Massachusetts Press, Amherst
Grob G (1973) Mental institutions in America: social policy to 1875. The Free Press, New York
Healy D (1999) The antidepressant era. Harvard University Press, Cambridge, MA
Hide L (2014) Gender and class in English asylums, 1890–1914. Palgrave Macmillan, London
Houston R (2000) Madness and society in eighteenth-century Scotland. Clarendon Press, Oxford

Hurren E (2012) Dying for Victorian medicine: English anatomy and its trade in the dead poor, 1834–1929. Palgrave Macmillan, Houndmills

Jones K (1993) Asylums and after: a revised history of the mental health services, from the early 18th century to the 1990s. The Athlone Press, London

Kirkbride T (1854) On the construction, organization, and general arrangements of hospitals for the insane, with some remarks on the treatment of insanity. Lindsay & Blackiston, Philadelphia

Kramer C (1998) A fool's paradise: the psychiatry of Gemüth in a Biedermeier asylum. Unpublished PhD thesis, University of Chicago, Chicago

Leckie J (2019) Colonizing madness: asylum and community in Fiji. University of Hawaii Press, Honolulu

MacDonald M (1981) Mystical Bedlam: madness, anxiety, and healing in seventeenth-century England. Cambridge University Press, Cambridge

Mahone S, Vaughan M (eds) (2007) Psychiatry and empire. Palgrave Macmillan, London

Malcolm E (1989) Swift's hospital: a history of St Patrick's Hospital, Dublin 1746–1989. Gill and Macmillan, Dublin

Marland H (2004) Dangerous motherhood: insanity and childbirth in Victorian Britain. Palgrave Macmillan, New York

Maudsley H (1867) Physiology and pathology of the mind. Appleton & Company, London

Maudsley H (1872) Is insanity on the increase? Br Med J 13(1):36–39

McCandless P (1979) "Build! build!": the controversy over the care of the chronically insane in England, 1855—70. Bull Hist Med 53(4):553–574

McGovern C (1986) Masters of madness: social origins of the American psychiatric profession. University of New England Press, Lebanon/Hampshire

Micale M (1995) Approaching hysteria: disease and its interpretations. Princeton University Press, Princeton

Micale M, Lerner P (eds) (2009) Traumatic pasts: history, psychiatry, and trauma in the modern age, 1870–1930. Cambridge University Press, Cambridge

Moran J (2019) Madness on trial: a transatlantic history of English civil law and lunacy. Manchester University Press, Manchester

Oppenheim J (1991) Shattered nerves: doctors, patients, and depression in Victorian England. Oxford University Press, Oxford

Paradis A (1994) L'asile québécois et les obstacles à la médicalisation de la folie (1845–1890). Can Bull Med Hist 11(2):297–334

Parle J (2007) States of mind: searching for mental health in Natal and Zululand, 1868–1918. University of Kwazulu-Natal Press, Durban

Parry-Jones W (1972) The trade in lunacy: a study of private madhouses in England in the eighteenth and nineteenth centuries. Routledge, London

Pelling M (1985) Healing the Sick Poor: social policy and disability in Norwich, 1550–1640. Med Hist 29(2):115–137

Philo C (2004) A geographical history of the institutional provision for the insane from Medieval times to the 1860s in England and Wales: the space reserved for insanity. Edward Mellen Press, Lewiston

Pick D (1989) Faces of degeneration: a European disorder, 1848–1918. Cambridge University Press, Cambridge

Pinel P (1800) Traité médico-philosophique sur l'aliénation mentale. Richard, Caille et Ravier, Paris

Porter R (1987) Mind-Forg'd Manacles: a history of madness in England from the restoration to the regency. The Athlone Press, London

Porter R (2004) Madness: a brief history. Oxford University Press, Oxford

Porter R, Wright D (eds) (2003) The confinement of the insane: international perspectives, 1800–1965. Cambridge University Press, Cambridge

Pressman J (1998) Last resort: psychosurgery and the limits of medicine. Cambridge University Press, Cambridge

Quétel C (2009) Histoire de la folie, de l'Antiquité à nos jours. Tallandier, Paris

Rothman D (1971) The discovery of the asylum: social order and disorder in the New Republic. Little Brown, Boston

Rothman D (1980) Conscience and convenience: the asylum and its alternatives in progressive America. Little Brown, Boston
Rush B (1812) Medical inquiries and observations upon the diseases of the mind. Kimber & Richardson, Philadelphia
Sadowsky J (2017) Electroconvulsive therapy in America: the anatomy of a medical controversy. Routledge, London
Schmiedebach H-P (2010) Inspecting Great Britain: German psychiatrists' views of British asylums in the second half of the nineteenth century. In: Roelcke V et al (eds) International relations in psychiatry: Britain, Germany, and the United States to World War II. University of Rochester Press, Rochester, pp 12–29
Scull A (1975) From madness to mental illness: medical men as moral entrepreneurs. Arch Eur Sociol 16:218–261
Scull A (1979) Museums of madness: the social organisation of insanity in nineteenth-century England. Allen Lane, London
Scull A (1993) The most solitary of afflictions: madness and society in Britain, 1700–1900. Yale University Press, New Haven
Scull A (2007) Madhouse: a tragic tale of megalomania and modern medicine. Yale University Press, New Haven
Scull A, MacKenzie C, Hervey N (1996) Masters of Bedlam: the transformation of the mad-doctoring trade. Princeton University Press, Princeton
Shepherd A, Wright D (2003) Madness, suicide and the Victorian asylum: attempted self-murder in the age of non-restraint. Med Hist 46(2002):175–196
Shorter E (1998) A history of psychiatry, from the era of the asylum to age of Prozac. Wiley, New York
Shorter E, Healy D (2007) Shock therapy, a history of electroconvulsive treatment in mental illness. Rutgers University Press, New Brunswick
Smith L (1999) 'Cure, comfort and safe custody': public lunatic asylums in early nineteenth-century England. Leicester University Press, Leicester
Smith L (2014) Insanity, race and colonialism: managing mental disorder in the post-emancipation British Caribbean, 1838–1914. Palgrave Macmillan, London
Smith L (2020) Private madhouses in England, 1640–1815: commercialized care for the insane. Palgrave Macmillan, London
Sposini F (2020) Just the basic facts: the certification of insanity in the era of the form K. J Hist Med Allied Sci 75:171–192
Sturdy H, Parry-Jones W (1999) Boarding-out insane patients: the significance of the Scottish system 1857–1913. In: Bartlett P, Wright D (eds) Outside the walls of the asylum: the history of care in the community, 1750–2000. Athlone, London, pp 86–114
Szasz T (1997) The manufacture of madness: a comparative study of the inquisition and the mental health movement. Syracuse University Press, Syracuse
Tomes N (1985) A generous confidence: Thomas Story Kirkbride and the art of asylum-keeping, 1840–1883. Cambridge University Press, Cambridge
Tuke D (1885) The insane in the United States and Canada. Reprinted by Nabus Press
Weisz G (2005) Divide and conquer: a comparative history of medical specialization. Oxford University Press, New York
Winston M (1994) The Bethel at Norwich: an eighteenth-century hospital for lunatics. Med Hist 38(1):27–51
Woodward J (1974) To do the sick no harm: a study of the British voluntary hospital system to 1875. Routledge, London
Wright D (1998) The certification of insanity in nineteenth-century England. Hist Psychiatry 9(35):267–290
Wright D (2001) Mental disability in Victorian England: The Earlswood asylum, 1847–1901. Oxford University Press, Oxford
Wu H (2021) Mad by the millions: mental disorders and the early years of the World Health Organization. MIT Press, Cambridge, MA

49. Forensic Psychiatry: Human Science in the Borderlands Between Crime and Madness

Eric J. Engstrom

Contents

Introduction	1274
Medical Jurisprudence and Penal Reform	1275
Alienism	1278
Degeneration Theory and Lombrosian Criminology	1282
Psychiatric Expertise in Foro	1285
The Psychiatrization of Danger: Prophylactic Assessments	1289
Feeble-Minded Children and Incorrigible Youth	1290
Psychopaths	1292
Clinical Assessment Tools	1293
Preventive Detention in National Contexts	1294
Judicial Gatekeeping	1297
Conclusion: Toward Therapies of Self-Governance and Patients' Rights	1299
Cross-References	1301
References	1301

Abstract

The chapter is a broad overview of the history of forensic psychiatry in the nineteenth and twentieth century. It tracks the emergence of forensic psychiatry from discourses on penal reform, alienism, and criminology. After considering the status of forensic experts in courtroom settings, it then describes the twentieth century turn to the prophylactic and therapeutic strategies that targeted the criminal lunatic and to the trend toward greater patients' rights. It emphasizes the dual realms in which forensic psychiatrists have operated: on the one hand, as expert witnesses before the court, subjected to public scrutiny at criminal trials (or coroner's inquests) where they pronounced defendants sane or insane; and on

E. J. Engstrom (✉)
Department of History, Humboldt University, Berlin, Germany
e-mail: engstroe@geschichte.hu-berlin.de

© The Author(s), under exclusive licence to Springer Nature Singapore Pte Ltd. 2022
D. McCallum (ed.), *The Palgrave Handbook of the History of Human Sciences*,
https://doi.org/10.1007/978-981-16-7255-2_94

the other hand, the "darker underbelly" of mostly invisible and often extralegal medical practices and administrative decision-making involved in the study, management, and rehabilitation of criminal lunatics.

Keywords

Forensic psychiatry · Prison reform · Alienism · Degeneration theory · Criminology · Expertise · Preventive detention

Introduction

Surveying the historical literature of the past several decades, it is easy to identify narratives that emplot the history of madness as stories of social marginalization and incarceration. For the most part, these narratives have operated explicitly as a critique of psychiatric science and practice, both inspired by and helping to inspire a broader anti-psychiatric movement in the 1960s and 1970s. This critique came to be associated with the work of authors such as Michel Foucault, Erving Goffman, Robert Castel, David Rothman, Andrew Scull, and others. And because it evolved against the backdrop of older, more hermetic, and too often hagiographic accounts of psychiatry's rise as a medical science, this tradition came to be described by scholars as "The Great Revision" (Micale and Porter 1994).

If nothing else, and although certainly not without its own pitfalls, this Great Revision has shown that the history of madness has been profoundly intertwined with evolving notions of crime and deviance. Whether attempting to understand the social dynamics leading to the marginalization and exclusion of criminal lunatics, the practices of police officers in apprehending them, the legal statutes applied in adjudicating them, the administrative protocols governing their commitment to psychiatric institutions, or even the medical techniques used to diagnose and treat them, historians have rightly underscored the normative categories that underpin and explain the phenomena of madness through the ages. To a remarkable degree, the history of madness can and has been written in terms of society's response – be it in the form of punishment, detention, management, therapy, or otherwise – to the lunatic's transgressions.

One of the consequences of this historiographic setting has been an especially keen interest in forensic topics. Sensational courtroom trials and legislative battles over criminal law reform have become the centerpieces of numerous articles and monographs about the history of madness. Such trials and debates have inspired historians to explore the often intractable problems, multivalent meanings, and ambiguous solutions that have characterized earlier encounters on the borderline between sanity and insanity. They have opened windows onto questions that, although confronted in very different ways throughout history, have lost none of their contemporary relevance: what exactly was the nature of the relationship between mental illness and crime? How could medical diagnoses be translated into legal frameworks? How did historical actors mediate between medical and legal discourse? Did psychiatrists – in their capacity as physicians – have simply to diagnose a person's medical condition? Or did their expertise extend further to an assessment of culpability or dangerousness? How could someone's past mental state, at the moment of the crime, be determined? These and other vexing questions were (and remain) hotly contested. Any plausible historical examination of them must necessarily attend not only to their enduring relevance, but also to the specifically contingent and quotidian contexts in which they were posed and answers sought.

In an effort to grasp the complexity of these issues, one could do worse than to study them in their many and evolving spatial contexts and configurations. As Michel Foucault demonstrated many years ago, notions of deviance and delinquency were decisively shaped by the regimes of power and knowledge embodied in the architecture and protocols of various institutional receptacles. Police stations, courtrooms, prisons, workhouses, asylum wards, etc., all exercised a profound influence over the perceptions of criminal deviance. But alone, none of these receptacles suffice to explain how criminal behavior was studied, defined, or dealt with. Nor can one assume that the respective aims and agendas of such institutions, bent as they generally were on punishing or normalizing deviance, always coincided or that moving between them involved seamless transitions within a larger disciplinary "archipelago" of modern society. In fact, an important part of the history of psychiatry as it relates to crime and mental disorder can be told as a story of the intersection and shifting jurisdictions of these institutions and of the sometimes dysfunctional relationships between them in the medicolegal borderlands. It is crucial to recognize that the place and impact of the human sciences throughout these borderlands has varied enormously, due not just to conceptual or methodological differences, but also to the different uses to which those sciences were put and the ends they served.

Medical Jurisprudence and Penal Reform

Any fuller understanding of the historical relationship between crime and madness must necessarily range beyond the narrower confines of the (trans)disciplinary specialty of forensic psychiatry to account for differing legal traditions and systems across Europe and North America. For legal frameworks have always legitimized and defined the scope of psychiatric practice and forensic reasoning (Eigen 2016; Hamlin 2019; Smith 1991). And those frameworks have differed markedly. For example, differing federal, regional, and municipal jurisdictions, and the specific disposition of the interfaces between them – not to mention ideological or political agendas – have decisively shaped the experience and remit of forensic practices. As a result, writing broader, synthetic histories of forensic psychiatry faces daunting challenges. While concepts like "born criminal" or "psychopath" certainly varied from one jurisdiction to another, their travel across borders has tended to be much smoother than that of legislatively crafted articles of code, let alone administrative protocols and courtroom cultures.

For understandable, albeit not quite so easily justifiable reasons, most historical scholarship has focused on issues related to *criminal* codes. Most prominently, the insanity defense and the adjudication of accountability have attracted the most attention. The insanity defense long predated the rise of medical jurisprudence in the late eighteenth and nineteenth centuries and has been the topic of numerous inquiries. Of significantly lesser concern to historians has been *procedural* justice, involving the narrower mechanics of courtroom deliberations and rules of evidence, as well as *administrative law*, including police law or poverty and health law (Bartlett 1999). Likewise historians of psychiatry have shown relatively less interest in *civil* codes and issues such as guardianship, marriage, divorce, child custody, workers compensation, or the adjudication of wills. And yet arguably, the adjudication of these issues has been just as important, if not more so than cases in criminal law (Moran 2019). Long ago, Michel Foucault (1978) noted how a broadening understanding of civil liability could migrate into criminal law, enabling notions of reduced criminal accountability. And more recently, Susanna L. Blumenthal (2016) examined ordinary civil cases in the United States, demonstrating how the contribution of the human sciences to the adjudication mental capacity has been at once decisive and ambiguous: On the one hand, their engagement in deliberations of legal accountability helped to construct the "default legal person" as the statutory embodiment of the responsible, law-abiding citizen. And yet those same sciences also threatened to deconstruct this normative legal personhood by exposing a "vast borderland between perfect sanity and total derangement."

Throughout Western history, societies and the legal statutes they brought forth have generally recognized that perpetrators who, at the time of the crime, were not of sound mind should not be held criminally responsible and punished (Robinson 1996; Watson 2011). The concept of *non compos mentis* extended beyond the insane to include others judged incapable of tending to their own affairs. As it applied to insanity in the early modern era, however, for the most part it identified a defect of reason. It was not until the eighteenth century that, in addition to intellectual defects, it came to be used in describing emotional and moral defects as well (Rollin 1996).

Of significantly more recent vintage, however, was the notion that, as a *medical* condition, madness could actually be the cause of a crime. Regardless of what punishment might be imposed, establishing whether or not defendants were criminally culpable could hinge on whether they were afflicted by mental illness. And with that understanding came the fundamental problem of medical jurisprudence – which persists to this day – of negotiating the interface between the medical symptoms of insanity on the one hand, and legal categories (such as free will, motivation, intent, or culpability) on the other. At stake along this interface – not least in terms of effective governance – was the "correct" allocation of criminal responsibility and punishment.

Contemporary notions of criminal responsibility can to this day be traced back to the eighteenth century Enlightenment. The ideals of the rule of law, individual liberty, social contracts, utilitarianism, progress, and rationality led to radical transformations in earlier systems of punishment. No one did more to spur change than the Italian jurist Cesare Beccaria. In his treatise *On Crimes and Punishments* (1764), he justified punishment as retribution for the willful breach of the social contract. But in order to eliminate the arbitrariness of the criminal justice systems in Europe's *anciens régimes*, he advocated that each type of crime be directly associated with a specific and clearly articulated form of punishment. Beccaria and his disciples insisted that the often arbitrary and corporal forms of traditional punishment needed to be reformed to make them more egalitarian and proportional to the crime. Fixed, legally codified punishments, imposed regardless of one's particular social rank, they argued, would not only help eliminate judicial corruption, but also enhance the prospects of convicts' rehabilitation and reintegration into society. And by relentlessly coupling crime and punishment, criminals would, as rational subjects, also come to recognize that it was in their own self-interest to obey the law and comport themselves accordingly (Wiener 1990). In England, these egalitarian, prophylactic, and rehabilitative aims were taken still further to new, utilitarian heights by Jeremy Bentham, who developed a moral arithmetic of punishment designed to discipline convicts and instill in them a sense of moral responsibility. By the early 1800s, the works of both Beccaria and Bentham were inspiring an entire generation of reformers devoted to Christian good works and/or the new moral science of penology. The zeal of figures like John Howard or Quaker philanthropists like Elisabeth Fry in the UK, Charles Lucas in France, Nikolaus Heinrich Julius in Germany, and Dorothea Dix in the United States unfolded in protracted disputes over moral responsibility and humanitarian assistance, often enough pitting retributive and utilitarian aims against one another.

Aside from its carceral tasks, the nineteenth century prison was also a factory and a school that aimed to rehabilitate convicts through work and moral instruction (O'Brian 1982). Mentally ill convicts unable to work or learn ran afoul of this prison regime: it elicited their mental deviance, made it visible, and punished it, often with solitary confinement (Walker 1968/1973). Initially, prison surgeons and chaplains had been well positioned to observe and respond to convicts' mental anguish. Growing concerns about hygiene, however, led to an expansion of prison medical care that spurred efforts to identify and study the mental condition of criminal lunatics, assess their suitability for punishment, and prevent suicides (Wiener 1990). These tasks were governed by standards of documentation and reporting that often became the basis of "scientific" study of various phenomena, from suicide and intelligence, to prison psychosis and simulation. Concepts such as "moral imbecility," "monomania," or "moral insanity" evolved in good part from the study of such incarcerated populations.

The amalgam of "criminal lunacy" expressed the ambiguous fusion of legal sanction and medical assessment that has been described as a medico-penal nexus (Moran 2014; Smith 1991). Certainly in the public imagination, crime and madness could overlap in diffuse ways, but experts tended to distinguish between the insane criminal and the criminally insane. In the case of insane criminals, physicians argued that madness had played either no or only a marginal role in their crimes and – providing their madness was cured – they could be tried, convicted, sentenced, and punished like any other offender. By contrast, the offenses of the criminally insane had been perpetrated on account of their madness, such that these individuals could in general *not* be tried, convicted, sentenced, and punished like other offenders. This distinction informed widespread nineteenth century debates on legal and penal reform.

All of this is to underscore, as Roger Smith (1991) has argued, that psychiatry became "occupationally associated more with the new specialty of penology than with the legal profession, and practitioners in penology and psychiatry, unlike many lawyers, shared a utilitarian rather than retributive attitude toward offenders." This distinction is crucial in understanding the dual realms in which nascent forensic psychiatrists operated: on the one hand, as expert witnesses in foro, i.e., before the court, where they pronounced defendants sane or insane and were subjected to public scrutiny; and on the other hand, the "darker underbelly" of mostly invisible and often extralegal medical practices and administrative decision-making involved in the study, management, and rehabilitation of criminal lunatics. In other words, "there was a sharp disjuncture between the legal discourse of mental state and the later administrative discourse of internment, whose categories were based on the dangerousness of the individual to society" (Goldstein 1998).

Alienism

Many of the Enlightenment and utilitarian ideals of these penal reformers were shared and actively promoted by early nineteenth century asylum superintendents, or so-called alienists. They too saw themselves as standard-bearers of rationality, individual liberty, and humanitarian reforms. They believed that, not unlike convicts, the mad too could be persuaded back to rational, self-interested behavior. The madman's deviance would be corrected not by physical punishment or heroic medical interventions, such as bleedings or purgings, but rather by the arguably milder psychological techniques of a "moral treatment" (Goldstein 1987). In the words of its most prolific advocate, the French alienist Jean-Ètienne Dominique Esquirol, moral treatment involved the "application of the faculty of intelligence and emotions in the treatment of mental alienation" (Kelly 2014). These alienists envisioned themselves as benevolence reformers and moral pedagogues, relying on their patriarchal authority as asylum superintendents to oversee an institutional therapeutic regime designed to return inmates to their "natural" and rational state of mind.

Alienists had often understood themselves to be liberating the mentally ill from both prisons as well as from the misconceptions of jurists or the general public. An essential rhetorical building block in the formation of alienism was the rebuttal of widely held beliefs, lasting well into the eighteenth century, that the insane were demonically possessed and that harsh inquisitorial methods (including beatings, exorcism, and the stake) were justified responses of local communities to their conditions. From the late eighteenth century, overcoming "fanatical" religious beliefs and preventing the mistreatment of the mad in their communities became key justifications for the establishment of mental asylums and potent manifestations of alienists' enlightened principles.

While early nineteenth century alienists and classical penologists shared a common set of Enlightenment values, they often found themselves in disagreement about the aims and effects of their respective institutional regimes. Much of the early development of nineteenth century alienism can be understood as a history of its demarcation from penal institutions and as an alternative to them. Early alienists were convinced that prisons were no place for the mentally ill and that they instead needed to be cared for in mental asylums. This conviction was accompanied by both an administrative and medical discourse: experts and government officials argued whether or how the institutions that housed mentally ill convicts could guarantee the dual aims of public safety and medical care. In addressing these issues, legal systems across Europe and beyond enacted legislation in the 1830s and 1840s that created a variety of institutional receptacles (Porter and Wright 2003).

For their part, mental asylums developed different institutional strategies that sought to reconcile these bifurcating aims. Some asylums preferred to concentrate convicts on special wards, while others insisted that individuals with criminal records be treated just like other patients, subjected to the same therapeutic regime, and distributed accordingly throughout the entire architectural space of the asylum. Often enough, however, the criminal lunatic simply remained confined in prisons and workhouses. In some cases, however, special institutions were created to detain them, such as the short-lived Rockwood Criminal Lunatic Asylum in Canada (1855), the asylum of Aversa (1876) near Naples in Italy, the Central Criminal Lunatic Asylum near Dublin in Ireland, or, most notoriously, the psychiatric prison hospital Broadmoor, which began admitting patients in the 1860s.

Beyond their medical ethos, nothing distinguished asylums from prisons more than the doctrine of "non-restraint." Increasingly from the middle of the nineteenth century, this doctrine became a litmus test of institutional reform used to distance asylums from their penal origins and to transform them into medical hospitals. As conceived and implemented by John Connolly and his supporters, non-restraint sought to banish all forms of mechanical coercion in the treatment of psychiatric patients. Advocates of non-restraint hoped not only to enhance the therapeutic prospects of their charges but also to improve the reputation of their own profession. At the same time, however, the implementation of non-restraint policies brought with it new and specific challenges, especially when it came to forensic issues and alienists' ambitions to diagnose and treat criminal lunatics.

The admission of criminals to psychiatric facilities was therefore not always greeted with unalloyed enthusiasm. Certainly, the early justification for building asylums had in good part been based the need to remove mentally ill convicts from an institutional regime of punishment and instead place them in a regime of medical treatment. But alienists also worried that the reputation of their therapeutic institutions would be tarred by association with criminals. As Oosterhuis and Loughnan (2014) have written: "Delinquent inmates undermined the ambition of psychiatrists to promote their field as a medical specialty, to organize lunatic asylums, like hospitals in general, as therapeutic institutions and thereby to dispel the association between mental institutions and houses of correction, detention centers, and prisons." Such concerns illustrate just how contentious the historical overlap of general and forensic psychiatry could be.

In legal discourse and in most of the reformed criminal codes of the nineteenth century, madness had remained a largely ill-defined concept that, to be recognized, had to rely on the ordinary, untrained eyes of citizens and public officials. But with the establishment of asylums and development of mental science, alienists developed more differentiated notions of madness based on their daily experiences with patients. Of particular forensic relevance were forms of madness in which rational faculties had *not* been entirely eclipsed. For example, Philippe Pinel's *manie sans délire* manifested itself in instinctual, emotional, or volitional, but *not* in cognitive impairment. Jean-Étienne Esquirol's *monomania* represented a kind of partial insanity characterized by more or less normal and reasonable behavior on the one hand, but also by irresistible urges and a reduced capacity for self-control on the other. Similarly in England in the mid-1830s, James Cowles Prichard popularized the notion of "moral insanity" to describe a non-delusional mental illness that, while leaving its victims rational faculties intact, had eroded their moral and ethical sensibilities. Pritchard's work thus refocused forensic attention on moral and affective considerations. Over time, these clinical findings came to be reflected in more refined courtroom testimony about madness and, as Joel Eigen (1991) has found, served to normalize the crimes of the mad by drawing sanity and insanity closer together.

All of these descriptions compounded far wider concerns about the integrity of human consciousness. Anton Mesmer's wildly popular "animal magnetism" had already disconnected cognition from consciousness. Franz Josef Gall's phrenology had carved up consciousness into discrete mental faculties (intellect, emotion, will). And developments in physiological psychology seemed to indicate that the human nervous system had a mind of its own.

Countering these disturbing insults to cognition, and above all bringing madness to heel for legal purposes, the so-called McNaughtan Rules were devised in 1843 and soon widely adopted throughout the British Empire and the United States (Eigen 1995; Loughnan and Ward 2014). The rules strictly limited what counted as exculpatory madness based on whether or not defendants knew what they were doing, or knew it to be wrong. The rules reasserted the boundary between sanity and insanity, ensconcing consideration of cognitive defects, but precluded consideration of partial insanity, as well as emotional or volitional disorders. For traditional jurists, the rules became nearly sacrosanct and attempts to revise them were viewed as assaults on the very "foundations of individual morality" (Rosenberg 1968). As a result, the rules effectively cemented a yawning and evermore contested rift between legal and medical discourse that lasted well into the twentieth century. As psychiatry evolved and more subtle borderline mental conditions were diagnosed, pressure grew on courts and legislatures to be more flexible and to recognize that mental illness (especially in cases involving irresistible impulses) could sometimes merely diminish defendants' responsibility rather than altogether extinguish it.

It was not just impaired cognition or moral fortitude, but the vagaries of consciousness itself that became the object of intense forensic scrutiny (Eigen 2003; Hacking 1995). Even more so than in the UK, in France and Germany the most vibrant forensic debates about culpability and free will in the 1870s and 1880s involved cases of somnambulism, hypnotic suggestion, and double consciousness. Assumptions about the presumed criminal proclivities of epileptics also had experts plumbing the depths of unconscious, autonomous behavior.

Although alienists' descriptions of partial madness eventually fell out of fashion, there arose no shortage of alternatives to replace them. In fact, latter nineteenth century casuistic practices – and the individualizing consequences they spawned – brought with them an explosion of new terminologies and taxonomies, some of which seemed to identify specifically exculpatory mental disorders. Especially prolific in this regard was the German psychiatrist Richard von Krafft-Ebing, whose textbooks on forensic psychiatry and sexual pathology cataloged an enormous range of deviant sexual behaviors. By the dawn of the twentieth century, the forensic expert's diagnostic toolbox was bursting with many more and similar kinds of mental deviance which often hewed to specific types of crime (homosexuality, homicidal mania, pyromania, kleptomania, necrophilia, querulousness, exhibitionism). Among many other things, such medical categories colonized a void opened up by the waning ability of either religious piety or rational self-interest to enforce social mores.

In the opinion of many alienists, one common, distinguishing feature of these new disorders was that laymen and jurists often failed to spot them. Defendants in court could exhibit no symptoms of madness, yet be severely mentally ill; and conversely, defendants who seemed eccentric or mad might well be of sound mind. And so mid-nineteenth century forensic experts were increasingly called upon to assess whether court defendants were simulating mental illness. Cases involving infanticide and postpartum depression provided an especially prolific and protean space for such evaluations (Jackson 2002; Marland 2004). Epilepsy in its so-called "masked" forms, which at the time was closely associated with impulsive and violent criminal behavior, posed especially daunting forensic challenges. Jurists were prone to assume that defendants would often feign mental illness to escape trial or punishment. But alienists countered that clinical expertise and scientific knowledge were required in order to diagnose these afflictions and only medical experts could marshal the conclusive evidence needed to reach determinations about defendants' behavior and state of mind.

Likewise, prison medical officials tended to take a skeptical view of prisoners who exhibited psychiatric symptoms. More often than not, they eschewed claims that prison life (especially solitary confinement) was harmful and were instead inclined to believe that convicts simulated their illness in order to avoid prison labor or to prompt transfer to psychiatric hospitals, from whence they might more easily escape. Alienists, however, were less suspicious. They argued that jumping to the conclusion that prisoners were simulating was harmful because it prevented their removal from the prison regime and delayed medical intervention. They believed that simulation was very rare and that, given their own broad experience observing the insane in their asylums, they could easily expose it.

If spotting the signs of criminal lunacy in prisons and courtrooms was fraught with empirical pitfalls, locating them in the general population was hardly less daunting. In addition to the efforts of alienists and pathologists, mid-nineteenth century moral statisticians were also applying the new quantitative techniques developed by Adolphe Quetelet to identify otherwise hard-to-discern traits of madness and criminal behavior. These inquiries, which were frequently sponsored or mandated by state agencies, posited new arrays of causality extending across domains of time, place, ethnicity, and moral disposition, documenting seasonal and diurnal fluctuations, as well as the effects of urban lifestyle, racial pedigree, or bastardy lineage. Purported rises in the rates of crime and madness spurred efforts not just to identify and parse both populations, but also to study etiological factors responsible for both group's deviance. The delineation of such groups most often relied upon data drawn a posteriori from already-segregated institutionalized populations. And in constructing them, statisticians played a decisive role in mapping empirically measurable individual traits/defects onto perceptions about the risks they posed to larger (usually national, ethnic, or religious) bodies politic. By 1900, these efforts had resulted a fitful, but inexorable shift from nineteenth century deterministic etiologies to twentieth century risk assessments.

Degeneration Theory and Lombrosian Criminology

The medicalization of criminal deviance, new statistical methodologies, and eventually Darwinian doctrines of human evolution all became important catalysts that drove latter nineteenth century theories of degeneration. Those theories were first articulated in 1857 by the French alienist Bénédict-Augustin Morel with the publication of his *Treatise on Physical, Intellectual and Moral Degeneration*. Undergirded by deeply religious notions of the human fall from divine grace, Morel developed a theory of hereditary degeneration, whereby the accumulation of morbid influences and their transmission along family lineages could, over the course of generations, lead the irreversible mental decline, infertility, and death. Morel's ideas resonated especially strongly in France, where concerns about national decline multiplied after the country's capitulation in the war against Germany in 1871. Following the publication of Morel's work and against the backdrop of the rapidly spreading influence of Darwinian ideas, many advocates – most notably Henry Maudsley in the UK and Emil Kraepelin in Germany – untethered Morel's degeneration theory of its religious moorings and posited degeneration as a natural phenomenon. Alienists had always attributed mental illness to heredity, but for most of the nineteenth century this rarely factored into decisions about institutionalization. By the early twentieth century, however, degeneration theory had become a common trope throughout Europe, infusing debates in a wide range of cultural and scientific fields, from psychiatry and criminology to literature and the arts. Degeneration theory enjoyed such broad purchase precisely because it appeared to provide a "natural" explanation for many of the social and moral anxieties facing early twentieth century elites. In an era of rapid urbanization, restive working classes, and a rising mass society, degeneration theory offered plausible explanations of everything from alcoholism, crime, prostitution, and nervous exhaustion to epilepsy, homosexuality, and falling fertility rates. Advocates of degeneration theory at once not only articulated such "problems" but also promoted sociomedical strategies designed to resolve them.

Certainly the most controversial advocate of degeneration theory was the Italian military physician and psychiatrist Cesare Lombroso (Gibson 2002). Although Lombroso entertained remarkably eclectic views about the etiology of crime, he became most renowned for his notion of the born criminal. He and his students – generally known as the positivist school of Italian criminology – argued that, in many cases, biology was destiny and that the deviance of criminals was a reflection of their inherently defective nature. Lombroso believed that Beccaria's philosophical defense of free will was outdated and that instead biological factors played a decisive role in criminal behavior. He hoped that by quantifying physical and psychological signs of degeneration he could transform the discipline of criminology into an empirical, statistical science. His notion of a born criminal marked a far more general shift in the locus of juridical reasoning from the *crime* as such to the body and mind of the *criminal* perpetrating the crime. The rallying cry of Lombroso and many other reformists was individualization. They demanded that the Beccarian criminal codes's inflexible alignment of specific crimes with specific punishments be radically reformed. Instead of just acknowledging crimes and their attendant penalties, the codes needed also to account for the body and mind of the accused. Punishment – like medical therapy – needed to be tailored to the individual, and its purpose redirected toward averting the danger that individuals posed to society. To these ends, every conceivable aspect of criminals' lives, past and present, became targets of intense scrutiny and taxonomic ambitions. These totalizing ambitions, compounded by a certain voyeuristic, bourgeois, and unsettled fascination with the transgressive behaviors of the underclasses, drove the science of criminology. The tools that Lombroso's science applied to elicit and delineate his "born criminal," as well as other criminal types, were mostly clinical, including anthropometric measurements, family and case histories, and psychological tests.

Lombroso's hotly contested doctrines were posited amidst older traditions of state medicine, medical jurisprudence, and criminal psychology. Emerging out of seventeenth and eighteenth century absolutist doctrines, state medical police in continental Europe were granted broad powers that included jurisdiction over issues involving public health and welfare, as well as forensic affairs. The tighter weave between the state and medical science on the continent meant that forensic psychiatry – as both academic discipline and courtroom practice – became more highly developed than in the Anglo-American world, where the adversarial judicial system spawned a "marketplace professionalism" (Mohr 1997). But the expansive continental notion of medical police was being eroded in the nineteenth century, partly in response to the growth of public hygiene as a medical specialty, but also as a consequence of the rise of mental medicine and alienism. In fact, many of the early professional organizations established in the mid-nineteenth century were explicit manifestations of alienist claims not only to the medical expertise of identifying, diagnosing, and treating mental illness, nor simply to the administrative expertise of institutionally managing large criminally deviant populations, but also to the forensic expertise traditionally associated with the fields of medical jurisprudence and criminal psychology. As a derivative of their alienist work, this early forensic psychiatry had evolved in tandem with the mental asylum and reflected its medico-psychological regime of moral treatment. Lombroso's starkly deterministic and biological convictions were especially anathema to these older, more voluntaristic and psychologically inclined forensic experts. As a result, Lombroso's impact on judicial proceedings seems to have been more limited than historians have often assumed. Even in Italy, most jurists rejected Lombroso's views and remained loyal to the classical, Beccarian school of penal reform.

Even in less traditional circles, Lombroso's doctrines became the object of intense controversy (Gibson 2014; Nye 1984). An opposing and quickly preeminent sociological or environmental school of criminology, relying heavily on neo-Lamarckian ideas about the inheritance of acquired characteristics, arose around Alexandre Lacassagne and Gabriel Tarde in France, as well as among some of Lombroso's own Italian colleagues. In the UK, Lombroso's doctrines were considered extremist and even in Germany, where there was considerable sympathy for his biological approach, much of his work was rejected as unscientific. Nevertheless, Lombroso's ideas spread widely and informed public debate on crime in the decades before World War One. In Italy, his supporters remained influential well into the mid-twentieth century. And even in French forensic discourse, where the logic of "social defense" was never as powerful, an enduring undercurrent persisted and even grew in influence in the second half of the twentieth century (Protais 2014).

The widespread reception of Lombroso's arguments was at least in part a consequence of a double standard that accompanied growing public awareness about insanity. On the one hand, this involved ethical qualms about the deleterious effects of punishment on offenders who were mentally ill. On the other hand, it was manifested in heightened concerns about the danger the mentally ill posed to tightening modern norms of respectability and socioeconomic cohesion. In attempting to reconcile these two concerns, reformers advocated greater consideration of mitigating circumstances in sentencing, moderation of excessive forms of punishment, the use of indeterminate sentencing, and de-penalization in favor of medical assessment and treatment. But the ostensible flexibility and "leniency" of such proposals was coupled with more invasive strategies designed to deal prophylactically with dangerous individuals – be they recidivists or simply potential criminals – not on the basis of what they had done, but rather on the basis of who they were and what they could do in the future. The promise of forensic psychiatry was that its practitioners could do this work of "prophylactic assignation" (Rose 1996) more effectively and responsibly than other groups, be they prison officials, judges, or the police.

Lombrosian criminology embodied a much wider shift at the end of the nineteenth century away from religious and moral discourses toward mechanistic and biological explanations of criminal behavior. At the time, like many other branches of medicine, psychiatry too sought to model its reasoning and research practices on those of the nomothetic natural sciences. But as Daniel N. Robinson (1996) has argued, the pivotal legal concept of *mens rea*, concerned as it is with terms like "judgement," "intelligence," "rationality" (i.e., with purposeful intent), had been an outgrowth of legal and moral thought rather than scientific or philosophical debates. The specter of determinism radically challenged the traditional understandings of free will that undergirded the entire notion of culpability and the rule of law, threatening to "de-moralize" criminality (Wiener 1990). Hence, ever since the early nineteenth century, the scientific and phrenological "atomization of the mind" had "predictably disorganizing effects on juridical and moral thought" (Robinson 1996). Whereas early penal reformer had relied on punishment and rationally guided self-discipline to repair criminals moral failings, in a post-Darwinian, positivistic world, new scientifically informed methods of prophylactic and therapeutic intervention now appeared necessary to correct their natural failings. One of the many and far-reaching consequences of this enormous shift was that perpetrators of such morally fraught crimes as suicide or infanticide, to a lesser extent homicide, and more recently homosexuality, were more likely to be seen as non-culpable.

Psychiatric Expertise in Foro

The emergence of alienism and criminology as "young" professions inevitably challenged the authority and practices of much "older" medical jurisprudence. Jurists were especially concerned that long-standing legal concepts like free will, intention, motive, responsibility, guilt, and retribution – anchoring, as they sometimes did, entire legal philosophies of crime and punishment – would be undercut by advances in mental science. Such concerns were neither new nor unfounded. Famously, the eighteenth century French physician and philosopher Julien Offray de La Mettrie had called for the resignation of all judges so that physicians might take their place as the only experts competent enough to replace them and to distinguish guilt from innocence. And of course equally famously, Immanuel Kant had insisted that forensic questions fell not within the jurisdiction of jurists or physicians, but rather of metaphysicians because they alone were able to assess whether offenders were in possession of their faculties of understanding and judgment when choosing to commit a crime.

Cognizant of such disagreements, historians have frequently crafted their narratives of the emergence of forensic psychiatry in terms of boundary disputes between medicine and law. And nowhere were these disputes enacted more dramatically than in the courtroom encounters between psychiatrists and jurists (Engstrom 2014). Ostensibly, psychiatrists were admitted to court in order to help render a decision. As experts, they were usually called not as eyewitnesses to a crime, but as post-facto observers of defendants' behavior. Their participation in trials could be voluntary, but it could just as easily be in response to unwanted court subpoenas. Their testimony drew on specialized – especially clinical – knowledge acquired over the course of their careers diagnosing and treating other psychiatric patients in general practice, mental asylums, or prisons. In general, the rules governing expert witness testimony (when they would be called upon and by whom, the scope of their evidence, the weight of their judgments in the legal decision-making process, etc.) varied widely between legal jurisdictions. The common law, adversarial systems in Britain and North America tended to pit experts against one another and their effectiveness usually depended on them convincing a lay audience of jurors and on withstanding cross-examination by opposing counsel. In the inquisitorial systems of continental Europe and South America, proceedings were led by judges and it was largely at their discretion that expert testimony was called upon and then either accepted or rejected in reaching a verdict. In the early nineteenth century, alienists' appearances in court were relatively rare and expert opinions were more likely to be recruited from the ranks of general physicians and state medical authorities. Only gradually did recourse to psychiatric expertise finally become an established norm by the early twentieth century.

Certainly one of the most common complaints voiced by psychiatrists was that their expertise was insufficiently accounted for, simply ignored, or even disdainfully dismissed in the courtroom. Professional journals and private correspondence are littered with lamentations about both judges and juries either mistrusting or dismissing psychiatric expertise. Although changes in procedural law had sometimes opened opportunities for greater use of psychiatric experts, neither juries nor judges – identified legally as the ultimate "fact finders" and arbiters of truth – were compelled to accept the expert opinions voiced by psychiatrists. In most respects, the authority (or lack thereof) of psychiatric expertise was not so much grounded in code, as it was in psychiatrists' standing in the public mind's eye, their courtroom performance, and in the reliability of their diagnostic and prognostic assessments. And by the turn of the twentieth century, public skepticism of that authority was growing rapidly.

Some observers believed that psychiatrists had only themselves to blame for their low standing in the eyes of judges, juries, and the general public. For the courtroom situation – especially in high-profile jury trials – had often made psychiatrists painfully aware of the shortcomings in their own medical science. The imprecision of psychiatric diagnoses contrasted sharply with that of other medical specialties and the natural sciences. Experts were often led astray by the stark divide separating psychiatric nosology, with its multitude of borderline disorders, and criminal codes that recognized no such distinctions. All too often they were inclined to fit the law to their science, rather than their science to the law. Overarching, general arguments that relied too heavily on disease-specific claims tended to fail unless they could be embedded in a plausible crime narrative that resonated with juries' and judges' notions of volition and rational decision-making (Finder 2006). When pressed to comment on defendant's criminal responsibility, forensic experts who preferred to stick to the medical facts of a case could appear evasive in the eyes of the court and the public, seeming to abdicate their responsibilities as expert witnesses. Furthermore, the semi-public forum of the courtroom lent itself to the exposure of intra-professional rivalries that cast an unseemly light on psychiatric practice. And in a more general sense, the consistently low esteem in which psychiatrists were held compromised trust in their forensic judgments in spite of the explosion of interest in nervousness and psychological themes at the turn of the century.

Aside from a host of self-inflicted liabilities, psychiatrists also saw numerous other factors that undercut their authority in the courtroom. For example, juries tended to discount psychiatric science out of concern for public safety. For if, based on psychiatric expertise, a defendant was found to be not accountable, then the prospect of the case being dismissed and the defendant walking free loomed large. In many such cases, jurors ignored psychiatric findings, preferring to see the mentally ill defendant convicted rather than set free. In addition, psychiatrists believed that their authority in court was undermined by the fact that criminal codes, in the event of acquittal, often made no provision for detaining defendants based on their current state of mind. As a consequence of these failings, some psychiatrists called on courts to impose measures to protect public security by having the acquitted defendants committed to an asylum or by transferring the defendant from the jurisdiction of the criminal court to that of civil or other courts. However, jurists objected to such calls on the grounds that in dispensing justice the courts could not assume the duties of the police.

Judges on the other hand were not necessarily any more inclined than juries to accept psychiatric evaluations. They had long been frustrated by the incompetence of many court-appointed general physicians. At the same time, because psychiatric assessments inherently involved judgments about human motivation, intent, and conduct, they represented a deep intrusion into what had traditionally been considered to lie within the purview of judges. This made it that much easier, compared with other kinds of technical expertise, for judges to dismiss psychiatric evaluations altogether. Especially in continental, non-adversarial legal jurisdictions, where it was first and foremost the judge's responsibility to call for an expert opinion on the mental state of a defendant, widespread suspicions that defendants simulated mental illness in order to escape conviction could frequently result in psychiatric experts never being called to the court in the first place. Jurists were especially critical of psychiatrists' ignorance of legal process and showed a growing preference for not just psychiatric experts, but for trained forensic psychiatrists, who were able better to understand and navigate the legal and procedural issues of criminal trials.

Nevertheless, in many cases, there emerged a common realm of collaboration and shared perspectives. As Tal Golan (2004) has emphasized, science and the law could be "mutually supporting belief systems." In Italy, jurists and positivists could put their differences aside in the face of "dangerous girls" whose accusations of incest threatened to undermine the honor of Italian families (Guarnieri 1998). In France, a *modus vivendi* emerged around an acknowledgment that medical expertise and therapeutic approaches could also be understood as forms of punishment and as means of protecting society more so than rehabilitating the criminal (Harris 1989). Similarly in the UK, the end of the nineteenth century saw a "period of rapprochement" when the gap between legal and medical attitudes to insanity narrowed significantly (Loughnan and Ward 2014). Svein Skålevåg has shown that in Norway, dating from the establishment of a government commission of forensic medicine in 1900, relations between psychiatrists and jurists were characterized by broad consensus. In the case of Switzerland, Urs Germann (2004) has shown how daily courtroom routines fostered pragmatic working relationships between jurists and psychiatric experts and enhanced the acceptance of expert testimony by psychiatrists and of medical explanations for criminal behavior. Increasingly, officials were prepared to accept medical evidence as a basis for reprieving convicted criminals or, better yet, avoiding their conviction in the first place. Collaboration with forensic experts also helped jurists to "outsource" cases that, although often involving serious offenses, were deemed incompatible with existing standards of jurisprudence. Evidence for this enhanced cooperation can also be seen in the proliferation of forensic societies in many countries throughout Europe. Furthermore, the emergence of a new literary genre of special forensic textbooks and practical guides helped psychiatrists navigate the pitfalls of the courtroom environment and to hone their performative skills as expert witnesses in foro – as scientists purveying facts rather than advocates arguing a case.

Even if psychiatrists and jurists negotiated modes of cooperation, as the twentieth century progressed, other groups (neurologists, psychologists, sociologists, social workers, etc.) disputed psychiatrists' claims to forensic expertise. For example, after 1900, psychiatric expertise was increasingly coming to be challenged by forensic psychologists (Rose 1985; Wolffram 2018). Like psychiatrists, they too were striving to apply their science of the individual to resolve pressing problems of crime, delinquency, and the efficient administration of justice. Responding in part to the sharp public criticisms directed at psychiatry and the courts, criminal psychologists exploited their experimental methodology to measure the mental abilities of defendants and witnesses alike. Their use of psychometric and statistical techniques attracted the attention of contemporary jurists, and were even used in the seminars of the prominent German criminal law reformer Franz von Liszt in Berlin. As Annette Mülberger (2009) has argued: "By putting findings obtained with the help of these research techniques in relation to problems in law, psychologists offered legal experts and judges a new way of arguing juridical matters."

And so the impact of psychiatric expertise in the courtroom remained ambiguous. Sometimes it influenced court decisions, at other times it was rejected outright, and on still other occasions it simply reinforced common sense assessments that had already tipped the judicial scales. The link between legal statute and courtroom decisions was hardly simple and direct. In fact, the outcomes frequently depended on local variables. For all the rhetoric at work parsing human behavior along voluntaristic/idealistic and deterministic/materialistic lines, in courtroom practice neither jurists nor psychiatrists were generally of one voice on such fundamental questions and, in fact, often spoke in mixed tongues. When it came down to the nitty-gritty and always intractable question of aligning mental disorders with criminal acts, inter- as well as intra-professional consensus was precarious. As Robert Nye (1984) has rightly argued, experts were not necessarily as transfixed by their own disciplinary logic as historians have usually assumed, but were instead also "sensitive to trends in public and political opinion."

Some historians have argued that forensic work was a boon to the development the psychiatric profession. They have claimed, for example, that forensic work helped sharpen psychiatric diagnosis or that the notoriety of public trials helped psychiatrists enhance their standing in the eyes of government and other professional groups or even spur demand for their services. And indeed, the acceptance of forensic expertise was often attributed to their role in the growth of knowledge about mental illness, which in turn was generously interpreted as a sign of the progress of science, and indeed of civilization itself. But there was also a downside to this work. Within the psychiatric profession itself, routine forensic work enjoyed little esteem and much of it was delegated to junior doctors as a burdensome chore. Forensic experts were easily identified with their doubly stigmatized patients. Furthermore, the drama of public trials could easily expose not just the fallibility of their diagnoses and theories about insanity, but also collegial discord. By the end of the nineteenth century, when salacious newspaper reports about courtroom proceedings became widespread, the sight of expert witnesses at loggerheads with each other engendered little confidence in their testimony and even less in their science. And so if nothing else, forensic psychiatry was a double-edged sword when it came to cultivating psychiatry's professional reputation and it was not uncommon to encounter ambivalent attitudes about both the opportunities and dangers of forensic testimony in foro.

This helps to underscore the fact that, more so than other branches of psychiatry, the development of forensic psychiatry has been inextricably entangled in public discourse. It has evolved in a symbiotic relationship to the *cause célèbres* and the advocacy movements that those cases often spawned. It has been on the heel of scandals and the visceral emotions they evoked – as much as of technological or conceptual innovations – that change has come, not just in the adjudication and diagnosis of criminal lunatics, but also in the formation of public attitudes about them and the crafting of public policy and legal statutes to deal with them. Above all, public discourse has been framed by new forms of media and literary genres, be it newspaper court reporting on the sometimes lurid details of a case, or often semi-autobiographic novels exploring the inner complexities of the human psyche and the traumas of institutionalization. Over time psychiatrists learned to appreciate and even exploit these new media more effectively for their own ends. But whether they liked it or not, expert forensic testimony in famous trials inevitably became the "true public face" (Gibson 2014) of psychiatry.

The Psychiatrization of Danger: Prophylactic Assessments

The early decades of the twentieth century witnessed a fundamental shift in the orientation of forensic psychiatry. The remit of forensic experts came to be defined much more broadly, expanding to include not just courtroom testimony, but also broader societal responsibilities of risk management. As much as the pessimistic implications of degeneration theory seemed to have dimmed psychiatrists' therapeutic hopes, it radically spurred their socio-hygienic, prophylactic interventions. New policy agendas emphasized mental or social hygiene, and "social defense" against deviant or harmful behavior. Much of the debate about criminality was moving past the politics of moral reform and being recast in terms of social hygiene and eugenic improvement. To the frequent, but varying extent that forensic experts contributed to these agendas, their conduct served as a model for political action. They laid claim to a role for their science in the administration of entire populations in the interests of national well-being and public security, although too often at the expense of patients' individual rights, privacy, and dignity.

The aim of these new forensic regimes was the management of dangerous individuals. Whereas for much of the nineteenth century experts had deliberated back and forth about the proper institutional receptacles for the *criminel-aliéné* versus the *aliéné-criminel*, increasingly after the turn of the century what took precedence was simply the management of the social dangerousness of both. And this demanded more effective institutional and administrative means. It is important to recognize that such risk management strategies were enabled in good part by a further disaggregation of institutionalized populations (Engstrom 2003; McCallum 2004). Psychiatric hospitals had increasingly sought to separate chronic patients from other acute, and often violent ones, building observation wards, cottage systems, and outpatient polyclinical facilities to better manage and treat their populations. Prisons too were devoting more resources to specialized wards for psychiatric inmates and juvenile detention facilities. And educational systems were introducing special classes and schools to accommodate feeble-minded children, striving to keep their charges off the streets and out of prisons or psychiatric hospitals.

What emerged from this "psychiatrization of danger" (Foucault 1978) was not just more refined divisions *within* psychiatric taxonomy, but also a refinement of its *outer* surfaces with otherwise "normal," i.e., non-pathological behavior. Much of the twentieth century was spent mapping this boundary-space and colonizing it with so-called borderline conditions. It was these liminal disorders that, in part, helped spur calls for the judicial recognition of reduced accountability and shifts away from punishment and towards treatment.

These trends manifested the fact that, in the early decades of the twentieth century, much of the debate about criminal deviance was moving past the politics of moral reform and being recast and expanded in terms of the management of dangerous individuals, social hygiene, and eugenic improvement. At the vanguard of these interventions (and to some degree the product of them) stood, in turn, juvenile delinquents. It was in this population of deviants – perceived to be at greatest risk, the least corrupted, and most amenable to rehabilitative interventions – that forensic experts envisioned their work's greatest potential.

Feeble-Minded Children and Incorrigible Youth

The expanding reach of nineteenth century child welfare and protection laws had helped spur greater judicial leniency around 1900, shifting emphasis away from punishment and toward prevention and various nonpunitive forms of resocialization. The day-to-day work of special courts for juvenile offenders served to raise a more holistic awareness of the environmental and familial vectors that influenced crime and delinquency and widened the scope of suspended or conditional sentencing, parole, and other non-carceral forms of rehabilitation. The key was to avoid prison sentences, albeit at the cost of widespread segregation in specialized institutions. Instead of trying defendants in court or sending them to jail, *diversionary strategies* sought to channel them into other more or less coercive therapeutic or educational facilities, for example by processing them through civil courts rather than criminal ones. For example in the UK, the Mental Deficiency Act (1913) allowed judges to commit convicted offenders to asylums or to guardianship. Furthermore, the creation of youth detention centers (Borstals) in the UK, Ireland, and several Commonwealth countries were examples of these enhanced efforts at social reintegration of corrigible offenders, diverting them away from punishment and toward education and treatment. Similar aims were also pursued by youth courts in Germany after their establishment in 1908, although there the influence of forensic psychiatry remained limited and well circumscribed within legal frameworks.

Emblematic of these developments was the juvenile court system established in Chicago in 1899. The court, which was mimicked across the country and overseas, represented an attempt to hew punishment more closely to defendants' mental conditions as well as to their social and familial milieus. The court and its chief justice, Harry Olson, entertained close contacts with the Eugenics Record Office at Cold Spring Harbor and established a psychopathic laboratory under the direction of William Hickson, not just to examine juvenile defendants forensically, but to study and develop "treatments" for them. Michael Willrich (1998) has described this new court-based regime of social governance as "eugenic jurisprudence." This involved a "synergy of state building and social theorizing" and the "aggressive mobilization of law and legal institutions in pursuit of eugenic goals" and the "socialization" of criminal law. The courts were "incorporating the therapeutic disciplinary techniques of psychiatry, medicine and social work into everyday judicial practice" (Willrich 2003).

Helping to facilitate these aims were new diagnostic techniques, most notably the measurement of intelligence. The association of crime with subnormal intelligence had long been a mainstay of criminological discourse. In many quarters, feeble-mindedness came to be associated with criminality, delinquency, and immorality. And because feeble-mindedness was usually viewed as a congenital disorder – rather than a temporary or episodic one like mental illness – it later became an especially fertile target for eugenic intervention. The Binet-Simon intelligence test helped quantitatively to undergirded claims *in foro* about feeble-mindedness, and became a useful tool for putting eugenic ideas into practical action. Henry H. Goddard, who was arguably one of the most prominent eugenicists in the United States, associated feeble-mindedness with a proclivity toward criminal behavior (males) and prostitution (females), especially in juveniles. Coining the term "moron," Goddard adapted Binet's techniques and paired them with the study of family lineages to recommend that many young delinquents, among others, be either sterilized or segregated from society. The legal right of states to sterilize the feebleminded was confirmed by the US Supreme Court in the infamous case of *Buck v. Bell* in 1927 (Cohen 2016). Thanks to the work of Goddard and others, the United States became a pioneer in the practices of eugenic sterilization and inspired eugenic initiatives in other countries as well. Although some 32 states implemented forced sterilization programs, there never evolved the kind of national program like the one implemented in the 1930s in Germany, where eugenic courts and medical experts collaborated to see countless thousands institutionalized, sterilized, and worse.

Above and beyond the promotion of scientific or racist agendas, pressures for more sweeping eugenic measures were also augmented by other more mundane forces. Rapid population growth, economic dislocation, and inadequate mental health funding contributed to a ballooning of institutionalized populations and spawned demands to reduce hospital expenses. In California, for example, involuntary sterilization was used to justify patient discharge and thus alleviate overcrowding (Braslow 1997). This reminds us that eugenically inspired involuntary sterilization, although attaining its radicalized and racist apogee in Germany in the 1930s, elsewhere not only predated and outlasted the Nazi regime, but could also be put to very different uses. Especially in the United States, Canada, and Scandinavia, sterilizations continued into the 1970s, although they were increasingly justified on social grounds rather than eugenic ones. In other words, political expediency and economic rationale, and not just psychiatric science or outright racism, helped to drive these policies.

Psychopaths

Calls for enhanced juridical flexibility vis-à-vis juveniles were echoed in other cases more closely associated with social defense doctrines, especially those involving so-called psychopaths (Walker 1968/1973). The roots of psychopathy stretched back into nineteenth century notions of degeneracy and pathological deviance. By the 1920s, psychopathy was increasingly being defined not as a disease, but as a constitutional personality disorder describing, among other things, unpredictable behavior, a lack of empathy and remorse, chronic delinquency and recidivism, and a resistance to traditional forms of psychiatric intervention. In other words, psychopathy represented a mental condition short of insanity that advocates insisted needed to be taken into account in legal proceedings. Although today many criminologists have all but abandoned the concept as scientifically and clinically useless – little more than "moral prejudice masked as diagnosis" (Schneider 2007) – it proliferated widely in mid-twentieth century scientific and public discourse, becoming synonymous will all manner of incorrigibly antisocial behavior. As Greg Eghigian (2015) has found, psychopathy became an "emblematic object for a paradigmatic form of twentieth-century prognostic governance, one stressing risk aversion, surveillance, and preventive intervention."

Perhaps nowhere was psychopathy's uptake so dramatic as in cases involving sexual violence. Especially in the United States, something approaching a sex crime panic swept across the country in the late 1940s, resulting in numerous sexual psychopath laws by the early 1960s. These laws targeted sexually violent male behavior as opposed to earlier and broader eugenic measures designed to prevent reproduction among "inferior stock." In part, the laws have been interpreted as blowback from World War II, which had heightened concerns about male violence, authoritarian and narcissistic personality types, and irrational impulses (Jones 2016). Concerns about sex offenders were echoed in other countries as well. Between 1939 and 1968 in Holland, some 400 sex-offender were castrated in pursuit of the ambiguous ends of curbing the sexual libido of offenders and preventing their ability to produce "inferior" progeny. The putatively "voluntary" treatment was never enacted into criminal or civil law, but instead governed by "informal protocol" (van der Meer 2014). In Britain, guidelines for the treatment of sexual offenders were published by prison authorities in 1949. And in 1959, psychopathy became enshrined in the Mental Health Act, where it was conceived to evolve from social conditions rather than psychoanalytically informed developmental problems. As Janet Weston (2017) has shown, forensic experts in England who advanced psychological explanations for sexual crimes (including prostitution and homosexuality) exerted surprisingly little influence over judicial procedures and sentencing, thanks partly to jury trials, but also to the resistance put up by defendants themselves.

Psychopathy's usefulness was in good part a measure of its own amorphous plasticity. This facilitated its swift appropriation beyond of psychiatry in areas such as criminology, psychology, public health, and the social sciences. The concept proved especially useful spurring sharper forms of intervention and legislative measures against recalcitrant and recidivist offenders. But as a *psychiatric* diagnosis, psychopathy ultimately "failed" because it was considered less a medical condition than a personality disorder. In postwar Scandinavia, doctors considered it to be "merely" a sign of social maladjustment (Bergenheim 2014; Parhi and Pietikäinen 2017). Coming to lack adequate somatic credentials, psychiatrists increasingly disputed that psychopathy held any exculpatory forensic significance. In the second half of the twentieth century, psychopathy was subsumed by the new-found nosologic category of antisocial personality disorder (ASPD).

Clinical Assessment Tools

Reliably identifying clinical signs of psychopathy was hardly less problematic than it had been for nineteenth century experts diagnosing moral insanity. The episodic quality of some mental disorders and the ever present specter of simulation made forensic diagnosis especially challenging. Over time, clinicians therefore developed a number of tools designed to enhance their clinical gaze. The limitations of early efforts in this direction, such as "projective" personality tests, like the Rorschach and Szondi tests, soon spawned calls for more reliable, structured techniques. Spurred on partly by efforts to rationalize the recruitment of soldiers in the twentieth century's World Wars, these methods soon escaped their clinical settings and were put to use in the service of more expansive biopolitical and normative ends. The Wechsler Adult Intelligence Scale (WAIS) was developed after World War I and began, from the 1930s, to displace the use of the Binet's methodology and to compensate for its biases. The Minnesota Multiphasic Personality Inventory (MMPI) was developed in the 1930s and 1940s to delineate personality types, probe the honesty of patient' responses, expose recalcitrant attitudes, and monitor treatment efficacy.

Such clinical technologies were *also* applied in forensic contexts and were soon complemented by more specific tools designed to address psycho-legal issues, especially the assessment of future behavior. To elicit signs otherwise invisible to standard clinical procedures, the psychiatrist Hervey Cleckley developed criteria for the diagnosis of psychopathy. And later the Canadian psychologist Robert D. Hare developed a Psychopathy Checklist (PCL) which became a popular research and clinical tool used to assess risk in penal populations. By the end of the century, competency and threat assessment tools and their acronyms (PCL, CST, CAST-MR, JSAT, HCR-20) had proliferated widely.

As clinical research tools, these techniques (checklists, questionnaires, standardized psychological and psychometric tests, structured interviews, rating scales, etc.) preformed clinical evidence, produced actionable data, and elicited otherwise invisible "symptoms" of dangerousness. In the process, they effectively served to delineate populations of adjudicatable and/or potentially dangerous individuals, thus underpinning the credibility of forensic science. In day-to-day clinical practice, these psychometric tools facilitated more rapid assessments and therapeutic interventions. They provided answers to pressing psycho-legal questions by helping explain past and current behavior and by predicting the parameters of future behavior.

But just how extensively these technologies were applied, remains to be studied. They sometimes proved difficult to apply across different forensic cultures and jurisdictions. For example, in Germany, where somatic traditions remained stronger than sociological or psychological ones, the evaluation of risk continued to rely on unstructured clinical assessments, whereas the use of structured guidelines remained uncommon (Müller-Isberner et al. 2000). Nevertheless, what these standardized assessment technologies *also* achieved was to help fuel an ensuing anti-psychiatric backlash of the 1960s and 1970s. As Shilling and Casper have shown, the psychometric testing regimes designed to plumb the depths of human personality came to be perceived by the public as coercive and a threat to individuals' privacy (Shilling and Casper 2015).

Preventive Detention in National Contexts

The mid-twentieth century witnessed a remarkable expansion of judges' abilities to impose alternative forms of sanction, including preventive detention on fully or partially irresponsible defendants who were considered to pose a threat to society. Holland's so-called Psychopath Laws (1928) introduced special restriction orders (*Terbeschikkingstelling*) that allowed convicted criminals – if they posed a threat to public order – to be placed involuntarily in psychiatric facilities, released on parole, or placed under psychiatric surveillance. Depending chiefly on psychiatric assessments, these orders could be renewed at regular intervals. In Italy, the new Rocco Penal Code of 1930 implemented administrative security measures for individuals thought to pose a threat to society. The code gave judges new powers to declare defendants who had been acquitted on grounds of insanity as "socially dangerous" and to have them institutionalized in an asylum for criminal lunatics for up to ten years. Belgium's influential Social Defense Act of 1930 and Germany's Law against Dangerous Habitual Criminals and on Rehabilitative and Preventive Measures (1933) introduced similar measures in cases of diminished responsibility, as did the Swiss Criminal code of 1937 and the UK's Homicide Act of 1957. Spain's

Dangerous and Social Rehabilitation Law (1970) was similarly based fundamentally on the principle of social defense, while at the same time – at least nominally – aiming to rehabilitate potentially dangerous individuals.

Initially in many jurisdictions, courts themselves had supervised preventive detention. But increasingly, jurisdiction over mentally ill acquittees was transferred to quasi-judicial/administrative bodies (security review boards, committees, tribunals, or penal courts) charged with reviewing continued detention primarily from the perspective of public safety. The verdicts of these bodies have been described as "insanity sentences" (Bloom et al. 2000) – analogous to and sometimes coupled with the criminal sentences – that were often enough perceived to put individuals in a kind of court-sanctioned double jeopardy. The usually mixed, transdisciplinary composition of these bodies (jurists, psychologists, psychiatrists, penal officials, laymen) formalized psychiatric participation, at once both ensconcing and potentially also diluting its input into posttrial decision-making processes.

These forensic regimes entailed a diversification of means involving the expansion and diffusion of prophylactic regulations that targeted social deviance. They reflects the ongoing and fluctuating tension between notions of punishment and social defense on the one hand, and efforts to treat and resocialize individuals on the other. However, the use of preventive detention was by no means always or everywhere lauded or even implemented. An adequate assessment of the historical significance of these extremely diverse forensic regimes is impossible without taking into account more specific national contexts. Their implications for everyone involved could differ markedly if they were embedded in liberal democratic, fascist, authoritarian, or socialist nation states. Several examples can help illustrate how national legislation and political circumstances could modulate these regimes.

France never enacted social defense legislation because of government concerns about cost and resistance among psychiatric practitioners (Protais 2014). It would be a mistake, however, to assume that practitioners were unable to avail themselves of alternative solutions. For example, although not duplicated elsewhere in France, a psychiatric wing was established in the local prison in Lyon after World War II, where psychiatrists drew heavily on psychoanalytic and phenomenological insights. A reform of the Penal Procedural Code in 1958 furthermore explicitly enjoined forensic experts to assess their examinee's dangerousness and prospects of rehabilitation. Similarly, during the 1950s and 1960s, forensic psychiatrists increasingly came to assess individuals presenting behavioral disorders as being legally culpable. And from the 1980s onward, the same assessments were even made of psychopathic and borderline personality disorders. This trend toward attributing greater responsibility to borderline patients was fed by psychotherapists working to reform asylum culture: for them, engaging patients on the assumption that they could take more responsibility for their actions, was a gateway to therapeutic treatment and cure. Furthermore, ever since measures to open up psychiatric hospitals began to be implemented in the 1960s, treating dangerous criminal patients in prisons came to be seen as a vital way of protecting the gains of institutional reform. The ensuing rise of victimology and the women's liberation movement in the 1970s only further spurred this trend away from attributions of irresponsibility.

Japan witnessed repeated legislative efforts to implement preventive detention, sometimes within the prison system, other times within the mental health system. And yet ultimately, these efforts failed (Nakatani 2000, 2006). The country's long tradition of domestic custody (*shitaku-kanchi*) was fundamentally transformed only after World War II when – unlike trends elsewhere – the country embarked on a major expansion of psychiatric institutions, which were used to house mostly involuntary patients. Furthermore, the relatively belated development of criminology in Japan, well after Lombroso's heyday, limited the influence of social defense arguments. The first attempt to implement preventive detention had been undertaken in the 1940s and was modeled along the lines of German *Sicherungsverwahrung*. A subsequent attempt was made in the 1960s after the stabbing of the American Ambassador Edwin O. Reischauer. But these efforts were thwarted, partly due to resistance among traditional jurists, but partly also due to psychiatry's critics and human rights advocates.

In Italy, there existed a long-standing legal and institutional tradition of forensic psychiatric hospitals (dubbed OPGs or *Ospedali Psichiatrici Giudiziari* in the 1970s) dating back to the last quarter of the nineteenth century which admitted defendants found to be socially dangerous (Gibson 2014; de Vito 2014). In the wake of criminological debates, there emerged a "double track system" (*sistema del doppio binario*) whereby responsible defendants were given retributive sentences and dangerous ones subjected to security measures. The Rocco Code (1930) then cemented this stark dichotomy, radically separating punitive and therapeutic measures. As a result, the forensic hospitals evolved within the criminal justice system, where they prioritized custody and public safety rather than therapy, thus perpetuated the carceral logic of nineteenth century institutions. These hospitals remained largely untouched by reforms in the 1970s and have only recently been closed and replaced with a residential system of forensic psychiatric care.

In the first half of the twentieth century in Argentina, an oligarchic regime collaborated with a medicolegal "state-within-a-state" dominated by positivist social reformers (Salvatore 2006). As in Europe, "dangerousness" and "social defense" became the watchwords of these biopolitical technocrats, who managed to entrenched themselves inside state institutions of social control, including the criminal justice system. What they conspicuously failed to achieve, however, were substantive revisions to the penal codes or judicial procedures. Instead, they had be satisfied with piecemeal changes and the exploitation of traditional legal instruments like pardons, suspended sentences, and conditional release. Within these niches evolved the "efficacious counterpoints" that helped manage social deviance and defuse social conflict, thereby ameliorating the otherwise repressive and exclusionary policies of the regime and helping stabilize its legitimacy.

In non-liberal, Eastern bloc countries, forensic psychiatry mostly remained a matter of courtroom expertise, while trends toward specific treatment and rehabilitation of criminals either stagnated or were retrograde before the final decades of the twentieth century. Dating back to the Zarist era in Russia, psychiatrists had unsuccessfully sought to assert that their expertise could better handle the dangerously insane than the police could, only themselves to be repeatedly rebuffed and their science debased by autocratic officials (Healey 2014). But after the Bolshevik revolution, new Soviet institutions enhanced the status of forensic psychiatry, professionalizing it and placing it on par with other medical specialties. New Soviet criminal and procedural codes in the 1920s ensured that police and the courts sought out forensic expertise to evaluate criminal suspects. These new responsibilities led to a blossoming in criminological research aspiring to help forge the archetypal New Soviet Man and Woman. By the mid-1930s, however, the Serbsky Institute had monopolized control over compulsory treatment, as forensic expertise was increasingly exploited and pressed into service inside Gulag labor camps, or simply eclipsed under Stalin's reign of terror by a secret police network of special psychiatric hospitals. In the post-Stalin era, forensic psychiatry enjoyed greater professional autonomy and recognition, and yet still continued to be compromised by political exploitation and used to help pathologize political dissidents as "socially dangerous." In Russia, forensic experts were "almost never asked to assess dangerousness and risk prediction" and there evolved neither specific regulations governing probation or parole, nor follow-up aftercare facilities (Ruchkin 2000).

On account of Germany's fraught twentieth century history, historians have long emphasized the importance of anthropological tropes in German criminology before 1933. And yet, German forensic experts had quickly debunked Lombroso's theories, and as late as the early 1930s, the country's preeminent criminologist, Gustav Aschaffenburg, maintained that criminal anthropology had achieved virtually nothing. Even the most tangible application of forensic expertise in Bavarian prisons in the 1920s was only nominally "biological," providing scientific cover for conservative and moralistic policies of the state's Catholic elites (Liang 2006). But the racist biopolitical agendas of the National Socialists quickly unlocked new opportunities to conduct research and implement policy. Newly erected Hereditary Health Courts had tens of thousands of mentally ill sterilized, although the sterilization of criminals was relatively rare (Wetzell 2006). The 1933 Law Against Dangerous Habitual Criminals established a dual track system of punitive and preventive measures that reformers had been demanding for decades. While during the Nazi era biopolitical security and prophylactic priorities dominated, the postwar era witnessed an expansion of the rehabilitative and "correctional imagination" (Eghigian 2015). In East Germany, forensic psychiatry was influenced by policies of ideological reeducation and the promotion of a "socialist personality." And the reform of the criminal code in 1968 eliminated court-ordered preventive detention altogether, which effectively refocused experts' priorities on forensic assessments and undercut nascent therapeutic efforts. In West Germany, preventive detention remained possible, with revisions to the criminal code in the 1970s allowing courts to enforce treatment if patients' "social prognosis" was favorable. After German unification in 1989, East Germany adopted the West German legal system.

Judicial Gatekeeping

In taking stock of these different national perspectives on preventive detention, it is important to recall the distinction drawn at the outset of this chapter between courtroom testimony on the one hand, and the "darker underbelly" of less visible administrative decision-making practices. In their study of forensic psychiatry in the UK, Loughnan and Ward (2014) found that while the influence of forensic psychiatry over post-conviction sentencing and treatment had grown, it had been curtailed in pre-conviction assessments of criminal responsibility in contested trials. This conclusion reminds us of the important gatekeeping function that courtroom proceedings have exerted over psychiatric expertise. One needn't go as far as Joel Eigen (2003) to insist that the "history of forensic psychiatry is the history of law" in order to recognize that the impact of shifting courtroom dynamics is crucial to any interpretation of forensic psychiatry's influence, or lack thereof.

Twentieth century developments in the United States help to underscore this point in two respects (Robinson 1996). First, in 1954, the US Court of Appeals for the District of Columbia Circuit articulated the so-called Durham Rule which expanded the scope of the insanity defense by allowing juries to find defendants not guilty by reason of insanity if they suffered from mental disease. The rule effectively acknowledged developments in the medical and behavioral sciences and thus enhanced the forensic influence of psychiatric expertise. Specifically, it recognized psychopathy and personality disorders as mental diseases, thereby overturning courts' long-standing reliance on the narrower McNaughten standard which hinged on defendants' ability to distinguish right from wrong. But in 1972, the same court overturned its rule, arguing that it gave too much power to psychiatrists and replacing it with the widely accepted guidelines set out by the American Law Institute (ALI) in its Model Penal Code published in 1962. And in the wake of the John Hinckley trial in 1982, the Insanity Defense Reform Act (1984) again restricted the scope of expert forensic testimony, leaving it to the jury to decide whether a defendant was insane or not and limiting expert testimony to assessments of defendants' mental abilities.

Second, not just the content of psychiatric expertise but also its quality came under judicial scrutiny and resulted in a shift in standards for the admission of forensic evidence (Golan 2004). For much of the twentieth century, judges in the United States had allowed evidence from expert witnesses only if it relied on generally established scientific practice. But by the end of the century, the erosion of confidence in scientific expertise contributed to judges assuming a more proactive role in assessing scientific methods. A new, so-called Daubert Standard (1993) raised the bar on judges gatekeeping responsibility when it came to the admissibility of evidence and weeding out "junk science." This evolution was especially consequential in a field as protean as psychiatry, where for much of the twentieth century generally accepted standards and methodologies could be particularly hard to come by. Among the consequences of these changes has been heightened scrutiny of the scientific integrity of tests used in forming expert opinions and calls for those tests to be subject to "validation studies" (Hamlin 2019).

Developments such as these lend broader credence to Loughnan's and Ward's (2014) conclusion that psychiatric claims to speak with scientific authority about criminal responsibility have been in a "state of perpetual negotiation" and psychiatric expertise at best "emergent and contingent."

Conclusion: Toward Therapies of Self-Governance and Patients' Rights

After the Second World War, the expansion of the welfare state and growing optimism about the potential to improve human behavior led forensic psychiatrists to place greater emphasis on the rehabilitation and treatment of mentally ill criminals. Increasingly these efforts sought to address not only outward behavior, but also to target the inner self and to affect changes in psychological attitudes and motivations that would allow individuals to function more effectively in the face of the challenges of life in modern societies. In place of early twentieth century interventions driven by anxieties about the moral decay and loss of community, priority now shifted toward greater responsibility for self-development and motivation. The aim of these efforts involved not just – or necessarily even primarily – the alleviation of suffering and injustice, but also the inculcation of certain norms of civic responsibility and self-discipline.

The key sites of these interventions have been, more often than not, the protean forensic regimes of juridical and administrative governance. North America and Europe have witnessed a proliferation of special, treatment-oriented courts (for substance abuse cases, juveniles, families, habitual criminals, domestic violence, etc.) that combine psychological and medical treatments with juridical compulsion in hopes of improving adjudicatory efficiency. The therapeutic rationale of these courts is manifest in more hybrid sentencing options (from involuntary hospital treatment and indeterminate sentencing or deferral to review boards, to conditional release, community supervision, or exoneration), sometimes stipulating that judges apply the "least restrictive" means necessary. Even more so than courtroom proceedings, administrative practices have proven to be especially amenable to psychiatric influence and have witnessed, in the words of Nikolas Rose (1996), an "infusion of heterogeneous 'psy'-judgements." Forensic responsibilities have evolved from crafting expert opinions and courtroom performances, toward treatment, prophylactic assessments, and resocialization. Psychiatrists' new role has become "less that of curing illness than of administering pathological individuals across an archipelago of specialist institutions" and instead engaging in the "prophylactic and preventative work of maximizing mental health." Forensic practitioners have come to operate across increasingly diverse domains of psychiatric intervention, for example civil commitment and competency hearings, evaluations in prisons, parole boards, police profiling, liasioning with general mental health programs and community aftercare services. This bleeding of forensic practices into broader extramural and extrajudicial domains has recently been described in terms of "forensic rule" and "forensic regimes" (Hamlin).

The evolution of these complex forensic regimes begs interpretation in the context of more general developments in the history of psychiatry and criminal justice systems. The emergence of new pharmacological treatments in the 1950s and the subsequent displacement of psychoanalysis by cognitive-behavioral therapies at the end of the twentieth century have radically transformed the therapeutic arsenal of psychiatrists. Likewise, their diagnostic practices have become evermore aligned with standardized psychiatric nosological systems like the American Psychiatric Association's Diagnostic and Statistical Manual of Mental Disorders (DSM) and the World Health Organization's International Classification of Diseases (ICD). Above all, the effects of psychiatric deinstitutionalization have markedly transformed the environment in which forensic experts operate. To cite only the most extreme example of the United States, headlong deinstitutionalization combined with the dragnet of retributive, "get tough," anti-crime policies (nourished by wider "lock 'em up" and "three strikes you're out" sentiments) have contributed to ever more individuals being "processed" through the criminal justice system rather than treated in mental health facilities, turning jails and prisons into the country's foremost sites of psychiatric intervention (Fuller Torrey et al. 2014).

The impact of these and other historical developments on post World War II forensic regimes and therapies still awaits sustained analysis on the part of historians of the human sciences. But it would be a mistake simply to assume that the effects of these new regimes and therapies have been strictly pejorative and hegemonic. If nothing else, these trends have involved a substantial displacement and repositioning of forensic expertise. Indeed, they have even led to calls for the development of legal criteria for insanity *independent* of psychiatric expertise. Concluding his study of the history of the insanity defense, Daniel Robinson (1996) has suggested that in clearcut cases of madness, where experts are on their most solid ground, their advice isn't really needed in the first place, and for far less obvious borderline cases it remains unclear whether it is better to rely on the often arcane and inaccessible evidence presented by psychiatric experts, or on nonspecialist evidence that is "empathically" accessible to juries and the general public. Whereas traditionally many interventions related to the deprivation of liberty required psychiatric assessments (insanity defense, civil commitment laws, guardianship, and competency provisions), recently the need for such assessments has been questioned and standards for doing without them have been developed (Slobogin 2015).

Furthermore, the emergence of the legal concept of informed consent after World War II, and more recently the rising disability rights movement since the 1970s, have helped to animate debate about patient rights, choice, and participation. As Felicity Callard et al. (2012) have noted, by the end of the twentieth century, an ongoing transition was underway from "an emphasis on the protection of society from people with mental illness towards the protection of the rights of people with mental illness within the framework of 'patient-oriented medicine'." This trend has been manifested in new national laws which have regulated involuntary treatment, and which, taken together, represent a watershed in the evolution of the legal rights of psychiatric patients. To cite only a few examples, Italy's so-called Basaglia Law (1978) established norms guaranteeing patients' dignity as well as their civil and political rights. The Danish Mental Health Act (1989) stipulated that psychiatric measures be minimally restrictive and based as much as possible (or in the case of psychosurgery absolutely) on patient consent, while strengthening patients' rights to object to treatment and to access information and counseling. The Polish Mental Health Act (1994) addressed the need to protect patient's civil rights and ensure informed consent. Following on such national legislation, the United Nations

Convention on the Rights of Persons with Disabilities (CRPD) (2006) stipulated that mental impairment not be grounds for denying patients' legal capacity and advocated a shift from the more traditional and paternalistic paradigm of patients' "best interests" toward a new paradigm stressing instead their "will and preference" (Callard et al. 2012).

Forensic science has been described as a "makeshift of plausibilities and poorly understood probabilities, a fabric of contingencies more than a well-ordered decision tree" (Hamlin 2019). If we accept this interpretation, we can't but expect that exploring the history of forensic psychiatry will continue to yield especially thorny entanglements. The phenomenon of madness has remained enduringly elusive and it has not been for want of trying that forensic experts have failed to fully "capture," disentangle, and master it. Hardly less elusive is the historical relationship between law and psychiatry. Neither discipline has been in a position to fully "occupy" or dictate its terms to the other; indeed, the two are incommensurable (Smith 1981). And historians of all stripes can be thankful for this incommensurability! For it has generated extraordinarily rich and prolific exchanges, opening historical windows onto evolving practices and attitudes about mental deviance far beyond the narrower disciplinary confines of forensic psychiatry.

Cross-References

▶ Asylums and Alienists: The Institutional Foundations of Psychiatry, 1760–1914
▶ The Madness of Women: Myth and Experience

References

Bartlett P (1999) The poor law of lunacy: the administration of pauper lunatics in mid-nineteenth-century England. Leicester University Press, London
Bergenheim Å (2014) Sexual assault, irresistible impulses, and forensic psychiatry in Sweden. Int J Law Psychiatry 37:99–108
Bloom JD, Williams WH, Bigelow DA (2000) The forensic psychiatric system in the United States. Int J Law Psychiatry 23:605–613
Blumenthal SL (2016) Law and the modern mind: consciousness and responsibility in American legal culture. Harvard University Press, Cambridge
Braslow J (1997) Mental ills and bodily cures: psychiatric treatment in the first half of the twentieth century. University of California Press, Berkeley
Callard F, Sartorius N, Arboleda-Flórenz J, Bartlett P, Helmchen H, Stuart H, Taborda J, Thornicroft G (eds) (2012) Mental illness, discrimination and the law: fighting for social justice. Wiley-Blackwell, West Sussex
Cohen A (2016) Imbeciles: the Supreme Court, American eugenics, and the sterilization of Carrie Buck. Penguin, New York
de Vito CG (2014) Forensic psychiatric units in Italy from the 1960s to the present. Int J Law Psychiatry 37:127–134
Eghigian G (2015) Unruly menace: a history of psychopathy in Germany. Isis 106:283–309
Eigen JP (1991) Delusion in the courtroom: the role of partial insanity in early forensic testimony. Med Hist 35:25–49

Eigen JP (1995) Witnessing insanity: madness and mad-doctors in the English court. Yale University Press, New Haven

Eigen JP (2003) Unconscious crime: mental absence and criminal responsibility in Victorian London. Johns Hopkins University Press, Baltimore

Eigen JP (2016) Mad doctors in the dock: defending the diagnosis, 1760–1913. Johns Hopkins University Press, Baltimore

Engstrom EJ (2003) Clinical psychiatry in imperial Germany: a history of psychiatric practice. Cornell University Press, Ithaca

Engstrom EJ (2014) Topographies of forensic practice in Imperial Germany. Int J Law Psychiatry 37:63–70

Finder GN (2006) Criminals and their analysts: psychoanalytic criminology in Weimar Germany and the first Austrian Republic. In: Becker P, Wetzell RF (eds) Criminals and their scientists: the history of criminology in international perspective. Cambridge University Press, New York, pp 447–469

Foucault M (1978) About the concept of the 'dangerous individual' in 19th-century legal psychiatry. Int J Law Psychiatry 1:1–18

Fuller Torrey E, Zdanowicz MT, Kennard AD, Lamb HR, Eslinger DF, Biasotti MC, Fuller DA (2014) The treatment of persons with mental illness in prisons and jails: a state survey. Treatment Advocacy Center. https://www.treatmentadvocacycenter.org/storage/documents/treatment-behind-bars/treatment-behind-bars.pdf. Accessed 30 Sep 2021

Germann U (2004) Psychiatrie und Strafjustiz. Entstehung, Praxis und Ausdifferenzierung der forensischen Psychiatrie am Beispiel der deutsch-sprachigen Schweiz, 1850–1950. Chronos, Zurich

Gibson M (2002) Born to crime: Cesare Lombroso and the origins of biological criminology. Praeger, Westport

Gibson M (2014) Forensic psychiatry and the birth of the criminal insane asylum in modern Italy. Int J Law Psychiatry 37:117–126

Golan T (2004) Laws of men and laws of nature: the history of scientific expert testimony in England and America. Harvard University Press, Cambridge

Goldstein JE (1987) Console and classify: the French psychiatric profession in the nineteenth century. Cambridge University Press, Cambridge

Goldstein JE (1998) Professional knowledge and professional self-interest: the rise and fall of monomania in 19th century France. Int J Law Psychiatry 21:385–396

Guarnieri P (1998) 'Dangerous girls,' family secrets, and incest law in Italy, 1861–1930. Int J Law Psychiatry 21:369–383

Hacking I (1995) Rewriting the soul: multiple personality and the sciences of memory. Princeton University Press, Princeton

Hamlin C (2019) Forensic facts, the guts of rights. In: Burney I, Hamlin C (eds) Global forensic cultures: making fact and justice in the modern era. Johns Hopkins University Press, Baltimore, pp 1–33

Harris R (1989) Murders and madness: medicine, law, and society in the fin de siècle. Clarendon, Oxford

Healey D (2014) Russian and soviet forensic psychiatry: troubled and troubling. Int J Law Psychiatry 37:71–81

Jackson M (ed) (2002) Infanticide: historical perspectives on child murder and concealment, 1550–2000. Ashgate, Aldershot

Jones DW (2016) Disordered personalities and crime: an analysis of the history of moral insanity. Routledge, Abingdon

Kelly B (2014) Custody, care & criminality: forensic psychiatry and law in 19th century Ireland. The History Press Ireland, Dublin

Liang O (2006) The biology of morality: criminal biology in Bavaria, 1924–1933. In: Becker P, Wetzell RF (eds) Criminals and their scientists: the history of criminology in international perspective. Cambridge University Press, New York, pp 425–446

Loughnan A, Ward T (2014) Emergent authority and expert knowledge: psychiatry and criminal responsibility in the UK. Int J Law Psychiatry 37:25–36

Marland H (2004) Dangerous motherhood: insanity and childbirth in Victorian Britain. Palgrave Macmillan, New York

McCallum D (2004) Personality and dangerousness: genealogies of antisocial personality disorder. Cambridge University Press, Cambridge

Micale M, Porter R (eds) (1994) Discovering the history of psychiatry. Oxford University Press, New York

Mohr JC (1997) The origins of forensic psychiatry in the United States and the great nineteenth-century crisis over the adjudication of wills. J Am Acad Psychiatry Law 25:273–284

Moran JE (2014) Mental disorder and criminality in Canada. Int J Law Psychiatry 37:109–116

Moran JE (2019) Madness on trial: a transatlantic history of English civil law and lunacy. Manchester University Press, Manchester

Mülberger A (2009) Teaching psychology to jurists: initiatives and reactions prior to World War I. Hist Psychol 12:60–86

Müller-Isberner R, Freese R, Jöckel D, Cabeza SG (2000) Forensic psychiatric assessment and treatment in Germany: legal framework, recent developments, and current practice. Int J Law Psychiatry 23:467–480

Nakatani Y (2000) Psychiatry and the law in Japan: history and current topics. Int J Law Psychiatry 23:589–604

Nakatani Y (2006) The birth of criminology in modern Japan. In: Becker P, Wetzell RF (eds) Criminals and their scientists: the history of criminology in international perspective. Cambridge University Press, New York, pp 281–298

Nye RA (1984) Crime, madness, and politics in modern France: the medical concept of national decline. Princeton University Press, Princeton

O'Brian P (1982) The promise of punishment: prisons in nineteenth century France. Princeton University Press, Princeton

Oosterhuis H, Loughnan A (2014) Madness and crime: historical perspectives on forensic psychiatry. Int J Law Psychiatry 37:1–16

Parhi K, Pietikäinen P (2017) Socializing the anti-social: psychopathy, psychiatry and social engineering in Finland, 1945–1968. Soc Hist Med 30:637–660

Porter R, Wright D (eds) (2003) The confinement of the insane: international perspectives, 1800–1965. Cambridge University Press, Cambridge

Protais C (2014) Psychiatric care or social defense: the origins of a controversy over the responsibility of the mentally ill in French forensic psychiatry. Int J Law Psychiatry 37:17–24

Robinson DN (1996) Wild Beasts & Idle Humours: the insanity defense from antiquity to the present. Harvard University Press, Cambridge

Rollin HR (1996) Forensic psychiatry in England: a retrospective. In: Freeman H, Berrios GE (eds) 150 years of British psychiatry – Volume II: The aftermath. Athlone, London, pp 243–267

Rose N (1985) The psychological complex: psychology, politics and Society in England, 1869–1939. Routledge, London

Rose N (1996) Psychiatry as a political science: advanced liberalism and the administration of risk. Hist Hum Sci 9:1–23

Rosenberg C (1968) The trial of the assassin Guiteau. University of Chicago Press, Chicago

Ruchkin VV (2000) The forensic psychiatric system of Russia. Int J Law Psychiatry 23:555–565

Salvatore RD (2006) Positivist criminology and state formation in modern Argentina, 1890–1940. In: Becker P, Wetzell RF (eds) Criminals and their scientists: the history of criminology in international perspective. Cambridge University Press, New York, pp 253–279

Schneider HJ (2007) Theorien der Kriminologie (Kriminalitätsursachen). In: Schneider HJ (ed) Internationales Handbuch der Kriminologie. De Gruyter, Berlin, pp 125–181

Shilling R, Casper S (2015) Of psychometric means: Starke R. Hathaway and the popularization of the Minnesota multiphasic personality inventory. Sci Context 28:77–98

Slobogin C (2015) Eliminating mental disability as a legal criterion in deprivation of liberty cases: the impact of the convention on the rights of persons with disabilities on the insanity defense, civil commitment, and competency law. Int J Law Psychiatry 40:36–42

Smith R (1981) Trial by medicine: insanity and responsibility in Victorian trials. Edinburgh University Press, Edinburgh

Smith R (1991) Legal frameworks of psychiatry. In: Berrios GE, Freeman H (eds) 150 years of British psychiatry, 1841–1991. Gaskell, London, pp 137–151 van der Meer T (2014) Voluntary and therapeutic castration of sex offenders in the Netherlands (1938–1968). Int J Law Psychiatry 37:50–62

Walker N (1968/73) Crime and insanity in England. 2 vols. Edinburgh University Press, Edinburgh

Watson KD (2011) Forensic medicine in Western society: a history. Routledge, Oxon

Weston J (2017) Medicine, the penal system, and sexual crimes in Engand, 1919–1960s. Bloomsbury, London

Wetzell RF (2006) Criminology in Weimar and Nazi Germany. In: Becker P, Wetzell RF (eds) Criminals and their scientists: the history of criminology in international perspective. Cambridge University Press, New York, pp 401–423

Wiener MJ (1990) Reconstructing the criminal: culture, law, and policy in England, 1830–1914. Cambridge University Press, Cambridge

Willrich M (1998) The two percent solution: eugenic jurisprudence and the socialization of American Law, 1900–1930. Law History Rev 16:63–111

Willrich M (2003) City of courts: socializing justice in progressive era Chicago. Cambridge University Press, Cambridge

Wolffram H (2018) Forensic psychology in Germany: witnessing crime, 1880–1939. Palgrave Macmillan, Cham

Psychiatry and Society

50

Petteri Pietikäinen

Contents

Introduction	1306
Part I: Mental Asylum as a Small Society	1308
Psychiatry in the Age of the Asylum	1308
"The Plea for the Silent": Patients in Mental Asylums	1310
Authority and Power in the Asylum	1313
Social Class and Therapy	1314
The Legacy of Asylums	1316
Part II: Marginality: Race and Politics	1318
Mental Health in Segregated Societies: The Case of Racial and Colonial Psychiatry	1319
Mental Health in a Segregated Society: The Case of the USA	1320
Protest Psychosis and Cultural Paranoia	1324
Alienation, Race, and Mental Illness	1326
Conclusion	1327
References	1329

Abstract

This chapter explores two themes that illustrate the close relations between psychiatry and society. The first one is the mental hospital as a small society, as an organized community in which madness was contained. This theme is examined in order to reveal how this principal institution of mental health care functioned as a mini-society that was separate from, and yet in close interaction with, the larger social organization. The main function of mental asylums was twofold: first, to protect society from the disruptions caused by the mentally ill and second, to help mental patients adjust to the norms and rules of larger society as well as to the norms and rules of the mental hospital. The second theme, "race" and marginality, draws on the experience of African Americans to explore the

---

P. Pietikäinen (✉)
History of Sciences and Ideas, University of Oulu, Oulu, Finland
e-mail: petteri.pietikainen@oulu.fi

© The Author(s), under exclusive licence to Springer Nature Singapore Pte Ltd. 2022
D. McCallum (ed.), *The Palgrave Handbook of the History of Human Sciences*,
https://doi.org/10.1007/978-981-16-7255-2_97

ways in which minorities have been treated and discussed by mainstream psychiatry in the United States. This theme delves into the role of psychiatric expertise enforcing social norms on the one hand, and conversely into psychiatry as a reflexive science that adapts itself to the changing norms and rules of society on the other. Psychiatric terms and theories have been incorporated into social and political ideas and practices, while racialist thinking has influenced the ways in which nonwhite minorities were treated as patients, clinical material, and subjects of medical experiments. Intrinsic to the discussion of both themes are the issues of social class, inequality, and marginality as exclusion.

Keywords

Social organization · Mental hospital as a small society · Race and racialist thinking in psychiatry · Marginality as exclusion

Introduction

Since psychiatry is the most social of all medical sciences, it is natural to speak of psychiatry and society in tandem. Psychiatry is social, because "madness" has been shaped by social forces and institutions, and madness itself has shaped societies and their cultures by eliciting from them institutional, legal, and other reactions. Mental illness has also impacted on the ways in which we understand ourselves, our fellow humans, and our social environment. For this reason, to study mental illness historically is to study the intellectual and socio-cultural changes that constitute varieties of human experience. In other words, the explanation for much of our behavior – both normal and abnormal – is linked to social contexts. We live within institutions and social structures that organize our lives. Our social environment includes parents, other relatives, friends, neighbors, colleagues, authority figures, the Church, the mass media, not to mention economic, legal, educational, or political institutions – as the poet John Donne put it in 1624, "no man is an island entire of itself."

Psychiatry enters the story of madness quite late. It was only during the nineteenth century that the explicitly medical idea of mental illness became firmly established in Western Europe and North America. This process was closely tied to the professionalization of mental medicine. Psychiatry was one medical specialty that emerged in tandem with the growth of the state and its various forms of governance. For centuries, "madness" had been a colloquial term signifying mental derangement that was traditionally discussed, explained, and treated by both medical men and "laymen," such as folk healers, village elders, and churchmen. From the late eighteenth century onward, madness was gradually replaced by "mental illness," a term denoting a medical or psychomedical disorder within the purview of medical experts called "psychiatrists" or "alienists" (for general histories of mental illness and psychiatry see Shorter 1997; Scull 2015; Pietikainen 2015).

This shift from madness to mental illness amounted to a massive medicalization of madness, which brought with it new explanations of insanity as well as new methods of treatment. As a new group of medical specialists, psychiatrists started to name and classify mental disorders, provide medical and moral treatment in mental hospitals, and explain the nature and causes of mental illness. During the nineteenth century, the picture of madness evolved from being heterogeneous and protean to rather more homogeneous and stable. By 1900, psychiatrists had more or less attained a monopoly in the management of madness. For this reason alone, they had become important agents of social power. Insanity was an unwelcome disturbance in the body politic, and psychiatrists were expected to make interventions that would minimize the social damage caused by people who had "lost their minds," thereby becoming totally or partly dysfunctional members of the polity (Castel 1988).

Social scientists have often seen mental illness as a form of deviance, "deviance" denoting a more or less consistent habit of breaking social norms. In this view, conforming to social norms and adjusting to the social environment are used to demarcate between mental illness and health. Sociologists are interested in social order and *disorder*, exemplified in criminality, rebelliousness, sectarian religiosity, sexual promiscuity, and madness. According to the moderate sociological view, social structures and processes influence the expression and representation of mental illness, whereas an extreme "sociologism" claims that mental illness is *produced* by social structures and processes. Because the mentally ill have caused social disorganization by deviating too much from the accepted rules, they have been stigmatized – their identity is "spoiled," as the sociologist Erving Goffman (1963) put it in his classic study. For Goffman, the negative attitude of community members and prejudices about deviancy paved the way for the increasingly common practice whereby oddly behaving individuals were seen as mentally deranged.

Historical explorations of the social and cultural factors of deviance are often illustrative and important. For example, for a white person, publicly to declare their atheism or antiracism was certainly a form of deviance in the southern United States in the early days of the civil rights movement. Yet, for the most part, people have not been pronounced mentally ill simply because their behavior has been deviant or unacceptable. There must have been some sort of radical change in the personality and behavior of individuals before they were regarded as mentally disordered. Moreover, it seems quite obvious that the onset of severe mental illnesses, such as schizophrenia or bipolar disorder (formerly known as manic depression), involves biological components. Mental illness is not a social construction that could be eradicated by simply changing the structures, values, and norms of society. Behind the historically and culturally varying signs and symptoms of madness, there is biology, the reality of nature. Right from the beginning, psychiatrists have recognized the importance of heredity in the onset of mental illness (Porter 2018).

What makes the study of mental illness a challenging endeavor is its tight entanglement with both biology and society, nature and culture. By looking closely at the social context of mental illness, one gains essential knowledge of, and insight into, the intricate ways in which the illness emerges and manifests itself, how the environment reacts to it, and, finally, how these reactions in turn effect the fate of the individual who is now considered "mad," "crazy," "insane," or – in today's parlance

– "mentally disordered." Some scholars have argued that the very purpose of psychiatry is to carry out the political goals of the government, including the control and standardization of its citizens (Foucault 2006; Castel 1988; Castel et al. 1982; Scull 1989). In the case of nondemocratic, authoritarian countries, such as the former Soviet Union (1917–1991) and today's China, this is undoubtedly true. The question of the political role of psychiatry in liberal-democratic countries is more complicated, but what is indisputable is that modern societies have employed policies and technologies that aim to combat behavior deemed antisocial and harmful to the social order. As a result, psychiatrists (and psychologists) have functioned as "architects of adjustment" (Napoli 1981): they have explained and managed serious mental disorders (psychoses), minor mental troubles (such as neuroses and what is now called depression), as well as social deviancy, like psychopathy, juvenile delinquency, labor protests, substance abuse, homosexuality, and even political protest. As one American psychiatrist put it in 1965, the "psychiatrist must truly be a political personage in the best sense of the word. He must play a role in controlling the environment which man has created" (quoted in Castel et al. 1982: 75).

This chapter explores two themes that illustrate the close relations between psychiatry and society. The first one is the mental hospital as a small society, as an organized community in which madness was contained. This theme is examined in order to reveal how this key institution of mental health care functioned as a mini-society that was separate from, and yet in close interaction with, the larger social organization. The second theme, "race" and marginality, focuses on the ways in which a specific minority, African Americans, was treated and discussed by mainstream psychiatry in the United States. This theme examines the role of psychiatric expertise in enforcing social norms on the one hand, and the status of psychiatry as a reflexive science that adapts itself to the changing norms and rules of society, on the other. Intrinsic to the discussion of both themes are the issues of social class, inequality, and marginality as exclusion.

Part I: Mental Asylum as a Small Society

From the early nineteenth century onward, mental illness became institutionalized in two fundamental ways. First, it was increasingly managed and explained within the institutional confines of the mental hospital. Second, a medical specialty called psychiatry or alienism was developed to manage and explain insanity. As a result, mental hospitals became small semi-autonomous societies that functioned as large Therapeutic Machines aiming to bring order to minds in disarray.

Psychiatry in the Age of the Asylum

Psychiatry first became an independent medical specialty in France (Goldstein 1987). The French medical community was at the forefront of medical science, and France was a pioneer in the development of medical science policy. In fact, France was the first nation in modern Europe to build a strong, centralized state administered by a corps of trained civil servants. Psychiatry in France, as in all of Europe, was tightly linked with the establishment of mental hospitals, or "asylums," as they were typically called in the nineteenth and early twentieth century. As its name suggests, it was designed to be a "refuge from the maddening world" (Gamwell and Tomes 1995: 9). Later, when institutional care devolved into custodial warehousing, the "insane asylum" became negatively connoted, prompting medical authorities to introduce less stigmatizing terms, such as "sanatorium" or "mental hospital."

In the early nineteenth century, when the French psychiatrist Philippe Pinel was developing a psychiatry based on rational, Enlightenment principles and moral treatment, we can see how he and other pioneers were inspired by the vision of the Therapeutic Society. In this better world of the future, the mentally ill would be taken care of, and may be even cured, by the new scientists and physicians of the soul. The idea and practice of early nineteenth-century moral treatment tried to implement this therapeutic vision using a new type of mental asylum where the "insane" would regain their senses and be restored to a form of citizenship required in a society bending toward political freedom and democratization. Some scholars have seen the mental asylum in early nineteenth-century France as a laboratory of democracy, where alienists tried to save the glimmer of sanity left in mental patients in order to restore their ability as social beings and autonomous persons (Gauchet and Swain 1999).

As in France (and Britain), the more systematic and publicly funded institutionalization of the mentally ill began in Germany during the first half of the nineteenth century (Kaufmann 1997). Enlightened German state bureaucrats were keen to turn traditional madhouses into modern mental asylums governed by both medical and administrative principles. Although German states only unified in 1871, the founding and functioning of German asylums was nevertheless linked to the consolidation of state power. This entailed extensive administrative reforms in local government, the penal system, education, religion, and medicine. To some extent, the German asylum system developed as a consequence of the secularization of Church properties after 1803. Monasteries, castles, and other ecclesiastic institutions were transformed into public mental hospitals. They were designed to be therapeutic establishments rather than purely custodial institutions that functioned as human warehouses, workhouses, or penitentiaries (Schott and Tölle 2006: 236–239). However, by mid-century, the early therapeutic optimism had dwindled away and alienists were gripped by a sense of disappointment and dissatisfaction.

In contrast to asylums, university hospitals were preoccupied with the scientific analysis of "clinical material" – that is, patients. Instead of focusing on patients' social milieu, as asylums had, they paid more attention to diagnostic techniques, classification, and laboratory research. Clinical psychiatrists and neurologists analyzed the patient's body and mind in hopes of finding keys to solve the puzzle of mental illness. Especially in Germany, clinics were top-notch medical institutions serving scientific, pedagogic, and educational purposes and providing medical students with practical opportunities to observe the mentally ill (Engstrom 2004). This project of turning alienists into medical scientists required a systematic and rigorous attempt at naming and classifying disease entities. To achieve this goal, psychiatrists began to look at the life of their patients through the narrow window of diagnosis. By the early twentieth century, psychiatric explanations of insanity in Germany were governed by the demands of diagnostic classification.

"The Plea for the Silent": Patients in Mental Asylums

The Age of the Asylum was launched with great therapeutic expectations in the early decades of the nineteenth century. Unfortunately, after 100 years, the optimism was replaced by doubt and pessimism. Not that there was no considerable progress in medicine during this era. On the contrary, society enjoyed better hygiene, better surgical techniques, and a better understanding of the role of cells and germs. The problem was that similar progress eluded mental medicine. As a matter of fact, it appeared to decline. Even if most psychiatrists were reluctant to admit it, even to themselves, by the early twentieth century, it was apparent that the asylum system was metaphorically in a straitjacket, as were some of its patients (Schott and Tölle 2006: 298). In short, the asylum system did not live up to its early expectations.

The influential French philosopher and historian Michel Foucault has called the asylum system a "gigantic moral imprisonment" (Foucault 2006: 511). Following Foucault's verdict, critical historians of the 1960s and 1970s tended to view mental hospitals as institutions of human misery, antitherapeutic black holes that sucked the patients inside and turned them into human wrecks. Asylums had soft cells, hard attendants, and staggering patients whose identity was stolen and whose minds were confused by the merciless machinery of the "system." We have read and heard about abuses in the "loony bin," "bughouse," "nuthouse," "booby hatch" (etc.), including mental and physical violence, mechanical and biochemical restraints, poor diet, dirt, noise, negligence, and an overall ambience of brutality and inhumanity. As the recent scholarship has demonstrated, to portray the asylum as some sort of snake pit is a gross overstatement, but in the public imagination asylums are still easily seen as colossal human warehouses, the dreadfulness of which stems from institutional insanity rather than from the mental derangement of patients.

In the early era of the asylum, the rules and practices of confinement and release were often ambiguous. There were rumors and stories about innocent people being incarcerated and certified as insane without a valid medical reason. It is true that families could place their relatives in private hospitals without anyone interfering. Occasionally the confined family member took the matter to court by pleading *habeas corpus*, a common law principle that requires a person under arrest to be brought before a judge or into court. But it could be exceedingly difficult to prove sanity from inside an asylum. If one tried too hard and passionately to convince the staff that "I'm sane, can't you see!" one's behavior could easily be regarded as the delusion of a disorganized mind. It was a tricky situation for the patient. Instead of trying to convince the staff with words, it was better to maintain composure and consistently behave in an organized and rational manner, a criterion which even "normal" people in the outside world found difficult to fulfill. Another option was to have close relatives or a spouse remove you from the asylum. This happened quite frequently to patients who had relatives who were willing to take responsibility for their care. Sometimes patients were more or less abandoned by their families, or they did not even have a family to care for them. If these patients did not get better within a reasonable amount of time, approximately one year, they sometimes became chronic cases and remained incarcerated for years or decades (Pietikainen 2015: 138).

Fortunately, and increasingly so the closer we get to our own time, there are published and nonpublished first-person accounts of what it was like to be an institutionalized mental patient. Judging by these patient narratives, life in an asylum was often neither very comfortable nor therapeutic. To many patients, confinement was a frightening experience. This was especially true of the more severely ill patients who were taken to the asylum against their will, or who were too confused or frightened to understand where they were, and why. It seems that, upon admission, many patients experienced at least a temporary worsening of their symptoms; they became depressed, anxious, fearful, and nonresponsive. This is how one British patient experienced her confinement in the 1950s:

> More weeks went by and I regained my strength, but I still could not remember who I was, nor how I came to be there, locked up in this vast strange building; sometimes sleeping in a long dormitory, sometimes in a small side room off a long corridor, at others in a padded cell on the floor. Always the doors were locked. One could tell when nurses passed by the jingle of keys at their waists. Never had I felt so frightened, lost, lonely or desolate. (*The Plea for the Silent*, 71)

Upon admission, many patients were uncertain about their whereabouts – Where am I? Why am I here? Who are these strange people who give me orders and push me around? Why doesn't anybody listen to me? As the late historian Roy Porter noted,

> hallucinations persecuting the sick person often came true in the asylum, full of strange, disorderly people, terrifying authority figures, and, not least [...] manacles and chains which brought to mind the torture engines so prominent in the imaginations of the demented. (Porter 1991: 198–199)

Such a picture is not the whole truth. In the patient records of Finnish mental hospitals, for example, there are letters from former patients who express gratitude for the treatment they received – "treatment" usually meaning rest and recuperation as well material security: the mostly indigent patients were given food, clothing, and shelter, which often meant a great deal to poor and abused mental patients (Pietikäinen 2020). Another example: in 1878, the American woman Anna Agnew was certified lunatic in Indiana and committed to an asylum. This followed repeated attempts to commit suicide and an attempt to kill one of her children. When the asylum engulfed her, she experienced a strong sense of relief. Her madness was acknowledged. Her family was safe. As she later wrote, "my unhappy condition of mind was understood, and I was treated accordingly." In this case, she received humane treatment (King 2002). Thus, not all patients experienced confinement negatively. At least some, such as the suicide-prone Anna Agnew, welcomed being institutionalized. The alternatives at least within the context of nineteenth-century society were hardly attractive. They included confinement in a welfare or penal institution, or wandering aimlessly in the community and relying on others for the necessities required for survival (Grob 1994: 102).

Recent studies have refuted the popular assumption that nineteenth-century asylums functioned primarily as dumping grounds or warehouses for misfits and undesirables. Obviously, families and local communities used confinement as a convenient way to get rid of difficult or unwanted individuals, but this was not the rule. Evidently, there was no evil intention or malicious power-mongering involved in the creation of the asylum system. The fate of institutionalized patients might very well have been sad and bleak, but in recent decades historians have devised a multidimensional picture of asylums and asylum life. Within this picture, Anna Agnew is almost as good an example of the asylum as is an anonymous chronic who languished for years or even decades on the back wards of mental institutions and unlearned all the social skills needed in the outside world. To understand the functional logic of the asylums, it is useful to see them as institutions that were designed to maximize normalcy and to maintain the therapeutic and patriarchal power that was legitimized by the asylum doctors' medical control of the patient's body and mind.

From the patients' point of view, a major problem with asylums was the lack of public monitoring. Governments sometimes failed to pass legislation to regulate and control mental hospitals, and even if such laws existed, there were often not enough dutiful administrators who systematically monitored whether hospitals actually followed the official regulations. Within asylums, supervision was also lax. Hospital boards were reluctant to interfere in the activities of the superintendent and his medical staff, and medical staff often neglected to supervise nurses and attendants. The lack of efficient supervision and control gave way to all sorts of violations. Violence and arbitrary discipline were often concealed and hushed up rather than brought out into the open. Officials were predisposed to protect the reputation of the hospital at all costs rather than to enter into public discussion of problems, such as the abuse of patients. Therefore, social control and consistent surveillance should not be considered a purely negative exercise of power. Deep inside the wards of asylums, the problem was often a lack of surveillance – at least from the point of view of powerless patients (Pietikainen 2015: 150).

There was one fundamental problem with mental asylums that remained more or less unresolvable: for the most part, patients did not feel comfortable there. And because they felt uncomfortable, there was a risk of their mental and physical condition deteriorating. In short, the longer one stayed in an asylum, the greater the chances of becoming institutionalized and being forever incapable of coping with the demands of everyday life. Quite often, patients were able to adjust to the social organization of the asylum, while their social adjustment to the society at large failed.

Authority and Power in the Asylum

The social structure of the mental asylum resembled the prevailing patriarchal European family. The superintendent was the venerated patriarch, whose relationship with the patients was very much like that between father and child. Together with his assistant physicians, nurses, and attendants, the patriarch-doctor offered protection, support, security, and consolation to his patients. Reciprocally, he expected his patients to submit to his authority and to the rules he had devised for the benefit of the whole hospital community. In the subordinate role, the patient was provided with the basic comfort of home, albeit without freedom, idleness, or luxury. If the patients defied the authority of the patriarch, it was his duty to punish them, for example, by constraining and confining unruly patients with the help of isolation rooms, straitjackets, and other restraints or by reducing their food or tobacco rations. Recalcitrant patients were treated very much like disobedient children, and their defiance became a symptom of their illness. A strictly observed hierarchy prevailed in the asylum (Pietikainen 2015: 137).

A guiding principle in the life of the asylum was regimentation. Through a mixture of instructions, guidance, and (if required) discipline, patients were coerced into adjusting to the miniature society that was the mental hospital. Life in the wards was regulated, repetitious, and uniform, and it was governed by rigid administrative and normative rules. One day in the asylum was very much like any other – changes and surprises were in principle unwelcome because they might have disturbed the order of the hospital, which in turned might have further disordered patients' minds. The law and order approach was justified by the assumption that the patients' paths to recovery required becoming conscious of the need to be in *control* of themselves. The main therapeutic idea was to strengthen the inmates' self-control and to reward responsible conduct. To achieve these goals, it was essential that life on the wards be structured and regulated. The conduct of the staff and the patients followed moral and religious principles. Through rigid structure and the administration of rewards and punishment, the medical management of madness resembled a carrot and stick approach. The ultimate goal of the psychiatric cure was to (re)integrate the recovered patient into the order of society (Engstrom 2004: 19–21).

By the early twentieth century, the therapeutic goal began to be formulated in terms of social adjustment, particularly in the USA, the most rapidly modernizing western nation. In contrast, in the more peripheral European countries such as Finland, mental illness was long understood as a social and moral problem rather than something medical. Those considered crazy, mad, or insane, lived in their local communities, and were taken care of by their family, community, and the local authorities – usually the more affluent farmers and the clergy. In developing countries that were almost wholly dependent on primary production, the social value and material security of individuals correlated very strongly with their ability to make a living and take care of themselves. Obviously, the mentally ill lacked this highly valued capacity for self-sufficiency and independent living (Pietikainen 2019: 20).

Governmental health policies played a leading role in the development of mental asylums in most European countries. The emergence of the notion of *public health* contributed to the construction of national mental health care systems that relied on large-scale institutions and the "mass-production" of mental health services. Without doubt, around the turn of the twentieth century, new asylums were established on a monthly if not weekly basis in different parts of Europe. Still, this "great confinement" – as Michel Foucault called it – did not necessarily mean that governments were eager to incarcerate and silence maladjusted "misfits." As a matter of fact, psychiatrists complained, and not without reason, that public authorities were inefficient in organizing mental health care. Administrators needed financial resources for the implementation of health care policy, but politicians were very often reluctant to "waste" taxpayers' money on the medical management of madness, because the mentally ill were considered to be inferior to the somatically ill. In many cases, the latter group could be restored to health and become productive citizens again. By contrast, the mentally ill were typically regarded as unproductive because they were incapable of working reliably or of doing anything "useful." As mental patients were perceived as debits on the social balance sheet, policy makers were not particularly sympathetic to their plight. They were more inclined to invest in general hospitals and in the treatment of the physically ill. Still, authorities and leading physicians had to face the fact that all the beds in mental hospitals were filled as soon as they became vacant, and new hospitals had to be built to meet growing demands for institutional care. After all, there were no viable alternatives – the "insane" could not be left to their own devices, for then they would become a burden to poor relief, the prison administration, and local communities (Pietikainen 2015: 152).

Seen through a political and societal lens, mental illness was associated with poverty and deprivation, and for this reason, it was a stigma that primarily tainted disadvantaged classes. In public mental hospitals, for example, class distinctions also affected the relationship between doctors and patients (Pietikainen and Kragh 2019). A great majority of asylum doctors came from the middle- or upper-middle classes, while the majority of patients were small landowners, urban and rural workers, servants, and craftsmen. This large socioeconomic and cultural gap between psychiatrists and patients made communication between them problematic: psychiatrists were inclined to equate the low level of education, and the "primitive" verbal utterances and behavior of patients with "feeblemindedness," and patients in turn often found it difficult to trust psychiatrists or sometimes even understand what they were saying. In their day-to-day lives, patients were more likely to communicate with the nurses and other staff members of the hospital, whose ways of talking and modes of behavior were closer to their own (Pietikainen 2019: 23).

Social Class and Therapy

Throughout much of the nineteenth century, most mental patients spent their days doing next to nothing, excluding daily rituals such as washing and eating. Later, especially in the early twentieth century, work became a more important aspect of asylum life. By that time, the ideas of utility and cost-effectiveness had become part and parcel of the social ethos in the western world. To keep people, mainly adults, in a hospital, sometimes for years, meant that these people were unproductive citizens and a heavy burden to society in that they did not work or study, and somebody had to pay their hospital bills. In Europe, with the exception of Britain (MacKenzie 2013), it was usually the state or municipalities that financed mental hospitals, and in the United States, there were both private and public hospitals, the latter owned by the states.

In the early decades of the twentieth century, public mental hospitals conformed to this utilitarian ethos and put patients to work when possible (Ernst 2015). Hospitals began to resemble therapeutic labor colonies, in which predominantly working-class patients toiled in the fields, barns, forests, laundry rooms, kitchens, stables, and weaving rooms. It was rarely obligatory for patients to work, but it was strongly encouraged and usually expected of those patients who were capable of doing so, and many were. Patients' mostly unpaid labor was justified by the supposed therapeutic effects of work on the mental condition of patients. Moreover, in many countries across Europe, more peaceful and easy-to-handle patients were also placed in family care, usually in farms near the hospital. There they helped with the farm work just like family members and hired farm hands, and this family-based community care was considered useful for patients who needed a sort of halfway house between leaving the hospital and finding a new home.

Work therapy began to be more systematically applied after the German asylum psychiatrist, Hermann Simon, developed what he called "active therapy" in the early decades of the twentieth century. He laid out its principles in his 1929 book (*Aktivere Krankenbehandlung in der Irrenanstalt*), which influenced psychiatrists throughout Europe. One of them was the chief physician of the Oulu Mental Hospital in Northern Finland. He began to apply Simon's "active therapy" to organize patients' work in a more systematic manner, and it had an energizing effect on the everyday life of the institution. When an increasing number of patients became engaged in different kinds of activities, it had both a soothing and invigorating effect on patients, especially the restless male patients. Another positive result of work therapy was that there was less need to use barbiturates, paraldehyde, and other drugs that made patients drowsy and groggy. Moreover, patients were no longer made to take prolonged baths for hours and hours to calm and pacify them. Last but definitely not least, the systematic employment of patients as a cheap labor force had beneficial effects on the hospital's finances.

It would certainly be an over-statement to claim that the underlying role of psychiatrists and medical authorities was to discipline and exploit mental patients by either engaging them in "work therapy" or simply neglecting them. Rather, what was at stake here concerned the basic responsibilities of institutional mental health care: (i) to offer shelter and provide treatment to patients, (ii) to protect society from the potential danger of mentally ill people, and (iii) to budget responsibly by using all available means – including the patients themselves – to make hospitals as economically self-sustainable as possible (Pietikainen 2019: 32).

Hopes for effective cures and quick discharge also encouraged asylum psychiatrists to introduce radical therapeutic interventions, especially psychosurgery, insulin coma therapy, and electroconvulsive therapy ("electroshocks"). In the 1940s, when it became quite apparent that these therapies often resulted in only short-term relief from symptoms rather than full recovery, psychiatrists had to rest content with the much less satisfactory goal of adjusting patients to institutional life *within* the asylum. To understand the widespread use of treatments that, from today's perspective, seem brutal and unnecessary, we need to be aware of the acute problem of overcrowding and therapeutic impotence that haunted institutional mental health care in all of the western world. Mental hospitals provided care not only to the mentally disordered, but also to those considered demented, psychopathic, epileptic and to those suffering from alcoholism and drug addiction. Restless, noisy, and violent patients were particularly difficult to manage before the introduction of modern psychotropic drugs in the mid-1950s. And it seems to be the case that the "difficult" patients were more easily singled out for psychosurgery, which was known to change the personality in a way that made the management of these patients easier (Meier 2015). What probably appealed to psychiatrists, superintendents, and hospital boards was the fact that lobotomized patients could often be discharged from hospital. A more rapid turnover of patients both reduced the costs and demonstrated to the world that mental health professionals were able to cure their patients or at least to alleviate their symptoms to the extent that they could live more independent lives, even if they could not earn their own living. In psychiatry, social recovery was a therapeutic achievement in itself. Yet, many institutionalized patients became so dependent on their attendant-guardians that upon discharge they could no longer function in the outside world. Unable to cope with the everyday challenges, they sought protection and shelter in well-ordered institutional life (Pietikainen 2015: 148, 262–264).

The criteria used to select potential lobotomy patients were influenced by the socio-economic, cultural, and psychological distance between the medical staff and the majority of patients. Since superintendents and other psychiatric authorities did not seem to expect much from their lower class patients, it was easier to select them for so-called "heroic" medical interventions. Conversely, middle-class and upper-middle-class patients were more likely to receive psychotherapy (Myers and Bean 1968: 105, 202–203). Quite predictably, the use of diagnostic categories has also been influenced by socio-cultural assumptions and prejudices. What is quite apparent is that, at least in the USA, lower-class patients have been diagnosed with psychopathy and related diagnoses more frequently than middle-class patients, who in turn have more often received the diagnoses of neurosis and manic-depression (Hollingshead and Redlich 1958: 220–233).

The Legacy of Asylums

In his influential book *Asylums* (1961), the Canadian-American sociologist Erving Goffman argued that mental hospitals are total institutions and that as such they are comparable to prisons. The book was partly based on his own experiences as a temporary field worker at St Elizabeths, a large, federally funded mental hospital in

Washington, DC, in the mid-1950s. Goffman defined total institution as a place of residence and work where a large number of similarly situated individuals cut off from wider society for an appreciable period of time, together lead an enclosed, formally administered life (Goffman 1961: xiii). Goffman argued that mental hospitals, prisons, boarding schools, and the military shape the inmates' identity and create for them a specific role that facilitates their adjustment to a regimented life. In his view, total institutions tended to rob individuals of their autonomy and sense of self. He observed how the "ill person" was socialized into the role of a mental patient. After the onset of the illness and the ensuing "prepatient phase," the patients were taken into the hospital, where they were subjected to an "institutional ceremony." In this ceremony, individuals became patients whose case histories were constructed in a way that justified incarceration and harsh treatment. The consequence of the institutional ceremony was the shrinkage of the patients' personality and the objectification of their life history.

Goffman wrote his book at a time when many public hospitals in the USA and elsewhere had long functioned as custodial institutions that were burdened by overcrowding, underfunding, and deplorable sanitary conditions. *Asylums* resonated widely, influencing mental health professionals and social critics across the western world. In West-Germany, Goffman's ideas contributed to reforms in mental health care during the 1960s and the 1970s (Schott and Tölle 2006: 208–209, 312). Goffman may have exaggerated the damaging aspects of life in mental hospitals, but his critical approach to asylums served the important purpose of provoking debates and promoting reforms.

Since the early 1960s, when Goffman's book was published, historical and sociological research on mental hospitals has challenged categorical claims about institutional care. An important conclusion to be drawn from these studies is that the asylum system was simultaneously many things. It was subject to constant negotiations involving different parties, such as families, poor relief administrators, police, hospital superintendents, and the patients themselves (Porter and Wright 2003: 4–5). Historians have begun to emphasize the active role of the family and local community in the management of madness. This is an important corrective to the earlier, "control-oriented" approach that focused on a "top down" history and on the power of psychiatrists and authorities. Historians have observed how, upon the admission of new patients, asylum doctors often just officially confirmed the certificate of madness already made by the families, local poor relief administrators, or other nonmedical community members. Thus, there is a scholarly need to study not only mental hospitals themselves but also the role of families and local communities – even in the Age of the Asylum, many mentally ill family members lived in their own homes or in the local homes for the elderly and disabled (Suzuki 2006; Gründler 2013).

From the late 1950s onward, so-called deinstitutionalization increasingly turned hospital-based institutional care into community-based (open) care. The therapeutic vision behind this transformation was laudable: let us keep the mentally ill in our communities rather than incarcerating them in large, inhumane institutions; and let us establish outpatient clinics, psychiatric wards in general hospitals, day hospitals, and social support networks, rather than locked wards, lobotomies, and mind-numbing existence in loony bins. What actually happened between 1960 and 2000 was in many jurisdictions the reverse of the "great confinement" of the Asylum Age: the whole structure of the asylum system was disassembled as the number of mental hospital beds was reduced to one tenth or so of what it had been during the heyday of large asylums. It is a contested issue whether national governments and US states had any serious intention of investing in mental health care beyond what was necessary to keep up the appearance of organized care. In the short term, closing down asylums was economically beneficial to public coffers, but to what extent were they prepared to *spend* money to create "community services" for mental patients, the great majority of whom were no longer sheltered and cared for in hospitals? Sadly, not much, and a solid infrastructure of mental health services is still either missing or underfunded in most western countries. As a consequence of deinstitutionalization, homelessness skyrocketed, for example, in the UK and the USA, and a growing number of mental patients ended up on the streets, in jails, or in homeless shelters. A leading figure in UK mental health politics, Baroness Elaine Murphy, "titled her account of community care between 1962 and 1990 'The Disaster Years'" (Taylor 2014: 116). To speak of the mental health care system is to speak of the most recent unfulfilled promise.

The American philosopher and psychologist William James long ago in 1908 pointed out the intrinsic problem with "total institutions" eloquently in a lecture he delivered at Oxford University: "Most human institutions, by the purely technical and professional manner in which they come to be administered, end by becoming obstacles to the very purposes which their founders had in view" (James 1909: 96). Following James, we can say that it was not so much that mental asylums developed in the wrong direction and lost their essential therapeutic purpose; rather, the problem lay in the very design of the asylum system. But as nobody could foresee this when the first modern mental hospitals were established in the early nineteenth century, one must be wary of blaming the "founding fathers" of the mental asylum for going astray. Besides, human history is largely a history of unintended consequences – we often cannot foresee, let alone control, the consequences of our actions. For this reason, judging previous generations by today's standards is anachronistic: right or wrong, they usually based their decisions on what was then considered solid knowledge and plausible assumptions about the proper treatment of mental illness.

Part II: Marginality: Race and Politics

In many societies of the world, there has been a clear-cut racial divide between ethnic groups, and this division has also affected psychiatry and mental health care. What this division has usually meant is that both governmental policies of racial segregation with attending oppression and the racialist beliefs and prejudices of ordinary people have influenced psychiatric theory and therapy. This was most conspicuous during the era of European colonialism, when psychiatry was closely involved in the racial health policy in African and Asian colonies.

What characterized the relations between race, society, and psychiatry from mid-nineteenth century to mid-twentieth century was the almost total naturalization of what is essentially a socio-cultural category, namely "race." The legitimacy of this category was seen to derive from nature – after all, races were deemed to be biologically different – whereas in reality it was a social construct based on prevailing beliefs, prejudices, and assumptions. What made "race" a deeply damaging category was that it was used as a divinely ordained and/or scientifically validated justification for social inequality, segregation, and sheer oppression of nonwhite peoples.

In the second half of this chapter, we will see how this enmeshing of biological and socio-cultural categories manifested itself in psychiatry. In the first section, the focus is mainly on apartheid era South Africa. Thereafter, the focus shifts to the USA and the ways in which the category of race influenced mental health care and psychiatric discourse on African Americans.

Mental Health in Segregated Societies: The Case of Racial and Colonial Psychiatry

Starting from the late nineteenth century, the most extreme form of racial division developed in South Africa during apartheid. Yet, contrary to what could have been expected, there was very little psychiatric abuse of the colored population in this era. This was not because defiant mental health professionals protected their black patients from government oppression, but simply because the psychiatric focus was very much on white patients and especially poor white men who were considered curable. Although white men only made up about 5% of the population, they occupied about 20% of the beds in mental hospitals. Thus, a considerable minority of the patients admitted to South-African mental hospitals were lower-class white (Afrikaaner) men, while the majority of "nonwhite" as well as female populations were simply neglected. Poor white men constituted the one social group that the apartheid government wanted to help and support (Jones 2014: 25, 41–42).

After World War II, the preferred apartheid policy was to confine the black population in reserves, areas that became known as *bantustans* or homelands. Blacks who were housed in overcrowded and underfunded mental hospitals received only custodial care, and these hospitals resembled warehouses or dilapidated prisons more than therapeutic institutions. Most of the institutionalized black patients were unemployed or unskilled workers, and therefore, much less valuable members of society than those who had jobs or whose skills were in greater demand, such as miners or agricultural and commercial workers. As in the USA, the stigmatizing diagnosis of schizophrenia was more widely used for black patients than for whites, who were given less severe diagnoses, such as depression or psychoneurosis (Jones 2014: 34).

Well documented, especially in the 1970s, are the human rights abuses that took place in mental health institutions where practitioners occasionally contributed to, or even facilitated, the mistreatment of black patients. But there is no evidence of practitioners giving blanket support to apartheid ideas and policies. Instead, relationships between psychiatrists, patients, and public officials were influenced by local circumstances, political forces, and international trends. When the era of apartheid ended in 1994 and the Truth and Reconciliation Commission (TRC) was established, the activities of the Medical Association of South Africa (the parent organization of the Society for South African Psychiatrists) were investigated. In its report, the TRC noted that the Medical Association had placidly accepted apartheid policies in medical practices. Other data compiled about the role of mental health practitioners demonstrated that (i) some practitioners participated in interrogations of political prisoners, (ii) some patients in mental institutions had died because of neglect or ill treatment, (iii) intrusive methods of treatment were employed, and (iv) racist language and a racist approach to the patients' needs were evident. Unfortunately, while state-sanctioned racism and official segregation have disappeared, mental health care has not really improved in postapartheid South Africa. Even racial discrimination has continued, as have custodial practices (Jones 2014: 177–182).

One of the most well-established and noncontroversial observations made by historians of psychiatry is that indigenous patients suffered under colonial rule. For example, in Australia, they are rarely noticeable in the records, and this may have been due to colonial policies of segregation and the indigenous people's removal and dispersal throughout Australia. However, indigenous inmates were not separated from European patients in the asylums, perhaps because – as a distinct minority – they were not considered a threat to the "white" identity of the institutionalized settlers. In New Zealand, "very few Maori patients were discharged from Auckland [Mental Hospital] when compared to European patients"; more than half of the Maori inmates died in the asylum (Coleborne 2010: 36–42, 111, 134). If white settlers sometimes felt uprooted in the colonies, there is hardly any doubt that indigenous people felt insecure and disempowered in an institution organized around principles, norms, and rules that were alien and probably frightening to ethnic peoples.

To take one more example, in colonial North Africa between the two world wars, French psychiatrists examined almost exclusively the normal and pathological mind of the indigenous population. They devised theories about the "primitive mentality" and "criminal impulsiveness" of North Africans. As if this was not degrading enough, even the "normal" consciousness of a Muslim was described as containing "a mixture of insanities in varying doses," as one French psychopathologist put it in 1908. If the more "docile" Tunisians were "dreamy" and prone to mysticism, the mind of the "aggressive" Algerians was presented as a threat to public safety. In addition, more invasive treatments, such as psychosurgery and shock methods, were first tested on Muslim patients (Keller 2007: 7–16, 108, 138).

What can be distilled from the avalanche of accounts of racialist psychiatry is that, in mental health care, race has long been an essential factor in the discovery (or, invention), explanation, diagnosis, and treatment of mental disorder. Within the western world, this was most apparent in the USA.

Mental Health in a Segregated Society: The Case of the USA

Due to the huge cultural influence of the USA, as well as modern communication technology, there is a growing global awareness of the racial injustices committed inside the criminal justice system and other public institutions as well. In the early

2020s, the Black Lives Matter movement has mobilized protest even in countries where there are hardly any "black" or African minorities. In light of global interest in most things American, and racial issues in particular, it is instructive to examine the role played by race in American psychiatry and to see how psychiatric understanding of race-specific mental health problems reflected deep currents of American society.

In the nineteenth and early twentieth century, medical authorities were inclined to link mental derangement with more advanced levels of civilization, and in this evolutionary framework, the more "primitive," "childlike," or "inferior" races, including Indians and African Americans, were portrayed as having underdeveloped nervous systems that made them largely immune to insanity and "endowed" them with higher tolerances for both emotional and physical pain. Accordingly, black Americans were more likely to be subjected to cruel and painful medical procedures, surgeries, and experiments. During the era of slavery, the slave owners had a convenient scientific justification for their violent punishments: blacks had an undeveloped sense of pain (Washington 2008: 58).

In 1905, the famous US psychologist G. Stanley Hall asserted that the "naturally cheerful Negroes" rarely suffer from melancholia or commit suicide. Thanks to their cheerful disposition and the attending greater tolerance for suffering, "Negroes" had a natural capacity to endure slavery that whites did not possess (Scott 1997: 14). And since the blacks were by and large "incapable" of going mad, there was no need to treat them in mental asylum either. This twisted logic also worked the other way around: slave owners were actually ensuring that blacks would not go mad. The medical rationale behind this popular idea was that African Americans could not stand the mental pressure of freedom. Some Southern physicians proclaimed that the "Negro" is a slave by nature, and there was even a separate diagnosis for slaves who suffered from a pathological desire for freedom: "drapetomania," literally "madness of the fugitive." In the Alabama Insane Hospital, one diagnosis given to a former slave in 1867 was "freedom" (Washington 2008: 143).

When African Americans displayed clear symptoms of "insanity," they were not necessarily admitted to mental institutions as patients. Some private asylums did not admit blacks at all, while others tried to keep black patients out of sight from white patients and visitors. State mental hospitals had more permissive policies, but they too placed blacks in segregated, inferior quarters or separate buildings. Moreover, as hospital funding for the treatment of poor people was generally provided at public expense, state officials often strove to save money by incarcerating mentally ill blacks in jails and almshouses. Thus, a policy of segregation that tarnished all of American society was practiced in mental health care too (Gamwell and Tomes 1995: 56).

Scientifically, the widely held assumption that the "black race" was inherently inferior was based on the doctrines of nineteenth-century racial biology and physical anthropology. Measurements of human skull had ostensibly demonstrated that the "Caucasian" (white) racial type had the largest skull and therefore the largest brain, while the black or African race had the smallest. According to this racialist and racist reasoning, due to their low intelligence, the destiny of the black race would remain on the lowest rung of the racial hierarchy (Barkan 1992). This prejudice – presented as scientific fact – indicated in turn that blacks, including African Americans, were incapable of autonomous living in complex industrial societies requiring educated work forces. Thus, they were in need of white care and supervision. An additional reason blacks had to be supervised was their inherent inability to control their animal-like sexual urges. This putative lack of self-control made "predatory" black men a threat to white society – not surprisingly, then, lynch mobs in the southern states often targeted black men who had allegedly sexually assaulted white women; indeed, rape became increasingly classified as a "Negro crime" (Freedman 2011). Black women in turn were assumed to be sexually aggressive seducers who were luring white men into immoral sexual relations (Washington 2008: 45).

As if this was not enough, African Americans were usual subjects of medical experimentation as well as clinical demonstration, even though this fact was no longer openly advertised in the medical community in the post-World War II era. In clinical psychiatry, imprisoned African Americans were selected as guinea pigs for psychosurgical experiments in the 1950s. The idea of employing psychosurgery to "cure" the black urban radicals surfaced in the 1960s, when racial tensions escalated to rioting and street violence in many urban areas. To provide an effective remedy to this social malady, white neurosurgeons developed different brain surgery techniques by experimenting on blacks, especially in the southern states. Some of the "aggressive" and "hyperactive" African Americans targeted for operation were children. The rationale behind these extremely invasive and often outrageously primitive methods was to excise the alleged seat of violence in the brain. The National Institute of Mental Health and the Law Enforcement Assistance Administration granted three surgeons more than a half million dollars to conduct brain research on urban "rioters" (Washington 2008: 286–287). On the other side of the color line, things looked quite different: in the segregated South, black men and boys faced not only the specter of psychosurgery, but also white men beating, torturing, and killing them with impunity.

The attitude of some medical authorities toward the "Negro" is revealed in a comment given by a public health physician in the 1940s: he suggested that instead of providing education to the blacks, they should be used as guinea pigs in research laboratories when they fall ill. In the 1960s, another expert referred to the misuse of black subjects by pointing out in a public lecture that it was cheaper to use blacks than cats for medical experiments, because the former were cheap and readily available (in prisons, for example) (Washington 2008: 10). Small wonder, then, that medical schools and hospitals "became firmly cemented into the African American consciousness as places of terror, violence, and shame, not of medical care" (Washington 2008: 114). One of the main problems of racialist health care in the USA was that the traditionally conservative medical community consistently ignored or discounted the structural reasons for the ill health of African-Americans, including poor housing, poor educational facilities, inadequate health care and nutrition, high infant mortality, inequalities in the labor market, and the overall effects of structural and institutional racism (Washington 2008: 152–153).

Not all white psychiatrists were racially prejudiced, however. In his book on Harlem's Lafargue Mental Hygiene Clinic, Gabriel N. Mendes sheds light on the history of antiracist psychiatry in New York. The main protagonist of Mendes' book is Fredric Wertham, a German-born Jewish senior psychiatrist of the New York City Department of Hospitals, who in 1946 founded an outpatient mental health clinic in Harlem together with two African Americans, writer and intellectual Richard Wright, whom he had befriended in the early 1940s, and Reverend Shelton Hale Bishop, an Episcopal priest. Already in the early 1930s, when he began his clinical work in New York, Wertham had noted the correlation between racism and mental disorders. Until 1958, when it had to close its doors due to Bishop's retirement, the death of some staff members, and lack of funding, Lafargue Mental Hygiene Clinic, located in the basement of Bishop's Episcopal parish house, provided an alternative to the racist mental health care system and challenged psychiatrists who ignored the lived experience and suffering of African American patients in Jim Crow America.

The fact that Wertham's Clinic did not receive any government or philanthropic funding and had to close down after operating only for 12 years shows that its emphasis on the relationship between racist oppression and mental problems found little favor among medical authorities, who were quite ill-disposed toward Wertham for his "calumnies" directed against the psychiatric establishment (Mendes 2015). Yet, as the historian Dennis Doyle has shown in his study of psychiatry in Harlem from the late 1930s to the late 1960s, the mental health needs of African Americans were not totally neglected by the policy makers in New York. The idea that blacks should have the same access to care as the white population was put in practice by the so-called "racial liberals" – psychiatrists, civil servants, and social reformers – who were determined to create race-neutral standards of care (Doyle 2016). By documenting attempts to overcome racial prejudices within mental health care system, Mendes' and Doyle's important studies paint a more nuanced picture of the relationship between race psychiatry and in the civil rights era.

By the early 1960s, the tectonic plates of social organization were shifting, and changes in race-related legislation were inevitable. Supported by liberal white students as well as church and trade union leaders, African Americans managed to win majority support for the passage of the 1964 Civil Rights Act and the 1965 Voting Rights Act. At the same time, urban unrest affected public discussions of race, one example being the popularity of "damage discourse": the African American family was described as pathological, both psychologically and socially. For more conservative academics, the outburst of "rioting" across the country led to the psychopathologization (personality assassination) of black radicals. Some social scientists and psychiatrists considered black nationalism a "psychosis," or an outgrowth of personality problems (Scott 1997: 110–111), while others presented black urban unrest as criminality. Many experts saw the culprit in what they imagined to be the disorganized black family. In particular, the problem was "matriarchy," that is, families without the presence of patriarchal fathers who would otherwise have acted as all-important authority figures, especially for wayward sons (Scott 1997: 147). This damage thesis was challenged by more radical thinkers, some of whom claimed that young African Americans who participated in the riots "were not known to be psychologically impaired or especially suffering from problems of masculine identity." In fact, juveniles who were arrested in the 1967 Detroit riot "were found by a psychological team to be less emotionally disturbed and less delinquent than typical juvenile arrestees" (Scott 1997: 170).

Related to the widespread imagery of damage during the civil right era was the more subtle "deprivation discourse" that described the lives of low-income families in terms of what was missing (Raz 2013). This liberal discourse was meant to be benevolent toward the poor, and it was accompanied by concrete programs and initiatives that assisted poverty-stricken black families. At the same time, it disclosed some deep-seated prejudices and preferences of white middle-class society. Yet, not all African Americans were poor and disadvantaged. In the 1950s, there was a sizable population of black businessmen, professionals, and so-called white collar workers in the USA, and together they constituted the "black bourgeoisie" (Frazier 1965). To disregard the differences in the socioeconomic status of African Americans was another way of seeing them as an amorphous and homogeneous mass, devoid of any individual traits.

Protest Psychosis and Cultural Paranoia

During the civil rights era in the 1960s, "angry black men" began to be diagnosed with schizophrenia, especially paranoid schizophrenia. This stigmatizing diagnosis was overwhelmingly applied to African American protesters, for example, in Michigan, where the Ionia State Hospital for the Criminally Insane confined black men for political rather than medical reasons (Metzl 2009). At the end of the 1960s, more than 60% of the asylum's inmates were black men, who were classified as "dangerous" and "paranoid." Many of these men hailed from impoverished neighborhoods in urban Detroit, a city powerfully shaped by racial tensions. In the 1960s, schizophrenia in the USA was transformed into a racialized illness characterized by aggression and volatility. To be sure, there were violent, agitated black men among the rioters, but to be angry and belligerent is not tantamount to being mentally ill – quite the contrary, if those who take part in rebellion represent an oppressed, often persecuted ethnic minority, they may have sound reasons for being angry, even paranoid. The use of psychiatric diagnoses and nomenclature for the denunciation of political enemies, movements (such as early feminist suffragettes and the workers' movement), and marginal groups has a long history, dating back at least to early twentieth-century crowd psychology, advocates of which provided psychomedical explanations of political phenomena and groups they disliked, be they feminists, socialists, minorities (such as the Jews), or whole nations, such as Russians or Chinese (Borch 2012).

Among psychiatric diagnoses, schizophrenia came to be closely associated with African Americans. In one large government study, researchers found that psychiatrists diagnosed schizophrenia in African American patients, especially men, "four times as often as in white patients" (Metzl 2009: x). Yet, the research team did not find any evidence that black patients would have been any more seriously ill than whites. The only explanation for the discrepancy between the black and white patients was race. Socio-political tensions and urban unrest encouraged the diagnostic application of schizophrenia to "difficult," angry black men. In 1968, two white psychiatrists (Walter Bromberg and Frank Simon) dubbed the kind of schizophrenia that allegedly afflicted these black men "protest psychosis." In their view, a number of black men had started to suffer from "delusional anti-whiteness" as well as "hostile and aggressive feelings" after they had joined the Nation of Islam, listened to speeches given by the black radical Malcolm X, or aligned with black militant groups aiming to topple existing white power structures. As a result, they manifested "paranoid projections" of racial antagonism between blacks and "Caucasians." According to these psychiatrists, militant black men were in need of psychiatric treatment, because their disorder posed a threat both to their own mental health and to the social order (of white America). Thus, delusions, hallucinations, and paranoia were literally caused by the black liberation struggle. Such reasoning was quite common in mainstream American psychiatry during the 1960s and 1970s. In some publications, black culture as a whole was described as a risk for mental health.

What conveniently escaped the clinical gaze of white psychiatrists was the violence committed by whites against the blacks during the civil right era. Some black activists and scholars pointed out that racism precipitated mental health problems in the ghetto, and they saw racism itself as a "mass psychosis," a delusion, and a pathological indication of a sick society (Metzl 2009: 101–102, 122). As the black Harvard professor Chester M. Pierce noted in 1969, to be black in the USA means to live closely with despair and hopelessness, to feel insecure all the time, and, in consequence, to become "psychologically terrorized" (Pierce 1969: 302–303).

What makes the study of the wide use of schizophrenia so intriguing is that the term and diagnosis was also used by the black civil rights leaders themselves. Obviously, they did not claim that the black psyche was inherently damaged or naturally prone to serious mental disorders. Rather, their intention was to reveal how racist white society produced mental suffering among African Americans (Grier and Cobbs 1969). By this line of reasoning, paranoid schizophrenia as a mental state was the result of black men's contradictory goals: on the one hand, they tried to conform to the norms and mores of white society, and on the other hand they resisted the injustices inflicted upon them by those same norms and mores. In this sense, schizophrenia was not only an illness; it was also a natural adaptation to a racist society. Martin Luther King used the term schizophrenia in just this sense when he referred to schizophrenia as "going on within all of us" in his famous "Unfulfilled Dreams" sermon in the spring 1968 (only a few months before he was murdered). This was a reference to a spiritual divide within the nation, but it was also an articulation of the conflict between the white and black America as well as within the black liberation movement. Until the end of the century, African American men were diagnosed with schizophrenia much more often than white men. In particular, the paranoid subtype of schizophrenia was a widely used diagnosis for black men in general and "belligerent," hostile, and angry black men in particular (Metzl 2009, x–xv).

While King represented a moderate, nonmilitant civil rights activism based on Christian ethics, for the radical groups and leaders, such as the Black Panthers and Malcolm X, the proper "cure" for schizophrenia was to dismantle white hegemony by almost any means necessary. The black radicals used the term "cultural paranoia" to refer to the African-Americans' healthy suspicion about the motives of white Americans – it was advisable to be mentally troubled and over-alert about white oppression and persecution (Malcom X was diagnosed with "prepsychotic paranoid schizophrenia" by the FBI). To the radicals, cultural paranoia helped blacks survive and fight back against white society. Thus, paranoid schizophrenia was a critique of white society as well as an element of black identity (Metzl 2009: 124–127; Grier and Cobbs 1969: 161). In short, to be paranoid was to resist racist society.

From the perspective of African American communities, what happened between the late 1950s and early 1970s was that "rioting" black men were easily diagnosed as being mentally disordered, while black women in inner cities were considered to be "inadequate" in their roles as mothers. As a social group, African Americans were regarded as illiterate laborers and, if they were male adults, potentially violent and a danger to the social order (Grier and Cobbs 1969: 135, 147). Obviously, what this emphasis on deprivation, deficiency, and disorder amounted to was a devaluation of African American culture and family life. For the black families themselves, the sense – and reality – of being unable to protect family members from economic insecurity, threats, humiliation, and harassment by law enforcement and other authorities constituted a heavy psychological burden (Grier and Cobbs 1969: 81).

Alienation, Race, and Mental Illness

In the nineteenth century, "alienation" had been a common term for mental illness, and psychiatrists working in mental asylums were typically called "alienists" (*aliéniste* in French) – they were experts in the medical treatment of mental alienation. Mental illness was understood as the insane persons' alienation or estrangement from themselves (their true identity) as well as from their social environment. In Europe and North America, the concept of alienation was revived in the 1950s, and it was now used more politically and culturally to signify estrangement from society's norms and values (Pietikainen 2007).

In the early 1950s, the idea of alienation was used in the framework of colonial racism by the Martinique-born French psychiatrist and political thinker Frantz Fanon. In his influential view, the alienation and mental illness of oppressed people were pathologies of "structural" or "institutional" racism, which manifested themselves in economics, education, the labor market, and institutions of law and order (police, legislatures, courts). This kind of racism was less overt and more subtle than the racism of individuals and groups, but for this very reason, it was also more difficult to eradicate. In racist societies, as Fanon pointed out, mental illness was an emphatically social problem, a result of social forces and power relations that, among other things, determined health and illness (Fanon 2008).

In racially segregated societies, such as the USA in the 1960s, many black authors saw alienation as the fundamental condition of the oppressed minority. While African Americans were part of American society, they were also kept on the margins, and this chronic marginalization created an internal divide which in turn resulted in an alienated state of mind. By the time the Civil Rights Act of 1964 came into force, African Americans had been second-class citizens for a century, and in the previous two centuries, most of them had been slaves (nationwide in 1860, there were almost four million enslaved African Americans compared to less than five hundred thousand free African Americans). Thus, to be profoundly alienated from the increasingly affluent white middle-class society was a "natural condition" for African Americans, most of whom had menial, low-paying jobs (Clark 1967: 17–21). The problem was aggravated by the fact that, in 1960, only 2.2 percent of American physicians were black, although African Americans constituted more than 10 percent of the population. This discrepancy was very visible in psychiatry, as white psychiatrists labeled oppressed blacks as "neurotic" and, in general, misdiagnosed African Americans due to their ignorance of black culture (Hare 1969: 45). Ultimately, the explanation for the increased diagnosis of schizophrenia among black American men can be found not in the individual psyches but in a society torn apart by civil rights issues (Metzl 2009: 94, 203). In other words, the alleged increase of schizophrenia among black men was mainly due to urban riots and the overall radicalization of African Americans – what was at stake here was the psychopathologization of black political protest. This is a reminder that medicine is closely linked to politics, power, and racial biases.

Conclusion

This chapter explored the dual aspect of psychiatry and society: in the first part, the focus was on the confinement of madness in mental asylums. The institutional solution to the medical and social problem of mental illness was to establish hospitals that functioned very much like mini-societies within the larger society of "normal" people. The main function of mental asylums was twofold: first, to protect the society from the disruptions caused by the mentally ill and second, to help mental patients adjust to the norms and rules of the larger society as well as to the norms and rules of the mini-society that was the mental hospital. If the patients could not be returned to society as re-adjusted or re-normalized persons, then they needed to become adjusted to the social order of the asylum. This was the design of the mental asylum, and it was the prevailing form of mental health care until the last third of the twentieth century, when so-called deinstitutionalization dramatically decreased hospitalized patient numbers (although the number of admissions each year remained high). Some patients became homeless and unprotected, and some ended up in nursing homes or, worse, in another total institution, the prison.

In the second part of the chapter, the focus was on the ways in which madness becomes part and parcel of social practices and cultural assumptions about race, citizenship, and family. As we have seen, racism influenced how the nonwhite minorities were treated as patients, clinical material, and subjects of medical experiments. At the same time, as the US case demonstrates, psychiatric terms and theories were incorporated into social and political discourse and practice, as a result of which psychopathological terms, such as schizophrenia, paranoia, and psychosis, were employed to denigrate minorities, depoliticize anger and resistance, and appeal to the moral sensibilities of the white majority (especially middle-class whites) in the name of equality and social justice. The latter kind of "damage imagery" was also employed by African Americans themselves, which showed how the language of pathology could be used for a variety of purposes: (i) to disparage one's opponent, (ii) to win sympathy and gain support for a disadvantaged minority, and (iii) to demonstrate how the disease of racism sickened society and how a sick society itself produced mental disturbances.

The notion of a sick (capitalist) society that became common in the 1960s links together the two parts of this chapter. In the USA, what could be called the New Left, including women's and gay liberation, as well as Black and Chicano Power, upset the guardians of traditional America, but eventually their struggle for political rights and recognition of their identities paid off, and overt racism, male chauvinism, and intolerance of sexual minorities became largely discredited (Hartman 2015). For its part, the American Psychiatric Association, under pressure from gay activists, removed homosexuality from its list of mental disorders in 1973. At the same time, mental patient organizations began to advocate the patients' right to humane treatment.

The important conclusion to be drawn here is that intrinsic in the arise of the New Left was a critical stance toward western society and its institutions, such as mental hospitals, prisons, and the rule of law. From the mid-1960s to the mid-1970s, "antipsychiatric" and critical views of psychiatry as an instrument of social control became part of leftist student movements, thereby influencing the attitudes of these future policy makers, academics, and public officials. What in particular united the critics was their assumption that psychiatry was a form of social control rather than a robust medical or psychological science. Critics of psychiatry had a damning view on the incarceration of "deviant" individuals in mental hospitals, but they also cast a shadow of doubt on the entire medical discourse on madness (Pietikainen 2015: 312–314).

New Left radicals questioned (re-)adjustment policies that aimed to return the mentally disordered and other deviants, such as homosexuals and drug addicts, to the conservative or reactionary social order. They also tended to regard the "alienation" of African Americans from the white society as a natural reaction to institutional racism. Radicals began to promote the idea that if you become too well adjusted to your social environment, it could be a sign of your mental pathology. According to critics, individuals were sane when they displayed political alertness, sexual permissiveness, and cultural liberalism and when they had emancipated themselves from the ossified normative system of their parents' generation. At the same time, anyone who uncritically accepted the norms, values, and demands handed down by authorities, parents, and educators risked becoming either mentally alienated or the Cheerful Robot memorably described by the radical American sociologist C. Wright Mills (Mills 1967: 171–173).

As demonstrated by more recent protest movements, including MeToo and Black Lives Matter, as well as the continued high prevalence of mental disorders, social pathologies still afflict even the most tolerant and liberal democratic societies.

Psychiatry, as part of the medical establishment, is just like any other social institution: it moves in the river of time and changes as the times change. But only with the benefit of hindsight are we able to better understand how the intricate and sometimes quite upsetting relations between psychiatry and society have evolved – and continue to evolve – into the new millennium.

References

Barkan E (1992) The retreat of scientific racism: changing concepts of race in Britain and the United States between the world wars. Cambridge University Press, Cambridge

Borch C (2012) The politics of crowds: an alternative history of sociology. Cambridge University Press, Cambridge

Castel R (1988) The regulation of madness: the origins of incarceration in France. Trans Halls WD. Orig French edn 1976. University of California Press, Berkeley

Castel R, Castel F, Lovell A (1982) The psychiatric society. Trans Arthur Goldhammer. Orig French edn 1979. Columbia University Press, New York

Clark K (1967) Dark ghetto. Harper & Row, New York

Coleborne C (2010) Madness in the family: insanity and institutions in the Australasian colonial world, 1860–1914. Palgrave Macmillan, Basingstoke

Doyle DA (2016) Psychiatry and racial liberalism in Harlem: 1936–1968. Rochester University Press, Rochester

Engstrom EJ (2004) Clinical psychiatry in Imperial Germany: a history of psychiatric practice. Cornell University Press, Ithaca

Ernst W (ed) (2015) Work, psychiatry and society, c. 1750–2015. Manchester University Press, Manchester

Fanon F (2008) Black skin, white masks. Trans Charles Lam Markmann. Orig French edn 1952. Pluto Press, London

Foucault M (2006) History of madness, ed Khalfa J, trans Murphy J, Khalfa J., Orig French edn 1961. Routledge, London

Frazier FE (1965) Black bourgeoisie, 1st edn 1957. Free Press, New York

Freedman EB (2011) "Crimes which startle and horrify": gender, age, and the racialization of sexual violence in white American newspapers, 1870–1900. J Hist Sex 20:465–497

Gamwell L, Tomes N (1995) Madness in America. Cultural and medical perspectives of mental illness before 1914. Cornell University Press, Binghamton

Gauchet M, Swain G (1999) Madness and democracy: the modern psychiatric universe. Trans Porter C. Orig French edn 1980. Princeton University Press, Princeton

Goffman E (1961) Asylums. Anchor Books, Garden City

Goffman E (1963) Stigma. Notes on the management of spoiled identity. Simon & Schuster, New York

Goldstein J (1987) Console and classify. The French psychiatric profession in the nineteenth century. Cambridge University Press, Cambridge

Grier WH, Cobbs PM (1969) Black rage. Jonathan Cape, London

Grob GN (1994) The mad among us. The Free Press, New York

Gründler J (2013) Armut und Wahnsinn: "Arme Irre" und ihre Familien im Spannungsfeld von Psychiatrie und Armenfürsorge in Glasgow, 1875–1921. Oldenbourg, Munich

Hare N (1969) Does separatism in medical care offer advantages for the ghetto? In: Norman JC (ed) Medicine in the ghetto. Appleton-Century-Crofts, New York, pp 43–50

Hartman A (2015) A war for the soul of America. University of Chicago Press, Chicago

Hollingshead AB, Redlich FC (1958) Social class and mental illness. Wiley, New York

James W (1909) A pluralistic universe. Longmans, Green & Co, New York

Jones TF (2014) Psychiatry, mental institutions, and the mad in Apartheid South Africa. Routledge, New York
Kaufmann D (1997) Aufklärung, bürgerliche Selbsterfahrung und die 'Erfindung' der Psychiatrie in Deutschland, 1770–1850. Vandenhoeck & Ruprecht, Göttingen
Keller RC (2007) Colonial madness. Psychiatry in French north Africa. University of Chicago Press, Chicago
King LJ (2002) From under the cloud at Seven Steeples, 1878–1885: the peculiarly saddened life of Anna Agnew at the Indiana Hospital for the Insane. Guild Press/Emmis, Zionsville
MacKenzie C (2013) Psychiatry for the rich. A history of Ticehurst private asylum. Routledge, London
Meier M (2015) Spannungsherde: Psychochirurgie nach dem Zweiten Weltkrieg. Wallstein Verlag, Göttingen
Mendes GN (2015) Under the strain of color: Harlem's Lafargue clinic and the promise of an antiracist psychiatry. Cornell University Press, Ithaca
Metzl JM (2009) The protest psychosis: how schizophrenia became a black disease. Beacon Press, Boston
Mills CW (1967) The sociological imagination, 1st edn 1959. Oxford University Press, Oxford
Myers JK, Bean LB (1968) A decade later: a follow-up of social class and mental illness. Wiley, New York
Napoli DS (1981) Architects of adjustment: the history of the psychological profession in the United States. Kennikat Press, Port Washington
Pierce CM (1969) Is bigotry the basis of the medical problems of the ghetto? In: Norman JC (ed) Medicine in the ghetto. Appleton-Century-Crofts, New York, pp 301–312
Pietikainen P (2007) Alchemists of human nature: psychological utopianism in Gross, Jung, Reich and Fromm. Pickering & Chatto, London
Pietikainen P (2015) Madness: a history. Routledge, London
Pietikainen P (2019) Pity the poor patient: the indigent mentally ill in late nineteenth and early twentieth-century Finland. In: Pietikainen P, Kragh JV (eds) Social class and mental illness in Northern Europe. Routledge, London
Pietikäinen P (2020) Kipeät sielut. Hulluuden historia Suomessa. Gaudeamus, Helsinki
Pietikäinen P, Kragh JV (eds) (2019) Social class and mental illness in northern Europe. Routledge, London
Porter R (ed) (1991) The Faber book of madness. Faber & Faber, London
Porter T (2018) Genetics in the madhouse: the unknown history of human heredity. Princeton University Press, Princeton
Porter R, Wright D (eds) (2003) The confinement of the insane. International perspectives 1800–1965. Cambridge University Press, Cambridge
Raz M (2013) What's wrong with the poor? Race, psychiatry and the war on poverty. The University of North Carolina Press, Chapel Hill
Schott H, Tölle R (2006) Geschichte der Psychiatrie. C.H. Beck, München
Scott DM (1997) Contempt and pity: social policy and the image of the damaged Black psyche, 1880–1996. The University of North Carolina Press, Chapel Hill
Scull A (1989) Social order/mental disorder. Anglo-American psychiatry in historical perspective. University of California Press, Berkeley
Scull A (2015) Madness in civilization. A cultural history of insanity, from the Bible to Freud, from the madhouse to modern medicine. Princeton University Press, Princeton
Shorter E (1997) A history of psychiatry. Wiley, New York
Suzuki A (2006) Madness at home: the psychiatrist, the patient, & the family in England, 1820–1860. University of California Press, Berkeley
Taylor B (2014) The last asylum. A memoir of madness in our times. Hamish Hamilton, London
The plea for the silent (1957) Christopher Johnson, London
Washington HA (2008) Medical apartheid: the dark history of medical experimentation on Black Americans from colonial times to the present. Anchor Books, New York

Early Child Psychiatry in Britain

51

Nicola Sugden

Contents

Introduction	1332
Doing the History of Child Psychiatry	1334
Institutional Records and Patient Case Files	1335
Legislation and Public Policy	1336
Practitioners' Accounts: Histories and Testimony	1336
Personal Papers and Correspondence	1337
Print and Media	1338
The Voice of the Child	1339
Tangled Roots: The Study of the Child and Child Insanity at the Turn of the Twentieth Century	1340
Child Psychiatry, Child Guidance, and Child Psychoanalysis Before the NHS	1346
Conclusion: Enduring Eclecticism; Contemporary Concerns	1351
References	1353

Abstract

This chapter surveys work in the history of child psychiatry to offer a history of early child psychiatry in Britain and suggest future research directions. It is intended as a general overview and as a guide to those embarking on research in this field. Accounting for the development of child psychiatry as a discipline reveals a deep and complex root network drawing from children's medicine, child guidance, psychoanalysis, education, social work, hospital psychiatry, and other institutional settings, necessitating a broad interpretation of "child psychiatry" to include the many theories and practices that took as their object the understanding and management of the child's mind. The gathering of practices under the "Child and Adolescent Mental Health Services" (CAMHS) umbrella in the late twentieth century belies an enduring eclecticism that has been a consistent feature

N. Sugden (✉)
Centre for the History of Science, Technology and Medicine, University of Manchester, Manchester, UK

© The Author(s), under exclusive licence to Springer Nature Singapore Pte Ltd. 2022
D. McCallum (ed.), *The Palgrave Handbook of the History of Human Sciences*,
https://doi.org/10.1007/978-981-16-7255-2_99

of child psychiatric theory, practice, and research in Britain. Across the period, children and childhood – both as empirical realities and as analytical categories – have posed particular questions to psychiatry and mental health care, especially in their ostensible vulnerability, liminality, and inaccessibility. Historians have approached this complex history by framing their analyses around specific psychiatric theories or hypotheses, prominent individuals or groups of practitioners, particular institutions, and the construction of diagnostic categories and disease subjects. These investigations have utilized case files, institutional records, past research and discussion papers, legislation and policy documents, oral history, and methods from a range of academic fields. The mind of the child remains a contested entity at the center of contemporary social issues, and the history of child psychiatry has a role to play in promoting children's welfare in the present and future.

Keywords

Britain · Childhood · Children · Psychiatry · CAMHS · Madness

Introduction

From the melancholia of ancient Greece to the neurodiversity of the twenty-first century, notions of extranormal dispositions or constitutions of the mind, brain, or soul (whether temporary or permanent) can be found throughout human history, personified variously in popular and scientific discourse by the madman, the lunatic, the schizophrenic, the hysterical woman, the unruly child, and others. Feared as dangerous, written off as incurable, embraced as prophesiers, celebrated as geniuses, and eulogized as tragic, these figures tell us as much – if not more – about contemporaneous entanglements of scientific knowledge, cultural values, and social norms as they do about lived experience. The shifting concepts and language of madness are generated through everyday practice, in a world thickly populated by people, practices, and things, all of them deserving of historical investigation. This rich history has proved fertile ground for historians over the last 70 years or so, yet within the broad and firmly established field of the history of madness child psychiatry and children as subjects of mental health care are understudied by trained historians (Evans et al. 2008; Taylor 2017). It is only recently that child psychiatry has become a major subject of historical research, in tandem with critical and interdisciplinary approaches that have disrupted progressivist narratives of psychiatry in twentieth century Britain. Although recent work has begun to make inroads into hitherto unexplored territory, far greater attention has been paid to the second half of the twentieth century than to the theories and practices of child psychiatry before its incorporation into the National Health Service (NHS). Children and childhood offer a particular set of challenges to the history of madness, including a relative paucity of primary sources, and ethical and interpretational quandaries that require the field to extend its scope and its armory. Nevertheless, there is an early history of child psychiatry to be told.

Reorienting the history of psychiatry to center children and childhood requires a broadening beyond any narrowly-defined notion of child psychiatry as a formal medical specialty. Although general psychiatry and the management of adult mental health have, of course, also been shaped by ideas and practices outside specialist services and institutions, child psychiatry has even more complex and eclectic origins. The mind of the child has been subject to investigation and intervention by a vast array of interests within a complex care ecology, incorporating maternity services and parenting advice, education, child guidance, hospital practice, psychology, psychoanalysis, pediatrics, primary care, the criminal justice system, and the care of young people with learning disabilities. The history of child psychiatry must seek to account for the emergence and development of child psychiatry and allied disciplines by examining these evolving interests and appreciating their place within a thick and changing constellation of child-oriented knowledges and practices over the last 150 years.

Historical expertise in the history of child psychiatry and children's mental health has begun to accumulate over the last 30 years, following the general turn towards child-centered histories of medicine in the 1990s typified in Roger Cooter's germinal edited collection on the history of child health and welfare (Cooter 1992). Historians have developed social histories of systems of health surveillance, management, and treatment in modern Britain, including child welfare (Hendrick 1994), the Schools Medical Service (Harris 1995), public health and social work (Welshman 1996), and child guidance (Stewart 2014). Childhood has been recognized as a significant analytical category and distinct object of study in the histories of medicine (Dwork 1993; Levene 2011) and of the psych-sciences (Herman 2003), and increasingly researchers are drawing on and contributing not only to the history of madness and the history of childhood, but also the History and Philosophy of Science, Science and Technology Studies, Medical Anthropology, Disability Studies, Ethnography, and Geography. Insights and concepts from Science and Technology Studies and the History and Philosophy of Science have been utilized to examine the construction of psychiatric diagnoses and the disease subjects to whom they are applied. Several scholars have traced the emergence and development of diagnoses that have come to be particularly associated with childhood, adolescence, or young adulthood, including autism (Evans 2017; Hollin 2014; Hollin and Pilnick 2015), ADHD (Smith 2012), and self-harm (Millard 2015). Historians of child psychiatry must engage with its relationship to psychology (Thomson 2006), social work (Long 2011), psychoanalysis (Sugden 2020), and the changing presence of the mind and psyche in general practice (Hayward 2015), and must also recognize the relevance of cultural histories of intellectual disability, "imbecility," and "idiocy" (McDonagh et al. 2018; McDonagh 2008). Finally, historians working in and around the history of child psychiatry have become increasingly sensitive to the responsibilities historians have in relation to contemporary social and political concerns (Hendrick 2003), including those rooted in socially constructed and historically contingent conceptions of childhood "as a distinct phase in the socio-cultural, economic and demographic life cycle" and the emergence of "normative 'standards' and wider narratives of (adult and child) perfectibility" (King and Taylor 2017).

The significance of this history is a question of doing justice to the past, present and future. Criticizing the adult-centric nature of much humanities research, Wall argues that "If the humanities focus in some way on 'the human,' including its meanings, diversities, constructions, and possibilities, then it would be curious to neglect the third of human beings who happen to be under the age of eighteen. This situation would appear all the more peculiar if the humanities are charged, as many argue, with challenging normative assumptions and investigating historically marginalized voices" (Wall 2013, p. 61). In recent decades, historians have begun to explore childhood as an analytical category, both ready for productive deployment in research and subject to its own conceptual history, and have sought to recover and incorporate children's lived experience of the past. This shift has begun to take hold in the history of madness. More than an intellectual project generating enriched analysis of the past, the history of child psychiatry is also urgent for the recognition and restitution of past injustices. The history of madness is rife with mistreatment and abuse, and it is incumbent upon researchers to confront this head-on and to consider how childhood – with its connotations of vulnerability and subjection to adult authority – has intersected with disempowering machinations of mental health "care" to harm and dehumanize. Nor is present mental health care in Britain a panacea for children, adolescents, and young people, whose needs and concerns too often remain unaddressed and underfunded. Wall defines "childism," the challenge he sets to the humanities, as "the effort not only to pay children greater attention but to respond more self-critically to children's particular experiences by transforming fundamental structures of understanding and practice for all." (Wall 2013, p. 21). By interrogating the origins and evolution of current understandings and practices, historians can enable their transformation. Research in the history of child psychiatry is crucial if young people in the present and future are to be seen neither as passive victims of overmedicalization nor merely as fragile barometers for the ill effects of modern life, but rather as active and complex agents whose minds and behaviors have troubled social and cultural norms and scientific research for centuries.

Doing the History of Child Psychiatry

The historian of child psychiatry faces several challenges in identifying and interpreting primary materials, and the availability and limitations of extant evidence has circumscribed research in the field. It is not always possible to reconstruct children's experiences or to recover their own voices. However, it is possible – indeed, urgent and productive – to center conceptions of children, childhood, and children's bodies and minds, and the invocation of those conceptions in the development and workings of bodies of knowledge, institutions, and understandings of psychiatric care. There is ample historical evidence for this project, and every reason to pursue it. Taking children and childhood as a central object of study, the history of child psychiatry raises new questions and generates new insights about mental health care in the past. The vast range and variety of relevant organizations and individuals has necessarily produced a disparate and widely dispersed tranche of archival materials for the historian to navigate, offering different directions for future research.

Institutional Records and Patient Case Files

In Britain (and elsewhere), historians' attention has gathered most intensely on the modern history of madness, in part due to the availability of extensive records from the asylums that proliferated in the nineteenth century and the mental hospitals that succeeded them and dominated institutional mental health care until the mid-twentieth century. Annual reports, patient casebooks, personal and professional correspondence, photographs, and documents from attached pathological laboratories as well as archaeological evidence have furnished historians with rich materials for studying asylum administration, tracing the evolving treatment and management of patients in relation to changing beliefs about the nature of mental illness, and reconstructing the everyday lives of patients, staff, and visitors. Children made up a vanishingly small proportion of asylum residents, often meriting only a footnote in asylum history. Yet Taylor's systematic investigation of five English asylums and their child inmates in the second half of the nineteenth century demonstrates that institutions developed distinct approaches to the treatment of children, following childhood insanity in and out of asylums and through local networks of care and recovering the stories of a handful of children who were resident in asylums (Taylor 2017). A Taylor-esque examination of the presence of children in other asylums across the same period and in the mental hospitals of the first half of the twentieth century is much needed. Though administrative records and published research papers describe the patterns of organization and treatment in the mental hospitals, details of the day-to-day lives of patients are often difficult to ascertain due to the temporary restriction of archive records containing personally identifiable information. It seems likely that historians' accounts of mental hospitals will expand over time to rival the literature on asylums as extant documentation becomes accessible and matches or surpasses the volumes available from the asylums. Evidence of child psychiatry in practice can also be found in surviving records from care settings that were not explicitly "psychiatric" and have not always been included in narrower definitions of child psychiatry. Patient case cards from the pre-NHS children's clinic at Paddington Green Children's Hospital, for example, are a window onto the kinds of complaints parents presented on behalf of their infant children at the intersection of psychiatric and somatic medicine (Sugden 2020). The history of children's medicine and pediatrics is itself understudied, and further exploration of children's hospitals and children's wards within general hospitals may prove illuminating for the history of child psychiatry. Furthermore, records relating to children living in or passing through specialist systems from orphanages and foster homes to reformatories and correctional facilities have offered significant insight into both the emergence and definition of "problem" behaviors and identities, and the practical application of psychiatric and psychological ideas in nonmedical settings.

To the potential sources of institutional records must also be added the archives, reports, press releases, and journals of professional organizations, associations, and learned societies. Wardle (1991) lists 16 organizations influential in the development of child psychiatric services in the twentieth century (including the Royal Medico-Psychological Association and its Section of Child Psychiatry; the Child Guidance

Council; the Association of Child Psychologists and Psychiatrists; and the Association for the Psychiatric Study of Adolescents), to which can be added groups working in psychoanalysis (the British Psychoanalytical Society), learning disability (e.g., Mencap, originally The National Association of Parents of Backwards Children and later The National Society for Mentally Handicapped Children), social work (e.g., the Association of Psychiatric Social Workers and British Association of Social Workers), charity work (e.g., the Children's Society; the National Society for the Prevention of Cruelty to Children), and other related fields. Include local and national government bodies and other organizations that have emerged in the subsequent 30 years (e.g., the Office of the Children's Commissioner) and the historian is faced with a formidable array of material – though one spread widely and unevenly and therefore challenging to synthesize.

Legislation and Public Policy

The most easily accessible sources are those which have been published in print and those which can be accessed remotely. UK legislation and policy documents are a matter of public record, and many have been digitized and made freely available. These records allow for the tracing of government priorities and ideals; they also project particular conceptions of childhood and child mental health whether directly (as in the outlining of the state's duties towards children in, e.g., Children Act 1948; Mental Health Act 1959), indirectly, or by omission (children and infants are almost entirely absent from the Lunacy Act 1890). Enacted legislation and public reports have, of course, been tempered by the debates informing them; supplementary documentation such as interim reports and evidence submitted to committees is usually preserved in the National Archives and frequently proves illustrative of the nuances of debate and changing values within and between governments and those seeking to influence them. Barrett (2019) and Wardle (1991) both cite key legislation most directly relevant to child psychiatry, though further work is needed to contextualize changes in law, including deeper examination of the various stimuli for, catalysts of, and forms of opposition to change, as well as comparison of how changes in law were implemented in different localities.

Practitioners' Accounts: Histories and Testimony

Histories written by practitioners of child psychiatry and allied disciplines reigned for much of the second half twentieth century (Cameron 1956; Kanner 1959; Walk 1964; Warren 1970; Hersov 1986; Wardle 1991), beginning in the postwar period when child psychiatrists began to think of themselves as constituting a distinct specialty with its own history to be located within the broader history of medicine. Usually delivered in the form of lectures and papers presented to learned societies and professional associations or their journals, these texts are often characterized by a keenness to describe child psychiatry as neglected or confused in the past; to emphasize recent "advances" and present the field as having recently become consolidated in some way; and to present the field in the present as being at a crucial crossroads or stage of development. Lectures and articles describing the "history," "development," or "state" of child psychiatry represent attempts by professionals to locate themselves and their work in time and place and, in so doing, to establish disciplinary boundaries through strategic inclusion and exclusion, aligning themselves with distinguished precursors and distancing from ostensible embarrassments of the past. These papers also tend to stake particular claims on the future, by promoting particular theoretical orientations, approaches to treatment, research directions, policy positions, and allocations of resources. Such texts are valuable as barometers for the contemporaneous concerns of leading practitioners and as evidence of the self-understanding and self-presentation of child psychiatrists both as individuals and as an emergent professional group. These accounts, then, are of value both as descriptions of the development of child psychiatry and when read as polemics about the state and direction of the field at the time they were written.

Practitioners also offer invaluable insight into the past in the form of direct testimony shared through oral history initiatives, including traditional one-on-one interviews and "witness seminar" formats that bring together groups of interested parties to discuss their experiences. It is promising that recent witness seminars in the history of psychiatry have represented a coming-together of practitioners and historians, whose relationship in the past has sometimes been tumultuous as critical historical accounts have been met with indignation as challenges to medical authority. Whether proceedings from witness seminars have been transcribed and presented in full (Graham et al. 2009) or published after summation and analysis mediated by historical expertise (Turner et al. 2015), they have captured stories that would surely otherwise have been lost to the historical record and provided stimulation for future research directions – as well as reminders that the categories "historian," "practitioner," and "service user" are not mutually exclusive (Turner et al. 2015). The next step for oral history initiatives in the history of child psychiatry must be to concentrate effort on diversifying participation by race, gender, and (dis)ability, and to ensure that the voices of nurses, therapists, social workers, and others working in allied disciplines are included. In the meantime, the seventieth anniversary of the founding of the NHS has furnished historians with material in the form of reflective pieces across many organizations in print and online, and the "NHS at 70" (2021) oral history project has collected relevant testimony from many practitioners and service users. New oral histories and direct testimonies are, of course, restricted to living memory and therefore provide material only about more recent developments.

Personal Papers and Correspondence

Many published collections of prominent practitioners' papers and correspondence are available and provide insight into their own professed beliefs and experiences. These selected and edited works (organized as they are around the life and work of an individual) generally omit the quotidian, incidental, and ephemeral remnants of medical practice – precisely the texts which might offer glimpses into the experiences and treatment of patients, illuminate care pathways, or demonstrate the links between different services. For example, the letters selected for publication by the editors and contributors of *The Collected Works of D.W. Winnicott* (Caldwell and Taylor Robinson 2016) are those which exemplify the pediatrician, psychoanalyst, and self-styled child psychiatrist's position on a given topic, contain particularly clear or poetic exposition of his theories, or demonstrate his interactions with fellow analysts and other prominent colleagues. These published examples are necessarily but a small portion of the vast surviving archive, where the majority of correspondence relates to referrals, requests for advice, and invitations to speaking engagements, interviews, or to write for publications. Usually each letter alone does not merit focused attention and analysis. Yet, these letters individually contain small details that build up into bigger pictures. Taken as a whole, such tranches of correspondence – which survive for many psychiatrists, psychologists, and psychoanalysts who worked with children – are a valuable source of information about the everyday realities of professional life in the care of children's mental health. They contain snippets of cases, capturing vestiges of children's lived experience and offering insight into the behaviors that parents, teachers, and doctors deemed problematic, the various treatments children were subject to, and the referral and care pathways they moved through. Often private correspondence reveals deeper layers to more public debates, capturing the related disputes, anxieties, or coordinated actions that preceded or ran in parallel with professional discussions; they also embody and therefore reveal connections between individuals and organizations, preserving the skeleton of a complex network of services for children. The Winnicott archives are a powerful reminder of the number and range of people and organizations concerned with child mental health in mid-twentieth century Britain: professional bodies (the Association for Child Psychology and Psychiatry; the Royal Medico-Psychological Association; British Paediatric Association); university departments of sociology, social science, social work, and psychiatry; charities (the National Association for Mental Health; the National Society for the Prevention of Cruelty to Children; the National Children's bureau); and local Children's Officers and Social Services Departments (Sugden 2020). Deeper engagement with these organizations and their histories – what conceptions of childhood they deployed; how they understood medical and psychiatric care; what contemporaneous issues came to the forefront of their work; and how these changed over time – will be crucial to understanding the scientific, social, and political milieu that catalyzed the professionalization and institutionalization of child psychiatry at mid-century.

Print and Media

Although an increasing proportion of children, adolescents, and young adults are identified as in need of support from Child and Adolescent Mental Health Services in Britain, the fact remains that a majority of children and families are not "known to" and have not interfaced with psychiatric services. And yet all have likely interacted with child psychiatry and psychology via books, news headlines, magazine articles, film, and television. Tracing the changing ways in which child psychiatric knowledge has been represented in print and on film is especially useful for exploring popular understandings of and engagement with ideas about the mind of the child, and – conversely – the influence of social and cultural representations on professional opinion and resulting changes to legislation and clinical practice. The history of child psychiatry has already benefitted from several studies that engage with cultural as well as scientific material: Shuttleworth has identified the co-emergence of an interest in the child's mind in literature as well as scientific writing in the late nineteenth century (Shuttleworth 2010); Urwin and Sharland's study of parenting advice notes increasing discussion of children's minds in the interwar period (Urwin and Sharland 1992); and the impact of James Robinson's short films "A Two-year-old Goes to Hospital" and "Going to Hospital with Mother" are widely cited as influencing changes to hospital practice in the 1950s by promoting the importance of infants' attachment to their mothers. News media is capable of drawing attention to injustice and scandal on the one hand, and fanning the flames of moral panic on the other, while consumption of popular culture – especially via modern technologies – is entangled and implicated in the emergence of novel pathological phenomena (e.g., gaming addiction) and novel experiences of mental health and illness, including within both supportive communities and dangerous online spaces (e.g., "pro-ana" websites). Greater historical engagement with these cultural milieus will enrich understandings of children's experiences.

The Voice of the Child

Within all of these materials, children's voices and experiences are notoriously difficult to coax out. A significant but fragile distinction must be sustained: that between "real" children physically present in the past, and the conceptions of children and childhood held and deployed by adults in the past. In many ways, the latter more easily lend themselves to historical study, since the majority of the historical record was created by adults. To research the history of children's medicine on the fringes of living memory is to work with traces of children's lives that are mediated by parents, doctors, teachers, social workers, and others; and documents are often anonymized, restricted, closed, or – in the case of many psychoanalytic case files – deliberately destroyed. The challenge of finding children's voices in the medical archives is multiplied by patient confidentiality and data protection, as well as the stigma often associated with mental illness: the material that is available deserves to be treated with sensitivity, and historians must also appreciate that past and present reticence and desires for privacy leave a silence in the archive whose exact shape and nature may be unknowable. Direct testimony from child patients is rare, and recollections from adults who were child patients is sometimes vague: as a former child wrote to D.W. Winnicott in 1966: "I consider I was fortunate to be your patient at Paddington Green donkeys years ago. All I remember is that I liked you! . . . I recall nothing of what you said to me" (Sugden 2020, p. 37). Some adults, though, do have vivid memories of their treatment as child patients, and historians as a whole could do more to systematically engage with the ethics of accessing and doing justice to these stories, as well as to broaden their conception of relevant firsthand evidence by considering non-pathological mental health and emotions alongside the illness captured in medical and psychiatric records. Here, resources from past and ongoing large-scale research projects could be of use: initiatives like the Mass Observation Archive (2021) and tools like those developed by Cohort and Longitudinal Studies Enhancement Resources (CLOSER) (2021) provide access to a wealth of data and insights about everyday life and health in twentieth-century Britain that could complement and contextualize the histories of the pathological versus the normal in children's mental health.

Tangled Roots: The Study of the Child and Child Insanity at the Turn of the Twentieth Century

The history of research and practices centered on children and their minds over the last 150 years is a story of branches growing apart only to meet, cross, and diverge once again. Child psychiatry is one of these branches, variously crisscrossing the worlds of child guidance, pediatrics, public health, education, psychology, psychoanalysis, and social work. Several accounts identify the period following the First World War as one where interest in generating knowledge about and practicing the regulation of children's bodies gave way to interest in their minds and emotions (Hendrick 1994; Urwin and Sharland 1992), with the first significant precursors to child psychiatry to be found in the Child Guidance Movement imported from the United States in the 1920s. Child Guidance was certainly crucial for the development of child psychiatry in Britain: it trained staff, established infrastructure, generated knowledge and discussion, influenced policy, and formed links with families, communities, and local services, normalizing the discussion of child development and the intercession of professional practitioners in family life. Yet a fascination with the child's mind, one that spurred research and debate and provided points of resonance that made Britain receptive to Child Guidance in the interwar years, was already apparent before the turn of the twentieth century.

References to and case reports of children displaying symptoms and behavior indicative of mental illness can be found intermittently throughout written history, as former RMPA President Alexander Walk emphasized in his address to the Association's Child Psychiatry section in 1964. Offering a "Pre-history of Child Psychiatry," Walk employed a laudably broad definition of the field and surveyed medical and philosophical literature from the previous 500 years for discussion of relevant cases, citing descriptions of epilepsy, sleep disturbance, dreams and nightmares, bedwetting, stammering, jealousy, hysteria, demonic possession, despondency, violence, uncontrollable behavior, schizophrenia, moral insanity, feeble-mindedness, idiocy, psychosis, overtaxed brain, inflammation of the brain, suicide, stealing, melancholia, mania, dementia praecox, adolescent insanity, neurasthenia, and juvenile general paralysis of the insane (Walk 1964). Walk's amble through history reveals an eclectic mix of "disorders," from common phenomena associated with childhood and youth (bedwetting; sleep disturbance), to physical diagnoses (inflammation of the brain), and modified descriptions of adult disorders (*juvenile* general paralysis of the insane) – a result of the uncoordinated researches informing the literature he cites. Walk drew few links with social and cultural context (though he did note that concerns about "scholastic overstrain" coincided with the expansion of educational opportunities for girls and women), but he did emphasize one turning point: he was clear in his identification of a general trend of multiplying case reports relating to children in the 1870s and 1880s (Walk 1964, p. 761).

The second half of the nineteenth century is precisely where Shuttleworth locates the "pre-history" of child psychiatry in Britain, based on a wide-ranging analysis of literary and scientific texts that reveals a newly widespread interest in the mind of the child. Between 1840 and 1900, Shuttleworth demonstrates, "the inner workings of the child mind became for the first time an explicit object of study across the cultural and disciplinary spectrum, from novels and autobiographies to psychiatric case studies" (Shuttleworth 2010, p. 2). Writers and researchers "shared preoccupations with the child, rather than a single definitive image of the child mind" (Shuttleworth 2010, p. 361), approaching their object with various techniques in their attempts to discern particular truths, whether poetic, social, evolutionary, psychological, or psychiatric. Escalating interest in the study of mental disorder in children is evidenced by the amount of space devoted to mental disorders of childhood in *The Journal of Mental Pathology and Psychological Medicine* (1848–1860, the first journal of its kind) which printed articles discussing the nature and prevalence of childhood mania, criminality, idiocy, suicide, and hysteria (Shuttleworth 2010, p. 28, 33). At the same time, anthropometric and observational data and early forays into experimental psychology were generating discussion about the development of children's senses, perceptions, and cognitive abilities. Heavily influenced by recapitulation theory (ontogeny recapitulates phylogeny), which "g[ave] the child an unprecedented position in the domain of knowledge, no longer an afterthought or irrelevance but the primary source of evidence and key to understanding in a range of disciplines," psychological and anthropological thinkers found in the child "an index of evolutionary development... an iteration of parental or species history rather than an entity in its own right"; a source of information about humanity in general and not children themselves (Shuttleworth 2010, pp. 344–345). Evolutionary and recapitulatory ideas underwrote psychiatric thought, too, manifesting in an emphasis on heredity, degeneration, and the tensions between children's undeveloped physical and mental state and the stresses of their modern environment: children's inheritance left them "doubly burdened" as "the carrier of primitive, animalistic passions, but also the attenuated nerves of an overdeveloped civilization and unbidden memories of the past" (Shuttleworth 2010, p. 353).

A series of publications around the turn of the century demonstrate some of the changing attitudes and new approaches towards mental illness and learning disability in children that followed this earliest period of interest in the child's mind. James Sully (1842–1923)'s 1896 book *Studies of Childhood* exemplifies the psychological approach prominent in the Child Study Movement at the turn of the century, as well as the gulf that already existed between child psychology as a field of research and early child psychiatric practice or the medical treatment of the child's mind. Sully was a psychologist who would go on to found the British Psychological Society. Well aware that the figure of the child had a rich literary and philosophical heritage, Sully cites Rousseau, Blake, Dickens, and Hugo in his introduction and positions his own work as their scientific corollary: "With the growth of a poetic or sentimental interest in childhood there has come a new and different kind of interest. Ours is a scientific age, and science has cast its inquisitive eye on the infant... we can now speak of the beginning of a careful and methodical investigation of child-nature, by men trained in scientific observation" (Sully 1896, pp. 3–4). And it was the "scientific observation" and description of the normal child – especially in relation to human evolution – that Sully concerned himself with in this book. A university professor rather than a medical man, his work had little to say about childhood madness, insanity, or idiocy. Rather, he sought to "trac[e] back [...] the complexities of man's mental life to their primitive elements in the child's consciousness," exploring simultaneously "the pathway both of the child-mind and the race-mind" (Sully 1896, pp. 7–8) – he sought the answers to more abstract questions than how to care for the mentally ill or learning disabled. It is significant, though, that many of the phenomena that Sully described in detail and from which he drew insight into children's thought processes were those which would later become diagnostically and therapeutically central to child psychiatric work, including imagination, play, art, questioning, fears, and lying.

In 1887, John Langdon Down (1828–1896), a former asylum physician, senior physician at the London Hospital, and founder and medical superintendent of Normansfield (a home for patients with learning disabilities), published *The Mental Affections of Childhood and Youth*, a book based on a series of his lectures. Although there is a brief discussion of mental illness (infantile mania; delusions; melancholy; moral insanity (Langdon Down 1887, pp. 91–96)), the book was devoted to the subject of "idiocy," "imbecility," and "feeble-mindedness" – though early on Langdon Down disavows the first of these terms: "I have no great liking for the term idiot. It is so frequently a name of reproach" (Langdon Down 1887, p. 5). Langdon Down categorized these "afflictions" by their supposed causes (congenital; accidental; developmental) and argued that "afflicted" children should be moved into institutions where they could be cared for appropriately. He emphasized that "basis of all treatment should be *medical*" (original emphasis), but "Medical, I mean in enlarged sense." The treatments he outlined were environmental, educational, and occupational: he advised that residents be kept warm, housed in spacious and well-ventilated bedrooms, and bathed regularly, and that they should be given a healthy diet, outdoor exercise, and physical training as well as moral and intellectual education and instruction in appropriate practical activities like gardening, farming, carpentry, and needlework. Corporal punishment should not be used, and each child should be "surrounded by influences of art and nature calculated to make his life joyous, to arouse his observation, and quicken his power of thought" (Langdon Down 1887, pp. 132–142). Landgon Down's vision of care for learning disabled children shares similarities with the moral management of asylum inmates pioneered by William Tuke at the York Retreat at the end of the eighteenth century and implemented in other institutions as the nineteenth century progressed; the expanding concept of "the medical" and the notion of environment and occupation as therapeutic in themselves would also resurface in the twentieth century. That Langdon Down set out this approach for children with learning disabilities in print almost a hundred years after the York Retreat was founded perhaps suggests that in general the treatment of these children at home and within other institutions may have been worse than that of children and adults in asylums at the same time. Though lauded for his work with learning disabled children and adults, Langdon Down must also be remembered – like so many of his fellows and colleagues – as a believer in race science. He freely shared his thoughts on "the remarkable resemblance of feeble-minded children to the various ethnic types of the human family": here was a take on recapitulation theory that equated learning disabled children with particular ethnic and national groups within a supposed racial and civilizational hierarchy (Langdon Down 1887, p. 7).

William W. Ireland (1832–1909)'s 1898 *The Mental Affections of Children: Idiocy, Imbecility, and Insanity* groups together childhood learning disabilities and mental illness, presenting these pathologies as closely related (he maintained that "Insanity in very young children is always accompanied or masked by idiocy") and as extreme examples of what he understood to be a fundamental difference between adults and children. Drawing on his experiences as medical superintendent of the Scottish Institution for the Education of Imbecile Children and as medical officer of Miss Mary Murray's Institution for Girls at Preston, Ireland emphasized difference in adults' and children's mental processes, casting the child as primitive and animalistic:

> Many of the mental processes which play so important a part in the life of the adult are either wanting or rudimentary in the child... The sanity of a child is something quite different from that of a man... the senses in the child are most acute: pain, disgust, deprivation, are resented with passionate keenness and provoke bursts of wrath or weeping... With a deficient experience and imperfect judgment children believe whatever is told to them, invent fictions, and make grotesque assertions; they are subject to vain terrors, and smile at real dangers. Some children are apt to indulge in acts of cruelty to animals and gloat over tales of bloodshed. (Ireland 1898, pp. 279–282)

Ireland offers an overview of types of insanity observed in children, including melancholia, suicide, mania, general paralysis, post-febrile insanity, and toxic insanity, but describes very little by way of treatment and management of such cases (Ireland 1898, pp. 294–306). His emphasis on the differences between adults and children represents another strand of late-nineteenth century medical and psychiatric thought: one that normalized the adult and medicalized childhood itself based on a conception of the child as incomplete, imperfect, and vulnerable to particular sicknesses of the mind and body.

This casting of children as a group of patients qualitatively different from adults and requiring special provision was encapsulated in the rise of the children's hospital in Victorian Britain. A range of specialist hospitals emerged over the course of the nineteenth century, organized around particular body parts (skin; chest; eye), by type of disease (nervous disorder; cancer; venereal disease), or associated with particular types of patient (women; children). This speciation represented the structuring of medical knowledge along new organizing principles; children's hospitals embodied a separation between adult and child patients and the nature of their illnesses (Sugden 2020). This gathering of children's medicine, precursor to pediatrics, included the appearance of a new figure in medicine: the children's doctor; a practitioner with special interest in and knowledge of children their health. And in, in the practices of children's medicine by children's doctors, more seeds of child psychiatric thought germinated. Leonard Guthrie (1858–1918), who spent most of his career at Paddington Green Children's Hospital, was one of the first physicians to publish a book about "nervous" children and their "disorders." Published in 1907, *Functional Nervous Disorders in Childhood* emphasized the importance of the internal emotional life of the child. Guthrie advocated for medical attention to be directed beyond children's bodies to their emotions, arguing that "Many nervous and other ailments are the outcome of the neurotic or emotional temperament, and all are aggravated thereby" (Guthrie 1907, p. 1), and sought in the child's mind causes of physical symptoms like asthma, chorea, and tics. Guthrie's view not only raised moral and emotional well-being to the same standing as physical health, but proposed a holistic worldview in which they were intertwined: "health and happiness go together," he wrote (Guthrie 1907, p. 7). He also invoked the preventative dimension of child psychiatry, presenting his book as intended to "emphasize the truism that the nervous child is the father of the neurasthenic adult" (Guthrie 1907, p. 1). This concern to attend to the mental health of children in order to secure the mental health of adults was felt with increasing urgency as national and imperial concerns about the health of Britain's population were sharpened by the crises of the First World War.

Children's Departments in other hospitals also produced specialists in children's mental health, especially as the nervous and anxious child became an object of fascination for parents, educators, and doctors. Hector Charles Cameron (1878–1958) served as physician in charge of the Children's Department of Guy's Hospital and published *The Nervous Child* in 1919. Like Guthrie, he promoted the centrality of the study of infancy and childhood to the work of preventive medicine and emphasized the existence and importance of a complex emotional life in even the youngest children. Cameron's writings draw on the fields that had turned their attention so intently towards the child at the end of the nineteenth century (he described the book as an introduction for those "seeking to apply there the teachings of Psychology, Physiology, Heredity, and Hygiene" to the care of children), but gone is the primitive creature of instinct from humanity's distant past, replaced by a being who is whole and individual: "The newborn baby has a personality of his own, and mothers will note with astonishment and delight how strongly marked variations in conduct and behaviour may be from the first. One baby is pleased and contented, another is fidgety, restless, and enterprising." These differences are reason for early skepticism about the strict sleeping and feeding schedules contemporaneously promoted by Truby King and others: for Cameron,

"A rigid routine in sleep is a good thing, but the routine belongs to the baby, not to the nurse... We may wreck everything by a blind adhesion to a too rigid scheme." Newly present, too, in Cameron's work is the mother and her close relationship to her child: this speaks to the new maternalism that came to characterize child-rearing and child care in the interwar period. As well as considering the relationship between doctors, mothers, and children, Cameron urged mothers not to blame themselves for difficulties breastfeeding, and advised bodily contact with the mother to calm children. In places Cameron's advice for the care of the nervous child is strikingly similar to Langdon Down's prescriptions for learning disabled children: he recommends management of the environment to keep children warm, expose them to fresh air, and reduce unnecessary stimuli, and suggests the provision of mental, moral, and physical training. Elsewhere, though, Cameron also advises the use (albeit sparingly) of sedatives like bromide and chloral (Cameron 1919, pp. 104–116). When considered as a text published during the beginnings of child psychiatry, one final aspect of Cameron's book seems especially prescient. His discussion of nervousness in older children begins with a consideration of the relationship between the doctor and his young patient: the doctor "must secure the child's confidence," whose treatment "calls for much insight":

> The child's confidence must be completely secured, and he must be encouraged to tell of all his sensations and of the reasons which prompt his actions. The nervous child has a horror of appearing unlike other children, and will suffer in silence. If his troubles are brought into the light of day with kindness and sympathy they will melt before his eyes. Even night-terrors are, as a rule, determined by the suppressed fears of his waking hours. (Cameron 1919, p. 135)

The attempt to include the child's perspective, the recognition of his sense of self and social world, and the suggestion that cultivating an understanding of his "sensations" and "reasons" is a marked change from the wild, violent, or hopeless children in the accounts of childhood insanity and learning disability that had begun to emerge 70 years earlier.

As exemplified in John Langdon Down and William W. Ireland's books cited above, proto-psychiatric texts on the child's mind frequently drew on the author's experiences working within nonmedical institutional settings for children. Schools, reformatories, and correctional facilities often employed or sought advice from physicians and psychologists with special interests in children's health and development. At the same time, they held populations of children who could be measured and observed with relative ease. Especially within correctional and specialist educational institutions, children's very presence was the result of, and a further cause for, scrutiny of their behavior and development. As particular behaviors and (dis) abilities became understood to be criminal, deviant, or otherwise lacking, they became "problems" ripe for medico-psychological investigation and intervention – in borstals and reformatories, this was increasingly the case after the 1940s as the public school-based model popular in the interwar period was supplanted by psychiatrically informed approaches including assessments, referrals, and treatment undertaken by psychiatrists (Jackson and Bartie 2014; Tebbutt 2020). Such nonmedical settings for the gathering of data and deployment of psychological and psychiatric ideas about children played a significant part in defining pathological behaviors and therefore shaping research problems (for example, defining and explaining child criminality and juvenile delinquency) around which child psychiatry could crystallize.

Shuttleworth credits the Victorians with "grant[ing] the child a new interiority, a complex subjectivity, complete with passions and traumas which defied easy analysis" (Shuttleworth 2010, p. 359). This turn to the mind of the child is borne out by studies of children as welfare recipients and as subjects of medical and parental authority. Levene's review and analysis of scholarship on the history of the "welfare child" identifies a shift from concerns about the body to concerns about the mind at the beginning of the twentieth century, an alteration in discourse and practice resulting from changing national priorities, increased scientific knowledge, fear of depopulation and degeneration, and the rise of the women's movement and maternalism in public policy (Levene 2006). Likewise Urwin and Sharland have demonstrated in their analysis of childcare literature in interwar Britain that there was a shift from a focus on bodies to a focus on minds, rooted in the influence of psychological theory and child guidance practices (Urwin and Sharland 1992). Analyses of texts about the psychological study of the child and the treatment and management of children suffering mental illness or learning disability, whether labelled as "nervous" or "anxious," "idiots" or "imbeciles," demonstrate the emergence of medical professionals who had cultivated a specialism in this area and their extension of knowledge derived from the study of the child to applications in medical and proto-psychiatric settings; later this knowledge percolated into nonmedical sites including correctional and educational institutions. Particularly after the turn of the century, the significance of the child's mind and inner emotional life was increasingly taken seriously and met with clinical response. As Cooter has argued, "In 1880 child health and welfare was not yet medicalised; instead it was a set of interests converging on the child. But by the 1920s child health and welfare was not only medicalised, it was serving as a powerful argument for extending the role of the state in health and welfare generally" (Cooter 1992, p. 12). It was in these circumstances that medical and psychiatric treatment of the child's mind expanded in hospitals and clinics, communities, and the home.

Child Psychiatry, Child Guidance, and Child Psychoanalysis Before the NHS

The period between the end of the First World War in 1918 and the establishment of the NHS in 1948 was one of acceleration and expansion of medical interest in the child's mind, laying ground for the speciation of child psychiatry and its integration into medical care and the British welfare state. Hypotheses and treatments were explored within psychiatric provision for children in hospitals, shaping the separation of child from adult psychiatry and determining the line of development it would follow in subsequent decades. Nascent concerns about child development found expression in the rise of the Child Guidance Movement, facilitated by philanthropic funding and an organizational model imported directly from the United States. The new Child Guidance Clinics brought together multidisciplinary teams of psychiatrists, psychologists, and psychiatric social workers (PSWs), took psychiatric knowledge and practice into communities and homes, and disrupted mutually exclusive notions of the normal and the pathological child under the banner of "maladjustment," extending psychiatric authority beyond acute and chronic cases to, potentially, any child. Psychoanalysts in Britain also turned their attention towards children at this time, with a small network of private practices and a children's department at the Institute of Psychoanalysis. Psychoanalytic ideas that found little traction in psychiatry were particularly influential in parenting, child care, and social work – another reminder of the interconnectedness of different disciplines and services and the multiplicity of approaches within them.

The interwar years in Britain saw the emergence of the mental hospital, as the old asylums began to reform and rename themselves. There were also new institutions, including the Maudsley, which opened as a mental hospital in 1923 and soon rose to prominence within British psychiatry, quickly becoming a leading center for teaching, training, and treatment. In 1948, it became home to the newly created Institute of Psychiatry under the charge of Aubrey Lewis, appointed as the first Chair of Psychiatry in Britain. At its opening, the Maudsley had a nominal children's department headed by D.W. Dawson, and analysis of surviving case and administrative records by Evans, Rahman and Jones provides insight into its workings and development through its first decades. An afterthought in the original plans for the hospital, the children's department initially saw a small number of children referred by friends or private doctors (Evans et al. 2008, p. 460). The children were physically examined for injuries, to ascertain their general constitution, and to identify any unusual sensory-motor reflexes; any past traumas or shocks were noted, as were signs of mental illness in relatives. They might be diagnosed with epilepsy, mental defect, neurosis, hysteria, or moral abnormality, or labelled with the colloquial terms "backward" or "nervous" (Evans et al. 2008, p. 456). This initial examination reveals the acceptance of a range of causes of mental disturbance and therefore the influence of multiple theories: from heredity to emotional disturbance or damage to the nerves caused by traumatic events. Mental illness was also to be differentiated from learning disability by the application of an intelligence test in most cases. Evans, Rahman, and Jones also note that dementia, mania, dementia praecox, and schizophrenic state – terms more usually reserved for adult patients – were among the diagnoses recorded for some postpubescent children, though they do not elaborate on this pattern (Evans et al. 2008, p. 457), and note some difference in patterns of treatment between adolescents and younger children, with the former thought to be better able to participate in and benefit from psychodynamic methods like free association and intensive psychotherapy that required cooperation and insight from the patient. This evidence of differential diagnosis and treatment by developmental stage is of significance to the history of adolescent psychiatry and its troubled existence both within child psychiatry and on the fringe of adult psychiatry (the emergence of specific adolescent units and departments and the organization of research and professional expertise around "the adolescent" was a postwar phenomenon, and one which ebbed and flowed (Turner et al. 2015, pp. 86–90)). More generally, the treatments employed at the Maudsley offer further evidence of the mix of approaches utilized in child psychiatry between the wars, from bed rest, fresh air, and bathing to hormonal treatments, sedatives, and anticonvulsants. Evans, Rahman, and Jones argue that this did not necessarily indicate etiological confusion: these treatments could be thought of as existing along a continuum, with psychoanalytic methods "merely regarded as more sophisticated ways to intervene in the child's unconscious life... sedatives and continuous baths could calm nerves and instinctive reflexes, while dream association and interpretation enabled the children to master unconscious impulses and drives" (Evans et al. 2008, p. 459).

The work of the child department at the Maudsley expanded rapidly, from 90 cases in 1924 to 432 in 1931 and 839 in 1935; at the same time, an increasing number of referrals came from London County Council agencies: the hospital was increasingly a major hub within a sociomedical network of organizations centered on the child, from child care committees and child guidance centers to schools and probation services. Under the surveillance and authority of this network, children were identified as mentally disturbed or learning disabled, and the Maudsley developed a crucial role in categorizing children's problems as mental, intellectual, or moral, and directing them to treatment, special education, or removal to an institution. This work required a shared language and conceptual framework: Evans, Rahman, and Jones argue that the Maudsley began to incorporate classifications and theories of child development that were used by local authority services, abandoning psychophysiology, evolutionary theory, and, later, individual psychological drives as the child care landscape came to focus on social and domestic rather than individual explanations (Evans et al. 2008, pp. 460–464). Further examination and comparative study of different child care services in this period might clarify how and why linguistic and conceptual convergence occurred in the sector, and what narrowing effect this might have had on the language and practice of child psychiatry in its formative years.

At the same time as a rising number of children received psychiatric care in hospital, an even greater number were brought under psychiatric authority and intervention by the rise of Child Guidance in Britain. Child Guidance was an international movement growing out of the Mental Hygiene Movement, which originated in the United States at the beginning of the twentieth century and promoted a preventive attitude towards mental health, turning medical and psychiatric attention towards infancy and childhood under the influence of the psychoanalytic and psychodynamic ideas. It is important to note that in interwar Britain, a number of clinics doing psychiatric work with children were established independently of the American influence and yet as part of the same wave of interest in child development and mental health. When the Tavistock Clinic opened in 1920 under the impetus of Hugh Crichton-Miller, a psychiatrist with psychodynamic leanings, its first patient was a child; children and families continued to be treated there alongside adults, and the Clinic developed its own training program for psychiatrists, therapists, and social workers working with children. In 1928, Margaret Lowenfeld

(1890–1973) opened her Children's Clinic for the Treatment and Study of Nervous and Difficult Children (later the Institute of Child Psychology), where she was a leading play therapy practitioner and theorist. Nevertheless, the majority of clinics doing psychiatric work with children were established under the auspices of the Child Guidance Movement, facilitated by American philanthropy, firsthand observation of how Child Guidance Clinics operated in America, and the wholesale implementation of the American organizational model and its psychiatrically-led team approach to cases. Stewart's historical analysis of Child Guidance in Britain between 1918 and 1955 charts the almost exponential spread of Child Guidance Clinics across the period, from a single clinic in London in 1927 to more than 60 by 1939 and over 300 in England and Wales by 1955 (Stewart 2014, p. 10). Children were referred to the clinics by their schools or parents (and, very occasionally, by primary care practitioners) with complaints such as stammering or bed-wetting. A team was assigned to each case, and diagnosis, advice, and treatment were given on the bases of in-person assessments at the clinic and home visits. Child Guidance in Britain was characterized by tension between the psychiatrists who led case conferences and the psychologists and psychiatric social workers who formed the rest of the team around the child. As Stewart notes, the American model firmly placed psychiatry in a position of prominence and this model was notionally instituted in Britain; yet British Child Guidance Clinics had closer links to the education system and its emergent practices of surveillance than to the medical establishment, and this favored the child psychologists and their valued expertise in cognitive development and psychological testing. The PSW's subordinate position was less hotly contested, though led to tension: the PSW had a closer relationship to and personal knowledge of children and their families developed through home visits, since families would encounter the psychiatrist and psychologist only in the Clinic itself (Stewart 2014). In other words, the structure of the Child Guidance team embodied a hierarchy (albeit an uneasy one) of knowledges and practices: the medical-psychiatric, the scientific-psychological, and the social-domestic. The rise of Child Guidance represented an expansion of state and medical authority over more families in ever more intimate ways. As Stewart emphasizes, under the influence of the preventive goals of mental hygiene, Child Guidance was concerned not with the "mentally deficient," delinquent, or mentally ill child, but with maladjustment in the "otherwise normal child." Psychological research was already concerned with the development and processes of mind and brain in the normal child; for psychiatrists, though, this concern represented an expansion of their attention from more severe and/or institutionalized cases posing difficulties in the present to milder symptoms that, it was believed, could lead to deeper problems in the future if left untreated. The interaction between psychiatric ideas and domestic life also became more intimate under the guise of Child Guidance, concerned as it was by the child's home environment, and particularly the relationships therein. Despite this, the British Child Guidance Movement did not, on the whole, employ the psychoanalytic leanings of its American progenitor: in Britain, psychoanalysis was institutionally isolated, making few inroads into hospitals and universities – and therefore psychiatric training and practice – except in the diluted form of the various psychotherapies deployed alongside other treatments. Since the psychiatrist was nominally at the head of the Child Guidance team, psychoanalysis was unlikely to be the prominent school of thought.

If British Child Guidance resisted the intrusion of psychoanalytic ideas into the realm of child psychiatry, it was not without a struggle from the British Psychoanalytical Society (BPS) who provided preferential and discounted analyses for PSWs and opened its own Child Department at the Institute of Psychoanalysis in 1941 to treat children from families who could not afford to pay for private analysis. Although it remained institutionally independent and made few formal inroads into medical and psychiatric training or hospital care, British psychoanalysis seems to have enjoyed an influence disproportionate to its numbers in the first half of the twentieth century, particularly in social work and education. The BPS had fewer than 100 members across this period (almost all of them resident and practicing in London) and only a small proportion of these practiced child psychoanalysis, which required additional training and accreditation. This handful of specialist child analysts in turn met few patients; even a preternaturally busy leading child analyst like D.W. saw no more than 40 or 50 new child patients each year (Sugden 2020).

Like psychologists, psychoanalysts had originally been interested in the mind of the child insofar as it might be the key to the mind of the adult; but they soon sought to exploit the therapeutic potential of the psychoanalytic setting for children (Sugden 2020). The development of child psychoanalysis in Britain was catalyzed by the arrival of two Viennese child psychoanalysts who began their careers on the Continent before moving to London in 1925 and 1938, respectively, and establishing themselves as preeminent members of the BPS: Melanie Klein (1882–1960) and Anna Freud (1895–1982). Klein and Freud inspired, analyzed, and trained the first generation of British child psychoanalysts; yet their presence and rivalry in the BPS also led to turbulence. Despite the paucity of psychoanalytic treatment for children, there was perhaps no more crucial theoretical question in British psychoanalysis by mid-century than the workings of the minds of babies and infants: Klein and Freud had radically opposing views on the infant psyche, which boiled over into a series of heated arguments at the BPS in 1941–1945.

Though internal strife plagued the BPS and formal links with and influence in medicine and psychiatry were not forthcoming, psychoanalytic thought was not entirely absent from child psychiatry, broadly defined. The psychoanalytic child was popular with parents and educators, having gained a foothold in their imaginations at the beginning of the interwar period, and analysts were able to make use of formal and informal opportunities like local lectures, training seminars, articles for specialist and general audiences, and radio broadcasts to share their interpretation of what went on in the child's mind. Some prominent individuals in education, child guidance, and social work had some training or interest in psychoanalysis, not least Susan Isaacs (1885–1948), who was the first Director of the Department of Child Development at the Institute of Education (IoE), University of London in 1933. Isaacs embodied the interrelation of different fields and professions that were focused on the child: she was a trained teacher with a degree in philosophy and postgraduate study in psychology who was also an active member of the BPS. Under her direction, the Child Care Course at the IoE became popular with teachers and social workers, and was infused with psychodynamic ideas – including, in the 1960s, lectures by D.W. (Sugden 2020).

Conclusion: Enduring Eclecticism; Contemporary Concerns

In Britain, the mind of the child and the subject of child "insanity" were of abiding concern from the middle of the nineteenth century. By the turn of the century, psychological and anthropometric studies of children and child development were well-established and informing discussion of the nature of heredity, humanity, and interiority as well as the more practical questions of parenting advice, education, and medical care. Notions of the importance of understanding and managing the child's mind, development, and environment were embodied in several texts at the turn of the century, and the early decades of the twentieth century witnessed emerging examples of specialist child psychiatric care, including the children's department at the Maudsley. In these contexts, a mixture of treatments with varying proposed mechanisms and associated with a host of ostensible etiologies of madness were in use, often within the same institution and even at the hands of the same individual practitioner. Through the interwar years, psychiatric care of the child was centered in institutions – mental hospitals; child guidance clinics; special schools – operating as part of a growing network of state and philanthropic services; increasing cooperation and coordination between these moving parts generated shared language and conceptions that shaped child psychiatry in its nascent form. Nonmedical settings, too, served as incubators of psychiatric thought and practice – especially institutions within the education and criminal justice systems. As the twentieth century wore on, some of the evolutionary and psychological influences of the late nineteenth century fell away, and psychodynamic ideas found more traction with social workers, educators, and parents than with psychiatrists and psychologists. A handful of independent clinics and centers continued to plough their own furrows. Psychiatric authority over the mind of the child was entrenched by the implantation of the American-model Child Guidance Movement in the 1920s; at the same time, several practitioners set up independent clinics and the British Psychoanalytical Society was turning towards the study and treatment of children. The term "child psychiatry" entered the everyday medical and psychiatric lexicon in the 1930s, though practitioners treating children with regard to their mental health did not necessarily identify themselves as "child psychiatrists." The gradual speciation and institutionalization of the field through mid-century manifested in the opening of further children's departments and units in psychiatric hospitals, the founding of a specific professional association, sections and committees within established organizations, and the creation of *The Journal of Child Psychology and Psychiatry*. Debates about the proper training for child psychiatry and the nature of its relation to adult psychiatry followed in the postwar era, as the nature of the field was contested by practitioners with different disciplinary backgrounds and theoretical orientations. Via these internal reflections, increasingly firm institutional foundations, and the coming of new generations of physicians with formal specialist training, "child psychiatry" crystallized into a discipline populated by people called "child psychiatrists." Initially little-changed by the formation and early development of the NHS, child psychiatry was impacted in the 1970s and 1980s by several critical reports in response to local failings of state services to adequately care for and protect children from abuse and mistreatment. After a series of reorganizations of the NHS and social services, the final decades of the century saw a gathering of child psychiatry and related services within the new umbrella of Child and Adolescent Mental Health Services (CAMHS), under frameworks intended to recognize and guide the involvement of multiple individuals and institutions in children's mental health, from General Practitioners (GPs), social workers, and schools to therapists, psychiatrists, and highly specialized inpatient units.

Historians are sometimes preoccupied with identifying change over time to the exclusion of finding resonances and points of convergence, yet it is often by seeking and establishing similarities that history can demonstrate its relevance to and impact on contemporary issues. Though the Child and Adolescent Mental Health Services of the late twentieth century look very different from the institutional care and Child Study Movement of the late nineteenth, there are some historical features that resonate with contemporary concerns and indicate, if not continuity across child psychiatry, then at least the enduring significance of its early history. Early investigations of the child's mind were influenced by heredity and eugenics, and the recent history of child psychiatry has seen the re-emergence of genetic and genomic research, reoriented now towards understanding risk factors and gene-environment interaction. It would seem that a longer history of genetics and child psychiatry is needed, one that accounts for continuities and changes between its eugenic origins and the technological promises for diagnosis and treatment in the postgenomic era. The late nineteenth century ostensibly saw the demise of restraint in institutions and calls for corporal punishment not to be used on mentally disturbed or learning-disabled children; yet types of restraint including pain-inflicting distraction techniques are still used on children in custody and secure facilities, sometimes leading to serious injury or death. Historians might provide insight as to how theories of children's mental state been used to justify or oppose such interventions across the twentieth century. The Tavistock Clinic was an early adopter of many approaches across the psych-sciences and a prominent training center after its founding in the 1920s. Now, at its centenary and as part of the Tavistock and Portman NHS Foundation Trust, its Gender Identity Development Service (opened in 1989) faces vitriolic criticism for providing life-saving gender-affirming treatment to transgender children. Historians of child psychiatry have a role to play in contextualizing the current controversy by analyzing the longer history of gender identity and psychopathology and how children and their mental health have been positioned as both at risk from and a danger to the social ordering of gender norms. The clearest and most generalizable link between past and present child psychiatry is the legacy of regional variation and the patchwork development of services that can still be traced back to centers of psychiatric expertise from mid-century, with specialist services often remaining in the same locality – or the same building – as their precursors, and concentrated more densely in London and the south-east.

If there is a single enduring feature of British child psychiatry, it is eclecticism. Even as some theories and treatments have disappeared (bed rest; bromide), novel approaches emerge (cognitive behavioral therapy; eye movement desensitization reprocessing) and find traction or rejection in different parts of the childcare ecology. With this in mind, the history of child psychiatry becomes less the story of how a particular medical specialism emerged and more the story of changing relationships between multiple services, practices, and conceptions of children and their minds. Future research will likely continue to multiply the history of British child psychiatry and to explore critical conceptual and methodological approaches. Productive avenues could include: problematizing childhood itself to consider changing conceptions of the life course, including the emergence of adolescence as a problematic developmental stage; according greater agency to children and their guardians and finding ways to understand why and how psychiatric care has been sought for children; and centering and reconstructing the complex pathways that patients follow to cut across the taxonomies of diagnoses, services, and disciplines that have served as convenient categories for the structuring of historical accounts. There is also room for the field to expand geographically, going beyond mainland Britain to extend recent histories of colonial psychiatry to consider the pathologization of certain child-associated behaviors as a colonizing practice, including the deployment of ideas of child-like-ness and childishness in relation to native and indigenous adults to diminish their autonomy and subject them to colonial authority. The history of child psychiatry is the history of scientific and cultural understandings of the child's mind and a group of practices and knowledges that have shaped and continue to influence childhood and adulthood, illness and health, in Britain and beyond. As a project, therefore, the history of child psychiatry demands an extensive and inclusive scope that includes reckoning with harm and abuse as well as care and treatment.

References

Barrett S (2019) From adult lunatic asylums to CAMHS community care: the evolution of specialist mental health care for children and adolescents 1948–2018. Rev Fr Civilis Britannique XXIV(3):1–16

Caldwell L, Taylor Robinson H (2016) The collected works of D.W. Winnicott. Oxford University Press, Oxford

Cameron HC (1919) The nervous child. Hodder & Stoughton, London

Cameron K (1956) Past and present trends in child psychiatry. J Ment Sci 102(428):599–603

Cohort and Longitudinal Studies Enhancement Resources (CLOSER) (2021) https://www.ucl.ac.uk/ioe/departments-and-centres/centres/cohort-and-longitudinal-studies-enhancement-resources-closer. Accessed 10 July 2021

Cooter R (1992) In the name of the child: health and welfare 1880–1940. Routledge, New York

Dwork D (1993) Childhood. Companion encyclopedia of the history of medicine, vol 2. Routledge, New York

Evans B (2017) The metamorphosis of autism: a history of child development in England. Manchester University Press, Manchester

Evans B, Rahman S, Jones E (2008) Managing the unmanageable: interwar child psychiatry at the Maudsley Hospital. Hist Psychiatry 19:454–475

Graham P, Minnis H, Nicolson M (2009) Witness seminar: the development of child and adolescent psychiatry from 1960 until 1990. Centre for the History of Medicine, University of Glasgow
Guthrie LG (1907) Functional nervous disorders in childhood. Hodder & Stoughton, London
Harris B (1995) The health of the school child: a history of the school medical service in England and Wales. Open University Press, Buckingham
Hayward R (2015) The transformation of the psyche in British primary care, 1870–1970. Bloomsbury, London
Hendrick H (1994) Child welfare: England 1872–1989. Routledge, London
Hendrick H (2003) Child welfare: historical dimensions, contemporary debate. Bristol University Press, Bristol
Herman E (2003) The Modern Social Sciences, Chapter 7. In: Psychologism and the Child. The Cambridge History of Science. Cambridge University Press, Cambridge, pp 649–662
Hersov L (1986) Child psychiatry in Britain – the last 30 years. J Child Psychol Psychiatry 27(6):781–801
Hollin G (2014) Constructing a social subject: autism and human sociality in the 1980s. Hist Hum Sci 27(4):98–114
Hollin GJS, Pilnick A (2015) Infancy, autism, and the emergence of a socially disordered body. Soc Sci Med 143:279–286
Ireland WW (1898) The mental affections of children: idiocy, imbecility and insanity. J. & A. Churchill, London
Jackson LA, Bartie A (2014) Policing youth: Britain, 1945–70. Manchester University Press
Kanner L (1959) The thirty-third Maudsley lecture: trends in child psychiatry. J Ment Sci 105:581–593
King S, Taylor SJ (2017) 'Imperfect children' in historical perspective. Soc Hist Med 30(4):718–726
Langdon Down J (1887) On some of the mental affections of childhood and youth: being the Lettsomian lectures delivered before the Medical Society of London in 1887, together with other papers. J. & A. Churchill, London
Levene A (2006) Family breakdown and the "welfare child" in 19th and 20th century Britain. Hist Fam 11(2):67–79
Levene A (2011) Childhood and adolescence. Oxford handbook of the history of medicine. Oxford University Press, Oxford, pp 322–337
Long V (2011) 'Often there is a good deal to be done, but socially rather than medically': the psychiatric social worker as social therapist, 1945–70. Med Hist 55(2):223–239
Mass Observation Archive (2021) http://www.massobs.org.uk/. Accessed 10 July 2021
McDonagh P (2008) Idiocy: a cultural history. Liverpool University Press, Liverpool
McDonagh P et al (eds) (2018) Intellectual disability: a conceptual history, 1200–1900. Manchester University Press, Manchester
Millard C (2015) A history of self-harm in Britain. Palgrave Macmillan, Basingstoke
NHS at 70 (2021) https://www.nhs70.org.uk/. Accessed 10 July 2021
Shuttleworth S (2010) The mind of the child: child development in literature, science, and medicine, 1840–1900. Oxford University Press
Smith M (2012) Hyperactivity: the controversial history of ADHD. Reaktion Books, London
Stewart J (2014) Child guidance in Britain, 1918–1955: the dangerous age of childhood. Routledge, London
Sugden N (2020) Winnicott's worlds: a history of psychoanalysis and childhood in Britain c.1920 – c.1975. Dissertation, University of Manchester
Sully J (1896) Studies of childhood. Longmans, Green and Co, London
Taylor SJ (2017) Child insanity in England, 1857–1907. Palgrave Macmillan, London
Tebbutt M (2020) Questioning the rhetoric of borstal reform in the 1930s. Hist J 63(3):710–731
Thomson M (2006) Psychological subjects: identity, culture, and health in twentieth-century Britain. Oxford University Press, Oxford

Turner J et al (2015) The history of mental health services in modern England: practitioner memories and the direction of future research. Med Hist 59(4):599–624

Urwin C, Sharland E (1992) From bodies to minds in childcare literature: advice to parents in inter-war Britain. In: In the name of the child: health and welfare 1880–1940. Routledge, London, pp 174–199

Walk A (1964) The pre-history of child psychiatry. Br J Psychiatry 110(469):754–767

Wall J (2013) Childism: the challenge of childhood to ethics and the humanities. In: The Children's table: childhood studies and the humanities. The University of Georgia Press, Athens

Wardle CJ (1991) Twentieth-century influences on the development in Britain of services for child and adolescent psychiatry. Br J Psychiatry 159(1):53–68

Warren W (1970) You can never plan the future by the past: the development of child and adolescent psychiatry in England and Wales. J Child Psychol Psychiatry 11:241–257

Welshman J (1996) In search of the "problem family": public health and social work in England and Wales 1940–70. Soc Hist Med 9(3):447–465

Human Experimentation and Clinical Trials in Psychiatry

52

Erika Dyck and Emmanuel Delille

Contents

Introduction	1358
Biology and Psychiatry	1359
A Psychopharmacological Turn	1365
Evidence-Based Medicine and Randomization: All Controlled Trials Are Not Randomized Controlled Trials (RCTs)	1368
Clinical Trials in Psychiatry	1371
Psychiatry, Experience, and Expertise	1372
Conclusion	1374
Cross-References	1376
References	1376

Abstract

This chapter examines the history of human experimentation in psychiatry, with a concentration on developments in the twentieth century. It explores bodily therapies that developed out of a desire to harness biological evidence to support physical interventions that might provide relief from mental symptoms. Some of these experiments helped to set the stage for another phase of experimentation characterized by drug trials, which introduced new methodological terminology and concepts. Double-blind trials, randomization, controlled trial methodology, and the interrogation of placebo effects changed the way that psychiatric experimentation unfolded in the second half of the twentieth century. This chapter shows that the new language of clinical trials created new moments of scientific

E. Dyck (✉)
Department of History, University of Saskatchewan, Saskatoon, Saskatchewan, Canada
e-mail: Erika.dyck@usask.ca

E. Delille
Department of Contemporary History, University Johannes Gutenberg, Mainz, Germany

Centre Marc Bloch, Berlin, Germany
e-mail: deem@cmb.hu-berlin.de

© The Author(s), under exclusive licence to Springer Nature Singapore Pte Ltd. 2022
D. McCallum (ed.), *The Palgrave Handbook of the History of Human Sciences*,
https://doi.org/10.1007/978-981-16-7255-2_93

optimism, and also generated unprecedented commercial opportunities. Despite the differences between clinical drug trials and physical interventions of the somatic era, the history of psychiatry reveals a legacy of experimentation aimed at diagnosing, classifying, and treating mental disorders. Although approaches, theories, and methods have changed, experimentation has remained a central part of the field. By considering how historians and scholars have interpreted this past, one can chart a longer relationship between experimentation and therapy that has characterized different moments of psychiatric history as well as galvanized responses against psychiatry for wielding undue power over people with mental illness.

Keywords

Somatic therapies · Psychopharmacology · Human experimentation · Chlorpromazine · Lobotomy · Randomized controlled trial · Placebo

Introduction

The history of psychiatry is a complicated story of human experimentation. While one might think of experimentation as a product of laboratory tests, clinical trials, and medical research, this book reveals how the field of psychiatry has explored different ways of identifying and managing madness. These different approaches to managing madness have also shaped our understanding of human experimentation, especially as one recognizes the blurry line that divides experimentation from therapy.

In many ways exploring different theories of causation and treatment represents a form of human experimentation. As other authors in this volume demonstrate, psychiatrists have adopted various approaches that rely upon different conceptualizations of madness and require different experimental approaches to justify a treatment modality, or in the case of anti-psychiatry an outright rejection of treatment or even pathologization. Early efforts in the nineteenth century to professionalize psychiatry coincided with the rise of the asylum and, consequently, custodial care. Large-scale institutionalization necessitated experimentation with a variety of ways of managing patient populations, while medicalizing madness and developing forms of care aimed at ameliorating unwanted behaviors. Moral therapy or work therapy in this context can itself be considered an experimental approach. Asylum-based care dominated the psychiatric landscape, consolidating the profession around principles of custodial care, disciplined environments, and structural-behavioral reforms. These systems of care relied upon systematic observations and statistical evidence that could be compared across sites; and they represented experiments in how to separate mental disorders from other disorders, and importantly, from normal behavior.

One might consider the age of the asylum as part of a longer legacy of human experimentation, but institutionalization also opened up new avenues for experimentation. Psychiatric patients were now clustered in dedicated facilities and became a concentrated population for medical research. Physical spaces of institutional wards divided patients according to shared traits in order to observe similarities and differences, across gender, age, or diagnosis. These closed communities provided opportunities to study human behavior in new ways, giving rise to new approaches to experimentation in psychiatry, including postmortem analysis that stimulated ideas about brain localization and a search for a biological basis for mental illness. More systematized observation also generated new categories of disease, classifications of distress, and standards for both diagnosis and treatment.

Historians of psychiatry have shown us that human experimentation has been a central feature in this field. Historian of early modern science, Larry Stewart argues that the focus on human bodies as objects of study was part of an epistemic shift that involved distinguishing humans from animals. The distinction grew out of Enlightenment thinking, suggesting that humans' capacity for rationality and conscious thought separated human bodies into both objects and instruments of observation (Stewart and Dyck 2017). Stewart points to Michel Foucault's proposition in *The Order of Things* (1970), where he argues that during the seventeenth century "to know nature now meant to observe it across a vast (artificial) table encompassing all of its myriad similarities and differences" (Stewart and Dyck 2017, 1). By the end of the nineteenth century the idea that human bodies were objects of experimentation was well integrated in biomedical research. Evolutionary ideas about the perfectibility of the human race, or the application of eugenics, appealed across scientific and social science fields, further exposing human bodies and minds to researchers' gaze (Paul 1998; Bashford 2014; Kevles 1998). Historians Stephen Casper and Delia Gavrus remind us that this culture of research permeated deeply into the mind sciences, giving rise to not only physiological studies of the structure of the brain but also the philosophical and theological understandings of the mind (Casper and Gavrus 2017, p. 5). Experimentation aimed at uncovering the workings of the human mind engaged research cultures across disciplines, stretching from the humanities, through the social sciences, and into medical science, all of which opened up new theoretical territory that fed into psychiatric classifications and treatments.

Biology and Psychiatry

In psychiatry the distinction between the normal and the pathological has invited close scrutiny of human bodies, behavior, thoughts, and brains. As historian Anne Harrington explains, the history of psychiatry is a history of a "troubled search for the biology of mental illness" (Harrington 2019). As Harrington and others have shown, this field is characterized by an enduring and in many ways unsatisfying set of attempts to bring empiricism to the mind sciences. Andrew Scull likewise explores this relationship between medical research and experimentation that attempted to rationalize behavior using biological principles. He explains that "linkages between mental symptoms and underlying tissue pathology served to reinforce the sense that biological research might help to uncover the aetiology of madness, but for the overwhelming bulk of mental illness, the hypothesized brain lesions remained as elusive as ever" (Scull 2015, p. 263). Despite the lack of definitive biological explanations, the very idea of a biological basis for madness instilled hopes that the experimental techniques applied to human bodies might relieve the suffering of the human mind. In the first half of the twentieth century Western psychiatry embraced biological models of psychiatry as the most promising avenue of research for ameliorating madness.

A number of scholars have pointed to a period between the end of the nineteenth century and the middle of the twentieth century when Western psychiatrists privileged biological explanations of mental disease. This moment has been variously described by historians in relatively critical terms: Anne Harrington calls it "biology in disarray," Andrew Scull explains it as a search for "desperate remedies," Elliot Valenstein calls it a period of "great and desperate measures," while Edward Shorter is the least critical, referring to it more neutrally as the era of the "first biological psychiatry" (Harrington 2019; Scull 2015; Valenstein 1986; Shorter 1997). Despite subtleties in their descriptions, these authors draw attention to a period in psychiatry when somatic therapies gained traction in the field. They all acknowledge how the focus on bodily therapies or physical interventions generated enthusiasm in the field, garnering several Nobel Prize awards for what contemporary observers recognized as major breakthroughs. Although these scholars and others writing about this period do not entirely agree about the legacy of this period, there is considerable consensus as to how Western psychiatry adopted these models of madness.

Human experimentation and bodily therapies went hand in hand with efforts to classify diseases and identify observable patterns in symptomatology. Harrington explains how linking mental diseases with physical causes stimulated research on these connections, pointing out that in 1897 "Richard von Krafft-Ebing [a German neurologist], oversaw an experiment to test for a possible causal link" between syphilis and general paralysis of insanity (GPI) (Harrington, p. 30). These German researchers injected 9 GPI patients with syphilis and kept them under observation to see whether they developed syphilitic symptoms, namely, fever, rash, and genital sores. According to Harrington, researchers knew that subjects could be affected by syphilis only once. Therefore, if these men injected with syphilis showed no symptoms, they must have already had the disease, or still be infected. The men did not develop symptoms, and the experiment was deemed a success. Regardless of how these men came to be subjects in the trial, the scientific evidence generated from this experiment demonstrated a verifiable link between syphilis and insanity. Attention could now focus on searching for other empirically observable relationships between germs, biological defects, and even brain lesions and maladapted behaviors and mental diseases.

In the wake of this observation, researchers searched for other relationships between biology and human behavior. Building on the momentum generated by the popular and attractive theories of evolution, degeneration, and eugenics, psychiatrists played their part in promoting studies of human behavior that had the potential to distinguish biological and inherited qualities from those accumulated through environmental and social contexts. Arbitrating the nature-nurture debate, human experimentation became crucial to the study of abnormal behavior, while the implications of that research went well beyond ameliorating individual symptoms and had the potential to influence human civilization in significant ways by clarifying where political interventions might fail, and biological ones were required.

The combined pressures of growing institutionalized populations and the desire to secure better cures for mental disease placed psychiatrists in a position to embrace human experimentation. As Dyck and Deighton (2017) have argued, the period between the 1930s and into the 1950s represented an important turning point in psychiatry. Historian Edward Shorter described it this way: "In the first half of the twentieth century, psychiatry was caught in a dilemma. On the one hand, psychiatrists could warehouse their patients in vast bins in the hopes that they might recover spontaneously. On the other, they had psychoanalysis, a therapy suitable for the needs of wealthy people desiring self-insight, but not for real psychiatric illness. Caught between these unappealing choices, psychiatrists sought alternatives" (Shorter 1997, p. 145). Psychiatrists embraced a number of radical bodily interventions, such as insulin shock therapy, malaria therapy, electro-shock or electroconvulsive therapy, lobotomies, and by the 1950s a host of pharmacological treatments. While these therapeutic innovations have been described as barbaric in hindsight, malaria therapy and lobotomies earned their innovators Nobel Prizes for pathbreaking research. Teaming up with neurologists and emphasizing a physical and increasingly scientific approach to mental health, psychiatry realigned itself with its biological roots after what Shorter describes as a brief hiatus with Freudian psychoanalysis from the turn of the century until after the Second World War (Shorter 1997, p. 145).

While some scholars suggest that this period helped propel psychiatry out of the doldrums of therapeutic failures, others maintain that these radical cures represent a culture of desperation and disillusionment within the field. Erika Dyck and Alex Deighton have argued that this desolate situation affected patients facing a lifetime in an institution, or alienation from their community, and it also affected psychiatrists whose reputation as medical specialists remained in jeopardy as their discipline seemed unable to match the triumphs and progress seen in other fields of medicine. These difficulties helped to pave the way for experimentation within mental hospitals, spurring hopes that by exploiting recent developments in biochemistry, neurology, and endocrinology, psychiatrists could both burnish their professional reputations and provide more sophisticated explanations for mental disorders. Within this culture of enthusiasm for experimental psychiatry and physical treatments, psychiatrists felt relatively free to use experimental methods to achieve therapeutic objectives within their institutions (Dyck and Deighton 2017).

Within this context, psychiatrists began borrowing techniques and theories from other parts of medicine, launching new kinds of experiments that were poised to change psychiatric practice. Scholars have explored these therapies in some detail, citing them as proof of radical experimentation and a trial and error approach to psychiatric research. Building on the theme of injecting patients with diseases to produce cures, or at least to observe reactions in an effort to diagnose them, a number of experiments involved the therapeutic use of other diseases. For example, Austrian physician Julius Wagner-Jauregg used malaria to treat patients with dementia paralytica and for that received a Nobel Prize in 1927. Wagner-Jauregg's approach involved injecting a patient with a known disease, in this case malaria, in order to produce an immune response, in this case a fever. Knowing that malaria was now treatable with quinine, Wagner-Jauregg reasoned that it was less risky to induce a treatable disease for the purposes of stimulating an immune response that had curative and rehabilitative effects on psychosis. The idea of causing or identifying an infection also inspired other research programs. Andrew Scull traced the notorious work of Dr. Henry Cotton at Trenton, New Jersey who in the 1920s theorized that psychosis was caused by "focal sepsis." To that end, Cotton performed hundreds of invasive surgeries on patients with psychotic disorders in an attempt to test his theory. His controversial approach resulted in unnecessary deaths and occasioned an inquiry further highlighting the controversial place of medical experimentation in the field of psychiatry (Scull 2005).

The theory that infection led to insights about pathological conditions was also applied to new technologies. The discovery of insulin in 1922 by Canadian physiologists Frederick Banting, Charles Best, John MacLeod, and biochemist J.B. Collip also earned international fame, and later a Nobel Prize for introducing a treatment for the fatal disease diabetes (Bliss 1982; Li 2003). Elsewhere, endocrinologists began applying theories drawn from this emerging field to psychiatry, further underscoring the desire at this time to harness the work of adjacent medical specialties to test the relationship with psychiatry (Evans and Jones 2012; Shorter and Fink 2010). Austrian neurophysiologist and psychiatrist Manfred Sakel hypothesized that using the newly approved insulin for psychotic patients and drug addicts, he could artificially put patients into a hypo-glycemic shock, causing the body to temporarily shut down and ultimately ameliorate psychotic symptoms (Shorter and Healy 2007). Despite criticisms of his work, Sakel claims that he successfully induced this early form of shock therapy to psychotic patients with strong results.

According to scholars Shorter and Healy, Sakel's insulin shock therapy helped to generate enthusiasm for other forms of shock therapy, ultimately including electro-convulsive or electro-shock therapy. This intervention, pioneered by Italian neurologist Ugo Cerletti, involved applying electrodes to the brain to induce a seizure. Like the preceding examples, Cerletti fixed his sights on psychotic symptoms, and took his cues from another hospital ward. In his case, Cerletti had observed that epileptic patients did not exhibit signs of psychosis, and he theorized that seizures associated with epilepsy may have a prophylactic or curative effect on psychosis. To test this hypothesis, he created an electro-convulsive machine designed to deliver electrical shock waves to a patient with the intention to cause a seizure. Initial results were positive and Cerletti's technique was picked up around the world, and later earned him a Noble Prize nomination.

While electro-shock therapies have become a popular symbol of this period of radical experimentation, lobotomies represent perhaps the darkest chapter of this era. Once again, borrowing from observations made in other areas of medicine, neurologists and anatomists continued to probe the brain for signs of abnormal growth, lesions, or physical clues that might explain aberrant behavior. Portuguese neurologist Egaz Moniz received a Nobel Prize in 1949 for his work in the 1930s developing the "burr-hole" technique for psychosurgery, which involved cutting white matter from the frontal lobe. Although Moniz is considered a pioneer in this field, American neurologist Walter Freeman, and to a lesser extent his research partner James Watt, popularized the technique in the United States, bringing popular and medical attention to his patented frontal lobotomy, which he pioneered at first using an ice pick as his cutting instrument. Historian Jack Pressman has provided a thorough and vivid examination of Freeman's efforts to champion the lobotomy as a humane intervention for psychotic patients, ultimately relieving them, Freeman believed, of unwanted psychotic symptoms and allowing patients to return home to families rather than face a lifetime in an institution (Pressman 1998). Joel Braslow has likewise examined the history of lobotomies, and other somatic therapies in California, highlighting the popularity of these techniques that appealed to families and psychiatrists as a radical, but necessary intervention (Braslow 1997). Physician-historian Mical Raz closely examined patient and family responses, showing a diversity of experiences to the Freeman-style lobotomy, further complicating our understanding of how to historically interpret this radical intervention as leap of scientific faith, or a humanitarian intervention, albeit a crude one (Raz 2015).

Raz and others reveal another side to this period of human experimentation in psychiatry, which draws attention away from the science of brains and behavior, and toward the social and cultural impact of these technologies. Scholars revisiting this history have recognized that this phase of experimentation set the stage for a backlash from patients and their families who later argued that these interventions were an abuse of power over marginalized, psychiatrized, and institutionalized people. Allan Beveridge explores ▶ Chap. 55, "Antipsychiatry: The Mid-Twentieth Century Era (1960–1980)" movements elsewhere in this volume, and some commentators have shown a clear link between the history of lobotomies, their depiction in popular media, and a growing set of protests against psychiatrists for unethical human experimentation.

Other scholars have stepped back from these specific treatments to argue that the longer history of experimentation in psychiatry culminated by mid-century in a cultural-medico consensus that mental diseases are in fact disorders of the brain. In other words, mental illness remains a medical – that is physical – disorder and its treatment and diagnosis should be the domain of bio-medicine. Psychologist Elliot Valenstein eventually describes this period as "blaming the brain," in a move away from social or environmental causes of distress (Valenstein 1998). Valenstein's critique centers on psychiatry, but also reminds readers that other domains of research had not lost sight of the social, cultural, and environmental factors in mental health and illness.

Alongside a growing interest in narrower physical approaches, psychiatrists also ventured into colonial medicine or comparative psychiatry, later described as ethnopsychiatry or transcultural psychiatry. These psychiatrists were particularly interested in cultural intersections with biology. This field, which tended to focus on colonialized subjects in places like Africa, South America, Asia, and among North American Indigenous populations, reflected both a colonial attitude toward subjugated populations, but also exhibited an intellectual curiosity about the relationship between biology, culture, and competing notions of civility as they pertained to mental disorder. Ethnopsychiatrists developed a more social and anthropological approach to understanding mental disorders, often without direct references to brain physiology, the genes, or any specific somatic etiology, thus challenging mainstream bio-medical discourse. Emil Kraepelin himself is considered the "father" of comparative psychiatry because of his travel and study in Dutch colonial psychiatric hospitals in Java at the end of the nineteenth century. Although he is more frequently remembered as the father of nosological categories like dementia praecox and manic-depressive illness, his observations were influenced by cultural comparisons and by the statistical analysis of social groups that would progressively underpin psychiatric epidemiology. Physicians, psychoanalysts, sociologists, and anthropologists started more formal collaborations in the 1930s and 1940s.

Psychiatric epidemiology examines the distribution of mental disorders within a population and emerged on the scientific scene during the second half of the twentieth century. An example of psychiatric epidemiology exists in the work of psychiatrist and anthropologist Alexander H. Leighton (1908–2007). Leighton launched his Stirling County Study in 1950s Nova Scotia, Canada to assess the effects of the sociocultural environment on the prevalence of psychiatric disorders. His work was contemporary with other pioneering epidemiological studies on chronic diseases such as the Framingham Heart Study in Massachusetts (1948). Both of these studies are ongoing, although the Framingham study is better known. The history of psychiatric epidemiology remains understudied, partly because it departed from mainstream psychiatry and depends on statistical indicators of illness rather than relying on clinical trials. Consequently, psychiatric epidemiology produces a different kind of scientific knowledge about mental health. It also reminds us of some of the alternative ways of studying mental disorders that have coexisted, including ones that embrace quantitative analysis, albeit without clinical trials and without a biological frame of reference (Dyck and Delille 2020).

Such cross-cultural approaches served to decenter mental health from a strictly Western point of view. According to psychiatrist Eric Wittkower, "cultural psychiatry concerns itself with the mentally ill in relation to their cultural environment within the confines of a given cultural unit, whereas the term transcultural psychiatry denotes that the vista of the observer extends beyond the scope of one cultural unit to another" (Wittkower 1970, p. 162). In the case of the Division of Social and Transcultural Psychiatry Research, founded by Wittkower and anthropologist Jacob Fried in Montreal at McGill University in 1955 (Delille 2018), the mental health consequences of acculturation, urbanization, and immigration (especially of displaced persons after the Second World War) were dominant themes. The 1950s and the 1960s therefore cannot be purely viewed as the ascent of biological psychiatry, as these social factors also demanded attention, particularly in the field of ethnopsychiatry. Historians of psychiatry have nonetheless often separated these narratives, without taking into account that even psychiatrists like Wittkower, with his classical training in physiology in German medicine, were interested in socio-cultural factors due to his own personal experience as an immigrant. Wittkower's personal experience as a transnational and transcultural figure shaped his approaches to the field of psychiatry during the 1960s. And Wittkower was not alone. Anthropologist and psychoanalyst George Devereux (Delille 2016), psychiatrist Henri Ellenberger (Ellenberger 2020), and others experienced life and research in different communities, bringing their perspectives to bear on ideas about how culture shapes experience, perception, and in turn mental health. In a post-war era of decolonization and growing awareness of indigenous sovereignty, acculturation into the dominant society captivated the attention of social researchers. Moreover, as multiculturalism emerged as a politically salient goal of non-assimilation, psychiatrists played a role in distinguishing culture from pathology. Social psychiatry relied on different instruments and methodologies, but it too engaged in human experimentation in the form of surveys and comparisons across different population groups. While early forays into ethnopsychiatry during the colonial period tended to exoticize racial and cultural characteristics, the field grew more critical and produced more sophisticated analyses of the prognosis of chronic mental disorders in cultural contexts.

On the basis of World Health Organization (WHO) studies, one can observe two tendencies. For some experts (Hopper et al. 2007), psychiatric recovery rates have been found to be better in the developing world (Cohen et al. 2008) than in industrialized countries, although other experts found no statistical evidence of a major difference. The controversy persists, revealing that quantitative data and evidence-based medicine are inseparable from interpretive frameworks, many of which are based on social, political, and cultural considerations. The variances in epidemiological distribution of mental illness are confounded by historical differences in access to medications and technologies. Psychiatric medications have been accessible in industrial countries for people diagnosed with a mental disorder. This access is less reliable or consistent in some geo-political regions, which alters the epidemiological landscape, but reinforces an idea that Western psychiatry is superior due to its capacity to more precisely measure the epidemiological distribution of illness. It also means that clinical trials, psychopharmaceutical products, and evidence-based medicine are not the dominant features of mental health programs everywhere. Widening the scope beyond Western psychiatry begins to challenge the primacy of clinical trials, pharmacological solutions, and the grip of the American Psychiatric Association's Diagnostic and Statistical Manual on the historical narrative of psychiatry.

A Psychopharmacological Turn

The auspicious introduction of chlorpromazine in psychiatry launched a psychopharmacological revolution that was a methodological game changer. Scholars and clinicians alike have drawn our attention to the 1950s as a pivotal period in the history of psychiatry, a moment that consolidated a professional landscape, introduced new ways of diagnosing mental disorders, and carved out new strategies for evaluating the efficacy of treatments. Some commentators suggest that without the embrace of psychopharmacology, governments may not have invested as heavily in the idea of deinstitutionalization, which in many respects transformed the mental health landscape and gave rise to new and experimental kinds of therapeutic communities. Some scholars contend that the introduction of psychopharmacological remedies modernized psychiatry, merging subjective and therapeutic interactions with statistical models and scientific principles, both of which rescued it from a cruder form of biological psychiatry as well what many considered to be ineffectual talk therapies. Psychiatrist and historian Gladys Swain (1987) emphasized the importance of this moment as early as the 1980s. Swain explained it as a paradox: the introduction of psychotropic drugs inaugurated a gradual and significant generalization of psychotherapies. She also maintained that the "classical" nosography elaborated during the nineteenth century in Western countries had become obsolete because of the global use of drugs, which are taken for a broad spectrum of ailments beyond the strictly mental health categories defined by Western authorities.

Other scholars more critically suggest that an increased focus on psychopharmacological solutions went hand in hand with a proliferation of disease categories, articulated most clearly in the American Psychiatric Association's Diagnostic and Statistical Manual, or DSM, which first appeared in 1952. Some of these critical commentators argue that the seismic shift toward psychopharmacology at mid-century, coupled with the emergence of the DSM as the authoritative catalogue of pathological disorders, paved the way for the development of Big Pharma, a politically powerful transnational force that has dominated the mental health arena ever since. Outspoken critics of this model have gone so far as to suggest that the pharmaceutical industry is more invested in the proliferation of mental disorders, and thus its own capital gains, than their amelioration.

While chlorpromazine may not be single-handedly responsible for all of these outcomes, its introduction in the early 1950s generated a wave of interest in rethinking the therapeutic landscape and perhaps even in re-articulating the language of mental health and illness. Reinterpreting the relationship between drugs and madness coincided with a questioning of the relationship between human experimentation and expertise. This period of reevaluation was not confined to psychiatry; the specter of clinical trials altered the relationship between the subject and observer, inspiring a host of critiques leveled at the notion of professional expertise, who could wield it, and who benefited or suffered under the new clinical gaze. Professional tensions between psychiatrists and psychologists intensified when drug therapies were at stake, while others examined these changes at the level of power using the lenses of philosophy, sociology, and history. Scholars asked penetrating questions about how it was that madness had become a commodity, or how psychiatrists had come to participate as political gatekeepers to citizenship with madness and mental disorder functioning as labels that threatened to restrict citizenship rights. It is no coincidence that the changes wrought in this period also coincided with the emergence of anti-psychiatry movements and scholarly critiques intent on rooting out the origins of psychiatric power and expertise over the pathologizing of human behavior.

Psychopharmacologist, psychiatrist, and author David Healy has argued that "the modern story of drugs and madness starts in 1950, with the synthesis of chlorpromazine, the first of the antipsychotics" (Healy 2002, p. 37). Healy, who has written extensively on this topic, provides readers with a detailed history of how psychiatrists eventually turned to pharmacological remedies out of a combination of frustration and pressure to modernize the field. His account in *The Creation of Psychopharmacology* offers a close reading of developments leading to the realization by French military surgeon Henri Laborit that chlorpromazine had the surprising effect of reducing stress, and stabilizing patients undergoing surgery (Healy, p. 79). Placing the body in a temporary and artificial state of "hibernation," Laborit experimented with the use of antihistamines on the nervous system in an effort to calm patients without sedation. Laborit recommended the name Largactil (chlorpromazine's eventual European trade name), and continued experimenting with it in surgery. A year later, according to Healy, Laborit gave it to his psychiatrist colleague, who ultimately found it unremarkable (Healy, p. 82). For the next 2 years chlorpromazine appeared in different experimental contexts, as an anti-emetic in surgical operations, and it bounced through different pharmaceutical houses, with limited interest. According to Healy, the next chapter of its story was rather auspicious.

> One day Heinz Lehmann, a German-Canadian [psychiatrist] who worked at the Verdun Hospital [Montreal, Quebec], read about chlorpromazine while taking a bath. The next day, he ordered supplies of the drug. He recruited a resident to help and gave chlorpromazine to seventy patients. He also gave it some nurses to study how it worked. (Healy 2002, p. 94)

Healy's description of how chlorpromazine came to the attention of the field of psychiatry is hardly a story reflective of precision, intent, and in many ways may not sound like a typical origin story for a drug now known for triggering a psychopharmacological revolution. However, this version of the story is nonetheless remarkable for what it reveals about the state of psychiatric experimentation in this period. The pace of research, the authority of the psychiatrist, the role of nurses, and the desire to test reactions on staff – all of these features are important hallmarks of a culture of experimentation that accompanied the emergence of the clinical trial in psychiatry, and its ascendancy as the gold standard in modern psychiatric research.

Lehmann's pioneering efforts at the Verdun Hospital, part of the Douglas Institute in Montreal in 1953, owed a lot to French psychiatrists who had already started using chlorpromazine in psychiatric wards in 1952, thanks to connections with Henri Laborit. The France-based psychiatrists did not yet know that the new molecule had a specific action; they began by combining its use with other drugs and therapies. The first physician who used it as the primary treatment was Pierre Deniker, then Professor Jean Delay's assistant in 1952. This application ultimately made them famous for observing the drug's sedative effect when used on psychotic patients. They were recognized with a Lasker Prize in medical science for this finding in 1957. It is very important to take into account that neither psychiatrists nor the public understood the specific action of chlorpromazine in 1952 and that Deniker and Mayer-Gross undertook no clinical trials at that time. The general understanding and the reception took several years: the first important scientific conference about the drug was organized in Paris in 1955, where the scientific community likely first became genuinely aware of this product as a new psychotic drug (Deniker and Mayer-Gross 1956).

While Healy and others have emphasized the historical importance of chlorpromazine for setting the field on a path toward psychopharmacological investments, chlorpromazine was itself not the product of clinical trials. Its introduction nonetheless helped to focus research energy on the development of psychotropic drugs, including other antipsychotic drugs, but also of other categories like anxiolytics and antidepressants. Massive investments in psychopharmacological solutions necessitated new methods of inquiry, and new ways of evaluating benefits and risks. Clinical trials eventually became the epicenter of this tectonic shift in psychiatry.

Pharmaceutical firms swelled in size and importance in this opportunistic marketplace. Ambitious psychiatrists also moved into this field as interlocutors between pharmaceutical solutions and prescribing physicians. One example was Swiss physician Roland Kuhn, who discovered the first antidepressant (imipramine) in 1957. His discovery led to fame and accolades for identifying a pharmaceutical intervention capable of ameliorating unwanted symptoms of persistent depression. Since his death in 2005, evidence has surfaced suggesting that his early patients were subjected to his experiments without their consent. According to historian Marietta Meier (2019) and her colleagues, Kuhn conducted trials without his patients' consent, and without following contemporary ethical guidelines. Moreover, he continued these trials for several decades after the introduction of the Nuremberg Code of Ethics in 1947, which had established legal restraints on human experimentation in medicine. Some of Kuhn's patients died in Munsterlingen hospital in the Swiss canton Thurgau as a direct result of Kuhn's experiments. The numbers of people involved are significant: Kuhn tested about 60 molecules on more than 1000 patients, including children. Therapeutic enthusiasm in the 1950s alone does not explain how these trials were allowed to continue (Weiss 2004). Kuhn maintained personal connections with pharmaceutical industries in Switzerland, suggesting that there may have been pressure, as well as financial and professional incentives, for Kuhn to oversee trials of new pharmaceuticals.

Evidence-Based Medicine and Randomization: All Controlled Trials Are Not Randomized Controlled Trials (RCTs)

Clinical trials in biomedicine began in earnest in the 1940s, culminating famously in the randomized controlled trial (RCT). The RCT involves dividing experimental subjects into two groups, assigned at random. One group receives the active intervention, the other receives a placebo or an alternative treatment. Neither the investigators nor experimental subjects are to know whether they are in the experimental group, or the placebo group, or have received a conventional therapy. Parts of this methodology have a longer history, including comparing therapeutic interventions or exploring placebos. But, as historian Edward Shorter writes, the RCT became "the gold standard in pharmaceutical trials, and its status was cemented during the mid-1950s" (Shorter 2011, p. 193).

In 1948, experimentation with streptomycin, an anti-tuberculosis drug, attracted international attention both for identifying an efficacious remedy for the deadly disease, and also for introducing a new method for measuring efficacy. Medical historians have devoured this story, showing that this method of testing, which involved comparing, standardizing, and repeating results across multiple testing sites, transformed modern biomedical research. Following on the heels of the successful streptomycin research, other drugs became the subject of examination that involved, according to Helen Valier and Carsten Timmerman, the "statistical technique of randomization, with new organizational techniques, such as the division of specialist labour, and central review and data collection, across multiple sites of study" (Valier and Timmerman 2008, p. 493). Scholars in Britain, Germany, the United States, and Canada have further argued that national state infrastructure seized upon this method, prioritizing it in funding initiatives and effectively institutionalizing the clinical trial and the RCT as the highest standard of evaluation.

American historian of medicine Harry Marks explains that, prior to the 1940s, using statistics in clinical research was relatively rare. Medical students were not even necessarily required to learn statistics, which was a method reserved for the domain of public health. As Marks and others have shown, the integration of statistics into experimental medicine came from British geneticist R. A. Fisher, who was primarily interested in cataloguing differences in plants (Marks 1997, p. 141). Biomedical researchers nonetheless picked up on Fisher's approach, which used statistical comparisons and randomization in order to mitigate bias. Marks explains that for Fisher, "the use of randomization ensured 'that neither our personal idiosyncrasies, consciously or consciously applied, nor our lack of judgement have entered into the construction of the two (or more) treatment groups and thus biased them in any way'" (Marks, p. 145). Fisher's influence upgraded the clinical trial methodology to the RCT by adding the randomizing feature as another characteristic safeguarding against bias (Marks, p. 145; Shorter 2011, p. 195).

For many researchers at this time, the controlled trial method meant comparing outcomes using two different drugs or forms of therapy. But, as Edward Shorter points out, it also gave rise to the notion of a placebo and its place in the comparative methodology. Shorter contends that the idea of a placebo also emerged in the 1940s, and attracted attention from psychiatrists, one of whom claimed: "the placebo is a specific psychotherapeutic device with values of its own." American physician Harry Gold went further to suggest that upon reflection, he believed that his patients often responded positively to a placebo: "I sleep better; my appetite is improved; my breathing is better…" leading Gold to conclude that "I think the placebo as a chemical device for psychotherapy has a definite place which cannot be filled by anything else in many cases" (Shorter 2011, p. 195).

The clinical trial offered biomedical research an innovative method that allowed for scaling up research while minimizing subjectivity. The timing was important. Clinical trials had occurred prior to World War II, but had been necessarily smaller, and at times regionally limited. State infrastructure combined with an international recognition of the need to increase ethical standards for human experimentation after the Nuremberg Trials brought the clinical trial under greater scrutiny. But clinical research was not uniform across medical fields, and psychiatry developed its own relationship with this emerging methodology.

Historian of medicine Othmar Keel (2011) has made a critical observation about the methodology of the randomized controlled trial. In *La medecine des preuves* Keel separates the logic of randomization and the use of placebo, casting doubt on the therapeutic aims of Big Pharma and the commercial objectives of a ballooning pharmaceutical industry. Randomization was introduced in the UK in the 1920s and 1930s, before becoming the "gold standard" in trial methodology. Additional criteria, like double blinding and placebos, came later. Unlike Edward Shorter's interpretation, Keel's analysis had nothing to do with the psychotherapeutic value of a placebo effect. Keel instead offers a radical deconstruction of the use of placebo by pharmaceutical corporations, resulting in a rigorous epistemological study of the "evidence-based medicine" of the twentieth century. Keel explains that all controlled trials are not in fact randomized, and the application of double blinds and placebo studies are inconsistently applied in many trials, despite meeting industry standards. There are a large variety of practices in use because private companies want first and foremost to sell drugs, and the protocols for trial methodology are a means to that end. Indeed, all controlled trials are not "randomized controlled trials" because companies do not always subscribe to randomization, and not all companies are held to the same standards. Here Keel distinguishes between trials organized by public agencies and those conducted by private companies, which pursue different objectives or are subject to different forms of oversight.

Secondly, Keel tackles the concept of placebo. He convincingly argues that the randomized controlled trial's validity is first based on features of randomization, and control over all variables, not on placebo. Randomization is a technique designed to control scientists' subjectivity and the arbitrary construction of patient groups: patients are assigned by chance to different groups. For Keel, the placebo is a completely separate component. The idea of the placebo, or the perceived psychological effect of a non-active intervention, emerged in the pharmaceutical industry after the Second World War. Pharmaceutical companies eventually used placebos, which included sugar pills or any substance with no therapeutic effect, as a trick to avoid comparing different drugs.

According to Keel, if the result of a psychotropic drug is better than a placebo after a controlled trial, it does not necessarily mean that the evaluation of the drug went through a process of randomization, and the new drug is not necessarily better than others already on the market. Randomization is key to this process, but regulatory agencies have limited or no capacity to oversee trials outside of the public jurisdiction, and randomization slows or restricts the introduction of new drugs to market, which is the aim of private pharmaceutical firms. Keel therefore advocates for multisite, public, and randomized trials. His recommendations require multilateral investments from regulators, governments, scientists, and pharmaceutical firms, investments that are neither the norm in the pharmaceutical sciences in general, nor in mental health in particular. However, despite these critiques it remains the case that clinical trials and human experimentation in psychiatry are not simply in the interest of scientists, physicians, philanthropists, or politicians, but also very much in the interest of private companies operating in competitive markets.

Clinical Trials in Psychiatry

Many scholars have written about the history of psychiatry over the course of the twentieth century, often characterizing its history as a story about the rise and fall of the psychiatric asylum, a story that often runs in tandem with an examination of psychiatry's enduring quest to define itself as a medical specialty. Caught between cultures of science and humanities, pulled toward science and fueled by philosophy, psychiatrists have the unenviable task of explaining human behavior and thoughts in ways that have historically defied mono-causal or even disciplinary traditions. For example, Harrington tracks this story through different intellectual, professional, and disciplinary efforts to search for the holy grail of mental illness (Harrington 2019). That is, she examines different intellectual pursuits as they tried to gain the upper hand in describing, categorizing, and ultimately treating mental illness over the twentieth century. In many ways, this is also a history of the emergence of different psy-professions and the scientific discourses they used to distinguish themselves from one another as they jockeyed for positions in what amounted to a race to discover an underlying logic to madness. Harrington is by no means alone. Edward Shorter has written extensively on this topic. He tips the scales in his interpretation toward the influence of biology, and the comparably more successful results from psychiatrists who more readily embraced the scientific principles borrowed from neurologists, biomedical scientists, and eventually, the statisticians. He contrasts these developments with the comparably ineffectual psychodynamic theories, whether Freudian, Jungian, or other talk-therapists who took their cues from an eclectic mix of disciplinary influences that included neurology and also philosophy and religion to theorize consciousness, memory, and experience in the clinical encounter.

Despite differences in kind, scholars tend to agree on certain milestone moments within this narrative. Enthusiasm for asylum-based care began to lose momentum by the end of the nineteenth century. Meanwhile exciting new theories of degeneration and evolution borrowed from biologists (and philosophers), breathed new life into the field of psychiatry as it shifted from that of a caretaking role, to one that might develop novel interventions by examining biological principles and applying them to mental diseases. Harrington and others trace the emergence of neuroscience, a stormy – but influential – affair with Freudian psychodynamic traditions, and then follows the immeasurable rise of faith in psychopharmaceutical promises, which shifts the focus to treatment and introduces a significant commercial angle to this field. Others have put their own spin on this history, but often arrive in a similar place. Sociologist Andrew Scull (2019), for example, has spent decades criticizing this history, exposing the internal debates within the broader field of brain sciences and therapists. The story of psychiatry, for Scull, is in fact a story of discontents

(Scull 2019). It is a story that has been told according to different epistemological approaches to understanding human behavior, and, as a result, its history is one of how different disciplines and professions have jockeyed for dominance in a field that continues to elude grand theories, and perhaps universal standards of evaluation.

But even with differences in interpretation, geographical focus, and temporal scope, historians often agree that the rise of Big Pharma in conjunction with the establishment of the *DSM* have contributed to what amounts to a madness industry. Harrington likewise suggests that the psychopharmaceutical promises have not resulted in a clear understanding of brain chemistry such that one can now definitely explain the relationship any better than proponents of earlier theories. And, upon reflection perhaps it is not surprising that the clinical trial did not have the same transformative effect on psychiatry as it did elsewhere in biomedicine. Thomas Ban, a psychiatrist, psychopharmacologist, and advocate for chlorpromazine in the 1950s, told David Healy in an interview in the 1980s "that whilst psychopharmacology had been responsible, and laudably so, for 'dragging psychiatry into the modern world', he reflected that in the process, psychiatry had become a laboratory focused research activity, with too little emphasis on making the results relevant to clinical conditions" (Tansey 1998, p. 79). While minimizing bias in drug trials is important, the history of psychiatry reminds us that experience, understanding, and communication is also a vital part of that clinical encounter.

Psychiatry, Experience, and Expertise

The field of psychiatry sat at a crossroads in the post-Second World War period. Throughout the Western world for much of the nineteenth and into the twentieth centuries, patients considered mad, insane, deviant, or disabled were segregated in asylums, as described elsewhere in this volume. These facilities became increasingly medicalized, shifting the language from asylum to hospital, and from insanity to mental disease or disorder; but the prognoses of the mad, whether they were called inmates or patients, remained grim. By the mid-twentieth century, reformers criticized these institutions for warehousing people who did not seem suited to work or participate as full citizens in a modern marketplace. Meanwhile, there was significant professional and disciplinary tension over who or what should be responsible for the care and management of individuals considered disordered.

By mid-century, Freudian psychoanalysis was the prevailing theoretical and clinical approach to explaining and treating the entire range of mental illnesses in the United States (Herzog 2016). At the same time, biological psychiatrists and somatic or bodily therapies attracted attention in the first half of the twentieth century, in part because they offered bold responses to decades of suffering, particularly for patients who could not be reached through talk therapies (Harrington 2019). Some psychiatrists reconnected the body with the mind during this period, ushering in radical interventions, such as insulin shock therapy, malaria therapy, electro-shock or electroconvulsive therapy, and lobotomies. By the 1950s some of these more aggressive interventions were abandoned or relaxed in favor of new pharmacological options. While these therapeutic innovations have been criticized, malaria therapy and lobotomies still earned their innovators Nobel Prizes for pathbreaking research, suggesting both that psychiatrists were keen to incorporate techniques from other fields, but were also reluctant to entirely isolate mind from body in their attempts to modernize (Shorter 1997; Valenstein 1986; Braslow 1997; Scull 2015).

Medical historian John Burnham has referred to the post-war period as the "golden age of medicine," a moment of unprecedented medical authority, largely stemming from the development of new medical technologies (Burnham 1982). New prescription drugs entered mainstream society at an unprecedented rate, launching a pill-popping phenomenon; gradually, taking pills became part of normal behavior, rather than an indication of abnormal behavior (Healy 2002; Tone 2009). Medical scientists and medical practitioners rose in esteem as new technological and pharmacological advancements promised to conquer an expanding list of complaints: pain, infection, menstruation, anxiety, depression, hypertensive disorder, alcoholism, and schizophrenia. It seemed that after the Second World War, armed with new ethical codes, new standards for evaluating the efficacy of medical treatments, and more sophisticated interdisciplinary and international cooperation, the clinical trial methodology was poised to transform modern medicine. Edward Shorter has described how pharmacology appealed to practitioners across methodological divides, softening professional tensions and allowing for this technology and its methodology to take root in a profession long searching for a unifying identity. For Shorter, "Pharmacotherapy spread like wildfire in these depraved settings [psychiatric facilities]. Initially there was little conflict between analysts and drug therapists because their spheres of influence did not overlap much" (Healy 2003, p. 144).

In North America, psychiatrists began publishing results of clinical trials in the 1950s. In 1954 Canadian-based psychiatrists in Saskatchewan published what some consider to be one of the earliest randomized, placebo-controlled trials of nicotinic acid, a study that they followed with another trial using d-lysergic acid diethylamide (LSD) (Shorter 1997; Dyck 2008; Hoffer and Osmond 1961a; Hoffer and Osmond 1961b). Despite being what might appear to be early converts to the RCT model, the investigators involved here, Humphry Osmond, John Smythies, and Abram Hoffer, instead became fierce opponents of the clinical trial. Hoffer at one point referred to the methodology as "double dummy," blinding both the observer and subject and learning very little in the process (Dyck 2008).

Far from being reductionist, Hoffer and Osmond argued that perceptions of the trials overcompensated for a desire to measure discrete effects, instead rendering experience useless, or of no enduring value in the therapeutic process. They argued that psychiatry needed to instead expand its conceptualization of experience and embrace environmental influences as fundamental to how we perceive ourselves, and how those perceptions effect communication. They borrowed the concept of *Umwelt* ("environment as experienced by an individual") from biologist and philosopher Jacob von Uexkull, who posited that both individual organisms and species as a whole have particular ways of experiencing the world (Bisbee et al. 2018, p. 413).

Perception, or the notion of a filtered experience of the world that prevents us from perceiving deeper truths, became a major theme in their research challenging the utility of the controlled trial. These researchers were not alone in their view that reducing psychiatric illness and its treatment to a series of symptoms and controlled reactions may offer a lucrative path for pharmaceutical products but did not fully appreciate a broader range of mental experiences, whether expressed as distress or not. Historian Justin Garson explains that during the 1950s attempts to model psychosis came in many forms, and some psychiatrists began exploring psychosis not simply as a cluster of unwanted symptoms, but as a window into the experience of madness (Garson 2017, p. 203). Garson shows that while an LSD-inspired model psychosis gained scientific attention, other forms of mimicry also emerged, giving rise to an amphetamine model-psychosis that gained popularity in the 1970s, supplanting the psychedelic model and further underscoring the role of dopamine in antagonizing psychotic symptoms (Garson, p. 204). These models of interrogation challenged the idea that madness could be studied, described, and treated using a series of precise controls, but instead threw open the doors of investigation to philosophical and phenomenological studies of experience that defied an RCT model.

Conclusion

The history of the clinical trial in psychiatry is in part a story about drugs or psychopharmaceuticals that deceptively suggests that mental illnesses stem only from faulty brain chemistry. The story is not altogether wrong, but it oversimplifies the place of clinical trials in psychiatry, and the way that the human brain has been studied and explained. It also potentially distorts our understanding of the interactions between psychoactive drugs and psychiatry in the context of human experimentation.

Since the nineteenth century modernizing agendas have propelled industrialization, commercialization, and along with it, facilitated and encouraged the massive global consumption of psychoactive substances (Dyck and Savelli 2021). Since the 1950s anti-psychotics, anti-depressants, and tranquilizers became a part of modern culture (Herzberg 2020; Tone 2009). In the 1960s, the popular British rock group, the Rolling Stones sang: "Mother needs something today to calm her down, and though she's not really ill, there's a little yellow pill" (*Mother's Little Helper, 1966*), signaling the mainstream acknowledgment of a psychopharmacological reach into everyday life, in this case through an anti-anxiety medication called Miltown.

Stepping back from the psychopharmacological turn in psychiatry, there remains a strong historical legacy of human experimentation and even clinical trials stretching back into the asylum era and through the early parts of the twentieth century with the development of shock therapies and lobotomies. Psychiatric interventions, including early drug experiments, were often justified as desperate measures or humanitarian interventions to provide relief from unwanted and distressing symptoms for a group of people whose mental states were considered to be chronic, unrelenting, and undesirable. Consent in such cases was not a simple matter of making informed decisions, but rather it was more often framed in a paternalistic context where psychiatrists were assumed to have their patients' best interest at the forefront of the experiment. These claims were especially compelling in cases where self-harm was a legitimate risk, or where a patient's inability to communicate clearly confounded the principles of providing free consent, or where a patient might be caught in the grip of a delusion that severed their connections to reality and put them and others in harm's way. Teasing apart the ethical and moral contexts faced by psychiatrists in this field is complicated, raising deeper questions about the relationship between human experimentation and therapeutic progress.

But the altruism required to support this interpretation is not always evident in the historical record. The history of psychiatric experiments is often presented as a dark chapter in medical history, peppered with examples of abuses of power over marginalized people. The crude attempts to alter brains, including with ice picks in the case of the lobotomy, or with electrical shocks, have become effective tropes and symbols of medical tyranny. Anti-psychiatrists seized upon cases of experimentation to justify rejecting psychiatry as a legitimate profession. Even less radicalized critics of psychiatry have harnessed the imagery of these early experiments to express how psychiatric power has real consequences for how people express different ways of thinking and being. The novel, turned film, *One Flew Over the Cuckoo's Nest* (Kesey 1962) exemplified this idea by bringing cinematic attention to a vivid display of psychiatry punishing noncompliant behavior using tools of the trade, which included talk therapy, shock therapy, and ultimately, a frontal lobotomy.

Clinical drug trials in psychiatry have not escaped this scrutiny, despite providing what appear to be less invasive interventions. Big Pharma has been the target of critics ranging from those who reject the medicalization of mental illness altogether, through to those who are frustrated with a treatment modality that merely softens symptoms while generating profits for companies that seem more addicted to the sale of mental health drugs than they are invested in the idea of curbing rates of mental illness (Healy 2003; Herzberg 2020).

The distinction between therapy and experimentation in psychiatry may be impossible to separate entirely, but it is also difficult to divide clinical trials from placebo. No experiment or therapy can effectively isolate the body and the mind, or neutralize the effects of talk therapy, group therapy, and any kind of psychotherapy. Even transcultural psychiatry challenges this biopolitical point of view because socio-cultural factors are involved in the patient's community life and world view.

In recent decades, the discourse of psychiatric experimentation has changed again. The controversy is not so much about whether psychiatry has become better at managing madness, but instead about scandals, legal responsibilities, and the application of ethics to psychiatric experimentation. Judgments are not confined to scientific laboratories or medical associations, but are reached in the court of public opinion, on social media, in the news, and in dramatized accounts of mental illness. In fact, the court of public opinion may even deliver appropriate judgments given that pharmaceutical drug trials control for variables of experience and that the marketplace can therefore facilitate real-life trials as people encounter pharmaceuticals in their diverse lives and consume them in a variety of ways that create experiences beyond those observable, or even imaginable, in a clinical trial.

Cross-References

▶ Antipsychiatry: The Mid-Twentieth Century Era (1960–1980)
▶ Asylums and Alienists: The Institutional Foundations of Psychiatry, 1760–1914
▶ The Mental Patient in History

References

Bashford A (2014) Global population: history, geopolitics, and life on earth. Columbia University Press, New York
Bisbee C, Bisbee P, Dyck E, Farrell P, Sexton J, Spisak J (2018) Psychedelic prophets: the letters of Aldous Huxley and Humphry Osmond. McGill-Queens University Press, Montreal/Kingston
Bliss M (1982) The discovery of insulin. McClelland & Stewart Ind, Toronto
Braslow J (1997) Mental ills and bodily cures: psychiatric treatment in the first half of the twentieth century. University of California Press, Berkeley
Burnham J (1982) American Medicine's golden age: what happened to it? Science 215(4539): 1474–1479
Casper ST, Gavrus D (eds) (2017) The history of the brain and mind sciences: technique, technology, therapy. University of Rochester Press, Rochester
Cohen A, Patel V, Thara R, Gureje O (2008) Questioning an axiom: better prognosis for schizophrenia in the developing world? Schizophr Bull 34(2):229–244
Delille E (2016) On the history of cultural psychiatry: Georges Devereux, Henri Ellenberger, and the psychological treatment of native Americans in the 1950s. Transcult Psychiatry 53(3): 392–411
Delille E (2018) Eric Wittkower and the Foundation of Montreal's transcultural psychiatry research unit after the Second World War. Hist Psychiatry 29(3):282–296
Deniker P, Mayer-Gross W (1956) Colloque International sur la Chlorpromazine et les Médicaments Neuroleptiques en Thérapeutique Psychiatrique, Paris 20–22. G. Doin, Paris
Dyck E (2008) Psychedelic psychiatry: LSD from clinic to campus. Johns Hopkins University Press, Baltimore
Dyck E, Deighton A (2017) Managing madness: Weyburn mental hospital and the transformation of psychiatric care in Canada. University of Manitoba Press, Winnipeg
Dyck E, Delille E (2020) Alternative therapies: psychedelic, primal scream, nude therapy, sociodrama. In: Pickren W (ed) Oxford encyclopedia of psychology. https://doi.org/10.1093/acrefore/9780190236557.013.629
Dyck E, Savelli M (2021) Crafting the modern via psychoactivity advertisements. Hist Pharm Pharmaceut Special Issue 63(1):80–96
Ellenberger H (2020) Ethnopsychiatry. Emmanuel Delille (ed). McGill-Queen's University Press, Montreal/Kingston
Evans B, Jones E (2012) Organ extracts and the development of psychiatry: hormonal treatments at the Maudsley Hospital 1923–1938. J Hist Behav Sci 48(3):251–276

Foucault M (1970) The order of things: an archaeology of the human sciences. Routledge, New York

Garson J (2017) A 'Model Schizophrenia': Amphetamine Psychosis and the Transformation of American Psychiatry. In: Casper ST, Gavrus D (eds) The History of the brain and mind sciences: Technique, Technology, Therapy. University of Rochester Press, Rochester, pp 202–228

Harrington A (2019) Mind fixers: psychiatry's troubled search for the biology of mental illness. W.W. Norton

Healy D (2002) The creation of psychopharmacology. Harvard University Press, Cambridge

Healy D (2003) Let them eat prozac. James Lorimer

Herzberg D (2020) White market drugs: big pharma and the hidden history of addiction in America. University of Chicago Press, Chicago

Herzog D (2016) Cold War Freud: psychoanalysis in the age of catastrophes. Cambridge University Press, Cambridge

Hoffer A, Osmond H (1961a) Double-blind clinical trials. J Neuropsychiatr 2:221

Hoffer A, Osmond H (1961b) In reply. J Neuropsychiatr 3:262

Hopper K, Harrison G, Janca A, Sartorius N (eds) (2007) Recovery from schizophrenia. A report from the WHO collaborative project, the international study of schizophrenia. Oxford University Press, Oxford

Keel O (2011) La Medecine des preuves. Les Presses de l'Universite de Montreal, Montreal

Kesey K (1962) One flew over the cuckoo's nest. Signet, New York

Kevles D (1998) In the name of eugenics: genetics and the uses of human heredity. Harvard University Press, Cambridge

Li A (2003) J.B. Collip and the development of medical research in Canada. McGill-Queens University Press, Montreal/Kingston

Marks H (1997) The progress of experiment: science and therapeutic reform in the United States, 1900–1990. Cambridge University Press, Cambridge

Meier M, Konig M, Tornay M, Klauser U (2019) Testfall Munsterlingen: klinische Versuche in der Psychiatrie, 1940–1980. Chronos, Zurich

Paul D (1998) The politics of heredity: essays on eugenics, biomedicine, and the nature-nuture debate. SUNY Press, New York

Pressman JD (1998) Last resort: psychosurgery and the limits of medicine. Cambridge University Press, Cambridge

Raz M (2015) The lobotomy letters: the making of American psychosurgery. Boydell & Brewer/ University of Rochester Press, Rochester

Scull A (2005) Madhouse: a tragic tale of megalomania and modern medicine. Yale University Press, New Haven

Scull A (2015) Madness in civilization: a cultural history of insanity from the bible to Freud, from the madhouse to modern medicine. Oxford University Press, Oxford

Scull A (2019) Psychiatry and its discontents. University of California Press, Oakland

Shorter E (1997) A history of psychiatry: from the era of the asylum to the age of prozac. John Wiley & Sons Inc., New York

Shorter E (2011) A brief history of placebos and clinical trials in psychiatry. Can J Psychiatr 56(4): 193–197

Shorter E, Fink M (2010) Endocrine psychiatry: soliving the riddle of melancholia. Oxford University Press, Oxford

Shorter E, Healy D (2007) Shock therapy: a history of electroconvulsive treatment in mental illness. University of Toronto Press, Toronto

Stewart L, Dyck E (eds) (2017) The uses of humans in experiment: perspectives from the 17th century to the 20th century. Brill/Rodopi Press, Leiden/Boston

Swain G (1987) Chimie, cerveau, esprit et societe. Paradoxes epistemologiques des psychotropes en medecine mentale. Le Debat 47:172–183

Tansey EM (1998) 'They used to call it psychiatry': aspects of the development and impact of psychopharmacology. In: Gijswijt-Hofstra M, Porter R (eds) Cultures of psychiatry. Clio Medica, Amsterdam

Tone A (2009) The age of anxiety: a history of America's turbulent affair with tranquilizers. Basic Books, New York

Valenstein E (1986) Great and desperate cures: the rise and decline of psychosurgery and other radical treatments for mental illness. Basic Books, New York

Valenstein E (1998) Blaming the brain: the real truth about drugs and mental illness. Simon & Schuster, New York

Valier H, Timmerman C (2008) Clinical trials and the reorganization of medical research in post-Second World War Britain. Med Hist 52(4):493–510

Weiss N (2004) No one listened to imipramine. In: Acker CJ, Tracy S (eds) Altering American consciousness: the history of alcohol and drug use in the United States, 1800–2000. University of Massachusetts Press, Amherst, pp 329–352

Wittkower E (1970) Transcultural psychiatry in the caribbean: past, present, and future. Am J Psychiatry 127(2):162–166

Colonial and Transcultural Psychiatries: What We Learn From History

53

Sloan Mahone

Contents

Introduction	1380
Transcultural Psychiatry and Its Histories	1381
The Universality of Mental Illness	1384
What Is "Culture-Bound" and What Is Not?	1387
Patients, Healers, and Healthcare/Traditional Healers	1388
Psychiatry and the "Colonial Mind"	1390
Conclusion	1398
Bibliography	1399

Abstract

Transcultural psychiatry as a discipline has a well-documented history and is now the subject of numerous retrospectives that chart the development and the shifting conceptual agendas of the field. In contrast, what we have come to think of as "colonial psychiatry" exists primarily as a historiographical category within which historians of medicine, psychiatry, or imperialism may contextualize the imposition of Western categories of "normal" psychology and deviance, race, and difference, as well as greater attention paid to the lived experience of colonialism and the politics of resistance. Ultimately, these disparate but entangled bodies of literature engage with larger questions of the universality of experience and expressions of suffering and distress amidst unequal power relations.

Keywords

Colonial psychiatry · Transcultural psychiatry · Ethnopsychiatry · History of mental illness · Global mental health · Colonialism

S. Mahone (✉)
University of Oxford, Oxford, UK
e-mail: sloan.mahone@history.ox.ac.uk

© The Author(s), under exclusive licence to Springer Nature Singapore Pte Ltd. 2022
D. McCallum (ed.), *The Palgrave Handbook of the History of Human Sciences*,
https://doi.org/10.1007/978-981-16-7255-2_95

Introduction

Transcultural psychiatry as a discipline has a well-documented history, often defined, redefined, and reinvented over the last several decades and largely from within. While it remains an active area of intellectual engagement, as opposed to a distinct set of clinical practices, it is now mature enough to be the subject of multiple retrospectives that endeavor to describe its birth, evolution, legacies, and futures. These are perhaps too numerous to describe in their entirety here, but it is worth giving some context to the field's efforts to reflect upon its own evolution and *raison d'etre*. The beating heart of transcultural psychiatry has always been and remains at McGill University in Montreal, the site of the first established intellectual center for what Raymond Prince termed the "outlandish matters" of interest in culturally situated psychological conditions or dissociative states related to, for instance, trance and spirit possession that elsewhere might be classed as schizophrenia (Prince 2000, 431–2). While Prince's description might be narrower than the broadest interests in comparative cultural psychiatry, his comment points to the difficulty in assessing one medical system of thought with the framework and definitions of another. This has proven to be an ever-shifting bar within transcultural psychiatry. However, the ambitions and idealism of the field are apparent in their attempts to address such research questions. It is worth reviewing some of these retrospectives in their own right for what they might tell us about both the internal and external politics, debates, and aspirations of the psychiatrists and anthropologists who make up the field (see for instance Littlewood 1990; Prince 2000; Bains 2005; Kirmayer 2007, 2013; Steinberg 2015; Delille 2016; and Delille and Crozier 2018).

This review will also engage with the precursors to transcultural psychiatry by looking at the vast body of work and long history of engagement of Western psychiatric practice within colonial territories. In contrast to its "transcultural" cousin, colonial psychiatry exists today primarily as a convenient historiographical category. This trend in historical analysis is concerned with the medical and intellectual history behind the implementation or rationale for colonial rule in tandem with both the experience of, and resistance to, colonial oppression. This sub-genre in historical writing comes forth from a vibrant scholarly interest in former colonial territories within the history of medicine and the history of psychiatry. These aim to uncover the dynamics of race, class, gender, and micro-politics as expressed through the imposition of conceptual frameworks that imposed perceptions of "otherness" on subject peoples. This is not a dynamic invented by colonialism. Rather, such intellectual biases and misuses of science and medicine typically occur where inequalities and marginalized groups exist.

Psychiatry as a medical specialty was practiced from the outset of colonial occupation, often as an outgrowth of prisons, and up until and beyond decolonization. However, early colonial era doctors were often only nominally trained in state-of-the-art psychiatric medicine, and they only ever held direct authority over a miniscule percentage of the population. Nonetheless, this ill-defined professional field often punched above its weight in terms of the enormous consequences felt by colonized populations both in terms of racist interpretations of their collective characters and in explicit policy directives that sought to undermine resistance to colonial rule. Despite a poor showing in terms of numbers, colonial psychiatrists, at least those who sought such acclaim, were imbued with a sense of authority on matters of race, culture, and modernization as they sent home to the metropole widely read medical reports from the "field."

The aim of this chapter is to engage with both transcultural psychiatry and colonial psychiatry in turn as entangled but contrasting intellectual enterprises and historical trajectories. Along the way, we will necessarily engage with intersecting disciplines such as the histories of anthropology, global mental health, and clinical medicine as well as incorporating both primary and secondary sources and addressing key methodological concerns.

Transcultural Psychiatry and Its Histories

In 1965, Eric Wittkower, from the Department of Psychiatry at McGill University, gave an opening address to the Ciba Foundation symposium on *transcultural psychiatry*. The event, held in London, attracted psychiatrists working around the globe and produced a volume of academic chapters alongside published commentary that followed the delivery of each conference paper (Wittkower in De Reuck and Porter 1965). Wittkower's task was to relate "recent developments in transcultural psychiatry," a field that he had helped to establish a decade before. Wittkower acknowledged the recent surge in interest in the field, but he also harkened back to the "true founder" of cultural or "comparative" psychiatry, Emil Kraepelin, who questioned how culture might influence the frequency or type of mental disorders experienced in places as different as Germany and Java. While it could be argued that travellers and doctors have always remarked upon the characters and psychologies of the unfamiliar peoples they encountered, Kraepelin's brief time in Java produced a paper that attempted to address larger questions of difference, universality, and the existence of what would ultimately come to be known as "culture-bound syndromes." Historian Holger Steinberg has suggested that Kraepelin's influence as a pioneer in transcultural psychiatry is negligible in comparison to how often he is cited as a "founder" (Steinberg 2015) but the document he produced remains an interesting historical marker to how psychiatrists operating within a world of empires attempted to make sense of difference in light of prevailing theories about race, degeneration, and ill-defined notions of "culture." Wittkower outlined for his more modern audience the myriad terms which routinely appeared under the umbrella of social psychiatry, including cultural, transcultural, cross-cultural, comparative, and international psychiatry (Wittkower and Rin 1965, 6–7). He stressed that transcultural psychiatry was primarily a field of research and theory, an observation that we might see as still accurate today. It should be noted, however, that the field has always maintained a degree of commitment to application whether as a means of evaluating or developing health services internationally or as an engagement with the mental health needs of immigrant communities (Prince 2000, 436; Lipsedge and Littlewood 1997; Fernando 1988). While some of the field's early approaches to cultural comparisons may seem outdated and exoticizing today, its proponents attempted to move beyond the racial determinism that held sway unchecked from much of the research output of colonial medicine. As much as transcultural psychiatrists worked to understand local variations in the appearance of mental illness and approaches to treatment, they were also asking larger questions in the hopes of moving the science behind psychiatry forward with cutting-edge research.

The Ciba Foundation symposium is interesting for the context in when it was produced – roughly a decade after the first newsletters on "transcultural research in mental health problems" appeared at McGill University. These first attempts to define a set of interests in psychiatry, anthropology, and culture and to report on interesting work happening throughout the world would evolve into a more substantial academic journal, *Transcultural Psychiatric Research Review*, and eventually today's *Transcultural Psychiatry* now with an emphasis on publishing original research. The CIBA papers were presented by leading experts in psychiatry and anthropology, such as Alexander Leighton and Margaret Mead, but more notably by next-generation leaders such as Thomas Adeoye (T.A.) Lambo and Pow Meng Yap who were from and worked in some of the regions under scrutiny. The broad discussion pointed to continued debates about the true definition of transcultural psychiatry, how it might differ from related fields of "cultural" psychiatry or "ethno"-psychiatry, and how it really ought to proceed in the world as either a distinct discipline or as a somewhat global or comparative turn to what was, in actuality, simply "psychiatry" (Ellenberger, [Review] 1968, 100). Throughout its history, transcultural psychiatrists have continued to critique the field from within. In 1974, psychiatrist Ajita Chakraborty penned a "challenge" to his colleagues with his article "Whither Transcultural Psychiatry" that both lauded the efforts of much important research but also called out the ethnocentric biases that characterized expressions of mental illness in non-white societies as "cultural," while the same observations and processes in white Western settings would be classified as "social psychiatry" or "family psychiatry" (Chakraborty 1974, 102–107). Chakraborty's essay takes as its starting point, a review of a prominent published collection *Transcultural Psychiatry* by Ari Kiev (1972). Both the book in question and its reviewer, according to Chakraborty, fell into the same traps including the failure to define what "transcultural psychiatry" actually is and the use of biased language that accepted uncritically the evaluation of non-Western groups as "natives" or primitive and thus worthy of a Western gaze without any suggestion that Western diagnostic processes are equally fraught with idioms of "culture." Chakraborty's critique is written from the perspective of an insider and a strong proponent of the ideals of what transcultural psychiatry could accomplish. Such internal critiques have continued to shape the field especially with the greater inclusion of authors from non-Western countries.

Historians today will look at these early collections as primary source material that may shed light on some of the nuances, tensions, and contradictions that preoccupied the discipline in its infancy. Decades later, Laurence J. Kirmayer of McGill's Division of Social and Transcultural Psychiatry would publish another brief retrospective stock-taking of the field – once again from within the pages of their own journal (*Transcultural Psychiatry*, 50, 1, 2013). Kirmayer's editorial looked back on the 1956 newsletter, the 1964 establishment of a scholarly journal concerned primarily with compiling observations from the field, and the shift in 1991 to the publication of original research at the crossroads of anthropology and psychiatry (Kirmayer 2013, 3). Kirmayer's editorial traces the intellectual development of transcultural psychiatry, including its evolution from "older ways" of understanding culture "as relatively isolated, distinct, self-contained, local worlds," an outdated preoccupation with the "exotic," and misunderstandings of culturally specific forms of communication and expression as pathological "culture-bound syndromes" (Kirmayer 2013, 4). This "colonial" way of thinking, mused Kirmayer, saw "culture" as something experienced by "others" as seen from the vantage point of a normative Western perspective. Thus, the earliest proponents of transcultural psychiatry were well aware of the intellectual baggage they had inherited from the wider set of colonial practices and literatures that tended to fetishize hierarchical notions of difference; however, resolving some of these dilemmas remained a work in progress.

A key turning point in some of these critiques of the field has focused on the tensions that existed between (or sometimes within) psychiatry and anthropology, including debates about how to modernize research practice. A watershed moment occurred in 1977 with the publication of an article entitled "Depression, Somatization, and the 'New Cross-Cultural Psychiatry'" by Arthur Kleinman which took as its starting point a strongly worded response to a published literature review of depressive disorders with its conclusion that there existed "insufficient evidence to support a prevalent view that depressive illness in primitive and certain other non-Western cultures has outstanding deviant features" (Singer 1975, p. 3). Kleinman, trained in both psychiatry and anthropology, argued that Singer's compilation of a wide body of literature represented an "old transcultural psychiatry" with its problematic data sets, imprecise or noncomparable definitions, and predictably flawed conclusions. Such research looked hard for "pure" forms of diseases, such as depression, rather than understanding diseases as "explanatory models" not as "things" (Kleinman 1977, 4). Kleinman called for further research along the lines of the "new" interdisciplinary approaches to the field which must begin with deeper ethnographic understandings of cultural context and local systems of belief before searching for universals based upon stringent Western categories and terminologies. Interestingly, the history and implications of this particular debate remain unsettled even today with new research by Donald McLawhorn revisiting the Singer-Kleinman debate as a means of assessing cultural psychiatry's theorization of Chinese thought and illness experience particularly around concepts of "somatization" (McLawhorn 2021).

In a 2007 *Textbook of Cultural Psychiatry* (Bhugra and Bhui, eds.), Kirmayer further outlined the field's historical trajectory, emphasizing that cultural psychiatry was driven primarily by "clinical imperatives" rather than "theoretical problems," but he also noted the long, troubled history of previous "taken-for-granted superiority of European civilization" (Kirmayer 2007, 5). Kirmayer offered an updated critique (albeit now more than a decade old) with cautions about the persistent legacy of colonial thinking within transcultural psychiatry as evidenced by the "continuing romance with exoticism, the de-contextualised view of mental-health problems and focus on culture-bound syndromes, efforts to reify and essentialize culture as individual traits, and the tendency to employ developmental hierarchies contrasting traditional and modern societies" (Kirmayer 2007, 13).

More recently, Emmanuel Delille has provided a tight history of cultural psychiatry as seen through the works of Georges Devereux and Henri Ellenberger, particularly in relation to their engagement with Native American communities in the 1950s (Delille 2016). Delille calls for an end to transcultural psychiatry's many origin stories, suggesting instead that the time has come to dig deeper methodologically within the copious sources that we now have in order to establish more nuanced histories of how decades of networks and rivalries actually operated. Although the vast corpus of published material provides some insights, Delille pushes his analysis further by reading the seminal texts alongside lecture notes, correspondence, and other institutional and personal archival holdings from ethnology, psychiatry, and related fields. Delille brings his expertise to bear with co-editor, Ivan Crozier, to a special issue of *History of Psychiatry* on "historicizing transcultural psychiatry," the result of an international workshop held in Paris in 2016 and organized around work on key figures in the development of the field. The result is a lively set of papers that, while focusing on individual actors, elucidate context-driven developments in a global network of ideas (Delille and Crozier 2018).

The Universality of Mental Illness

The many origin stories of transcultural psychiatry have elucidated the problems inherent in applying universal definitions and dominant Western approaches to local experiences, beliefs, and systems. They have also flagged the difficult legacy left by the intellectual output of deeply politicized colonial era research – all of which has prompted further debates about the legitimacy of a truly "transcultural" approach to psychiatry whether one sees this as a largely theoretical or applied area of research. Psychiatrist Jatinder Bains, from the University of Sydney, notes that from its inception, transcultural psychiatrists were concerned with questions of the "universality" of mental illnesses or related presumptions about the existence of the same mental illnesses everywhere but with the expectation of sometimes drastically different modes of expression. He frames the rise of transcultural psychiatry as an expression of the "optimism" expressed by humanitarian and social science approaches to mental health – a postwar global engagement that sought to address the aftermath of the Second World War by applying expertise in how culture and environment interacted with personality types and mental illness in light of global migration, rapid social change and the tensions associated with interconnectedness (Bains 2005, 141).

Such questions predate the evolution of transcultural psychiatry as a research field, harkening back to Kraepelin and those before him who attempted to confirm or deny the presence of Western-defined illnesses such as depression or schizophrenia across cultures or to assess how certain illnesses, if common to everyone, might be understood or expressed differently in different regions of the world. The approach to such questions can vary significantly across disciplines, from anthropology to neuroscience. Psychological anthropologist Erika Bourguignon (1924–2015), a specialist in studies of trance states, shamanism, and spirit possession wrote extensively about how societies institutionalize locally understood and accepted means of experiencing altered states of consciousness. Bourguignon opens her comparative case study of Joaõ, an *Umbandista* from Saõ Paulo, and a case of "multiple personality disorder" in New York City with reference to the psychoanalytic principle of a "psychic unity of mankind" (Bourguignon 1989, 371). This proposed that the value of making comparisons in experiences of illness and healing was to uncover those cultural institutions that might be "out in the open" and "culturally implemented" in one group but are "repressed" or pathologized in another. The work of Bourguignon and others not only prompt questions about the nature of universality, or "psychic unity," but also shine a light on the social construction of categories of illness, definitions of the "normal and abnormal," and the dynamic relationship between patients and healers.

In her 1989 article, Bourguignon presents two cases of "alternate personalities": that of "Mrs. G," a black woman in her 40s from New York City alongside her aggressive and slovenly alternate personality, "Candy," and "Joaõ," described as "a shy and mild-mannered mulatto 33-year-old" from Saõ Paulo with his alternate, a provocative female, "Margarida" (Bourguignon 1989, 373). The two personal illness stories unfold in remarkably similar ways. Their most significant differences come from the methods by which these periods of distress are mediated. Mrs. G/Candy's case is represented by a psychotherapist's clinical report, and Joaõ/Margarida is described within an American anthropologist's ethnographic account of a Brazilian Umbanda possession cult. Joaõ is understood as a medium for "Margarida" as well as for several other spirits who visit him, a fact corroborated by other Umbandistas. Mrs. G., in contrast, reacts hostilely to her therapist's diagnosis that she possesses "Candy" as a split-off part of her own deeply troubled personality. The comparative case study prompts a number of questions about the universality of many illness experiences with further questions about the efficacy of some approaches to diagnosis and treatment. Joaõ, Bourguignon writes, "turned to the spiritual-therapeutic institution of Umbanda specifically in order to learn how to give expression to these personalities" (Bourguignon 1989, 376). In contrast, Mrs. G.'s relationship with her therapist was oppositional and distressing as his approach could only see her "multiple personality" as a severe pathology or defect even amidst the psychiatric profession's own contentious disagreements as to whether the controversial diagnosis of multiple personality disorder truly existed.

Philosopher of science, Ian Hacking has given a historical perspective to what became an apparent "epidemic" of multiple personality diagnoses (MPD) in American psychiatry in the 1980s. Even as the dissociative state came to be understood as a consequence of the trauma of sustained childhood sexual abuse, there remained a ferocious debate within the field about the "reality" of the disorder (Hacking 1995). Bourguignon also notes the history of childhood abuse in many multiple personality cases and notes that the nature of treatment often takes a "punitive" turn when psychiatrists are skeptical of their patients' claims. Depending upon one's personal orientation, the similarities in the cases of Mrs. G and Joaõ could suggest that Umbandista possession is a culturally formulated coping strategy employed to address an underlying dissociative state such as MPD. Alternatively, the case comparison may suggest that there could be a possession like experience at play for Mrs. G. with the "recognition of the patient's subjective experience of alien intrusion and discontinuity of identity." While Bourguignon's dual case study allows us to engage with these questions of context and terminology, she concludes that the similarities observed in Mrs. G. and Joaõ remain cursory as the theoretical understanding of the illness experience and its therapeutic response remain radically different in practice (Bourguignon 1989, 383).

While psychiatry, anthropology, and even philosophy might grapple with the nature or existence of specific disorders, others continued to question the usefulness of looking for universals at all. For some, such questions were themselves exercises in bias. In 1973, psychiatrist E. Fuller Torrey picked up the general point of universality in his think piece, "Is Schizophrenia Universal?" (Torrey 1973). Torrey's contention was that no genuine foundational research base existed to answer the universality question. Rather, proponents of one stream of thought or the other merely substantiated their own agendas with selective lists of citations from dubious sources such as colonial era racially charged research studies. The very same question, however, was taken up a decade later by Jablensky and Sartorious who, harkening back once again to Kraepelin's aspirations to find common threads in understanding schizophrenia (or *dementia praecox*), heralded the feasibility of global comparisons aided in great part by the multi-site research capacities of the World Health Organization (Jablensksy and Sartorious 1988). In 1994, the *British Journal of Psychiatry* published a "revisiting" of the universality question which outlined the core motivation behind such studies. If universality can be determined, the authors noted then the likelihood of a biological etiology for mental illness can help to validate modern diagnostic categories (Patel and Winston 1994). The power behind who may define what constituted "universal" categories of illness was also noted by the authors in their assessment of the international criteria for classifying mental disorders and their symptoms. Within a team of 47 leading psychiatrists in charge of developing the *International Classification of Diseases* manual (ICD-10, WHO 1992), only 2 were from an African country. Thus, the biases inherent in the field for decades remained apparent as illnesses that did not conform to Western descriptions were considered to be cultural masking of "real" Western illnesses (Patel and Winston 1994, 438).

There is surely no "final word" on the question of universality (of schizophrenia or of mental diseases overall), but anthropologists Tanya Luhrmann and Jocelyn Marrow present a nuanced collection of case studies that attempt to put the problematic and scattered diagnosis of schizophrenia into deep discussion by way of 12 case studies that reveal the lived experience and social context for schizophrenia (Luhrmann and Marrow 2017). The ways in which one's social environment and life circumstances may affect the outcome of the illness, there has been little work to explain how this occurs at home and amidst personal lives and relationships. Even as the diagnosis, crafted in the West and debated in global forums, remains imprecise, the reality of suffering caused by this experience of illness is vast. "There is no such thing as schizophrenia, and this is its ethnography" writes Luhrmann. This may yet be the most critical approach to understanding the truly universal burden of mental illness.

What Is "Culture-Bound" and What Is Not?

Embedded within, or perhaps adjacent to, the question of universality, is the historical description of psychiatric syndromes termed "culture-bound." The term was coined by psychiatrist Pow Meng Yap in a paper he published in 1951 ("Mental diseases peculiar to certain cultures"). As the various writings from the "new" transcultural psychiatry attest, there has been an intellectual collapse of this category as it becomes shorthand for the exotic behaviors and illnesses of more primitive peoples. Ivan Crozier has reintroduced Yap's article as a classic text by placing it within the conceptual framework in which Yap and his contemporaries worked which included questions of the universal validity of Western psychiatric categories even as he relied on them himself (Crozier 2018). As Yap engaged with the questions that also vexed his colleagues, he also called for greater research efforts in non-Western settings so that the dominant diagnostic system in place could be re-examined. In effect, Yap called for a revalidation of Western psychiatry using the additional expertise of non-Western sources. The literature on "culture-bound syndromes" from psychiatry (colonial and transcultural) and from anthropology and history is vast. Much of it has focused on a few classic cases such as *koro* (extreme anxiety that one's penis is shrinking or retracting), *latah* (an extreme startle response), and various forms of possession. However, the most famous example remains *amok*, thought to be unique to the Malays and characterized by a sudden outburst of homicidal rage that could result in multiple murders without apparent enmity or motive (Ellis 1893). Recent work within the history of psychiatry has reassessed not the validity of amok as a psychiatric disorder, but the enormous field of commentary and analysis *amok* has generated. Imai, Ogawa, Okumiya, and Matsubayashi (2019) identified 88 English language articles on *amok* in a systematic literature review and traced the shifting meanings ascribed to it. The authors outline a long trajectory of meanings from the possible ancient etymology of the word *amok* to increasingly psychologized and medicalized understandings of both *amok* and the Malay character. The authors conclude that such systematic reviews of disease categories help us understand how to frame modern diagnostic categories and to approach illness categories overall with more caution.

The World Health Organization's efforts to develop a global classification system for mental disorders are taken up by Harry Yi-Jui Wu in *Mad by the Millions: Mental Disorders and the Early Years of the World Health Organization*, a historical treatment of postwar techno-idealism of modern institutions and networks of expertise (Wu 2021). The ambition behind the WHO's push for a global social psychiatry project maps onto the same debates around universality and the efficacy of cross-cultural comparisons that have continued to engage the field. Wu traces the proposals and politics of the WHO's engagements with its member states as well as the scale of collaborative networking that took place within a long series of international platforms, symposia, and meetings. However, such engagements remained largely within the purview of Western-trained elite experts. The authority given to Western medical interventions and to systems of classification continues to raise questions about the applicability of dominant diagnostic frameworks that fail to acknowledge significant expertise in countries where "traditional" healers are the first port of call for a majority of the population.

Patients, Healers, and Healthcare/Traditional Healers

While late-colonial and postcolonial efforts to engage with a meaningful epidemiology of mental disorders often operated at scale, there were also many examples of deeply contextualized ethnographic work on mental health and illness taking place throughout the world. Anthropologists and psychiatrists, including some who were dually trained, have produced groundbreaking foundational studies on systems of regionally specific psychotherapies and systems of diagnosis and healing. Margaret J. Field, a psychiatrist/anthropologist working in Ghana (formerly the Gold Coast), produced several key articles about concepts of mental ill health, as well as an extensively documented description of healing shrines that addressed local concerns about witchcraft, depression, schizophrenia (to use the author's biomedical definitions), and "security" (Field 1960).

Ursula M. Read and Solomon Nyame conducted research on mental health in Ghana with a particular emphasis on the roles played by families. They have elucidated the precarity with which the mentally ill and their families must navigate means of treatment amidst the pressures of daily life (Read and Nyame 2019). The series of intimate interviews conducted demonstrate the degree to which daily life (the "mundane" tasks of cooking, eating, washing, and making a living) is disrupted when mental illness is present. The authors stress that these quieter daily moments are understudied amidst a research field that has focused on comparing explanatory models of disease and healing practices. As a result, the findings have a stronger "feel" for the universal as a presentation of local circumstances with universally understood expressions of distress or struggle.

A special issue of the journal *Social Science & Medicine* (v. 15B, 1981) presented a series of provocative and groundbreaking papers organized around "Causality and Classification in African Medicine and Health" which incorporated both individual ethnographic accounts of practice and studies of the broader framework of systems, taxonomy, and medical pluralism. Detailed inquiries into the many facets of health-seeking behavior within communities elucidate both the internal order at play within countries or regions and the multiplicity of options at work as both patients and healers operate in dynamic and sometimes contrasting healing methods. The means of studying such systems without the polarizing language of "traditional" and "modern" remains a challenge and source of reflection even today. Violet Nyambura Kimani's work on plural healthcare in central Kenya and J. Kimpianga Mahaniah's paper on health in Kinshasa, Zaire (Democratic Republic of Congo), both outline the multidimensional characters of medical treatments available to the population, a dynamic fostered both by consumer choices and the demands caused by population influx to cities where a stable health infrastructure may be lacking. Harriet Ngubane, author of the ethnography *Body and Mind in Zulu Medicine: An Ethnography of Health and Disease in Nyuswa-Zulu Thought and Practice* (1977), notes the pluralistic nature of medical care in rural South Africa with a particular emphasis on the contrasts in the doctor/patient relationship when engaging with both "Western-trained doctors" and "indigenous healers." In particular, Ngubane highlights the significant barriers faced by patients when moving between systems such as negative attitudes of medical doctors toward previous engagement with local healers, poor communication about a diagnosis or treatment, and obstacles related to available time or funds to travel to hospitals (Ngubane 1981). The themes presented in these still relevant early collections address the social dynamics, in their many forms, of health decision-making and ways of accessing care. Key theoretical works such as Arthur Kleinman's *Patients and Healers in the Context of Culture* (1980) have provided conceptual frameworks for understanding the illness experience by presenting both ethnographic detail and theoretical models which may be considered for other social contexts. Increasingly, ethnography and oral history work have been incorporated into not only works of transcultural psychiatry and medical anthropology but also as first steps in the design of health interventions.

In 1986, anthropologist Murray Last and sociologist and traditional healer Gordon Chavunduka co-edited a first-of-its-kind volume, *The Professionalisation of African Medicine*, which evolved from an international conference held in Botswana. The collection today has less resonance as a description of comparative modern healing practices, but it charts the early efforts of healers in African countries to organize and professionalize their services through dialogue with other systems of care and the development of professional associations. The volume is notable for its engagement with social scientists and practitioners working on and within the continent. The questions and problems raised by participants mirror those asked by transcultural psychiatrists including points of definition and terminology, methods for collaborations, and how and why one might compare ways of healing.

Researchers from anthropology, public health, and medicine (including psychiatry) continue to study the modern practices of traditional healers as well as the experiences of patients seeking treatment from local specialists. Attempts to find meaningful ways to collaborate for the good of the patient remain in the forefront of healthcare settings in low-to-middle-income countries (LMICs) especially with the realization that access to modern hospitals may be limited for a large segment of the population. A recent review essay analyzed existing studies that assessed the perceptions of both traditional healers and biomedical practitioners toward collaboration. While the breadth of the review was limited and strongly tilted toward Africa, it showed increasing openness to finding common ground despite contrasting ideas about disease causation and a skepticism about the efficacy of some medications.

Some previously held assumptions from the medical field have been given more nuance with this new research. Most notably, it has become clear that with mental health treatments in particular, many patients will opt for the services of a traditional healer even when they have access to modern hospital-based treatment (Green and Colucci 2020). This is perhaps not so surprising as people have always made the most of a whole host of options available to them. As the authors here note, more research is needed to understand the full range of intangible and hard-to-quantify benefits a patient may receive from local healers.

Psychiatry and the "Colonial Mind"

The breadth of exciting work carried out by decades of research in transcultural psychiatry has nonetheless had to set itself apart from the sorts of colonial practices that came under fire from critics like Frantz Fanon. To some extent, the field, particularly in the early years, has had a dual burden of developing its own research trajectories while addressing the impact of deeply biased work that, despite obvious flaws, made it into the mainstream of psychiatry and medicine. Those who have endeavored to write the history of transcultural psychiatry have presented a general trajectory of the key figures and research trajectories of what constitutes a coherent set of professional principles and practices – even as they were debated from within. In contrast, historians of colonial psychiatry are often interested in the dynamic taking place on the peripheries of psychiatry. These might the use of psychiatric ideas and language within the politics of imperialism and rebellion, the intersection of psychiatry with healthcare systems more broadly, and the histories of related disciplines such as anthropology, psychology, and education. As such, histories of colonial mental "asylums" are very often about something *else*. The primary sources for these studies vary widely; however, the archival holdings of colonial states, while potentially extensive, must be assessed critically and cross-referenced with other types of sources as much as possible. Quite simply, many of the sources that purport to offer psychological interpretations of individuals or groups tend to shed more light onto the proclivities, prejudices, and anxieties of colonial practitioners and the governments they served.

Megan Vaughan's groundbreaking work on colonially contrived "idioms of madness" was the first to historicize the fallacies of colonial constructions of African madness. By focusing on European doctors' preoccupations with categorizing African normality (and abnormality) in the context of what was seen as the rapid imposition of "civilization," Vaughan describes the "double othering" that occurred as doctors reacted with alarm to African delusions that incorporated symbols that did not belong to them such as claims to European royalty or the unlikely ownership of material luxuries (Vaughan 1983). Vaughan's article, alongside a later chapter appearing in her book of essays, *Curing their Ills*, and article on "the madman and the migrant" by anthropologists John and Jean Comaroff (1987) have paved the way for a now vibrant field of scholarship on the discursive and material histories of "madnesses" that have their roots in colonialism, inequalities, and resistance.

While the scholarship in this area has grown from the 1980s onward, the history of psychiatry against the backdrop of colonialism was, in effect, first elucidated by those who were living it. The writings of revolutionary thinkers such as Frantz Fanon, Octave Mannoni, and Aimé Césaire provide both experiential and philosophical contemporary accounts of the impact of colonial occupations and oppression on colonized peoples although their approaches vary significantly. Fanon is perhaps the exemplar of the critique of "schools" of psychiatric ideology that exerted their influence throughout colonized Africa. His multiple polemics, including *The Wretched of the Earth*, *Black Skin, White Masks*, and *A Dying Colonialism*, include chapters that are explicit in their accounts of the devastating effects of psychiatry's racism within systems of oppression. Richard Keller has given the most in-depth account of Fanon's writings within the context of a history of colonial medicine. In his longue durée account, *Colonial Madness: Psychiatry in French North Africa* (2007), Keller traces discourses from within France and French psychiatry that transformed ideas of the Muslim character within the colonial imagination of the Maghreb. Fanon has been the subject of myriad biographical works and critiques, including newer work by Gibson and Beneduce (2017) which contextualizes Fanon's largely untranslated works in psychiatry (in contrast to his more famous polemical writing) in the sphere of his anticolonial politics.

While historians have called upon the works of Michel Foucault to frame the viciously unequal power dynamics inherent in colonial medical systems, frequently the analysis stops there. More nuanced histories of specific colonial contexts continue to show that relationships of power always include more resistance, agency, and even collaboration than a theory of "biopower" alone can articulate. Fanon's writings take explicit issue with the construction of a medico-psychiatric knowledge base emanating from both North African and East African "schools" of psychiatry, commenting that the racist pseudo-intellectualism of published doctors like J.C. Carothers in Kenya easily made its way into medical education, colonial government policy, and institutions like the World Health Organization as authoritative teaching on race (Fanon, *Wretched*, 1968 [original in French, 1961] 303). As Fanon, originally from Martinique, practiced as a psychiatrist in Blida Psychiatric Hospital in Algeria, he became deeply disillusioned by and critical of the institutionalized racism inherent in the psychological characterizations of colonized peoples, who were also, at their core, wounded by the burden of colonial occupations. The characterization of the "Muslim personality" in Algeria as steeped in criminality easily parallels J.C. Carothers' writings on the "African mind" which identified the African maladaptation to modern civilization to the extent that forms of rebellion were classed either as pathology or as a regression to savage impulses (McCulloch 1995; Mahone 2006). Carothers, with only a diploma in psychological medicine gained later in his career, is perhaps seen as the most notorious example of the abuses of a "colonial psychiatrist" as his key writings from Kenya were instrumental in rationalizing the violent colonial response to upholding rule and stifling the Mau Mau rebellion with a rhetoric that turned aspects of widespread concentration camps into centers for psychological "rehabilitation" (McCulloch 1995).

At close glance, it becomes clear that the "East African school" of psychiatry consisted primarily of Carothers himself with the international status afforded him due to his World Health Organization commissioned monograph, *The African Mind in Health and Disease* (WHO 1953), and the subsequent analysis of Kenya's struggle for independence in the government report, *The Psychology of Mau Mau* (Nairobi 1954). The East African school was not a movement in any formal way. Its importance stems from the degree to which the writings of its few psychiatrists were cited by contemporaries and supported by powerful institutions despite the clear racial politics at play. This is not to say that Carothers was entirely marginal in his ideas. Chloe Campbell's study of the eugenics movement's intellectual production from Kenya, *Race and Empire: Eugenics in Colonial Kenya*, traces how a small but vocal medical intelligentsia, publishing in the 1930s in journals within and outside of Kenya, came to influence race and eugenic science beyond Kenya's settler colony borders (Campbell, Manchester 2007; Mahone in Mahone and Vaughan 2007). Despite the virulent criticism of writers like Carothers, the published material they produced created a body of racist science that would seep into policy, education, science, and psychological medicine as well as popular thinking around matters of race for those who were already receptive to its message.

Despite the significant objections to some of his most outrageous claims, including the contention that the mind of the African was similar to that of a "lobotomized" European, Carothers continued to be included in far more progressive international research symposia comprising the foremost anthropologists and (transcultural) psychiatrists of the late 1950s and early 1960s. Nigerian psychiatrist Thomas Adeoye Lambo's critiques of Carothers and his ilk predate those of Fanon despite having to accommodate Carothers' participation in the international network of burgeoning transcultural psychiatrists (Lambo, ed., Conference Report, Abeokuta, 1961). Lambo developed groundbreaking innovations to the provision of psychiatric care through his Aro Hospital project in Abeokuta, Nigeria. The material reimagining of what effective care might look like worked in tandem with tearing down the deeply ingrained racism in psychiatry as theorized by Carothers. In his study of "cultural factors in paranoid psychosis," Lambo begins with a call to dismantle the dangerous scientific thinking prevalent in psychiatric work emanating from some of the medical voices coming from the prejudices of the colonial system. Lambo specifically called out Carothers' work as *reductio ad absurdum* (Lambo 1955, 239–40). Most interestingly, Lambo's hard-hitting critiques of "colonial psychiatry" eschew the philosophical polemics that Fanon would express famously in *The Wretched of the Earth*. Instead, he co-opted the scientific authority of Carothers as an "expert" in African psychiatry by characterizing such writings as failures in medical and scientific terms. Lambo's legacy has been well-acknowledged largely through the works of Jonathan Sadowsky, *Imperial Bedlam: Institutions of Madness in Colonial Southwest Nigeria* (Berkeley 1999), and Matthew M. Heaton, *Black Skin, White Coats* (2013). Heaton, in particular, offers an analysis of the beginnings of Nigerian psychiatry and its role in decolonizing the medical discipline on a global stage and within the broader context of Lambo's enormous contribution to transcultural psychiatry (Heaton 2013). Matthew Heaton's essay in the *History of Psychiatry* special issue on "historicizing transcultural psychiatry" (2018) offers an incisive analysis of T.A. Lambo's career with a non-Western-centric point of departure. In contrast to many other works in this area, Heaton's insightful overview moves past the significant preoccupation with the field's impact by suggesting that we should be looking more at the "politics of the 'transcultural' than the practice of 'psychiatry'" (Heaton 2018, 315). Indeed, Lambo was one of a handful of African psychiatrists who bridged the colonial and postcolonial contexts, and he was and is recognized as an early leader in an emerging global mental health movement – uniquely from the vantage point of an Africa-based clinician. Lambo's task, according to Heaton, was nothing less than a "project of decolonizing psychiatric knowledge and practice" (Heaton 2018, 316). While Lambo has arguably achieved this conceptually and historically in many ways, the provision of mental healthcare in Nigeria has continued to suffer from a lack of investment and prioritization postindependence, a reality that continues to impact many low- to middle-income countries today. While this holds true in terms of a "modern" infrastructure for psychiatry, it would be wrong to imply that mental healthcare is absent across the continent. Indeed, Lambo's innovation within the Western model was to incorporate the expertise of traditional healers within the psychotherapeutic environment of the Aro Village hospital scheme, thus assuring that treatment embraced local understandings of illness and wellness.

Like his contemporaries, Lambo considered questions of the universality of mental illnesses like schizophrenia, but rather than theorizing whether or not schizophrenia existed in Africa, he differentiated symptomatology based upon urban and largely Western-educated populations and rural and largely illiterate populations. Thus, "rural Africans" tended to express delusions associated with schizophrenia in terms of witchcraft or supernatural causes – factors that earlier colonial psychiatrists failed to recognize due to a "willful desire not to see them" alongside assumptions that witchcraft beliefs were part and parcel of the primitive collective delusions of a backward society. Lambo also noted particular pathologies of modern Western life which seemed to be on the rise, particularly among housewives, in the newly developed housing estates in Great Britain. Those affected exhibited increased anxiety with hysterical features and depression. The impact of this social insecurity caused by the alienation, boredom, and isolation of this new mode of living was termed "suburban neurosis" (Lambo 1959). Although this particular article is not among the most prominent of his writings, it is interesting for the way in which Lambo rewrites the racist assumptions of colonial medicine that saw Africans breaking down when introduced too rapidly to "civilization." Lambo's brief article entitled "Rapid Development Can Threaten Mental Health," published in the *International Journal of Social Work*, reoriented the discussion to address the process of rapid urbanization, the dearth of social welfare provision. In other words, poor mental health was linked to development, but in terms of rampant inequality and not the innate capacities or collective psychoses of Africans.

J.C. Carothers' most notorious theories such as the childlike qualities of adult Africans and the "idleness" of Africans' frontal lobe are repeated in most contemporary and modern critiques of the psychiatric practices he represented and the general tone of the work emanating from colonial Africa. However, psychiatrist Femi Oyebode, in a general history published in the *International Review of Psychiatry*, looked at more casual examples of what constituted "evidence" in Carothers' assessments of the psychological and personality shortcomings of African peoples. In a 1951 article that described the "unreliability" of Africans, Carothers drew up a list of examples that he personally observed in Africans in his employ. This compilation of apparent evidence included "my cook often does not tell me he wants more firewood until the latter is quite finished…the houseboys cannot put furniture back level with the wall…the egg-boy brings eggs on Mondays and Thursdays. He failed to come last Thursday on account of the rain, so he did not come again till Monday, though there was nothing to prevent his coming on Friday" (Oyebode 2006, 323). That privileged settler colonists in Kenya who were accustomed to engaging with "house boys" and "egg boys" held such views is not surprising, but as Oyebode points out, this laundry list of petty racist grievances appeared in one of the prominent medical journals of its day without apparent question or protest. That such statements could constitute "evidence," Oyebode asserts, indicates "how far these beliefs were part of widely held assumptions in Western society about Africans but also about other subject peoples in the colonial period" (Oyebode 2006, 324). Indeed, it is striking that although there were numerous published criticisms of Carothers' grander claims, his work was not thrown out in its entirety as lacking scientific credibility. Rather, ongoing work of others continued to sidestep some of his claims but cherry-picked other observations as a contribution to the general corpus of important medical work on Africa.

While T.A. Lambo is widely recognized for his extraordinary impact on the international stage, he had important African contemporaries (although not in great numbers). Yolana Pringle has authored a recent study of Ugandan psychiatry in the era of decolonization, a central figure of which was psychiatrist Stephen Bosa of Butabika Hospital in Kampala (Pringle 2019). Pringle's aim is to situate Ugandan psychiatry into the dual settings of a newly independent country (in a decolonizing continent) amidst a surge in transnational concerns and networks in global mental health. Uganda, like other states, participated in the overarching debates concerned with imagining what psychiatry's true impact could and should be in the process of rebuilding a nation. Pringle's work is a welcome addition, as is Heaton's, on the history of African psychiatry with African practitioners at the center. This is an area of scholarship worthy of development as other prominent figures such as Sudanese psychiatrist Tigani El Mahi well known in their own countries but left out of a broader historiography.

Saul Dubow charts the institutionalization of scientific racism underpinning the rise of apartheid and race relations in modern South Africa, looking well beyond psychiatry to interrogate a broad spectrum of scientific and social science disciplines for the ways they have entrenched intellectual racism into the mainstream (2008). While seemingly cut-off from the rest of sub-Saharan Africa, similar work has come out of studies of psychiatry amidst South Africa's race politics before, during, and after apartheid. Many of these are article length but seminal in their own right. Harriet Deacon has written about "madness and race" within Robben Island's early manifestation as Cape Colony's "lunatic asylum" (Deacon 1996). Sally Swartz, emeritus professor of psychology at the University of Cape Town, has also documented the history of the treatment of insanity in the Cape (Swartz 2015). Swartz has more recently penned a Fanon-inspired historico-psychoanalytical study of the role of "ruthlessness" in psychic development and how this has resonance not only with historic trauma of past colonialisms but also in modern protest movements, such as those felt across university campuses amidst attempts to "decolonize" ubiquitous mainstream institutional racisms visible in curricula and memorialization in material culture such as statues (Swartz 2018).

Despite some dominance in the historical field, colonial psychiatry was not only an Anglo enterprise in Africa. Recent work by Katie Kilroy-Marac interrogates the lasting legacy of the Fann Clinic in Senegal and its postcolonial engagement with transcultural psychiatry in the 1960s and 1970s. Kilroy-Marac's study presents an ethnography where theory and history meet praxis. The Fann Clinic has an important place in the history of psychiatric treatment in Francophone Africa which stands in stark contrast to the more custodial colonial hospitals in British East Africa. Dr. Henri Collomb and colleagues trained Senegalese psychiatrists, and the clinic was deeply engaged with the local contexts of "culture," including locally produced knowledge of healing practices. Kilroy-Marac's ambitions move past a strict institutional history of the hospital, relying instead on memory, imagination, and even "hauntings" from postcolonial Senegal (Kilroy-Marac 2019). Kilroy-Marac's fascinating ethnography offers a confession upfront as the ethnographer admits that she is following a trail of sightings of Dr. Collomb's ghost roaming the hallways of the modern Fann Clinic, a sensitive metaphor for remnants and echoes of the past. Collomb's innovative work has been explored in Alice Bullard's work, but her comparative piece on Frantz Fanon and Collomb's dual inquiries in 1956 into the efficacy of personality tests for Africans is particularly insightful. Bullard shows these early efforts to refine psychiatric practice in African settings, using the diagnostic tools of psychiatry, to provide culturally meaningful and locally appropriate interventions. These studies, Bullard notes, "demonstrate the difficulties confronted by doctors intent on overcoming racism, who are compelled to work within broader structures that enforce racism" (Bullard 2005, 226). These exercises in the application of personality tests give some perspective on early attempts to put transcultural psychiatry's theoretical corpus to work in practice. The efforts of psychologists to standardize knowledge and measurement can be seen as well in Erik Linstrum's study, *Ruling Minds: Psychology in the British Empire* (2016). Linstrum captures how the discipline of psychology worked in practice to measure and modernize colonial subjects through "minds," "tests," and "experts," illustrating the benefits of drilling down into the history of a single discipline with its multilayered approaches and logics. Both Bullard and Linstrum's work shows that there is an operational history to psychiatry and psychology as the disciplines engaged with empire, a context that moves us beyond simply citing the racism and abuses of the colonial mindset.

New work by Claire Edington (based on her doctoral research) also challenges the notions of a hegemonic colonial presence with her study of how the Vietnamese engaged with, and altered, the French colonial asylum (Edington 2019). Edington illustrates the many entanglements and contradictions inherent in the asylum which was embedded within the local community in ways that made it negligible as a locus of strict control. Patients moved in and out of the asylum and continued to engage with their communities, families, and local approaches to mental health treatment. This porous environment, with Edington's focus on the totality of the social lives of patients, offers a refreshing take on histories of hospitals. Edington's is not the first work to engage more fully with such an approach. Catharine Coleborne's study, *Madness in the Family: Insanity and Institutions in the Australasian Colonial World, 1860–1914*, stands out for its innovative use of letters between asylum doctors and the families of patients in colonial Australasia in the nineteenth century. Coleborne's history is less about a "colonial psychiatry" than it is about the intimacy of distress within communities, households, and marriages (Coleborne 2010). The letters are enormously revealing for painting a picture of bleakness among a class of colonial life and particularly among women. The focus on "patient" is almost irrelevant in Coleborne's study as the domino effect of struggle and tragedy for entire families comes into view through the hospitalization of one family member.

While histories of colonial psychiatry have often taken a single hospital as their focus, the chief aim has always been to interrogate the broader social, political, or discursive field. The nature of psychiatric care, confinement, and terminology lends itself to asking broader questions about nineteenth-century India or twentieth-century Vietnam. Waltraud Ernst has contributed much to the field in terms of her own research path and as an editor of collections on transnational medical networks. Ernst's early institutional history *Mad Tales from the Raj: Colonial Psychiatry in South Asia, 1800–58* (2010) concerns itself with the treatment of the British insane in India particularly in light of how the colonial setting reframed questions of respectability and prestige. Burma is most certainly an understudied region within the history of psychiatry, but Jonathan Saha has produced a single article that exposes very familiar British concerns about maintaining control over a subject population and those deemed insane, an approach best characterized as both "overwhelming indifference and excessive intervention" (Saha 2013, 407). Lawrence E. Fisher has produced an innovative ethnographic and sociological account of how "madness" is perceived and experienced in multiple settings (within and outside of the mental hospital) in Barbados in the 1970s (Fisher 1986). The study is not as well-known as it should be in some circles, perhaps as its methodologies place it outside of the frame of much of the historical work on psychiatry and its colonial legacies.

The myriad regional case studies that make up this historiographical field have led to an interest in comparative and thematic collections in book form and special issues of journals. Two volumes organized broadly by the framework of empire include *Colonialism and Psychiatry* (Bhugra and Littlewood 2001) and *Psychiatry and Empire* (Mahone and Vaughan 2007). Together they cover a range of work on colonial territories allowing for nuanced intellectual histories specific to place. Broad colonial approaches to the "native mind" are shown to be present in most colonial territories, but how this played out in the Dutch East Indies, Fiji, or Trinidad provides glimpses of local agency and action (see Hans Pols, Jacqueline Leckie, and

Roland Littlewood in Mahone and Vaughan 2007). Alice Bullard's analysis of the transition from "colonial psychiatry" to "transcultural psychiatry" was enacted upon *both* Senegal and France an important departure from assumptions that it is only former colonies that become transformed states (Bullard 2007).

Further edited volumes such as *Transnational Psychiatries: Social and Cultural Histories of Psychiatry in Comparative Perspective, C. 1800–2000* (Ernst and Mueller 2010) move beyond the comparative case study approach to present a more thematic entanglement of material and discursive interconnections and, in some cases, *dis*connections. Central to such discussions, and a core theme in this Palgrave volume, is a historicized approach to historiographical and methodological concerns and practices that must accompany our forays into the colonial archive. Roland Littlewood and Simon Dein compiled a collection of original papers from the field with an aim to reassessing canonical and influential texts of theory and practice from both cultural psychiatry and its relation, medical anthropology (Littlewood and Dein 2000). The volume introduces a discussion for this set of diverse texts, placing each into some historical context and highlighting core themes or legacies they represent. The juxtaposition of myriad ethnographic and psychiatric theoretical field studies (many of them now outdated despite their iconic status) allows for a thematic rather than easily digested chronological reading of the field – a useful reminder that transcultural psychiatry represents multiple contexts, intellectual trajectories, and historical groundings. A co-edited collection by Hans Pols and Harry Yi-Jui Wu open up regions that have not received as much attention but offer much by way of comparison. Their special issues (parts I and II) offer insights from Peru and Chile, the Dutch East Indies (Pols' area of specialization), and Maoist China (Pols and Wu 2019). This growing area of scholarship is simply too vast to describe in its entirety, but the vibrancy of the field can be seen in these imaginative collections, often highlighting the work of younger scholars.

The history of psychoanalysis intersects with the histories of colonialism and anthropology in a variety of contexts but perhaps most significantly in India as outlined by Christiane Hartnack (2001) whose work illustrates the tensions and remakings of imported intellectual systems of thought despite broader claims to the universality of Freudian principles (Hartnack 1990). The ambitious collection *Unconscious Dominions: Psychoanalysis, Colonial Trauma, and Global Sovereignties* (Anderson et al. 2011) engages both with the psychoanalytic turn within colonial states and as a unique form of analysis that has dominated twentieth-century thought. The editors re-historicize traces of the psychoanalytic by "bringing the history of psychoanalysis into colonial focus and employing this colonized psychoanalysis for purposes of postcolonial critique" (Anderson et al. 2011, 3). The stellar contribution of the volume is to illustrate not how one thing was enacted within the other, but how necessarily co-dependent colonial expansion and psychoanalytic thought could be. Such interdependence might take the form of individual psychoanalysts' deeply constructed musings on the colonial subject, or in some cases, psychoanalysis framed the conditions by which the colonial "elite native" might rise to a level of "civilization" such that they could be deemed complex enough to be subjects of psychoanalysis – a status that also rendered them closer to citizen than subject

(Anderson et al. 2011, 7). The establishment of the Indian Psychoanalytical Society in 1922, comprising both Indian and British practitioners, created, to some extent, a domain that sought to control intellectual output in ways that were culturally and politically relevant. Bengali psychoanalyst Girindrasekhar Bose is the subject of works by both Hartnack (in Unconscious Dominions, 2011) and Shruti Kapila (in Psychiatry and Empire, 2007) whose work on what she terms the "forgotten encounter between Freud, Jung and India" similarly charts how discourse and practice shifted amidst competing ideas about the rationality or irrationality of religion.

Psychoanalysis makes a stronger showing in India than in Africa; however, South Africa provides the rather extraordinary example of a white psychoanalyst's engagement with a black healer in the 1930s in Johannesburg. The result of Sachs' encounters with black patients in a mental hospital in Pretoria, his analytical training in Germany, and his friendship/rivalry with "John Chavafambira" in a Johannesburg slum, resulted in the 1937 publication of *Black Hamlet* (republished in 1947 as *Black Anger*) which purported to be a psychoanalytic reading of an "African Negro." As a source document, the book relates a detailed picture of interwar, urban South Africa, as well as the tense, sometimes poignant, and often problematic encounter between two formidable individuals in both collaboration and competition. *Black Hamlet* should be read alongside Saul Dubow's in-depth piece, "Wulf Sachs's *Black Hamlet*: A Case of 'Psychic Vivisection?'" (1993). As Dubow points out, the book is today both a fascinating ethnography of the tensions of South Africa in this period and a troubled document in its own right that throws light on how we might engage methodologically with a primary source (including comparisons to the changes made to the 1947 publication) that are steeped in bias and decontextualized claims.

Conclusion

Finally, the recent push to engage with a global mental health will soon have origin stories of its own. A starting point for this (future) analysis might be the overview by three anthropologists who take not individual localities and ideas, but "global mental health" itself as their object of inquiry. This 2019 special issue offers to problematize global mental health by way of three thematic threads or "genealogies" that are worth listing here: (1) the qualitative turn of global health away from earlier international health and development, (2) networks and social movements, and (3) diagnostically and metrics-driven psychiatric imperialism, reinforced by pharmaceutical markets (Lovell et al. 2019). The anthropological study of psychiatry (as an ethnography rather than anthropologists critiquing psychiatric practice) has produced groundbreaking work such as Tanya Luhrmann's *Of Two Minds* (2000) about the schism in American psychiatry as it grapples with psychodynamic and biomedical (organic disease) schools of thought. However, for our purposes and harkening back to Matthew Heaton's call to understand the "politics of the transcultural," a genealogical approach to a global system mental health research, policy, and practice is particularly useful. In their introduction, the authors include a brief history of the global milestones that have come to be recognized as important calls to action in global mental health; these include the World Health Organization's report *Mental Health: New Understanding, New Hope* (2001) which illustrated the scale of the mental health "treatment gap" particularly in low- to middle-income countries as well as a call from the Mental Health Group of *The Lancet* in 2007 to "scale up" mental health services as a matter of urgency in order to address the global burden of mental ill health. Lovell, Read, and Lang's genealogy further complicates the origin story of global mental health by parsing the historical understanding of the influence of institutions like the WHO with pointed reminders to look deeper at the World Bank, donor relationships, and the role of big pharma. Individual papers that make up the special issue recover familiar ground with new ways of conceptualizing longstanding debates from both colonial era and transcultural psychiatry. Nicolas Henckes presents "schizophrenic infrastructures" as a way of "insisting" on the "materiality of the professional and scientific networks behind schizophrenia" (Henckes 2019). Henckes stresses the global nature of the concept of schizophrenia throughout the twentieth century but looks deeper at the structures that were in place to inform how and when such diagnoses were employed. Roberto Beneduce closes this collection with a philosophical reflection on how we might reframe the problems caused by the imposition of a Western hegemonic approach to mental health. This incisive articulation of the histories, tropes, entanglements, and hauntings of how mental health concerns are poorly served by structural inequalities that remain hidden behind many disguises, from diagnostic categories to unequal collaborations, and euphemistic terminologies that have had their day such as the "professionalization" of healers still suggests an ambivalence to "traditional" expertise (Beneduce 2019). Psychiatric diagnosis, Beneduce suggests, often "names a disorder and at the same time hides another issue, difficult or impossible to evoke." The evolution of this vast area of inquiry, at its best, continues to search for ways to articulate, with some degree of "psychic unity," what we mean, what we feel, and how we might alleviate distress and suffering wherever we find it.

Bibliography

Anderson W, Jenson D, Keller R (2011) Unconscious dominions: psychoanalysis, colonial trauma, and global sovereignties. Duke University Press, Durham

Bains J (2005) Race, culture and psychiatry: a history of transcultural psychiatry. Hist Psychiatry 16(2):139–154

Beneduce R (2019) "Madness and despair are a force": global mental health, and how people and cultures challenge the hegemony of western psychiatry. Cult Med Psychiatry 43:710–723

Bhugra D, Littlewood R (2001) Colonialism and psychiatry. Oxford University Press, Oxford

Bourguignon E (1989) Multiple personality, possession trance, and the psychic unity of mankind. Ethos 17(3):371–384

Bullard A (2005) The critical impact of Frantz Fanon and Henri Collomb: race, gender, and personality testing of North and West Africans. J Hist Behav Sci 41(3):225–248

Bullard A (2007) Imperial networks and postcolonial independence: the transition from colonial to transcultural psychiatry in Mahone and Vaughan. Psychiatry Emp:197–221

Campbell C (2007) Race and empire: eugenics in colonial Kenya (Studies in imperialism (Manchester, England)). Manchester University Press, Manchester

Carothers JC (1953) The African mind in health and disease: a study in ethnopsychiatry, World Health Organization. Monograph series, no. 17. World Health Organization, Geneva

Carothers JC (1954) The psychology of Mau Mau. Printed at the Govt. printer, Nairobi

Chakraborty A (1974) Whither transcultural psychiatry? Transcult Psychiatr Res Rev 11(2): 102–107

Coleborne C (2010) Madness in the family: insanity and institutions in the Australasian colonial world, 1860–1914. Palgrave Macmillan, Basingstoke

Comaroff J, Jean (1987) The madman and the migrant: work and labor in the historical consciousness of a South African people. Am Ethnol 14(2):191–209

Crozier I (2018) Introduction: Pow Meng Yap and the culture-bound syndromes. Hist Psychiatry 29(3):363–385

De Reuck AVS, Porter R (1965) Ciba foundation symposium: transcultural psychiatry. Little, Brown and Company, Boston

Deacon H (1996) Madness, race and moral treatment: Robben Island lunatic asylum, Cape Colony, 1846–1890. Hist Psychiatry 7(26):287–297

Delille E (2016) On the history of cultural psychiatry: Georges Devereux, Henri Ellenberger, and the psychological treatment of Native Americans in the 1950s. Transcult Psychiatry 53(3): 392–411

Delille E, Crozier I (2018) Historicizing transcultural psychiatry: people, epistemic objects, networks, and practices. Hist Psychiatry 29(3):257–262

Dubow S (1993) Wulf Sachs's Black Hamlet: a case of 'psychic vivisection'? Afr Aff 92(296):519

Dubow S (1995) Scientific racism in modern South Africa. Cambridge University Press, Cambridge

Edington C (2019) Beyond the asylum: mental illness in French colonial Vietnam (Studies of the Weatherhead East Asian Institute, Columbia University). Ithaca, New York

Ellenberger H (1968) Review: transcultural psychiatry, Ciba Foundation Symposium. Can Psychiatr Assoc J 13(1):100

Ellis WG (1893) The amok of the Malays. J Ment Sci 39(166):325–328

Ernst W (2010) Mad tales from the Raj: colonial psychiatry in South Asia, 1800–58 (Anthem South Asian studies). Anthem Press, London

Ernst W, Müller T (2010) Transnational psychiatries: social and cultural histories of psychiatry in comparative perspective, c.1800–2000. Cambridge Scholars, Newcastle

Fanon F (1963) The wretched of the earth. MacGibbon & Kee, London

Fanon F (1968) Black skin, white masks. MacGibbon & Kee, London

Fernando S (1988) Race and culture in psychiatry. Croom Helm, London

Field MJ (1960) Search for security: an ethno-psychiatric study of rural Ghana. Northwestern University Press, Evanston

Fisher L (1986) Colonial madness: mental health in the Barbadian social order. Rutgers University Press, New Brunswick

Gibson N, Beneduce R (2017) Frantz Fanon, psychiatry and politics (Creolizing the canon), London

Green B, Colucci E (2020) Traditional healers' and biomedical practitioners' perceptions of collaborative mental healthcare in low- and middle-income countries: a systematic review. Transcult Psychiatry 57(1):94–107

Hacking I (1995) Rewriting the soul: multiple personality and the sciences of memory. Princeton University Press, Princeton

Hartnack C (1990) Vishnu on Freud's desk: psychoanalysis in colonial India. Soc Res 57(4): 921–949

Hartnack C (2001) Psychoanalysis in colonial India. Oxford University Press, Delhi

Heaton M (2013) Black skin, white coats: Nigerian psychiatrists, decolonization, and the globalization of psychiatry, New African histories series, Athens

Heaton M (2018) The politics and practice of Thomas Adeoye Lambo: towards a post-colonial history of transcultural psychiatry. Hist Psychiatry 29(3):315–330

Henckes N (2019) Schizophrenia infrastructures: local and global dynamics of transformation in psychiatric diagnosis-making in the twentieth and twenty-first centuries. Cult Med Psychiatry 43:548–573

Imai H, Ogawa Y, Okumiya K, Matsubayashi K (2019) Amok: a mirror of time and people. A historical review of literature. Hist Psychiatry 30(1):38–57

Jablensky A, Sartorius N (1988) Is schizophrenia universal? Acta Psychiatr Scand 78(S344):65–70

Keller R (2001) Madness and colonization: psychiatry in the British and French Empires, 1800–1962. J Soc Hist 35(2):295–326

Keller R (2007) Colonial madness: psychiatry in French North Africa. University of Chicago Press, Chicago

Kiev A (1972) Transcultural psychiatry. The Free Press, New York

Kilroy-Marac K (2019) An impossible inheritance: postcolonial psychiatry and the work of memory in a West African clinic, Oakland

Kimani VN (1981) The unsystematic alternative: towards plural health care among the Kikuyu of central Kenya. Soc Sci Med 15B:333–340

Kirmayer L (2007) Cultural psychiatry in historical perspective. In: Bhugra, Bhui (eds) Textbook of transcultural psychiatry. Cambridge University Press, Cambridge

Kirmayer L (2013) 50 years of transcultural psychiatry. Transcult Psychiatry 50(1):3–5

Kirmayer L (2020) Toward a postcolonial psychiatry: uncovering the structures of domination in mental health theory and practice. Philos Psychiatry Psychol 27(3):267–271

Kleinman AM (1977) Depression, somatization and the "new cross-cultural psychiatry". Soc Sci Med 11(1):3–9

Kleinman A (1980) Patients and healers in the context of culture: an exploration of the borderland between anthropology, medicine, and psychiatry. University of California Press, Berkeley

Lambo TA (1955) The role of cultural factors in paranoid psychosis among the Yoruba tribe. J Ment Sci 101:239–266

Lambo TA (1959) Rapid development can threaten mental health. Int J Soc Work 2(3):30–32

Lambo TA (ed) (1961) First pan-African psychiatric conference, Abeokuta, Nigeria. Government Printer, Ibandan

Lambo TA (1962) The importance of cultural factors in treatment (with special reference to the utilization of the social environment). Acta Psychiatr Scand 38:176–182

Last M, Chavunduka G (1986) The professionalisation of African medicine. Manchester University Press, Manchester

Linstrum E (2016) Ruling minds: psychology in the British empire. Massachusetts, Cambridge

Littlewood R (1990) From categories to contexts: a decade of the 'new cross-cultural psychiatry'. Br J Psychiatry 156(3):308–327

Littlewood R, Dein S (2000) Cultural psychiatry and medical anthropology: an introduction and reader. Athlone, London

Littlewood R, Lipsedge M (1997) Aliens and alienists: ethnic minorities and psychiatry, 3rd edn. Routledge, London

Lovell AM, Read UM, Lang C (2019) Genealogies and anthropologies of global mental health. Cult Med Psychiatry 43(4):519–547

Luhrmann T (2000) Of two minds: the growing disorder in American psychiatry. Picador, London

Luhrmann T, Marrow J (2017) Our most troubling madness: case studies in schizophrenia across cultures. University Press Scholarship Online, Oakland

Mahaniah JK (1981) La structure multidimensionnelle de guerison a Kinshasa, capitale du Zaire. Soc Sci Med 15B:341–349

Mahone S (2006) The psychology of rebellion: colonial medical responses to dissent in British East Africa. J Afr Hist 47(2):241–258

Mahone S, Vaughan M (2007) Psychiatry and empire, Cambridge imperial and post-colonial studies series, Basingstoke

McCulloch J (1995) Colonial psychiatry and 'the African mind'. Cambridge University Press, Cambridge

McLawhorn D (2021) The neurasthenic-depression controversy: a window on Chinese culture and psychiatric nosology. Routledge, London

Ngubane H (1977) Body and mind in Zulu medicine: an ethnography of health and disease in Nyuswa-Zulu thought and practice. Academic Press, London

Ngubane H (1981) Aspects of clinical practice and traditional organization of indigenous healers in South Africa. Soc Sci Med 15B:361–365

Oyebode F (2006) History of psychiatry in West Africa. Int Rev Psychiatry 18(4):319–325

Patel V, Winston M (1994) 'Universality of mental illness' revisited: assumptions, artefacts and new directions. Br J Psychiatry 165(4):437–440

Pols H, Wu H (2019) Psychology and psychiatry in the global world: historical perspectives. Hist Psychol 22(3):219–224. (4):289–368

Prince R (2000) Transcultural psychiatry: personal experiences and Canadian perspectives. Can J Psychiatr 45:431–437

Pringle Y (2019) Psychiatry and decolonisation in Uganda. United Kingdom, London

Read U, Nyame S (2019) "It is left to me and my God": precarity, responsibility, and social change in family care for people with mental illness in Ghana. Africa Today 65(3):3–28

Sachs W (1937) Black Hamlet: The mind of an African Negro revealed by psychoanalysis. G. Bles, London

Sadowsky J (1999) Imperial bedlam: institutions of madness in colonial southwest Nigeria. University of California Press, Berkeley

Saha J (2013) Madness and the making of a colonial order in Burma. Mod Asian Stud 47(2): 406–435

Singer K (1975) Depressive disorders from a transcultural perspective. Soc Sci Med 9:289–301

Smith L (2014) Insanity, race and colonialism: managing mental disorder in the post-emancipation British Caribbean, 1838–1914. Hampshire, Basingstoke

Steinberg H (2015) Emil Kraepelin's ideas on transcultural psychiatry. Aust Psychiatry 23:531–535

Swartz S (2015) Homeless wanderers: movement and mental illness in the Cape Colony in the nineteenth century. Ebook Central, Claremont

Swartz S (2018) Ruthless Winnicott: the role of ruthlessness in psychoanalysis and political protest. Routledge, London

Torrey EF (1973) Is schizophrenia universal? An open question. Schizophr Bull 1(7):53–59

Vaughan M (1983) Idioms of madness: Zomba lunatic asylum, Nyasaland, in the colonial period. J South Afr Stud 9(2):218–238

Vaughan M (1991) Curing their ills: colonial power and African illness. Polity, Cambridge

Wittkower E, Rin H (1965) Recent developments in transcultural psychiatry. Ciba Symp Transcult Psychiatry:4–25

Wu H (2021) Mad by the millions: mental disorders and the early years of the World Health Organization. MIT Press, Cambridge

Yap PM (1951) Mental diseases peculiar to certain cultures: a survey of comparative psychiatry. J Ment Sci 97(407):313–327

Geriatric Psychiatry and Its Development in History

54

Jesse F. Ballenger

Contents

Introduction	1404
Ancient Entanglements of Mental Infirmity and Aging	1404
The Dementia Conundrum in European Psychiatry Through the Early Twentieth Century	1405
The Geriatric Imperative	1408
Psychiatry, Dementia, and the Social Transformation of Aging	1410
The Biomedical Deconstruction of Senility and the Establishment of Geriatric Psychiatry	1413
Conclusion	1416
References	1416

Abstract

Although medical concern with mental disorder in old age can be found in every society since antiquity, only in the second half of the twentieth century did geriatric psychiatry take shape as a distinct professional specialty within medicine, by the 1980s replete with its own professional organizations and journals. This development occurred across the countries of North America and Europe in the decades following World War II, as modern Western welfare states confronted the unprecedented challenges of rapidly aging societies – particularly the specter of a dramatic increase in the prevalence of dementia. Thus, the relatively recent development of geriatric psychiatry as a specialty is connected to the much longer history of ideas in medicine and society about mental health and aging. Since antiquity, physicians have noted that the aged are particularly susceptible to mental infirmity and debated whether this should be thought of as disease or as part of the aging process itself. This debate has continued in various forms down to the present. Disentangling the relationship between aging and disease to

---

J. F. Ballenger (✉)
College of Nursing and Health Professions, Drexel University, Philadelphia, PA, USA
e-mail: Jfb83@drexel.edu

© The Author(s), under exclusive licence to Springer Nature Singapore Pte Ltd. 2022
D. McCallum (ed.), *The Palgrave Handbook of the History of Human Sciences*,
https://doi.org/10.1007/978-981-16-7255-2_89

develop an age-appropriate approach to therapeutics remains the core problem of all geriatric medicine, including geriatric psychiatry.

> **Keywords**
>
> Geriatric psychiatry · Gerontology · Aging and mental health · Dementia · Alzheimer's disease · History of psychiatry

Introduction

Although medical concern with mental disorder in old age can be found in every society since antiquity, only in the second half of the twentieth century did geriatric psychiatry take shape as a distinct professional specialty within medicine, by the 1980s replete with its own professional organizations and journals. This development occurred across the countries of North America and Europe in the decades following World War II, as modern Western welfare states confronted the unprecedented challenges of rapidly aging societies – particularly the specter of a dramatic increase in the prevalence of dementia. Thus, the relatively recent development of geriatric psychiatry as a specialty is connected to the much longer history of ideas in medicine and society about mental health and aging. Since antiquity, physicians have noted that the aged are particularly susceptible to mental infirmity and debated whether this should be thought of as disease or as part of the aging process itself. This debate has continued in various forms down to the present. Disentangling the relationship between aging and disease to develop an age-appropriate approach to therapeutics remains the core problem of all geriatric medicine, including geriatric psychiatry.

Little has been written about geriatric psychiatry as a medical specialty. What literature there is consists almost entirely of short historical sketches by geriatric psychiatrists of the origin and development of their specialty, usually limited to the single national context of its author. Historians have written considerably more on the history of ideas about mental illness and aging, though this is still an area that has gotten relatively little attention compared to other topics in the history of psychiatry. This chapter will connect these disparate strands of scholarship to show that the emergence of geriatric psychiatry can be understood in the context of broader changes in medical ideas and social attitudes about aging and mental health. Given that some key points rest on a slender body of scholarship, this claim must remain tentative, less asserting firm conclusions than pointing to fruitful areas for further historical inquiry.

Ancient Entanglements of Mental Infirmity and Aging

It is a common misconception that, in premodern societies, old age was virtually nonexistent and little notice was taken of medical problems associated with it because life expectancy at birth was likely less than 40 years in every society before 1900, and as low as 20–30 in ancient Greece and Rome. But life expectancy is the average number of additional years a person can expect to live *at any given age*. The very low life expectancy *at birth* in premodern societies was largely a product of very high infant and child mortality. If a person lived beyond childhood, they stood a good chance of living to adulthood, and an adult stood a decent chance of living into old age. As will be described later in this chapter, it is certainly true that modern societies have seen an increase in longevity since the late nineteenth century and a dramatic aging of the population since the mid-twentieth century with an associated rising prevalence of dementia. But old age and all of its associated challenges, including dementia, were far from unknown in premodern societies.

Archaeologist Martin Smith used evidence from written historical sources, forensic analysis of ancient burial sites, and ethnographic accounts of premodern populations that had persisted in isolation until encounters with modern society in the twentieth century to create models of the age structure for several different forms of premodern society. Extrapolating from these models with current dementia rates, he argues that dementia has most likely posed a significant challenge to human societies for thousands of years. For example, he estimates that out of Imperial Rome's population of about 450,000 in 1 C.E., more than 9,500 would likely have had dementia (Smith et al. 2017). However rough such estimates may be, they raise interesting and important questions for historians about how societies in the past understood and treated people with dementia. How was dementia understood? What sort of arrangements were made for the care of people with dementia?

Historical scholarship has produced little work on how people with dementia were cared for in premodern societies. But scholars have noted that descriptions of mental infirmity associated with old age that we would likely call dementia appeared sporadically in literary, legal, and medical texts in all ancient societies (Boller et al. 2007; Papavramidou 2018). In general, none of these texts sought to disentangle dementia from the broad array of debilitating physical and mental losses associated with aging (Ballenger 2017). Though some authors, most famously Cicero in *De Senectute* (44 B.C.), argued that mental deterioration might be delayed through regimes of diet, exercise, and intellectual stimulation, all concurred that it must be understood as one of the calamities of growing old. As the Roman playwright Terence put it in a throwaway line in his second century C.E. comedy *Phormio*, "*s*enectus ipsa morbus est" – old age is itself a disease. This basic idea that dementia was part of a group of infirmities that were inevitable at some point in old age remained the dominant view in Western medicine into the twentieth century, even as physicians identified and described the specific pathophysiological changes in the brain that marked it.

The Dementia Conundrum in European Psychiatry Through the Early Twentieth Century

The isolation of dementia as a specific condition afflicting the elderly begins with the effort of French alienist Phillippe Pinel to establish a rational taxonomy of mental disorders. In his widely influential book on the classification of mental disorders published in 1798, Pinel identified dementia as one of the four basic types of mental disorder. His concept of dementia included a broad range of symptoms and behavior that all related to the loss of psychosocial competence. Throughout the nineteenth century, psychiatrists gradually narrowed this concept of dementia to only the impairment of intellectual abilities. In this way, noncognitive symptoms were increasingly viewed as mere epiphenomena of dementia, or as indications of separate disorders not necessarily associated with aging. The growing importance of clinical-pathological correlation led late-nineteenth-century psychiatrists to categorize different kinds of dementia according to the pathological changes that could be observed in brains at autopsy such as cortical atrophy, arteriolosclerosis, and strokes (Berrios 1995).

By the last quarter of the nineteenth century, many observers thought psychiatry was lagging behind other areas of medicine (Rosenberg 1992). A succession of dramatic discoveries in microbiology had increasingly allowed medicine to define discrete disease entities associated with specific pathologies, and occasionally to even develop effective interventions. Louis Pasteur's rabies vaccine in 1885 was widely celebrated in the mass media and established a widespread popular belief (at least in the USA) that modern medicine should be characterized by constant progress (Hansen 2009). The only genuine achievement psychiatry could point to along these lines was general paresis – one of the most common diagnosis of patients in the mental asylums that were growing to massive size across Europe and North America. General paresis was first described as an entity associated with a specific brain pathology in 1822 by French physician Antoine Bayle. German psychiatrists linked it to syphilitic infection in the 1850s, but this remained controversial until the early twentieth century when researchers identified syphilis spirochetes at autopsy. Austrian Julius Wagner-Jauregg began treating paretic patients by deliberately infecting them with malaria, thus inducing fever sufficient to kill the spirochetes. In 1927, he was awarded the Nobel Prize for this work (Scull 2015).

In this context, German psychiatrist Emil Kraepelin led a group of researchers investigating senile dementia and similar conditions, hoping to establish it as a second major mental disorder for which a clear basis in brain pathology could be demonstrated. By 1910, they seemed to have succeeded brilliantly in this. In 1906, Kraepelin's protégé Alois Alzheimer presented a paper to a group of German academic psychiatrists meeting in Tübingen that described the case of Auguste Deter, a 51-year-old woman with progressive dementia correlated at autopsy with general cortical atrophy, numerous senile plaques, and neurofibrillary tangles. Alzheimer concluded that "on the whole, it is evident that we are dealing with a peculiar, little-known disease process... We must not be satisfied to force it into the existing groups of well-known disease patterns" (Alzheimer 1906).

In 1910, based on this and a few more cases that had been published since the initial report, Kraepelin created the eponym "Alzheimer's disease" in his massive and influential *Ein Lehrbuch der Psychiatrie,* then in its eighth edition. The term was to distinguish the relatively rare cases of presenile dementia, occurring in people younger than 65 from the much more common occurrence of dementia in older people. Kraepelin made this distinction despite the fact that Alzheimer's disease and senile dementia appeared to have identical clinical symptoms, course, and pathological features. Age of onset seemed to be the only factor Kraepelin used to differentiate the cases. This classification quickly generated confusion about the relationship between Alzheimer's and senile dementia, and how either or both of them were related to aging. In a long 1911 paper that provided a more detailed discussion of the 1906 case and several others, Alzheimer himself seemed to call the distinction into question:

> "As similar cases of disease obviously occur in late old age, it is therefore not exclusively a presenile disease, and there are cases of senile dementia which do not differ from these presenile cases with respect to the severity of the disease process. There is then no tenable reason to consider these cases as caused by a specific disease process. They are senile psychoses, atypical forms of senile dementia. Nevertheless, they do assume a certain separate position so that one has to know of their existence." (Alzheimer 1911)

As historian of psychiatry German Berrios concludes, it seems that in both the 1906 and 1911 papers, Alzheimer intended nothing more than to point out that senile dementia could sometimes occur in a relatively young person (Berrios 1995).

Why then did Kraepelin create the separate entity of Alzheimer's disease? Over the decades, a number of theories have been advanced: that by establishing Alzheimer's disease as a second example of a mental illness with clearly defined brain pathology, he would beat back the rising tide of psychoanalysis (Torack 1979); that he was motivated by competition with other departments in Europe, most notably that of Arnold Pick in Prague, to add prestige to his own group (Amaducci et al. 1986); that he sought to justify the cost of Alzheimer's expensive pathology lab in Munich (Myfanwy and Isaac 1987); or that he did so in the honest conviction that age of onset was legitimate reason to make a distinction between the two, at least until more decisive evidence could be produced (Beach 1987). All of these explanations are plausible enough, and none are mutually exclusive. But more historical research on Kraepelin and Alzheimer is needed before any explanation can be regarded as conclusive.

For now it seems most plausible to take Kraepelin's decision at face value. It made sense to him to categorize as a disease the rare cases of dementia occurring at an age when it would be unexpected, as against the more commonplace occurrence of dementia in old age. This reflected the continued assumption, deeply embedded in Western medicine and society, that physical and mental deterioration were common, if not universal in old age. That this remained true in Germany during this period could be seen in the institutional structure of psychiatric care, in which patients with dementia were typically admitted to state-run asylums for chronic, incurable patients. These institutions were located in rural areas that provided custodial care in isolation from the university clinics that conducted advanced psychiatric research. This two-tiered structure of psychiatric care made it difficult for Kraepelin and Alzheimer, working at a university clinic in Munich, to pursue their work on dementia because they had trouble finding cases to study. This difficulty could be seen in Alzheimer's now-famous first case. Alzheimer admitted Auguste Deter in

1901 to the state asylum in Frankfurt where he worked before joining Kraepelin in Munich. But after joining Kraepelin in Munich, he continued to follow her case, and twice blocked her husband from transferring her to a cheaper but more distant asylum, presumably so he could more easily arrange to have postmortem pathological brain specimens sent to him after her death. For his part, Kraepelin had begun a broad study of psychosis in older patients with the aim of classifying all disorders arising in patients after the age of 45 in order to more precisely demarcate the full range of psychiatric disorders, but quickly abandoned the project because of a lack of patients in this group at the university clinic (Engstrom 2007).

In the event, Kraepelin's distinction between Alzheimer's presenile and senile dementia persisted for decades, leaving dementia as a marginal category. Although Alzheimer and Kraepelin established a correlation between a clinical syndrome and well-defined brain pathology – an accomplishment that was and remains a rare achievement in psychiatry – dementia retained a liminal status in psychiatry. If thought of as a disease – Alzheimer's – it was a rare presenile condition, not a major disease. But if thought of as the much more common senile dementia, it seemed less like a disease than an extreme variant of normal aging – and hence much less relevant to psychiatrists like Alzheimer and Kraepelin who were interested in advancing psychiatric knowledge, not in addressing the problems of aging per se.

The Geriatric Imperative

Modern geriatrics emerged out of the Paris School of clinical medicine in the early nineteenth century. It encouraged the method of clinical-pathological correlation that psychiatrists like Alzheimer and Kraepelin pursued. But more importantly for the history of geriatrics, it also made the aging body itself an important locus of scientific interest. Emphasizing careful clinical pathological correlation as the means of establishing rigorous disease classification and diagnosis, and working in two large public hospitals – the Bicêtre and Salpêtrière – with large populations of elderly patients, French physicians began to identify particular degenerative changes in body tissues that they argued characterize old age as a physiologically distinctive stage of life. In this new view, the aged body was not just more susceptible to various diseases. It was itself different than the normal adult body, undergoing processes of deterioration that physicians had to take into account in order to properly diagnose and treat disease in the aged (Haber 1986; Katz 2019).

As we have seen, this was not an entirely new idea. Modern medical science was simply confirming the ancient idea that old age was itself a disease. But the Paris physicians took this a step further by suggesting that degenerative changes in the aging body meant that diseases manifested differently in older people. Thus the Paris School encouraged specific research on older people and the process of aging. By the mid-nineteenth century, at least 15 physicians had published monographs focused on these ideas about aging and disease. Particularly influential were a series of lectures by the preeminent neurologist Jean-Martin Charcot, first published in 1861 and translated into English in 1881 as *Clinical Lectures on the Diseases of Old Age*.

Charcot made the case that "the special study of the diseases of old age cannot be contested.... We have come to recognize in reality that the pathology presents its difficulties, which cannot be surmounted except by long experience and profound acquaintance with its peculiar characteristics" (quoted in Haber 1986, p. 74).

The call for attention to the particularity of the medical problems of aging was taken up with great vehemence by the Austrian-born American physician Ignatz Leo Nascher, who in 1909 coined the term "geriatrics" to describe "the same field in old age that is covered by the term pediatrics in childhood, to emphasize the necessity of considering senility and its diseases apart from maturity and to assign it a separate place in medicine" (Nascher 1909). Just as pediatrics was needed to establish standards of diagnosis and therapy appropriate to the developing body of infants and children, Nascher argued that geriatrics was necessary to guide physicians in the treatment of the elderly. But Nascher had to acknowledge that there were important differences. Where pediatrics could establish clear standards of health as children developed health and intelligence, geriatrics must struggle to differentiate "normal" processes of deterioration from disease. Ultimately, Nascher argued that "it is impossible to draw a sharp line between health and disease in old age. With every organ and tissue undergoing a degenerative change which affects the physiological functions, it is a matter of personal opinion to determine at which point the changes in the anatomic features and physiologic functions depart from the normal changes of senility and to what degree" (quoted in Haber 1986, p. 76). Moreover, where physicians looked upon the young as worthy patients who would clearly benefit from medical research and treatment, Nascher thought it natural that they would find the elderly abhorrent, too decrepit and infirm to profit from advances in medicine. "While the dependence of the child arouses sympathy," he wrote, "in the aged the repugnance aroused by the disagreeable facial aspect and the idea of economic worthlessness destroys the sympathy we bestow upon the child and instills a spirit of irritability if not positive enmity against the helplessness of the aged" (quoted in Haber 1986, p. 77). Nascher's many articles in medical journals and his 1914 textbook of *Geriatrics* emphasized the inevitability of increasing physical and mental infirmity in old age, and the futility of treatment. The most medicine could hope for was to avoid making this inevitable decline worse. Perhaps because of Nascher's intensely negative views of aging and the elderly, it took several decades to establish geriatrics as a medical specialty.

In the UK, a more optimistic version of geriatrics developed in the 1940s and 1950s around the work of the physician Marjory Warren. In 1936 she was put in charge of medical services for a large hospital that included some long-neglected workhouse wards. These wards were overcrowded with chronically ill old people, most of whom had never been diagnosed, and many of whom suffered from iatrogenic complications. After conducting a systematic assessment of each patient, she found that many of these chronic conditions were in fact treatable. With rehabilitative services, mobility aids, and supportive community services many older patients were able to return home. Warren became an energetic advocate of the approach she developed, speaking widely in the UK and abroad about the importance of a comprehensive approach to the medical care of the elderly. She began training nurses and medical undergraduates in comprehensive geriatric medical care, and began to garner praise and emulation in the USA and Europe (Evans 1997; Mulley 2012).

Geriatric medicine developed in the context of increasing recognition of aging itself as a significant problem for social policy in the modern state. Policy experts noted that the demographic transition and improvements in public health led to increased survival into old age, making the elderly an increasingly large proportion of the population. At the same time, industrialization and urbanization created new challenges for people trying to continue living independent, meaningful lives as they aged (Haber and Gratton 1994; Thane 2005). Governments and major philanthropic organizations began to invest in research on aging in the late 1930s. Most notably, the Josiah Macy Foundation in the USA convened an impressive array of leaders in the biological, medical, and social sciences to participate in a conference and contribute to an influential multidisciplinary handbook called *The Problems of Aging*, published in 1939 (Achenbaum 1995). During the 1940s and 1950s, the organizational infrastructure of the aging field took shape in the USA and the UK as some of the most important organizations, government agencies, and research institutes were formed, including the American Geriatrics Society (1942), the Gerontological Society of America (1945), and the British Geriatrics Society (1947). The nascent National Health Service in Britain established geriatric consultancy services shortly after its founding in 1948 (Achenbaum 1995; Evans 1997).

Psychiatry, Dementia, and the Social Transformation of Aging

Meanwhile, aging was beginning to attract more serious attention within American and British psychiatry. In large part, this was a matter of necessity as elderly patients with dementia had become a disproportionately large share of the patient population of the sprawling, government-run mental hospitals in both countries. These hospitals had been optimistically created in the nineteenth century as therapeutic institutions on which the professional authority of psychiatry rested. But because psychiatry regarded senile dementia as incurable, its rising prevalence undermined the therapeutic environment mental hospitals were supposed to provide. Worse, because the overall population of society was aging, the problem was bound to get worse – a demographic avalanche that would bury the mental hospital as a viable therapeutic institution (and the professional legitimacy of psychiatry along with it). Many leaders in psychiatry argued that dementia was not properly understood as a mental illness, so some other institutional arrangement should be found to provide care for such patients. This would eventually occur, albeit with woeful problems, when the modern nursing home industry was created in the wake of deinstitutionalization in the 1960s and 1970s. But other psychiatrists embraced the challenge of dealing with senile dementia as a pressing psychiatric problem and laid the foundations for geriatric psychiatry as a specialty (Ballenger 2000; Hilton 2015).

In the UK, psychiatrists challenged the widely held idea that all mental disturbances in older people could be reduced to the irremediable effects of aging and worked to establish services for effective treatment and prevention of mental illness in old age. In 1955, Martin Roth published an extensive study of mental disorders in old people and argued that there were five types – not all of which should be regarded as incurable. Nick Corsellis conducted pathological studies that provided strong evidence that Alzheimer's disease and senile dementia were a single entity, separate from normal aging. Felix Post's work at the Bethlem Maudsley Royal Hospital in London showed that some mental illnesses in the old could be effectively treated. Though these initiatives did not enjoy wide acceptance among national policymakers and psychiatric leaders, they did win followers in health districts across the country who worked to establish therapeutic, rehabilitative, and social support services for the elderly in about half of the health services districts in the country (Hilton 2015).

In the USA, psychiatric concern for the elderly developed in a way that reflected the strength of psychoanalytic and psychodynamic psychiatry. In the mid-1930s, American psychiatrists, led by David Rothschild of the Worcester State Hospital, developed a new theory of dementia that sidestepped the nosological confusion created by Kraepelin's distinction between Alzheimer's disease and senile dementia, and created a more optimistic framework for approaching the proliferation of older people with dementia in the state mental hospitals. Rothschild's publications in the early 1930s followed Alzheimer's methods of correlating clinical dementia with histological study of brain pathology, but he became more interested in the number of cases that did not fit the model of dementia as a brain disease. Around the same time, German researchers conducted a large-scale study that confirmed problems with the correlation between brain pathology and dementia. A significant minority of patients clinically diagnosed with dementia showed little pathology at autopsy, while others with little to no dementia were found at autopsy to contain significant numbers of plaques and tangles. Rothschild theorized that this could be best accounted for by a differing ability among individual people to compensate for brain pathology. In his view, age-associated dementia was not simply the inevitable result of a deteriorating brain, but a dialectical process between the brain and the psychosocial context in which an aging person was situated. Problems in a person's personality structure, experience of emotional trauma, and loss of family and social support contributed at least as much to the development of dementia as the biological processes within the brain that produced plaques and tangles (Ballenger 2000).

This emphasis on psychosocial factors over brain pathology in the etiology of dementia brought age-associated dementia into the ambit of mainstream American psychiatry, and provided a logical basis for making meaningful therapeutic interventions. From the 1930s through the early 1960s, there was a growing number of articles about dementia in the two most prestigious psychiatric journals in the USA, the *American Journal of Psychiatry* and the *Archives of Neurology and Psychiatry*. Many of these articles enthusiastically reported on the successful use of therapies that had previously been considered inappropriate for aged patients, such as psychotherapy, ECT, and drug treatments. But the enthusiasm of this literature probably said more about how badly clinicians wanted to find effective treatments than about their actual efficacy. Few of these articles were RCTs, and initial positive results were seldom replicated when more careful studies began to be conducted in the 1970s (Ballenger 2000).

But the psychodynamic model of dementia did more than provide a rationale for desperate therapeutic endeavors in state hospitals. It also provided a framework for understanding the total experience of aging in the post-World War II USA. During the 1940s and 1950s, US psychiatrists writing about dementia largely ceased the investigation of brain pathology. They also abandoned attempts to establish carefully bounded disease entities, instead gathering Alzheimer-type dementia, cerebral arteriosclerosis, and functional mental disorders together into a broad concept of senile mental deterioration, whose pathological hallmarks were not to be found in the brain, but in modern social relations. The cause of senile mental deterioration was not the aging brain, but a society that stripped the elderly of any meaningful role in life through mandatory retirement and social isolation. Suffering intense stigma and deprived any meaningful social role, it was not surprising that older people experience mental deterioration. "In our present social set-up, with its loosening of family ties, unsettled living conditions and fast economic pace, there are many hazards for individuals who are growing old," he wrote. "Many of these persons have not had adequate psychological preparation for their inevitable loss of flexibility, restriction of outlets, and loss of friends or relatives; they are individuals who are facing the prospect of retirement from their life-long activities with few mental assets and perhaps meagre material resources" (Rothschild 1947, p. 125). Other psychiatrists pushed this line of thinking even further, arguing that social pathology was in fact the cause of brain pathology. "Senility as an isolable state is largely a cultural artifact and that senile organic deterioration may be consequent on attitudinal alterations," Maurice Linden and Douglas Courtney argued in Linden and Courtney 1953, thought they acknowledged that it would be difficult to find evidence to support this (Linden and Courtney 1953, p. 912). But David C. Wilson was more confident, arguing that it was simply a matter of waiting for "laboratory proof" to support what was clear in clinical experience: the "pathology of senility is found not only in the tissues of the body but also in the concepts of the individual and in the attitude of society." Wilson listed social deprivations that afflicted the elderly, and argued that "factors that narrow the individual's life also influence the occurrence of senility. Lonesomeness, lack of responsibility, and a feeling of not being wanted all increase the restricted view of life which in turn leads to restricted blood flow" (Wilson 1955, p. 905). To mid-century Amercian psychodynamic psychiatrists, it seemed that the psychosocial losses of aging that could account for the constricted blood vessels of the aged brain (Ballenger 2000).

By bringing together the symptoms of dementia with broader concerns about the loss of status, social isolation, and stigma that often accompanied aging in modern society, the psychodynamic approach to senile mental deterioration gained currency far beyond professional psychiatry. It was an especially useful idea for social gerontologists and their allies who were working to transform retirement from a dreaded loss imposed on the elderly to a satisfying and desired culmination of life by making it more financially secure and socially meaningful. For social gerontologists, the high prevalence of senile mental deterioration among the elderly was an indictment of society's failure to meet the challenge of aging (Ballenger 2006).

The psychodynamic approach to dementia resonated with the concept of individual "adjustment" to aging that was important in the emerging field of social gerontology. The aging process inevitably challenged the individual with disruptions such as the end of paid employment, changes in living arrangements, or the death of friends and family. Failure to properly adjust to these disruptions resulted in social isolation and mental deterioration. But individuals could also adjust successfully, forming new relationships and discovering new interests and activities to replace those that were lost (Pollak and Heathers 1948; Donahue and Tibbitts 1957; Havighurst and Albrecht 1953; Friedmann and Havighurst 1954; Tibbitts and Donahue 1960). Although the individual had ultimate responsibility to adjust to the challenges of aging, prominent gerontologists like Robert J. Havighurst argued that "in modern America the community must carry the responsibility of creating conditions that make it possible for the great majority of older people to lead the independent and emotionally satisfying lives of which they are capable" (Havighurst 1952, p. 17). Society's responsibility toward the elderly was not just altruistic. In a society like the USA with a rapidly aging population, failure to help older people adjust would result in a disastrous increase in senility. According to Jerome Kaplan, an advocate for recreation programs for the elderly, "with the number of people who are over 65 increasing significantly each year, our society is today finding itself faced with the problem of keeping a large share of its population from joining the living dead – those whose minds are allowed to die before their bodies do" (Kaplan 1953, p. 3). It was thus incumbent upon society to develop programs to support older people and provide them with meaningful activities.

The psychodynamic model of dementia thus provided support for social gerontology's agenda for transforming old age. By the end of the 1970s, much of this agenda had been accomplished in the USA and other affluent countries. Pension and social security entitlements had been greatly expanded, though not to an equal extent for all older people; age discrimination was mostly outlawed; the stigmatization of aging was increasingly challenged in popular and professional discourse; and effective advocacy groups were formed to make the aging a visible and potent political force in US society (Calhoun 1978; Haber and Gratton 1994). At the same as these social and policy changes heightened expectations for a good old age, deinstitutionalization increased the burden that dementia posed to families in the community. All of this combined to make senile dementia an increasingly devastating prospect by the 1970s (Ballenger 2006).

The Biomedical Deconstruction of Senility and the Establishment of Geriatric Psychiatry

The success of the psychodynamic model *beyond* psychiatry created a contradiction that necessitated change *within* psychiatry. The expansive concept of senility that had currency within psychodynamic psychiatry and gerontology from the 1940s to the 1960s no longer seemed appropriate in the new era of aging that was emerging in the 1970s. "Ageism" replaced "adjustment" as the key term in social gerontology for a more aggressive and politicized generation of gerontologists and aging activists. The term ageism was coined by psychiatrist and gerontologist Robert Butler in 1968 to describe the "process of systematic stereotyping of and discrimination against people because they are old, just as racism and sexism accomplish this with skin color and gender" (Butler 1975, p. 12). One of the worst aspects of ageism, according to Butler, was the belief that the process of aging entailed inevitable physical and mental decline. Butler and other American gerontologists argued that "senility" was neither a medical diagnosis nor a useful sociological concept, but a "wastebasket term" applied to any person over 60 with a problem. Worse, it rationalized the neglect of those problems by assuming that they were inevitable and irreversible. "'Senility' is a popularized layman's term used by doctors and the public alike to categorize the behavior of the old," Butler argued. "Some of what is called senile is the result of brain damage. But anxiety and depression are also frequently lumped within the same category of senility, even though they are treatable and often reversible." Because both doctors and the public found it so "convenient to dismiss all these manifestations by lumping them together under an improper and inaccurate diagnostic label, the elderly often did not receive the benefits of decent diagnosis and treatment" (Butler 1975, pp. 9–10).

Butler did not discount the reality of irreversible brain damage, as had an earlier generation of psychodynamic psychiatrists. Rather, he argued that the refusal to systematically distinguish the various physical and mental disease processes from each other and from the process of aging itself was a manifestation of the ageism that kept society from taking the problems of older people seriously. In this view, all of the physical and mental deterioration commonly attributed to old age was more properly understood as the product of disease processes distinct from aging. Acting on this assumption, a group of clinical psychiatrists, neurologists, neuropathologists, and biochemists who entered the field in the late 1960s and 1970s worked to recast dementia as a number of disease entities distinct from aging, chief among them Alzheimer's disease.

An essential part of the transformation of dementia was to establish a clearer understanding of its relationship to brain pathology. As noted above, British psychiatrist Martin Roth began this work in the 1950s. In the mid-1960s, he was joined by neuropathologist Bernard Tomlinson, who developed precise methods of counting plaques and tangles in autopsied brain tissue, and psychologist Gary Blessed, who developed quantitative scale to assess the severity of dementia. In a series of heavily cited articles, the group established statistically significant correlations between the number of plaques found in the brain and the degree of dementia. This, they argued in an article in *Nature,* refuted a central claim of the psychodynamic model. "Far from plaques being irrelevant for the pathology of old age mental disorder, the density of plaque formation in the brain proves to be highly correlated with quantitative measures of intellectual and personality deterioration" (Roth et al. 1966, p. 110). They argued that brain pathology differentiated 90% of the cases of dementia in their study from the normal controls. They chalked up the remaining

10% to pathological associations with dementia that were not yet recognized. The authors concluded that psychodynamic psychiatry's claim that equally severe pathological destruction could be found in normal old age as found in senile or arteriosclerotic dementia was groundless. Although a small number of researchers continued to question the relative importance of brain pathology (e.g., Kitwood 1987), Roth, Tomlinson, and Blessed were widely regarded as having definitively proved that there was a strong and probably causal relationship between brain pathology and dementia (Katzman and Bick 2000).

The other essential step was to clear up the nosological confusion that had surrounded dementia since Kraepelin created the term Alzheimer's in 1910. Part of this involved establishing clear distinctions between progressive, age-associated dementias such as Alzheimer's disease and reversible dementias produced by treatable conditions. In 1980, the US National Institute on Aging convened a task force which published widely cited guidelines listing dozens of medical conditions that could produce dementia-like symptoms that physicians needed to rule out before making a diagnosis of dementia (National Institute on Aging 1980).

But the main thing was redefining the relationship between Alzheimer's disease and senile dementia. In 1976, neurologist Robert Katzman published an editorial in the *Archives of Neurology and Psychiatry* arguing that since both the clinical course and pathological features of Alzheimer's presenile and senile dementia were essentially identical, the distinction ought to be dropped. Crucially, Katzman called the unified entity Alzheimer's disease rather than senile dementia to emphasize that it was not something that should be regarded as one of the natural and inevitable changes associated with aging. Conceptualized this way, Alzheimer's disease went from being a rare condition to a major killer. Katzman estimated that there were as many as 1.2 million people with Alzheimer's in the USA, and between 60,000 and 90,000 deaths per year – making it the fourth leading cause of death in the USA (Katzman 1976). In 1978, Katzman and like-minded colleagues enlisted the support of several directors of institutes within the US National Institutes of Health for a major workshop intended to resolve these issues and encourage new researchers from a variety of fields to begin research on the disease (Katzman et al. 1978). By the end of the decade, there was a clear consensus that Alzheimer's disease and senile dementia were a unified disease entity distinct from the process of normal aging.

This reformulation of Alzheimer's was politically savvy, allowing those researchers, policymakers, and family members who wanted to raise public awareness and government funding to make a persuasive case that Alzheimer's disease constituted a dire threat to society that must be fought. When Robert Butler became the founding director of the National Institute on Aging (NIA) in 1974, he became perhaps the most forceful advocate for Alzheimer's research. Following the disease-specific lobbying strategy that had worked well for other institutes within the NIH, Butler made Alzheimer's the focal point of research in the NIA. The strategy succeeded brilliantly, and by the end of the 1980s Alzheimer's had become one of the most widely feared diseases and the NIA's budget for research on it increased by more than 800% (Fox 1989). Though the global interactions of the Alzheimer's research community have not been well studied, the networks of biomedical researchers that elevated Alzheimer's disease to a major public issue in the USA were transnational, and nongovernment organizations were created around 1980 in the USA, Canada, the UK, and eventually more than 100 countries to raise public awareness and support for research.

Conclusion

The rise of Alzheimer's disease as a major public issue, and the transformation in attitudes toward aging that it reflected, created the necessary context for the emergence of geriatric psychiatry. Although neurology came to be seen as the medical specialty with primary responsibility for research and diagnosis of dementia, psychiatry retained a central role (Grossberg and Lake 1998). More importantly, elevating Alzheimer's disease to a major issue attracted unprecedented public attention and funding to the mental health needs of older adults. In the 1970s and 1980s, departments and programs in geriatric psychiatry were created in many medical schools, and professional associations were founded to advance the specialty like the European Association of Geriatric Psychiatry (1973), the American Association of Geriatric Psychiatry (1978), and the International Psychogeriatric Association (1982). By the 1980s, geriatric psychiatry was firmly established as a full-fledged medical specialty committed to proper diagnosis and treatment of the unique manifestations of mental illness in older people.

References

Achenbaum WA (1995) Crossing frontiers: gerontology emerges as a science. Cambridge University Press, Cambridge, New York

Alzheimer A (1906) A charceteristic disease of the neocortex. Translated by Katherine Bick. In Luigi Amaducci, Bick, and Ginacarlo Pepeu. (ed.), *The early story of Alzheimer's disease: translation of the historical papers by Alois Alzheimer, Oskar Fischer, Francesco Bonfiglio, Emil Kraepelin, Gaetano Perusini.* Liviana Press, Adova, Italy, 1987

Alzheimer A (1911) On certain peculiar diseases of old age. Translated by Forstl, H., & Levy, R. (1991). Hist Psychiatry, 2(5 Pt 1)

Amaducci LA, Rocca WA, Schoenberg BS (1986) Origin of the distinction between Alzheimer's disease and senile dementia: how history can clarify nosology. Neurology 36(11):1497–1499

Ballenger JF (2000) Beyond the characteristic plaques and tangles: mid-twentieth century U.S. psychiatry and the fight against senility. In: Whitehouse P, Maurer K, Ballenger JF (eds) Concepts of Alzheimer disease: biological, clinical and cultural perspectives. Johns Hopkins University Press, Baltimore

Ballenger JF (2006) Progress in the history of Alzheimer's disease: the importance of context. J Alzheimer's Dis 9:1–9

Ballenger JF (2017) Dementia: confusion at the borderlands of aging and madness. In: Eghigian G (ed) Routledge history of madness and mental health. Routledge, New York, pp 297–312

Beach TG (1987) The history of Alzheimer's disease: three debates. J Hist Med Allied Sci 42: 327–349

Berrios GE (1995) Dementia: clinical section. In: Berrios GE, Porter R (eds) A history of clinical psychiatry: the origin and history of psychiatric disorders. Athlone Press, London, pp 34–51

Boller F, Bick K, Duyckaerts C (2007) They have shaped Alzheimer disease the protagonists, well known and less well known. Cortex 43(4):565–569

Butler RN (1975) Why survive? Being old in America, 1st edn. Harper & Row, New York

Calhoun RB (1978) In search of the new old; redefining old age in America, 1945–1970. Elsevier, New York

Donahue WT, Tibbitts C (1957) The new frontiers of aging. University of Michigan Press, Ann Arbor, Michigan

Engstrom E (2007) Researching dementia in imperial Germany: Alois Alzheimer and the economies of psychiatric practice. Cult Med Psychiatry 31(3):405–413

Evans JG (1997) Geriatric medicine: a brief history. Br Med J 315(7115):1075

Fox PJ (1989) From senility to Alzheimer's disease: the rise of the Alzheimer's disease movement. Milbank Q 67(1):58–102

Friedmann EA, Havighurst RJ (1954) The meaning of work and retirement. University of Chicago Press, Chicago

Grossberg GT, Lake JT (1998) The role of the psychiatrist in Alzheimer's disease. J Clin Psychiatry 59(Suppl 9):3–6

Haber C (1986) Geriatrics: a specialty in search of specialists. In: Van Tassel DD, Stearns PN (eds) Old age in a bureaucratic society: the elderly, the experts, and the state in American history. Greenwood Press, New York, pp 66–84

Haber C, Gratton B (1994) Old age and the search for security : an American social history. Indiana University Press, Bloomington

Hansen B (2009) Picturing medical progress from Pasteur to polio: a history of mass media images and popular attitudes in America. Rutgers University Press, New Brunswick

Havighurst RJ (1952) Social and psychological needs of the aging. Ann Am Acad Pol Soc Sci 279:11–17

Havighurst RJ, Albrecht RE (1953) Older people, 1st edn. Longmans, Green, New York

Hilton C (2015) Psychiatrists, mental health provision and 'senile dementia' in England, 1940s–1979. Hist Psychiatry 26(2):182–199

Kaplan J (1953) A social program for older people. University of Minnesota Press, Minneapolis

Katz S (2019) Charcot's O+lder women: bodies of knowledge at the interface of aging studies and women's studies. In: Cultural aging. University of Toronto Press, Toronto, pp 37–52

Katzman R (1976) Editorial: the prevalence and malignancy of Alzheimer disease. A major killer. Arch Neurol 33(4):217–218

Katzman R, Bick KL (2000) Alzheimer disease: the changing view. Academic, San Diego

Katzman R, Terry RD, Bick KL (1978) National Institute of Neurological and Communicative Disorders and Stroke., National Institute on Aging., & National Institute of Mental Health (U.S.)

Kitwood T (1987) Explaining senile dementia: the limits of neuropathological research. Free Assoc 10:117–138

Linden M, Courtney D (1953) The human life cycle and its interruptions: a psychologic hypothesis. Am J Psychiatr 109:906–915

Mulley G (2012) A history of geriatrics and gerontology. Eur Geriatric Med 3(4):225–227

Myfanwy T, Isaac M (1987) Alois Alzheimer: a memoir. Trends Neurosci 10:306–307

Nascher IL (1909) Geriatrics. NY Med J 90:358–359

National Institute on Aging, C. T. F (1980) Senility reconsidered. Treatment possibilities for mental impairment in the elderly. Task force sponsored by the National Institute on Aging. JAMA 244(3):259–263

Papavramidou N (2018) The ancient history of dementia. Neurol Sci 39(11):2011–2016. https://doi.org/10.1007/s10072-018-3501-4

Pollak O, Heathers G (1948) Social adjustment in old age; a research planning report. Social Science Research Council, New York

Rosenberg CE (1992) The crisis in psychiatric legitimacy: reflections on psychiatry, medicine, and public policy. In: Exploring epidemics and other essays in the history of medicine. Cambridge University Press, New York

Roth M, Tomlinson BE, Blessed G (1966) Correlation between scores for dementia and counts of 'senile plaques' in cerebral grey matter of elderly subjects. Nature 209(18):109–110

Rothschild D (1947) The practical value of research in the psychoses of later life. Dis Nerv Syst 8: 123–128

Scull A (2015) Madness in civilization : a cultural history of insanity, from the bible to Freud, from the madhouse to modern medicine. Princeton University Press, Princeton

Smith M, Atkin A, Cutler C (2017) An age old problem? Estimating the impact of dementia on past human populations. J Aging Health 29(1):68–98. https://doi.org/10.1177/0898264315624905

Thane P (2005) The 20th century. In: Thane P (ed) A history of old age. J. Paul Getty Museum, Los Angeles, pp 265–300

Tibbitts C, Donahue WT (1960) Aging in today's society. Prentice-Hall, Englewood Cliffs

Torack RM (1979) Adult dementia: history, biopsy, pathology. Neurosurg 4(5):434–442

Wilson DC (1955) The pathology of senility. Am J Psychiatry 111:902–906

Antipsychiatry: The Mid-Twentieth Century Era (1960–1980) 55

Allan Beveridge

Contents

Introduction	1420
R. D. Laing (1927–1989) and His Colleagues	1422
David Cooper (1931–1986)	1426
Laing's Other Colleagues	1429
Thomas Szasz (1920–2012)	1430
The Myth of Mental Illness	1431
The Manufacture of Madness	1432
Commentaries	1433
Erving Goffman (1922–1982)	1434
Commentaries	1436
Postscript to *Asylums*	1438
Michel Foucault (1926–1984)	1439
Franco Basaglia (1924–1980)	1443
Concluding Remarks	1445
Judgments	1447
References	1448

Abstract

The antipsychiatry movement of the 1960s continues to attract analysis and commentary. Its story has been told in memoirs, biographies, and oral histories; it has undergone scholarly scrutiny; it has been the subject of newspaper articles and documentaries; it has been portrayed in films, novels, and plays; and it has lent itself to a great deal of mythologizing. Its legacy continues to be contested. For some commentators, it represents a period when mainstream psychiatry was triumphantly revealed as an authoritarian arm of the state. Further, the writings of antipsychiatrists, in this view, eloquently and courageously uncovered the hypocrisies and iniquities of Western society. For others, the antipsychiatrist movement was a temporary and rather ineffective protest that was very much a product of its

A. Beveridge (✉)
Royal College of Physicians of Edinburgh, Edinburgh, UK

© The Author(s), under exclusive licence to Springer Nature Singapore Pte Ltd. 2022
D. McCallum (ed.), *The Palgrave Handbook of the History of Human Sciences*,
https://doi.org/10.1007/978-981-16-7255-2_91

time. Its leading lights put forward a highly romanticized view of madness which proposed that the mentally ill should remain untreated, a stance, according to this view, that was morally and clinically indefensible. More recently, historians have sought to provide a more nuanced picture which takes account of both not only the idealism of the movement, but also its failings. This chapter examines the key players in the movement: RD Laing, David Cooper, and their colleagues; Thomas Szasz; Erving Goffman; Michel Foucault; and Franco Basaglia. It concludes by assessing the significance and legacy of the antipsychiatry movement, and the reasons for its rise and fall.

Keywords

Antipsychiatry · RD Laing · David Cooper · Thomas Szasz · Erving Goffman · Michel Foucault · Franco Basaglia

Introduction

The antipsychiatry movement of the 1960s continues to attract analysis and commentary. Its story has been told in memoirs, biographies, and oral histories; it has undergone scholarly scrutiny; it has been the subject of newspaper articles and documentaries; it has been portrayed in films, novels, and plays; and it has lent itself to a great deal of mythologizing. Its legacy continues to be contested. For some commentators, it represents a period when mainstream psychiatry was revealed as an authoritarian arm of the state. Further, the writings of antipsychiatrists, in this view, eloquently and courageously uncovered the hypocrisies and iniquities of Western society. For others, the antipsychiatrist movement was a temporary and ineffective protest that was very much a product of its time. Its leading lights put forward a highly romanticized view of madness which proposed that the mentally ill should remain untreated, a stance, according to this view, that was morally and clinically indefensible. More recently, historians have sought to provide a more nuanced picture which takes account of both not only the idealism of the movement, but also its failings.

The period, however, is still a difficult one to characterize: Does one focus on key individuals or on wider social movements? Does one concentrate on particular localities or the wider global picture, which takes in America, Holland, Italy, and Britain? Does the antipsychiatric movement represent a revolution in the approach to the mentally ill, or are there continuities with the past? And, indeed, what does the term, "anti-psychiatry" mean, and did all the leading players associated with the movement accept the label? Indeed, as Crossley (2006) points out, many of the ideas of the key theorists changed dramatically over time, so it is difficult to discern, let alone outline, a coherent and consistent "anti-psychiatry" discourse.

It may be helpful, at this stage, in order to orientate the reader, to provide a broad-brush account of the development of antipsychiatry. It should be acknowledged from the start that there has been criticism of psychiatry since its inception and that this has continued to this day. This chapter concerns with one particular era, the 1960s and 1970s, which can be seen as a specific manifestation of a rather longer tradition of critique. Its immediate origins were in the publication of several books at the beginning of the 1960s. In 1960, the Glasgow psychiatrist, R.D. Laing (1960), published *The Divided Self*, which brought an existential perspective to the understanding of disturbed mental states such as schizophrenia. In *The Myth of Mental Illness*, published in 1961, the American psychiatrist, Thomas Szasz (2003), argued that there was no such thing as mental illness, that it was a man-made invention. In 1961, the Canadian sociologist, Erving Goffman (1968), brought out *Asylums*, in which he argued that much of the so-called "mad" behavior of psychiatric patients was a product of being incarcerated in an institution that stripped them of their self-respect. The French philosopher, Michel Foucault (1967), published his doctoral thesis, *Folie et deraison,* in 1961, which was abridged and translated into English as *Madness and Civilisation* in 1965. In it, Foucault upended the standard history of psychiatry, as a tale of benign progress, and claimed the mentally ill had enjoyed comparative freedom until what he called "The Great Confinement" of the seventeenth century. As well as books, there were radical experiments aiming to reform the mental hospital. In 1962, the South African psychiatrist, David Cooper, set up Villa 21, a residential unit at Shenley Hospital in Hertfordshire, England, in which he sought to introduce a more democratic structure by challenging rigid roles of "doctor," "patient," or "nurse." In the early 1960s, the Italian psychiatrist, Franco Basaglia, was introducing radical changes to the public mental hospital at Gorizia in northeast Italy with the ultimate aim of its eventual closure. Basaglia was the charismatic leader of the Democratic Psychiatric Movement, which promoted "alternative psychiatry" in Italy in the 1960s and 1970s.

As the 1960s progressed, the antipsychiatric movement became involved with the burgeoning counterculture; with the arts, particularly film and theater; and with the New Left. Laing (1967) became a counterculture guru, attracting large audiences on his tours of American university campuses, and writing the cult classic, *The Politics of Experience and the Bird of Paradise*, in which he portrayed madness as a mystical and potentially self-healing journey. He and his colleagues also set up Kingsley Hall in London as a different kind of psychiatric hospital where mentally troubled people could stay, free from coercion or compulsory medication. In 1967, "The Dialectics of Liberation" conference took place at the Roundhouse in North London. This event focused on the nature of violence and the possibility of liberation and included presentations from Herbert Marcuse, Stokely Carmichael, Allen Ginsberg, Gregory Bateson, and others, including Laing and Cooper. In the conference volume, Cooper presented these therapists as "anti-psychiatrists." Despite Laing's objections, and despite very few other therapists ever identifying themselves as antipsychiatrists, the term caught on in the media and among scholars of mental health. Cooper (1967) had, in fact, used the term in his book of the previous year, *Psychiatry and Anti-Psychiatry.*

In the late 1960s, the antipsychiatric movement started to unravel. Colleagues fell out with each other and pursued different directions. Laing went to Ceylon (now Sri Lanka) to study with the country's holy men; Cooper followed a more political path.

Media interest waned, and antipsychiatry was no longer seen as a vital part of the zeitgeist. In Italy, Basaglia and colleagues persevered with their goal to transform and close down the mental hospitals. Szasz, who never considered himself to be an antipsychiatrist, continued to write prolifically about what he saw as the abuses of psychiatry. For some, the demise of the antipsychiatry movement was a triumphant vindication of the values and methods of mainstream psychiatry; for others, it marked the defeat of a kinder and more humane way of responding to the mentally disturbed. A middle way was taken by the Irish psychiatrist and commentator, Anthony Clare, who felt many of the approaches of the antipsychiatrists, such as attending to the patient as a person, had been absorbed by orthodox psychiatry. Indeed, the patients' movement, with its challenge to medical authority and demand for basic human rights, could be seen as a direct heir of antipsychiatry.

That then is a very simplified and brief account of the antipsychiatric movement. We will now go on to consider the leading participants in more detail.

R. D. Laing (1927–1989) and His Colleagues

The main figures in the British Anti-Psychiatry scene were R. D. Laing, David Cooper, Aaron Esterson, Leon Redler, Morton Schatzman, and Joseph Berke. Most commentators consider Laing to be the most prominent figure in the antipsychiatry movement. He was the earliest to publish on the subject, and his writings were the most eloquent and best-selling. He achieved world fame (or notoriety) in the 1960s and was considered one of its significant cultural figures. Michael E. Staub (2011: 63) states: "By 1965 Laing was internationally heralded as the counterculture's principal psychiatrist." The American writer, Angela Carter, observed: "I suppose that R.D. Laing's *The Divided Self* was one of the most influential books of the sixties – it made madness, alienation, hating your parents... it made it all glamorous" (Quoted in Staub 2011: 64). Peter Sedgwick (1982: 67) stated: "it was R.D. Laing who dominated the scene longest, as arch-seer and prophet-in-chief." Nick Crossley (1998), in his survey of the British Anti-Psychiatry movement, allots Laing prime place. However, Oisin Wall (2018) maintains that the emphasis on Laing underplays the role of his colleagues. Likewise, Adrian Chapman (2016) feels that David Cooper, a close colleague of Laing's, has suffered unjustifiable neglect.

Before considering Laing in more detail, it is worth considering his core ideas. He maintained the following:

1. The experience of madness is understandable. Existential phenomenology provides a means of seeing psychosis through the eyes of the sufferer. (Laing (1960) *The Divided Self*)
2. Psychosis is meaningful if one relates it to the disturbed family communications which the patient has endured. (Laing and Esterson (1964) *Sanity, Madness and the Family*)

3. Madness is a reasonable response to so-called "sane" society. In effect, it is society that is mad. (Laing (1967) *The Politics of Experience and The Bird of Paradise*)
4. Psychosis is a journey of self-discovery, which can bring about enlightenment, both spiritual and personal to the voyager. (Laing (1967) *The Politics of Experience and The Bird of Paradise*)

Ronald David Laing was born in 1927 in Glasgow (Beveridge 1998, 2011). After graduating in medicine at Glasgow University in 1951, he worked at the Glasgow and West of Scotland neurosurgical unit at Killearn, near Loch Lomond, where he met Joe Schorstein, a leading neurosurgeon, whom Laing was to later describe as "my spiritual father." Schorstein was the son of a Viennese Rabbi and was very well read in the literature of European Existential philosophy. He served to greatly expand Laing's knowledge of continental thinkers. Laing and Schorstein were part of a philosophical discussion group, made up of doctors, philosophers, and writers, which held regular meetings in Glasgow.

Laing's time in Glasgow was seminal to the evolution of his thought. However, this has often been overlooked by biographers, who have been mystified that the early Laing was so well versed in Existential philosophy, coming as he did from that supposed backwater of civilization, the West of Scotland. In their book, *The Eclipse of Scottish Culture*, Beveridge and Turnbull (1989) have demonstrated that there was, in fact, a lively culture of philosophical debate in Glasgow during this period. Participants were deeply engaged with European existentialist thought, a tradition which Laing drew on heavily in his first book, *The Divided Self*.

After serving in British Army psychiatric units in England, Laing's first non-military posting was at Gartnavel Royal Hospital, where the superintendent was the kindly, if somewhat eccentric, Angus MacNiven, who was wary of the recently introduced psychotropic medication. At Gartnavel, Laing participated with senior colleagues, John L. Cameron and Andrew McGhie, in a project they were conducting that was later dubbed the "Rumpus Room" experiment. The back wards of Gartnavel were overcrowded and understaffed. The researchers posed the question: To what extent was the behavior of the patients in these wards a consequence of their living conditions? Twelve patients were chosen to spend part of the day in a large comfortably furnished room with a high nurse to patient ratio. When it was seen that the patients' behavior improved markedly, the researchers surmised that this was due mainly to nurses being able to form deeper, more meaningful relationships with the patients.

Laing's clinical experiences at Gartnavel and in the British Army informed his first book, *The Divided Self*. In the book, Laing's (1960) sought "to make madness, and the process of going mad comprehensible." He drew on the work of existentialist philosophers, such as Kierkegaard, Sartre, Buber, and writers, such as Franz Kafka and William Blake. Laing contended that conceiving of a mentally ill person as a malfunctioning, biological mechanism served to dehumanize them.

A celebrated passage from the book exemplifies Laing's approach. He quotes an extract from Kraepelin, in which the German clinician presents a schizophrenic patient to his colleagues. Kraepelin provides a summary of the patient's speech and behavior and goes on to assert that the patient makes no sense, that, in fact, he is demonstrating all the signs and symptoms of schizophrenia.

In contrast, Laing tries to understand what is actually happening with the patient. He contends that the young man was protesting against being exhibited like a creature in a zoo, and that his response can be seen as meaningful, if this is taken into account. Laing (1960: 31) writes:

> Now it seems clear that this patient's behaviour can be seen in at least two ways... One may see his behaviour as "signs" of a "disease"; one may see his behaviour as expressive of his existence.

The Divided Self was published in 1960 after Laing had moved to London to train in psychoanalysis at the Tavistock. In 1961, he completed *The Self and Others* (Laing 1961), which looked at the interpersonal aspects of psychosis. His 1964 book *Sanity, Madness and the Family*, which he cowrote with fellow Glasgow medical graduate, Aaron Esterson, located the origins of schizophrenia in the context of the disturbed family communications to which the patient had been subjected during their formative years (Laing and Esterson 1964). Although Laing later claimed that he never said that the family caused schizophrenia, this book created an impression that parents were somehow responsible for their child's breakdown. The book was criticized by academic psychiatrists, who observed that there was no control group, and that the authors had based their work on their own subjective feelings about the families they interviewed, rather than on standardized rating procedures. Anthony Clare (1980) complained: "their study is a clear case of having your classifying cake and eating it; they disagree with the whole idea of the concept of schizophrenia yet one crucial factor that is common to the families in question is that such a diagnosis has been applied to a family member" (p. 112). Daniel Burston (1996) remarked that Laing and his colleagues came to the somewhat disappointing conclusion that the patterns of disordered communication they observed in the families of schizophrenics were often present in normal families as well.

Laing's departure from orthodox psychiatry was completed in 1967 with the publication of *The Politics of Experience and The Bird of Paradise* (Laing 1967: 50). In an arresting polemic, Laing wrote:

> From the moment of birth, when the stone-age baby confronts the twentieth century mother, the baby is subjected to these forces of violence, called love, as its mother and father have been, and their parents and their parents before them. These forces are mainly concerned with destroying most of its potentialities. The enterprise is on the whole successful. By the time the new human being is fifteen or so, we are left with a being like ourselves. A half-crazed creature, more or less adjusted to a mad world. This is normality in our present world.

It was clear that Laing was now appealing to the counterculture gallery of the 1960s. He was on what Sedgwick (1982) has called "the Radical Trip." Laing's writings resonated with the postwar student generation, especially in the States, where his works were eagerly bought and read. Laing himself became a star attraction at lectures and conferences throughout the latter half of the 1960s. In his talks, he grouped the psychotic patient with the criminal and the political dissident in a coalition of oppressed bearers of an authentic statement about the human condition (Clare 1990). Crossley (1998) maintains that there was a thematic link between Laing and the counterculture. Both were concerned with such issues as alienation; love, personal relationships, and self-development; and the critique of science and technology.

In *The Politics of Experience and the Bird of Paradise*, Laing also depicted the experience of psychosis as a voyage of self-discovery that could lead to spiritual enlightenment. This echoed the mindset of the counterculture, and it also had literary precedents in the writings of the Romantics. Nevertheless, Laing's latest work went down badly with mainstream psychiatry, who saw it as an irresponsible attempt to glamourize mental disorder. Although the book was a best seller in its day, it ultimately served to diminish Laing's reputation as a serious commentator on psychiatric matters.

During this period, Laing and his colleagues set up the Philadelphia Association. Founded on 8 April 1965, it was initially a charitable operation devoted to the creation of therapeutic communities for people in crisis. It advocated that a psychotic breakdown was not a symptom of genetic abnormality or physical disorder, but an existential crisis that was potentially a breakthrough to a more authentic way of being (Burston 1996). Laing and his colleagues also set up Kingsley Hall, in London, as a refuge for the mentally distressed who did not want pills or institutionalization. It was based on the belief that madness was a potentially self-healing journey, and that if sufferers were provided with a supportive and nurturing environment, without compulsion or physical treatments, they would recover. During the 1960s, Kingsley Hall attracted visitors from around the world, as well as celebrities, poets, rock stars, misfits, and former psychiatric patients (Burston 1996: 86).

The end of the 1960s saw Laing retreating to Ceylon (since 1972, Sri Lanka) to live among the country's religious community, before returning to Britain 2 years later. Thereafter, his star fell and his literary output was sporadic and insubstantial, though the autobiography of his early years, *Wisdom, Madness and Folly* (Laing 1998), was thoughtful, interesting, and as ever, somewhat self-serving. Burston (1996: 102) judged that "Laing was an unwitting accomplice in his own downfall."

Laing continues to divide opinion. One view, mainly held by psychiatrists, is that he had a brief period of fame in the 1960s when protesting against the establishment was fashionable, but that his views on schizophrenia were irresponsible in that they blamed parents for the illness of their offspring and proclaimed that medication should not be prescribed. In this view, Laing's later descent, hastened by excessive alcohol consumption, seemed to confirm that the man was, all along, an unstable impostor and that his theories had no relevance to contemporary approaches to mental health. The alternative view, mainly held by nonpsychiatrists, is that, during a period when the mentally ill were hidden away in institutions and largely forgotten, he brought attention to their plight. Laing asserted that psychiatric patients were people too, and that, if one only had a little patience and compassion, their apparently perplexing behavior could be understood. From this perspective, Laing's subsequent demonization by the psychiatric establishment demonstrates that it is immovably in thrall to a biological model of mental illness and to maintaining its power.

David Cooper (1931–1986)

> Our experience originated in studies into that predominant form of socially stigmatised madness that is called schizophrenia. Most people who are called mad and who are socially victimised by virtue of that attribution (by being "put away," being subjected to electric shocks, tranquillizing drugs, and brain-slicing operations, and so on) come from family situations in which there is a desperate need to find some scapegoat, someone who will consent at a certain point of intensity in the whole transaction of the family group to take on the disturbance of each of the others, in some sense, suffer for them. In this way the scapegoated person would become a diseased object in the family system and the family system would involve medical accomplices in its machinations. The doctors would be used to attach the label "schizophrenia" to the diseased object and then systematically set about the destruction of that object by the physical and social processes that are termed "psychiatric treatment."

Thus David Cooper (2015) wrote in the introduction to "The Dialectics of Liberation" in 1967. The passage gives a good idea of Cooper's antipsychiatry sentiments, as well as his fervent and polemical style. Cooper coined the term "anti-psychiatry" in his eponymous book, published in 1967, and later used the term to apply to others, including R.D. Laing, who rejected the label. In fact, Adrian Laing's (1994) claim that there was only ever one antipsychiatrist, and that was David Cooper, has some validity. Cooper kept faith with the revolutionary political principles of antipsychiatry after others, like Laing, had abandoned the political for the spiritual. Cooper's writings express the central tenets of antipsychiatry and do so in a dramatic, uncompromising, if often strident fashion. As Chapman (2016) points out, Cooper has been comparatively neglected in historical accounts of antipsychiatry. In part, this may be because his hectoring style of writing lacked the elegance of Laing's, and was frequently simplistic and sloganeering. In *Zone of the Interior*, Clancy Sigal's (2005) main character, Dr. Willie Last, who hails the madman as the new proletarian and the harbinger of revolution, is often taken as a satirical portrait of Laing, but, as Chapman convincingly argues, he is much closer in spirit to Cooper than Laing.

David Cooper was born in 1931 in Cape Town, South Africa. He graduated in medicine from the University of Cape Town in 1955. In his home country, Cooper was involved in underground resistance to the Apartheid regime. He spent time in China undergoing political education, probably in the mid-1950s. Later, he traveled to Cuba in the early days of the revolution. He trained in psychiatry in England in the 1950s and held several hospital appointments, during which time he met R.D. Laing. Cooper left the National Health Service (NHS) in 1966 for private practice and political activism.

While a senior registrar in his last post in the NHS, Cooper attempted to run a therapeutic community at Villa 21 at Shenley Hospital in Hertfordshire, a project which has been examined by Wall (2018). Cooper went further than the earlier proponents of the therapeutic community: In Villa 21, the staff did not wear uniforms and ate from the same plates as the patients. Wall traced two former residents of Villa 21, whom he calls "Adam" and "Ben." Their testimony helps to bring alive the day-to-day reality of Villa 21. Interestingly, they each had opposing views of the place. Adam found Dr. Cooper caring and thought his stay had improved his mental health, whereas Ben felt he survived his time in Villa 21 *in spite of* his treatment, rather than *because* of it. Both agreed that patients were expected to "perform" madness, especially if there were visitors to the unit. Indeed, Adam became a "star" patient. As Wall (2018: 81) astutely observes:

> In many ways this left the residents in a classic double-bind position: on the one hand they were at the hall to work through their madness, with the intention of emerging on the far side; but on the other hand they had been granted access to this exciting community by virtue of their overt craziness and there was the constant threat that they might, in Clancy Sigal's words, "go down and come up straight."

However, Chapman considers that Cooper should be given credit for conducting the first experimental antipsychiatric community and noted that it took place in an N.H.S. psychiatric hospital. Further, it was Cooper who persuaded the hospital authorities to allow the project to go ahead. Cooper believed the function of the standard psychiatric hospital was to dehumanize people, and this was one of the key beliefs of antipsychiatry.

In 1965, Cooper was involved along with others in setting up the mental health charity the Philadelphia Association (PA). However, he showed little interest in the subsequent establishment of Kingsley Hall and left the organization in 1971 in protest. The PA initiated a formal training program in psychotherapy which, from Cooper's point of view, illustrated its failure to recognize the wider political nature of mental distress.

Cooper was involved in the Antiuniversity of London (1968–1971), a radical educational project that offered inexpensive courses without entrance requirements. The teachers included Laing, Joseph Berke from the P.A., as well as the writer Alexander Trocchi, the anthropologist Francis Huxley, the artist Jeff Nutall, and the feminist Juliet Mitchell. The Antiuniversity aimed to question the division between teacher and student, and to focus on personal and political development.

Later, Cooper met Franco Basaglia at a conference in Portugal in 1974 and was impressed by his Psichiatria Democratica movement. The Portugal conference foreshadowed the setting up of the International Network of Alternatives to Psychiatry (INAP) in 1975, with which Cooper was involved. Cooper came to believe that the future of radical psychiatry lay beyond the hospital.

Cooper's early writing was concerned with two of France's great cultural icons: Jean-Paul Sartre and Michel Foucault, both of whom were to be important to the antipsychiatry movement. With Laing, Cooper wrote *Reason and Violence*, a highly condensed version of Sartre's *Critique of Dialectical Reason* (Laing and Cooper 1964), the first volume of which would not be translated until 1976. Wall (2018: 122) judges that *Reason and Violence* was "an important and undervalued work in the history of anti-psychiatry... it laid out many of the philosophical underpinnings of the anti-psychiatric project and, more specifically, their counter-cultural turn." In contrast, Sedgwick (1982: 78) thought it was "so compressed as to be virtually incomprehensible... and resembling a precis for private study." Cooper also wrote an introduction to the first English translation of Foucault's *Madness and Civilisation*. Madness, Cooper (2001) claimed, represents "some sort of lost truth." Chapman considers that Cooper played an important part in introducing English readers to the work of Foucault and Sartre.

In 1967, Cooper brought out his first book of original writing, Psychiatry and Anti-psychiatry (Cooper, 1967), which gives an account of his evolution from psychiatrist to antipsychiatrist. It also details his Villa 21 experiment. He began by stating that a critically aware psychiatrist had to resist being "numbed or engulfed by the institutionalising process of formal training and day-to-day indoctrination in the teaching hospital" (Cooper 1967: ix). The major problem area in psychiatry, he claimed, was schizophrenia:

> What I have attempted to do in this monograph is to take a look at the person who has been labelled schizophrenic in his actual human context and to enquire how this label came to be attached, and what it signifies for the labellers and labelled. (Cooper 1967: ix)
> This is a study of one mode of social invalidation... First, a person is progressively made to conform to the inert, passive identity of invalid or patient... second, the process whereby almost every act, statement, and experience of the labelled person is systematically ruled invalid according to certain rules of the game established by his family, and later by others, in their efforts to produce the vitally needed invalid-patient. (Cooper 1967: x.)

Cooper maintained that the violence of psychiatry was on a spectrum which included mass-extermination, and he made frequent references to the Nazis throughout the book. He drew on double-bind theory in accounting for how one member of the family came to be labeled schizophrenic. He held that the mother was especially culpable in this process.

His next book was *The Death of the Family* (Cooper 1971). Wall (2018) believes that this book is underrated, for it succeeded in bringing together the different strands of the antipsychiatric views of the family. In the opening pages, Cooper (1971: 4) writes:

> The power of the family resides in its social mediating function. It reinforces the effective power of the ruling class in any exploitative society by providing a highly controllable paradigmatic form for every social institution. So we find the family form replicated through the social structures of the factory, the union branch, the school (primary and secondary), the university, the business corporation, the church, political parties and governmental apparatus, the armed forces, general and mental hospitals, and so on.

Wall contends that Cooper was attempting to disrupt what he saw as the mystifying and invalidating strategies, not only of the hospital but also those of the family, the police, the school, and the university – all the institutions that, the antipsychiatrists believed, sought to limit the range of mental life. Wall (2018: 80) observes: "On the face of it, these were grandiose, and even deluded ambitions."

In his 1980 book, *The Language of Madness*, Cooper (1980: 23) underlined his radical perspective on mental illness, declaring: "*all delusion is political statement ... and all madmen are political dissidents.*" For Cooper, madness was both a resistance to and a sign of the repressive nature of the family. Cooper, at times, clearly associated madness with the purity of childhood. He held a very romanticized view of the madman as a truth-teller. For him, Antonin Artaud was such a figure.

As Chapman (2016) concedes, Cooper's sexual politics were extremely dubious, and he was attacked by Elaine Showalter (1987) in *The Female Malady* for abusing the power invested in his status as a therapist. He also concedes that Cooper's linking of madness with the wider social contexts could lack nuance and tend toward sloganeering. However, he felt that his strength lay in his insistence on the need to understand the symptoms of the individual, not only in terms of individual psychology or family dynamics, but also in terms of broader social, institutional, and economic forces.

Laing's Other Colleagues

Aaron Esterson grew up in Glasgow, where he studied medicine, qualifying in 1951. He worked in several British psychiatric hospitals between 1954 and 1962, latterly at Napsbury Hospital. There he ran a unit for women with schizophrenia and carried out research which formed the basis of *Sanity, Madness and the Family*, which he cowrote with Laing. In the early 1960s, he was a research associate at the Tavistock Institute. In November 1962, Joe Berke, a young American medical student, contacted RD Laing for the first time. According to Wall (2018: 72–73), it was "an event which was of key importance to the establishment of anti-psychiatry in the international counter-culture." Berke was connected with various luminaries in the American counterculture, such as Timothy Leary, Allen Ginsberg, and Michael Hollingshead, all of whom he subsequently introduced to Laing. Berke and his colleague Leon Redler had both studied medicine in New York. Both had worked with Maxwell Jones at Dingleton Hospital, before moving to Kingsley Hall. Morton Schatzman arrived in London in 1967, having also studied medicine at New York. All of these antipsychiatrists were also involved with the Philadelphia Association.

Berke was the therapist who looked after Kingsley Hall's most famous resident, Mary Barnes, and together they wrote a book entitled *Mary Barnes. Two Accounts of a Journey through Madness* (Barnes and Berke 1971). Her story has been much retold, becoming a *cause celebre* for the antipsychiatry movement and the basis of a play by David Edgar in 1979. At Kingsley Hall, she was allowed to regress, smear feces on the wall, refuse to wash or eat, and had to be fed from a baby's bottle. Her case raised the question of how much freedom a person should be granted, especially if their behavior impinges on the well-being of others. Other residents certainly objected to the foul smells emanating from her room. It also posed the question of whether it is therapeutic to allow or encourage a mentally person to "regress" – could it actually make them worse?

By the end of the 1960s, tensions were mounting in the antipsychiatry group, and in the early 1970s it collapsed altogether. In December 1966, Laing moved out of Kingsley Hall, leaving the day-to-day operations to Redler, Schatzman, and Berke. Kingsley Hall disintegrated into chaos by the end of the decade, and even Laing admitted that it had not been a great success. For many, its failure illustrated the limits of a noninterventionist approach to mental illness. In its defense, it did anticipate and possibly inspire subsequent attempts to treat the mentally ill outside the institution. In 1968, Esterson fell out with Laing after the latter assaulted him. Esterson, who was alarmed, both by Laing's increasingly erratic behavior and by the lack of order at Kingsley Hall, resigned from the Philadelphia Association. In 1970, Berke and Shatzman left the Philadelphia Association to establish the Arbours Association. Laing left London in 1971 to spend a year in Ceylon, studying Eastern religion. Cooper resigned from the Philadelphia Association in 1971 to pursue revolutionary politics in Argentina.

Wall (2018: 90) offers an explanation for the breakup of the antipsychiatry group: "Unlike the stable and socially mandated authority of the hospital, the countercultural network that the anti-psychiatrists moved into was plagued by unstable, manipulative, and charismatic leaders – Laing and Cooper soon became classic examples of this." Ingleby (1998) felt Laing, in particular, had a talent for falling out with everyone.

Thomas Szasz (1920–2012)

Thomas Szasz contrasts with the other figures considered here. He was on the political Right, whereas most of the other antipsychiatrists were on the Left. He was a professor of psychiatry, while Laing and Cooper worked, for the most part, outside mainstream psychiatry. Unlike them, he was never immersed in the counterculture and, in fact, rejected its philosophy. According to Staub (2011), what distinguished Szasz from other antipsychiatrists was his complete lack of interest in the theory that the cause of mental illness lay within the family and the notion that asylums could drive people mad. He resisted the idea of reforming psychiatry as he felt there was no point in trying to fix what he regarded as an evil system. Szasz had a disdainful attitude toward other antipsychiatrists, particularly Laing, about whom he made disparaging personal remarks. Szasz fiercely rejected the label "antipsychiatrist" and did not see himself as part of the antipsychiatry movement, nor did he work collaboratively with any of its members. Unlike Laing, Cooper, and Basaglia, all of whom died at a comparatively young age, Szasz lived to his 90s and had a successful professional and writing career long after the antipsychiatric era ended. Here we will only be concerned with those aspects of his career which intersect with the antipsychiatry period.

Originally from Hungary, Szasz emigrated to the United States and trained at the Chicago Institute for Psychoanalysis in the 1950s, at a time when psychoanalysis held sway in American psychiatry (Szasz 2004). He later moved to Syracuse, New York, where he became a professor of psychiatry. In 1961, he published *The Myth of Mental Illness* (Szasz 2003), in which he argued that there was no such thing as mental illness, that it was a man-made invention. In a prolific writing career, he went on to expand on this idea, asserting that the madman was responsible for everything he said and did. Szasz claimed that hallucinations were "disowned self conversations" and delusions were "stubborn errors or lies," both of which were "created" by the patient and could be "stopped" by them (Szasz, 2002). He argued that the term "schizophrenia" did not represent a genuine disease but was a "sacred symbol of psychiatry," invented by psychiatrists to allow them to lock up their patients. He held that "symptoms" hid and expressed what was really going on in human interactions. He inveighed against what he saw as the medicalization of "the problems of living." He protested against compulsory detention and treatment, and against the insanity defense. In 1969, he made common cause with Scientologists to fight against compulsory incarceration. Szasz maintained that psychiatrists should not be involved with the courts and that the notion of an insanity defense was untenable. People should take full responsibility for their actions, and to claim that mental illness led them astray was a denial of responsibility. It was simply an attempt to avoid blame, and psychiatrists seemed only too keen to collude with this maneuver.

Szasz's philosophy of freedom was a key aspect of his thought and led him to consider suicide as the ultimate act of free will. A libertarian from the political right, Szasz argued against state intervention and was opposed to the ethos of Britain's NHS. He favored a type of private practice in which the patient and doctor agreed a voluntary contract.

The Myth of Mental Illness

In *The Myth of Mental Illness*, Szasz (2003) maintained that disease or illness can affect only the body, and therefore there can be no such entity as mental illness. He claimed that "Mental illness" was a metaphor. Minds can be "sick" only in the sense that jokes are "sick" or economies are "sick" (Szasz 2003: 267). Psychiatric diagnoses, he argued, are stigmatizing labels, masquerading as medical diagnoses and applied to people whose behavior annoys or offends others. He claimed:

> Those who suffer from or complain of their own behavior are usually classified as "neurotic"; those whose behavior makes others suffer, and about whom others complain, are usually classified as "psychotic." (Szasz 2003: 267)

Szasz stated that mental illness was not something a person *has* but was something they *do* or *are*. Since, in his opinion, there was no such thing as mental illness, there was, therefore, no need for psychiatric treatment. Szasz distinguished between "voluntary" and "involuntary" psychiatry and described the latter: "Typical involuntary psychiatric interventions are commitment or measures carried out under threat of commitment, and psychiatric 'diagnoses' and 'treatments' imposed on persons by parents, schools, courts, military authorities, and other social and governmental agencies" (Szasz 2003: 260). He maintained that psychiatric treatment against people's will was indefensible and an attack on individual autonomy, concluding: "There is no medical, moral, or legal justification for involuntary psychiatric interventions. They are crimes against humanity" (Szasz 2003: 268).

In fact, Szasz argued psychiatry was not even a medical speciality. It was not concerned with mental illness; rather psychiatrists dealt "with personal, social, and ethical problems in living" (Szasz 2003: 262). In place of orthodox psychiatry, Szasz put forward his notion of therapy based on game theory. He maintained that:

> Personal conduct is always rule-following, strategic, and meaningful. Patterns of interpersonal and social relations may be regarded and analysed as if they were games, the behaviour of the players being governed by explicit or tacit game rules (Szasz 2003: 268).

The therapist's role was to help the client to elucidate the inexplicit game rules.

Szasz contended that powerful institutional forces maintained the tradition of keeping psychiatric problems within the conceptual framework of medicine. The problem of "mental illness," he advised, must be recast and redefined in "a morally explicit science of man," because human behavior was fundamentally *moral* behavior (Szasz 2003: 263).

Szasz concluded by stating:

> ... the concept of mental illness... undermines the principle of personal responsibility, the ground on which all free political institutions rest. For the individual, the notion of mental illness precludes an enquiring attitude towards his conflicts which his "symptoms" at once conceal and reveal. For a society, it precludes regarding individuals as responsible persons and invites, instead, treating them as irresponsible patients. (Szasz 2003: 262)

The Manufacture of Madness

In this book, Szasz (1971) put forward his thesis that the contemporary belief in mental illness and the consequent "persecution" of mental patients was similar to the belief in witchcraft and the persecution of witches in the Middle Ages. He compared "messianic Christianity" with what he called "messianic psychiatry" and held that, in modern times, medicine had replaced theology; the psychiatrist, the inquisitor; and the insane, the witch. As a result, the "persecution" of mental patients in the present day had replaced the persecution of heretics in the past. Szasz stated that his book was the history of psychiatry "from its theoretical origins in Christian theology to its current practices couched in medical rhetoric and enforced by police power."

He wrote:

> ... it is necessary to explain and justify situations where individuals are "treated" without their consent and to their detriment. The concept of insanity or mental illness supplies this need. It enables the "sane" members of society to deal as they see fit with those of their

fellows whom they can categorise as "insane". But having divested the madman of his right to judge what is in their own best interests, the people – especially psychiatrists and judges... have divested themselves of the corrective restraints of dialogue... In this medical rejection of the Other as a madman, we recognise, in up-to-date semantic and technical garb, but underneath it remarkably unchanged, his former religious rejection as a heretic. (Szasz 1971: xvi)

Szasz introduced the concept of the "Therapeutic State," which he saw as a collaboration between psychiatry and government to counter any actions, behaviors, or thoughts which they deemed unwelcome or disruptive by reclassifying them as symptoms of mental illness, and thus in need of "treatment." The advocates claimed they were trying to help their fellow man but, in Szasz's view, were blind to the fact people might not want their help and would prefer to be left alone. They believed themselves to be patients' allies, when they were actually their adversaries. If the patient refused help, this was a sign of mental illness and legitimized the need for compulsory treatment.

Szasz went on to describe what he called "Institutional Psychiatry," which he contrasted negatively with "Contractual Psychiatry."

The principle problem in psychiatry has always been, and still is, violence: the threatened and feared violence of the "madman", and the actual counter-violence of society and the psychiatrist against him. The result is the dehumanisation, oppression, and persecution of the citizen branded "mentally ill"... The best, indeed the only, hope for remedying the problem of "mental illness" lies in weakening – not strengthening – the power of Institutional Psychiatry. Only when this peculiar institution is abolished will the moral powers of uncoerced psychotherapy be released. Only then will the potentialities of Contractual Psychiatry be able to unfold – as a creative human dialogue unfettered by institutional loyalties and social taboos... (Szasz 1971: xvii)

Szasz held that the "Institutional" psychiatrist was a bureaucratic employee, who was paid by a private or public organization, rather than by the patient. In contrast, the "Contractual" psychiatrist was a private entrepreneur, who was paid directly by the patient. He concluded

... there are, and can be, no abuses *of* Institutional Psychiatry, because Institutional Psychiatry *is* itself, an abuse... just as the Inquisition was the characteristic abuse of Christianity, so Institutional Psychiatry is the characteristic abuse of Medicine. (Szasz 1971: xxv)

Commentaries

In *The Myth of Mental illness,* Szasz advised that we should judge clinicians by what they *do*, not what they *say*. In the case of Szasz, this is especially pertinent. How did the man, who said there was no such thing as mental illness and claimed that psychiatric treatment was an abuse of human rights, actually deal with his patients? We can gain some answer to this question from American clinicians who knew Szasz and his work. Fuller Torrey (2019) speculates that it is possible that Szasz never diagnosed or treated any patients with schizophrenia and that he never had to take care of any patient who needed to be detained in hospital. He quotes Szasz as saying: "I have never, ever given drugs to a mental patient." Torrey's speculations are borne out by Szasz (2004) himself, who records in his memoir that as part of his training, he was advised to spend time in a mental hospital treating patients with serious psychiatric illness, but that he refused to go. Torrey concludes that Szasz's statements about schizophrenia are "fatuous" (Torrey 2019: 99). Allen Frances (2019) writes that he was initially impressed when he read Szasz as a trainee, but later found Szasz's theories were clinically impractical. He concludes: "Schizophrenia is only a 'myth' to those with no clinical experience or life experience" (Frances 2019: 172). Frances recalls asking Szasz what he would do if his son was hearing voices telling him to kill himself: Would he favor compulsory detention? Szasz said he would intervene: "I am a father first and a protector of human rights second" (Frances 1971: 170). It is difficult to see this as anything other than hypocrisy: advising others to do what one would not do oneself.

Sedgwick (1982: 153) was perturbed that Szasz advocated that doctors should not concern themselves with the social causes of illness and that they should only see patients who were able to pay their fees:

> An organically defined medical science, blind to the most obvious connections between social environment and personal ailment: an equally individualistic psychoanalytic framework, available only for those patients who are well enough (and well-off enough) to pay cash: such are Szasz's positive therapeutic ideals.

The philosopher Jennifer Radden (2019: 249) judges that expecting people to take responsibility for their symptoms seems "inhuman and unrealistic." Staub (2011) criticized Szasz for refusing to accept that mental illness might be real. Wilson (2019) felt that Szasz's emphasis on individualism denies the real nature of mental illness and the suffering of patients and their families, and serves to stigmatize psychiatrists. He does concede that Szasz highlighted the dangers of arbitrary incarceration and of the unnecessary creation of new categories of mental disorders, and that he did puncture the scientific pretensions of some mental health professionals. Staub (2011) felt that Szasz's writing had a great influence on the American legal system, and that his ideas helped to promote and support gay rights and feminism.

Erving Goffman (1922–1982)

Erving Goffman was a Professor of Anthropology and Sociology at the University of Pennsylvania (Burns 1999). His main connection with the antipsychiatric movement is his 1961 book, *Asylums*, in which he offered a sociological critique of the mental institution, based on his observational work in St Elizabeth's Hospital in Washington (Goffman 1968). Goffman went undercover and presented himself as an assistant to the athletic director. Had he posed as a patient, he might have gained greater insight into their lives, but it would have restricted his mobility. The book contributed to the growing feeling that mental hospitals were places of repression, rather than therapy. As Dimitri N. Shalin (2013) has observed, the parallels Goffman drew between concentration camps and mental hospitals rang true to the generation that witnessed the rise of totalitarian states, experienced the horrors of World War II, and became involved with the civil rights movement. Goffman showed how patients had their dignity and self-worth stripped away in order to make them conform with the hospital system. He also showed how patients developed strategies to beat the system and, indeed, survive it.

Asylums was divided into four essays. The first essay outlined the characteristics of "total institutions." The second dealt with "the moral career of the mental patient" and detailed their journey from the status of prepatient to inpatient. The third looked at the underlife of the institution and was subtitled "A study of ways of making out in a mental hospital." The last essay was concerned with the "medical model" of a hospital, especially the doctor-patient relationship.

Goffman (1968: 11) began by defining what he meant by a "total institution": "A total institution may be defined as a place of residence and work where a large number of like-situated individuals, cut off from the wider society for an appreciable period of time, together lead an enclosed, formally administered round of life." Throughout the book, he repeatedly compared the asylum to the Nazi concentration camps.

Goffman (1968: 130) coined the term, "the betrayal funnel," to describe what he saw as the conspiratorial relationship between the relatives of patients and staff to get the patient admitted to hospital. He described how the newly admitted patient was processed by undergoing what he called "a series of abasements, degradations, humiliations, and profanations of self. His self is systematically, if unintentionally mortified" (Goffman 1968: 24). On admission, the inmate was stripped of their clothes and belongings. Personal details about the inmate were collected in files which all the staff could read. Inmates needed to request permission from staff to carry out basic activities. New patients were robbed of their "accustomed affirmations" and were "subject to a rather full set of mortifying experiences: restriction of free movement, communal living, diffuse authority of a whole echelon of people" (Goffman 1968: 137). However, the inmate could undermine the system by making "secondary adjustments" or developing "angles" to procure things they wanted. Goffman (1968: 268) writes:

> When existence is cut to the bone, we can learn what people do to flesh out their lives. Stashes, means of transportation, free places, territories, supplies for economic and social exchange – these apparently are some of the requirements for building up a life.

In Goffman's scheme, the nature or reality of mental illness was barely considered. Instead, he contended that the patient had indulged in "situational improprieties," which psychiatrists then recast as mental illness. He argued: "Stigmatization as mentally ill and involuntary hospitalization are the means by which we answer these offences against impropriety" (Goffman 1968: 268). If the inmate expressed disaffection with the system, the asylum staff saw it as evidence of illness. Goffman held that much of the apparently odd behavior displayed by patients was, in fact, their method of dealing with the oppressive nature of the institution in which they found themselves incarcerated.

After the patient had been in the ward for a certain amount of time, staff would want to see "improvement" in their condition. This led staff to encourage "proper" conduct and to express disappointment when a patient was not making progress. However, some symptoms, like muteness and apathy, might be tolerated because they made the ward routine easier. If the patient behaved, they were rewarded with better conditions; if not, they were punished by having privileges withdrawn. As Goffman observed:

> Mental patients can find themselves in a special bind. To get out of the hospital, or to ease their life within it, they must show acceptance of the place accorded them, and the place accorded them is to support the occupational role of those who appear to force this bargain. (Goffman 1968: 335)

Commentaries

Goffman's *Asylums* has had a mixed reception: Some hailed it as exposing the oppressive nature of mental institutions, while others accused it of painting them in an unfair and unnecessarily bleak light. On the positive side, Seamus MacSuibhne (2011) judged: "Goffman's key role was in humanising patients and drawing attention to patterns of interaction that dehumanised them." In like manner, Tom Burns (2020) wrote.

> Despite the book's uncompromising message, *Asylums* was readily accepted by the profession. Its warning of the malevolent effects of total institutions, confirmed by regular scandals, became a central driver of the deinstitutionalization process that dominated the following decades. Goffman's revelations about the power structure in hospitals informed the development of multidisciplinary working.

However, Goffman also had several critics. Matthew Gambino (2013) observed:

> A closer look at the changes transforming St. Elizabeth's around the time of Goffman's research reveals a far richer portrait of institutional culture. Group therapy, psychodrama, art and dance therapy, patient newspapers, and patient self-government – each of which debuted at the hospital in the 1940s and 1950s – provided novel opportunities for men and women to make themselves heard and to take their fate into their own hands... surviving documents suggests that those who participated found their involvement rewarding and empowering.

Gambino judges that Goffman's failure to appreciate fully the capacities of patients led him to underestimate the importance of these developments. Similarly, Staub (2011) observed that *Asylums* made no mention of the improvements that were taking place in asylums at the time. In his opinion, the book achieved its goal of disparaging the mental hospital by ignoring the changes which were already underway. Shalin (2013) agrees, stating that what was missing in Goffman's early work was any acknowledgment that psychiatric treatment might benefit patients, that it achieved anything other than pacifying relatives and flattering the psychiatrists' inflated egos. Goffman also overstated the extent to which the state-run mental hospitals catered to the involuntarily committed patients. While Goffman claimed that voluntary patients were only a tiny percentage, they actually amounted to half the patients admitted to St. Elizabeth's during this period (Shalin 2013). As for Goffman's notion of the "funnel of despair" – the supposed conspiratorial relationship between relatives of patients and staff – Shalin (2013) noted: "Studies in the late sixties showed that three-quarters of patients discharged from mental institutions did not feel betrayed."

In a comprehensive review of the research which tested Goffman's claims, Weinstein (1982) concluded that his work was deeply flawed. The social situation of mental patients was really quite different from the portrait drawn by Goffman. In contrast to his claim that patients suffered a "mortification of self," this was not borne out by quantitative research which found that most patients' self-esteem improved or stayed the same 1 month after admission. In addition, Goffman's contention that mental hospitals "converted" patients and changed patients' conceptions of themselves to the hospital's conception of them was also inaccurate. Patients were not powerless and impotent, but often responsible participants in their treatment.

Goffman's characterization of the asylum as a "total institution" analogous to other institutions, such as concentration camps, was considered to be spurious and overdrawn. Institutions are not homogenous, and even among asylums there were many differences. The mental hospital was not a "closed system" apart from the rest of society. Research has suggested that Goffman paid too much attention to the negative side of asylums, and too little to the therapeutic and rehabilitative aspects. The large proportion of voluntary patients during the period Goffman was studying and the active participation of patients in their treatment belie his assertions. Weinstein (1982) judged that Goffman did not convincingly establish that mental hospitals were coercive and tyrannical and that patients suffered from abandonment and loss of rights. Goffman's picture of the patients entertaining a relentlessly negative attitude toward the asylum was not corroborated by researchers. Indeed contemporary and more recent research has found that the majority of patients had positive attitudes. The majority wanted to enter the hospital. Goffman's contention that patients were admitted for reasons unrelated to mental illness, such as social factors, was again shown to be unfounded. Weinstein (1982) concludes:

> The bleak picture painted by Goffman of the social situation of mental patients derives mainly from his use of the total institution model. He places mental hospitals in the same category as prisons, concentration camps, monasteries, orphanages, and military organizations...
>
> Goffman's view of asylums is not so much wrong as it is one-sided. He focused on the negative and debilitating characteristics of mental hospitalization without giving adequate attention to the therapeutic effects.

Critics were also concerned that Goffman did not seem to accept the reality of mental illness. Shalin (2013) wrote: "it seems apparent with the passage of time that Goffman downplayed the medical dimension of mental disorder." Shalin pointed out that Goffman placed "mental illness" and "sickness" in quotation marks to convey his disparaging attitude to psychiatry. Mental illness for him was a social construct that colluding others successfully impose on a victim. Sedgwick (1982) complained that Goffman denied the reality of psychiatric symptoms by insisting that they were always dependent on the context in which they occurred – that they were "situational improprieties," not evidence of insanity. Behavior considered pathological inside the asylum might be considered perfectly normal outside it. Goffman claimed that: "I know of no psychotic misconduct which cannot be matched precisely in everyday life by the conduct of persons who are not psychologically ill nor considered to be so!" (quoted in Sedgwick 1982: 45). Sedgwick concedes that, while context plays a part, it is not the full explanation of someone's utterances or behavior; mental illness is of crucial importance too.

For Siegler and Osmond (1971), the chief shortcoming of Goffman's work was that he analyzed mental hospitals without regard for the concept of mental illness and was extremely unclear as to how inmates ended up in hospital. As they write:

> The official function of the mental hospital is to treat psychiatric illness, but its true function appears to be to subdue, degrade, and humiliate the people who are confined there, so they will be easier to control... Goffman has managed to conjure up something that is worse than a concentration camp, a total institution in which the inmates live in a frightful exile for no reason.

Postscript to *Asylums*

Shalin (2013) examined a paper entitled "The Insanity of Place," which Goffman (1971) wrote some years after the publication of *Asylums*. He found that Goffman's views had changed dramatically. When he referred to mental illness, he no longer placed the term in quotation marks. Goffman now acknowledged that there was an organic dimension to mental illness. The supposed collusion between the doctor and the relative against the patient – "the funnel of despair" – was now between the doctor and the patient against the carer. Goffman no longer valorized patients as abused human beings suffering from conspiratorial designs. The author's sympathies were now with families forced to endure manifestly disturbed members whose behavior disrupted domestic life. Shalin (2013) observes:

> Situational improprieties are framed here in a starkly negative light, with no romanticizing of the rebellious tactics celebrated in *Asylums*. The scourge of families, disruptive behavior has little to do with quest for freedom and a good deal with insanity. The anxious tone in IP [Insanity of Place] contrasts with the ironic discourse of *Asylums*. The author pictures would-be patients as seriously impaired individuals who are overdue for institutionalization, indulged by their therapists, and tragically hurtling towards their fate.

Shalin puts Goffman's change of viewpoint in context. Goffman had a difficult relationship with his wife who suffered from mood swings for many years and underwent psychiatric treatment. She eventually killed herself in 1964. Shalin concludes that, while Goffman's experience of his troubled marriage might have influenced his antipathy to psychiatrists and the views expressed in *Asylums*, it definitely influenced the perspective he adopted when he came to write "The Insanity of Place."

Michel Foucault (1926–1984)

The French philosopher Michel Foucault's work has encompassed a wide range of subjects, such as madness, knowledge, sexuality, power, and punishment. His main link with the antipsychiatry movement is his book, *Madness and Civilisation* (Foucault 1967). His doctoral thesis was published in 1961 as *Folie et deraison*, which he subsequently abridged as *Histoire de la Folie a l'age Classique* in 1964. It was translated in this form into English as *Madness and Civilisation* in 1965. RD Laing had recommended the book to Tavistock, the English publisher, reporting that it was "an exceptional book… with a thesis that thoroughly shakes the assumptions of traditional psychiatry" (quoted in Foucault 2006). David Cooper (2001: vii) provided the foreword to the English translation, calling it "a remarkable book." Thomas Szasz quoted approvingly from it in *The Manufacture of Madness*, and Franco Basaglia cited *Madness and Civilisation* as an influence.

Foucault's book basically presented the history of psychiatry from an antipsychiatric perspective. Where previous histories had told of the benign progress of psychiatry with psychiatrists as the heroes developing increasingly more humane and effective treatments, Foucault argued that the mad enjoyed comparative freedom until the seventeenth century, which ushered in what he called the "Great Confinement" of large numbers of the mentally ill.

In an interview he gave to a French journalist, Foucault provided a useful summary of the book:

> Madness only exists in society… It does not exist outside of the forms of sensibility that isolate it, and the forms of repulsion that expel or capture it. Thus one can say that from the Middle Ages up to the Renaissance, madness was present within the social horizon as aesthetic and mundane fact; then in the seventeenth century – starting with the confinement [of the mad] – madness underwent a period of silence, of exclusion. It lost the function of manifestation, of revelation, that it had in the age of Shakespeare and Cervantes (for example, Lady Macbeth begins to speak the truth when she becomes mad), it becomes laughable, delusory. Finally the twentieth century collars madness, reduces it to a natural phenomenon, linked to the truth of the world. From this positivist expropriation derive both the misguided philanthropy that all psychiatry exhibits towards the mad, and the lyrical protest that one finds in poetry from Nerval to Artaud, and which is an effort to restore to the experience of madness the profundity and power of revelation that was extinguished by confinement. (as quoted in James Miller 1993: 98)

Foucault sought to establish how "madness" was first divided from "reason." He held that the emergence of psychiatry, instead of alleviating the plight of the mentally ill, had resulted in their imprisonment and silencing. In the Middle Ages, he claimed, madness was regarded as a vice. Subsequently, in the Renaissance, the mad, by now a feature of "everyday life," were regarded not only as a threat to society, but also as a grim example that reason was fragile and could be easily overturned. Despite this, the Renaissance also witnessed the phenomenon of the "Ship of Fools," whose passengers were the mentally deranged, gliding down the rivers of the Rhineland and Flemish canals. Although Foucault conceded that this was mainly a literary and artistic invention, he insisted that some ships actually existed. He claimed that the boats took madmen from town to town, where they led an "easy wandering existence" and were "allowed to wander in the open countryside" (Foucault 1967: 8). He judged that the "Classical Age" (the seventeenth and most of the eighteenth century) reduced "to silence the madness whose voices the Renaissance had just liberated" (Foucault 1967: 38). He wrote:

> Madness was thus torn from that imaginary freedom which still allowed it to flourish on the Renaissance horizon. Not so long ago, it had floundered about in broad daylight: in *King Lear*, in *Don Quixote*. But in less than a half-century, it had been sequestered and, in the fortress of confinement, bound to Reason, to the rules of morality and to their monotonous nights. (Foucault 1967: 64)

From the seventeenth century onward, Foucault maintained, society's response to the mad was to incarcerate them. He argued that lazar houses for lepers, which had lain empty for two or three centuries, were used to exclude not only the mad, but also poor vagabonds and criminals (Foucault 1967: 7). At this stage, according to Foucault, the mad were not differentiated from the criminal and the idle, and all were confined in the same institution. He argued that the aim of confinement was not to cure the sick, but rather it was motivated by "the imperative of labour" (Foucault 1967: 46). All were put to work, as the bourgeois authorities prized industry. During the eighteenth century, the mad were differentiated from the rest of society's rejects and confined separately in asylums. Foucault (1967:58) suggests this was because the mad "distinguished themselves by their inability to work and to follow the rhythms of collective life."

He asserted that The Age of Reason ended the dialogue of the sane and insane. He writes:

> As for a common language, there is no such thing; or rather, there is no such thing any longer; the constitution of madness as a mental illness, at the end of the eighteenth century, affords the evidence of a broken dialogue, posits the separation as already effected, and thrusts into oblivion all those stammered, imperfect words without fixed syntax in which the exchange between madness and reason was made. The language of psychiatry, which is a monologue of reason *about* madness, has been established on the basis of such a silence. (Foucault 1967: x–xi)

In a chapter entitled "The Birth of the Asylum," Foucault mocked traditional histories of psychiatry which saw the advent of psychiatry and the creation of the asylum as proof of the humane progress in the treatment of the mentally ill. He declared that, prior to this, others had also advocated separating the mad from the criminal and idle.

He writes:

> We can see how the political critique of confinement functioned in the eighteenth century. Not in the direction of a liberation of the mad; nor can we say that it permitted a more philanthropic or a greater medical attention to the insane. On the contrary, it linked madness more firmly than ever to confinement... (Foucault 1967: 227)

Foucault then went on to consider two heroes of the traditional history of psychiatry: Samuel Tuke and Philippe Pinel. He turned the traditional history on its head and maintained that their asylums oppressed the mentally ill just as much as before, but in a more subtle and sophisticated way. In traditional histories of psychiatry, the establishment of the York Retreat is portrayed as a significant advance in the treatment of the mentally ill. In response to the abuses of patients at the York Asylum, Samuel Tuke, a Quaker tea merchant, founded the Retreat, which was initially run without doctors. Instead of physical restraints, the asylum staff favored a type of "moral treatment," whereby inmates were to be treated with kindness, and the provision of meaningful occupation. Foucault detected something sinister in this approach, arguing that physical restraint had been replaced by self-restraint. The patient was expected to police himself so that he behaved with decorum in public.

Foucault writes:

> The obscure guilt that once linked transgression and unreason is thus shifted; the madman as a human being originally endowed with reason, is no longer guilty of being mad; but the madman, as a madman, and in the interior of that disease of which he is no longer guilty, must feel morally responsible for everything within him that may disturb morality and society, and must hold no one but himself responsible for the punishment he receives. (Foucault 1967: 246)

Foucault pointed out that the Retreat was founded on Quaker principles, which particularly favored putting the inmates to work as part of its program of moral treatment. He maintained that Tuke's project needed to be reevaluated. Although historians claimed that the Retreat had liberated the insane and provided a therapeutic milieu, for Foucault (1967: 247) these were: "only justifications. The real operations were different. In fact, Tuke created an asylum where he substituted for the free terror of madness the stifling anguish of responsibility." Foucault suggested that the existence of madness was judged on what was visible to the observer, how the individual acted, not on their innermost thoughts. He declared:

> Madness no longer exists except as *seen*... The science of mental disease, as it would develop in the asylum, would always be only of the order of observation and classification. It would not be a dialogue. (Foucault 1967: 250)

Philippe Pinel also occupies an important place in traditional histories of psychiatry. He was the man, who, in the middle of the French Revolution, liberated the mad from their chains and brought in his own version of moral treatment to their care. Foucault was having none of it, writing "the asylum becomes, in Pinel's hands, an instrument of moral uniformity and of social denunciation" (Foucault 1967: 259). Further, he contended: "The asylum is a religious domain without religion, a domain of pure morality, of ethical uniformity" (Foucault 1967: 257). According to Foucault, everything was organized so that the madman would be aware that he inhabited a world in which he was watched, judged, and condemned. The asylum functioned as a "juridical space," where the doctor judged and sentenced the patient. There emerged, what Foucault termed, "the medical personage." The doctor assumed a dominant role in the asylum, but this was not through his knowledge of medical science, but rather because he was thought to possess a kind of non-specialized wisdom.

Foucault ended his historical narrative with the advent of psychoanalysis, which he found had some potential:

> Freud went back to madness at the level of its *language*, reconstituted one of the essential elements of an experience reduced to silence by positivism… he restored in medical thought, the possibility of a dialogue with unreason. (Foucault 1967: 198)

Ultimately, however, Foucault concluded psychoanalysis had failed to communicate with the mad and failed to understand the language of madness. For him, the voice of the mad could only be heard fleetingly:

> Since the end of the eighteenth century, the life of unreason no longer manifests itself except in the lightning-flash of works such as those of Holderlin, of Nerval, of Nietzsche, or of Artaud – forever irreducible to those alienations that can be cured, resisting by their own strength that gigantic moral imprisonment which we are in the habit of calling… the liberation of the insane by Pinel and Tuke. (Foucault 1967: 278)

James Miller (1993: 103) persuasively contends that *Madness and Civilisation* "is not really about the mental illness at all – it is rather, about the philosophical value accorded to the lives, utterances, and works of artists and thinkers conventionally deemed "mad."" Foucault thought that the rudimentary experience of madness was potentially accessible and that its language was preserved in some of the major works of European artists. Antonin Artaud, the French actor, artist, and playwright, was a particular influence on Foucault. Artaud developed a psychotic illness and spent many years in a psychiatric hospital. For Foucault, he was a "figure of *daimonic* heroism, an artist who embodied a new way of knowing" (Miller 1993: 96). As Ian Hacking (2006: vi) observed, Foucault, at least in earlier versions of his book, saw the mad in romantic terms: "A romantic fantasy lurks here, the purity of the possessed, those who not only speak the truth in paradox, like the fools in Shakespeare, but are also themselves the truth." Sengoopta (1999) points out that Foucault's almost romantic idealization of Renaissance ideas of madness has been questioned seriously by historians, especially his notion that the mad were part of everyday life.

Sengoopta further observes that Foucault's postulation of an undifferentiated Great Confinement of the mad and bad all over Europe during the seventeenth century has also been challenged by historians. Roy Porter (1990) demonstrated how inapplicable the notion of the Great Confinement was to Britain. During this period, there were only a small number of institutions, and the mad who were incarcerated were not mixed with other deviant groups. Historians have also shown that, contrary to Foucault's assertion, the "Ship of Fools" never existed (Middlefort 1980). Sengoopta asks if moral treatment was bad, as Foucault claims, what were psychiatrists to do?

In many ways, the historians' objections to *Madness and Civilisation* are beside the point for the purpose of this chapter. Rather we are interested in how it was perceived by and its influence on the antipsychiatrists. The central tenets of Foucault's book chimed with many of them. The notion that psychiatry, borne of the Enlightenment, was not a benign, humane enterprise, but one that oppressed the insane, echoed their own views. Foucault's attempt to locate the voice of the mad was also a concern of the antipsychiatrists, notably Laing, who held that technical psychiatric terminology served to distance the clinician from the patient and to prevent them from hearing what their patients were actually saying. Like Foucault, Laing and Cooper tended to have a romanticized view of madness. Like him, they were intrigued with Artaud. Early on, Laing had read and been greatly inspired by Artaud's essay on Van Gogh, in which he argued that society could not cope with its visionaries and creative geniuses and, instead, locked them up in asylums. Cooper (1967: 33) claimed that: "Artaud saw too much and spoke too much of the truth. He had to be cured." On a more general level, Sedgwick (1982) contended that Foucault's book accorded well with the late 1960s opposition to authority, medical or otherwise, and the celebration of craziness.

Franco Basaglia (1924–1980)

Franco Basaglia was the charismatic leader of a radical movement promoting "alternative psychiatry" in Italy in the 1960s and 1970s (Donnelly 1999; Foot 2015). He led *Psichiatria Democratica* or the Democratic Psychiatric Movement, which had a wider social base and wider influence than the antipsychiatric currents elsewhere in the world (Crossley 2006). He had contact with both, R.D. Laing and David Cooper, and was also influenced by Foucault's *Madness and Civilisation* and Goffman's *Asylums*, as well as the work of Maxwell Jones on therapeutic communities.

Basaglia became the head of a provincial public mental hospital at Gorizia in northeast Italy near the Yugoslav border. He initially tried to humanize the institution and create a climate where the patient was treated as an individual, rather than as an object of the medical diagnostic gaze. Basaglia and his colleagues drew on phenomenology and existential psychiatry to try and achieve this. He argued that, in approaching the patient, the clinician should place the diagnosis "in brackets" to avoid any "pre-formed value judgment." He stressed the importance of the patient telling their life story and uncovering how their past experiences, especially of being institutionalized, shaped their current feelings and behavior.

Borrowing from Goffman, he argued that much of what seemed to be the symptoms of mental illness was actually the result of institutionalization. To democratize the hospital, Basaglia and his colleagues introduced a type of therapeutic community. Daily assemblies for patients and staff were introduced, and the general aim was to destabilize the institutional regime. Basaglia believed that the creation of the therapeutic community was only the first stage in a process of challenging the current psychiatric system. It was certainly an advance over the custodial hospital, he conceded, but it was still part of the institution whose "contradictions" would have to be confronted eventually if the patients were ever to recover their "subjecthood" and capacity to act with autonomy. He and his colleagues wanted to provoke the patients into what they called "institutional rage" against the mental hospital. The ultimate goal was to move to the next stage: Rather than making the asylum more therapeutic, it should be closed all together. Basaglia held that psychiatrists should engage in political activism to change society's relation to the mentally disordered.

Basaglia (1968) gave an account of the Gorizia experiment in his book, *L'istituzione negata (The Institute Negated)*. According to Michael Donnelly (1999: 271), the book:

> ... came to serve as a manifesto for the growing political ambitions of alternative psychiatrists. It also struck a wide resonance in Italian society at large, quickly becoming a key text in the radical students' and workers' mobilisations that swept Italy during 1968–69.

A national organization was formed, *Psichiatria Democratica*. According to Basaglia's biographer, John Foot (2015: 390): "It was an extraordinary time, of revolutionary language accompanied by real reform... Hierarchies were broken down and subverted. Asylums became centers of change and hope." In an essay entitled "Breaking the circle of control," Basaglia (1981: 184) gave a good summary of his key ideas. Referring to the dismantling of the asylum at Gorizia, he writes:

> Many of these changes were parallel to those wrought by the "therapeutic community" movement in England and elsewhere; but the underlying aim was far more radical. The humanization of asylum life was not seen as an end in itself, but only the first step; the ultimate goal was the abolition of the asylum itself.

Basaglia felt that the very existence of the asylum embodied a contradiction inherent in the nature of psychiatry. It was trying to reconcile two opposing roles: therapy and custody. Psychiatry was compromised by its role in controlling deviance and maintaining public order. But this dichotomous approach was also fundamental to psychiatry's place as a social institution. He averred:

> It was necessary to look beyond the asylum, at the role psychiatry had in society at large: for psychiatric diagnoses were rooted in the prevailing moral order, which defined normality and abnormality in its own rigid terms; and it was the class system itself which gave rise to the fact that the "lower orders" made up the bulk of the psychiatrist's cases... We had to go beyond the world of the mental hospital to confront the madness of "normal" life... their common origin in class divisions and the unequal distribution of power. (Basaglia 1981: 85)

Basaglia (1981: 192) concluded "When the mentally ill are no longer segregated... we are forced to recognise their peculiarities and at the same time to discover our own: for 'normality' can be just as much of a distortion as madness."

Foot considers that Basaglia's political analysis of the origins of mental disorder could be simplistic, for example, Basaglia's claim that the mad were the poor, the victims of class and capitalism. Was this actually true? Although mental illness afflicts the lower socioeconomic groups disproportionately, all social classes are affected. Basaglia, like David Cooper, was guilty, at times, of applying unsubtle and unnuanced political theory to the complexities of mental illness.

Basaglia began a second pilot experiment in 1971, when he was invited by the provincial administration of Trieste to reorganize the local mental health services, then centralized in a 1200 bed hospital. Physical constraints were abolished, as was ECT. Medications were used but, according to Basaglia (1981: 188), "solely in order to facilitate the development of relationships and never as a treatment in their own right." Once again, Basaglia's aim was to move care from the hospital to the community. He stated that his community teams were not like traditional CMTs because they still retained the sanction of confinement, whereas in Trieste, they had abolished the asylum and thus the threat of incarceration. He also set up a range of cultural projects designed to mix ex-patients with community residents and both practically and symbolically break down barriers between the institution and extramural society. Basaglia left Trieste in 1979 to take over psychiatric services in Rome, and the surrounding Lazio region, but died suddenly the following year.

Donnelly judges that Basaglia's legacy is ambiguous. He inspired the sweeping mental health reforms enacted in 1978, the so-called Basaglia Law, but its implementation was patchy. Basaglia's model was not easily sustainable or repeatable. Foot points out that the Italian experience was much more than just the life story of Basaglia, remarkable as his contribution was: It was a mass movement. He concedes that the movement was often marked by sectarian disputes, wild, exaggerated claims, and the use of inflammatory language. He also acknowledges that patients and relatives often paid a heavy emotional and human price for the move from the institution to the community. Despite this, Foot (2015: 392) concludes that the Italian experience was unique because the asylums were closed by the people working inside them:

> This was a collective "no". And this "no" changed the world. It was unacceptable to treat human beings in the way they were being treated – without rights, without autonomy, without the possibility of using knives and forks, without their own hair, without any control over their own treatment, without freedom.

Concluding Remarks

Rather than seeing the antipsychiatry movement as representing a radical rupture with the past, several commentators have pointed to the continuities with previous developments in psychiatry. Catherine Fussinger (2011) compared the approach of antipsychiatrists to the postwar psychiatric experiments with therapeutic communities and found several similarities. Mathew Thomson (1998) maintained that the roots of British antipsychiatry lay in the aftermath of the Second World War when there was a reorientation of the psychological sciences and their position within society. Beveridge (2011) detailed the early influences on Laing, which included his clinical years in Glasgow. Laing was to describe himself as a "conservative revolutionary," by which he meant he was carrying on the tradition of the older Glasgow psychiatrists, such as Angus MacNiven, the Superintendent of Gartnavel Royal Hospital, who were skeptical of the new physical treatments such as psychosurgery and medication, and who saw their role as protecting their patients from harm.

Although one can trace continuities in their approach to mental illness between the antipsychiatry movement and earlier developments in psychiatry, as a form of *protest* the antipsychiatry movement differed from previous protests in that it was primarily led by psychiatrists, rather than those from outside the profession. It was located in a specific time – the 1960s and early 1970s – and formed a significant part of the zeitgeist, embracing both the counterculture and the politics of the New Left. According to Crossley (1998), what made antipsychiatry unique was its radicalism. Unlike previous critics of psychiatry, antipsychiatrists not only did question psychiatric treatment and practice, but they also questioned the very basis of psychiatry itself: its purpose, its concept of mental illness, and the very distinction between sanity and insanity. In addition, antipsychiatrists extended their gaze beyond the psychiatric hospital to the wider society, which they castigated for the huge psychological and emotional damage it wreaked on its citizens. Burns (2020) asked what made antipsychiatry so celebrated in the 1960s? He suggested:

> ... the answer may lie in the confidence of the baby boomer generation who were emerging with enormous optimism and energy from post-war austerity. The civil rights movement and anti-Vietnam protests in the USA mirrored the student protests of the late 1960s Europe. A generation was maturing that had seen their parents' authority discredited and felt emboldened to reject it. Psychiatry straddles science, moral values, and social aspirations... it is not surprising that the rebellion first erupted into psychiatry.

Crossley (2006) intriguingly speculates that if mainstream establishment psychiatry in Britain had been more open to the ideas of Laing and his colleagues, there may never have been an antipsychiatry movement. By blocking dissent, the establishment fueled protest.

Duncan Double (2006) compared the views of the various antipsychiatrists and outlined two groups. The first group, which held that mental illness did not exist, included Szasz. We could add Goffman to this group. The second group held that mental illness was a reaction to life and included Laing, Basaglia, and Foucault. Double subdivided the second group: Laing related madness to interpersonal behavior, particularly within the family, while Basaglia and Foucault related it to wider societal factors. Double (2002) observed: "The essence of anti-psychiatry derives from the sense in which psychiatry itself is regarded as part of the problem."

Several commentators have noted that the leading figures were all men. Chapman (2016) writes: "Sigal's *Zone of the Interior* provides a sharp, satirical critique of antipsychiatry's failure to address matters of gender, and, specifically, how the appeal of madness – its 'sexiness' – gets mixed up with the physical appeal of distressed women to male anti-psychiatrists." Jones (1998) pointed to antipsychiatry's unproblematized attitude to gender roles and observed that, like much of the 1960s counterculture, the movement was "very largely boy's stuff, a guy thing." But, on the other hand, it is important to recognize that some feminists, such as Juliet Mitchell (1974), praised Laing's concentration on the plight of mentally ill women and their role in the family.

The antipsychiatric movement as a cultural force waned dramatically in the early 1970s. Wall (2018) links its demise to the wider picture that saw the diminishing influence of the counterculture generally. According to Jones (1998): "The very term, 'anti-psychiatry' though coined by David Cooper, was essentially a media term... It served as a flag of convenience for the media." Jones observed, just as the antipsychiatric movement was catapulted to worldwide attention by the media, once the media lost interest, its decline was inevitable:

> The mediatisation of its arch proponents signalled both the *splendours* and the *miseres* of anti-psychiatry as a cultural movement... The very newsworthiness and chic-ness of the 'anti' in the late 1960s and early 1970s offered a window of opportunity for change which the predecessors... had never enjoyed... Yet the accompanying risk was that once anti-psychiatry was no longer current and newsworthy, then the views of the anti-psychiatrists would no longer carry cultural and political clout...

Judgments

Wall (2018: 160) judged that, "The anti-psychiatrists' position was wildly romantic and seemed to flatly ignore the difficult realities of mental illness." He found it shocking because the practitioners surely knew that schizophrenia often brought great pain to the sufferers and their families. John Wing criticized the anti-psychiatrists' view that the family was implicated in the causation of schizophrenia because, he contended, this model labeled, scapegoated, and stigmatized every bit as much, if not more so, than the medical model it was seeking to replace (Quoted in Clare 1980: 194–195). Staub (2011) observed that the antipsychiatrists could not resolve the relationship between the individual and society, a problem, he conceded, that had perplexed generations of psychiatrists for decades before.

However, Foot (2015: 40–41) claimed: "Since the end of the 1970s, with some notable exceptions, the work of anti-psychiatrists... has been systematically misrepresented, ridiculed and/or excised from history. Anti-psychiatry has become a term of abuse." Staub (2011: 6) agreed: "Derogatory declarations about anti-psychiatrists have gone unchallenged for decades." Jones (1998) remarked: "To reduce anti-psychiatry to a set of *depasse* ideas and concepts misses the dimension of the movement which was a carnivalesque celebration of the symbolic inversion of medical authority and established legitimacy, and to underestimate the force and freshness of the 'anti-psychiatric moment' in western culture." In a generally critical essay, Digby Tantam (1999), nevertheless, praised the antipsychiatrists' challenge to psychiatric diagnosis, which he felt led to the subsequent refinement of psychiatric diagnosis and made psychiatrists more sensitive to labeling their patients. Tom Burns (2013) judged that the positive legacy of antipsychiatry lay in the development of rehabilitation, in the closure of the asylums, and in their plea that the psychiatrist needed to attend to the patient's view of the world. Nasser (1995) suggested that the antipsychiatry movement led to the later development of advocacy groups, while Crossley (1998) felt that the work of Laing and his colleagues ignited the user movement. Although differing on certain points of theoretical approach and not attracting the same all-consuming media attention as the antipsychiatrists in their heyday, the Critical Psychiatry movement of our time has continued the tradition of questioning the assumptions of mainstream psychiatry.

References

Barnes M, Berke J (1971) Mary Barnes: two accounts of a journey through madness. MacGibbon and Kee, London
Basaglia F (1981) Breaking the circuit of control. In: Ingleby D (ed) Critical psychiatry. The politics of mental health. Penguin, Harmondsworth, pp 184–192
Basglia F (1968) L'instituzione negate. Einauda, Torino
Beveridge A (1998) R.D. Laing revisited. Psychol Bull 22:452–456
Beveridge A (2011) Portrait of the artist as a young man. The early writings and work of RD Laing, 1927–1960. Oxford University Press, Oxford
Beveridge C, Turnbull R (1989) The eclipse of Scottish culture. Inferiorism and the intellectuals. Polygon, Edinburgh
Burns T (1999) Erving Goffman. Routledge, Oxon
Burns T (2013) Our necessary shadow. The nature and meaning of psychiatry. Allen Lane, London
Burns T (2020) A history of antipsychiatry in four books. Lancet Psych 7(4):312–314
Burston D (1996) The wing of madness. Harvard University Press, Cambridge, MA
Chapman A (2016) Re-coopering anti-psychiatry: David Cooper, revolutionary critic of psychiatry. Crit Radic Soc Work 4(3):421–432. https://doi.org/10.1332/204986016X1473688814636
Clare A (1980) Psychiatry in dissent. Controversial issues in thought and practice, 2nd edn. Tavistock, London
Clare A (1990) Ronald David Laing 1927–1989: an appreciation. Psychol Bull 14:87–88
Cooper D (1967) Psychiatry and anti-psychiatry. Tavistock, London
Cooper D (1971) The death of the family. Allen Lane, London
Cooper D (1980) The language of madness. Penguin, Harmondsworth
Cooper D (2001) Introduction. In: Foucault M (ed) Madness and civilization. A history of insanity in the age of reason. Routledge Classics, London, pp vii–viii
Cooper D (ed) (2015) The dialectics of liberation. Verso, London (First published, 1968)
Crossley N (1998) R.D. Laing and the British anti-psychiatry movement: a socio-historical analysis. Soc Sci Med 47:877–889
Crossley N (2006) Contesting psychiatry. Social movement in mental health. Routledge, London
Donnelly M (1999) Franco Basaglia (1924–1980). In: Freeman H (ed) A century of psychiatry. Mosby, London, pp 270–272
Double D (2002) The history of anti-psychiatry: an essay review. Hist Psychol 13:231–236
Double D (ed) (2006) Critical psychiatry. The limits of madness. Palgrave, London
Foot J (2015) The man who closed the asylums. Franco Basaglia and the revolution in mental health care. Verso, London

Foucault M (1967) Madness and civilization. A history of insanity in the age of reason. (trans: Howard R). Tavistock Publications, London

Foucault M (2006) History of madness. (trans: Murphy J and Khalfa J). Routledge, London

Frances A (2019) The myth and reality of mental illness. In: Haldipur CV, Knoll JL IV, v d Luft E (eds) Thomas Szasz. An appraisal of his legacy. Oxford University Press, Oxford, pp 169–176

Fuller Torrey E (2019) Schizophrenia: sacred symbol or Achilles heel? In: Haldipur CV, Knoll JL IV, v d Luft E (eds) Thomas Szasz. An appraisal of his legacy. Oxford University Press, Oxford, pp 98–103

Fussinger C (2011) "Therapeutic community", psychiatry's reformers and anti-psychiatrists: reconsidering changes in the field of psychiatry after Word War 11. Hist Psychol 22:146–163

Gambino M (2013) Erving Goffman's asylums and institutional culture in the mid-twentieth century United States. Harv Rev Psychiatry 21(1):52–57

Goffman E (1968) Asylums. Essays on the social situation of mental patients and other inmates. Pelican Books, Harmondsworth (originally published, 1961)

Goffman E (1971) The insanity of place. In: Goffman E (ed) Relations in public: microstudies of the public order. Basic Books, New York, pp 335–390

Hacking I (2006) Preface. In: Foucault M (ed) History of madness. Routledge, London, pp ix–xii

Ingleby D (1998) The view from the North Sea. In: Gijswijt-Hofstra M, Porter R (eds) Cultures of psychiatry and mental health care in postwar Britain and the Netherlands. Rodopi, Amsterdam, pp 295–314

Jones C (1998) Raising the anti: Jan Foudraine, Ronald Laing and anti-psychiatry. In: Gijswijt-Hofstra M, Porter R (eds) Cultures of psychiatry and mental health care in postwar Britain and the Netherlands. Rodopi, Amsterdam, pp 283–294

Laing R D (1960) The divided self. A study of sanity and madness. Tavistock Publications, London

Laing RD (1961) The self and others. Tavistock, London

Laing RD (1967) The politics of experience and the bird of paradise. Penguin, Harmondsworth

Laing A (1994) R. D. Laing. A biography. Peter Owen, London

Laing RD (1998) Wisdom, madness and folly. The making of a psychiatrist, 1927–1957. Canongate, Edinburgh (originally published, 1985)

Laing RD, Cooper D (1964) Reason and violence. A decade of Sartre's philosophy 1950–1960. Tavistock, London

Laing RD, Esterson A (1964) Sanity, madness and the family. Penguin Books, London

MacSuibhne S (2011) Erving Goffman's *Asylums* 50 years on. B J Psych 198:1–2

Middlefort HE (1980) Madness and civilisation in early modern Europe. In: Malamont B (ed) After the reformation: essays in Honour of J.H. Hexter. University of Pennsylvania Press, Philadelphia, pp 247–265

Miller J (1993) The passion of Michel Foucault. Harper Collins, London

Mitchell J (1974) Psychoanalysis and feminism. Penguin, Harmondsworth

Nasser M (1995) The rise and fall of anti-psychiatry. Psychol Bull 19:743–746

Porter R (1990) Foucault's great confinement. Hist Hum Sci 3:47–54

Radden J (2019) Rights, responsibilities, and mental illness: a chronology of the Szasz decades. In: Haldipur CV, Knoll JL IV, v d Luft E (eds) Thomas Szasz. An appraisal of his legacy. Oxford University Press, Oxford, pp 237–255

Sedgwick P (1982) Psycho politics. Pluto Press, London

Sengoopta C (1999) Michel Foucault (1926-1984). In: Freeman H (ed) A century of psychiatry. Mosby, London, pp 245–248

Shalin DN (2013) Goffman on mental illness: *Asylums* and "the insanity of place" revisited. Symbol Inter 37(1):122–144

Showalter E (1987) The female malady. Women, madness and English culture, 1830–1980. Virago Press, London

Siegler M, Osmond H (1971) Goffman's model of illness. B J Psych 119:419–424

Sigal C (2005) Zone of the interior. Pomona Books, Hebden Bridge (first published in 1976)

Staub ME (2011) Madness is civilization. When the diagnosis was social, 1948–1980. Chicago University Press, Chicago
Szasz T (1971) The manufacture of madness. A comparative study of the inquisition and the mental health movement. Routledge and Kegan Paul, London
Szasz T (2002) The meaning of mind: language, Morality and Neuroscience. Syracuse University Press, Syracuse
Szasz T (2003) The myth of mental illness. Foundations of a theory of personal conduct. Harper Perennial, New York (originally published in 1961)
Szasz T (2004) An autobiographical sketch. In: Schaler JA (ed) Szasz under fire. The psychiatric abolitionist faces his critics. Open Court, Chicago and La Salle, pp 1–28
Tantam D (1999) RD Laing and anti-psychiatry. In: Freeman H (ed) A century of psychiatry. Mosby, London, pp 202–208
Thomson M (1998) Before anti-psychiatry: "mental health" in wartime Britain. In: Gijswijt-Hofstra M, Porter R (eds) Cultures of psychiatry and mental health care in postwar Britain and the Netherlands. Rodopi, Amsterdam, pp 43–60
Wall O (2018) The British anti-psychiatrists. From institutional psychiatry to the counter culture. Oxon, Routledge, pp 1960–1971
Weinstein RM (1982) Goffman's *Asylums* and the social situation of mental patients. Ortho Psych 11(4):267–274
Wilson S (2019) Study on the Szaszophone: theme and variations. In: Haldipur CV, Knoll JL IV, v d Luft E (eds) Thomas Szasz. An appraisal of his legacy. Oxford University Press, Oxford, pp 5–11

Part IX

History of Economics

History of Thought of Economics as a Guide for the Future

56

Dieter Bögenhold

Contents

Introduction: The Economy and Society Viewed Academically	1454
Divergencies and Pluralism in Economics	1457
The Decoupling of Economics and History	1461
Problems and Failures in Economics	1466
Conclusion	1467
References	1468

Abstract

Causes for the distinct and growing separation of the academic domains of economics and sociology are ongoing processes of specialization and fragmentation. Thanks to the multiplication of publications and knowledge in economics, degrees of specialization have emerged. One of the great paradoxes in economics is the existence of mainstream economics, which is taught to undergraduate students and dominates textbooks, side by side with fresh and provocative new contributions, which enter the arena and become established by public and academic debate, being awarded prestigious prizes in the process. The chapter tries to draw a few lines of development in economics oscillating between continuity and change. Especially, the interplay between different domains in social sciences is discussed as fields of tension and cooperation between economics, sociology, history, and psychology.

Keywords

History of economic thought · Interdisciplinary studies · Social sciences · Economics · Psychology · History · Sociology · Imperialism of economics · Social scientification

D. Bögenhold (✉)
Department of Sociology, Universitat Klagenfurt, Klagenfurt am Wörthersee, Austria
e-mail: Dieter.Boegenhold@aau.at

© The Author(s), under exclusive licence to Springer Nature Singapore Pte Ltd. 2022
D. McCallum (ed.), *The Palgrave Handbook of the History of Human Sciences*,
https://doi.org/10.1007/978-981-16-7255-2_1

Introduction: The Economy and Society Viewed Academically

The history of thought of economics can and must be seen through different lenses. The recent introductory article is almost guided by a scientific background of a *sociology of economics*. Sociology deals with an analysis of different elements of thought and tries to decode them by acknowledging different zones of time and space which act as institutional determinants of social and economic processes of change. Economy and society take place in different settings and can be treated in a singular when thinking this bonding in an abstract way. When trying to think economy and society more in an applied way with concrete applications in different locations and times it must be thought in a plural as economies and societies. Practically economies differ across the globe and along historical times and theoretical reflections try to take those diversications into account. Therefore discussion in this chapter intentionally tries to integrate views by history, sociology, and psychology into a discussion of the major lines of development of economics. This can be summarized as sociology of economics.

An economy is never isolated in a vacuum but always exists in a field of social coordinates determined by space and time as well as in a related framework of culture and other institutions. This is why one of the major theorists in economics and sociology included "economy" and "society" (Weber [1921] 1972) in the title of his book, already turning it into a kind of manifesto because the economy and society are mentioned in the same breath. The "and" then gives us the idea of a peaceful coexistence between two spheres, which have since evolved to become separate domains of scientific treatment. Following Max Weber ([1921] 1972), Parsons and Smelser (1956) expanded on the topic in the 1950s with their book *Economy and Society* (1956), explaining that the two had become increasingly separated in the early decades of the twentieth century. More specifically, they stated that only a few authors who were competent in sociological theory have a "working knowledge of economics, and conversely ... few economists have much knowledge of sociology" (Parsons and Smelser 1956).

Causes for the distinct and growing separation of the academic domains of economics and sociology are ongoing processes of specialization and fragmentation. Thanks to the multiplication of publications and knowledge in economics, which can be separated into management sciences and pure economics, degrees of specialization have emerged. The two subjects expanded in horizontal and vertical directions, giving rise to a wealth of subdisciplines which were related to their own new universes of discourse with their own research organizations, networks, journals, curricula, career paths, and publication routines. A process of growing academic fragmentation could be observed which permanently generated new islands of academic activity and knowledge while bridges between them were often nonexistent or invisible. Ultimately, academic development in the twentieth century may be classified as a process of ongoing vulcanization within the landscape of economic and social sciences (Wallerstein et al. 1996).

With this ongoing cellular division within the economic and social sciences, an intrinsic autism emerged through which individual units became their own worlds for and in themselves. Max Weber had discussed this very matter more than one hundred years ago in his famous lecture on "Science as an Occupation" ("Wissenschaft als Beruf" ([1919] 1988, 588), in which he said that individual authors only have the feeling of excellence and exclusivity in science when referring to the smallest niches of specialization. In more recent times, too, individuals are increasingly unaware of the interrelationships and resonances between disciplines and they "become slaves of their discipline's approach" (Hunt and Colander 2011, 19).

In sociology, too, by the second half of the twentieth century, some authors were starting to reject historical tendencies toward differentiation. One of the major new positions was modern system theory, where the economy was treated as a subsystem of society (Luhmann 1988). Alternatively, "new economic sociology" was emerging (Granovetter and Swedberg 1992; Smelser and Swedberg 2005), the starting point for fostering new lines of thought and creating several new research societies following the credo of socioeconomics, which was a very common label at the end of the nineteenth century. Those new ambitious strands of work did not only reflect upon the original ideas of socioeconomics but they also tried to acknowledge and integrate emerging social network analysis and historical research as well as contemporary cross-national and institutional perspectives in order to argue against the so-called "imperialism of economics" (Lazear 2000, Becker 2010 for economics, and Granovetter 1993, 2017 for sociology).

Discussing the relationship between society and the economy, thus, inevitably leads to a discussion of sociology and economics. Practicing this principle ultimately clears the way to speaking about academic subjects as unique bodies of knowledge since the academic subjects themselves (economics, sociology, history, psychology, etc.) are very fragmented. What undergraduates learn as textbook knowledge at the beginning of their studies is very much a stereotype of a discipline since many different applications and interpretation exist in reality. For example, sociology is a bit like a patchwork quilt with more than 60 different research committees within the International Sociological Association which do not only have different universes of discourse, newsletters, and career paths but also their very own lifeworlds and social rules. The textbook *The Structure of Sociological Theory* (Turner 1998) has 35 chapters, each dealing with a different theoretical, sociological approach. This underlines the fact that there is not a single, unique sociology as an academic discipline but an immense heterogeneity of different, often competing, paradigms which provide a broad cosmos of different islands of knowledge, so much so that it is often difficult to say what the concrete and pure content of sociology is at all.

The same observation is valid for economics. The question for the concrete domain of economics has a long tradition and the often quoted statement by Jacob Viner "economics is what economists do" (quoted in Barber 1997, 87; Bögenhold 2010) has been quoted very often. Looking at topics of economists activities shows that content of what economists do has been shifted and extended several times. No clear lines of demarcation between economics and surrounding academic subjects exist, although economics has expanded its domains clearly. Especially the expansion by Gerry Becker (1993, 2010) in direction of explanation of family behavior or diverse other fields of human activities (addiction, sports, and restaurant visits) has

Fig. 1 The spectrum of academic domains in the social sciences. (Author's own representation)

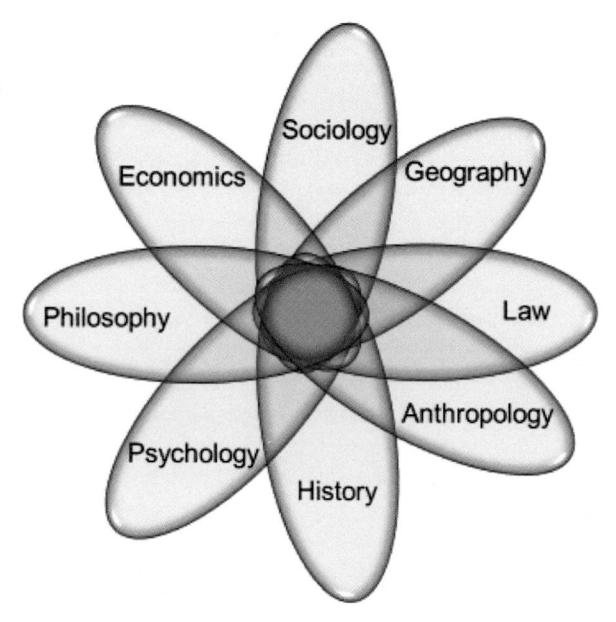

changed and enlarged the terrain of economics (see Herfeld 2012). However, it is difficult to define clearly where one academic domain starts and where its competencies end and where and when another academic domain takes over clear responsibility.

Since no such clear frontiers for the landscape of social sciences exist, the task to define the spheres has remained being difficult and did not reach further than earlier tautological describtions provided by Viner. During the last 150 years academic areas have not only multiplied the number of areas and labels but also internally deepened with the emergence of diverse new curricula, final degrees, associations, and journals. In the same time the number of publications increased exponentially (Fig. 1).

Historically, the emergence of modern economics is closely related to the ascent of neoclassical economics, which evolved on the foundations of marginal utility theory established in the 1880s and 1890s. Economists started to try and establish a new approach to economics which should be *theoretical* and *universal*. "Genuine" or "pure" economics, as Walras ([1874] 1954) described it, was the credo behind a new way of doing economics in imitation of the natural sciences, namely with clear methods and the ambition to arrive at precise laws. The idea was to enhance discussions of economic affairs by making them apply to modern capitalism at a universal level, independently of whether economists were reflecting on the economies of Belarus, Belgium, Bolivia, or Botswana in the eighteenth, nineteenth, twentieth, or twenty-first centuries; in other words, they should apply to economies regardless of their concrete culture and historical setting.

In order to achieve a higher level of abstraction, the involvement of mathematics was considered to be essential. Jevons, one of the originators of marginal utility theory alongside Walras and Menger, wrote in the introduction to his seminal book: "It is clear that Economics, if it is to be a science at all, must be a mathematical science" (Jevons 1871, introduction). In fact, economics has become increasingly associated with "mathiness" (Romer 2015) and several systematic studies describe the historical process of integrating mathematics into economics (McCloskey 1987; Weintraub 2002; Morgan 2012). In an attempt to reduce numerous developments within economics in the twentieth century to one common denominator, the birth and subsequent development of neoclassical economics was the most important factor. Economies are mostly observed and treated on a general and universal level as if they existed in a vacuum which has no institutions and no contextual time-space framework. "Pure economics" was a program for abstractness which was a paradigm for most of the twentieth century. Critically, authors qualified this paradigm as a kind of academic religion (Nelson 2001). In parallel, a variety of new fields of application emerged, including industrial economics, labor economics, agricultural economics, household economics, or the economics of age or culture, underlining the general trend toward specialization and differentiation. However, if we are speaking about *mainstream economics* and if we take a look at the principal topics in introductory courses for students, neoclassical economics is still alive and kicking.

Economics has increased its size in terms of involved people, topics, journals, and societies but splitted also between economics on one side and business administration and management on the other side. It is sometimes difficult to say where the border between those two camps runs concretely.

As already mentioned, the semantic bracket of the economy and society has a long tradition, but the earlier combination of economics, philosophy, sociology, history, and behaviorism has broken up. Philosophy disappeared as soon as economics claimed to be a positive science which is to be proved by data. Sociology was increasingly lost by economics, which led to society disappearing as well. Social institutions and their inherent dynamics, culture, and questions of social stratification as well as the behavior of different social classes were no longer on the agenda. The trappings of history and society were abandoned for abstract theorizing in economics, which increasingly became a kind of monodiscipline without overlaps.

Divergencies and Pluralism in Economics

Ideas of homo oeconomicus are already discussed since early decades in twentieth century, partly under the label of heterodox economics (Dorfman 1946–1959). In order to achieve high applicability, neoclassical mainstream economics operated with the theorem of *homo oeconomicus*. Homo oeconomicus is a type of actor who is rational and profit-seeking without showing changing preferences and emotional considerations. All human attrbutes are reduced to an economic calculation of maximization. A further assumption of neoclassical thought is that all actors share the same information: Knowledge packages in economy are equally distributed to all members of society. Finally, economies are perceived as being in a state of equilibrium (Hodgson 1994).

Of course, each of these three assumptions leads to different forms of discussion and related criticisms. Modern economies cannot be conceptualized adequately if the unequal distribution of knowledge is not taken into consideration, a fact that was centrally addressed by Hayek (1945) in the mid twentieth century. In fact, information asymmetries are driving engines for sources of innovation, new markets, and growth.

One of the great paradoxes in economics is the existence of mainstream economics, which is taught to undergraduate students and dominates textbooks, side by side with fresh and provocative new contributions, which enter the arena and become established by public and academic debate, being awarded prestigious prizes in the process. Psychologist Herbert Simon (1982), for example, received a Nobel Memorial Prize in economic sciences for his critique of *homo oeconomicus* with the introduction of his famous concept of *bounded rationality* (Bögenhold 2016), which had a major contribution to make to decision theory. His principle research question is: "How do human beings reason when the conditions for rationality postulated by neoclassical economics are not met?" (Simon 1989, 377). The concept of bounded rationality takes into account that, first, agents often act in ways which could be characterized as nonrational behavior driven by emotions; second, the use of bounded rationality emphasizes the fact that access to information is limited if the same bits of information are not shared which would be necessary to decide between the alternatives; and third, even in a situation where information is shared equally, human beings are characterized by cognitively diverse and also limited skills. Simon reflects upon traditional economic theory, which deals with conceptions of the "economic man" who is "economically engaged" and in the same sense also "rational."

Hodgson (2012, 46) has identified several Nobel laureates in economics since the 1970s as being very critical of the concept of the rational egoistic man, among other things. The list includes Herbert Simon, Douglass North, Amartya Sen, Daniel Kahneman, and George Akerlof, but can be extended to Richard Thaler (1994, 2016) as well. Each of these economists clearly rejected the idea of rational behavior as followed for decades before. Vane and Mulhearn (2005) could even show that Nobel Prizes in economics were increasingly awarded for contributions with a strong sociological or psychological emphasis, a tendency which has been labeled the growing "social scientification" of economics (Bögenhold 2018).

According to Kahneman, social action must be interpreted as a kind of choice between alternatives and, therefore, social action is difficult to predict since human beings often act intuitively and are driven by emotions: "The central characteristic of agents is not that they reason poorly but that they act intuitively. And the behavior of these agents is not guided by what they are able to compute but what they happen to see at a given moment" (Kahneman 2003, 1469). Daniel Kahneman (2012) also distinguished between experience and memory, while the most recent Nobel laureate Richard Thaler (1994, 2016) continued working here more explicitly and broadened behavioral economics. In some way one may include here also the recent Nobel prize recipients Milgrom and Wilson (The Royal Swedish Academy of Sciences 2020) who investigated in new auction formats by employing game theory and microeconomics to explain collective behavior on auction which is also based upon thought suggested by Richard Thaler.

Douglas North (1990, 1991), who was awarded the Nobel Memorial Prize for his achievements in institutional economics, wrote a chapter on behavioral assumptions in a theory of institutions and argued: "Although I know of very few economists who really believe that the behavioral assumptions of economics accurately reflect human behavior, they do (mostly) believe that such assumptions are useful for building models of market behavior in economics and, though less useful, are still the best game in town for studying politics and the other social sciences. I believe that these traditional behavioral assumptions have prevented economists from coming to the grips with some very fundamental issues and that a modification of these assumptions is essential to further progress in the social sciences. The motivation of these actors is more complicated (and their preferences less stable) than assumed in received theory. More controversial (and less understood) among the behavioral assumptions, usually, is the implicit one that the actors possess cognitive systems that provide true models of the worlds about which they make choices . . . " (North 1990, 17).

It was increasingly argued that economics had to show an interest in openness toward behavioral and cognitive approaches (Akerlof 2007; Akerlof and Kranton 2000; Akerlof and Shiller 2009) in order to map economic phenomena and their developments more realistically, which was a kind of credo of "new economics." Akerlof (2007) not only complained in his *Presidental Address* at the conference of the *American Economic Association* that the issue of human motivation was too much absent in theory in (macro) economics but also that topics as social norms and values in general as well as many questions and matters in sociology were not taken into account. Neither the distribution of information nor the formation of preferences was really considered appropriately (Kahneman and Tversky 2013); neither were institutional constraints for decision-making acknowledged (Ariely 2010). Behavioral economics can be done at different levels, either at the level of individual actors or groups and their ways of decision-making, or it may be employed at a macro-cultural level which are the animal spirits as firstly used by J.M. Keynes ([1936] 1964, 161) and later broadly discussed by Akerlof and Shiller (2009). Animal spirits mean a kind of spontaneous optimism or persimism toward the evaluation of future developments and the question of doing investments or purchases. It is clear when referring to those terms economics takes perspectives by different academic domains and their tool boxes.

All of these shifts toward a stronger acknowledgment of motivation in macroeconomics is not only connected with an emphasis on behavioral aspects but also with the need to consider sociological competencies: "Sociology has a further concept that gives an easy and natural way to add those norms to the utility function. Sociologists say that people have an ideal for how they should or should not behave. Furthermore, that ideal is often conceptualized in terms of the behavior of someone they know, or some exemplar whom they do not know" (Akerlof 2007, 10). It is especially the social context seen by Akerlof (2007) which gives a frame to social action and its learning processes. Last but not least, religion is also considered to be one of the tools to socialize individual's economic behavior: "Sociology is dense in examples of people's views as to how they and others should behave, their joy when they live up to those standards, and their discomfort and reactions when they fail to do so" (Akerlof 2007, 10). Acknowledging sociology helps to understand consumption processes, including their inherent preferences for choices which are sometimes hidden but almost always in contrast to those abstract utility functions as used in economics. In other words, "sociology gives motivations for consumption that are very different from the reasons for it in the life-cycle model" (Akerlof 2007, 15). In wording of Bowles (1998) consumption practices are based upon endogenous preferences which are socially learned and not part of the human DNA.

It is remarkable that Nobel Prize winner in economics Akerlof not only refers to the relevance of social norms but also to sociology as a pertinent academic discipline in general to deal adequately with social behavior. Here, in fact, we see the opposite of Nobel laureate Becker's imperialism, which tries to operationalize behavior and every form of social activity in categories of utility maximizing. Akerlof (2007) discusses sociologists like Goffman and Bourdieu as if no border existed between economics and sociology. He clearly acknowledges what sociologist DiMaggio (1994) had expressed earlier. Namely "the starting point of any discussion of life styles and consumption patterns must be the work of Thorstein Veblen and Pierre Bourdieu" (DiMaggio 1994, 458).

Looking at different topics reveals considerable thematic analogies in neighboring disciplines, which should be analyzed and explored in order to see how the contours and borders of the academic landscapes are changing and in which directions developments are evolving (Rosenberg 2018). Ultimately, sociology is concerned with the question as to what people do and why they do it in the way they do in a special social setting. As already mentioned, economics has separated into management and business administration on the one hand and pure economics including micro- and macroeconomics, economic policy, and a growing apparatus of econometrics on the other. It is so interesting to see the overlaps between the individual subjects. One historian, Robert William Fogel (1993), was awarded a Nobel Prize in economics for his work on economic and social history. Three psychologists, too, received the Nobel Prize in economics (Simon in the year 1978; Kahneman in the year 2002; Thaler in the year 2017) for establishing and then expanding on behavioral economics and one sociologist, at least, was shortlisted as a candidate (Mark Granovetter in 2015). This illustrates that Viner's old slogan (see above) that economics is what economists do has really outlived its usefulness since the boundaries are shifting. What is more, definitions do not only arise from the inside but are also proposed from the outside, namely by prestigious committees and their authority. This further underlines the blurred boundaries between these disciplines and others which are not listed in the quartet in Fig. 2. These overlaps are also in permanent state of transition and reconfiguration.

Scientific progress is often contingent and never rational, in the sense that it follows arithmetic rules of combinations. The "market" for ideas is neither efficient

Fig. 2 Fields of tension and cooperation between economics, sociology, history, and psychology. (Author's own representation)

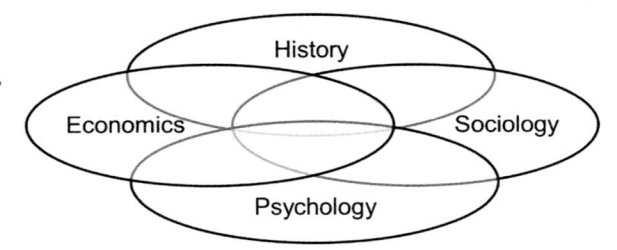

nor perfect. Academic progress also proceeds by fits and starts, related to a series of mistakes by which intellectual resources are wasted, and, as a consequence, there are indeed intellectual gems lying around unexploited, waiting for someone to pick up (Collins 2002). Especially as economics has attempted and is still maintaining interesting shifts in the direction of psychology or sociology, this kind of academic poaching should be noticed (Granovetter 2017).

In sociology, especially, the works of Jon Elster (1983, 1999) and Randall Collins (1992, 2005) focus on emotions and issues of rationality. Both took up elements of discussion which can already be found in the classics by Georg Simmel, Max Weber, Norbert Elias, or Ervin Goffman. In the *American Sociological Association* and in the *International Sociological Association*, research committees on emotions were established as recently as the 1980s. All of these diverse ongoing activities within various academic branches can be bundled under the flag of "bounded rationality" as well. However, Swedberg (2015, 2016) compared sociology critically with the cognitive sciences, explaining that "[s]ociologists have failed to address a number of topics that are important to theorizing, and that cognitive scientists have already been working on for several decades. .... Cognitive scientists have also developed some important insights in other areas where sociologists are active but have not been particularly innovative. Studies of meaning, memory and emotions are some examples of this" (Swedberg 2016, 18–19).

The Decoupling of Economics and History

In order to enhance the universal applicability and theoretical depth of economic theories, economics had to strengthen its abstractness and reduce its concrete links to specific times and spaces. The solution was to discuss the economy *in abstracto* rather than *in concreto* and to construct clean and proper models of the functioning of a capitalist economy. The twentieth century made substantial progress in elaborating and consolidating a new foundation for economics, mostly as a neoclassical conception, which is still, for the most part, textbook knowledge. Very much of this new way and form of economics can be studied when even looking at the cosmos of Chicago economics which included in itself so much of twentieth-century state of economics (see Emmett 2010). In parallel, different applications multiplied, from household economics to agricultural or industrial economics, welfare economics, public choice theory, and sport or transport economics, which broadened the terrain of economics and provided new specializations. So, the conclusion to draw may be that we have entered "*The Age of the Applied Economist*" (Backhouse and Cherrier 2017) at the same time as elegant theoretical reasoning has increasingly lost its former historical nexus and foundation. In other words, we have experienced a tricky situation in economics which shows, on the one hand, increased specialization, abstractness and mathiness, and ongoing tendencies toward processes of an increased pluralization of economics (Hodgson et al. 1992; Schabas 1992; Davis 2016) on the other.

The more complex economics proved to be, the smaller the real terrain of neoclassical theory became, although the general image of economics, especially when looking at it from the outside, still retains the dominance of neoclassical orthodoxy. However, mainstream economics is also, in itself, fragmented and always changing (Cedrini and Fontana 2017). The twenty-first century looks back on this scientific period of development, differentiation, and consolidation as a feature of the twentieth century. The link which was formerly maintained by economics and other domains in the social sciences, especially to philosophy, was exchanged against new links to econometrics and mathematics (Chichilnisky 2017). In other words, economics has started to forget history (Hodgson 2001). The history of economic theory has been abolished or relegated to other disciplines (philosophy of science or theory of knowledge). One could argue with Hodgson that "prowess with formal technique has replaced the broader intuitive, methodological and historical intellectual grounding required of the great economist. Such qualities were emphasized and personified by both Alfred Marshall and John Maynard Keynes. Today, economists are no longer systematically educated in economic history, the philosophy of science or the history of their own discipline" (Hodgson 2007, 19).

Historical analyses and observations fall into two different categories at least. One topical area is concerned with the history of economic thought as an academic discussion of the scientific evolution of and change in economics. The other topical area is concerned with (material) social and economic history, which means trying to analyze patterns of real historical change in the economy and society. The *Cliometric Society* especially takes care of these circumstances and necessities to use history as a crucible to examine economic theory in order to deepen our knowledge of how, why, and when economic change occurs (Haupert 2016, 4).

Goldin (1995) discussed the relevance of both Nobel recipients Fogel and North and underlined the importance of their work for economics: "What is it then that makes economic historians, such as Robert Fogel and Douglass North, unique among economists? It is not that they study the past, use historical data, exploit the past for natural experiments, use a particular methodology, are open to the ideas from other disciplines, or find lessons in the past for developing countries. Rather, it is all of these plus one indispensable ingredient. Economic historians study economies over the long term. The evolution of economies is their particular niche" (Goldin 1995, 207) (see also Hodgson 2017 for a discussion of North in particular).

In some way those historians have a very clear proximity to representatives of old or new institutional theory, which, ever since the first battle of methods between

Gustav Schmoller and Carl Menger (Schumpeter 1926; Swedberg 1990; Bögenhold, 2008; Louzek 2011), has been part of old and new institutional economics. What Boulding (1971) reported already more than 50 years ago, namely that modern graduates have rarely read a piece which is older than 10 years, is still true (Roncaglia 2014). The half-life knowledge of publications is getting shorter and shorter. The number of publications on the history of economics has also been declining compared to the total number of publications in economics, while a large body of these publications are concentrate in just five highly specialized journals (Marcuzzo and Zacchia 2016, 36).

Joseph A. Schumpeter dealt very systematically with methodological questions of the composition of different perspectives and academic tools in his "*History of Economic Analysis*" (Schumpeter 1954). Schumpeter argues not only in favor of economic history as rendering a service to economic theory, but also in favor of "a sort of generalized or typified or stylized economic history" (Schumpeter 1954, 20), which includes institutions like private property, free contracting, or government regulation. According to Schumpeter, there are multiple reasons to study history, ranging from pedagogical advantages and new ideas to new insights into the ways of the human mind. Regarding the *pedagogical* advantages, he argues that, for students, it is very difficult to approach a field without knowing how it relates to a specific historical time. For a thorough understanding, historical background is required: "Scientific habits or rules of procedure are not merely to be judged by logical standards that exist independently of them; they contribute something to, and react back upon, the logical standards themselves" (Schumpeter 1954, 5). Finally, economics deals in particular with a unique historical process. Scientific progress must be interpreted as a permanent process of overwriting previous knowledge with new statements and new understanding in order to arrive at a more appropriate version of theoretical reasoning (Bögenhold 2014).

In a different context it was Robert Merton (1987) who used his formulation of strategic research materials, which involve the significance of historical data to provide a sense of development. Just reading a contemporary piece in economics (or any other social science) does not allow one to get the full sense of that piece, if one does not understand the context of creation. Ideas, it was said, have their own history; telling the story of an idea's development was "internal" or "absolutist" history (Emmett 2003, 533).

Human beings do not have clear knowledge about the future since the future is, in principle, unknown and uncertain to varying degrees. In contrast, we know the past as a collection of facts, which may provide ideas about the courses of development and, consequently, about future possibilities when interpreted systematically. As Boettke puts it, "[t]he use of intellectual history instrumentally follows both from the idea that all that is important in the past is *not* necessarily contained in the present, and the idea that mining the past might offer concepts which point the way to more productive theory construction today. Following this path we may find dead-ends in current trends of thought which force us to reconsider the earlier moment of choice and then imagine the path that could have been followed instead. . . . . But reading an old work in economics is not unlike watching a silent film or news clips of an old baseball game... .." (Boettke 2000).

In philosophy it was already Georg Wilhelm Friedrich Hegel ([1837] 2001) who claimed, very convincingly, that an academic subject must always be taught and interpreted in the light of its own history. So, philosophy would always include the history of philosophy as well. Hegel's philosophy of history is a neglected academic area concerning the necessary contextualization of knowledge to provide a more adequate working compass. Kurz (2016) indicated that not only is it important to remember that the huge changes in the economy over the last few centuries have also changed our view of the economy and society (Kurz 2016, 3), but that the history of economic thought is also changing. Each generation writes its own history, new knowledge is always created and each generation is "keen not only on being original but on being perceived as such. But each generation also searches for meaningful progenitors so it can share in their renown and brilliance" (Kurz 2016, 2).

If we are really serious that economics is part of social sciences (Marchionatti and Cedrini 2017), we have to look for the links between economics and neighboring fields in the social sciences; anthropology, sociology, (social-)psychology, and, of course, history are some of those neighbors, always providing fruitful ideas and connections to prevent economics from becoming sterile and too isolated. The credo of the historical school was that researchers should dive into the ocean of economic history with its manifold historical details in order to gather so many details that they may be generalized (Shionoya 2001, 2005).

Douglas North (1990) was among those prominently awarded economists who rejected the abstract and universal approach of economics by reintroducing (new) institutional economics which stresses upon the context of specific economies within their changing societies. He emphasized that abstract "all are alike" approaches for the working of modern capitalism are sterile and extremely decontextualizing because they do not acknowledge the specific historical "rules of the game" (North 1990; Baumol 1990) and also the broad institutional variety of existing institutions in capitalism.

Cultures along with their corresponding times and spaces – and this brings us full circle to the discussion introduced at the beginning – bring with them different considerations of individual rationality. Culture can be seen as an analytical variable that is indicative of different constellations of norms and corresponding behavior (North 1990; Jones 2006). Accordingly, culture operates as a framework of and for behavior and is a factor that represents real – as opposed to abstract – economies and societies. The historian David Landes succinctly summed up this kind of statement as follows: "Culture makes almost all the difference" (Landes 2000, 2). Assuming this to be true, then one conclusion must be that not merely sociology but also historical science are vitally important for the adequate examination of socioeconomic processes. Economic historians consistently stress the tremendous importance of "cultural factors in economic growth" (Cochran 1960) and, following on from this, conclude that the "really fundamental problems of economic growth are non-economic" (Buchanan and Ellis 1955, 405).

Although, in the view of Sombart, theory is the prerequisite to any scientific writing of history. Postulating "no theory – no history!" (Sombart 1929, 3), he warned, on the other hand, against doing theory without history: "The economic theorist moved in an unreal, abstract world. He concentrated his attention upon the exchange operations of 'economic men.' He failed to reap the abundant harvest offered by the manifold variety of actual life, and thus deprived the economic historian of indispensable material" (Sombart 1929, 8).

"Understanding instititutional diversity" (Ostrom 2005) was the new credo and academic aim for institutional economics. Elinor Ostrom who received the Nobel award for economics as first female person in history in the year 2009 did her work explicitly interdisciplinary. When Ostrom is working on social capital which is rooted in sociology and which goes back to Georg Simmel and his ideas of *crossing of social circles* (Simmel 1908) and to Granovetter's *"The Strength of Weak Ties"* (she refers to the *"The troika of sociology, political science and economics"* (Ostrom and Ahn 2009a, b). The wording of a troika is exactly the program of converging disciplines of which contemporaries are increasingly convinced that is necessary. Reflections on social capital show also that beside financial and human capital social dimensions are acknowledged as a domain of capital which are usually beyond the scope of (pure) economics. Ostrom refreshes also thought which was already worked out by socioeconomist and historian Karl Polanyi ([1944] 2001) in the mid of the twentieth century when he added the dimensions of reciprocity and redistribution as key principles of market integration in addition to (core) market principles with (market) offer and demand (Bögenhold 2007).

The fact that Elionor Ostrom received the prestigious Nobel prize as first women so late in history of economics underlines also the dominance of men in economics and its history (Becchio 2019; Marcuzzo and Rosselli 2008). In parallel, also the role of gender was increasingly encountered as an important research topic in economics (Goldin 1992; Sevilla 2020; Naz and Bögenhold 2020) which explored social and economic diversity rather than just operating with the axiom of an agent.

Another way of practicing economics was the integration of developmental perspectives such as globalization (Stiglitz 2002, 2006) and human capabilities in context with ethics and inequalities (Sen 1987, 1992) which is by choosing the topic already a specific heterodox way of doing economics. Sen's capability approach provides a challenging social evaluation of people's living and working conditions which is asking for degrees of relative consent by participants (Naz 2016) which are taken by their own biographies and life experiences rather by statistical data of policy advisers.

It is also very interesting that the highly cited book by economist Thomas Piketty (2014) *Capital in the Twenty-First Century* which is focusing at wealth and income inequalities since the early twentieth century till recent times has inspired many conferences worldwide to take social and economic inequalities as general conference slogan. However, since times of Karl Marx and Max Weber social stratification and inequalities have been primarily a topic under the umbrella of sociology and history while it has started to merge with economics. Even here we find processes of disciplinary exchange and convergence.

Problems and Failures in Economics

Mainstream economics is, also, in itself fragmented and always changing (Cedrini and Fontana 2017). Former links between economics and other domains of social science, especially to philosophy, were exchanged for new links to econometrics and mathematics. Mainstream economics became increasingly associated with abstractness and formalism (Lachmann 1950, 1975), which went hand in hand with an ongoing trend that meant that even the history of economic theory was forgotten. One of the main takeaways of reasoning here is that, most recently, many substantial concepts from psychology, history, and sociology have been taken up by economists and incorporated into their body of knowledge without their really or fully being informed by the original sources. This theft of ideas could be seen negatively, or, indeed, positively as new interdisciplinary domains and synergies emerge. In fact, from a perspective of philosophical economics, one can speak of an ongoing social-scientification of economics (Bögenhold 2010), which is increasingly incorporating ideas brought forth by neighboring social science disciplines. Hedström et al. (2009) present a collection of those topics and research lenses.

The demand by the *Gulbenkian Commission on the Restructuring of Social Sciences* (Wallerstein et al. 1996) to "open the social sciences" should be taken seriously. As a consequence of those processes of the simultaneous multiplication and fragmentation of academic knowledge, new frontiers of academic organization (must) evolve: "What seems to be called for is less an attempt to transform organizational frontiers than to amplify the organization of intellectual activity without attention to current disciplinary boundaries. To be historical is after all not the exclusive purview of persons called historians. It is an obligation of all social scientists. To be sociological is not the exclusive purview of persons called sociologists. It is an obligation of all social scientists. Economic issues are not the exclusive purview of economists. Economic questions are central to any and all social scientific analysis. Nor is it absolutely sure that professional historians necessarily know more about historical explanations, sociologists more about social issues, economists more about economic fluctuations than other working social scientists. In short, we do not believe that there are monopolies of wisdom, nor zones of knowledge reserved to persons with particular university degrees" (Wallerstein et al. 1996, 98).

The social sciences can be regarded as an orchestra with different instruments and different roles. Psychology, history, and sociololgy can certainly play a crucial part in that orchestra. Not only the *Gulbenkian Commission on the Restructuring of Social Sciences* (Wallerstein et al. 1996), but also the first *Social Science Report* by UNESCO (1999) pointed to the problem that academic competences are often defined using an exclusive terminology: "Disciplines are classified under either the one (for example, economics, sociology, political science, as social sciences) or the other (for example, psychology, anthropology and linguistics, as human sciences)" (UNESCO 1999, 12).

Despite the need for specialization in academic training, transdisciplinary attempts are also necessary in order to increase the potential of insights: "There is no doubt that disciplinary separations are part of the scientific endeavour and have a clear heuristic and educational value. It is also obvious that a competent social scientist is a person with a high level of training and expertise in one of the core disciplines, without which he/she cannot cross, with relevance and usefulness, disciplinary frontiers, to cooperate with other specialists. However, at the cutting edge of science, in advanced research, interdisciplinarity or transdisciplinarity is required, combining theories and methods from different disciplines according to the nature of the research" (UNESCO 1999, 12). The conclusion that has been reported so far across different platforms of science management is that "the future is cross-disciplinary" and "social science is central to science" overall (Campaign for Social Science 2015).

Following Akerlof's (2020) argumentation, we should be much more liberal and tolerant and much less apodictic toward the question as to the right way of doing economics: too often those ways have changed, too random is that matter which is just at the forefront of truth and power of definition, and too visible are the visible shortcomings and misconceptions in economics which show time and again that economics is far from being a hard science like physics. Akerlof writes: "The norms regarding how economics should be done should call for flexibility of methodology—instead of insistence on methodological purity that might be perfect for some important problems, but leaves other problems and other approaches outside the domain of economic research. Historically, those paradigms—norms for how economic research should be done, and also for what constitutes 'economic research'—have developed out of an evolutionary process" (Akerlof 2020, 416). The coronavirus pandemic and its unpredictable disaster and its disastrous effect on for the world economy coupled with the obvious interdisciplinary and global interrelations have revealed our limited powers to forecast and explain developments when small unknown issues enter the stage because they were excluded from initial calculations. The same admission of failure happened with the global economic crisis in 2007–2008 which was, in itself, the same plea to open up social sciences in order to arrive at a discussion and theoretical orchestration which is more adequate.

Conclusion

The reciprocal integration of economy, society, and culture (Granovetter 2017) and the increasingly visible overlappings of psychology, history, sociology, and economics must be better acknowledged in academic reflections of a science of science so that disciplinary authorities will be defined accordingly. Of course, there are also immanent consequences for public funding and political decisions for the use of research money which are always in change and in a continuous and historically changing battle over public funding (Solovey 2020). Social-sciencation is an ongoing process which we can acknowledge in diverse examples in economics and which has serious implications for a new division of the academic landscape.

References

Akerlof GA (2007) The missing motivation in macroeconomics. Am Econ Rev 97(1):5–36
Akerlof GA (2020) Sins of omission and the practice of economics. J Econ Lit 58(2):405–418
Akerlof GA, Kranton RE (2000) Economics and identity. Q J Econ 115:715–753
Akerlof GA, Shiller RJ (2009) Animal spirits: how human psychology drives the economy, and why it matters for global capitalism. Princeton University Press, Princeton
Ariely D (2010) Predictably irrational: the hidden forces that shape our decisions. Harper Perennial, New York
Backhouse RE, Cherrier B (2017) The age of the applied economist: the transformation of economics since the 1970s. Hist Politi Econ 49(Supplement):1–33
Barber WJ (1997) Reconfigurations in American academic economics. A general practitioner's perspective. Daedalus 126(1):87–103
Baumol WJ (1990) Entrepreneurship: productive, unproductive, and destructive. J Polit Econ 98(5):893–921
Becchio G (2019) A history of feminist and gender economics. Routledge, New York
Becker GS (1993) The economic way of looking at behavior. J Polit Econ 101(3):385–409
Becker GS (2010) Economic imperialism. Relig Lib 3(2). https://www.acton.org/events
Boettke PJ (2000) Why read the classics in economics? George Mason University, Unpublished paper
Bögenhold D (2007) Polanyi. In: Weir RE (ed) Encyclopedia of American social class, vol II. Greenwood Press, Westport, pp 620–621
Bögenhold D (2008) Economics, sociology, history: notes on their loss of unity, their need for re-integration and the current relevance of the controversy between Carl Menger and Gustav Schmoller. Forum Soc Econ 37(2):85–101
Bögenhold D (2010) From Heterodoxy to Orthodoxy and Vice Versa: Economics and Social Sciences in the Division of Academic Work. T Am J Econ Sociol
Bögenhold D (2014) Schumpeter as a universal social theorist. Atl Econ J 42:205–215
Bögenhold D (2016) Bounded rationality. In: Ritzer G, Rojek C (eds) Encyclopedia of sociology, 2nd edn. Blackwell, London. https://doi.org/10.1002/9781405165518.wbeos0787
Bögenhold D (2018) Economics between insulation and social-Scienciation: observations by a sociology of economics. J Econ Issues 52(4):1125–1142
Boulding K (1971) After Samuelson, who needs Adam Smith? Hist of Politi Econ 3(2):225–237
Bowles S (1998) Endogenous preferences: the cultural consequences of markets and other economic institutions. J Econ Lit 36(1):75–111
Buchanan NS, Ellis HS (1955) Approaches to economic development. Twentieth Century Fund, New York
Campaign for Social Science (CfSS) (2015) The business of people. Social science over the next decade. Sage, London
Cedrini M, Fontana M (2017) Just another niche in the wall? How specialization is changing the face of economics. Camb J Econ 42(2):427–451
Chichilnisky G (2017) Mathematical economics. Edward Elgar, Cheltenham
Cochran TC (1960) Cultural factors in economic growth. J Bus Hist 20(4):515–530
Collins R (1992) Sociological insight. An introduction to non-obvious sociology. Oxford University Press, Oxford
Collins R (2002) The sociology of philosophers. Harvard University Press, Cambridge
Collins R (2005) Interaction ritual chains (2. print., and 1. paperback print). Princeton University Press, Princeton
Davis JB (2016) Economics imperialism versus multidisciplinarity. Hist Econ Ideas 24(3):77–94
DiMaggio P (1994) Social stratification, life-style, and social cognition. In: Grusky DB (ed) Social stratification in sociological perspective. Race and gender. Westview Press, Boulder, pp 458–468
Dorfman J (1946–1959) The economic mind in American civilization, vol 5. Viking Press, New York

Elster J (1983) Sour Grapes. Studies in the subversion of rationality. Cambridge: Cambridge University Press
Elster J (1999) Alchemies of the Human Mind: Rationality and the Emotions. Cambridge: Cambridge University Press
Emmett RB (2003) Exegesis, hermeneutics, and interpretation. In: Samuels WJ, Biddle JE, Davis JB (eds) A companion to the history of economic thought. Blackwell, Oxford, pp 523–537
Emmett RB (ed) (2010) The Elgar companion to the Chicago school of economics. Edward Elgar Publishing, Cheltenham
Goldin C (1992) Understanding the gender gap: an economic history of American women. Oxford University Press, New York
Goldin C (1995) Cliometrics and the Nobel. J Econ Perspect 9(2):191–208
Granovetter MS (1993) The nature of economic relationships. In: Swedberg R (ed) Explorations in economic sociology. Russell Sage Foundation, New York, pp 3–41
Granovetter MS (2017) Society and economy: framework and principles. The Belknap Press of Harvard University Press, Cambridge, MA
Granovetter M, Swedberg R (1992) *The sociology of economic life*. Boulder: Westview Press
Haupert M (2016) History of cliometrics. In: Diebolt C, Haupert M (eds) Handbook of cliometrics. Springer, Berlin/Heidelberg, pp 3–32
Hayek FA (1945) The use of knowledge in society. Am Econ Rev, September 35(4):519–530
Hedström P, Bearman PS, Bearman P (eds) (2009) The Oxford handbook of analytical sociology. Oxford University Press, Oxford
Hegel GWF (2001) The history of philosophy. Batoche Books, Kitchener
Herfeld C (2012) The potentials and limitations of rational choice theory: an interview with Gary Becker. Erasmus J Philos Econ 5(1):73–86
Hodgson GM (1994) The return of institutional economics. In: Smelser NJ, Swedberg R (eds) The handbook of economic sociology. Princeton University Press, Princeton, pp 58–76
Hodgson G (2001) How economics forget history. Routledge, New York
Hodgson G (2007) Evolutionary and institutional economics as the new mainstream? Evol Inst Econ Rev 4(1):7–25
Hodgson GM (2012) From pleasure machines to moral communities: an evolutionary economics without homo economicus. University of Chicago Press, Chicago
Hodgson G (2017) Introduction to the Douglass C. North memorial issue. J Inst Econ 13(1):1–23
Hodgson GM, Mäki U, McCloskey D (1992) Plea for a pluralistic and rigorous economics. Am Econ Rev 82(2):25. 60
Hunt EF, Colander DC (2011) Social science. An introduction to the study of society. Allyn & Bacon, Boston et al.
Jevons WS (1871) Theory of political economy. McMillan, London
Jones EL (2006) Cultures merging. A historical and economic critique of culture. Princeton University Press, Princeton
Kahneman D (2003) A perspective on judgment and choice: mapping bounded rationality. Am Psychol 58(9):697–720
Kahneman D (2012) Thinking, fast and slow. Penguin, London
Kahneman D, Tversky A (2013) Prospect theory: an analysis of decision under risk. In: Handbook of the fundamentals of financial decision making: part I. World Scientific Publishing, Singapore/Hackensack, pp 99–127
Keynes JM ([1936] 1964) The general theory of employment, interest, and money [1936]. Harcourt, San Diego
Kurz HD (2016) Economic thought. A brief history. Columbia University Press, New York
Lachmann LM (1950) Economics as social science. S Afr J Econ 18:233–241
Lachmann LM (1975) Makroökonomischer Formalismus und die Marktwirtschaft. J.C.B. Mohr, Tübingen
Landes D (2000) Culture makes almost all the difference. In: Harrison LE, Huntington SP (eds) Culture matters. How values shape human progress. Basic Books, New York, pp 2–13

Lazear EP (2000) Economic imperialism. Q J Econ 115(1):99–146
Louzek M (2011) The battle of methods in economics: the classical Methodenstreit—Menger vs. Schmoller. Am J Econ Sociol 70(2):439–463
Luhmann N (1988) Die Wirtschaft der Gesellschaft. Suhrkamp, Frankfurt
Marchionatti R, Cedrini M (2017) Economics as social science economics imperialism and the challenge of interdisciplinary. Routledge, London
Marcuzzo MC, Rosselli A (2008) The history of economic thought through gender lenses. In: Bettio F, Verashchagina A (eds) Frontiers in the economics of gender. Routledge, London/New York, pp 3–20
Marcuzzo MC, Zacchia G (2016) Is history of economics what historians of economic thought do? Hist Econ Ideas 24(3):29–46
McCloskey D (1987) Econometric history. Macmillan, London
Merton RK (1987) Three fragments from a sociologist's notebooks: establishing the phenomenon, specified ignorance, and strategic research materials. Annu Rev Sociol 13(1):1–29
Morgan M (2012) The world in the model: how economists work and think. Cambridge University Press, Cambridge
Naz F (2016) Understanding human well-being: how could Sen's capability approach contribute? Forum Soc Econ 16(3):316–331
Naz F, Bögenhold D (2020) Unheard voices: women, work and political economy of global production. Palgrave Macmillan, London
Nelson RH (2001) Economics as religion: from Samuelson to Chicago and beyond. Pennsylvania State University Press, University Park
North DC (1990) Institutions, institutional change and economic performance. Cambridge University Press, Cambridge
North DC (1991) Institutions. J Econ Perspect 5(1):97–112
Ostrom E (2005) Understanding institutional diversity. Princeton University Press, Princeton
Ostrom E, Ahn T-K (eds) (2009a) Handbook of social capital: the troika of sociology, political science and economics. Edward Elgar, Cheltenham
Ostrom E, Ahn TK (2009b) What is social capital. *Social capital: reaching out, reaching in.* In: Ostrom E, Ahn T-K (eds) Handbook of social capital: the troika of sociology, political science and economics (2009). Edward Elgar, Cheltenham, pp 17–35
Parsons T, Smelser NJ (1956) Economy and society: a study in the integration of economic and social theory. The Free Press, Glencoe
Piketty T (2014) Capital in the twenty-first century. Harvard University Press, Cambridge, MA
Polanyi K (2001) The great transformation: the political and economic origins of our time [orig. 1944]. Beacon Press, Boston
Robert William Fogel (1993) was awarded a Nobel Prize in economics for his work on economic and social history in the year 1993.
Romer PM (2015) Mathiness in the theory of economic growth. Am Econ Rev 105(5):89–93
Roncaglia A (2014) Should the History of Economic Thought be Included in Undergraduate Curricula? Economic Thought 3(1)1–9.
Rosenberg A (2018) Philosophy of social science. Routledge, New York
Schabas M (1992) Breaking away: history of economics as history of science. Hist Polit Econ 24(1): 187–203
Schumpeter JA (1926) Gustav von Schmoller und die Probleme von heute. Jahrbuch für Gesetzgebung, Verwaltung und Volkswirtschaft im deutschen Reich. Duncker and Humblot, Leipzig, pp 337–388
Schumpeter JA (1954) History of economic analysis. Oxford University Press, Oxford
Sen A (1987) On ethics and economics. Basil Blackwell, Oxford
Sen A (1992) Inequality reexamined. Oxford University Press, Oxford
Sevilla A (2020) Gender economics: an assessment. Oxf Rev Econ Policy 36(4):725–742
Shionoya Y (ed) (2001) The German historical school: the historical and ethical approach to economics. Routledge, London

Shionoya Y (2005) The soul of the German historical school: methodological essays on Schmoller, Weber and Schumpeter. Springer, New York

Simmel G (1908) Die Kreuzung sozialer Kreise. In: Simmel G (ed) Soziologie. Untersuchungen über die Formen der Vergesellschaftung. Duncker & Humblot, Berlin, pp 305–344

Simon HA (1982) Models of bounded rationality. MIT Press, Cambridge

Simon HA (1989) The scientist as a problem solver. In: Klahr D, Kotowsky K (eds) Complex information processing. How do human beings reason. Erlbaum, Hillsdale, pp 375–398

Smelser NJ, Swedberg R (eds) (2005) The handbook of economic sociology. Princeton University Press, Princeton

Solovey M (2020) Social science for what?: battles over public funding for the other sciences at the National Science Foundation. MIT Press, Boston

Sombart W (1929) Economic theory and economic history. Econ Hist Rev 2(1):1–19

Stiglitz J (2002) Globalization and its discontents. W. W. Norton, New York

Stiglitz J (2006) Making globalization work. W. W. Norton, New York

Swedberg R (1990) The new 'battle of methods'. Challenge 33(1):33–38

Swedberg R (ed) (2015) Theorizing in social science: the context of discovery. Stanford University Press, Stanford

Swedberg R (2016) Before theory comes theorizing or how to make social science more interesting. Br J Sociol 67(1):5–22

Thaler RH (1994) Quasi rational economics. Russel Sage Foundation, New York

Thaler RH (2016) Misbehaving. The making of behavioral economics. W.W. Norton, New York/London

The Royal Swedish Academy of Sciences (2020) The prize in economic sciences 2020. Stockholm. Download: www.nobelprize.org/uploads/2020/09/press-economicsciences2020.pdf

Turner JH (1998) The structure of sociological theory. Wadsworth Publishing, Belmont

UNESCO (ed) (1999) World social science report. UNESCO, Paris

Vane HR, Mulhearn C (2005) The Nobel memorial laureates in economics. An introduction to their careers and main published works. Edward Elgar, Cheltenham

Wallerstein I et al (1996) Open the social sciences. Report of the Gulbenkian commission on the restructuring of the social sciences. Stanford University Press, Stanford

Walras L ([1874] 1954) Elements of pure economics. Richard D. Irwin, Homewood

Weber M (1972) Wirtschaft und Gesellschaft [orig. 1921]. J.C.B. Mohr, Tübingen

Weber M (1988) Wissenschaft als Beruf [1918–19]. In: Weber M (ed) Gesammlte Aufsätze zur Wissenschaftslehre. Tübingen, J.C.B. Mohr, pp 524–555

Weintraub RE (2002) How economics became a mathematical science. Duke University Press, Raleigh

Classical Political Economy

57

Heinz D. Kurz

Contents

Introduction	1474
Method and Content of Classical and of Marginalist Economic Theory	1475
Homo mercans, *Homo laborans*, and *Homo inventivus*	1478
Three Grand Orders of Men: Landlords, Workers, and Capitalists	1479
Information Asymmetries and the Banking Trade	1480
Money and Currency	1481
The Classical Surplus Approach to the Theory of Value and Distribution	1483
Necessaries Versus Luxuries	1484
"Natural" Versus "Market Price"	1485
Quantities of "labor embodied"	1486
The "fundamental law of distribution"	1487
Scarce Natural Resources	1489
Technical and Organizational Change	1490
The "law of population"	1493
Foreign Trade	1494
The Role of Government and the State, Taxation, and Public Debt	1496
Conclusion	1498
References	1498

Abstract

This chapter informs about the method, analytical structure, and content of classical political economy, represented most prominently, in Britain, by authors such as Adam Smith and David Ricardo. It focuses attention on what is common to them especially in the theory of value and distribution. These economists explain all property incomes (in particular the rents of land and profits) in terms of the surplus product that obtains after all necessary means of subsistence, or wages, in the support of workers and all means of production used up in the course of production have been deducted from gross output levels. In conditions

H. D. Kurz (✉)
Graz Schumpeter Centre, University of Graz, Graz, Austria
e-mail: heinz.kurz@uni-graz.at

of free competition, the no-rent part of the surplus is distributed as profits at a uniform rate on all capitals (consisting of means of subsistence and of production) in the economy. The prices that support the given distribution of income are called "natural" prices or "prices of production." Profits are the main source of saving alias investment, that is, the accumulation of capital, which together with technical progress and the growth of population decides about the pace at which the economy develops and grows.

Keywords

Accumulation of capital · Development and growth · Division of labor · Foreign trade · Free competition · Income distribution · Long-period method · Population growth · Prices · Profits · Public debt · Rents · Surplus · Taxes · Technical progress · Value · Wages

Introduction

Karl Marx coined the concept of "classical political economy" in *A Contribution to the Critique of Political Economy*. Representatives of it were William Petty in Britain and Pierre Le Pesant de Boisguilbert in France in the seventeenth century up until David Ricardo in Britain and Simonde de Sismondi in France at the beginning of the nineteenth century (see Marx [1859] 1970: 52). In his *Theories of Surplus Value*, Marx referred to the classical political economists as including the Physiocrats, Adam Smith and Ricardo, who sought "to grasp the inner connection of the phenomena" under consideration ([1861–3] 1988: 358). In volume I of *Capital*, he contrasted classical political economy and "vulgar economy," which is said to deal with "appearances only" (Marx [1867] 1954: 85 n.). Marx called Ricardo "the last great representative" of classical political economy (ibid: 24), a view the early Joseph A. Schumpeter ([1912] 1954: 62–7) explicitly shared. Marx considered prominent authors including John R. McCulloch and John Stuart Mill, often regarded as main representatives of British classical political economy, to be part of its decline.

Interpreters from Edwin Cannan (1893) to Mark Blaug (1987, 2008) and Denis O'Brien (1975, 2004) entertained a different concept of "classical political economy." They saw it to refer to pre-marginalist analysis in the period roughly from the mid-eighteenth to the mid-nineteenth century. In this view, it was an early and primitive version of demand-and-supply analysis, with the focus on production and supply, and consumption and the demand side still under- or even undeveloped. The alleged "shortcoming" involved was overcome, it was contended, by elaborating marginal utility theory in the second half of the nineteenth century. The idea underlying this perspective was that as a scientific subject the discipline progressed from its early beginnings to its modern constructions, involving the elaboration of ever more sophisticated, rich, and coherent versions of demand-and-supply theory. In this view, there was only a single kind of economic analysis that provided us with a more and more thorough and correct understanding of the economic phenomena under consideration. However, as we shall see, this view is difficult to sustain. Classical and marginalist economics differ in important respects – the former is not simply an early and rude precursor of the latter. This is discussed in some detail in Walsh and Gram (1980), Garegnani (1984), and Kurz (2016a, 2018). See also Kurz and Salvadori (1995, 2015) and Faccarello and Kurz (2016, especially vol. III).

It deserves to be mentioned that marginalist theory is also known under the name of "neoclassical economics," a term coined by Thorstein Veblen in 1900. Veblen responded to Alfred Marshall's claim that marginalist theory consisted simply in the further cultivation of the seed planted at the time of the classical authors, thus the name *neo*classical. However, Veblen stressed that the neoclassical school had very little in common with other schools, especially the classical one. In substance, the concept of marginal utility had already been anticipated by the German economist Karl Heinrich Rau in 1833, whose work inspired, among others, Hermann Heinrich Gossen, who is often but wrongly credited with having invented the concept; see Kurz (2016b: 255–8).

Method and Content of Classical and of Marginalist Economic Theory

Clearly, demand and supply play some role in any kind of economic analysis, classical, marginalist, Marxist, Austrian, or other. The question is: precisely *which* role? This question can be answered in very different ways, a case in point being the classical authors versus the marginalists (see on this Kurz 2016a: chaps 2 and 4). These differences have their roots in fundamentally different methodological outlooks on the subject. Most importantly, the classical economists took the socioeconomic system as they experienced it, stratified in social classes – workers, landowners, and capitalists – who perform different roles in the process of the production, distribution, and use of commodities and the wealth of a nation. In the tradition of Aristotle's *zoon politicon* (ζώον πολιτικόν), individuals are seen as social beings whose motivations, aspirations, capabilities, and so on are largely shaped by society or the milieu from which they come.

Another characteristic feature of the classical authors is their *objectivist* point of view. This was most effectively expressed when William Petty opted in favor of a "physician's outlook" on economic problems and decided to express himself "in terms of *Number, Weight,* or *Measure* ... and to consider only such Cases, as have visible Foundations in Nature, leaving those that depend upon the mutable Minds, Opinions, Appetites, and Passions of particular Men, to the Consideration of others" (Petty [1690] 1986: 244). Interestingly, the alternative he described fits surprisingly well marginal utility theory, which, together with marginal productivity theory, builds the two pillars of marginalism. Similarly, James Mill, a friend of Ricardo's and the father of John Stuart Mill, stated: "The agents of production are the commodities themselves ... They are the food of the labourer, the tools and the machines with which he works, and the raw materials which he works upon" (Mill 1826: 165). Production, these authors insisted, is a process of "productive consumption," in which means of production and means of subsistence of workers have to be "destroyed" in order to get the commodities needed and wanted. The amounts that have to be destroyed reflect the "difficulty" and "cost" of production and inform about the productivity of the economic system.

The focus of attention in classical political economy is on the coordination of economic activities via interdependent markets within a system of the *social division of labor*. Which conditions does an economy have to meet in order to reproduce itself? When will it develop and grow? When will it stagnate or shrink? In this analysis, the issues of socioeconomic *reproduction* and *development* assume center stage. The approach is systemic and general, looking at the economy in its entirety and its interrelated parts and seeking to understand its "law of motion" (Karl Marx). The main problem is the dynamic behavior of the system: an investigation of its static properties is only a step toward this goal. Important elements in this colossal painting of socioeconomic life are as follows::

- The factors affecting the pace at which capital accumulates
- The determinants of population growth
- The impact of technical progress induced by competitive conditions on economic growth and income distribution
- The role of the scarcity of renewable resources as well as the exhaustion of depletable resources in all this
- The conflict over the distribution of income between workers and the propertied classes and between capitalists and landowners
- The role of money and the banking sector in facilitating economic transactions, but also in endangering the stability of the system
- Foreign trade as an important channel to deepen the division of labor and to raise labor productivity
- The ways and means government has to influence the course of things

In contrast, marginalist authors start from the needy individual. Its behavior is analyzed in terms of Robinsonades, contemplating the production and consumption of Robinson Crusoe (before he met Friday) in Daniel Defoe's novel with the same title. According to Lionel Robbins (1932), (marginalist) economics studies "human behaviour as a relationship between ends and scarce means which have alternative uses." *Homo oeconomicus*, economic man, dominates the stage. Marginalism endorses *methodological individualism*, which does not take society as one encounters it, but is intent on reconstructing it in terms of the interaction of self-seeking individuals. Because of the postulated infinity of man's wants and wishes and the finiteness of the means to satisfy them, the analysis revolves around the concept of the *scarcity* of goods and services and the options available to *Homo oeconomicus* to make the best of it. Within this framework, social relations may be relatively unimportant and economic interaction weak.

Depending on the set of givens of the theory – preferences of agents, their initial endowments of goods and means of production, and the set of technical alternatives from which they can choose to produce the various goods – an *equilibrium* may exist in which all markets clear simultaneously. A market is understood as the confrontation of the collective demand of and the collective supply for a particular good or service. The theory seeks to determine the set of relative prices, at which equilibria in all markets obtain simultaneously, that is, in all markets, the quantities demanded equal the quantities supplied. Depending on circumstances, in such an economy-wide equilibrium, several agents (in the extreme: all of them) remain in a state of autarky and only a few (none) get involved in what a commentator once called "a little trading on the side." As can be shown, social cohesion vanishes entirely when one takes the spatial dimension of economic activity into account in the simplest case possible. Then the competitive price mechanism can explain neither the emergence of spatial economic concentration nor extensive trade streams. In fact, with constant returns to scale, economic activity will be evenly distributed across a homogeneous plain, carried out by autarkic units of production and consumption. There is no society in any meaningful sense.

This chapter is based analytically on the most advanced version of classical political economy, which we owe to Sraffa (1951, 1960) and people elaborating on his contribution. Space limitations do not allow for a full exposition of all of its fascinating details and a comprehensive treatment of the subject matter. At the same time, an attempt will be made to be faithful to what major classical authors actually have written. This applies especially to Ricardo, because we owe him important insights into the working of the economic system and corrections of the doctrine of Adam Smith. The common analytical core that can be discerned in the economic investigations of the classical authors consists essentially in their explanation of all property incomes (rents, profits, and interest) in terms of the *surplus product* that obtains for a given system of production in use and given real wages. However, apart from this, there exist noteworthy differences between their analyses, which, alas, can only be touched upon in passing. For a discussion of similarities and differences among the classical economists, see, inter alia, Garegnani (1984), Kurz and Salvadori (1995, 1998, 2015), Marshall ([1890] 1920), Kurz (2016a), and Faccarello and Kurz (2016).

The difference between the classical and the marginalist approach especially to the theory of value and distribution – the core of their analyses – is well expressed in the distinction between *political economy* and *economics*, the former being used for the classical and the latter for the marginalist school of thought. At the time of the classical economists, the separate discipline *sociology* did not yet exist. The study of social relationships and interaction predates the proper foundation of a separate discipline and was an integral part of political economy. The need for a separate discipline was felt only after marginalism and *Homo oeconomicus* had conquered economic studies and largely purged them of sociological themes and concepts, such as social class, power, and domination.

Homo mercans, *Homo laborans*, and *Homo inventivus*

Understanding the economic world presupposes understanding human beings, man's nature and disposition, his innate characteristics, urges and desires, physical, mental, and emotional faculties, and so on. David Hume (1711–1776) in *A Treatise of Human Nature* (1739) developed a naturalistic view of man and opposed philosophical rationalism by arguing that human behavior is often governed by passion and not reason. Adam Smith in *The Theory of Moral Sentiments* (1759) argued that moral judgment is nothing innate to man: it is rather the result of a dynamic interaction and learning of people. Observing others and reflecting upon their judgments of one-self and others makes one aware of oneself and of how one is perceived. The natural desire to achieve "mutual sympathy of sentiments" with others shapes peoples' habits and eventually their norms of behavior and conscience, which is the faculty that constrains self-interest. The way this is effectuated is via an "impartial spectator" – the "man within the breast" – whose approval individuals seek. While in the *Theory of Moral Sentiments* Smith developed a theory of the roots of peoples' moral behavior, in *An Inquiry into the Nature and Causes of the Wealth of Nations* (1776), he focused attention on the economic sphere in which self-interest prevailed. A major concern of his was whether and to what extent competition was able to hold self-interest in check.

There has been a controversy about whether Smith's two major works are contradicting each other, known as "Das Adam Smith Problem." There is now widespread agreement, though, that this is not the case. The two works rather deal with different features of human nature and the particular situations in which they come to the fore.

In order to survive, humans have to consume and to produce. Smith erects his political economy upon a philosophical-cum-empirical anthropology. Man is endowed with faculties and motives that are conducive to association, cooperation, and competition. There is, Smith writes, "a certain propensity in human nature ... to truck, barter, and exchange one thing for another" (WN I.ii.1). But man is not only able to do so, he is also dependent upon it: "In civilized society he stands at all times in need of the cooperation and assistance of great multitudes, while his whole life is scarce sufficient to gain the friendship of a few persons" (WN I. ii.2). From this Smith concludes in a famous passage of *The Wealth* that

> man has almost constant occasion for the help of his brethren, and it is in vain for him to expect it from their benevolence only. He will be more likely to prevail if he can interest their self-love in his favour, and shew them that it is for their own advantage to do for him what he requires of them. Whoever offers to another a bargain of any kind, proposes to do this. Give me that which I want, and you shall have this which you want, is the meaning of every such offer; and it is in this manner that we obtain from one another the far greater part of those good offices which we stand in need of. (WN I.ii.2)

To this he adds: "It is not from the benevolence of the butcher, the brewer, or the baker, that we expect our dinner, but from their regard to their own interest. We address ourselves, not to their humanity but to their self-love, and never talk to them of our own necessities but of their advantages." (WN I.ii.2)

He even sees the division of labor which propels material well-being as rooted in the propensity under consideration: "it is this same trucking disposition which originally gives occasion to the division of labour." (WN I.ii.3)

Smith's analytical construction thus rests essentially on the following premises:

- The market allows the natural faculties of man to take full effect and therefore is the appropriate social institution of organizing economic affairs.
- Functioning markets accommodating an ever-deeper social division of labor are the key to a growing labor productivity and hence real income per capita, which according to Smith expresses the wealth of a nation.

The types of economic agents populating Smith's world are the *Homo mercans* and *Homo laborans*, but also the *Homo inventivus*. As he emphasizes: "the desire of bettering our condition ... comes with us from the womb, and never leaves us till we go into the grave." (WN II.iii.28) These agents accumulate capital, expand markets, intensify the division of labor, and bring about "improvements" – innovations – in each and every sector of the economy. They invent tools and machines to abbreviate the drudgery of work and increase the productivity of labor.

Three Grand Orders of Men: Landlords, Workers, and Capitalists

The classical authors distinguished between "three grand orders of men," or social classes, landlords, workers, and capitalists. This classification was not decided in terms of a single dimension or criteria only, that is, which kind of property the respective order of men possessed: land and natural resources, labor power or industrial, commercial, and financial capital. According to Smith, there is another dimension of great importance – the access to information and knowledge. *Landlords* receive a revenue (rent) that "costs them neither labour nor care, but comes to them ... independent of any plan or project of their own." This makes them indolent and "renders them too often, not only ignorant, but incapable of that application of mind which is necessary in order to foresee and understand the consequences of any publick regulation" (WN I.xi.p.8). *Workers* are worse off. The worker's "condition leaves him no time to receive the necessary information, and his education and habits are commonly such as to render him unfit to judge even though he was fully informed." The worker is most in danger of being manipulated: "In the publick deliberation, therefore, his voice is little heard and less regarded, except upon some particular occasions, when his clamour is animated, set on, and supported by his employers, *not for his, but their own particular purposes*" (WN I.xi.p.9; emphasis added). Best informed in economic and political matters are *merchants* and *master manufacturers,* who "during their whole lives ... are engaged in plans and projects" and who therefore "have frequently more acuteness of understanding than the greater part of country gentlemen" (WN I.xi.p.10). Possessed of a "superior knowledge of their own interest," they are the source of economic development. Their selfishness may, however, be detrimental to the interests of the other classes and society at large, because they are keen "to narrow the competition [and raise] their profits above what they naturally would be, to levy, for their own benefit, an absurd tax upon the rest of their fellow-citizens." Smith adds:

> The proposal of any new law or regulation of commerce which comes from this order, ought always to be listened to with great precaution, and ought never to be adopted till after having been long and carefully examined, not only with the most scrupulous, but with the most suspicious attention. *It comes from an order of men, whose interest is never exactly the same with that of the publick, who have generally an interest to deceive and even to oppress the publick, and who accordingly have, upon many occasions, both deceived and oppressed it.* (WN I.xi.p.10; emphasis added)

Those who are better informed and knowledgeable may benefit from their superior understanding at the cost of their customers, consumers, and workers. While Smith does not use the terms, what he refers to is asymmetric information that gives rise to moral hazard and adverse selection (see below). Smith deplored the "wretched spirit of monopoly" (WN IV.ii.21) that seeks to reap above normal profits not by innovations but by limiting competition.

Information Asymmetries and the Banking Trade

Information asymmetries play a particularly important role in the banking and financial sector. Bankers, Smith stressed, are often willing to take risks, knowing that in case of failure, the potential costs of their decisions will be borne by others. Investment projects whose expected profitability is abnormally high are, as a rule, also more risky. However, as recurrent financial crisis illustrate, this fact is time and again ignored. People fall victim to "irrational exuberance" (Alan Greenspan). Smith's respective observations read like a commentary on the most recent bursting of a financial bubble. With a hypertrophic expansion of the banking business, bankers "can know very little about [their debtors]" and are inclined to give credit to "chimerical projectors." These would employ the money "in extravagant undertaking, which ... they would probably never be able to compleat, and which, if they should be compleated, would never repay the expence which they had really cost." (WN II.ii.77)

Chimerical investors are willing to offer high rates of interest to banks because they expect very high profits from their "extravagant" undertakings and, should these fail, do not intend to pay back the debt. The "sober and frugal debtors," who "might have less of the grand and the marvellous, [but] more of the solid and the profitable," would, on the contrary, after careful calculation, be willing to pay only a lower rate of interest. Banks can therefore be expected to go for the former and not for the latter. This involves *adverse selection* that transfers a great part of the capital of a country "from prudent and profitable, to imprudent and unprofitable undertakings" (WN II.ii.77). Therefore Smith opted in favor of regulating the banking trade to avoid a serious systemic risk. "The obligation of building party walls, in order to prevent the communication of fire," he emphasized, "is a violation of natural liberty, exactly of the same kind with the regulations of the banking trade which are here proposed." (WN II.ii.94) Good government has to contain the dark sides of selfishness and instead stimulate diligence, industry, and creativity. The regulatory task, Smith was clear, is a permanent obligation: it is a race between the cleverness of legislators and the cunning of business people.

According to the classical economists, the relationships between different classes and groups of people were not generally harmonious, as is frequently maintained, but often characterized by conflicts reflecting antagonistic interests. What precisely were the sources of these conflicts, did they endanger prosperity and economic development, and what could be done to mitigate their destructive effects?

Money and Currency

Ricardo began his professional career as a stockjobber at the London Exchange and monetary issues therefore played an important role in his life. He is typically portrayed as a representative of orthodox monetary views and of the quantity theory of money, according to which the price level is proportional to the quantity of money in the economy. This interpretation is not faithful to what Ricardo wrote and fails to recognize that his monetary theory developed in close correspondence with his theory of value and distribution. Ricardo's monetary analysis is first and foremost a response to the inflationary tendencies in Britain at the time of the Napoleonic wars and involves a fierce criticism of the Bank of England, a private institution (up until 1946), that allowed its governors and directors to reap huge profits, which, in Ricardo's view belonged to the public.

In February 1816, Ricardo published proposals for a secure currency (see *Works* IV: 43–141) and presented anew his "Ingot Plan." He proposed a return to the Gold Standard by rendering bank notes convertible not into specie (coins) but into bullion (gold ingots). This implied the demonetization of gold in domestic circulation and was expected (i) to allow Britain to continue to use paper money as the actual means of payment; (ii) to reduce the need for gold reserves held by the Bank of England and reduce the rise in the value of gold; and (iii) to curb the huge profits pocketed by the representatives of the Bank. The House of Commons decided on a plan for the gradual return to note convertibility in bullion, starting in early 1820 and ending in May 1821 at the pre-1797 parity. During this period, Ricardo's plan was implemented, but immediately after the old parity had been restored, the Bank of England decided to return to note convertibility in coin. This led to huge profits pocketed by its directors, who in anticipation of the move had accumulated large amounts of gold. These they now sold to the Bank at very favorable terms – thus exemplifying the kind of self-enrichment Ricardo was keen to abolish.

In 1823, Ricardo developed a plan for the establishment of a National Bank, which was published posthumously in February 1824 (see *Works* IV: 271–300). His plan had first taken shape in 1815 and made public in the first edition of the *Principles*

(*Works* I: 361–3). Of the two operations that the Bank of England performed – issuing paper currency and advancing loans to merchants and investors – the former should be entrusted to independent commissioners. These were supposed to act as bankers to the government, but are "totally independent of the control of ministers." This would not thwart the provision of the economy with money, but "in a free country, with an enlightened legislature" (*Works* I: 362), it would transfer a part of the profits of the Bank to the national Treasury and thus to the public.

In the *Principles*, Ricardo pointed out the important role of a standard of value in terms of which the causes of changes in the prices of commodities were supposed to be identifiable. After some deliberation, he decided to take gold as standard because he felt that it met reasonably well the requirement of being an "invariable standard of value" that was produced at all times with roughly always the same amount of labor needed directly and indirectly per ounce. On the one hand, gold was a commodity like any other commodity, whose value was regulated like that of other commodities by the amount of labor expended in its production (see below). On the other hand, gold served as money under the gold standard and as such was not a commodity. The "only use" of the standard, Ricardo insisted, "is to regulate the quantity, and by the quantity the value of a currency" (*Works* IV: 59). If the state coins money and charges a seigniorage for coinage, "the coined piece of money will generally exceed the value of the uncoined piece of metal by the whole seignorage charged" (*Works* I: 353). Hence the value of gold (of a given weight and fineness) and the value of money will differ and the difference will depend on the quantity of money provided. An ideal monetary system Ricardo defined in the following way: "A currency is in its most perfect state when it consists wholly of paper money, but of paper money of an equal value with the gold which it professes to represent." (*Works* I: 361) Hence the quantity of paper money in circulation "should be regulated according to the value of the metal which is declared to be the standard" (*Works* I: 354). This does not require paper money to be convertible in specie to secure its value. It suffices that "paper might be increased with every fall in the value of gold, or, which is the same thing in its effects, with every rise in the price of goods" (*Works* I: 354). According to Ricardo, the increase in the price level during the suspension of the convertibility of bank notes between 1797 and 1821 was caused by printing too much money and disregarding the role of the monetary standard.

Deleplace (2015: 355) concluded: "Ricardo's concept of monetary standard ... had a revolutionary content, which put it far ahead of its time." According to Bonar (1923: 298), Ricardo's Ingot Plan "was to be the euthanasia of metal currency." This is expressed too strongly, because in Ricardo's view, gold was to preserve its role as domestic monetary standard and as bullion) to serve as the means of settlement of international debts.

Ricardo advocates the purchasing power theory of exchange rates and the theory of a gold currency including the mechanism that is seen to bring about an equalization of the balance of payments. In the Bullion controversy, which generated important insights into the functioning of a monetary system without convertibility, Ricardo joined the "bullionists" who were in favor of a swift return to the gold standard. An increase of the domestic relative to the foreign price level via the flow of commodities and capital engenders a falling external value of the domestic currency and thus prompts a tendency towards the parity of its purchasing power at home and abroad.

The Classical Surplus Approach to the Theory of Value and Distribution

The core piece of classical political economy is how it tackles the problem of value and distribution. This involves explaining the sharing out of the product among various claimants (workers, capitalists, and landowners) in terms of wages, profits, and rents, and determining the system of relative prices that supports this distribution. Close scrutiny shows that the unifying element in the analyses of different classical authors is that they all explain the *general rate of profits* in the economy, the *rents* paid to the proprietors of the different types of land, and the *ordinary or "natural" prices* ruling in markets *at a given time and place* in terms of the following givens or independent variables (see Sraffa 1951, 1960):

1. The gross outputs of the various commodities produced during a year.
2. The set of methods of production actually employed in producing these outputs, where these methods reflect the current state of technological knowledge in the economy.
3. The amounts of the various qualities of land available in the economy to be used in production.
4. The real wage rate (or, in the case of heterogeneous labor, the real wage rates) of workers in terms of a given bundle of commodities workers can afford with their money wages.

We may illustrate the classical *surplus approach* to value and distribution in terms of an exceedingly simple example. Assume that there is only a single commodity, wheat alias corn. It is used as a consumption good that feeds people and as a production good needed in the production of itself (as seed). Assume further that there is only a single quality of land available that exists in abundance. Landowners competing for tenants who cultivate the land bid the rent down to zero so that the problem of the rent of land need not concern us here. There is only one kind of labor and the wheat wage rate per unit of it is taken to be given. A numerical example may clarify the main ideas. Assume that altogether 100.000 tons of wheat are being produced during a year by 200.000 workers, each of which receives a wage in terms of wheat at the beginning of the year to feed himself and his family that amounts to 0.3 tons per year. Total wages paid annually thus equal $200.000 \times 0.3 = 60.000$ tons of wheat. Assume that the seed that has to be put up with at the beginning of the year equals 20.000 tons of wheat. Wages and seed are for simplicity taken to equal total *cost of production* and thus amount to $60.000 + 20.000 = 80.000$ tons of wheat. They constitute the physical *capital*, consisting of means of production (seed) and means of sustenance in the support of workers and their families (wages), that has to be advanced at the beginning of the *period of production*, which is supposed to be a year. The *surplus product*, which in our case will be appropriated as *profits* by capital owners, is equal to 20.000 tons of wheat. The ratio of profits and capital invested gives the *rate of profits*, which in the example is

$$\text{Rate of profits} = \text{Profits}/\text{Capital} = 20.000/80.000 = 0.25 \text{ or } 25 \text{ per cent}$$

Necessaries Versus Luxuries

Wheat in our example is what the classical authors called a "necessary," because it is indispensable in the reproduction of the workforce and therefore also in the reproduction of the social product as a whole. Other commodities that the classical authors considered necessaries include clothes, dwellings, etc., or "wage goods" more generally, and commodities needed in their production, such as coal and iron. These commodities were needed directly or indirectly in each and every line of production, whereas "luxuries" were not, being pure consumption goods enjoyed by the propertied classes, capitalists and landowners, or means of production needed in the production of other luxuries (e.g., race horses).

The different role of luxuries in the economic system can be exemplified by extending our numerical example to include whiskey, a good consumed by the well-off, whereas workers cannot afford it. Whiskey is produced by means of wheat. Assume that 10.000 tons of the surplus of wheat in the above numerical example are used to produce 5.000 hectoliters of whiskey. Assume also that of the 10.000 tons 6.000 tons are used to pay 20.000 workers, each of which gets the same wage per unit of labor employed as in wheat production, that is 0.3 tons. The remaining 4.000 tons of wheat are processed into whiskey.

In conditions of free competition, capital employed in the whiskey industry will yield the same rate of profits as capital employed in wheat production, that is, 25 per cent. This implies that the price per hectoliter of whiskey, p_w, has to adjust relative to the price per ton of wheat, p_c, so that this condition is met. In wheat production, the price of the aggregate output equals

$$100.000 p_c = (1+r)\, 80.000 p_c \tag{1}$$

and in whiskey production, it equals

$$5.000 p_w = (1+r)\, 10.000 p_c. \tag{2}$$

From Eq. (1), we get the already known $r = 0.25$. By plugging this in Eq. (2), we can determine the price ratio of the two commodities,

$$p_w/p_c = 2.5. \tag{3}$$

That is, 1 hectoliter of whiskey is worth 2.5 tons of wheat.

The important message of this little illustration of the classical approach to value and distribution is this: The set of data or independent variables 1–4 suffice to determine the general rate of profits, r, and the relative competitive price of whiskey, p_w/p_c. No other data are needed. As Sraffa (1960) has shown, this holds true also in more general cases, with many commodities, several means of production, wage goods and luxuries, scarce land, and so on, on which more below.

"Natural" Versus "Market Price"

The price ratio determined in (3) reflects what Adam Smith and David Ricardo called "natural," "normal," or "ordinary prices," and Ricardo, Robert Torrens, and Karl Marx called "prices of production." The characteristic feature of these prices is that they reflect the permanent and systematic forces at work in competitive conditions and cover costs of production of the various commodities plus a *uniform* rate of profits on the capitals invested. "Actual" or "market prices" are in addition subject to a multiplicity of "accidental" and "temporary" factors interfering with the fundamental forces. By their very nature, market prices defy an explanation of sufficient generality. The classical economists' therefore focused attention almost exclusively on prices of production.

A uniform rate of profits reflects the working of competitive forces. The natural price, Smith defined, is

> the central price, to which the prices of all commodities are continually gravitating. Different accidents may sometimes keep them suspended a good deal above it, and sometimes force them down even somewhat below it. But whatever may be the obstacles which hinder them from settling in this center of repose and continuance, they are constantly tending towards it. (Smith, WN I.vii.15)

The "gravitation" of market prices toward natural prices, or rather their "oscillation" around them, is caused by the profit-seeking behavior of capitalists. Ricardo expounded:

> While every man is free to employ his capital where he pleases, he will naturally seek for it that employment which is most advantageous; he will naturally be dissatisfied with a profit of 10 per cent., if by removing his capital he can obtain a profit of 15 per cent. This restless desire on the part of all the employers of stock [capital], to quit a less profitable for a more advantageous business, has a strong tendency to equalize the rate of profits of all, or to fix them in such proportions, as may in the estimation of the parties, compensate for any advantage which one may have, or may appear to have over the other. (Ricardo, *Works* I: 88–9)

Changes in the employment of capital and labor across the economy are taken to bring about a tendency toward a uniform rate of profits. Financial capitalists, Ricardo argued, who are possessed of "a circulating capital [i.e. liquid funds] of a large amount" lend capital to industrial capitalists. Due to this "floating capital," profit rate deviations are reduced more rapidly. On the basis of this presumption, Ricardo felt entitled to abstract altogether from the "temporary effects" produced by "accidental causes" and focus on "the laws which regulate natural prices, natural wages and natural profits, effects totally independent of these accidental causes" (*Works* I: 89–92).

Quantities of "labor embodied"

Up until now, "labor values" or quantities of "labor embodied" in the different commodities have not been mentioned. Yet what was later called the "labor theory of value" is typically, albeit with little justification, considered the linchpin of classical political economy. According to it, commodities exchange in proportion to the amounts of labor needed directly and indirectly in their production: Expressed in terms of a famous example adapted from Adam Smith, if altogether 20 units of labor are needed to catch a beaver and 10 units to hunt a deer, then 2 deer should be worth, or exchange for, 1 beaver. Ricardo is considered as the most ardent advocate of the labor theory of value. The truth, however, is that he took the labor embodied principle only as a makeshift solution to a problem, the complexity of which he failed to fully master. He actually spoke "of labour as being *the foundation of all value*, and the relative quantity of labour as *almost exclusively* determining the relative value of commodities" (*Works* I: 20; emphases added). Quantities of labor embodied alone, he insisted, do not explain fully correctly relative natural prices. The reason is that it does not only matter how much labor is needed altogether, but also how this labor is employed over time, that is, when it is employed in the course of the production of a commodity. Assume that in order to catch a beaver, one has first to produce a trap, which takes 18 units of labor, followed by 2 units to collect the animal. In order to hunt a deer, one has first to produce a spear, which takes 3 units, followed by 7 units to trace and kill the animal. Beaver production obviously needs relatively more indirect labor (spent on producing a means of production or capital good), whereas deer production needs relatively more direct labor (spent on tracing and culling the beast). In competitive conditions, the wages paid to workers in early periods of time engaged in producing means of production have to be discounted forward at the ruling rate of profits. Since in our example beaver production needs not only absolutely more labor, but also relatively more indirect labor than deer production, a beaver will be worth more than two deer, as the labor theory of value implies.

The deviation of relative competitive prices from labor values can also be illustrated in terms of our wheat-whiskey example. Let v_c be the labor value of a ton of wheat and v_w the labor value of a hectoliter of whiskey. Then the following equation describes in labor units wheat production:

$$100.000 v_c = 200.000 + 20.000 v_c \tag{4}$$

Gross output of wheat has a labor value of $100.000 v_c$, which is made up of 200.000 units of direct labor and $20.000 v_c$ units of indirect labor "embodied" in the seed capital put up with. Solving the equation with respect to v_c gives

$$v_c = 2.5,$$

that is, 2.5 man years are needed altogether to produce a ton of wheat (2 man years are needed directly and 0.5 years indirectly via the used up input of wheat in wheat production.) As regards whiskey production, the following labor value accounting applies:

$$5.000 v_w = 20.000 + 4.000 v_c. \tag{5}$$

Plugging $v_c = 2.5$ into Eq. (5) and solving it for v_w gives

$$v_w = 6.$$

The ratio of the two labor values equals

$$v_w/v_c = 6/2.5 = 2.4. \tag{6}$$

Comparing Eqs. (3) and (6) shows that relative competitive prices deviate from relative labor values: In price terms, whiskey is more expensive relatively to wheat than in labor value terms – 2.5 vs. 2.4. The reason for this is that whiskey is produced with relatively more indirect labor, incorporated in the capital good wheat, than wheat itself. Put differently, the ratio of wheat input to direct labor input in whiskey production is larger than in wheat production: 4.000 wheat/20.000 labor = 1/5 > 20.000 wheat/200.00 labor = 1/10. The difference between 2.5 and 2.4 expresses the compound interest effect of discounting forward wages paid in a more or less distant past.

Ricardo was clear that the labor theory of value does not fully correctly determine relative prices, but he felt that it provided an approximation that was good enough to adopt it as a makeshift solution. In the *Principles of Political Economy*, he therefore developed his argument *as if* the labor embodied principle was strictly true. To be clear, Ricardo (like Smith) was not of the opinion that labor is the only source and substance of value, as Marx later contended.

If commodities are produced by means of commodities, and if we wish to discuss a system with numerous commodities, prices (but also labor values) could only be determined by solving a system of simultaneous equations; see Sraffa (1960) and Kurz and Salvadori (1995). The mathematics needed were not yet at the disposal of the classical economists (and Marx), who therefore attempted to tackle the problem in terms of the tools available to them and by invoking simplifying assumptions.

The "fundamental law of distribution"

According to Ricardo, the "principal problem in Political Economy" (*Works* I: 5) consisted in establishing the "laws" that regulate the distribution of the product between capitalists, workers, and landowners in a dynamic socioeconomic setting, in which capital accumulates, the population grows, the scarcity of some natural resources increases, there is technical progress, and there is foreign trade.

In the above, the real wage rate is taken to be given. How is its level determined? According to Adam Smith, there is a conflict over the distribution of income. The "common wages of labour," he stated, depend "every where upon the contract usually made between those two parties, *whose interests are by no means the same.*" He continued: "The workmen desire to get as much, the masters to give as little as possible. The former are disposed to combine in order to raise, the latter in order to lower the wages of labour" (WN I.viii.11; emphasis added). Smith opined that

> It is not, however, difficult to foresee which of the two parties must, upon all ordinary occasions, *have the advantage in the dispute, and force the other into compliance with their terms.* [1] The masters, being fewer in number, can combine much more easily; and [2] the law, besides, authorises, or at least does not prohibit their combinations, while it prohibits those of the workmen. ... [3] In all such disputes the masters can hold out much longer. A landlord, a farmer, a master manufacturer, or merchant, though they did not employ a single workman, could generally live a year or two upon the stocks which they have already acquired. Many workmen could not subsist a week, few could subsist a month, and scarce any a year without employment. (WN I.viii.12; emphasis added)

Because of reasons [1]–[3], the bargaining position of the "labouring poor" is weak and they are often bound to accept the conditions dictated by employers in the "dispute" over wages. "Masters," Smith stressed, "are always and every where in a sort of tacit, but constant and uniform combination, not to raise the wages of labour above their actual rate. To violate this combination is every where a most unpopular action, and a sort of reproach to a master among his neighbours and equals." He added: "We seldom, indeed, hear of this combination, because it is the usual, and one may say, the natural state of things which nobody ever hears of." (WN I.viii.13) It is only with a rapid accumulation of capital and thus a swift growth in the demand for "hands" that masters would break the combination and bid up wages in order to increase their work force.

Ricardo confirmed Smith's point of view in terms of his "fundamental law of distribution." The law implies that for a given system of production actually in use there is an inverse relationship between the share of wages and the general rate of profits: "The greater the portion of the result of labour that is given to the labourer, the smaller must be the rate of profits, and vice versa" (*Works* VIII: 194). For a given system of production, wages and the rate of profits cannot rise simultaneously: one class can gain only at the cost of some other class. The harmonious view of society was naïve and ignored the constraint binding changes in the distributive variables.

The law can be exemplified with regard to the above numerical example with two sectors. Assume that the yearly wage rate happens to be larger, that is, 0.32 instead of 0.3 tons, then total wages paid in wheat production amount to 64.000 tons of wheat and total wheat capital employed is 84.000 tons. The surplus distributed as profits is correspondingly smaller and equals 16.000 tons. The rate of profits is lower and equals 16.000/84.00, that is, just a little more than 19 per cent instead of the previous 25 per cent. Calculating the price of whiskey in terms of wheat in the new situation gives 2.476: whiskey is now relatively cheaper, because due to the lower rate of profits, the compound interest effect is smaller.

When wages absorb the entire surplus, there are no profits and the rate of profits is nil. In this case, the ratio of the two prices is equal to the ratio of the two labor values: $(p_w/p_c)|_{r\,=\,0} = (v_w/v_c)$. From this it follows that the labor theory of value explains relative prices fully correctly in a profitless economy (which, however, is of little interest to Ricardo). They also do so in the extremely special case in which the production of all commodities exhibits the same input proportions. Beyond these cases, the labor theory of value is at best approximately true, as Ricardo thought it was. Hence, while important as a makeshift solution, the labor theory of value can hardly be called a linchpin of classical political economy.

Scarce Natural Resources

Up until now we have assumed that natural resources play no perceptible role in the argument. But this is, of course, not true in actual life. Smith had argued in accordance with the Physiocrats that the ground rent paid to the owners of land and natural resources was an expression of the "fertility of nature." Nature was seen to cooperate with man for free and thus increase his or her productivity. This is also the reason why Smith contended that agriculture was more productive than manufacturing, a view Ricardo strongly opposed: rent, he argued, expressed rather the "niggardliness" of nature! If land of the best quality and location were available in unlimited quantity, there could be no ground rent, because cost-minimizing producers would always be able to meet society's need for wheat by cultivating only the best quality of land. However, since in actual fact this kind of land is not available in abundance and will become *scarce* at some point as production increases, it is necessary to meet effectual demand by simultaneously cultivating also inferior lands, which exhibit larger unit costs of production, or by cultivating the best quality of land more intensively, which is also only possible at rising unit costs. As a result, returns fall either extensively or intensively, leading to *extensive* or *intensive rents*.

If, for example, demand is large and cultivation is extended to qualitatively less fertile plots of land, production costs per quarter of wheat will be higher. In order for the larger quantity to be brought forth, the wheat price will have to rise. The higher price enables the owners of the superior quality of land – which continues to produce at lower unit cost – to collect a rent from their tenants, which is just large enough to result in equal costs (inclusive of rent) on both qualities of land. In this new situation, no rent is paid on the inferior land, which is not scarce and which represents what later was called "marginal" land in the given situation. Ground rent is therefore a *differential rent* attributable to differences in production costs per quarter of wheat. To Ricardo, trained in the financial business and in financial mathematics, the connection between the annual rent per hectare of a piece of land of given quality m, q_m, and the land price per hectare, p_m, was clear. If one discounts all future annual rent payments at the prevailing interest rate i in order to get their so-called *present* or *capital value*, one arrives at the formula for eternal rent: $p_m = q_m/i$. If the lease happens to be £100 and the interest rate 5 per cent (or 0.05), the price per hectare of that land will be £2.000.

With society's growing need for wheat, and abstracting from technological progress, the unit cost of wheat would rise, as would its price, the money wage rate (to keep real wages constant) and ground rents on all cultivated lands. Therefore, for a given real (and a rising money) wage rate and an ever smaller fertility of marginal land, there would necessarily ensue a falling tendency of the rate of profits for producers in agriculture and, via the mobility of capital and the ensuing process of gravitation of prices toward their new normal levels, in the economy as a whole. This was Ricardo's explanation of the *tendency of the rate of profits to fall* in the hypothetical case in which there was no technological progress. But since there typically *is* technological progress, Ricardo insisted: "it is difficult to say where the limit is at which you would cease to accumulate wealth, and to derive profit from its employment." (*Works* IV: 179). The widespread view (see, for example, Rostow 1990: 34, 87; Blaug 2009; Solow 2010) that Ricardo saw the stationary state lurking around the corner therefore cannot be sustained. It mistakes Ricardo's method of *counterfactual reasoning* – what would happen, if there was no technical progress, but capital accumulates and population grows? – for a statement about actual economic development. Yet Ricardo was no Horseman of the Apocalypse as his intellectual counterpart, Thomas Robert Malthus ([1829] 1989), who saw mankind forever exposed to misery and deprivation. Despite clear evidence to the opposite, Ricardo is frequently taken to share Malthus' pessimism. Nothing could be further from the truth. (We come back to this in section 14 below.)

His findings prompted Ricardo to invest large parts of the fortune he had gained at the London Stock Exchange on the occasion of the defeat of the Napoleonic troops in the Battle of Waterloo in 1815 to buy land and become one of England's wealthiest landowners. He understood: should the accelerating accumulation of capital and the ensuing shortage of lands following Waterloo lead to an increase of the lease on the land in the numerical example above to £180 and at the same time the interest rate (as a result of a declining rate of profits) fall to 3 per cent (or 0.03), the price of land would rise from £2.000 to £6.000. No bad deal at all! While Ricardo rejected Smith's theory of rent, he confirmed the latter's dictum that landlords "love to reap where they never sowed" (WN I.vi.8): when land is getting more and more scarce, ground rents and land prices will rise. Landlords are the happy beneficiaries of a development of the economy to which they contributed nothing.

Technical and Organizational Change

If capital accumulates and the population grows, more and more of the less fertile lands have to be cultivated. Should the real wage rate remain constant, the rate of profits must necessarily fall. Since capital accumulation depends on the rate of profits – in a fully classical spirit, Marx was to call the rate of profits "the stimulus of capitalist production and condition as well as driver of accumulation" – both fall or rise together. With a falling tendency of it, the system tends towards a *stationary state*: economic growth comes to a standstill. Ricardo saw technical and organizational change as a factor that countervails the niggardliness of nature. For a thorough analysis of Ricardo's theory of accumulation and growth, see Salvadori (2020).

To Adam Smith, the division of labor was the most important engine of economic growth and the lever increasing labor productivity and per capita income. In his view, the division of labor starts within firms, then extends to a division of labor between firms, regions, and eventually countries. The social division of labor is the driving force behind globalization. It (i) yields gains from specialization, (ii) saves time that is lost in changing from one task or job to another, and, most importantly, (iii) promotes the development of useful machines. Labor power is replaced by machine power, and production gets mechanized – a process for which there is no end in sight.

Smith anticipated the emergence and rise of a sector today known as Research and Development (*R&D*). Those employed in it "are called philosophers or men of speculation [i.e. scientists], whose trade it is, not to do any thing, but to observe every thing; and who, upon that account, are often capable of combining together the powers of the most distant and dissimilar objects. In the progress of society, philosophy or speculation becomes, like every other employment, the principal or sole trade and occupation of a particular class of citizens." (WN I.i.9) New economically useful knowledge is getting systematically generated, enabling "improvements" in production and organization. Two centuries prior to the emergence of the concept of "knowledge society," Smith had already explicitly identified the "quantity of science" as the foundation of society's productive powers.

According to him, the engine of the growth of wealth was capital accumulation. It engendered a virtuous circle: by increasing the extent of markets, capital accumulation facilitated an ever-deeper social division of labor, which is accompanied by rising labor productivity and as a consequence higher profits and incomes more generally. This leads in turn to further capital accumulation, further productivity growth, and so on and so forth, following an incessantly upward spiral. Capital accumulation is both a source and an effect of the continual transformation from within which the market system experiences. The process under consideration is characterized by *circular and cumulative causation*: accumulation and innovation feed on themselves.

Ricardo deepened the analysis of technical change and its impact on the economy. New methods of production replace old ones and new commodities and the methods to produce them enter the system. If technical change concerns the production of "necessaries" or wage goods or capital goods needed directly or indirectly in the production of these, then for a constant general rate of profits the real wage rate will increase. As early as in the *Essay on Profits* of 1815, Ricardo stressed that "it is no longer questioned" that improved machinery "has a decided tendency to raise the real wage of labour" (*Works* IV: 35). This is possible without a fall in the general rate of profits, because improved machinery reduces the quantity of labor needed directly and indirectly in the production of the various commodities: it reduces "the sacrifices of labour" (*Works* IV: 397).

Clearly, those who introduced improved machinery typically had no interest in increasing real wages, Ricardo in fact called machines "mute agents of production," because different from workers they do not ask for higher real wages and better working conditions. Machines were often explicitly introduced in order to keep the aspirations of workers at bay. However, the profit-seeking efforts of manufacturers generated effects these had not intended and could not possibly have foreseen: by accelerating capital accumulation, the growth of the demand for labor power increased, which exerted an upward pressure on wages.

If technical change concerned "luxuries," consumed by the propertied classes, the general rate of profits would not change, given the real wage rate; only the prices of luxuries would fall relative to those of other commodities and would eventually perhaps become affordable even to workers.

According to Ricardo and those following in his footsteps, technical progress typically reduces the amount of labor needed directly and indirectly in producing the various commodities. He put forward a rich typology of different types or forms of technological progress (direct labor saving, indirect labor or capital saving, land saving, and so on) and insisted that these forms may affect employment, wages, profitability, the rents of land, and other important economic variables in very different ways. In the chapter on machinery, added to the third edition of the *Principles* (1821), Ricardo discussed a particular form of technical progress, which, he insisted, "is often very injurious to the interests of the class of labourers" (*Works* I: 388). The case under consideration is "the substitution of machinery for human labour," which reduces the gross product. It is accompanied by an increase in labor productivity and in the capital-to-output ratio, and thus a decrease in the maximum rate of profits corresponding to wages that are hypothetically equal to nil. The kind of technical progress under consideration is both labor saving and (fixed) capital using. It leads to what was later called "technological unemployment," which will exert a downward pressure on real wages, viz. its injurious effect on workers. Interestingly, Marx was to take Ricardo's above form of technical progress to dominate capitalist development, reflected in a growing "organic composition of capital" and thus a falling maximum rate of profits, which, Marx was convinced, will eventually also force the actual rate of profits to fall.

Foreign trade, the classical economists insisted, increases the set of commodities and methods of production to which a country has access via imports bought with exports, and therefore can be expected to affect income distribution and relative prices via the channels mentioned above. Ricardo also had a clear understanding of *induced technical change*: A newly invented machine, for example, may not be adopted by cost-minimizing producers, because at the given wage rate and prices, it would not be profitable to do so. It has been born into an environment that is inimical to it, which is reflected in extra costs its employment would incur. Joseph A. Schumpeter would call it an "invention" that would not (immediately) become an "innovation." However, with the accumulation of capital and the growth of population, wages and prices are bound to change and may eventually render the invention worth adopting (see Kurz 2015: section 7).

Interestingly, Ricardo even reflected upon the limiting case of a *fully automated system of production* and its consequences: "If machinery could do all the work that labour now does, there would be no demand for labour. Nobody would be entitled to consume any thing who was not a capitalist, and who could not buy or hire a machine." (*Works* VIII: 399–400). He thus anticipated a trend toward automation that advanced economies are presently experiencing and the potentially serious problems regarding employment and income distribution it poses.

The "law of population"

According to a widespread opinion, a centerpiece of the doctrines of the classical economists is the "law of population," which was most forcefully advocated by Malthus (1798). The law implied that any increase of real wages above bare subsistence spurs population growth, which then drives real wages down again: a rise in real wages is self-defeating. Pronouncements of this mechanism are already to be found in earlier authors, including Smith, who argued that a growing division of social labor and the increase in labor productivity it entails could outweigh population growth and allow for modestly rising real wages. Later critics from Ricardo to Marx accused Malthus of vastly underrating the effect of technical change; he was in fact a technological pessimistic.

Ricardo's point of view in this regard is particularly interesting, because he and Malthus were in close contact with one another and discussed matters of political economy on a fairly regular basis. While Ricardo assumed a given and constant real wage in one part of his economic analysis, he abandoned it in another one: He distinguished between (a) the determination of the rate of profits, the rents of land and relative prices in *given* economic circumstances, that is, at a given time and place and given real wages and (b) the movement of all distributive variables, including wages, and prices in *changing* circumstances, that is, over time, when capital accumulates, there is technical progress and the population grows. In the former case, Ricardo saw the rate of profits and relative prices as fully determined in terms of the given system of production and the given real wages. For an essentially tactical reason, he was prepared to come partly Malthus's way by assuming the law of population, because then the real wage rate could be taken as a magnitude fixed at the subsistence level. This rendered the explanation of profits residually in terms of the surplus product (see The Classical Surplus Approach to the Theory of Value and Distribution above) a good deal easier and should have prevented Malthus from escaping Ricardo's logic.

When Ricardo in his theory of capital accumulation and economic development then turned to a system continually in motion and incessantly transformed from within, it was clear to him that real wages could no longer be assumed to be fixed: they were variable and reflected inter alia the impact of innovations on the one hand and the growing scarcity of some natural resources on the other. Ricardo therefore explicitly distanced himself from Malthus's law of population. He stressed the historically and socially contingent dimensions of the natural wage (*Works* I:

96–7) and that "population may be so little stimulated by ample wages as to increase at the slowest rate – or it may even go in a retrograde direction" (*Works* I: 169). "Better education and improved habits" may break the connection between population and necessaries (*Works* II: 115). Workers may get "more liberally rewarded" and thus participate in the sharing out of the surplus product (*Works* I: 48). If such a constellation prevailed for a longer period of time, a sort of ratchet effect may obtain: higher real wages become customary and define a new level of "natural" wages.

Therefore the concept of "natural wages" in Ricardo is specified with reference to the wealth of a society and the growth regime it just experiences. It does most certainly not reflect a given and constant real wage rate, as numerous interpreters wrongly contended. Ricardo felt the need to replace the real (that is, commodity) wage rate by a share concept, or "proportional wages" (Sraffa 1951: lii), that is, "the *proportion* of the annual labour of the country ... devoted to the support of the labourers" (*Works* I: 49; emphasis added). As we have seen, this share may contain a part of the surplus product. It was on the basis of this wage concept that Ricardo asserted his fundamental proposition on distribution: the rate of profits depends inversely on proportional wages (see also Gehrke 2011).

It should thus be clear that the classical surplus explanation of profits applies both in a regime, in which the law of population holds, and in one, in which it doesn't. Ricardo was, of course, mainly interested in the latter.

Foreign Trade

The mercantile system of monopolies, import restrictions and export promotion, Adam Smith argued, led to a misallocation of resources, stunted economic dynamism, and entailed undesirable distributional effects. It stood in opposition of the liberal principles of "equality [in the sense of equal rights], liberty, and justice." Smith was especially critical of the mercantilist promotion of cities and thus of the manufacturing sector and of foreign long-distance trade. This went against the "natural course of things," in which agriculture is the first sector to develop, followed by industry and cities in tandem with domestic trade, and foreign commerce only in a final stage. As we have heard already in the above, he supposed agriculture to exhibit the highest added value on the (mistaken) ground that in it "nature labours along with man" and "costs no expense," a view Ricardo showed to be untenable. But foreign trade, although it offered opportunities for higher profits, also harbored higher risks and greater insecurity for capital investment. The risk-averse capital owner therefore preferred to invest at home and, in pursuing his own advantage there, provided for higher domestic employment and income. Smith concluded: "He is . . . led by an invisible hand to promote an end which was no part of his intention."

Smith was an eloquent advocate of free trade and what today is called "globalization." But his advocacy was tied to an important condition: *all* countries and parties involved should benefit from free trade and not just some of them, which points once more to the importance of good government. The mercantile policy was, on the contrary directed toward a highly asymmetric result, with firms like the East India Company making huge profits at the cost of both the majority of foreign and domestic population. Smith was one of the fiercest critics of this company.

Ricardo regarded Smith's explanation of how countries specialize based on absolute advantages in production costs for goods as incomplete. Assume, he argued, one country could produce all goods at lower costs than the rest of the world. Then, initially, only this country would export goods, which foreign countries would import. However, according to David Hume's *price-specie flow mechanism*, this would lead to a flow of gold (the money commodity) from the rest of the world to the country. A redistribution of the amount of money would then imply that prices in the country rise and those abroad fall. At some point, some commodities abroad would become cheaper than in the country, so that the absolute cost advantage gets reversed and (some of) the other countries can now export the commodities under consideration. Which commodities are concerned?

Ricardo answered this question in terms of the *principle of comparative advantage*. He exemplified the principle in terms of the trade of cloth and wine between England and Portugal. Assume that in Portugal 90 h of labor are needed to produce a bale of cloth and 80 h to produce a cask of wine. In England, it takes 100 h for cloth and 120 h for wine. Therefore, Portugal possesses an absolute advantage with respect to both products, and with respect to wine also a comparative (relative) advantage: the cost differential for wine (80 vs. 120) is larger than for cloth (90 vs. 100). Correspondingly England faces an absolute disadvantage with respect to both products but a comparative advantage with respect to cloth. For Portuguese producers, it is worthwhile to specialize in the production and export of wine while importing cloth from England, where the English absolute disadvantage is comparatively small.

We may explain Ricardo's important principle, which Paul A. Samuelson called both "true" and "nontrivial," in another way, drawing attention to the involved possibility of *arbitrage*, meaning here the exploitation of differences in relative prices in the two countries involved. Assume that the two countries have their own currencies, which are supposed to be nonconvertible – Portugal the Portuguese Real (R) and England the Pound ($£$). Assume that the money prices of the quantities of cloth and wine in the two countries are proportional to the quantities of labor spent in producing them, and assume for simplicity that the numbers are the same, the only difference being that now, instead of Portuguese and English labor, we have R and $£$ (Table 1).

The reader will quickly see that trade would be favorable to merchants of both countries. (In the following, we abstract for simplicity from transport costs.) Take the case of an English merchant. He may buy for 100 $£$ cloth, ship it to Portugal, and sell it there for 90 R. With this sum of money, he may then buy wine from a Portuguese wine grower and get altogether $90/80 = 9/8$ units of wine. This quantity of wine he then ships to England and sells it for $9/8 \times 120£ = 135£$. He thus yields a profit of $(135 - 100) £ = 35 £$, or a rate of profit of 35 per cent on an investment of 100 $£$. (Notice that the English merchant can use the same ship to export and import goods from and to England.) A similar consideration applies to a Portuguese merchant.

Table 1 Different prices for different goods

Price in Reals (Portugal) and Pounds (England) of a given quantity of		
	Cloth	Wine
In Portugal (Real)	90	80
In England (Pound)	100	120

The case under consideration shows that while commodities are exported and imported, the currencies of the two countries do not cross borders: they stay in the countries of origin; there are no flows of money into and out of a country.

What applies to specialization between countries or regions also applies to trade between people. The happy message of Ricardo's finding is this: whoever is inferior to another person in any kind of productive activity can nonetheless become involved in a division of labor that is mutually beneficent. In this way, Ricardo added an important verse to Adam Smith's hymn of praise on the division of labor.

The Role of Government and the State, Taxation, and Public Debt

Smith redefined the roles of the private and of the public sector. According to him, the government should only assume tasks that private agents are either incapable of carrying out at all or cannot do so as well as the government or can do it only at a higher cost. Once the legitimate tasks of the state are fixed, the means to finance them must be decided. According to Smith, the maxim to follow is that the private sector should not be burdened with excessive taxation.

Smith's remarks on this matter are frequently interpreted as a plea for a "minimal" or "night watchman state." This interpretation cannot be sustained. The *Wealth of Nations* includes an impressive set of tasks for the state but also a severe criticism of the hypertrophic mercantilist state. Smith proposed to transform the old authoritarian state into a modern constitutional and achievement-oriented state that reacts appropriately to the changing challenges it is confronted with. He recognized, for instance, that the division of labor could have negative side effects: the devaluation of artisanal skills and the replacement of adult with child labor. He called for state-financed elementary school education to cushion the negative impact of this development. He listed other responsibilities of the state, including the administration of justice, policing, and national defense; the provision of infrastructure to facilitate the movement of people and commodities; and the organization of large-scale projects in the general interest. In light of historical experiences – especially the introduction of paper money in France at the beginning of the eighteenth century and the ensuing Mississippi Bubble – Smith also advocated, as we have already heard, regulating the unstable banking sector, since "those exertions of the natural liberty of a few individuals, which might endanger the security of the whole society, are, and ought to be, restrained by the laws of all governments." And while he considered paper money on a par with technical progress, because it allowed a society to save on the costly provision of gold and silver, he warned that the commerce and industry of a country "cannot be altogether so secure, when they are thus, as it were, suspended upon the *Daedalian wings of paper money.*" According to Greek mythology, Daedalus was a gifted craftsman who built wings of wax and feathers with which he and his son Icarus escaped from the island of Crete after having been imprisoned by Minos. But hubris – or "irrational exuberance"? – made Icarus ignore his father's warnings: he got too close to the sun, the wax in his wings melted, and he plunged into the sea and died.

Taxes, Smith proposed, should be proportionally equal, and thus addressed both the *ability-to-pay principle* (that taxation should be based on income) and the *equivalence principle* (that taxation should be based on the benefits experienced by the taxpayer as a result of government activity).

In the *Principles*, Ricardo devoted a substantial space and a great deal of attention to taxation and especially the problem of tax incidence and the impact of taxes on the pace of capital accumulation and economic growth. He insisted: "There are no taxes which have not a tendency to lessen the power to accumulate. All taxes must either fall on capital or revenue" (*Works* I: 152). However, he added, the burden of a tax is not necessarily borne by whoever pays it. This he then illustrated with reference to a number of cases involving both direct and indirect taxes. For example, on the premise that workers are paid a subsistence wage, a tax on wages could not be borne by workers: nominal wages would rise leaving real wages constant and the tax would accordingly be shifted to capitalists. A similar reasoning applies to the case in which a tax is laid on wage goods or "necessaries." Taxes on "luxuries" on the contrary "fall on those only who make use of them" (*Works* I: 205).

In full accordance with his doctrine that (differential) rent does not enter the price of commodities, Ricardo insisted that "A tax on rent would affect rent only; it would fall wholly on landlords, and could not be shifted to any class of consumers" (*Works* I: 173). A tax on profits would increase the prices of the products: "if a tax in proportion to profits were laid on all trades, every commodity would be raised in price" (*Works* I: 205). Depending on the consumption patterns of the different classes of society, this would affect their respective members differently. A rise in the price of wage goods would again entail a corresponding adjustment of nominal wages: "Whatever raises the wages of labour, lowers the profits of stock; therefore every tax on any commodity consumed by the labourer, has a tendency to lower the rate of profits" (*Works* I: 205), and, as a consequence, the rate of capital accumulation.

The classical authors were not principally opposed to public debt. Smith, for example, advocated taking up loans in order to invest in public infrastructure, and all authors were clear that exceptional circumstances, such as natural catastrophes or wars, necessitated financing the costs via public debt. This was necessary not least because the tax base at the time in which they wrote was still fairly small. For example, no general income tax existed as yet. However, the classical economists were suspicious of political spendthrifts who had variously managed to ruin entire nations by wasting other people's money in conspicuous luxury consumption and belligerent adventures. When by the time of the end of the Napoleonic wars England had amassed a huge public debt, Ricardo proposed to pay it off by a once and for all tax on property (capital, land, money). Schumpeter put forward a similar proposal after World War I designed to abolish the public debt of Austria.

Conclusion

The chapter has provided a short account of classical political economy, focusing attention on the writings of British economists, especially Adam Smith, David Ricardo, and those elaborating on their contributions up until modern times. The themes covered include the method of analysis adopted, which focuses on the long period, and the approach chosen, which revolves around the determinants of the social surplus generated in the economic system; its distribution among different kinds of agents, workers, capitalists, and land owners; and the use to which it is put, that is consumption and investment or capital accumulation. The areas dealt with include the problems of value and distribution, foreign trade, socioeconomic development and growth, population dynamics, technical progress, and the role of the government and the state. Several misapprehensions in the secondary literature on the classical authors are pointed out. Most importantly perhaps, it is argued that the classical authors were no strict advocates of the labor theory of value, but had recourse to it as a kind of makeshift solution because they lacked the necessary analytical tools to cope properly with the problem of the determination of value and distribution in a system in which commodities are produced by means of commodities. This lacuna was made good only in the twentieth century, which provided the classical approach with a solid foundation: it gave it new vigor through rigor. It turned out that in many respects the classical authors' intuitions were basically correct and showed the way to a fruitful investigation of the laws governing the production, distribution, and utilization of the wealth of a nation. It also became clear that the classical authors were not early and incoherent advocates of the later marginalist or neoclassical or demand-and-supply approach, as their critics contended. They rather had elaborated a fundamentally different analysis of economic phenomena of genuine significance. Studying classical political economy and the contributions of its modern representatives is therefore not a sign of a purely historical and antiquarian interest: it exposes one to modern economics of a nonmainstream variety.

Acknowledgment The author of this chapter is most grateful to the *Munich Social Science Review* (MSSR), New Series, for granting permission to adapt the paper on "Classical Political Economy" published in MSSR, vol. 2/2019, pp. 121–155, for the *Palgrave Handbook*.

References

Aspromourgos T (1995) On the origins of classical economics. Distribution and value from William Petty to Adam Smith. Routledge, London\New York

Aspromourgos T (2009) The science of wealth. Adam Smith and the framing of political economy. Routledge, London\New York

Bailey S (1825) A critical dissertation on the nature, measures, and causes of value; chiefly in reference to the writings of Mr. Ricardo and his followers. Hunter, London. Reprint 1967, New York, A.M. Kelley

Blaug M (2009) The trade-off between rigor and relevance: Sraffian economics as a case in point. History of Political Economy, 41(2):219–47

Blaug M (1987) British classical economics. In: Eatwell J, Milgate M, Newman P (eds) The new Palgrave. A dictionary of economics, vol 1, pp 434–445

Bonar J (1894) Classical economics. In: Palgrave's Dictionary of political economy, vol I

Bortkiewicz L von (1907) Zur Berichtigung der grundlegenden theoretischen Konstruktion von Marx im 3. Band des Kapital, Jahrbücher für Nationalökonomie und Statistik, 34: 319-335.

Cannan E (1893) A history of the theories of production and distribution in english political economy from 1776 to 1848. Perceval & Co, London. Reprint 1967, New York: A. M. Kelley

Cantillon R ([1755] 1964) Essai sur la nature du commerce en général, London: Fletcher Gyles. Edited, with an English translation, by H. Higgs as Essay on the nature of trade in general, London 1931: Macmillan. Reprint 1964, New York, A. M. Kelley

Dobb M (1972) Theories of value and distribution since Adam Smith. Ideology and economic theory. Cambridge University Press, Cambridge

Faccarello G, Kurz HD (eds) (2016) Handbook on the history of economic analysis. Three vols. Vol. I: Great economists since Petty and Boisguilbert; Vol. II: Schools of thought in economics; Vol. III: Developments in major fields of economics. Edward Elgar, Cheltenham\Northampton

Garegnani P (1984) Value and distribution in the classical economists and Marx. Oxford Economic Papers 36:291–325

Gehrke C (2011) "Price of wages": a curious phrase. In: Ciccone R, Gehrke C, Mongiovi G (eds) Sraffa and modern economics, vol I. Routledge, London, pp 405–422

Gehrke C (2012) Rent, as share of produce, not governed by proportions. In: Levrero ES, Palumbo A, Stirati A (eds) Sraffa and the reconstruction of economic theory, volume 3: Sraffa's legacy. Interpretations and historical perspectives. Palgrave Macmillan, London

Gehrke C, Kurz HD (1995) Karl Marx on physiocracy. The European Journal of the History of Economic Thought 2(1):54–92

Gehrke C, Kurz HD (2006) Sraffa on von Bortkiewicz: reconstructing the classical theory of value and distribution. History of Political Economy 38:91–149

Kurz HD (2016a) Economic thought: a brief history. Columbia University Press, New York

Kurz HD (2016b) German and Austrian schools. In: Faccarello G, Kurz HD (eds) Handbook on the history of economic analysis, Vol. II: Schools of thought in economics, Cheltenham\Northampton, pp 252–273

Kurz HD (2018) Klasik Politik Ekonomi. In: Eren AA, Kirmizialtin E (eds) Iktisat Sosyolojsisi. Kurucu Düsünürler ve Iktisat Okullari Özelinde Bir Çalisma. Heretik, Ankara, pp 83–126

Kurz HD, Salvadori N (1995) Theory of production. A long-period analysis. Cambridge University Press, Cambridge

Kurz HD, Salvadori N (eds) (1998) The elgar companion to classical economics, two vols. Edward Elgar, Cheltenham and Northampton

Kurz HD, Salvadori N (eds) (2015) The elgar companion to David Ricardo. Edward Elgar, Cheltenham and Northampton

Malthus TR (1798) An essay on the principle of population, as it affects the future improvement of society with remarks on the speculations of Mr. Godwin, M. Condorcet, and other writers. J. Johnson, London

Malthus TR ([1820] 1989) Principles of political economy, Variorum edition, 2 vols, edited by John Pullen, Cambridge University Press, Cambridge

Marshall A ([1890] 1920). Principles of economics. 1st ed. 1890, 8th ed. 1920. Reprint, reset (1977), Macmillan, London

Marx K ([1867] 1954) Capital. A critique of political economy, vol. I. Progress Publishers, Moscow

Marx K ([1859] 1970) A contribution to the critique of political economy. International Publishers, New York

Marx K ([1861-63] 1988) Economic manuscript of 1861–63. A contribution to the critique of political economy ["Theories of Surplus Value"], in K Marx, Frederick Engels: collected works, vol. 30, International Publishers, New York

Mill J (1826) Elements of political economy. 3rd revised and corrected edition. Baldwin, Cradock and Joy, London. Reprint 1965, AM Kelley, New York

Mill JS ([1848] 1965) Principles of political economy with some of their applications to social philosophy, Toronto: University of Toronto Press.

Neumann, J. von (1937), Über ein mathematisches Gleichungssystem und eine Verallgemeinerung des Brouwerschen Fixpunktsatzes, *Ergebnisse eines mathematischen Kolloquiums*, 8: 73–83. English translation as A Model of General Economic Equilibrium. Rev Econ Stud, 13: 1-9.

O'Brien DP (1975) The classical economists. Oxford University Press, London

O'Brien DP (2004) The classical economists revisited. Princeton University Press, Princeton and Oxford

Pasinetti LL (1977) Lectures on the theory of production. Macmillan, London

Petty W ([1690] 1986) Political Arithmetick, London: Robert Clarel. Reprinted in: C. Hull (ed.) (1899), The economic writings of Sir William Petty, vol. I. Reprint (in one volume) 1986. A. M. Kelley, New York pp. 232-313.

Ricardo D (1951–73) The works and correspondence of David Ricardo, 11 vols, edited by Piero Sraffa with the collaboration of Maurice H. Dobb, Cambridge University Press, Cambridge

Roncaglia A (2001) The wealth of ideas. A history of economic thought. Cambridge University Press, Cambridge

Rostow WW (1990). Theories of economic growth from David Hume to the present. Oxford University Press, New York

Salvadori N (2020) Ricardo's theory of growth and accumulation. A modern view. In The Graz Schumpeter Lectures, Routledge, London.

Schumpeter JA ([1912] 1954) Epochen der Dogmen- und Methodengeschichte, in: M Weber (ed) Grundriß der Sozialökonomie, Tübingen: J.C.B. Mohr. English translation as Economic doctrine and method: a historical sketch, London: Allen & Unwin.

Smith A (1759 [1976]) The theory of moral sentiments, edited by DD Raphael, AL Macfie, The Glasgow edition of the works and correspondence of Adam Smith, Oxford: Oxford University Press

Smith A (1776 [1976]) An inquiry into the nature and causes of the wealth of nations, 2 vols, edited by RH Campbell, AS Skinner, WB Todd, The Glasgow edition of the works and correspondence of Adam Smith, Oxford University Press, Oxford.

Solow RM (2010). Stories about economics and technology. The European Journal of the History of Economic Thought, 17(5):1113–126

Sraffa P (1951) Introduction. In D. Ricardo (1951–73), vol. I.

Sraffa P (1960) Production of commodities by means of commodities. Cambridge University Press, Cambridge

Torrens R (2000) Collected works of Robert Torrens, 8 vols, edited and introduced by Giancarlo de Vivo. Thoemmes Press, London

Walsh V, Gram H (1980) Classical and neoclassical theories of general equilibrium: historical origins and mathematical structure. Oxford University Press, New York and Oxford

Neoclassical Economics: Origins, Evolution, and Critique

58

Reinhard Neck

Contents

Introduction	1502
What Is Neoclassical Economics?	1503
Predecessors	1505
Antoine Augustin Cournot (1801–1877)	1506
Heinrich Hermann Gossen (1810–1858)	1508
The Founding Fathers	1512
William Stanley Jevons (1835–1882)	1512
Léon Walras (1834–1910)	1516
Carl Menger (1840–1921)	1519
The Development of Neoclassical Economics: From Heterodoxy to Mainstream Microeconomics	1520
Partial Equilibrium Analysis	1521
Welfare Economics	1522
Properties of a General Economic Equilibrium	1523
Alternative Market Structures	1524
Macroeconomics: Economic Crisis as a Challenge for Neoclassical Economics	1525
The Great Depression	1525
The Answer: John Maynard Keynes (1883–1946)	1526
Keynesianism and Monetarism: Macroeconomics after World War II	1530
Conclusion	1535
References	1536

Abstract

This chapter provides an overview of neoclassical economics. The term is explained and contrasted with heterodox alternatives. The historical origins of neoclassical economics are presented, emphasizing some forerunners (Antoine Augustin Cournot, Heinrich Hermann Gossen) and discussing the three

---

R. Neck (✉)
Department of Economics, Alpen-Adria-Universität Klagenfurt; Karl Popper Foundation Klagenfurt; Kärntner Institut für Höhere Studien; and CESifo, Klagenfurt, Austria
e-mail: reinhard.neck@aau.at

© The Author(s), under exclusive licence to Springer Nature Singapore Pte Ltd. 2022
D. McCallum (ed.), *The Palgrave Handbook of the History of Human Sciences*,
https://doi.org/10.1007/978-981-16-7255-2_5

"founding fathers" of the English, Lausanne, and Austrian Schools (William Stanley Jevons, Léon Walras, and Carl Menger). A brief account of a few later contributions to neoclassical economic theory is given, and its ascent to become the dominant approach to economic theory is sketched. Economic crises and the lack of an adequate explanation for involuntary unemployment are identified as the main weaknesses of neoclassical economics. The alternative of Keynesian macroeconomics is presented, showing how macroeconomics dealt with these shortcomings. We conclude that neoclassical economics is still the dominant approach in economic theory, due to its superior mathematical methodology and its ability to embed alternative approaches into its system of theories as well as its ability to adapt its doctrines to accommodate elements of alternative theories.

Keywords

History of economics · Neoclassical economics · General equilibrium · Partial equilibrium · Microeconomics · Macroeconomics · Marginalism

Introduction

Seen from a bird's-eye view, the history of economic theory can be roughly divided into two periods: about one hundred years (1776–1870) dominated by the "classical school" (Adam Smith, David Ricardo, John Stuart Mill, and Karl Marx, among others) and the current period since 1870 with the emergence, rise, heyday, and ensuing dominance of the "neoclassical school." Although this picture is necessarily very superficial, it is not completely wrong when the state of the scientific discipline is evaluated according to the number and quality of the publications written during these years. The scientific literature in economics, more specifically contributions to the most respected journals and books, is dominated by research which can be characterized as following neoclassical methodological principles. Even the (numerous) critics of the neoclassical paradigm and "heterodox" economists start their investigations mostly by chafing at neoclassical economics. This demonstrates that a survey of neoclassical economics is an indispensable part of a handbook of the history of economic science. That said, it is difficult to choose from the wealth of contributions in this field. In this chapter, we concentrate on a few contributors to neoclassical economics and a main (partial) contender, namely Keynesianism.

The next section attempts to clarify the meaning of the term "neoclassical economics" and is followed by discussions of some of the predecessors of neoclassical economics (especially Cournot and Gossen), an overview of its founding fathers Jevons, Walras, and Menger as representatives of the three "schools" (English, Austrian, and Lausanne) and a sketch of a few later important developments, controversies, and criticisms, including Keynesianism as the main competitor (or fellow traveler) of neoclassical economics. The last section concludes by looking both backward and a little bit forward. Due to the huge number of economists who may be classified as neoclassical economists, a discussion of only a small selection of them must suffice. For more details, consult encyclopedias like the *New Palgrave Dictionary of Economics*, books about the history of economic doctrines like the classics Schumpeter (1954) and Blaug (1997) or the more recent three-volume handbook by Faccarello and Kurz (2016).

What Is Neoclassical Economics?

When economists talk to each other about "neoclassical economics," a strange aporia occurs, similar to what St. Augustine (*Confessiones* XI, 17) faced when asked, "What is time? If no one asks me, I know what it is. If I wish to explain it to him who asks, I do not know." ("Quid est enim tempus? Si nemo ex me querat, scio; si quaerenti explicare velim, nescio.") As for neoclassical economics, in addition to the presence of different (and partially mutually exclusive) attempts to define or explain this term, we have to distinguish between adherents and critics of the "neoclassical school" or paradigm, with the latter usually embellishing their justifications with some accusation of how unrealistic, unethical, etc. it is and why it contradicts common sense or good social science.

Looking into textbooks of economics, we find that sometimes neoclassical economics is identified with developments starting around 1870 and called "marginalism," an approach to analyzing economies that emerged with the use of calculus, and sometimes it has been equated with the dominant ("orthodox") economic science until today. Although neither of these interpretations is totally beside the point, it does not tell the whole story. Digging a little bit deeper and consulting leading histories of economic thought, we may distinguish between the use of mathematics as an instrument of neoclassical economic analysis and some basic assumptions (or axioms, or methodological prescriptions) on which neoclassical economics is founded.

For instance, in a widely used, brief, though precise, textbook of the history of economic thought, Sandelin et al. (2014) write:

> The neoclassical breakthrough is often dated to the 1870s. A characteristic feature of neoclassicism is its use of marginal concepts – such as marginal utility, marginal cost and marginal revenue – to determine the behaviour that drives the market forces of supply and demand. [...] the neoclassical economists generalized it [viz. the marginal principle] to a universal principle of rational economic behaviour. [...] The fundamental assumptions of utility and profit maximization were probably one reason why many early neoclassicists were inclined to use mathematics as a tool. Beginning to follow the methods of natural scientists, some of them considered their discipline to be an exact science like mechanics or other parts of physics. (pp. 40–42)

Somewhat more specifically, Roll (1992) differentiates neoclassical from classical economics (represented by economists such as Adam Smith, David Ricardo and Karl Marx) by emphasizing exchange instead of production as being central for markets, focusing on a subjective instead of an objective approach, starting from individuals instead of from the structure of a society, and doing (or purporting to do) scientific "economics" instead of classical "political economy." Samuels (1990) adds to this a list of different (and sometimes contradictory) meanings:

> (1) the subjective marginal utility theory of the 1870s and beyond; (2) the economics of Alfred Marshall; (3) the work of twentieth century writers working in the tradition, or mold, established by Marshall and some others, most notably Léon Walras; (4) some combination of the foregoing; (5) work under the aegis of the (Samuelsonian) neoclassical synthesis of microeconomic price and resource allocation theory with Keynesian macroeconomics. (p. 3)

It is interesting to note that the origin of the term "neoclassical economics" does not stem from an insider of this "school of thought" but from Thorstein Veblen (1900), a founding father of American institutionalism, in reference to Marshall's economics. For this and the subsequent evolution of the term, see Aspromourgos (1986, 2018). Veblen's text motivated Tony Lawson (2013) to link neoclassical economics to his ideas of social ontology, according to which "neoclassical" refers to economic approaches which recognize the importance of an evolutionary science but which remain in a "closed system ontology" (which roughly means that they deal inadequately with change and complexity). Although this seems to be a touch far-fetched, it even became the subject matter of an extensive body of secondary literature; see Morgan (2015, 2016), among others. Another possibility is to equate neoclassical economics with mainstream economics (e.g., Dequech 2007-2008) as opposed to several strands of heterodox economics, the latter comprising such diverse approaches as post-Keynesian, Marxian, classical (post-Ricardian) economics and many others. In view of the nature of the term "neoclassical" as a weasel word used by friend and foe alike, it is no wonder that Colander (2000) declared the death of neoclassical economics, albeit only in reference to its name but not its content.

Here, we will use the term akin to Arnsperger and Varoufakis (2006) without, however, taking on the negative implications drawn by these authors. Thus "neoclassical economics" refers to all work resting on the following three methodological principles:

- *Methodological individualism:* Neoclassical economists base their explanations on the behavior of individual agents. Alternatively, in the parlance of the structure-agency debate in sociology, their explanations start from agency within a given structure, which itself has to be explained by the (possibly past) actions and interactions of individual agents. Contrary to Arnsperger and Varoufakis, most neoclassical economists do not believe in this principle as a (hidden) "meta-axiom" in the sense of a dogma of faith; instead, they regard it as a method to explain social and, in particular, economic phenomena. Methodological individualism is fully consistent with ontological, political, and ethical individualism but does not imply any of them (and is not implied by any of them). There is a vast body of literature on methodological individualism in economics, sociology, and philosophy; see Udehn (2001) or, more recently, Zahle and Collin (2014) and Di Iorio (2015). Although the emergence of neoclassical economics with its

individualistic methodology may be explained by the flourishing liberalism of the late nineteenth century, it has become a methodological principle in its own right, independent of any political connotations.
- Instead of Arnsperger and Varoufakis' notion of methodological instrumentalism, the term *methodological preferentialism* is preferred: Neoclassical economists assume that individual agents (and they alone) have consistent (noncontradictory) preferences, which do not depend on their beliefs or on their resources (endowments). They aim to act in accordance with these preferences as far as possible, i.e., within the constraints given to them. While Becker and Stigler (1977) interpret the essence of methodological preferentialism *De gustibus non est disputandum* to mean that tastes neither change capriciously nor differ essentially between individuals, as a methodological principle for neoclassical economics, this is too restrictive. The intuitive justification for this principle is the idea that every human being attempts to do what they consider to be best for them in any particular situation, given their state of information in that situation. In a wider sense, it can be interpreted as the assumption of rational behavior or *homo oeconomicus* or rational choice; optimization is one form of such behavior but not the only one. The principle of methodological preferentialism has since found many applications in fields other than economics like sociology or political science (see Kirchgässner 2008). Like methodological individualism, methodological preferentialism is not without alternatives: methodological holism for the former and other forms of human behavior such as those outlined by Max Weber (1921/1922) for the latter.
- Methodological equilibration or (better) the *methodology of equilibrium analysis:* Equilibrium is a term that economists imported from the leading science of the nineteenth century, namely physics. It denotes a state of a system where, in the absence of external influences, there is no tendency for the system to change its position or its path. In neoclassical economics, in many cases (but by far not in all), it means the equality of (planned) supply and demand or, more generally, the equality of planned and actual values of variables. Even more generally, it can mean the same as the "solution concept" for a model or a class of models (for example, in game theory). This general meaning seems less controversial as it is an absolute requirement for a model to be solved; however, there may still be objections to model analyses in economics. More opposition from critics of neoclassical economics has arisen from the idea that equilibrium is assumed to emerge in a coordination process (especially one that is "automatic"), sometimes with normative implications. However, this is not an idea which has been embraced by all neoclassical economists.

Predecessors

Already during the period dominated by classical economics, in the eighteenth century and first two-thirds of the nineteenth, there were a few economists who developed ideas that later became the cornerstones of neoclassical economics.

Here the focus is on just two of them, Cournot and Gossen, but they had several contemporaries predating the great founders of neoclassicism, like Jules Dupuit (1804–1866) in France, who developed a theory of consumers' surplus, demand, and utility in the context of engineering. Likewise, some forerunners in the use of mathematical methods in economics should be mentioned here, such as Daniel Bernoulli (1700–1782) and Johann Heinrich von Thünen (1783–1850), who examined optimal decisions under uncertainty (Bernoulli 1954) and topics in spatial economics (von Thünen 1826).

Antoine Augustin Cournot (1801–1877)

Antoine Augustin Cournot was born in 1801 into a middle-class family in Gray in the French Département of Haute-Saône in the Bourgogne-Franche-Comté region. After attending school in Gray and Besançon, in 1821 he entered the École Normale Supérieure in Paris. In 1823, he graduated from the Sorbonne in mathematics and was immediately employed by the former Napoleonic Marshal Gouvion-Saint-Cyr as an advisor to the marshal himself and tutor to his son, a position that gave him enough spare time to continue his studies in various fields and obtain a doctorate in science (mechanics and astronomy) in 1829. Mentored by mathematician Poisson, Cournot became professor of mathematics in Lyon in 1834 but soon left this position for administrative posts as Recteur in Grenoble and later Dijon and as inspector general of public education. In 1862, he retired and returned to Paris but continued his research until his death in 1877. For more details on his life, see Moore (1905) and Martin (2016).

Antoine Augustin Cournot was a mathematician, philosopher, historian, and economist. He can even be regarded as an early sociologist (Leroux 2019). A highly prolific writer, he published books in all four main fields, translated works on astronomy and mechanics from English into French and edited and annotated a book by the great Swiss mathematician Leonhard Euler. The full range of his own scientific writings can be seen in his *Collected Works* (Cournot 1973–2010). His work in mathematics included an elementary textbook on calculus, a book on the theory of functions and infinitesimal calculus, and another on the correspondence between algebra and geometry as well as books and papers on probability. He wrote a history of modern ideas, one about the history of the fundamental ideas of science, and a book on public education in France. Among his philosophical books, there is a book on epistemology and one on the philosophical systems of his day. His first book on economics using mathematical methods, especially calculus (Cournot 1838), is particularly interesting. Unfortunately, this book was not well received either by his fellow economists or by mathematicians (cf. Theocharis 1990). In order to gain acceptance for his ideas, he wrote two other, less technical books on the same topic, the last one, putting his work in the context of the history of economic doctrines, appearing in the year of his death in 1877, which also were not well received.

Several reasons account for the lack of success of his mathematical work relating to economics. First, the mathematics used was relatively elementary and, hence, not of interest to mathematicians. Second, economics at that time was overwhelmingly nonmathematical and even well-educated economists simply did not understand either the mathematics or Cournot's revolutionary approach to economics. It was not until the "founding fathers" of neoclassical economics, especially Jevons and Walras, developed their own mathematical versions of price theory that Cournot became better known among economists. But there is also a personal reason for the unfortunate fate of his book: Cournot's life was marred by weak eyesight and a gradual loss of vision. In the last few years of his life, he was nearly blind and could not do any mathematical work; he needed helpers to read (nonmathematical) papers aloud and then write down and edit his texts so that they could be published. This may explain why he did not elaborate on some of his work, although highly original and creative, sufficiently well to make it attractive to its audience.

As the two later books did not contribute much to the edifice Cournot created, we shall concentrate here on his 1838 book, which is the only one that has been translated into English, edited by Irving Fisher in 1897 (2nd edition 1927). For a modern interpretation of Cournot's text, see Shubik (2018). In the preface, Cournot tries to correct economists' misconceptions about the potential of mathematics for their field of study, especially their belief that mathematics is only useful for numerical calculations:

> I propose to show in this essay that the solution of the general questions which arise from the theory of wealth, depends essentially not on elementary algebra, but on that branch of analysis which comprises arbitrary functions, which are merely restricted to satisfy certain conditions. (Cournot 1927, p. 4)

With typical modesty, he points toward the limited practical value of mathematical economic theory, stating:

> [...] I believe that there is an immense step in passing from theory to governmental applications; [...] and I believe, if this essay is of any practical value, it will be chiefly in making clear how far we are from being able to solve, with full knowledge of the case, a multitude of questions which are boldly decided every day. (Cournot 1927, p. 5)

Chapters I and II of Cournot (1927) outline the concepts of wealth and value, relating the latter to what is now known as a relative price. Chapter III develops a theory of exchange, including the special case of exchange equations derived from arbitrage conditions. Chapter IV contains what is probably the first mathematical formulation of the "law of demand," with demand for a good generally being a negatively sloped function of the price of that good. In Chaps. V–IX, Cournot develops a theory of markets, starting with monopoly (including monopoly with taxation) and proceeding to competition among producers and unlimited competition. It is remarkable that he did not start from perfect competition (anonymous competition among many market participants without any of them influencing the price), which was what most economists did before him and many still do after him, but from the cases of no competition (monopoly) and competition among the few, i.e., an oligopoly.

Cournot's theory of monopoly is virtually the same as can be found in any textbook of microeconomics today, an enormous achievement for a writer in the 1830s.

However, Cournot's accomplishments are even greater than establishing the mathematical theory of monopoly. This becomes clear when his theory of competition is examined and compared with the modern theory of games. Cournot's solution of the duopoly (competition between two producers) implicitly uses the fundamental concept of noncooperative equilibrium introduced more than 100 years later by John F. Nash (1951). This concept, which is sometimes rightly also called the Cournot-Nash equilibrium (though more often only the Nash equilibrium), considers a situation of independently optimizing (e.g., profit-maximizing) agents who mutually influence each other and do not cooperate; under some conditions, such a situation may result in an equilibrium. Cournot analyzed both a stable and an unstable equilibrium and derived the first-order conditions for such an equilibrium, assuming quantity to be the strategic variable of the agents. Although Bertrand (1883), in a book review of Cournot and Walras, is credited with having developed a parallel analysis with price as a strategic variable, this case is already in Cournot, as are concepts like the elasticity of demand (usually attributed to Marshall) and key concepts of the theory of the firm (Morrison 2003). Microeconomic textbook presentations of conditions like "marginal cost equals marginal revenue" for a monopoly as well as "marginal cost equals price" for perfect competition, including an examination of second-order conditions, were first developed by Cournot. Although even the high priest of neoclassical economics, Alfred Marshall, acknowledged Cournot's achievements, albeit in a more clandestine way (Shubik 2018, p. 2422), this author is still very much underrated in the literature on the history of economics.

The last three chapters (X–XII) of Cournot's treatise are less elaborate, but his attempts at obtaining results for "social income," an aggregate of the incomes or outputs of an economy, are brave digressions into the (future) field of macroeconomics and use rudimentary comparative statics methods to examine the effects of a change in price and consumption of a good on social income. Altogether, Cournot has to be regarded as a pioneer in mathematical economics, one of the fathers of modern microeconomics and, given his broad interests shown by his publications in diverse fields, a role model of a scientist, overcoming a physical handicap in an admirable way.

Heinrich Hermann Gossen (1810–1858)

If a book states in its preface:

> I believe I have accomplished for the explanation of the relations among humans what a Copernicus was able to accomplish for the explanation of the relations of heavenly bodies. I believe that I have succeeded in discovering the force, and in its general form also the law of the effect of this force, that makes possible the coexistence of the human race and that governs inexorably the progress of mankind. And just as the discoveries of Copernicus have made it possible to determine the paths of the planets for any future time, I believe that my

discoveries enable me to point out to man with unfailing certainty the path that he must follow in order to accomplish the purpose of his life.

Whether I have erred in this belief will depend on whether, like the discoveries of Copernicus, my arguments have the power of convincing others of their validity. When they will have thus proved themselves valid, a Kepler, a Newton may then succeed in working out more precisely the laws of the force that moves humanity! (Gossen 1983, p. cxlvii)

then we must expect it to reveal the most fundamental ideas about the laws of human society and, hence, to be an opus that will be eagerly read by social scientists, radically changing their way of thinking about the subject of their science. Indeed, the book by Gossen (1854) can be regarded as providing a cornerstone of modern (micro)economic analysis, the theory of utility, in mathematical form, which even today (at least in the neoclassical camp) is the starting point of textbooks and lectures about this subject all over the world. The book also contains several ideas and results that, much later, were rediscovered by various neoclassical economists.

Rarely have the staunch beliefs of the writer of a scientific work been ignored in such a cruel way as happened to this author. Apart from a few mentions in relatively obscure works, Gossen's book was not appreciated at all by his contemporaries in economics. Several years later, he withdrew the book from circulation. Ironically, the men who are now (rightly) generally considered to be the founding fathers of neoclassical economics, William Stanley Jevons and Léon Walras, discovered that the essence of some of the ideas they found independently was already in Gossen's forgotten book. These two men were kind enough to pay tribute to their unfortunate predecessor, and Walras (1885) wrote an article about Gossen after some research into his life.

All that is known about Gossen's biography is reported in Nicholas Georgescu-Roegen's extensive introduction to the English translation of Gossen's only book (Gossen 1983); see also Hayek (1991), Niehans (2018), or Kurz (2016) for other perspectives on Gossen's life and work. Heinrich Hermann Gossen was born in 1810 in Düren (then French, now German), the son of a tax collector, and attended grammar school, where he acquired all of his mathematical knowledge. On the orders of his father, he studied law and political science at the universities of Bonn and Berlin and started work as a civil servant in Cologne in 1834. As he disliked public service, his performance in this profession was far below average, and failing to obtain a promotion, he resigned in 1847. He then continued his studies in Berlin and founded an insurance firm, which did not turn out to be very successful either. He spent his final years living with his two sisters. Disappointed by the failure of his book and with his health deteriorating, he eventually died of tuberculosis in 1858 in Cologne.

Gossen's book of 1854 is by no means an easy read: It is not subdivided into chapters; it contains a lot of mathematical stuff, sometimes with boring numerical examples; and both the mathematical approach and the liberal policy prescriptions were contrary to the then prevailing views of the dominant German Historical School. Despite these shortcomings, reading the original, or the English translation, is still worthwhile today as it expresses several original ideas not contained in the works of more widely read authors. Most notable are Gossen's First and Second Laws. The First Law postulates that marginal utility decreases, or:

> The magnitude of pleasure decreases continuously if we continue to satisfy one and the same enjoyment without interruption until satiety is ultimately reached. (Gossen 1983, p. 6)

Here, in the context of the maximization of lifetime utility, Gossen introduces what later became known as marginal utility, thus foreshadowing the intertemporal optimization of an objective function and the idea of the allocation of time. In his mathematical treatment, he assumed linear marginal utility and did not mathematically formulate or solve the problem of maximizing utility over time. A reformulation of the First Law, which relates it to economics proper and which is sometimes called Gossen's Third Law, with the First Law referring more specifically to pleasure (utility) in view of limitations of time (Hayek 1991, pp. 364ff.), is:

> With the increase in that quantity, the value of each additional atom must decrease steadily until it sinks to zero. (Gossen 1983, p. 35)

While the general substance of the First Law was already known by utilitarian philosophers such as Jeremy Bentham and even earlier, this is not true of Gossen's Second Law, which was his own original idea and which he formulated as follows:

> In order to maximize his total pleasure, an individual free to choose between several pleasures but whose time is not sufficient to enjoy all to satiety must proceed as follows: However different the absolute magnitudes of the various pleasures might be, before enjoying the greatest pleasure to satiety he must satisfy first all pleasures in part in such a manner that the magnitude of each single pleasure at the moment when its enjoyment is broken off shall be the same for all pleasures. (Gossen 1983, p. 14)

In modern terminology, this states that the optimal allocation of resources requires the equality of marginal utilities in all different activities. It is a necessary condition for utility maximization, which "initiated the 'marginal revolution' in the theory of value" (Niehans 2018, p. 5391) or price theory.

In addition to this condition for the single individual, Gossen analyzed exchange and production by labor as methods of increasing "enjoyment." For bilateral exchange, the Second Law took the form of the equality of marginal utilities between individuals for each good, assuming cardinal measurement and the interpersonal comparability of utility. This later became, without these assumptions, the equality of marginal rates of substitution between all pairs of goods. Exchange in a market under given prices led to Gossen's formulation of the "banner of the marginal revolution" (Niehans 2018, p. 5329):

> Man obtains the maximum of life pleasure if he allocates all his earned money E between the various pleasures and determines the e [E and e are different amounts of money, RN] in such a manner that the last atom of money spent for each pleasure offers the same amount of pleasure. (Gossen 1983, p. 109)

Gossen also introduced a distinction between consumption goods ("objects of first category"), complementary goods ("objects of second category"), and "objects of third category," the latter indicating the means of production, which has a flavor of Menger's later distinction between goods of different orders. His theory of production is rudimentary, but there is an important consideration of (in modern terminology) labor supply, relating labor to disutility within his utilitarian framework. Although an explicit theory of market equilibrium is absent in Gossen's book, implicitly, the suggestion that excess demand (supply) leads to increasing (decreasing) prices points toward the assumption of a stable market equilibrium. Another interesting point is Gossen's conviction that utility, or the intensity of preferences, is measurable, coupled with his proposals to perform such measurements.

The last part of Gossen's book deals with his program for economic and social policy. Apart from a proposal for land reform, which involves public ownership of land, his ideas are extremely liberal (or, in US terminology, libertarian) in the sense of Adam Smith's "invisible hand" (without quoting him). In the context of proposals for monetary reform, his formulation of the quantity theory of money, which he relates to his First and Second Laws on the use of money by the individual, is worthy of note. This well-known theory comes with a sketch of a theory of demand:

The final result of a change in the quantity of the commodity serving as money is as follows:

- The prices of all objects offered for sale change proportionately to this change in quantity.
- The quantity of the commodity serving as money held by every individual for the direct satisfaction of pleasure – and, hence, also the resulting total life pleasure – varies closely with a change in the price of that commodity, which is in inverse proportion to a change in its quantity. (Gossen 1983, pp. 224 f.)

Gossen's political views are part of his metaphysical and even religious view of the world as a creation of God with some kind of preestablished harmony. If everybody acts in accordance with the laws of human nature (as discovered by him) and the "other" laws of nature, this will result in maximal welfare because it is in accordance with God's plans. In his concluding section, he states:

> Given the mode in which the other laws of nature function, the Creator knew how, through the law of decrease in pleasure, to cause man – the creature to whom He granted the greatest self-determination – to use this freedom only in the most desirable way for the benefit of the whole universe. [...] He made egoism the sole and irresistible force by which humanity may progress in the arts and science for both its material and intellectual welfare. (Gossen 1983, p. 299)

Thus Gossen's normative position concurs with what is often said to be a consequence of the subjectivism and individualism of neoclassical economics. However, in Gossen's case, it follows from his combination of a positivistic belief in natural-law-like social laws and religious belief in a benevolent God as the creator of the world including its laws. If one of these presuppositions is removed, totally different normative conclusions are possible. For example, if we do not believe in the laws as being equivalent to natural ones, there could be possibilities for governments to improve on the results of laissez-faire. So there is no strict relation between Gossen's scientific achievements – his utility theory – and his normative recommendations.

The Founding Fathers

Most books on the history of economics state that, starting in the early 1870s, neoclassical economics emerged independently from the ideas of three economists working independently of each other who went on to found their own schools: William Stanley Jevons with the English School, Carl Menger with the Austrian School, and Léon Walras with the Lausanne School. Although this is not completely wrong and will form the framework of this chapter, too, some qualifications are necessary. First, as the previous section has shown, there were important economists who formulated key ideas of neoclassical economics well ahead of the founding fathers. Second, although there are remarkable differences between the three which must not be overlooked (Jaffé 1976), none of them founded a "school" proper. They had students and followers, but the further evolution of neoclassical economics occurred through interaction among economists belonging to any of the three groups, and even the founding fathers themselves corresponded with each other. Finally, the direct influence of the three on the development of neoclassical economics was less important than that of other economists: Jevons' influence was greatly surpassed by that of Alfred Marshall; Menger had few direct students, in contrast to later generations of economists in the Austrian School; and Walras' successor, Vilfredo Pareto, was at least as influential as the founder himself.

The following subsections look at each of the economists briefly, including an attempt to characterize their specific contributions to neoclassical economics. Due to the availability of their works and a vast body of secondary literature, we confine ourselves to some general remarks on their lives and the main works relevant to the development of the neoclassical paradigm in economics.

William Stanley Jevons (1835–1882)

Among the three giants, the claim to primogeniture is due to William Stanley Jevons, as is clear from the paper (Jevons 1866) he already read before the Economic Science and Statistics Section of the British Association for the Advancement of Science in 1862. Moreover, Jevons was the most prolific writer of the three despite his short life span; he was a real polymath, publishing extensively in areas like philosophy, statistics, mathematics, meteorology, astronomy, engineering, and even music. For instance, he constructed a mechanical computer (the "logic piano") to derive logical conclusions from premises. To some, his hypothesis of sunspots being responsible for business cycles, though somewhat dubious, is his best known idea. His textbook on logic (not formal logic but Aristotelian) sold many copies, was widely used by students, and was regarded as the best introduction to the field for many years.

William Stanley Jevons was born in Liverpool in 1835. His parents were middle-class people with strong interests in science and engineering, arousing in their gifted son an early passion for these areas. He studied mathematics and chemistry at University College London, where he became heavily influenced by the French mathematician Augustus De Morgan, who gave the most advanced lectures in England at that time (Maas 2018). In 1853, Jevons became a gold assayer for the newly established mint in Sydney, Australia, a position that provided him with a good salary – essential in the light of his father's business going bankrupt – and allowed him enough time to pursue his interests in experimental science, on which (especially in meteorology) he contributed several well-received papers. He also wrote a social survey of Sydney based on observations obtained by examining especially the poor workers' quarters of the city.

Upon returning to England in 1859, Jevons resumed his studies at University College London, specializing in political economy, and graduated there. After a few not so well accepted publications on logic, chemistry, and economics, his breakthrough came in 1865 with his book on *The Coal Question*, a thorough statistical analysis which is still of interest today, not only for economic historians but also for environmental economists. It opened the door for him to become professor of logic and Cobden lecturer on political economy at Owens College in Manchester the following year. In this position, he published his two most important works, in economics (Jevons 1871) and the philosophy of science (Jevons 1874) as well as several other books and articles. In 1876, he returned to University College London as professor of political economy but resigned in 1880 due to the high teaching load. In the last years of his life, Jevons wrote about economic policy and particularly social policy problems. In 1882, he drowned while swimming near Hastings on the south coast of England.

Although several other economics publications by Jevons are also well worth reading, his opus magnum in terms of economic theory is undoubtedly *The Theory of Political Economy* (Jevons 1871; the final fifth edition was edited by his son, Herbert Stanley Jevons, in 1957). In the preface, William Stanley Jevons stated clearly the double objective of this book: first, to base economics on what he considered psychological foundations, that is, on the needs and wants of the consumers, and second, to use mathematical methods to obtain insights into economic problems. In his words:

> In this work I have attempted to treat Economy as a Calculus of Pleasure and Pain, and have sketched out, almost irrespective of previous opinions, the form which the science, as it seems to me, must ultimately take. I have long thought that as it deals throughout with quantities, it must be a mathematical science in nature if not in language. [...] The Theory of Economy thus treated presents a close analogy to the science of Statical Mechanics, and the Laws of Exchange are found to resemble the Laws of Equilibrium of a lever as determined by the principle of virtual velocities. (Jevons 1871, pp. viif.)

This statement can be interpreted as the neoclassical research program in a nutshell: As consumption is the ultimate goal of all economic activities – an insight already gained by Adam Smith – the study of the economy has to start with demand and with the consumer or household. Demand rather than supply ultimately explains economic relations and production, in contrast to the Classical School of Smith, Ricardo, and John Stuart Mill, and also has to be explained by the needs and wants of households (and individuals). Bentham's philosophy of utilitarianism had an influence on Jevons, as he himself admitted. Methodologically, the emphasis is on using mathematics because of the analogy between physical and economic systems, and the advances made by physics and engineering in the nineteenth century point toward promising uses of similar methods in the social sciences and economics in particular.

In the preface to the second edition, Jevons severely criticizes the then prevailing classical economists Ricardo and Mill (the latter, in particular, frequently throughout the book as well) and distances himself from them and the "English School" which were completely ignorant of the insights of authors such as Cournot and Gossen, which he only discovered after the first edition of his book had been published (Jevons 1957, pp. xlivf.). He also refers to his congenial neoclassical colleague, Léon Walras (whom he called his friend and with whom he corresponded repeatedly), and his central idea of "rareté" (scarcity) to explain economic value. As a sign of his (and neoclassical economists') break with the Classical tradition, he favors the label "economics" instead of the classical term of "political economy" (Jevons 1957, p. xiv). In his characteristic modesty, he pays tribute to predecessors like Dupuit, Cournot, and Gossen, stating that "it is evident that novelty can no longer be attributed to the leading features of the [his] theory" (ibid., p. xxxviii). This seems to be somewhat exaggerated and belittles his achievement of independently (re-)inventing the results of his predecessors and presenting them in a more accessible way than they did – at the age of 36 (or, if we consider the first presentation, 27).

As for the use of mathematics in economic analysis, Jevons collected a list of predecessors and contemporaries, which was continued after his death by his wife, his son, and Irving Fisher. He repeatedly emphasized that

> [...] all economic writers must be mathematical so far as they are scientific at all, because they treat of economic quantities, and the relations of such quantities, and all quantities and relations of quantities come within the scope of mathematics. Even those who have most strongly and clearly protested against the recognition of their own method, continually betray in their language the quantitative character of their reasonings. (Jevons 1957, p. xxi)

This strong plea for mathematization, formalization, and quantification may seem superfluous today when economic theory is regarded by many as an application of mathematics, but in the nineteenth century this claim was regarded as exotic by the overwhelming majority of economists.

The structure of Jevons' book is as follows: He starts with a brief introduction on the scientific character of economic theory, arguing again for the use of mathematics. In Chapter II, a "theory of pleasure and pain" is sketched, resulting in what was later called diminishing marginal utility (Gossen's First Law). Instead of marginal utility, he uses the somewhat misleading term "(final) degree of utility" (in contrast to "total utility"). The equality of marginal utility for different goods as a characteristic of the household's optimum (Gossen's Second Law) is justified verbally using a marginalist argument, without a link to the optimization calculus behind it. A rudimentary (verbal) extension to a dynamic utility calculus is also provided. Instead of the notion of "household" for an economic agent, he uses "trading bodies" for market participants, which may include individuals and groups of individuals, as well as households and firms.

Next, Jevons moves onto a theory of exchange, the topic of Chapter IV, which is probably the best and most important part of the book. He defines the notoriously obscure word "value" as the ratio of exchange, introducing the idea of a relative price, which then became the central tenet of neoclassical price theory. In contrast to Cournot's procedure, Jevons starts from a perfect market (and, tacitly, perfect competition) and states the "law of one price" (he calls it the "Law of Indifference") for such markets, characterizing an equilibrium in such a market by the condition that the relative price of each pair of goods exchanged must equal the inverse of their marginal utilities (in modern terms: their marginal rates of substitution). Problems with indivisible goods, the indeterminacy of bilateral trade, and others are discussed briefly, contrary to the claims of some critics (Peach 2018). Polemicizing against John Stuart Mill, Jevons stresses the gains from trade (international or otherwise) due to the voluntary nature of any exchange (under the conditions assumed).

Chapters V, VI, and VII deal with the markets for the factors of production and their prices: labor – wages, land – rent and capital – interest, respectively. The secondary literature is divided over the interpretation of what Jevons' theory really means, but when looking at what his son wrote (in Appendix I) and the preface to the 2nd edition (Jevons 1957), it should be clear that he changed his position from one marred by traditional Classical doctrines to a truly neoclassical marginal productivity theory (albeit without an explicit production function). Each factor of production contributes to the product and receives its reward according to the evaluation of the output, on the one hand, and its own contribution, on the other; there is no difference between these factors with respect to the relation between the amount of their use in production and their reward, which is finally driven by the utility attributed to the output by the consumers.

From the concluding remarks (Chapter VIII), Jevons' formulation of the economic problem as the subject of his book is remarkable:

> The great problem of Economy may, as it seems to me, be stated thus: *–Given, a certain population, with various needs and powers of production, in possession of certain lands and other sources of material: required, the mode of employing their labour so as to maximise the utility of the produce.* (Jevons 1871, p. 255, italics in original)

This statement clearly foreshadows Robbins' famous definition of the economic problem. It clearly formulates a research program for neoclassical economics. Notwithstanding the incomplete execution of this program by Jevons himself, it is fair to credit him with the honor of having gone several steps along the path prescribed by that program. The clear exposition of his theory is also to his credit, which distinguishes it from his predecessors.

Léon Walras (1834–1910)

Marie-Esprit Léon Walras was born in Évreux in the Département Eure, France, in 1834. His father Antoine Auguste Walras was a teacher of philosophy and rhetoric and an educational administrator. Although he did not hold a truly academic position, he was not only a philosopher but also an economist, and he had a significant influence on his son Léon (as the latter was known). The *Collected Works in Economics* of Walras father and son (Walras and Walras 1987–2005) were even edited together to stress the direct intellectual link between the two. Although having obtained a Bachelor's degree, Léon twice failed to enter the École Polytechnique, dropped out of the École des Mines, and then tried to find his place in French society as a writer of novels, a journalist, a clerk in a railway office, the director of a bank for cooperatives, a newspaper editor, a public lecturer, and a bank employee after his bank went into liquidation. He did not obtain an academic qualification in economics and hence never found an academic position in France.

However, under pressure from his father, he committed himself by solemn oath to continuing his father's endeavors in economics, in particular by following his policy prescriptions (which included the aims of nationalizing land and abolishing income taxes) and strengthening them scientifically with the help of mathematics. Indeed, Léon remained faithful to this obligation and wrote many publications on these and other topics of economic policy and social reform. He even applied twice (unsuccessfully) for the Nobel Peace Prize for having delivered a scientific answer to the so-called social question. He also considered his theoretical work, which is of interest in the context of neoclassical economics, as the foundation for his applied and policy writings (Baranzini 2016). Strongly opposed to the laissez-faire positions of his contemporary French economists in the tradition of Jean-Baptiste Say, Walras as a social reformer (or even socialist) can be seen as a heterodox economist during his lifetime (Gallois 2011) – an irony of fate in view of the frequent criticism of neoclassical theory as "neoliberal mainstream."

In spite of his lack of academic credentials as an economist, Léon Walras was lucky to obtain a position at the University of Lausanne, Switzerland, after he gave a presentation at a conference there in 1860. Ten years later, after having proved his worth as an academic economist, he was recommended by an influential politician from the Canton Vaud and appointed tenured full professor in Lausanne in 1871. Although he retired from his professorship for health reasons, he was productive as a scientific writer almost until the end of his life. The *Collected Works* contain several other theoretical and many applied economic books and articles, but we will concentrate on his most important work, that is, his *Éléments d'Économie Politique Pure*, of which four (or, if the posthumous definitive edition is counted separately, five) editions appeared. It is also the only book that was translated into English, the definitive edition appearing as late as 1954 in a translation by William Jaffé (Walras 1954), with a translation of the third edition (by Donald A. Walker and Jan van Daal) published in 2014. The reason for the new translation lies in a strange scientific dispute over Walker's claim that the third edition should be considered the final and best one due to Walras' diminishing intellectual powers after its appearance (see De Vroey 2017 and the references given there). Not in a position to judge this highly contested hypothesis, we will stick to Jaffé's position and his translation of the definitive edition here.

The first edition of *Éléments* appeared in two parts, in 1874 and 1877. In later editions, the order of presentation was changed and several substantial modifications were made. Part I positions the pure theory within Walras' research program of theoretical and applied economics, emphasizing social wealth as the main subject matter of pure theory, which is defined as all things that are "rare," by which he means scarce and useful. This term had already been used many times by Walras' father; later in the text and elsewhere, he also uses it as being synonymous with "marginal utility." Part II analyses the bilateral exchange of two commodities and determines the equilibrium, a situation where the effective offer (his notion for supply) and the effective demand are equal. Utility curves are introduced, the conditions for the maximization of utility are derived, and the relationship between utility and demand is discussed, including a sketch of the derivation of the demand curve from utility maximization. He stresses the importance of "rareté" as the cause of the price of the good, or the value in exchange.

Part III of the definitive edition contains what Walras is famous for, an analysis of general equilibrium, which is a situation where all markets in an economy are simultaneously in equilibrium. As a first step toward an analysis of general equilibrium, he extends the results of Part II to the case of three commodities and then presents the conditions for the general m-commodity market geometrically and algebraically. Money is introduced as "numéraire," a standard of value, but its main function as a medium of exchange, which was included rudimentarily in the first edition, is treated in much more detail in Part VI from the second edition onward. The conditions for a general equilibrium express the equality of the quantities offered to those demanded and determine the prices of the commodities. These are relative prices (price ratios) or prices expressed in terms of the "numéraire." As the m equations are not independent due to the condition that the total value of social wealth must be equal to the total value of the commodities offered (an aggregate budget constraint, later termed Walras' Law), $m-1$ equations serve to determine $m-1$ prices and, hence, according to Walras, allow a solution to be calculated.

In the situation of a disequilibrium, when offer and demand do not coincide for one or several goods, Walras proposes a so-called "tâtonnement" (groping, trial-and-error) process, which he models in a similar way to what happens at an auction: An auctioneer calls a set of randomly chosen prices for each commodity. The market participants choose their bids and let the auctioneer know. For goods with excess demand (demand greater than offer), the auctioneer increases the price and decreases it for goods where there is excess supply. This procedure continues until all markets attain their equilibrium. It was clear to Walras that this "tâtonnement" process is merely a notional device to show that an adjustment process where the price increases on excess demand leads to a general equilibrium. Further topics in the central Part II of the *Éléments* include an analysis of the effects of changes in prices as a result of changes in utility and in the quantities possessed (his Law of the Variations of Commodity Prices), the fact that the general equilibrium remains unchanged under some additional assumptions if commodities are redistributed among traders (his Theorem of Equivalent Redistributions of Commodity Holdings) and the refutation of previous price theory, especially the labor theory of value and Say's utility theory.

Part IV provides a theory of production, in which production uses as factors of production land, labor, and capital. He distinguishes capital (in the broader sense), which, as income, may be material and immaterial and includes all factors of production owned by the agent concerned, from capital proper, which is an industrial product (commodities produced as factors of production). He introduces the entrepreneur as an agent who organizes and implements the production and sales process; the entrepreneur may receive positive or negative (loss) profits in contrast to incomes from the production factors (rent for land, interest for capital, and wages for labor). The theory of general equilibrium is then extended to a market system with production using the same mathematical apparatus as in Part III. Extensions to markets with capital formation and credit are the subject of Part V while Part VI presents a theory of money, with a general-equilibrium investigation into different forms of money (bimetallic standard, fiduciary money, and others) and of foreign exchange. Here, in his pure theory with perfect competition, the economic content is close to the classical dichotomy, separating the monetary sphere from the real one, while dynamics and disturbances are relegated to applied theory.

Part VII contrasts his (neoclassical) theory of production based on the theory of marginal productivity with the physiocratic and classical theories of explanations of prices and income from factors of production. The final Part VIII deals with the effects of measures introduced by the state in the form of price fixing and taxation as well as of monopoly on the general equilibrium. All of these generalizations were later taken up and extended by other authors working within the general-equilibrium paradigm of neoclassical microeconomic theory.

It must be stressed that the analysis of the interdependences of markets in Walras' general equilibrium theory assumes, as given, the initial resources, production technology, the preferences of the economic agents, and their initial endowments. Perfect competition is also assumed throughout. These assumptions are still made in many general equilibrium models today, often without detailed discussion. Walras had attempted to complement his treatise on pure economics (general equilibrium theory) with two more, on social economics (theory of the distribution of wealth) and applied economics (theory of the production of social wealth). This was only partially implemented in his late writings, but he was well aware that his "pure" theory did not provide an approximation to reality. Rather, it is something of a

Platonic "idea" in a metaphysical sense; it may thus be criticized as a form of Model Platonism as elucidated by Hans Albert (1998).

Carl Menger (1840–1921)

Carl Menger has the unique position in the history of economic doctrines of being a founder of both mainstream neoclassical economics and a school, the Austrian School, which, at present, is regarded by many as heterodox. Although this is mainly due to later developments within this "school," namely those emanating from Ludwig von Mises (1881–1973) and Friedrich August von Hayek (1899–1992), here it is necessary to concentrate on the neoclassical elements of Menger's theory; the Austrian School proper is dealt with in another contribution to this Handbook. Critics of neoclassical economics denounce the Austrian School for its radical economic liberalism (or libertarianism), which is in marked contrast to most other heterodox economic schools, such as Marxist, Post-Keynesian, or ("old") institutionalist economics, which are more ready to accept or support government interventions in markets. But again, as with Walras, the case for Menger being a promoter of libertarianism is not strong, to say the least; this characteristic is due to Mises and Hayek but not to the earlier members of the Austrian School (for this issue and the following text, see Neck 2014).

Carl Menger was born in 1840 in Neu-Sandec in Galicia (now Nowy Sącz in Poland), which, at the time, was part of the Habsburg Monarchy and, after 1867, the Austrian or "Cisleithanian" half thereof. He studied law and economics at the universities of Prague and Vienna and obtained his Doctor of Law from the University of Cracow. After some years as a journalist, he started doing research in economics, resulting in his first and most important book (Menger 1871). Although it was called a "first part," there never was a second one, despite Menger living until 1921 and retiring prematurely in order to devote more time to completing the planned four-volume treatise; instead, his son Karl, a mathematician, edited a posthumous second edition of the 1871 book with considerable additions, which appeared in 1923.

The book counted as his "Habilitation" thesis, giving him the right to teach and supervise doctoral students at the University of Vienna, and later leading to the positions of Associate (1873) and Full Professor (1879) at that university. Moreover, in 1876 Menger was appointed private lecturer to the emperor's son Crown Prince Rudolf (who committed suicide in Mayerling, Lower Austria, in 1889). Fortunately, these lectures have survived in the form of shorthand notes by the Crown Prince, providing us with valuable information about Menger's (and, for that matter, also Rudolf's) positions on economic policy issues (Streissler and Streissler 1994). Thanks to his services in this capacity and advice given to the monarchy's government in monetary affairs, he was conferred the title of "Hofrat" (privy councillor) and made a member of the "Herrenhaus," the House of Lords.

Among Menger's further academic work, his book on the methodology of social sciences and economics (Menger 1883) is most important. It launched the so-called

"Methodenstreit" (Battle of Methods) with the German Historical School of Economics, and with Gustav Schmoller in particular. Schmoller's (1883) criticism of Menger was followed by a harsh rebuttal (Menger 1884), and both sides continued the battle without coming to even a minimal consensus. In addition, Menger worked on capital theory, money, and currency reform. He retired from his chair in 1903 and died in 1921.

In Austria, Menger had a number of followers, making up the second generation of the Austrian School of Economics, although he did not agree with all of their theoretical positions. Among them, Eugen von Böhm-Bawerk (1851–1914) and Friedrich von Wieser (1851–1926) are best known. A third generation, not directly influenced personally by Menger, includes Ludwig von Mises and Joseph A. Schumpeter (1883–1950), among others. Menger also exerted considerable influence on nineteenth century economists abroad, especially in Italy, the Netherlands, the USA, and the UK.

Carl Menger's main contribution to economic theory was his subjective value theory, which he developed after critically contesting the classic explanation of prices by production costs as explained especially by Adam Smith and David Ricardo (and implicitly Karl Marx and his labor theory of value, although Menger, in contrast to Böhm-Bawerk, did not take position directly against him). It starts with the individual and the individual's needs, which, together with the constraints the individual is confronted with, form the basic driving force of economic decisions. According to Menger, the value or price of what he calls first-order goods (consumer goods) is determined directly by the decisions of individuals and especially their needs while the value of so-called higher-order goods (goods with derived demand, especially capital goods) is derived from the demand for first-order goods and, hence, indirectly from consumers' valuations. This creates a chain of valuation processes related to the subjective needs of individuals.

Menger's subjective theory of value, coupled with his more elaborate price theory under monopoly and under competition, is the main reason that led to his being regarded as one of the three fathers of neoclassical economics, together with William Stanley Jevons and Léon Walras. On the other hand, Menger and other Austrian School economists departed from the mainstream of neoclassical economics by linking supply to subjective valuations as well (the notion of opportunity costs having been created by Menger's follower, Friedrich von Wieser), by regarding mathematics to be of limited use in economics and by emphasizing dynamic processes instead of concentrating on states of (partial or general) equilibrium. Being the first of the three fathers of neoclassical economics to explain prices primarily by subjective valuations, Menger, nevertheless, was strongly influenced by German economists of his time, although it seems that he was not aware of Gossen's work (or of Cournot's).

The Development of Neoclassical Economics: From Heterodoxy to Mainstream Microeconomics

The evolution of neoclassical economic theory in the 100 years since the death of its last founding father can be characterized by three main characteristics: The neoclassical paradigm gradually began to make inroads into academic economics, the "three schools" amalgamated, and higher mathematics was increasingly used for analysis. The first characteristic has resulted in the modern theory of microeconomics, laid down in (elementary to advanced) textbooks, taught in courses at universities worldwide, and dominating the research literature in the leading (English language) journals. There is a consensus among a large group of economists in many countries about the truth and adequacy of the results of neoclassical microeconomic theory (although less so about macroeconomics). The second characteristic, in fact, debunks the myth of the "three schools." Already the founding fathers corresponded with each other about matters of economic theory, and their contemporaries and followers did so at an increasing pace, supported by the rise in communication technologies. The only exception is the split of the Austrian School from the neoclassical mainstream, but this is a development dating from the second half of the twentieth century (and from outside Austria). Finally, the increasing use of mathematics, as also reflected in the curricula of economics departments, has led to the discovery of many theoretical insights and even (at least in the case of game theory) to new fields of mathematics. It is difficult for the "Gallic village" of heterodox economics to have its voice heard by neoclassical economists now representing the mainstream, although at least macroeconomics developed a respected alternative, Keynesianism, and in times of crisis, alternatives to neoclassical economics are more fashionable than otherwise.

It is not possible to sketch the development of neoclassical economics over the last hundred plus years in a single article; moreover, there are several excellent books on this topic, from the textbooks and handbooks of economic doctrine to more specialized surveys. What can be done here is to present a few glimpses of milestones in this evolution, however subjective and superficial this might be. We think that the most interesting and important work has been done in the following four fields.

Partial Equilibrium Analysis

The Walrasian theory of general equilibrium was initially not well received by economists, especially in the English-speaking world, due to both the relatively high degree of mathematical knowledge it requires and the fact that it was written in French. A more practical way of approaching economic problems was and still is partial equilibrium analysis, the analysis of the market for a single good without considering other markets, except under some "ceteris paribus" conditions. The doyen of this type of investigation was undoubtedly Alfred Marshall (1842–1924), who became very influential as professor of political economy at the University of Cambridge from 1885 to 1908, the most prestigious chair in the UK. At least as important for his fame is his textbook *Principles of Economics* (Marshall 1890), in which he presented an overview of the microeconomic theory of his time and included many of his own results. Virtually the entire theory of costs and production in today's elementary and intermediate textbooks comes from this book, as do notions like the elasticity of demand, consumers' surplus, substitution, and long and short periods. Although Marshall also uses mathematics to derive his results, the level is considerably lower than with Walras and the book is much easier to read than Walras' *Éléments* (even in the excellent English translation by Jaffé) – in Marshall's book, there are nearly no mathematical formulae but many descriptive illustrations and practical examples. Generations of economics students and practitioners have used this book, which saw eight editions published during its author's lifetime, followed by a carefully arranged ninth variorum edition published under the auspices of the Royal Economic Society in 1961. Of course, abstaining from taking the interdependence of markets into account has its price: Some conclusions drawn by Marshall are rather far-fetched, and the economic laws he states stand on shaky ground, as do his policy conclusions. Nevertheless, the book is still worth reading today, not only for historians of economics but also for business people who want a lively introduction to the field.

Welfare Economics

In spite of economists' frequent, explicit claims to avoid value judgments, neoclassical economics has developed an extensive normative apparatus for its theoretical toolbox. Its general-equilibrium version comes from Vilfredo Pareto (1848–1923), Léon Walras' successor to the chair of economics at Lausanne. Pareto was the son of an Italian Marchese and a French woman; he published in both languages. After graduating in mathematics, physics, and engineering and studying economics on his own, Pareto contributed a great deal to economic theory (among others, Pareto 1906), applied economics, and statistics; later in life, he also wrote sociological books, but his best known concept was the idea of Pareto optimality or allocative efficiency. A state of an economy which has this property is such that it is impossible to increase the utility of any of its agents without reducing the utility of at least one other member. This is a weak criterion because also rather undesirable states fulfill it: For example, a situation where one agent has all of the goods of the economy and all the others have none is Pareto efficient if the rich agent does not want to donate anything to the poor. Pareto showed that under some additional conditions, the general equilibrium of an economy with perfect competition is efficient in the above sense, a result that later became known as the First Fundamental Theorem of Welfare Economics and is frequently (though not quite rightly) used as a justification for the optimality of a free market economy.

Another way of introducing welfare considerations came from Arthur Cecil Pigou (1877–1959), who succeeded Alfred Marshall as professor in Cambridge. In Pigou (1920), he developed the concept of externalities, that is, the positive or negative effects of an agent's economic actions (in consumption or production) that benefit or harm another agent without being rewarded or sanctioned by the market. Environmental problems are usually caused by such (mostly negative) externalities. Pigou proposed subsidies and taxes (the "Pigovian tax") for positive and negative externalities, respectively. Another solution was later proposed by Ronald H. Coase (1910–2013) based on negotiations between the originator and the person affected to internalize the effect by payments to either party. Thus, very early onward, neoclassical economists dealt with one of the world's most pressing problems nowadays, developing the subdiscipline of environmental economics.

An important controversy which lasted for several decades was the so-called Socialist Calculation Controversy about the possibility of an efficient allocation of resources in a socialist economy, i.e., an economy with private ownership of capital abolished and with central government planning of the economy. Ludwig von Mises opened this debate in 1920 by explicitly denying this possibility due to socialism's purported lack of rational accounting. Socialist Polish economist Oskar R. Lange (1904–1965) countered by constructing a model of a socialist quasi-market economy, where central planning is carried out by managers sworn in to set prices equal to marginal costs, as is done in a market economy under perfect competition. In this way, according to Lange, capitalist profit maximization and, hence, allocative efficiency could be preserved in a socialist economy as well. Later contributions introduced the problems of imperfect information and different incentive structures in both systems as arguments for or against socialism. Eventually, the controversy was decided empirically by the breakdown of "real socialism," the Eastern block and the Soviet Union in 1989/1991. For neoclassical theory, the debate showed how centralized, in fact, the Walrasian model of general competitive equilibrium theory is – perhaps it is even better suited as a model for a socialist rather than for a capitalist economy.

Properties of a General Economic Equilibrium

In line with Walras, neoclassical economists investigated the fundamental properties of a general equilibrium covering its existence, uniqueness, and stability. Walras was content to assert that if there are as many exchange equations as prices to be determined, then a general equilibrium of an economy exists. In the 1920s and 1930s, however, some problems arose with free goods (goods without a positive price) that could not be solved by counting equations. In a mathematical colloquium in Vienna, which gathered economists, mathematicians, and some philosophers from the Vienna Circle, the mathematical statistician Abraham Wald (1902–1950) raised the issue of the lack of existence of a general equilibrium and supplied a partial answer. General answers defining the conditions for the existence of a general equilibrium were given by Kenneth J. Arrow (1921–2017), Gérard Debreu (1921–2004), and Lionel W. McKenzie (1919–2010) using advanced topological methods (cf. Düppe and Weintraub 2014).

The problems as to whether there is only one equilibrium or more (uniqueness) and whether an economy returns to an equilibrium after having left it were also studied but with less success. It turned out that fairly restrictive assumptions were needed to demonstrate these properties. An even more devastating result is the Sonnenschein-Mantel-Debreu theorem (cf. Kirman 2016). These authors showed that the only conditions imposed on aggregate excess demand by the usual assumptions about individuals in general-equilibrium theory are the continuity of the aggregate excess demand function in prices, Walras' Law, the homogeneity of degree zero (only relative prices matter for the individual's decision), and aggregate demand going to infinity when the price goes to zero (nonsatiation). As there are economies with multiple and unstable equilibria fulfilling these conditions, we cannot expect to find reasonable conditions for the uniqueness and stability of a general equilibrium of an economy, which raises serious doubts about the scientific value of general equilibrium theory.

Alternative Market Structures

Perfect competition and monopoly may represent extreme types of markets, but other ones are more frequent in reality. Restaurants, for instance, usually have a menu with more or less variation: Chinese, Italian, French food, etc. They are confronted with the demand curve of lovers of the different categories of food and can opt to change their prices even if their competitors (suppliers of related but not identical food) do not do this. Based on this observation, Edward H. Chamberlin (1899–1965) and Joan M. Robinson (1903–1983) created the theory of monopolistic or imperfect competition according to which competing firms, within limits, behave like monopolists because they can, through product differentiation, tie a segment of customers to them. The standard formalization of this market was developed by Dixit and Stiglitz (1977); it is now a workhorse model with many applications.

Oligopoly is another possible market structure: competition among the few, not the many. Cournot presented an early model of a duopoly; in his case, the roles of the competitors are symmetric, and both use either prices or quantities as choice variables. An alternative is the asymmetric solution introduced by Heinrich von Stackelberg (1905–1946): Here a "leader" has the advantage of deciding on and implementing their strategy while the other agent, the "follower," has to act upon the decision of the leader. Both the Cournot solution and the Stackelberg solution were adopted by game theorists as solution concepts for a strategic game. Game theory itself, although extending far beyond economics, can also be interpreted as a spin-off of neoclassical economics because it was invented thanks to the collaboration of an economist (Oskar Morgenstern, 1902–1977) and a mathematician (John von Neumann, 1903–1957); it assumes that each player (agent) tries to maximize their own utility.

Imperfect competition, oligopoly, and other market structures are topics relating to industrial organization, a branch of microeconomics which deals with the effects of different market structures on the behavior and performance of an economy. It is an active area of research which has resulted in many insights into markets going far beyond those investigated before World War II. Especially in connection with research on imperfect information and the resulting incentive structures, the theory of industrial organization has provided a new view on the market economy as well as a more differentiated view of the functioning of economic systems than the dichotomy of a market versus a planned economy. The same is true for the theory of the firm, which has been greatly modified by taking issues of information and incentives into account going far beyond the black box view held by the first generation of neoclassical economists in the nineteenth century.

Macroeconomics: Economic Crisis as a Challenge for Neoclassical Economics

The Great Depression

After World War I and the economic turbulences in its aftermath, in the 1920s ("the Roaring Twenties"), the economies of most industrialized countries boomed. Some economists (notably Irving Fisher) even thought that the scientific understanding of an economy had reached such a high standard that crises could no longer occur. This was attributed to the dominant economic doctrine of the day, which was one or another version of neoclassical economic theory. Although there were still alternative ways of doing economics, such as American Institutionalism or the German Historical School, both of which rejected methodological individualism and deductive reasoning (especially mathematics), these schools gradually lost their adherents and became minority programs within economics. The normative consequences of neoclassical economics, in the sense of advocating laissez-faire policies, became predominant ideologies in (at least academic) economics to a similar extent as in the era of prewar liberalism. Nevertheless, economic policy makers did not shy away from intervening in the economy, and first attempts at developing scientific theories of economic policy appeared in the literature (Braun 1929; later: Bresciani Turroni 1942).

Against this background, the onset of the Great Depression, an economic crisis that lasted from 1929 to 1939, came as a severe shock, not only due to its severity (a decrease in production and income and an increase in unemployment to more than 25 percent of the labor force in most industrialized countries at the height of the crisis in 1932–1933) but also to its long duration. It started with a stock market crash on the New York Stock Exchange in October 1929 and spread rapidly throughout the United States, to Europe, and the rest of the world. The fall in stock prices caused by their speculative overvaluation led to panics and excessive debt, a loss in consumer confidence, bank runs, the bankruptcy of banks and manufacturers, massive layoffs of workers, and a self-enforcing vicious circle of downturn in all major economies. Although there is some consensus among economic historians that misguided economic policies contributed to the extent of the Great Depression (Whaples 1995), the question as to which alternative(s) could have avoided its spread to the global economy is still subject to disagreement.

High, increasing, and seemingly endless unemployment was the most catastrophic feature of the Great Depression. This came at a time when unemployment insurance was underdeveloped or even nonexistent and labor force participation was much lower than today: When a breadwinner lost their job, this often meant a loss of income for the entire family. Those affected by this situation suffered from poor mental health and feelings of being worthless, which very frequently led to depression and, in many cases, to suicide. An alternative reaction was a loss of confidence in ruling elites and in democracy and a tendency to join extremist political parties such as the Fascists in Italy and other countries, the Nazis in Germany, the Communists in many countries, or the Anarchists in Spain. In 1933, when the Nazi Party came into power supported by the masses of unemployed and desperate people in Germany on the back of promises to bring back full employment (and national greatness after the lost war), its leader Adolf Hitler quickly installed a brutal and bloody dictatorial system. By some devious form of covered public debt and forced labor, the regime managed to increase production (not least of military hardware) and restore nearly full employment by 1936, which contributed greatly to its popularity (in addition to its uncontested propaganda). As Hitler's politics led directly into World War II with its millions of casualties, it is maybe not false to claim that the Great Depression was the main cause of this humanitarian and political catastrophe.

Neoclassical economics was not in a position to provide satisfactory answers to the challenge posed by the Great Depression. At the climax of the Great Depression, Arthur Cecil Pigou (1933) published a book entitled *The Theory of Unemployment*, in which he claimed, based on neoclassical economic theory, that unemployment as a phenomenon of disequilibrium could vanish rather quickly, without much intervention by the government, with the help of the equilibrating forces of decreasing prices, wages, and interest rates. The economic policy recommendation that could be drawn from this analysis (although not actually drawn explicitly by Pigou himself) would be to wait until such an equilibrium materialized. In view of the enormous economic and political costs of the slow adjustment of actual market economies toward equilibrium, it is no wonder that this neoclassical answer to the global crisis appeared inadequate, to say the least.

The Answer: John Maynard Keynes (1883–1946)

The most effective answer to the Great Depression challenge came from one of the most important economists of all times, John Maynard Keynes. He was the oldest child of Florence Ada Keynes, a local social reformer, and John Neville Keynes, lecturer and later Registrary (the senior administrative officer) at the University of Cambridge, himself a cultured liberal, philosopher, and economist who outlived his famous son by 3 years. John Maynard Keynes was top of his class at school and excelled during his studies of philosophy, mathematics, history, and economics at King's College, Cambridge, with Alfred Marshall being his teacher and Pigou his college friend. In 1905, he graduated in mathematics and started working for the India Office of the British government the following year. There he worked on his PhD dissertation on the theory of probability, published as Keynes (1921), and prepared his first economic publication on India in the *Economic Journal* (Keynes 1909; see also Keynes 1913), the journal of the Royal Economic Society, of which he became editor in 1911 and secretary of the society 2 years later, holding both positions until shortly before his death in 1946.

In 1909, he started lecturing in economics at Cambridge University; he never became full professor. Later on, in addition to his editorial and lecturing obligations, Keynes actively worked as an economic advisor to the British government, notably during World War I, in the peace negotiations afterward (heavily criticizing the Versailles Treaty: Keynes 1919; also 1922), during the Great Depression, and in World War II (especially during the Bretton Woods negotiations about the postwar international monetary system). Although he was a member of the Liberal Party and was offered political posts several times, he never accepted any. His political independence is also evident from his critical essays on economic policy and, besides his opposition to the peace treaty in particular, his proposals for monetary reform (Keynes 1923), his critical attitude toward laissez-faire policies (Keynes 1926) and the UK's return to the gold standard (1925).

During the 1920s and later, Keynes worked intensively on problems of monetary theory and policy. He proposed a reform of the UK's monetary system after the war (Keynes 1923) and castigated then treasurer Winston Churchill for the return to the gold standard at a parity that later turned out (as predicted by Keynes) to be too high for the postwar economy of the UK (Keynes 1925). The result of his theoretical investigations in this field was the two-volume book *A Treatise on Money* (Keynes 1930). When the Great Depression was at its worst point and Pigou proclaimed "the" neoclassical answer to it, Keynes felt deeply annoyed by this inadequate reaction to the social, economic, and political catastrophe which, even in the democratically consolidated UK, resulted in angry groups marching with Oswald Mosley's fascists and Wal Hannington's communists. Keynes decided to work out (together with a circle of eminent Cambridge economists, including Richard F. Kahn and Joan M. Robinson; see Kahn 1984) an alternative to neoclassical theory. He heavily criticized what he called the "Classics," meaning both classical and neoclassical economists, and formulated a "general theory" in his opus magnum *The General Theory of Employment, Interest and Money* (Keynes 1936), a book that was tremendously successful in both academia and politics. It can be regarded as the cornerstone of modern macroeconomic theory.

Although there are many, partly contradictory, interpretations of what Keynes' "vision" was of how the economy worked, several of his new concepts were widely and rapidly accepted, giving birth to abundant theoretical and empirical research following the publication of the *General Theory*. These include the (macroeconomic) consumption function, the multiplier, and the marginal efficiency of capital and liquidity preference, among others. The hypothesis of the consumption function may serve to illustrate a decisive point in Keynes' *General Theory*. He stated (Keynes 1936):

> We [...] define what we shall call the *propensity to consume* as the functional relationship χ between Y_w, a given level of income in terms of wage-units, and C_w the expenditure on consumption out of that level of income, so that
> $C_w = \chi(Y_w)$ [...] (p. 90).
>
> [...] the propensity to consume is a fairly stable function so that, as a rule, the amount of aggregate consumption mainly depends on the amount of aggregate income [...] (p. 96)

These passages show a few key elements of Keynes' theory, which were elements of virtually every subsequent "Keynesian" theory:

- Keynes dealt with *aggregates*, that is, variables relating to the entire economy such as aggregate consumption, aggregate income, aggregate production (output), aggregate demand and supply, etc., and not with variables relating to single households or firms, separate goods or markets. This contrasts with those versions of methodological individualism that always first call for the microeconomic foundations of aggregate relationships. Although Keynes argued for his consumption function by referring to the psychology of individual consumers, it later turned out that his original consumption function was not easily reconcilable with the neoclassical core concepts of demand and supply (Clower 1965). While neoclassical demand and supply of a good depend primarily on the (relative) prices of the good under consideration and of other goods, aggregate consumption and other components of aggregate demand depend on aggregate quantities such as real income or, in turn, aggregate demand.
- Perhaps the most crucial concept in Keynes' theory is *aggregate demand*. According to Keynes' principle of effective demand, it is the aggregate demand of consumers, firms, the government, and foreign countries (the rest of the world) which determines output and employment in an economy. If this is true, we must look for insufficient demand (not only for goods and services, but also for labor = employment) if we want to explain mass unemployment such as that prevailing in the Great Depression. Although price movements may be responsible for the development of aggregate variables to a certain extent, aggregate demand (which is a real variable, i.e., essentially a quantity) is the prime mover of an economy.
- Keynes developed a theory of involuntary unemployment. He defined it as follows (Keynes 1936, p. 15; entire sentence in italics in the original):

> Men are involuntarily unemployed if, in the event of a small rise in the price of wage-goods relatively to the money-wage, both the aggregate supply of labour willing to work for the current money-wage and the aggregate demand for it at that wage would be greater than the existing volume of employment.

In more modern terminology, this can be expressed to mean that involuntarily unemployed workers would be willing to work at current nominal wages but could not find a job due to insufficient demand for labor and for goods and services. Keynes notes that the absence of involuntary unemployment is consistent with "full" employment, which is a state of affairs where all unemployment is either frictional (including seasonal) or voluntary (i.e., in neoclassical terms, where the marginal disutility of labor is larger than the prevailing real wage).

- Keynes also gave an alternative definition of involuntary unemployment, which, according to him, "amounted to the same thing" (ibid., p. 26):

   > [...] a situation in which aggregate employment is inelastic in response to an increase in the effective demand for its output. Thus Say's law, that the aggregate demand price of output as a whole is equal to its aggregate supply price for all volumes of output, is equivalent to the proposition that there is no obstacle to full employment.

   Say's Law, which is usually abbreviated to say "supply creates its own demand," is equivalent to assuming the equality of aggregate demand and aggregate supply at prevailing prices. Keynes denies the validity of this "basic postulate" of neoclassical economics – sometimes considered as being equivalent to the assumption of general equilibrium, and sometimes merely as an accounting identity or an aggregate budget condition like Walras' Law – and attacks the "Classics" for their dogmatic adherence to it. His "general theory" includes the "neoclassical" one as a special case under the (exceptional) assumption of full employment.

- This theoretical approach of Keynes had profound economic policy consequences. If we regard the possibility of enduring involuntary unemployment, this amounts to a market failure, in particular if it takes on the form of mass unemployment as in the Great Depression. It means that an unregulated system of markets, as advocated by laissez-faire neoclassical economists, cannot escape from this situation; hence, below full employment is a (temporary, at least) state of equilibrium, a so-called unemployment equilibrium. Keynes and later Keynesians stressed this aspect of his theory and called for an adequate theory to explain such a situation and for remedies to cure this evil. The basic idea of Keynesian policy prescriptions in such a situation is the concept of a countercyclical stabilization policy. According to this idea, the government (or, more precisely, the public sector) should step in and support the insufficient demand of private agents. In a recession or a crisis, the government should reduce taxes and/or increase public expenditures to increase aggregate demand until involuntary unemployment is wiped out. On the other hand, in a boom with a lack of an adequate labor supply and inflation, the task of the government should be to increase taxes and/or decrease public expenditures to repay public debt accrued during a recession. In addition to these measures of fiscal (or budgetary) policy, the central bank can support them by increasing the quantity of money or reducing its interest rate in order to induce firms to invest and consumers to consume more. After World War II, policy makers followed these principles more or less closely, with considerable success during the period of rebuilding the Western economies until the 1960s.

Keynes' *General Theory* was received with great interest by economists, first in the UK and then all over the world. A quick reaction came from John R. Hicks (1937), who, in an extended book review, created what was later called the IS-LM model and can be found in most macroeconomics textbooks. In this paper, he identified the consumption function with the multiplier, considered it and the marginal efficiency of investment (investment depending on the rate of interest) and the liquidity preference (money demand depending on the rate of interest due to the speculation motive) theories to be the key innovations in the *General Theory* and put them together in a model with a market for goods and services and a money market. Implicitly, a bonds market (a market for financial assets) complemented these two to form an aggregate general equilibrium model. Franco Modigliani (1944) augmented it with a labor market (with two versions: Keynesian and (neo-) classical) and an aggregate production function. In this form, it became the ultimate workhorse for macroeconomists and the core of macroeconomics textbooks until far into the post-World War II period – and for many undergraduate economics students even to the present day. It is characteristic for the dominance of neoclassical equilibrium thinking that Keynes' ideas were thus quickly embedded into that theoretical framework. The master himself (Keynes 1937) did not object to this interpretation of his views.

Shortly after having published the *General Theory*, Keynes had a heart attack. After recovering, he returned to academic and consultant work that was driven mostly by the necessities of dealing with the problems of the imminent and then actual Second World War. He served in various capacities for the British government and was the main UK negotiator in the negotiations on the international monetary system, which eventually led to the Bretton Woods arrangements. The enormous amount of work that Keynes shouldered during these years caused deteriorating health, and he finally suffered a fatal heart attack in 1946. He was survived by both of his parents and his wife; he had no children. Toward the end of his life, he became Baron of Tilton, taking his seat in the House of Lords, in addition to receiving several other honors. The Royal Economic Society honored Keynes by editing the 30-volume *Collected Writings of John Maynard Keynes* (finished 1989). Several biographies were written about him, for instance, by his successor as editor of the *Economic Journal*, Roy Harrod (1951), by the main editor of the Collected Writings, Donald Moggridge (1980), or the three-volume biography by Robert Skidelsky (1983–2000).

Keynesianism and Monetarism: Macroeconomics after World War II

As mentioned above, after the war Keynesianism became the dominant paradigm in macroeconomic theory and policy in most Western countries, alongside neoclassical theory as the basis for microeconomics. The split in economic theory and the relationship between microeconomics (dealing with individual agents and their interactions) and macroeconomics (dealing with aggregates) are largely due to the impact of Keynes on the course of mainstream theory, but not every facet can be covered here. Instead, the focus is on those aspects that are relevant for attempts to clarify the relationship between the two areas and their pertinence to neoclassical economics. For a detailed study of the history of macroeconomics, see De Vroey (2016).

At first sight, it may seem strange that there are two fields of economic theory in the first place: Is there not only one economy? Nearly no microeconomist would deny that it is possible to aggregate some firm- or household-specific variables over an economy, and most macroeconomists would agree to the fact that aggregate variables are somehow related to the behavior of households and firms. Nevertheless, microeconomics and macroeconomics are separate topics of courses in universities all over the world, and specialization in one of these areas is accepted as the legitimate choice of scholars in academia everywhere. Hence, what are the relations between these two fields, and how do they relate to the ideas of neoclassical and Keynesian economics?

The first attempt at reconciling neoclassical microeconomics with Keynesian macroeconomics can be identified in the so-called neoclassical synthesis, a term coined by Paul Anthony Samuelson (1915–2009) in the third edition of his textbook *Economics* (1955). Samuelson was the creator of modern mathematical neoclassical microeconomics in his *Foundations of Economic Analysis* (Samuelson 1947). Here he showed that the use of the mathematical concepts of maximization, equilibrium, and stability could elucidate a great number of economic problems by providing "operationally meaningful theorems" which serve to derive empirically testable hypotheses. Although this book, which was based on the author's PhD dissertation at Harvard University, also contained some mathematical formulations of Keynesian models, its main contribution was to provide a (perhaps "the") mathematical foundation for neoclassical microeconomics. However, as becomes clear from several of Samuelson's macroeconomic and policy papers, he also broadly embraced most of Keynes' theoretical ideas on unemployment as well as his prescripts on fiscal and monetary policy.

The neoclassical synthesis is Samuelson's version of bridging the gap between neoclassical and Keynesian thinking. Basically, for Samuelson, neoclassical microeconomic theory holds under the conditions of full employment while Keynesian macroeconomics is the theory appropriate to situations where the resources of the economy (including labor) are not used to full capacity. A related (though not congruent) interpretation says that neoclassical economics is the theory for the long run while Keynesian economics is valid for the short run. Through Samuelson's textbook, which became the best-selling economics textbook of all times, this interpretation became dominant from the late 1940s to the end of the 1960s and remained respectable afterward, in spite of its simplicity and lack of theoretical consistency. This fact is mainly due to the empirical success of applying Keynesian policies on the macroeconomic level and using neoclassical theory for explaining microeconomic phenomena.

More sophisticated analyses emerged when authors such as Don Patinkin (1956) and Robert W. Clower (1965) attempted to reconcile the Keynesian message with the Walrasian general-equilibrium version of neoclassical price theory. Patinkin's work showed that many problems arise when Keynesian monetary theory is confronted with neoclassical microeconomics and, in addition to providing a model of the labor market under neoclassical and Keynesian conditions, opened up many fruitful directions for macroeconomic research. Clower, in turn, started from the observation that the Keynesian consumption function is incompatible with the model of the consumer's decision in Walrasian theory, the latter assuming (relative) prices, including wages, as given to consumers and their income as determined endogenously by them while the former treats income (for instance, quantity of labor times wage) as the main determinant of consumption (and given to them). Clower's solution was to interpret Keynesian economics as disequilibrium theory, analyzing a situation of quantity restrictions for consumers and producers. Leijonhufvud (1968), Barro and Grossman (1976), and Malinvaud (1977), among others, also followed this avenue of research. The main problem with this disequilibrium approach was to explain its required assumption that prices and wages are sticky, which limited the acceptance of the fixed-price models. Clower (1967) proposed a cash-in-advance constraint to explain quantity restrictions for consumers; nominal rigidities (sticky wages and prices) are the consequences of the cost to firms of changing prices in models by Taylor (1979) and Calvo (1983), among others.

The "Keynesian consensus," the belief that Keynes' theory and policy prescriptions, which were supported by the IS-LM (investment = saving, liquidity demand = money supply) and AD-AS (aggregate demand = aggregate supply) textbook models, lasted until the end of the 1960s. At that time, a phenomenon occurred that seemed to contradict Keynesian macroeconomics, namely stagflation, a situation with simultaneously low or negative growth in output, high unemployment (stagnation), and high inflation. Stagflation first appeared in the UK and then, in the early 1970s, in other Western countries. Keynesian macroeconomists usually considered demand side developments as causes of adverse development, which caused low growth and high unemployment when negative (insufficient demand) and high inflation when positive (excess demand). This gave rise, in their view, to a trade-off between inflation and unemployment, as expressed by the negative relation between the rates of unemployment and inflation, graphically shown by the Phillips curve (Phillips 1958): If you want to reduce unemployment, you will have to take (expansionary) measures and will have to put up with more inflation. Stagflation, however, is a phenomenon that cannot occur when the Phillips curve is valid, and it entails unpleasant policy consequences: If you fight unemployment by expansionary fiscal or monetary policies, this will increase what is already high inflation; if you combat inflation, you will get even higher unemployment. The trade-off turns into a choice between inevitably increasing one of two evils.

Although the "Keynesian consensus" was strong during the first three decades after World War II, it was far from being universally accepted by all economists. The main opposition came from those whom Keynes called the "Classics," i.e., neoclassical economists. During Keynes' lifetime, his most outspoken critic was Friedrich August Hayek, a prominent member of the Austrian School, who had become professor at the London School of Economics (LSE) already at the beginning of the 1930s, seconded by the chairman of LSE's economics department, Lionel

Robbins, whose definition of economics in his *Essay* (Robbins 1935) is a strong, succinct formulation of the neoclassical program:

> Economics is a science which studies human behaviour as a relationship between ends and scarce means which have alternative uses. (p.16)

Hayek, who was a personal friend of Keynes' despite their severe dissent over matters of economics, criticized several positions of Keynes in detail but did not provide an overall criticism of the *General Theory*. Pigou, whom Keynes had attacked directly, wrote some pieces (Pigou 1950) against the *General Theory*, which did not, however, exert any lasting influence, apart from the so-called "Pigou effect" (which, in fact, was first discovered by Austrian Harvard professor Gottfried Haberler) as a mechanism to restore full employment equilibrium without policy intervention. According to this idea, a fall in the price level during a depression would lead to an increase in real wealth and, as a consequence, to an increase in consumption (and possibly investment) sufficient to boost employment.

A more effective "counter-revolution" to the "Keynesian revolution" came from a group of economists who saw themselves in the tradition of the neoclassical Chicago School. Their generally acknowledged frontman was Milton Friedman, who – like Hayek – received the Nobel Prize in the 1970s. Friedman changed his arguments several times during his long and successful academic career but had the permanent aim of demonstrating the superiority of rules over discretionary economic policy, a doctrine already propagated by Chicago economist Henry C. Simons (1936): Discretionary policies react to the current state of the economy while rules prescribe long-term commitments for policy makers. Friedman first argued – in contrast to Keynes – that monetary policy was more effective at influencing the economy than fiscal (budgetary) policy. Due to this argument, the entire economic movement was given the label Monetarism. Friedman tried to demonstrate this in empirical studies in which the predictive power of a monetary model in the tradition of the quantity theory of money was greater that a Keynesian one (Friedman and Meiselman 1963). This did not remain uncontested; see Poole and Kornblith (1973) on this controversy.

Although, according to his theory, monetary policy is highly effective, monetarists argue against its use for controlling the economy over the business cycle in a discretionary way. One reason for this is the long and variable lags in its effects, which often lead to procyclical outcomes of such activities (Friedman 1961). Friedman presented several versions of monetary theories, such as a theory of the demand for money (Friedman 1956) or a model purportedly summarizing his thinking about money and the macroeconomy (Friedman 1970). By developing and empirically testing a consumption function which made consumption dependent on a long-term concept of income (permanent-income hypothesis; Friedman 1957), he provided an argument about the ineffectiveness of fiscal policy with respect to stabilizing the economy.

The most effective blow to the "Keynesian consensus," however, came from Friedman's presidential address to the American Economic Association in 1968 (Friedman 1968). Here he developed the theory of the natural rate of unemployment as the neoclassical counterpart to Keynesian involuntary unemployment. The key innovation of this theory was the explicit introduction of expectations to the Phillips curve relation and the proposition that in the long run there is no trade-off between inflation and unemployment; instead, increasing aggregate demand by expansionary monetary or fiscal policies to reduce the rate of unemployment below its natural level leads to permanent and accelerating inflation. This theory, some components of which were simultaneously discovered by Edmund S. Phelps (1967), was very well suited to explaining the accelerating inflation of the 1970s and is, thus, one of the rare examples of a theoretical hypothesis developed ahead of the empirical event supporting it. Although the introduction of an aggregate supply curve in addition to the aggregate demand curve of the Keynesian textbook model also does a major part of the job of explaining the stagflation episodes of the 1970s and 1980s, these events led to the demise of the "Keynesian consensus" and a revival of neoclassical theory in the form of monetarism.

An even more radical version of monetarism is New Classical Macroeconomics with its policy ineffectiveness theorem. Using the twin assumptions of rational expectations (that agents know the "true" model of the economy) and continuous general equilibrium, Robert E. Lucas (*1937) and others (particularly Sargent and Wallace 1976) showed that even systematic stabilization policies following a contingent rule (for instance, derived as an "optimal" one) can have no effect on real quantities such as output or employment. Economic agents react to such systematic policy rules and change their behavior to counteract policy makers, which changes the structure of the economy in the sense that a new model of it is required (Lucas 1976). In addition, such policies are time inconsistent: Once implemented at the beginning of the planning horizon, in later periods policy makers do not continue following them because it is no longer optimal for them to do so (Kydland and Prescott 1977). These theoretical developments fundamentally changed the content of mainstream academic research in macroeconomics; the "old-fashioned" Keynesian way of arguing is now the resort of heterodox currents (such as Post-Keynesian Economics), although some politicians and authors of textbooks still use and believe them.

DSGE models (dynamic stochastic general equilibrium models) characterize the current standard of macroeconomic research (for an introduction, see Sbordone et al. 2010; for an example, Smets and Wouters 2003; for techniques, Canova 2007). In most cases, they are based on the assumption of rational expectations, assume general equilibrium in markets of goods and services (but often not in the labor market) or movements on a path toward such equilibrium, use assumptions about representative households and firms, and are calibrated (instead of estimated econometrically) to numerical empirical data. Two main classes of DSGE models prevail: (strictly neo-) classical ones (including real business cycle models, initiated by Kydland and Prescott 1982) and New-Keynesian ones, which typically assume monopolistic competition and wage (and often price) stickiness to explain involuntary unemployment. Although this class of models is subject to frequent criticism as being too rigid for applied work on empirical forecasts and policy recommendations (e.g., Stiglitz 2018, Fair 2012), it shows considerable flexibility with respect to several of its assumptions. For instance, New-Keynesian DSGE models use monopolistic instead of perfect competition; rational expectations can be replaced relatively easily by other behavioral assumptions (e.g., De Grauwe 2012); there are DSGE models with heterogeneous agents (Hommes 2021); and econometric estimation (including Bayesian econometrics) can support numerical calibration methods, etc. Hence, while neoclassical orthodoxy in general and DSGE modeling in particular came under fire after the Great Recession (the financial and economic crisis of 2007–2009) and the Covid-19 recession since 2020, it would be premature to write off these methods, especially in view of the not fully satisfactory alternatives available so far.

Conclusion

To summarize, the development of neoclassical economics has been traced from its predecessors to the present, with a special emphasis on its main current competitor, Keynesian macroeconomics. Neoclassical economics achieved the transition from obscure heterodoxy to mainstream economic theory and has retained this position for over a century. The main reason for this is the fact that it explains, in a theoretically consistent way, the demand side of markets by utility theory and the prices of goods in a system of markets, in contrast to previous classical economics, which concentrated on the supply side of the economy. Another reason for the success of neoclassical economics is its sound methodological foundation through mathematical reasoning and – at least to some extent – empirical evidence in favor of some of its theoretical predictions. On the other hand, neoclassical economics failed to explain how economic crises disturb the general equilibrium other than by introducing exogenous events and, even more disturbingly, has failed to explain the long duration of some of these crises. Keynesianism provided an alternative that could explain such crises and even propose advice to governments regarding measures to combat them. But Keynesianism, too, has its theoretical problems, the most severe one being the need to rely on wage and price stickiness, which, in the medium and long run, is not a plausible assumption.

Another asset of neoclassical economics is its ability to accommodate modifications of some of its assumptions without abandoning the key elements of its methodology. For instance, the obviously unrealistic assumptions of perfect competition, utility maximization by all agents, and permanent general equilibrium could and can be modified, and the modification can be embedded in the edifice without demolishing its base. Imperfect competition, monopoly, and oligopoly were explained within the neoclassical paradigm, and the theory of games can even be judged to be a success, not only for oligopoly price theory but also as a general theory of human interactions under conflicting interests. Behavioral economics successfully replaced the strict maximization hypothesis for economic agents and led to empirically and experimentally backed theories of human behavior with far-reaching consequences for economic theory without compromising on methodological individualism as a central tenet of neoclassical economics and the rationality of human behavior in particular. More difficulties have arisen from attempts to modify the assumption of general equilibrium, but at least with the alternative concept of temporary equilibrium, the Keynesian challenge of involuntary unemployment could also be countered, albeit imperfectly. A deeper problem arises from the inability of general equilibrium theory to develop economically meaningful conditions for the uniqueness and stability of equilibria and especially from the Sonnenschein-Mantel-Debreu theorems. So far, most neoclassical economists have reacted to this problem with benign neglect or by simply assuming a unique and stable equilibrium or by resorting to partial equilibrium analysis.

Probably the most important reason for the lack of success of alternative theories is the methodological dearth of heterodox economic theories. Mathematical modeling, which formulates its assumptions unambiguously and derives its results by irrefutable deductive logic, is one asset of neoclassical economists (and an important barrier to entry into the scientific community) which, when acquired as part of economists' human capital, serves to lend reputation and scientific credibility to their arguments. If we dare look into the future of economic theory, this will probably grant a long life to – at least undogmatic modified variants of – neoclassical economics. Economic scientists are therefore advised to take up relevant criticisms of neoclassical economics and analyze them using the same instruments as the best neoclassical economists do. This may be the best way forward to more satisfactory theories "of man in the ordinary business of life," as Alfred Marshall put it for the subject of economics.

Acknowledgments Parts of this chapter have been reprinted/adapted with permission from Neck, R. (2014) On Austrian Economics and the Economics of Carl Menger. Atlantic Economic Journal 42, 217–227 (2014). https://doi.org/10.1007/s11293-014-9422-6

References

Albert H (1998) Modell-Platonismus. Der neoklassische Stil des ökonomischen Denkens in kritischer Betrachtung. In: Albert H Marktsoziologie und Entscheidungslogik Zur Kritik der reinen Ökonomik. Mohr Siebeck, Tübingen, pp 108–142
Arnsperger C, Varoufakis Y (2006) What is neoclassical economics? The three axioms responsible for its theoretical oeuvre, practical irrelevance and, thus, discursive power. Panoeconomicus 53: 5–18
Aspromourgos T (1986) On the origins of the term "neoclassical". Camb J Econ 10:265–270
Aspromourgos T (2018) Neoclassical. In: The new Palgrave dictionary of economics, 3rd edn. Palgrave Macmillan, London, pp 9382–9383
Baranzini R (2016) Marie-Esprit-Léon Walras. In: Faccarello G, Kurz HD (eds) Handbook of economic analysis, Edward Elgar, Cheltenham, vol I, p 245–261
Barro and Grossman (1976) Money, employment and inflation. Cambridge University Press, Cambridge, UK
Becker GJ, Stigler GS (1977) De gustibus non est disputandum. Am Econ Rev 67:76–90
Bernoulli D (1954) Exposition of a new theory on the measurement of risk. Econometrica 22: 23–36. Original 1738: Specimen theoriae novae de mensura sortis. Commentarii Academiae Scientiarum Imperialis Petropolitanae 5:175–192

Bertrand J (1883) Review of: Théorie mathématiques de la richesse sociale par L Walras et Recherches sur les principes mathématiques de la théorie des richesses par AA Cournot. Journal des Savants 67:499–508

Blaug M (1997) Economic theory in retrospect, 4th edn. Cambridge University Press, Cambridge, UK

Braun MS (1929) Theorie der staatlichen Wirtschaftspolitik. Deuticke, Leipzig

Bresciani Turroni C (1942) Introduzione alla politica economica. Einaudi, Torino

Calvo GA (1983) Staggered prices in a utility-maximizing framework. J Monet Econ 12:383–398

Canova F (2007) Methods for applied macroeconomic research. Princeton University Press, Princeton

Clower RW (1965) The Keynesian counter-revolution: a theoretical appraisal. In: Hahn FH, Brechling FPR (eds) The theory of interest rates. Macmillan, London, p 103–125. First published as: Die Keynesianische Gegenrevolution: Eine theoretische Kritik. Schweizerische Zeitschrift für Volkswirtschaftslehre und Statistik 99: p 8–31

Clower RW (1967) A reconsideration of the microfoundations of monetary theory. Western Economic Journal 6:1–8

Colander (2000) The death of neoclassical economics. Journal of the History of Economic Thought 22:127–143

Cournot AA (1838) Recherches sur les principes mathématiques de la théorie des richesses. Hachette, Paris. Also in Cournot (1973–2010), vol. VIII

Cournot AA (1927) Researches into the mathematical principles of the theory of wealth. Macmillan, New York. English translation of Cournot (1838)

Cournot AA (1973-2010) Oeuvres completes, 11 vols. Vrin, Paris

De Grauwe P (2012) Lectures on behavioral macroeconomics. Princeton University Press, Princeton

Dequech D (2007-2008) Neoclassical, mainstream, orthodox, and heterodox economics. Journal of Post Keynesian Economics 30:279–302

De Vroey M (2016) A history of macroeconomics from Keynes to Lucas and beyond. Cambridge University Press, Cambridge, UK

De Vroey M (2017) Review of: DA Walker and J van Daal (eds and trans) Léon Walras, Elements of theoretical economics: or the theory of social wealth, Cambridge University Press, Cambridge 2014. Journal of the History of Economic Thought 39(special issue 1):144–147

Di Iorio F (2015) Cognitive autonomy and methodological individualism. The interpretative foundations of social life. Springer, Cham

Dixit AK, Stiglitz JE (1977) Monopolistic competition and optimum product diversity. Am Econ Rev 67:297–308

Düppe T, Weintraub E (2014) Finding equilibrium. Arrow, Debreu, McKenzie and the problem of scientific credit. Princeton University Press, Princeton

Faccarello G, Kurz HD (eds) (2016) Handbook of economic analysis, 3 vols. Edward Elgar, Cheltenham

Fair RC (2012) Has macro progressed? J Macroecon 34:2–10

Friedman M (1956) The quantity theory of money – a restatement. In: Friedman M (ed) Studies in the quantity theory of money. University of Chicago Press, Chicago, pp 3–21

Friedman M (1957) A theory of the consumption function. Princeton University Press, Princeton

Friedman M (1961) The lag in effect of monetary policy. J Polit Econ 69:447–466

Friedman M (1968) The role of monetary policy. Am Econ Rev 58:1–17

Friedman M (1970) A theoretical framework for monetary analysis. In Gordon RJ (ed) Milton Friedman's monetary framework. A debate with his critics. University of Chicago Press, Chicago pp. 1–62

Friedman M, Meiselman D (1963) The relative stability of monetary velocity and the investment multiplier in the United States, 1897–1958. In: Commission on money and credit, stabilization policies. Prentice-Hall, Englewood Cliffs, pp 165–268

Gallois N (2011) L'école française contre Walras, économiste hétérodoxe. L'Economie politique 51(3):7–32
Gossen HH (1854) Entwickelung der Gesetze des menschlichen Verkehrs, und der daraus fließenden Regeln für menschliches Handeln. Viehweg, Braunschweig
Gossen HH (1983) The laws of human relations and the rules of human action derived therefrom. MIT Press, Cambridge MA. English translation of Gossen (1854), with introductory essay by N Georgescu-Roegen
Harrod (1951) The life of John Maynard Keynes. WW Norton, New York
Hayek FA (1991) Hermann Heinrich Gossen. In: Bartley WW III, Kresge S (eds) The trend of economic thinking. Essays on political economists and economic history. Collected works of F.A. Hayek. Liberty Fund, Indianapolis, pp 352–372
Hicks JR (1937) Mr Keynes and the "classics": a suggested interpretation. Econometrica 5:147–159
Hommes C (2021) Behavioral and experimental macroeconomics and policy analysis: a complex systems approach. J Econ Lit 59:149–219
Jaffé W (1976) Menger, Jevons and Walras de-homogenized. Econ Inq 14:511–524
Jevons WS (1866) Brief account of a general mathematical theory of political economy. J Stat Soc Lond 29:283–287. Also Appendix III in Jevons (1957)
Jevons WS (1871) The theory of political economy. Macmillan, London
Jevons WS (1874) The principles of science: a treatise on logic and scientific methods. Macmillan, London
Jevons HS (ed) (1957) The theory of political economy by W. Stanley Jevons, 5th edn. Kelley & Millman, New York
Kahn RF (1984) The making of Keynes' general theory. Cambridge University Press, Cambridge, UK
Keynes JM (1909) Recent economic events in India. Econ J 19:51–56. Also in vol XI of Keynes (1971–1989)
Keynes JM (1913) Indian currency and finance. Macmillan, London. Also vol I of Keynes (1971–1989)
Keynes JM (1919) The economic consequences of the peace. Macmillan, London. Also in vol II of Keynes (1971–1989)
Keynes JM (1921) A treatise on probability. Macmillan, London. Also vol VIII of Keynes (1971–1989)
Keynes JM (1922) A revision of the treaty. Macmillan, London. Also vol III of Keynes (1971–1989)
Keynes JM (1923) A tract on monetary reform. Macmillan, London. Also vol IV of Keynes (1971–1989)
Keynes JM (1925) The economic consequences of Mr. Churchill. Hogarth Press, London. Also in vol IX of Keynes (1971–1989)
Keynes JM (1926) The end of Laissez-faire. Hogarth Press, London. Also in vol IX of Keynes (1971–1989)
Keynes JM (1930) A treatise on money. Vol 1: The pure theory of money. Vol 2: The applied theory of money. Macmillan, London. Also vols V and VI of Keynes (1971–1989)
Keynes JM (1936) The general theory of employment, interest and money. Macmillan, London. Also vol VII of Keynes (1971–1989)
Keynes JM (1937) The general theory of employment. Quarterly Journal of Economics 51: 209–223. Also in vol XIV of Keynes (1971–1989)
Keynes (1971–1989) The collected writings of John Maynard Keynes. Macmillan, London
Kirchgässner G (2008) *Homo oeconomicus*. The economic model of behaviour and its applications in economics and other social sciences. Springer, New York
Kirman A (2016) General equilibrium theory. In: Faccarello G, Kurz HD (eds) Handbook of economic analysis. Edward Elgar, Cheltenham, vol III p 236–253
Kurz HD (2016) Hermann Heinrich Gossen. In: Faccarello G, Kurz HD (eds) Handbook of economic analysis, Edward Elgar, Cheltenham, vol I, p 196–202
Kydland FE, Prescott EC (1977) Rules rather than discretion: the inconsistency of optimal plans. J Polit Econ 85:473–491

Kydland FE, Prescott EC (1982) Time to build and aggregate fluctuations. Econometrica 50: 1345–1370

Lawson T (2013) What is this 'School' called neoclassical economics? Camb J Econ 37:947–983. Also in: Morgan J (2015) p 30–80

Leijonhufvud A (1968) On Keynesian economics and the economics of Keynes. Oxford University Press, New York

Leroux R (2019) Antoine-Augustin Cournot as a sociologist. Palgrave Macmillan, London

Lucas R (1976) Econometric policy evaluation: a critique. In Brunner K, Meltzer A (eds) The Phillips Curve and Labor Markets. Carnegie-Rochester Conference Series on Public Policy 1, Elsevier, New York pp 19–46

Maas H (2018) Jevons, William Stanley (1835–1882). In: The new Palgrave dictionary of economics, 3rd edn. Palgrave Macmillan, London, pp 7115–7128

Malinvaud E (1977) The theory of unemployment reconsidered. Basil Blackwell, Oxford

Marshall A (1890) Principles of economics. Macmillan, London. Variorum edition 1961

Martin T (2016) Cournot (A). In: Kristanek M (ed) L'Encyclopédie philosophique. https://encyclo-philo.fr/cournot-a

Menger C (1871) Grundsätze der Volkswirthschaftslehre. Braumüller, Vienna. English translation: Menger C (1976) Principles of economics. New York University Press, New York

Menger C (1883) Untersuchungen über die Methode der Socialwissenschaften und der Politischen Oekonomie insbesondere. Duncker & Humblot, Leipzig. English translation: Menger C (1996) Investigations into the method of the social sciences. Libertarian Press, Grove City, PA

Menger C (1884) Die Irrthümer des Historismus in der deutschen Nationalökonomie. Hölder, Vienna

Modigliani F (1944) Liquidity preference and the theory of interest and money. Econometrica 12: 45–88

Moggridge DE (1980) John Maynard Keynes. An economist's biography. Macmillan, London

Moore HL (1905) The personality of Antoine Augustin Cournot. Q J Econ 19:370–399

Morgan J (2015) What's in a name? Tony Lawson on neoclassical economics and heterodox economics. Camb J Econ 39:843–865

Morgan J (ed) (2016) What is neoclassical economics? Debating the origins, meaning and significance. Routledge, London

Morrison CC (2003) Augustin Cournot and neoclassical economics. Atl Econ J 31:123–132

Nash J (1951) Non-cooperative games. Ann Math 54:286–295

Neck R (2014) On Austrian economics and the economics of Carl Menger. Atl Econ J 42:217–227

Niehans J (2018) Gossen, Hermann Heinrich (1810–1858). In: The new Palgrave dictionary of economics, 3rd edn. Palgrave Macmillan, London, pp 5390–5394

Pareto V (1906) Manuale di Economia politica con una introduzione alla scienza sociale. Società Editrice Libraria, Milan. English variorum edition: Pareto V (2014) Manual of political economy. Oxford University Press, Oxford

Patinkin (1956) Money, interest and prices: an integration of monetary and value theory. Row, Peterson and Company, Evanston IL

Peach T (2018) Jevons as an economic theorist. In: The new Palgrave dictionary of economics, 3rd edn. Palgrave Macmillan, London, pp 7106–7115

Phelps ES (1967) Phillips curves, expectations of inflation and optimal unemployment over time. Economica NS 34:254–281

Phillips AW (1958) The relation between unemployment and the rate of change of money wage rates in the United Kingdom, 1861-1957. Economica NS 25:283–299

Pigou AC (1920) The economics of welfare. Macmillan, London

Pigou AC (1933) The theory of unemployment. Macmillan, London

Pigou AC (1950) Keynes's general theory – a retrospective view. Macmillan, London

Poole W, Kornblith EBF (1973) The Friedman-Meiselman CMC paper: new evidence on an old controversy. Am Econ Rev 63:908–917

Robbins L (1935) An essay on the nature and significance of economic science, 2nd edn. Macmillan, London

Roll E (1992) A history of economic thought, 5th edn. Faber and Faber, London

Samuels WJ (1990) Introduction. In: Hennings K, Samuels WJ (eds) Neoclassical economic theory. Kluwer, Boston, pp 1–12

Samuelson PA (1947) Foundations of economic analysis. Harvard University Press, Cambridge MA

Samuelson PA (1955) Economics. An introductory analysis, 3rd edn. McGraw-Hill, New York

Sandelin B, Trautwein HM, Wundrak R (2014) A short history of economic thought, 3rd edn. Routledge, London

Sargent TJ, Wallace N (1976) Rational expectations and the theory of economic policy. J Monet Econ 2:169–183

Sbordone AM et al (2010) Policy analysis using DSGE models: an introduction. Federal Reserve Bank of New York Economic Policy Review 16:23–43

Schmoller G (1883) Zur Methodologie der Staats- und Sozialwissenschaften. Jahrbuch für Gesetzgebung, Verwaltung und Volkswirthschaft im deutschen Reiche 7:239–258

Schumpeter JA (1954) History of economic analysis. Allen & Unwin, London

Shubik M (2018) Cournot, Antoine Augustin (1801–1877). In: The new Palgrave dictionary of economics, 3rd edn. Palgrave Macmillan, London, pp 2416–2424

Simons HC (1936) Rules versus authorities in monetary policy. J Polit Econ 44:1–30

Skidelsky R (1983–2000) John Maynard Keynes. 3 vols. Macmillan, London

Smets F, Wouters R (2003) An estimated dynamic stochastic general equilibrium model of the Euro Area. J Eur Econ Assoc 1:1123–1175

Stiglitz JE (2018) Where modern macroeconomics went wrong. Oxf Rev Econ Policy 34:70–106

Streissler EW, Streissler M (eds) (1994) Carl Menger's lectures to Crown Prince Rudolf of Austria. Edward Elgar, Cheltenham

Taylor J (1979) Staggered wage setting in a macro model. Am Econ Rev 69(2):108–113

Theocharis RD (1990) A note on the lag in the recognition of Cournot's contribution to economic analysis. Can J Econ 23:923–933

von Thünen JH (1826) Der isolirte Staat in Beziehung auf Landwirthschaft und Nationalökonomie. Friedrich Perthes, Hamburg

Udehn L (2001) Methodological individualism. Background, history and meaning. Routledge, London

Veblen T (1900) The preconceptions of economic science. Q J Econ 14:240–269

Walras L (1954) Elements of pure economics or the theory of social wealth. Allen & Unwin, London. English translation of: Walras L (1926) Éléments d'économie politique pure ou théorie de la richesse sociale, definitive edition, R Pichon, Paris

Walras (1885) Un économiste inconnu: Hermann-Henri Gossen. Journal des Économistes 30: 68–90. Also in Walras and Walras (1987–2005) vol IX, p. 311–330

Walras A, Walras, L (1987–2005) Auguste et Léon Walras: Oeuvres économiques complètes, 14 vols. Economica, Paris

Whaples R (1995) Where is there consensus among American economic historians? The results of a survey on forty propositions. J Econ Hist 55:139–154

Weber M (1921/1922) Wirtschaft und Gesellschaft. Mohr, Tübingen. English translation: Weber M (1968) Economy and society. An outline of interpretive sociology. Bedminster Press, New York

Zahle J, Collin F (eds) (2014) Rethinking the individualism-holism debate. Essays in the philosophy of social science. Springer, Cham

The Austrian School and the Theory of Markets

59

David Emanuel Andersson and Marek Hudik

Contents

Introduction	1542
The Birth of the Austrian School and the *Methodenstreit*	1543
The Debate on Economic Planning and the Character of Knowledge	1544
Competition in Austrian Economics	1547
The Great Divergence	1548
Institutions Reexamined	1550
Entrepreneurship	1553
Austrian Methodology	1555
Conclusions	1557
References	1558

Abstract

The Austrian school of economics emerged as a distinct research program in the 1870s in opposition to the German Historical School. In its early days, it was part of the neoclassical mainstream. The origin of several enduring concepts, such as marginal utility, opportunity cost, and subjective preferences, can be traced to early Austrian contributions. Austrian economists such as Eugen von Böhm-Bawerk and Ludwig von Mises were also among the earliest critics of Marxist economics and central planning, arguing that Marx's labor of theory of value is erroneous and that a socialist economy is impossible. In the 1930s, the Austrian school diverged from mainstream economics. The new rift was between the Austrian approach, which analyzes markets as institutionally

D. E. Andersson (✉)
IBMBA Program, College of Management, National Sun Yat-sen University, Kaohsiung City, Taiwan
e-mail: davidemanuelandersson@cm.nsysu.edu.tw

M. Hudik
Faculty of Business Administration, Prague University of Economics and Business, Prague, Czechia

Center for Theoretical Study, Charles University, Prague, Czechia

© The Author(s), under exclusive licence to Springer Nature Singapore Pte Ltd. 2022
D. McCallum (ed.), *The Palgrave Handbook of the History of Human Sciences*,
https://doi.org/10.1007/978-981-16-7255-2_6

channeled entrepreneurial discovery processes, and the mainstream approach with its focus on models of market equilibria. Within the Austrian school, there was also a deepening of its underlying subjectivism to encompass not only consumer preferences, but also the subjectivity of knowledge and expectations. In the twenty-first century, Austrian concepts and theories remain influential in social theories that emphasize the role of institutions in human development.

Keywords

Austrian school · *Methodenstreit* · Economic planning · Market process · Spontaneous orders · Entrepreneurship · Institutions · Methodology

Introduction

The Austrian school of economics is a remarkable case in the history of ideas. It emerged in the 1870s in opposition to the historical approach that was dominant in German-speaking Europe. It soon became a part of the mainstream in economics, where it remained until World War II. In this period, its leading representatives, such as Carl Menger, Friedrich von Wieser, Eugen von Böhm-Bawerk, Ludwig von Mises, and Friedrich Hayek, made contributions that became the basis of the new economic orthodoxy. At the same time, the school resisted the move of the discipline toward formalization and quantitative methods. Consequently, it later found itself outside the economic mainstream, where it has remained until this day. Nevertheless, some Austrian ideas became influential in other disciplines. In particular, Hayek's contributions are recognized in political philosophy, while Israel Kirzner's work is one of the most influential theoretical foundations for contemporary research in entrepreneurship.

This chapter provides a brief history of the Austrian school from its birth in the late nineteenth century in the cosmopolitan capital of the Austro-Hungarian Empire up to the present day. It highlights its distinguishing features vis-à-vis the dominant paradigm, and shows how their salience changed over time. In its early stage, the Austrian school emphasized marginalism, opportunity cost, methodological individualism, and subjectivism – concepts which are explained in the following section. As these aspects were incorporated into mainstream economics, other characteristics of the Austrian school came to the fore; in particular its realism, its process-based approach, and its analysis of the character of human knowledge.

Many other scholars have discussed the main traits of the Austrian school, such as Backhouse (2000), Boettke (1996), Caplan (1999), Salerno (2002), and Huerta de Soto (2008). This chapter differs from earlier treatments by presenting these traits in a more historical context. In particular, it shows how key ideas originated in debates with other schools of thought, and in some cases, how traits that were originally specifically Austrian were incorporated into standard economics, thereby ceasing to be associated with Austrian economists. Consequently, other aspects of the Austrian approach came under the spotlight. While the essential characteristics of the Austrian school are relatively constant, their relative importance varies over time. The focus is on the Austrian analysis of the markets, which limits the scope of this chapter. Important topics that are omitted include the Austrian analysis of business cycles and capital theory (see Butos (1994), Horwitz (2000), and Garrison (2001) for summaries of Austrian contributions to macroeconomic theory).

This chapter is organized as follows. Section "The Birth of the Austrian School and the *Methodenstreit*" focuses on the birth of the Austrian school and the characteristics that distinguished it from the German historical school. Section "The Debate on Economic Planning and the Character of Knowledge" summarizes the "economic calculation debate," which highlighted the role of markets in coordinating economic activities and laid the foundations for later Austrian contributions to entrepreneurship theory, institutional analysis, and the theory of spontaneous order. Section "Competition in Austrian Economics" presents the Austrian view on competition. Section "The Great Divergence" explains what caused the divergence of the Austrian school from mainstream economics in the postwar era. Section "Institutions Reexamined" discusses the Austrian view of institutions, while section "Entrepreneurship" focuses on entrepreneurship. Section "Austrian Methodology" discusses Austrian methodology and its varieties. The final section concludes.

The Birth of the Austrian School and the *Methodenstreit*

In 1871, Carl Menger published the book *Principles of Economics*. With this book, he aimed at improving the ideas of the German historical school – the then-dominant school of economic thought in the German-speaking countries. Indeed, he dedicated his book to Wilhelm Roscher, one of the school's dominant representatives, who later gave Menger a positive mention in his own work (Roscher 1874). The reaction from the rest of the German historical school was less favorable. Gustav Schmoller, an influential member of the younger generation, attacked Menger's work in his writings. In one of his articles, Schmoller pejoratively referred to the Mengerian approach to economics as the "Austrian school," so as to distinguish it from the German historical school. Thus the label – and the school – was born.

The exchange between Menger and Schmoller came to be known as the *Methodenstreit* – a dispute over methods. From this debate, some of the main characteristics of the Austrian school emerged. Schmoller advocated an empirically based approach, which emphasized historical relativity at the expense of theoretical universalism. He focused on collective phenomena, such as social classes, and analyzed their historical development. He further argued that collective phenomena – although existing in the minds of acting individuals – cannot be analyzed in terms of individual behavior. In contrast, Menger formulated a theoretical approach that highlights how universal laws govern the behavior of individuals. He constructed a theory based on "marginalism," which focuses on the principle that individuals make decisions at the margin. Additionally, he adopted invisible-hand explanations of institutions (Menger [1883] 2009). In the case of the institution of money, Menger demonstrated how an institution can emerge as the unintended consequence of a multitude of individual actions. Hayek (1952) is a restatement of – and elaboration upon – this aspect of Mengerian theorizing.

In the literature on the history of economic thought, Menger is primarily remembered, along with Leon Walras and William Stanley Jevons, as one of the founding fathers of the marginalist approach in economics. All three agreed on the importance of the concept of marginal utility for economic analysis. However, their contributions differed substantially. The crucial difference is that for Walras and Jevons, marginal utility represented a starting point for developing a quantitative theory of markets. By contrast, Menger's interest was not in quantitative theory. He primarily focused on a careful definition of concepts and their mutual relationships. This essentialism also characterizes the contributions of some of Menger's followers (see e.g., Linsbichler 2017).

Building on Menger's contributions, the next generations of the Austrian school developed the theory of value further. Friedrich von Wieser (1914) introduced the concept of opportunity cost and coined the term marginal utility (*Grenznutzen*); Eugen von Böhm-Bawerk ([1884] 1890) laid the foundations of the modern theory of capital and interest with his focus on the time dimension and his notion of the roundaboutness of production processes (that said, the Austrian theory of capital later diverged from the mainstream – cf. Hayek 1941; Lachmann 1977; Lewin 1999). Böhm-Bawerk ([1896] 1898) also formulated an influential critique of Marx's economic theory. Other contributions followed. Ludwig von Mises ([1912] 1980) developed the Austrian theory of money and business cycles. He also developed and reinterpreted aspects of Menger's methodology (Mises, [1933] 1981). In particular, Mises pointed out that marginalism serves as the basis of not only economic action, but also other types of actions. He used the term "praxeology" to denote this general theory of action.

Many of these Austrian contributions were integrated into standard economic theory: Menger, Böhm-Bawerk, and Wieser stand next to Jevons, Walras, Marshall, Pareto, and Edgeworth as the founders of neoclassical economics. The methodological basis of this synthesis rests on Mises's idea of a universal theory of action, which was restated and popularized by Robbins (1932). The new orthodoxy was marginalist, subjectivist, and individualist. Therefore, the characteristics that initially defined the Austrian school ceased to be the distinguishing features of the school, and debates with other schools in economics were reduced to disagreements about specific issues within the type of shared theoretical framework that exists within any discipline. This view was also expressed by Mises ([1933] 1981, p. 214) and Hayek (1968, p. 52). Nonetheless, the Austrian school did not disappear as other of its characteristics came to the fore.

The Debate on Economic Planning and the Character of Knowledge

The first and second generations of Austrians defined some of the crucial characteristics of the school, such as marginalism, opportunity cost, and methodological individualism. Another debate, which helped to specify and highlight additional features of the school, was the debate on economic planning. This debate began in 1920 with Mises' article "Economic Calculation in the Socialist Commonwealth" (Mises, [1935] 1975) and continued until the 1940s. Mises argued that market prices were indispensable for the ability of producers and consumers to engage in rational economic calculation. The impossibility of calculation in nonmarket systems of governance thereby became a central concern of Austrians (Rothbard, [1962] 1970). Combined with the works of Coase (1937), Williamson (1985), and Grossman and Hart (1986), Mises' insight represents an essential aspect of the modern Austrian theory of the firm (Bylund 2015; Foss 1994).

The debate on the rationality of central planning proceeded in German up until the late 1920s, when it was taken up by English-speaking economists. The leading supporters of socialism accepted Mises' argument regarding the indispensable role of prices in economic calculation (Lange 1936; Lerner 1934), but they believed that prices could be determined without markets and private ownership. A crucial counter-argument was raised by Hayek, who pointed out that in order to set prices in the socialist economy, planners would have to possess relevant information about tastes and technologies. However, this information is not available in a ready-to-use form. Instead, economic agents act upon fragmented pieces of knowledge that are subjective, incomplete, and may be impossible to centralize. Hayek's argument inspired a vast stream of research in both Austrian (Kirzner 1973; Rizzo and Whitman 2019) and mainstream (Hurwicz 1969; Grossman and Stiglitz 1976) economics. Fritz Machlup (1973) developed the most elaborate theory of knowledge within the Austrian camp.

Hayek's own analysis of the knowledge problem focused on the limits of human reason as regards its ability to accumulate a sufficiently large amount of knowledge for the purpose of designing and controlling complex systems. For him, socialism exemplified the "fatal conceit" that ignored the limits of reason in its attempt to plan society as a whole. Hayek subsequently attempted to formulate a methodology to study complex phenomena such as the market (Hayek 1955, 1967). He argued that in the case of complex phenomena, detailed explanations and specific predictions might not be feasible, and the best we can achieve are *explanations of principle* and *pattern predictions*. Building on the insights of Adam Smith, Adam Ferguson, and Carl Menger on the spontaneous emergence and functioning of various institutions, and borrowing from disciplines such as biology, linguistics, and law, he showed that decentralized systems could perform better than centralized ones. Hayek's insights inspired modern research on bounded (Frantz and Leeson 2013) and evolutionary (Smith 2007) rationality, as well as complexity science (Arthur 2010; Bowles et al. 2017; Vriend 2002).

The calculation debate revealed some additional aspects of the Austrian approach. Some economists in the socialist camp argued that consumers are not able to make rational choices about goods and services (Dobb 1933). Against this view, Austrians advocated consumer sovereignty (Mises 1949). For them, consumer sovereignty was merely an aspect of their subjectivism – the positive and normative relevance of individuals' preferences, knowledge, and beliefs. As Hayek ([1955] 1964, p. 52) put it, the notion of "objective needs" refers to "somebody's views about what the people ought to want." Newer Austrian literature elaborates on the idea of consumer sovereignty in critiques of new forms of paternalism associated with the work of Richard Thaler and other behavioral economists (Rizzo and Whitman 2019).

Yet another aspect of the Austrian school that emerged from the debate on economic planning was its dynamic view of markets. Socialists argued that capitalism has an inevitable tendency toward monopolization. Therefore, for them, the choice was not between competitive capitalism and socialist planning, but between monopoly capitalism and socialist planning (Persky 1991). They argued that the textbook model of competition, with a large number of small sellers, is inapplicable to contemporary capitalist economies. Although early Austrians contributed to the textbook model of competition (Böhm-Bawerk, [1884] 1890), later generations realized that it misses many important aspects of real-world competition, such as innovative entrepreneurship. The next section describes the Austrian view of competition in more detail.

On a more general level, the calculation debate focused on two generic institutional settings, "capitalism" and "socialism." This is understandable, given the political and social circumstances of the period. Nevertheless, after 1989, such debates became outdated, and the focus shifted to forms of governance other than pure markets and pure central planning. More recent research has focused on the issues of self-governance in the case of missing or incomplete markets due to common ownership, externalities, or public goods (Candela and Geloso 2018; Leeson 2007; Skarbek 2014). Here Austrian build on the contributions of Buchanan, Coase, and the Ostroms (Aligica et al. 2019; Boettke and Lopez 2002). A closely related literature uses Austrian insights to analyze legal and political systems (Leeson 2011 2014; Piano 2019). The acknowledgment of the existence of institutional diversity distinguishes this modern Austrian literature from the mainstream and positions the school closer to institutional economics (Hodgson 2019; Hudik and Bylund 2021). Nonetheless, a trademark of the Austrian approach is pessimism regarding government as opposed to bottom-up solutions to social problems, including the problems associated with economic development and transitions (Coyne 2008; Powell 2014).

The key methodological principle underlying Austrian analyses of government is behavioral symmetry (Boettke, Coyne, and Leeson 2007). According to this principle, the motivations and cognitive capacities of political and market actors do not differ. If market actors are assumed to be self-interested and fallible, this should also hold true for political actors. People are the same whether they act as buyers, sellers, legislators, or bureaucrats. Behavioral symmetry is violated when the government is assumed to be benevolent and omniscient, while market participants are self-interested and fallible. In behaviorally asymmetric models, governments are assumed to provide solutions to market failures. Once the behavioral symmetry is acknowledged, the possibility of government failure emerges. Under these settings, it is unclear whether the government improves or worsens market outcomes in the case of market failures, and a comparative institutional analysis is called for.

Competition in Austrian Economics

One of the distinguishing features of Austrian economics is its theory of competition. There are two types of approaches to theorizing about market competition: one treats competition as a "noun" or a state of affairs, the other interprets it as a "verb" or a process (McNulty 1967). Mainstream economics develops the concept of competition as a state of affairs. Here competition is understood as the absence of monopoly power (Stigler 1957). The idea of competition as a process has been developed by Schumpeter (1942) and Hayek (1948; 2002 [1968]), although its origin can be traced back to Adam Smith (McNulty 1967).

Hayek (2002 [1968]) characterizes competition as a discovery procedure. He emphasizes that competition is useful when its results cannot be predicted. Hence, competition is a tool to address questions to which we do not know the answer, such as "what is the most efficient way of producing a given good?" or "what is the most productive allocation of a given resource?" By contrast, the conventional model of competition as a state of affairs typically assumes well-informed agents. In Hayek's view, the assumption that everything relevant is known results in uninteresting and useless models of competition (Hayek 1945).

Later contributions developed this process-based approach to markets further (cf. Ikeda 1990; Thomsen 1992; Machovec 1995; Boettke and Prychitko 1994). They typically highlight differences between the theory of the market process and the mainstream approach, including its modern elaborations and refinements. The process view focuses on the moves toward or away from equilibrium, whereas the mainstream view is concerned with the properties of an equilibrium as such, without analyzing how it was achieved in the first place. According to the Austrians, the crucial agent behind all market processes is the entrepreneur.

Although entrepreneurship was discussed by Menger (1871) and Mises (Mises 1949), the Austrian theory of entrepreneurship is primarily associated with the work of Israel Kirzner (1973). According to Kirzner, entrepreneurs are akin to arbitrageurs who are alert to profit opportunities stemming from price differences. Their activities lead to the equilibration of markets, although equilibrium is thought of as a hypothetical state that may never be achieved in reality. Modern theories of entrepreneurship build on Kirzner's theory (1973), as well as on Mises' insights about market processes (Foss and Klein 2012). But it is only against the backdrop of an imperfect institutional structure that entrepreneurship becomes possible, as is shown later.

One of the Austrian critiques of the mainstream approach to markets, especially in its Walrasian form, is that it includes very few variables: It focuses on prices and quantities while treating technology, consumer preferences, and resource endowments as exogenous factors that affect prices and quantities. Nonetheless, it ignores the roles of social norms, customs, traditions, legislation, or governance structures. In short, the standard mainstream model lacks institutional detail. Until the development of modern contract theory, the mainstream model of competition only considered two generic institutions: the market and the firm. By contrast, Austrians have been concerned with the emergence and functioning of various institutions since Carl Menger. The following section presents the Austrian view of institutions.

The Great Divergence

To understand the postwar development of Austrian economics, it is necessary first to note certain features of mainstream neoclassical economics as it evolved from the 1950s onward. Three developments had especially far-reaching impacts on what postwar economics departments taught and did research on:

1. The split of mainstream economics into micro and macro, which reflected the success of Keynes's general theory (1936).
2. The formalization of mainstream economics, which shifted the focus toward issues that could be addressed with mathematical tools.
3. The development of econometrics, which over time became the preoccupation of economists concerned with empirical research. Econometric studies, as a rule, demonstrate a greater interest in aggregates than is common among Austrians.

In the wake of the Great Depression, Keynes's theory engendered policy recommendations such as deficit spending that seemed to offer a way out of economic crises. These crises were, according to Keynes, often caused by a vicious circle of self-reinforcing pessimism among investors and other market actors. Keynes argued that there is a need for a different way of reasoning when theorizing about macrophenomena than the traditional way of just building models based on conventional micro-approaches. This had the lasting consequence of splitting mainstream education and research into its two major subdisciplines of microeconomics and macroeconomics.

The second development had little in common with Keynes's original approach, which was more institutional and less mathematical than typical postwar economics. Because Keynes's institutional specificity and subjective approach to expectations and probability theory were difficult to integrate into formal models, later macroeconomists reformulated Keynes's original theory in favor of a more mathematical and institutionally poor approach that was to become known as the "neoclassical synthesis."

A key focus in Keynes's original theory, which he shared with Knight (1921) as well as some later Austrians (Lachmann 1957), was on the irremediable uncertainty of the shape of the future economy and, in particular, the economy in the distant future. This had the important implication that the distant future is unstructured in the present and thus not amenable to forecasts based on probabilities derived from past or present economic data, although such data can sometimes be used to predict the near future. Thus Austrian, original Keynesian, and Knightian approaches are entirely at variance with the mainstream treatment of probability theory, whereby the future is often treated as structurally equivalent to the past and present.

The increasing formalization of economics began in American universities almost immediately after World War II (Blaug 2003; Weintraub 2002). Many of the newly appointed professors in US economics departments had a background in mathematics or physics rather than in social science. They thus had a strong preference for mathematical models over the type of verbal reasoning that had been the mainstay not only of pre-war economics but which remained the standard approach in related social science disciplines such as political science and sociology. The new breed of economists began to dominate, although less formal approaches were not completely stamped out: seminal books and articles by Coase, Williamson, North, Alchian, Demsetz, and Schelling have had an enormous influence on many postwar economists. Although the Austrian school is not inherently anti-mathematical (Hudik 2015), its dominant representatives chose not to adopt formal tools of analysis. It did not manage to remain relevant to the mainstream (Foss 2019). Unlike many behavioral and neoinstitutional economists, Austrians did not present their findings as realistic deviations from a mainstream benchmark model.

The third trend, which was toward a more prominent role for econometric methods in empirical research, reflected a shift toward extensive use of data and empirical testing. As in the case of formalization, Austrians were not part of this development. There were several reasons for this. First, many Austrians were influenced by Mises's view that economic theory is not subject to empirical testing. Instead, according to Mises, empirical analysis presupposes theory. Moreover, empirical findings have limited generalizability: they hold only for specific times and places as the world is in constant flux. Additionally, Hayek objected to the use of statistics for studying social structures as it "deliberately and systematically disregard[s] the relationships between the individual elements" (Hayek, [1955] 1964, p. 61). Although neither Mises nor Hayek denied the usefulness of statistics to address certain questions, these were not the questions that they were interested in. Other Austrians followed in their footsteps (Leeson 2020).

This lack of interest in statistics and econometrics does not mean that Austrians are uninterested in empirical evidence. However, this evidence comes in different forms. Austrians highlight the role of analytical narratives:

> Borrowing from sociology and anthropology, economics may employ survey, interview, and participant observe techniques to glean new empirical knowledge from its subjects (the narrative) to be analyzed in light of aprioristic rational choice theory (the analytic), leading to analytically rigorous but institutionally rich examinations (Leeson and Boettke 2006, p. 263).

The underlying assumption behind this method is that many economic phenomena are heavily influenced by idiosyncratic factors (Hudik and Bylund 2021). Therefore, the search for generally valid quantitative relationships has only limited value.

The 1950s and 1960s were not receptive to these ideas. These were the years of high theory. It culminated in the mathematician Gerard Debreu's (1959) highly abstract version of general equilibrium theory, where all practical considerations regarding knowledge, expectations, or institutions were pushed aside in favor of a timeless and spaceless mathematical model of perfectly efficient and stable markets. While this turned out to be an interlude of limited duration, as evidenced by the increasing proportion of economists focusing on investigating empirical phenomena (rather than theoretical models) in narrow domains such as a specific market or industry, it did nevertheless lead to a parting of ways between the mathematical mainstream and more verbal alternative schools, including the Austrian school among many others. It is thus against this backdrop that the divergence of Austrian and mainstream economics after World War II must be interpreted.

Institutions Reexamined

In the early twenty-first century, it is easy to forget that economics and the other social sciences once shared a preoccupation with institutions. Even Menger, the alleged opponent of institutional approaches, was an institutionalist by postwar standards. Other economists of his time were even more institutional.

But it is really the continued focus on institutions that offers a compelling explanation of why Hayek seemed more relevant to the scholarship of political philosophers and political scientists than to postwar economists. It could even be argued that Hayek may have chosen noneconomists as his main audience so as to preserve the relevance of his ideas in a scholarly context, in spite of the increasing irrelevance of his economic theories to postwar economists. To economists such as Debreu, Samuelson, and others of their ilk, Hayek must have seemed like an anachronism that could be ignored. This is all the more ironic since Hayek was considered one of the leading economic theorists in the pre-World War II era, having devoted himself to business cycle theory, price theory, and the pure theory of capital (Hayek 1941).

The most enduring legacy of Hayek's contribution to social science is arguably his theoretical framework of the social world as consisting of spontaneous and planned orders, or what are sometimes called *cosmos* and *taxis*.

As Moroni (2014) shows, this is an elaboration of two strands of thought with their origin in the Scottish Enlightenment. One was first associated with Adam Ferguson and Carl Menger, and is preoccupied with the spontaneous or unplanned order of evolving human institutions. Ferguson ([1767] 1995) claimed that such orders "are made with equal blindness to the future; and nations stumble upon establishments, which are indeed the results of human action, but not the execution of any human design." But it was Menger who gave the fullest expression of this view when he presented his theory of the spontaneous institutionalization of money:

> It is not impossible for media of exchange, serving as they do the commonweal in the most emphatic sense of the word, to be instituted also by way of legislation, like other social institutions. . . . Putting aside assumptions which are historically unsound, we can only come fully to understand the origin of money by learning to view the establishment of the procedure, with which we are dealing, as the spontaneous outcome, the unpremeditated resultant, of particular individual efforts of the members of a society, who have little by little worked their way to a discrimination of the different degrees of saleableness in commodities. (Menger 1892, p. 38)

Societies may thus adopt institutions, not because a ruler or government decided to adopt them on the basis of their real or apparent attractiveness, but instead in a gradual fashion as the result of decentralized trial-and-error experimentation with different institutional options. Hayek (1988) claims that the spread of various institutions is an evolutionary process with unequal selection propensities. Institutions spread through conquest or voluntary imitation, and their perceived relative success influences the likelihood of long-term retention. As a general tendency, those institutions that are more conducive to the survival and growth of human societies have a greater tendency to proliferate and survive over long periods of time than those that are less supportive of what Hayek sometimes calls "the Great Society" (from Adam Smith) and sometimes "the Open Society" (from Karl Popper). And it is here that Hayek makes the connection between the spontaneous order of evolving institutions and the spontaneous order of the market *within* an evolved institutional structure:

> It is in the *ius gentium*, the law merchant, and the practice of the ports and fairs that we must chiefly seek the steps in the evolution of law which ultimately made an open society possible. Perhaps one may even say that the development of universal rules of conduct did not begin within the organized community of the tribe but rather with the first instance of silent barter when a savage placed some offerings at the boundary of the territory of the tribe in the expectation that a return gift would be made in a similar manner, thus beginning a new custom (Hayek 1979, p. 82).

Hayek believed that the most successful evolved institutional structures were spontaneous orders themselves, with the market order as the most important institutional framework not only for survival, but for civilization and economic development more generally. So we have the long-run spontaneous order of an evolving institutional structure, which eventually gives rise to the "Great Society." And this society consists of the spontaneous order of the market, as well as other such orders, notably science (Polanyi 1962) and democracy (diZerega 1989).

The idea of the market as a spontaneous order was first articulated by Adam Smith with his "invisible hand" metaphor (Smith 1776). What Hayek and later Austrian economists did was primarily to trace out the implications of the classical economists' institutionally rich understanding of how markets work. The conception of the market as a spontaneous order consists of a few distinguishing features:

1. It is a social order that has emerged through human interaction, but it has not been designed by anyone in particular.
2. It does not have an overriding aim, unlike planned orders such as profit-seeking firms and other organizations.
3. The market order coordinates the actions of goal-directed individuals or firms, some of which may have contradictory goals.

Central to all spontaneous orders, whether markets or not, is the existence of a *systemic resource*. The accumulation of a systemic resource is evidence of system-specific success. Such accumulation also shows other participants in the same order ways in which success can be achieved. The "accumulators" – whether individuals or organizations – become benchmarks for prospective imitators. The systemic resource of the market order is money, while in the democratic order it is votes. It is indirect exchange that makes the accumulation of money (the systemic resource) possible, but the exchange of goods for money also generates an order-specific signal, the market price.

Hayek's favored term for the spontaneous order of the market is *catallaxy* (Whately 1832). The rules that govern it are the rights and liabilities associated with the creation, protection, and exchange of property rights over scarce resources. Money signaling not only provides guidelines for successful business strategies. It also disseminates exchange ratios – money prices – that approximate underlying resource scarcities within interconnected networks. Indeed, it is through decentralized exchanges that market prices come into being. Prices disseminate localized subjective knowledge to other market participants (Hayek 1945). In their most advanced instantiations, the resulting price-coordinated networks may encompass millions of anonymous traders.

As mentioned earlier, the market order or *catallaxy* is not the only spontaneous order. A contemporary of Hayek, Michael Polanyi, conceived of science in a way that is similar and yet distinct from markets (Polanyi 1962). In the case of science, the systemic resource is fuzzier but still functional, and consists of the best cross-disciplinary signals of reputation. The counterpart to money would then be a vaguer measure such as citation-weighted publications in the long run, or prestige-weighted publications in the short run.

Table 1 offers a comparison of the three most important spontaneous orders in a capitalist liberal democracy, which are the *catallaxy*, democracy, and science (Andersson and Taylor 2012). Success in catallactic markets is measured in money. In a democracy, it is measured in vote tallies and in science, in publications and citations.

Table 1 Systemic attributes of three spontaneous orders

Attribute	*Catallaxy* (Markets)	Democracy	Science
Systemic resource	Money	Votes	Citation-weighted publications
Systemic resource distribution	Roughly equal to highly skewed	Equal (for insiders)	Extremely skewed
Revealed preferences	Continuous (willingness to pay)	Discrete (one person, one vote)	Continuous (e.g., article downloads; citation of others)
Aggregation of preferences	No (individual choice)	Yes (collective choice)	No (individual choice)
Typical *taxis*	Business firm	Political party	University or institute
Effect of replacing *cosmos* with *taxis*	Central planning	Autocracy	Dark ages

Source: Adapted from Andersson and Taylor (2012)

The spontaneous order framework has been one of the most fertile theoretical research programs within postwar Austrian economics. It has also affected other schools of thought within economics. In particular, the influential neoinstitutional school of Coase, North, and Williamson has implicitly adopted many of the underlying assumptions of Hayek's institutional turn after 1945. In particular, North's distinction between formal institutions (laws) and informal constraints (norms) closely mirrors notions of a slowly evolving institutional structure as Hayek presents them in his two last treatises (Hayek 1979, 1988). And the preoccupation with transaction costs in Coase (1937) and with opportunistic behavior in Williamson (1985) would not be possible without the original Hayekian insight that market participants have limited subjective knowledge when interacting with others (Hayek 1937).

Entrepreneurship

The spontaneous order approach is a framework for thinking about institutions. Some institutions affect the market as a whole and are thus part of the slowly evolving *cosmos*. Other institutions are instrumental. They have the purpose of furthering the specific goals of a *taxis*, such as a profit-seeking firm. But institutions only become important in a world with bounded knowledge and limited information about market opportunities and constraints. Institutions are rules of behavior that can be formal, as in the case of laws and regulations, or informal, as in the case of conventions and norms (North 1990). Institutions structure economic (i.e., de facto) property rights (Barzel 1989).

Institutions only become meaningful as an analytical focus if we assume imperfect knowledge since perfect knowledge implies perfectly defined and communicated property rights. The phenomenon of entrepreneurship is similar to institutions in this sense. If we assume perfect knowledge and zero transaction costs, all gains from trade can be instantaneously realized, and there are no profit opportunities that have not already been discovered.

The early Austrian economists recognized the importance of entrepreneurs in the economy, but they were never the main focus of their treatises. Nevertheless, Menger was aware of the importance of the entrepreneur when he wrote that

> [t]he process of transforming goods of higher order into goods of lower or first order, provided it is economic in other respects, must also always be planned and conducted, with some economic purpose in view, by an economizing individual. The individual must carry through the economic computations of which I have just been speaking, and he must actually bring the goods of higher order, including technical labor services together (or cause them to be brought together) for the purpose of production. (Menger [1871] 1994, pp. 159–160).

Menger is approaching a theory of entrepreneurship in the above passage, but his reference to an "economizing individual" differs from Israel Kirzner's (1973) later distinction between economizing and entrepreneurial behavior. Schumpeter's (1934) entrepreneurship theory is more rigorous in its definitions, but its reliance on Walrasian equilibrium as a point of departure marks it as a "hybrid theory" rather than a pure Austrian one (Rothbard 1987).

It was Kirzner who developed a recognizably Austrian theory of entrepreneurship by elaborating on certain features of Austrian theorizing that are especially evident in "Economics and Knowledge" (Hayek 1937) and "The Use of Knowledge in Society" (Hayek 1945). Hayek alludes to the existence of a market function that effects greater market coordination in these two articles, without ever mentioning that this function is carried out by the entrepreneur. Kirzner's seminal book, *Competition and Entrepreneurship* (1973), is arguably an attempt to formulate a theory that fills in the gaps in Hayek's two path-breaking papers (Hayek 1937 1945).

Unlike Schumpeter (1934), Kirzner (*op. cit.*) does not refer to equilibria as "circular flows" that can be improved upon as new innovations enter the economy. Kirzner's conception reflects the postwar understanding of equilibrium as a perfect-knowledge stable state. In Kirzner's framework, this market equilibrium exercises a "gravitational pull" on economic action, although it is never attained. In spite of superficial similarities, the Austrian notion of equilibrium differs from its counterpart in conventional price theory (Hudik 2018, 2020). In the Austrian analysis, equilibrium refers to a (hypothetical) real-world situation where individuals' plans are fully coordinated (Hayek 1937). By contrast, equilibrium in mainstream price theory is an analytical tool that does not refer to anything real. The aim of this tool is to derive all implications that result from an exogenous change, such as technological shocks or policy interventions (Machlup 1958). In the mainstream approach, equilibrium is inseparably connected to comparative statics, where properties of equilibria are studied but there is no room for market processes or entrepreneurial activities.

In the modern Austrian understanding of the market as a process, the market is never entirely coordinated, but it is always moving in the direction of greater coordination. And it is the entrepreneurs who fulfill the role of being the coordinating agents. Kirzner (1973) approaches entrepreneurship in its most basic form as a problem of interlocal arbitrage.

The basic idea is that there are two unconnected local markets. They sell the same good in each market, but at different local market prices. An alert individual notices the price discrepancy and is able to buy it at a lower price in the cheaper market and sell it at a higher price in the dearer one. This is then a source of pure entrepreneurial profit. This profit will attract imitators. A new integrated market with an intermediate price is the final outcome of this entrepreneurial arbitrage process. According to Kirzner, entrepreneurial action hence integrates previously separated markets, equalizing prices. The new market price contains more economic knowledge, since it integrates the dispersed local knowledge of more individuals than previously.

Arbitrage is, of course, not the only type of entrepreneurship. Kirzner (ibid.) distinguishes between three types: arbitrage, speculation, and innovation. He interprets speculation as a type of intertemporal arbitrage, whereby the entrepreneur "observes" a later expected higher price and a present lower one. Innovation is the most complex type of entrepreneurship; an alert entrepreneur discovers a greater output price than the sum of the prices of all inputs that contribute to the production of the output.

Like Schumpeter (1934), Kirzner (1973) makes a clear separation between the entrepreneur and the capitalist. The pure entrepreneur has no need for resources; the only thing he/she needs is *alertness*, which denotes the ability to observe existing profit opportunities that other market participants have failed to notice. If the realization of the profit opportunities requires resources, it then becomes the economic role of capitalists to lend the means to acquire them, for which they receive interest as factor income. Entrepreneurship is thus the only source of profits in a pure market economy. In Kirzner's conceptualization of a pure market economy, there are no barriers to entry, and lending is asset-neutral. This theory thus shares some of the hallmarks of neoclassical models of competitive markets.

When entrepreneurs bridge markets, they eliminate knowledge gaps. According to Kirzner (ibid.), each such coordinative action brings the market a little bit closer to a fully coordinated perfect-knowledge market. But because there will always be changes to factors such as technology, consumer preferences, and resource availabilities, which Kirzner, like his neoclassical contemporaries, treats as exogenous variables, this (hypothetical) end state remains elusive in real-world markets. Thus, the entrepreneur-driven tendency toward equilibration does not imply equilibrium.

In modern Austrian theories of the market process, it is the focus on the role of entrepreneurs and, additionally, the effects of institutions on entrepreneurial opportunities and incentives which constitute the greatest divergence from the neoclassical mainstream. With the long-standing focus on local knowledge, it is but a short step to endow entrepreneurs with the key role of initiating changes that reflect their spatio-temporally specific knowledge of market conditions. Thus, the focus on market agents' imperfect knowledge generates many of the most distinctive features of the Austrian school: the importance of property rights and other institutions, the role of entrepreneurs, and the subjectivity of not only preferences but also of knowledge and expectations.

Austrian Methodology

Austrians traditionally emphasize methodological issues. This emphasis distinguishes them from mainstream economists who typically leave methodological debates to specialists. Many representatives of the Austrian school, including Menger, Mises, Hayek, Machlup, Rothbard, and Lachmann, wrote extensively on method. Although their views shared some basic characteristics, such as methodological individualism, subjectivism, and the rejection of inductive methods, they emphasized and elaborated upon different aspects of their shared Austrian starting point. These differences in emphasis stemmed from differences in their research focus, and they often allowed Austrians to reach out to various allies outside the Austrian camp.

Arguably, the most influential statement of the Austrian method is by Mises ([1933] 1981; 1949; [1962] 1978; [1957] 1985), who highlighted the role of theory in interpreting empirical observations. In his view, theory cannot be tested by experience because it has a similar character as logic and mathematics. It is a lens that exists prior to experience, and it enables the theorist to grasp historical facts. Mises further argued that the a priori character of the theory is implied by the fact that individuals' minds have a common logical structure. This common structure allows for the possibility of a general theory of human action – praxeology. Although Mises' methodological views may seem bizarre to a reader trained in mainstream economics, they did not differ dramatically from the views of classical economists such as Nassau William Senior or Jean-Baptiste Say. They also formed the basis of the early neoclassical methodology as formulated by Lionel Robbins. Within the Austrian school, Mises's views were further elaborated and reinterpreted by Rothbard (1956, 1957, 1970, 1973, 1976), but all Austrians built on Mises's methodology in one way or another.

While Mises' methodology centered on the concept of human action, Hayek focused more on social interactions and the coordination of human activities (O'Driscoll 1977). The study of interactions required addressing empirical questions about how people acquire information (Hayek 1937; Hudík 2011). This observation led Hayek to develop an institutional analysis that was inspired by Menger and the philosophers of the Scottish Enlightenment such as Adam Smith and Adam Ferguson. In line with Menger and Mises, Hayek ([1955] 1964) emphasized methodological individualism, subjectivism, and the theoretical method. His essential point was that the opposite methodological views – methodological collectivism, objectivism, and historicism – stem from "scientism," which Hayek described as a slavish imitation of the method of the natural sciences. Later he modified the definition of scientism as an imitation of what is mistaken for the method of natural science. He objected that scientism did not recognize the limits of human reason in the study of the type of complex phenomena that all social sciences have to deal with.

Under the influence of Popper, Hayek adopted methodological monism, i.e., the view that the methods of the natural and social sciences are in principle the same. However, he maintained that these methods have to be adjusted for the study of complex phenomena in disciplines such as economics and evolutionary biology. Therefore, according to Hayek, the crucial distinction is not between natural and social sciences but between disciplines studying simple and complex phenomena. He attempted to develop a methodology to account for such phenomena using Popperian language (Hayek 1955; 1967). Hayek's proposed method for studying complex phenomena later inspired the new field of complexity economics (Axtell 2016; Rosser 2012; Koppl 2010).

Ludwig Lachmann was similar to Hayek in some respects, but paid more attention to the implications of subjective expectations. Like Hayek, Lachmann focused on the coordination of human activities. But his emphasis on the subjective character of expectations led him to contend that disequilibria are ubiquitous. Nonetheless, he also highlighted the guiding role of institutions in complex environments, as well as their capacity to make the expectations of market participants converge under ideal circumstances (Lachmann 1977). Like Mises, Lachmann was inspired by Max Weber's conception of "ideal types" (Lachmann 1971). Lachmann's methodology has been most amenable to hermeneutical analyses and thus furthest away from the neoclassical program based on rational choice, although there have been attempts to bridge the two approaches (Koppl and Whitman 2004). The Lachmannian approach was further developed and applied by Lavoie (2011), Prychitko (1994), Lewin (1994), Chamlee-Wright (2008), and Storr (2019a). A notable methodological preference of this group of Austrians is the use of participant observation and other qualitative approaches usually associated with anthropology and qualitative sociology. This approach also has many commonalities with the theorizing of Shackle (1952), Loasby (1999), and heterodox schools of thought such as Post-Keynesianism, old institutionalism, and evolutionary economics.

While Lachmannian methodology is arguably furthest away from mainstream economics, the methodological views of Fritz Machlup represent the most direct bridge between the Austrian school and mainstream price theory. Machlup embraced an eclectic array of theoretical approaches, ranging from phenomenal sociologists such as Alfred Schutz to Chicago school economists such as Milton Friedman. In his many contributions, he elaborated on various Austrian topics, such as the rejection of naïve empiricism (Machlup 1955) and the differences between natural and social sciences (Machlup 1961). Nevertheless, he also offered sympathetic interpretations of some concepts and theories common in mainstream price theory, including equilibrium (Machlup 1958), marginalism (Machlup 1967), and *homo oeconomicus* (Machlup 1970).

For Austrians, the methodology of economics is a central concern. Austrian approaches are continually reexamined and contrasted with developments in other schools of thought in economics and other disciplines (e.g., Caldwell 1982; Dekker 2016; Linsbichler 2017). In this way, Austrians attempt to attain a deeper understanding of what economics is all about.

Conclusions

Austrian economics has been shaped and reshaped in its ongoing debate with other schools of thought within economics. Early Austrians highlighted the role of an economic theory based on marginalism, subjectivism, and opportunity costs. They shared these characteristics with other variants of neoclassical economics. Nonetheless, the mainstream approach later adopted econometrics as their preferred tool for studying empirical regularities, which Austrians rejected as inadequate for an understanding of the connections that structure complex phenomena. The Austrian approach highlights the importance of factors that are difficult to quantify or aggregate for the purposes of econometric analysis. Consequently, Austrians emphasize methods such as historical narratives and case studies. Moreover, they traditionally pay careful attention to methodological issues (Sima 2019).

Unlike mainstream neoclassical economics, Austrians did not adopt formal modeling. Although there have been attempts to formalize some aspects of Austrian theory (Hudik 2020; Littlechild and Owen 1980), most Austrian work has remained informal. The advantage of such an approach is that it encourages economists to focus on real-world problems rather than on mathematical puzzles. Yet at the same time it makes communication with the rest of the economic profession unnecessarily difficult. Some argue that Austrian economics would gain influence by adopting a more formal approach (Foss 2019; Hudik 2015), while others suggest that Austrians should cultivate their relationship with disciplines other than economics, particularly social anthropology, cognitive psychology, political philosophy, and phenomenological sociology (Boettke 2019; Storr 2019b).

The distinguishing feature of the Austrian school is its emphasis on bottom-up institutional solutions to social problems. This emphasis combines Austrian insights on spontaneous orders, institutions, entrepreneurship, and the boundedness of human reasoning. By addressing real-world problems, Austrians continue to provide valuable contributions to our understanding of markets and economic development processes.

References

Aligica PD, Boettke PJ, Tarko V (2019) Public governance and the classical-liberal perspective: political economy foundations. Oxford University Press, Oxford, UK

Andersson DE, Taylor JA (2012) Institutions, agglomeration economies and interstate migration. In: Andersson DE (ed) The spatial market process. Emerald, Bingley, pp 233–263

Arthur WB (2010) Complexity, the Santa Fe approach, and non-equilibrium economics. Hist Econ Ideas 18(2):149–166

Axtell RL (2016) Hayek enriched by complexity enriched by Hayek☆. In: Boettke PJ, Storr V (eds) Revisiting Hayek's political economy. Emerald, Bingley, pp 63–121

Backhouse R (2000) Austrian economics and the mainstream: view from the boundary. Q J Austrian Econ 3(2):31–43

Barzel Y (1989) Economic analysis of property rights. Cambridge University Press, Cambridge, UK

Blaug M (2003) The formalist revolution of the 1950s. J Hist Econ Thought 25(2):145–156

Boettke PJ (1996) What is wrong with neoclassical economics (and what is still wrong with Austrian economics). In: Foldvary F (ed) Beyond neoclassical economics. Edward Elgar, Cheltenham, UK

Boettke PJ (2019) What is right about Austrian economics? In: D'Amico DJ, Martin AG (eds) Assessing Austrian economics. Emerald, Bingley, pp 125–137

Boettke PJ, Lopez EJ (2002) Austrian economics and public choice. Review of Austrian Economics 15(2–3):111–119

Boettke PJ, Coyne CJ, Leeson PT (2007) Saving government failure theory from itself: recasting political economy from an Austrian perspective. *Constitutional Political Economy, 18*, 127–143

Boettke PJ, Prychitko DL (eds) (1994) The market process. Edward Elgar, Aldershot

Bowles S, Kirman A, Sethi R (2017) Retrospectives: Friedrich Hayek and the market algorithm. J Econ Perspect 31(3):215–230

Butos WN (1994) The Hayek-Keynes macro debate. In: Boettke PJ (ed) The Elgar companion to Austrian economics. Edward Elgar, Aldershot, pp 471–477

Bylund PL (2015) The problem of production: a new theory of the firm. Routledge, Abingdon

Caldwell BJ (1982) Beyond positivism: economic methodology in the twentieth century. Routledge, New York

Candela RA, Geloso VJ (2018) The lightship in economics. Public Choice 176(3–4):479–506

Caplan B (1999) The Austrian search for realistic foundations. South Econ J 65(4):823–838

Chamlee-Wright E (2008) The structure of social capital: an Austrian perspective on its nature and development. Rev Polit Econ 20(1):41–58

Coase RH (1937) The nature of the firm. Economica 4(16):386–405

Coyne CJ (2008) After war: the political economy of exporting democracy. Stanford University Press, Stanford

Debreu G (1959) The theory of value: an axiomatic analysis of economic equilibrium. Yale University Press, New Haven

Dekker E (2016) The Viennese students of civilization: the meaning and context of Austrian economics reconsidered. Cambridge University Press, New York

diZerega G (1989) Democracy as a spontaneous order. Crit Rev 3(2):206–240

Dobb M (1933) Economic theory and the problems of a socialist economy. Econ J 43(172):588–598

Ferguson A ([1767] 1995) *An essay on the history of civil society.* In: Oz-Salzberger, F (ed), Cambridge University Press, Cambridge, UK

Foss NJ (1994) The theory of the firm: the Austrians as precursors and critics of contemporary theory. Review of Austrian Economics 7(1):31–65

Foss NJ (2019) Austrian economics: a tale of lost opportunities. In: D'Amico DJ, Martin AG (eds) Assessing Austrian economics. Emerald, Bingley, pp 111–124

Foss NJ, Klein PG (2012) Organizing entrepreneurial judgment: a new approach to the firm. Cambridge University Press, Cambridge, UK

Frantz R, Leeson R (eds) (2013) Hayek and behavioral economics. Springer, Heidelberg

Garrison RW (2001) Time and money: the macroeconomics of capital structure. Routledge, London

Grossman SJ, Hart OD (1986) The costs and benefits of ownership: a theory of vertical and lateral integration. J Polit Econ 94(4):691–719

Grossman SJ, Stiglitz JE (1976) Information and competitive price systems. Am Econ Rev 66(2): 246–253

Hayek FA (1937) Economics and knowledge. Economica 4(13):33–54

Hayek FA (1941) The pure theory of capital. The University of Chicago Press, Chicago

Hayek FA (1948) *Individualism and Economic Order.* University of Chicago Press, Chicago

Hayek FA (1945) The use of knowledge in society. Am Econ Rev 35(4):519–530

Hayek FA (1952) The counter-revolution of science: studies in the abuse of reason. Collier-Macmillan, London

Hayek FA (1955) Degrees of explanation. Br J Philos Sci 6(23):209–225

Hayek FA ([1955] 1964) The counter-revolution of science: studies on the abuse of reason. The Free Press of Glencoe, Collier-Macmillan, London

Hayek FA (1967) The theory of complex phenomena. In: Hayek FA (ed) Studies in philosophy, politics and economics. Taylor & Francis, Abingdon

Hayek FA (1968) Economic thought VI: the Austrian school of economics. In: Sills DL, Merton RK (eds) International encyclopedia of the social sciences. New York, Macmillan

Hayek FA (1979) Law, legislation and liberty. University of Chicago Press, Chicago

Hayek FA (1988) The fatal conceit: the errors of socialism. University of Chicago Press, Chicago

Hayek FA ([1968] 2002) Competition as a discovery procedure. Q J Austiran Econ 5(3): 9–23

Hodgson GM (2019) Austrian economics is still not institutional enough. In: D'Amico DJ, Martin AG (eds) Assessing Austrian economics. Emerald, Bingley, pp 101–110

Horwitz S (2000) Microfoundations and macroeconomics: an Austrian perspective. Routledge, London

Hudík M (2011) Why economics is not a science of behaviour. J Econ Methodol 18(2):147–162

Hudík M (2015) Mises and Hayek mathematized: towards mathematical Austrian economics. In: Bylund P, Howden D (eds) The next generation of Austrian economics: essays in honor of Joseph T. Salerno. Mises Institute, Auburn, pp 105–122

Hudík M (2018) Equilibrium analysis: two Austrian views. Cosmos+Taxis 6(1–2):1–10

Hudík M (2020) Equilibrium as compatibility of plans. Theor Decis 89(3):349–368

Hudík M, Bylund PL (2021) Let's do it Frank's way: general principles and historical specificity in the study of entrepreneurship. J Inst Econ:1–16. https://doi.org/10.1017/S1744137421000205

Huerta de Soto J (2008) The Austrian school: market order and entrepreneurial creativity. Edward Elgar, Cheltenham
Hurwicz L (1969) On the concept and possibility of informational decentralization. Am Econ Rev 59(2):513–524
Ikeda S (1990) Market process theory and "dynamic theories" of the market. South Econ J 57(1):75–92
Keynes JM (1936) The general theory of employment, interest and money. Macmillan, London
Kirzner IM (1973) Competition and entrepreneurship. University of Chicago Press, Chicago
Knight FH (1921) Risk, uncertainty, and profit. Houghton Mifflin, Boston
Koppl R (2010) Some epistemological implications of economic complexity. J Econ Behav Organ 76(3):859–872
Koppl R, Whitman DG (2004) Rational-choice hermeneutics. J Econ Behav Organ 55(3):295–317
Lachmann LM (1957) Capital and its structure. Sheed, Andrews and McMeel, Kansas City
Lachmann LM (1971) The legacy of max weber. The Glendessary Press, Berkley
Lachmann LM (1977) Capital, expectations, and the market process: essays on the theory of the market economy. Sheed, Andrews, and McMeel, Kansas City
Lange O (1936) On the economic theory of socialism: part one. Rev Econ Stud 4(1):53–71
Lavoie D (2011) The interpretive dimension of economics: science, hermeneutics, and praxeology. Rev Austrian Econ 24(2):91–128
Leeson PT (2007) An-arrgh-chy: the law and economics of pirate organization. J Polit Econ 115(6): 1049–1094
Leeson PT (2011) Government, clubs, and constitutions. J Econ Behav Organ 80(2):301–308
Leeson PT (2014) Anarchy unbound: why self-governance works better than you think. Cambridge University Press, Cambridge, UK
Leeson PT (2020) Economics is not statistics (and vice versa). J Inst Econ 16(4):423–425
Leeson PT, Boettke PJ (2006) Was Mises right? Rev Soc Econ 64(2):247–265
Lerner AP (1934) Economic theory and socialist economy. Rev Econ Stud 2(1):51–61
Lewin P (1994) Knowledge, expectations and capital. The economics of Ludwig M. Lachman: attempting a new perspective. Adv Austrian Econ 1:233–256
Lewin P (1999) Capital in Disequilibrium: the role of Capital in a Changing World. Routledge, London
Linsbichler A (2017) Was Ludwig von Mises a conventionalist? A new analysis of the epistemology of the Austrian School of Economics. Palgrave Macmillan, Basingstoke
Littlechild S, Owen G (1980) An Austrian model of the entrepreneurial market process. J Econ Theory 23(3):361–379
Loasby B (1999) Knowledge, institutions and evolution in economics. Routledge, London
Machlup F (1955) The problem of verification in economics. South Econ J 22(1):1
Machlup F (1958) Equilibrium and disequilibrium: misplaced concreteness and disguised politics. Econ J 68(269):1–24
Machlup F (1961) Are the social sciences really inferior? South Econ J 27(3):173–184
Machlup F (1967) Theories of the firm: Marginalist, behavioral, managerial. Am Econ Rev 57(1):1–33
Machlup F (1970) Homo Oeconomicus and his class mates. In: Natanson M (ed) Phenomenology and social reality. Springer, Dordrecht, Netherlands, pp 122–139
Machlup F (1973) The production and distribution of knowledge in the United States. Princeton University Press, Princeton
Machovec FM (1995) Perfect competition and the transformation of economics. Routledge, London
McNulty P (1967) A note on the history of perfect competition. J Polit Econ 75(August):395–399
Menger C (1871) Grundsätze der Volkswirtschaftslehre. Wilhelm Braumüller, Vienna
Mengers C ([1883] 2009) Investigations into the method of the social sciences, with special reference to economics. Ludwig von Mises Institute, Auburn
Menger C (1892) On the origin of money. Econ J 2(6):239–255
Moroni S (2014) Two different theories of two distinct spontaneous phenomena: orders of actions and evolution of institutions in Hayek. Cosmos + Taxis: Studies in Emergent Order and Organization 1(2):9–23

North DC (1990) Institutions, institutional change and economic performance. Cambridge University Press, Cambridge, UK
O'Driscoll G (1977) Economics as a coordination problem: the contributions of Friedrich a. Hayek. Sheed Andrews and McMeel, Kansas City
Persky J (1991) Retrospectives: Lange and von Mises, large-scale enterprises, and the economic case for socialism. J Econ Perspect 5(4):229–236
Piano EE (2019) State capacity and public choice: a critical survey. Public Choice 178(1):289–309
Polanyi M (1962) The republic of science: its political and economic theory. Minerva 1(1):54–73
Powell B (2014) Out of poverty: sweatshops in the global economy. Cambridge University Press, New York
Prychitko DL (1994) Ludwig Lachman and the interpretative turn in economics: a critical inquiry into the hermeneutice of the plan. Adv Austrian Econ 1:303–319
Rizzo MJ, Whitman G (2019) Escaping paternalism: rationality, behavioral economics, and public policy. Cambridge University Press, Cambridge, UK
Robbins LC (1932) An essay on the nature and significance of economic science. Macmillan, London
Roscher W (1874) Geschichte der National-Oekonomik in Deutschland. R. Oldenbourg, Munich
Rosser JB Jr (2012) Emergence and complexity in Austrian economics. J Econ Behav Organ 81(1): 122–128
Rothbard MN (1956) Toward a reconstruction of utility and welfare economics. In: Sennholz M (ed) On freedom and free Enterprise. D. Van Nostrand, Princeton, pp 224–262
Rothbard MN (1957) In defense of "extreme Apriorism". Southern Econ J 23(3):314–320
Rothbard MN ([1962] 1970) Man, economy, and state: a treatise on economic principles. Nash, Los Angeles
Rothbard MN (1970) *Power and Market: Government and the Market.* Menlo Park, CA: Institute for Humane Studies
Rothbard MN (1973) Praxeology as the method of the social sciences. In: Natanson M (ed) Phenomenology and the social sciences, vol 2. Northwestern University Press, Evanston, pp 311–339
Rothbard MN (1976) Praxeology: the methodology of Austrian economics. In: Dolan EG (ed) The foundations of modern Austrian economics. Sheed Andrews, Kansas City, pp 19–39
Rothbard MN (1987) Breaking out of the Walrasian box: the cases of Schumpeter and Hansen. Rev Austrian Econ 1(1):97–108
Salerno JT (2002) The rebirth of Austrian economics – in the light of Austrian economics. Q J Austrian Econ 5(4):111–128
Schumpeter JA (1934) The theory of economic development. Harvard University Press, Cambridge, MA
Schumpeter JA (1942) Capitalism, socialism and democracy. Harper & Brothers, New York
Shackle GLS (1952) Expectation in economics. The University Press, Cambridge, UK
Sima J (2019) Austrian school identity and unavoidable trade-offs in its long-term progress. In: D'Amico DJ, Martin AG (eds) Assessing Austrian economics. Emerald, Bingley, pp 61–68
Skarbek D (2014) The social order of the underworld: how prison gangs govern the American penal system. Oxford University Press, New York
Smith A (1776) An inquiry into the nature and causes of the wealth of nations. W. Strahan and T. Cadell, London
Smith VL (2007) Rationality in economics: constructivist and ecological forms. Cambridge University Press, New York
Stigler GJ (1957) Perfect competition, historically contemplated. J Polit Econ 65(1):1–17
Storr VH (2019a) Ludwig Lachmann's peculiar status within Austrian economics. Rev Austrian Econ 32(1):63–75
Storr VH (2019b) On the status of Austrian economics. In: D'Amico DJ, Martin AG (eds) Assessing Austrian economics. Emerald, Bingley, pp 81–88
Thomsen E (1992) Knowledge and prices: a market-process perspective. Routledge, London

von Böhm-Bawerk E ([1884] 1890) Capital and interest: a critical history of economic theory. Macmillan, London
von Böhm-Bawerk E ([1896] 1898) Karl Marx and the close of his system: a criticism. T. Fisher Unwin, London
von Mises L ([1912] 1980) The theory of money and credit. Liberty Fund, Indianapolis
von Mises L ([1933] 1981) Epistemological problems in economics. New York University Press, New York
von Mises L ([1935] 1975) Economic calculation in the socialist commonwealth. In Hayek, F. A. (ed), Collectivist economic planning, Kelley Publishing, Clifton, pp 87–130
von Mises L (1949) Human action: a treatise on economics. Yale University Press, New Haven
von Mises L ([1957] 1985) Theory and history: an interpretation of social and economic evolution. The Ludwig von Mises Institute, Auburn/Washington, DC
von Mises L ([1962] 1978) The ultimate Foundation of Economic Science: an essay on method. Sheed Andrews and McMeel, Kansas City
von Wieser F (1914) Theorie der gesellschaftlichen Wirtschaft. Mohr Siebeck, Tübingen
Vriend NJ (2002) Was Hayek an ace? South Econ J 68(4):811–840
Weintraub ER (2002) How economics became a mathematical science. Duke University Press, Durham
Whately R (1832) Introductory lectures on political economy. B. Fellowes, London
Williamson OE (1985) The economic institutions of capitalism. Free Press, New York

Joseph Schumpeter and the Origins of His Thought

60

Panayotis G. Michaelides and Theofanis Papageorgiou

Contents

Introduction	1564
The General Economic Framework	1564
On Technological Change and Business Cycles	1567
On Individualism	1571
On the Future of Capitalism	1575
On Credit and Banking	1578
Conclusions	1580
References	1581

Abstract

Joseph Alois Schumpeter's *oeuvre* has been heavily cited in various theoretical traditions. However, many aspects of his work have been less widely discussed or even unexplored. In this context, the origins of his thought should be traced back in several theoretical traditions and schools of economic thought. This chapter argues that several ideas or elaborations of various theoreticians from different schools of thought such as the Austrian social democrat Emil Lederer, the Austro-Marxist Rudolf Hilferding, the American institutionalist Thorstein Veblen, the French sociologist Gabriel Tarde, and the members of the GHS Max Weber and Werner Sombart can be traced in the Schumpeterian doctrine.

Keywords

Schumpeter · Lederer · Hilferding · Veblen · Tarde · Weber · Sombart

P. G. Michaelides (✉)
Laboratory of Theoretical and Applied Economics and Law, National Technical University of Athens, Athens, Greece
e-mail: pmichael@central.ntua.gr

T. Papageorgiou
University of Patras, Patras, Greece
e-mail: thpapag@upatras.gr

© The Author(s), under exclusive licence to Springer Nature Singapore Pte Ltd. 2022
D. McCallum (ed.), *The Palgrave Handbook of the History of Human Sciences*,
https://doi.org/10.1007/978-981-16-7255-2_8

Introduction

Despite the fact that – apart from Keynes – Schumpeter was probably the only truly great economist of the twentieth century, his affinities have not been widely discussed in the literature so far (Kessler 1961, 334; see also Michaelides et al. 2010). The purpose of this chapter is to analyze the influence of certain economists of various origins on the Schumpeterian oeuvre. More specifically, Schumpeter's similarities with certain theoreticians such as the Austrian social democrat Emil Lederer, the Austro-Marxist Rudolf Hilferding, the American institutionalist Thorstein Veblen, the French sociologist Gabriel Tarde, and the members of the GHS Max Weber and Werner Sombart are considered.

In this chapter, Schumpeter's views are claimed to be heavily affected by the socioeconomic environment and its subsequent theoretical context of the time, given the fact that the choice of the research questions is rather determined by the definite historical context and circumstances (Blümle and Goldschmidt 2006) and the fact that a "chief merit" of the Schumpeterian work was the fact that he gave "an individual and partly original shape" to ideas previously borrowed from others (Suranyi-Unger 1931, 319). Of course, Schumpeter undoubtedly not only read widely but also comprehended what he read (Samuels 1983, 6). Schumpeter aimed at scrutinizing every argument valuable to him (Shionoya 2008, 5). Thus, Schumpeter was perceived as a controversial writer (Allen and Rostow 1994, 165) and his works were interpreted in a contrasting way (Taylor 1955, 12–22), failing to construct a dominant theory for the functioning of capitalism, as John Maynard Keynes did.

In this spirit, the present chapter draws from previous relevant works by the first author and his colleagues, namely Michaelides and Milios (2005, 2009, 2015), Michaelides and Theologou (2010), Michaelides et al. (2010, 2011), Vouldis et al. (2011), Papageorgiou and Michaelides (2016), and Papageorgiou et al. (2013), although an appreciable amount depends on other significant contributions (Haberler 1950; MacDonald 1965; Swedberg 1989; Andersen 1991; Streissler 1994; Chaloupek 1995; Diebolt 1997, 2006; Shionoya 1997, 2005; Ebner 2000; Hodgson 2001, 2003; Allgoewer 2003).

The reminder is as follows: section "The General Economic Framework" provides the general economic framework; section "On Technological Change and Business Cycles" revolves around technological change, business cycles, and determinism; section "On Individualism" analyzes individualism; section "On the Future of Capitalism" focuses on the future of capitalism; section "On Credit and Banking" focuses on the role of credit; and finally section "Conclusions" concludes.

The General Economic Framework

Joseph Alois Schumpeter was born in 1883 at the Austrian part of Moravia. He was appointed assistant professor at the University of Czernowitz in 1909; between the years 1911 and 1919 he became full professor in Graz (1911–1919) in the field of political economy, being also an exchange professor at Columbia University (1913–1914). He was affiliated with the German Socialization Commission in 1918, and in 1919 he became minister of finance in the social democratic government (Haberler 1950, 346). Next, in 1921, he was appointed president of Biederman Bank, and in 1924 he became full professor in Germany in the University of Bonn. The years between 1932 and 1950 – year of his death – he was appointed professor at Harvard University, while serving as president of the American Economic Association. Schumpeter's oeuvre incorporated a wide range of topics, such as the dynamics of economic and social evolution (e.g., *Theory of Economic Development*, 1912, and *Business Cycles*, 1939), the assimilation of economic, sociological, and political aspects with reference to capitalism (e.g., Capitalism, Socialism, and Democracy, 1942), and the history of economic analysis (e.g., Economic Doctrine and Method, 1914, and History of Economic Analysis, 1954). He was fond of developments in mathematical economics and econometrics of his time, but also he opted for a universal social science consisting of economic sociology and economics (Shionoya 2004). Eugene von Boehm-Bawerk was a key personality for Schumpeter's life, while "the experience of those early years in Vienna never really left him" (Harris et al. 1951, 89).

Even though Schumpeter could not fit into a single "school of thought," he criticized aspects of the neoclassical approach in economics and mainly equilibrium theorizing and rationality. The economists of the classical school made an effort to develop general laws to explain the economic process. The marginalists shifted attention to individual choices, still trying to extract "laws" for both the individual and the economy, considering political economy as a strict science established on exact laws similar to the ones of the natural sciences: As Warlas (1954 [1874], 81; see also Papageorgiou and Michaelides 2016, 3) argued "pure theory in economics is a science which resembles the physio-mathematical sciences in every respect."

The basis of neoclassical economics of the time, the conception of static equilibrium, elaborated by Leon Walras, was partly rejected by Schumpeter, in favor of "development," a notion implying more unstable and evolving paths for the economic process. He argued: "I felt very strongly that this was wrong, and that there was a source of energy within the economic system which would of itself disrupt any equilibrium that might be attained" (Schumpeter 1989 [1937], 166). As Rosenberg (1994, 44–45) argued: "Schumpeter committed himself to the historical analysis of this process of mutation as an alternative to the equilibrium paradigm." Through the rejection of the Walrasian equilibrium he wanted to give substance to certain theories of his time (Arena 1992; Heilbroner 1998). According to Schumpeter: "Development is the distinct phenomenon entirely foreign to what may be observed in the circular flow or in the tendency towards equilibrium." Schumpeter noted, in the first Japanese edition of his Theory of Economic Development, that he purposed at creating "a theoretic model of the process of economic change in time . . . to answer the question how the economic system generates the force which incessantly transforms it" (Clemence 1951, 158–159). A circular flow process, which, without innovative activities, leads to a stationary state, is described in the book. The stationary state is, for Schumpeter, a Walrasian equilibrium, which was only applicable to a stationary process, despite taking account of the interdependences of economic variables. Schumpeter (1928) made it clear that "stationary flow" is a theoretical abstraction and can be used only as a starting point for the analysis.

Emil Lederer had been described by Schumpeter (1954, 884) as the "leading academic socialist of Germany in the 1920's." He studied law and economics at the University of Vienna. Among his friends or classmates there were Ludwig von Mises, Joseph Schumpeter, Otto Bauer, and Rudolf Hilferding. He taught at the University of Heidelberg while also giving lectures as guest professor at Tokyo Imperial University. Lederer, just like Schumpeter, considered the notion of equilibrium as being inadequate to analyze accurately an economic system, noting that if equilibrium were to have any sense, the data must have been fixed so that "the inherent or observed tendencies towards change would have to be ignored." In that spirit, "the idea of economic equilibrium can be effectively applied under a static system, but such a system is based on assumptions that remove it from most of the problems that have to be dealt with in actual practice" (Lederer 1938, 78). Nevertheless, the examination of a static system can be useful in analyzing the short run effects and tendencies when many of the dynamic factors can reasonably be treated as rigid. Lederer argued that the static system should be defined in the narrowest sense, i.e., assuming zero growth rates for population and capital, given that the static system ought to serve as a reference point for comparison and "the accidental inclusion of one or more elements of the dynamic system creates confusion in which it is difficult to distinguish the essentials of a static system and the consequences of disturbances from the outside" (Lederer 1938, 86). Schumpeter followed the same logic for the explanation of the mechanism that launches the system in motion, previously being into a state of immobility.

For Lederer static equilibrium as indicated by the nonappearance of idle factors "comes from the attitude of the laissez faire school, which invested the economic system with a harmony that is entirely unjustified within the dry and precise framework of the static system" (Lederer 1938, 81). In practice, it is crucial to "consider a longer period, with the changes that may normally be expected to occur within it. In that case the concept of static equilibrium has no meaning. That is why the concept of moving equilibrium was developed in its place" and "this moving equilibrium means a system of 'disturbances'" (Lederer 1938, 91), which being merged crop a dynamic system where any regressive movements do not prevent any further progress (see also Michaelides et al. 2011, 177).

Gabriel Tarde (1843–1904) was engaged in legal studies becoming *Juge d' instruction*. He studied the human nature being especially interested in the explanation of human motivation. According to Tarde, the drivers of social evolution are incorporated by successful initiatives (Taymans 1950) bringing about the struggle between routine and innovation, notions that mark the circular flow. Similarly to Tarde, Schumpeter inaugurated his analyses with a thesis which, excluding any innovative activities, caused a situation of stationary state. The stationary state is relevant only for a stationary process, i.e., one which is the outcome of adaptation to powers operating on it. For Schumpeter, economic development engaged spontaneous and discontinuous change in the channels of flow, disturbances of equilibrium, permanent alternation, and displacement of the previously existing state of equilibrium.

Schumpeter acclaimed Pareto for eliminating the concept of "utility" and argued that the maximization of rationality was an unrealistic story (Freeman and Louçã 2001). Also, for Schumpeter the entrepreneur is not spotted by any rationality on her hedonism; there is not any sense of rationality from the part of the entrepreneur in his "characteristic motivation of the hedonist kind"; "hedonistically the entrepreneur would be irrational" (Schumpeter 1983 [1934], 92). Finally, Schumpeter opposed Austrian psychologism arguing that "as saturation increases the demand for further food would decline and, as a result, saturated individual is only willing to pay a decreasing price for every additional quantity...Why is such an explanation given?" (Schumpeter 1908, 68).

Veblen was born in 1857 and lived in the Gilded Age of American capitalism, being definitely influenced by the general economic context. The world of young Veblen resembled greatly to a self-sufficient household economy, greatly contrasted to the nineteenth-century American capitalism (O'Donnell 1973, 200–201), and the seeds for his radical thought could be found there. He attacked neoclassical economics on the point of the purely individualistic basis of its theory: "An adequate theory of economic conduct, even for social purposes cannot be drawn in terms of the individual simply – as is the case in the marginal-utility economics – because it cannot be drawn in terms of the underlying traits of human nature simply" (Veblen 1909, 629). Meanwhile, "the human material with which the inquiry is concerned is conceived in hedonistic terms; that is to say, in terms of a passive and substantially inert and immutably given human nature. The psychological and anthropological preconceptions of the economists have been those which were accepted by social sciences some generations ago" (Veblen 1898, 389).

On Technological Change and Business Cycles

According to Schumpeter ([1934] 1983, 66) economic development encompassed the following cases "1. The introduction of a new good . . . or a new quality of a good. 2. The introduction of a new method of production . . . 3. The opening of a new market . . . 4. The conquest of a new source of supply . . . 5. The carrying out of the new organisation of any industry." The entrepreneur is the bearer of change (Schumpeter, [1934] 1983, 79–80). Schumpeter clearly discriminated the process of economic development from growth due to the gradual expansion in population and capital rises in productivity based on innovation are the prime drivers for economic development. In *Socialism, Capitalism, and Democracy* Schumpeter ([1942] 1950, 81ff) related market structure to the launching of innovations and more precisely pointed the significant role of large oligopolistic firms toward technical innovation. Thus, according to Schumpeter, technology is the foundation of economic evolution and comes out as the composition of new combinations. Fluctuations refer to three different sources, "namely: external factors (i.e. changes in commercial policy, diseases, changes in gold production because of new discoveries, revolutions and disasters), growth (i.e. changes in economic data which occur continuously in the sense that the increment/decrement per unit of time can be currently absorbed by the system without perceptible disturbance) and innovation (i.e. the historic and irreversible change in the way of doing things and more specifically changes in production functions which cannot be decomposed into infinitesimal steps)" (Konstantakis and Michaelides 2017, 17).

Hilferding defines economic development as "expansion of production [. . .] in general can be attributed to the opening of new markets, the establishment of new branches of production, the introduction of new technology, and the expansion of needs resulting from population growth" (Hilferding 1910, 258) and "the introduction of new machinery, the assimilation of related branches of production, the exploitation of patents" (Hilferding 1910, 123–124). Moreover, Hilferding in an identical to Schumpeter way, argued that: "[O]nce a combination has come into existence as a result of economic forces it will very soon present opportunities for the introduction of technical improvements in the process of production" (Hilferding 1910, 197).

The great combinations are "obliged to introduce these [technical] improvements, for otherwise there is a danger that some outsider will use them in a renewed competitive struggle [. . .]. [I]n this case technical improvements mean an extra profit, which is not eliminated by competition" (Hilferding 1910, 233). It is this technical superiority that permits the monopolistic firms to preserve their position: "The corporation can thus be equipped in a technically superior fashion, and what is just as important, can maintain this technical superiority" (Hilferding 1910, 123). As a result, "An industrial enterprise which enjoys technical and economic superiority can count upon dominating the market [. . .] can increase its sales, and after eliminating its competitors, rake extra profits over a long period, which more than compensate it for the losses sustained in the competitive struggle" (Hilferding 1910, 191). This technical superiority enables the monopolist firms to reproduce themselves "these technical advantages, once achieved, in turn become powerful motive for forming combinations where purely economic factors would not have brought them about" (Hilferding 1910, 197). "The corporation can thus be equipped in a technically superior fashion, and what is just as important, can maintain this technical superiority" (Hilferding 1910, 123).

For Lederer, economic development consists of: "the opening up of new markets, the manufacture of new products, and improved methods of production in the broadest sense of the term" (Lederer 1938, 230). Just like Schumpeter, Lederer explicitly analyzed technical development as the differentiating aspect of a real dynamic system; technical change is crucial, since technical development is more likely than other sources of alteration to generate sudden change, not easily absorbed in a harmonious process of readjustment and adaptation (see Lederer 1931, 112; 1938, 89). Technical development is, thus, responsible for "the extensive ups and downs in production that are typical of our modern capitalist process" (Lederer 1938, 90).

For Schumpeter, innovation is a significantly distinct phenomenon from invention: "innovation is possible without anything we should identify as invention, and invention does not necessarily induce innovation, but provides of itself [. . .] no economically relevant effect at all" (Schumpeter 1939, 84). Also, Schumpeter argued that the social process that produces innovations is clearly distinct "economically and sociologically" from the social process that produces inventions (Ruttan 1959, 597). Schumpeter discriminates innovation, which is endogenous, from invention, which is exogenous, in the following terms: "innovation is endogenous to the system, but is finally determined by the entrepreneurial function, that unique capacity to make combinations" (Freeman and Louçã 2001, 59).

Tarde advanced the searching for the laws of invention and individual innovation even further. He distinguished between practical and theoretical inventions: The latter consisting of hypotheses, philosophical systems, mythological conceptions, and scientific discoveries, and the former from verbal innovations (neologisms), political innovations, ritual innovations, military innovations, industrial innovations, artistic, judiciary innovations, and literary innovations. Practical innovations appear after the theoretical ones. Another differentiation was the one between inventions unable to be substituted and others able to be substituted: Only the theoretical, such as scientific, discoveries are unable to be substituted, the practical, such as industrial, ones are often substituted by others (Tarde 1902, 4).

Tarde stressed the fact that inventions are the outcome of a new combination of previously existing resources and concepts, i.e., from imitation. "Every machine consists of old tools, old methods, combined in a different way" (Tarde 1902, 5). Invention is the outcome of logic and teleology: a mixture of judgment, reasoning, deduction, and adaptation. Of course, "at the source of a new invention there is something else than just combined imitations of prior inventions. There is the main originality of this combination" (Tarde 1902, 6). Without this, "there would be no change in the channels of the flow (from the same to the same) no change in the production function (from equations to equations)" (Taymans 1950, 618). The forces that innovate are new given that they are implanted upon the old ones (Tarde 1902, 1).

Schumpeter ([1942] 1950, 65), in a Tardian spirit, wrote that "development consists primarily in employing existing resources in a different way, in doing new things with them, irrespective of whether those resources increase or not." In that spirit, he labeled the accomplishment of new combinations "enterprise" and the individuals who accomplished them "entrepreneurs." However, "entrepreneurs" cannot simply do this when they encounter new tasks given that while in the habituated paths their own experience and ability is sufficient, when encountered with innovations, they are in need of guidance (Schumpeter [1934] 1983, 79–80). Put it differently, in the circular flow they are obliged to swim with the stream, while having to try to swim against the stream if they wish to alter its channel (Prendergast 2006, 255). Thus, "it follows that novelty needs to be forced upon the majority of economic agents, as progress in general is basically a result of force and confrontation" (Ebner 2006, 504). Following Scott's (1998, 104) argumentation: The entrepreneur is, by definition, the man who acknowledges that the new combination is made. Thus, he should be distinguished from the capitalist (who bears the risk) and from the inventor (who has the ideas), although it is possible for one person to have in parallel all these three identities, at the same time. Apparently, based on the previously mentioned Schumpeterian analysis, it could be argued that nobody would probably care too much for the use of "invention," instead of "innovation" as related to the foremost reason for evolution, since this is exactly what Tarde meant by the term "invention." In any case, the term "innovation" emerged in Schumpeter's oeuvre in 1927 (Taymans 1950, 616).

In Veblen's oeuvre, the agent of change is to be found, principally, in technology. Technology in the Veblenian framework is "a joint stock of knowledge derived from past experience, and is held and passes on as an invisible possession of the community at large" (Veblen 2001 [1921], 28). Technological advance "is an affair of the community, not a creative achievement of individuals" (Veblen 1990 [1914], 103). Technology is of social and cumulative character, while technological change "is always in process of change" and is "held and carried forward collectively" (Veblen 1990 [1914], 103). Put differently, "the technological system is an organisation of intelligence, a structure of intangibles and imponderables, in the nature of habits of thought" (Veblen 1967 [1923], 280).

In a Schumpeterian vein, economic change is not continuous, because "innovations are not evenly distributed over time, but appear if at all discontinuously in groups, swarms or clusters" (Schumpeter 1939, 223). Innovations tend to cluster given the fact that every time something totally new and untried has been accomplished, much less effort is needed not only to do the same thing again but also to do similar things in different ways. The clustering nature of technical change brings about the business cycle and its phases; the business cycle is defined as the wave-like movement, which is typical to industrial change. The cyclical pattern of the economic process is summarized by Elliot (1993, 14): "development occurs through a cyclical process' and as a result 'cyclical fluctuations are no barrier to economic growth and recessions are not necessarily indicators of capitalism failure or breakdown."

According to Tarde (1902, 1): "social transformations are explained by the individual initiatives which are imitated, I do not say that invention, successful initiative, is the only acting force, nor do I say that it is actually the strongest force, but I say that it is the directing, determining, and explaining force." This argument reminds strongly of Schumpeter, for whom development is mostly the result of innovation, i.e., "the outstanding fact in the economic history of capitalist society" (Schumpeter 1939, 61).

Large firms, for Schumpeter, become conducive to technological progress: "There are superior methods available to the monopolist which either are not available at all to a crowd of competitors" (Schumpeter 1954, 101). In the same vein, "the perfectly bureaucratized giant industrial unit [. . .] ousts the small or medium-sized firm" (Schumpeter 1954, 134). In this spirit, the large firm is able to attract superior "brains" and to set up an array of practices to protect their risk-bearing investments (Schumpeter [1942] 1950, 110). Hilferding expressed a similar view: "the corporation can install new technology and labour saving processes before they come into general use, and hence produce on a large scale, and with improved, modern techniques, thus gaining an extra profit, as compared with the individually owned enterprise" (Hilferding 1910, 123–124). Consequently, "[T]he largest concerns introduce the improvements and expand their production" (Hilferding 1910, 233).

Summarizing, "a corporation [. . .] is able, therefore, to organize its plant according to purely technical considerations, whereas the individual entrepreneur is always restricted by the size" (Hilferding 1910, 123, emphasis added). In this context, Hilferding expressed the thesis, known as "Hilferding's Hypothesis," that "the size and technical equipment of the monopolistic combination ensure its superiority" (Hilferding 1910, 201), which has, in general terms, significant similarities with "Schumpeter's Hypothesis" arguing that: "large firms with considerable market power, rather than perfectly competitive firms were the 'most powerful engine of technological progress' (Schumpeter 1954, 106).

On Individualism

Schumpeter's methodology is rooted in his version of economic sociology and defines the relationships between individual behavior and collective entities (Festre and Garrouste 2008, 365, 381). But "who is the bearer of change in the capitalist process" for Schumpeter? On the one hand, Schumpeter's innovations are introduced by entrepreneurs, driven by the "dream and will to found a private kingdom," the "joy of creativity," and the "will to conquer" giving to the argument an anthropomorphic and animistic character (Papageorgiou and Michaelides 2016, 14). Schumpeter focuses on the individual as the pioneer of economic change, "the bearer of innovation," reflecting his adherence to "the Schumpeterian version of methodological individualism" (Papageorgiou and Michaelides 2016, 14). By this term, Schumpeter acknowledges methodological individualism as meaning "just that one starts from the individual in order to describe certain economic relationships" (Schumpeter 1908) or to use evolutionary terms, the individual may be acknowledged as the agent of evolution. Methodological individualism is a term coined to Schumpeter resembling greatly to Max Weber's individualistic ideal-type approach (see also Michaelides and Milios 2009). Firstly, according to Weber's approach, a unique historical figure that had "charisma," i.e., magical powers is the origin of any ideal type, a type of individual that may paralleled with Schumpeter's entrepreneur, who differs from the social norm. Secondly, Weber's ideal type may be interpreted focusing mainly on her institutional characteristics, "the personality of a man in society comprises dispositions both of a more private and temperamental kind, and of a more public and institutional kind. Only certain individuals are disposed to weep during the death-scene in Othello" (Watkins 1957, 40). The institutional dimension of Weber's ideal type seems to coincide with "the institutional constraints that influence individual action in the works of Schumpeter" (Papageorgiou and Michaelides 2016, 15).

In Schumpeterian terms, methodological individualism is "the approach in which determination of economic phenomena, for instance, values and prices, surplus value and profits can be traced to individual decisions and choices" (Schumpeter 1954, 888–889). Schumpeter made a sharp distinction between political and methodological individualism since the two concepts share nothing in common. The first refers to the "freedom of people to develop themselves and to take part in well-being and to follow practical rules. The second just means that one starts from the individual in order to describe certain economic relationships" (Schumpeter 1908, 90–91). Individualism for Schumpeter is substantially different from its neoclassical definition: Individual is the origin for all economic phenomena in the latter, whereas for the former it can be used as analytical tool for certain economic relationships, especially entrepreneurship. Individual is not considered as homo economicus that aims at satisfying her needs through maximizing utility, but as a "human being who has social and psychological motives, as well" (Papageorgiou and Michaelides 2016, 15).

According to Schumpeter, individualist initiative was indispensible for social evolution and economic development. In fact, Schumpeter (1910, 51) had already argued that the "herd of consumers" needed to be "mastered and guided" by the "leading personalities" of the sphere of production. In this context: "[L]eadership [...] does not consist simply in finding or creating the new thing but in so impressing the social group with it as to draw it on in its wake" (Schumpeter 1912, 88). Put differently, entrepreneurship is "essentially a phenomenon that comes under the wider aspect of leadership" (Clemence 1951, 254–255). For Schumpeter, entrepreneurial profits are means to achieve further ends and, thus, "entrepreneurship is driven by motivations that are alien to the rationalist foundations of capitalist civilisation" (Ebner 2006, 504).

For Schumpeter (1983 [1934], 74, 75) "enterprise is called the carrying out of the new combinations and the one that carries them out is the entrepreneur." In other words, entrepreneurship becomes the "ultimate cause" of capitalist development, since "the mechanisms of economic change in capitalist society pivot on entrepreneurial activity" (Schumpeter 1951 [1947], 150) as opposed to the manager, the capitalist, or industrialist "who merely may operate an established business" (Schumpeter 1983 [1934], 74). The entrepreneur is "rational" when he accomplishes the new plans, though he is in no sense rational in his "characteristic motivation of the hedonist kind"; "hedonistically the entrepreneur would be irrational" (Schumpeter 1983 [1934], 92). The motivation is both social and psychological: the "will to found a private kingdom," a "dynasty" inspired by the ideal of the medieval lordship, encompassing "from spiritual ambition down to mere snobbery," the "will to conquer," the "will to prove superior to others, to succeed for the sake of success itself and finally the intrinsic motive of getting things done," "the joy of creativity," and thus entrepreneurship is not a "vocation," since "everyone is an entrepreneur only when he carries out new combinations" (Schumpeter 1983 [1934], 77–78). Entrepreneurs are differentiated from common people to the "spiritual constitution" (Schumpeter 1911, 163 cf., 142–143). Considerable resistance is been found by the entrepreneur from his surroundings; people are set in their path as a "railway embankment in the earth" (Schumpeter 1983 [1934], 84). He is a "man of action" ready to get into "energetic action" (Schumpeter 1911, 132). The entrepreneur comes out of the existing economic reality; he "appears out of nowhere."

For Tarde, the innovator/inventor does not behave under a solid pattern of methods. The adaptive behavior of the innovator consists of a reaction to a given set of conditions proceeding by a causal relationship determined and described by theory (Taymans 1950, 619). Every invention is a combination of two, or more elements. There is not but a single line, a unique series of inventions carried out by a logical deduction: there are, starting from each invention, millions of possible paths, but only some of them are materialized (Tarde 1902, 10). It is important to consider all possibilities that led to an abortion, so as to avoid committing the mistake of believing in single-linear formulas of social evolution. Furthermore, evolution and change are "made possible through individual invention based on repetition: [. . .] social transformations are explained by the individual initiatives which are imitated, I do not say that invention, successful initiative, is the only acting force, nor do I say that it is actually the strongest force, but I say that it is the directing, determining, and explaining force" (Tarde 1902, 1).

Thanks to the imitative diffusion, the superior individual is working solely for the collectivity where she belongs. Furthermore, the main part of the individual tends to socialize (Tarde 1902, 11). Will the necessity of superior individuals decline? Tarde's (1902, 11) reply is negative given that the easier inventions are the ones that emerge first, providing a reason to why there are inventions (innovations) that appeared simultaneously in the past, independently the one from the other in different parts of the world. These discoveries, according to Tarde, "are made by individuals and these discoveries are interconnected and philosophically interwoven, with other individuals" (Michaelides and Theologou 2010, 366).

Veblen argued that socioeconomic evolution must be regarded as a substantial unfolding of life (Veblen 1897, 137) where "it is primarily the social system that would preserve or develop the capacity for change, not significantly the human genotype" (Veblen 1990 [1914], 18). His understanding of the nature of capitalism is expressed in the following quotation: "The economic life history of the individual is a cumulative process of adaptation of means to ends that cumulative change as the process goes on, both the agent and his environment being at any point the outcome of the last process" (Veblen 1898, 391). The evolutionary process takes place through the habits of thought, conventions, and institutions given that habits both endure and adapt in line with "changes in material facts." The material means of life take the form of "prevalent habits of thought, and it is as such that they enter into the process of industrial development" (Veblen 1898, 375).

A major shift in Schumpeter's work could probably be traced in *Capitalism, Socialism, and Democracy,* where Schumpeter seems to argue against his previous individualistic stance. There he stresses the importance of the effects of social milieu on individual action: "we know that every individual is fashioned by the social influences in which he grows up. In this sense he is the produce of the social entity or class and therefore not a free agent" (Schumpeter 1931, 286). Veblen seems to be closer to the mature views of Schumpeter. The shift in Schumpeter's approach could not be irrelevant with his move to the USA, where holistic and institutional approaches were dominant. In this latter schema, the entrepreneur looses her importance, and as a result the process of economic development comes to an end. Schumpeter (1949, 51) admitted that "the entrepreneurial function need not be embodied in a physical person and in particular in a single physical person." Of course, Schumpeter still stressed the importance of individual entrepreneurs, although in a different institutional setting; for example, a production engineer in the research and development department of a large firm.

Hilferding believed that "monopoly capitalism" inaugurated a distinction between the entrepreneur (i.e., the head of the firm's managerial staff) and the capitalist (i.e., the owner or principal creditor). He argued that the new form of corporation: "converts what has been an occasional, accidental occurrence in the individual enterprise into a fundamental principle; namely, the liberation of the industrial capitalist from his function as industrial entrepreneur" (Hilferding 1910, 107). The entrepreneur is in charge for the use of capital in the course of production, while the capitalist is the person who advances or lends his capital and bears the risk. On the other hand, Hilferding identified another "personality" part of the firm's management staff, who has analogous tasks to those of an (innovative) manager, and whose goal is to make new ideas work properly: "The separation of capital ownership from its function also affects the management of the enterprise. [T]hey [the managers] will develop firm's plants, modernize obsolete installations, and engage in competition to open up new markets" (Hilferding 1910, 126). Thus, it "separates management from ownership and makes management a special function" (Hilferding 1910, 347).

According to Schumpeter, the difference between the manager and the entrepreneur is to be found in the following difference: "carrying out a new plan and acting according to a customary one are things as different as making a road and walking along it" (Schumpeter 1912, 85). Put differently, "surmounting this opposition is always a special kind of task which does not exist in the customary course of life, a task which also requires a special kind of conduct" (Schumpeter 1912, 87). In his mature works, Schumpeter observed the rise of a "collective" type of entrepreneurship: "[the] entrepreneurial function need not to be embodied in [. . .] a single physical person. Every social environment has its own ways of filling the entrepreneurial function [. . . it] may be and often is filled co-operatively. With the development of the largest-scale corporations this has evidently become of major importance: aptitudes that no single individual combines can thus be built into a corporate personality; on the other hand, the constituent physical personalities must inevitably to some extent, and very often to a serious extent, interfere with each other" (Schumpeter 1949, 260–261).

In a similar vein, Hilferding concluded that a form of hostility appeared between capitalist owners and managers, the former being motivated by short-term profit maximization, the latter motivated by long-run strategies based on innovation: "the separation of capital ownership from its function also affects the management of the enterprise. The interest which its owners have in obtaining the largest possible profit as quickly as possible, their lust for booty, which slumbers in every capitalist soul, can be subordinated to a certain extent, by the managers of the purely technical requirements of production. More energetically than the private entrepreneur they will develop the firm's plant, modernise obsolete installations, and engage in competition to open up new markets" (Hilferding 1910, 126).

Schumpeter integrated the functions of Hilferding's "innovative manager" to those of the entrepreneur and related them with the "spirit of capitalism." He, thus, attributed the role of innovation to the entrepreneur who, without being a capitalist, "might exist only in the framework of the capitalist regime" (Michaelides and Milios 2015, 139).

Thus, the separation of roles came out of the predominant type of firm structure. The roots of this idea could be traced back to Hilferding, who first presented the "liberation of the industrial capitalist from his function as industrial entrepreneur" (Hilferding 1910, 107).

On the Future of Capitalism

Regarding the future of capitalism Schumpeter famously asked himself: "Can capitalism survive? No, I do not think it can. The thesis I shall endeavour to establish is that the actual and prospective performance of the capitalist system is such that its very success undermines the social institutions which protect it, and inevitably creates conditions in which it will not be able to live and which strongly point to socialism as the heir apparent" (1975 [1942], 61). Nonetheless, he argued that there were three main reasons for this (see also Papageorgiou and Michaelides 2016, 16).

First, the role of entrepreneurs was becoming outmoded, they are replaced by trained specialists responsible for technological innovation; technological innovation is increasingly rationalized and transformed into a routine procedure. In this path, entrepreneurs were transformed to wage earners and lost their motive to innovate. Second, small entrepreneurs, who are the defenders of capitalism, were weakening, being absorbed by the big business, thereby transferring ownership to shareholders. In addition, "capitalism produces intellectuals and other groups that have a vested interest in social unrest" (see also Papageorgiou and Michaelides 2016, 16). Capitalism is "constitutionally" unable to produce "emotional attachment to the social order"; it produces "almost universal hostility" to itself (Schumpeter 1975 [1942], 143, 145). This hostility is not only "external" but also internal, rooted deeply in the subconscious of the small entrepreneur, given that at this phase of capitalism the "bourgeois is unheroic to the point of the comic and does not even dare to say boo to a goose" (Schumpeter 1975 [1942], 138). Business is increasingly routinized, becoming an issue in the hands of professional managers, in parallel with innovation.

Third, the institutional framework of capitalism was staggering. The most likely scenario is a movement toward "increasing bureaucratization of economic life, coupled with an increasing dominance of the labour interest" (Schumpeter 1951 [1946], 203). According to Schumpeter, this augmenting reliance on government would lead to "guided capitalism" and finally to "state capitalism" which is a notion very closely connected to socialism, one might just as well call it precisely this (Schumpeter 1951 [1946], 204). This demise of capitalism takes place despite the cultural and economic achievements of capitalism; "modern medicine, modern painting and the modern novel are among its creations, a culture which is rationalistic, pacifist and pre-scientific" (Schumpeter 1975 [1942], 126).

Sombart's conception of *Vergeistung* of the big firm seems to move in parallel with the effects of Schumpeterian theoretical construction of the solidification of the big firm. Sombart saw two trends in the economic process regarding the evolution of relationships between different enterprises (Chaloupek 1995, 135n): firstly, augmenting specialization in the production of goods and services and secondly, increased concentration of production in enterprises of augmented size. In the second path of the economic process Sombart saw the process of *Entseelung* (de-animation) and *Vergeistung* (spiritual reification), which are regarded to be a result of capitalism: "rationalization goes hand in hand with reification in which individual animal spirits ('soul') are replaced by abstract concepts." The evolution of capitalism was just a special case, which was transformed "from a community of lively individuals tied to each other by personal relationships into a system of artfully designed interdependent work performances which are executed by functionaries in human shape" (Sombart 1927, 895). Capitalist enterprise consisted of three (sub)-systems: "(i) administration, (ii) accounting and (iii) production called 'instrumental system'" (Sombart 1927, 901n).

Consequently, leadership – as expressed by qualities such as the function and personality of the entrepreneur, being the typical motor of economic development in capitalism for Sombart (see Prisching 1996) – had been substituted by a bureaucratic apparatus. In this way, bureaucratization went on parallel with a growing perfection of methods, leading to the limitation of economic fluctuations (Sombart 1927, 680n). Production for the shake of profit gave gradually way to the production for shake of the population (Sombart 1927, 1015). An example was given through the case of the state operating the railways. Thus, "the community will extend its area of control. Free enterprise is replaced by semi-public entities" (Sombart 1925, 26).

The "bureaucratisation of economic life" (Schumpeter [1942] 1950, 206) simplifies the transition to a socialist but "bureaucratic apparatus" by establishing new modes of managerial responsibility and selection that "could only be reproduced in a socialist society" (Schumpeter [1942] 1950, 206–207), and defined socialism as "an institutional arrangement that vests the management of the productive forces with some public authority" (Schumpeter [1942] 1950, 113). For Schumpeter, this new mode of economic organization, "the growth of the great combines or the rise of trustification, is the final stage of capitalism [. . .A]fter that, it would resolve itself into socialism" (TeVelde 2001, 24). Capitalism's "tendency toward self-destruction [. . .] may well assert itself in the form of a tendency toward retardation of progress" (Schumpeter [1942] 1950, 162). For Schumpeter, the various factors "make not only for the destruction of the capitalist but for the emergence of a socialist civilization [. . .]. The capitalist process not only destroys its own institutional framework but it also creates the conditions for another" (Schumpeter [1942] 1950, 162). Schumpeter claimed that "things and souls are transformed in such a way as to become increasingly amenable to the socialist form of life." The capitalist "economic process tends to socialize *itself* – and also the human soul" and so "the technological, organizational, commercial, administrative, and psychological prerequisites of socialism tend to be fulfilled more and more" (Schumpeter [1942] 1950, 168, 219). These factors also "point to socialism as the heir apparent" (Schumpeter [1942] 1950, 61). Furthermore, predictions go that far as to say that "private enterprise will lose its social meaning through the development of the economy and the consequent expansion of the sphere of social sympathy. [. . .] Society is growing beyond private enterprise [. . .] That too is certain" (Schumpeter 1918, 131). Thus, "the hour [of socialism] will come" (Schumpeter 1918, 131). The view that Schumpeter expressed on socialism in 1918 seems surprising, given its contradiction to texts published before 1918, arguing in a very conservative or even monarchist narrative, praising Catholic–conservative and upper nobility views (Blumenthal 2008, 647–648). This shift could be related to the revolution and the rise of socialism in 1918 offering a chance to Schumpeter to update his political profile into a progressive way. For Schumpeter the epoch that he was living was a transitional phase of capitalism, which was the unique result of the coexistence of "two different epochs," i.e., capitalism and absolutism. Current capitalism existed alongside feudal remnants, with the bourgeoisie being subject to the power of imperialist autocracy. A transitional "social regime could be named imperialist capitalism to differentiate it from anti-imperialist pure capitalism, a theoretical construct designating a hypothetical gradual countertrend extrapolated into the future" (Michaelides and Milios 2015, 136).

For Hilferding the distinction between entrepreneurs and capitalists paved the way to socialism: "finance capital puts control over social production increasingly into the hands of a small number of large capitalist associations, separates the management of production from ownership and socializes production to the extent that this is possible under capitalism" (Hilferding 1910, 367). This "facilitates enormously the task of overcoming capitalism [...] it is enough for society, through its conscious executive organ – the state conquered by the working class – to seize finance capital in order to gain control [...] of production" (Hilferding 1910, 367). And "since all other branches of production depend upon these, control of large-scale industry already provides the most effective form of social control" (Hilferding 1910, 367). Hilferding clarified that "the tendency of finance capital is to establish social control of production, but it is an antagonistic form of socialization, since the control of social production remains vested in an oligarchy." And that "the response of the proletariat to the economic policy of finance capital [. . .] cannot be free trade, but only socialism" (Hilferding 1910, 366), but socialism seen as "the organization of production, the conscious control of the economy not by and for the benefit of capitalist magnates but by and for society as a whole, which will then at last subordinate the economy to itself" (Hilferding 1910, 366–367).

For Hilferding, the extinction of free competition came in the following way: "the basis of this association is the elimination of free competition among individual capitalists by the large monopolistic combines" (Hilferding 1910, 301). Thus, "it is also clear that monopolistic combines will control the market" (Hilferding 1910, 193). And it was Schumpeter himself who recognized that Hilferding's core vision "is interesting and original" (Schumpeter 1954, 881). Apparently, Hilferding has influenced importantly Schumpeter. Common feature of the two theoreticians is the emphasis stressed on the domination of the market by large monopolistic enterprises and the "incompatibility" of such kind of organizations with technological change. Regarding the issue of imperialism, even though there is a clear a divergence, Schumpeter's analysis seems to have been sketched responding to Hilferding's elaborations or put differently, acting as a "mirror image." In addition, both authors started from a similar theoretical point, namely, the separation of roles between the entrepreneur and the capitalist, the big enterprise of "trustified capitalism," and thus arrived at the same conclusion: the inevitability of socialism.

For Veblen outcome of technological change is the struggle between the two basic classes, a struggle taking place also in their habits of thought respectively. On the one hand, there are the business people along with their "pecuniary habits"; "their ulterior end sought is an increase of ownership not industrial serviceability" (Veblen 1975 [1904], 24) and the class of businessmen is interested in making "the disturbances of the system large and frequent" since their profit appears in the "conjunctures of change." The outcome of the struggle is unknown posing a "definite limitation towards the prediction for the advent of socialism" – at least in the mature works of Veblen (Papageorgiou and Michaelides 2016, 12). Business interests are dissociated from the interests of the community; "business principles" are principles of property; they are corollaries under ownership; older than the machine industry although their full development is to be found in the machine era. On the other side, the engineers and the "industrial and mechanical employments" are governed from the values of workmanship and serviceability. Thus, mechanical efficiency is their primary habit and technological progress is the result of their developed labor (Papageorgiou and Michaelides 2016, 12). The evolution of the capitalist system is the outcome of the relation between them, described as strive and conflict.

The similarity of the routinization of innovation and business becoming more and more an affair for professionals could entail that the operation of capitalism is synonymous with routine and goes along with a period that the scale of business is increased, highlighting the need for entrepreneurial planning going hand by hand with a subsequent boom in the need for hiring managerial workforce, a period also dominated by the prediction of a weakening capitalism and the rise in socialism in 1918.

On Credit and Banking

Schumpeter (1983 [1934], 74) pointed out that "the 'new combination of means of production' and credit" were the "fundamental phenomena of economic development." Additional purchasing power is provided to the entrepreneur via credit enabling him to foster development "granting credit in this sense operates as an order on the economic system to accommodate itself to the purposes of the entrepreneur" (Schumpeter 1983 [1934], 107). Schumpeter pointed the importance "of credit means of payment created *ad-hoc*, which can be backed neither by money in the strict sense nor by products already in existence" (Schumpeter 1983 [1934], 106). Credit carries out the functions of "enabling the entrepreneur to withdrawn producers' goods which he needs from their previous employments, by exercising a demand for them, and thereby to force the economic system into new channels" (Schumpeter 1983 [1934], 106).

Hilferding conceived paper money as different from legal tender "which emerges from circulation as a social product," and from "credit money which is a 'private affair', not backed by the government" (Schumpeter 1910, 66), where money can be substituted by a promise to pay. Hilferding argued that credit originated as an outcome of the altered function of money as a means of payment. Thus, a purchase not followed by direct payment "means that one capitalist has enough surplus capital to wait for payment for the purchaser, the money due is credited" and "money is [. . .] merely transferred" (Hilferding 1910, 82). Nevertheless, when a promissory note functions as a means of payment, "money capital has been saved, and this type of credit is called 'circulation credit'" (Hilferding 1910, 83).

Increases in production were believed to act as a simultaneous expansion of circulation and "the enlarged circulation process is made possible through an increase in the quantity of credit money" (Hilferding 1910, 83). However, circulation credit does not "transfer money capital from one productive capitalist to another; nor does it transfer money from other (unproductive) classes to the capitalist class" (Hilferding 1910, 87); that role is played by "capital credit." This form of credit comprises a transfer of money to those who make use of it as money capital, i.e., for the objective of purchasing the elements of productive capital (Michaelides and Milios 2015, 141). Credit "puts money into circulation as money capital in order to convert it into productive capital" (Hilferding 1910, 88), extending the scale of production while the scale of circulation is enlarged using previously idle money.

Lederer's view is similar to the one expressed from Schumpeter that anyone who wants to act as an entrepreneur chasing profit must raise funds, the provision of which is a role of the capitalist (Vouldis et al. 2011, 445). He shared Schumpeter's view of credit as necessary for economic expansion: "the introduction of a new process of production can only be held up by the absence of extra means of payment" (Lederer 1938, 224; see also Lederer 1925). Moreover, many nongrowing enterprises would have to contract in the absence of access to credit (Lederer 1938, 230). Economic activity is not financed by past savings but only from access to new credit (or new savings), which is identical to the creation of supplementary production capacity (Lederer 1930, 514). For Lederer, additional credit provides "the fuel without which any dynamic power would spend itself very quickly" (Lederer 1936, 156).

To sum up, the determination of credit by its demand side is given specific attention by all Schumpeter, Lederer, and Hilferding and is considered as necessary for the functioning of capitalism. Schumpeter viewed credit as inseparably linked to entrepreneurial activity and the introduction of innovations. Credit "enable[s] the entrepreneur to withdraw producers' goods which he needs from their previous employments, by exercising a demand for them, and thereby to force the economic system into new channels" (Schumpeter 1983, [1934], 106). The capitalist provides credit or alternatively funds which are outcome of previously successful innovative initiatives and entrepreneurial profit. The financial risk is been born by the capitalist, while the entrepreneur risks his job and his reputation and, given that capital utilization is solely the diversion of the factors of production to new paths, the capitalist has some power to delineate new paths to production. In this vein, Schumpeter (1983, [1934], 74) argued that "new combinations of means of production" and "credit" are the "fundamental phenomena of economic development" and that "fresh opportunities arise of expanding production through credit" (Schumpeter 1983, [1934], 230). Lederer focused on the importance of innovation in increasing the demand for credit, since technical improvements are the main reasons that feed entrepreneurial wants for borrowing: "heavy demands on the credit market are therefore only likely to arise as the result of sudden prospects of large profits, created in particular by the opening up of new markets, the manufacture of new products, and improved methods of production in the broadest sense of the term. But . . . technical progress . . . may be regarded as the main cause of the demands for credit which arise" (Lederer 1938, 230; see also Vouldis et al. 2011, 446).

Conclusions

Despite the fact that Schumpeter was among the greatest economists of the twentieth century, his affinities have not been widely discussed in the literature so far. It is argued that Schumpeter's views were strongly influenced by the socioeconomic environment of the time, since Schumpeter not only read widely but also comprehended what he read (Samuels 1983, 6). In this chapter, the work of Joseph Schumpeter was analyzed with reference to other economists that influenced him such as the Austrian social democrat Emil Lederer, the Austro-Marxist Rudolf Hilferding, the American institutionalist Theorstein Veblen, the French sociologist Gabriel Tarde, and the members of the GHS Max Webber and Werner Sombart.

Schumpeter rejected partly the fundamental conception of standard neoclassical economics, the conception of static equilibrium – a conception elaborated by Leon Walras – in favor of "development," a more unstable and evolving path of the economic process. In the same vein, Lederer acknowledges the concept of equilibrium as inadequate to analyze correctly an economic system, noting that in order to have meaning, data must have been fixed so that "the inherent or observed tendencies towards change would have to be ignored" (Lederer 1938, 78). For Schumpeter, enterpreneurial hedonism is in no way rational and the entrepreneur is in no sense rational. Veblen accused neoclassical economics because of the purely individualistic foundation of its theory.

In the Schumpeterian system, technology is the foundation of economic evolution and comes out as the composition of new combinations. Fluctuations refer to three different sources, i.e., external factors, growth, and innovation. For Hilferding the great combinations must introduce technical improvements, given that in any other case there is a danger that an outsider will make use of them in order to make extra profits, which are not removed by competition. Furthermore, Lederer notably acknowledged technical development as the aspect that differentiates a real dynamic system; technical change is crucial, because technical development is more likely to generate sudden change that is not easily absorbed in a harmonious path of readjustment and adaptation. Innovations for Schumpeter tend to cluster because every time something totally new and untried has been accomplished, much less effort is needed to do the same thing again but also to do similar things in different ways. For Tarde individual invention based on repetition enables evolution and change, while social transformations are explained by the imitation of individual. Thanks to the imitative diffusion the superior individual is working solely for the collectivity where she belongs. A major shift in Schumpeter's work could probably be traced in *Capitalism, Socialism, and Democracy* where Schumpeter himself seems to argue contrary to his previous individualistic views by pointing the significance of the social context on individual action.

For Hilferding the entrepreneur is responsible for the use of capital in production, while the capitalist is the person who advances or lends his capital bearing the risk. However, Hilferding identified another "personality" who belongs to the firm's management staff, having similar responsibilities to those of an (innovative) manager, and whose role is to make new ideas work properly. Schumpeter incorporated the modernizing functions of Hilferding's "innovative manager" to the ones of the entrepreneur and related them with the "spirit of capitalism."

Schumpeter argued instead that there were three main reasons for the demise of capitalism: firstly, the role of entrepreneurs was becoming out of date, substituted by trained specialists who were more and more responsible for technological innovation. At the same time, technological innovation is being rationalized into a routine procedure. Second, small entrepreneurs, who are the defenders of capitalism, were weakening, being absorbed by big business, thereby transferring ownership to shareholders. Thirdly, the institutional environment of capitalism was staggering. The most probable scenario is a transition toward augmenting bureaucratization of economic life, leading gradually to "guided capitalism" and finally to "state capitalism," a notion which is very closely related to socialism; one might just call them identical. Sombart's conception of *Vergeistung* of the big firm seems to move in parallel with the effects of Schumpeterian theoretical construction of the solidification of the big firm coinciding also with the effects of the Schumpeterian "automatisation." Sombart saw two trends in the economic process: firstly, increased specialization in the production of goods and services and secondly, increased concentration of production in enterprises of augmented size. In the second path of the economic process Sombart saw the process of *Entseelung* (de-animation) and *Vergeistung* (spiritual reification), which he considered to be resulting from capitalism. Schumpeter had expressed before 1918 very conservative or even monarchist views, adopting Catholic-conservative and upper nobility arguments and in that sense, his view on socialism expressed in 1918 seems surprising. This change could be attributed to the rise of socialism in 1918 giving Schumpeter a chance for updating his political profile into a more progressive direction. For Hilferding the difference between capitalists and entrepreneurs opened the way to socialism. The control of finance capital over social production and the increasing concentration of finance capital into the hands of a small number of large capitalist associations facilitate enormously the overcoming of capitalism. The tendency of finance capital is to establish social control of production, but it is an antagonistic form of socialization, since the control of social production remains vested in an oligarchy.

References

Allen R, Rostow W (1994) Opening doors: the life and work of Joseph Schumpeter. Transaction, New Brunswick

Allgoewer E (2003) Emil Lederer: business cycles, crises, and growth. J Hist Econ Thought 25: 327–348

Andersen E (1991) Schumpeter's Vienna and the schools of thought, Smaskrift Nr. 70, Aalborg University, Aalborg, August

Arena R (1992) Schumpeter after Walras: "economie pure" or "stylised facts"? In: Lowry T (ed) Perspectives on the history of economic thought, vol VIII. Edward Elgar, Aldershot

Blumenthal K (2008) Economic theorist and 'entrepreneur of popularisation': Schumpeter as finance minister and journalist. Eur J Hist Econ Thought 15(4):641–672

Blümle G, Goldschmidt N (2006) From economic stability to social order: the debate about business cycle theory in the 1920s and its relevance for the development of theories of social order by Lowe, Hayek and Eucken. Eur J Hist Econ Thought 13(4):543–570

Chaloupek G (1995) Long term economic perspectives compared: Joseph Schumpeter and Werner Sombart. Eur J Hist Econ Thought 2(1):127–149

Clemence RV (1951) Essays of economic topics of J. A. Schumpeter. Kennikat Press, Port Washington

Diebolt C (1997) La théorie de la sous-consommation du cycle des affaires de Emil Lederer. Econ App 50:27–50

Diebolt C (2006) Progrès technique et cycles économiques dans la pensée allemande de l'entre-deux-guerres: l'apport d'Emil Lederer. Working paper no. 9, Association Française de Cliométrie, Paris

Ebner A (2000) Schumpeter and the 'Schmollerprogramm': integrating theory and history in the analysis of economic development. J Evol Econ 10:355–372

Ebner A (2006) Institutions, entrepreneurship, and the rationale of government: an outline of the Schumpeterian theory of the state. J Econ Behav Organ 59(4):497–515

Elliot J (1993) Schumpeter's theory of economic development and social change: Exposition and assessment. Int J Soc Econ 12(6/7):6–33

Festre A, Garrouste P (2008) Rationality, behavior, institutional, and economic change in Schumpeter. J Econ Methodol 15(4):365–390

Freeman C, Louçã F (2001) As time goes by: from the industrial revolution to the information revolution. Oxford University Press, Oxford

Haberler G (1950) Joseph Alois Schumpeter: 1883–1950. Q J Econ 64:333–372

Harris, Haberler SE, Leontief GW, Mason ES (1951) Professor Joseph A. Schumpeter. Rev Econ Stat 33(2):89–90

Heilbroner R (1998) The worldly philosophers. Penguin, New York. [2000]

Hilferding R (1910) Finance Capital. Routledge, London. [1981]

Hodgson GM (2001) How economics forgot history: the problem of historical specificity in social science. Routledge, New York

Hodgson GM (2003) Schumpeter's 'entrepreneur' in historical context. Adv Austrian Econ 6:267–270

Kessler M (1961) The synthetic vision of Joseph Schumpeter. Rev Polit 23:334–355

Konstantakis KN, Michaelides PG (2017) Does technology cause business cycles in the USA? A Schumpeter-inspired approach. Struct Change Econ Dyn 43:15–26. ISSN 0954-349X

Lederer E (1925) Konjuktur und Krisen. In: Grundriss der Sozialoekonomik. J.C.B. Mohr, Tübingen

Lederer E (1930) Ort und Grenze des zusätzlichen Kredits. Archiv für Sozialwissenschaft und Sozialpolitik 63(3):513–522

Lederer E (1931) Technischer Fortschritt und Arbeitslosigkeit. J.C.B. Mohr, Tübingen

Lederer E (1936) Developments in economic theory. Am Econ Rev, Supplement, papers and proceedings of the forty-eighth annual meeting of the American Economic Association, March 26(1):151–160

Lederer E (1938) Technical progress and unemployment. King and Son, Geneva

MacDonald R (1965) Schumpeter and Max Weber – central visions and social theories. Q J Econ 79 (3):373–396

Michaelides PG, Milios JG (2005) Did Hilferding influence Schumpeter? Hist Econ Rev 41:98–125

Michaelides PG, Milios JG (2009) Joseph Schumpeter and the German historical school. Camb J Econ 33(3):495–516

Michaelides PG, Milios JG (2015) The Schumpeter-Hilferding Nexus. J Evol Econ 25(1):133–146

Michaelides PG, Theologou K (2010) Tarde's influence on Schumpeter: technology and social evolution. Int J Soc Econ 37(5):361–373

Michaelides PG, Milios JG, Vouldis A, Lapatsioras S (2010) Heterodox influences on Schumpeter. Int J Soc Econ 37(3):197–213

Michaelides P, Milios J, Vouldis A, Lapatsioras S (2011) Emil Lederer and Joseph Schumpeter on economic growth, technology and business cycles. Forum Soc Econ 39(2):171–189

O'Donnell LA (1973) Rationalism, capitalism, and the entrepreneur: The views of Veblen and Schumpeter. Hist Political Econ, 5(1):199–214

Papageorgiou T, Michaelides P (2016) Joseph Schumpeter and Thorstein Veblen on technological determinism, individualism and institutions. Eur J Hist Econ Thought 23(1):1–30

Papageorgiou T, Katselides I, Michaelides P (2013) Schumpeter, commons, and Veblen on institutions. Am J Econ Sociol 72(5):1232–1254

Prendergast R (2006) Schumpeter, Hegel and the vision of development. Camb J Econ 30(2):253–275

Prisching M (1996) The entrepreneur and his capitalist spirit – sombart's psycho- historical model. In: Werner S (ed) (1863–1941) Social Scientist, Vol. II. Marburg, Metropolis, pp 301–30

Rosenberg N (1994) Joseph Schumpeter: radical economist. In: Shionoya Y, Perlman M (eds) Schumpeter in the history of ideas. University of Michigan Press, Ann Arbor, pp 41–57

Ruttan W (1959) Usher and Schumpeter on invention, innovation, and technological change. Q J Econ 73:596–606

Samuels W (1983) The influence of Friedrich von Wieser on Joseph A. Schumpeter presidential address history of economics society, May 1982. J Hist Econ Thought 4(2):5–19

Schumpeter J (1908) Das Wesen und der Hauptinhalt der theoretischen Nationalëokonomie. Duncker and Humblot, Berlin

Schumpeter JA (1910) Die neuere Wirtschaftstheorie in den Vereinigten Staaten. Schmollers Jahrbuch fuer Gesetzgebung. Verwaltung und Volkswirtschaft im Deutschen Reiche 34:1–52

Schumpeter J (1911) Theorie der wirtschaftlichen Entwicklung. Dunker & Humbolt, Leipzig

Schumpeter J (1927 [1951]) Die sozialen Klassen im ethnisch homogen Milieu [Social classes in an ethnically homogenous environment]. Archiv für Sozialwissenschaft und Sozialpolitik 57:1–67, in Schumpeter (1951)

Schumpeter JA (1918) Die Krise des Steuerstaates, Zeitfragen aus dem Gebiet der Soziologie, vol 4. Leuschner and Lubensky, Graz and Leibzig

Schumpeter J (1928) The instability of capitalism. Econ J 38:361–386

Schumpeter J (1931) The present state of economics or on systems, schools and methods. Kokumin Keizai Zasshi (J Econ Bus Admin) 50(5):679–705. Cited in: http://www.schumpeter.info/text2_1.htm

Schumpeter J (1939) Business cycles: a theoretical, historical and statistical analysis of the capitalist process. McGraw-Hill, New York

Schumpeter JA ([1942] 1950) Capitalism, Socialism and Democracy. Harper & Row, New York

Schumpeter JA (1949) Economic history and entrepreneurial history. In: Clemence R (ed) Essays: Joseph Schumpeter. Transaction Publishers, New Brunswick

Schumpeter J (1951 [1946]) Capitalism. In: Clemence RV (ed) Essays: on entrepreneurs, innovations, business cycles and the evolution of capitalism. Harvard University, Cambridge, MA, pp 184–205

Schumpeter J. (1951 [1947]) The creative response in economic history. In: Clemence RV (ed) Essays: on entrepreneurs, innovations, business cycles and the evolution of capitalism. Harvard University, Cambridge, MA, pp 221–231

Schumpeter J (1954) History of economic analysis. Oxford University Press, New York

Schumpeter J (1975 [1942]) Capitalism, socialism, and democracy. Harper and Brothers, New York

Schumpeter J (1983 [1934]) The theory of economic development. Harvard University Press, Cambridge, MA

Schumpeter J (1989 [1937]) Preface to the Japanese edition of Theorie der wirtschaftlichen Entwicklung. In: Essays. Transaction, New Brunswick, pp 165–168

Scott MF (1998) A new view of economic growth. Oxford University Press, New York/Oxford

Shionoya Y (1997) Schumpeter and the idea of social science: a meta-theoretical study. Cambridge University Press, Cambridge

Shionoya Y (2004) Scope and method of Schumpeter's universal social science: economic sociology, instrumentalism, and rhetoric. J Hist Econ Thought 26(3):331–347

Shionoya Y (2005) The soul of the German historical school: methodological essays on Schmoller, Weber and Schumpeter. Springer, New York

Shionoya Y (2008) Schumpeter and evolution: an ontological exploration. In: Shionoya Y, Nishizawa T (eds) Marshall and Schumpeter on evolution: economic sociology of capitalist development. Edward Elgar, Cheltenham

Sombart W (1925) Prinzipielle Eigenart des modernen Kapitalismus, Grundriss der Sozialoekonomik, vol IV, Abteilung/ 1, Teil, J.C.B. Mohr, Tuebingen

Sombart W (1927) Der moderne Kapitalismus, vol III/1,2: Das Wirtschaftsleben im Zeitalter des Hochkapitalismus, Duncker and Humblot, Munich/Leipzig

Streissler EW (1994) The influence of German and Austrian economics on Joseph Schumpeter. In: Shionoya Y, Perlman M (eds). Schumpeter in the history of ideas, University of Michigan Press, Ann Arbor, pp 13–38

Suranyi-Unger T (1931) Economics in the twentieth century. Norton, New York

Swedberg R (1989) Joseph A. Schumpeter and the tradition of economic sociology. J Inst Theor Econ 145:508–524

Tarde G (1902) L'invention considerée comme moteur de l'évolution sociale. Société de Sociologie de Paris:1–12. June 11 (also in: Revue Internationale de Sociologie, vol 7, pp 561–574)

Taylor OH (1955) Schumpeter's history of economic analysis. Rev Econ Stat 37(1):12–22

Taymans A (1950) Tarde and Schumpeter: a similar vision. Q J Econ 64(4):611–622

TeVelde RA (2001) Schumpeter's theory of economic development revised. In: The future of innovation studies. Eindhoven University of Technology, the Netherlands, 20–23 Sept 2001

Veblen T (1897) Review of Max Lorenz, Die Marxisasche Socialdemokraae. J Polit Econ 6(1): 136–137

Veblen T (1898) Why is economics not an evolutionary science? Q J Econ 12(3):373–397. Reprinted in Veblen (1919)

Veblen T (1904) The theory of business enterprise. Charles Scribners, New York. Reprinted (1975). Augustus Kelley, New York

Veblen T (1909) The limitations of marginal utility. J Polit Econ 17(9):620–636

Veblen T (1914) The instinct of workmanship, and the state of the industrial arts. Augustus Kelley, New York. Reprinted (1990) with a new introduction by Murphey, M. G. and a 1964 introductory note by Dorfman, J. Transaction Books, New Brunswick

Veblen T (1967 [1923]) Absentee ownership and business enterprise in recent times: the case of America. Beacon, Boston

Veblen T (2001 [1921]) The engineers & the price system. Batoche Book, Kitchener

Vouldis A, Michaelides P, Milios J (2011) Emil Lederer and the Schumpeter-Hilferding-Tugan-Baranowsky Nexus. Rev Polit Econ 23(3):439–460

Warlas L (1954 [1874]) Elements of pure economics or the theory of social wealth. Routledge Library Editions, London, reprinted 2003

Watkins JWN (1957) Historical explanation in the social sciences. Br J Philos Sci 8(2):104–117

Learning from Intellectual History: Reflection on Sen's Capabilities Approach and Human Development

61

Farah Naz

Contents

Introduction	1586
Learning from the Intellectual History of Economic Thought	1588
Ethical Blind Spot in Neoliberal Economics	1590
Sen and the Revival of Ethical Traditions in Economic Thought	1591
An Overview of Sen's Capabilities Approach (CA)	1593
Capability Approach and Development Economics	1595
Capability Approach and Human Development	1598
Sen's Capability Approach and Feminist Concerns About Gender Inequalities	1601
Concluding Remarks	1606
References	1607

Abstract

The end of the Second World War, and the subsequent creation of independent states from colonies, marked the beginning of serious interest among scholars and policy makers in the field of development economics. Though there is agreement among scholars that the ultimate goal of development is to increase the wellbeing of individuals, there is disagreement on the fundamental question of whether economic growth automatically leads to human wellbeing. The mainstream approaches to economic development often equated development with economic growth. However, policies based on this approach failed to solve the problem of poverty and inequality in developing countries. Instead, there were negative social consequences such as the widening of inequality and the weakening of the social fabric in many developing states. As a result of growing inequality and poverty, scholars began to question development policies that focused narrowly on economic growth as an indicator of development. Consequently, there was a gradual shift from growth-oriented economic development approaches to the

F. Naz (✉)
Department of Sociology and Criminology, University of Sargodha, Sargodha, Pakistan
e-mail: farah.naz@uos.edu.pk

© The Author(s), under exclusive licence to Springer Nature Singapore Pte Ltd. 2022
D. McCallum (ed.), *The Palgrave Handbook of the History of Human Sciences*,
https://doi.org/10.1007/978-981-16-7255-2_7

alternative human development agenda. During the 1980s, important contributions to development theory improved scholars' understanding of the concept of poverty and its measurement. The emergence of Sen's capability approach, which later provided the theoretical foundation for the human development paradigm in economics, was among these conceptual advancements that shed a new light on conceptualizing poverty, inequality, and development. This chapter draws on Amartya Sen's capability approach and recent developments in capability literature, to provide a critical evaluation of the development processes.

Keywords

Capability approach · Gender · Development

Introduction

The economic and social development of developing countries had not been a topic of academic interest, as such, until the end of the Second World War and of the colonial system in the late 1940s. The end of the Second World War, and the subsequent creation of independent states from former colonies, marked the beginning of serious interest among scholars and policy makers in the field of development economics. An appropriate conceptual framework was required to formulate development strategies and design economic policies for the newly independent states. As a field of economics, this body of emerging knowledge intended to prescribe policy and develop tools for the social and economic transformation of the former colonies. The development paradigm during this time period focused on economic growth and modernization as a means of economic development. It was widely believed that economic growth would ultimately eliminate income and social inequalities.

Thus, the tool kit of development economists during the 1950s contained conceptual frameworks such as "big push" and "balanced growth." The prevailing development strategy during the 1950s advocated for large investments and industrialization as an engine for growth and development. The 1960s, however, were a time for the doctrine of "economic dualism," in which the backward agricultural sector was conceived as an important sector for development to take off. A gradual shift in emphasis took place, and the debate regarding balanced and unbalanced growth was a major concern in economic circles during this time. Nonetheless, gross national product (GNP) growth was still considered as the main goal of economic development. However, by the 1970s, the failure of growth-oriented development strategies led to a re-evaluation of the economic and social development process.

Considering the level and scale of inequality and many associated issues, GNP as objective of development was dethroned by the mid-1970s. The twin objectives of growth and poverty alleviation have changed the conceptual landscape of the development paradigm. New development strategies, which focused on the relationship between variables such as education, nutrition, health, fertility, infant mortality, and the birth rate, broadened the horizons of development thinking. "Redistribution with growth" and "basic needs" were proposed as alternative development strategies. However, the heavy debt of developing countries blocked the development process in the 1980s – this is often cited as "the lost development decade" – and the focus of development policy shifted towards adjustment and stabilization issues (Thorbecke 2019).

In retrospect, it does appear that the dominant approaches to economic development often equated development with economic growth. The International Bretton Woods Institution has played a vital role in pushing these approaches of economic development through various policy instruments such as the 1980s' structural adjustment programs. However, such policies failed to solve the problem of poverty and inequality in developing countries. Instead, the social consequences of these policies, such as widening of inequality and weakening of social fabric in many developing states, have been reported (UNDP 2011; Stiglitz et al. 2009; Dalziel et al. 2018) in literature. As a result of growing inequality and poverty, scholars began to question such development policies that focused narrowly on economic growth as an indicator of development. Consequently, there was a gradual shift from growth-oriented economic development approaches to the alternative human development agenda.

During the 1980s, important contributions to development theory improved scholars' understanding of the concept of poverty and its measurement. The emergence of Sen's capability approach, which later provided the theoretical foundation for the human development paradigm in economics, was among these conceptual advancements that shed a new light on conceptualizing poverty, inequality, and development. Other important theoretical contributions during the 1980s that are worth highlighting include human development approaches and the neo-institutional framework. The capabilities approach (CA), developed in 1985, was a seminal contribution, which provided a useful framework for conceptualizing poverty. CA was a comprehensive and operationally useful approach for poverty analysis. Economic growth was longer the sole criterion for development. CA brought individuals back to the center of development goals.

Sen's (1999a) interdisciplinary capabilities approach has challenged the commodity-based understanding of poverty. This approach is a combination of elements from philosophy and economics, and provides an alternative approach to previously dominant economic frameworks. Poverty and development are understood in terms of a broader humanist perspective rather than the maximization of utility. According to Sen (1999a), poverty is the deprivation of one or more basic capabilities. These capabilities are indispensable for people to achieve at least the minimum functioning required to live their desired life. Sen's (1999a) CA provides the theoretical basis for the human development approach and broadens the scope of the human development approach by paying attention to factors such as economic shortfalls and gender inequalities along with monetary data.

The capability approach takes the achieving of capabilities as both the end and the means of development (Gasper 2002). The limitation of development frameworks that rely on GNP in conceptualizing poverty is evident from the fact that basic material needs cannot be fulfilled without other capabilities like education, healthcare, and political and civil rights (Sen 1999a). This broader and contextual approach to poverty is in contrast with the neoliberal approach advanced by the Washington consensus. In contrast to the previous income-based approaches, the human development approach seeks contextualized data from developing countries which reflect individuals' capabilities and interpretations of freedoms ascribed to life by the people themselves. According to Sen (1999a), through the human development paradigm, a *"more illuminating picture can be obtained from information on aspects of life in these parts of the world"* (p. 99). The pioneers of the human development framework, Sen, Martha Nussbaum, and Mahbub-ul-Haq, were inspired by the ideas of Aristotle, pointing out that *"wealth is not the good we are seeking, for it is merely useful for the sake of something else"* (Haq 1995, p. 13).

The same perspective is reflected in the ideas of classical political economists and philosophers such as Adam Smith, Immanuel Kant, and Thomas R. Malthus (Sen 1989; Nussbaum 2011; Haq 1995). This confusion over means and ends is still relevant in development economics, whether it is the mainstream or the heterodox tradition. Though there is a considerable level of agreement among scholars that the ultimate goal of development is to increase the wellbeing of individuals, there is disagreement on the fundamental question of whether economic growth automatically leads to the improvement of human wellbeing. This chapter draws on Amartya Sen's capability approach (a key contribution in the field of development economics) to provide some insight into the meaning of growth and development.

Learning from the Intellectual History of Economic Thought

According to Aristotle, economics is a field of practical philosophy that is embedded in a predetermined framework of social norms. Thus, in ancient philosophy, using the word moral with economics would have been redundant (Götz 2015). Even during the Middle Ages, the view on economics was quite restricted, and economics was often approached from the domestic and moral perspective. Profitability or macroeconomic linkages were not reflected in economic deals. However, the eighteenth century *cameralists* and *physiocrats* took commerce and national accounting as an important dimensions of the economy. This new discourse replaced the Aristotelian discourse of an economy guided by morals, and the order of the household was replaced by the principles of productivity. Thus, economy acquired a new meaning – a global, dynamic system capable of growth.

The idea that economics is a nonnormative object of study independent of moral imperatives acquired its "oracular authority" during the mercantilist period. Towards the end of nineteenth century, Kantianism emerged as the dominant paradigm in moral philosophy, ultimately divorcing moral from political economy (Götz 2015). Although morality was an important component of the early political economy, there are few references to morality in contemporary theory. Morality is treated as a matter of individual preferences that lie outside the scope of public validation (Sayer 2000). From the perspective of a contemporary observer, this shift from normative to positive economics is a paradigm shift. Like economic behavior, moral validation is also excluded from consideration in economics. This split in normative and positive economics led to the rationalization of values, thus expelling the moral questions from positive economics (Sayer 2001).

Economists in the twentieth century were often cynical about normative and evaluative judgments in economics. They were keen to promote value-free economics. Lionel Robbins (1935) famously argued: *"Economics deals with ascertainable facts; ethics with valuations and obligations"* (Robbins 1935, pp. 148–9). Consequently, neoliberal economic theory has evolved in a way that has created a sharp distinction between economic deliberation and normative values (Morris 2010). Neoliberal capitalism not only provides the analytical framework of economic organization, it also provides the overriding ethical context in which social and economic relations are organized. However, rising inequality and the growing economic insecurity among the masses at a global scale have not only deteriorated trust in the market economy, they have also raised many questions about the legitimacy of the dominant economic paradigm – it is perceived to privilege economic elites. Therefore, neoliberal economics is often critiqued for endorsing egoism that ignores ethical formation for the pursuit of material goals. Sen argued against this kind of moral skepticism in *Collective Choice and Social Welfare* (1970).

The ethical foundation of the neoliberal economic paradigm is built on the model of *homo economicus* individuals: individuals driven by self-interest. However, this, as argued in the rest of the chapter, is not a true reflection of human nature. The *homo economicus* model works only under restrictive conditions, since the exclusive focus on utility maximization of subjective material preferences fails to acknowledge other human emotions such as altruism, trust, wellbeing, and happiness. Sen (2002) rightly doubts the model.

> This narrow view of rationality as self-interest maximization is not only arbitrary, it can also lead to serious descriptive and predictive problems in economics (given the assumption of rational behavior). In many of our actions, we evidently do pay attention to the demands of cooperation. . . . Indeed, within the narrow view, there are ongoing challenges in explaining why people often work together in interdependent productive activities, why public-spirited behavior is often observed (from not littering the streets to showing kindness and consideration to others), or why rule-based conduct standardly constrains narrowly self-seeking actions in a great many contexts. (Sen 2002, pp. 23–4)

Although in principle economic behavior and moral disposition might be separated, in practice both are deeply intertwined, and ethical concerns unavoidably infiltrate even a good deal of both positive and normative economics. Though this "pursuit of self-interest" model is widely accepted, there is enough evidence to support the involvement of other emotions in economic transactions. Individual also commit themselves to altruistic goals (Alvey 2011). Human sentiments cannot be divided on the poles of egoism and altruism, rather, there is a range of im/moral sentiments such as benevolence, gratitude, compassion, pride, shame, envy, selfishness, vanity, sense of justice, prudence, and propriety. Economic processes are also affected by all these emotions (Sayer 2007).

If we look back at the classical political economy of the eighteenth and early nineteenth centuries, Adam Smith – a founding father of modern economics and advocate of commercial society – himself believed that happiness is not necessarily linked with wealth and other material things. Instead, he believed that the continuous and uninterrupted pursuit of material goods undermines people's tranquility and happiness. Although people in commercial societies are more content than in some other forms of societies, this is not due to material things. There are other factors such as relative liberty and a sense of security that help people in commercial societies to act more virtuously and enjoy rewarding relations with friends and family. Adam Smith's economic theory is embedded in a broader theory of moral order and sentiments (Sayer 2004). According to Smith, insecurity and dependence were a great evil of pre-commercial societies that caused misery in these societies. In commercial societies, money or wealth is not a source of tranquility in itself, rather, an enhanced sense of security and liberty are the source of tranquility. Economic exchange, like all other social interactions, is affected by the actor's sense of justice (as Adam Smith emphasized in his book *The Theory of Moral Sentiments*).

Existing studies in behavioral economics also provide support for the position that people are inclined towards values such as altruism and fairness. This calls for a serious reconsideration of the history and ethical foundations of neoclassical economics, which reduces economic goals to individual material gain. *Homo economicus* is expected to respect laws and the right to individual property, and his actions are guided by economic rationale rather than moral norms. Financial incentives and material gains have precedence over norms and values in economic transactions. This indicates an ethical blind spot in neoclassical economics. Market efficiency declines without commitments to values such as honesty, trust, and fairness. In case we agree on these insights, we need to draw a moral boundary around free markets and we must create a space for some agreed upon common goods as desirable goals in neoclassical economics.

Ethical Blind Spot in Neoliberal Economics

Historically, economics emerged as a branch of moral philosophy, and to restore this relation, we need ethical reflection in economic reasoning. Liberal theory in its economic guise turns moral and political values into subjective preferences (Sayer 2000). However, the utility-maximizing *homo economicus* individual is a mere ideological construction. Every economic activity has a constitutive moral assumption. People operate within layers of reality, for example, gender norms, organizational systems, and community values are realities, which might shape the lives of individuals in a variety of ways. Individuals have the capacity to reflect on these different dimensions of their lives, and they also have the capacity to support or undermine one position or another (Kofti 2016). Economic exchange might have other goals beside profit, such as acquiring social prestige, enhancing one's social capital, or maintaining group solidarity (Naʾre 2011). Economic exchange has both a moral and an economic logic. Therefore, it is futile to argue for giving priority to economic logic over moral logic and vice versa as they are different expressions of the same "kernel of human relationships" (Thompson 1961, p. 38).

Individuals are ethical and evaluative beings (Sayer 2011, p. 142), and their economic behavior, as well as practices, are intertwined in the social sphere (which is governed by social norms and values.) Every social relation is governed by a set of morals giving a sense of rights and responsibilities to an actor. The same is true of economic relations; to enact economic roles, one must have some idea about what is expected or proper behavior (Sayer 2007). Thus, economic behavior, on the one hand, is affected by human evaluations and sentiments and, on the other hand, it also affects human evaluation, sentiments (Sayer 2000), and external values that are about responsibility, distribution, and rights (Bolton and Lasser 2013).

There is a call from within economics to humanize economics by reviving the ethical tradition of economics. Consequently, moral dimensions of economic life are subject to reflection and constant negotiation in modern societies. There are two polar views about morally guided economic action: One is dismissal and the other is an exaggerated scope for morality in economic activities. However, this bipolar view of economic actions curtails the possibility of hybrid economic actions. Economic systems need to be re-evaluated and steered where it is deemed appropriate to do so (Sayer 2000). This reinterpretation may deal with overall economic relations, or it could be related to a specific aspect. One focuses on reviving the overall economic system, while the other wants to improve the system by focusing on the existing limitations.

Sen and the Revival of Ethical Traditions in Economic Thought

As Putnam notes (2002), in contrast to the contemporary approach of mainstream economics, values play an important part in the writings of Sen, and his work can best be interpreted as a revival of the classical economic tradition. Sen's major contribution to the history of economic thought is that he has recovered an integrated approach to ethics, ontology, and economics that characterized the classical political economy of Adam Smith. Sen highlighted the contradiction between the "non-ethical character" of the discipline cherished by mainstream economics and the historical evolution of the subject as an offshoot of ethics (Sen 1987).

In the construction of his economic framework, Sen was inspired by Adam Smith's understanding of human behavior (Eiffe 2010). Sen has been apprehensive about the distance between ethics and economics, and has contended that "modern economics has been substantially impoverished by the distance that has grown between economics and ethics" (Sen 1987, p. 7). In his own work, he focused on the moral dimension of economic analysis, which is a key feature of the Cambridge economic tradition. He suggested a multidimensional approach to the study of economic behavior and rationality. Sen was deeply concerned with the moral and philosophical dimensions of economic analysis. Sen advocates a close connection between moral philosophy and economics.

Human behavior is complex and characterized by a variety of emotions. Sen criticized mainstream economic theory on the grounds that it provides a poor description of human behavior. The self-interest pursuit approach of rational choice theory (which provides a basis for utility theory in macroeconomic analysis) assumes that self-interest is dominant over all other motivations in economic agency. Another important feature of rational choice theory is the internally consistent unique preference ordering, which is explained in terms of utility function. The utility function, which provides explanations of individual preferences, is an analytical tool that explains how preferences determine choices.

However, Sen notes that human behavior is affected by conflicting motivations, goals, and values that are not necessarily consistent with each other. Furthermore, preference order is also dependent on external factors, and actual human behavior is an outcome of various competing motivations. Various motivations such as conventional rule following, social commitment, and moral imperatives prevent individuals from choosing the welfare optimizing alternatives. For example, women's desires in traditional societies are shaped by existing institutional arrangements that drive a wedge between personal choice and personal welfare (Sen 1987). Their desires are not shaped by their individual self-interest, rather they have a different understanding of their individual self. This different way of thinking about the self curtails the possibility of vocal protest against gender inequalities at the household level. Thus, one must be cautious when using the term "preference," to state whether one means actual choices or personal welfare.

Sen concludes that this internal consistency approach of utility theory is conceptually flawed (Sen 2002), and self-interest on its own lacks explanatory power to account for all the relevant reasons for economic choice. Conflict of preferences, goals, or values may occur during economic exchange. According to Sen, the situation of unique ordering that is presented in mainstream economic theory is just one possibility among many others which may or may not occur. According to Sen (2002, p. 8), rational choice theory "has denied room for some important motivations and certain reasons for choice, including some concerns that Adam Smith had seen as parts of standard 'moral sentiments' and Immanuel Kant had included among the demands of rationality in social living (in the form of 'categorical imperatives')."

According to Sen (1996):

> Welfare economics is a major branch of 'practical reason'. There are no good grounds for expecting that the diverse considerations that are characteristic of practical reason, discussed among others, by Aristotle, Kant, Smith, Hume, Marx, or Mill, can, in any sense, be avoided by taking refuge in some simple formula like the utilitarian maximization of utility sums, or a general reliance on optimality, or going by some mechanical criterion of technical efficiency or maximization of the gross national product. (Sen 1996, p. 61)

Sen (1985) criticizes the mainstream approaches of development economics that equate wealth with welfare. He argues that income is not a good indicator of wellbeing, but rather it should be measured in terms of a set of achieved functioning and capabilities. Functioning refers to what an individual is, whereas capabilities are the set of potential functionings. Sen's ideas about wellbeing have some similarities with Pigou's (1912, 1920) and Sidgwick's (1883) ideas about wealth and welfare. Both agree that wealth and welfare may not necessarily coincide but the point of departure in their thought is that they equated wellbeing with utility, whereas Sen rejects utility as a measure of happiness and desire satisfaction.

Sen critiqued the scalar measure of wellbeing used in mainstream welfare economics. He advocated that actual preferences are not the real basis for the evaluation of wellbeing; freedom to choose is also important. He proposed a multidimensional approach to wellbeing. This line of argument is also closely linked with the capabilities approach (CA) of Amartya Sen, which is not only of theoretical interest but also has practical relevance for policy design (Crocker and Robeyns 2010). In CA, Sen is concerned with the ontological description of a space for assessing wellbeing. However, his focus is more on the ontological description rather than on an ethical prescription. CA provides a general framework for the evaluation of wellbeing or justice, and not a full-fledged ethical theory. He proposed the expansion of human capabilities as an objective of economic development, a topic on which Sen has written extensively.

An Overview of Sen's Capabilities Approach (CA)

Although Sen acknowledge the importance of resources to wellbeing, he strongly believes that material resources are a means and not an end. Therefore, he proposes that the expansion of human capabilities or potential function should be the main goal of development. In introducing the idea of development as an expansion of human capabilities, Sen (1989) has opened up the possibilities of envisioning development as a means of human flourishing. Based on this conception of development, Sen (1983) argued that the development process must be concerned with capabilities. Capabilities refer to the potential opportunities that individuals have to achieve certain functionings, the ability of the individual to live the life that they have reason to value. Taking human capabilities as the main currency for measuring development makes it possible to have a more realistic assessment of development and the effectiveness of economic growth in achieving this goal. CA does not assume that economic growth automatically increases human wellbeing.

In the first volume of the *Handbook of Development Economics*, Amartya Sen (1988) looks at the relationship between development and economic growth; though this relationship is important, it is also a source of confusion. The capability approach has provided a theoretical basis for many development paradigms, for example, CA facilitated and reinforced the human rights-based approach to development as well as feminist economics and development. CA provides a normative framework for the evaluation of human wellbeing. This approach is built on the interrelated concepts of functionings, capabilities and freedoms, and agency. Capability is defined as "the various combinations of functionings (beings and doings) that the person can achieve. Capability is, thus, a set of vectors of functionings, reflecting the person's freedom to lead one type of life or another ... to choose from possible livings" (Sen 1992, p. 40).

According to Sen (1987), capability is a positive freedom that entails both instrumental and intrinsic value. Functionings and capabilities are closely interconnected. Functioning refers to the "various things a person may value doing or being" (Sen 1999a, p. 75). Capabilities, the ability of an individual to choose specific functionings, depend upon the functionings that were achieved by the individual earlier in his or her life. For example, good nourishment, health, and education are directly needed functionings for capabilities (Gandjour 2008). According to Sen (1992), functionings such as physical health and the absence of poor nourishment are important for people's ability to choose their desired life. There are other functionings, such as mental health, which are very important for the individual ability to do something. Mental health problems can indirectly affect the achieved functionings of an individual, for example, an individual who is suffering from some kind of mental disorder might negatively perceive the existing opportunities, and thus fail to avail themselves of those potential opportunities (Gandjour 2008). Whereas agency is defined as the ability to pursue goals that one values and has reason to value, an agent is "someone who acts and brings about change" (Sen 1999a, p. 19).

In order to illustrate the relationship between capability, functioning, and agency, Sen (1983) has given the simple but interesting example of a bicycle. A bicycle is a commodity that has many characteristics, such as it enables one to play or commute from one place to another. So we can say that a bicycle is a commodity that provides an individual with the capability to achieve functionings such as play or mobility. But if an individual is unable to ride a bike, the simple possession of a bike will not ensure play or mobility. So what matter most is the capability and the freedom of the individual, and not the possession of material goods. The link between material things and human goals is crucial. CA carefully differentiates between means and ends, both at a theoretical and an empirical level. Commodities are just a means to enlarge the set of human capabilities and have mere instrumental value. However, the conversion of material goods into functioning depends on other factors such as having the ability and freedom to do so. Therefore, income and consumption are not valid measure of capabilities. As opposed to the conventional approach of development analysis, CA differentiates between means and ends, and pays attention to the factors that help the conversion of resources into capabilities. Thus, capabilities are more important than functioning and have intrinsic value. Two individuals might have the achieved the same functionings but with different sets of capabilities.

In CA, functionings and capabilities intercede between utility and welfare. Sen (1990) trailed Aristotle's idea that material things are not an end in themselves, but that they are means towards some ends. However, it is also true that in some cases they might fail to serve as a means. Sen critiqued traditional development economics on the grounds that the focus is on national product, aggregate income, and the total supply of particular goods. He instead highlighted the role of entitlements of people and the capabilities these entitlements engender (Sen 1983). He argued to shift the focus of development economics from traditional measures of growth, such as aggregate income, to individual entitlements. Sen (1983) defines entitlements as a set of different commodity bundles that an individual can have in a society by using all the rights and opportunities that he or she has. Entitlements that include material consumption and other resources serve as a source to produce wellbeing. Entitlements help to generate capabilities, which in turn enlarge individual choices in society through increased participation. However, the web between the various sets of entitlement, capabilities, and functionings is highly complex and tangled.

Sen (1993) notes that capabilities and functionings are better indicators of the quality of life. Functionings refer to the existing condition of people. There are various functionings that an individual can achieve in a society. For example, health conditions, happiness, income, and sustenance are some of the functionings that an individual can achieve in a society. The functionings that one achieves are the outcome of the choices that one has previously made. The standard of living indicates a set of various opportunities that a person can avail themselves of in order to gain diverse achievements in life. The standard of living for an individual is likely to increase with the increase in the sets of available functionings in life (Jasek-Rysdahl 2001). However, functionings are not the sole indicator of standard of living, rather, capabilities also play a vital role to better conceptualize that. As mentioned earlier, capabilities are a set of possible functionings that are accessible to an individual that he or she is free to choose from (Sen 1993). Capabilities and entitlements have a two-way relationship; whether an individual has some capabilities or not depends upon his or her entitlements in society, and the entitlements of an individual depend upon individual capabilities.

Capability Approach and Development Economics

As a theoretical framework, CA contradicts the mainstream approaches in development economics and challenges many foundational concepts of welfare economics. Although most approaches of development economics agree that the goal of development is to improve human conditions, these approaches mainly focus on the maximization of utilities by satisfaction of preferences. Income is used as a proxy for utility in these approaches. In CA, Sen emphasizes the central role of capabilities or potential functionings for measuring wellbeing, as freedom to choose is also important along with the availability of choices. Sen (1993) criticize mainstream economic theory for exclusively focusing on actual preferences, while ignoring the freedom to choose, in the measurement of wellbeing. Freedom to choose is a very important concept in Sen's capabilities framework, which is consistent with his multidimensional perspective on wellbeing. Departing from utilitarianism and rational choice theory, Sen in his later writings on CA laid emphasis on the centrality of moral values. In CA, he intends to promote human freedom as a means of individual wellbeing. Freedom to choose in a given situation is also a basic requirement for human development and is one of the basic elements required to achieve the goal of social justice. According to Sen (1993),

> The freedom to lead different types of life is reflected in the person's capability set. The capability of a person depends on a variety of factors, including personal characteristics and social arrangements. (p. 33)

Capabilities include both positive as well as negative freedoms. Positive freedom refers to the ability of a person to choose whatever he or she desires to choose, whereas negative freedom refers to a situation where a person, despite having choices, is not in a position to choose the desired end due to other constraints. For example, a women who has a legal right to file a domestic violence complaint might lack the required infrastructure to reach law-enforcing agencies. In this case, she has a negative freedom, as her freedom is constrained by the existing infrastructure of society. Negative freedom is related with the process aspect of freedom, which entails "autonomy of choice" and "non-interference" (Naz 2016). Thus, positive freedom, also known as opportunity freedom, helps to bridge the gap between empowerment and agency (Jasek-Rysdahl 2001). Sen has differentiated between agency and wellbeing freedom as well (Qizilbash 2005). He believes that individuals are responsible agents who can choose between various ends. They can either choose to act or refuse to do so.

According to Sen (1999a), an agent is,

> someone who acts and brings about change, and whose achievements can be judged in terms of her own values and objectives, whether or not we assess them in terms of external criteria (p. 19).

However, there is a link between the agency and capabilities of an individual. It is only possible to make choices and act responsibly if one has the possibility to do so. Therefore, in order to promote wellbeing, we have to design policies that create opportunities and open more choices for people. Thus, freedom and social welfare are not only an important means to promote wellbeing, but they also play an instrumental role in promoting economic development. As Martins (2007) has convincingly argued, freedom is not just a desirable ethical goal, but also an ontological constituent of reality. Amartya Sen has acknowledged diversity and pluralism both from a theoretical, as well as a practical perspective. On the one hand, he defends capitalism as the best system to provide freedom and, on the other hand, he also criticizes its theoretical foundations and monolithic dimension. His approach reconciles ethics and capitalism, and he argues that economic exchange takes place within a social and moral framework (Sen 1987). In Sen's work, moral and material issues are disentangled without preferring one or the other (Rajapakse 2015). In his work *Poverty and Famine* (1981), Sen claimed that the immediate cause of famines is not crop failure; rather, malfunctions of the moral economy are the most proximate cause of famines.

The term moral economy was first used in the eighteenth century, but after the publication of *The Moral Economy of the English Crowd in the Eighteenth Century* (Thompson 1971), it has become popular. Despite its increasing popularity, the moral economy concept itself remains undertheorized and elusive due to the multiple meanings associated with the term. Moral economy states the rights and responsibilities of individuals and institutions towards others. Historically, the concept of moral economy was applicable to societies where there were few markets and economic activity was largely regulated by means of moral norms. However, moral norms are also present in modern capitalist economies (Sayer 2000). The relationship between economy and society is the focal point of moral economy. Exclusion of moral dimensions from any rational analysis of the economy or market is no longer accepted. The market is socially embedded and mediated by institutions, individuals, and communities. There is a constant struggle below the calm surface of the capitalistic system. There is a call from within economics to pay attention to the micro foundation of economics (Bolton and Lasser 2013). The dominant approaches to economic development often equate development with economic growth. However, social consequences of policies resulting from the application of these approaches have been acknowledged (UNDP 2011; Stiglitz et al. 2009; Dalziel et al. 2018) in existing literature.

Thus, the utility-maximizing *homo economicus* is a mere ideological construction. Without creative capacity and the moral commitment of individuals, the market can hardly survive. Human beings have the capacity to reflect and reason about norms, inequalities, and obligations. There are two popular streams in literature that try to conceptualize moral economy. One stream focuses on the systems of provisions that develop outside the market (such as through informal practices). According to this analysis, economic practices are embedded in social and institutional structures, moral obligations, and societal norms. For example, given the vulnerable nature of human beings, norms regarding responsibilities for the care of the elderly, sick, and children are a case in point (Sayer 2004). There are two ways to fulfill economic responsibilities for other people, one is through the provision of unpaid services and the other is through purchasing services from the market either privately or through the tax system. The norms can also be influenced by economic organization. Increased mobility due to economic globalization has led to a change in family systems (from joint to nuclear), and elderly care has become marketized. However, there are also cultural norms and social structures that are not easily influenced through economic organization. Economic responsibilities for and towards others need serious consideration (Sayer 2004). In *Poor Economics* (2011), Banerjee and Duflo convincingly argue that under conditions of extreme poverty, it is necessary to deliver various services below the market price.

> —we have already seen that in many instances government intervention is necessary precisely when, for some reason, the free market cannot do the job. For example, many parents may not end up immunizing their children or giving them deworming pills, both because they do not take into account the benefit this would have for others and because of the time inconsistency problems (p. 254)

It is clear that some responsibilities are institutionalized while others are subject to contestations and open to interpretation. Helping friends and family members depend on, in Smith's terms, the moral sentiment of actors and reciprocal relationships (Sayer 2004). The second stream of literature on moral economy is interested in the norms and values per se. This interpretation of moral economy is not so concerned with the production and distribution of resources; rather, its interest lies in the economy of morals. Thus, the second stream focuses on the study of morals, and economy has become a stranger in the nascent field of "moral anthropology" (Zigon 2007).

Moral economy has close ties with the subject of political economy. Moral economic approaches are not only interested in a grounded understanding of global political economy processes, but they also historicize everyday power relations. Moral economy is particularly interested in analyzing various dimension of social reproduction, such as political culture, norms, and expectations of the various groups of people. The implicit notion in these two perspectives is that the market alone is not capable of solving all problems in an interconnected world. Morality is very important for the people, because it is related to the things that people value in their lives which seriously affect their overall sense of wellbeing. Normative rationales matter for actors. The moral dimension is unavoidable in economic relations too. Economic relations are also shaped by moral norms such as rights, duties, and entitlements. Morality has its own intrinsic value, regardless of whether there are penalties for bad conduct or not. We want to do good to enhance wellbeing and avoid harm to others (Sayer 2004).

However, there is a possibility that in some cases ethical and economic evaluations of an activity may come into conflict with each other. In some cases, what is good for one person has to be compromised for economical imperatives. This might be a likely scenario in societies in which capitalist forms of organization are more dominant (Jessop 2002). Values are overridden by economic rationale. Profit is not an end, rather it is a means towards an end (Sayer 2004). Many heterodox economists have proposed various alternatives to income-based measures of growth and development.

Capability Approach and Human Development

Since the foundational work of CA in the 1980s by Amartya Sen and Martha Nussbaum, the framework has continued to evolve and become richer. CA has been used across disciplines as a theoretical framework with different methods and for a variety of purposes. As a theoretical framework, CA is different from other mainstream approaches of development economics. The human development paradigm used the capability approach as a normative framework (Haq 1995). Human development is sometimes used synonymously with capability approach, but the two are distinct from each other. However, core elements of both approaches overlap. Human development is the application of CA in the field of development, whereas the scope of CA extends beyond development. In the first Human Development Report (HDR), Haq defined human development as:

A process of enlarging people's choices. The most critical ones are to lead a long and healthy life, to be educated and to enjoy a decent standard of living. Additional choices include political freedom, guaranteed human rights and personal self-respect—what Adam Smith called the ability to mix with others without being 'ashamed to appear in public'. (UNDP 1990, p. 10)

The concept of human development presented in the Human Development Report was much broader than the one widely accepted in conventional development economics. Human development is a human-centered approach of development that differentiates between economic means and human ends. For instance, poverty is defined as capability deprivation that people experience in their life, rather than lack of income. Human development uses multidimensional measures of development and challenges existing structures of power. Reconceptualizing the problems and concepts such as poverty, rights, and cultural identities is a major contribution of the human development paradigm. The flourishing of human beings has been reprioritized as the major development goal. Consequently, the information base was shifted from commodities to individuals. Human development is a complex framework, which requires new metrics to measure its progress. Measuring human development raises many questions, both at a conceptual and a practical level. For example, which capabilities to include and how to assign weightage to different capabilities is arguably a difficult task. Sen helped Haq to develop the Human Development Index (HDI) as a measure of human development that combines education, health, and a "decent standard of living" as the three important dimensions of human development. The indicators that are used in the HDI observe functioning as a proxy for capabilities. Although the HDI is reductionist and does not fully capture the complex concept of human development, Sen explains "we have to see the human development index as a deliberately constructed crude measure, offered to rival the GNP" and that would "serve to broaden public interest" in a fuller accounting of human progress (Sen 2003, p. x). Thus, the major contribution of the HDI is the shift in development focus from growth to a human-centered approach of development.

Another important agenda of human development is agency (the ability of the individual to set and choose their desired goals in life). According to Sen, human development is concerned with "how human agency can deliberately bring about radical change through improving societal organization and commitment" (Sen 2003, p. vii). Agency and freedom are fundamental goals of the human development paradigm. People can push for social change through collective and individual actions by exercising their freedoms. The capability approach has been an effective tool in redirecting development theory and practice towards broader human-centered goals beyond human development. There are novel problems in economics that have become subject to contestation, such as feminist contestations of gendered economic roles and unequal division of labor. The capability approach was also influential in the development of the feminist economics that arose parallel with the human development paradigm. Both approaches complemented each other with concepts and tools.

Feminist economists have uncovered gender biases in the discipline of economics and called for economics to extend the study of economic life beyond the formal economy (Nelson 2010). They have challenged underlying gendered normative assumptions regarding economic activities and how value is assigned to various economic activities. In her book *Women's Role in Economic Development* (1970), Ester Boserup pointed out that economic development processes in postcolonial economies have had an implicit gender bias, which in turn has reinforced unequal gender relations. She has effectively demonstrated that a variety of productive activities are performed by women. However, in the market economy, only those activities which are performed within the ambit of market transactions and paid labor are considered productive activities. Care work that is responsible for the provision of human life is mostly unpaid and performed by the women within the private sphere of the household. Though essential to the proper functioning of market economies, care work, such as the fulfilment of basic needs and the provision of support to the elderly, sick, children, and nonhuman beings, is often unacknowledged and performed outside the market (Schildberg 2014). Feminists demand to put such provisioning of human life at the center of economic debate, rather than focusing on individual choices made under conditions of scarcity. However, this would require not only a fundamental shift in the way economic rationality is defined, but also in the way society and the economy is organized. Scholars have highlighted the gender biasness in basic postulates, economic models, and desired research methods of mainstream economics. Furthermore, they argue that care is not just unpaid work, rather it is a system of social relationships, based on ethics and values which acknowledges human interdependence and their vulnerabilities (Tronto 1993).

A general tendency in mainstream literature is to support the popularized notion of a split between economic and social life. Social life is considered a domain of life that is laden with values and ethics, whereas economic life is considered to be guided by its own mechanical and impersonal rules. The discussion of ethics and markets is also divided along the same dichotomous lines. Conservative free market economists consider business ethics to be an oxymoron, while others acknowledge that there is a possibility for the application of at least some sort of rationalist and individualist ethic. In contrast, critics from the political left believe that capitalism and ethics are not compatible with each other so they want to build a new and more humane economic system by destroying the impersonal capitalist machine. They claim that capitalism is a means to institutionalize greed. However, the intrinsically mechanical and antisocial view of economic life curtails the possibilities of imagining the idea of "care ethics."

Insight from feminist economics provides us a useful analytical lens to explore the relationship between care ethics and markets. Feminist economists have challenged the traditional understanding of gender and economic behavior. They argue that the image of an automaton economic man is misleading (Ferber and Nelson 1993). All human beings, as argued by Nelson (2010), are individuated and connected in relationships. There is vast literature (Folbre and Nelson 2000; Herzberg 1987; Kusnet 2008) that provides empirical evidence to support the statement that individuals do not enter the market without their feelings, values, ethics, and sociality.

Feminist approaches recognize interdependencies between market, firms, state, and households. Families are also important economic units and relations are directed by a combination of factors such as power, convention, and moral commitment. Considering the patriarchal structure, gendered division of resources, and unpaid work, the family can be categorized as an immoral economy (Mies 1997). Feminist economics has highlighted how commodification, particularly that of labor power, has resulted in the loss of self-respect and standards for women. They have discussed commodification and ethics of care in relation to the commodification of these services. Critical evaluation of development processes by feminist economists has raised question about the effectiveness of mainstream development theories to enhance economic and social wellbeing. Furthermore, they have also challenged the underlying assumptions about what constitutes a good implicit in orthodox neoliberal development theories.

Sen's Capability Approach and Feminist Concerns About Gender Inequalities

Feminist scholarship and the capabilities approach have overlapping commitments in the area of gender inequality and poverty. A critical evaluation of the development process by feminist economists has raised many questions about the potential of neoliberal policy-led development processes to enhance economic and social wellbeing. A feminist analysis of the development processes offers useful insight to rethink the mainstream economic development paradigm. CA and associated participatory methods in development studies have highlighted the gender biases and inequalities embedded in human capital and basic need approaches. Sen also emphasized the importance of CA to study gender inequalities, because human capabilities are not just created but in some cases are also negatively affected by development policies, markets, and other social arrangements. The flexible and human-centered development approach proposed by Sen highlighted the different obstacles faced by women in overcoming poverty and inequality. As argued by Beneria and Sen (1981), development is not a gender-neutral linear process that guarantees improvement in overall quality of life; rather it is an uneven and disruptive process which has a differential impact on men and women.

According to Sen (1999a, p. 108), women in developing countries have a different experience of poverty due to the gendered nature of capabilities deprivation. The capabilities approach, which provides a multifaceted understanding of inequalities, has the potential to address feminist concerns and questions (Robeyns 2003). Feminism shares with Sen's capability approach a core belief that acknowledges people as the end of development. Therefore, economy should be geared towards meeting the needs of people (Floro 2016). Feminist economics has challenged the core assumptions of utilitarianism and recognizes the role of nonmarket economic exchange, such as unpaid care work and social reproduction, in flourishing human capabilities.

Sen has also criticized welfare economics for its exclusive focus on income (Sen 1985, 1987, 1992, 1993, 1995, 1998) and argued that the focus should be on capabilities and freedom. Sen claimed that income is not an end in itself, rather it should be used to achieve higher ends such as enhancement of individual functionings and capabilities of people (Pressman and Summerfield 2000). Amartya Sen's work on capabilities has provided an alternative framework of development, which incorporates the goal of gender equality. However, Sen has not given a complete list of desired capabilities, rather, he has cited some fundamental capabilities, such as basic liberties, freedom of movement, freedom of association, and freedom of occupational choice, against a background of diverse opportunities such as positions of responsibility in political and economic institutions; income and wealth; and the social bases of self-respect (Naz 2016). Thus, development is about the expansion of these capabilities. Human life can be enriched by expanding the range of choices through provision of basic capabilities, such as to be healthy and well nourished, to be knowledgeable, and to participate in communal life.

This line of Sen's work has the potential to address overarching feminist concerns about gender inequality. In his book *Inequality Reexamined*, Sen has argued that

> the question of gender inequality ... can be understood much better by comparing those things that intrinsically matter (such as functionings and capabilities), rather than just the means [to achieve them] like ... resources. The issue of gender inequality is ultimately one of disparate freedoms (Sen 1992, p. 125).

However, Sen has not proposed a definite list of capabilities. Sen's capability approach is a normative tool that is not directly applicable to the study of gender inequalities. In the words of Robeyns (2003), "it is not a mathematical algorithm that prescribes how to measure inequality or poverty, nor is it a complete theory of justice" (p. 64). Rather, it provides a general framework that could be useful to access inequalities. Due to the underspecified nature of CA and the lack of a defined list of capabilities, there are methodological challenges in the practical application of CA. There is an ongoing discussion in existing literature regarding the application of CA in empirical research. How to choose the most valuable capabilities? Whether one should adopt a subjective or an objective method to choose the most valuable capabilities is a serious epistemological issue that needs to be sorted. However, despite these methodological challenges, many scholars have used CA as an analytical tool in their studies (Pelenca et al. 2015; Zimmermann 2006).

The major question is how to measure capability and which capabilities are most relevant. Sen has emphasized that the set of individual capabilities must lead to the freedom to live the life that a person has reason to value. This involves a normative evaluation of people's capabilities, and Sen criticized a utility-based evaluation of individual wellbeing. However, he also acknowledged the importance of material resources, as inequality of economic resources may generate other inequalities in capabilities and functioning. In case of gender inequalities in functioning and capabilities, Agarwal (1994) has argued that "the gender gap in the ownership and control of property is the single most critical contributor to the gender gap in economic wellbeing, social status, and empowerment" (p. 1455). To analyze gender inequality, one must pay attention to the inequalities in resources which cause gender inequalities in capabilities and functionings. According to Robeyns (2003), Sen's capability approach provides a useful framework for the analysis of gender inequalities for three important reasons: Firstly, CA is an ethically (or normatively) individualistic theory, which implies that individuals are the unit of normative judgment. Whereas ontologically CA is a not an individualistic theory, and it acknowledges that women's wellbeing can be subsumed under the household or the community. As compared to other standard wellbeing approaches, CA acknowledges the intra-household inequalities in access to resources. Secondly, CA pays attention to people's functioning and capabilities in both market and nonmarket settings.

This is important for gender inequality research, as women spend most of their time outside the market. Thirdly, CA is important for the gender-related assessments of wellbeing and disadvantage, because it acknowledges human diversity such as race, age, ethnicity, gender, sexuality, and geographical location. According to Sen

> Investigations of equality – theoretical as well as practical – that proceed with the assumption of antecedent uniformity . . . thus miss out on a major aspect of the problem. Human diversity is not a secondary complication (to be ignored or to be introduced "later on"); it is a fundamental aspect of our interest in equality. (1992, p. xi)

Notwithstanding the relevance of CA to the study of gender inequalities, it also has one major limitation caused by the underspecified nature of CA, which might render it vulnerable to androcentric interpretations and applications. Within the existing literature on CA, there is also an ongoing debate among scholars on whether to define a basic universal list of central human capabilities or to have a public debate to draw a list of desired capabilities (Claassen 2011). Martha Nussbaum (2003) has emphasized the need to validate one definite list of valuable capabilities to apply the capability approach for the better understanding of gender inequalities. In order to tackle these intellectual and legitimacy issues in the selection of relevant capabilities for gender inequality assessment, Robeyns (2003, 2005) suggested five criteria to draw such lists: first, the criterion of explicit formulation; second, the criterion of methodological justification; third, the criterion of sensitivity to context; fourth, the criterion of different levels of generality; and fifth, the criterion of exhaustion and nonreduction.

Since the 1980s, Amartya Sen and Martha Nussbaum's foundational work on CA has expanded enormously, encompassing contributions in wide-ranging fields. The capability framework is comprised of a core set of concepts, and its applications are found in a variety of disciplines, such as economics and ethics (Robeyns 2009), poverty (Alkire et al. 2015), and gender (Sen 1990; Razavi 1996; Nussbaum 2000; Robeyns 2003; Walker et al. 2014). Progress on furthering the application of CA to gender inequality analysis is desired, both on empirical as well as theoretical fronts. Having a proper understanding of gender inequalities and underlying capabilities deficit would help to deal with them. Since the debate on CA is still evolving and the scope of analysis has been extended, further conceptual debates have emerged on the theoretical and conceptual fronts. According to Robeyns (2017), it is important to rearticulate CA by taking into account the way CA has evolved in literature.

Emphasis on individual and collective agency is another important and overlapping area of interest between CA and feminist scholarship. Feminist approaches to development advocate for the removal of institutional barriers and power hierarchies to ensure gender equality. Collective agency resides in groups, which can shape development contours through collective action. Both Sen and Nussbaum have highlighted the limitation of aggregating different life experiences into a single metric of satisfaction (assuming that all individuals have the same preferences) (Nussbaum 2011). Sen recognizes the distinctiveness of the individuals and therefore CA has been criticized by some scholars for its emphasis on individualism. Frances Stewart and Séverine Deneulin (2002) argued that CA was too individualistic to accommodate collective action that can play a pivotal role in countering dominant market forces.

To overcome the problem of individualism in CA, Evan's (2002) notion of collective capabilities has received much attention from capability scholars. There is an ongoing debate in recent literature about the possibility and role of collective capabilities, social structures, and groups (Robeyns 2005) for human development. Peter Evans has closely linked the notion of collective capabilities with collective action, thus creating the possibility to envision the role of collectives in enhancing the lives of people. Literature on collectivities has espoused debate on the allegedly individualistic stance of CA. Stewart and Deneulin (2002) criticized that due to its methodological individualism, CA does not recognize the importance of irreducible social goods.

In her intellectually stimulating review article, Leßmann (2020) has summarized various strands of the capability literature on collectivity, and also presented many suggestions for how Sen's capability approach can accommodate collectives. Collectivity is a very broad concept and collectivities range from small face-to-face groups to larger societies. Collectivities can be defined both from the internal and external point of view. Collectives can be formed through the deliberate choice of their members, such as self-help groups. These are called internally defined collectivities. Collectivities that are formed on the basis of ascription by others based on shared characteristics are called externally defined collectivities. Member of these groups are often born into such collectivities, for example, membership of ethnic groups is nonvoluntary. The notion of collectivity has its connections with CA, because the structural features of both these collectivities affects the wellbeing and agency of their members. Sen has provided a distinction between these interrelated but distinct concepts. In order to explain wellbeing and agency goals, Sen has referred to Adam Smith's notion of sympathy and commitment.

Caring for someone specific is an act of sympathy that aims at wellbeing. Leßmann (2020) states that commitment is a case of agency. It is related to other people and sometimes it is even in conflict with personal wellbeing (Sen 1987). Thus, the notion of commitment or agency draws a line between personal choice and personal welfare; they might not necessarily coincide with each other (Sen 1999b). According to Sen (1999b), there is a close connection between agency, commitment, and the group. Groups are often the locus of individual commitment. In the words of Sen, "people are committed to comprise 'families, friends, local communities, peer groups, and economic and social classes.'" (Sen 1999b, p. 85). Sen linked commitment with identity and acknowledged multiple identities and commitments. Commitments or agency goals are wide-ranging and not tied to one specific aim. On the other hand, wellbeing goals are more specific. Stewart (2005) highlighted the importance of living together in a group and posits that groups and membership of groups affect people's sense of wellbeing. Groups also affect the values and choices of group members. Due to their engagement in a collectivity, the members of a group acquire new collective capabilities or functioning bundles that help them to acquire a life that they have reason to value individually. According to Ibrahim (2006), collective capabilities are generated through a process of collective action, and there is a close link between collective capabilities and the exercise of human agency.

These interpretations of collective capabilities provide space to envision the interaction between individual and collective capabilities within a capability framework. According to Sen (2002, p. 41), "we sometimes act as a member of a group (e.g. 'we voted for our candidate') without seeing it as primarily an individual act (e.g. 'I voted for our candidate')." In this case, the unit of agency in choosing an electoral candidate is broader than the individual action of casting a vote. In his work on famine, Sen (1982) concludes that the occurrence of famine can be prevented through the collective action of citizens through media and legislature. According to Schmid (2007), Sen's notion of commitment beyond personal choice is a reflection of shared group goals. These goals are hard to reduce as individual goals. In the words of Schmid (2007), "Togetherness is irreducible" (p. 221). However, this line of argument goes against feminist demands to value women as individual people rather than subsuming their identity under the household.

This notion of the irreducibility of social goods is used as a feminist critique of CA that is often associated with methodological individualism. As mentioned earlier, Robeyns (2008) responds to this critique by introducing the notion of ethical individualism in this debate. She argues that CA can be ethically individualist without embracing ontological and methodological individualism. Thus, the notion of ethical individualism builds a connection between CA and feminist demands of valuing women as individuals. Qizilbash (2014) furthers this debate and differentiates between methodological individualism and normative or moral individualism. Taking this approach, Qizilbash (2014) concluded that the capability approach underlines the significance of individual choice rather than attaching intrinsic value to social forms or culture. The core of Qizilbash's (2014) argument is that CA preserves "core elements of normative individualism." CA recognizes collectives as structures within which individuals are embedded.

CA has provided an alternative framework for development that shifts the focus of development from material wealth to the development of human capabilities. The focus of development has been shifted away from material consumption to the enhancement of the overall wellbeing of people. Wellbeing is broad and multidimensional, and ranges from receiving care at different stages of life to having access to food, decent work, safe water, and active participation in community life. Thus, CA has created the possibility to incorporate nonmarket work such as social reproduction, and care work, which is mostly performed by women, into the discussion of the economic processes. Mainstream development economics, which is built on the implicit assumption that work is defined only in terms of pay or profit (Beneria et al. 2016), does not acknowledge certain activities and experiences that are of greater concern to women, such as the provision unpaid of care work within families and communities.

Care provisioning, meeting the care needs of young, sick, disabled, and elderly members of society, though a very important element of societal wellbeing, is often neglected. Many feminist scholars have underlined the contribution of women's unpaid work to social reproduction (Floro 2019). Both paid and unpaid care work performed by women is important for sustaining human life and promoting wellbeing. However, women's care work has rarely been acknowledged in mainstream theories of economic development. Although much attention has been paid to the capital accumulation and formulation of developing countries, the maintenance and generation of the labor force required for this process has been assumed as a natural given. Consequently, associated daily unpaid care work performed by family members remains invisible in official statistics (Folbre 2006). However, the twin health and economic crisis, COVID-19, accentuates the relevance of care work and human-centered development frameworks. This is a prime example of the failure of the public health system, which later resulted in devastating health, economic, and societal outcomes across the globe. COVID-19 has badly affected the constituent elements of human development such as income, health and, education. As projected in a UNDP report (2020), it is not just a health emergency, rather, it is an unfolding of the human development crisis, which has already started to affect various dimensions of economic and social development in unprecedented ways. These outcomes cannot be directly linked with gross national income or geographical locale within the global North or South. Instead, social determinants of health such as gender, poverty, physical environment, race, and ethnicity have shaped the health outcomes of the global pandemic (Abrams and Szekler 2020). In these times of crisis, exploring alternative approaches to growth and development has perhaps become more urgent and starkly visible than ever. There are many important points from feminist economic research, such as the integration of social provisioning and care, which can serve as guideposts for transforming the future development goals.

Concluding Remarks

The neoliberal economic development paradigm has evolved in a way that has created a sharp distinction between economic deliberation and normative values. *Homo economicus* is expected to respect laws and the right to individual property, and his actions are guided by economic rationale rather than moral norms. The general assumption in the orthodox theory of economic development is that economic growth will automatically enhance the quality of life for everyone. In this model, financial incentives and material gains are given precedence over norms and values in economic transactions. However, the economy is an interdependent social system, and economic exchange has both a moral and an economic logic. Amartya Sen presented CA as an alternative to neoclassical theory of economic development. He attempted to shift the focus in the field of economics and development studies from an exaggerated emphasis on growth towards the issues of personal wellbeing, agency, and freedom. Sen offered many compelling arguments for going beyond the notion of utility and welfare when it comes to judging personal wellbeing or human development (Clark 2005). CA, which provides a representative, individualistic, and multifaceted understanding of inequalities, has the potential to address development challenges in the contemporary world. Unless development pathways are revaluated with conscious awareness about underlying societal mechanisms, strategies such as those aiming at the improvement of overall human wellbeing might not be effective.

References

Abrams EM, Szekler SJ (2020) COVID-19 and the impact of social determinants of health. Lancet Respir Med. https://doi.org/10.1016/S2213-2600(20)30234-4

Agarwal B (1994) The gender and environment debate: lessons from India. In: Arizpe L, Stone MP, Major DC (eds) Population and environment: rethinking the debate. Westview Press, Boulder, pp 87–124

Alkire S, Foster J, Seth S, Santos E, Roche JM, Ballon P (2015) Multidimensional poverty measurement and analysis. Oxford University Press, New York

Alvey JE (2011) Ethics and economics, today and in the past. J Philos Econ V(1):5–34

Banerjee AV, Duflo E (2011) Poor economics: a radical rethinking of the way to fight global poverty. Public Affairs, New York

Beneria L, Sen G (1981) Accumulation, reproduction, and "women's role in economic development": Boserup revisited. Signs J Women Cult Soc 7(2):279–298

Beneria L, Berik G, Floro M (2016) Gender, development and globalization: economics as if all people mattered. Routledge, New York

Bolton SC, Lasser K (2013) Work, employment and society through the lens of moral economy. Work Employ Soc 27(3):508–525

Boserup E (1970) Woman's role in economic development. St. Martin's Press, New York

Claassen R (2011) Making capability lists: philosophy versus democracy. Polit Stud 59:491–450

Clark DA (2005) Sen's capability approach and the many spaces of human well-being. J Dev Stud 41:1339–1368. https://doi.org/10.1080/00220380500186853

Crocker DA, Robeyns I (2010) In: Morris CW (ed) Amartya Sen. Cambridge University Press, Cambridge, pp 40–59

Dalziel P, Saunders C, Saunders J (2018) Wellbeing economics: capabilities approach to prosperity. Palgrave Macmillan, Cham

Eiffe FF (2010) Amartya Sen reading Adam Smith. Hist Econ Rev 51(1):1–23. https://doi.org/10.1080/18386318.2010.11682153

Evans PB (2002) Collective capabilities, culture and Amartya Sen's development as freedom. Stud Comp Int Dev 37(2):54–60. https://doi.org/10.1007/BF02686261

Ferber MA, Nelson JA (eds) (1993) Beyond economic man: feminist theory and economics. University of Chicago Press, Chicago

Floro MS (2016) Feminist approaches to development. In: Ghosh J, Kattel R, Reinert E (eds) Elgar handbook of alternative theories of economic development. Edward Elgar Publishing Ltd., Cheltenham, pp 416–440

Floro MS (2019) Feminist economist's reflections on economic development: theories and policy debates. In: Nissanke M, Ocampo JA (eds) The Palgrave handbook of development economics: critical reflection on development and economics. Palgrave Macmillan, Cham, pp 61–108

Folbre N (2006) Measuring care: gender, empowerment, and the care economy. J Hum Dev 7(2): 183–199

Folbre N, Nelson JA (2000) For love or money – or both? J Econ Perspect 14(4):123–140

Gandjour A (2008) Mutual dependency between capabilities and functionings in Amartya Sen's capability approach. Soc Choice Welf 31:345–350. https://doi.org/10.1007/s00355-007-0283-7

Gasper D (2002) Is Sen's capability approach an adequate basis for considering human development? Rev Polit Econ 14(4):435–461. https://doi.org/10.1080/0953825022000009898

Götz N (2015) Moral economy': its conceptual history and analytical prospects. J Glob Ethics 11(2):147–162

Haq M (1995) Reflections on human development. Oxford University Press, New York

Herzberg F (1987) One more time: How do you motivate employees? Harv Bus Rev September–October:109–120

Ibrahim S (2006) From individual to collective capabilities: the capability approach as a conceptual framework for self-help. J Hum Dev 7(3):397–416. https://doi.org/10.1080/14649880600815982

Jasek-Rysdahl K (2001) Applying Sen's capabilities framework to neighborhoods: using local asset maps to deepen our understanding of well-being. Rev Soc Econ 59:313–329

Jessop B (2002) The future of the capitalist state. Polity, Cambridge

Kofti D (2016) Moral economy of flexible production: fabricating precarity between the conveyor belt and the household. Anthropol Theory 16(4):433–453

Kusnet D (2008) Love the work, hate the job: why America's best workers are more unhappy than ever. John Wiley & Sons, Hoboken

Leßmann O (2020) Collectivity and the capability approach: survey and discussion. Rev Soc Econ. https://doi.org/10.1080/00346764.2020.1774636

Martins N (2007) Ethics, ontology and capabilities. Rev Polit Econ 19:37–53. https://doi.org/10.1080/09538250601080768

Mies M (1997) Do we need a new "moral economy"? Can Woman Stud 17(2):12–20

Morris CW (2010) Ethics in economics. In: Morris CW (ed) Amartya Sen. Cambridge University Press, Cambridge, pp 40–59

Na're L (2011) The moral economy of domestic and care labour. Sociology 45(3):395–416

Naz F (2016) Understanding human well-being: how could Sen's capabilities contribute? Forum Soc Econ. https://doi.org/10.1080/07360932.2016.1222947

Nelson JA (2010) Care ethics and markets: a view from feminist economics. Tufts University, Medford

Nussbaum M (2000) Women and human development. The capability approach. Cambridge University Press, Cambridge

Nussbaum MC (2003) Capabilities as fundamental entitlements: Sen and social justice. Fem Econ 9(2–3):33–59

Nussbaum M (2011) Creating capabilities: the human development approach. Harvard University Press, Cambridge

Pelenca J, Bazilec D, Cerutid C (2015) Collective capability and collective agency for sustainability: a case study. Ecol Econ 118:226–239. https://doi.org/10.1016/j.ecolecon.2015.07.001

Pigou AC (1912) Wealth and welfare. Macmillan, London

Pigou AC (1920) The economics of welfare. Macmillan, London

Pressman S, Summerfield G (2000) The economic contributions of Amartya Sen. Rev Polit Econ 12:89–113

Putnam H (2002) The collapse of the fact/value dichotomy and other essays. Harvard University Press, Cambridge

Qizilbash M (2005) Sen on freedom and gender justice. Fem Econ 11:151–166. https://doi.org/10.1080/13545700500301551

Qizilbash M (2014) Are modern philosophical accounts of well-being excessively 'individualistic'? Int Rev Econ 61(2):173–189. https://doi.org/10.1007/s12232-014-0204-x

Rajapakse N (2015) Bringing ethics into the capitalist model: Amartya Sen's approach to economic theory and financial capitalism. Revue LISA/LISA e-journal. https://doi.org/10.4000/lisa.8233

Razavi S (1996) Excess female mortality: an indicator or female insubordination? A note drawing on village-level evidence from South Eastern Iran. Politeia 12(43–44):79–96

Robbins L (1935) An essay on the nature and significance of economic science, 2nd edn. Macmillan, London

Robeyns I (2003) Sen's capability approach and gender inequality: selecting relevant capabilities. Fem Econ 9(2–3):61–92. https://doi.org/10.1080/1354570022000078024

Robeyns I (2005) The capability approach: a theoretical survey. J Hum Dev 6(1):93–117. https://doi.org/10.1080/146498805200034266

Robeyns I (2008) Sen's capability approach and feminist concerns. In: Comim F, Qizilbash M, Alkire S (eds) The capability approach: concepts, measures and applications. Cambridge University Press, Cambridge, pp 82–104

Robeyns I (2009) The capability approach. In: Peil J, Van Staveren I (eds) The handbook of ethics and economics. Edward Elgar Publishing, Cheltenham/Northampton, pp 39–46

Robeyns I (2017) Wellbeing, freedom and social justice: the capability approach re-examined. Open Book Publishers, Cambridge

Sayer A (2000) Moral economy and political economy. Stud Polit Econ Spring:79–103

Sayer A (2001) For a critical cultural political economy. Antipode Why Things Matter to People: Social Science, Values and Ethical Life. Cambridge University Press 33:687–708

Sayer A (2004) Moral economy. Department of Sociology, Lancaster University, Lancaster. http://www.comp.lancs.ac.uk/sociology/papers/sayer-moral-economy.pdf

Sayer A (2007) Moral economy as critique. New Political Econ 12(2):261–270

Sayer A (2011) Why things matter to people: social science, values and ethical life. Cambridge University Press

Schildberg C (2014) A caring and sustainable economy. A concept note from a feminist perspective. http://library.fes.de/pdf-files/iez/10809.pdf. 08 Dec 2019

Schmid HB (2007) Beyond self-goal choice: Amartya Sen's analysis of the structure of commitment and the role of shared desires. In: Peter F, Schmid HB (eds) Rationality and commitment. Oxford University Press, pp 211–226

Sen A (1970) Collective choice and social welfare. Holden-Day, San Francisco

Sen A (1981) Poverty and famine: an essay on entitlement and deprivations. Clarendon Press, Oxford

Sen AK (1982) Choice, welfare and measurement. Basil Blackwell, Oxford

Sen A (1983) Development: which way now? Econ J 93:745–776

Sen A (1985) Commodities and capabilities. North-Holland, Amsterdam

Sen A (1987) On ethics and economics. Basil Blackwell, Oxford

Sen A (1988) The concept of development. In: Chenery H, Srinivason TN (eds) Handbook of development economics, vol 1. Elsevier Science Publishers B.V., Amsterdam, pp 9–26

Sen A (1989) Development as capability expansion. J Dev Plan 19:41–58

Sen A (1990) Gender and cooperative conflict. In: Tinker I (ed) Persistent inequalities. Oxford University Press, New York, pp 123–149

Sen A (1992) Inequality reexamined. Oxford University Press, Oxford

Sen A (1993) Capability and wellbeing. In: Nussbaum M, Sen AK (eds) The quality of life. Oxford University Press, Oxford, pp 30–53

Sen A (1995) The political economy of targeting. In: van de Walle D, Neat K (eds) Public spending and the poor. John Hopkins University Press for the World Bank, Baltimore, pp 11–24

Sen AK (1996) On the foundations of welfare economics: utility, capability, and practical reason. In: Farina F, Hahn F, Vannucci S (eds) Ethics, rationality, and economic behaviour. Clarendon Press, Oxford

Sen A (1998) Mortality as an indicator of economic success and failure. Econ J 108(446):1–25
Sen A (1999a) Development as freedom. Oxford University Press, Oxford
Sen A (1999b) Rational fools: a critique of behavioural foundations of economic theory. In: Sen AK (ed) Choice, welfare, and measurement, Paperback edn. Harvard University Press, Cambridge, pp 84–106
Sen A (2002) Rationality and freedom. Harvard University Press, Cambridge
Sen AK (2003) Sraffa, Wittgenstein, and Gramsci. J Econ Lit 41:1240–1255
Sidgwick H (1883) The principles of political economy. Macmillan, London
Stewart F (2005) Groups and capabilities. J Hum Dev 6(2):185–204. https://doi.org/10.1080/14649880500120517
Stewart F, Deneulin S (2002) Amartya Sen's contribution to development thinking. Stud Comp Int Dev 37(2):61–70. https://doi.org/10.1007/BF02686262
Stiglitz J, Sen A, Fitoussi J (2009) Report by the Commission on the Measurement of Economic Performance and Social Progress. https://www.insee.fr/en/information/2662494. Accessed 25 Dec 2020
Thompson EP (1961) The long revolution (part II). New Left Rev I(10):34–39
Thompson EP (1971) The moral economy of the English crowd in the eighteenth century. Past Present 50:76–136
Thorbecke E (2019) The history and evolution of the development doctrine, 1950–2017. In: Nissanke M, Ocampo JA (eds) The Palgrave handbook of development economics: critical reflection on development and economics. Palgrave Macmillan, Cham, pp 61–108
Tronto J (1993) Moral boundaries. A political argument for an ethics of care. Routledge Veil, New York/London
UNDP (2011) Human development report: sustainability and equity. Palgrave Macmillan, New York
UNDP (1990) Human Development Report 1990. New York: Oxford University Press
UNDP (2020) COVID-19 and human development: assessing the crisis, envisioning the recovery. http://hdr.undp.org/en/hdp-covid. Accessed 9 Jan 2021
Walker J, Berekashvili N, Lomidze N (2014) Valuing time: time use survey, the capability approach, and gender analysis. J Hum Dev Capab 15(1):47–59
Zigon J (2007) Moral breakdown and the ethical demand: a theoretical framework for an anthropology of moralities. Anthropol Theory 7(2):131–150
Zimmermann B (2006) Pragmatism and the capability approach: challenges in social theory and empirical research. Eur J Soc Theory 9:467–484. https://doi.org/10.1177/1368431006073014

Part X

History of Ethnography and Ethnology

History of Ethnography and Ethnology: Section Introduction

Bican Polat

Abstract

This chapter is a section introduction and provides a brief analytic summary of each chapter included in the section. The chapters introduce the history of ethnography and ethnology against the backdrop of a wide range of contexts, from nation-building efforts and imperial projects to transnational specimen exchange networks and foundation-sponsored research initiatives. The section as a whole aims to explore the development of ethnological knowledge from a transnational perspective, situating the singularity of each particular ethnographic tradition with an eye toward cross-border flows and global entanglements.

Keywords

Ethnology · Ethnography · Transnational history · Global intellectual history

The emergence of ethnographic and ethnological knowledges in the Age of Enlightenment was one of the key events in the history of the human sciences, a watershed that indelibly racialized their future development. The expansion of European trade and colonization from the Age of Discovery onwards had led to various kinds of asymmetrical encounters between European and non-European peoples. As various parts of the globe were integrated into the capitalist economy during these encounters, written accounts about the non-European world proliferated in Europe. From early travelogues to more systematic writings based on geographical expeditions, knowledges accumulated over time soon gave rise to a formal discourse that expanded Europe's horizons beyond its geographical boundaries and ethnocentric limits. Integration of these knowledges into existing epistemic structures resulted in critical transformations in the European human sciences ranging from the birth of

B. Polat (✉)
Tsinghua University, Beijing, China
e-mail: bicanpolat@mail.tsinghua.edu.cn

comparative studies of language and religion to the advent of the stadial theories of history. Over the course of the eighteenth century, "ethnography" thus came to refer to empirical studies of particular peoples and their traditions while "ethnology" was taken to signify the general and comparative study of humanity in all its particularities (Vermeulen 2015)

This section covers the history of ethnographic and ethnological knowledges in Europe and beyond from the second half of the eighteenth century to the interwar period. The individual chapters of the section examine the development of distinct knowledge traditions against the backdrop of a wide range of contexts, from nation-building efforts and imperial projects to transnational specimen exchange networks and foundation-sponsored research initiatives. Taken together, these chapters thus explore the transformation of ethnographic and ethnological discourses with a focus on cross-border flows and global entanglements, paying equal attention to developments in and interchanges between "central" ethnographic traditions that arose in English- and German-speaking countries and "peripheral" traditions that evolved in countries such as Japan and Russia.

Ildikó S. Kristóf situates the Hungarian tradition of ethnography within a broader cultural context, examining how it developed in the late eighteenth century at the intersection of the cultural-political discourses that arose when Europeans were gaining political control over non-Western cultures. Kristóf's chapter tracks the emergence of "world ethnography" or "universal ethnography" in the Kingdom of Hungary between 1760 and 1830, identifying the scholarly networks that forged it on the basis of textual resources imported from a primarily Germanic geographical and natural history tradition. According to Kristóf, three principal networks paved the path for a Hungarian style of world ethnography from the 1760s onwards: (1) the Jesuit missionaries based at the Academy of Trnava who used their almanacs for publicizing geographical and ethnographic knowledge gained from their overseas activity in Africa, America, and Asia, (2) the Lutheran scholars in Bratislava/Breslau who imported a wide variety of Western European travelogues and accounts of discoveries to the Kingdom of Hungary and began publishing a number of geography books from the 1810s onwards, and (3) a loosely organized circle of individual pastors who served the Reformed Church of Hungary and were involved in the translation into vernacular Hungarian of principal Western European scientific works in natural history, geography, and the emerging science of world ethnography.

Kristóf pays attention to the Eurocentric bias that seeped into the print literature of the period, documenting a wide diversity of influences that informed how those texts represented non-European peoples, ranging from the discursive legacies of the classical Greek and Roman tradition to the age-old rhetorical stereotypes used for depicting and classifying aboriginal peoples (such as demonization, hierarchization, animalization, barbarization, and exoticization). She also highlights the political significance of this cultural-intellectual movement by placing it within the context of the Hungarian national awakening. Underlining the transnational underpinnings of this movement (which was spearheaded primarily by Hungarian Protestant scholars trained in the German universities such as Göttingen, Jena, and Halle), Kristóf's contribution shows how the authors and publishers of the period self-consciously engaged in a scientific revolt against the existing Catholic, imperial, and absolutist science of the Habsburg monarchy and strove to create vernacular scientific discourses modelled on the Protestant trends prevailing in German universities. It was from within this intellectual milieu that anthropological discourse in Hungary emerged in the form of world ethnography and exerted its influence on the first generation of Hungarian fieldworkers who flourished in the aftermath of the 1848 Revolution.

Brooke Penaloza-Patzak, likewise, adopts a transnational approach in her chapter, reconstructing the disciplinary development of early ethnology within the context of trans-Atlantic networks that were being built during the second half of the nineteenth century. She explores the development and institutionalization of anthropology in the United States and German and Austro-Hungarian Empires with a focus on the specimen exchange networks formed by highly heterogenous groups of scholars and a wide diversity of ethnographic object collections. To highlight the interdisciplinary character of knowledge production in this period, her account begins by identifying the specific contributions of several different fields of expertise to the creation of a shared disciplinary perspective. In pre-twentieth century anthropology, the raw material for ethnographic analysis was often brought together by practitioners who drew on their linguistic and cultural knowledge about a particular region and expertise in navigation and trading. Objects obtained by these agents of cultural exchange were then analyzed by other practitioners trained in the natural sciences and medicine, including Adolf Bastian (1826–1905) and Franz Boas (1858–1942), who sought to subject those specimens to an empirical analysis and make inferences about their significance for the scientific study of human societies. Penaloza-Patzak shows how the knowledge produced within these interprofessional encounters was increasingly mobilized through the work of scholarly societies that were established in major European and US metropoles such as St. Petersburg, Halle, Berlin, Vienna, Philadelphia, and Washington, DC, especially from the 1860s onwards. The transactions and other kinds of publications produced by these societies served as tools for initiating scholarly traffic across national borders while also paving the path for transnational specimen exchange networks long before the creation of dedicated ethnographic departments and museums in the 1880s.

Based on both quantitative and qualitative methods, Penaloza-Patzak illustrates the role of Berlin museum's specimen collections from the Pacific Northwest in the creation of an exchange network that linked prominent practitioners in Berlin, Vienna, and the United States. Making use of records relating to correspondence, publications, and specimen exchanges, she thus explores the interpersonal and institutional alliances that practitioners including Bastian and Boas helped forge across national and disciplinary borders.

João Leal also explores cross-national transfer of anthropological knowledge in his chapter, with a particular attention to how such epistemic transfers intersected with the nation building process. His contribution tracks the historical development of Portuguese ethnography from the 1870s to the 1960s in relation to both the country's national political concerns and the theoretical trends that prevailed in Central European anthropology. Leal begins his analysis by defining Portuguese anthropology as a "nation-building anthropology," asserting that it was primarily developed through studies of rural folk tradition (*Volkskunde*) rather than studies of other peoples (*Völkerkunde*), the ethnographic style that evolved in France and England. He then goes on to show how Portuguese ethnography began its journey with the goal of constructing ethno-genealogical arguments that would buttress a civic and territorial discourse on national identity. Leal identifies four distinct phases in the development of this ethnographic nationalism and shows how in each phase Portuguese anthropologists selectively integrated within their ethnographies the theoretical models that originated in Central Europe (and later in the United States). He thus illustrates how central anthropological theories were filtered by Portuguese ethnographers to address shifting national concerns over the course of a series of exercises in the "ethnographic imagination" of the nation.

In the 1870s and 1880s, Portuguese ethnographers imagined the nation as a homogeneous cultural entity with no internal divisions and began constructing an ethno-genetic discourse to link the country's existing folk traditions to remote ethnic origins. During this first phase, ethnographers drew on Max Müller's comparative mythology and pre-evolutionist diffusionist theories to restore the ancient character of Portuguese culture. In the 1900s, ethnographic representations began to portray Portugal as a more heterogeneous entity, whose internal diversity and regional differences were studied from a more diversified research perspective, which included traditional material culture, folk art, and the social and economic organization of rural communities. The earlier romantic enthusiasm for the Portuguese folk was replaced in this phase with an evolutionist approach that framed folk culture as the expression of the country's fall from past glory. Evolutionism allowed Portuguese ethnographers to establish an ideological equation between the non-European "savages" and the Portuguese peasantry, providing the rhetorical tools that helped classify its customs as "primitive" or "barbarian." In the 1910s and 1920s, a new optimism about Portugal's destiny started to take root in ethnographic writings, giving rise to representations of folk culture as the manifestation of the nation's genius. Ethnographers began to project an artificially homogeneous image of the country during this time and described different local and regional realities as the minor and mutually interchangeable expressions of Portugal's "essence." Leal notes that the pattern of relationship that existed between center and periphery in the earlier two phases changed during this phase, whereby the previous openness toward theoretical trends in Central European anthropology was replaced with an insular attitude. Theory was considered irrelevant as the earlier ethno-genetic models of explanation were supplanted with a strong descriptive orientation that focused on celebrating (rather than explaining) the Portuguese folk.

During what Leal considers the fourth phase of Portuguese anthropology, which spanned from 1930 to 1960, the main anthropologists then active portrayed the country as a complex reality encompassing both unity and diversity and qualified its "national character" as paradoxical – a stance epitomized in Jorge Dias' work that integrated then-obsolete diffusionism with the culture and personality school of American anthropology. Leal shows how this syncretic theoretical orientation enabled Dias and his contemporaries to reconstitute a pluralist ethno-genealogy of the nation while also allowing them to redefine national identity in terms of the presumed particularities of the Portuguese temperament. Based on an analysis of long-term trends, Leal's chapter thus illustrates how central anthropological theories were selectively appropriated by Portuguese ethnographers with respect to their usefulness in addressing national concerns, also paying attention to the novel meanings and uses these theories acquired as they were transformed to comply with local demands.

Sergei Alymov also explores the intersections between nationalism and ethnographic theory in his chapter, but instead of examining the cross-national transfer of long-term trends between center and periphery, he focuses on the evolution of a single concept against the backdrop of arguably the most politically punctuated peripheral tradition of ethnography. Alymov reconstructs the development of the concept of ethnos from its origins in the turbulent political context of the final years of the Russian Empire through its transmutations within the Soviet scientific establishment to its gradual disappearance following the Perestroika. His reconstruction begins with the earliest articulations of this concept within the nationalist agendas of Ukrainian ethnographers such as Fedor K. Volkov and Nikolai Mogilyansky, who deployed it to denote the ethnic-national units living under the sovereignty of the Russian Empire and naturalize their differences in a positivist research framework. Inspired by the late nineteenth century vision of anthropological science developed at the École d'Anthropologie in Paris, these Saint Petersburg academics defined anthropology as a natural science devoted to investigations of human anatomy and types in general and ethnology as the study of particular races on the basis of their physical, linguistic, and cultural traits. The concept of ethnos was first formulated within this pre-Soviet intellectual milieu, designating a theoretical category that allowed for inquiries into the differences between the Great and Little Russians and the particularities of a distinct homogenous Ukrainian type.

This naturalistic category was then incorporated by Sergei Rudenko into his biosocial program where it was elaborated on the basis of correlations drawn between cultural and physical types. Alymov notes how the Great Break (1929) hampered further development of the theory of ethnos in its biosocial form and prevented ethnographers from weaving links between social structures and features investigated in physical anthropology. The Marxist ethnographic literature developed during Stalinism only allowed for the formulation of "sociological" or historical concepts to explain society and demanded autonomy of social laws from natural laws. Alongside the Rudenko's program, Alymov also tracks the development of an alternative interpretation of the ethnos concept through the work of Sergei M. Shirokogoroff, who strove to reconfigure the role of ethnology within a humanistic framework, defining it as a unifying discipline that combined ethnography (study of culture), physical anthropology, and linguistics. Shirokogoroff sought to explain the laws that governed the growth and interaction of different ethnoses by investigating their social and material cultures and ecological adaptations while at the same time studying the continuity of physical types within particular groups in spite of linguistic and cultural assimilation.

Alymov's analysis highlights the uneasy relationship the Soviet ethnos-thinking had with the official theory of the nation that was supported by Stalin until the 1950s, while also exploring the debates that took place in the 1960s when a more relaxed intellectual climate enabled both a revival of interest in the theory of ethnos and a new synthesis between the natural and the social sciences. This new phase in the history of Soviet ethnography saw an increased emphasis on the notion of ethnogenesis, the study of ethnic origins of modern peoples. Influential scholars of the period such as Yulian V. Bromley and Lev N. Gumilyov agreed on the need to study ethnoses as the historically formed communities of people bound together by their geography, language, and culture, while at the same time clashing over whether to reconstruct their genesis in terms of social or natural processes. At any rate, in the 1970s and the 1980s, the ethnos-thinking was expanded to include new formulations based on theoretical advances (such as cybernetic theory) and methodological innovations (such as integration of demographic data on marriages to demonstrate the statistical validity of endogamy as the defining feature of the ethnos and its stability). Against the naturalistic tendencies gaining foothold in this period, Mikhail Kryukov and his colleagues promoted the notion of "ethnic consciousness," while also shifting the substantive focus of the field toward investigations of the Chinese ethnos in a multivolume historical study. During the last years of the Soviet Union, other important intellectual battles were fought between the proponents of the ethnos theory and the Marxists experts on the national question, who strove to downplay the importance of ethnic traditions and maintain the primacy of "the social" over "the ethnic." Despite the latter's ideological efforts, problems of ethnic identity were increasingly raised during the Perestroika and eventually led to what Alymov calls the unmaking of the ethnos theory in the aftermath of the Union's dissolution. Placing this intellectual demise within the context of the political conflicts of the post-Soviet "ethnic renaissance," Alymov's chapter attributes the fall of the ethnos theory to the inability of its long durée/diachronic perspective to deal with nascent ethno-political problems and concludes its century-long history by noting its eventual replacement by a synchronic research agenda, which applied constructivist and instrumentalist approaches to the study of ethnicity.

Shingo Iitaka's chapter focuses on the history of another peripheral ethnographic tradition that had however reached a global significance due to its entanglement with an imperial agenda. Iitaka examines the development of Japanese ethnography in Micronesia between 1914 and 1944, when this vast constellation of small islands was under the administrative control of the Empire of Japan. His contribution tracks the production of ethnographic knowledge by a loose network of Japanese merchants, navy officers, colonial administrators, professional scholars, and amateur writers in conformity with the administrative objectives set by the South Seas Government (*Nan'yō-chō*), the government agency charged with the task of managing the Japanese colonies in Micronesia until its dissolution at the end of the Pacific War (1941–1945).

The Japanese administration in Micronesia militarized the occupied islands and encouraged immigration from Japan while also promoting economic development through exploitation of the islands' natural and human resources. However, it did not put in place any organized research programs comparable to the ones launched by the Germans and the Americans. Iitaka shows how Japanese ethnographies of

Micronesia were produced in a practical research context directly shaped by colonial policies, which were designed to satisfy the administration's informational demands. A series of surveys on customary laws began to be conducted from 1925 onwards in order to provide the Japanese jurists with knowledge to adjudicate Micronesian lawsuits. Compiled as *Reports of Customary Laws in the South Sea Islands*, these surveys made available a wide range of practical information, such as the origins and legends of kin groups, the hierarchies of traditional chiefs, and comparisons of kinship categories between the Japanese and the Micronesians. These surveys were followed by studies of land tenure that aimed to identify and manage land ownership among individuals and kin groups. From the 1930s, the earlier surveys of custom and land tenure left their place to more systematic ecological and ethnological expeditions that were funded by the South Seas Development Company (*Nan'yō Kōhatsu*), which controlled the islands' sugarcane industry. During this time, Micronesian ethnographies were disseminated through the South Sea Islands Cultural Society (*Nan'yō Guntō Bunka Kyōkai*) and its publication outlet *Nan'yō Guntō*. Conducted by scholars trained in important metropolitan institutions (such as Tokyo Imperial University, Kyoto Imperial University, and the Institute of Pacific Relations), these investigations helped generate knowledge central to the administration of the Micronesian colonies while also allowing for the application of European academic perspectives to the study of practical administrative problems. Commissioned by the *Nan'yō-chō*, the interwar Japanese ethnographies brought the theoretical categories of British functionalism and German legal studies to bear on problems of interpreting the local data generated through surveys of land and customs.

The focus of these Micronesian ethnographies soon expanded to include a wider set of problems, ranging from the vanishing songs and dances of the islands to the rapid social changes indigenous groups faced due to the influx of Japanese immigrants. Iitaka's rich account pays attention to the nuanced approaches Japanese investigators adopted to respond to the colonial situation. In addition to examining the works produced by jurists such as Zennosuke Nagakawa and anthropologists such as Ken'ichi Sugiura, he examines the critical contributions made to the ethnographic imagination of Micronesia by artists such as Hisakatsu Hijikata, who not only produced rich illustrations of the material culture of the islands but also taught Micronesians wood-carving skills during his stay in Palau.

Dennis Bryson's chapter shifts the focus towards the development of a "central" ethnological tradition between the two world wars. Bryson examines the formation of the culture and personality approach in interwar American anthropology at the intersection of foundation-sponsored research and the style of ethnographic inquiry forged by Franz Boas. Bryson's reconstruction of this noted school of anthropology highlights the two key factors that played an important role in its consolidation from the 1920s onwards: a new series of investigative themes promoted by the nascent federal research bodies (such as the Social Science Research Council and National Research Council which were heavily funded by the Rockefeller philanthropies) and the investigative trends evolving in Boasian circles, especially through the work of Ruth Benedict, Margaret Mead, and Edward Sapir.

Key to the development of foundation-sponsored research in the interwar period were two influential administrators, Robert S. Lynd and Lawrence K. Frank, who guided the research agendas of the SSRC and the Rockefeller boards, respectively. The new epistemic vision set by these administrators involved the promotion of interdisciplinary social science research and the systematic inquiries into a new series of topics, including child development and the relationship between culture and personality. To achieve the goals promoted by these interdisciplinary research initiatives, the Boasians elaborated the notion of culture they had received from their mentor and synthesized it with the notion of personality, in the process articulating a more holistic and integrated approach to culture. Bryson's account explores both the similarities and differences in the perspectives adopted by Benedict, Mead, and Sapir while they were striving to bring their ethnological expertise to bear on problems of explaining how culture influenced the formation of personality.

Articulating the notion of cultural configuration, Benedict expanded the Boasian idea of culture as an integrated and patterned totality and elaborated on how individuals were "conditioned" by the cultural patterning of their group. However, she was less interested in examining the specifics of the conditioning process with respect to child-rearing and educational methods – a theme taken up by Mead in a series of ethnographies that integrated psychodynamic interpretations of personality formation with the Boasian approach to cultural totalities. It was primarily through a set of theoretical writings published by Sapir during the 1920s and 1930s that this culturalist outlook was expanded to include problems about understanding the relationship of the individual to society. According to Sapir, the individual was not a passive recipient of cultural beliefs and practices but responded to them in idiosyncratic ways, sometimes resisting cultural patterns and even shaping them.

Bryson's contribution combines a close attention to the writings produced by the culture and personality advocates with an analysis of overarching categories prevalent in American social science during the interwar years. According to Bryson, the social scientists' increased concern with the culture concept from the 1920s onwards and their attempt to integrate it within studies of personality were part and parcel of a social engineering agenda that sought to reform the American population by way of fostering normalized and cooperative personalities. He further emphasizes how this hybrid investigative style responded to then existing tensions in American society (including increasing economic inequality, conflict between traditionalist and modernist cultural trends, ascendance of individualistic and materialist values) in a novel way: reconfiguring the social as cultural pattern, this anthropological movement helped replace a view of historically articulated structures of power and domination within society with one that focused on cultural beliefs and practices.

References

Vermeulen HF (2015) Before Boas: the genesis of ethnography and ethnology in the German Enlightenment. The University of Nebraska Press, Lincoln/London

Before Fieldwork: Textual and Visual Stereotypes of Indigenous Peoples and the Emergence of World Ethnography in Hungary in the Seventeenth to Nineteenth Centuries

Ildikó S. Kristóf

Contents

Introduction: Archives as a "Field" for a Historian of Anthropology	1622
Euro-American Eyes, Indigenous Eyes, and Hungarian Eyes: Questions of Research Methodology	1623
Global Ethnography in Hungary: An Entangled History	1624
Three Examples of the Representation of Indigenous Peoples: Demonization, Hierarchization/Barbarization, and Exoticization	1629
Local Contexts: Jesuits, Lutherans, and Calvinists as Agents of World Ethnography	1634
Notions of Geography: Notions of Hierarchy – Examples of Siberia and the Arctic Region	1640
Conclusion or the Afterlife of Stereotypes: Orientalism or Cultural Colonialism?	1645
References	1646

Abstract

This chapter explores the emergence of a discourse on world ethnography in the Kingdom of Hungary between the late seventeenth and the early nineteenth century. The author regards archives as a "field" for a historian of anthropology and elaborates on three main points: first, the principal agents of the so-called "world ethnography" in local, Jesuit, Lutheran, and Calvinist contexts; second, the respective historical sources that resulted from their work, i.e., missionary accounts, travelogues, (school)books of geography, and (school)books of natural history; and third, cultural stereotypes occuring in both texts and images, and relating to non-European indigenous peoples, for example, those of America, Asia, and Oceania. Examining the rise of global ethnography in Hungary as an entangled history, this chapter presents three detailed examples of the representation of indigenous peoples: demonization, hierarchization/barbarization, and exoticization. Demonstrating the Eurocentric background of Enlightenment ideas

---

I. S. Kristóf (✉)
Institute of Ethnology, Hungarian Academy of Sciences, Budapest, Hungary

© The Author(s), under exclusive licence to Springer Nature Singapore Pte Ltd. 2022
D. McCallum (ed.), *The Palgrave Handbook of the History of Human Sciences*,
https://doi.org/10.1007/978-981-16-7255-2_107

like that of *savagery – barbarism – civilization*, the chapter analyzes stereotypes relating to American Indians, Asian peoples (especially, the Chinese and the Samoyed), Polar peoples (the Greenland Inuit and the Sámi), and the Aborigins of Australia and Oceania.

Keywords

History of anthropology · History of ethnography · Representation of indigenous peoples · Cultural colonialism · Cultural history of Hungary · Visual stereotypes · Demonization · Barbarization

Introduction: Archives as a "Field" for a Historian of Anthropology

The emergence of world ethnography – an ethnography of universal scope – as part of the history of sociocultural anthropology can be studied in archives and rare book collections. These places constitute the "field" for a historian of anthropology. The following example illustrates very well how and in what particular ways. *Geographiae Particularis Epitome*, a geography textbook published in Vienna in 1727 and also familiar in the Kingdom of Hungary, surveyed the characteristics of the then-"known" parts of the world, i.e., the *four* continents. The main characteristics of each continent were provided in brief Latin phrases which were included in a formalized visual structure. The structure apparently served as an instrument of *ars memoriae* (or *ars memorativa*, the art of memory) to help the reader to imagine and also to memorize the faraway, non-European lands and regions of the world. This art of memory was based on the visual shape of the lands, as well as on certain ideas and concepts that readers – European students – could/should associate with them. According to the text, the continent(s) of America could be imagined as a set of *two triangles*, while Asia could be visualized as a *Tartar (or Tatar) riding a horse*. The horse was supposed to be imagined in a sitting position so that the head of the Tatar was Anatolia and the right hindleg of the horse was the Kamchatka Peninsula. The creature lay across the entire continent of Asia, and the parts of his body were to be used to memorize the different regions and countries there (*Geographiae* 1727).

This example tells us that the European geography literature before 1848 was characterized by a certain conceptual, ideological – and frequently anthropomorphic – approach (Stagl 1995, 155–170). The Kingdom of Hungary as part of the Habsburg Empire between 1697 and 1918 was no exception. This, however, is not a well-researched aspect of the cultural history of that country. This chapter will demonstrate in this study that an archival approach to the history of ethnography is worth attempting in Hungary even if it impacts on both the temporal (chronological) and material (spatial) aspects of the survey that could be written about it. In fact, it would also impact on the possible research methodology to be employed.

Euro-American Eyes, Indigenous Eyes, and Hungarian Eyes: Questions of Research Methodology

The *history of travel* and the *history of anthropology* have fertilized one another in various conceptual and methodological ways in Europe since the 1980s/1990s. Being originally the products of ambiguous sociohistorical processes, sometimes colonialism itself, they first had to identify their roots to move closer to their own subjects, namely non-European indigenous peoples. From the 1970s, a number of Euro-American researchers started criticizing the Eurocentric orientation in anthropology, history, historiography, etc. from a perspective which was more and more frequently called *postcolonial* (Clifford and Marcus 1986; Pratt 1992; Thomas 1994). Somewhat later, indigenous thinkers and scholars – Māori, Sámi, Native American, and other scholars – themselves started instructing their European colleagues to identify how Euro-American culture historically gained a dominant position over indigenous cultures and how it provided (and in many cases still provides) a master discourse for the representation of the latter (Said 1978; Mihesuah 1998; Fixico 2003; Sz. Kristóf 2007, 2017c). Embedded in the broader cultural-political discourses of a Eurocentric interpretation of otherness in the nineteenth century – an interpretation which the emergent field of anthropology itself shared – "cultural areas" were constructed with peoples classified as "primitive" and "civilised" according to rather rigid, evolutionary, and imperial/colonial – but also earlier, Christian and hierarchical – views. Anthropologists such as Mary Louise Pratt started to regard and analyze those views as products of "imperial eyes" (Pratt 1992). Underlying the relevance of such an examination is the question of whether there existed only Euro-American, and especially Western European (British and French) "eyes", to depict non-European indigenous peoples (Said 1978), or whether there were also other, Central European/Central Eastern European "eyes" with their own respective interpretive potential. Do we find particular interpretive "eyes" in the German, Austrian, Russian, or, for that matter, Hungarian traditions of ethnography/anthropology?

The present chapter will discuss the "Hungarian eyes" only, but is indebted, however, to the approach of certain scholars working within other European anthropological traditions, especially the Spanish, Portugal, and German ones (Cañizares-Esguerra 2006; Rubiés 2007; Vermeulen 2015; Sárkány and Somlai, 2003, Sárkány 2018). As for the archival material relating to Hungary, it relies on the author's own research, carried out in various archives in Budapest (Hungary) in recent years: the University Library of the Eötvös Loránd University, a descendant of and successor to the old library of the Jesuit academy in Nagyszombat/Trnava/Tyrnau (in present-day Slovakia), founded in 1635; the Library of the Hungarian Academy of Sciences, founded in 1826; and the National Széchényi Library. Additional research was conducted in the Herzog August Bibliothek in Wolfenbüttel (Germany) and the Ethnological Collections of the Georg-August-Universität in Göttingen (Germany).

Global Ethnography in Hungary: An Entangled History

How did the field of anthropology – and especially its particular, earlier branch called "world ethnography" or "universal ethnography" – emerge in the Kingdom of Hungary before 1848? Figure 1 demonstrates the visual manifestation of such a global ethnography from the year 1805. It is a testament to the presence of a detailed knowledge of the indigenous peoples living in the "known" parts of the world, such as Greenland (1), Unalaska (2), Virginia (3), Patagonia (4), and Tierra del Fuego (5). The richly illustrated *Bilderbuch* was published in Vienna, having come out in a quadrilingual edition in German, French, Latin, and Hungarian (Bertuch 1805). As the pictures tell us, world ethnography was conceived in that period as a basically *descriptive* field. It seems to have been used as such not only by the author of the album, Friedrich Justin Bertuch (1747–1822), a German publisher and scholar of natural history from Weimar, but a little bit all over Europe. In that period of the late Enlightenment and early Romanticism, the field of anthropology appears to have used not so much an analytical, but a literally descriptive discourse attempting to identify the *external* elements of non-European cultures, i.e., the costumes, skin color, hairstyle, body ornamentation, forms of habitation, arms and tools of the indigenous peoples of the world, etc.

In the following, three main points on the emergence of such a discourse in the Kingdom of Hungary will be discussed. Who were the principal agents of "world ethnography" at the turn of the nineteenth century? What kind of historical sources do we have to shed light on their work? How would certain characteristics of their work relate to later ethnographic research on particular continents and their indigenous peoples, for example, America and/or Asia in the long durée?

These questions can only be answered by putting them into a broader perspective and by identifying more general sociocultural contexts for early anthropology in European history. The relevant general sociocultural contexts are the following. It is the period between approximately 1760/1770 and 1830 that the field of anthropology and world ethnography started emerging on our continent, in Western Europe as well as in the Central Eastern European region, including the Kingdom of Hungary (Marshall and Williams 1982; Thomas 2003; Liebersohn 2006; Lüsebrink 2006; Vermeulen 2015). The historical contexts were obviously different in many respects in the Western and Eastern parts, but there were some factors – social, political/ideological, and also scholarly/methodological factors – that seem to have been similar or even common, and without which we would not have this field at all.

A variety of primary sources and the related religious and/or scholarly discourses demonstrate that a huge, mosaic-like stock of knowledge of the indigenous peoples of Asia, the American continent(s), the Pacific area, and also the "the North" – or the "Arctic region" as it was called in the age – appeared in Europe between the 1760 and 1830. This knowledge was fairly detailed, it was rather ethnographic in character, and it also reached the Central Eastern region of Europe. The *Bilderbuch* by Bertuch noted above depicted a whole series of indigenous peoples from Asia, for example (see Fig. 2). The book consisted of images of peoples from India (1) and Siberia (2) – in the latter case the male figure is a Yakut, the female a Chukchee – in

Fig. 1 Peoples of America and Oceania. Fridrich Justin Bertuch: Természethistóriai képeskönyv az ifjúság hasznára és gyönyörködtetésére [Illustrated book of natural history for the use and amusement of the youth]. Bécs, 1805. (Courtesy of the Library of the Hungarian Academy of Sciences)

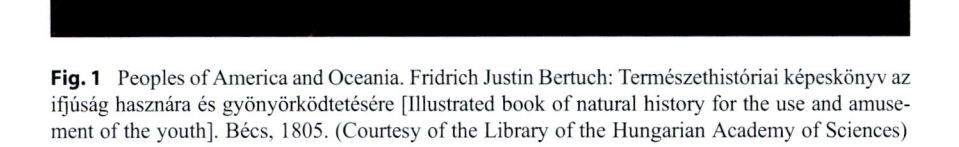

addition to Kalmyk Tatars (3) and Arabs (4). As the *Bilderbuch* demonstrates, the ethnographic knowledge of the period was mostly preserved in textual and visual/figural forms, but also in the form of objects. The latter were related to a number of

Fig. 2 Peoples of Asia. Fridrich Justin Bertuch: Természethistóriai képeskönyv az ifjúság hasznára és gyönyörködtetésére [Illustrated book of natural history for the use and amusement of the youth]. Bécs, 1805. (Courtesy of the Library of the Hungarian Academy of Sciences)

social/institutional contexts that tended to transform themselves in the period from private sociocultural milieus to common/collective ones, for example, from privately owned collections of aristocrats to museums and cabinets of study in the various universities of Europe. The surveys of excellent cultural historians, such as

Justin Stagl, Paula Findlen, and Han Vermeulen, have described those seventeenth- and eighteenth-century cultural trends identifiable in Western and certain Central European countries (Findlen 1994; Stagl 1995; Te Heesen and Spary 2001; Vermeulen 2015).

As for the appearance of such knowledge in the Kingdom of Hungary, this history has not yet been written. As for the period discussed here, the advent of ethnography and anthropology seems to have been embedded in the cultural program of the Enlightenment, just like in Western European countries, and it also appears to have been related to the discourses of what was called *useful literature* – travelogues, accounts of discoveries, geographies, natural histories, etc. – just like in Western Europe (Harbsmeier 1994; Stagl 1995; Rubiés 2007; Kontler 2001, 2014; Gurka 2003, 2010). Together with those discourses, the emergence of domestic and global ethnography in Hungary also seems to have been related to a certain extent to late eighteenth-century political science (Tóth 2015; Bodnár-Király 2017).

As for its conceptual background, the historical forms of an anthropological/ ethnological knowledge would seem to have been conveyed as well as infiltrated by a number of significant historical-philosophical ideas of the age, first and foremost by the Enlightenment concepts of *linearity* and *stadiality* according to which human societies evolve in history (Harris 1968; Stocking 1987; Sárkány and Somlai 2003; Bödeker et al. 2008). An illustration of such a linear and stadial development of the societies of the world can be seen in Fig. 3. This is an engraved image from a Hungarian edition of the work of Georg Christian Raff (1748–1788) which was originally entitled *Naturgeschichte für Kinder* (A Natural History for Children). The author was a professor of geography and natural history in the lyceum of Göttingen, and his work saw no fewer than three Hungarian translations in the period between 1779 and 1846 (Sz. Kristóf 2011, 2013). An idea of stadiality and graduality is discernible from the depiction of non-European societies in the images in the original book as well as in editions in different languages, among them, Hungarian. By means of the layout and the direction of reading, going from top to bottom and from left to right, Fig. 3 explains to the viewer how human societies are expected to develop from an imagined stage of *nature* to that of *savages* (gatherers), and from that to the stage of *non-European pagan civilizations*.

Other images in the book reveal similar pictorial stereotypes, a particular *ars memorativa* of the late eighteenth and early nineteenth century. Figure 3 clearly outlines a chain of successive stages of progress, a sequence of an imagined linear social formation. According to this pattern, the path of evolution starts at simple gathering societies, which consist of half-naked, fruit-picking "East Indians" and other simple fishing and hunting societies, such as the American Indians, the South African "Hottentots" or Khoisans, and the Northern European Sámi ("Lapps"), and it leads to a "semi-civilised" phase. This stage is represented by the depiction of China, for example: in the nineteenth-century editions of Raff's textbook, the images show ornate pagodas and richly decorated clothes and robes worn by tea-harvesting Chinese figures, as in Fig. 3. Through the apparently "less developed" societies of Eastern Europe that live closer to nature and animals, the evolution is shown to reach its culmination in the hard-working, industrialized

Fig. 3 The development of human societies. Raff György Keresztély: Természet História Gyermekek' számára. Második magyarítás. Ford. Vajda Péter [Natural history for children. Second Hungarian translation. Translated by Péter Vajda]. Kassa, 1837. (Courtesy of the National Széchenyi Library, Budapest)

Western European civilization with silkworm farming, whaling, sugar production, regular houses, ships, and weapons. The idea behind such representations is the lineage of an imagined progress from "savagery" to "barbarism," then "semi-barbarism" or "semi-civility," and, finally, "civilisation." Raff's images and visual scenes attempted to fix this lineage in the memory and thinking of the students and readers of the era. Most probably, Raff's textbook incorporated the available progressive theories of the age, such as those originating at Göttingen: that of Arnold Hermann Ludwig Heeren (1760–1842), August Ludwig von Schlözer (1735–1809) or, perhaps, William Robertson (1721–1793) etc. Raff's editors extracted and schematized some already simplified, vulgarized versions of those theories and organized them into easy-to-memorize arrangements, scenes, and images. The book does not specify which theories were used, nor does it mention any authors. However, with the various translations of *Naturgeschichte*, the images covered most of the European countries in the period, the Kingdom of Hungary included, so they had the potential to spread this stadial view of societies and cultures all over the European continent (Sz. Kristóf 2011, 2013).

Such ideas and visual arrangements seem to have exerted a long-lasting impact on the way(s) in which non-European indigenous peoples were imagined and represented in Europe, including Central Eastern Europe and Hungary. This is why the images of indigenous peoples always have an *entangled* history. They have always been embedded in the main political/ideological discourses as well as the representational conventions of the age. They never existed without such a discursive context. Any history of anthropology should account for those contexts, it should explore the social and also the conceptual frames of the representations in which those images were embedded. It should endeavor to shed light on their historical and local sociocultural background. A series of postcolonial examinations as well as critical, analytical, and reflexive studies have been carried out in that direction in travel history, for example, since the early 1990s (e.g. Barker, Hulme and Iversen 1994; During 1994; Thomas 1994; Cañizares-Esguerra 2006; Rubiés 2007). It is also worth applying their lessons to the Kingdom of Hungary.

Three Examples of the Representation of Indigenous Peoples: Demonization, Hierarchization/Barbarization, and Exoticization

As regards the sociocultural embeddedness of the representation of indigenous peoples, the Kingdom of Hungary was no exception. Let us see three examples, three different ways or sociocultural strategies, of representation from there.

A number of sixteenth- and seventeenth-century printed works available in the Kingdom of Hungary provided a demonized description and visual depiction of the American Indians. They paralleled and confirmed the views of the Catholic missionaries trained primarily at the Jesuit academy in Nagyszombat/Trnava/Tyrnau. Figure 4 shows a diabolical representation created like this by the Flemish artist Theodor de Bry (1528–1598) for a travelogue by the Italian Girolamo Benzoni (1519–1570) in the 1590s. The engraving illustrates very well how the Caribbean Indians were represented as heathens adoring the demons and/or the devil itself, according to Christian theological concepts. The works of De Bry, Benzoni, and many other early explorers of America were available in the old library of the Jesuit academy in Nagyszombat/Trnava/Tyrnau and in many other institutional and private libraries in seventeenth-century Hungary (Benzoni 1644; Sz. Kristóf 2012b, 2016a, b; Groesen 2008). Similar diabolical representations existed for Asia, for example, of the Chinese (see Fig. 5).

At the other end of the period discussed here around 1835, the Australian Aborigines were depicted as intriguing and mysterious, but rather brutish, savage creatures by

Fig. 4 American Indians diabolized. Hieronymus Benzoni: Historiae Antipodum sive Novi Orbis, qui vulgo Americae et Indiae Occidentalis nomine usurpatur [A history of the Antipods or, the New World which is commonly called America and West India]. Mattheus Merian, Frankfurt am Main, 1644. (Courtesy of the University Library of Eötvös Loránd University, Budapest)

Pál Almási Balogh (1794–1867), a Hungarian physician. Almási Balogh had never traveled to Australia, but he most probably read and used the texts and images from a series of contemporary Western European travelogues, which were available in his private library. His manuscript, entitled *Az ember Ausztráliában* (Man in Australia) and written in Hungarian, is held today in the University Library of the Eötvös Loránd University, Budapest. It consists of a total of 19 pages, and it is divided into the following units or sections: Specificities of the physical structure and living environment of the Australian Aborigines (untitled), "Religion" (approximately two-and-a-half pages), "Dwelling places" (approximately two pages), "Lifestyle" (approximately one page), "Marriage" (about five-and-a-half pages), "Superstitions" (about two-and-a-half pages), "Inclinations" (about two pages), "Clothing" (less than one page), and finally "Language" (slightly more than one page) (Sz. Kristóf 2014b: 129–130). Thus, Almási Balogh compiled a full general ethnography, a so-called ethnographic "profile" (Service 1958, 1963) of the inhabitants of the fifth continent.

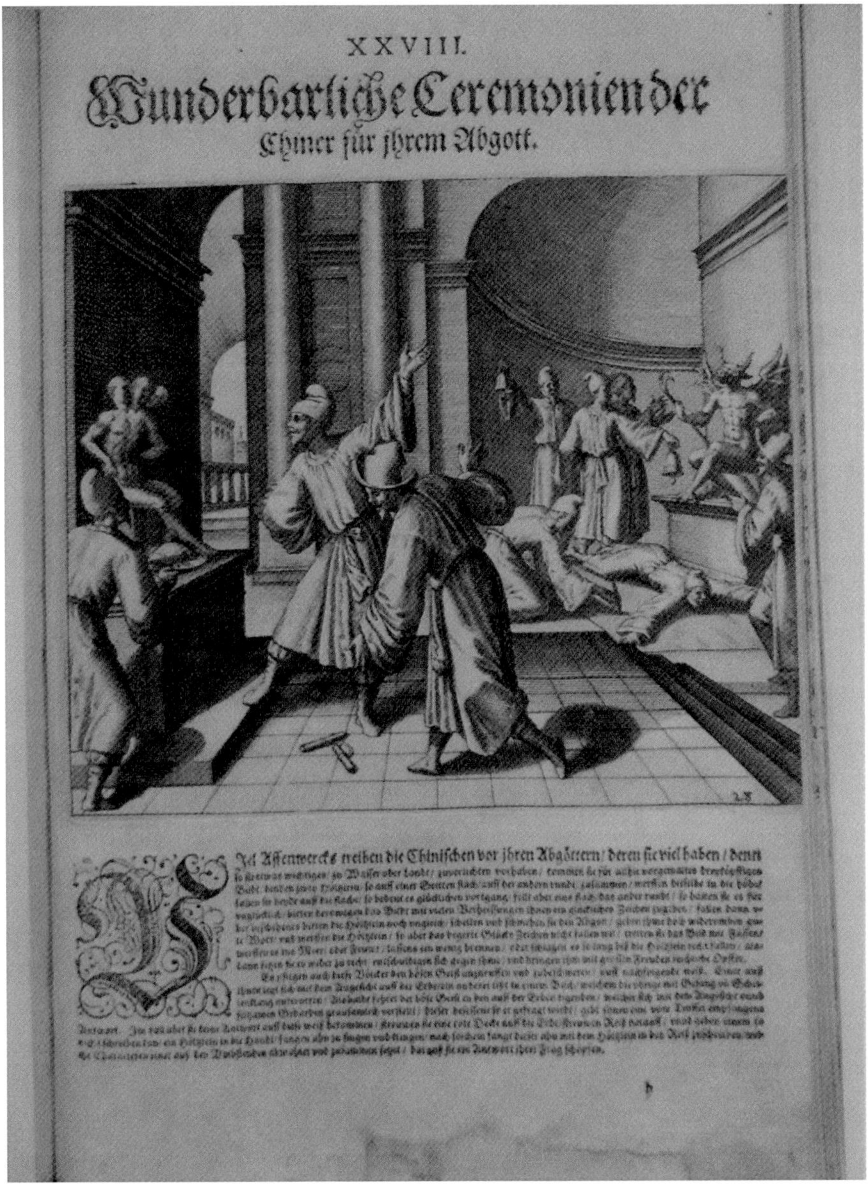

Fig. 5 Chinese rituals diabolized. Jan Huyghen van Linschoten: Ander Theil der Orientalischen Indien Von allen Völckern Insulen Meerporten fliessenden Wassern und anderen Orten.. [The other part of East India, about all the peoples, islands, harbours, running waters and other places..] hrsg [edited by] Hans Dieterich and Hans Israel von Bry. Johan Saur, Franckfurt am Meyn, 1599–1601. (Courtesy of the University Library of Eötvös Loránd University, Budapest)

A microphilological analysis reveals that the Hungarian physician actually compiled his manuscript from the works of James Cook (1728–1779), Allan Cunningham (1791–1839), David Collins (1756–1810), and other British and French travelers and scholars. One of his most important sources was a contemporary French collection of illustrated travelogues which were edited by Jules Dumont d'Urville (1790–1842) and saw several editions in French and German during the first half of the nineteenth century. Almási Balogh had two copies of this book (in German) in his private library (Sz. Kristóf 2014b). The works that he relied on as scholarly literature also impacted the way he spoke about the Aborigines. His manuscript conveyed to its Hungarian readers a version of the early Western *colonial approach* that interpreted the indigenous peoples as inferior in both physical and cultural aspects and treated them rather – although not altogether – disdainfully. Figure 6 is a drawing of an Aboriginal from the album edited by Dumond D'Urville and in the possession of Almási Balogh, which was designated "scary" for European – and Hungarian – eyes. However, Almási Balogh attributed the alien cultural characteristics of the Australian natives to the local environment, climate, and hardships of life, and he introduced "Benilong," i.e., Woollaware Bennelong (ca. 1764–1813), the famous elder of the Eora (Koori) people, to a Hungarian audience in long paragraphs (Sz. Kristóf 2014b).

Between the two ends of the chronology discussed here, the Greenland Inuit have been rather exoticized by the Reverend Mihály Dobosy (1780–1853), Hungarian translator of a mission account by David Cranz (1723–1777), a Moravian pastor. The original work of the latter entitled *Historie von Groenland* (History of Greenland) was published in German in Barby, Sweden, in 1765, and the Hungarian translation came out in Buda in 1810. Cranz's original work depicted the Inuit both textually and visually as kind of innocent children of nature, and the Hungarian adaptation preserved that idealizing tone very well. It even added to it in exoticizing "the North" and its clever indigenous people (Cranz 1765; Cranz 1810; Sz. Kristóf 2016a).

Mihály Dobosy was a highly educated pastor in the countryside in the service of the Reformed (Calvinist) Church, who knew and used both German and French. In 1805 he spent a year at the University of Göttingen, and this experience had a great significance for him as well as for the history of anthropology in Hungary. As he says in the *Foreword* of his translation of Cranz, he was so influenced by the geography courses he took in Göttingen and certain "objects" he saw in the university's natural history cabinet that he decided to establish this novel "ethnographic geography" approach after returning to Hungary. And he did so. After his return, he began to translate and became an excellent interpreter of the new scholarly literature of the era, published especially in German, on travel and geographical and ethnographic discoveries. His attitude toward the Greenland Inuit was rather positive: He expressed his gratitude to Providence that he was able to "encounter" such simple and innocent people (Sz. Kristóf 2016a). Figure 7 is an image of the *kayak* and *umiak* of the Inuit of Greenland from the Hungarian translation of Cranz.

Demonization, hierarchization, barbarization, and exoticization – these were the principal ways of religious/colonial interpretation and representation of cultural otherness in early modern Europe, including the Kingdom of Hungary. They were

Fig. 6 An Aborigin of Australia. Jules Dumont d'Urville et al. (hrsg.): Malerische Reise um die Welt. Verfaßt von einer Gesellschaft Reisender und Gelehrter. Ins Deutsche übertragen... von Dr. A. Diezmann, Band II. [Picturesque voyages around the world. Made by a society of travelers and scholars, volume II]. Leipzig: Industrie-Comptoir (Baumgärtner), 1835. (Courtesy of the University Library of Eötvös Loránd University, Budapest)

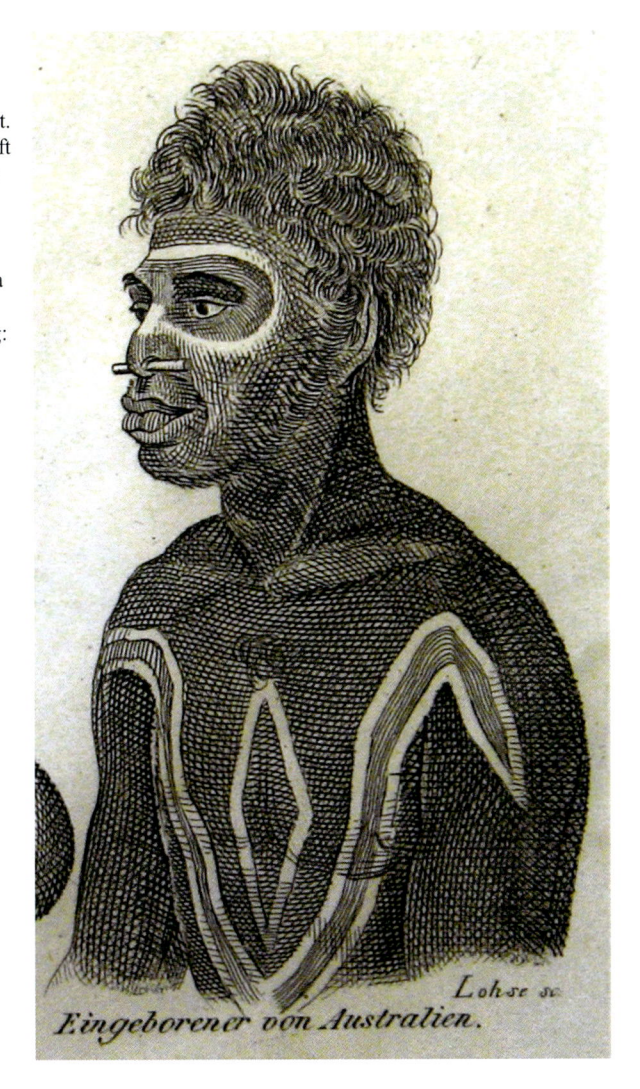

widespread and rather international: No actual colonies were needed for them to appear. They seem to have arrived in Hungary – a country without colonies – mostly from Western Europe together with the translations and adaptations of early ethnographic/anthropological works. They were, so to say, "armchair representations," i.e., textual and visual stereotypes that could be accessed *without any fieldwork*, without having even traveled to the lands described – as was the case with the Reverend Mihály Dobosy and Dr. Med Pál Almási Balogh, both noted above.

There existed a *multiplicity* of local contexts and interpretive filters in the reception of non-European indigenous peoples in the Kingdom of Hungary. They are discussed in the following.

Fig. 7 An umiak and a kayak of the Inuit of Greenland. Grönlánd históriája, melyben ez a' tartomány lakosival együtt leíródik és a' természeti-históriára sok jegyzések tétetnek. Ford. Dobosy Mihály [History of Greenland in which this land is described together with its inhabitants and natural history is discussed abundantly]. Landerer Anna, Buda, 1810. (Courtesy of the Library of the Hungarian Academy of Sciences)

Local Contexts: Jesuits, Lutherans, and Calvinists as Agents of World Ethnography

It appears that there were always specific, *local* cultural-political filters, built on the specific local cultural-political milieus that constituted the closer context in which and through which the larger international representations were received and adopted (recreated or adapted) in the eighteenth–/early nineteenth-century Hungary. The local characteristics of the reception, the *local appropriation*, of the early ethnographic/anthropological knowledge are themselves worth considering (on the term *appropriation*, see Chartier 1989, 1992, 1995).

Let us first see who the principal agents of the emerging world ethnography were in Hungary and what kind of historical/archival sources are available on them.

Considering the period between 1760/1770 and 1830, at least three groups of such agents could be identified that worked more or less consciously to import a global ethnographic/anthropological lore in Hungary.

The first group was that of the Jesuits. The Jesuit missionaries educated in Nagyszombat/Trnava/Tyrnau tended to support the Habsburg governments in Vienna, but they had their own independent interests as a powerful denomination as well as a well-organized province (the so-called Austrian province). The main church and the academy of the Jesuit order in Nagyszombat/Trnava/Tyrnau as a Central-Eastern European center for missionary formation were advertised in the local almanacs of the academy as early as the second half of the seventeenth century. Founded in 1635, the Jesuit academy used its almanacs to publicize both the overseas – mostly American and Asian – activity of the missionaries and the new geographical/ethnographic knowledge resulting from that activity (Sz. Kristóf 2014a).

The second group of agents consisted of Lutheran scholars in Pozsony/Bratislava/Breslau (in present-day Slovakia). During the 1810s, they gathered around a knowledgeable superintendent called János Kiss (1770–1846), and they produced, among other printed works, a number of geography books of global scope. For example, János Tomka-Szászky (1692–1762) published an *Introductio in orbis antiqui et hodierni geographia, in duos tomos divisa, quorum prior continens [. . .] Europam, posterior Asiam, Africam, et Americam* in 1777 (Tomka-Szászky 1777). An engraving next to the title page was an allegorical representation of the four continents as females. In the foreground, Europe was dressed in classical Greek attire, Asia was wearing a turban, and Africa and America were smaller creatures wearing feathered headdresses in the background. The image suggested a particular idea of stereotypization and hierarchization of the continents.

It is also worth mentioning in this place that it was especially a group of Protestant – especially, Lutheran – scholars who made contact with the German naturalist and traveler Alexander von Humboldt (1769–1859) during the 1790s–1810s. They started studying his early works and came under the influence of his natural history. Humboldt visited the Kingdom of Hungary on two occasions, in 1797 and 1811 (Sz. Kristóf 2017a, b, 2018b).

The third group of agents involved in the transmission of Western European ideas in the Kingdom of Hungary included individual pastors serving the Reformed Church of Hungary. This was not so much an organized group of scholars as much as a loosely organized circle of churchmen with scholarly interests in natural history, geography, and the emerging field of world ethnography. Rev. József Fábián (1762–1825) lived in the Hungarian town of Veszprém, for example. He was one of the translators of *Naturgeschichte für Kinder* by Georg Christian Raff noted above in 1799. Rev. Mihály Dobosy lived in another town, Szentes, in Csongrád County. He translated the *Historie von Groenland* by David Cranz. We should also include Pál Almási Balogh, the Calvinist physician mentioned above. He was the private physician to Count István Széchényi (1791–1860) and Governor Lajos Kossuth (1802–1894), and he compiled the first ethnographic profile of the Australian Aborigines in Hungary around 1835 (on Raff: Sz. Kristóf, 2011, 2013; on Cranz and Dobosy: Sz. Kristóf, 2016a; on Almási Balogh: Sz. Kristóf, 2014b).

As regards the scholarly activity of the three groups in relation to global ethnography and its sources, the following points can be made.

The Jesuits produced numerous manuscript accounts (*relationes*) of their missionary work, and they compiled several *synopses* of world history. The almanacs published by the academy in Nagyszombat/Trnava/Tyrnau, and some other Jesuit academies in the region in the seventeenth and eighteenth centuries include an important stock of knowledge based on both *relationes* and *synopses* tied to America, Africa, and Asia (Sz. Kristóf 2014a). Having worked as a missionary among the Moxo Indians in today's Bolivia between 1753 and 1767, Ferenc Éder Xavér S.J. (1727–1772) wrote up the history of his work (in Latin), including a detailed ethnographic description of the Moxos (Boglár and Bognár 1975; Sz. Kristóf 2012a). The almanac of the academy, published in Kassa/Košice/Kaschau (in present-day Slovakia) in 1745, contained a long description (in Latin) of the costumes of the Crimean Tatars. This description was based on an account by P. Du Bois and another French Jesuit missionary working in the region (Sz. Kristóf 2014a, 215). Many of the longer Latin texts, called *dissertationes* (treatises) and included in the almanacs of Nagyszombat/Trnava/Tyrnau as everyday readings in the seventeenth century, provided stereotypical representations, sometimes age-old stereotypes in the depiction of aboriginal peoples living in faraway lands. They told, for example, of the so-called *homo sylvestris* (man of the woods). As Fig. 8 shows, the latter was imagined as a wild, hairy half-animal, half-human creature. It was quite popular in the Jesuit descriptions of some non-European (Asian and American) "pagans" (*ethnici*) that the travelers and missionaries either had not yet seen or had only caught sight of for a few minutes. The image is an illustration of the work of Gaspar Schott S.J. (1608–1666), entitled *Physica curiosa* and published in Würzburg in 1662. A copy of this work came into the possession of the Jesuits of Nagyszombat/Trnava/Tyrnau, and was quite widely used (Sz. Kristóf 2014a, 220–224). Beyond *homo sylvestris*, a legion of headless, dog-headed, and other kinds of monsters (*monstra*) were discussed in the readings of the Jesuit academy almanac. Parallel to ancient Greek and Roman historical narratives, such monsters were located in Central Asia, Scythia, and/or India (Sz. Kristóf 2014a, esp. 218–224).

Moreover, a certain kind of demonizing representation was also present in the early library of the Nagyszombat/Trnava/Tyrnau Jesuits, as mentioned above. This literature, mostly illustrated travelogues, was linked not only to America, but also to certain parts of Asia. For example, the diabolical images of the Flemish – Protestant – artist, Theodor de Bry himself, formed part of the collection of books of the Jesuits. Together with the diabolical representation of America discussed above, the latter had in his possession the travel account of a Dutch merchant, Jan Huygen van Linschoten (1563–1611), from the years 1599–1601. The images of that book show various ways in which the Chinese were imagined to regularly adore demons in their religious ceremonies (Linschoten 1599–1601; Sz. Kristóf, 2012b, 60–61) (see Fig. 5 above). It seems certain that diabolical lore about Asia was also part of the heritage of the Nagyszombat/Trnava/Tyrnau Jesuits.

In sum, the Jesuits in Hungary appear to have preserved much of the legacy of the classical Greco-Roman tradition in their representation of alien, non-European

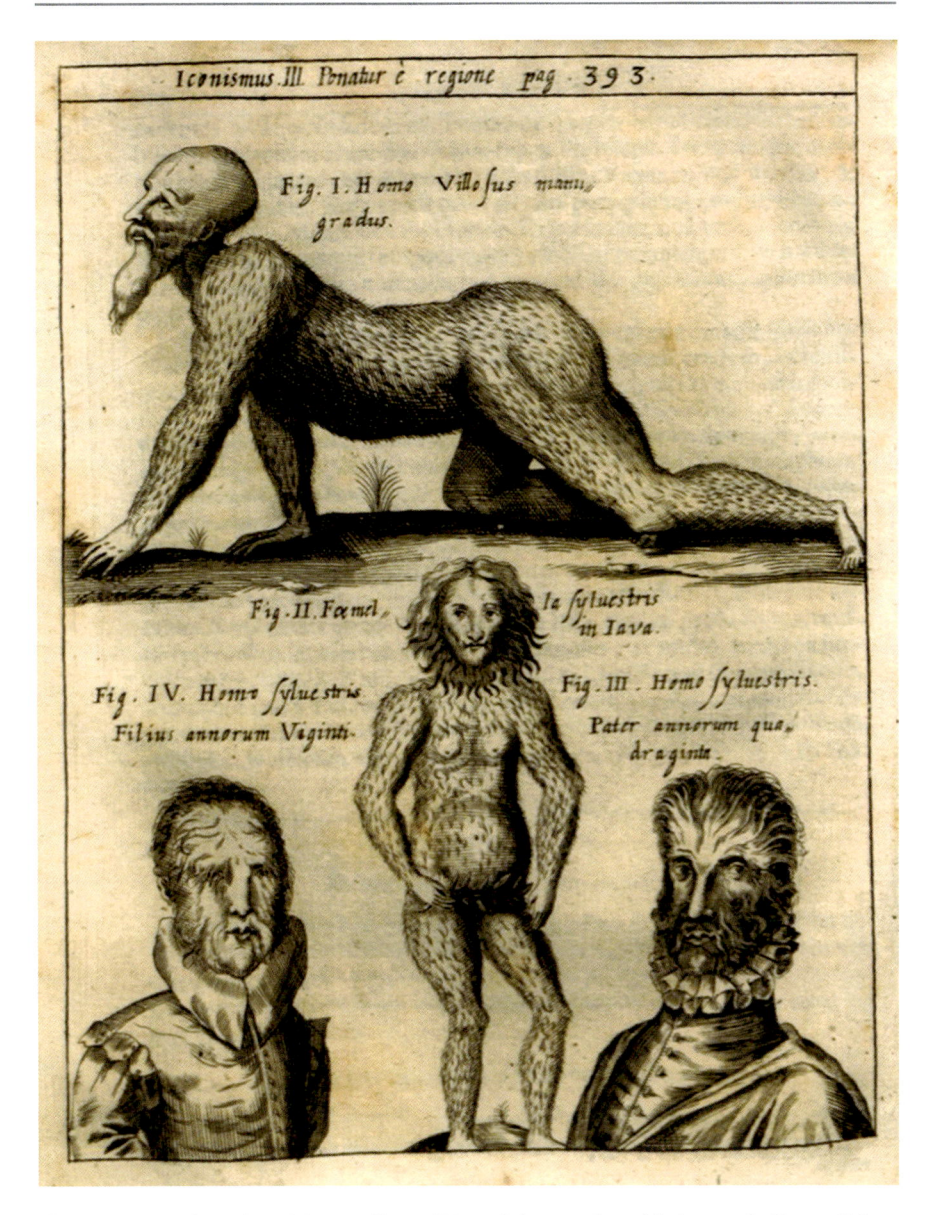

Fig. 8 Homo sylvestris and homo pilosus [Man of the woods and hairy man]. Gaspar Schott, Physica curiosa sive mirabilia naturae et artis [Remarkable physics or the wonders of nature and arts]. Herbipoli, 1662. (Courtesy of the University Library of Eötvös Loránd University, Budapest)

peoples. They also imported and adapted a great deal of that of the so-called Age of Discovery (sixteenth to eighteenth centuries). "Monsterizing" and/or demonizing descriptions of indigenous peoples are to be found in their writings well into the eighteenth century. One has to admit, however, that the discursive contexts and the interpretation in which those descriptions are embedded would seem to be increasingly marked by doubt and rationalization as time goes by and one approaches the eighteenth century. In any case, a certain amount of diabolical/demonological lore related to non-European indigenous peoples also infiltrated the geography books of the Catholic denomination published in the Kingdom of Hungary during the eighteenth century (Sz. Kristóf 2014a, 2018b, 355–369).

The case of the Protestants and the ethnographic/anthropological lore that they conveyed in the Kingdom of Hungary in the period under discussion was different in many respects. The Protestants did not do as much missionary work at that time as could be expected, but they engaged in an enormous *translation* effort. They translated a large number of foreign scholarly works into vernacular Hungarian. It was they – both Lutherans and Calvinists – who imported eighteenth-century Western European travel writing into Hungary and adapted it: the works of Joachim Heinrich Campe (1746–1818), James Cook (1728–1779), Jean-François Galaup de La Pérouse (1741–1788), George Macartney (1737–1806), and so on.

One of the most remarkable features of the Protestant translations – and missing in the case of the Jesuits – was a strong *Germanophile* attitude. The translations were done either from a German text and/or the idea of the cultural transfer itself originated in one or another German scholarly context, for example, that of the German universities of the late eighteenth century. The universities at Jena and Halle and especially Göttingen drew Hungarian Protestant students in that period (Szögi 2001). The translations thus also reveal a *political* dimension to and implication of the importation of Western European travel writing, ethnography/anthropology, and natural history into the Kingdom of Hungary. Subject to the Habsburg Empire since the end of the Ottoman wars (1699), Hungary was expected to integrate into Austria politically, administratively, and culturally. The aim of the Hungarian Protestant authors and publishers of the translations in question was admittedly to *subvert* the existing Habsburg – absolutist, Catholic, imperial, and partly also Jesuit – culture and scholarship and to lead a *scholarly revolt* against them. The Hungarian translators and publishers were in fact pastors, scholars, educated noblemen, etc. deeply involved in what was called the Hungarian national awakening movement (ca. 1795–1848). Their final goal, expressed by a good number of them in their writings, was to create *Hungarian scholarship* as such, i.e., a scholarly discourse in the vernacular language which had never existed before. To this end, they imported foreign – especially German and Protestant – scholarly works and translated them into Hungarian. Apart from translations, replicas of foreign works were also made, such as *A' Világ ritkaságai* (*The Rarities of the World*), translated from German by the Hungarian lawyer Ferenc Farkas (1785–ca. 1844) and published in Pest in 1807. This book can be considered a rather simplified replica of Bertuch's *Bilderbuch* mentioned above, which contained numerous images and texts of ethnographic/anthropological interest (Farkas 1807). One of its engravings, for example, depicts the Great Wall of China with "Tatar riders" racing about in the lower section of the image (see Fig. 9). This picture was apparently intended to represent a cultural and also civilizational opposition between the settled and nomadic peoples of the ("known") world, which was depicted quite frequently in the period under discussion.

Fig. 9 The Great Wall of China and Tartar riders. Ferenc Farkas: A' Világ ritkasági avagy a természet és a mesterség remekjei I [The rarities of the world or, the masterpieces of nature and arts, volume I]. Hartleben, Pest, 1807. (Courtesy of the Somogyi Library, Szeged)

It is especially important to bear in mind that the Hungarian translators wanted to create a vernacular scholarly discourse and they did so by means of Protestant patterns, originating primarily at German universities. This is a significant feature of the emergence of various scholarly fields in general in the Kingdom of Hungary during the late eighteenth and early nineteenth centuries and that of anthropology/

ethnology in particular. A long series of works reveal a characteristically *mixed* discourse in various scholarly fields, which came into being in the period under discussion: texts written in the vernacular (Hungarian), but structured in foreign – mostly German – frames and patterns of thought. This was also the case of an emerging natural history and anthropology/world ethnography. The translations of travelogues by David Cranz and Captain James Cook and natural histories, like that of Georg Christian Raff, were based, together with many others, on German originals, themselves translations in some cases. It is this particular combination of linguistic and political aspects, not sufficiently known and/or appreciated as yet in the history of scholarship, that was characteristic *locally* in Hungary in the period under discussion. A final example, the editions of the *Voyages* of James Cook in Hungary show that there were at least four different attempts to translate them around the turn of the nineteenth century. Three of the four were based on German texts (originally translated from English), and three of the four were done by Hungarian Protestant translators (Sz. Kristóf 2011, 2013, Forthcoming).

In sum, the main sources available for early anthropology/world ethnography in the Kingdom of Hungary in the period under discussion were missionary and travel accounts, geography (school)books, and natural history (school)books, resulting from the interests and work of the different local denominations. They conveyed religious as well as scholarly ideas above all from the German principalities, as well as from England, France – and in the case of the Jesuits – Italy, Spain, and Portugal, to Hungary. The transfer of such ideas was made through the mediation of primarily Latin and German languages and texts.

Notions of Geography: Notions of Hierarchy – Examples of Siberia and the Arctic Region

Let us examine more closely some of the ideas that were brought to Hungary and adapted there. It is worth analyzing how they relate to some more or less imaginary characteristics of the non-European lands and peoples. In the case that follows, Asia – especially, Siberia – and some of its indigenous peoples will be considered.

A widely known anthropomorphic map, included in the *Cosmographia* by Sebastian Münster (1488–1552) and well known in the Kingdom of Hungary in the period, depicted the continent of Europe as a queen in 1570. *Europa regina* ("Queen Europe") was represented as a female aristocrat lying across the continent and treading on Western Asia. The political symbolism indicated by the parts of her body was also elaborated in the text tied to the image. The map and the text together suggested that the Western parts/countries of Europe were superior in sociocultural value – i.e., civilization – to the Eastern ones. Even the Eastern European parts/countries were shown as superior within that hierarchy of values to those constituting the hem of the attire of "Queen Europe," i.e., "Moscovia," "Scythia," and the lands which were to be found directly under her feet, i.e., "Asia," "Tartaria," etc. (Münster 1570). As mentioned at the beginning of this chapter, such politically inspired body symbolism could reveal contemporary or even earlier forms of classification and means of memorization of geography and human societies than those that formed part of the linear and stadial approach characteristic of the Age of Enlightenment. The geographies and natural histories of the era testify to a long history and wide distribution of such a hierarchical and anthropomorphic way of thinking. In general, Europe was depicted as a powerful queen – or, in other cases, a triumphant virgin – in early modern European scholarly/geography literature, while Asia tended to be represented as a wild savage, for example, a Tatar riding a horse. This was also the case in the Kingdom of Hungary. However, the figure of the Tatar had a special significance for Hungarians, due to several – and in earlier times truly devastating – Tatar (Mongol) attacks against the Kingdom of Hungary (such as in 1241–1242, 1694, and 1717), the memory of which was kept alive by historians and other scholars. Accordingly, the illustrated album entitled *A' Világ ritkaságí* (*The Rarities of the World*) and published by Hungarian Ferenc Farkas (in Pest, 1807) emphasized a gap between the nomadic, "wild" peoples of Asia and the settled, urbanized, and "civilised" ones there. This sharp contrast was confirmed and even further developed in the geography books of the era.

If we take a closer look at those geographies, we find that each always has a particular section on Asia, more specifically, on Siberia. These sections reveal interesting details about the prevailing ideas and concepts behind the classification of the indigenous peoples living there. There were two basic types of geography book in the Kingdom of Hungary around the turn of the nineteenth century. On the one hand, there were shorter, synopsis-like works and, on the other, more detailed, more narrative *Weltbeschreibung*-like (description of the world-like) ones. Among the latter, the content of the Jesuit geographies – such as that of the *Geographica Globi Terraquei Synopsis*, the most popular geography in Catholic Hungary in the eighteenth century – was divided according to empires and religious administrative structures, such as provinces and mission areas. An edition of this geography, published in the Hungarian town of Győr in 1746, sorted contemporary knowledge of Northern Asia according to a division between the so-called "Muscovite" empire and "Tartaria" or "Tartaria Magna," i.e., Siberia (*Geographica* 1746). These geographies tended to structure the knowledge they conveyed according to then-existing political, religious, and sometimes also cultural – or even imaginary – groups or units. They briefly discussed – or, rather listed – the indigenous peoples living in the region. For example, *Introductio in orbis antiqui et hodierni geographia* by János Tomka-Szászky, published in Pozsony/Bratislava/Breslau and Kassa/Košice/Kaschau in 1777 and noted above, listed several of the Finno-Ugric peoples living in the "ancient kingdom of Tartaria," which formed part of the Russian Empire in Tomka-Szászky's time and was then organized into the prefecture of Kazan. Tomka-Szászky mentioned those Finno-Ugric people rather briefly, but he also referred to the possible relationship between their language and Hungarian: "Praefectura Kazany: regnum olim Tartariae. Inhabitant praeter Tartaros et Russos, *Cseremissi*, *Mordvini*, *Csuvassi*, *Wotiaci*, *Permiaci*, *Siriaeni*, et *Wogules*, qui eiusdem dicuntur esse originis, cuius Hungari, et lingue cognatione ad eosdem accedere" etc. (Tomka-Szászky 1777, 752, my emphasis).

Just like the *Introductio* by Tomka-Szászky, the geography books of the synopsis type, published mostly in Latin, tended to provide a rather brief characterization of the indigenous peoples inhabiting a particular administrative unit (of an empire). Instead of describing them, they frequently stated that these peoples were pagan and that they believed in "demons" or the "devil" itself. The other type, the *Weltbeschreibung*-like geographies, however – which were not only published in Latin, but also in German and Hungarian toward the end of the period discussed – offered incomparably more details of the lifeways and customs of the indigenous peoples. They seem to have led toward the voluminous world ethnographies of the turn of the century.

As regards eighteenth-century printed literature in general, one finds that it conveyed certain stereotypical representations of non-European indigenous peoples, especially those subjected to the great empires in America and Asia. Some of those representations appear to have survived in the approach of such nineteenth-century fieldworkers as ethnographers, anthropologists, linguists, and geographers. The same holds true for the Kingdom of Hungary. The case of Siberia and "the North" is especially revealing in this respect (Mészáros et al. 2017). This will be discussed in the following, final part of the paper.

Among the cultural stereotypes – or ways of othering – that survived into the nineteenth century are those that originate in the descriptive ethnography of the early modern era, for example. This kind of ethnography, as discussed above, focused on the external, material elements of culture. The historical representation of the Samoyed (a Siberian, linguistically Uralic people) is a good example. An engraving, attached to an Asian travelogue by Jan Huyghen von Linschoten and produced by Theodor De Bry, depicted the Samoyed and their material culture as very *simple*, even *miserable*, for European eyes (see Fig. 10). Linschoten's travelogue was published in Frankfurt between 1599 and 1601, as mentioned above, but its picture of the extreme simplicity of the people of Siberia seems to have enjoyed a long afterlife among generations of geographers, linguists, and ethnographers to come (Linschoten 1599–1601, XLI). Accordingly, the *diabolical* representation, the alleged idolatry of non-European indigenous peoples, was widely used as a means of description and cultural/religious classification, independently of the denominations to which the authors belonged. The diabolical representation also found expression, as we have seen, among the Asian images of Theodor De Bry. The Samoyed themselves were depicted as "adoring their idol" (Linschoten 1599–1601, title page).

One finds among the most widespread cultural stereotypes another (and similarly unjust) historical image of the so-called "Arctic peoples" being *ugly* and *disgusting*. János Kis, a Hungarian Lutheran author, claimed in his *Természet tsudái* (Wonders of Nature), published in Pozsony/Bratislava/Breslau in 1808, that the "Arctic peoples" (*"poláris emberek"* in Hungarian) are indeed "ugly" and "dirty," that they stink, that they regularly live under the ground, that the Samoyeds are cannibals, etc. (Kis 1808). Rev. János Kis had never visited "the North," not even Siberia: He had read these stereotypes in contemporary printed geographies and repeated them in his own work. This kind of "uglifying" representation of the peoples of the North seems to go back to the early modern lore of *climates*, permeating European geography and

XLI.
Eygentliche Contrafaytung der wilden Samuiten genannt. 6.

Jefe Leut genannt Samuiten, funden wir auff dem Fußveften Landt an dem ort den man Waygat nennet/ waren faft anzufehen wie Wilden/ als wir aber mit jnen zu reden kamen/ befandt es fich/ daß es fittfame ehrerbietige Leut waren/ fie find von ftatur klein/ mit breitem flachen Angeficht/ kleinen Augen/ die Knie ftehen jnen außwerts den Beinen/ fie tragen lange Haar/ flechten diefelbige in einen Zopff/ binden fie hinden zufammen. Ihre Kleydung find rohe Fell/ darmit fie von dem Haupt biß auff die Fußfolen bekleidet find/ fie hatten jhre Schlitten bey fich/ vnnd je an einem Schlitten zween Damhirfch vorgefpannt/ welche fo fchnel mit einer oder zwo Perfon daß auff fort fprungen/ daß jnen kein Pferdt gleich lauffen mag.

Fig. 10 Samoyeds of Siberia. Jan Huyghen van Linschoten: Ander Theil der Orientalischen Indien Von allen Völckern Insulen Meerporten fliessenden Wassern und anderen Orten. [The other part of East India, about all the peoples, islands, harbours, running waters and other places..] hrsg [edited by] Hans Dieterich and Hans Israel von Bry. Johan Saur, Franckfurt am Meyn, 1599–1601. (Courtesy of the University Library of Eötvös Loránd University, Budapest)

natural history in the seventeenth century, Hungary included. It can be found later in the linear and stadial view of the development of human societies, characterizing the philosophy of history of the Enlightenment. According to the latter, the "Arctic peoples," together with other simple societies, represent the *first stage* of human social development, the stage of "savages" (Sz. Kristóf, 2014a, 210–211; Harbsmeier 1994, 2001). This "stage" was not altogether despised; it was sometimes even idealized and exoticized in European – and Hungarian – literature. Certain characteristics were, however, attributed to it, among which material simplicity and physical ugliness were the two most common stereotypes.

As for the possible influence of those ancient hierarchical and Eurocentric images on the nineteenth-century fieldworkers, one should keep in mind that the knowledge that they – travelers, ethnographers, geographers, engineers, etc. – managed to acquire about Asia, for example, and other faraway, non-European lands was first formed by schoolbooks. And schoolbooks like geography books, books on natural history, travel accounts, etc. abounded with age-old cultural stereotypes. Stereotypical representations of indigenous peoples surrounded Hungarian travelers and fieldworkers both in schools and in academic contexts. They must have impacted their way of thinking. And it is most certain that those travelers and fieldworkers reacted to and reflected on them both in empirical research contexts and in their scholarly writings. An interesting and very important research question for the future would be to learn how the kind of conceptual, ideological, often Eurocentric representations of the indigenous peoples of Siberia and "the North" shaped the minds of our pioneer anthropologists, such as Antal Reguly (1819–1858), János Jankó (1868–1902), Bernát Munkácsi (1860–1937), and others. We should know more about how these peoples influenced their interests and scholarly thinking and how these scholars (most probably) transformed themselves in the field. Preliminary thoughts in this direction are included in a research report by our international team (Mészáros et al. 2017).

There are some truly solid stereotypes which seem to have lived a considerably long life and which originated in books on geography and world ethnography from the late eighteenth and early nineteenth centuries. Let us see another example from the cultural history of Hungary in closing. The so-called "Finno-Ugric–Turkic debate" (ca. 1860s–1880s) related to the schoolbook representation and cultural stereotypes of certain Siberian/Northern indigenous peoples at the turn of the nineteenth century. As regards the origin and the linguistic relatives of Hungarian, the debate was supported by an abundant philology of the description of those peoples, for example, in geography books. The anti-Ugric party, represented by scholars like Turcologist Ármin Vámbéry (1832–1913), expressed its dislike of – and even contempt for – the Finno-Ugric languages with quite similar ideas as those implied in the "uglifying" representation of the peoples of the North. The pro-Turkic party argued that the Finno-Ugric/Sámi peoples, championed by Finno-Ugric linguists such as József Budenz (1836–1892) and ethnographers such as Pál Hunfalvy (1810–1891), "stank," "smelled like fish," etc. The debate was broadened into a public discourse toward the end of the nineteenth century, and the way of life of the Sámi, the Finns, and other Finno-Ugric peoples was often represented as all too simple, miserable, and despicable. It was contrasted with that of the Turkic peoples, which was held in much higher regard. Thus, a possibility of an ancient linguistic and ethnic kinship with the Turkic peoples was considered glorious, glamorous – and thus desirable. Kinship with the Finno-Ugric peoples, however, was viewed as rather shameful by the pro-Turkic party, scholars, and noblemen alike (Pusztay 1977). It is worth mentioning that the debate also appears to have taken on denominational/cultural dimensions. The possibility of a closer Hungarian-Sámi linguistic kinship was considered all the more lamentable and deplorable by the Protestant protagonists of the debate in that the idea was originally proposed in 1770 by János Sajnovics (1733–1785), a Jesuit scholar (Sajnovics 1770; Kontler 2013; Kontler and Aspaas 2015). The chief defenders of the Ugric-Hungarian kinship – Budenz and Hunfalvy – were, however, Lutherans (of German descent).

Conclusion or the Afterlife of Stereotypes: Orientalism or Cultural Colonialism?

It would be extremely important to ascertain exactly how the works of early geography and world ethnography discussed above exerted their influence on the first (and maybe also the second) generation of Hungarian fieldworkers. This is a little researched area so far. It would also be interesting to learn how these fieldworkers' interpretation of the indigenous peoples changed as they actually met them in person in America, Asia, etc. during the second half of the nineteenth century. Were indigenous people still viewed and classified according to the hierarchical, Eurocentric stereotypes of earlier times or not? Did the indigenous peoples themselves react to those views in one way or another? And, finally, one should ask: To what extent can we consider those views and stereotypes versions of "Orientalism" (in the sense of the Palestinian-American scholar, Edward W. Said) and/or were these views and stereotypes certain forms of "cultural colonialism" (in the sense of the British anthropologist, Nicholas Thomas) (Said 1978; Thomas 1994)? A thorough examination of writings by Hungarian fieldworkers from these aspects would contribute to a better understanding of the late nineteenth-century afterlife as well as the overall impact of European Enlightenment anthropology and the earlier world ethnography of a truly *long durée*.

Acknowledgements This research was funded by a four-year grant from the National Research, Development, and Innovation Office for the years 2016–2020: No NKFIH 11957, *A tudományos tudás áramlásának mintázatai Magyarországon, 1770–1830 [The Circulation of Scholarly Knowledge in Hungary, 1770–1830]*. Earlier versions of this text were presented at two conferences: firstly, at the *Representations of Indigenous Peoples of the Asian Peripheries of the Russian Empire (Northern and Inner Asia) in the Legacies of Travelers from Austro-Hungary* workshop held at the Institut für Kultur- und Sozialanthropologie at the University of Vienna on 22–23 February 2017; and secondly, at the biennial *Staying, Moving, Settling* conference of the European Association of Social Anthropologists (EASA) conference in the History of Anthropology Network (HOAN) session held in Stockholm on 16 August 2018. A considerably shorter version of this study appeared as Sz. Kristóf (2019). The English of the paper was revised by Thomas A. Williams.

The chapter is based on the author's own research, which was conducted in various archives in Budapest, Hungary, such as the University Library of Eötvös Loránd University which is a descendant and successor of the ancient library of the Jesuit academy of Nagyszombat/Trnava (in today's Slovakia), founded in 1635, the Library of the Hungarian Academy of Sciences, founded in 1826, and the National Széchenyi Library, Budapest.

References

Ablonczy B (2016) Keletre, magyar! A magyar turanizmus története [Hungarians, eastward ho! The history of Hungarian Turanism]. Jaffa, Budapest. [In Hungarian]

Almási Balogh P (cca. 1835) Az ember Australiában [Man in Australia]. Manuscript, University Library of Eötvös Loránd University, Budapest. [In Hungarian]

Benzoni H (1644) Historiae Antipodum sive Novi Orbis, qui vulgo Americae et Indiae Occidentalis nomine usurpatur. [A history of the Antipods or, the New World which is commonly called America and West India] Mattheus Merian, Frankfurt am Main (Copy of the University Library of Eötvös Loránd University, Budapest) [In Latin]

Bertuch F J (1805) Természethistóriai képeskönyv az ifjúság használatára és gyönyörködtetésére. [Illustrated book of natural history for the use and amusement of the youth] Bécs (Copy of the Library of the Hungarian Academy of Sciences). [In Hungarian]

Bödeker H E, Büttgen P, Espagne M (eds) (2008) Die Wissenschaft vom Menschen in Göttingen um 1800. Wissenschaftliche Praktiken, institutionelle Geographie, europäische Netzwerke [The science of man in Göttingen around 1800. Scientific practices, institutional geograpy, European networks], Göttingen [In German]

Bodnár-Király T (2017) Államleírás és a „statisztika elmélete" a 18–19. század fordulóján [The description of the state and the „theory of statistics" in the turn of the 18th and 19th centuries]. Századok [Centuries] 151:971–986. [In Hungarian]

Bodrogi T (1997) A néprajzi érdeklődés kialakulása [The emergence of ethnographic interest]; Vadak, természeti népek, primitívek [Savages, natural peoples, primitives]. In: Bodrogi T (ed) Mesterségek, társadalmak születése [the birth of arts and societies]. Fekete Sas, Budapest, pp 9–35. [In Hungarian]

Boglár L, Bognár A (1975) Éder X. Ferenc leírása a perui missziókról a XVIII.századból [the description of Ferenc Éder X. of the Peruvian missions in the 18th century]. Ethnographia LXXXVI 1:181–192. [In Hungarian]

Cañizares-Esguerra J (2006) Puritan Conquistadors. Iberianizing the Atlantic 1550–1700. The Stanford University Press, Stanford

Chartier R (1989) Le monde comme représentation [The world as a representation]. Annales E.S.C. 6: 1505–1520 [In French]

Chartier R (1992) Laborers and voyagers: from the text to the reader. Diacritics 22(2):49–61

Chartier R (1995) Forms and meanings: texts, performances, and audiences from codex to computer. The University of Pennsylvania Press, Philadelphia

Clifford J, Marcus GE (1986) Writing culture. The poetics and politics of ethnography. The University of California Press, Berkeley

Cranz D (1765) Historie von Groenland enthaltend Die Beschreibung des Landes und der Einwohner... insbesondere die Geschichte der dortigen Mission der Evangelischen Brueder zu Neu-Herrnhut und Lichtenfels, Mit acht Kupfertafeln und einem Register [History of Greenland containing the description of the land and the inhabitants...particularly the history of the mission of the evangelical brothers of Neu-Herrnhut and Lichtenfels there, with eigth engravings and an index]. Heinrich Detlef Ebers, Barbÿ, Weidmanns Erben, Leipzig (Copy of the Somogyi Library, Szeged). [In German]

Cranz D (1810) Grönland históriája, melyben ez a' tartomány lakosival együtt leíródik és a' természeti-históriára sok jegyzések tétetnek. Fordította és egynéhány világosító jegyzésekkel bővítette Dobosy Mihály Vaiszlói Réfor. Prédikátor. Grönland mappáival és rajztáblákkal

[History of Greenland in which this land is described together with its inhabitants and natural history is discussed abundantly. Translated and enlarged with enlightening notes of Mihály Dobosy, preacher of the Reformed Church in Vajszló. With maps of Greenland and pictures]. Landerer Anna, Buda (Copy of the Somogyi Library, Szeged). [In Hungarian]

Dumont d'Urville J et al (eds) (1835) Malerische Reise um die Welt. Verfaßt von einer Gesellschaft Reisender und Gelehrter. Ins Deutsche übertragen... von Dr. A. Diezmann [Picturesque voyages around the world. Made by a society of travelers and scholars. Translated in German by Dr. A. Diezmann, volume II]. Band II. Industrie-Comptoir (Baumgärtner), Lepizig [In German]

During S (1994) Rousseau's patrimony: primitivism, romance and becoming other, In: Barker F., Hulme P, Iversen M. (eds) Colonial discourse/postcolonial theory. Manchester – New York, p 47–71

Farkas F (1807) A' Világ ritkaságai a'vagy a Természet és Mesterség remekjei, I. köt. [The rarities of the world or, the masterpieces of nature and art, volume I]. Hartleben, Pest (Copy of the Somogyi Library, Szeged, Hungary). [In Hungarian]

Findlen P (1994) Possessing nature: museums, collecting, and scientific culture in early modern Italy. The University of California Press. London, Berkeley

Fixico DL (2003) The American Indian mind in a linear world. American Indian studies and traditional knowledge. Routledge, New York and London

Geographiae (1727) = Geographiae Particularis Epitome...Ex Auctoribus recentioribus collecta [An outline of a geography relating to particular places...collected from various authors]. Viennae Austriae (Copy of the Somogyi Library, Szeged, Hungary). [In Latin]

Geographica (1746) = Geographica Globi Terraquei synopsis [a geographical synopsis of the world], Győr. Copy of the University Library of Eötvös Loránd University, Budapest. [In Latin]

Gurka D (2003) Reflexiók és iniciatívák. Az Európán kívüli világ (re)prezentációja a göttingai egyetemen [Reflections and initiatives. The (re)presentation of the world outside of Europe in the university of Göttingen]. Magyar Filozófiai Szemle [Journal of Hungarian Philosophy] 3: 341–357. [In Hungarian]

Gurka D (2010) Göttingen dimenziói. A göttingeni egyetem szerepe a szaktudományok kialakulásában [the dimensions of Göttingen. The role of the university of Göttingen in the emergence of the disciplines], Gondolat Kiadó, Budapest. [In Hungarian]

Harbsmeier M (1994) Wilde Völkerkunde. Andere Welten in deutschen Reiseberichten der Frühen Neuzeit [Wild Ethnology. Other worlds in early modern German travelogues], Campus Verlag, Frankfurt, New York (Historische Studien, Band 12). [In German]

Harbsmeier M (2001) Stimmen aus dem äußersten Norden. Wie die Grönländer Europa für sich entdeckten [Voices from the farthest North. How the natives of Greenland explored Europe]. Jan Thorbecke Verlag, Stuttgart (Fremde Kulturen in alten Berichten). [In German]

Harris M (1968) The rise of anthropological theory. Thomas Y Crowell Company, New York

Kis J (1808) Természet' tsudái, országok' nevezetességei és nemzetek' szokásai, mellyek külömbféle munkákból öszveszedegettettek [Wonders of nature, monuments of countries and customs of nations, gathered from various works]. Pozsony. [In Hungarian]

Kontler L (2001) William Robertson and his German audience on European and non-European civilisations. Scott Hist Rev LXXX:63–89

Kontler L (2013) Distances celestial and terrestrial. Maximilian Hell's Arctic expedition of 1768–1769: contexts and responses in scholars in action. In: Holenstein A, Steinke H, Stuber M (eds) The practice of knowledge and the figure of the savant in the 18th century II. Brill, Leiden, pp 721–750

Kontler L (2014) Translations, histories, enlightenments: William Robertson in Germany 1760–1795. Palgrave Macmillan, Basingstoke

Kontler L, Aspaas PP (2015) Before and after 1773: central European Jesuits, the politics of language and discourses of identity in the late eighteenth century Habsburg monarchy. In: Almási G, Subarić L (eds) Latin at the crossroads of identity. Brill, Leiden, pp 95–118

Liebersohn H (2006) The travelers' world. Europe to the Pacific. Harvard University Press, Cambridge, MA/London

Linschoten J H von (1599–1601) Ander Theil der Orientalischen Indien Von allen Völckern Insulen Meerporten fliessenden Wassern und anderen Orten [The other part of East India, about all the peoples, islands, harbours, running waters and other places..].. hrsg [edited by] Hans Dieterich and Hans Israel von Bry. Johan Saur, Franckfurt am Meyn (Copy of the University Library of Eötvös Loránd University, Budapest). [In German]

Lüsebrink H-J (ed) (2006) Das Europa der Aufklärung und die außereuropäische koloniale Welt [Enlightenment Europe and the non-European colonial world]. Wallstein Verlag, Göttingen (Das achtzehnte Jahrhundert Supplementa) [In German]

Marshall PJ, Williams G (1982) The great map of mankind. British perceptions of the world in the age of enlightenment. Dent J. M. & Sons Ltd, London/Melbourne/Toronto

Mészáros C, Krist S, Bashkuev V, Bělka L, Hacsek Z, Nagy Z, Sántha I, Sz. Kristóf I (2017) Ethnographic accounts of visitors from the Austro-Hungarian monarchy to the Asian peripheries of Russia and their contribution to the development of systematic ethnological studies in the monarchy: preliminary results and research perspectives. Acta Ethnographica Hungarica 62(2): 465–498

Mihesuah DA (ed) (1998) Natives and academics. Researching and writing about American Indians. The University of Nebraska Press, Lincoln/ London

Münster S (1570) Cosmographia [Cosmography]. Basel (Copy of the Somogyi Library, Szeged). [In Latin]

Pratt ML (1992) Imperial Eyes. Travel Writing and Transculturation, Routledge/London/New York

Pusztay J (1977) Az „ugor-török háború" után. Fejezetek a magyar nyelvhasonlítás történetéből [After the „Ugric-Turkic war." Chapters from the history of Hungarian linguistic comparison]. Magvető, Budapest. [In Hungarian]

Raff Gy K (1837) Természet História Gyermekek' számára. Második magyarítás. Ford. Vajda Péter [Natural history for children. Second Hungarian translation. Translated by Péter Vajda]. Kassa. [In Hungarian]

Rubiés J-P (2007) Travellers and cosmographers. Studies in the history of early modern travel and ethnology. Ashgate Variorum, Aldershot, Hampshire/Burlington

Said EW (1978) Orientalism. Pantheon Books, Random House Inc., New York

Sajnovics JSJ (1770) Demonstratio idioma Ungarorum et Lapponum idem esse [a demonstration of how the language of the Hungarians and Lappons is the same]. Tyrnaviae. Copy of the University Library of Eötvös Loránd University, Budapest. [In Latin]

Sárkány M (2018) Etnográfia és szociográfia [Ethnography and sociography]. In: Tóth P P (ed) A magyar szociográfia a 20–21. században [Hungarian sociography in the 20th-21st century]. Gondolat Kiadó, Budapest, p 89–105. [In Hungarian]

Sárkány M, Somlai P (2003) A haladástól a kontingenciáig. Vázlat a szociokulturális evolúció változó elméleteiről [from progress to contingency. A sketch of the changing theories of sociocultural evolution], Szociológiai Szemle [sociological review] 13(3): 3–26. [In Hungarian]

Schott G (1662) Physica curiosa sive mirabilia naturae et artis [remarkable physics or, the wonders of nature and arts]. Herbipoli. Copy of the University Library of Eötvös Loránd University, Budapest. [In Latin]

Service E (1958) A profile of primitive culture. Harper & Brothers, New York

Service E (1963) Profiles in ethnology. Harper & Row, New York/London

Stagl J (1995) A history of curiosity. The theory of travel 1550–1800. Harwood Academic Publishers, Chur etc (Studies in Anthropology & History, vol. 13)

Stocking GW Jr (1987) Victorian anthropology. Free Press, New York

Sz. Kristóf I (2007) „Kié a hagyomány és miből áll? Az Indigenous Studies célkitűzései a jelenkori amerikai indián felsőoktatásban" [Whose is tradition and what does it consist of? The aims of indigenous studies in American Indian higher education]. In: Wilhelm G (ed) Hagyomány és eredetiség. Tanulmányok [Tradition and authenticity. Studies]. Néprajzi Múzeum, Budapest p 153–172. [In Hungarian]

Sz. Kristóf I (2011) The uses of natural history. Georg C. Raff's Naturgeschichte für Kinder (1778) in its multiple translations and multiple receptions. In: Adams A, Ford P (eds) Le livre demeure [The book persists]. Studies in book history in Honour of Alison Saunders. Droz, Genève p 309–333

Sz. Kristóf I (2012a) The uses of demonology. European missionaries and native Americans in the American Southwest (17–18th Centuries). In: Szőnyi Gy E, Maczelka Cs (eds) Centers and peripheries in European renaissance culture. Essays by East-central European Mellon Fellows. JATEPress, Szeged

Sz. Kristóf I (2012b) Missionaries, monsters, and the demon show. Diabolized representations of American Indians in Jesuit libraries of 17th and 18th century upper Hungary. In: Kérchy A, Zittlau A (eds) Exploring the cultural history of continental European freak shows and "Enfreakment". Cambridge Scholars Publishing, Newcastle upon Tyne, pp 38–73

Sz. Kristóf I (2013) Domesticating nature, appropriating hierarchy: the representation of European and non-European peoples in an early-nineteenth-century schoolbook of natural history. In: Demski D, Sz. Kristóf I, Baranieczka-Olsewska K (eds) . Competing Eyes. Visual Encounters with Alterity in Central and Eastern Europe. L'Harmattan, Budapest, pp 40–66

Sz. Kristóf I (2014a) Local access to global knowledge: Historia naturalis and anthropology at the Jesuit University of Nagyszombat (Trnava), as transmitted in its almanacs (1676–1709). In: Almási G (ed) A divided Hungary in Europe, Study Tours and intellectual-religious relationships, vol 1. Cambridge Scholars Publishing, Newcastle, pp 201–228

Sz. Kristóf I (2014b) The representation of the Australian aborigines in text and picture: Dr. med. Pál Almási Balogh (1794–1863) and the birth of the science of anthropology in Central Europe/Hungary. Caiana Revista academica de Historia del Arte y Cultura Visual de Centro Argentino de Investigadores de Arte (CAIA) (Buenos Aires), No 5, Segundo semestre, dossier special: Csúri P, García Ferrari M (eds) „Ciencia y Cultura Visual": 126–140

Sz. Kristóf I (2016a) „Terepmunka" a Terepmunka kora előtt: Dobosy Mihály, a grönlandi inuitok és az etnológia (antropológia) születése a 19. század eleji Magyarországon [„Fieldwork" before the age of fieldwork: Mihály Dobosy, the Greenland Inuit and the birth of ethnology (anthropology) in Hungary in the beginning of the 19th century]. Ethno-lore XXXII: 1–23. [In Hungarian]

Sz. Kristóf I (2016b) „Indi legendi & scribendi usum mirantur." Egy kulturális sztereotípia múltjáról és régi magyarországi előfordulásáról [„Indi legendi & scribendi usum mirantur." About the stereotypical representation of America in Hungary in the early modern period]. In: Nyerges J, Verók A, Zvara E (eds) MONOKgraphia. Tanulmányok Monok István 60. születésnapjára [MONOKgraphia. Studies for the 60th birthday of István Monok]. Kossuth Kiadó, Budapest, p 435–446. [In Hungarian]

Sz. Kristóf I (2017a) Alexander von Humboldt és a Podmaniczkyak. Mozaikok a földrajz és az egyetemes néprajz 19. századi tudománytörténetéhez [Alexander von Humboldt and the Podmanitzkys. new findings in the history of geography and ethnology in the 19th century]. In: Gurka D (ed) A báró Podmanitzky család szerepe a 18–19. századi magyar kultúrában [The role of the Podmanitzky family in Hungarian culture in the 18th/19th Century). Gondolat, Budapest, p 25–40. [In Hungarian]

Sz. Kristóf I (2017b) Alexander von Humboldt és Magyarország: Egy romantikus természettudós jelentősége a magyarországi egyetemes néprajzi érdeklődés kibontakozásában [Alexander von Humboldt and Hungary. The importance of a Romantic scientist in the rise of interest in world ethnography in Hungary]. Századok [Centuries], 151, 5: 987–1006. [In Hungarian]

Sz. Kristóf I (2017c) "No Visitors Beyond This Point:" Rules of conduct for tourists in Native American reservations and their cultural-political contexts in the USA In: Réka M. Cristian, Andrea Kökény and György E. Szőnyi eds. Confluences: essays mapping the Manitoba-Szeged partnership. JATEPress: Szeged, p 123–137

Sz. Kristóf I (2018a) Amerika und seine UreinwohnerInnen in den ungarischen Kalendern des 17. Jahrhunderts: David Frölich vs. die Jesuiten. [America and its indigenous people in Hungarian almanachs in the 17th century]. In: Herbst K-D, Greiling W (eds) Schreibkalender und ihre Autoren in Mittel-, Ost- und Ostmittel-Europa (1540–1850) [Almanachs (Schreibkalender) and their authors in Central-, East- and East-central Europe (1540–1850)]. Edition Lumière, Bremen, pp 355–369. [In German]

Sz. Kristóf I (2018b) Alexander von Humboldt and Hungary: national identity and the emergence of modern sciences. In: Semsey V (ed) National identity and modernity 1870–1945. Latin America, southern Europe, east Central Europe. KRE, L'Harmattan, Budapest, pp 391–406

Sz. Kristóf I (2019) The emergence of world ethnography in Hungary before 1848: agents and sources. In: Gurka D (ed) Changes in the image of man from the enlightenment to the age of romanticism: philosophical and scientific receptions of (physical) anthropology in the 18–19th centuries. Gondolat Kiadó, Budapest, pp 207–223

Sz. Kristóf I (Forthcoming) Tahiti in Hungary: the reception of the voyages of James Cook and the emergence of cultural anthropology in hungary. In: Krász L (ed) Sciences between tradition and innovation – Historical perspectives/Wissenschaften zwischen tradition und innovation – historische Perspektiven. Praesens Verlag, Wien

Szögi L (2001) Magyarországi diákok németországi egyetemeken és főiskolákon 1789–1919 [Hungarian students in German universities and high schools, 1789–1919]. Budapest. [In Hungarian]

Te Heesen A, Spary E C (eds) (2001) Sammeln als Wissen. Das Sammeln und seine wissenschaftsgeschichtliche Bedeutung [Collecting as knowing/collection as knowledge. The practice of collecting and its meaning in the history of science]. Wallstein Verlag, Göttingen („Wissenschaftsgeschichte"). [In German]

Thomas N (1994) Colonialism's culture. Anthropology, travel and government. Polity Press, Cambridge

Thomas N (2003) Discoveries. The voyages of captain cook. Penguin Books, Allen Lane/London, etc

Tomka-Szászky J (1777) Introductio in orbis antiqui et hodierni geographia, in duos tomos divisa, quorum prior continet...Europam, posterior Asiam, Africam, et Americam [Introduction in the geography of the antique and the modern world, divided in two parts, from which the first one contains...Europe, then Asia, Africa and America]. Posonii et Cassoviae (Copy of the Somogyi Library, Szeged). [In Latin]

Tóth G (2015) Bél Mátyás, a történész [Matthias Bel, the historian]. In: Békés E, Kasza P, Lengyel R (eds) Humanista történetírás és neolatin irodalom a 15–18. századi Magyarországon [Humanist history writing and Neo-Latin literature in Hungary in the 15th–18th century]. MTA BTK Történettudományi Intézet, Budapest p 157–167 (Convivia Neolatina Hungarica 1). [In Hungarian]

van Groesen M (2008) The representations of the overseas world in the De Bry collection of voyages (1590–1634). Brill, Leiden, Boston

Vermeulen H (2015) Before boas. The genesis of ethnography and ethnology in the German enlightenment, Critical studies in the history of anthropology. The University of Nebraska Press, Lincoln/London

Wolff L, Cipolloni M (eds) (2007) The anthropology of the enlightenment. The Stanford University Press, Stanford

Scientists and Specimens: Early Anthropology Networks in and Between Nations and the Natural and Human Sciences

64

Brooke Penaloza-Patzak

Contents

Introduction	1652
Who Were Early Anthropologists?	1655
Gathering Places: Early Special Interest Societies	1662
Ethnographic Museums and Society Membership: A Transnational Case Study	1665
Ethnographic Departments and Museums	1666
National Societies, International Memberships	1667
National Interests and the International Congress of Americanists	1669
Personal Correspondence and Institutional Object Exchange	1669
Conclusion	1674
References	1676

Abstract

Merging a transnational approach to the history of science with methods from museum anthropology, this chapter traces the emergence and expansion of early anthropology networks rooted special interest societies and the natural sciences. The focus here is on the interplay between practitioners and the objects – from publications to specimens – with which they worked during a singularly seminal and transient time in which the nucleus of anthropological activity migrated from scientific societies, to museums and then universities. The chapter also presents a detailed case study of the people and things that linked early anthropology in the USA and German and Austro-Hungarian Empires. Adolf Bastian (1826–1905), inaugural director of Berlin's ethnological museum and now considered a father of German anthropology, was a key figure in these developments, as was his one-time staff member Franz Boas (1858–1942), himself now thought of as a father of modern US anthropology. Zeroing in on their engagements in the

---

B. Penaloza-Patzak (✉)
Department of History and Sociology of Science, University of Pennsylvania, Philadelphia, PA, USA
e-mail: penaloza.patzak@univie.ac.at

© The Author(s), under exclusive licence to Springer Nature Singapore Pte Ltd. 2022
D. McCallum (ed.), *The Palgrave Handbook of the History of Human Sciences*,
https://doi.org/10.1007/978-981-16-7255-2_104

subfield of Americanist anthropology, however, among others the scramble for Northwest Coast artifacts, World's Columbian Expedition in Chicago, and the International School for American Archeology and Ethnology (ISAAE) in Mexico, the chapter retraces how Americanist networks grew to transcend the activity and influence of both.

Keywords

Interdisciplinary histories of science · Disciplinary coalescence · Anthropology · Scholarly societies · Museums · Specimen exchange · Scientific tours · Transnational science

Introduction

The study of humankind began as a hobby pursued by, among others, physicians, philologists, geographers, geologists, military officers, explorers, and antiquarians. The process by which this subject of inquiry came into being as a scientific field, in the sense of practitioners' shared general objectives and notions of what constituted data, not to mention methods for procuring that data, was profoundly shaped by the technological advances that distinguished the second half of the nineteenth century. With electrical telegraphy presenting humans the world over with an unprecedented potential for speed in communication, and steam-powered ships and trains driving travel to become faster and more affordable than ever before, the advent of modern globalization had arrived. These were favorable circumstances for the development of early anthropology, in particular the phenomenon of network formation that forms the focus of this chapter, with special attention to the mobilities of people and things, which played so central a role in the initiation and maintenance of those networks. This era proceeded from what Vermeulen describes as the late eighteenth century "conceptualization of ethnology" (Vermeulen 2015) and would eventually give way to the institutionalization of anthropology in the university context. Before proceeding, a note on terminology: The aims and methods of research conceived as anthropological in nature varied by region and changed over time. This in mind, the term "anthropology" used throughout this contribution should be understood as implicating a range of practitioners and practices engaged in the study of humanity in the broadest sense, including, but not limited to, archaeological, sociocultural, linguistic, and biological inquiries.

Three circumstances broadly characterized international relations from the eighteenth to the early twentieth century and were likewise formative in the evolution of modern anthropology: Imperialism, the rise of the modern nation state, and a therewith related rise in identification with ones' nation, otherwise known as nationalism. In recent years, nationalism and isolationism have again been on the rise. In 2015, member states of the European Union clashed in bitter disagreement over the acceptance and freedom of movement of asylum seekers from non-member nations, in 2017 the USA announced its withdraw from the Paris Agreement, and in 2020 the

United Kingdom officially withdrew from the European Union in the move popularly known as "Brexit." These and other current events lend new urgency to research into how and why transnational scientific networks develop, flourish, ebb, and are reconfigured over time. It is becoming more and more common for historians to approach the development of scientific disciplines as a transnational or global, rather than national, phenomenon. Still, it is important to remember the pervasive impact of the nation as a social, ideological, and economic entity on the development of scientific disciplines. This includes anthropology, which was so fundamentally informed by an assumption of white Euro-American preeminence and the otherness of those who did not fall within these categories. While there were certainly those who took up studies within their own and other local cultures (folk studies, *Volkskunde*), practitioners by and large trained their gazes on non-white non-Euro-Americans. In this sense, early anthropology was an almost intrinsically transnational enterprise, bringing together highly heterogeneous groups of individuals with diverse interests and stakes in the production of anthropological knowledge, whose interactions were shaped by unequal power structures.

As this suggests, European and American countries have especially long histories as initiators of anthropological investigation. By the early twentieth century, however, this branch of science had begun to spread, likewise taking root in precisely those areas that had been characterized as "other" by the "industrialized, urbanized, and literate European world of the nineteenth century" (Krotz 2006, 88). This research advances a nuanced understanding of the role of transnational networks in that spread, emphasizes the subjectivity of claims regarding central and peripheral national traditions, considers how they came to be considered as such, and tracks the capacity of these perceptions to change over time. People and things linking the German, Austro-Hungarian, and US American traditions form the focus here, but it is essential to remember that the networks they populated were by no means constricted by that constellation of national borders. Indeed, without exception they extend far beyond those borders, to England, France, Mexico, Brazil, India, Japan, China, Taiwan, Russia, and beyond.

Early transnational networks in anthropology tended to be rooted in special interest scientific societies that sprouted from more venerable scholarly societies. Two examples are the Imperial Russian Geographic Society, founded in St. Petersburg in 1845 and which included a division of ethnography from the outset (Vermeulen 2015), and the Berliner Gesellschaft für Anthropologie, Ethnologie und Urgeschichte (BGAEU, Berlin Society for Anthropology, Ethnology and Prehistory, 1869), which began as a subsection of the Gesellschaft Deutscher Naturforscher und Ärzte (society of German natural scientists and physicians, 1822). In many instances, these older organizations had already established a tradition of publication exchanges with their cognates in other nations, and affiliates with an interest in anthropology were able to avail themselves of these extant networks to initiate exchange with other similarly curious interlocutors. One particularly well-organized and far-reaching organ for scholarly exchange was the Smithsonian Institution's International Exchange Service (Fig. 1). Established in 1849, the International Exchange Service operated out of Washington, D.C. and transmitted US-produced

Fig. 1 "Smithsonian International Exchange Program Commences." International Exchange Service employees convey crates of scholarly and government documents from their workrooms in the Castle cellar to a horse-drawn wagon waiting on its north façade. Courtesy of the Smithsonian Institution Archives. Image # MAH-13318

scientific journals and other printed matter free of charge to scientific societies the world over. In the wake of the US congress's 1879 establishment of the Smithsonian-affiliated Bureau of Ethnology (later Bureau of American Ethnology, BAE), the International Exchange Service would play a pivotal role in the dissemination of Washington, D.C.-based and -backed anthropological research in particular.

From the 1850s on, the European capitals of first London, then Paris, and eventually Vienna, followed by Philadelphia, Chicago, and St. Louis in the US, began hosting international expositions, also known as World's Fairs. Later on in the chapter, we will demonstrate how these events represented an important venue for the convergence of and exchange between an international body of early practitioners, facilitating the longer-term expansion of transnational networks. Ethnographic departments, museums, and their specimen collections were also central coordinates in the formation of transnational relations, not to mention the methodological and epistemic development of the field (Penaloza-Patzak 2018a, b, 2020), and this contribution places special emphasis on the role of specimen exchanges in the codification of these networks.

This overarching aim here is to begin to account for the manifold and interlacing interests, institutions, interlocutors, and interactions that contributed to the constellation of practices and epistemic frameworks that constituted anthropology. Inherent in this undertaking is recognition that, as Ash put it, "neither the sciences themselves nor the societies and cultures in which they are practiced are closed systems" (Ash 2008). This being the case, we also look farther back in history, and to the practitioners and networks, which populated the academic institutionalization and "globalization" of the discipline. In order to do this, an emphasis will be placed on local contexts, resources, and disciplinary orientations, and how and why these facilitated or impeded specific network configurations. What emerges is a complex portrait of a discipline that was practiced differently in different local and national contexts, but was to a great extent unified by a few underlying convictions and questions. Of the former: Human culture was part of the natural world, thus well suited to analysis via natural science concepts and techniques, and in order to establish its credibility this new field need necessarily progress from a natural science perspective. Of the latter: How had human culture come into being in its countless variations across the globe? And what role did the natural world play in the development of culture?

Who Were Early Anthropologists?

Throughout the mid to late nineteenth century, ethnographic objects began to be extracted from broader encyclopedic collections and brought together into discrete collections intended to serve anthropological inquiry. This phenomenon resulted in the world's first public ethnographic divisions, departments, and museums opening up in metropolitan centers across Asia, Europe, and the Americas. Some of the most prominent among these included the Peter the Great Museum of Anthropology and Ethnology in St. Petersburg, the Museo Nacional de Antropología, in Madrid, Rijks Ethnographisch Museum in Leiden, the Neues Museum ethnographic department in Berlin, the Musée de l'Ethnographie in the Trocadero, and the Smithsonian Institution United States National Museum in Washington, D.C. (see Table 1).

It was within these spaces that the discipline of anthropology began to coalesce in the 1880s. Up until this time there was no standardized training for those interested in practicing anthropology, and the field of knowledge drew on the material and intellectual contributions of individuals active in a broad range of professions. Recognizing the extent of this range is crucial in reconstructing these networks and the dynamics at play therein. There were autodidacts such as Edward Burnett Tylor (1832–1917) in Oxford and Norwegian-born Johan Adrian Jacobsen (1853–1947); Physicians such as Paul Broca (1824–1880) in Paris, Daniel Garrison Brinton (1837–1899) in Philadelphia, and Rudolf Virchow (1821–1901), Adolf Bastian (1826–1905), and Karl von den Steinen in Berlin; Geologists such as John Wesley Powell (1834–1902) in Washington D.C., and Ferdinand von Hochstetter (1829–1884) and Franz Heger (1853–1931) in Vienna; Zoologists such as Enrico Giglioli in Florence, and geographers like Franz Boas (1858–1942) in New York

Table 1 Early ethnological and anthropological societies, journals, and public museums: An overview, 1788–1910

Year	Organization
1788	Royal Batavian Society of Arts and Sciences, Section for Geography and Ethnology, Batavia
1799	Société des Observateurs de l'homme, Paris
1836	Kunstkamera, Ethnographic Museum, St. Petersburg (now Peter the Great Museum of Anthropology and Ethnography)
1837	Rijks Japansch Museum, Leiden (now at Museum Volkenkunde)
1839	Société Ethnologique de Paris, Paris
1841	*Mémoires de la Société Ethnologique*
1842	American Ethnological Society, New York City
1845	*Transactions of the American Ethnological Society*
1843	Ethnological Society of London, London
1848	*Journal of the Ethnological Society of London*
1845	British Museum, Department of Antiquities, Ethnological Gallery
1847	Musée royal d'Armures, d'Antiquités et d'Ethnologie, Brussels, (now at Musée Art et Histoire)
1851	Kongelige Etnografiske Museum, Copenhagen (now at Nationalmuseet)
1859	Société d'Anthropologie de Paris, Paris
1859	*Bulletins et Mémoires de la Société d'Anthropologie de Paris*
1859	Société d'Ethnographie, Paris
1862	Königliche Ethnographische Sammlung (today Museum Fünf Kontinente), Munich
1863	Anthropological Society, London
1864	*Journal of the Anthropological Society of London*
1866	*Archiv für Anthropologie. Zeitschrift für Naturgeschichte und Urgeschichte des Menschen* (international)
1866	Museu Paranaense, Curitiba, Brazil
1866	Peabody Museum of American Archaeology and Ethnology, Cambridge
1869	Museum für Völkerkunde Leipzig, Leipzig (now Grassi Museum für Völkerkunde)
1869	Berlin Gesellschaft für Anthropologie, Ethnologie und Urgeschichte, Berlin
1869	*Zeitschrift für Ethnologie*
1869	Museo di Storia Naturale, Antropologia e Etnologia, Florence
1870	Anthropologische Gesellschaft Wien, Vienna
1870	*Mitteilungen der Anthropologischen Gesellschaft Wien*
1870	Deutsche Gesellschaft für Anthropologie, Ethnologie und Urgeschichte, national
1870	Società Italiana die Antropologia e di Etnologia, Florence
1871	*Archivio per l'Antropologia e la Etnologia*
1871	Royal Anthropological Institute, London
1871	*Journal of the Anthropological Institute of Great Britain and Ireland*
1872	Magyar Nemzeti Múzeum, Ethnographic Department, Budapest (now at Néprajzi Múzeum)
1873	Smithsonian Institution, Division for Anthropology, Washington, D.C.
1873	American Museum of Natural History, Division of Anthropology, New York City
1875	Museo Nacional de Antropología, Madrid

(continued)

Table 1 (continued)

Year	Organization
1875	Neues Museum Berlin, Department for Anthropology, Ethnology and Prehistory (now at Berlin Ethnological Museum)
1875	Gesellschaft Deutscher Naturforscher und Ärzte, Division for Ethnology and Geography, and Division for Anthropology and Prehistory, Leipzig
1876	Naturhistorisches Museum, Department for Anthropology, Ethnology and Prehistory, Vienna (now at Weltmuseum)
1878	Königliches Zoologisches und Anthropologisch-Ethnographisches Museum, Dresden (now Museum für Tierkunde)
1878	Musée d'Ethnographie du Trocadéro, Paris (now at Musée de l'Homme)
1879	Anthropological Society of Washington, Washington, D.C.
1888	*American Anthropologist*
1879	Bureau of (American) Ethnology, Washington, D.C.
1880	*Annual Report of the Bureau of Ethnology to the Secretary of the Smithsonian Institution*
1879	Museum für Völkerkunde, Hamburg (now Museum am Rothenbaum)
1882	American Ass. for the Advancement of Science, Anthropological Section, Washington, D.C.
1884	University Museum, Oxford (now Pitt Rivers Museum)
1884	Jinruigaku no tomo (Friends of Anthropology), Tokyo
1886	*Jinrui Gakkai Hokoku* (Reports of the Anthropological Society)
1887	Museu Arqueològic i Antropològic, Buenos Aires (now Museo de La Plata)
1888	*Internationales Archiv für Ethnographie*, international
1898	Ethnographic Museum, Cairo
1889	Magyar Néprajzi Társaság (Hungarian Ethnographical Society), Budapest
1890	*Ethnographia*
1894	Columbian Museum, Department of Anthropology, Chicago (now Field Museum)
1901	Musée d'ethnographie, Geneva
1901	Temporary Commission for the Survey of Traditional Customs in Taiwan, Taiwan
1902	American Anthropological Association, New York City
1910	Geological Survey of Canada, Division of Anthropology, Ottowa
1910	Museo Nacional de Arqueología, Historia y Etnografía, Mexico City (now Museo Nacional de Antropología)

(Barth 2005; Vermeulen 2015; Huang 2017; Kuwayama 2017; Song 2017; Penaloza-Patzak 2018b)

City; Indigenous guides and interpreters such as George Hunt (1854–1933) in Fort Rupert, Canada; Archeologists like Albert Grünwedel (1856–1935) in Berlin; and missionaries such as US-based Henry Voth (1855–1931) and Reverend John Roscoe (1861–1932) in Uganda; Government agents such as Israel Powell (1836–1915) in British Columbia, and business men like Carl Hagenbeck (1844–1913) in Hamburg; Political activists Waldemar Bogoraz (1865–1936) and Lev Sternberg (1861–1927) in Siberia; and even couples who worked collaboratively, such as Felix (1854–1924) and Emma von Luschan (née von Hochstetter 1864–1931) and Eduard Seler

(1849–1922) and Caecilie Seler-Sachs (1855–1935) in Berlin; and Waldemar Jochelson (1855–1937) and Dina Jochelson-Brodskaya (1862–1941) in Siberia.

Conceived as an empirically based natural science dedicated to the investigation of human psychology and culture as enmeshed in the natural world, early anthropology was a fundamentally interdisciplinary endeavor in which knowledge creation was reliant on several fields of expertise. Consider, for instance, a few links in the chain of individuals and interests that bore ethnographic material from the field into the museum. Between 1881 and 1883, nearly 7,000 objects collected from along Canada's Pacific Northwest Coast were sent to the ethnographic department at the Neues Museum in Berlin, Germany. In order to understand why and how, we need to go back a number of years earlier, to the moment when Adolf Bastian, a widely traveled German national and ships' physician who had assembled a number of smaller ethnographic collections, met William Healey Dall. Dall was a US American naturalist who had taken part in several expeditions to the Pacific Northwest, and would go on to become the Honorary Curator of Mollusks at the Smithsonian Institution in Washington D.C. Dall, possessed of a salvage mentality now considered characteristic of his time (Gruber 1970), was convinced that Native American cultures would soon succumb to the purported civilizing influence of Euro-American culture, and impressed upon Bastian the urgent need to collect whatever possible material vestiges of the Pacific Northwest Coast cultures.

By 1876, Bastian had been appointed director of the anthropology, ethnology, and prehistory collections at Neues Museum in Berlin. With Dall's injunction still in mind, he set out to procure as large as possible a collection from the Pacific Northwest. First, he needed an experienced collector. Carl Hagenbeck, owner of the eponymous Hamburg zoo who had gained recognition in anthropological circles for his *Völkerschauen*, recommended his agent Johan Adrian Jacobsen. Jacobsen, a one-time whaler who had by that point been bringing people, animals, and object collections to Europe on Hagenbeck's behalf for several years, accepted the commission. He spent the next three-and-a-half-years in the Pacific Northwest assembling a collection of objects that, once in Berlin, amounted to some 6,730 accessions (Hatoum 2015, Glass and Hatoum manuscript). Bastian considered Jacobsen's expedition and resulting collection a resounding success. This was in part thanks to George Hunt, who had worked as Jacobsen's guide and interpreter during his time in Kwakwaka'wakw territory. Hunt leveraged his connections with local community members to help Jacobsen assemble a collection of unusual antiquity, and Jacobsen noted more than once that Hunt was the "best expert" he had "ever met" (Penaloza-Patzak 2018b).

Once the objects had arrived at the Berlin museum, Albert Grünwedel, one of Bastian's directorial assistants, oversaw the process of accessioning whereby the ethnographic objects were translated into scientific data. A specialist in Asian studies with a background in classical philology and archaeology, Grünwedel was assisted by Eduard Seler and Franz Boas, who had been taken on as temporary staff to help with the Jacobsen collection. Seler had a background in botany and mathematics, and was working toward his doctorate in philology with a dissertation on Mayan conjugation. Boas, with a doctorate in physics, was working toward his Habilitation in geography with a work that drew together geographic and anthropological research he had recently conducted on Baffin Island. In the years to come, these two would come to number among the world's most prominent Americanists, the latter even collaborating with Hunt to build a collection for the American Museum of Natural History (AMNH) in New York City, which would rival Jacobsen's, but at this juncture both were little more than newcomers to this emerging field.

This is just one of countless examples of the heterogeneity of practitioners involved in bringing together the raw material that was to serve as the basis for anthropological studies. Some, such as Hunt and Jacobsen might be characterized as classic "go-betweens" (Schaffer et al. 2009; Raj 2016), agents of intercultural exchange who combined linguistic, cultural, or interpersonal knowledge about a region with expertise in navigation or trading, without whom it would have been impossible to assemble, much less document and transport, an ethnographic collection of any magnitude or scientific quality. Others, like as Bastian, Seler, and Boas, came from backgrounds in the natural sciences and medicine, and were looking to understand what man-made objects could tell them about humankind and its relation to nature. Some like Grünwedel had trained in the humanities and brought a historical perspective to object analysis. In a sense they were all go-betweens, coming from diverse cultural and disciplinary backgrounds and making different material and intellectual contributions to the unique interpretations of anthropology, which were coming into being in different museums in different cities.

Epistemic frameworks and research orientations tended to vary from one nation, and indeed even city and institution, to the next. These were to a considerable degree guided by the leading figures in each domain: museum directors and governing members of scholarly societies who benefitted from the support of institutional infrastructures and government funding. Consider, for example, some of the major figures in the USA and German and Austro-Hungarian empires: In 1915 Franz Boas, who had by that point made a name for himself as a pioneer in modern US anthropology, credited John Wesley Powell, Brinton, and Frederick Ward Putnam (1839–1914) with shaping early US anthropology (Boas 1915). Powell ascribed to a unilinear notion of cultural evolution driven by democratization and technologization popularized by the likes of Lewis Henry Morgan (1818–1881) and Gustav Klemm, and his 1877 *Introduction to the Study of Indian Languages* – a workbook for the field – set a precedent in the study of Native American linguistics. After attending some university-level courses, Powell served as a cartographer with the Union Army, and was appointed director of the newly founded BAE in 1879. This organization was responsible for gathering the material relating to Native American populations collected by the US Department of Interior in the course of its geological surveys, and transmitting this to the Smithsonian Institution. In Boas's estimation, Powell had brought order to the US government's various strands of anthropological research as well to the process of collecting the data, which would be up to the next generation of practitioners to analyze (Boas 1915).

Brinton, who had taken a medical degree, served as a Union surgeon, and practiced civilian medicine for a few years before retiring to focus on anthropological research, shared Powell's views on the primacy of language for ethnographic research. In 1884 he was named Professor for Ethnology and Archaeology at the Philadelphia Academy of Natural Sciences, the oldest institution of its kind in the USA, and he did the groundwork for US recognition of anthropology as a scientific field by fostering its recognition in the academy. Brinton's work was guided by a conviction in the basic "psychical unity" of mankind, and considered schemas like Morgan's too mechanical to do justice to the complex processes involved in cultural development; at the same time, however, he maintained an unwavering belief that humankind was comprised of separate races with varying capacities to attain a Euro-American-based notion of ultimate civilization.

Whereas Powell and Brinton primarily focused their research on language, Putnam considered material culture the most essential resource for the study of culture, and carved out spaces for anthropology in the museum and university contexts analogous to those created by Powell and Brinton within the government and academy, respectively. As a young man, he had studied zoology at Harvard University under Louis Agassiz, but left before earning his degree. In 1864 Putnam was named director of the newly established Peabody Museum in Salem, Massachusetts (now Peabody Essex Museum), and curator at the Peabody Museum of Archaeology and Ethnology at Harvard in 1874. Over the course of the next 20 years, he would continue in the former position while at the same time taking part in the establishment of other outposts for anthropology, including the Columbian Museum (today Field Museum) in Chicago and AMNH in New York City. In his capacity at the Peabody and as a professor at Harvard, Putnam organized private funding for field work, directed archaeological digs, prepared some of the USA's first university-trained anthropologists, and secured a place for anthropology in the museum and university by implementing exacting methods and standards for documentation.

Anthropology in the USA came together under the direction of a motley group: A Civil War veteran turned geologist, a physician, and a zoologist turned archaeologist. The leading figures in Germany, where there was a clear prevalence of physicians, were in that sense more homogeneous (Ranzmaier 2011; Feest 1995). In addition to Adolf Bastian, Rudolf Virchow, who was also based in Berlin, also warrants special recognition for his role in the development of early networks in anthropology. Virchow received his medical degree from Friedrich-Wilhelms University and was appointed a full pathologist at Charité in 1846. Following his involvement in the March Revolutions of 1848, Virchow was forced to resign from the hospital and leave Berlin, but returned in 1856 to assume directorship of the Institute for Pathology at Charité, and accept the chair for Pathological Anatomy and Physiology at Friedrich-Wilhelms-Institut. An early proponent of microscopic pathology, Virchow's approach to anthropology is best elucidated via analogy to his contribution to postmortem studies: He introduced the method of individual and microscopic examination of each body part and organ – an empirical, inductive analysis of the history of interrelated units comprising an organism – which continues to serve as the foundation of contemporary practice (Penaloza-Patzak 2018b). Virchow's scientific perspective was pervaded by his insistence on observation-based analysis, a factor which also resulted in his vocal opposition to what he considered Darwin's as yet unproven, though increasingly accepted, theory of biological evolution as the projection of this onto analyses of human culture.

In brief elaboration on the sketch of Bastian already begun, it suffices for the task at hand to simply assert that his approach to ethnology complemented Virchow's to physical anthropology and prehistoric archaeology, and by the end of the nineteenth century the former's prolific publication activity and position as director of Berlin's ethnographic museum had made him one of the most visible proponents of German anthropology. Like Virchow, under whom he had studied, Bastian was also a vocal skeptic of Darwin's evolutionary theory and its extension to sociocultural studies, instead asserting cultural evolution to be a matter of deterministic parallel development. Like Virchow's method for autopsy, and similarly couched in biological terms, Bastian intended his museum's extensive material collections to serve as the raw data, which would lead from specific instances to the more general principles guiding cultural development. He considered each object a constitutive element in an aggregate data set that would eventually serve as the basis for future anthropological research of a statistical and "comparative genetic" nature, which would be ruled by "induction regulated by deduction" (Bastian 1900; Penaloza-Patzak 2018b).

The configuration of anthropology within the *Vielvölkerstadt* of the Austro-Hungarian Empire was yet different again, with two distinct traditions sprouting from each of the Empire's urban centers: a German-speaking tradition based in Vienna, and a Hungarian-speaking tradition based in Budapest. While the latter lays claim to one of Europe's earliest dedicated ethnographic collections (1847) and university chairs for anthropology (1881), the focus here will be on anthropology in Vienna, which by virtue of linguistic propinquity and the long and complicated history of rivalry between Austria and Germany presents a more intriguing comparison with the Berlin-based tradition. Some 550 km south-east from Berlin as the crow flies, the field of anthropology in Vienna was predominantly populated by geologists with an interest in prehistoric – or paleo – ethnology, which was considered a subfield of prehistory. In Vienna, as in the USA, research was primarily focused on cultures within the nation's own borders (Ranzmaier 2013; Feest 1995).

One of the most influential figures in Vienna-based anthropology was German-born Ferdinand von Hochstetter (1829–1844), a naturalist who had taken part in the Austrian Imperial Navy's first circumnavigation of the globe, known as the Novara Expedition (1857–59), and was later appointed professor of geology and minerology at the Technische Hochschule. In 1876 Hochstetter was named director of what would become the k.k. Naturhistorisches Hofmuseum (NHM), including a new department for anthropology and ethnology. He based his organizational concept for the museum on Darwin's evolutionary model, with the anthropology and ethnology collections to be divided into three categories: crania and skeletal material from throughout history and the world over; prehistoric European artifacts; and artifacts from primitive, ancient civilized American, and eastern and south Asian peoples. The idea was to signal parallels between departments, drawing physical anthropology into conversation with zoology, the prehistory collection animating the relationship of geology to history, and the ethnographic collection serving as part pendant to the prehistory collection and part illustration of "primitive" culture

(Penaloza-Patzak 2018b). Another important figure in Viennese anthropology was geologist Ferdinand Leopold von Andrian-Werburg (1835–1914) who, as we will discuss shortly, was instrumental in the extra-museal development of Vienna-based anthropology.

As Berlin began to gain more and more international recognition as a hub of anthropological activity, the tendency toward a more local focus among Vienna-based practitioners relegated that tradition to a more peripheral position on the international stage. Franz Heger, who had studied geology under Hochstetter at the Technische Hochschule and began working as his assistant in the royal collections in 1877, would go on to become a central figure in the internationalization of Vienna-based anthropology. Initially engaged to catalogue the ethnographic, prehistoric, and anthropological holdings, Heger was so taken with this work that he shifted his attention from geology to the new field of anthropology. When Hochstetter died in 1884, Heger was named director of the ethnology department, and despite or because of Andrian-Werburg's decision to focus AGW-affiliated research on Austria-Hungary and the Balkans, used his position at the museum as a starting point for cultivating relations with international colleagues and working to expand the department's non-European collections.

Gathering Places: Early Special Interest Societies

As the synopses imply, research of an explicitly anthropological nature first began to be structured and coordinated at the institutional, municipal, and national levels in the mid-nineteenth century. Up until that time, research tended to be carried out as an extension of practitioners' day jobs, and in relative isolation from other enthusiasts. Scientific societies thus posed a vital forum for exchange between early practitioners and the crystallization of disciplinary aims. The core membership of these societies was comprised of local- and nationally based practitioners and enthusiasts and select international and honorary members, but with society meetings often open to all, these became indispensable venues for obtaining news of the field and points of contact for local and visiting practitioners. Bearing in mind that we have as yet too little knowledge of the histories of too many non-Euro-American anthropological traditions to cite a definitive date or geographic location at which this field of inquiry entered into the purview of the world's learned societies, Table 1 presents a provisional overview of societies, journals, and museums which contributed to the coalescence of early anthropology, and offers a point of reference and departure for the following analysis of early network formation.

As these founding dates and locations intimate, early anthropology was almost inevitably anchored to the exploratory or expansionist endeavors of the nations within or on behalf of which it was practiced. Thus it was that the Dutch established the Royal Batavian Society of Arts and Sciences Section for Geography and Ethnology in 1788 in the city now known as Jakarta at a time when it was under colonial rule; and the Temporary Association for the Investigation of Taiwanese Old Customs was established by the Japanese government in 1901 shortly after winning the Sino-Japanese War (Vermeulen 2015; Song 2017). Intercultural encounter was an elemental aspect of national efforts toward expansionism and imperialism, of exploration, and of attempts to seek, establish, and maintain, not to mention relinquish, control. Physicians, geologists, and geographers were generally among the essential staff on any given foreign expedition, and alongside their work in these capacities, oftentimes collected observations and data of an anthropological nature as well.

The reports, visual depictions, and specimens they brought back from these trips awakened among Europeans an intensified awareness of themselves as such, and curiosity about how people lived outside of Europe. From the late seventeen hundreds up until the mid-nineteenth century, competing colonial interests within Europe and abroad led to intense rivalries between the United Kingdom and France, and Germany and Austria-Hungary in particular, with the USAs entering into this constellation in the early twentieth century. This era also ushered in the widespread establishment of national academies of sciences and their associated publications. The former became an invaluable tool for facilitating intellectual exchange between societies, as did key figures such as Benjamin Franklin (1706–1790), Jean-Baptiste Lamarck (1744–1892), Johann Gottfried von Herder (1744–1803), Georg Forster (1754–1795), and Alexander von Humboldt (1769–1857) whose travel and activity spanned and linked research in different nations.

By the mid-1860s, print media was flourishing and technological advancements in railway and steamship travel revolutionized movement throughout Europe and beyond. Scientific societies and institutions in, among others, St. Petersburg, Halle, London, Paris, Vienna, Philadelphia, and Washington, D.C. entered into reciprocal exchange agreements for society transactions and other scholarly publications. They became points of convocation for resident practitioners as well as those visiting from other lands, and therewith fostering in a more international sense scholarly communities, providing an intellectual, institutional, and economic infrastructure for the advancement of anthropology. In addition to Bastian, Dall, and Hochstetter, other well-traveled figures in early anthropological circles included Armand de Quatrefages (1810–1892), Morgan, Ferdinand von Hochstetter, Tylor, and Otis T. Mason (1838–1908).

During this era, there was a pervasive conviction among practitioners regarding the epistemic and legitimizing gains to be made from the appropriation of methods and concepts from the natural sciences, zoology, and botany in particular. At the same time, practitioners also assumed a relationship between problems in geography and ethnology on the one hand, and prehistory and anthropology on the other (Penaloza-Patzak 2018b). This state of affairs is reflected in the formulation of, for instance, the Royal Batavian Society of Arts and Sciences Section for Geography and Ethnology and the GDNÄ divisions for ethnology and anthropology. As noted with regard to Dall's collection-making advice to Bastian, many considered Euro-American culture the apex of civilization, and ranked other cultures according to the degree to which they approximated the same. Attitudes such as these reflected the projection of unilinear notions of biological evolution advanced by, among others, Darwin, on to the study of culture, and exerted a marked influence on the direction of anthropological research, classification, and exhibition while at the same time helping to justify Euro-American attempts to dominate the rest of the world's peoples. There were of course exceptions to this tendency, including the tradition that flourished in Berlin under Bastian (Bastian 1884), and ideological differences on this point had an unmistakable impact on the development of and inclusion in international networks.

Up until the mid-nineteenth century, scholarly societies were the primary point of convocation for local and traveling practitioners and enthusiasts of anthropology. Short-lived though they were, the Société des Observateures de l'Homme and Société Ethnologique de Paris were among the first of their kind, and the former began publishing its *Mémoires* in 1841. Three years later, Albert Gallatin (1761–1849), Edward Robinson (1794–1863), and Henry Schoolcraft (1793–1864) founded the American Ethnological Society (AES) in New York City, and began publishing the society's *Transactions* in 1864. These publications promoted and disseminated the work of society-affiliated scholars, constituted tools for the initiation and maintenance of scholarly exchange across municipal and national borders, and supplied a precedent and basis for the formation of subsequent transnational journals, such as the *Archiv für Anthropologie* and *Internationales Archiv für Ethnographie*, and specimen exchange networks.

The AES, a gentlemen's social club modeled on the Société Ethnologique de Paris, was dedicated to promoting what it deemed the "most important and interesting branch of knowledge, that of Man and the Globe he inhabits." With 152 members by 1845, the bulk of foreign corresponding and honorary members who the society elected were based in the United Kingdom (23), more specifically London (19), which had recently become home to an anthropological society of its own. The French Empire and German Confederation were matched as home to the next largest groups of foreign members (14 each), with nearly half of the members within the latter residing in Berlin (6), most notably von Humboldt, Jacob Grimm, Richard Lepsius, and Leopold Ranke. The Austrian Empire was, in contrast, home to just two elected members, both in Vienna and one of whom was the American consul (AES 1845).

In 1869 Bastian, Virchow, and others cofounded the aforementioned BGAEU and its *Zeitschrift für Ethnologie* (*ZfE*, journal for ethnology). In 1870, Andrian-Werburg helped found the Anthropologische Gesellschaft Wien (AGW, anthropological society of Vienna) and its *Mittheilung der Anthropologischen Gesellschaft in Wien* (*MAGW*, report of the anthropological society in Vienna), and later on that year the Deutsche Gesellschaft für Anthropologie, Ethnologie und Urgeschichte (DGEAU, German society for anthropology, ethnology and prehistory) was established. AGW officials elected to maintain a certain distance from the latter, which was intended as an umbrella organization to link German-language efforts in the field, and underscored their sovereignty by electing pathologist Carl von Rokitansky (1804–1878) as the AGW's first president. This appointment was intended to rival the acclaim of Virchow and Bastian and as a direct challenge to Berlin-based anthropology more broadly. Like Andrian-Werburg, Hochstetter and others at the AGW, Rokitansky was a vocal supporter of Darwin (Ranzmaier 2013; Gingrich

2005; Müller 1871), and enthusiastic in the application of biological evolutionary principles to the study of culture.

Differences in the status accorded to Darwin by the BGAEU and AGW demonstrate the extent to which society leaders' scientific beliefs impacted the memberships they extended. Darwin was named one of the AGW's first foreign honorary members in 1872 (followed 1 year later by Ernst Haeckel), but was never accorded anything more than corresponding membership to the BGAEU, and this some 80 years after the society was founded (*Mittheilungen der Anthropologischen Gesellschaft in Wien* 1872; Zeitschrift für Ethnologie 1877). By the mid-1880s, the disconnect between the internationally oriented program Heger was developing at Vienna's NHM and the AGW's focus on anthropology within the *Vielvölkerstadt* had rendered anthropology in the Austro-Hungarian capital less cohesive than in, for instance, Berlin, where the anthropological society and ethnographic museum were under the same leadership and shared intricately intertwined research agendas. This, alongside the AGW's calculated remove from German-based anthropology, initially consigned practitioners in Austria-Hungary to a more peripheral position within the trans-European and transatlantic networks, which practitioners began initiating in this era. This notwithstanding, Heger's sustained efforts throughout the 1890s and early twentieth century, in particular his decision to host the 1908 International Congress of Americanists (ICA) at the NHM, ultimately succeeded in drawing Austrian-based practitioners into more sustained conversation with their international colleagues.

Upon its establishment in 1879, the Washington D.C.-based BAE was, as mentioned, initially tasked with transmitting data on Native American populations from the Department of Interior to the Smithsonian Institution. Soon however, the Bureau became a research entity in its own right. The Anthropological Society of Washington (ASW) was likewise established in 1879, BAE Director Powell as its founding president, and soon-to-be Smithsonian ethnology curator Otis T. Mason its secretary. In 1888 the ASW began issuing a journal, today known as *American Anthropologist*, and the society's influence soon superseded that of the AES, the former seat of US-based anthropology, in New York City. Thus, it was that the foremost venues for anthropology in Washington, D.C. were, as in Berlin, overseen by the same few individuals, and as a result of a certain extent of consistency in their aims and practices quickly attained a higher degree of visibility and recognition than was the case with, for example, Vienna.

Ethnographic Museums and Society Membership: A Transnational Case Study

In the early nineteenth century, objects of an ethnographic nature began to be culled from older, in many cases imperial, collections of natural and manmade curiosities and exhibited in dedicated ethnographic departments and museums. By the 1880s several such institutions had been established in cities across Europe, Russia, and the Americas. These object collections became local centers for anthropological research and popular education alike and, as the ease of travel increased, also became points of pilgrimage for practitioners interested in learning how the new field of anthropology was conceived and practiced in different cities. They embarked on national and international museum tours, with a stay in any given city generally entailing curator-led introductions to local museum and library collections, formal and informal gatherings with area practitioners, and attendance at local society meetings. Upon return home, practitioners' museum reports were published and disseminated via local institutional or society publications, and were at times even picked up for translated and reprint in publications further afield (Boas 1887; Bahnson 1888; Dorsey 1899). This form of information finding and networking became a powerful vehicle for not only smaller-scale intellectual and material exchange but the broader coalescence of the field (Penaloza-Patzak 2018b). The following presents a microstudy of transnational networks as they were initiated and spread between practitioners in Berlin, Vienna, and the USA.

Ethnographic Departments and Museums

The earliest public ethnographic departments and museums in the North German Confederation were established in Munich, Leipzig, Berlin, Dresden, and Hamburg. Following unification in 1871, Berlin, which had up until that point served as the capital of Prussia, was named capital of the new empire. In 1886, the ethnographic department within Berlin's Königlichen Museen (KMB) was transferred to a dedicated building, the Königliches Museum für Völkerkunde (KMV). Directed by Bastian, this institution was the chief destination for ethnographic acquisitions made in the name of the empire, and soon gained special recognition among international practitioners for the magnitude of its collections, which included extensive holdings from the Americas.

A decade before in Philadelphia, material collected from North Pacific indigenous communities, which was on prominent display at the Centennial Exposition World's Fair, sparked what has been dubbed a "scramble for Northwest Coast artifacts" (Cole 1985). By the mid-1880s, Bastian's museum had assembled one of the world's premier collections of objects from the Pacific Northwest. The museum's oldest material from the region was collected during Captain Cook's third voyage (1776–1780), and extended to include objects brought back from the second and third circumnavigations of the Prussian merchant ship Prinzeß Louise (1829–1834), as well as the previously Jacobsen collection, which comprised some 7,000 objects (Hartmann 1973; Bolz and Sanner 1999; Glass and Hatoum manuscript). These holdings would prove key in initiating a lively exchange network between practitioners working in Berlin and North America.

In 1866 the Peabody Museum of American Archaeology and Ethnology was established in Cambridge, the first of its kind in the US context. A few years later in 1873, a division for anthropology was established at the AMNH in New York City. The Smithsonian Institution, which had begun accepting ethnographic and anthropological material collected during government surveys as early as 1858, finally established its own division for anthropology in 1883 at the US National Museum (USNM). Otis T. Mason was curator of ethnology within that division, which was expanded into a department in 1897. Whereas most European ethnographic museums concentrated on collection budgets and activity on non-local material, the situation in North America was quite different. The express focus at the Peabody, AMNH, and Smithsonian was material documenting indigenous North and South American cultures. By the mid-1880s, the USNM had already gained a certain level of international recognition, and certainly more so than the AMNH, which had yet to appoint an anthropology curator; however the AMNH laid claim to a more extensive collection of highly desirable Pacific Northwest holdings, including some 500 objects of considerable antiquity, which the museum had purchased from Canadian Indian Agent Israel Powell, and some four thousand objects purchased from US Naval Lieutenant George T. Emmons (Willmott 2006; Glass 2011).

Hochstetter established the first ethnographic department in Vienna in 1876, but it wouldn't be open to the public at the new Naturhistorisches Museum (NHM, natural history museum), and this under the directorship of Heger, until 1889. With Habsburg interests in the Americas dating back to the Viceroyalty of New Spain, and more recently underscored by the ascendancy of Austrian Archduke Maximillian to the Mexican throne (1864–67), Austria-Hungary had a complicated history of American involvements. Among Vienna's most valuable ethnographic holdings at the time included Pacific material brought back to Europe by Cook and Johann Natterer's (1787–1843) collections made during the Austrian Brazil Expedition (1817–1835). Today the former comprises the world's second largest surviving Cook collection and, alongside the Cook material in Berlin, highlights the role of early collections in awakening among Europeans an awareness of non-European cultures, entwining European capitals via their various stakes in these collections, and stoking colonial ambition (Kaeppler and Stone 2011, Feest 2011).

National Societies, International Memberships

By the early 1890s, Bastian and Virchow, J.W. Powell and Otis T. Mason, and Andrian-Werburg and Heger were recognized as among the most prominent figures working in anthropology in their respective nations. As the differences in status accorded by the BGAEU and AGW have shown, society membership can be a profitable tool for gauging transnational engagements and establishing where individual practitioners were particularly active or recognized. As mentioned, early foreign membership in the AES manifested a clear bias toward England, but by the 1880s the USA's first anthropological society had all but disbanded, and held little sway over the direction of US anthropology. In its stead, the Smithsonian Institution and ASW in Washington, D.C. had become the new loci of anthropological research. The ASW named its first German and Austro-Hungarian honorary and corresponding members in 1883. Of special note among the former were Bastian, Virchow, Virchow's former student Ernst Haeckel, and A.B.. Meyer, director of the Zoologisches und Anthropologisch-Ethnographisches Museum in Dresden.

Of the ASW's German memberships, it is no surprise that Bastian and Virchow were chosen as representatives of German-based anthropology; tactically speaking, it would have been foolish to ignore the men at the helm of that empire's anthropological society and collection. Still, the more theoretical dimensions of their work were far from consistent with the trend in Washington, D.C., where Powell, and the ASW community more generally, were nothing if not sure about the rationality of applying unilinear evolutionary principles to analyses of human culture. This would emerge as one of the most basic points of departure between the Berlin and Washington, D.C.-based knowledge systems, and one which would prove the most difficult to reconcile. Still, Berlin and Washington, D.C.-based luminaries shared a common, indeed epochal, dedication to amassing the raw data that they imagined would fuel future anthropological study, and each took a vested interest in the other if for no reason more than the potential he represented to expand their own ethnographic holdings via exchange (Penaloza-Patzak 2018b).

The BGAEU began naming US-based corresponding and honorary members in 1872, the first of these being Belgian-born archaeologist Charles Rau (1826–1887) at the Smithsonian. Bastian spent 1875–1876 collecting in the Americas. On his way back to Berlin, he stopped in Philadelphia to have a look at the Centennial Exhibition (Fischer 2007) and in Washington, D.C., where he met with the director of the Smithsonian and inquired into the possibility of establishing a reciprocal exchange program (Penaloza-Patzak 2018a, b). Bastian likely came into contact with Dall in one of these two places, and this acquaintance alongside the Pacific Northwest ethnographic material on display at the Centennial Exhibition instilled within him the sense of urgency that would go on to fuel the Jacobsen collection. Over the course of the next 5 years, the BGAEU elected several more US-based members who were in one way or another affiliated with the Smithsonian or had contributed notable ethnographic collections to the Centennial Exhibition. Among these included Powell, George M. Wheeler, and F.V. Hayden (two other former US geological surveyors, 1842–1905 and 1829–1887, respectively), and German-born polar explorer Emil Bessels (1847–1888).

For all of Vienna's relative proximity, the BGAEU extended markedly less memberships to scholars based there, a circumstance in large part due to the ideological and organizational dissimilarities between the two traditions. The first honorary membership was extended to Hochstetter in 1881 (nearly 10 years after the first US-based members were elected), and the second to Heger over 10 years later in 1893. As mentioned, the Vienna-based AGW extended few foreign memberships in its early years. Like the AES in New York City, the majority of early foreign members elected to the ASW hailed from England and France. By 1883, however, this trend began to shift, and the number of German-based AGW members began to break even with the number of those working in France. In 1884 Bastian and Powell were both named honorary members, and by 1886 the list had expanded to include, among others, Virchow, Brinton, and Rau.

The early prevalence of memberships accorded by the anthropological societies of New York City, Berlin, and Vienna to practitioners in London and Paris underscores the extent to which these two cities were an early nexus for forerunners in the development of the field. This perception changed as the century wore on. By the mid-1880s, the constellations of participation within the anthropological societies in Berlin, Washington, D.C., and Vienna show that foreign membership conferrals began to proceed less from aspirational affiliations and more from matter-of-fact contact made in the course publication exchanges, society meetings, museum tours, and research visits (Goldstein 1994; Nichols and Parezo 2017; Penaloza-Patzak 2018b). Comparative analysis of these memberships from year to year moreover indicates that they also progressed in a largely reciprocal fashion, with the BGAEU leading the group both in terms of the early date at which it began conferring foreign memberships and number conferred.

National Interests and the International Congress of Americanists

Another organization that merits special consideration within the tangle of anthropological interests that converged on the Americas in the late nineteenth and early twentieth centuries is the International Congress of Americanists (ICA), established in Nancy, France in 1875. With an invitation issued to "All persons engaged in the study of America ... and the ethnographical writings on the races of America," the impulse for this interdisciplinary affair dated back to Alexander von Humboldt's time spent writing up the findings from his American travels in Paris (Fletcher 1913). While the authority to host annual meetings was initially restricted to European nations, the subject matter awakened broad appeal far beyond those borders, and twenty-nine countries in total from, among others, North, Central, and South America, sent delegates to the ICA inaugural congress. In 1881 Powell became the first to submit a report on US-based Americanist activity to the ICA. This document, which closed with a call for future meetings in the USA and the remark that "the [Americanist] field has already been extensively cultivated [by US scholars] and its magnitude recognized," stands as early indication that the theoretical, methodological, and material contours of anthropology in and of the Americas were becoming increasingly contested themes on the stage of European and European-American scientific relations well before the turn of the century (Penaloza-Patzak 2018b).

Personal Correspondence and Institutional Object Exchange

Correspondence, publication, and specimen exchanges constituted the mainstays for initiating, maintaining, and advancing the individual bonds between practitioners and institutions that comprised transnational networks. Analysis of these three mediums for exchange, or combinations thereof, in conversation with one another affords panoramic overviews of the arc of international engagements between constellations of practitioners, as well as detailed glimpses into specific episodes in those trajectories. It also offers a great deal of insight into the dynamics informing how these relationships developed and were conducted, and how this changed over time in response to, among others, economic and national or international political circumstances. Take, for example, Table 2, which presents the quantitative record of Franz Boas's correspondence with colleagues in Berlin from 1880–1923. This data provides a point of departure for establishing how a junior scientist who started out in a precarious position in Berlin maintained contact with, and eventually became deeply embedded within, German anthropology while at the same time functioning as a formative figure in US anthropology.

Close attention to material exchange yields an intimate glimpse into the mechanics of exchange that shaped these networks. Regarded in conjunction with Table 3, which presents the history of material transactions realized by US institutions with institutions in Germany and Austria, we can begin to plot the exact contours of the roles that Boas and other fellow practitioners played in the coalescence of transnational exchange networks. Of special note here is the high correlation between the frequency of correspondence exchanged between individuals on one hand, and of specimens exchanged between institutions with which those individuals were affiliated on the other, a circumstance which of course underscores the fundamental interdependence of interpersonal and interinstitutional engagements. Likewise significant is the evidence these tables present regarding the potential of individual practitioners to shape scientific exchange on an international scale. Boas spent more than the first decade of his career moving from one short-term appointment to the next, first in Germany and then in the USA. As Table 3 demonstrates, the scope of longer-term exchange networks initiated while he worked at each of these institutions outlived his tenure at any single one of them, and posed a vital contribution to the formation of an internationally geared branch of Americanist anthropology.

Before relocating to the USA, Boas spent a year training in the Neues Museum ethnological collections under Albert Grünwedel, and Grünwedel and Bastian were his primary Berlin-based correspondents during his first years working in anthropology. Shortly after moving to the USA, Boas, who had yet to secure steady full-time employment, brokered the first explicitly ethnographic exchange to take place between the KMV and the USNM in Washington, D.C. Bastian had laid the foundations for this exchange during his 1876 meeting with the then director Joseph Henry (1797–1878), but it wasn't until 1887 that Boas's sustained presence in Washington made it possible to successfully negotiate such a transaction, and this using his own recent Pacific Northwest Coast collection as the Berlin contribution (Penaloza-Patzak 2018a).

Following the conclusion of that exchange, Bastian's communications to Boas dwindled, while communications from Bastian's assistant Grünwedel increased. A subset of letters Grünwedel and Boas exchanged during this time are especially interesting for what they can tell us about the branching of their correspondence, what had up until that time been one-on-one, into a network. The trajectory began with Boas courting Grünwedel's help in organizing an exchange between the ethnology department for the 1893 World's Columbian Exhibition (WCE) in Chicago (where Boas was at that point working as assistant to department director Fredrick Ward Putnam) and the KMV. Putnam's vision for the department was to depict the "material and moral progress of American civilization" since the time of Columbus's arrival in the so-called new world (Hinsley 2016). Grünwedel,

Table 2 Correspondence received by Boas from colleagues in Berlin and Vienna, 1880–1923[a]

Correspondent	1880–1887	1888–1892	1893–1897	1898–1902	1903–1907	1908–1913	1914–1918	1919–1923	Total
A. Bastian (KMV)	13	5	6	3	–	x	x	x	**27**
A. Grünwedel (KMV)	10	17	1	3	1	2	–	–	**34**
E. & S. Seler (KMV)	–	3	7	6	8	2	5	2	**33**
K. von den Steinen (KMV)	–	–	3	5	14	1	5	6	**34**
F. & E. von Luschan (KMV)	2	1	6	3	1	3	9	6	**31**
F. Heger (NHM)	–	3	5	2	7	18	7	6	**48**
TOTAL	**25**	**29**	**28**	**22**	**31**	**26**	**21**	**20**	**201**

[a]Sample based on surviving correspondence housed at the American Philosophical Society, Philadelphia, PA. X = deceased

Table 3 US Ethnographic Transactions with Institutions in the German and Austro-Hungarian Empires, 1887–1909[a]

Year	Institutions	Individuals	Type	Material
1887	USMN / KMV	Boas / Bastian	Exchange	Canadian items / North Pacific coast items
1893	WCE / KMV	Boas (Putnam, Holmes) / Seler	Exchange	Papier mâché casts, Honduras / Plaster casts, Guatemala
1893	WCE / NHM	Boas (Putnam, Holmes) / Heger	Exchange	Papier mâché casts, Guatemala / Polynesian and Central African items
1895	KMV / AMNH	Seler / AMNH	Gift	Seler exped., Mexican items
1895	USNM / NHM	Dall / Heger	Gift	Archeological items
1898	AMNH / KMV	Boas / unknown	Exchange	"Eskimo" items / "White Nile" items
1898	AMNH / König. Mus., Dresden	Boas / Meyer	Exchange	Alaskan items / items from Dutch New Guinea
1899	AMNH / KMV	Boas / von den Steinen	Gift	Jesup exped., Pacific Northwest items
1899	Peabody / KMV	Putnam / KMV	Gift	Plaster casts, Copan
1900	AMNH / KMV	Saville / Bastian	Unknown	Oaxacan tomb lintel
1902	AMNH / KMV	Boas / Grünwedel	Exchange	Plaster casts, Vancouver Island / Plaster casts, British Columbia, busts
1902	AMNH / NHM	Boas / Heger	Gift	Plaster casts
1903	AMNH / KMV	Boas / Stumpf	Gift	Phonographic recordings
1903	AMNH / KMV	Boas / von den Steinen	Sale	Cumberland Sound Inuit items
1904	USNM / KMV	Holmes / Seler	Unknown	Plaster casts
1905	AMNH / KMV	Saville / Seler	Exchange	Plaster casts
1905	Columbian Mus. / KMV	Dorsey / von den Steinen	Exchange	Wyman coppers / Brazilian items
1905	Columbian Mus. / KMV	Dorsey / von den Steinen	Sale	North American baskets
1905	USNM / KMV	Holmes / Grünwedel	Exchange	Plaster casts and other items / unknown
1906	AMNH / KMV	Boas / von den Steinen	Sale	Comer collection items
1906	AMNH / KMV	Boas / Seler	Gift	Botanical specimens
1909	AMNH / KMV	Saville / Seler	Unknown	Cayapa collection

[a]Based on records housed at the respective institutions, see Penaloza-Patzak 2018a, b for more detailed information and analysis

preoccupied with other matters, forwarded Boas's request to Africanist Felix von Luschan, who ultimately forwarded it to Eduard Seler, the museum's newly appointed Americanist.

A specialist in pre-Columbian cultures, Seler was enthusiastic about an exchange that had the potential to bring more American material into the Berlin collection. So too was Franz Heger at the NHM in Vienna, and both agreed to contribute ethnographic material to the WCE on the condition that their institutions were compensated in kind with American material once the Chicago fair was over. As Table 3 indicates, the 1893 exchanges initiated a surge of transactions that would cement relations between these nations for more than a decade to come. It was as a result of these transactions that Putnam was brought into correspondence with Heger in Vienna and Seler and Bastian in Berlin, and that Heger was able to actualize the NHM's first explicitly ethnographic exchange with a US museum – therewith gaining footing within the landscape of transnational exchange. At the close of the fair, William H. Holmes (1846–1933) at the BAE was named curator of ethnology at the new Columbian Museum, which was built to house the collections assembled in Chicago for the fair. Holmes consequently assumed responsibility for concluding the exchanges Boas initiated with Seler and Heger, and therewith inherited a precedent of exchange with the KMV and NHM for the fledgling museum in Chicago.

The exchanges Seler and Boas organized for the WCE were the first of many collaborations between the two in years to come. Following Boas's appointment as Assistant Curator of Ethnology at the AMNH in New York City, their alliance would bring about a veritable flood of transactions between the AMNH and KMV, and provide the framework for a robust network of practitioners linking the ethnographic departments at their respective museums. In 1897, George Dorsey (1868–1931), who had studied under Putnam at Harvard and received the first US Ph.D. in anthropology, took over Holmes's position at the Columbian Museum while Holmes returned to the Smithsonian as head curator of Anthropology. Holmes carried the relationships with the ethnographic departments in Berlin and Vienna to his new position in Washington, D.C., while Dorsey was initiated into the exchange network by virtue of his new role as curator of the Chicago collections. The practice of specimen exchange thus fostered relations between practitioners and institutions in different countries while at the same time propagating itself via everchanging interpersonal and interinstitutional alliances and affiliations.

Between 1911 and 1914 the International School for American Archeology and Ethnology (ISAAE) in Mexico, initiated by Boas and with Seler as its first sitting director, brought a transnational group of Americanist anthropologists into closer cooperation than ever before (Penaloza-Patzak 2018b). Negotiations to establish the school had begun as early as 1905, but it wasn't officially called into being by a committee of delegates from Columbia University, Harvard, the University of Pennsylvania, Germany, Mexico, and France until the 1910 International Congress of Americanists in Mexico City. The ISAAE's directorate implemented a multi-institutional and multinational funding system intended to free their research program from constraints set forth by any given sponsor, and in exchange for their financial support sponsors could send one student per semester to train at the ISAAE and received a portion of material collected in the course of field research.

In April 1914, US forces occupied Veracruz, aggravating the already tense atmosphere resulting from the ongoing Mexican Revolution, and gave rise to

ISAAE sitting director Alfred M. Tozzer's (1877–1954) swift departure from Mexico. The school and its collections were placed in the care of former ISAAE director Manuel Gamio (1883–1960), Alfonso Pruneda (1879–1957), and George Charles Marius Engerrand (1877–1961), all of whom were Mexican nationals. Two months later, in June of 1914, an even larger-scale conflict diverted ISAAE affiliates' attentions away from their work and collections in Mexico. The assassination of Archduke Franz Ferdinand, heir to the Austro-Hungarian throne, occasioned a cat's cradle of international alliances, which by the end of August had likewise drawn Russia, Germany, France, and the UK into the First World War. Although the US maintained an official stance of neutrality until 1917, and efforts to sustain exchange between US, German, and Austro-Hungarian practitioners continued up until that point, the field of Americanist anthropology – in particular at the government-affiliated USNM and BAE in Washington, D.C. – became increasingly politicized. When the USA entered the war on the side of the allies, this gradual suffocation of international exchange turned into to a period of virtually suspended interaction. As the war came to an end, some US-based practitioners made attempts to reconnect with colleagues working in Central Powers nations, but these reconfigured networks operated on very different terms, necessitating new modes and means of exchange (Penaloza-Patzak 2021).

Conclusion

By 1910 the number of specimen exchanges sent shuttling between Americanist networks in the USA and Germany had already begun to slow. The vestiges and inheritors of the anthropological networks, which had come together during the mid- and late nineteenth century coalescence of anthropology to facilitate the exchange of knowledge and material in the museum had begun migrating to other, in large part university, institutional structures. This was an artifact of the more general ebbing of the museum age. Museum collections remained important tools for research and instruction, but there was a fundamental shift in field priorities from the amassment of specimen collections to the professionalization of anthropology. Many of the disciplinary aims and methods, which had been developed in the course of collections-based work in museums, were now codified. The age of large-scale government and privately funded collection was winding to an end, more and more universities were establishing anthropology programs of their own, and it was this setting which was considered most conducive to the training of future specialists. During the earliest decades of the museum age, practitioners who came from a wide range of disciplinary backgrounds had little recourse to formal training and developed fluency in anthropology in the course of their work. University instruction introduced a potential for instructional standardization unknown in the museum era, and one which the initiated increasingly equated with legitimacy and professionalism. The practice of ethnographic specimen exchange continued to play a role in cementing transnational research into the 1910s, but increasingly became a by-product, rather than a focus, of those efforts.

In the 1880s and 1890s, practitioners and practitioners in the making toured foreign museums to learn how anthropology was carried out elsewhere. Once university-based anthropology became institutionalized around the turn of the twentieth century, Boas, Seler, and other university-based anthropologists began encouraging students to round-out their training via research stays with colleagues within their professional networks at foreign universities, students began supplanting specimens as the primary vehicle of exchange (Penaloza-Patzak 2018b). Temporarily upended by the First World War, elements of this mobility persisted well into the interwar period, by which point they became part of a wholly different – and less equal – economy of exchange.

By the onset of World War I, the USA was coming into its own as an economic and political power, yet Berlin, where anthropologists traced a scientific fascination with the Americas back to at least as far back as Alexander von Humboldt's early nineteenth century expeditions, still stood as an indisputable center for Americanist research. The war politicized this position, and marked a crisis of sorts in the field. In the interest of vaulting themselves into place as masters of a field many considered theirs by birthright, groups of US-based practitioners in Washington, D.C. and elsewhere sought to gain purchase over Americanist anthropology by, for example, ostracizing German and Austrian colleagues and appropriating the ICA during the war-induced vacuum of European influence (Penaloza-Patzak 2018b).

Boas' initial aspiration had been to foster international cooperation by refashioning the ICA into an international ethnological congress with an American subdivision. This initiative failed to resonate, however, with many Washington D.C.-based colleagues, who hoped to merge the ICA with the Pan-American Conference (the latter dedicated to establishing independence from Europe by encouraging commercial, social, economic, military, and political cooperation between North, Central, and South American nations) and therewith promote a US – and at the very most pan-continental – strain of Americanist anthropology (Penaloza-Patzak 2018b). Upon USA's entry into the war, conventional postal communication between US practitioners and those in Central Powers nations became all but impossible and the Smithsonian's International Exchange Service ceased shipments to Germany and Austria-Hungary. By the end of the 1910s, the relative ease with which prewar cooperations with German and Austrian colleagues and museums had been accomplished was a world away.

The First World War did not extinguish relations between anthropologists in the USA, Germany, and Austria, nor did it stop the international circulation and exchange of individuals, correspondence, publications, and specimens. Still, the generational transformations within the field of anthropology and widespread financial hardship that beset Europe during and in the wake of the war brought about a fundamental reconfiguration of the interests and aims of the networks, which had previously drawn those fields together. Even once peace was reestablished, organizations such as the International Research Council put forth concerted effort to bar German and Austrian scientists from participating in the postwar reconstruction of international science, an intervention which had a an immediate and pronounced effect on the intellectual and material resources of anthropologists working in former

Central Power nations. In the months immediately following the war, it became clear that economic conditions were such that German and Austrian anthropologists, societies, and institutions lacked the means to conduct and publish research as they had in the prewar years, much less participate as equal partners in the production of an internationalist strain of Americanist anthropology (Penaloza-Patzak 2021).

Nationally oriented histories of anthropology and other scientific disciplines offer critical insight into how intranational circumstances impact the ways in which science could and does come into being and is practiced in different countries. Disciplinary histories also have their place. Both, however, offer insight into merely one facet of what are actually transnational and transdisciplinary histories. They are in essence first steps in the direction of understanding scientific disciplines as in relation to global dynamics and constellations of questions, methods, and theories rooted in several other disciplines and which may stand in seemingly little relation to one another from a contemporary point of view. Such is the case with early anthropology.

Close attention to transnational and transdisciplinary networks offers unique insight into the complex dynamics at play in the development of scientific disciplines, and reveals disciplinary development as phenomena inherently shaped by a highly contingent body of global factors. The contours defining the study of human kind in different cities – even linguistically and geographically close-lying cities like Vienna and Berlin – had the potential to vary greatly. Social and political alliances, ideological orientations, and economic and material resources were all highly influential on the type of anthropology that could be and was carried out at any given institution at any given time. These local orientations and circumstances in turn, provide a great deal of insight into the dynamics at play in how and why individuals chose to initiate and maintain contact with and between specific institutions and societies in other nations, and how these contacts developed into broader networks connecting practitioners in different countries by way of shared disciplinary perspectives, concerns, and material resources.

References

American Ethnological Society (1845) Transactions of the American ethnological society 1. Bartlett & Welford, New York

Ash M (2008) Forced migration and scientific change after 1933: steps toward a new approach. In: Scazzieri R, Simili R (eds) The migration of ideas. Watson Publishing International, Sagamore Beach, pp 161–178

Bahnson C (1888) Ueber ethnographischen Museen mit besonderer Berücksichtigung der Sammlungen in Deutschland, Oesterreich und Italien. Mittheilungen der Anthropologischen Gesellschaft in Wien 18:109–164

Barth F (2005) Britain and the commonwealth. In: Barth F, Gingrich A, Parkin R, Silverman S (eds) One discipline, four ways: British, German, French and American anthropology. University of Chicago Press, Chicago, pp 3–57

Bastian A (1884) Einige Blätter zur Kolonial-Frage. Ferd. Dümmlers Verlagsbuchhandlung Harrwitz und Gossmann, Berlin

Bastian A (1900) Völkerkunde und Völkerverkehr, unter seiner Rückwirkung auf die Volksgeschichte. Weidmannsche Buchhandlung, Berlin
Boas F (1887) Review of Dr. Kristian Bahnson on European ethnographical museums. Science 10: 245–246
Boas F (1915) Frederic Ward Putnam. Science 42:330–332
Bolz P, Sanner H (1999) Indianer Nordamerikas, die Sammlung des Ethnologischen Museums Berlin. G+H Verlag, Berlin
Cole D (1985) Captured heritage: the scramble for Northwest Coast artifacts. University of Washington Press, Seattle
Dorsey GA (1899) Notes on the anthropological museums of Central Europe. Am Anthropol 1(3): 426–474
Feest C (1995) The origins of professional anthropology in Vienna. In: Rupp-Eisenreich B, Stagl J, Acham K (eds) Kulturwissenschaften im Vielvölkerstaat: zur Geschichte der Ethnologie und verwandter Gebiete in Österreich, ca. 1780 bis 1918. Böhlau, Vienna, pp 113–131
Fiedermutz-Laun A (1970) Der kulturhistorische Gedanke bei Adolf Bastian. Systematisierung und Darstellung der Theorie und Methode mit dem Versuch einer Bewertung des Kulturhistorischen Gehaltes auf dieser Grundlage. F. Steiner, Wiesbaden
Fischer M (2007) Adolf Bastian's travels in the Americas (1875–76). In: Fischer M, Bolz P, Kamel S (eds) Adolf Bastian and his universal archive of humanity. G. Olms, Hildesheim, pp 191–206
Fletcher AC (1913) A brief history of the international congress of Americanists. Lancaster, The New Era Printing Company
Gingrich A (2005) From the nationalist birth of Volkskunde to the establishment of academic diffusionism: branching off from the international mainstream. In: Barth F, Gingrich A, Parkin R, Silverman S (eds) One discipline, four ways. University of Chicago Press, Chicago, pp 76–93
Glass A (2011) Objects of exchange: social and material transformation on the late nineteenth-century Northwest Coast. Yale University Press, New Haven
Glass A, Hatoum R (Manuscript, volume in preparation for publication) From British Columbia to Berlin and back again: Jacobsen's Kwakwaka'wakw collection across three centuries. In: Baglo C (ed) Johan Adrian Jacobsen: trader of traditions
Goldstein D (1994) "Yours for science" the Smithsonian Institution's correspondents and the shape of scientific community in nineteenth-century America. Isis 84(4):573–599
Goodrum MR (2016) The beginnings of human palaeontology: prehistory, craniometry and the 'fossil human races'. Br J Hist Sci 49(3):387–409. https://doi.org/10.1017/S0007087416000674
Gruber JW (1970) Ethnographic salvage and the shaping of anthropology. Am Anthropol 72(6): 1289–1299
Hatoum R (2015) The Berlin Boas Northwest Coast collection: a challenging vocabulary for cultural translation. In: Etges A, König V, Hatoum R, Brüderlin T (eds) Northwest Coast representations: new perspectives on history, art, and encounters. Reimer, Berlin, pp 27–66
Hinsley CM (2016) Anthropology as education and entertainment. Frederic Ward Putnam at the world's fair. In: Hinsley CM, Wilcox DR (eds) Coming of age in Chicago: the 1893 world's fair and the coalescence of American anthropology. University of Nebraska Press, Lincoln, pp 1–77
Huang S (2017) Convergence and divergence between Taiwanese and US anthropologies. Asian Anthropol 16(3):181–189. https://doi.org/10.1080/1683478X.2017.1346963
Ildikó P, György P (2011) Hungary/Magyar köztársaság, introduction: a brief history and current state of physical anthropology in Hungary. In: Márquez-Grant N, Fibiger L (eds) The Routledge handbook of archaeological human remains and legislation. An international guide to laws and practice in the excavation and treatment of archaeological human remains. Routledge, New York, pp 185–202
Kaeppler AL, Stone S (2011) Holophusicon – the Leverian museum: an eighteenth-century English institution of science, curiosity, and art. ZKF Publishers, Altenstadt
Krotz E (2006) Mexican anthropology's ongoing search for identity. In: Ribiero GL, Escobar A (eds) World anthropologies: disciplinary transformations within systems of power. Berg, Oxford, pp 87–109

Kuwayama T (2017) Japanese anthropology, neo liberal knowledge structuring, and the rise of audit culture: lessons from the academic world system. Asian Anthropology 16(3):159–171. https://doi.org/10.1080/1683478X.2017.1346891

Mitteilungen der Anthropologischen Gesellschaft in Wien 2 (1872)

Müller F (1871) Ueber die Verschiedenheit des Menschen als Rassen und Volks-Individium. Mittheilung der anthropologischen Gesellschaft Wien 1(22):247–267

Nichols C, Parezo NJ (2017) Social and material connections: Otis T. Mason's European grand tour and collections exchanges. Hist Anthropol 28(1):58–83

Parkin R (2005) The French-speaking countries. In: Barth F, Gingrich A, Parkin R, Silverman S (eds) One discipline, four ways: British, German, French and American anthropology. University of Chicago Press, Chicago, pp 157–251

Penaloza-Patzak B (2018a) An emissary from Berlin: Franz Boas and the Smithsonain Institution, 1887–88. Mus Anthropol 41(1):30–45. https://doi.org/10.1111/muan.12167

Penaloza-Patzak CB (2018b) Guiding the diffusion of knowledge: the transatlantic mobilization of people and things in the development of US anthropology, 1883–1933. PhD diss, University of Vienna

Penaloza-Patzak B (2020) Capital collections, complex systems. Vienna, Berlin, and ethnographic specimen exchanges in transnational fin de siècle scientific networks. In: Ash M (ed) Science in the metropolis. Vienna in transnational context, 1848–1918. Routledge, New York, pp 152–171

Penaloza-Patzak B (2021) Friends in deed: allies in the interwar struggle for 'German' science and art. Academies and World War I: The Aftermath [special issue] Acta Historical Leopoldina 78:xx

Raj K (2016) Go-betweens, travelers and cultural translators. In: Lightman B (ed) A companion to the history of science. Wiley Blackwell, Chichester

Ranzmaier I (2013) Die Anthropologische Gesellschaft in Wien und die akademische Etablierung anthropologischer Disziplinen an der Universität Wien, 1870–1930. Böhlau, Vienna

Schaffer S, Roberts L, Raj K, Delbourgo J (2009) The brokered world: go-betweens and global intelligence, 1770–1820. Watson Publishing International, Sagamore Beach

Song P (2017) Anthropology in China today. Asian Anthropol 16(3):228–241. https://doi.org/10.1080/1683478X.2017.1356573

Vermeulen HF (2015) Before Boas. The genesis of ethnography and ethnology in the German enlightenment. Lincoln, University of Nebraska Press

Vermeulen HF (2019) Ethnographie, ethnologie und anthropologie im 18. und 19. jahrhundert: einheit, vielfalt und zusammenhang. Mitteilungen der Berliner Gesellschaft für Anthropologie. Ethnologie Urgeschichte 40:91–117

Willmott C (2006) The historical praxis of museum anthropology: a Canada-US comparison. In: Harrison J, Darnell R (eds) Historicizing Canadian anthropology. UBC Press, Vancouver

Zeitschrift für Ethnologie 8 (1877)

Center and Periphery: Anthropological Theory and National Identity in Portuguese Ethnography

65

João Leal

Contents

Introduction	1680
Portuguese Anthropology Between 1870 and 1970: An Overview	1681
Theoretical Influences in Portuguese Anthropology	1688
Anthropological Theory and National Identity	1690
Conclusion	1694
References	1695

Abstract

While in some central European countries and the USA, anthropology developed as a research on non-Western societies and cultures, in peripheral countries of Europe, the discipline was mainly concerned with the study of local folk tradition. Focusing on the Portuguese case, this chapter examines the relationship between these two anthropological traditions. It argues that this relationship was a complex one. While some theoretical references originating from the center were assimilated at a reasonable pace by Portuguese anthropologists, others only became relevant when they were already obsolete in the center; others still were completely absent in the historical development of Portuguese anthropology. The reasons for such a situation are related to the "nation building" character of Portuguese anthropology. The relationships established with the anthropological

---

A shorter version of this chapter was published in the volume *Anthropologie de la Méditerranée. Anthropology of the Mediterranean* (Leal 2001). Besides a more detailed approach, the present version has also revised and updated some of the arguments presented in 2001. Thanks are due to the editors of *Anthropologie de la Méditerranée. Anthropology of the Mediterranean* – especially to Dionigi Albera – for the permission to republish this revised version of the chapter.

J. Leal (✉)
CRIA, Universidade Nova de Lisboa, Lisboa, Portugal
e-mail: joao.leal@fcsh.unl.pt

© The Author(s), under exclusive licence to Springer Nature Singapore Pte Ltd. 2022
D. McCallum (ed.), *The Palgrave Handbook of the History of Human Sciences*,
https://doi.org/10.1007/978-981-16-7255-2_102

paradigms originating at the center were based on their potential usefulness to Portuguese anthropology as an exercise in the "ethnographic imagination" of the nation.

Keywords

History of anthropology · Ethnology · National identity · Portugal

Introduction

Two distinct research orientations are generally accepted in the history of the Western anthropological tradition. One of them – corresponding to the German concept of *Völkerkunde* (the study of [other] peoples) – prevailed in some "core" European countries, such as Great Britain and France, and in the USA. Here, anthropology developed as a discipline centered on the study of non-Western societies and cultures. The other – corresponding to the German concept of *Volkskunde* (the study of [one's own] people) – was adopted in the "peripheral" countries of Europe. According to *Volkskunde*, anthropology should be concerned with the study of rural folk traditions. While *Völkerkunde* had several links with the colonial projects of European empires, *Volkskunde* was mostly associated with "nation building" in peripheral European countries. Besides major differences in terms of their object and ideological background, these two traditions of anthropological research also presented several methodological and theoretical differences.

Some of them have already been reasonably well identified. However, much less is known about the similarities between these two distinct anthropological traditions, particularly about the influence that developments taking place in one field had on the other and about the nature of that influence. This chapter aims at examining, from an historical perspective, the relationships between one peripheral tradition of *Volkskunde*, the Portuguese ethnographic tradition, and the central anthropological tradition.

One example of the classical approach to relationships between anthropological center and periphery can be found in John Davis' introduction to his classic *People of the Mediterranean* (1977). There, the author equates peripheral anthropology with nationalism and theoretical anachronism:

> In some countries the work of providing a scientific basis for nationalist claims took on such symbolic significance that anthropology ceased to be a developing academic activity altogether but was rather fossilised so that a contemporary ethnographer from France, or England, or America, carrying the very latest lightweight intellectual machine guns in his pack, may be suddenly confronted by a Tylorian or Frazerian professor, a Japanese corporal from the jungle, to wage a battle only he knows is still on (Davis 1977: 3–4).

Anachronism is also one of the topics addressed by João Pina Cabral in his approach of the history of Portuguese anthropology. After considering "the preservation of anachronistic scientific theories and methods" (1991: 13; *own translation*) a major trait of Portuguese academia until the 1974 revolution, he affirms that, in the specific field of anthropology, the 1940s marked "the beginning of a tendency for academic anachronism that was only to disappear after the end of the dictatorial regime" (Cabral 1991: 27; *own translation*). Important anthropologists, such as Leite de Vasconcelos (Cabral 1991: 27) or Jorge Dias (Cabral 1991: 34–35), are given as examples of this anachronistic orientation of Portuguese anthropology.

Obviously we can speak of a certain desynchronization between core anthropological theory and peripheral ethnological traditions, which became more evident – not after the 1940s, as Pina Cabral argues – but between the 1910s and the 1930s (see Schippers [1995] for the interwar period). In theoretical terms, while central anthropology moved away from historicist perspectives to a synchronic analysis of "primitive" cultures, *Volkskunde* continued to privilege an approach to peasant cultures as the products of unique ethnic histories. And while fieldwork *à la Malinowski* became the dominant methodology in the study of "primitive" cultures, short fieldwork trips and extensive survey remained the most usual tools for studying peasant cultures. Cosmopolitan anthropology and local ethnology – to use the terminology proposed by Tomas Gerholm (1995) – evolved according to distinct logics and did not interact often. However, a number of researches has suggested that there is need for a more complex approach to the articulation between center and periphery in the history of European anthropology (Herzfeld [1986]; Vermeulen and Alvarez Roldan [1995]; Ó Giolláin [2004]).

This chapter follows that suggestion. Its objective is to take a closer look at the dialectics between peripheral ethnographic nationalism and central anthropological theory in the Portuguese ethnographic tradition. The chapter starts by presenting an overview of the history of Portuguese anthropology from 1870 to 1970, with special emphasis on its "nation-building" characteristics. It then addresses the main theoretical influences that prevailed in the different periods of the historical development of Portuguese anthropology: from Müllerian comparative mythology and evolutionism to diffusionism and North American culturalism. It tries to show how these influences were strongly dependent on their potential usefulness to Portuguese anthropology as an exercise in the "ethnographic imagination" (Anderson 1991 [1983]) of the nation. The concluding section of the chapter argues that the relationship between Portuguese anthropology and central anthropological paradigms should be seen as a continuous and mutable exercise of selection and reinterpretation of theoretical trends based on a nationalist criterion. Rather than anachronism, the key words for analyzing such an exercise are syncretism (or hybridization) and bricolage (Lévi-Strauss 1962).

Portuguese Anthropology Between 1870 and 1970: An Overview

From a historical point of view, Portuguese ethnographic tradition presents some paradoxical characteristics, when compared to other European countries (for a more detailed analysis of the history of Portuguese anthropology between 1870 and 1970, see Leal [2000a, 2006]).

In fact, the choice between "anthropology" and "ethnology," between *Völkerkunde* and *Volkskunde*, has been frequently explained – as noted before – as a choice related to the political and ideological circumstances of nineteenth-century Europe. "Anthropology" developed in countries that ruled over a colonial empire, while "ethnology" developed in countries that had a classical national problem, i.e., countries in which a process of national autonomy and/or independence had started. Following Stocking (1982), one could thus trace a clear separation between "anthropologies of empire-building," prevailing especially in France and England, and "anthropologies of nation-building," prevailing in the majority of the remaining European countries.

This distinction is somewhat paradoxical when applied to Portugal. In fact, despite the existence of an empire and the absence of a national problem in the classical sense, Portuguese anthropology emerged and developed, from 1870 until 1960, as a "nation-building anthropology," i.e., as an anthropology that not only favored the study of rural folk tradition but also conducted that study as part of a search for Portuguese national identity. Indeed, despite the existence of a prior tradition of physical anthropology, often framed in racialist terms, the first signs of a consistent socio-anthropological interest in non-European cultures and societies only developed in the late 1950s, when Jorge Dias (1907–1973) – the most important Portuguese anthropologist of the twentieth century – conducted (together with his wife Margot Dias [1908–2001]) a thorough study of the Makonde of northern Mozambique (Dias 1964; Dias and Dias 1964, 1970). Until then, and for almost 80 years, the intellectual interests of Portuguese ethnographers and anthropologists were mostly focused on Portuguese folk culture.

It is possible to find a twofold explanation for this paradox, which, as we shall see, is more apparent than real. The absence of an anthropological tradition of "empire-building" in Portugal must be related – as Rui Pereira has argued (1998: XI) – to the weakness of Portuguese colonialism. Exerted by a peripheral country itself strongly dependent on central European powers, particularly Great Britain, Portuguese colonialism was, until the late 1950s, marginal to the great tendencies and concerns of Portuguese economical, and political life, having an intermittent and discrete impact on the national ideological and cultural scene. It was only on the eve of the anti-colonial wars in the Portuguese colonies that such a situation drastically changed. The "scientific occupation" of the colonies, namely through the mobilization of the social sciences, was then defined as a priority of the colonial politics of Salazar's regime, thus creating a more favorable environment for the anthropological study of African cultures. Simultaneously, anthropology and other social sciences were introduced in the curriculum of the old Colonial School, whose name was changed to University Institute for Social Sciences and Overseas Politics (*Instituto Superior de Ciências Sociais e Política Ultramarina*).

The simultaneous predominance of a "nation-building" anthropology can be explained by the central role played by questions related to Portuguese national identity in Portuguese intellectual life over the last 150 years (Lourenço 1978). Despite being considered as one of the "old, continuous nations" (Seton-Watson 1977) of the West – and in that sense as a country without the classical national problem present in other peripheral European countries which have adopted the "nation-building" anthropological perspective – Portugal has been consistently regarded by its intelligentsia as a "problematic" country. According to Lourenço (1978), this problematic nature of the country, as seen by most of its intellectuals, derives from a persistent feeling of "ontological frailty." Based on a sort of historical hyperconsciousness regarding the nation's past greatness, represented by the Portugal of the Discoveries, this feeling has been nurtured by the contrast between that glorious past and an irrelevant present: Portugal, once one of the first nations of Europe, had been reduced to a marginal place among Western nations. This feeling of "ontological frailty" was repeatedly brought to the surface by a series of "traumatic" historical events. Among them was the occupation of the country by the foreign troops of Napoléon Bonaparte between 1802 and 1820, which led the royal family to escape to Rio de Janeiro (Brazil) and was followed by Brazil's independence in 1822; the awareness of Portuguese backwardness relative to cultivated Europe following the introduction of railway connections to Europe in the late nineteenth century; and the English Ultimatum (1890), which, by imposing English sovereignty over the territories of contemporary Zimbabwe and Zambia, drastically limited Portuguese colonial claims in Africa. In the twentieth century, the disappointment with the Republic established in 1910, that led to the military coup of 1926, and 40 years of dictatorial regime from 1926 to 1974, reinforced these feelings of national "frailty" among the elites.

It is in this context that the predominance of an ethnographic and anthropological tradition of "nation-building," similar to that prevailing in European countries with a classical national problem, can be understood. In accordance with the model proposed by Anthony Smith (1991), the perception of Portugal as a problematic country led to a need for reinforcing the civic and territorial discourse on national identity – characteristic of European "old continuous nations" – with ethno-genealogical arguments, in order to confer greater strength upon a debilitated "national existence." Besides a common territory and the rights and duties shared by its citizens, the strength of the nation was also to be measured in terms of the deeper social ties created by a common culture and a shared ethnic ancestry.

It is possible to identify four main periods in that process of formation and consolidation of Portuguese anthropology as a "nation-building anthropology" or, to quote Benedict Anderson (1991 [1983]), as a series of exercises in the "ethnographic imagination" of the nation. These periods are: the 1870s and 1880s; the turn of the century; the first three decades of the twentieth century; and finally, the period that extends from 1940 to 1960.

There are, in each of these periods, different protagonists and research objectives, and also different ways of addressing the link between folk culture and national identity.

The major Portuguese anthropologists of the 1870s and 1880s, the period that corresponds to the emergence of anthropology as a discipline in Portugal, were Teófilo Braga (1843–1924), Adolfo Coelho (1847–1919), Consiglieri Pedroso (1851–1910), and Leite de Vasconcelos (1858–1941), who were also leading figures in Portuguese intellectual (and political) life. Teófilo Braga was the founder of a national tradition of history of literature and a central figure in the republican movement that led to the establishment of the Portuguese Republic in 1910. Following the Republican revolution, he was appointed the head of the provisional government, and, in 1915, he was elected President of the Republic. Pedroso was also active in the republican movement and was elected MP in the 1880s. He was professor of Universal History at Lisboa University and one of the Presidents of the Geographical Society. Adolfo Coelho was the initiator of pedagogy in Portugal and Leite de Vasconcelos was the founder of the National Museum of Archaeology and a key figure in the establishment and initial development of archaeology in Portugal.

The main research areas of Portuguese anthropology in the 1870s and 1880s were folk literature and traditions. Whereas folk poetry and folk tales were the most representative fields of research within folk literature (see Coelho [1985]; Braga [1987]; Pedroso [1910]), the study of folk traditions focused on a heterogeneous set of topics, running from popular beliefs and "superstitions" to rural festivals and *rites de passage* (see Coelho [1993a]; Pedroso [1988]; Braga [1985]; Vasconcelos [1986]). Despite their emphasis on collecting folk literature and traditions from "genuine" oral sources, these ethnographers rarely did fieldwork and their main informants were individuals with a rural background who lived in Lisboa. Portugal was seen as a homogeneous cultural entity, without internal boundary lines or divisions and the prevailing interpretation of folk culture was basically ethno-genealogical (Smith 1991). As in other nation-building anthropologies (Wilson [1976]; Herzfeld [1986]; Handler [1988]; Ó Giolláin [2004]), folk literature and traditions were seen as the contemporary survivals of the most remote and distinctive ethnic origins of the country. Portugal, as a nation, far from being the accidental outcome of a series of more or less recent political and military events, was seen as the product of remote ethnic origins, of which folk literature and traditions were the remains. The discussion of these remote ethnic origins gave rise to different approaches, but one of the most important – for instance in the works of Braga and Vasconcelos – viewed the Lusitanians as the main ancestors of Portuguese folk culture (Leal 2000a: 63–82). Anthropologists were also interested in assessing the "ethnic psychology" of the Portuguese, and folk literature and traditions were seen – especially in Braga's and Coelho's works – as gatekeepers to the national soul.

By the turn of the century, the most important Portuguese anthropologists were Adolfo Coelho – the only ethnographer of the 1870s and 1880s still active then – and Rocha Peixoto (1866–1909), who died prematurely in 1909 and whose work was characterized by an innovative sociological approach to folk culture. Portuguese anthropology became a more plural field, and topics such as traditional material culture, folk art, and the social and economic organization of rural and fishing communities were integrated in the research agenda. Fieldwork became a major tool for collecting information on these diverse aspects of Portuguese folk culture and the country's internal diversity started to be addressed. Portugal was now seen as a more heterogeneous entity unfolding in a multiplicity of regional and/or local realities. In fact, despite its ethnic homogeneity, Portugal – as several authors have stressed (Ribeiro 1963 [1945], Mattoso 1985) – was characterized, up until the 1970s, by the diversity of its rural culture, both in geographical and in sociocultural terms. Coelho's and Peixoto's writings were central in acknowledging this internal diversity of the country. Although they did not address it in a systematic way, they nevertheless proposed a more complex image of the country than the one prevailing in the 1870s and 1880s.

This period was also characterized by a series of events, perceived as "traumatic" to national pride – particularly the English Ultimatum (1890) which drastically limited Portuguese colonial claims on Africa – that gave rise to a series of skeptical reflections on the nation's viability, built around the idea of "national decline." Portuguese anthropology was strongly influenced by these tendencies and, rather than asserting national identity – as in the 1870s and 1880s – it questioned the viability of Portugal as an independent nation. From a romantic token of national identity, folk culture was turned into an expression of the country's decline. For instance, in some programmatic essays written in the 1890s by Adolfo Coelho (1993b, c), the study of the consequences of national decline among the Portuguese people was considered one of the main tasks of anthropological research. At the same time, in the analysis of some specific areas – such as traditional pottery, vernacular architecture (Peixoto 1967a, b), or folk theatre (Coelho 1993d) – negative judgments replaced the romantic enthusiasm for Portuguese folk culture. Thus, writing about traditional pottery, Peixoto stated that it was characterized by "a narrow inspiration and an immovable dependence on traditionally inherited forms. We find [Portuguese pottery] immutably prehistoric" close to "some of the crudest types of primitive potteries" (1967a: 103; *own translation*), with "resources incomparably inferior to those of many populations considered barbarians" (1967a:112; *own translation*). As to Coelho, he wrote on the symptoms of the "ethnic illness" of the Portuguese people in equally disappointed terms. According to him, the Portuguese were characterized by an "almost constant spirit of hesitation," a "progressive inability to work," the "predominance of selfish feelings over communitarian ones," an "excessive imitative spirit," a "frequent moral insanity," and "pessimism, hypochondria and social fatalism" (1993b: 692–693; *own translation*).

In the wake of the establishment of the Republic and throughout the 1910s and 1920s, optimism about the nation's destiny returned to Portuguese cultural life. Portuguese ethnography was now viewed as a crucial undertaking to reunite the Portuguese people with their homeland and to "nationalise" Portuguese life (see Ramos [1994] for a more detailed discussion of the nationalist tendencies prevailing in Portuguese cultural life in the 1910s and 1920s). This nationalist ethnography – of which Vergílio Correia (1888–1944), Luís Chaves (1888–1975), and Pires de Lima (1908–1973) were some of the most prominent representatives – had a strong descriptive bias and was mainly focused on the study of folk art: a composite group of objects, running from traditional pottery to folk architecture and traditional clothing. Based on short researches "excursions" to the countryside, it emphasized an artificially homogeneous image of the country, in which different local and regional realities were considered minor and mutually interchangeable expressions of the nation's "essence." This ethnography lacked theoretical ambitions: rather than explaining Portuguese folk culture, its aim was to celebrate it as an expression of the genius of the nation. Theoretical analysis was thus replaced by "literary" descriptions loaded with nationalist rhetoric and complemented with photographs and "artistic" drawings.

This kind of ethnography was to become, from the 1930s onwards, the official ethnography of Salazar's dictatorial regime – the Estado Novo (literally the "New State") (see Alves 2013). In fact, under António Ferro – the director of the SPN/ SNI (Secretary for National Propaganda/Secretary for National Information) – folk culture became a major tool for the nationalist disciplining of Portugal. Folklore groups were strongly supported by the regime and a national Museum of Folk Art was inaugurated in Lisboa (Oliveira 2019). Ferro also launched a national contest – *A Aldeia mais Portuguesa de Portugal* (literally "The Foremost Portuguese Village of Portugal") – aimed at awarding rural virtues as tokens of the nation's *grandeur*. Together with the Empire, folk culture also became a major tool for the international affirmation of Portugal's dictatorship: this was a country of content peasants (and happy African "natives"), united under the "benevolent" tutelage of Salazar.

As the Estado Novo turned ethnography into propaganda, other anthropologists were seeking alternative ways of studying Portuguese peasant cultures. This was mostly the case of Jorge Dias, the main Portuguese anthropologist of the twentieth century. Fascinated by Portuguese peasant cultures, Dias took his PhD in ethnology (*Volkskunde*) at Munich University and, after his return to Portugal, founded the first research center in Portuguese anthropology – the *Centro de Estudos de Etnologia Peninsular* (Centre for the Study of Peninsular Ethnology). He was later to become Professor of Anthropology at Coimbra University, Lisbon University, and ISCSPU (University Institute for the Study of Social Sciences and Overseas Politics), and was the founder of the National Museum of Ethnology (initially *Museu de Etnologia do Ultamar* [Overseas Museum of Ethnology]). He was the author of such important works as the monograph he wrote on Rio de Onor (1953a) – a mountain community in NE Portugal – "Os Elementos Fundamentais da Cultura Portuguesa" ("The Fundamental Elements of Portuguese Culture") (1953b) – an essay dedicated to the study of Portuguese national culture – and, together with his wife, Margot Dias, he wrote a comprehensive monograph on the Makonde of northern Mozambique (Dias 1964; Dias and Dias 1964, 1970) – which was the first anthropological study of an African ethnic group produced in Portugal. As General Secretary of the CIAP (*Commision Internationale des Arts et Traditions Populaires*) from 1954 to 1957, he also played an important role in the development of the field of European Ethnology, aimed at bringing together the different European national traditions of study of folk culture. His contribute was particularly relevant – as Rogan (2013, 2015) has stressed – for the affirmation of an inclusive approach to *Völkerdunke* and *Volkskunde* – or to folklore and anthropology – viewed not as distinct disciplines, but as two aspects of the same undertaking.

Following the launching of the *Centro de Estudos de Etnologia Peninsular*, Dias formed a team, that, besides his wife Margot Dias, included Ernesto Veiga de Oliveira (1910–1990), Benjamim Pereira (1928–1920), and Fernando Galhano (1904–1995). The main focus of the team was the study of Portuguese peasant cultures and societies, structured around four main areas: (a) the study of mountain rural communities of northern Portugal (Dias 1948a, 1953a); (b) the study of vernacular architecture and traditional agricultural technologies, such as ploughs, granaries, wind- and water-mills (see, for instance, Dias 1948b; Dias, Oliveira and Galhano 1963; Oliveira and Galhano 1992); (c) the study of popular feasts and other cyclic celebrations (Oliveira 1984; Pereira 1973); and (d) global studies of Portuguese culture (Dias 1953b, 1955, 1960).

While Jorge Dias did intensive fieldwork in two rural villages in northern Portugal (Rio de Onor and Vilarinho da Furna), the main methodological tool used by Dias and his colleagues was extensive survey. However, unlike their predecessors – whose research was focused on particular areas of the Portuguese countryside – their frequent field trips covered the whole country and provided the most comprehensive and detailed image of rural Portugal ever produced. They were thus able to challenge the artificially homogeneous image of the country that was prevalent in Estado Novo ethnography, and address Portuguese folk culture as a reality encompassing both unity and diversity.

A significant part of their work was directed at the systematic study of the factors that showed the diversity of the country. Based on the model proposed by the geographer Orlando Ribeiro (1911–1997) (Ribeiro 1963 [1945]), Jorge Dias distinguished three major cultural areas in Portugal: Mediterranean Portugal, in the southern part of the country; Transmontano Portugal (from Trás-os-Montes, literally meaning "behind the mountains," one of the provinces of NE Portugal) in the northeast; and Atlantic Portugal in the northwestern part of the country. But where Ribeiro had stressed the importance of natural conditions to this threefold division of the country, Dias emphasized the importance of its ancient ethnic roots, expressed in the existence of distinct sociocultural configurations. While Lusitanian roots were prevalent in Transmontano Portugal, the northeast was seen as strongly influenced by the Swabians and other German speaking peoples that had settled the Iberian Peninsula in the fifth century, and Mediterranean Portugal had an important Roman and Arab cultural background. Each one of these areas was also connected to distinct types of rural settlement and material culture and to different patterns of social and family organization (Dias 1955, 1960). While the nuclear family prevailed in the south, the extended family was dominant in the northeast, and in the NW Portugal, women had an important role in family and social life. Social organization also presented some significant contrasts. While northeast Portugal was characterized by strong traditions of communitarian social organization, in the NW, practices of mutual help were widespread, and in Mediterranean Portugal, due to the prevalence of latifundia, social cohesion was weaker.

Due to the influence of Ribeiro and Dias, this emphasis on the diversity of the country was also the driving force of other studies of Portuguese folk culture developed in the 1950s and 1960s by the ethnomusicologist Michel Giacometti (1909–1990) (Giacometti and Graça 1960–1970) and by the modernist architects who, between 1957 and 1961, conducted the *Inquérito à Arquitetura Popular em Portugal* (National Survey of Vernacular Architecture) (VVAA 1961). Leite de Vasconcelos – who between the late 1880s and the 1930s had redirected his research to archaeology and only returned to anthropology after his retirement – was also an important author in the mapping of local and regional differences of rural Portugal (Vasconcelos 1936, 1942).

Together with his interest in the diversity of Portuguese folk culture, Dias also thematized its unity. His best known essay on the subject is "Os Elementos Fundamentals da Cultura Portuguesa" ("The Fundamental Elements of Portuguese

Culture") (Dias 1953b), which has remained, until today, a very influential essay among Portuguese intellectual circles committed to a contemporary reflection about "what it is to be Portuguese" (Leal 2000a: 99–104)). Throughout the essay, Jorge Dias proposes a description of the "Portuguese temperament" based on previous approaches to Portuguese ethnic psychology. Besides the essays written on the topic by nineteenth century Portuguese anthropologists, Dias was mostly influenced by the work developed at the beginning of the twentieth century by the poet Teixeira de Pascoaes (1877–1952), who proposed *saudade* (longing, nostalgia, homesickness) as the key concept for understanding Portuguese national identity.

Teixeira de Pascoaes viewed *saudade* as a feeling that combined "the desire of the being or thing that is loved, with the pain of its absence" (Pascoaes, 1986 [1912]; *own translation*). Associating a physical and a spiritual element, an orientation towards the past with an orientation towards the future, the feeling of *saudade* was considered by Pascoaes to be a structuring theme of Portuguese culture and the most original trait of Portuguese ethnic psychology. One of the main arguments used by the poet to support this idea related to the allegedly untranslatable character of the word *saudade* (on the invention of *saudade*, see Leal [2000b]).

Pascoaes' nationalist approach to *saudade* played an important role on Dias' approach to Portuguese national character. According to Dias, the Portuguese personality was a very complex and paradoxical one, being based on a series of conflicting psychological traits. It combined, for instance, a remarkable capacity of adapting to different surroundings – allegedly expressed in a process of colonization, through assimilation or miscegenation, distinct from that of other European countries – with a strong capacity for keeping its own character. Other oppositions present in the Portuguese character were a strong capacity of dreaming versus a powerful will of action; an intrinsic goodness versus a propensity towards violence and cruelty; a strong feeling of individual freedom versus the importance of powerful values of solidarity; a lack of sense of humor combined with an intense irony. Following Pascoaes, Dias viewed *saudade*, with its allegedly unique merging of the lyricism of the dreamer, the obstinacy of the man of action and a strong fatalism, as the most powerful expression of this paradoxical "national character."

In the context of several trips he made to Brazil in the early 1950s, Dias also became interested in thematizing the cultural continuities between Portugal and Brazil, and coined the concept of a Luso-Brazilian cultural area, thus giving a geopolitical tone to his reflections on the unity of Portuguese folk culture (Leal 2021).

Theoretical Influences in Portuguese Anthropology

These distinct periods of Portuguese anthropology were linked to diverse theoretical influences (the only exception being the first decades of the twentieth century, when, as was pointed out earlier in this chapter, a mainly descriptive ethnographic approach lacking in theoretical ambition became dominant in Portuguese anthropology).

In the 1870s and 1880s, Portuguese anthropology was strongly influenced by Max Müller's comparative mythology and by pre-evolutionist diffusionist theories, running from Benfey's diffusionism to Lenormant's "turanianism" (see Cocchiara [1981], for a discussion of these different theories; on comparative mythology see, for example, Dorson [1968: 160–186]; Schremp [1983]; Carroll [1985]; Belmont [1986: 93–120; and Kippenberg [2002: 36–50]). Evolutionism was the dominant theoretical influence in Portuguese anthropology at the turn of the century. While evolutionary theories had already exerted some influence on the previous generation of ethnographers, in particular on Consiglieri Pedroso (Leal 1988), after the 1890s, its significance in Portuguese anthropology became hegemonic. Finally, in the 1930s and 1940s, Jorge Dias and his team were deeply influenced by German diffusionist theories, with which Jorge Dias had come in contact during his PhD studies in Germany, and by the North American school of "Culture and Personality," especially by Ruth Benedict's work, that Dias probably knew after his first trip to the USA, in the early 1950s. One of the major expressions of his culturalist "conversion" is his participation in the 1952 Wenner-Gren Symposium, whose proceedings were published in *An Appraisal of Anthropology Today* (Tax et al. 1953). Besides some individual participations on the debates, Dias also presided, together with Margaret Mead, over one of the sections of the Symposium (Leal 2021: 51). The influence of North American culturalism is also evident in Dias' monograph on Rio de Onor (1953a), in which he uses the opposition between Dionysian and Apollonian cultures – established by Ruth Benedict in her famous book *Patterns of Culture* (1934) – to characterize the spiritual orientation of local folk culture.

The different theoretical influences prevalent in Portuguese anthropology from the 1870s to the 1960s were associated with distinct patterns of relationship between center and periphery.

During the 1870s and 1880s, albeit some temporal *décalage*, Portuguese anthropology was well up-to-date with the main anthropological central theories. Comparative mythology and pre-evolutionist diffusionism, although dominant, coexisted with an open attitude towards evolutionism, then at its peak. Portuguese anthropologists were well read in several languages, were aware of the main anthropological production of the period, were affiliated to several international scientific societies, and some of their papers were published in England and in France. One of the most interesting examples of this cosmopolitanism of nineteenth-century Portuguese anthropology is Consiglieri Pedroso. His anthology of folktales (1882) was published in London by the Folklore Society (founded, among others, by Tylor and Lang) even before its Portuguese edition. At the same time, Pedroso corresponded regularly with several foreign anthropologists and was a member of various international scientific societies (see Leal 1988). Fascinated by the Russian tradition of folk studies, he even travelled to Saint Petersburg and Moscow. The case of Leite de Vasconcelos must also be mentioned. Being a prolific and versatile author, whose interests covered such different areas as ethnography, archaeology, and dialectology, he took his PhD in dialectology in Paris, where he presented his *Esquisse d'une Dialectologie Portugaise* (An Essay on Portuguese Dialectology) (Vasconcelos 1901).

Portuguese anthropologists at the turn of the century kept in touch with some of the main trends within the discipline. Evolutionist theories – although already in crisis (Stocking 1995) – were still important, and their influence was crucial in the work of Rocha Peixoto and Adolfo Coelho. The latter was also aware of the work of Marcel Mauss, Émile Durkheim, and Franz Boas, although he made only a limited use of their theories.

Finally, between 1940 and 1960, the work of Jorge Dias and his team can be analyzed according to two distinct patterns. One of them, associated with diffusionism, is linked to a certain theoretical anachronism. Indeed, since the 1920s, diffusionism, although still relatively important in German anthropology, had lost ground both in the USA and in the United Kingdom. Despite this declining influence, it was nevertheless central to Dias' research on Portuguese rural culture and to his historicist concerns about the ethnic origins of Portugal's internal diversity. The other pattern, expressed in the influence of the North American school of "Culture and Personality," shows Dias' concern with a theoretical *aggiornamento* and a desire to coming closer to the leading theories in a great international center. This *aggiornamento* was nevertheless selective: the impact of British functionalism, the other major theoretical tradition in the international anthropology of the time, was very weak in the work of Jorge Dias and his team. In fact, although Dias considered his monograph on Rio de Onor (1953a) as being close to a functionalist approach *à la Malinowski*, the actual traces of such an influence – both in the internal organization of the book and in its generic theoretical approach – were rather irrelevant. One would have to wait until the 1960s and the publication of Dias' monographs on the Makonde of northern Mozambique, for such an influence to become visible in Portuguese anthropology (functionalism was by then, of course, the basic reference in Africanist anthropology). The contribution of these monographs to the international affirmation of Dias among British Africanist anthropologists must be stressed. That is the reason why Max Gluckman, Meyer Fortes, M. G. Marwick, and John Beattie participated in the volumes of *In Memoriam Jorge Dias* (VVAA 1974), thus paying a posthumous tribute to Dias' anthropological achievements.

The foregoing discussion shows that there was a variable pattern in the relationship between periphery and center in the Portuguese case. An open attitude towards the center seemed to prevail, with the exception of first decades of the twentieth century. However, this attitude was selective and its outcomes were very diverse. Some theoretical references originating from the center were assimilated at a reasonable pace, as was the case with comparative mythology in the 1870s and 1880s, with evolutionism at the turn of the century, and with the "Culture and Personality" school in the decades following 1940. Others only became relevant when they were already obsolete in the center; this was the case of diffusionism from the 1940s onwards. Finally, others were absent, as was the case of British functionalist tradition until its Africanist appearance in the work of Dias in the 1960s.

Anthropological Theory and National Identity

The reasons for such a situation are manifold. But it can be argued that one of the main explanations is related to the "nation-building" character of Portuguese anthropology. The relationship established with central anthropological paradigms was, in every period, based on their potential usefulness to Portuguese anthropology as an exercise in the "ethnographic imagination" of the nation.

Thus, the influence of comparative mythology and pre-evolutionist diffusionism in the 1870s and 1880s was linked to the prevalence of an ethno-genetic discourse on national identity. Portuguese ethnographers of that period, in resorting to comparative mythology or to pre-evolutionist diffusionist theories, were mainly interested in using them as "scientific" tools with which to support the ancient and unique character of Portuguese culture. Conversely, the lack of enthusiasm for evolutionism was linked to its inadequacy for these exercises in the ethno-genealogy of the nation. More concerned with humankind in general and universal stages of development than with particular histories of specific cultures, evolutionism was of limited use to the first generation of Portuguese anthropologists.

Its later success at the turn of the century was linked to the skeptical approach to national identity that prevailed at the time, which, as was pointed out, gave rise to an interpretation of folk culture as an expression of national decline. It is against this background that evolutionism emerged as a major theoretical reference: the ideological and moral equation it established between the non-European "primitive" and the European peasant became an important rhetorical tool to characterize the condition of the Portuguese in a negative way The Portuguese peasantry was considered to be close to the "savage state" and its customs were recurrently classified as "primitive" or "barbarian" (on the comparison between "primitives" and "peasants," see Stocking [1987: 186–237]).

The lack of theoretical ambition that characterized Portuguese anthropology from the 1910s to the 1930s was a result of the emergence of a nationalist ethnography with a strong descriptive orientation, whose main concern was celebrating rather than explaining "the people." Theory, any theory, especially if coming from abroad, was seen as an obstacle to that celebration of the Portuguese folk. However, it must be noted that both the descriptive orientation and the nationalist tone prevailing in Portuguese ethnography of that period mirrored more general tendencies at work in other European countries with strong *Volkskunde* traditions (Schippers 1995). The same applies to the emphasis given to the study of folk art (see Klein and Widbom [1994], for the case of Sweden).

The impact of diffusionism on the work of Jorge Dias must be seen as a result of his concern with the systematic characterization of the main cultural areas of Portuguese folk culture. Diffusionism became therefore fundamental for the reconstitution of a pluralist ethno-genealogy of the nation which bestowed on the Lusitanians, the Romans, the Germanic "tribes" that settled the Iberian Peninsula after the fall of the Roman empire, and the Arabs, the role of civilizing heroes of Portuguese culture. His use of the diffusionist concept of cultural area to characterize the continuities between Portuguese and Brazilian folk cultures is also revealing. What was at stake was the nationalist affirmation of the transnational relevance of Portuguese culture, linked to its historical role in European expansion. As for the "Culture and Personality" school, its influence on the work of Dias must also be understood in the context of the nationalist logic of Portuguese anthropology. The writings of Ruth Benedict and the North American studies of "national character" provided scientific support for a project focused on the definition of national identity based on the alleged particularities of the Portuguese temperament. The absence of functionalism in that period should also be seen in this context: there was nothing in functionalist theory susceptible to recycling by an anthropology whose main objective was the construction of national identity. Functionalism, with its vehement denial of conjectural history – whether evolutionist or diffusionist – and its emphasis in society as opposed to culture, was useless to Portuguese anthropology of that period.

To sum up, the dialogue established by Portuguese anthropologists with central anthropological theory depended on the usefulness of the theoretical models of the latter to the exercises in the "ethnographic imagination" of the nation by the former.

The relationship between Portuguese anthropology and central anthropological paradigms should thus be seen as a continuous and mutable exercise of selection and reinterpretation of theoretical trends based on a nationalist criterion. This exercise has strong similitudes with syncretic and hybridization processes studied by anthropologists. Syncretism, for instance, has been defined as "an 'interlock' of religious and cultural elements of different origins into a situation of contact" (Rudolph 2004: 82). As to hybridization, it has been characterized as a "process which allows for the simultaneous co-existence (or combination) of forms and voices, but also to their mutual blending and transmutation" (Kapchan 1993: 304). Similarly, Portuguese anthropology between 1870 and 1970 can also be viewed as an "interlock" based on the "co-existence" of different "forms and voices" characterized by their "different origins" and by processes of "blending and transmutation" through which central anthropological theories were locally recycled and transformed, acquiring new meanings in the process.

Orvar Löfgren has defined nationalism as a "gigantic do-it-yourself kit." According to him, nationalism is "an international ideology imported for national ends" (1989: 8):

> a kind of check list: every nation should have not only a common language, a common past and destiny, but also a national folk culture, a national character or mentality, national values, perhaps even some national tastes and a national landscape (...), a gallery of national myths and heroes (and villains), a set of symbols, including flag and anthem, sacred texts and images (1989: 8–9).

This definition equates the work of nation-building with *bricolage* (Lévi-Strauss 1962). The relationship between Portuguese ethnography and central anthropological paradigms should be seen from this perspective: as a work of bricolage in which materials produced in different contexts were used and recycled in a context that was distinct from the original, sometimes leading to an unpredictable outcome. The Portuguese anthropologist, rather like the *bricoleur* described by Lévi-Strauss,

> Excited by his project (...) has to turn back to an already existent set made up of tools and materials to consider and reconsider what it contains and, finally and above all, to engage in a sort of dialogue with it, and, before choosing between them, to index the possible answers which the whole set can offer to his problem (Lévi-Strauss 1962: 37; *own translation*).

Central anthropological theories, the tools and materials in Lévi-Strauss's definition, not only acquire new characteristics in the course of this operation, but, because they become part of a new set with a distinct purpose (the construction of national identity), undergo surprising and sometimes bizarre changes of meaning.

There are a number of examples that illustrate this argument particularly well. One of them relates to the way comparative mythology was used by nineteenth-century Portuguese anthropologists. The comparativism advocated by Max Müller aimed at linking European folk literature and traditions to their initial source: primordial Indo-European thought. According to Müller, a certain set of traditions should be compared, first of all, with identical sets, located in the same synchronic layer, but belonging to distinct geographic-cultural places. This process was aimed at identifying a core that was shared by these different sets of traditions, separating local additions from common Indo-European roots. Once identified, this common core should be compared with the more archaic Indo-European tradition in order to elucidate the original meaning of beliefs and practices that had in the meantime lost their primordial transparency. In each comparative stage, the objective was not the identification of particularities but rather the search for similar or related elements. In the Portuguese case, as in other nineteenth-century European anthropological traditions, these theoretical priorities were subverted by the nationalist appropriation of comparative mythology. Rather than being used to interpret Portuguese folk traditions by reference to the primordial Indo-European tradition, comparative mythology was frequently applied to assert the singularities of Portuguese culture within the Indo-European context. The final outcome of this application of comparative mythology was an emphasis on particularities, rather than on shared references.

A particularly good example of this trend is Consiglieri Pedroso's interpretation of the Portuguese folk traditions of *mouras encantadas* (literally "enchanted Moorish women") (1988: 217–227). Pedroso starts by comparing *mouras encantadas* to similar representations in other Indo-European traditions, such as Germanic *nixen*, Slavic *rusalki*, English *lac-ladies*, or Greek *naiada*. But from such a comparison, Pedroso only retains the singularity of the Portuguese *mouras encantadas*, which he describes as "one of the most poetic creations of the Portuguese folk tradition" (1988: 218; *own translation*).

A further example of the unexpected results of this filtering of central anthropological theories by a nation-building anthropological tradition can be found at the turn of the century in the combination of evolutionism with theories of national decline. Evolutionism regarded progress as the main force behind human history and tended to look suspiciously at any references to degeneration. These were considered to be inspired by the "degenerationist" narratives typical of the biblical paradigm. Primitive condition should not be seen as the result of a downfall, but rather as the outcome of an incomplete process of ascension. Resorting to evolutionism as a frame of reference able to support a negative vision of Portuguese folk culture, the Portuguese ethnographers of the turn of the century, not only ignored evolutionist's resistance to issues such as degeneration and decline but also brought together – in a tour de force of which probably they were not aware – two completely distinct narratives about the causes of primitive condition.

Finally, a third example concerns the combination of "Culture and Personality" theory and ethnic psychology in the work of Jorge Dias. Indeed, one could say that Benedict's views were mainly used by Dias as a scientific varnish for the updating and expanding of previous literary reflections on the Portuguese national character which, as was pointed out earlier in this chapter, viewed *saudade* as a token of Portuguese national identity. Based on North American studies of national character, which he explicitly quotes in the introduction, Jorge Dias' essay can thus be seen as an attempt at providing credibility to the project of defining Portuguese national identity by reference to an ethnic essence of psychological nature. In the context of Portuguese culture, this project did not have a "scientific" nature and some its more significant developments had taken place in fields such as literary essays or poetry.

All these examples clearly demonstrate the heterodox consequences of the peripheral appropriation of central orthodoxy. Transferred from the contexts where they were originally produced, central theories were not passively absorbed but actively appropriated, so that they could serve the needs of an anthropological project whose leitmotif was the construction of national identity.

Conclusion

To sum up, rather than being exclusively a history of "Tylorian" or "Frazerian Professors" and "Japanese Corporals" – as John Davis (1977) defended – the history of Portuguese anthropology is also a history of "Müllerian" and "Benedictian Professors." And, if sometimes – although not always – those "Professors" may look like "Japanese corporals," it is because the battles they were fighting were not the great wars of central anthropological theory, but the small cultural wars of nation-building. Difficult to understand from a central point of view, those battles were, nevertheless, meaningful enough for those who fought them.

This is one of the reasons why anachronism seems to be a problematic category for the analysis of peripheral traditions of "nation building" anthropology. Their protagonists were, on the contrary, strongly involved with the predicaments of their time and their choice of central anthropological theories did not necessarily result from ignorance but from the priority given to political and analytical agendas strongly imbedded in the present.

In the process, at least in the Portuguese case, they were able to provide a comprehensive picture of folk culture, ranging from folk traditions and literature, to folk art and material culture, and peasant family and social organization. The country they mapped has undergone deep and enduring transformations. But if these "worlds we have lost" (Laslett 1965) are still accessible to us, it is because of their researches.

In architectural theory, a distinction is made between the concept and the project. The concept can be good but its realization can be mediocre (or bad), and vice-versa. It might be that the concept underlying "nation building" anthropologies was an arguable one. But it gave rise to a series of realizations which are indispensable to our contemporary understanding of the lives and predicaments of the peasants who constituted, until the 1960s, the majority of the European population. In their own particular ways, these "nation building" anthropologists were able to give this "peasant majority" a voice.

References

Alves V (2013) Arte popular e nação no Estado Novo. A política folclorista do Secretariado de Propaganda Nacional [Folk art and nation under the Estado Novo. The folkloristic policies of the National Secretary for propaganda]. Imprensa de Ciências Sociais, Lisboa

Anderson B (1991 [1983]) Imagined communities. Reflections on the origin and spread of nationalism. Verso, London

Belmont N (1986) Paroles païennes. Mythe et folklore [Pagan words. Myth and folklore]. Éditions Imago, Paris

Benedict R (1934) Patterns of culture. Houghton Mifflin, New York

Braga T (1987 [1883]) Contos tradicionais do povo português [traditional folktales of the Portuguese people]. Publicações Dom Quixote, Lisboa

Braga T (1985 [1885]) O povo português nos seus costumes, crenças e tradições [The customs, beliefs and traditions of the Portuguese people]. Publicações Dom Quixote, Lisboa

Cabral JP (1991) Os Contextos da Antropologia [The contexts of anthropology]. Difel, Lisboa

Carrol M (1985) Some third thoughts on Max Müller and solar mythology. Arc Eur Soc XXVI:263–290

Cocchiara G (1981 [1952]) The history of folklore in Europe. Institute for the Study of Human Issues, Philadelphia

Coelho A (1985 [1879]) Contos populares portugueses [Portuguese folktales]. Publicações Dom Quixote, Lisboa

Coelho A (1993a [1880]) Materiais para o estudo das festas, costumes e crenças populares portuguesas [materials for the study of Portugues feasts, customs and beliefs]. In: Obra etnográfica [Ethnographic works], vol. I. Publicações Dom Quixote, Lisboa, pp 277–372

Coelho A (1993b [1890]) Esboço de um programa para o estudo antropológico, patológico e demográfico do povo português [Proposals for the anthropological, pathological and demographic study of the Portuguese people]. In: Obra Etnográfica [Ethnographic works], vol I. Publicações Dom Quixote, Lisboa, pp 681–701

Coelho A (1993c [1896]) Exposição etnográfica portuguesa. Portugal e ilhas adjacentes [Portuguese ethnographic exhibit. Portugal and the Atlantic islands]. In: Obra Etnográfica [Ethnographic works], vol I. Publicações Dom Quixote, Lisboa, pp 703–736

Coelho A (1993d [1910]) Cultura e analfabetismo [Culture and illiteracy]. In: Obra etnográfica [Ethnographic works], vol II. Publicações Dom Quixote, Lisboa, pp 251–299

Davis J (1977) People of the Mediterranean. An essay in comparative social anthropology. Routledge & Kegan Paul, London

Dias AJ (1948a) Vilarinho da Furna. Uma aldeia comunitária [Vilarinho da Furna. A communitarian village]. Instituto para a Alta Cultura, Porto

Dias AJ (1948b) Os arados portugueses e as suas prováveis origens [Portuguese ploughs and their probable origins]. Instituto para a Alta Cultura, Porto

Dias AJ (1953a) Rio de Onor. Comunitarismo agro-pastoril [Rio de Onor. Agro-pastoril communitarism]. Instituto para a Alta Cultura, Porto

Dias AJ (1953b) Os elementos fundamentais da cultura portuguesa [The fundamental elements of Portuguese culture]. In: Proceedings of the international colloquium on Luso-Brazilian studies. Nashville, pp 51–65

Dias AJ (1955) Algumas considerações acerca da estrutura social do povo português [Some observations on the social structure of the Portuguese people]. Rev Ant 3(1):1–20

Dias AJ (1960) Tentamen de fixação das grandes áreas culturais portuguesas [An essay on Portuguese cultural areas]. In: Estudos e ensaios folclóricos em homenagem a Renato de Almeida [Folclore studies and essays in tribute to Renato de Almeida]. Rio de Janeiro, pp 431–454

Dias AJ (1964) Os Macondes de Moçambique. Aspectos históricos e económicos [The Makonde of Mozambique. Historjcal and economical aspects], vol I. Junta de Investigações do Ultramar, Lisboa

Dias AJ Dias M (1964) Os Macondes de Moçambique. Cultura material [The Makonde of Mozambique. material culture], vol II. Junta de Investigações do Ultramar, Lisboa

Dias AJ Dias M (1970) Os Macondes de Moçambique. Vida social e ritual [The Makonde of Mozambique. Social and spiritual life], vol III. Junta de Investigações do Ultramar, Lisboa

Dias AJ Oliveira EV Galhano F (1963) Sistemas primitivos de secagem e armazenagem de produtos agrícolas. Os espigueiros portugueses [Portuguese granaries]. Instituto de Alta Cultura, Porto

Dorson R (1968) The British folklorists. A history. Routledge & Kegan Paul, London

Gerholm T (1995) Sweden: central ethnology, peripheral anthropology. In: Vermeulen HA, Roldan AA (eds) Fieldwork and footnotes. Studies in the history of European anthropology. Routledge, London, pp 159–170

Giacometti M Graça FL (1960–1970) Antologia da música regional portuguesa [Anthology of Portuguese regional music]. Arquivos Sonoros Portugueses, Lisboa (5 LPs)

Handler R (1988) Nationalism and the politics of culture in Quebec. The Wisconsin University Press, Madison

Herzfeld M (1986) Ours once more. Folklore, ideology and the making of modern Greece. Pella, New York

Kapchan D (1993) Hybridzation and the marketplace: emerging paradigms in folkloristics. Wes Fol 2–4:303–326

Kippenberg H (2002) Discovering religious history in the modern age. Princeton University Press, Princeton

Klein B, Widbom M (eds) (1994) Swedish folk art. All tradition is change. Harry Abrahams Publishers, New York

Laslett P (1965) The world we have lost. Routledge, London

Leal J (1988) Prefácio [Preface]. In: Pedroso C (ed) Contribuições para uma mitologia popular portuguesa e outros escritos etnográficos [Contributions to a Portuguese folk mythology and other ethnographic works]. Publicações Dom Quixote, Lisboa, p 13–40

Leal J (2000a) Etnografias portuguesas (1870–1970). Cultura popular e identidade nacional [Portuguese ethnographies (1870–1970). Folk culture and national identity]. Publicações Dom Quixote, Lisboa

Leal J (2000b) The making of 'saudade'. National identity and ethnic psychology in Portugal. In: Dekker T, Helsloot J, Wijers C (eds) Roots and rituals. The construction of ethnic identities. Het Spinhuis, Amsterdam, pp 267–287

Leal J (2001) 'Tylorean professors' and 'Japanese corporals': anthropological theory and national identity in Portuguese ethnography. In: Albera D, Blok A, Bromberger C (eds) Anthropologie de la Méditerranée. Anthropology of the Mediterranean. Maisonneuve et Larose, Paris, pp 645–662

Leal J (2006) Antropologia em Portugal. Mestres, percursos, transições [Anthropology in Portugal. Masters, pathways, transitions]. Livros Horizonte, Lisboa

Leal J (2021) Os anos 'brasileiros' de Jorge Dias" [The 'Brazilian' years of Jorge Dias]. In: Silva AT (ed) Cartas do Brasil. Correspondência de antropólogos e folcloristas brasileiros para Jorge Dias (1949–1972) [Letters from Brazil. Correspondence from Brazilian anthropologists and folklorists to Jorge Dias (1949–1972)]. Etnográfica Press, Lisboa, pp 47–70

Lévi-Strauss C (1962) La Pensée sauvage [The savage mind]. Plon, Paris

Löfgren O (1989) The nationalization of culture. Eth Eur XIX:5–24

Lourenço E (1978) Da literatura como interpretação de Portugal [Literature and the interpretation of Portugal]. In: O labirinto da saudade. Psicanálise mítica do destino português [The labyrinth of saudade. Mythical psichoanalysis of the Portuguese destiny]. Publicações Dom Quixote, Lisboa, pp 85–126

Mattoso J (1985) Identificação de um país. Ensaio sobre as origens de Portugal [Identification of a country. Essay on the origins of Portugal]. Editorial Estampa, Lisboa

Ó Giolláin D (2004) Locating Irish folklore. Tradition, modernity, identity. Cork University Press, Cork

Oliveira EV (1984) Festividades cíclicas em Portugal [Cyclical celebrations in Portugal]. Publicações Dom Quixote, Lisboa

Oliveira A (2019) Herança de António Ferro. O Museu de Arte Popular [António Ferro's legacy. The Museum of Folk Art]. Caleidoscópio-DGPC, Lisboa

Oliveira EV, Galhano F (1992) Arquitetura tradicional portuguesa [Traditional Portuguese architecture]. Publicações Dom Quixote, Lisboa

Pascoaes T (1986 [1912]) O espírito lusitano ou o saudosismo [The Lusitanian spirit and *saudosismo*]. In: Botelho A Teixeira A (eds). Filosofia da saudade [The philosophy of saudade]. Imprensa Nacional-Casa da Moeda, Lisboa, pp 21–35

Pedroso Z (1882) Portuguese folktales. The Folklore Society, London

Pedroso Z (1910) Contos populares [Portugueses folktales]. Vega, Lisboa

Pedroso Z (1988 [1878–1882]) Contribuições para uma mitologia popular portuguesa [Contributions to a Portuguese folk mythology]. In: Contribuições para uma mitologia popular portuguesa e outros escritos etnográficos [Contributions to a Portuguese folk mythology and other ethnographic works]. Publicações Dom Quixote, Lisboa, pp 83–302

Peixoto R (1967a [1900]) Etnografia portuguesa. Indústrias populares. As olarias do Prado [Portuguese ethnography. Folk industries]. In Obras [Complete works], vol I. Câmara Municipal da Póvoa do Varzim, Póvoa do Varzim, pp 89–132

Peixoto R (1967b [1904]) A Casa portuguesa [The Portuguese house]. In: Obras [Complete works], vol I. Câmara Municipal da Póvoa do Varzim, Póvoa do Varzim, pp 153–165

Pereira B (1973) Máscaras portuguesas [Portuguese masks]. Junta de Investigações do Ultramar, Lisboa

Pereira R (1998) Introdução à reedição de 1998 [Introduction to the 1998 reedition]. In: Dias J Os macondes de Moçambique [The Makonde of Mozambique], vol I. Comissão Nacional para a Comemoração dos Descobrimentos-Instituto de Investigação Científica Tropical, Lisboa

Ramos R (1994) A Segunda fundação (1890–1926) [The second foundation (1890–1926)]. Círculo de Leitores, Lisboa

Ribeiro O (1963 [1945]) Portugal, o Mediterrâneo e o Atlântico [Portugal, the Mediterranean and the Atlantic]. Sá da Costa, Lisboa

Rogan B (2013) Sigurd Erixon on the post-war international scene, international activities, European ethnology and CIAP from 1945 to the mid1950s. Arv 69:89–152

Rogan B (2015) A remarkable congress and a beloved general secretary: CIAP/SIEF, Arnhem 1955 and Jorge Dias. Etn 19(3):567–576

Rudolph K (2004) Syncretism: from theological invective to a concept in the study of religion. In: Leopold AM, Jensen JS (eds) Syncretism in religion. A reader. Routledge, New York, pp 65–85

Schippers T (1995) A history of paradoxes: anthropologies of Europe. In: Vermeulen HE, Alvarez Roldan AA (eds) Fieldwork and footnotes. Studies in the history of European anthropology. Routledge, London, pp 234–246

Schrempp G (1983) The re-education of Friederich Max Müller: intellectual appropriation and epistemological antinomy in mid-victorian thought. Man 18:90–110

Seton-Watson H (1977) Nations & states. An enquiry into the origins o/nations and the politics of nationalism. University Paperbacks, Methuen/London

Smith A (1991) National identity. Penguin Books, London

Stocking G (1982) Afterword: a view from the center. Ethnos 47:72–86

Stocking G (1987) Victorian anthropology. The Free Press, New York

Stocking G (1995) After Tylor. British social anthropology 1888–1951. Athlone, London

Tax S, Eiseley L, Rouse I, Voegelin C (eds) (1953) An appraisal of anthropology today. International Symposium on Anthropology of the Wenner-Gren Foundation. University of Chicago Press, Chicago

Vasconcelos JL (1901) Esquisse d'une dialectologie portugaise [An essay on Portuguese dialectology]. Aillaud, Paris

Vasconcelos JL (1936) Etnografia portuguesa [Portuguese ethnography], vol II. Imprensa Nacional-Casa da Moeda, Lisboa

Vasconcelos JL (1942) Etnografia portuguesa [Portuguese ethnography], vol III. Imprensa Nacional-Casa da Moeda, Lisboa

Vasconcelos JL (1986 [1883]) Tradições populares de Portugal [Portuguese folk traditions]. Imprensa Nacional-Casa da Moeda, Lisboa

Vermeulen J, Alvarez Roldan AA (eds) (1995) Fieldwork and footnotes. Studies in the history of European anthropology. Routledge, London

VVAA (1961) Portuguese folk architecture. Sindicato Nacional dos Arquitetos, Lisboa

VVAA (1974) In memoriam António Jorge Dias, 3 vols. Instituto de Alta Cultura – Junta de Investigações do Ultramar, Lisboa

Wilson W (1976) Folklore and nationalism in modern Finland. Indiana University Press, Bloomington

Making and Unmaking of Ethnos Theory in Twentieth-Century Russia

66

Sergei Alymov

Contents

Introduction	1700
Key Characters	1702
Fedor Volkov and the Politics of Ukrainian Identity	1703
The Concept of Ethnos and the Teaching of Anthropology	1704
The Development of Volkov's Methodology by Sergei Rudenko	1705
Marxism and the End of Biosocial Theory in the Soviet Union	1706
Shirokogoroff As Ethnographer and Theorist	1707
A New Rapprochement of Natural and Human Sciences in the USSR	1709
The Revival of Ethnos and the Questioning of Stalin's Theory of Nation	1710
Refining Ethnos Theory at the Academy in the 1970s – Early 1990s	1713
Nationalities Study in the USSR: Nation Versus Ethnos/"Scientific Communists" Versus Ethnographers	1717
Perestroika and the Decline of Ethnos in Russian Academia	1720
Conclusion	1721
References	1723

Abstract

The chapter is a historical account of the development of the concept of ethnos in twentieth-century Russia. The first formulations of ethnos as a subject matter of ethnography as a science appeared in the 1910–1920s. Their authors were anthropologists educated in Paris and Saint-Petersburg. They considered ethnography as a part of anthropology, which they believed to be a natural science. N.M. Mogilianskii and F.K. Volkov studied East Slavic anthropology and were activists of Ukrainian nationalism. They considered physical anthropology a key to differentiating ethnoses, and that linguistic, cultural, and physical anthropology maps were instrumental in this regard. S.M. Shirokogoroff produced a complicated theory of ethnos centered on ecological adaptation of ethnoses and their relations with neighbors.

S. Alymov (✉)
Institute of Ethnology and Anthropology, Russian Academy of Sciences, Moscow, Russia

The concept of ethnos and the biosocial theory behind it was inadmissible for Soviet Marxism and was forbidden during the period 1930–1950. The experimental period after Stalin's death made it possible to criticize his definition of nation as well as enabled a rapprochement of natural sciences and humanities. The theory of ethnos thrived during the late Soviet period, although its main theorists – Iu.V. Bromlei and L.N. Gumilev – sharply criticized each other. They were also challenged by experts in national question. The Perestroika and ethnic conflicts of the 1990s made ethnographers turn to constructivism and instrumentalism, so the concept of ethnos was marginalized in academia.

Keywords

Ethnicity · Ethnos · Theory of ethnos · Nation · Nationalism · the USSR · Russia · Russian anthropology · Soviet ethnography · History of anthropology

Introduction

This chapter gives a historical account of the theory of ethnos in the Russian Empire and Soviet Union. Unlike late Soviet era narratives (Bromlei 1986; Vainshtein and Kryukov 1986), it does not approach the development of the idea as a teleological progressive process. It neither engages in polemics against the theory (Filippov 2010), nor evaluates its epistemological qualities against other theories of ethnicity (Banks 1996: 17–23). The approach of the present chapter is a historicist one. Although the text concentrates on the main figures' discussions, the author tries to provide as much intellectual, political, and social background as possible, following the model of "multiple contextualization" (Kuklick 1998). It makes clear that the thinking of pioneers of ethnos cannot be understood without taking into account their political personae and transnational influences and connections (Anderson and Arzyutov 2019).

The chapter follows the rises and falls of the concept in the Russian Empire and the Soviet Union in chronological order. The first generation of ethnos theorists were a group of like-minded scholars, preoccupied with establishing ethnography and anthropology in Russian universities and, besides, acting as political activists at the time of the Russia's Great Revolution and the Civil War. On the one hand, they followed the well-established tradition of Russian ethnography in defining the subject matter of the discipline in terms of nationality (Knight 1998; Vermeulen 2015). On the other, they brought in a distinctively new emphasis on physical anthropology and positivistic standards of knowledge.

The second half of the story is played out in a radically different milieu, i.e., that of the Soviet 1960s. The death of Stalin in 1953 and the denunciation of his "cult" in 1956 started a revision of political and intellectual legacies of the whole Stalinist period. The revival of ethnos had a lot to do with the reestablishment of the biosocial theorizing, banned in the early 1930s. Another central intrigue of this period is uneasy relations between the Soviet ethnos-thinking and Stalin's theory of nation. The first emerged in a deliberate opposition to the second. So, it is necessary to follow the development of the Soviet ethnos-thinking among ethnographers with an eye on their relations with their immediate rivals, i.e., experts on the theory of nation and national relations from the discipline of scientific communism.

The intellectual tradition that produced ethnos theory was formed around such institutions as the Department of Geography and Ethnography of Saint Petersburg University, the Russian Anthropological Society of Saint Petersburg University, the Russian Museum, and the Museum of Anthropology and Ethnography (Kunstkamera, or MAE) of the Academy of Sciences. The main features that characterized their thinking were:

1. A training in natural sciences and to an extent a shared positivistic idea of biosocial laws that govern society as a "natural" phenomenon
2. An interest or training in physical (biological) anthropology
3. A connection to the discipline of geography and sometimes geographical determinism
4. Borrowings from contemporary French and German anthropology
5. A vision of anthropology as an umbrella natural science of "man" that stemmed mainly from the French tradition of anthropology. Ethnography was seen as one of its subdisciplines.

The idea of ethnos as a subject matter of ethnography as a discipline was suggested for the first time, in Russia, by the ethnographer and museum curator Nikolai Mikhailovich Mogilianskii (1871–1933) in his article "Ethnography and its Tasks" (Mogilianskii 1908). In 1916 Mogilianskii published an essay "On the Subject Matter and Tasks of Ethnography" with the following definition of ethnos:

The ἔθνος [ethnos] concept — is a complex idea. It is a group of individuals united together as a single whole by several general characteristics. [These are:] common physical (anthropological) characteristics; a common historical fate, and finally a common language. These are the foundations upon which, in turn, [an ethnos] can build a common worldview [and] folk-psychology – in short, an entire spiritual culture (Mogilianskii 1916: 11).

After 1916, the five core elements of Mogilianskii's definition (a single collective identity; a physical foundation; a common language; a common set of traditions or destiny; and a common worldview) would appear in successive descriptions of Russian and Eurasian ethnos theory for the next 100 years. A full-fledged "theory of ethnos" was developed by the Russian émigré ethnographer Sergei Shirokogoroff who came from the same intellectual milieu of Saint-Petersburg academics. In the first book-length monograph on the topic, he included many of the same attributes:

[An] ethnos is a group of people, speaking a common language who recognize their common origin, and who display a coherent set [kompleks] of habits [obychai], lifestyle [uklad zhizni], and a set of traditions that they protect and worship. [They further] distinguish these [qualities] from those of other groups. This, in fact, is the ethnic unit – the object of scientific ethnography. (Shirokogoroff 2010: 16)

There was one characteristic that Mogilianskii shared with his older friend and teacher Fëdor Kondratievich Volkov [Khfider Vovk]: their Little Russian/Ukrainian origins and active involvement in the Ukrainian national movement and politics. The fact that this program was conceived in ethnic-national terms made these anthropologists particularly mindful of ethnic divisions while their scientific anthropological outlook contributed to the way they naturalized these differences. The appearance of "ethnos thinking" should be considered not as an invention of pure scientists, but in the political context of the turbulent last years of the Russian Empire, replete with national parties and movements at the age of collapsing empires and rising nation-states.

Key Characters

Anthropologist, archaeologist, and ethnographer Fedor Kondratievich Volkov (1847–1918), or Vovk, was educated at the departments of natural sciences of the faculty of physics and mathematics at the universities of Odessa and Kiev. As a result of increasing persecutions of the Ukrainian movement, in which he took an active part, Volkov left the Russian Empire. In 1887, after a peripatetic period involving many cities and countries of residence, he finally settled in Paris, where he attended lectures of leading French anthropologists, including Paul Broca's disciple Léonce Manouvrier, and Paul Topinard, and was on the editorial board of the journal L'Anthropologie. In 1905 he received a master's degree in natural sciences for his dissertation, Skeletal Variations of Feet among the Primates and Races of Man, under the supervision of Ernest-Théodore Hamy. After the 1905 Revolution Volkov returned to Russia, and in 1907 was appointed as curator at the Russian Museum and started teaching at Saint Petersburg University. He died in 1918 on his way from Petrograd to Kiev. In March 1918, several months before his death, he was elected the head of the department of geography and ethnography at Kiev University (Franko 2000).

Nikolai Mikhailovich Mogilianskii was born in 1871 in Chernigov in Malorossia [Little Russia, now Ukraine]. In 1889 he entered the natural sciences division of the Saint-Petersburg University where he listened, among others, to the lectures of the anthropologist and geographer Eduard Petri and the anatomist Petr Lesgaft. In 1894 he went abroad to continue his education in Paris. During his stay there he studied anthropology at l'École d'Anthropologie under Léonce Manouvrier, Gabriel de Mortillet, Charles Letourneau, and others. In Paris he became close friends with Fedor Volkov who influenced him as a more experienced anthropologist and compatriot. Upon returning to Saint-Petersburg Mogilianskii became a professional anthropologist and ethnographer. He worked in the Emperor Alexander III's Russian Museum until 1918. He also lectured in anthropology and geography in several educational institutions. After the Bolshevik revolution Mogilianskii moved to Kiev where he held high posts in the government of independent Ukraine under getman (a military commander in Eastern and Central Europe) Pavlo Skoropadskii.

In 1920 he immigrated to Paris. In 1923 he moved to Prague where he resumed his teaching and research. Mogilianskii died in Prague in 1933.

Sergei Mikhailovich Shirokogoroff (1887–1939), the youngest of the three, was born in Suzdal', an ancient town in the Central Russia. From 1907 to 1910 he audited courses at the Faculté des Lettres, University of Paris and l'École d'anthropologie de Paris and read extensively anthropological literature. Upon his return to Russia in September 1911 he enrolled at Saint Petersburg University, The Division of Physics and Mathematics, where he could have heard the lectures of V.K. Volkov. During his Saint Petersburg years, he was active in anthropological circles and even was appointed Head of Department of [Physical] Anthropology of the Museum of Ethnography and Anthropology (MAE) in 1917. He was also involved in the work of the Russian Geographical Society's Commission for Making Ethnographic Maps of Russia, headed by Volkov. In 1912, 1913, and 1915, he and his wife Elizaveta conducted their field research among the Evenki in Transbaikal region. In 1917 Shirokogoroff, like the other protagonists of the story, left Petrograd, but went eastward. He spent the years 1918–1922 mostly in Vladivostok, which at this time changed hands between the revolutionary Red Army and the counterrevolutionary Whites. The Shirokogoroffs never returned to Russia. Sergei Mikhailovich worked at a number of Chinese universities, including Sun Yat Sen University in Canton (Guăngzhōu; 1927–1930), and the Tsinghua University in Beijing (1930–1937). He died in Beijing in 1939 (Anderson and Arzyutov 2019; Anderson 2019; Arzyutov 2019).

Fedor Volkov and the Politics of Ukrainian Identity

In the early period of his life Volkov was influenced by Ukrainian historians Nikolai Kostomarov (1817–1885), Volodimir Antonovich [Włodzimierz Antonowicz] (1834–1908), and other intellectuals who had laid down the basis of Ukrainian studies and Ukrainian nationalism. If the first one represented the Romantic definition of Ukrainian identity, the second one rendered it with positivistic arguments of physical anthropology (Alymov 2019: 84–88).

When Volkov returned to the Russian Empire during the First Russian Revolution in 1905–1907, he encountered a thriving Ukrainian community in Saint Petersburg. After the declaration of civil liberties and the convening of the first parliament (Duma) (1906), Ukrainian nationalists could legally engage in public politics. They published journals, had their representatives in the First State Duma, and vividly discussed the question of national determination and autonomy. In 1906 the newspaper Ukrainskii vestnik (Ukrainian Herald) published Volkov's article "Ukrainians from the Anthropological Point of View." Discussing various "ethnic indicators," he claimed that "the successes of somatic anthropology [...] urged [scholars] to look for other, more lasting ones, which happen to be purely physical indicators like the colour of bones, hair and eyes, proportions and forms of various parts of the body and, predominantly, its skeleton" (Volkov 1906: 418). Volkov argued that they all showed a similar pattern of geographic variation along a northeastern-southwestern axis from a comparatively short, blond, long-headed type to the brachycephalic population of tall stature, dark hair and eyes and a straight and narrow nose that he believed to be "the main Ukrainian type."

The final, classic version of Volkov's studies of Ukraine were published in the second volume of a rich and well-illustrated edition, "The Ukrainian People in its Past and Present," published in Saint Petersburg by Maksim A. Slavinskii (1868–1945), the same journalist who edited Ukrainskii vestnik. Their conclusions were the following:

1) 1) The Ukrainian people on the entire territory is distinguished by a range of common ethnographic characteristics, which leaves no doubt that it constitutes an ethnic unity that definitely stands out among other Slavic peoples.
2) The Ukrainian people preserved in its everyday way of life a considerable number of vestiges from the past, proving that it had not undergone very deep ethnic influences from outside, and that in spite of an eventful history it developed its ethnographic characteristics consistently and quite uniformly.
3) As all other peoples, it was exposed to a certain extent to external ethnographic influences and assimilated some alien forms, but not to a degree that could alter its main ethnographic characteristics and deny its common Slavic type.
4) In particular of its ethnographic way of life, the Ukrainian people manifests the closest similarity with its Western neighbours – Southern Slavs, such as Bulgarians and Serbs, as well as Romanians, who remain a quite Slavic people ethnographically. Poland was the main conduit of cultural diffusion from the European West.
5) In their most ancient form, the ethnographic characteristics of Belarusians and Great Russians are close if not identical to those of the Ukrainians. (Volkov 1916: 647)

This conclusion conveys the central intuition of ethnos-thinking, i.e., that cultural, linguistic, and physical anthropology's borders laid on the geographical map must coincide, thus revealing geographical outlines of ethnoses. The project of making ethnographical maps of this kind was launched with Volkov's active participation, but was scrapped because of the war and revolution (Alymov and Podresova 2019).

The Concept of Ethnos and the Teaching of Anthropology

Another important aspect that helps understand the formation and significance of the concept of ethnos was the position of ethnography and anthropology in the university. The beginning of the First World War stimulated the authorities to look for an alliance with scientists who, from their side, were also willing to cooperate in the war effort. In 1915, under the newly appointed liberal minister of popular enlightenment Pavel Ignatiev, a draft of a new University Charter was sent to university councils for discussion. This and other bureaucratic procedures continued until the Revolution, and the charter was never approved. Nevertheless, it triggered a round of debates about the academic teaching of anthropology/ethnography and its status as a natural or human science. In fact, Mogilianskii's article "On the Subject Matter and Tasks of

Ethnography" – which contained the definition of ethnos – was his motion in this debate.

Mogilianskii was emphatic about the distinction between the history of culture that had as its subject matter human culture in general, and ethnography that dealt with ethnos and its specific features. He suggested establishing two departments – anthropology and ethnography – at the faculty of natural sciences, and a department of history of culture at the faculty of history and philology. Mogilianskii followed the opinion of Volkov who suggested the establishment of an Anthropological institute with departments of physical anthropology, prehistoric anthropology, and ethnography. The model for this institute was the École d'Anthropologie in Paris, the only place where, according to Volkov, anthropological sciences were taught "in their entirety" (Volkov 1915: 102). Both Volkov and Mogilianskii adhered to the late nineteenth century French vision of anthropology and ethnology which, however, had developed in a rather peculiar way. The term "anthropology" was used to denote "a natural science devoted to 'positive' investigations into human anatomy, the variety of human physical types, and 'man's place in nature'" (Williams 1985: 331) while ethnology was usually defined as the study of races (Conklin 2013: 53). Mogilianskii's understanding of ethnic differences echoed this definition. He subscribed to Paul Broca's definition of anthropology as a "science that studies the human group in its entirety, its details, and its relations to nature" (GARF R-5787-1-93: 2). Tribes and peoples were defined as "lesser units" within a few large racial groups that "differ from each other by secondary characteristics." As an example, he cited the visible physical differences between a tall, blond, and blue-eyed Norwegian and a brown, dark-eyed, and dark-haired Portuguese, both of whom would be classified within a single "white race" (GARF R-5787-1-93: 4).

Mogilianskii continued to teach anthropology and write about Ukrainian matters in emigration. In Paris, in 1921, he wrote unpublished article "Ukraine and Ukrainians" in which he attempted to integrate ethnography, history, physical anthropology, and current politics into an inclusive characterization of an "ethnic type." There he also referred to Volkov's conclusions as decisive evidence that proved the difference between Great and Little Russians and the existence of a distinct homogeneous Ukrainian type (GARF R-5787-1-34: 9–11).

The Development of Volkov's Methodology by Sergei Rudenko

Mogilianskii died in exile, and his post-1917 writings remained for the most part unpublished and inaccessible to readers in the USSR. That was also the case with Sergei Shirokogoroff's writings. Nevertheless, the idea of ethnos and Volkov's methodology, with its complex investigation and mapping of data from physical anthropology, as well as from material and spiritual culture, was followed by the generation of his pupils who stayed in the Soviet Union.

The most prominent among them was Sergei Ivanovich Rudenko (1885–1969), an Ukrainian born in Kharkov. He studied anthropology in Saint Petersburg with Volkov and spent a year in 1913–1914 attending classes at the École d'Anthropologie in Paris and working in Léonce Manouvrier's laboratory. Undoubtedly, it was his book, *The Bashkirs: An Ethnological Monograph*, that established Rudenko as one of the leading Russian anthropologists. It was published in two volumes: The Physical Type of Bashkirs (1916) and The Way of Life of Bashkirs (Byt bashkir) (1925) (Rudenko 1916, 1925). This book was written under the obvious influence of Volkov's methodology and reflected the model Volkov suggested in his writings on the "Ukrainian People in its Past and Present." Volkov distinguished three groups of Ukrainians: northern, middle, and southern Ukrainians whose dialectal and cultural borders roughly coincided with those of anthropological types. The correlation between "types," ascertained on the basis of physical anthropology, linguistics, and cultural traits was the issue that also intrigued Rudenko. In his book he distinguished three major cultural types of Bashriks (eastern, south-western, and northern) and concluded that they retained their most ancient "pure" Turkish cultural forms in the eastern type.

At the First Turkological Congress in 1926, Rudenko gave a paper titled "The Current State and Next Tasks of the Ethnographical Studies of the Turkish Tribes," in which he presented an ambitious research program and made a series of theoretical observations characteristic of the Volkov school. He claimed that it was possible to speak of a physical type that was characteristic for the Turks and which manifested itself most vividly in the Kazakh-Kyrgyzes. Rudenko proposed to "determine the geographical distribution of the individual cultural (бытовые) elements and their combinations in the closed biological units that we call ethnic groups" (Rudenko 1926: 86). This study was to reveal the "provincial and regional groupings" that presumably coincided with the peculiarities of a physical type and dialects. His presentation ended with a reference to exact scientific methods and biological metaphors: "In order to succeed in developing our knowledge about the biology of human societies, the life of ethnic groups, and the factors which influence their lives, in order to clarify the evolution of the human culture, we must switch from dilettantism to precise scientific investigation (Ibid: 88)."

Marxism and the End of Biosocial Theory in the Soviet Union

Rudenko's grandiose program was doomed. It was formulated right before the Cultural Revolution and the "Great Break," which shook the life of the whole country in 1929. Among other disruptions, such as the restructuring and the Bolshevik's "takeover" of the Academy of Sciences, there came a firm philosophical dictate that social laws should be shown to work independently of natural laws. Within ethnography, this placed a taboo on any direct reference to the social structures being linked to biological processes. As historian Mark Adams has observed, this was epitomized by the emergence of a new pejorative term "biologizirovat'" (to biologize). He further reflected that "no field that linked the biological and the social survived the Great Break intact" (Adams 1990: 184).

Valerian Aptekar', a fervent proponent of Marxism and student of the linguist and archaeologist academician N. Ya. Marr, started a campaign against ethnology as science. He claimed that "culture" and "ethnos" – two central concepts of ethnology – were construed as natural, metaphysical, or biological substances with their own immanent forces. Cultures and ethnic groups in ethnological discourse are endowed with biological, chemical and physical characteristics, thus portraying social processes as analogous with those in organic and even nonorganic nature. He also insisted that ethnos in ethnological discourse stood for a thinly disguised race. Although Aptekar"s critique was to an extent justified, his accusations of ethnographers in racism were hardly adequate given the distinction that they made between two concepts since the times of Eduard Petri. Nevertheless, the huge ideological turn of the late 1920s to the early 1930s led to a devastating critique of "bourgeois" science, including purges of many prominent ethnographers, and the creation of a Marxist ethnographic literature that used only "sociological" or historical concepts (Slezkine 1991).

Rudenko's suggestions about correlations between cultural and physical types as well as formulations like "biology of human societies" became ideological anathema. Rudenko was arrested in the summer of 1930 in Bashkiria, but there is no direct evidence that the repressions against him were related to his scientific views. The researcher was named in the so-called "academic case" against the "All-People's Union for the Revival of Russia" – an organization fabricated by the Soviet secret police (OGPU) to deal with politically conservative academics. Rudenko was charged with the squandering of resources during his expeditions. Any further development of the theory of ethnos with its biosocial implications became impossible in the Soviet Union during Stalinism.

Marxism had its own vocabulary for speaking about "nationalities question." It comprised the central notion of nation and a set of terms for designating "prenational" identities, such as tribe and nationality (*narodnost'*). These terms related to the precapitalist social forms, which were arranged in what Francine Hirsch calls "the Marxist timeline of historical development" from primitive society to socialism. The goal of the Soviet state was to assist them in this development through "state-sponsored evolutionism" to which ethnographers also subscribed (Hirsch 2005: 7–10). Nation was defined by Stalin in 1913 as "an historically evolved, stable community of people, which is united by a common language, territory, economic life, and a psychic individuality manifested in a common culture" (cited in: Meissner 1976: 58).

Shirokogoroff As Ethnographer and Theorist

Unlike Volkov and Mogilianskii, Shirokogoroff created an elaborate theory of ethnos, which produced not only definitions, but a quasi-scientific narrative of the growth and interaction of ethnoses. According to Shirokogoroff himself, the first attempt to formulate his theory was partly due to the first expedition to Tungus-speaking population of Zabaikal'e (Transbaikal region). Shirokogoroff and his wife Elizaveta set out to Zabaikal'e in search of "pure" Tungus-speaking tribes to document their vanishing languages and cultures in a classical "salvage ethnography" style. The realities on the ground – creolized groups with mixed spoken languages – were far from his clear-cut preconceptions. This made Shirokogoroff concentrate mostly on anthropological measurements. He also was a meticulous student of material culture, an interest that is related to the central theme of his theory, namely, ecological adaptation and the role of material culture in the resilience of ethnoses. As David G. Anderson concludes, "His painstaking physical anthropological work was intended to illustrate the continuity of physical types within the groups in spite of linguistic and cultural assimilation. From his first fieldwork he developed the counterintuitive idea that a demographically sparse, hunting culture could define the ethnic landscape of half of a continent" (Anderson 2019: 234).

In 1922 and 1923 Shirokogoroff published his ideas in a book intended as an introduction to ethnology for students of the Far Eastern University. Among the reasons for the hasty publication of the brochure he also pointed at his willingness "to abolish the gap between so called humanistic and natural sciences," the project that his ethnology was to accomplish (Shirokogoroff 2010: 9–11). Unlike Volkov and Mogilianskii, Shirokogoroff considered ethnography as humanities discipline that studies material, social, and spiritual culture of ethnoses, drawing on psychology, linguistics, sociology, and "technology." But Shirokogoroff's scheme of the division of sciences was similar to the one of his older colleagues, only with ethnology as a unifying discipline, integrating the study of culture (ethnography), physical anthropology, and language and "seeking to establish the laws, to which the life of individual ethnoses obey. Like biology, which is a science of life in general, ethnology is a science of ethnoses as forms in which humanity lives and develops, <...> hence, ethnology is the pinnacle of the study of man" (Shirokogoroff 2010: 34). Shirokogoroff's version of laws was his equations of "ethnic equilibrium" that related the territory of ethnos, the density of its population and the level of its culture. They also took into consideration the strength of neighboring ethnoses.

As David G. Anderson points out, Shirokogoroff's theory "to contemporary readers seems to combine the anthropogeography of Ratzel, with a concern over performed ethnic boundaries anticipating those of Frederik Barth," although neither he, nor anybody else managed to operationalize his quasi-scientific formulas. It is also important to keep in mind the political dimension of Shirokogoroff's thought. As Dmitry V. Arzyutov has shown, his lively interest in politics took more practical application in Vladivostok, where he was deeply involved with "Non-Socialist Movement," uniting anti-Bolsheviks of the region. He published a series of political pamphlets in which he envisioned a kind of non-party "national movement," expressing a "national will," a concept which bears resemblance to his ethnos' adaptation strategies. His political and scholarly works were written in the situation of great chaos and in a region torn in struggle between several superpowers. His imaginative visions of ethnic equilibrium and nations united in one will, probably, served as a solace in the midst of a highly unbalanced reality (Arzyutov 2019). Meanwhile, his émigré status and right-wing politics were enough for making him persona non grata, although his ethnogenetic hypothesis was occasionally cited by Soviet anthropologists. It was only toward the late Soviet period that his writings were cited as a precursor of the Soviet theory of ethnos.

A New Rapprochement of Natural and Human Sciences in the USSR

The term "ethnos" reappeared in ethnographers' publications soon after the end of the Great Patriotic War. It was the period of the publication of Stalin's article "The National Question and Leninism" (1949) and "Marxism and Problems of Linguistics" (1950) which, along with Stalin's famous 1945 toast to the Russian people, signaled the strengthening of the tendency toward Soviet-Russian patriotism. Pavel Ivanovich Kushner (Knyshev) (1889–1968) reintroduced ethnos into post-war Soviet ethnography in an influential book "Ethnic Territories and Ethnic Borders" (1951). His definition of ethnos acknowledged both history and geography – and ignored physical form: "Ethnic phenomena distinguish the everyday life [byt] of one people from another. The set of such special markers include differences in language, material culture, customs, beliefs, etc. The sum-total [sovokupnost'] of such specific differences in everyday lives of peoples, preconditioned by the history of those peoples, and the effect of the geographical environment upon them is called 'ethnos' (Kushner (Knyshev) 1951: 6)." It is telling that after Stalin's smashing of Marrism, Rudenko was able to return to his reflections on ethnos. In an unpublished sketch "Etnos and Culture" he defined etnos as a people [narod] or a group [narodnost'] demonstrating all the characteristics of a nation and differing from the latter by the "presence of the commonality of the somatic origin of its members, which is not a requirement for a nation" (SPF ARAN 1004-1-40: 1). The standard definition, of course, lacking "commonality of somatic origin," appeared in the first postwar textbook of ethnography: "For a correct understanding of the tasks of ethnography it is important to specify the meaning which this branch of historical science put across the notion of "a people" (ethnos). Peoples are historically formed groups of people, bound together by the commonality of the territory of their formation, their language and culture" (Tolstov et al. 1957: 10).

As the study of ethnic origins of modern peoples (ethnogenesis) became a high priority under the new director of the Institute of Ethnography Sergei Pavlovich Tolstov (1907–1976), a close cooperation between physical anthropologists and ethnographers was encouraged. Physical anthropological measurements could ascertain degrees of homogeneity and diversity among speakers of certain linguistic groups as a sort of independent measure of ethnogenetic progress (Debets et al. 1952: 28–29). The leading physical anthropologist Valerii P. Alekseev (1929–1991) epitomized this resumption of a multidisciplinary approach by the new generation. His doctoral dissertation "The Origins of the Peoples of the Eastern Europe" used craniological research to balance arguments about ethnogenesis. This book partially "rehabilitated" Volkov's views on the anthropological distinctiveness of Ukrainians (Alekseev 1969: 164). Later he hypothesized that the earliest period of ethnogenesis harks back to the Paleolithic age, when "a certain correlation of the borders of primitive populations, cultural traditions and primal languages" existed (Alekseev 1982: 52).

Viktor A. Shnirel'man observed, in the 1960s there was a renewed interest in and enthusiasm for linking human behavior to genetic heredity (Shnirel'man 2011: 252–80). The search for a new synthesis between the social and natural sciences was proclaimed by no other than the president of the Academy of Sciences, Mstislav V. Keldysh (1911–1978). In his speech at the general meeting of the Academy in October 1962 he declared: "We cannot leave the social sciences with the task of developing themselves [in isolation]. There is no clear-cut division between the social, natural, and technical sciences" (Keldysh 1962: 6).

Both major ethnos theorists of the 1960s – Yulian Vladimirovich Bromlei (1921–1990) and Lev Nikolaevich Gumilev (1912–1992) – were influenced by this intellectual atmosphere. Bromlei published his first "ethnographical" article in coauthorship with Alekseev, and his first theoretical article titled "Ethnos and endogamy" speculated about the role of inmarriages in "stabilizing" ethnos. Gumilev was more ingenious in this regard, drawing inspiration from a wide range of disciplines, including ecology and earth sciences, genetics, and biophysics. The strongest influences on his thinking about ethnicity, as Mark Bassin argues, came through two outstanding Soviet geneticists, Nikolai Vladimirovich Timofeev-Resovskii (1900–1981) and Mikhail Efimovich Lobashev (1907–1971), the latter one also being a pioneer of ethology in the USSR (Bassin 2016: 31–32).

The Revival of Ethnos and the Questioning of Stalin's Theory of Nation

The 1960s in the Soviet academia were a time of experimentation and debates that were encouraged from above and tried the limits of the mandated freedom. In 1963, the Soviet Academy of Sciences staged a wide-ranging debate on methodological issues in the humanities and social sciences (Markwick 2001: 156). Academicians Pëtr N. Fedoseev (1908–1990) and Iurii P. Frantsev (1903–1969) wrote a sort of instruction manual for de-Stalinization, which encouraged social scientists, including ethnographers, to rewrite sociological and historical laws and to embark on interdisciplinary research (Akademiia nauk SSSR 1964: 16, 37).

This long hoped-for measure resulted in a string of publications that, among other things, sought to challenge Stalin's definition of nation. Early attempts to criticize Stalin's definition among ethnographers took place already in 1955. The party cell of the Institute of Ethnography lambasted one of ethnographer and demographer Viktor I. Kozlov's papers as revisionist and accused him of reviving Kautsky's idea that personal national identities (or, in Soviet language, national self-consciousness) constitute the only characteristic of nationhood (TsGAM P7349-1-13: 10–11). Nevertheless, in 1958 Kozlov and philosopher S.T. Kaltakhchan initiated another attempt to revise Stalin's definition of nation (Kozlov 1995: 24). In 1966–1968 the journal "Questions of History" organized a discussion of the concept of nation. Its participants suggested some modifications of Stalin's definition, including the idea to put national self-consciousness on the list of characteristics of nation, but no definitive conclusions of the discussion were made (Meissner 1976: 65–71).

In the mid-1960s these early attempts were followed by articles of leading ethnographic authorities, such as Sergei A. Tokarev and Nikolai N. Cheboksarov. These articles consolidated a classification of ethnic groups according to major historical epochs or formations (primitive, slave-owning feudal, capitalist, and socialist societies). They debated the foundations or common features like kinship, language, culture, or integrated economy, which consolidated ethnic groups of different epochs. According to these theories, nation was the most evolved type of ethnic group, demonstrating maximal integration on all levels of Stalin's list of common features (Cheboksarov 1967; Tokarev 1964). Philosopher Yuri I. Semenov contributed to this development by coining the term "social organism" as a universal historical agent, a notion that could serve as an umbrella term for all social collectives from clans to nations (Semenov 1966). The term was picked up by Kozlov in his definition of ethnic group as "a social organism which forms on a certain territory out of groups of people who possessed or developed a common language, common cultural characteristics, social values and traditions, and a mixture of radically varied racial components" (Kozlov 1967: 111). Semenov also managed to publish a rather harsh critique of Stalin's theory of nation, denying it any originality and exposing the definition as a plagiarism from Karl Kautsky and Bruno Bauer (Semenov 1967; Semenov 1994). The stage seemed to be set for the revival of organicist and, concurrently, intellectualist vision of ethnos offered by Iulian Bromlei.

Iulian V. Bromlei (1921–1990), a son of historian of Antiquity and a grandson of the famous theater theoretician and director K.S. Stanislavski, was trained as historian of medieval Croatia. Since 1958 he served as a secretary of the Department of History of the Academy of Sciences, so he was well aware of the discussions among historians, philosophers, and ethnographers of the 1960s. In January 1966 he was appointed director of the Institute of Ethnography and found himself in a position where he was forced to adjudicate the raging theoretical debates in order to earn respect among his peers. The early years of his career were marked by a controversy, triggered by his article "Ethnos and Endogamy (Bromlei 1969). There he claimed that endogamy – the tendency for members of one group to prefer to marry partners of their own group – was a "mechanism of ethnic integration." This direct reference to a biological foundation to ethnicity quickly got the new director into trouble. The head of the Department of the Near and Middle East, Mikhail S. Ivanov (1909–1986) started a campaign of attacks against his boss. Ivanov claimed that if ethnoses are "stabilized" by endogamy this not only negates the Marxist formations of Bromlei's thinking, but makes ethnos a biological category (Anon. 1970: 89). During a special meeting that was called to discuss the article, almost all members of the Institute rose to speak in support of the new director (Anon. 1970). Having consolidated his intellectual and administrative authority, Bromlei published his first and most well-known monograph "Ethnos and Ethnography" (1973). That was a theoretical tract which defined ethnos as the main subject matter for ethnography and delimited its borders against related disciplines: "... ethnos <...> can be defined as a historically formed community of people who share relatively stable characteristics of culture (including language) and psychic setup, an awareness of their unity and distinction from other similar units" (Bromlei 1973). This definition was meant to differentiate between linguo-cultural groups/identities that do not need to be politically, economically, and territorially united (ethnicos in Bromlei's terms) and "ethno-social organisms" that constitute states or forms of pre-state political organization, i.e., tribes.

Lev N. Gumilev (1912–1992), a son of two great Russian poets, Anna Akhmatova and Nikolai Gumilev, followed a different path to his version of ethnos theory. Having served more than 13 years in Stalin's prisons, he was never fully accepted to the Soviet academia. In an article titled "A Biography of a theory, or a Self-Obituary" he described his interest in the rhythms of World History, personal contacts with the peoples of Central Asia and Siberia, and V.I. Vernadskii's idea of biochemical energy as major influences on his thinking (Gumilev 1988). He published his first article, introducing ethnos, in 1965 – a full 2 years before Bromlei's first published intervention (Gumilev 1965). Like Bromlei and other ethnos theorists, Gumilev believed that there can be no single characteristic such as language, culture, or common origin that defines ethnoses. He also subscribed to Engel's concept of the forms of matter and strictly differentiated between biological and social forms of matter with separate laws governing each. But if for Bromlei and his colleagues, ethnic groups were just another type of social groups, Gumilev insisted that all ethnic phenomena and processes "belonged strictly to the natural realm" (Bassin 2016: 28). The laws that govern them were different from social laws, governing the development of "societies" through the range of Marxian social formations. Gumilev insisted that he discovered the natural laws, governing the life cycles of ethnoses. As Mark Bassin explains, "Gumilev viewed the ethnos as a biological organism, and like all organisms it developed through a fixed process of growth, maturation, and decline" (Bassin 2016: 54–55). These life cycles, originally driven by some cosmic energy and genetic mutation among "passionate" individuals, in Gumilev's terminology were called "ethnogenesis," a term introduced by D.A. Koropchevskii at the turn of the twentieth century and commonly used by Soviet ethnographers to indicate the study of the formation of ethnoses.

Gumilev's radical reinvention of biosocial science and his pretension of being a founder of truly scientific ethnology in the USSR could not have passed unnoticed by Bromlei and other theorists from the Institute of Ethnography. There followed a long-lasting polemic, in which Gumilev was predictably lambasted for biologizing. During Perestroika he was somewhat vindicated by soaring popularity of his writings and a palpable failure of Soviet nationality policy and doctrines. Gumilev controversially linked the strained ethnic tensions in the crumbling Soviet federation to Bromlei's misguided theories. Bromlei retaliated by labeling Gumilev's distinction of "passionate" and "subpassionate" peoples as covert racism (Vainshtein 2004: 624–27). But to fully appreciate the role of "official" ethnography and ethnos theory in intellectual life and politics of the late Soviet period, one must consider the development of the theory per se, and the configuration of the whole field of ethnic studies and expertise within which this theory functioned.

Refining Ethnos Theory at the Academy in the 1970s – Early 1990s

Anthropologist Sergei Sokolovsky suggested a periodization of the history of Soviet ethnography, in which he branded the 1950–1980s as a scholastic period (Sokolovskii 2012). Indeed, after the publication of Bromlei's "Etnos and Ethnography" (1973), the lion's share of theoretical efforts of Soviet ethnographers was directed at refining and reformulating the concept of ethnos. Some of these efforts were aimed at finding factors that contributed to the stability of ethnoses in a way similar to Bromlei's endogamy. The second option was to try to apply the theory to historical or anthropological material. Finally, many scholars speculated about characteristics that distinguish ethnoses from one another and from other social groups. The following review will consider only the most remarkable works in all three directions and try to evaluate the outcomes with which these enquiries met the crisis and the breakup of the Soviet Union.

In 1972, Sergei A. Arutyunov and Nikolai N. Cheboksarov published an article titled "The Transmission of Information as a Mechanism of Existence of Ethnosocial and Biological Groups of Humanity." Arutyunov (b. 1932) was an expert on the culture and archaeology of Japan and Beringia, Cheboksarov (1907–1980) was a leading specialist on physical anthropology of China and South-Eastern Asia and head of the Department of Foreign Asia, Australia, and Oceania. In his lectures on China at Moscow State University, Cheboksarov had cited the works of Shirokogoroff and introduced his listeners to the concept of ethnos already in the early 1950s (Pimenov 2015: 115). Arutiunov and Cheboksarov observed that the ethnos theory was lacking an explication of a "mechanism of connections that account for the wholeness of the very existence of an ethnic group" (Arutiunov and Cheboksarov 1972: 11). The scholars theorized that this mechanism was simply the communication of the members of ethnic groups, the density of which is thicker inside the group than with members of other groups. They envisioned that, if there was a way to register and map all forms of oral and written communication between people (which, they admitted, was most likely impossible even in the future), the clusters of high density would coincide with ethnic groups and the "zones with low density – with borders between them" (Arutiunov and Cheboksarov 1972: 19). The density of "info-contacts" generally increases from tribe of primitive society to modern nations. Important clarification concerning the difference between nations and ethnoses (although nations were considered by authors as a form of ethnos) was the reliance of nations predominantly on synchronic contacts, while ethnoses are maintained mostly by diachronic contacts, i.e., cultural traditions, transmitted through generations (the distinction, equivalent to Bromlei's ethnoses and ethnicoses). Biosocial imagination played its part in this cybernetic version of ethnos theory. The authors relied on Valerii P. Alekseev's works which upheld the "population concept of race" as well as Bromlei's endogamy; they also considered ethnoses as hierarchical structure of endogamous populations.

The tendency of grounding the reality of ethnos in demographical statistics of marriages permeates the history of the theory from its early years. There were, however, two important discussions that form a logical sequence: Bromlei's theoretical intervention, which led to the consolidation of his authority, was followed by Zoia P. Sokolova's (b. 1930) attempt to demonstrate the statistical validity of endogamy as the ethnos' defining feature. Being a leading authority on the ethnography of Khanty and Mansi, for decades she had collected archival marital statistics of Khanty and Mansi, available since the eighteenth century. In the early 1990s she published a monograph "Endogamous Area and Ethnic Group" followed by an article "Endogamy and Ethnos." She believed that her Khanty and Mansi material was conclusive enough to elevate the status of endogamy from an "auxiliary" feature or a prerequisite for ethnos to its full-fledged characteristic. She estimated that for an ethnos to be "stable" (like the Khanty from the eighteenth to nineteenth century), it has to have 85–90% of endogamous families, and the level of endogamy below 85% (among the Manci) signals the process of its "erosion" (Sokolova 1992).

Sokolova's article was discussed in three issues of "Etnograficheskoe Obozrenie" (Ethnographic Review) in 1992 by ethnologists, physical anthropologists, demographers, and archaeologists. Although most comments were to an extent positive, the spirit of an epoch has clearly changed, and the reception was far from unanimous. Two theorists of younger generation, Viktor A. Shnirelman and Sergei V. Sokolovsky, flatly rejected the central thesis of the article. Shnirelman blamed ethnos theory for not taking into account "active role of ethnic consciousness" and provided an example of the Tlingit of Alaska: they married mostly outside their group, but "preserved themselves as ethnos" (Shnirel'man 1992). Sokolovsky, who himself studied statistics of marriages, claimed that ethnos' endogamy was a statistical result of endogamy of dems (i.e., small relatively isolated local populations), as much as endogamy of Africa, France, or Eastern Siberia (Sokolovskii 1992).

The most ambitious research project that applied formulations and methods of ethnos theory to empirical material was, undoubtedly, the six volume ethnic history of the Chinese, published by Mikhail V. Kryukov and Nikolai N. Cheboksarov in collaboration with historians and philologists Vladimir V. Maliavin (b. 1950), Leonard S. Perelomov (1928–2018), and Mikhail V. Sofronov (b. 1929). The volumes were published between 1978 and 1993 and covered the whole span of Chinese history since the earliest Paleolithic findings till the beginning of the twentieth century. The "ethnic history" was a quite all-encompassing concept: "our science defines ethnic history as a process of consequent changes of all features that characterize peoples of a certain region" (Kryukov et al. 1987: 5–6). Following this definition, each volume contained a detailed account of Chinese political, social, and cultural history with a breadth of topic ranging from the analysis of archaeological findings and climate condition of the region to the subtleties of the Confucian worldview or the development of Chinese theater. The books also featured polemic with Gumilev's interpretation of "cycles" of Chinese ethnos (Kryukov et al. 1979: 7–8, 279–80).

The authors of the series arrived at their own conception of ethnic history, quite distinct from Gumilev's cyclicism or Bromlei's schematic theorizing. The kernel of their approach was an analysis of "ethnic consciousness," which they conceived as identity that "focuses in itself the most important features of an ethnos, concentrating its values" (Kryukov et al. 1979: 277). In a retrospective theoretical reflection, published in the last volume, the authors openly attacked Stalin's definition of nation and the ubiquitous "triad" of tribe-nationality-nation. They claimed that since the 1970s Soviet ethnographers engaged in revisiting this definition, relegating such characteristics as language, common territory of psychic make-up to the status of conditions of the formation of an ethnos (Kryukov et al. 1993: 6). Ethnic self-consciousness gradually appeared as the most important and to an extent the only relevant defining feature of an ethnos. It "reflected" all objective cultural characteristics of the group, but selected only those that separated the ethnos from others, thus providing it with a worldview and self-understanding. Ethnographers can acknowledge the formation of the ethnos only when its identity and self-designation are fully formed, which happened for the "ancient Chinese" in seventh to sixth centuries B.C. and for the modern Chinese (the Han) in tenth to thirteenth centuries A.D (Kryukov et al. 1984: 292–3).

The cybernetic version of ethnos theory was also considered by Kryukov and coauthors a failed attempt to reanimate "the still-borne triad," since it was still connected to the sequence of Marxist formations, whereas "typology of ethnic groups must be constructed on ethnic features per se, the most important of which is ethnic consciousness" (Kryukov et al. 1993: 7–9, 377). Scaling up their conclusions on the world history level, Kryukov and coauthors distinguished two periods of "explosions of ethnicity," which chronologically roughly coincided with Classical Antiquity (and Ancient Chinese period) and Modernity. These "explosions" were caused not directly by the modes of production, but by broadly defined social factors such as elimination of clan and estate distinctions and increased social mobility. Two epochs produced different types of identities. The first one was based on an ethnocentric worldview, which relegated others to the status of subhumans, the second gave rise to a "relational" ethnicity, which acknowledged the unity of humanity, differentiated by certain cultural characteristics (Kryukov et al. 1993: 316–19, 375–9). It is important to note that for all of the emphasis Kryukov and other Soviet ethnographers put on identity, they never gave up on the idea of defining "objective" criteria of ethnic differentiation and were faithful materialists in considering ethnic consciousness a "reflection" of objective characteristics of ethnoses. In 1988 Kryukov himself initiated another attempt at creating an objective classification of ethnoses and sub-ethnoses. He suggested a program of research that would employ three main defining features of ethnoses – identity, language, and endogamy – to create a consistent ethnic classification. He believed this would be possible using quantitative linguistic criteria for differentiating between language and dialect, along with mapping endogamous collectives and collecting ethnic names (Kryukov 1988). Nevertheless, the discussion that followed Kryukov's article revealed widely different opinions concerning feasibility of this program, which did not arrive at empirical implementation anyway.

The third trend in ethnos-thinking, i.e., reflection on particular characteristics that differentiate ethnoses, reveals the paradox that ethnos theorists were perfectly aware of: no cultural feature or characteristic can be used as universal criteria for such differentiation. Language, religion, race, common origin, material and spiritual culture, etc., might acquire function of ethnic differentiation, but none of them squarely coincide with ethnic borderlines. A ethnos theorists Viktor I. Kozlov penned a series of articles about certain "features" in its relations to ethnos: "Ethnos and Economy" (1970), "Ethnos and Territory" (1971), and "Ethnos and Culture" (1979). The conclusions of these articles were similar: while in some cases these "elements" can conduct the function of ethnic differentiation, none of them are indispensable markers of ethnic specificity. According to Bromlei, "ethnically specific" culture was the subject matter of ethnography. But in practice, as Kozlov noted, ethnographers often committed a logical mistake, substituting an "ethnic/national culture" for the culture of an ethnos/nation, while in reality only a slight part of the latter is ethnically specific or unique (Kozlov 1979: 81).

Soviet ethnographers had a concept of "ethnic self-consciousness," roughly equivalent to English ethnic identity, which could take the place of a universal marker of ethnoses. Its increasing importance seems to be one of the most important tendencies of Soviet ethnos-thinking of the 1970–80s. It must be remembered that ethnic/national identity was not included in Stalin's list of characteristics of nation. Bromlei and other theoreticians of ethnos routinely gave credit to Pavel I. Kushner for introducing this notion into Soviet ethnography (Bromlei 1973: 96). In the 1960s, Cheboksarov and Kozlov elevated the epistemological status of ethnic self-consciousness (expressed in ethnic endonym) to the level of the most important characteristic of ethnos (Cheboksarov 1967; Kozlov 1967). At the same time, Bromlei and other theorists were quick to add that ethnic self-consciousness "as any form of consciousness – a secondary phenomenon, derivative from objective factors" (Bromlei 1973: 110). This logic had been in place already in P.I. Kushner (Knyshev)'s famous article about ethnic self-consciousness among the people of the Russian-Ukrainian border. During the fieldwork he came to conclusion that ethnic identity is the only marker that can reveal the identity of culturally similar groups, but, at the same time, recommended ethnographers to study all material and spiritual culture as ethnic "identifiers" (Kushner (Knyshev) 1949).

The importance of ethnic self-consciousness was increasingly visible in ethnos theories by the end of Soviet period for several reasons. Firstly, Bromlei observed that "with the technical and social progress the sphere of ethnic sides of culture shrinks. <...> [ethnic specificity] moves into the sphere of behavior and spiritual culture, including social consciousness" (Bromlei 1983: 140). Secondly, during the Perestroika he had to account for what he called "ethnic paradox," or the "growth of ethnic consciousness," which seemed against all odds to the standard official notion of "dying out" of ethnic contradictions. His explanation of the growing nationalism was for the main part structural. He mentioned such factors as unequal social and economic development of Soviet republics, unfair centralized distribution of resources among them, as well as ethnic nepotism and "parasitic attitude" of certain republican elites, misguided educational politics, etc. But there were "subjective" factors as well: some writers, politicians, historians, and ethnographers fed nationalism by lengthening the historical record of their nation and demanding privileges for their languages (Bromlei 1988: 124–127, 170–175). The emphasis on "subjective" factors of nationalism and active role of ethnic activists would become central concern for the next generation of ethnologists, who rejected ethnos theory in favor of the concept of ethnicity. But before turning to the consequences the new expert role of post-Soviet ethnologists had on their theory, one has to look at the structure of the field of ethnic and nationalities study in the USSR and theoretical issues that this structure stimulated.

Nationalities Study in the USSR: Nation Versus Ethnos/"Scientific Communists" Versus Ethnographers

As Dmitry Arzyutov has shown, Shirokogoroff used the terms ethnos and nation interchangeably, but in different contexts: ethnos figured in his ethnological works, while nation was reserved for political pamphlets. He envisioned nation as a kind of social organism, animated by popular will and not divided into political parties – a vision characteristic for right-wing conservatism and populism (Arzyutov 2019). This understanding of nation was, of course, unacceptable for Soviet nation studies, which relied on Marxist class analysis and Stalin's definition of nation. The central intuition of this approach was a historical nature of nation, formed simultaneously with capitalism as a result of national bourgeoisie's drive to unify national markets and production. How did ethnos fit into this worldview that put ultimate emphasis on "objective" criteria and class interests?

Ethnographers were not alone in the field of the study of nations. Historians of Soviet nationality politics and philosophers were safeguarding the Marxist orthodoxy. They were employed at such ideological strongholds of the CPUS as The Institute of Marxism-Leninism, the Academy of Social Sciences of the Central Committee of CPSU, etc. By the 1960s there had been achieved a kind of division of labor between historical materialism and social scientists: the former elaborated Marxist theory, the latter did empirical research, based on that theory. Usually the two camps coexisted smoothly, but when social scientists, like Bromlei, encroached on the territory of high theory, some counter-reaction was to be expected.

On 30 July 1985 a historian and head of the sector of theory of nation and national relations of the Institute of Marxism-Leninism Mikhail Ivanovich Kulichenko (1926-?) wrote a letter to the newly elected General Secretary of the Communist Party Mikhail S. Gorbachev. He signaled about major deficiencies in the field of nation studies: "The heart of the mistakes comes to the following: nation and all things national, including the relations between peoples, all their values are reduced to ethnic factors" (ARAN. F. 457. op. 1. D. 772. L. 8). Some two decades ago, lamented Kulichenko, "monographs about the most difficult aspects of lives of nations and international relations were published without a single mention of 'ethnos' or ethnic." Today, on the contrary, ethnographers even opine that they can do without "nation" since it is only a sub-variety of ethnos. The problem, as Kulichenko stressed, was not only terminological. The ethnic as a substance and category is in itself a conservative matter, not related to class and political life. Socialist nations and their values, on the contrary, are so much transformed by socialism that concentration on ethnic matters leads to deficient conclusions. To compromise Bromlei, Kulichenko even cited a positive review by American political scientist Martha Brill Olcott (b. 1949), who approved of Bromlei's "depolitisation" of the field as well as his "assertion of the primacy of psychological basis of social conduct" (ARAN. F. 457. Op. 1. D. 772. L. 11).

Ethnographers and their leader Bromlei were described by Kulichenko as a powerful community which outcompeted a few scattered experts on national question. In this particular case he was right: his attack was not successful. The letter was directed to the Academy of Sciences to the head of its social sciences division, Petr N. Fedoseev. He called a commission of high rank academicians to consider the case. Their verdict encouraged the intensification of research on national relations, but evaluated Kulichenko's accusations ungrounded and uncomradely (ARAN. F. 457. Op. 1. D. 772. L. 2–3).

For a reader not aware of the intricacies of Soviet Marxism, the discussion may sound purely scholastic. Nevertheless, behind it stood several decades of theorizing and institutional development, which seem to have led almost inevitably to certain tension between the two "camps." As has been said, soon after Stalin's death in 1953, and especially after the denunciation of the cult of his personality in 1956, there started a revision of his political and intellectual legacies. The definition of nation was not an exception. The first such attempt was a discussion that took place at the Division of History of the Academy of Sciences in November 28, 1957. The speaker was an Armenian writer and poet Khoren G. Adzhemian (1907–1968). Since 1944 he had spoken at a number of historical discussions defending highly unorthodox views combining Russian nationalism and imperialism (Tikhonov 2013). This time he was attacking Stalin's theory of nation head on. He flatly denied the idea that nations arise with capitalism and classified nations according to the epoch of their formation as ancient nations (Egyptians, Indians, Greeks, Chinese, etc.), feudal (French, English, Russians, etc.), capitalist (North Americans), and socialist ("new" nations of the USSR such as Yakuts, Nenets, etc.) ones. The formation of a nation has to do with the creation of their culture, art, and "self-awareness," rather than with economic and territorial unity. Excluding the later "materialist factors," but adding "physiognomic characteristic" (and explicitly ignoring the fear of "biologizing"), he came up with a radically "idealist" conclusion: "we consider indispensable the following national attributes: national self-awareness, understanding of their national identity by every person. Bound together by a single culture which includes language, art, knowledge of their own history, land, up and down sides, traditions, customs, etc., nation is a one whole organism" (Iurganov 2020: 155). Adzhemian also suggested a non-standard use of the term "ethnos" to signify (instead of nationality) all pre-national forms of ethnic identities (Iurganov 2020: 148).

This discussion was buried in archives and led to no visible result. This becomes more understandable if one compares it to the discussion of the same issue that took place on the pages of the journal "Voprosy istorii" (Issues of History) in 1966–1968. The discussion started in 1966, simultaneously with the establishment of the Commission for the study of national relations. It was headed by academician Evgenii M. Zhukov (1907–1980), whose deputy at the Division of historical science of the Academy of sciences was Bromlei. The formation of the commission was a response to the critique of the lack of research in this field, expressed at the 23th congress of the CPSU (1966) (ARAN. F. 1731. Op. 1. D. 72. L. 98). As has been said, the discussion itself did not produce any breakthroughs, because it centered around Stalin's four characteristics of nation and suggestions of slight modification of their list (the most debatable categories were national psychology and state). Nevertheless, several articles featured innovations that tended to converge ethnos and nation. For example, M. Dzhunusov came up with the definition of nation as a "social-ethnic organism" and "the highest form of ethnic collectivity." As Vasilii F. Filippov pointed out, a few years later Bromlei would use similar term – ethno-social organism – to designate people with common ethnic identity who form a nation state or a tribe (Filippov 2010: 67).

Taking stock of the debates of the 1960s and 1970s, Kulichenko underlined three points: the necessity to stress the historical and transient nature of nations, the necessity to realize the hierarchy of their characteristics, and to clarify the idea of national consciousness and psychology (Kulichenko 1983: 65). As a leading expert on the theory of nation and national relations, Kulichenko was the most outspoken opponent of such constructions as "ethno-social organism" and the whole tendency, most pronounced among ethnographers, to think of nations as the most developed form of ethnic collectives. He defined "the ethnic" as "formed in remote ages: language, deeply-rooted forms of material culture, folklore, folk art, traditions, customs, rites, specific characters of consciousness and psychology, "i.e. the most conservative and essentially classless part of the national [life]" (Kulichenko 1983: 20). The principal theoretical point that Kulichenko made time and again in his publications was preeminence of "the social" over "the ethnic" (Kulichenko 1981: 50). He admitted that ethnic was also "social" in a broad sense, but the latter for him was inseparably associated with the division of classes. Even elements of ethnic traditions, languages, etc., were so transformed during the formation of nations under capitalism that their existence inside national culture cannot be perceived as the direct continuation of their pre-national development. In his books Kulichenko cited Bromlei positively, but disputed with V.I. Kozlov and P.I. Kushner (Knyshev). While the latter was routinely credited by ethnographers as the first one to point to the importance of "ethnic self-consciousness" as a marker of ethnos, for Kulichenko, Kushner incorrectly "asserted the role of *national* self-consciousness in determining *ethnic* differences" (Kulichenko 1981: 98, author's italics).

Experts on nation and national question from the Institute of Marxim-Leninism maintained the primacy of "the social" over "the ethnic" for a number of reasons. Professionally, their job consisted in providing a consistent narrative of "the blossom and convergence of nations" of the USSR and the making of a "new historical community – the Soviet people," using party documents and statistics as their sources. The role of ethnic traditions and native languages was systematically downplayed by them because this new community was based on Soviet ideology and practices which, as they claimed, played a central role in the development of socialist nations. Another leading expert, Eduard A. Bagramov, describes the reluctance of Brezhnev and Soviet leadership to tackle the national question. The official dogma stated that the question had been "solved," and the limits of the allowed heterodoxy was to dispute whether blossoming or convergence was the leading tendency of the epoch (Bagramov 2003: 55). From time to time experts wrote memos to the authorities, but they were used only to provide a line or two for another speech at a plenum of the Central Committee. Nevertheless, scientific communists and ethnographers cooperated within the only academic interdisciplinary body in this field – the Scientific Council for the Study of National Question. Since the mid-1970s Bromlei was the head of the council, Kulichenko and Bagramov were his deputies. So, fighting off Kulichenko's attack, Bromlei had all reasons to point to the fact that Kulichenko had never raised any concerns about the activity of the Council. Moreover, Kulichenko's book was published in a series, edited by Bromlei, and he holds a copy of it with the author's grateful inscription (ARAN. F. 457. Op. 1. D. 772. L. 24–25).

Perestroika and the Decline of Ethnos in Russian Academia

Things were rapidly changing with the coming of Perestroika. National issues now could be openly discussed. At the end of 1987, the Institute of Marxism-Leninism held a meeting to discuss Bagramov's paper "National Question under Contemporary Conditions." Bagramov compiled a list of urgent problems: a visible imbalance of representatives of "titular nationality" in national republics' governments should be amended, schooling in local languages must be attended, the promotion of the Russian language should be less aggressive, etc. Other issues, discussed by Bagramov and colleagues at this meeting included nepotism of "national" elites, the widespread Islamization, inadequate expertise, lack of any governmental body that ruled nationality politics, etc. The main question, though, was the sharp and widespread growth of nationalism, which already led to mass demonstrations in Yakutsk, Alma-Ata, and Nagornyi Karabagh. Solomon I. Bruk (1920–1995) Bromlei's deputy from the Institute of Ethnography put the issue of nationalism in no uncertain terms: "national relations and their sharpness, at least in the nearest future, will increase. It is connected to the fact that the more a nation develops, the thicker, so to say, is her cultured layer, the intelligentsia, especially creative one, is growing, the more there are such problems, more competitive situations and more tensions" (RGASPI. F. 71. Op. 2. D. 350. l. 62).

In 1988 the discussions continued at the Institute of Ethnography under the auspices of Bromlei's Council. Scholars from the Academy of Sciences and from party's institutions voiced widely different opinions concerning the merits and future of the hierarchical structure of national and autonomous republics (Bromlei 1989; Korshunov and Cheshko 1989). The amount of problems generated within this structure can be seen from the statistics, provided by Mikhail N. Goboglo: 55 million people lived outside of "their" ethnic republics or, like Russian Germans, did not have them at all (Guboglo 1989: 27). Suddenly these "minorities" found themselves in jeopardy vis-à-vis "titular" nationalists. In the midst of this unexpected "national revival" ethnographers tried to make sense of the events as well as find their place as experts. Soon the leading voice in these discussions belonged to Bromlei's deputy and soon-to-be the next director of the Institute of Ethnography, Valerii A. Tishkov (b. 1941). He would also be an initiator and the main driving force behind the unmaking of the ethnos theory and the development of Russian post-Soviet ethnology in general.

In a series of publications in 1989 Tishkov criticized the ideological style of Soviet expertise and the "backwardness" of ethnos theory. Ironically, in his critique he was on the same page with Kulichenko, although for different reasons: "The understanding of nation as a type of ethnic community of capitalist and socialist epochs leads to the admission of its ethnic homogeneity, while the discrete character of nations, especially in modernity, is more than obvious" (Tishkov 1989). The dominant approach to ethnicity as "objective social groups," in his opinion, left Soviet social scientists unaware of the "mythopoetic mobilizing potential" of the "national factor." He defined nation as "a substance of spiritual culture and collective consciousness, an intragroup concept, not something defined from without by scholars or politicians." Tishkov sharply criticized not only ethnos theory, but also Soviet system of national-territorial republic, passport registration of nationality, and the hierarchy of nations and nationalities, etc. (Tishkov 1989: 10).

Tishkov's most pronounced critique of ethnos theory can be found in his English-language book "Ethnicity, Nationalism and Conflict in and after the Soviet Union. The Mind Aflame" (1997). Its methodological angle was to criticize primordialism of Shirokogoroff, Gumilev, and Bromlei as misguided and dangerous theory. The dangers of primordialism for him were revealed in the post-Soviet time, as it was propagated by nationalist intellectuals. His book "was intended inter alia as a critical analysis of the overly exclusive role played by post-Soviet intellectuals and political actors who are overloaded with the legacy of obsolete knowledge and who frequently lack the ability of self-reflection" (Tishkov 1997: xv). Tishkov himself adhered to "methodological individualism" and constructivism, seeing ethnicity as "a part of the repertoire that is calculated and chosen consciously by an individual or a group in order to satisfy certain interests and to achieve certain goals" (Tishkov 1997: 11). Constructivism and instrumentalism that underline the active pole of individual activists (or, in Tishkov's terms, "ethnic entrepreneurs") was arguably more adequate methodology for interpreting ethnic conflicts. It also suited the analysis of Soviet nation building – the enterprise not less "constructivist" than the clash for power among post-Soviet politicians and intellectuals. Thus, the "ethnic processes" of ethnos theorists were transformed by Tishkov into top-down view of history, in which the main actors were political and national elites, "no more than 5% elite elements <...> who can produce and impose any myth and project for the rest of populace" (Tishkov 1997: 297). To avoid this, Tishkov recommended that the category of ethnos "should be removed from public and, probably, academic discourse" (Tishkov 1997: 21; see also Cheshko 1994).

Conclusion

The sketch of a history of the ethnos concept presented above provides us with some general insights into the history of (Russian) anthropology and ethnology. The birth of the concept should be seen in the context of institutionalization of anthropology and the dispute over its place among sciences. As Nathaniel Knight reminds, the concept was also formulated in evident opposition to evolutionism (Knight 2019). Although the idea of progress was not alien to Shirokogoroff and especially to Mogilianskii, their view of history as an arena of the struggle of ethnoses was a view that have come naturally to these scholars, observing the bloodshed of the First World War and the collapse of European empires.

The victory of Marxism as a dominant and increasingly intolerant ideology of the Soviet Union made the pursuit of this kind of theory impossible. Its incompatibility with the Marxist ideology was not only on the political ground. Soviet Marxism adhered to a rather radical form of social determinism, which made geographical or biological factors unacceptable in interpreting human history (a subfield of which ethnography was proclaimed to be in the 1930s). The Bolsheviks created a unique "affirmative action empire," which consisted of more than fifty national republics and autonomous regions with their "title" nationalities, written languages, and intelligentsias, in some cases constructed from scratch (Martin 2001). This kind of "state evolutionism" obviously did not need the concept of ethnos with its semi-biological semimystical driving forces and life-cycles. Besides, although the socialist state "promoted ethnic particularism," it was not the particularism that early ethnos theorists espoused. Mogilianskii, following Volkov, viewed nation as united by identical physical and cultural characteristics of its members. Shirokogoroff's nation was more of a collective will and a shared mindset. But both abhorred Marxist theory and Bolshevik practice.

How then can one explain the reemergence of the theory, seemingly predestined to the dustbin of history, at least in the USSR? As has been shown, the discontent with Stalin's theory of nation was pronounced in the first years after the dictator's death. Ethnographers gradually, but steadily moved to more inclusive understanding of their field. Ethnos provided them with identifiable "subject matter" and respected theory, which allowed them to consolidate under one disciplinary title researches as widely different as the study of hominization of apes and the statistics of interethnic marriages in the USSR. In the 1960–70s the "biologization" of the social was not such ideological anathema, and even radical divergence of ethnos from Marxist social science by Gumilev was not approved, but tolerated. As Serguei Oushakine showed, Bromlei and especially Gumilev provided a language for nationalism when they "distilled" ethnos from the socio-political realm (Oushakine 2009: 86–95). They also offered fresh approaches to explaining history not as class struggle, but as creation of elitist groups of "passionate" or advanced individuals (Vershovtsev and Petryashin 2021). This, of course, should not obscure radical differences in their worldviews. Bromlei was loyal to official internationalism and Soviet nationality politics to the last days of his life (Bromlei 1989), quite unlike Gumilev, whose celebration of ethnic particularism became attractive for Russian, Eurasian, and Turkic nationalists alike (Bassin 2016: 177–305).

In the last years of the Soviet Union the debate about the ethnic or social character of nation led to the conflict between the leading theorist of ethnos and the leading theorist of nation. While Kulichenko's insistence on the capitalist origin of nations and inappropriateness of speaking about them as ethnic communities can be read as a constructivism of sorts, it was the dogmatism of the scientific communism's studies of "nationality question" that provided ethnos theory some of its charms. The notion of ethnos applied to all ethnic groups, thus at least rhetorically equalizing small-numbered ethnic groups and multimillion nations, the quality particularly attractive for those, who were well aware of the pitfalls and inconsistencies of Soviet nationality politics.

The turbulent realities of post-Soviet "ethnic renaissance" forced ethnologists to turn away from the longue durée historical research to the minutiae of current ethno-political issues and conflicts. Ethnos theory was obviously rather poorly equipped for such tasks. The critique of ethnos and synchronic research agenda went hand in hand with the growing dissatisfaction with Soviet ethnography's identity as historical science. But this turn also mirrored long-term processes in the discipline. Looking back at the history of the ethnos theory after 1945, one can see it as a slow, but constant growth of the importance of ethnic identity among characteristics, defining ethnos. The final turn of post-Soviet ethnological theorists to altogether constructivist and instrumentalist approaches seems to be an ironically consistent conclusion to a century of ethnos thinking.

Acknowledgments This chapter is published in accordance with the research plan of the Institute of Ethnology and Anthropology, RAS. The research for this chapter benefited from the Economic and Social Research Council (ESRC ES/K006428/1) "Etnos: A life history of the etnos concept among the Peoples of the North" (2013–2017). Some parts of the text appeared in print in the following publications: Alymov S.S., Anderson D.G., Arzyutov D.V. Etnos Thinking in the Long Twentieth Century. In David G. Anderson, Dmitry V. Arzyutov and Sergei S. Alymov (eds), Life Histories of Etnos Theory in Russia and Beyond. Cambridge, UK: Open Book Publishers, 2019, https://doi.org/10.11647/OBP.0150 (pp. 21–76); Alymov S.S. Ukrainian Roots of the Theory of Etnos. In David G. Anderson, Dmitry V. Arzyutov and Sergei S. Alymov (eds), Life Histories of Etnos Theory in Russia and Beyond. Cambridge, UK: Open Book Publishers, 2019, https://doi.org/10.11647/OBP.0150 (pp. 77–144).

Archives

ARAN	Archive of the Russian Academy of Sciences
RGASPI	Russian State Archive of Social and political History
TsGAM	Central State Archive of Moscow
SPF ARAN	Saint Petersburgh Branch of the Archive of the Russian Academy of Sciences
GARF	State Archive of the Russian Federation

References

Adams MB (ed) (1990) The wellborn science: Eugenics in Germany, France, Brazil, and Russia. Oxford University Press, New York/Oxford

Akademiia nauk SSSR (1964) Istoriia i sotsiologiia [History and sociology]. Nauka, Moscow. (In Russian)

Alekseev VP (1969) Proiskhozhdenie narodov Vostochnoi Evropy [The Genesis of the peoples of the Eastern Europe]. Nauka, Moscow

Alekseev VP (1982) O samom rannem etape rasoobrazovaniia i etnogeneza [On the earliest period of the Formation of Races and Ethnoses]. In: Bromlei YA, Kubbel LE, Pershits AI (eds) Etnos v doklassovom i ranneklassovom obshchestve [Ethnos in preclass and early class society]. Nauka, Moscow, pp 32–54

Alymov SS (2019) Ukrainian Roots of the Theory of Etnos. In Anderson DG, Arzyutov DV, Alymov SS (eds) Life Histories of Ethnos Theory in Russia and Beyond. Open Book Publishers, Cambridge, UK, pp 77–144

Alymov SS, Podresova SV (2019) Mapping Etnos: the geographic imagination of Fedor Volkov and his students. In: Anderson DG, Arzyutov DV, Alymov SS (eds) Life histories of ethnos theory in Russia and beyond. Open Book Publishers, Cambridge, UK, pp 145–202

Anderson DG (2019) Notes from His "Snail's Shell": The Siberian fieldwork of Sergei M. Shirokogoroff and the groundwork for ethnos thinking. In: Anderson DG, Arzyutov DV, Alymov SS (eds) Life histories of ethnos theory in Russia and beyond. Open Book Publishers, Cambridge, UK, pp 323–377

Anderson DG, Arzyutov DV (2019) The Etnos Archipelago. Sergei M. Shirokogoroff and the Life History of a Controversial Anthropological Concept. Curr Anthropol 6(60):741–773

Anon. (1970) Obsuzhdenie stat'i Iu. V. Bromleia Etnos i endogamiia [The discussion of Iu.V. Bromlei's article "Ethnos and Endogamy"]. Sovetskaia etnografia 3:86–103

Arutiunov SA, Cheboksarov NN (1972) Peredacha informatsii kak mekhanizm sushchestvovaniia etnosotsial'nykh i biologicheskikh grupp [Transmission of information as a mechanism of existence of ethno-social and biological groups]. Rasy i narody [Races and Peoples] 2:8–30

Arzyutov DV (2019) Order out of Chaos: political intrigue and the anthropological theories of Sergei M. Shirokogoroff. In: Anderson DG, Arzyutov DV, Alymov SS (eds) Life histories of ethnos theory in Russia and beyond. Open Book Publishers, Cambridge, UK, pp 278–324

Bagramov EA (2003) Natsional'naia problematika prezhde i teper' (sub'ektivnye zametki) [Ethnic and national studies now and then (subjective notes)]. In: Kozlov SI (ed) Akademik Iu. V. Bromlei i otechestvennaia etnologiia. 1960–1990e gody [Academician Iu.V. Bromlei and Russian Ethnology. The 1960–1990s.]. Nauka, Moscow, pp 47–86

Banks M (1996) Ethnicity: anthropological constructions. Routledge, London/New York

Bassin M (2016) The Gumilev Mystique: biopolitics, Eurasianism, and the construction of community in modern Russia. Cornell University Press, Ithaca

Bromlei IV (1969) Etnos i ėndogamiia [Ethnos and endogamy]. Sovetskaia etnografiia [Soviet Ethnography] 6:84–91

Bromlei IV (1973) Etnos i etnografia [Ethnos and ethnography]. Nauka, Moscow

Bromlei IV (1983) Ocherki teorii etnosa [Essays in the theory of ethnos]. Nauka, Moscow

Bromlei IV (1986) Teoriia etnosa [Theory of ethnos]. In: Bromley IV, Shtrobakh G (eds) Svod etnograficheskikh poniatii i terminov [A Dictionary of ethnographic concepts and terms], vol 2. Nauka, Moscow, pp 41–53

Bromlei IV (1988) Natsional'nye protsessy v SSSR: v poiskakh novykh podkhodov [The National processes in the USSR: in search of new approaches]. Nauka, Moscow

Bromlei IV (1989) O razrabotke natsional'noi problematiki v svete reshenii XIX partkonferentsii [On developing of ethnic and national studies according to the decisions of the 19th party conference]. Sovetskaia etnografiia [Soviet ethnography] 1:4–18

Cheboksarov NN (1967) Problemy tipologii etnicheskikh obzhnostei v trudakh sovetskikh uchenykh [The problems of typology of ethnic groups in the works of Soviet scholars]. Sovetskaia etnografiia [Soviet Ethnogr] 4:94–109

Cheshko SV (1994) Chelovek i etnichnost' [Man and ethnicity]. Etnograficheskoe obozrenie [Ethnogr Rev] 6:35–49

Conklin AL (2013) In the Museum of man: race, anthropology, and Empire in France, 1850–1950. Cornell University Press, Ithaca

Debets GF, Levin MG, Trofimova TA (1952) Antropologicheskii material kak istochnik izucheniia voprosov etnogeneza [Physical anthropology material as a source for the study of ethnogenesis]. Sovetskaia etnografiia [Soviet Sthnography] 1:22–35

Filippov VR (2010) "Sovetskaia teoriia etnosa". Istoriograficheskii ocherk ["The Soviet theory of ethnos": a historiographic essay]. Institut Afriki RAN, Moscow

Franko O (2000) Fedir Vovk – vchenii i gromads'kii diiach [Fedir Vovk as a scholar and activist]. Vidavnitstvo Evropeis'kogo universitetu, Kiev

Guboglo MN (1989) Natsional'nye gruppy i men'shinstva v sisteme mezhnatsional'nykh otnoshenii v SSSR [Ethnic groups and minorities in the system of ethnic relations in the USSR]. Sovetskaia etnografiia [Soviet Ethnogr] 1:26–41

Gumilev LN (1965) Po povodu predmeta istoricheskoĭ geografii: (Landshaft i etnos): III [On the subject matter of historical geography (landscape and ethnos): III]. Vestnik Leningradskogo universiteta [Herald of Leningrad University] 3(18):112–120

Gumilev LN (1988) Biografiia nauchnoi teorii ili Avtonekrolog [A biography of a theory, or an obituary of oneself]. Znamia [Banner] 4:202–216

Hirsch F (2005) Empire of nations: ethnographic knowledge and the making of the Soviet Union. Cornell University Press, Ithaca

Iurganov AL (2020) O pervom opyte destalinizatsii v filosofskom ob'iasnenii "natsional'nogo voprosa" [On the first attempt at destalinization in the philosophical discourse on "national question"]. Filosofskii zhurnal [Philosophical Jpurnal] 1(13):138–157. https://doi.org/10.21146/2072-0726-2020-13-1-138-157

Keldysh MV (1962) Stroitel'stvo kommunizma i zadachi obshchestvennykh nauk: (Rech' na Obshchem sobranii AN SSSR) [The Building of communism and the tasks of social sciences: a discourse at the meeting of the Academy of Science of the USSR]. In: Stroitel'stvo kommunizma i obshchestvennye nauki. Materialy sessii Obshchego Sobraniia Akademii Nauk SSSR 19–20 oktiabria 1962 goda [The Building of communism and social sciences. Materials of the meeting og the Academy and Sciences. October 19–20, 1962]. Izdatel'stvo AN SSSR, Moscow, pp 5–8

Knight N (1998) Science, empire, and nationality: ethnography in the Russian Geographical Society, 1845–1855. In: Burbank J, Ransel DL (eds) Imperial Russia: new histories for the empire. Indiana University Press, Bloomington, pp 108–141

Knight N (2019) Epilogue: why Etnos (still) matters. In: Anderson DG, Arzyutov DV, Alymov SS (eds) Life histories of ethnos theory in Russia and beyond. Open Book Publishers, Cambridge, UK, pp 389–402. https://doi.org/10.11647/OBP.0150

Korshunov AM, Cheshko SV (1989) Obzor vystuplenii [A review of speeches]. Sovetskaia etnografiia [Soviet Ethnogr] 1:18–25

Kozlov VI (1967) O poniatii etnicheskoi obshchnosti [On the Notion of ethnic group]. Sovetskaia Etnografiia [Soviet Ethnogr] 2:100–111

Kozlov VI (1979) Etnos i kultura [Ethnoc and culture]. Sovetskaia etnografia [Soviet Ethnogr] 3:71–86

Kozlov VI (1995) Problematika etnichnosti [Ethnic studies]. Etnograficheskoe obozrenie [Ethnogr Rev] 4:39–55

Kryukov MV (1988) Etnos i subetnos [Ethnos and subethnos]. Rasy i narody [Races Peoples] 18:5–21

Kryukov MV, Maliavin VV, Sofronov MV (1979) Kitaiskii etnos na poroge srednikh vekov [The Chinese ethnos on the eve of the middle ages]. Nauka, Moscow

Kryukov MV, Maliavin VV, Sofronov MV (1984) Kitaiskii etnos v srednie veka (VII–XIII A.D.) [The Chinese ethnos in the middle ages (the 7th–13th centuries AD)]. Nauka, Moscow

Kryukov MV, Maliavin VV, Sofronov MV (1987) Etnicheskaia istoriia kitaitsev na rubezhe srednevekov'ia i novogo vremeni [An ethnic history of the Chinese on the eve of modernity]. Nauka, Moscow

Kryukov MV, Maliavin VV, Sofronov MV, Cheboksarov NN (1993) Etnicheskaia istoria kitaitsev v XIX – nachale XX veka [An ethnic history of the Chinese in the 19th – early 20th centuries]. Nauka, Moscow

Kuklick H (1998) Speaking with the dead. Isis 1(89):103–111

Kulichenko MI (1981) Rastsvet i sblizhenie natsii v SSSR: Problemy teorii i metodologii [The Blossoming and the convergence of nations in the USSR: problems of theory and methodology]. Mysl, Moscow

Kulichenko MI (1983) Natsiia i sotsial'nyi progress [Nation and social progress]. Nauka, Moscow

Kushner (Knyshev) PI (1949) Natsional'noe samosoznanie kak etnicheskii opredelitel' [National self-consciousness as an ethnic indicator]. Kratkie soobshcheniia Instituta etnografii [Brief Reports of the Institute of Ethnography] AN SSSR 8:3–9

Kushner (Knyshev) PI (1951) Etnicheskie territorii i etnicheskie granitsy [Ethnic territories and boundaries]. Izdatel'stvo AN SSSR, Moscow

Markwick RD (2001) Rewriting history in Soviet Russia: the politics of revisionist historiography, 1956–1974. Palgrave, Basingstoke

Martin T (2001) The affirmative action empire: nations and nationalism in the Soviet Union, 1923–1939. Cornell University Press, Ithaca

Meissner B (1976) The Soviet concept of nation and the right of national self-determination. Int J 1: 56–81

Mogilianskii NM (1908) Etnografiiia i ee zadachi [Ethnography and its tasks]. Ezhegodnik Russkogo antropologicheskogo obshchestva [Russian Anthropological Society Annual] 3:1–14

Mogilianskii NM (1916) Predmet i zadachi ètnografii' [Subject matter and tasks of ethnography]. Zhivaia starina [Living Antiq] 25:1–22

Oushakine SA (2009) The patriotism of despair: nation, war, and loss in Russia. Cornell University Press, Ithaca

Pimenov VV (2015) Moia professiia – etnograf [Profession – ethnographer]. Avrora, Moscow

Rudenko SI (1916) Bashkiry: Opyt ètnologicheskoi monografii. Chast' 1: Fizicheskii tip Bashkir [The Bashkir: an ethnological monograph. Part 1: physical type of the Bashkir]. Tipografiia Iakor', Petrograd

Rudenko SI (1925) Bashkiry. Opyt ètnologicheskoi monografii. Chast' 2: Byt Bashkir [The Bashkir: an ethnological monography. Part 2: Bashkir's lifestyle]. Gosudarstvennaia tipografiia imeni I. Fedorova, Leningrad

Rudenko SI (1926) Sovremennoe sostoianie i blizhaishie zadachi etnograficheskogo izucheniia turetskikh plemen' [Contemporary state and nearest tasks of ethnographic study of the Turkish tribes]. In: Pervyi Vsesoiuznyi tiurkologicheskii s'ezd (Stenograficheskii Otchët) [The first Turkology congress (the stenographical report)]. Obshchestvo obsledovaniia i izucheniia Azerbaidzhana, Baku, pp 77–88

Semenov II (1966) Kategoriia "sotsial'nyi organizm" i eë znachenie dlia istoricheskoi nauki [The Category of "social organism" and its meaning for the historical science]. Voprosy istorii [Quest Hist] 8:88–106

Semenov II (1967) K opredeleniyu poniatiia "natsiia" [Towards defining the notion of "nation"]. Narody Azii i Afriki [The Peoples of Asia and Africa] 4:86–102

Semenov II (1994) O razlichii mezhdu dokazatel'stvami Ad veritatem i Ad hominem, o nekotorykh momentakh moei nauchnoi biografii i epizodakh iz istorii sovetskoi etnografii i eshche raz o knige N.M. Girenko "Sotsiologiia plemeni" [On the Difference between Ad veritatem and Ad hominem arguments, some moments of my scientific biography, episodes from the history of Soviet ethnography, and one more time of the book "The Sociology of Tribe" by N.M. Girenko]. Etnograficheskoe obozrenie [Ethnographic Review] 6:3–19

Shirokogoroff SM (2010) Etnos. Issledovanie osnovnykh printsipov izmeneniiz etnicheskikh i etnographicheskikh iavlenii [Ethnos. The Study of main principles of changes in ethnic and ethnographic phenomena]. Kafedra Sotsiologii MGU, Moscow

Shnirel'man VA (1992) Obsyzhdenie stat'i Z.P. Sokolovoi [The discussion of the article by Z.P. Sokolova]. Etnograficheskoe obozrenie [Ethnogr Rev] 3:85–90

Shnirel'man VA (2011) Porog tolerantnosti. Ideologiia i praktika novogo rasizma [The threshold of tolerance. Ideology and practice of new racism]. Novoe literaturnoe obozrenie, Moscow

Slezkine Y (1991) The fall of Soviet ethnography, 1928–38. Curr Anthropol 32(4):476–484

Sokolova ZP (1992) Endogamiia i etnos [Endogamy and ethnos]. Etnograficheskoe obozrenie [Ethnogr Rev] 3:67–78

Sokolovskii SV (1992) Obsyzhdenie stat'i Z.P. Sokolovoi [The Discussion of Z.P. Sokolova's article]. Etnograficheskoe obozrenie [Ethnogr Rev] 3:91–93

Sokolovskii SV (2012) Proshloe v nastoiashchem rossiiskoi antropologii [The past in the present of Russian anthropology]. In: Elfimov AL (ed) Antropologicheskie traditsyi, stereotipy, paradigmy

[Anthropological traditions, stereotypes, paradigms]. Novoe literaturnoe obozrenie, Moscow, pp 78–108

Tikhonov VV (2013) Kak "malen'kie lyudi" tvorili bol'shuyu istoriyu: fenomen "malen'kogo cheloveka" i ego rol' v poslevoennykh ideologicheskikh kampaniiakh v sovetskoi istoricheskoi nauke [How the rank-and-file people made big history: the phenomenon of the rank-and-file and its role in the post-WWII ideological campaigns in Soviet historical science]. Istoria i istoriki: istoriograficheskii vestnik [History and historians: the historiographical herald]. 2011–2012. Nauka, Moscow, pp 108–124

Tishkov VA (1989) O novykh podkhodakh v teorii i praktike mezhnatsional'nykh otnoshenii [On new approaches in the theory and practice of interethnic relations]. Sovetskaia Etnografiia [Soviet Etnogr] 5:3–15

Tishkov VA (1997) Ethnicity, nationalism and conflict in and after the Soviet Union: The Mind Aflame. Sage Publications, London

Tokarev SA (1964) Problema tipov etnicheskikh obshchnostei (k metodologicheskim problemam etnografii) [The problem of the types of ethnic communities [towards the methodological problems of ethnography]. Voprosy filosofii [Quest Philos] 11:43–53

Tolstov SP, Levin MG, Cheboksarov NN (1957) Ocherki obshchei etnografii. Obshchie svedeniia, Avstraliia i Okeaniia, America, Africa [Essays in general ethnography. General notions, Australia, Oceania. America, Africa]. Izdatel'stvo AN SSSR, Moscow

Vainshtein SI (2004) Iulian Vladimirovich Bromlei: chelovek, grazhdanin, uchenyi Iulian Vladimirovich Bromlei: person, citizen, scholar. In: Tishkov VA, Tumarkin DD (eds) Vydaiushchiesia otechestvennye etnologi i antropologi XX veka [Outstanding Russian ethnologists and anthropologists of the 20th century]. Nauka, Moscow, pp 608–627

Vainshtein SI, Kryukov MV (1986) Sovetskaia etnograficheskaia shkola [Soviet ethnographic school]. In: Bromlei IuV, Shtrobakh (eds) Svod etnograficheskikh poniatii i terminov, vol. 2. [A dictionary of ethnographic concepts and terms, vol. 2]. Nauka, Moscow, pp 114–124

Vermeulen HF (2015) Before Boas: the genesis of ethnography and ethnology in the German enlightenment. University of Nebraska Press, Lincoln

Vershovtsev D, Petryashin S (2021) Teorii etnosa vs. marksizm: istoriia mimeticheskogo soprotivleniia [Theories of ethnos vs. Marxism: a history of mimetic resistance], in print

Volkov FK (1906) Ukraintsy v antropologicheskom otnoshenii [Ukrainians from anthropological point of view]. Ukrainskii vestnik [Ukrainian Herald] 7:418–426

Volkov FK (1915) Antropologiia i ee universitetskoe prepodavanie [Anthropology and its university course]. Ezhegodnik Russkogo Antropologicheskogo obshchestva [Russ Anthropol Soc Annu] 5:99–107

Volkov FK (1916) Etnograficheskie osobennosti ukrainskogo naroda [Ethnographic characteristics of Ukrainian people]. In: Volkov FK, Grushevskii MG, Kovalevskii MM, Korsh FE, Krymskii AE, Tugan'-Baranovskii MI, Shakhmatov AA (eds) Ukrainskii narod v ego proshlom i nastoiashchem [The Ukrainian people in its past and present]. Tip. tov-va Obshchestvennaia pol'za, Petrograd, pp 455–647

Williams EA (1985) Anthropological institutions in nineteenth-century France. Isis 76(3):331–348

Assessing Ethnographic Representations of Micronesia Under the Japanese Administration

67

Shingo Iitaka

Contents

Introduction	1730
Japanese Anthropology/Ethnology and Micronesia	1732
Development of Anthropology and Ethnology in Modern Japan	1732
Development of Japanese Ethnography in Micronesia	1735
Ethnographers Commissioned by *Nan'yō-chō*	1740
Ethnographic Imaginations in the Journal *Nan'yō Guntō*	1744
Legacy of Japanese Ethnography in Micronesia	1746
Conclusion	1750
Cross-References	1752
References	1752

Abstract

This chapter examines how Japanese officials, scholars, and amateur writers represented Micronesia under the Japanese administration (1914–1944). It also investigates the contemporary meanings of these ethnographies in local societies. Japanese scholars expedited Micronesia, both in the earlier stages of administration and after it was well established. In the late 1930s, both professional scholars and amateur writers, commissioned by the South Seas Government, produced ethnographic representations of Micronesia. Hisakatsu Hijikata, an artist who conducted art education for indigenous children, left ethnographies of the Palau Islands, and Satawal in the Yap Islands group. Applied sciences appeared in the late 1930s, such as Ken'ichi Sugiura's anthropology of land tenure, which was influenced by British functionalism, and Zennosuke Nakagawa's comparative study of family laws and laws of inheritance, which was influenced by German legal studies. A local journal, *Nan'yō Guntō*, first published in 1935 by the South Sea Islands Cultural Society, contained ethnographic articles as well as propaganda news. These ethnographic representations were certainly born at the

---

S. Iitaka (✉)
Faculty of Cultural Studies, University of Kochi, Kochi, Japan

intersection of administrative demands, interests in "primitive" peoples in Japan's colonies, and academic influence from Europe in the humanities and social sciences. The ethnographies written by the Japanese in the pre-Pacific War era were not well known and were not fully returned to local societies in the postwar era. However, some ethnographies, such as those by Hijikata, were rediscovered as valuable in the era of nation-building in Micronesia.

Keywords

Micronesia · Palau · Japan · Ethnography · Ethnology · Empire · Administration · Colonialism

Introduction

After the crisis of representation raised by Edward Said, colonial ethnographies – here roughly defined as the representation of people under colonial situations – by European and North American anthropologists came under severe criticism as products of colonial administration. Ethnographies by the Japanese also developed with Japan's modern nation-building and its expansion; the Empire of Japan incorporated inner colonies such as Okinawa and Hokkaido in the early Meiji era, and later expanded to overseas colonies and territories, including Taiwan, Korea, Micronesia, and Manchuria. However, these prewar ethnographies written by the Japanese were not critically assessed in the 1980s, as critical theories from the West could not be applied directly to Japan. It required a different perspective to reflect on the ambivalent positionality of modern Japan, which was regarded by the West as a mysterious Oriental country, but which recognized Asian and Pacific neighbors as colonial subjects under the Empire of Japan (Sakano 2005: 499–503). The reluctance to assess prewar ethnographies was also due to the historical processing of the events. Japan's defeat in the Pacific War and the sudden loss of its overseas territories had created a discontinuity in historical consciousness between the prewar era and the postwar era.

In the prewar era, the origin of Japanese interactions with Japan's Asian and Pacific neighbors as part of a long-term historical process was the topic of much discussion. Some discussions, such as the theory that Japanese and Koreans had a common ancestor as well as the theory of a southern origin of the Japanese people, provided an ideological underpinning for the expansion of the Japanese Empire. In marked contrast to these discourses centering on the interaction among the people of the Asia-Pacific region, the myth of Japan as a homogeneous nation developed during the postwar era (Oguma 1995). Japanese citizens started to believe that the population of the Japanese Islands had been homogeneous ever since the dawn of history. Ainu, Okinawans, and those with different ethnic origins were regarded as atypical Japanese, even if they had lived in Japan and held Japanese citizenship. While Japanese anthropologists in the post-Pacific War era started different types of fieldwork in the former territories of the Empire of Japan after the 1965 removal of the ban on Japanese voyages abroad, the political context in which the prewar ethnographies were produced was overlooked. The attitude of these scholars seems to correspond to the collective amnesia about Japan's former overseas territories, which was closely connected to the invention of the myth of a homogeneous Japan. Furthermore, they hesitated to mention the ways in which Japan's prewar anthropologists had been involved in the making of national policies, especially during the Pacific War, such as the pacification of ethnic groups in such frontier colonies as Manchuria and the declaration of the benefits of ethnological research to the Japanese administration and occupation forces (Nakao 2016).

With the passage of time, some Japanese anthropologists and historians began to critically assess prewar ethnographies. In the late 1990s, their discussion points included the differentiation of disciplines in modern Japan, the construction of the Japanese identity by encountering colonial others, the utility of applied science under the Japanese administration, and the development of ethnographies before and during the war (e.g., Shimizu 1999; Sakano 2005). In the 2010s, a critical investigation of prewar ethnographies developed to the next stage. Peculiar historical and political contexts in modern Japan, which produced ethnographies both at home and in the colonies, were further investigated. Both historians and anthropologists contextualized ethnographies from the fields where ethnographic data were collected (Nakao 2016; Sakano 2019). Katsumi Nakao extensively visited former overseas territories of Japan, corresponding to the Greater East Asia Co-Prosperity Sphere, and conducted interviews with local populations. At home, he also interviewed former Japanese residents and those who were directly familiar with the anthropologists who had conducted fieldwork during the Japanese administration era. Toru Sakano, setting up an analytical scheme of an "island" where Japanese scientists met each other and colonial others, investigated colonial encounters – the way various agencies from both Japan and Palau encountered and interacted at the Palau Tropical Biological Station (*Palao Nettai Seibutsu Kenkyūjo*: パラオ熱帯生物研究所) (Sakano 2019). Katsuhiko Yamaji investigated the development of ethnography, focusing on continuity from the prewar to the postwar era (Yamaji 2011). Yuko Mio developed a comparative perspective. Her research project focused on the historical consciousness in Taiwan and Micronesia, where liberation from the Empire of Japan was followed by new administrations in the postwar era (Mio 2021). It also investigated how colonial legacies, including ethnographies written by the Japanese, were rediscovered and appropriated by the local population in the postwar era.

This chapter examines how Japanese officials, scholars, and amateur writers represented Micronesia (*Nan'yō Guntō*: 南洋群島) under the Japanese administration (1914–1944), with a focus on the ethnographic representations of Palau, which were produced between 1935 and 1945. Compared to the ethnographies of Taiwan, Korea, and Manchuria, ethnographies of Micronesia are fewer in number and thus have received scant attention; however, they include important works relevant to the development of ethnographies in Japan. While many Micronesian ethnographies by Japanese scholars of that time were influenced by German ethnographers such as Augustin Krämer and British anthropologists of functionalism such as Bronisław Malinowski, the administrative demands in Japan's *Nan'yō Guntō* also stimulated the peculiar development of ethnographic investigation. Akitoshi Shimizu's extensive work on the history of ethnographies in modern Japan already included a historical analysis of Micronesian ethnographies (Shimizu 1999). Matori Yamamoto also diachronically reviewed the development of Pacific Studies from the prewar to the postwar era (Yamamoto 2005). Rather than restricting the scope of investigation to those aspects of the history of the discipline that are irrelevant to colonial history, this chapter pays attention to the colonial agencies – the various agencies under the colonial situation, both from suzerain states and colonized societies – who took the responsibility of producing ethnographies under the tense situation leading up to the Pacific War and their interrelationship with the field. Here, the author avoids the dichotomized viewpoints between the colonizer and the colonized and pays attention to the entanglement of various colonial agencies. Previous studies held similar perspectives (e.g., Nakao 2016; Sakano 2019), yet this chapter focuses not only on professional scholars but also on amateur writers to investigate the entire colonial situation, which contributed to the development of ethnographic imagination at the edge of the Empire of Japan. In concrete terms, objects of investigation include those scholars who were commissioned by the South Seas Government (*Nan'yō-chō*: 南洋庁) for the research of land and customs in the late 1930s, as well as the activities of the South Sea Islands Cultural Society (*Nan'yō Guntō Bunka Kyōkai*: 南洋群島文化協会), which published a local journal titled *Nan'yō Guntō*. This chapter also investigates the contemporary meanings of these ethnographies in local societies. Ken'ichi Sudo has already assessed the significance of the ethnographies by Hisakatsu Hijikata (土方久功, 1900–1977) in contemporary Micronesia (Sudo 1991: 40). Here, the author highlights a discussion about how to return their history to local populations by translating Japanese ethnographies as well as archiving Japanese colonial documents.

Japanese Anthropology/Ethnology and Micronesia

Development of Anthropology and Ethnology in Modern Japan

Before delving into the development of ethnography in Micronesia, this study will provide a brief history of Japanese anthropological and ethnological interests since the Meiji era. Shōgorō Tsuboi (坪井正五郎, 1863–1913) pioneered anthropological investigation during the earliest stage of modern nation-building in Japan (Yamaji 2011). He formed a small voluntary association named *Friends in Anthropology* in 1884, which later developed into the *Anthropological Society of Tokyo* (東京人類学会) in 1886. The society issued a journal entitled *The Bulletin of the Tokyo Anthropological Society*. The title of the journal changed to *The Journal of the Anthropological Society of Tokyo*, and then to *Anthropological Science*, which continues to be published by the Anthropological Society of Nippon (日本人類学会). At the time that Tsuboi established the society, physical anthropology, archaeology, racial studies, ethnological studies, and folklore were not separate disciplines. Even after their fields split into specialized disciplines, Japanese physical anthropologists and cultural anthropologists had organized joint meetings well into the 1990s.

In contemporary Japan, cultural anthropology, rather than ethnology, is commonly used to refer to the subdivision of general anthropology, while the term "ethnology" was commonly used in the prewar era.

Tsuboi's research interests included the origin of the people who lived in the Japanese Islands. He insisted that Korpokkur, smaller people who appear in the Ainu folk tales, had lived in the Japanese Islands before the Ainu. He also insisted that Korpokkur were different from the people of the *Jōmon* period. Although many archaeologists and anthropologists disagreed, his claims stimulated discussions over the origin of the Japanese people. Tsuboi arranged the Academic Anthropic Pavilion (学術人類館) at the fifth National Industrial Exhibition held in Osaka in 1900. Because the pavilion exhibited peoples from Okinawa, Hokkaido, and Taiwan, his attempt at comparing these peoples was severely criticized. At the same time, Tsuboi was the first cultural relativist in modern Japan, who had a comparative perspective of the different ethnic groups (Yamaji 2011: 19). In fact, Tsuboi taught the anthropological course at Tokyo Imperial University and trained anthropologists such as Akira Matsumura (松村瞭, 1875–1936) and Ryūzō Torii (鳥居龍藏, 1870–1953), who conducted field research in Japan's overseas territories through comparative lenses. Matsumura joined an expedition to Micronesia right after the Japanese Navy's occupation in 1914. Torii conducted extensive fieldwork all over Asia, including Taiwan, Korea, Northeast China, Southwest China, Mongolia, the Kuril Islands, and Sakhalin. Torii is now considered one of the pioneer Japanese ethnologists who conducted extensive fieldwork overseas.

Ethnographic inquiry in modern Japan, deriving from physical anthropology and racial studies, began to develop in the late 1920s, both in the metropolis and in colonies of the Empire. In the metropolis, ethnological interest in the origin of the Japanese in comparison with the Ainu, Okinawan, and other Asian peoples was growing in the journal *Minzoku* (民族: ethnos/nation), which was first published in 1925 and included articles from ethnologists, historians, sociologists, and linguists (Yamaji 2011: 27). Cross-cultural studies developed through the establishment of the *Japanese Society of Ethnology* (*Nihon Minzoku Gakkai*: 日本民族学会) in 1934 and its bulletin, the *Japanese Journal of Ethnology* (*Minzokugaku Kenkyū*: 民族学研究), which was first published in 1935 (Yamaji 2011: 29). At that time, ethnos and ethnology had political connotations (Sakano 2005: 408). Ethnic issues, represented by the establishment of the State of Manchuria (満州国) and the expansion of Nazism in Europe, attracted wide attention in Japanese academic circles. The development of ethnology in Japan paralleled the tense political situation, which eventually led to the Second World War. Masao Oka (岡正雄, 1898–1982) was the central figure who led the development of Japanese ethnology between the 1920s and the 1940s. While he learned history-oriented ethnology from Wilhelm Schmidt at the University of Vienna, he was also influenced by the tense political situations both in Japan and Europe.

> Since the socio-political circumstances in the Society [Japanese Society of Ethnology] were rapidly changing during the late 1930s, and ethnologists were to be entirely involved in the wartime situation of the 1940s……., it is not appropriate to interpret, in static terms, the

theoretical trends of specialist ethnologists at the metropolitan centre in the early years of the Society. Oka, for instance, had already voiced scepticism about the conventional history-oriented ethnology he learned in Vienna, when he travelled through the Balkans in 1933 (Shimizu 1999: 150) (square brackets added by the author).

Oka appealed to government officials and thus the Ethnic Research Institute (*Minzoku Kenkyūjo*: 民族研究所) was established in 1943 under academic mobilization in wartime.

In the colonies, a specific interest in different cultures developed in both academia and the administration. Although extensive research was not conducted during the earlier stage of the administration in Korea, the Government-General of Korea (*Chōsen Sōtokufu*: 朝鮮総督府) initiated more vigorous research on traditional customs after the March 1st Movement in 1919, which demanded independence from Japan. Tomoe Imamura (今村鞆, 1870–1943), a police officer in residence at local police substations, collected ethnographic materials both through his work and through his involvement in folklore society in Korea. There were also researchers temporarily hired by the Government-General of Korea, such as Chijun Murayama (村山智順, 1891–1968), who left extensive reports on Korean folk religion. Photographs taken during his fieldwork in Korea are now accessible online through Keio University. Keijō Imperial University was established in Korea in 1924. Takashi Akiba (秋葉隆, 1888–1954), who studied anthropology in London under the influence of Malinowski and Radcliffe-Brown, worked in the Law and Literature Department of the university. He trained young ethnologists such as Seiichi Izumi (泉靖一, 1915–1970) and Kiichirō Ando (安藤喜一郎) (Nakao 2016: 224). Akiba's publications included the *Study on Korean Shamanism* (朝鮮巫俗の研究), co-authored by his colleague, Chijō Akamatsu (赤松智城, 1886–1960) (Asakura 2011: 131). Although it is unclear to what degree these studies about Korean folk religion contributed to the administration, ethnographies on the Korean people developed as the administrative demands grew for collecting further local information after the 1919 independence movement (Nakao 2016: 107).

In Taiwan, ethnographic research was conducted during the earlier stage of the administration, since Taiwan was an unknown world to Japan. Studies of Taiwan's indigenous peoples developed after the earlier expeditions by Kanori Inō (伊能嘉矩, 1867–1925) and Ryūzō Torii. The Government-General of Taiwan (*Taiwan Sōtokufu*: 台湾総督府) organized the Provisional Taiwanese Custom and Practice Research Committee (臨時台湾旧慣調査会) in 1901, led by Santarō Okamatsu (岡松参太郎, 1871–1921) from Kyoto Imperial University, in order to investigate the administrative laws under the Qing dynasty, as well as the customary laws of the Han and indigenous peoples. When Taihoku Imperial University was established in 1928, Nenozō Utsushikawa (移川子之蔵, 1884–1947), who had studied cultural anthropology at Harvard University, opened the Institute of Ethnology (*Dozoku-jinshugaku Kenkyūshitsu*: 土俗人種学研究室) in the Faculty of Literature and Politics. Although this minor institute produced only one Japanese graduate, Tōichi Mabuchi (馬淵東一, 1909–1988) (Shimizu 1999: 135), an extensive survey of Taiwan's indigenous peoples was conducted between 1930 and 1932, which resulted in huge volumes titled *The Formosan Native Tribes: A Genealogical and Classificatory Study* (台湾高砂族系統所属の研究). The description in the volumes was highly technical. Their intensive and elaborate research was made possible because the Japanese ethnographers used the information from the census and customary surveys conducted by the administration as well as the reports written by police officers in residence at local police substations. In this sense, the administrative process and the ethnographic writing in Taiwan were closely connected (Nakao 2016: 83).

Development of Japanese Ethnography in Micronesia

An evaluation of Japanese ethnography in Micronesia by American anthropologists stated that there were no organized research programs by the Japanese comparable to the *Südsee-Expedition*, a scientific expedition to German-administered Micronesia and Melanesia led by Georg Thilenius from 1908 to 1910, and the Coordinated Investigation of Micronesian Anthropology (CIMA) conducted under the earlier United States administration.

> In some respects, the Coordinated Investigation of Micronesian Anthropology resembled the earlier BAE [Bureau of American Ethnology] and PES [Philippine Ethnological Survey] research efforts; it was assumed, or hoped, that ethnographic data would be of value for administration of dependent peoples. In Micronesia, the German Südsee Expedition between 1908 and 1910 had been a similar effort, but most of the results of that investigation were published long after German had lost the islands to Japan in 1914 and thus were of no administrative use. While some anthropological work was done by individual Japanese researchers......, there was no organized research program comparable to the Südsee Expedition or CIMA efforts during the Japanese colonial era between the two world wars (Kiste and Falgout 1999: 26) (square brackets added by the author).

Nevertheless, following the trends in other overseas colonies, the Japanese ethnographic investigation of Micronesia developed over the course of the Japanese administration, especially after the establishment of *Nan'yō-chō* in 1922, to meet the administrative demands.

Naval Administration Era

After Japan declared war against Germany under the Anglo-Japanese Alliance, its Navy occupied the German-held territories of Micronesia in October 1914. These islands – Micronesia north of the equator, except for Guam and the Gilbert Islands – came to be known in Japan as *Nan'yō Guntō* (South Sea Islands), which were administered by the Empire of Japan for about 30 years. Japan's Navy administered the area from 1914 to 1919. During the naval administration era, natural scientists from Japanese universities visited the islands, but their research perspectives were not well defined. Since it was not certain whether the international community would allow the Empire of Japan to administer Micronesia permanently, the administrative demands placed on these scientists were not clear.

Two months after the Navy's occupation in 1914, Japanese scholars from various fields of natural science, such as medicine, botany, zoology, oceanography, agriculture, and forestry, made expeditions to Micronesia to collect general information on the area, as well as to investigate the utility of the natural resources on the islands (Sakano 2005: 356). Physical anthropologists such as Akira Matsumura, influenced by Tsuboi at Tokyo Imperial University, joined the earlier expedition, and Matsumura later published *Contributions to the Ethnography of Micronesia* (Matsumura 1918). Scholars from the humanities and social sciences were not invited because their works were not claimed as applied sciences useful for the administration. The disciplines of the humanities and social sciences, which blurred into physical anthropology and the science of races, were not specialized in Japan at that time (Sakano 2005: 18; Yamaji 2011: 19). Thus, fieldwork methods to investigate indigenous cultures and societies had not yet been developed. Matsumura contributed his work to a journal published by the Anthropological Institute of Tokyo University. It was a physical anthropological record with many photographs of Micronesian peoples, most of which were taken from the front and side. The report also covered topics such as "clothing and personal adornment," "food and other articles," "dwellings and household utensils," "navigation and fishing," "implements and weapons," and "decorative patterns" (Matsumura 1918). Matsumura excused the limitation of his research as follows.

> We spent sixty-four days on the voyage, covering 11,200 nautical miles. Thus, we were mostly at sea, spending only a very short time on land, and in fact only a few hours in the case of some of the islands. Under these circumstances, it was extremely difficult to undertake anything like exhaustive researches and collections. Happily, however, Mr. J. Shibata, Assistant in the Anthropological Institute of the University, and Dr. K. Hasebe were among our party, the former to undertake archaeological researches and collect ethnographical objects, and the latter to study physical anthropology. Mr. Shibata, in particular, was always with me, rendering me valuable assistance not only during our voyage but after our return to Tokyo, for which I desire to express my warm thanks. (Matsumura 1918: 2)

The national inspection, of which Matsumura was a part, was hasty and not well organized. Although Japanese scientists had not learned about the research methods for the field investigation, the photography was employed by Tsuboi's successors. Like Matsumura, Ryūzō Torii also used a camera in the field. Their photographic plates are now stored at the University Museum at the University of Tokyo. Both Matsumura and Torii's photographs were digitized and made accessible via the online database of the museum.

During the Naval administration, ethnographic information on Micronesia was collected through Japanese residents engaging in the trade of copra, made of coconuts on the islands. When the Navy officials needed information on the islands, Japanese merchants stationed in Micronesia were valuable and useful informants. The reports compiled by the military personnel included the names of these Japanese merchants, such as Koben Mori (森小弁, 1869–1945), a pioneer immigrant who moved to the Chuuk Islands at the end of the nineteenth century. Military officials compiled ethnographies for themselves. Kichitarō Tōgō (東郷吉太郎, 1867–1942), the second naval commander, compiled an ethnography titled *Natural Features of South Seas Islands* (Shimizu 1999: 138). Shizuo Matsuoka (松岡静雄, 1878–1936), a younger brother of Kunio Yanagita (柳田国男, 1875–1962) who was the founder of Japanese folklore, was stationed in Pohnpei Island during the occupation as a Major of the Imperial Navy. In *Ethnography of Micronesia* (ミクロネシア民族誌), published in 1927, Matsuoka referred to German ethnographies by John Kubary and Augustin Krämer. The former is a Polish naturalist who worked at the Museum Godeffroy in Hamburg. The latter joined the *Südsee-Expedition* conducted by the Museum of Ethnology in Hamburg.

Matsuoka also used ethnographic data collected from questionnaires distributed through *Nan'yō-chō* in later years. The ethnography was expected to be useful for the administration, but the extent of its contribution to the administration is unclear. Shimizu evaluated Matsuoka's ethnography as follows.

> His questionnaire......is simply a list of items to be inquired about, with neither interpretations nor notes. Nevertheless, it was unique as a comprehensive questionnaire for filed research compiled privately by a Japanese anthropologist. As far as the method of data-collection is concerned, Matsuoka may be compared with those 'armchair' anthropologists in the West who collaborated with 'amateur ethnographers' living in the colonies. He was also parallel to Western 'armchair' anthropologists in that his interest in anthropology was integrated with his inquiries into the classic literature, in his case that of ancient Japan. (Shimizu 1999: 139)

Nan'yō-chō Era

After an international agreement on Japan's administration in the area by the Versailles Treaty, civil administration was implemented in Micronesia in 1919. *Nan'yō-chō* was established in Koror, Palau, in 1922, and thus Japan started to hold mandates over Micronesia under the League of Nations. The Japanese administration in this region continued even after Japan withdrew from the League of Nations in 1935. A naval officer became the Governor-General of *Nan'yō-chō* in 1943 during the ongoing Pacific War. The government suspended its functions in April 1944, as the entire staff was converted into paramilitary personnel and evacuated to the jungle.

The characteristics of the Japanese administration in Micronesia included the promotion of immigration, economic development, and militarization. Immigration from Japan and its territories had been promoted in the Marianas since the early 1920s and in the rest of the area since the early 1930s. In 1935, the total number of Japanese immigrants reached 51,861, which outnumbered Micronesians. The Empire of Japan explicitly exploited the natural and human resources of Micronesia. A large number of Japanese immigrants engaged in the sugar cane industry in the Mariana Islands. Some immigrants cultivated new fields on the larger volcanic islands of Palau and Pohnpei for agricultural use. Others engaged in fisheries of bonito, tuna, and pearls. Phosphate mining on Angaur, Palau, also became a major industry. Micronesians were exploited as cheap laborers and were involved in harsh work, such as mining. After Japan withdrew from the League of Nations in 1935, administrative policies, such as the cultural assimilation of indigenous Micronesians and Japanese immigration to the islands, were reinforced. At the end of the administration, Micronesia became a foothold where military personnel either proceeded to or withdrew from the southern war fronts of Melanesia.

During the *Nan'yō-chō* era, humanities and social sciences professionals were required to conduct practical research, since *Nan'yō-chō* and its related organizations needed local ethnographic knowledge for administrative purposes. The South Seas Court of Justice (*Nan'yō-chō Hōin*: 南洋庁法院) conducted a series of surveys on customary laws in Micronesia in comparison with Japanese civil family laws. The jurists from the court, as investigation commission members, conducted a survey after 1925. Their works were compiled as *Reports of Customary Laws in the South Sea Islands* (*Nan'yō-chō* 1939), which included technical descriptions of Micronesian customary laws: the origins and legends of kin groups, the hierarchies of traditional chiefs, and comparisons of kinship categories between the Japanese and the Micronesians. Even though ethnography at the time kept the informants' names anonymous in principle, the report explicitly mentioned the personal names of Micronesian informants. The jurists at the court used the reports as handbooks to solve Micronesian lawsuits.

When Japanese immigrants in Micronesia became prominent in number, *Nan'yō-chō* began a survey of land to enclose public land that was not used by Micronesians, although it included common land. This survey started in Saipan as early as 1923, when the number of Japanese who engaged in the sugar cane industry increased. It continued on other islands until 1932. In 1933, a survey to identify titles among individuals and kin groups in Micronesia began. While the German administration had partly conducted a land survey in Pohnpei and the Mariana Islands, the modern notion of land ownership was not familiar to Micronesians. The land policy under the Japanese administration introduced arguments over land that have continued to the present. In the late 1930s, jurists and ethnologists took part in the research on land and customs in Micronesia. Zennosuke Nakagawa (中川善之助, 1897–1975) from Tohoku Imperial University was a jurist commissioned by *Nan'yō-chō* to arrange the research items for the survey. Ken'ichi Sugiura (杉浦健一, 1905–1954) from Tokyo Imperial University was employed temporarily by the Local Affair Division of *Nan'yō-chō* to conduct surveys of land tenure under the administration. Hisakatsu Hijikata, an artist and ethnographer who was fluent in local languages, supported those scholars (Nakao 2016: 228).

The South Sea Islands Cultural Society, established in 1935 as a branch of the Local Affairs Division of *Nan'yō-chō*, published a journal titled *Nan'yō Guntō*, to which professional scholars such as Hijikata and Sugiura as well as amateur writers living in Micronesia contributed their articles. The editor of the journal, Masaaki Noguchi (野口正章), published a couple of books on Palau, which contained ethnographic descriptions of the local society in transition (Noguchi 1941a, b). In the later Japanese administrative era, government-controlled companies and organizations actively engaged in research activities to implement national policies. The South Seas Development Company (*Nan'yō Kōhatsu*: 南洋興発), which managed the sugar cane industry in the Mariana Islands, provided rich funds for the joint field research. Supported by the company, Nobuhiro Matsumoto (松本信廣, 1897–1981) from Keio University organized an expedition to Micronesia and New Guinea in 1937, which was joined by leading ethnologists and archaeologists. Various material objects collected in New Guinea appeared in pictorial records (Society of the South 1937, 1940). Keio University now stores approximately 2000 material objects from Melanesia, most of which were originally owned by the South Seas Development Company. Famous Uli figures, carved objects from New Ireland, Bismarck Archipelago, were originally collected by Isokichi Komine (小嶺磯吉, 1866–1934), a Japanese resident who engaged in trading, shipbuilding, and copra production (Yamaguchi 2015). The South Seas Development Company also funded an expedition to Pohnpei in 1940, which was joined by ecologists led by Kinji Imanishi (今西 錦司, 1902–1992) from Kyoto Imperial University. The result was published as a volume titled *Ponape: An Ecological Study* (Imanishi 1944). After this experiment on the volcanic island of Micronesia, Imanishi's group expedited to the Greater Khingan (大興安嶺), a wider volcanic mountain range, most of which was located in Manchuria (Imanishi 1952).

By this time, the field investigation by Japanese scientists had become more professional and specialized. Professional anthropologists, archaeologists, and ethnologists conducted fieldwork from the perspectives of their disciplines. At the same time, their investigation was influenced by administrative concerns, such as promotion of Japanese immigrants. Shimizu evaluated Imanishi's expedition to Pohnpei as follows.

> In 1940, just before the Pacific War, a group of young ecologists, led by Imanishi Kinji, made a survey of Pohnpei and left a comprehensive record of their findings. Being free from preoccupation with salvage anthropology, they vividly recorded their observations on the life of Pohnpeians and Japanese settlers. (Shimizu 1999: 141).

Even for the ecologists conducting research on the volcanic island, the conditions in which Japanese settlers lived were also a matter of concern.

During the last years of the administration, ethnographies of Micronesia were involved in the Southern Expansion Doctrine *(Nanshin-ron:* 南進論), under which the Empire of Japan was expanding from the Inner South Seas (*Uchi-Nan'yō*: 内南洋), equivalent to Japan's *Nan'yō*, to the Outer South Seas (*Soto-Nan'yō*: 外南洋). The publications by the Institute of the Pacific (*Taiheiyō Kyōkai*: 太平洋協会), a research organization established in 1938 to install national policies, included *Great South Seas: Its Culture and Its Soil* (1941), *Pacific Ethnology* (1943), and *Pacific Sphere: Its People and Its Culture* (1944). These books covered a wider range of the Pacific area from Micronesia to Southeast Asia and Melanesia. The authors included specialists from *Nan'yō-chō*, as well as scholars conducting research in Micronesia (e.g., Nakagawa 1941; Sugiura 1941; Hijikata 1993: 1–84). Yoshitarō Hirano (平野 義太郎, 1897–1980), a Greater Asianist who converted from Marxism, organized some ethnologists contributing their articles to these volumes (Shimizu 2013: 51–57). Military officials were attached to the research division of the South Seas Colonization Corporation (*Nan'yō Takushoku*: 南洋拓殖), a government-controlled corporation established in 1936. Among others was Fukashi Kamijō (上條深志), a Captain of the Imperial Navy, who utilized the data collected by the Navy during the earlier occupation era and compiled ethnographies of Palau and Yap (Kamijō 1938, 1939). The Ethnic Research Institute, established in 1943 through Oka's petition to government officials, covered the entire area of the Japanese Empire. Its members included Sugiura and Ichirō Yawata (八幡一郎, 1902–1987), both from Tokyo Imperial University, who were in charge of the investigation of Japan's *Nan'yō* (Nakao 2016: 333; Shimizu 2013: 29).

In addition to these ethnographic investigations conducted in close relationship with the administration, there were other ethnographic works distanced from *Nan'yō-chō* and its related organizations. In 1932, Tadao Yanaihara (矢内原忠雄, 1893–1961), a scholar of colonial studies from Tokyo Imperial University, was required to investigate Micronesia by the Institute of Pacific Relations, an international organization established in 1925 to provide a forum for discussion among the Pacific Rim nations. Yanaihara published *Pacific Islands under Japanese Mandate* in 1935 (Yanaihara 1935) and its English version in 1940 from the Oxford University Press. He pointed out that indigenous Micronesians faced rapid social changes with the influx of Japanese immigrants. Although he critically assessed the administrative problems, he generalized the changes as a transition under the induced monetary economy (Yanaihara 1935: 356).

Kiichirō Ando, who studied ethnology under Takashi Akiba from Keijō Imperial University, was also distinctive. In 1933, he published *Culture and History in the South Seas* (南洋風土記), in which kinship and customary laws in Micronesia were investigated under the influence of Malinowski (Nakao 2016: 224). Although most scholars came to Micronesia either as specialists commissioned by *Nan'yō-chō* or as members of expedition teams from the metropolis, Ando seemed to approach local society by himself, although he acknowledged the assistance of higher officials of *Nan'yō-chō* in his book. Some medical doctors working at local hospitals also had ethnographic interests in Micronesia. Those stationed in Palau formed a voluntary association named the Palau Folklore Society (*Palao Minzoku Sadankai*: パラオ民俗瑣談会). The society published a bulletin named *Female Chiefs* (*Joshū*: 女酋), hinting at the matrilineal and hierarchical society in Micronesia. Hijikata, who became close to medical doctors during his stay in Palau, contributed an article with rich illustrations to the bulletin (Totsuka 1931).

Ethnographers Commissioned by *Nan'yō-chō*

Although a university was not established in Micronesia under the Japanese administration, ethnological or anthropological inquiry, following the trends in other colonies, developed in the latter half of the 1930s when Hijikata, Sugiura, and Nakagawa joined the survey of Micronesian land and customs. While Nakagawa and Sugiura were from the imperial universities in mainland Japan, Hijikata lived in Micronesia. Nakagawa was influenced by the Comparative Law in Germany, while Sugiura was influenced by British anthropology (Shimizu 1999: 141; Nakao 2016: 226). Their academic backgrounds matched the administrative demands of *Nan'yō-chō*. Hijikata, knowledgeable on indigenous societies and fluent in local languages, supported Nakagawa, Sugiura, and other scholars from mainland Japan (Nakao 2016: 219).

Hijikata moved to Palau in 1929, 3 years after he graduated from the Tokyo School of Fine Arts (*Tokyo Bijutsu Gakkō*: 東京美術学校). *Nan'yō-chō* commissioned him to teach indigenous children how to carve wooden boards in elementary schools for Micronesians (*kōgakko*: 公学校). He realized that the best way to do so was to follow the decorations on the beams of the traditional meeting houses containing various legendary and historical stories in Palau. While the Japanese administration never gave the Micronesians the same legal status as Japanese citizens, it culturally assimilated indigenous peoples. Micronesian children were required to attend elementary schools, in which they received limited education, including basic Japanese, ethics admiring the Emperor of Japan, and vocational training. Hijikata's engagement in art education was part of the education. The trained children entered a contest (*hinpyōkai*: 品評会) in which their works were appraised to win prizes. Longing for primitive cultures away from civilization, Hijikata left Palau in 1931 and moved to Satawal, a coral atoll at the easternmost end of the Yap islands group. After returning to Palau in 1939, he supported research activities by professional scholars, including those of Nakagawa and Sugiura. He also inspired Atsushi Nakajima (中島敦, 1909–1942), who edited textbooks for Micronesian children and later published literary works based on his life in Palau. Hijikata himself left remarkable ethnographic works on Palau and Satawal. His ethnographies were translated into English in the 1990s and have been highly praised in both Japan and Micronesia (Hijikata 1993, 1995, 1996, 1997). Hijikata's diary was also published in Japanese (Sudo and Shimizu 2010–2014).

As a graduate of art school, Hijikata left extensive sketches of and essays on material cultures in Micronesia. He also collected material cultural artifacts during his stay in Palau and Satawal and produced artworks of paintings and carvings in which the nature and peoples of Micronesia were the main subjects. Hijikata briefly returned to Japan in 1939 to hold an exhibition of artifacts collected in Micronesia at the Tokyo office of the South Sea Islands Cultural Society as well as at the anthropology course at the School of Science of Tokyo Imperial University. The exhibition attracted and inspired anthropologists, archeologists, and linguists working in Japan's *Nan'yō*. The collection was donated to the university, and later in the postwar era, transferred to the National Museum of Ethnology, established in 1974. Ichirō Yawata, an archeologist from the anthropology course of Tokyo Imperial University, introduced Hijikata's work to academic circles and wrote explanatory notes on the collection in an academic journal as well as in a pictorial record (Sudo 1991: 39).

While Micronesian society changed drastically with the influx of Japanese immigrants, Hijikata had a strong longing for primitivism. The material cultural artifacts he collected and sketched were a crystallization of Micronesian traditions that had avoided being destroyed amid rapid social changes. Palauan traditional meeting houses (*bai*) with rich carvings and paintings inspired his creative activity and his instruction of indigenous children in carving. He described *bai* as a library in a pre-literate society, which stored legendary figures, ancestors, moral stories, and historical events (Hijikata 1993: 54). In addition to works related to material cultures, Hijikata left various ethnographic descriptions of Micronesian cultures and societies. He often referred to German ethnography written by Krämer to describe the Palauan social structure and kin group. His descriptions were accurate, and thus highly praised by Japanese ethnologists and anthropologists who visited Micronesia in both the prewar and postwar eras. Hijikata was skeptical of applying a theoretical model to an indigenous society, thus ignoring local perspectives (Sudo 1991: 40). It was important for him to establish a concrete understanding of local perspectives by living among Micronesians. He wrote in his diary in August 1939 that matters for investigation set up by Nakagawa for the study of land and customs were too abstract for the study of native people (Nakao 2016: 220). At the same time, Hijikata recorded the impact of the administration in his diary (Sudo and Shimizu 2010–2014). He saw arguments over land between Japanese immigrants and local Palauans in Koror, where the population grew rapidly in the 1930s. On the remote island of Satawal, he witnessed the attending of Micronesian children at schools, the drafting of laborers for phosphate mining in Angaur, and the conflicts between islanders and Japanese merchants.

Ken'ichi Sugiura, in the early 1930s, contributed extensively to two different journals with homophonous names in Japanese, namely, *Folklore* (*Minzokugaku*: 民俗学) and *Ethnology* (*Minzokugaku*: 民族学), which led Japanese anthropology and ethnology at that time. Sugiura was influenced by Fritz Graebner, a German ethnologist known for his development of the culture circle (*Kulturkreis*) theory. He also participated in a community survey promoted by Kunio Yanagita in remote domestic villages (Izumi 1954: 72). In the late 1930s, however, Sugiura became a devoted reader of Malinowski and Radcliffe-Brown, and quickly absorbed functionalism in British anthropology. He was then prepared to conduct ethnographic research in Micronesia. Sugiura first visited Micronesia in 1937 as a member of the expedition to Micronesia and New Guinea, which was funded by the South Seas Development Company (no author 1937). With Tomoaki Nakano (中野朝明), a junior in the religious studies course at Tokyo Imperial University, Sugiura conducted ethnological research in some of the villages in Palau, where he continued the survey of land and customs after officially being commissioned by *Nan'yō-chō*. When he was entrusted with research in Micronesia in 1938, he visited a wider range of areas, including Palau, Yap, Pohnpei, and Chuuk. His publication list tells us that he investigated Micronesia from various academic perspectives: diffusionism toward material cultures, a perspective in religious studies toward folk beliefs, and a functionalist view toward land tenure (Shimizu 1999: 141). His ethnographic description of the Palauan social structure included a concrete geographical description of villages, the ranking of chief titles, the connections between the kin group and the chief title, and the network of kin groups. Here, Sugiura relied on Krämer's ethnography, as did Hijikata. Clearly, German ethnography was an important reference point for Japanese ethnographers working in Micronesia. At the same time, the British anthropology of functionalism was the theoretical background of his work.

Due to the wider coverage of fields and topics, Sugiura's ethnographic works were deemed to be extensive but nonspecific. It has also been pointed out that he did not provide critical comments on the administrative impact on indigenous societies (Shimizu 1999: 143). He declared the utility of applied anthropology for the administration at the beginning of the report on land tenure in Palau and Pohnpei, although the ethnographic description seemed too technical and professional, with rich folk terms and a complicated discussion of social structures (Sugiura 1944). Referring to Malinowski's influence on Japanese ethnographers, Shimizu evaluated Sugiura's ethnography on Micronesian land tenure as follows.

> ......among the works of the contemporary Japanese anthropologists, Sugiura's work on land tenure was the most comparable to Malinowski's idea of 'practical anthropology'. In contrast to Malinowski......, however, Sugiura lacked a critical attitude both to the colonial administration and to his fellow anthropologists; he could only be a remote counterpart of Malinowski in Japan. In spite of this limitation, it is an interesting fact that the Micronesian colony brought up a unique type of field anthropology distinct from those of Taiwan and Korea. (Shimizu 1999: 143)

It is unclear how his ethnographic description was appreciated and utilized by the administration (Shimizu 1999: 142). There was an inconsistency between what he declared at the beginning of the ethnography and what he actually described within it. Nonetheless, the political context of his research is clear. *Nan'yō-chō* conducted land surveying and recording, and thus compiled the Land Registration Record (*Tochi-daichō*: 土地台帳). When the United States administration in the 1970s initiated hearings to determine the titles, the written record became an authoritative reference point even though it contained inappropriate registrations. The confusion surrounding the registration of land has continued to this day, and there are numerous arguments over land in Micronesia. At that time, Sugiura observed the confusion caused by the introduction of the modern notion of land ownership in Micronesia, where kin groups held rights over land and the matrilineal system complicated claims over land (Sugiura 1944: 266).

Sugiura's field research was assisted by Hijikata, who had rich knowledge about Palauan culture and language. Nakagawa also expressed his thanks to Hijikata, claiming that Hijikata was kind enough to correct many mistakes in the folk terms he recorded in his field notes (Nakagawa 1941: 90). However, their ethnographic representations resulted in what Hijikata recognized as an inappropriate way to represent indigenous cultures (Sudo 1991: 40). Sugiura analyzed the ethnographic materials of land tenure in Micronesia from the perspective of British functionalism. He also referred to the administrative categories of land to discuss social change. Such theorization and reduction by administrative demand contradicted Hijikata's style. Nakagawa, in charge of arranging the research items, was convinced that German ethnographies attached too much importance to the descriptions of material cultures, and thus did not assign enough space for sociological descriptions (Nakagawa 1941: 89). Nakagawa's style also contradicted Hijikata's style, which focused on material cultures and stuck to indigenous perspectives. At the same time, Hijikata started to describe Palauan society in a more rigid way in the late 1930s, likely influenced by the perspectives of Nakagawa's comparative law and Sugiura's anthropology of functionalism (Nakao 2016: 222).

Ethnographic Imaginations in the Journal *Nan'yō Guntō*

In addition to the professional scholars and intellectuals who engaged in the survey of Micronesian land and customs in the late 1930s, there were other artists, novelists, and amateur writers who left behind descriptions of Micronesia in both true stories and works of fiction. In addition to Atsushi Nakajima, who was close to Hijikata, there were some famous writers and artists who portrayed Micronesia. Among these was Tatsuzō Ishikawa (石川達三, 1905–1985), the first winner of the Akutagawa Prize. He visited Palau, where his brother engaged in the cultured-pearl industry, and published *Record of a Chigger Mites Island* (赤虫島日誌) in 1943 (Sakano 2019: 170). Artists such as Toshi Maruki (丸木俊, 1912–2000) and Atsushi Someki (染木煦, 1900–1988) were also inspired by their visits to Japan's *Nan'yō*. Someki's book included rich illustrations of Micronesian artifacts (Someki 1945). Maruki stayed in Micronesia for 6 months in 1940 and met Hijikata in Palau (Takizawa 2008). Although she is famous as an artist who recorded the misery of the atomic bomb in Hiroshima, she also left extensive sketches of the peoples and cultures of Micronesia.

Masaaki Noguchi was an amateur writer and editor. He wrote a couple of books about Japan's *Nan'yō*, especially about Palau (Noguchi 1941a, b), although his name and works are obscure. He was originally from the Gotō Islands in Nagasaki Prefecture and worked as a teacher. Although the details of his personal history are unknown, Noguchi moved to Tokyo and edited a journal titled *Nan'yō Guntō*, which was published by the South Sea Islands Cultural Society (Kawamura 2001: 5). Under the Local Affairs Division of *Nan'yō-chō*, the society arranged cultural activities to promote administrative policies. He first worked at the Tokyo office and later moved to Palau in 1939 when the society's editorial department relocated to Koror. In Palau, he became the chief editor of the journal. Without being restricted by academic perspectives, Noguchi wrote frankly about the ongoing social changes in Micronesia. Further, despite his position in a society controlled by *Nan'yō-chō*, Noguchi sometimes wrote comments criticizing the administration.

Noguchi's main concerns were the trends of living among the Japanese immigrants in Micronesia, including the conditions of immigrant colonies, his daily life at the office, interactions with the officers of the South Seas Governments, the liveliness at the construction of *Nan'yō* Shrine (*Nan'yō Jinja*: 南洋神社) in 1940, the shortage of goods due to the war, and others. His essays also embodied Japan's Southern Expansion Doctrine. He evaluated Japan's *Nan'yō* highly and deemed it to be a strategic point for further expansion to the South, as well as the only tropical zone under the Empire of Japan with rich raw materials (Noguchi 1941a: 16–17). In this sense, his writing was not an ethnographic description, but an editorial commentary on the development of the Empire of Japan. At the same time, what he wrote accidentally contained vivid descriptions of the local society in transition. At this point, Noguchi is praised as an amateur ethnographer. He had the opportunity to work with scholars commissioned by *Nan'yō-chō* in the late 1930s, such as Hijikata who contributed many articles, poems, and illustrations to *Nan'yō Guntō*. Noguchi published a booklet that provided an outline of Micronesian artifacts collected at the South Seas Government Exhibition Room (*Nan'yō-chō Bussan Chinretsujo*: 南洋庁物産陳列所), in which Hijikata was involved after returning to Palau from Satawal (Noguchi 1940). In this booklet, Noguchi mentioned Hijikata's contribution to art education in Palau. Noguchi was also on close terms with jurists from the South Seas Court of Justice, including Hikorokuro Okuno (奥野彦六郎, 1895–1955), who wrote an article on Palauan kin groups in the postwar era (Okuno 1950).

Likely influenced by those intellectuals familiar with Micronesian cultures and societies, Noguchi investigated indigenous confusion over authority, even though he kept imperial eyes on the islands. Based on his interview with a pioneer Japanese merchant familiar with local affairs, who had moved to Palau during the German administrative era, Noguchi described disputes between a Palauan traditional leader and a new elite appointed as a headman by *Nan'yō-chō*. Hijikata and Sugiura reconstructed the traditional Palauan chieftainship, but they wrote little about the influence of the administration. Noguchi's description was specific, as his description included the personal name of the Palauan appointed as a headman, who was at a loss on how to deal with the traditional chief in his village (Noguchi 1941b: 157). Noguchi also used direct speech sentences to describe the Palauan's distress, although the speech would not be sourced from the transcript of his interview with the Palauan but was a product of his imagination based on his interview with the old Japanese merchant. Given the fact that Noguchi was affiliated with the South Sea Islands Cultural Society under *Nan'yō-chō*, his critical view of confusing circumstances set up by the administration must have been challenging. However, his description fell into a colonial discourse justifying the administration when he said that the concerned Palauan new elite was encouraged by the Japanese officials and thus wrestled with administrative works. He also claimed that, as a result, the traditional chief lost motivation to demonstrate his leadership (Noguchi 1941b: 164–165). Noguchi's perspective was distorted, but his description anticipated what American anthropologists in the postwar era found in their investigations when the United States administration introduced democratic elections to Micronesia (e.g., Force 1960). The relationship between traditional chiefs and elected leaders became a major discussion point among American anthropologists, especially among those conducting political anthropological research in Palau and Pohnpei. Petersen wrote about this tension as follows.

> There may be a relationship between anthropologists' perceptions of indigenous social organization and their views concerning the relative merits of introducing American representative democracy. For instance, to the extent that Force and Force interpreted indigenous Palauan society as highly stratified and ruled by powerful chiefs, they praised the introduction and ultimate success of American patterns of government. Useem and Fischer, on the other hand, emphasized the basic democratic character of traditional Micronesian societies, and doubted the good such American institutions could bring to the Palauans and Pohnpeians. (Petersen 1999: 172)

Similar to Noguchi's descriptions, the journal *Nan'yō Guntō* held an ambivalent character. It was a propaganda journal that was first published in 1935 and continued until the end of 1943. Noguchi served as the chief editor and issuer of the journal during most of the time he was in Palau. The journal extensively advertised the policies arranged by *Nan'yō-chō*, especially in the tense situation leading up to and during the Pacific War. The journal contained rich articles by local Japanese residents, scholars from metropolises, and officials from *Nan'yō-chō* and the central government. Some articles involved the administration, while others shied away from administrative concerns or even had critical comments on the administration, as Noguchi did. The journal also contained a few comments by Micronesians, although it is not clear whether the editors actually used their voices. For example, their comments seem to be exploited to justify the cultural assimilation policies of the Japanese Empire. Pictures of Micronesians' voluntary labor were intentionally inserted to promote the full mobilization of the public in the tense situation.

Hijikata extensively contributed articles and illustrations to *Nan'yō Guntō* in 1940 and 1941, when he temporarily worked at the South Seas Government Exhibition Room. The contribution might have been part of Hijikata's duty because *Nan'yō Guntō* was published by the South Sea Islands Cultural Society, a branch of the Local Affairs Division of *Nan'yō-chō*, which was in charge of the surveying of land and customs. The cover design of the journal changed at the beginning of 1940 when Hijikata's illustration, depicting coconut trees and a summerhouse at the seashore, appeared on the cover. The new cover page must have evoked a typical image of *Nan'yō* for Japanese readers. Other illustrations Hijikata contributed to the journal included Micronesian figures and landscapes, as well as images from the legends depicted on the beams of *bai* in Palau. His poems were also released with the illustrations. Some were from his life in Satawal, while others were from his life in Palau at the time. He wrote a series of reports on Palau that covered wider topics from marriage and divorce to games and plays. These reports were later included in his collected works (Hijikata 1993). Sugiura also contributed a short article about Palauan songs and dances, which he did not mention in other formal reports or academic journals. The South Sea Islands Cultural Society had asked him to conduct the research, as it aimed to record the vanishing performing arts in Micronesia (Sugiura 1940: 58). It was paradoxical that the society engaged in salvaging traditional cultures in Micronesia, while it assumed the responsibility of promoting colonial policies under *Nan'yō-chō*, thus, contributing to the rapid social change.

Legacy of Japanese Ethnography in Micronesia

Micronesian ethnographies written under the Japanese administration were not merely historical but were rediscovered in the post-Pacific War era by the Japanese, Americans, and Micronesians. With the introduction of the United States administration after the Pacific War, most of these ethnographies written in Japanese sank into oblivion as English became a new administrative and official language in Micronesia. Few ethnographies were referred to by American researchers who worked under the United States administration. Exceptions included the English version of Yanaihara's work, often referred to as a study of administrative impacts on the indigenous society, that the United States administration dealt with in the postwar era. Sugiura's study of land tenure was not referred to, even in the study of land tenure, conducted at an earlier stage of United States administration by Shigeru Kaneshiro (Kaneshiro 1958), a Japanese-American who worked as a district anthropologist in Palau (1950–1951) and Yap (1951–1957) (Kiste and Marshall 1999: 472–473). Only a few pages of Sugiura's study of land tenure were translated by DeVerne Reed Smith into English in 1986 (Smith 1986). However, it was not well known to the Western academic world, as the journal in which the translation appeared was only published in Japan. In a retrospective assessment of American anthropology in Micronesia (Kiste and Marshall 1999), postwar Micronesian ethnographies by Japanese anthropologists were often mentioned, but most of the prewar ethnographies by the Japanese were not referred to, except for a few works such as Hijikata's translated ethnographies.

Some scholars, who had conducted research in Micronesia during the Japanese administration, published their work in the post-Pacific War era (e.g., Okuno 1950; Sugiura 1949). However, these publications did not serve as main discussion points in academic circles, as scholars hesitated to confront the legacy of the Japanese Empire in former overseas territories. When Japanese anthropologists began conducting fieldwork again in Micronesia in the 1970s, they regarded ethnographies written in Japanese in the prewar era as important references. However, they were mainly concerned with extracting the ethnographic knowledge of Micronesia rather than assessing the colonial context in which the ethnographies were produced. At best, prewar ethnographies were listed in bibliographical works detailing prewar studies of overseas territories of the Empire of Japan (Hatanaka 1979). In 1987, Iwao Ushijima (牛島巖) and Ken'ichi Sudo (須藤健一), two leading Japanese oceanists in the post-Pacific War era, edited a volume titled *Cultural Uniformity and Diversity in Micronesia*. The volume was a compilation of Pacific Studies in postwar Japan with contributions from linguists, archeologists, and anthropologists. Mac Marshall, a leading anthropologist of Oceania studies in the United States, rated the volume highly because of the rich ethnographic data included and its broad range of inquiry. At the same time, he commented that Japanese anthropologists should have included theoretical discussions and have focused on the legacy of the Japanese administration by taking advantage of their access to historical documents.

> There is here a great wealth of descriptive field data, often in impressive detail, but the authors rarely engage in major contemporary issues in anthropological theory and debate. In fact, much of the book strikes this reviewer as old-fashioned, reflecting emphases within anthropology in an earlier era. For example, there is considerable discussion of material culture and oral history, and speculation about possible historical connections based on rather superficial resemblances. None of the contributors examines the economic, political or social changes that have profoundly altered Micronesian societies since the end of the First World War. This is particularly unfortunate since our Japanese colleagues are in a position to mine the Japanese language documentary sources from the 40-year period of Japan's

administration of the islands, to chronicle and analyse the transformations that occurred during those decades. (Marshall 1988: 358–359)

With the exception of Machiko Aoyagi (青柳真智子), who examined Modekngei, a syncretic religion arising in Palau under the Japanese administration, other anthropologists mostly focused on kinship and social organization in traditional societies. Aoyagi's book on Modekngei was published in Japanese in 1985 and later in English in 2002 (Aoyagi 2002). Her study is significant as it is a rare ethnographic investigation of the Japanese anthropologist about the cultural transformation in Japan's *Nan'yō*. In the preface of her ethnography, Aoyagi mentioned the initial difficulties with conducting research about the new religion because Modekngei followers were reluctant to provide answers to her questions.

> At the beginning of my years of study, every time I met with such refusal, I felt that I was refused because I was a Japanese. But later I realized that this did not seem to be the real reason for their refusal. Different from religions that propagate their religious doctrines through missionaries, the Modekngei religion did not consider it necessary to construct a systematic doctrine, and the followers themselves often did not know even the names of the gods they worshipped. (Aoyagi 2002: vi)

Since Medekngei was reported as an anti-Japanese religious movement, Aoyagi was worried about her positionality as a Japanese person. Although she concluded that she faced refusal at the earlier stage of her research not because of her nationality but because of the character of the religion itself, it was inferred from her doubt that postwar Japanese anthropologists had to be conscious of their positionality as burdened with the old Japanese Empire. They even found that elderly informants who had been educated at the elementary schools set up by the Japanese administration would speak to them in fluent Japanese. Thus, Japanese anthropologists are inevitably confronted with the legacy of the Japanese Empire during their fieldwork in Japan's former overseas territories (Iitaka 2019: 232).

During the nation-building era in Micronesia, some ethnographies were considered valuable for learning about past Micronesian societies. German ethnographies are often referred to for reconstructing traditional Micronesia prior to the intervention of colonial administrations. Krämer's ethnographies were translated into English, which intellectuals in Oceania widely read now. The Belau National Museum keeps the exhibition space for Krämer's work. It also holds an exhibition space for Hijikata. Hijikata's ethnographies were translated into English in the 1990s by the Sasakawa Peace Foundation, Japan. Palauans rated Hijikata highly as an artist who contributed to the development of the storyboard, a woodcarving of traditional legends originally from the beams of *bai*. Storyboard became a famous genre of art in postwar Palau. Hijikata's diary, open to the public in recent years in the original Japanese text (Sudo and Shimizu 2010–2014), includes a rich description of the ongoing social changes in Palau and Satawal at that time. Part of the diary is included in his collective works as miscellaneous topics. In Koror, Palau, where both *Nan'yō-chō* and a large-scale Japanese settlement were established, traditional Palauan land, which had belonged to the local community, was sold, lent out, or divided among Palauan kin groups.

The sites where the *bai* (meeting house) and *diangel* (canoe house) are located are also considered to be a part of the community land. However, as I have already mentioned, these official residences have been sold or rented out. Thus these were registered as private property; or they may not be registered now but are treated as such. It is said that the *bai* of Meketii [a meeting house in Koror] is now divided into half and half: one half is controlled by the Idid [the highest-ranking clan in Koror] and the other half is controlled by the Ikelau [the second highest-ranking clan in Koror]....... (Hijikata 1993: 256) (square brackets added by the author)

Japanese ethnologists or anthropologists in the prewar era adhered to academic perspectives such as British functionalism or were restricted by administrative demands; therefore, they did not record a vivid description of the ongoing social changes in their ethnographies. However, fragmentary descriptions remain in miscellaneous records including Hijikata's unpublished manuscripts. Such descriptions are further to be found and evaluated as ethnographic representation in order to investigate social change under the administration.

Amateur writers' descriptions, such as Noguchi's, also have similar potential, although their perspectives contained colonialist views. *Nan'yō Guntō* was reprinted in the 2000s and the 2010s in the original Japanese text, but availability was limited to those in Japanese academic circles. While prewar ethnographic representations by the Japanese have their origins in colonialism, contemporary readers can make use of them to critically investigate the colonial situation at that time. Nonetheless, prewar Micronesian ethnographies by the Japanese have not been fully returned to local society. Many of them, written in Japanese and not translated into English or local languages, are still in the possession of the Japanese, especially Japanese scholars in the fields of history, anthropology, and area studies.

Some resources of Japan's *Nan'yō* were archived at the Yanaihara Tadao Archive at the University of the Ryukyus Library, which offers online digitized materials collected by Yanaihara during his research trips to Micronesia. The National Museum of Ethnology arranged the Ethnology Research Archives, which include materials collected and written by Hijikata and Sugiura. However, these materials are still limited, scattered, and not easily accessible to foreigners without Japanese competence. In contrast, various documents and photographs recorded under the United States administration in the post-Pacific War era are available in the Trust Territory of the Pacific Archives. Both professional scholars and common Micronesians can access these archives online. Using social networking services, some Micronesians discussed whose figures and landscapes were captured in the photos. Here, colonial documents become objects interpreted and reinterpreted by the local population. This process owes a great deal to the United States librarians and archivists who worked to "return history" to the Micronesians through the archives (Peacock 2002). Peacock, paying attention to the political context of possessing colonial materials by the institution in the former suzerain state, evaluated the role of the Trust Territory of the Pacific Archives.

> Again, the University of Hawai'i may be criticized for holding unique materials that more properly belong in Micronesian archives. The fact is that the library was initially reluctant to

accept any materials other than the microfilm......My predecessor, the distinguished bibliographer and curator Renée Heyum, did not want to take on this material, as the Pacific Collection had, at that time, not ventured into visual holdings.......University librarian John Haak, however, felt strongly that this valuable resource would be a most significant addition to the Pacific Collection, and given the uncertainties of preservation measures available at that time in the region, Heyum put aside her opposition. I hope that this anecdote illustrates the true nature of the library's role in regard to the Trust Territory Archives. Its goal was then and is now to work to preserve the records of the American colonial period in Micronesia and to ensure that these records are available to the people of the islands, in their national archives and through images caught and conveyed on the Internet. (Peacock 2002: 127–128)

Micronesian ethnographies written under the Empire of Japan are not merely administrative or anthropological products. They are also appropriated by local populations to learn about their own societies. Therefore, it is necessary to archive colonial documents and ethnographic materials in order to improve accessibility for Micronesians as well as Western scholars. Confining these documents only to Japanese academics may lead to an exploitation of knowledge. Here is a thought-provoking comment about Atsushi Nakajima's short story titled *Mariyan*, modeled on a Palauan woman named Maria Gibbons, who was from the highest-ranking clan in Koror and whom Nakajima and Hijikata knew well. When the novel was translated in 2014, her descendants offered comments, including the following by one of her sons.

My comments to the story are very brief. Why did the author write the story? What were his reasons or purpose? The ending of the story was abrupt and incomplete which I believe is unfair to the readers and an injustice to Mariyan.

Finally, my personal feeling about the story is pretty much in agreement with the details about Mariyan's life as described (character, interests, status, religion, etc.). I witnessed the same during my childhood which I am proud of and admired.

The story indeed brought back good memories and made me feel closer to Mariyan once again. (Kawaji 2014: 76)

Even though this is a comment on a fictional account based on Nakajima's experience in Palau, the questions raised in this reflection can also be raised about colonial ethnography. The accountability of ethnography to the local population has been widely discussed after the crisis of representation in the 1980s. Japanese postwar anthropologists conducting their fieldwork in Japan's former overseas territories have the responsibility to assess the colonial contexts of prewar ethnographies and thus present their implications to the local population.

Conclusion

As investigated above, ethnographies of Micronesia were advanced in the late 1930s, both by administrative demands and academic concerns. Those who took responsibility for representing Micronesia included professional scholars as well as amateur writers. Scholars commissioned as researchers of Micronesian land and customs by *Nan'yō-chō* seized the opportunity to advance their research activities in territories overseas. This trend corresponds to that in the metropole, such as the establishment of the Ethnic Research Institute and the Institute of the Pacific. The fact that Sugiura was a member of the former institute and that Hijikata, Sugiura, and Nakagawa contributed to publications by the latter institute exemplifies the correspondence between the metropole and the colony. The local journal *Nan'yō Guntō*, published by the South Sea Islands Cultural Society, was also a locus in which amateur writers such as Noguchi had the chance to write frankly about local societies, while professionals such as Hijikata contributed short articles for the general public. Compared to the postwar ethnographies written by American anthropologists, these prewar ethnographies written by the Japanese were not known to the Western academic world and local societies in Micronesia, except for a few works such as Hijikata's ethnographies which were translated into English.

The research activities surrounding Hijikata, Sugiura, Nakagawa, and Noguchi in the late 1930s indicate that ethnographies of Micronesia were written at the intersection of administrative demands, interests in "primitive" peoples in Japan's colonies, academic influence from Europe in the humanities and social sciences, and the colonial gaze of the public. Modern Japan, being gazed at by the West, sought to obtain subjectivity to objectify colonial others in Asia and the Pacific (Sakano 2005: 5). Despite the different perspectives among officials, scholars, and amateur writers, they likewise had the chance to construct themselves as subjects turning their gaze upon "primitive" others who were either to be preserved in ethnographies or represented as colonial objects waiting for civilization. These ethnographic representations in the late 1930s took place under a broader political context. The survey of land and customs was required in accordance with the influx of Japanese immigrants into Micronesia. *Nan'yō Guntō* was published to promote the Southern Expansion Doctrine to the general public. At the same time, ethnographic representations were discordant with the administration. Hijikata's dissatisfaction with rigid survey items, Sugiura's highly specific discussion of Micronesian social structure, and Noguchi's critical comments on the administrative policies tell us that there was intimacy and distance between the administration and ethnographic representations.

In addition, those who wrote Micronesian ethnographies were not independent but interacted with each other from different positions, as is evident from the relationship among Hijikata, Sugiura, Nakagawa, and Noguchi. This suggests that ethnographic representations under the colonial situation are not grasped appropriately by restricting the research subject only to the academic world. Ethnographies in Micronesia in the late 1930s were products of a contact zone, in which various colonial agencies, including administrators, immigrants, professional researchers, amateur writers, and local populations met and interacted with one another. Rather than depicting academic history preserved in a clean room, relationships among colonial agencies, including ethnographers, must be observed concretely.

Finally, attention should be paid to the fact that the agency of the local population was not activated under the Japanese administration. While the quality of Hijikata's ethnographies was supported by his close relationship with indigenous people, other ethnographies were written under the support of Japanese informants like Hijikata.

Few Japanese scholars could communicate in local languages. Japanese linguists compiled Palauan legends told by a Palauan who studied at the *Tenrikyō* (天理教: a Japanese new religion) school in Nara, but published them only in Japanese (e.g., Miyatake 1932). Since the Japanese language had stronger influence as an administrative language, ethnographers did not need to acquire competence in local languages to collect ethnographic materials. Japanese residents or local populations who had learned how to speak Japanese provided information in the field. In turn, the voices of the Micronesians might not have been reflected well. When the Japanese language lost its presence with the collapse of the Japanese Empire, the prewar ethnographies faced difficulty in acceptance by Americans and Micronesians. The way Micronesians might appropriate the prewar Japanese ethnographies in the future, as the Palauans praised Hijikata's works in the era of nation-building, should be further investigated.

Cross-References

▶ Center and Periphery: Anthropological Theory and National Identity in Portuguese Ethnography
▶ Making and Unmaking of Ethnos Theory in Twentieth-Century Russia

References

Aoyagi M (2002) Modekngei: a new religion in Belau, Micronesia. Shinsensha, Tokyo
Asakura T 朝倉敏夫 (2011) 植民地期朝鮮の日本人研究者の評価—今村鞆・赤松智城・秋葉隆・村山智順・善生永助 [Evaluation of Japanese researchers in colonial Korea: Tomoe Imamura, Chijō Akamatsu, Takashi Akiba, Chijun Murayama, and Eisuke Zenshō]. In: Yamaji K 山路勝彦 (ed) 日本の人類学—植民地主義、異文化研究、学術調査の歴史 [Japanese anthropology: colonialism, the study of other cultures, and history of academic research]. Kansei Gakuin University Press, Nishinomiya, pp 121–150. [In Japanese]
Force RW (1960) Leadership and cultural change in Palau. Chicago Natural History Museum, Chicago
Hatanaka S (comp) (1979) A bibliography of Micronesia compiled from Japanese publication, 1915–1945, Research Institute for Oriental Cultures, Gakushuin University, Tokyo.
Hijikata H (1993) In: Endo H (ed) Collective works of Hijikata Hisakatsu: society and life in Palau. Sasakawa Peace Foundation, Tokyo
Hijikata H (1995) In: Endo H (ed) Collective works of Hijikata Hisakatsu: gods and religion of Palau. Sasakawa Peace Foundation, Tokyo
Hijikata H (1996) In: Endo H (ed) Collective works of Hijikata Hisakatsu: myths and legends of Palau. Sasakawa Peace Foundation, Tokyo
Hijikata H (1997) In: Sudo K (ed) Collective works of Hijikata Hisakatsu: driftwood: the life in Satawal Island, Micronesia. Sasakawa Peace Foundation, Tokyo
Iitaka S (2019) Positionality of East Asian anthropologists in Pacific Studies: comments on Guo Peiyi. Jpn Rev Cult Anthropol 20(2):227–235
Imanishi K 今西錦司 (1944) ポナペ島—生態学的研究 [Ponape Island: an ecological study]. Shōkō Shoin, Tokyo. [In Japanese].
Imanishi K 今西錦司 (1952) 大興安嶺探検—1942年探検隊報告 [Expedition to the Greater Khingan: report of the 1942 expedition]. Mainichi Newspapers, Tokyo. [In Japanese].

Izumi S 泉靖一 (1954) 故杉浦健一教授と人類学・民族学 [The late prof. Ken'ichi Sugiura and his anthropological and ethnological research] 民族學研究 [Jpn J Ethnol] 18(3):266–272. [In Japanese]

Kamijō F 上條深志 (1938) パラオ島誌 [Ethnography of Palau]. Nan'yō Shinpo, Koror. [In Japanese]

Kamijō F 上條深志 (1939) ヤップ島誌 [Ethnography of Yap]. Nan'yō Takushoku, Koror. [In Japanese]

Kaneshiro S (1958) Land tenure in the Palau Islands. In: Tobin J, Fischer J, Emerick R, Mahoney F, Kaneshiro S (eds) Land tenure patterns: Trust Territory of the Pacific Islands volume 1. Office of the High Commissioner, Trust Territory of the Pacific Islands, Guam, pp 294–332

Kawaji Y (ed) (2014) Nakajima Atsushi's "Mariyan" and Maria Gibbon, its inspiration. Minatonohito, Kamakura

Kawamura M 川村湊 (2001) 野口正章『外地』解説 [Notes on Masaaki Noguchi's "Overseas territories"]. In: Kawahara I 河原功 (comp), Kawamura M 川村湊, Shirakawa Y 白川豊 (eds) 日本植民地文学精選集 南洋群島編 2 [Selected works of literatures in Japan's colonies: South Sea Islands II]. Yumani Shobo, Tokyo, pp 1–6. [In Japanese]

Kiste RC, Falgout S (1999) Anthropology and Micronesia: the context. In: Kiste RC, Marshall M (eds) American anthropology in Micronesia: an assessment. University of Hawai'i Press, Honolulu, pp 1–9

Kiste RC, Marshall M (eds) (1999) American anthropology in Micronesia: an assessment. University of Hawai'i Press, Honolulu

Marshall M (1988) Reviewed work: Ushijima, Iwao and Ken'ichi Sudo (eds) Cultural uniformity and diversity in Micronesia. Senri Ethnological Studies, no. 21. National Museum of Ethnology, Osaka, 1987. J Polyn Soc 97(3):358–360

Matsumura A (1918) Contributions to the ethnography of Micronesia. J Coll Sci Imp Univ Tokyo 40:1–174

Mio Y (ed) (2021) Memories of the Japanese Empire: comparison of the colonial and decolonization experiences in Taiwan and Nan'yō Guntō. Routledge, London

Miyatake M 宮武正道 (1932) パラオ島の伝説と民謡 [Legends and songs in the Palau Islands]. Tōyō Minzoku Hakubutukan, Nara. [In Japanese]

Nakagawa Z 中川善之助 (1941) 中部カロリン群島に於ける家族と姓族 [House (im) and mother-sib (ainang) in the middle of Caroline Islands]. In: Institute of the Pacific 太平洋協会 (ed) 大南洋—文化と農業 [Great South Seas: its culture and its soil]. Kawade Shobo, Tokyo, p 87–172. [In Japanese]

Nakao K 中生勝美 (2016) 近代日本の人類学史—帝国と植民地の記憶 [History of anthropology in modern Japan: memory of the Japanese Empire and its colonies]. Fukyosha Publishing Inc., Tokyo. [In Japanese]

Nan'yōchō 南洋庁 (ed) (1939) 南洋群島々民旧慣調査報告書 [Reports of customary laws in the South Sea Islands]. Nan'yōchō, Koror. [In Japanese]

no author (1937)「南の會」第一回南洋群島民族學調査團旅程 [Itinerary of the first ethnological expedition to Nan'yō Guntō by the Society of the South]. 史学 [The historical science] 16(3): 110–116. [In Japanese]

Noguchi M 野口正章 (1940) 群島の島民と其の文化 [Micronesian islanders and their material cultures]. South Sea Islands Cultural Society, Koror. [In Japanese]

Noguchi M 野口正章 (1941a) 今日の南洋 [South Seas today]. Sakagami Shoin, Tokyo. [In Japanese]

Noguchi M 野口正章 (1941b) パラオ島夜話 [Night tales on Palau]. Kensetsusha Shuppanbu, Tokyo. [In Japanese]

Oguma E 小熊英二 (1995) 単一民族神話の起源—「日本人」の自画像の系譜 [The myth of the homogeneous nation: a genealogy of "Japanese" self-images]. Shinyōsha, Tokyo. [In Japanese]

Okuno H 奥野彦六郎 (1950) ミクロネシアにおける「同生地族」の形成 [The interaction of consanguinity and locality in the formation and "ramification" of kin-groups in Micronesia] 民族學研究 [Jpn J Ethnol] 14(3):200–210. [In Japanese]

Peacock KM (2002) Returning history through the Trust Territory Archives. In: Jaarsma SR (ed) Handle with care: ownership and control of ethnographic materials. University of Pittsburgh Press, Pittsburgh, pp 108–129

Petersen G (1999) Politics in postwar Micronesia. In: Kiste RC, Marshall M (eds) American anthropology in Micronesia: an assessment. University of Hawai'i Press, Honolulu, pp 145–195

Sakano T 坂野徹 (2005) 帝国日本と人類学者—1884–1952年 [The Empire of Japan and Japanese anthropologists: 1884–1952]. Keisō Shobo, Tokyo. [In Japanese]

Sakano T 坂野徹 (2019) 〈島〉の科学者—パラオ熱帯生物研究所と帝国日本の南洋研究 [Japanese scientists on islands: Palau Tropical Biological Station and South Seas Studies under the Empire of Japan]. Keisō Shobo, Tokyo. [In Japanese]

Shimizu A (1999) Colonialism and the development of modern anthropology in Japan. In: Bremen JV, Shimizu A (eds) Anthropology and colonialism in Asia and Oceania. Curzon Press, Richmond, pp 115–171

Shimizu A 清水昭俊 (2013) 民族学の戦時学術動員—岡正雄と民族研究所、平野義太郎と太平洋協會 [Ethnology and wartime academic mobilization: Oka Masao and Ethnic Research Institute, Hirano Yoshitaro and Institute of the Pacific]. In: International Center for Folk Culture Studies 国際常民文化研究機構 (ed) 第二次大戦中および占領期の民族学・文化人類学 [Ethnology and cultural anthropology during World War II and the occupation]. International Center for Folk Culture Studies, Yokohama, p 17–82. [In Japanese]

Smith DV (1986) Land tenure of the South Sea Islanders by Ken'ichi Sugiura I–III. 太平洋学会誌 [J Pac Soc] 29:156–165, 31:152–226, 32:139–196

Society of the South 南の会同人 (1937) ニューギニア土俗品圖集 南洋興発株式会社蒐集 上巻 [Material cultures in New Guinea: collection by the South Seas Development Company I]. Nan'yō Kohatsu, Tokyo. [In Japanese]

Society of the South 南の会同人 (1940) ニューギニア土俗品圖集 南洋興発株式会社蒐集 下巻 [Material cultures in New Guinea: collection by the South Seas Development Company II]. Nan'yō Kohatsu, Tokyo. [In Japanese]

Someki A 染木煦 (1945) ミクロネジアの風土と民具 [Natural features and fork materials in Micronesia]. Shōkōshoin, Tokyo. [In Japanese]

Sudo K (1991) The Micronesian studies by Hisakatsu Hijikata, ethnographer. In: Setagaya Art Museum 世田谷美術館 (ed) 土方久功展—南太平洋の光と夢 [The Hisakatsu Hijikata exhibition: light and dream of Micronesia]. Setagaya Art Museum, Tokyo, pp 36–40

Sudo K 須藤健一, Shimizu H 清水久夫 (eds) (2010-2014) 土方久功日記I–V [The diary of Hisakatsu Hijikata I–V], 国立民族学博物館調査報告 [Senri ethnological reports] 89, 94, 100, 108, 124. National Museum of Ethnology, Suita. [In Japanese]

Sugiura K 杉浦健一 (1940) 南洋群島島民の歌謡・舞踊・採集項目 [Research items for Micronesian songs and dances]. 南洋群島 [Nan'yō Guntō] 6(6):58–64. [In Japanese]

Sugiura K 杉浦健一 (1941) 民族学と南洋群島統治 [Ethnology and the administration of the South Sea Islands]. In: Institute of the Pacific 太平洋協会 (ed) 大南洋—文化と農業 [Great South Seas: its culture and its soil]. Kawade Shobo, Tokyo, pp 173–218. [In Japanese]

Sugiura K 杉浦健一 (1944) 南洋群島原住民の土地制度 [Land tenure of South Sea Islanders]. 民族研究所紀要 [Bull Jpn Soc Ethnol] 1:167–350. [In Japanese]

Sugiura K 杉浦健一 (1949) ミクロネシヤ原住民の政治組織—民主的なパラオ島民と封建的なポナペ島民 [Micronesians' traditional polities: democratic Palauans and feudalistic Pohnpeian Humanities]. 人文 [Humanities] 3(1):28–52. [In Japanese]

Takizawa K 滝沢恭司 (ed) (2008)「南洋群島」美術年表 [A chronological table of art history in the South Sea Islands]. In: Machida City Museum of Graphic Arts and Tokyo Shinbun 町田市立国際版画美術館・東京新聞 (eds) 美術家たちの「南洋群島」[The South Sea Islands and Japanese artists: 1910–1941]. Tokyo Shinbun, Tokyo, pp. 150–164. [In Japanese]

Totsuka K 戸塚皎二 (ed) (1931) 女酋 [Joshū] 2. Palao Minzoku Sadankai, Koror. [In Japanese]

Yamaguchi T 山口徹 (2015) ウリ像をめぐる絡み合いの歴史人類学―ビスマルク群島ニューアイルランド島の造形物に関する予察 [Historical anthropology of entanglement surrounding 'Uli' figures: review of carved objects from New Ireland, Bismarck Archipelago]. 史学 [The historical science] 85(1/2/3):401–439. [In Japanese]

Yamaji K 山路勝彦 (2011) 日本人類学の歴史的展開 [Historical development of Japanese anthropology]. In: Yamaji K 山路勝彦 (ed) 日本の人類学―植民地主義、異文化研究、学術調査の歴史 [Japanese anthropology: colonialism, the study of other cultures, and history of academic research]. Kansei Gakuin University Press, Nishinomiya, p 9–73. [In Japanese]

Yamamoto M (2005) Pacific islands studies in Japan. 亞太研究論壇 [Asia-Pacific Forum] 30:76–95. Center for AsiaPacific Area Studies, Academia Sinica

Yanaihara T 矢内原忠雄 (1935) 南洋群島の研究 [Pacific islands under Japanese mandate]. Iwanami Shoten, Tokyo. [In Japanese]

Archives/Database

Digital archive of early photographs taken in eastern Asia and Micronesia by Japanese anthropologists, University Museum, University of Tokyo 東京大学総合研究博物館 東アジア・ミクロネシア古写真資料画像データベース. http://umdb.um.utokyo.ac.jp/DJinruis/torii_catalogue/hajime.php?link=hajime. Accessed on 20 Mar 2021

Ethnology Research Archives 民族学研究アーカイブズ. http://nmearch.minpaku.ac.jp/. Accessed on 20 Mar 2021

Image archive of the Trust Territory of the Pacific Islands. http://libweb.hawaii.edu/digicoll/ttp/ttpi.html. Accessed on 20 Mar 2021

Image database of colonial documents at the Yanaihara Tadao Archive, University of the Ryukyus Library 琉球大学付属図書館矢内原忠雄文庫植民地関係資料. http://manwe.lib.uryukyu.ac.jp/yanaihara/. Accessed on 20 Mar 2021

Photo collection of Murayama Chijun 村山智順所蔵写真選. http://web.flet.keio.ac.jp/~shnomura/mura/index.htm. Accessed on 20 Mar 2021

The Anthropological Society of Nippon 日本人類学会. https://anthropology.jp/. Accessed on 20 Mar 2021

Boasian Cultural Anthropologists, Interdisciplinary Initiatives, and the Making of Personality and Culture during the Interwar Years

68

Dennis Bryson

Contents

Part I: Introduction	1758
Part II: Personality and Culture and the Boasians	1760
Franz Boas: The Founding Figure	1760
Ruth Benedict: The Cultural Patterning of Personality	1762
Margaret Mead: Culture, Temperament, and Personality	1765
Edward Sapir: The Individual As Carrier, Interpreter, and Elaborator of Culture	1771
Part III: Conclusion	1779
References	1780

Abstract

This chapter will examine the manner in which a group of major American cultural anthropologists came to play a key role in formulating and promoting the personality and culture approach in the United States during the 1920s and 1930s. It will thus focus on the group of Boasian cultural anthropologists – namely, Columbia University anthropologist Franz Boas, and, perhaps more significantly, his students Edward Sapir, Ruth Benedict, and Margaret Mead – who created and elaborated the emerging field, in part as the result of their participation in the interdisciplinary initiatives of the Social Science Research Council (SSRC) and the Rockefeller philanthropic organizations during the interwar years. Accordingly, the chapter will deal with various projects, networks, and seminars and conferences sponsored and fostered by the Rockefeller foundations and the SSRC in which the Boasian anthropologists participated; the role of Lawrence K. Frank and Robert S. Lynd in fostering SSRC and Rockefeller-sponsored initiatives will be indicated. The trajectory of the culture concept as elaborated by the Boasian anthropologists will also be examined. The chapter will

---

D. Bryson
Department of American Culture and Literature, Faculty of Humanities and Letters, İhsan Doğramacı Bilkent University, Ankara, Turkey
e-mail: dennis@bilkent.edu.tr

argue that this group, with the encouragement of Frank and Lynd, viewed personality and culture as an interdisciplinary endeavor that would integrate the social sciences. It would also, they hoped, elaborate an approach to reform and social engineering oriented toward altering the cultural patterns associated with child rearing and education. It was thus affiliated with what has been called by J. Meyerowitz the "biopolitics of child rearing." The chapter will offer a critical assessment of this approach.

Keywords

Personality and culture · Social engineering · Interdisciplinary initiatives · Social Science Research Council (SSRC) · Rockefeller philanthropy · Ruth Benedict · Margaret Mead · Edward Sapir

Part I: Introduction

This chapter will examine the manner in which a group of major American cultural anthropologists came to play a key role in formulating and promoting the personality and culture approach in the United States during the 1920s and 1930s. More specifically, it will focus on the group of Boasian cultural anthropologists – namely Columbia University anthropologist Franz Boas, and, perhaps more significantly, his students Edward Sapir, Ruth Benedict, and Margaret Mead – who created and elaborated the field, in part at least, as the result of their participation in the interdisciplinary initiatives of the Social Science Research Council (SSRC) and the Rockefeller philanthropic organizations during the interwar years. (The SSRC was funded by and worked closely with the Rockefeller boards during the interwar years; thus, during the 1920s, it was closely affiliated with the Laura Spelman Rockefeller Memorial [LSRM], a Rockefeller foundation, which sponsored major initiatives in the social sciences during this decade [Fisher 1993].)

This chapter will demonstrate that the Boasian formulations, sponsored and amplified by the foundation and SSRC interdisciplinary initiatives, represented the key inspiration for the emergent field that came to be known as personality and culture. To be sure, these cultural anthropological formulations and philanthropic/ organizational initiatives were closely intertwined with each other during the 1920s and 1930s. Thus, as key administrators of the SSRC and LSRM, respectively, Robert S. Lynd and Lawrence K. Frank encouraged the elaboration of the personality and culture approach as a means both to integrate and unify the social sciences and to promote social engineering and reform. (Lynd and Frank were good friends as well as intellectual allies; they had known each other since the 1910s, when both were ambitious young men living and working in New York City. Frank took a position with the LSRM in 1923 and continued to work for Rockefeller philanthropic organizations until 1936. Lynd was the "permanent secretary" of the SSRC from 1927 to 1931. For more on the two, see Smith 1994: 120–158; Bryson 2002). Lynd and Frank advanced personality and culture by planning and organizing interdisciplinary conferences, such as those held in Hanover, New Hampshire, in 1930 and 1934, as well as by means of other initiatives, such as the organization of SSRC committees focused on personality and culture. Two Boasian cultural anthropologists, Mead and Sapir, were very much involved in these initiatives, which fostered networks of cultural anthropologists with other social scientists and specialists. Meanwhile, from the 1920s on, Boasians such as Sapir, Benedict, and Mead elaborated the notion of culture that they had received from their mentor Boas; synthesizing this idea with that of personality, they arrived at a more holistic, integrated concept of culture. By 1930, under the rubric of personality and culture, the SSRC and the Rockefeller foundations, with the active encouragement of Frank and Lynd, had taken up and were fostering the Boasian agenda by providing various intellectual forums and outlets for its articulation (see especially Bryson 2002, 2009, 2015).

The social scientists, clinicians, and administrators involved in elaborating the personality and culture approach during the interwar years responded to the tensions and discontents of this turbulent period – which included accelerated cultural change and conflict, intergenerational strife, and roller coaster swings between the seeming prosperity of the 1920s and the economic collapse of the 1930s – by formulating a special version of social engineering and reform. Dismayed by what they saw as the unraveling of the interrelationship of the individual to society and culture, they proposed that American culture be reoriented and reconstructed. Generally, their chief means to achieve this would be to reconstruct the cultural patterns of child rearing and education in order to foster healthy, cooperative, and friendly personalities. Frank, who was placed in charge of the Laura Spelman Rockefeller Memorial's program in child study and parent education in 1923, embraced an early, perhaps even prototypical, version of this approach with great enthusiasm, exclaiming that with better and more enlightened practices of child nurture, "at last man may take charge of his own destiny" (Frank 1939: 5). With the increasing influence of psychoanalysis among social scientists during the 1930s, the personality and culture approach to social engineering came to focus on the psychodynamics of personality formation within cultural settings. In consequence, the new approach came to acquire the label personality and culture – or culture and personality, as some preferred to put it. Again, Frank played a leading role in promoting the revised approach, now as an officer for the General Education Board (a Rockefeller philanthropic foundation) – and again he expressed utopian expectations for this approach, declaring that it could usher in a new era of "cultural self-determination, social volition, and group responsibility" (Frank 1936: 343). Historian Joanne Meyerowitz (influenced by the French theorist Michel Foucault) has aptly characterized the new social engineering agenda embraced by scholars working in personality and culture as the "biopolitics of child rearing." According to Meyerowitz, such scholars "called for ... a liberal form of biopolitics that would reconfigure a group's behavior and health by reshaping the personalities of its members ... The way to enhance the quality of the population was not through selective breeding of so-called races but through selective nurturing of certain cultural traits" (Meyerowitz 2010: 1059).

In dealing with the history of personality and culture during the interwar years, this chapter will begin by briefly noting the contribution of Boas, who anticipated the personality and culture approach in his writings and pronouncements of the late 1910s and the early 1920s. Instructively, he offered his mentorship and support to the three cultural anthropologists who played the largest role in advancing personality and culture during the interwar years – Benedict, Mead, and Sapir. These anthropologists developed their own perspectives on personality and culture, and worked on varied projects within this field. Sapir developed a programmatic agenda stressing the individual and his or her idiomatic relationship to culture; he also emphasized, in a manner congruent with the SSRC approach to the cross-fertilization of the social sciences, how personality and culture could integrate and perhaps provide unity to the social sciences. In contrast, Benedict stressed the impact of the cultural configuration and its patterns on the individual; significantly, she was less concerned with an interdisciplinary agenda during the interwar years in comparison to Sapir and Mead. Mead, who was more intellectually aligned with Benedict than with Sapir in important ways, did extensive fieldwork as an anthropologist, much of it pertinent to personality and culture; thus, in a number of studies she examined the development of children and youth along with other issues relevant to the formation of personality in varied cultural settings. All of these anthropologists were concerned with social engineering, though Benedict and Mead were perhaps more explicit than Sapir in embracing such an agenda. As suggested above, the chapter will deal with the various SSRC and Rockefeller-sponsored interdisciplinary initiatives and projects in which Mead and Sapir were involved; as will be demonstrated below, these initiatives had a large impact on their approaches to personality and culture. Most importantly, modes of social engineering, particularly those associated with the biopolitics of child rearing, were associated with the SSRC and Rockefeller initiatives.

Part II: Personality and Culture and the Boasians

Franz Boas: The Founding Figure

The story of the formulation of the personality and culture approach most appropriately begins with Franz Boas (1858–1942), who has been universally acknowledged as the founding figure of American cultural anthropology. Embracing a "historical particularist" approach to cultural anthropology, he eschewed universal schema of social evolution – in which each human society was to proceed on a unilinear path from savagery to barbarism to civilization – and instead focused on the manner in which each culture developed in its own way. His overall approach to culture – and, more specifically, his holistic orientation to cultures and his formulations on such issues as cultural relativity, plurality, and determinism vis-à-vis human behavior – has had a large impact not only among American anthropologists, but on social scientists and the public more generally. Significantly, Boas was an out-spoken opponent of "scientific" and other modes of racism, and he came to advocate a cultural relativist approach – a perspective emphasizing that cultures should be understood on their own terms and not prejudged by the values and standards of Western observers. Nevertheless, Boas did not elaborate a systematic theory of culture and its various aspects; rather he formulated his ideas in scattered writings over the years in ways that were not always consistent. (For more on Boas and his culture concept, see Stocking 1982: 195–233; Patterson 2001: 45–50). It fell to his students – which included such distinguished anthropologists and cultural observers as Alfred Kroeber, Zora Neale Hurston, Melville Herskovits, Ruth Benedict, Margaret Mead, and Edward Sapir – to elaborate the culture concept and other aspects of Boas's approach to cultural phenomena. Instructively, by 1920, Boas was clearly moving in the direction of the personality and culture approach. Moreover, during the 1920s and 1930s, several of his students – Benedict, Mead, and Sapir – came to elaborate a concept of culture heavily inflected by the personality and culture approach. In doing so, they played a key role in launching this approach as a major social scientific perspective.

By the late 1910s and certainly by 1920, Boas's orientation in anthropology was shifting in ways that clearly anticipated the personality and culture approach. Even before then, in the 1890s and early 1900s, Boas had become concerned with a key aspect of the field that was later to emerge as personality and culture: namely, human growth and development within varying social contexts. Thus, he initiated studies of the development of various groups of children, including those from immigrant, African American, and affluent white native-born backgrounds. Instructively, he found that the rate of physical and mental development among individual children varied in ways that often deviated from standard growth curves. Moreover, the social environment in which children grew up influenced their development (Patterson 2001: 48–49). Boas also addressed issues pertinent to personality and culture when, in 1917, he gave a lecture to Barnard students denouncing the conformity that came to prevail in the aftermath of the United States' entry to World War I. Thus, he exhorted students to resist pressures to conform in the face of the intense emotional atmosphere created by American involvement in the war. Boas suggested that psychological and cultural understandings and, more generally, an anthropological outlook, would help students to do this. Along such lines, he indicated that the students would need to develop empathy with the nonconformist, to transcend "the fetters that the past imposes upon us," and to be willing to critically assess "our social structure" as well as "attitudes accompanied by strong outbursts of emotion" (Boas quoted by Brick 2006: 90).

Boas demonstrated an even greater shift toward the culture and personality outlook, and indeed in his anthropological orientation more generally, in his article "Methods of Ethnology," which was published in the *American Anthropologist* in 1920. By then, Boas had become dissatisfied with both the evolutionary and diffusionist approaches, and he noted approvingly that anthropologists were coming to focus on "cultural history" as "the dynamic phenomena of cultural change" (Boas 1920: 314). In order to advance the new emphasis, anthropologists would need to address the "problem of the relation of the individual to society." Instructively, as Boas observed, "The activities of the individual are determined to a great extent by his social environment, but in turn his own activities influence the society in which he lives, and may bring about modifications in its form" (316). As Stocking (1976: 22) has noted, the 1920 article demonstrated a reorientation in Boas's approach to anthropology more generally, as Boas expressed in the piece a concern with such phenomena as a culture's "inner development," "acculturation," and the "interdependence of cultural activities," as well as with the relationship of the individual and society. Moreover, according to Stocking, the article formulated "a change in emphasis between the central components of Boasian analysis – from [cultural] 'elements' to 'processes' and 'patterns' – in the context of a simultaneous shift in analytic perspective from the diachronic to the synchronic ... [T]he focus was to be on the 'dynamic changes in society that may be observed at the present time' [in Boas words]" (ibid.). Precisely how Boas's students elaborated the new emphases will be examined below.

Ruth Benedict: The Cultural Patterning of Personality

Ruth Benedict (1887–1948) is justifiably considered one of the major figures of the "classical" personality and culture school (Piker 1994). She stressed in her work the manner in which culture shaped personality as well as the implications for social engineering of such a perspective. Trained by Boas in the early 1920s, she became a close friend and associate to two other major figures involved in the formulation of personality and culture: Margaret Mead (who was her lover for a time) and Edward Sapir. Thus, as Banner (2003) and Handler (1986) have noted, Benedict engaged over the years in significant intellectual exchanges with Mead and Sapir, sharing and elaborating ideas with them in the emerging field of personality and culture (though Sapir became increasingly critical of Benedict's work in the 1930s). Benedict taught at Columbia University in the Anthropology Department for about two decades. Embracing various aspects of the culture concept formulated by Boas, she came to focus on the idea of cultural configuration and to apply this idea to personality and culture. In doing so, she came to emphasize certain aspects of the Boasian approach while de-emphasizing others. Thus, she came to stress the idea of culture as an integrated, patterned totality, as Boas had come to do, but she was somewhat less interested in the individual and the specific aspects of his or her relation to culture. Compared to her friends and colleagues Mead and Sapir, Benedict displayed relatively little interest in child-rearing practices and, more generally, in the specific processes by means of which the individual adopted and responded to the cultural patterns of his or her group. Thus, while she asserted that the individual was "conditioned," "shaped," and "moulded" by the group's cultural patterns – and that most individuals would simply accept this patterning "because of the enormous malleability of their original endowment" (Benedict 1934a: 254) – it was not until 1938 that she examined some of the specifics of the conditioning process with regard to child-rearing and educational methods in her writings. Indeed, she never became very interested in psychoanalytic concepts and the psychodynamics of personality formation (see, for example, Heyer Young 2005: 189–192). Nevertheless, Benedict's cultural configurational approach, in emphasizing how cultural patterns shaped the individual's potentials and behavior, had major implications for personality and culture.

Unlike Mead and Sapir, Benedict was not involved in the initiatives in personality and culture promoted by Lawrence K. Frank and Robert S. Lynd under the auspices of Rockefeller philanthropy and the Social Science Research Council. To be sure, Benedict was concerned, as were Frank, Lynd, and Mead, with a mode of social engineering geared toward the reorientation and rebuilding of culture. Her goals and methods paralleled in some ways, but also differed, from those advanced by this group. The contrast with Frank is especially instructive. Stressing the importance of childhood and youth, Frank wanted to see child-rearing and educational practices altered so that sane, cooperative, well-adjusted personalities would be produced. Benedict, on the other hand, was less interested in promoting a rather conformist vision of a society filled with sociable and well-adjusted individuals than in promulgating tolerance toward "abnormal" individuals such as homosexuals. (She believed that "abnormal" individuals were unable to adjust to the group's prevailing cultural patterns due to congenital temperamental factors.) Moreover, her vision of social engineering was not primarily geared toward reorienting child-rearing practices, but toward the conscious direction of change in the group's cultural configuration as a whole. Her overall aims were to promote the understanding and tolerance of abnormality as well as to ameliorate the problems created by traditionally sanctioned but antisocial patterns of behavior, such as excessive egotism, rivalry, and conformity, which she saw as prevalent in contemporary American culture. Hence, she indicated that a new version of normality might be formulated and perhaps disseminated by means of an educational program (she did not provide specific details on this program) (Benedict 1934a: 271, 274). As Benedict exclaimed:

> No society has yet achieved self-conscious and critical analysis of its own normalities and attempted rationally to deal with its own social process of creating new normalities within its next generation. But the fact that it is unachieved is not therefore proof of its impossibility. It is a faint indication of how momentous it could be in human society. (Benedict 1934b: 77)

In *Patterns of Culture* (1934a), her best-known work, Benedict provided a detailed formulation of her approach to culture – an approach that owed much to Boas's concern with the "genius of culture," as he himself suggested in his introduction to her book (xxi). In developing her approach, Benedict examined in some detail the cultures of three "primitive" societies – the Zuni, Dobuans, and Kwakiutl – in order to discern the fundamental "direction," "drives," patterns, and values of these cultures. In doing so, she depicted the "cultural configurations" of these groups. She found Gestalt psychology and the writings of the German thinkers Wilhelm Dilthey, Oswald Spengler, and Friedrich Nietzsche (especially the latter's treatment of the distinction between the Dionysian and Apollonian ethos) to be of special assistance in her efforts at interpreting and analyzing the cultural configurations of the three societies. Benedict's basic argument was that cultures, as with languages with regard to phonemes, selected a relatively limited number of behaviors and interests from the vast array ("a great arc") of potential behavioral traits and human inclinations – and, moreover, that each culture patterned and elaborated this selection as it moved toward increasing integration as a cultural configuration – "discarding those types of behavior that are uncongenial" along the way. According to Benedict, these processes tended to proceed in a "non-rational and subconscious" manner, as did linguistic processes (Benedict 1934a: 24, b: 72). In elaborating her approach, Benedict came to encompass social institutions under the rubric cultural configuration – and thus came to see them in terms of the culture's direction, drives, motives, traits, and patterns, rather than in terms of their relations within articulated structural totalities. Benedict thus proceeded with the reconfiguration of the social as the cultural. Accordingly, as she put it:

> The significant sociological unit, from this point of view … is not the institution but the cultural configuration. The studies of the family, of primitive economics, or of moral ideas need to be broken up into studies that emphasize the different configurations that in instance after instance have dominated these traits. (Benedict 1934a: 244)

Notwithstanding her previous neglect of the field, by 1938 Benedict had begun to exhibit some interest in child-rearing practices in her research and publications. In an article published that year in *Psychiatry*, "Discontinuities and Continuities in Cultural Conditioning," Benedict dealt with child rearing in some detail, though she was especially concerned with the discontinuities that characterized cultural expectations vis-à-vis the individual during different phases of the life cycle (Benedict 1938; Heyer Young 2005: 179–181). Benedict continued to address child-rearing practices in her work on the study of "culture at a distance" with the Office of War Information during World War II and, later, with the Research on Contemporary Cultures group during the Cold War years. Benedict thus came to recognize the need for the study of child-rearing methods in order to delineate the "national character" of enemies and friends during the 1940s. However, she still did not demonstrate any great interest in the psychoanalytic approach, and she remained fundamentally concerned with how methods of child rearing fit in with other aspects of the group's cultural configuration (Mead 1974/2005: 57–75). Thus, in examining the swaddling of infants in several Central and Eastern European groups, Benedict focused on the diversity of swaddling practices and on the messages communicated and the values assigned to these practices by these groups (Benedict 1948). To be sure, in her book on Japanese culture, *The Chrysanthemum and the Sword* (1946), she did deal with such psychoanalytically oriented themes as weaning, toilet training, and sexual expression in the chapter titled "The Child Learns." But the emphasis was upon how such child-rearing methods contributed to the emergence of a dualistic outlook in the culture of the Japanese: a sense of ease and appreciation of life, including sensual life, derived from their relatively easy and privileged childhood, on the one hand, and the acceptance of the restraints and pressures applied to them as adults, on the other (Benedict 1946: 286–296).

In summary, Benedict elaborated a personality and culture approach closely affiliated with social engineering. Building on themes initially formulated by Boas, she came to stress the patterning of cultural configurations and, going beyond Boas in significant ways, the manner in which these patterns could fashion the personalities of individuals. Some individuals, she came to recognize, might not quite fit into the cultural patterns offered to them by the groups that they lived in – and, elaborating on Boas's suggestions regarding cultural relativity, she advocated tolerance toward such individuals. Along such lines, she wanted to challenge "customary opinions and causes" as well as "absolutist philosophies" and to formulate "a more realistic social faith, accepting as grounds of hope and as new bases for tolerance the coexisting and equally valid patterns of life which mankind has created for itself from the raw materials of existence" (Benedict 1934a: 278). Notwithstanding such critical and humanistic sentiments, in focusing on the cultural configuration as the key object to be known as well as the primary terrain for social engineering and reform, Benedict's version of personality and culture reconfigured the social as culture and thereby failed to articulate in a fully adequate way the manner in which social institutions and structures had a major impact on society, culture, and personality.

Margaret Mead: Culture, Temperament, and Personality

Like her friend and mentor Ruth Benedict, Margaret Mead pioneered in the study of personality and culture during the interwar years. As with Benedict, she also played a major role in elaborating the Boasian culture concept and displayed a lively concern in her work with social engineering as cultural reform and reorientation. Unlike Benedict (at least Benedict before the late 1930s), however, Mead was very interested in children and youth in diverse cultures, as we will see below. Moreover, and not coincidently, Mead was very involved in Rockefeller and SSRC-sponsored initiatives in personality and culture during this period. Indeed, it was in no small part due to her participation during the 1930s in such initiatives as the Hanover Seminar on Human Relations and the SSRC's Subcommittee on Cooperative and Competitive Habits that Mead came to play such a major role in formulating the personality and culture approach. Her friendship with foundation executive Lawrence K. Frank as well as her intellectual association with Yale social psychologist John Dollard were fostered by these initiatives – and both relationships provided significant inspiration for Mead's work in personality and culture. Mead's fieldwork experiences were also important in encouraging her to elaborate her perspective on personality and culture. Thus, in planning and conducting her fieldwork, Mead pioneered in the study of children and adolescents in various "primitive" societies; in doing so, she was especially concerned with how the young were fashioned by, but also responded to, the culture of their group. Accordingly, Mead elaborated a mode of social engineering aimed at revising the cultural practices associated with raising and educating the young in modern American society. Along such lines, as a prominent public intellectual during much of the twentieth century, Mead wrote a number of widely read books and articles for popular magazines (and made frequent pronouncements for popular consumption in other venues) providing Americans with advice on problems associated with the care and education of the young, family life, gender roles, and other pressing cultural issues. Mead's work in personality and culture was also influential within the discipline of anthropology; indeed, her influence in anthropology went beyond the field of personality and culture and related topics to encompass issues such as innovative fieldwork methods, including the use of photography (Sullivan 1999).

It was during the 1920s that Mead became engaged in the study of cultural anthropology and that she moved to launch her career in this field. In the United States, the 1920s was a decade characterized by significant social and cultural turbulence and innovation – including the clash of traditional and "modernist" cultural trends, intergenerational conflict, the large-scale emergence of mass consumer society, and rampant economic inequality. American youth were especially affected by such cultural change and conflict, and Mead came to address their problems and lives, along with those of the young in "primitive" cultures, in her work during this decade. For Mead, all of this began in the fall of 1920, when she transferred from a provincial Midwestern college (as she saw it) to Barnard College, a women's college closely associated with Columbia University in New York City. A course given by Franz Boas during her senior year at Barnard introduced Mead to the study of cultural anthropology. After finishing an MA in psychology at Columbia University, she went on to pursue her doctorate in anthropology at Columbia under the guidance of Boas and his assistant Benedict.

It was Boas who sent Mead to study adolescence in Samoa on her first fieldwork expedition (Mead 1972: 127–128). Moreover, it was Boas who suggested, as he indicated in his foreword to the book that resulted from Mead's Samoan fieldwork, her *Coming of Age in Samoa*, that a focus on "the way in which the personality reacts to culture" should become a key issue for anthropological study (Boas in Mead 1928/2001: xxii). (To be sure, as scholars have noted, this issue was in fact not the focus of her first book; see, for example, Brick 2006: 94.) Most importantly, in *Coming of Age in Samoa* Mead addressed, setting a pattern that she would follow in her future work, problems characterizing contemporary American culture – and, taking cues from the "primitive" culture that she had studied, proposed reforms which would help Americans deal with these problems. Accordingly, she suggested that Americans learn from the communal style of raising children in Samoa and de-intensify the relationship between parents and children – a relationship that had been fraught with neuroses and "complexes" in the hothouse emotional atmosphere of the isolated middle-class nuclear family. Most importantly, Mead proposed that Americans provide children and youth with an "education for choice" – one which would not insist on imposing the values and beliefs of parents and teachers on the young but rather foster tolerance and open-mindedness. According to Mead, the young "must be taught how to think, not what to think" (Mead 1928/2001: 169). Thus, children and youth should be able to choose their own standards and ways of living in accord with their individual interests, gifts, and "temperamental types" in "our heterogeneous, rapidly changing civilization" (169–170).

A key focus of Mead's work during the interwar years was the idea of culture. In *Coming of Age in Samoa* and in the widely read books that followed, including *Growing Up in New Guinea* (1930) and *Sex and Temperament in Three Primitive Societies* (1935), Mead played a major role in popularizing the notion of culture. She also elaborated the culture concept in publications aimed at more academic circles, such as the volume that she edited and to which she contributed, *Cooperation and Competition among Primitive Peoples* (1937). In all of this work, Mead elaborated on the notion of culture as formulated by Boas and Benedict. Accordingly, as Mead developed her approach to culture as the "patterns of life" worked out by the group (Mead 1928/2001: 11), she came to emphasize how culture shaped the lives of the individual members of this group. Moreover, she came to insist on seeing various cultures as more or less integrated wholes. Nevertheless, Mead came to believe that the influence of culture on the individual might be limited by constitutional, seemingly biological, temperamental factors. Furthermore, under the influence of the British social anthropologist A. R. Radcliffe-Brown, Mead became concerned with the relationship of culture to the group's "social system" and "social structure." During the 1930s, these issues became salient for her work on personality and culture, especially with regard to such issues as the relationship of "social personality" to temperament and of "character formation" to social structure.

As Mead became increasingly involved in reconfiguring the social as culture during the 1930s, a kind of slipperiness came to characterize her idea of culture. Thus, in her contributions to *Cooperation and Competition among Primitive Peoples*, Mead tended to equate culture with Radcliffe-Brown's "social system" and with "society," using these terms interchangeably and even using the term culture more frequently than the latter two terms. (Hence, she tended to write of "Arapesh culture," "Dakota culture," and so on, rather than of Arapesh society or the Dakota social system.) In seeing societies and social systems as cultures, Mead elaborated a somewhat vague but all-embracing notion of culture. Thus, it seemed to Mead, culture included all aspects of social life, including family and kinship networks, technology and economic arrangements, art, religious life, political institutions, and so on. Nevertheless, for Mead, culture was triumphant, in a sense, over other aspects of social life – thus, the "cultural emphases" of the group vis-à-vis competitive, cooperative, or individualistic practices would prevail over the techniques, economic methods, and/or the material environment. Along such lines, according to Mead, societies were competitive, cooperative, or individualistic in emphasis not primarily as the result of the scarcity or abundance of natural resources or their techniques of production, but as the consequence of such aspects of their culture as the "educational process" or "the social structure prescribed by culture" (Mead 1937/1961: 13, 459). The implications of such a move were problematic, as will become evident below.

To be sure, Mead's attempt to synthesize the Boasian culture concept with Radcliffe-Brown's approach to social system and structure might arguably be a valid and illuminating way to approach the "primitive" societies studied by anthropologists. But what of modern market-oriented societies in which, according to economic historian and anthropologist Karl Polanyi (1944/2001), the self-regulating market of the liberal economic system has become seemingly dis-embedded from social relations and culture, and, as such an apparently autonomous system, has come to be prevail over other aspects of sociocultural life? In such societies, land and labor – inextricably intertwined as these are with all other aspects of social life – become, along with money, "fictitious commodities," while the individual comes to focus his or her economic activities on the pursuit of private gain. Of course, Mead was critical of modern market-oriented civilization and its economic ethos. Thus, she was critical of the economic model that posited the individual competing against others under conditions of scarcity – that is, the "hypothesis which regards man as primarily a craving organism who will compete with other organisms only to satisfy a desire for an object of which there is a limited supply" (Mead 1937/1961: 15). For Mead, cultural factors, not human nature or instincts, fostered cooperative or competitive behavior, as noted above. Moreover, in arguing for the prevalence of culture over the economic, Mead was engaged in formulating a critique of the modern economic system – a system in which narrow materialistic values and interests had come to prevail to the detriment of wider social, cultural, and artistic values and concerns. Nevertheless, she was unable to thematize the manner in which the advent of liberal market-oriented society transformed the relationship of the economic arrangements to society and culture – setting up the former as a seemingly self-regulating market system that assumed dominance over other social and cultural forms. Instructively, Mead came to advocate "effective, responsible intervention in our own culture" (Mead 1962: 121), largely on the microcultural level of child-rearing and educational practices – that is, on the terrain of the "biopolitics of child rearing," to use Meyerwitz's phrase – rather than an overall agenda of economic planning, such as that advocated by Polanyi (1944/2001).

Inspired by her participation in the Hanover Seminar on Human Relations during the summer of 1934 as well as by her work with the SSRC Subcommittee on Cooperative and Competitive Habits the following fall and winter, Mead increasingly came to adopt a psychoanalytic perspective to the problem of character formation within cultural and social structural settings. She thus came to focus on the educational process – child-rearing practices, the child's relations to its parents, and the like – by means of which the individual's "cultural character" was fashioned. Influenced by Dollard's neo-Freudian psychoanalytic approach (as well as by the perspectives of Erich Fromm and Karen Horney), Mead focused on "the relationship between character formation and the way in which the developing individual learns to cope with his impulse structure on the one hand, and with the institutions of society, on the other" (Mead 1962: 128). To be sure, Mead had encountered Freud in the past, even as an undergraduate student at Barnard College, and she was familiar with Malinowski's effort to apply Freud's approach to the Trobriand Islanders as well as the former's critique of the universality of the Oedipus complex (Mead 1931). (At Barnard, Mead had taken a course on the psychological aspects of culture from a Freudian perspective from sociologist William F. Ogburn [Mead 1972: 111]. She had read Malinowski's *Sex and Repression in Savage Society* – in which he investigated the relevance of Freud's formulations to the Trobriand Islanders – and she referred to this book in her 1931 article on primitive children. Most likely, she read the book before completing *Coming of Age in Samoa*.) Nevertheless, Mead assigned special significance with regard to her adoption of a psychoanalytic approach to her conversations with Dollard during the Hanover Seminar and its aftermath. From Dollard, she noted, she learned the importance of studying the life history of the individual over a period of years; such an approach would take into account how the "socially relevant biological factors" were culturally fashioned so that the "new organism" (the child) became "added to the group" (Mead 1937/1961: 1–2, 483–484). Along such lines, Mead examined issues pertinent to the psychoanalytic account in her work, including feeding and weaning, sphincter control, relations between children and parents, and other caregivers as well as with siblings, modes of sexual expression, and the like (see, for example, Mead 1962: 118–120). Scholars such as Maureen Molloy (2008) have claimed that Mead's understanding of psychoanalysis was relatively superficial, but the influence of it on her work was clearly great.

Even as Mead worked out her psychoanalytically oriented approach to character formation under the influence of Dollard in the mid-1930s, she was involved in the elaboration of a very different perspective to the problem of personality and culture – one based on the study of seemingly inborn temperaments and associated psychological types. From early 1933 on, Mead worked on formulations of her "squares hypothesis" – initially in the context of intense conversations with anthropologists Reo Fortune and Gregory Bateson in the field in New Guinea (Sullivan 2004a). Mead came to be especially concerned with the male/female axis in temperamental traits in a document outlining her squares hypothesis in March 1933; those with "male" temperaments, she claimed, tended to be paternalistically possessive while those with "female" temperaments tended to be maternal and caring (Banner 2003; Sullivan 2004a, b). Mead suggested in this document that biological men and women could possess either form of temperament; indeed, she claimed, men and women demonstrated both temperaments in equal numbers. She also elaborated on the distinction between the introverted, self-centered "fey" temperamental type and the extroverted, matter-of-fact "Turk" type. Though Mead felt that she and her associates had made a "tremendous discovery" (Mead 1972: 220), Mead did not publish an explicit account of her squares hypothesis during this period. She was persuaded by Benedict and Boas that adequate evidence for her theory was lacking. Moreover, her theory of innate psychological temperaments might be misused for unsavory political purposes (see, for example, Sullivan 2004a, b). (To be sure, Mead slightly qualified her theory of the biological/genetic basis of temperamental difference. Thus, she endorsed the assumption that "there are different temperamental differences between human beings which if not entirely hereditary at least are established on a hereditary base very soon after birth" [Mead 1935/1963: 284].)

In any case, Mead's theory of temperamental types, or at least important aspects of this theory, did inform her book *Sex and Temperament in Three Primitive Societies* (1935/1963). Mead believed this to be her "most misunderstood book" (Preface to the 1950 edition), and no doubt the manner in which her hypothesis of innate temperament was rather obliquely expressed in the book had much to do with this. Mead had started to write this book during the 1934 Hanover Seminar; there she had met the Columbia endocrinologist Earle T. Engle, whose formulations on hormones, sexuality, and gender had a significant influence on her. In conversations with Mead, Engle noted that genetic, hormonal, and embryological factors acted together in complex ways to influence the individual's sexual development; indeed, as one scholar has suggested, Engle "had decided that biologists could no longer precisely state what constituted a man or a woman" (Banner 2003: 350). Such a position, challenging the rigid modern Western male/female dichotomy, was elaborated in Mead's *Sex and Temperament*. While modern Western cultures assigned standardized temperamental traits to the "social personalities" of men and women – so that individuals born as biological females were supposed to be maternal and submissive, and individuals born as biological men were supposed to be aggressive and dominating – temperamental traits associated with sex were assigned to men and women in very different ways in the three New Guinea cultures that Mead examined in her book. According to Mead, among the Arapesh, both women and men were nurturing, cooperative, and gentle, while, in contrast, among the Mundugumor, both women and men were aggressive, relatively unmaternal, and sexually assertive. Meanwhile, among the Tchambuli, women were impersonal and managing, while men tended to be dependent and emotional. Mead's overall point was that temperamental traits rigidly dichotomized as male and female in Western civilization, while representing inborn constitutional aspects of temperament, could nevertheless be assigned to men and women very differently in other cultures. Thus ideas about what constituted maleness and femaleness varied greatly in different cultures.

Mead was especially concerned with individuals who deviated as the result of temperamental factors from the group's standardized social personality – or the group's standardized sets of personality dichotomies, such as those regarding the male/female distinction. Thus, some men might be endowed with seemingly feminine temperamental traits, while some women might possess seemingly male traits – and, as a result, come to be considered deviants, most notably homosexuals, in their societies. On the issue of homosexuality, Mead's thinking was intricate and somewhat confused, generally calling for tolerance, but also advocating the eradication of conditions that she believed to cause homosexuality, such as the rigid sex-stereotyping of individuals. In any case, the call for breaking down the rigid "sex-dichotomy of temperament" and thereby for tolerance for the fundamental temperamental differences of individuals was perhaps the dominant theme of her book (see especially Molloy 2008: 126–130). Along such lines, Mead called for a mode of social engineering geared toward creating "educational institutions [which would] develop to the full the boy who shows a capacity for maternal behavior [as well as] the girl who shows an opposite capacity that is stimulated by fighting against obstacles" (Mead 1935/1963: 321).

Mead embraced "cultural relativism" in her work. For Mead, this did not amount to an ethical relativism that proclaimed all cultural values and practices to be of equal validity; rather, she believed "that any item of cultural behavior must be understood in relation to the culture from which it comes, and that, torn out of its context, it is meaningless" (Mead 1937/1961: vi). Mead thus believed that the aim of anthropology was to "understand other possible ways of life, other solutions which are open to human beings," thus shedding light on "the different conditions of personal and social life under which human beings can survive and flourish, and also what prices must be paid by some members of society that others – either of different sex, age, status, temperament, or intellectual endowment, may have a full development" (quoted by Brick 2006: 107). To be sure, Mead was critical of various aspects of the "primitive" societies that she studied, even as she endeavored to understand them as integrated wholes. Thus, she was critical of the Samoans' inability to appreciate personal individuality as well as of their tendency to regiment human relations (Mead 1928/2001). She was also critical of the Manus' "Puritan" penchant to overvalue work and the acquisition of property, their strong sense of guilt, and their rigidity in "subduing...sex life to meet supernaturally enforced demands" (Mead 1930/2001: 124). Nevertheless, she hoped that modern Americans could learn from these cultures and perhaps adopt variations of their more positive traits. But while Mead could be critical of the modern American economic system and its ethos – with its poverty, overwork and alienation of workers, and economic instability, as well as its emphasis on aggressiveness, competiveness, and possessiveness (Brick 2006: 94, 95) – her approach to social engineering did not go far in addressing such issues. Rather, Mead focused her social engineering agenda on culture and, more specifically, on disparate cultural practices and traits (notwithstanding her proclamations on the importance of approaching cultures in a holistic fashion). As Molloy has observed with regard to Mead's focus on culture:

> This "culture" is sufficiently vague and non-located that it can be seen as the source of our difficulties .... It can be seen as "the system," an impersonal force that, because we can break it down to its components, may be amenable to change. Because it is made up of values, attitudes, behaviors, and rituals, we can frame change without hurting anyone or challenging fundamental property relations. (Molloy 2008: 132–133)

Edward Sapir: The Individual As Carrier, Interpreter, and Elaborator of Culture

Perhaps Boas's most brilliant student, Edward Sapir (1884–1939) elaborated the Boasian culture concept and the personality and culture approach by stressing the agency of the individual vis-à-vis culture. Significantly, to an even greater degree than Margaret Mead, Edward Sapir flourished within the world of SSRC and Rockefeller-sponsored conferences, seminars, and committees. Sapir was not only comfortable with interdisciplinary intellectual deliberation and discussion within such venues, but he was greatly inspired by them. Indeed, it was within such contexts that, encouraged by his friend and intellectual ally Harry Stack Sullivan, he came to call for an alliance between cultural anthropology and psychiatry in order to further a more comprehensive vision of social science. In the 1930s, Sapir came to describe this vision as "social psychology," and he elaborated this vision in a series of programmatic essays on the theoretical underpinnings of his proposal for allying cultural anthropology and psychiatry. Though he was more low-keyed in his advocacy of social engineering than Benedict and Mead, nevertheless Sapir's approach to personality and culture did have clear implications for such a project.

Sapir hoped that the new interdisciplinary approach that he was formulating would transcend the fragmentation of the social sciences by focusing on the personality of the individual. To be sure, the individual personality would need to be considered within its social and cultural settings, but within these settings, Sapir emphasized, the individual personality was to be viewed as the carrier, interpreter, and elaborator of culture. Along such lines, Sapir rejected a reified approach to culture – one which thematized culture as a system abstracted from the individual and his or her personality and behavior – while stressing the individual's idiosyncratic perspective on cultural beliefs and practices. (Sapir seems to have initially elaborated his critique of the reified notion of culture in his response to an article by Alfred Kroeber on the "superorganic." See Sapir 1917.) Instructively, for Sapir the idea of "society" receded from the purview of his interdisciplinary approach; by the 1930s, he made it clear that he was not primarily interested in the relationship of the individual to society. In doing so, he contributed to the reconfiguration of the social as the cultural (to be sure, as the latter was embedded in and enacted by individual personalities). Most importantly, he formulated a distinctive vision of the personality and culture approach – one that was to be influential in future years.

Sapir did not explicitly embrace social engineering as the rational direction of culture in the manner of Mead and Benedict. Nor did he write popular books that included proposals for cultural reform, as the two had. As he explained in a letter to Benedict in 1925, he had "no desire to save or help save humanity"; he would content himself with his erudite linguistic endeavors, leaving the salvation of the world to the likes of Mead and Benedict (cited in Mead 1959: 180). Perhaps more significantly, as Handler (1986) has suggested, Sapir's recognition of the unconscious cultural patterning of behavior led him to be skeptical of conscious efforts to engineer cultural patterns; according to Sapir, those studying culture might come up with useful schemes for "the medicine of society," but for the most part, it would seem, cultural patterns would have to be accepted (cited by Handler 1986: 151). Notwithstanding such sentiments, during the 1930s Sapir did move toward aligning himself with a social engineering approach. Thus, by embracing the psychiatric perspective on the significance of the individual personality, Sapir elaborated a therapeutic vision that had important implications for social engineering and cultural reorientation. Along such lines, in dealing with child development and personality formation, as he did in some of his work of the 1930s, Sapir elaborated a perspective with a close affinity with what Meyerowitz has described as the "biopolitics of child rearing."

The son of Jewish immigrants from German Pomerania, who had brought their young son Edward to the United States in 1890, Sapir grew up in modest circumstances on the Lower East Side of New York City. Receiving a scholarship to attend Columbia University, he did well in his studies, eventually coming to pursue the study of linguistics and anthropology under Franz Boas (King 2019: 137–138). The latter came to recognize Sapir as his most brilliant student, and Sapir followed in the footsteps of his mentor by specializing in the study of Native American (as well as Germanic and Semitic) languages. Sapir continued to work on Native American languages as the chief of the Division of Anthropology of the Geological Survey of the Canadian National Museum in Ottawa, Canada. A professional breakthrough came to Sapir in 1925, when he was hired by Fay-Cooper Cole of the University of Chicago to teach anthropology and linguistics at this institution. Sapir was brought to Chicago with LSRM funding – and, on his arrival, he was quickly introduced to the world of SSRC and Rockefeller-sponsored initiatives in interdisciplinary social science. Thus, while at Chicago, he became involved in the Institute for Juvenile Research and the Local Community Research Committee; moreover, during these years, he came to represent the American Anthropology Association as a member of the SSRC's board of directors (Darnell 1986: 161; SSRC 1934).

Most importantly, Sapir participated in such major initiatives as the American Psychiatric Association's First and Second Colloquia on Personality Investigation, organized by Harry Stack Sullivan in the late 1920s (the first event sponsored by the LSRM in conjunction with the American Psychiatric Association [APA], the second sponsored by the SSRC as well as by the APA and the LSRM); the SSRC's conference on acculturation and personality, held as part of the 1930 Hanover Conference; the SSRC's Advisory Committee on Personality and Culture (1931–1934); and the Seminar on the Impact of Culture on Personality, held at Yale University in 1932–1933. (Sapir left Chicago for a position at Yale in 1931, but he continued his work with the SSRC and Rockefeller-sponsored projects. He also worked with the National Research Council during 1930s, helping to launch the Council's Committee on Personality in Relation to Culture in 1935.) As indicated above, to a large degree, it was within the context of these conferences, seminars, and committees that he formulated his special approach to personality and culture.

As Sapir elaborated his thinking on issues pertinent to personality and culture during the interwar years, he came to believe that psychiatry would provide cultural anthropology, and the social sciences more generally, with the key to understanding personality. He had started to become interested in psychiatry in the late 1910s as the result of his efforts to cope with the mental illness of his first wife. Accordingly, he became interested in the writings of Jung and discussed the latter's theory of psychological types with a colleague at a 1924 conference (see Mead 1972: 124). Most importantly, his interest in psychiatry was given major impetus when he met psychiatrist Harry Stack Sullivan – who had formulated a neo-Freudian approach to "interpersonal relations" within the field of psychiatry – at a conference in Chicago in 1926. The two had a long conversation at the conference, and the personal friendship and intellectual alliance between the two that grew out of this lasted until Sapir's death in 1939. During the late 1920s and the 1930s, Sapir and Sullivan participated in a series of conferences and initiatives – generally backed by SSRC and funded by Rockefeller philanthropy – dealing with the emerging field of personality and culture. For Sapir, these initiatives provided important early venues in which to formulate his view on the relationship of psychiatry and anthropology to the field of personality and culture. Thus, at the 1928 colloquium on investigating personality, Sapir proposed bridging the gap between psychiatry and such social sciences as anthropology and sociology by acknowledging that both personality, on which psychiatry focused, and society, the focus of sociologists and anthropologists, dealt with "systems of ideas." According to Sapir:

in the last analysis there is no conflict between the concept of "culture" and the concept of "personality".... I would say that what really happens is that every individual acquires and develops his own "culture" and that "culture," as ordinarily handled by the student of society, is really an environmental fact that has no psychological meaning until it is interpreted by being referred to personalities or, at the least, a generalized personality conceived as typical of a given society. (APA 1929: 79)

In the 1930s, Sapir continued to formulate his perspective on personality and culture while participating in a series of SSRC initiatives in this field – including the sessions on acculturation and personality held as part of the 1930 SSRC Hanover Conference (Sapir chaired and contributed to these subconference sessions); the Advisory Committee on Personality and Culture of the SSRC; and the Yale Seminar on the Impact of Culture on Personality, which Sapir directed (with the assistance of John Dollard). As he participated in these efforts, Sapir came to elaborate an interdisciplinary outlook for the social sciences – one quite congruent with the aim of the SSRC to establish collaboration and cross-fertilization within these disciplines. (Instructively, Sapir wrote several proposals during the early 1930s, which he submitted to the SSRC and other agencies, outlining community studies which would include detailed studies of the life histories of individuals by teams of specialists from various disciplines. None of these interdisciplinary projects were actually initiated, but they may have had influence on other projects in future years. See Bryson 2009: 365–366, 380–381, 2015: 91.) As noted above, Sapir believed that the coupling of anthropology, with its focus on cultural patterns, and psychiatry, with its stress on the personality of the individual, would provide the "mother science" needed to integrate the social sciences. He described the emerging science as "social psychology" in several of his key papers from this period – but he also came to use the rubric "personality and culture" in referring to this field. Indeed, it was the latter phrase that scholars generally came to utilize in characterizing Sapir's project (Singer 1961; Darnell 1986; Piker 1994).

Sapir elaborated his approach to personality and culture in a number of programmatic statements published during the 1930s. In one of the more important of these statements "Cultural Anthropology and Psychiatry" (1932), Sapir dealt in some detail with how he believed that the two fields could collaborate with and complement each other in order to develop a comprehensive social scientific outlook. To be sure, psychiatry, with its origins in medicine, focused on the individual and aimed "to diagnose, analyze, and, if possible, cure those behavior disturbances of individuals which [demonstrate] serious deviations from the normal attitude of the individual toward his physical and social environment" (Sapir 1932: 143) – while cultural anthropology was concerned with the group and its culture, the latter consisting of "the generalized forms of action, thought, and feeling ... in their complex interrelatedness" (140). Nevertheless, according to Sapir, by coming to understand that both psychiatry and anthropology focused on individuals and their interactions and behavior within social and cultural contexts, one could get beyond a reified notion of culture as a superorganic entity imposed on individuals in an impersonal manner, on the one hand, and beyond a medical approach to the individual viewed simply as an organism, abstracted from society and culture, on the other (143–147).

Sapir suggested that the new field resulting from the coupling of psychiatry and anthropology would provide the basis for an inclusive science of human behavior (see also Sapir 1934b: 199: Sapir 1939: 183–186). Instructively, the emerging interdisciplinary field would be able to deal with issues pertinent to cultural change; by transcending superorganic notions of culture, it would lead "to the more dynamic study of the genesis and development of cultural patterns because these cannot be realistically disconnected from those organizations of ideas and feelings which constitute the individual" (Sapir 1932: 147). More particularly, according to Sapir, psychoanalysis, by focusing on the manner in which culturally patterned "systems of ideas and feelings" had an impact on the mental disorders experienced by individuals, could contribute to the new interdisciplinary perspective – though, to be sure, Sapir had major reservations with regard to the manner in which psychoanalysts had appropriated aspects of cultural anthropology and its findings (144–145, 148–150). (Sapir continued to exhibit interest in implications of the work of Jung, Alfred Adler, and even Ernst Kretschmer for the study of personality during the 1930s [see 1934a], but increasingly Sapir came to see psychoanalysis as, in Darnell's words [1986: 175], "the only systematic method to allow examination of individual behavior.") Significantly, by the early 1930s, Sapir had come to believe that the concept of society would not play a key role in his proposed interdisciplinary social science. According to Sapir, the true locus of culture was not society – a mere "cultural construct" in any case – but in individuals and their interrelations. As Sapir put it:

> The true locus of culture is in the interaction of specific individuals and, on the subjective side, in the world of meanings which each one of these individuals may unconsciously abstract for himself from his participation in these interactions. (1932: 151)

While Sapir did not deny, and indeed at times stressed, that culture fashioned the individual and his or her personality, his assignment of the "locus" of culture to individuals and the realm of "interpersonal relations," rather than to society and its organization, had problematic implications. To be sure, Sapir's elaboration of a therapeutic humanist perspective, in which the individual was not simply a pawn of the superorganic realm of culture but could develop his or her own idiosyncratic version of culture and even exert some degree of creativity with regard to culture, can be seen – and often has been seen by scholars such as Darnell (1986) – as a significant achievement. Moreover, Sapir's emphasis on the psychiatric approach to "personality as a reactive system which is in some sense stable or typologically defined for a long period of time, perhaps for life" – as well as his psychoanalytically oriented concern with how "concepts of behavior equivalences, such as sublimation, affective transfer, rationalization, libido and ego relations" establish "invariance of personality" – certainly contributed a distinctive and engaging perspective to personality and culture (Sapir 1934a: 165). Instructively, however, while Sapir attributed a high degree of stability and endurance to personality – and would allude to such phenomena as "personality organization" (1932: 162), "personality structure" (1939: 174), and "psychic structure" (1934b: 201) in his writings of the 1930s – he seemed relativity uninterested in attributing lasting organization and structure to the sociopolitical realm. Thus, in declining to provide substance and reality to society, Sapir obscured the manner in which organized and relatively long-lasting social structures, especially those manifested in the political economic systems of specific societies, interacted with culture in order to shape the subjectivities of the human beings living in these societies.

On such issues, it is instructive to compare Sapir's perspective to that of the major late-twentieth-century American anthropologist Michelle Rosaldo (1984). For Rosaldo, cultural meanings and scenarios shape the sense of selfhood and the affects experienced by persons in all societies – but the sociopolitical organization of these societies has a major impact on the cultural patterns and practices prevailing within them. According to Rosaldo, the very sense of being an individual self, with a sense of interiority and the potential for being described by an array of terms for specific personality traits and motives, is characteristic of certain modes of non-egalitarian sociopolitical organization. Along such lines, Rosaldo stressed "the theoretical point that relates lives of feeling to conceptions of the self, as both of these are aspects of particular forms of polities and social relations. Cultural idioms provide the images in terms of which our subjectivities are formed, and, furthermore, these idioms themselves are socially ordered and constrained" (150). (Instructively, Rosaldo initially presented her paper at an SSRC conference held in 1981.)

Sapir's affinity to the biopolitics of child rearing was demonstrated in another important programmatic essay of the 1930s, "The Emergence of the Concept of Personality in the Study of Culture" (1934b). Significantly, the article was based on a paper given at a conference on child development sponsored by the National Research Council in June 1933. Sapir elaborated on the issue that he had dealt with in his 1932 essay, again calling for the collaboration of cultural anthropology and psychiatry in order to advance his interdisciplinary agenda (and again labeling this agenda "social psychology"). In more detail than the earlier article, Sapir focused on the concept of personality in the 1934 piece; he was especially interested in the genesis of personality as the child encountered and appropriated aspects of the cultural patterns presented to it. As in the previous article, Sapir was critical of reifying approaches to culture, and he stressed the importance of individual personality for understanding culture and its changes. Sapir also made suggestive comments pertinent to social engineering in this essay. According to Sapir, "The interests connected by the terms culture and personality are necessary for intelligent and helpful growth because each is based on a distinctive kind of imaginative participation by the [scientific] observer in the life around him" (1934b: 196). Thus, from a cultural perspective, such an observer might be concerned with behavior associated with conscience and values, as transmitted to the child by the parents and other adults, while from a personality perspective, the observer might focus on "behavior as self-expressive" (197). Along such lines, Sapir indicated that the study of the genesis of personality would be fruitful for reevaluating the cultural materials studied by anthropologists; some issues, such as certain of the qualitative aspects of the father–child relationship in various cultures, would come to be stressed, while others, such as the child's membership (or lack thereof) in a clan, would be de-emphasized. Most importantly, Sapir came to stress the importance of studying child development in the essay. The focus should be on the "culture-acquiring child," whose "personality definitions and potentials" were involved in the effort "from the very beginning to interpret, evaluate, and modify every cultural pattern, sub-pattern, or assemblage of patterns that it will ever be influenced by" (205). Accordingly, Sapir proposed that a study be made on the manner in which children acquired cultural patterns, from birth to perhaps the age of 10; such a study would stress the relevance of cultural patterns for the development of personality.

A complex and nuanced thinker, Sapir was critical of various aspects of American society and culture during the interwar years. If he did not, as suggested above, adequately address the relationship of "personality" to the sociopolitical organization of society, he nevertheless formulated trenchant critiques of American economic life. Thus, in a major essay "Culture, Genuine and Spurious" (1924), Sapir criticized the "spurious" culture that he believed to characterize modern American industrial civilization. According to Sapir, this civilization no longer permitted individuals to participate in culture in the way that many primitive peoples had been able to in their "genuine" cultures. While a genuine culture was unified, yet "richly varied," and "inherently harmonious, balanced, and self-satisfactory," inviting individual engagement and creativity, a spurious culture was fragmented, lacking in harmony and integration, hypocritical, and fostered passivity and conformity among its members (90–93). Thus, in the spurious culture of contemporary America, workers, such as the "telephone girl," were simply cogs in the machine-age industrial system, confined to manipulating technical procedures without being able to meet their "spiritual needs" for meaningful activity (92). On the other hand, Native Americans had been able to actively participate in their genuine cultures, even within the most humble of activities, in a way that provided them with spiritual enrichment. Instructively, gesturing toward a Boasian approach to cultural relativity, Sapir noted that while both advanced civilizations and those at a "lower level" could achieve genuine culture, it would be easier for the latter to attain such culture. In a somewhat different vein, Sapir continued to elaborate a critical perspective on American economic civilization during the 1930s. Thus, in a 1939 essay he criticized the notion of "economic man" – the view that "man" "is endowed with just those motivations which make the known facts of economic behavior in our society seem natural and inevitable" (Sapir 1939: 172) – that had come to prevail in academic economics in the United States. He also condemned in this essay the maldistribution of income in this country, even suggesting that it could cause mental health problems for Americans (186–193).

A few years after the publication of his essay on genuine and spurious culture, Sapir again criticized an aspect of what he believed to be the spurious culture of contemporary America. This time, however, he was not primarily concerned with a kind of cultural spuriousness that blocked individual cultural engagement and creativity, but with individual activities that were disconnected with significant and, to his way of thinking, valid cultural norms associated with traditional beliefs, such as the distinction between love and lust (a distinction stressed by his friend Harry Stack Sullivan [Handler 1986: 146]). No doubt having Margaret Mead and her popular book *Coming of Age in Samoa* in mind, Sapir condemned in his "Observations on the Sex Problem in America" (1928) the sexual experimentation that he saw as being rampant in some circles during the 1920s. Although Sapir himself was critical of the constraints that had been placed on sexual practices by American culture, he believed that some of his contemporaries had gone too far in rejecting the repressive restrictions of the past. Thus, they had come to embrace "promiscuity" and "free love," and this had been especially damaging for women (as summarized by Meyerowitz 2010: 1066). Sapir also condemned homosexuality. As he put it, "The cult of the 'naturalness' of homosexuality fools no one but those needing a rationalization of their own problems" (as quoted in ibid.). In general, for Sapir, the pursuit of sexual experience as an end in itself or as a means to self-realization, seemed narcissistic and self-defeating (Handler 1986: 144–149). Instructively, in spite of Sapir's emphasis on individual cultural participation and creativity, he believed that culture and custom exercised a force on individuals that could not, and should not, be ignored. He thus believed that individual cultural engagement and creativity could flourish by means of self-discipline, not by simply rejecting cultural formulations and norms (148). (According to Handler, Sapir was specifically thinking of artists in making such assertions [148], but, in some ways, as indicated above, these could for him be applied to people more generally.) In condemning what he considered as sexual promiscuity, especially when engaged in by women, and in condemning homosexuality, Sapir departed from the more tolerant outlook promulgated by Mead and especially by Benedict. In this sense, he was not altogether willing to adhere to the notions of cultural relativity upheld by his fellow Boasians.

In formulating his personality and culture viewpoint, Sapir was influenced by the Boasian approach to culture, especially as it was elaborated from 1920 on, as well as by Sullivan's neo-Freudian perspective and by psychiatric concepts and theories more generally; undoubtedly, the interdisciplinary initiatives of the SSRC and the Rockefeller philanthropic organizations (as well as of the NRC) also had an important intellectual impact on him by providing him with venues and networks within which to develop and express his ideas. (Instructively, key administrators/social scientists of these organizations, in particular Frank and Lynd, clearly found Sapir's approach quite congenial. See Lynd 1939; Frank 1948.) Sapir advanced his approach to personality and culture by means of a series of programmatic papers published in the 1930s; by means of his participation in conferences, seminars, and committees during the interwar years; and by teaching courses on such topics as the "psychology of culture" during these years. He did not actually engage in ethnographic fieldwork oriented toward personality and culture, and he wrote no ethnographic studies fleshing out his approach. Nevertheless, he had a large impact on the field, exerting a major influence on many of the students and junior colleagues with whom he worked; indeed, a number of these would become prominent figures within personality and culture over the years. Thus, Sapir's vision of the field significantly influenced not only Ruth Benedict and Margaret Mead, but such scholars as Weston La Barre, Irvin Child, John Whiting, Otto Klineberg, Scudder Mekeel, Irving Hallowell, and Clyde Kluckhohn (see especially Darnell 1986; Kluckhohn 1944). While Sapir's vision of personality and culture perhaps exaggerated the role of the individual – and, more especially, of the individual's personality – in relation to culture and society, it had the advantage of viewing culture as an assemblage of practices enacted by individuals in group contexts, as opposed to a reified set of rules and schemes imposed on people in an impersonal manner. It also offered useful suggestions for dealing with cultural and social change – while, at the same time, advancing a therapeutic-biopolitical approach to social engineering, one that focused on promulgating individual agency and activity within cultural settings.

Part III: Conclusion

This chapter has attempted to demonstrate how the encounter of a group of Boasian cultural anthropologists with SSRC and foundation interdisciplinary initiatives led to the elaboration and consolidation of the personality and culture approach during the interwar years. This approach came to flourish in the United States during the middle decades of the twentieth century (see, for example, Singer 1961). Indeed, it gained special prominence with the "national character" studies aimed at the enemies and allies of the United States during World War II and the Cold War; instructively, Benedict and Mead were both quite involved with these studies. While the SSRC and the foundations attempted to advance personality and culture in order to coordinate and unify the social sciences and to orient them toward social engineering, it is important to note that these organizations did not, on their own, originate the ideas and perspectives associated with personality and culture. Rather, they took up, developed, and amplified formulations initially elaborated by the Boasians and other practitioners of the human sciences. Thus, the SSRC and the foundations did not simply impose a set of ideas and outlooks on researchers and scholars; the relationships that evolved were interactive and mutually enhancing, not simply hegemonic.

To be sure, there were problematic aspects of the personality and culture perspective that emerged during the interwar years. For one thing, the personality and culture approach entailed a reconfiguration of the social as culture and, more specifically, as cultural patterns and practices. With the new emphasis on culture – an emphasis that could, according to Meyerowitz (2010: 1057), even be characterized as an "epistemic shift in social thought" – notions of society and social structure receded into the background, while ideas of culture and cultural patterns came to the fore. (A striking example of this was the manner in which, in both scientific and popular discourse, the term "culture" came to replace "society"; both scientists and laypeople thus came to talk about "cultures" rather than "societies.") Along such lines, researchers in personality and culture came to focus on the realm of the micropractices of raising children, educating them, the relation of children to other children in the school and neighborhood, family life, and the like. In other words, they came to focus on the domain of interpersonal relations, generally within face-to-face communities. Increasingly, the overarching structures and systems that more or less dominated, but extended well beyond, small-scale groups and communities seemed to recede from the purview of researchers – a problem that would become especially salient when dealing with modern complexly-organized societies. Thus, the approach elaborated by personality and culture researchers tended to focus on cultures and cultural configurations, but generally not on the overarching structures of societies.

Furthermore, it is important to note the close affiliation of personality and culture and its social engineering agenda – particularly as this agenda took the form of the biopolitics of child rearing – with the process of normalization (see, for example, Foucault 2007). To be sure, advocates of the personality and culture approach such as foundation officer Lawrence K. Frank were not primarily concerned with disciplining and normalizing the behavior and attitudes of the individual – but rather with disseminating norms of mental and physical health, optimal growth and development, cooperation, and adjustment throughout culture. The issue was thus not so much that specific individuals were sick, mentally ill, and maladjusted in various cultures – and therefore that they needed to be treated and/or disciplined – but that the cultures, considered as more or less functioning wholes, could be sick, disordered, and disintegrating, and therefore in need of reorientation and reconstruction (see especially Frank 1936). Moreover, investigators of personality and culture, such as Benedict and Mead, stressed cultural relativity and proclaimed the importance of tolerance, individual choice, and resistance to conformist pressures. Nevertheless, in spite of their embrace of such positions, proponents of and researchers in personality and culture continued to see some kinds of cultural patterns, behaviors, affects, and personalities as healthy, harmoniously integrated, well-adjusted, and normal, while others continued to be seen as unhealthy, unintegrated, maladjusted, and abnormal. Thus, even as researchers and administrators generally counseled tolerance and understanding, as advocates of the mental health and social well-being of populations, they retained the prerogative to assess and judge cultural configurations and patterns, as well as to assess and judge the forms of personality associated with these cultural patterns and environments. Such a prerogative might at times have humanly oppressive implications, in spite of the researchers' embrace of tolerance and cultural relativity.

References

APA (1929) Proceedings. First colloquium on personality investigation held under the auspices of the American Psychiatric Association Committee on the Relations with the Social Sciences; December 1–2, New York City, Baltimore

Banner LW (2003) Intertwined lives: Margaret Mead, Ruth Benedict, and their circle. Knopf, New York

Benedict R (1934a/1989) Patterns of culture. Houghton Mifflin, Boston

Benedict R (1934b) Anthropology and the abnormal. J Gen Psychol 10:59–79

Benedict R (1938) Continuities and discontinuities in cultural conditioning. Psychiatry 1:161–167

Benedict R (1946/2005) The chrysanthemum and the sword: patterns of Japanese culture. Houghton Mifflin, Boston

Benedict R (1948) Child rearing in certain European countries. In: Mead 1959:449–458

Boas F (1920) Methods of ethnology. Am Anthropol 22:311–321

Brick H (2006) Transcending capitalism. Cornell University Press, Ithaca

Bryson D (2002) Socializing the young: the role of foundations, 1923–1941. Bergin and Garvey, Westport, CT

Bryson D (2009) Personality and culture, the Social Science Research Council, and liberal social engineering: the Advisory Committee on Personality and Culture, 1930–1934. J Hist Behav Sci 45:355–386

Bryson D (2015) Mark A. May: scientific administrator, human engineer. Hist Hum Sci 28:80–114

Darnell R (1986) Personality and culture: the fate of the Sapirian alternative. In: Stocking 1986, pp 156–183

Fisher D (1993) Fundamental development of the social sciences. University of Michigan, Ann Arbor

Foucault M (2007) Security, territory, population. Trans. G. Burchell. Palgrave Macmillan, New York

Frank LK (1936) Society as the patient. Am J Sociol 42:335–344

Frank LK (1939) Forces leading to child development viewpoint and study. Lawrence K. Frank papers, National Library of Medicine, History of Medicine Division, Modern Manuscripts Collection, Ms. C 280b, box 11. Bethesda, MD

Frank LK (1948) Personality and culture: the psychocultural approach. Interstate Printers and Publishers, Danville, IL

Handler R (1986) Vigorous male and aspiring female: poetry, personality, and culture in Edward Sapir and Ruth Benedict. In: Stocking 1986, pp 127–155

Heyer Young V (2005) Ruth Benedict: beyond relativity, beyond pattern. University of Nebraska Press, Lincoln

King C (2019) Gods of the upper air. Doubleday, New York

Kluckhohn C (1944) The influence of psychiatry on anthropology in America during the past one hundred years. In: One hundred years of American psychiatry, American Psychiatric Association, Columbia University Press, New York, pp 589–617

Lynd RS (1939) Knowledge for what? Princeton University Press, Princeton

Mandelbaum DG (ed) (1949) Edward Sapir: culture, language and personality: selected essays. University of California, Berkeley

Mead M (1928/2001) Coming of age in Samoa. Perennial, New York

Mead M (1930/2001). Growing up in New Guinea. Perennial, New York

Mead M (1931) The primitive child. In: Murchison C (ed) A handbook of child psychology. Clark University Press, Worcester, pp 669–687

Mead M (ed) (1935/1963) Sex and temperament in three primitive societies. William Morrow, New York

Mead M (ed) (1937/1961). Cooperation and competition among primitive peoples. Beacon Press, Boston

Mead M (ed) (1959) An anthropologist at work: the writings of Ruth Benedict. Houghton Mifflin, Boston

Mead M (1962) Retrospects and prospects. In: Gladwin T, Sturtevant WC (eds) Anthropology and human behavior. Anthropological Society of Washington, Washington, DC, pp 115–149

Mead M (1972) Blackberry winter: my earlier years. Touchstone, New York

Mead M (1974) Ruth Benedict: a humanist in anthropology. Columbia University Press, New York

Meyerowitz J (2010) "How common culture shapes the separate lives": sexuality, race, and mid-twentieth-century social constructionist thought. J Am Hist 96:1057–1084

Molloy MA (2008) On creating a usable culture: Margaret Mead and the emergence of American cosmopolitanism. University of Hawai'i Press, Honolulu

Patterson TC (2001) A social history of anthropology in the United States. Berg, New York

Piker S (1994) Classical culture and personality. In: Bock P (ed) Handbook of psychological anthropology. Greenwood Press, Westport, CT, pp 1–17

Polanyi K (1944/2001) The great transformation. Beacon Press, Boston

Rosaldo M (1984) Toward an anthropology of self and feeling. In: Shweder R, LeVine R (eds) Culture theory: mind, self, and emotion. Cambridge University Press, New York, pp 137–157

Sapir E (1917) Do we need a "superorganic"? Am Anthropol 19:441–447

Sapir E (1924) Culture, genuine and spurious. In: Mandelbaum 1949, pp 78–119

Sapir E (1928) Observations on the sex problem in America. Am J Psychiatr 8:519–534
Sapir E (1932) Cultural anthropology and psychiatry. In: Mandelbaum 1949, pp 140–163
Sapir E (1934a) Personality. In: Mandelbaum 1949, pp 164–171
Sapir E (1934b) The emergence of the concept of personality in a study of cultures. In: Mandelbaum 1949, pp 194–207
Sapir E (1939) Psychiatric and cultural pitfalls in the business of getting a living. In: Mandelbaum 1949, pp 172–93
Singer M (1961) A survey of culture and personality theory and research. In: Kaplan B (ed) Studying personality cross-culturally. Row, Peterson and Co., New York, Evanston, IL, pp 9–90
Smith MC (1994) Social science in the crucible. Duke University Press, Durham
Social Science Reseach Council (SSRC) (1934) Dicennial report, 1923–1933. SSRC, New York
Stocking GW (1976) Introduction. In: Stocking (ed) American anthropology. 1921–1945. University of Nebraska Press, Lincoln, pp 1–74
Stocking GW (1982) Race, culture, and evolution: essays in the history of anthropology. University of Chicago, Chicago
Stocking GW (1986) Malinowski, Rivers, Benedict and others: essays on culture and personality. University of Wisconsin, Madison
Sullivan G (1999) Margaret Mead, Gregory Bateson, and Highland Bali: fieldwork photographs of Bayung Gede, 1936–1939. University of Chicago, Chicago
Sullivan G (2004a) A four-fold humanity: Margaret Mead and psychological types. J Hist Behav Sci 40:183–206
Sullivan G (2004b) Of feys and culture planners: Margaret Mead and purposive activity as value. In: Janiewski D, Banner LW (eds) Reading Benedict, reading Mead. Johns University Press, Baltimore, pp 101–114

Part XI
Gender and Health in the Social Sciences

Gender and Health in the Social Sciences: Section Introduction

69

Meagan Tyler and Natalie Jovanovski

Abstract

This chapter is a section introduction on the history of gender and health in the social sciences, highlighting significant changes to the conceptualization of both concepts for over a century. By tracing the broad social scientific histories underlying these concepts, this introductory chapter highlights various shifts from structural to agentic theories over the years. It is these shifts, we argue, that have resulted in significant debates between theorists about the best ways to understand gender and health, which the authors in this section of the handbook discuss. This introductory chapter provides an outline of some of the key arguments in the social sciences, tracing the significance of *Our Bodies, Ourselves* to the Women's Health Movement in the West (Chapter: Our Bodies, Ourselves: 50 Years of Education and Activism); the decolonization of methodologies in research with Indigenous Peoples in Canada (Chapter: Reclaiming Indigenous Health Research and Knowledges as Self-Determination in Canada); the criticisms of the social ecological approach in addressing men's violence against women (Chapter: The Limits of Public Health Approaches and Discourses of Masculinities in Violence Against Women Prevention); the psychopathologization of women through unquestioned gender norms (Chapter: The madness of women: Myth and Experience); the gendered normalization of women's food restriction practices throughout Western history (Chapter: Feminine Hunger: A Brief History of Women's Food Restriction Practices in the West); and the analysis of industries that profit from the sexual violence of women and girls (Chapter: Systems of Prostitution and Pornography: Harm,

M. Tyler (✉)
RMIT, Melbourne, VIC, Australia
e-mail: Meagan.Tyler@rmit.edu.au

N. Jovanovski
University of Melbourne, Melbourne, VIC, Australia
e-mail: jovanovskin@unimelb.edu.au

Health, and Gendered Inequalities). Taken together, it is hoped that readers continue these debates in the social sciences and continually challenge and build on existing understandings of gender and health.

Keywords

Gender · Sex roles · Health · Social sciences · Interdisciplinary

As the various sections of this handbook show, a broad umbrella of sociologists, anthropologists, psychologists, political scientists, critical race theorists, and women's and gender studies scholars (among others) have contributed to the scientific study of society and developed ways to explain how and why some social phenomena occur. An interest in gender, health – and, more recently gender and health – has developed in a wide variety of ways to advance an ever-changing understanding of issues surrounding physical, psychological, and social well-being and inequality. Social scientists have played a particularly important role in shaping discussions around gender and health; two topics have frequently overlapped in the literature and have both been informed by changes to social theory. Due to theoretical and paradigmatic shifts in the social sciences, the two concepts of "gender" and "health" have often faced significant changes and been subject to debates and tensions along the way. In this introduction we offer some key points regarding these tensions to help set the scene for the diverse chapters that follow.

The term gender – described more commonly in early literature, as "sex role" – has been a particularly malleable concept in the social sciences. Used in sociology by structural functionalist writers, "sex roles" were originally seen as a way of "structuring American society along ... sex lines" (Edwards 1983, p. 387). Sociologists, such as Talcott Parsons, spoke of the structural importance of men and women having complimentary social roles, for instance, male as "head" or breadwinner and female as homemaker (Parsons and Bales 1955), and the on the role of the nuclear family in reinforcing social cohesion. However, as Barbara Risman and Georgiann Davis (2013, p. 735) note: "Few social scientists were concerned with issues of sex and gender before the middle of the twentieth century. The field has literally exploded in the last several decades." This "explosion" coincided with the rise of social movements in North America and other parts of the West, such as the New Left, Black Power, Women's Liberation, and the gay and lesbian rights movement. At the same time, social scientific analyses started to focus on sex roles through the lens of power and domination. According to Tyler (2017), social scientific theories in the 1960s and 1970s shifted from seeing sex roles as a cohesive social force (e.g., functionalism) to a set of habitualized behaviors made to look "natural" (e.g., social constructionists), or socially reinforced forms of male domination and female subordination (e.g., feminist and conflict theoretical analyses). While there were critiques of the notion of sex roles for its rigid focus on individual socialization (rather than as a structural force), some feminist theorists tried to retain the term as an important way of explaining the hierarchical pattern of socalization that results in women's sex-based subordination (e.g., Delphy 1984).

Despite arguments about the usefulness of the term "sex roles," particularly among radical and materialist feminist theorists, a paradigmatic shift, especially apparent from the 1980s onward, resulted in what some have referred to as the "postmodern turn" in the social sciences and academia more broadly (Susen 2015). The increasing dominance of poststructural and postmodern perspectives – broadly focused on subjectivities, the decentralized nature of power, and "petite" (as opposed to grand) narratives – saw the term "sex roles" largely replaced with the concept of "gender." This reconceptualization arguably shifted the focus away from the oppressive structure of "sex roles," as they affect the sex classes, and onto an analysis that positions gender as performative and even playful (Butler 1990; West and Zimmerman 1987). There were also important challenges in the 1980s and 1990s to notions of gender that failed to account for the nature of intersecting structures of oppression, in particular around issues of racism (e.g., Collins 1990; Crenshaw 1989; Lorde 1984). These changes and challenges in understandings of "gender" have, unsurprisingly, resulted in ongoing tensions and debates between theoretical perspectives in the social sciences, some of which are reflected in the differing foci of the chapters in this section.

Similar shifts have occurred in the social scientific research on "health." Structural functionalist perspectives, such as those advanced by Parsons (1951), initially focused on health from the context of the "sick role," where the role of the physician served as a gatekeeper to the patients' "sanctioned deviance," curing them of their illness and thereby restoring social order. By the 1960s and 1970s, conflict theorists returned to the work (Marx et al. 1972) and other identity-based social movements, to analyze the power of health professionals – and health systems more broadly – on the health and well-being of marginalized groups. This social scientific shift from structural functionalism to conflict theoretical perspectives also resulted in understandings of racism in health care (e.g., Gee and Ford 2011; Williams 1996; Willie et al. 1973), the effects of capitalism on workers' health (e.g. Scambler 2009), and the way that understanding women's health is constrained by male-dominated biomedical perspectives (e.g., Boston Women's Health Book Collective 1970; Oakley 1993). The Boston Women's Health Book Collective, and their iconic feminist text *Our Bodies, Ourselves*, is one example of feminist resistance to the health system during the time (see: Agigian and Sanford 2021; this section), but other movements were also influenced by a focus on power and domination in relation to colonialism that still influences the health outcomes of marginalized groups today (see Bourassa et al. 2021; this section). While the postmodern turn in the academy in the 1980s and 1990s resulted in questioning notions of selfhood, illness, and well-being, (Hodgkin 1996; Lupton 2008) the rise of neoliberalism and market-driven reforms were changing the material realities of health systems around the world (McGregor 2008). The shift to health as an individual responsibility and illness as a potential personal failing, rather than a result of structural inequality, became ever more pervasive (Lupton 1993; Minkler 1999).

Many of the ideas presented in this section of the handbook have emerged from one or more of these themes and theoretical traditions, or been spurred by critiques of them. Indeed, as demonstrated in the chapters that follow, gender and health have frequently coalesced in the social scientific and broader public health literature – in a myriad of ways – and been used to explain social patterns and inequities in relation to health.

We introduce our section of the handbook with a chapter from one of the original members and co-founders of the iconic Boston Women's Health Collective, Wendy Sanford, discussing the groundbreaking multi-edition book *Our Bodies, Ourselves* (OBOS). In collaboration with sociologist and executive director of *OBOS Today*, Amy Agigian, Sanford traces 50 years of OBOS's successes and challenges, including their more recent work advancing intersectional understandings of women's health. The chapter begins with reflections on the grassroots origins of OBOS, where Sanford discusses the power of a small group of women sharing their frustration about male physicians who failed to see women as experts of their own experiences. The women sought to empower other women by producing a text that demystified the world of medicine and the female body, penning the first edition of OBOS in 1970. In their chapter, Agigian and Sanford argue that while OBOS was groundbreaking for centering women's experiences of their bodies, and influential to generations of public health and social scientific researchers, a more intersectional approach to women's health activism is needed in the twenty-first century. They show how, through prism of race and sexuality, OBOS has managed to continue their vital work. Perhaps one of the most important social scientific legacies that OBOS has contributed to is the understanding that health should be viewed through the lens of social justice.

Social scientific approaches to health have also used a social justice lens to challenge methodological orthodoxies derived from colonialist research practices. In their chapter on decolonizing discussions of health, Bourassa and colleagues trace the negative impact of colonialism on the health and well-being of Indigenous Peoples in Canada (such as First Nations, Inuit, and Metis Peoples). They argue that while Canada boasts a universal health care system, stark disparities continue to exist between Indigenous and non-Indigenous People in Canada. Understanding the sociocultural context of these differences requires an understanding of Eurocentric research methods that have been used to engage Indigenous Peoples. Replacing the "helicopter research" approaches used by government agencies throughout Canada – approaches that failed to engage communities in decision-making – Bourassa and colleagues discuss the importance of Indigenous research methodologies (IRM), which involve a process of self-determination whereby Indigenous Peoples have control over the data that is produced and how their data will be used to address their health issues and disparities.

Like Bourassa and colleagues, other critical public health scholars have drawn on social scientific theories to address health inequities. They have also, however, identified times where public health has inconsistently adopted social scientific concepts, with potentially detrimental consequences. Bob Pease's chapter on constructions of masculinity in public health, for example, points to a disjunct between feminist social scientific perspectives and epidemiological approaches in the prevention of men's violence against women. While feminist social scientific perspectives see gender as a system of power contributing to men's violence against women, the epidemiological approach found most frequently in public health sees masculinity as a variable among many. Pease criticizes the prevailing socio-ecological model used in public health, an approach informed by biological ecology, as the most appropriate method of understanding violence against women and argues that "masculinity" must be viewed as more than just a variable among many, or as a social determinant of health. Instead, he argues that public health must not lose sight of patriarchy as a harmful system and making masculinity visible as the system of power in discussions of the prevention of men's violence against women.

While the role of masculinity is often described as a variable in public health, the role of femininity has been a persistent undercurrent in psychological constructions of women's madness. In her chapter on gendered constructions of mental health throughout Western history, Jane Ussher exposes the myths surrounding women's madness by taking a critical look at the power of medical and psychiatric experts and their "regimes of truth." Specifically, Ussher looks the way that women have been pathologized throughout Western health discourses and judged to be driven by their "raging hormones" and other "hysterical" behaviors tied to their female reproductive anatomy. She argues that diagnoses such as hysteria and anorexia nervosa were typically found in women who were from wealthy, middle-class homes that confined them to a life of domestic servitude, and could more accurately be described as a reasonable response to women's experiences of inequality than a reflection of some internal pathology.

Similar themes involving femininity and self-harm are addressed in Natalie Jovanovski's chapter, which takes a feminist sociological look at the history of women's food restriction practices in the West. Jovanovski defines gender as a set of norms, roles, and stereotypes that order the sex classes hierarchically, positioning women in a subordinate social class to men. She argues that women's relationships with food and eating have been informed by these rigid prescriptions of femininity, and that examples of "feminine hunger" – or acceptable forms of gendered self-laceration – can be found all throughout Western history. Jovanovski shows that what may seem like disparate examples of female hunger throughout history are actually connected by a common thread of women's continued subordinate social status, or the notion that it is normal and, indeed, socially acceptable for women to go without. She ends her chapter with a call for action; one that propels the women's health movement to focus on women's right to food, eating, and nourishment.

Social and political science analyses on health have also challenged the industries responsible for reinforcing women's sexual subordination. As Meagan Tyler and Maddy Coy discuss in the last chapter of this section, it is important to consider all forms of sexual exploitation in terms of health and gendered inequalities. This is important not only because of the extremely negative psychological and physical health outcomes for women and girls in prostitution but also because of the extraordinary overrepresentation of marginalized groups (e.g., Indigenous women and girls, migrant women and girls, and women and girls in poverty). Furthermore, these issues of health and well-being need to be conceptualized in a more holistic way, through forms of structural analysis, to understand the ways in which systems of prostitution feed on inequalities of capitalism, patriarchy, white supremacy, and colonialism. This suggests that certain industries, such as the global sex trade, are both a cause and consequences of intersecting inequalities and pose a threat to well-being and status of women and girls in multifaceted ways.

Together, all these chapters show the depth and breadth of differing arguments and approaches to issues of gender, health – and gender and health – in the social sciences today, through the prism of varied key issues: from colonial oppression to mental illness, and from eating disorders to men's violence against women and prostitution. We hope that readers get a sense of the immense value to be had in continuing these debates and challenging existing understandings.

References

Boston Women's Health Collective (1970) Women and their bodies. New England Free Press
Butler J (1990) Gender trouble: feminism and the subversion of identity. Routledge, New York
Collins PH (1990) Black feminist thought. Routledge, New York
Crenshaw K (1989) Demarginalizing the intersection of race and sex: a black feminist critique of antidiscrimination doctrine, feminist theory, and antiracist politics. Univ Chic Leg Forum 1 (1989):139–167
Delphy C (1984) Close to home: a materialist analysis of women's oppression. University of Massachusetts Press, Cambridge, MA
Edwards A (1983) Sex roles: a problem for sociology and women. Australia and New Zealand Journal of Sociology 19(3):385–412
Gee G, Ford C (2011) Structural racism and health inequalities: old issues, new directions. Du Bois Review 8(1):115–132
Hodgkin P (1996) Medicine, postmodernism, and the end of certainty. Br Med J 313(7072):1568–1569
Lorde A (1984) Sister outsider: essays and speeches. Crossing Press Feminist Series, Berkeley
Lupton D (1993) Risk as moral danger: the social and political functions of risk discourse in public health. Int J Health Serv 23(3):425–435
Lupton D (2008) A postmodern public health? Aust N Z J Public Health 22(1):3–5
Marx K, Engels F, Tucker R (1972) The Marx-Engels reader. Norton, New York
McGregor S (2008) Neoliberalism and health care. Int J Consum Stud 25(2):82–89
Minkler M (1999) Personal responsibility for health? A review of the arguments and the evidence at century's end. Health Education and Behaviour 26(1):121–141
Oakley A (1993) Essays on women, medicine and health. Edinburgh University Press, Edinburgh
Parsons T (1951) The social system. The Free Press, Glencoe
Parsons T, Bales R (1955) Family, socialization, and interaction process. Free Press, Glencoe
Risman B, Davis G (2013) From sex roles to gender structure. Current Sociology Review 61 (5–6):733–755
Scambler G (2009) Capitalists, workers and health: illness as a 'side-effect' of profit-making. Soc Theory Health 7(2):119–128
Susen S (2015) The 'postmodern turn' in the social sciences. Palgrave Macmillan, New York
Tyler M (2017) Sex roles. In: Turner B, Kyung-Sup C, Epstein C, Kivisto P, Outhwaite W, Ryan M (eds) Encyclopaedia of social theory. Wiley-Blackwell, (online) ISBN: 9781118430873
West C, Zimmerman D (1987) Doing gender. Gend Soc 1(2):125–151
Williams D (1996) Racism and health: a research agenda. Ethn Dis 6(1–2):1–8
Willie C, Kramer B, Brown B (eds) (1973) Racism and mental health: essays. University of Pittsburgh Press, Pittsburgh

Our Bodies, Ourselves: 50 Years of Education and Activism

70

Amy Agigian and Wendy Sanford

Contents

Introduction: Our Bodies, Ourselves: A Founder's Perspective (Wendy Sanford)	1792
Remaking the Book, Widening Our Lens	1794
OBOS Translations and Adaptations	1795
Beyond the Book	1796
Coda	1797
Our Bodies Ourselves: A Scholar's Perspective (Amy Agigian)	1798
How *Our Bodies, Ourselves* Changed the USA and the World	1798
Conclusion: We Need Our Bodies Ourselves Today	1800
Endnotes	1801
References	1801

Abstract

Our Bodies Ourselves has been at the forefront of the women's health movement for half a century, both reflecting and shaping that movement. This chapter summarizes that history, including how its mission and distribution have evolved over time. Co-authored by a founder of Our Bodies Ourselves, and the current director of Our Bodies Ourselves Today, this chapter includes both first person narrative and critical analysis. It explores the impact of the book and the organization that wrote and updated it, in the USA and globally. Selling millions of

---

(***Our Bodies Ourselves Founders:** Ruth Davidson Bell Alexander, Pamela Berger, Ayesha Chatterjee, Vilunya Diskin, Joan Ditzion, Paula Doress-Worters, Nancy Miriam Hawley, Elizabeth MacMahon-Herrera, Pamela Morgan (1949–2008), Judy Norsigian, Jamie Penney, Jane Kates Pincus, Esther Rome (1945–1995), Wendy Sanford, Norma Swenson, Sally Whelan, Kiki Zeldes)

A. Agigian (✉)
Department of Sociology and Criminal Justice, Suffolk University, Boston, MA, USA
e-mail: aagigian@suffolk.edu

W. Sanford
Our Bodies, Ourselves, Newton, MA, USA

© The Author(s), under exclusive licence to Springer Nature Singapore Pte Ltd. 2022
D. McCallum (ed.), *The Palgrave Handbook of the History of Human Sciences*,
https://doi.org/10.1007/978-981-16-7255-2_28

copies, and translated and adapted into 33 languages, Our Bodies Ourselves is one of the most influential books of the New Left social revolutions that begin in the 1960s. It has always combined three critical strands of a new and empowering approach to women's health: (1) clear information, as scientifically accurate as possible, accessibly presented, (2) women's personal stories, and (3) political analysis of the medical care system and of sexual relationships.

Keywords

Abortion · Sexuality · Feminism · Global · Women's health movement · Social movements · Consumer movements · LGBT · Women · Consciousness raising · Health care systems · Activism · Writing and revision

Introduction: Our Bodies, Ourselves: A Founder's Perspective (Wendy Sanford)

Our book contains real material about our bodies and ourselves that isn't available elsewhere, and we have tried to present it in a new way – an honest, humane, and powerful way of thinking about ourselves and our lives. We want to share the knowledge and power that come with this way of thinking...For some of us it was the first time we had looked critically, and with strength, at the existing institutions serving us. The experience of learning just how little control we had over our lives and bodies, the coming together out of isolation to learn from each other in order to define what we needed, and the experience of supporting one another in demanding the changes that grew out of our developing critique – all were crucial and formative political experiences for us. We have felt our potential power as a force for political and social change...Finding out about our bodies and our bodies' needs, starting to take control over that area of our life, has released for us an energy that has overflowed into our work, our friendships, our relationships with men and women, for some of us our marriages and our parenthood...Learning to understand, accept, and be responsible for our physical selves, we are freed of some of these preoccupations and can start to use our untapped energies. Our image of ourselves is on a firmer base, we can be better friends and better lovers, better people, more self-confident, more autonomous, stronger and more whole. (OBOS 1973: Preface)

In May of 1969, as second-wave feminism was gaining momentum in the Boston area, several future authors of "Our Bodies Ourselves" met at an early women's liberation conference at a local women's college. We were white, mostly in our twenties, and college educated. We were also able-bodied, cis-gendered, and primarily heterosexual, though we had yet to learn to identify ourselves in these ways. Most of us were married to men; many of us had small children. In a workshop about women and health care, called "Women and Their Bodies," we shared personal stories about having babies, being sexual, and having abortions (illegal at that time). As we revealed our own experiences, we were struck by how little we knew about how our bodies worked. We called ourselves "the doctors group," and decided to create a list of doctors who were respectful of women.

Unfortunately, we soon realized that none of our own (male) physicians could be included on that list: they were condescending, sexist, and unwilling to share their knowledge. Concerned, challenged and angry, we understood we would have to do our own research on women's health, reproduction, and sexuality, and teach each other what we needed to know. We were determined to be fully informed in the doctor's office or clinic as well as with sexual partners – "between the sheets," as we wrote at the time.

We taught ourselves about birth control, natural childbirth, masturbation, orgasm, and more. We spoke openly with each other about our lives, many of us for the first time. We felt joy in this, and a sense of power and connectedness. We realized that often we knew more about ourselves than our doctors did. We came to understand that we had been blaming ourselves for problems that many women faced – postpartum depression, for example, or the frustrations of male-centered sex – problems rooted in lack of information and the social and political effects of sexism. Through our developing feminist lens and our own lived experience, broadened through consciousness raising (CR), we began to embrace a woman-centered view of ourselves and to awaken to the full implications of having been socialized in a sexist, patriarchal society. We were one of innumerable feminist groups embracing consciousness raising as a method of bringing women's experience and wisdom together to generate resistance and change.

We began to offer free "Women and their Bodies" courses to local women, weaving together three critical strands of a new and empowering approach to women's health: (1) clear information, as scientifically accurate as possible, accessibly presented, (2) women's personal stories, and (3) political analysis of the medical care system and of sexual relationships. We affirmed the women who wanted to be mothers, and those who did not; we insisted that women should have the tools to make these choices. We had no training in nursing or medicine. This was an asset enabling us to do the work of demystification of medicine, health, and medical care. We read the medical literature as ordinary women.

As word of the "Bodies" course spread, several groups in other cities contacted us, asking for our materials. We decided to put our knowledge into an accessible format that other women could use, first as single topic pamphlets, and then as a book. In 1970, we published a 193-page book on stapled newsprint, with the non-profit, socialist New England Free Press (BWHBC 1970). While we first called the book "Women and Their Bodies," we soon changed its title to "Our Bodies, Ourselves: A Course By and For Women" to emphasize women's right to take full ownership of our bodies (BWHBC 1971). We saw the book, not as an end in itself, but as a tool to help women gather, form CR groups and take charge of their lives. News of the book spread through word of mouth, radical bookstores, and ads in the Whole Earth Catalog, and it quickly became an underground success, selling some 230,000 copies.

In 1972, 12 of us met to rewrite and enlarge the book in order to include the stories of the many women who had written us in response to the newsprint *Our Bodies, Ourselves*. For this new edition, we heatedly debated the politics of turning to a profit-making, mainstream press. We decided that commercial publication would allow the book to reach an even wider readership of women. We then formally incorporated as the nonprofit Boston Women's Health Book Collective, resolving to take no personal profit from the Collective's work. After we had negotiated to make discounted copies available to health clinics and colleges that served women, and to secure funding for a Spanish-language translation of the book, Simon & Schuster published its first edition of "Our Bodies, Ourselves: A Book By & For Women" in 1973 (BWHBC 1973).

Appearing in that edition for the first time was the historic "In America they Call Us Dykes," written by "A Boston Gay Collective." Lesbian women were an unseen and oppressed minority in 1973, sometimes even within the women's movement. We felt it was crucial that the book include a chapter on lesbians. Since none of us identified as lesbian at that point, we invited a local group of lesbians to write for the book. The women insisted on full editorial control: we were to change nothing they wrote, even if it shocked us. "In America they Call Us Dykes" turned out to be an historic document of love and liberation by and for lesbians. Many graying lesbians today (present author included) remember that chapter as both electrifying and freeing us.

The next edition of "Our Bodies, Ourselves," released in 1976 (BWHBC 1976), after abortion became legal in the USA, landed on the New York Times Best Seller list, where it remained for almost 3 years. Despite several right-wing efforts to ban *Our Bodies, Ourselves* from US schools and libraries for its bold sexuality and opposition to capitalist medicine (Sanford and Doress 1981), the book continued to reach more and more women.

Remaking the Book, Widening Our Lens

The preface and first chapter ("Our Changing Sense of Self") of the 1973 edition reflect our heady excitement about breaking out of roles expected of white, college-educated women at that time, and experiencing what was, for us, a new collective sisterhood. Women of color and working-class health activists soon challenged us to look beyond our own privileged perspective, and to reach beyond our personal health and sexuality concerns: to include freedom from sterilization abuse as a reproductive right, for example, and the right of all women, regardless of income, to expert and well-funded maternal care. In time, we formed joint projects with women of color activist groups like the former National Black Women's Health Project (now the Black Women's Health Imperative), and, later, Sistersong. Amigas Latinas en Acción pro Salud (ALAS), a Latina women's health education project (later called the Latina Health Initiative), initially worked out of the Our Bodies Ourselves (OBOS) office as a sister organization and grew into a core program of OBOS. In the 1980s, several women from Boston Self Help, a disability rights organization in Boston, worked with us to create a new chapter on Body Image. These projects challenged us to increase our knowledge and understanding of interrelated issues across race, class, and physical ability, and to widen our perspectives.

Our critique of the profit-making medical care system in the USA also expanded and became more complex. We realized that doctors often did not pay attention to what research deemed to be "best" practices. As we learned more about how studies and clinical trials were conducted, we saw how inadequate or dangerous and finally even fatal the failure to observe best practices could be. We became advocates for women's right to be informed about the risks of medical treatments and procedures, including drugs, medical devices, and surgical interventions. We became early, active and vocal critics of the pharmaceutical industry, and learned from others already engaged in this struggle.

Over the course of nearly five decades, we updated, revised and re-envisioned the original "Our Bodies, Ourselves" multiple times (OBOS n.d.-a "The Nine U.S. Editions"). The book eventually sold four and a half million copies worldwide. Each new edition reflected changing medical information, new ways of understanding the social and political factors that shape women's lives, the expanding work of women's health activists from diverse countries and ethnic groups around the world, and the experiences of a wider diversity of women – including, in the twenty-first century, gender diverse, and trans women.

A particular methodology made these continuing changes possible. Each chapter was written by a small group of women's health activists – in the earliest editions, primarily founders and circulated among diverse readers, from lay women to professionals. Those of us with skills as editors worked with the resulting drafts to make sure the text was consistent, clear, jargon-free, and expressed in an affirming, accessible, woman-to-woman voice. We pursued a fundamental determination to document every statement we made in the informational components of our books, and moreover, to keep copies of this essential scientific and technical information on file in case, as lay women, we were ever challenged. The writing model we developed persisted into later editions, with an ever-widening circle of contributing authors from emerging women's health projects around the country and the world. Over the years and editions, many hundreds of people contributed to the books, bringing their diverse experiences, expertise, and skills to our work of creating accurate health content grounded in women's lives. We created other books along the way, bringing our multi-faceted lens to several life-experiences of women, from parenthood to menopause to aging (OBOS n.d.-c "Publications").

The ninth – and most likely final – edition of "Our Bodies, Ourselves" was published in 2011 (BWHBC 2011). The Library Journal named the 2011 edition one of the best consumer health books of the year. Also in 2011, Time magazine recognized "Our Bodies, Ourselves" as one of the best 100 nonfiction books (in English) since the founding of Time in 1923. In 2012, the US Library of Congress included the original "Our Bodies, Ourselves" in Books that Shaped America, an exhibit of 88 nonfiction and fiction titles "intended to spark a national conversation on books written by Americans that have influenced our lives."

OBOS Translations and Adaptations

From early on, the book's popularity triggered a wave of foreign translations. In 1973, we used funds provided through our contract with Simon and Schuster to hire

Latina women to translate the book directly into Spanish. "Nuestros Cuerpos, Nuestras Vidas (NCNV)," was published in 1977, with about 50,000 copies distributed to US clinics and health centers, and eventually to Peace Corps workers and Latin American activists (BWHBC 1977). Other early translations were published by commercial presses in Europe, but, as early as 1974, women's groups in other countries began approaching us and asking for permission not only to translate but also to adapt "Our Bodies, Ourselves" to meet the needs of women in their own cultures and communities. Later in the twentieth century and moving into the twenty-first, global groups developed resources ranging from full-length books, to pamphlets, to – in Nigeria – an HIV-AIDS sticker campaign in Yoruba on the local canoe transport system used for travel between home and market. In the Latina Health Initiative, an integral program of OBOS (first known as ALAS), women from 23 different groups and 19 countries in Latin America and the Caribbean replaced the early Spanish-language translation of *OBOS* with a culturally relevant adaptation, also called "Nuestros Cuerpos, Nuestras Vidas." Published in 2000, by Seven Stories Press, NCNV reflected the experiences and needs of Latinas in the USA and across Latin America (BWHBC 2000; Shapiro 2005). In 2001, OBOS formalized the Our Bodies Ourselves Global Initiative, which provided support to and worked closely with women's groups adapting the book (OBOS n.d.-b "Global Initiative").

By 2020, nearly 50 years after its first publication, women's groups have created resources based on "Our Bodies, Ourselves" in 33 languages (including Tibetan, Russian, and Armenian), reaching millions of women around the world (OBOS n.d.-b "Global Initiative"). The most recent project, "Corps Accord: Guide de Sexualité Positive," was completed in 2019 by a Quebecoise women's group (La CORPS Féministe 2019). A new edition came out in France in 2020 (Collectif NCNM 2020). New adaptations are in development in Brazil and Morocco.

Beyond the Book

Over these 50 years, the organization initially known as the Boston Women's Health Book Collective, and later as Our Bodies Ourselves (OBOS) has engaged in activism, advocacy, and collaboration with other social change and social justice nonprofit groups and movements across the globe (Davis 2007; Wells 2010). After the first decade, many founders stepped away to pursue careers and other book projects. A small team of founders and staff (hired to meet the demands resulting from the success of the book) worked together to build the infrastructure of the organization, carry out the day-to day-work, and enable the blockbuster book to evolve into continuing programs that would carry the work into the future. During its peak years, OBOS had 8–15 paid staff members and consultants, a thriving board of directors and advisory board, an active founders group, generous donors, and an extensive array of interns and volunteers. The original founders group expanded twice, to include six long-time staff members who played pivotal roles in the organizational and programmatic growth that made OBOS a vibrant nonprofit.

Over time, we transformed much of our office, by then in Somerville, Massachusetts, into a Women's Health Information Library and opened it to the public. Eventually this library, which documented our research for each edition of *Our Bodies, Ourselves*, contained more than 5000 books and 100,000 documents in files. We developed a Women and Health Documentation Center Network for linkage and exchange with similar women's health centers in Brazil, Malaysia, Mexico, and Chile – a thrilling project at the dawn of wide-spread computer and Internet use. Dissatisfied with the medicalized, elitist, often sexist indexing languages available for cataloging materials on women and health, we developed our own "Women and Health Thesaurus," asserting that "the way in which OBOS collects, organizes, and presents material is openly political, with a frankly articulated critical, feminist, activist perspective." Further into the Internet age, we created a health information website that reached over a half a million people each month, a widely-read blog that examined the politics of women's health, and a website exploring the complex issues of global commercial surrogacy. Throughout, we advocated actively – at federal and state levels in the USA and the press – on women's reproductive justice causes, from tampon, contraceptive, and breast-implant safety, to menstrual equity, abortion rights, commercial reproduction issues, and licensure of Certified Professional Midwives (CPMs).

In 2018, due to financial pressures, OBOS transitioned to a volunteer-driven model. Archives for the administrative, programmatic, and organizational work of OBOS are held at Harvard University's Arthur and Elizabeth Schlesinger Library on the History of Women in America (Schlesinger n.d.). The Information Center files went to Harvard University's Countway Medical Library (OBOS 1980–2000). The Boston University Medical Library took the approximately 5000 books in the OBOS library, although they did not retain all of them.

While the organization still engages in activism, provides some support to global groups, and maintains a legacy website, it is no longer able to create and maintain ever-changing health content. In response to these changes, Our Bodies Ourselves Today (OBOS Today) emerged – a dynamic sister initiative housed at Suffolk University's Center for Women's Health and Human Rights. OBOS Today is working to build a world-class online platform that will be accessible to all and worthy of the Our Bodies Ourselves name. It will provide women and girls with the most timely, trustworthy, critical, and inclusive information about our health, sexuality, and well-being. It is powered in part by feminist faculty members and Boston-area university students, many of them young women of color, as well as a broad network of feminist health and sexuality experts. With the blessing of the OBOS founders and board, and access to the organization's broad and deep networks of feminist experts, OBOS Today aims to become an intergenerational, dynamic, go-to source for women and girls, people who are trans, gender-nonconforming, and/or intersex, and all those who care about us.

Coda

It is been half a century since the early founders of Our Bodies, Ourselves first met. Many of us are now in our seventies and eighties; two have died. We are grateful for our life-long connections with each other, and determined – amidst the dangerous shift to the right in the USA and many parts of the world – to continue defending and expanding women's right to control our bodies. Finally, we marvel how even the definition of "woman" has changed since we began our work. The afterword we contributed to the 2014 resource, *Trans Bodies, Trans Selves* (Erickson-Schroth 2014), reflects some dimensions of this changing awareness: "We, a group of cisgender women, now know that we can no longer say 'a woman's body' and mean only one thing. One person's body may have a penis and testicles, and be a woman's body. Another person's body may have breasts and a clitoris, and be a man's body. The revolutionary point is that we can name our gender identity for ourselves and rightfully expect respect and recognition. 'Our bodies, ourselves' grows in meaning daily."

Our Bodies Ourselves: A Scholar's Perspective (Amy Agigian)

How *Our Bodies, Ourselves* Changed the USA and the World

Our Bodies, Ourselves gave voice to, profoundly shaped, and helped to spread and sustain the US women's health movement. In fact, historian Linda Gordon calls *Our Bodies, Ourselves* "the American left's most valuable written contribution to the world (Gordon 2008)." In continuous print for almost 50 years, hundreds of thousands of copies have been distributed to women's centers and clinics, assigned in universities and medical schools, and become part of US culture. *Our Bodies, Ourselves* empowered countless individual women with health information and feminist perspectives on their bodies, their relationships, their sexuality, their reproduction, and of course their health. It has affected the course of countless women's lives, careers, and birth experiences. It centered women's bodily autonomy and sexual pleasure, while it promoted relationship equality and kindled political activism in the women's health movement. It changed health services and policy (Black & Neuhauser 2006), as well as academic disciplines (Clawson 1998; Zola 1995: 6). It intersected with movements that transformed sexual politics in the USA, including women's liberation, reproductive rights, patient rights, and LGBTQ rights.

When *OBOS* emerged, there was nothing else like it. Prior to the women's health movement, women's bodily realities were not discussed factually in public or in print; those who tried to do so faced censure and condemnation. The names of women's body parts could not be published in the newspaper and were not spoken aloud. Ordinary people were so unfamiliar with sexual and health information that such references were typically considered obscene. No nonprofessional women wrote and published about their own personal, embodied experiences. Only "experts," i.e., doctors (more than 90% of whom were male), were able to publish about women's bodies.

One reason the book was so effective is that it embodied and was a means to spread, the Consciousness Raising method so central to the US women's liberation movement. CR was a feminist process that elevated and validated women's lived experiential knowledge and cast women as experts in our own lives. At the same time, it created solidarity and self-awareness of women as an oppressed group deserving of liberation. CR was also a method that could be, and was, embraced by groups of women all over the world (e.g., Farah 1991).

The longevity and impact of *Our Bodies, Ourselves* rests in part on its powerful use of language. The book draws in its readers with a warm and plural authorial "We." This "OBOS Style" (Wells 2010: 10) invites the reader to identify not only with the other women whose voices shape the book, but also to experience herself as a collaborator, or potential collaborator, with the authors. The book's widely collaborative authorship, sometimes engaging up to 500 writers, readers, and editors for a single edition (Bonilla 2005: 1), "brought the knowledge of experts and advocacy groups to the text; it helped the collective construct a relationship between personal experience of the body and disciplinary medical knowledge" (Wells 2010: 10).

Some scholars have argued that *Our Bodies, Ourselves'* unique, feminist, and collective language practices constructed a new conception of a woman's body. In this view, the book "is revolutionary because it is a body co-authored by readers, via their own self-explorations and because it is a personal, whole, and storied body that represents a departure from traditional, atomized, and depersonalized medical depictions" (Emmons 2012: 208). Others point to its power to transform women's understandings of our lives (e.g., Tuana 2006).

The "We" of the contributors, authors, and editors evolved over the years as OBOS became more intentional and committed to making the book inclusive in terms of race, class, sexual orientation, and many of those not initially included in the text (Bonilla 2005; Lindsey 2005). The multiple voices in all versions of *Our Bodies, Ourselves* enabled the book to absorb changes in the women's movement for nearly five decades, while retaining its core story – women's personal and bodily self-knowledge, challenges to medical and scientific expertise, and feminist solidarity could empower women and change the world in which we live.

The women of the OBOS organization participated in energized, supported, and learned much from the global women's health movement (e.g., Bogic 2018: 222). They worked with many global and country-based feminist groups on women's health issues such as sexuality, gender-based violence, human rights in childbirth, contraception, and abortion. OBOS founders participated in historic women's conferences, including those held in Mexico City (1975), Houston (1977), Copenhagen (1980), Nairobi (1985), Cairo (1994), Beijing (1995), and Guadalajara (2002), as well as scores of other national and international conferences and meetings.

Further, each of the 33 existing translation/adaptations represents a women's health group or groups that worked with OBOS to publish the material most important to their local and national contexts. The process of creating the adaptation-translations strengthened feminist network building in the countries and regions among people involved in all aspects of women's health. These national women's health movements were and are typically among the most progressive and feminist movements/groups in a given country, whether or not they worked on translations/adaptations of *Our Bodies, Ourselves*. They are often anti-war, anti-fundamentalist, anti-neoliberalism, pro-human rights, and pro-environmentalist

(Gordon 2007). These national and international women's health networks make up the global women's health movement, parts of which work to influence women's health-related policy at the UN, the WHO, and other international forums.

True to its activist values, the BWHBC made a decision early on to make sure all foreign royalties stayed in the country where they were earned, and/or to compensate women there who worked on the book, and/or to make copies available more easily and cheaply to those who could otherwise not afford it. Similarly, their contract with their publisher specified that only local women's groups could translate the book, adapt it, and/or publish books "inspired by" *Our Bodies, Ourselves*. By restricting any personal enrichment that might have come from authoring such a popular volume, the OBOS founders expressed their deep commitment to feminist social justice.

The OBOS organization and the women's health movement in general also profoundly challenged the US medical establishment (Ruzek 2007). Feminist health activists resisted paternalism, insisting instead on honesty and shared decision-making. They dramatically increased the number of women physicians. They insisted on women's right to contraception and abortion. They fought for women's inclusion in health research and an end to forced and coerced sterilization. They brought discussions of breast cancer and other previously taboo women's health issues into the open. And they won at least incremental improvements in safety on hormonal contraception and other medical concerns for women (Norsigian 2019). The OBOS organization has been credited with helping to pave the way for the creation of the US/NIH Office of Research on Women's Health, the Women's Health Department in Vienna, Austria, and other entities within the powerful institutions that shape our lives (Norsigian 2019). While such gains perpetually need to be re-fought against patriarchal and corporate interests, even partial achievement of the feminist women's health agenda points to the possibility of enduring transformations.

Conclusion: We Need Our Bodies Ourselves Today

While the work of the OBOS organization has had a tremendous impact both in the USA and globally, the problems that necessitated its creation persist. The women's health movement has faced continual backlash, whether against women's health, reproductive rights, our critiques of the profit motive in medicine, or feminist sexuality. These challenges are not going away, in the USA or internationally.

Women the world over still need accurate, easily accessible, woman-centered, culturally appropriate information about our health and well-being, including our reproductive and sexual health and rights. Women continue to struggle for access to health care that addresses our needs for sexual and reproductive rights and justice, and for a powerful voice in the policies that affect our health and our lives. Women still need a trustworthy source of feminist, reality-based health, reproduction, and sexuality information that grounds these issues both in diverse women's lived experiences and in our political contexts.

This is especially important in our current political moment. Around the world, women face the proliferation of misogynistic, anti-science, free-market and religious fundamentalisms that threaten our health and our futures. Deregulation is increasing toxic exposures and exacerbating climate chaos. In the USA, the government is attacking reproductive health and rights, obstructing access to abortion, defunding Planned Parenthood, and implementing abstinence-only education not just for the young but also for adults as a substitute for contraception. At the global level, we are seeing successful USA attempts to turn back the clock on hard-won feminist gains. One stunning example is the US threat to veto a UN resolution condemning rape as a war crime until language agreeing to provide "timely reproductive and sexual health services" to survivors of sexual violence in conflict was removed (Nichols 2019).

Our Bodies, Ourselves originally addressed the problem of lack of access to women-centered information about women's health and sexuality. Today we have a glut of information, but much of it is scattered, inaccurate, biased, exploitative, profit-driven, and with a far too narrow definition of women's health. Those who fall outside of the most privileged strata by virtue of their race, class, age, immigration status, gender identity, disability, or access to the Internet are often least able to find relevant and trustworthy information about our health.

Our Bodies Ourselves Today, an online resource based at Suffolk University in Boston, is aiming to meet this need. Building on the formidable legacy of *Our Bodies, Ourselves*, it is nourished by the experience, networks, and blessing of the OBOS founders and board.

It would be hard to overstate the transformative social impact of *Our Bodies, Ourselves* (Morgen 2002). The world has been profoundly changed by the power of the women's health movement. Both the book and the activist work of the OBOS organization remain critical touchstones of this transformation. The challenge for oncoming generations is to sustain and build on these gains, and to democratize and institutionalize them even more for generations to come.

Endnotes

Links to Preface and Chapter 1 of the 1973 edition of Our Bodies, Ourselves:

Preface: https://bit.ly/1973Preface
Chapter 1: https://tinyurl.com/obos73-chp1

References

Black N, Neuhauser D (2006) Books that have changed health services and health care policy. J Health Serv Res Policy 11(3):180–183
Bogic A (2018) Translation and feminist knowledge production: the Serbian translation of our bodies, ourselves. Alif 38:203
Bonilla ZE (2005) Including every woman: the all-embracing "we" of our bodies, ourselves. NWSA J 17(1):175–183

Boston Women's Health Book Collective (BWHBC) (1973) Our bodies ourselves: a book by and for women. Simon & Schuster, New York

Boston Women's Health Book Collective (BWHBC) (1976) Our bodies ourselves: a book by and for women, Revised and expanded. Simon & Schuster, New York

Boston Women's Health Book Collective (BWHBC) (1977) Traducido al español por Raquel Scherr-Salgado y Leonor Taboada. *Nuestros cuerpos, nuestras vidas: un libro por y para las mujeres*. Boston Women's Health Book Collective, Somerville. Translated into Spanish by Raquel Scherr-Salgado and Leonor Taboada. Our bodies, our lives: a book by and for women

Boston Women's Health Book Collective (BWHBC). (1980–2000) Subject files, 1980–2000. H MS c261. Harvard Medical Library, Francis A. Countway Library of Medicine, Boston. https://id.lib.harvard.edu/ead/med00085/catalog

Boston Women's Health Book Collective (BWHBC) (2000) Nuestros Cuerpos, Nuestras Vidas: La guia definitiva para la salud de la mujer latina. Seven Stories Press, New York

Boston Women's Health Book Collective (BWHBC) (2011) Our bodies, ourselves. Simon & Schuster, New York

Boston Women's Health Book Collective (BWHBC) (2019) *Corps accord: guide de sexualité positive*. Traduction et adaptation La CORPS Féministe. Les éditions du remue-ménage, Montréal

Boston Women's Health Collective (1970) Women and Their Bodies (first edition). Boston: New England Free Press. ASIN: B07JR2S5YJ

Boston Women's Health Collective (1971) Our Bodies, Our Selves (First Edition). Boston: New England Free Press https://www.nefp.online/our-bodies-ourselves

Clawson D (ed) (1998) Required reading: sociology's most influential books. University of Massachusetts Press, Boston

Collectif NCNM (2020) Notre corps, nous-mêmes: Nouvelle édition. Françoise Laigle, Paris

Davis K (2007) The making of our bodies, ourselves: how feminism travels across borders. Duke University Press, Durham

Emmons KK (2012) Review of *our bodies, ourselves and the work of writing*. Lit Med 30(1):207–213

Erickson-Schroth L (2014) Trans bodies, trans selves: a resource for the transgender community. Oxford University Press, Oxford

Farah N (1991) Our bodies, ourselves: the Egyptian women's health book collective. Middle East Rep 173:16–25

Gordon L (2008) Translating our bodies, ourselves: the feminist health manual's message has evolved as its impact has spread globally. The Nation, 29 May 2008. https://www.thenation.com/article/translating-our-bodies-ourselves/

Lindsey ES (2005) Reexamining gender and sexual orientation: revisioning the representation of queer and trans people in the 2005 edition of our bodies, ourselves. NWSA J 1:184

Morgen S (2002) Into our own hands: the women's health movement in the United States, 1969–1990. Rutgers University Press, New Brunswick

Nichols M (2019) Bowing to U.S. demands, U.N. waters down resolution on sexual violence in conflict. Reuters World News. 23 Apr 2019. https://www.reuters.com/article/us-war-rape-usa/bowing-to-us-demands-un-waters-down-resolution-on-sexual-violence-in-conflict-idUSKCN1RZ27T

Norsigian J (2019) Our bodies ourselves and the women's health movement in the United States: some reflections. Am J Public Health 109(6):844–846

Our Bodies Ourselves (OBOS) (n.d.-a) *Our bodies, ourselves*: the nine U.S. editions. Retrieved from https://www.ourbodiesourselves.org/publications/the-nine-u-s-editions/

Our Bodies Ourselves (OBOS) (n.d.-b) OBOS global initiative. Retrieved from https://www.ourbodiesourselves.org/global-projects/global-initiative/

Our Bodies Ourselves (OBOS) (n.d.-c) Publications. Retrieved from https://www.ourbodiesourselves.org/publications/

Ruzek S (2007) Transforming doctor-patient relationships. J Health Serv Res Policy 12(3):181–182

Sanford W, Doress P (1981) *Our bodies, ourselves* and censorship. Libr Acquisit Pract Theor 5: 133–142

Schlesinger Library on the History of Women in America (n.d.) Boston women's health book collective: research guide. Radcliffe Institute for Advanced Studies. Harvard University. https://guides.library.harvard.edu/schlesinger_bwhbc

Shapiro ER (2005) Because words are not enough: Latina re-visioning of transnational collaborations using health promotion for gender justice and social change. NWSA J 17(1):141–172

Tuana N (2006) The speculum of ignorance: the women's health movement and epistemologies of ignorance. Hypatia 3:1

Wells S (2010) Our bodies, ourselves and the work of writing. Stanford University Press, Stanford

Zola IK (1995) Shifting boundaries: doing social science in the 1990s – a personal odyssey. Sociol Forum 10(1):5–19

Reclaiming Indigenous Health Research and Knowledges As Self-Determination in Canada

71

Carrie Bourassa, Danette Starblanket, Mikayla Hagel, Marlin Legare, Miranda Keewatin, Nathan Oakes, Sebastien Lefebvre, Betty McKenna, Margaret Kîsikâw Piyêsîs, and Gail Boehme

Contents

Introduction	1806
The Canadian Context	1807
Indigenous Health Disparities and Social Determinants of Health	1809
History of Indigenous Research in Canada	1813
Closing the Gap with Indigenous Research: Building Community Capacity	1816
Reclaiming Indigenous Research and Knowledges as Self-Determination	1820
Conclusion	1826
References	1827

C. Bourassa · D. Starblanket · M. Hagel (✉) · M. Legare · N. Oakes
University of Saskatchewan, Regina, SK, Canada
e-mail: carrie.bourassa@usask.ca; danette.starblanket@usask.ca; mikayla.hagel@usask.ca; marlin.legare@usask.ca; Nathan.Oakes@usask.ca

M. Keewatin
University of Saskatchewan, Regina, SK, Canada

All Nations Hope Network, Regina, SK, Canada
e-mail: miranda.keewatin@usask.ca

S. Lefebvre
Laurentian University, Sudbury, ON, Canada
e-mail: sy_lefebvre@laurentian.ca

B. McKenna
University of Regina, Regina, SK, Canada

M. Kîsikâw Piyêsîs
All Nations Hope Network, Regina, SK, Canada
e-mail: kisikawpiyesis@allnationshope.ca

G. Boehme
File Hills Qu'Appelle Tribal Council, Fort Qu'Appelle, SK, Canada
e-mail: gboehme@fhqtc.com

© Crown 2022
D. McCallum (ed.), *The Palgrave Handbook of the History of Human Sciences*,
https://doi.org/10.1007/978-981-16-7255-2_33

Abstract

The establishment and implementation of Indigenous research teams can be key to finding solutions to closing the gaps in Indigenous health outcomes. However, research within Indigenous communities can be challenging at times because of the ways past research has been conducted in these communities as well as past traumas related to colonial history. Research by colonial institutions, in Canada and elsewhere, have left Indigenous communities with feelings of mistrust, with no obvious results from their participation in those research projects (Boyer BB, Dillard D, Woodahl EL, Whitener R, Thummel KE, Burke W. Clin Pharmacol Ther 89:343–345, 2011). Despite this, Indigenous Peoples have been able to reclaim the research process through resilience and self-determination. They have begun collaborative work with research partners in order to implement relevant research processes that work for them using Indigenous Research Methodologies. These strong partnerships built on mutual respect, reciprocity, responsibility, and relevance are key to advancing meaningful research for Indigenous communities. This chapter demonstrates how Indigenous research teams must not only partner with Indigenous communities but take direction from them. This includes the ability to administer (including hosting their own funds), direct, and guide the research project. This is a process of reclaiming research and is a strong act of self-determination. Indeed, the research community has now begun adapting to these processes including honoring Indigenous worldviews when undertaking research with Indigenous communities.

Keywords

Ceremony · Community Research Advisory Committee (CRAC) · Health Disparities · Indigenous Peoples (First Nation, Métis, Inuit) · Community-Based Research (CBR) · Indigenous research · Indigenous Research Methodologies (IRM) · Ownership, Control, Access, and Possession (OCAP) · Resilience · Respect · Relevance · Responsibility · Reciprocity · Self-determination · Traditional Knowledge

Introduction

Indigenous Peoples have utilized their Knowledge bases to function as organized societies since time immemorial. The ways in which these organized societies used and exercised their Knowledges can be beneficial to the way Traditional Knowledges are understood. In the context of this discussion, Indigenous Knowledge is seen as a tangible and unique entity that exists within a person. There is no pan-Indigenous approach to Knowledge as each community utilizes unique teachings, and as such, each Indigenous Knowledge system is viewed as an integral and valid component of Indigenous culture and research. Knowledge is a spirit that accompanies us throughout life, and therefore, the importance and sacredness of this term is recognized in this piece of literature through the capitalization and pluralization of "Knowledge(s)" when in connection with Indigenous culture. It is imperative the research community changes the way they see the relevance of these Knowledges, which can change the health outcomes of Indigenous Peoples. Reclaiming Knowledges through Indigenous Research Methodologies (IRM) and having control of the data produced in research *with* and at the direction of Indigenous communities has been key to working toward closing the gap in health issues and disparities that are experienced by Indigenous Peoples in Canada – including HIV/AIDS, dementia, aging in place, and lack of technology in at-home care – and it is the context of Canada that is focused on in this chapter.

The Canadian Context

The expansion of settler society imposed colonial systems on the Indigenous Peoples in what is now Canada (which includes First Nations, Inuit, and Métis Peoples) by the British Empire. For centuries, Indigenous Peoples residing within Canada have been subjected to racist policies that have tried to destroy and assimilate them (Bourassa et al. 2004). Since 1876 the *Indian Act* (IA) has imposed a legal definition of who is a "status Indian" in Canada. According to the Government of Canada, a status Indian is the legal status of a person who is registered as an Indian under the *Indian Act* in Canada (Government of Canada 2020). It has affected both Indigenous men and women across Canada for generations; however, it has particularly impacted women. Between 1876 and 1985, if a status Indian woman (those women who were legally status according to the government and on the *Indian Act* registry) married a non-status man (including a non-status Indian man, a Métis or an Inuit man) then she lost her Indian status and so did her children. Further, up until 1960, if a person was not a "status Indian," they were not allowed on the reserve. This was part of the government's official assimilation policy from 1876–1973. The same policy did not apply to status Indian men, who could marry non-Indian women and retain their status. In fact, their non-Indian wife would gain status under the *Indian Act* and so would their children. This sexist policy subsequently had intergenerational effects on families. The roles of Indigenous women changed greatly. Patriarchal models of governance were imposed on communities where women had great authority and respect. They often owned the lodges and made key decisions. Women were highly honored as the givers of life and well-respected in the communities. Everyone had important, yet equitable, roles to play and those roles were valued, honored, and respected by all (Kubik et al. 2009). The impacts of the *Indian Act* changed the way women's roles as Knowledge Keepers were seen within Indigenous communities and this loss of identity facilitated the effects of intergenerational trauma.

Furthermore, racist policies specifically designed to assimilate and destroy Indigenous cultures were put in place, giving rise to institutions such as Residential Schools, which left in their wake destruction and intergenerational trauma that is still felt to this day (Aguiar and Halseth 2015). Over the course of 100 years, more than 150,000 Indigenous children were placed in Residential Schools, removing them from the influence of their cultures and languages in an effort to assimilate them into the Canadian colonial system. It is estimated that more than 6,000 children died as a result of the school system (Truth and Reconciliation Committee 2015). Another policy that has disrupted the livelihoods of Indigenous people is the "Sixties Scoop" (or 60s Scoop). The "Sixties Scoop" is the action of mass removal or "scooping" of Indigenous children from their homes, families, and communities through the 1960s. This action led to subsequent adoption into predominantly non-Indigenous, middle class families across Canada and the United States. This action left many adoptees with a loss of cultural identity. The separation of the adoptees continues to affect those who have experienced this action in Indigenous communities to this day (Government of Canada 2017). The removal of children through Residential Schools and the 60s Scoop had devastating effects on families. It also played a role in further marginalizing women and changing their roles in their communities. The intergenerational effects of the loss of culture(s) and breakdown of families from the abuse suffered in Residential Schools and in foster homes for generations are being seen now in communities. Although the effects of intergenerational trauma are seen within Indigenous community today, it must be noted that Indigenous communities are resilient as there are numerous protective factors like culture and Traditional Knowledges that provide strength and healing (Bourassa et al. 2017a; Bourassa et al. 2015).

Despite the fact that Canada is one of the most advanced countries in the world, boasting a universal healthcare system, there are major social and health disparities that Indigenous Peoples face. These include lowered life-expectancy, increased rates of chronic conditions such as diabetes and heart disease, and increased youth suicide rates. These conditions might be attributed to less-than-average household incomes, barriers in accessing healthcare, loss of traditional ways of knowing, and intergenerational trauma that is rooted in the colonial policies such as Residential Schools and the 60s Scoop. To this day, only minor changes have been made at the federal level to introduce or change policies that would benefit Indigenous Peoples by improving their quality of life while increasing their access to culturally safe and accessible healthcare (Richmond and Cook 2016). In the past, Indigenous communities were encouraged to participate in research studies that boasted to provide the knowledge they needed to increase their quality of life. However, a history of research being done *on* Indigenous communities and Peoples rather than *with* them (and at their direction) has been seen in the past. This has included nutritional experiments performed on children attending Residential Schools to test the efficacy of supplements as meal replacements, which was conducted at the expense of starvation and nutritional deficiency for the children (MacDonald et al. 2014). It has also included a common type of research conducted in Indigenous communities known as "helicopter research," whereby research institutions and government agencies undertake research in communities without fully engaging or involving the community in decision making, leaving with the data and not reporting back to the community upon completion (Boyer et al. 2011; Foulks 1989). This exploitation has hurt the way communities see research and, subsequently, may be affecting health outcomes (Ferreira and Gendron 2011).

However, a wave of resilience and self-determination has spread through Indigenous communities. By taking control, Indigenous communities can decide if research is beneficial to them and dictate how the research is to be undertaken. Mutually reciprocal relationships are built with researchers, who take directives from the community in the development of the research projects, their implementation, data collection and analyses, and knowledge dissemination. Indigenous Research Methodologies (IRM) have become standard when engaging in research *with* and *at* the direction of Indigenous Peoples and communities. This has given rise to the developments of new research and data analysis methods, which are inclusive, relevant, reciprocal, responsible, and respectful. Indigenous Peoples have also become the rightful stewards of their data, deciding who can access the data, how it can be utilized, and how it will be disseminated.

Utilizing IRM has resulted in many advances in various fields of health research that include but are not limited to HIV/AIDS, dementia, and aging in place. The introduction of technology to assist in aging in place has helped to bridge the gap in the health disparities experienced by Indigenous Peoples. An increasing number of Indigenous communities are engaging in research in new ways and using new technologies that essentially serve to deconstruct the colonial impacts that have attempted to break them down. This includes keeping Indigenous older adults in the community, as these are the individuals who carry Knowledge that is so highly valued within community.

Indigenous Health Disparities and Social Determinants of Health

Health and wellness are two public health terms that are often used interchangeably but have been defined differently. Both of these terms will be used in this chapter to understand the important role that the social determinants of health play in Indigenous health disparities. The World Health Organization (WHO) defines health as "a state of complete mental, physical, and social well-being and not as only the absence of illness or disease" (WHO 2019). Comparatively, the National Wellness Institute (NWI) defines wellness as "a conscious, self-directed and evolving process of achieving full potential; a multidimensional and holistic, encompassing lifestyle, mental and spiritual well-being, and the environment; and is positive and affirming" in its nature (National Wellness Institute n.d.). An important expansion of the WHO definition of health is that the health of individuals as well as communities is influenced multifactorially by physical, economic, and social environments (WHO 2010). It is important to note that the unique contextual factors of health and wellness also contribute to the mental health of Indigenous Peoples, and are directly linked to the way they feel, think, and act. The overarching context of a person's life and how they interact with it is what determines and influences the outcome of their health (WHO 2010). When speaking in terms of community, the internal balance of an individual is not only implicated by body, mind, and spirit alone, but external factors must be considered; this includes family, home setting, community, built environment, land resources, and culture(s) (First Nations in BC Knowledge Network

2016). These definitions of health and wellness are fundamental to understanding how the contextual determinants of health are impacting Indigenous Peoples in Canada today.

Health disparities can be defined as areas of health burdens where healthy prevention strategies can be applied to achieve optimal health for socially marginalized groups (Centers for Disease Control and Prevention 2018). The social determinants of health, or extraneous influences affecting a person's accessibility to achieving holistic health, may help to explain such high rates of ill health among Indigenous Peoples living in Canada. The health disparities between Indigenous Peoples and non-Indigenous Peoples in Canada are alarming to health professionals and researchers alike, and the gap between these two groups continues to grow (Reading and Wien 2009). Indigenous Peoples in Canada face chronic health conditions, which are often compounded by one another. Common chronic health conditions include diabetes, heart disease, vascular disease, and pulmonary disease to name a few. Many of these comorbidities are related to increases in a range of cognitive issues such as dementia (Petrasek MacDonald et al. 2018, p. 13). In part, these chronic health conditions have led to a lowered life expectancy for Indigenous Peoples; one that is much lower than that of non-Indigenous Peoples (Frideres 2016; Bourassa 2011). The Centers for Disease Control and Prevention also acknowledge that health disparities promote social and health inequity and that this occurrence is the result of both historic and contemporary unequal distribution of political, economic, social, and natural resources (Centers for Disease Control and Prevention 2018). Health disparities of Indigenous Peoples in Canada are also linked to a combination of contextual social determinants of health. In order to better understand how to improve health through Indigenous research, it is important to call into question why such discrepancies still exist. Factors such as food insecurity, obesity, smoking, lower graduation rates, and insufficient housing can all be attributed to the explanation as to why Indigenous communities self-report poorer health outcomes than non-Indigenous communities (Statistics Canada 2018).

Utilizing upstream thinking, the social determinants affecting health have the possibility to not only affect diverse dimensions of health, but they also affect a person's access or availability to relieve or ameliorate such inequalities (Reading and Wien 2009). The social determinants of Indigenous health can be thought of as an environmentally inclusive timeline that can be classified into three broad categories; proximal, intermediate, and distal. Proximal determinants of health include the physical and social environment, as well as individual health behaviors and characteristics; intermediate determinants include available resources, community infrastructure, and systems performances and capacities; and distal determinants include political, social, historical, and economic entities (Reading and Wien 2009). It is important to continually examine and assess the health and well-being of a population in order to ensure equitable access to achieving health for all; this is especially true with a nation as diverse as Canada. Despite Canada's reputation as a healthy and socially progressive nation, large health discrepancies remain between Indigenous and non-Indigenous Canadians.

Proximal social determinants involve factors such as home stability, suicide, substance abuse, access to good nutrition and health, and commercial tobacco use. The Indigenous population in Canada is expanding rapidly; the overall percentage of Canada's Indigenous population in 1996 was at 2.8% increasing to 3.8% in 2006 and 4.9% in 2016 totaling today 1,673,785 individuals. Between the years of 2006 and 2016, the percentage of the Indigenous population increased by 42.5%, making this population increase four times greater than the rest of the Canadian population (Statistics Canada 2018). Indigenous children accounted for an estimated 46% of the Indigenous population under the age of 25 in 2012 (Center for Suicide Prevention 2013).

Intermediate determinants are those which include available resources, community infrastructure, and systems performances and capacities such as judicial and legal issues, access to good education, access to adequate housing, and access to health services. On-reserve housing funding policies are the responsibility of the Federal Government of Canada, which then becomes the responsibility of the First Nation communities once allocated. In Canada, First Nations are the fastest growing segment of the Canadian population while in tandem with being the largest portion of the population who experience low quality housing (Joseph 2018). Poorly structured, deteriorating, and unsanitary housing conditions are largely responsible for an increased spread of infectious and respiratory diseases, chronic illnesses, low quality nutrition, physical injuries, mental health issues, and community violence (National Collaborating Centres for Public Health 2017). It is believed that remote First Nations communities experience a lack of economic development, which includes housing, where if addressed and supported in an effective and healthy manner may ameliorate health issues linked to low socioeconomic status of community members (Reading and Wien 2009). First Nations schools operated under an outdated Band Operating Funding Formula since 1987, which were supposed to include essential technology, language immersion programs and educators, sports and recreation, and student data management systems and libraries (Assembly of First Nations 2019). According to the *Indian Control of Indian Education Policy in 1972*, the Federal Government has successively failed to provide enough funding so that educational learning environments and systems can improve the overall learning outcomes for First Nations communities (Assembly of First Nations 2019). Education, literacy, social status, and physical environments are essential social determinants of health according to the National Collaborating Center for Aboriginal Health, which are also linked to upstream social issues that plague the future such as lack of employment opportunities, low-income status, insecure housing, nutritional deficiencies, and overall reducing quality of life outcomes (Cauchie 2017).

Furthermore, non-Indigenous healthcare is funded by provincial and territorial governments outlined in the *Canada Health Act (1984)*, and still today the *Indian Act, 1876* outlines Indigenous health services and supports as the Federal Government's responsibility (Nolan et al. 2017). Post-colonial policies affects health services available on the reserve by limiting the provision of quality services, vast isolated geographic reserve locations, lack of qualified practitioners, underdeveloped infrastructure, lack of adequate funding, high rates of discrimination, lack of culturally sensitive practices, lack of effective health surveillance, and lack of self-governance (Brown et al. 2016). The fact that there is a dichotomy of funded health services between urban versus reserve communities for Indigenous Peoples across Canada may also explain the vast discrepancies in health disparities and health outcomes. Indigenous Peoples are dealing with policies and a healthcare system that functions under Western-Eurocentric views, which lacks culturally sensitive practices, and this may be a major determining factor for the willingness of Indigenous Peoples to access health services. Along with inequitable outcomes to health, Indigenous Peoples living on reserve must bear less access to physicians and other healthcare services. For example, Indigenous Peoples are more likely to have come in contact with a community nurse than with a family doctor, where the rest of the Canadian population is more likely to have come into contact with a family doctor than with a community nurse (Statistics Canada 2015). Every Canadian deserves equitable access to health services in order to ensure all have equal opportunity to access the care needed to achieve holistic health.

Distal determinants are those which include political, social, historical, and economic entities such as family destabilization and altered child rearing environments. Since first contact and colonization the holistic model for community living and family structure for Indigenous Peoples in Canada has been extremely destabilized as the relocation of families and communities into reserves interrupted healthy Traditional Knowledge systems until they were eliminated from Indigenous lifestyle (Ross 2017). Residual colonial impacts from the past have been resonating intergenerationally in the poor health of Indigenous Peoples up to today. An environment that is central to Indigenous philosophies and present to every individual as a human being is the concept of family. Family is an important concept to include as a social determinant of health because all Indigenous family health disparities are affected and perpetuated by distal, intermediate, and proximal determinants of health; the consequences of assimilation events have eradicated culture, eroded traditional values, and destabilized the structure of Indigenous families (Center for Suicide Prevention 2013). If a family becomes destabilized, a parent's communication, ethics, traditions, beliefs, and values have a stronger chance of being influenced by the negative intergenerational trauma that will then be distributed to children. This perpetuated cycle can have long-lasting effects on family cycles and child development. Today's adults are the result of yesterday's childhood experiences and these correlate to how Indigenous Peoples' health is partly the result of their experiences growing up. This is imperative to understand because many Indigenous parents today are victims of intergenerational trauma, which has completely altered the structure of traditional parenting skills as well as adapting to contemporary parenting skills (Aguiar and Halseth 2015).

The disparities in health exist on the basis of race in Canada and this has been mentioned in various reports including the Truth and Reconciliation Commission (TRC), the Royal Commission on Aboriginal Peoples (RCAP), The United Nations Declaration on the Rights of Indigenous Peoples (UNDRIP), and the National Inquiry into Missing and Murdered Indigenous Women and Girls (MMIWG).

These are not about blame or shame; they are about understanding collectively that there is a burden of ill health for Indigenous Peoples. Understanding the context of how inequitable conditions and healthcare services are more than just barriers, but also about the systemic racism found within these systems is key. Indigenous Peoples who do not have a physician or do not wish to go to the hospital because they fear stigma and racism are faced with inequitable conditions. Indigenous Peoples have pushed through with resilience and are seeking ways in which to break the barriers and reclaim their Knowledges and health. Self-directed health research is emerging in Indigenous communities as one of the means to break these barriers in reclaiming Knowledges and health.

History of Indigenous Research in Canada

The history of Indigenous research in Canada is dominated by settlers' appropriation, exploitation, and commercialization of Indigenous Peoples' Knowledges. The United Nations Educational, Scientific, and Cultural Organization defines Indigenous Peoples' Knowledge as:

> ["T]he understanding, skills, and philosophies developed by societies with long histories of interaction with their natural surroundings. For rural and Indigenous Peoples, local Knowledge informs decision-making about fundamental aspects of day-to-day life." (United Nations Educational, Scientific, and Cultural Organization 2018)

Indigenous Peoples' Knowledges encompass everything from languages, understandings, beliefs, worldviews, and values. These Knowledges are passed on from the Elders and Knowledge Keepers, who themselves received it from their Elders and Knowledge Keepers. It is transmitted in the form of storytelling, ceremonies, prayers, song, and dance (Kovach 2009). Knowledge is gathered through experiences and relationships, both human and nonhuman. It shines through in cultures and ceremonies, and is highly respected (Bruchac 2014). In this sense, it is acknowledged that Indigenous People have been gathering Knowledges for thousands of years through research, and this has contributed to the wealth of Knowledges that have been passed down and, in fact, continues to be passed down to future generations (Wilson 2008).

As European settlers began to arrive in North America, a colonial system was put in place which, in part, attempted to eradicate Indigenous Peoples and their ways of knowing. With this came colonial policies which severely restricted Indigenous Peoples and their rights to participate in any traditional activities such as ceremonies and speaking their languages. This was further perpetuated by the introduction of Residential Schools, which took 150,000 Indigenous children and placed them in institutions with the sole purpose of erasing any trace of Indigenous cultures, including the erasure of their languages – the lifeblood of their cultures. The process of assimilation of Indigenous children into the Canadian colonial system included forcing them to speak English and replacing their Traditional beliefs, cultures and languages with Christianity. During the Residential Schools system, up to 6,000 children lost their lives. Subsequently, this system has been considered cultural genocide under the United Nations guidelines (Truth and Reconciliation Committee 2015). During hearings regarding the effects of Residential Schools on Indigenous Peoples, it came to light that several research programs had been performed *on* the children in the schools. One such research program was a nutrition experiment. Between the years of 1942 and 1952, the effects of nutritional supplements on malnourished children were observed. Groups of children were denied proper nutrition and were instead given supplements in order to see if these would be sufficient to sustain health (MacDonald et al. 2014). Not only were the methods unethical, the research was performed without consent of the participants (who were underage) and with parents unaware of the situation (MacDonald et al. 2014). Such types of research were commonplace in these systems.

The Canadian government continued to use various pieces of legislation and policy to initiate and undertake questionable research (MacDonald et al. 2014). Consequently, the research community utilized similar methodologies, undertaking research that would be considered highly unethical by today's standards (MacDonald et al. 2014). In years to follow, research teams with no knowledge or connection to Indigenous Peoples and their communities advanced on communities and undertook western scientific research that did not benefit the Indigenous Peoples they were studying. This became known as "helicopter research," which refers to researchers coming in-and-out of communities without exercising any level of research ethics relative to the communities' needs and rights. Furthermore, the results of these types of studies were never shared with the communities, although most were published in scientific journals (Ferreira and Gendron 2011).

There were several high-profile research studies that are often used as prime examples of helicopter research performed in Indigenous communities, and the negative impacts they had on the communities. In 1979, the Center for Research on the Acts of Man conducted a study regarding alcohol use in the community of Barrow, Alaska, the home of the Inupiat people. The results of this study had major impacts both in the scientific world and with the public, and further perpetuated stereotypes in popular media. Dr. Foulks, a member of the research team that administered the survey to the community, wrote a retrospective analysis 10 years later admitting that while he felt their research intentions were "noble," the study encountered political and ethical dilemmas. He stated that "difficulties might have been avoided had [the researchers] obtained better insight into the community's beliefs regarding the nature of 'the problem,' and had [they] been better able to ensure more total community participation in deciding how the results of the study were to be used" (Foulks 1989). In the 1980s, a study on rheumatoid arthritis was conducted on the Nuu-Chah-Nulth First Nation in British Columbia, Canada. Upon completion of the study, samples that had been collected were shared among other researchers to conduct further studies, some genetic in nature. While the sharing of samples was not uncommon at the time, the Nuu-Chah-Nulth community saw this unconsented sharing of samples as a breach of trust and subsequently asked for the samples to be returned to them (Boyer et al. 2011). Eventually Indigenous communities grew to mistrust and reject these forms of helicopter research, including the data that came from them. It became evident that change was needed.

Until recently, research within Indigenous communities has resulted in broken relationships with non-Indigenous researchers and institutions. Part of the reason for the level of distrust that Indigenous communities feel is due to their traditional and even Sacred Knowledges being taken for granted and distorted by researchers. Such negative implications are the result of this area of research having an absence of laws to protect this information as well as absence of recognition for Indigenous community rights. Therefore, a relatively new mandate for research *with* Indigenous communities is surfacing known as OCAP®, which designates Indigenous communities the right to Ownership, Control, Access, and Possession of all Knowledge protection (FNIGC 2019). According to the highlights of the Royal Commission on Aboriginal Peoples (RCAP) Report (1996), Indigenous Peoples have historically encountered strenuous and unhealthy relationships with non-Indigenous governments and academic institution personnel (Archives Canada 2016; Government of Canada 2010). To date, academic institutions are now placing efforts to improve the quality of engagement and expanding the knowledge base of academic and non-academic research by adopting the second draft of Tri-Council Policy Statement (TCPS) on Ethical Conduct of Research Involving Human Participants: Section 6: Research Involving Aboriginal Peoples (Government of Canada 2008). The TCPS is a set of Research Guidelines prepared by the Ethics Office of the Canadian Institutes for Health Research to ensure researchers and their institutions convey culturally safe research when involving Indigenous Peoples. These are integral guidelines as they provide specific and contextual considerations to make when engaging with Indigenous communities during research.

Indigenous community representatives and researchers have been in support of creating space for ethical engagement in order to protect the Knowledges that make their way from within the community to published literature and academic reports, because as Indigenous Peoples it is imperative to protect and respect oral intergenerational tradition and Knowledges. Willie Ermine, an Indigenous Knowledge Keeper, has established his title in academia by dedicating his work and publications toward promoting ethical practices with Indigenous Peoples and promoting Indigenous thought. He also has established a title in communities by working extensively with Indigenous Knowledge Keepers throughout his career (Wîcihitowin 2019; Working Better Together 2015). According to Ermine (2007), it is imperative to follow a moral procedure of respect to receive oral traditions, as most of the teachings come from the heart of Indigenous societies and families, and thus, researchers must understand where boundaries lie so that protocols are followed to avoid dishonoring Indigenous Knowledges and Traditions (Ermine 2007). He also states that the spirit of deeply rooted Indigenous Knowledges in its existence must not be violated by the actions of external human communities, and moral considerations are key to improving transcultural engagement and knowledge translation. It is a danger for colonial systems to create the image of Indigenous Peoples because with this, the colonial systems also take advantage of using the information to manipulate the Knowledge as a form of control to suit colonial interests (Ermine 2007).

Various reports have pushed for change over the decades. In 1996 the RCAP put forward a number of recommendations related to conducting research in Indigenous communities with Indigenous Peoples. In 2013 the United Nations Special Rapporteur made a report on Canada's relationship with Indigenous Peoples. A few years later, in 2015 the Truth and Reconciliation Commission (TRC) addressed research objectives in their 94 Calls to Action. None of these, however, have been taken seriously or have been acted on by the government of Canada or much of the research community (Jewell and Mosby 2019).

In 2014 the United Nations Special Rapporteur James Anaya argued that,

> It is difficult to reconcile Canada's well-developed legal framework and general prosperity with the human rights problems faced by Indigenous [P]eoples in Canada, which have reached crisis proportions in many respects. Moreover, the relationship between the federal Government and Indigenous [P]eoples is strained, perhaps even more so than when the previous Special Rapporteur visited Canada in 2004, despite certain positive developments since then and the shared goal of improving conditions for indigenous [P]eoples. (Anaya 2014; 6)

Canada is in an era of reconciliation. The TRC has worked hard to correct some of the historic wrongs; however, the disparities between Non-Indigenous Canadians and Indigenous Peoples remain (Reading and Wien 2009).

Various reports have pushed for change over the decades. In 1996 the RCAP put forward a number of recommendations related to conducting research in Indigenous communities with Indigenous Peoples. In 2014 the United Nations Special Rapporteur, James Anaya, made a report on Canada's relationship with Indigenous Peoples and stated:

> It is difficult to reconcile Canada's well-developed legal framework and general prosperity with the human rights problems faced by Indigenous [P]eoples in Canada, which have reached crisis proportions in many respects. Moreover, the relationship between the federal Government and Indigenous [P]eoples is strained, perhaps even more so than when the previous Special Rapporteur visited Canada in 2004, despite certain positive developments since then and the shared goal of improving conditions for indigenous [P]eoples. (Anaya 2014; 6)

A year later, in 2015, the Truth and Reconciliation Commission (TRC) addressed research objectives in their 94 Calls to Action. None of these, however, have been taken seriously or have been acted on by the government of Canada or much of the research community. Canada is in an era of reconciliation. The TRC has worked hard to correct some of the historic wrongs; however, the disparities between Non-Indigenous Canadians and Indigenous Peoples remain.

Closing the Gap with Indigenous Research: Building Community Capacity

Indigenous research methodologies are evolving practices in how researchers are to approach and engage communities. They are focused on the needs and wants of the communities, ensuring direction from the communities. In other words, Indigenous research methodologies address research in a manner that allows the researcher to serve the communities in the capacity that is needed by the communities. Indigenous research emboldens Indigenous Knowledges. In an Indigenous-led research project, the research process begins with ceremony led by Elders and Knowledge Keepers. Communities often employ a research advisory committee of community members, Elders, and Knowledge Keepers as a means of involving the community in every step of the process. Ceremonies are addressed from a holistic perspective, which means considering the components of mental, emotional, physical, and spiritual aspects. The spiritual aspect should not be forgotten in the process of the work that is being done within the community (Bourassa 2018). When addressing communities, it is important to consider that ceremony can happen in many different ways. For example, a project is often started with ceremonies and offerings to ask for spiritual guidance from the ancestors to lead the project in a good way. The significance of doing things in a "good way" is demonstrated across Indigenous cultures as the expression connotes the importance of incorporating and honoring Indigenous Knowledges and spirituality in everything a person does (Flicker et al. 2015). The expression is used in Anishinaabe Teachings of the Seven Grandfathers that embody wisdom, love, respect, bravery, honesty, humility, and truth. In the context of research within Indigenous communities, research in its entirety, must be done in a good way to highlight the importance and honor the connections between the spiritual and physical world. This can be done through a sweat lodge ceremony, a pipe ceremony, and/or a sharing circle or even in a simple prayer as it cannot be forgotten that some communities are Christian and it is essential to honor all belief systems. The Elders and the communities will direct the ceremonies that need to take place. Moreover, ceremonies often occur throughout the duration of the project and particularly at the end of the project. Often this will be in the form of a feast to celebrate and includes knowledge dissemination to the community members.

With the growing interest in Indigenous research, systems and principles had to be put in place in order to protect the interest and Knowledges of Indigenous Peoples. Historically, Indigenous Knowledges were used to further the colonial system's interests, and thus, require guidelines on how to proceed with the research, and how to govern the data that was collected during the research. The ways in which Indigenous data is to be collected and disseminated must be ethically governed; as such, organizations such as the First Nations Information Governance Centre (FNIGC) put forth principles such as ownership, control, access, and possession (OCAP®) and provide relevant training around Indigenous research. OCAP® is generally utilized in all areas of genuine Indigenous research and the training is offered through and is a registered trademark of the FNIGC (FNIGC 2019; Long and Dickason 2016). It should be noted that for Inuit and Métis communities, they are now developing their own community data protocols. Some communities who have not yet developed these data protocols may adapt the OCAP® guidelines but it is important for researchers to be aware of these developments (Bourassa 2018).

Another set of principles that are important when performing research with Indigenous communities are the Four Rs: Respect, Reciprocity, Relevance, and Responsibility. The 4 Rs were developed in 1991 as a means to assist Indigenous students and academics traverse the educational climates in a culturally relevant manner. Kirkness and Barnhardt developed these tools to assist teachers and professors in decolonizing educational curricula and to improve educational outcomes for Indigenous students (2019). The intention of the 4 Rs are as follows: Respect First Nations cultural practices, include education into curricula that is Relevant to First Nations experiences and worldviews, maintain Reciprocal relationships with First Nations students and educational stakeholders, and practice Responsibility through participating in Indigenous cultural practices and actively seeking Indigenous Knowledges (Kirkness and Barnhardt 2001). The 4 Rs have since been adapted to apply to Indigenous research. In the context of research, the Four Rs apply in a similar but distinct way to Indigenous research practices and in any research capacity that involves Indigenous Peoples, Indigenous Knowledges, Indigenous lands, Indigenous communities, or any research protocol that may include Indigenous perspectives and interests. These 4 Rs of research apply to both non-Indigenous and Indigenous researchers when conducting research practices that include Indigenous Peoples (Truth and Reconciliation Canada 2015). According to Indigenous researcher Shawn Wilson, an Opaskwayak Cree academic and the author of the book *Research is Ceremony: Indigenous Research Methods*, the Respect component of the 4 Rs applies to Indigenous research through the researcher expressing humility and gratitude when approaching First Nations communities with the intent of beginning or continuing research (Wilson 2008). If a research organization is to succeed in conducting research in collaboration with Indigenous populations and developing relationships with the communities, regardless of if the organization is Indigenous-based or not, the 4 Rs of Indigenous research must be studied and set into practice. Research projects must be conducted with the 4 Rs as a priority in the actual purpose of the project's outcome. The community that is a collaborator or co-researcher of the research needs to direct the researcher in every aspect of the project from creating the initial research agreements, deciding deliverables, recruiting participants, data collection and analysis, and evaluating culturally safe research methods. It is the responsibility of every research organization and individual to practice these 4 Rs to create better research outcomes with Indigenous Peoples and to better serve the communities through purposeful and relevant research.

Respect is also one of the Seven Grandfather Teachings of the Anishinaabe and is generally regarded as the base of ethical, culturally safe research (McGregor et al. 2018). In addition to the daily practices that one may show respect to other individuals, researchers must also display respect to Indigenous research participants and co-researchers by acknowledging where their research data comes from and the dedication required to protect and cherish Indigenous Knowledges and Indigenous data (Wilson 2008). Reciprocity is the concept of maintaining mutually beneficial relationships by working with the communities and displaying Respect for the communities put into practice. In order for Indigenous research to be successful it must be reciprocal in nature in that it benefits both the researcher and the needs of the communities involved. This applies to ethical research practices such as returning the data collected and analyzed to the community and directing research with the best interests of the community at the forefront and cultural practices such as offering tobacco to Elders and sharing their own stories to research stakeholders being respected (Kovach 2009). In addition, it is important to ensure that communities have the opportunity to participate in the research process from start to finish or to the ability that they feel they wish to and that includes data analysis. It is the Responsibility of the researcher to maintain relationships with Indigenous communities, just as it is the researcher's responsibility to commit to accountability toward the research stakeholders. This includes but is not limited to a strict adherence to research agreements made with communities, protecting their data in a responsible, attentive manner, and to ensure that Indigenous research is not misused or exploited in any way, shape, or form (Wilson 2008). Finally, researchers must ensure that the research conducted is Relevant to the Indigenous communities and their needs. This not only applies to the relevancy of the research to Indigenous needs as a whole but also to the specific needs of the communities themselves. This is why it is integral to the success of Indigenous research that it is directed by community members who have a great understanding of the needs and barriers that exist within their communities. Research under the 4 Rs is not to be done just for the sake of science or for the gain of the researchers, it must be done with a purpose for the community (McGregor et al. 2018).

Research done *by* Indigenous Peoples and *with* Indigenous communities has been key in closing the gap of health disparities that is found within these communities. By taking control of the research, communities are able to dictate what areas of interest are, how the research will be conducted, and how the data will be utilized and disseminated. As mentioned, IRM is key to achieving these goals. An example of an Indigenous research methodology that is often used is Community-Based Research (CBR). CBR suits the research needs of Indigenous communities as it recognizes the community as a unit of identity; this helps the community build on strengths and resources found within the community, helps integrate reciprocal Knowledge, action, and learning, and facilitates collaborative partnerships within the communities and with agencies outside the communities (LeVeaux and Christopher 2009; Lewis and Boyd 2012). Furthermore, using CBR helps involve all partners equally in the research process from its inception and development, participant recruitment, data collection and analysis, and knowledge dissemination with purpose of using the knowledge gained and turning it into action toward improving health, quality of life, and eliminating health disparities (Minkler and Wallerstein 2003). The creation of partnerships is a very important concept of research with Indigenous communities; a level of trust must be established and fostered between the research organization and the community. A good partnership between researcher and community will have negotiated values, framework, methodologies, ownership, and dissemination in order to do research in a good way (Ball and Janyst 2008). There is strong evidence that CBR does in fact impact and improve health outcomes of Indigenous Peoples by reducing the health disparities that are evident in those communities. Firstly, CBR brings the community together, and incorporates Knowledges, etiology, and Traditional Ways of Knowing into empirical science-based methods to directly benefit the community in their own areas of interest. Secondly, being directly involved in the process does in itself improve health as community members feel empowered in taking control of their own health and the outcomes of the research that is being conducted (Buchanan et al. 2007).

When Indigenous communities develop their research project using CBR, it is important for them to use research methodologies that align with their Traditional Ways of communicating Knowledges (Kovach 2009). Firstly, by using IRM such as storytelling, Indigenous communities are able to use ways that are traditional to them to gather data. This is powerful in that it not only empowers the communities when doing research, but also uses relatable ways that helps build trust between the communities and the researcher(s). Indeed, it is not uncommon for Western research methodologies that align themselves with Indigenous worldviews to be adapted. Phenomenology as a methodology aligns itself with Indigenous worldviews in that methods such as sharing circles and one-on-one interviews allow for participants to share their living experiences with the researchers as storytelling is an important way of transmitting Knowledge with Indigenous Peoples (Wilson 2001). Furthermore, CBR allows for the research to center itself around ceremony, which is also an important part of Indigenous ways. This can include, but is not limited to, gifting tobacco in the form of tobacco ties, using traditional medicines, smudging, sweat lodges, and having Elders involved to guide both the ceremonies and the research processes (Flicker et al. 2015). As research in Indigenous communities continues to grow and move forward, researchers and communities become involved in the development of new research methods and methodologies to best suit the research needs.

Indigenous communities have begun to take ownership of their research, with a desire to learn and teach more than ever before. Researchers have started using Indigenous understandings to reintroduce historical contexts and traditional methods into contemporary society. Such examples include Bourassa et al. who looks at parenting in traditional society (Bourassa et al. 2017b) and Auger et al. who examine how self-determination comes from incorporating traditional healthcare practices back into community (Auger et al. 2016).

Reclaiming Indigenous Research and Knowledges as Self-Determination

Indigenous Peoples have embraced the concept of taking control of their research. Ensuring Indigenous Peoples are able to utilize IRM in their research projects naturally incorporates their Knowledge systems into the research. Indigenous Research and Indigenous Knowledges are two areas that are of primary importance to Indigenous Peoples and their communities. When Indigenous Peoples as a collective enact their own research agenda, it is an expression of self-determination. Researchers must learn how IRM can benefit the work they do. Below is a representation of how communities are reclaiming Indigenous research and Knowledges as expressions of self-determination.

Morning Star Lodge (MSL) is a community-based Indigenous health research lab located in Regina, Saskatchewan, Canada and operates through the University of

Saskatchewan. MSL is a health research lab and being Indigenous led, it centers around the concept of mentorship. Each individual who is associated with MSL, whether a research assistant, a student, Elder, Knowledge Keeper, a community partner, community researcher, or community member, each individual holds a unique gift and set of skills that are essential to the team. Everyone is encouraged to share their gifts and skills with the rest of the team, in that, each team member is equal and there is no hierarchy. Elders and ceremony are at the forefront of the research. The teachings that are part of the Indigenous Research Methodologies being utilized include full moon ceremonies, sweat lodge ceremonies, naming ceremonies, sunrise ceremonies, tobacco tie teachings, tobacco tie offerings, medicine teachings, plant teachings, land-based learning, feasts, and sharing circles among a wide array of other potential teachings.

MSL practices the 4 Rs in every research consideration that requires action within Indigenous communities. This includes but is not limited to data collection, participant recruitment, data analysis, and mandating community-driven research. Morning Star Lodge practices the 4 Rs toward community partners by keeping culturally safe research practices and ceremonies remain in the forefront of the organization's priority. Firstly, all research projects go through ceremony to begin the project in a good way. All MSL research projects are also directed by the members of a Community Research Advisory Committee (CRAC) to conduct research with Indigenous communities in a way that is respectful, relevant to the needs of the communities, and maintains a reciprocal relationship with the community members and the communities. Morning Star Lodge also practices the 4 Rs by taking the accountability of the research data with the level of seriousness and care that the communities' data deserves. Not only is this data well-protected but is also brought back into the communities in the form of knowledge dissemination techniques such as educational workshops, symposiums, videos, and written materials that are distributed into communities by the CRAC members. All research decisions made by MSL are made with and at the direction of the community members that the lab serves while maintaining an open and receptive channel in which community members can express the needs of their communities as well as the types of research that would benefit the pressing matters that may affect their communities. Not only does this allow for Morning Star Lodge to conduct research that suits the needs of the communities, but it is also a useful vehicle to maintain a mutually beneficial relationship with up to 11 different Indigenous communities in one Tribal Council and several communities in urban settings and across Canada.

As many of the research projects that MSL conducts with community takes place within the File Hills Qu'Appelle Tribal Council (FHQTC) in Southern Saskatchewan, it was integral that a strategy be implemented to ensure that the research directions and results remained under the sovereignty within the communities that the research takes place in. In order to address these research needs within the community and ensure that the research is community directed, the CRAC is comprised of community-members and health directors representing the 11 communities of the FHQTC and was formed in 2016 to direct Morning Star Lodge with all research aspects. The CRAC meets with Morning Star Lodge several times per year in order to direct the research team on the progression of their research projects, address the research needs of their communities, disseminate knowledge within their communities, assist Morning Star Lodge in data analysis, and consult with Morning Star Lodge on the implications of research results to their respective communities (Dieter et al. 2018).

Another organization that Morning Star Lodge implements the 4 Rs of Indigenous health research with is the All Nations Hope Network (ANHN). All Nations Hope Network is a network of Indigenous agencies, organizations, and individuals whose goal is to provide better health outcomes for Indigenous populations in Saskatchewan through the treatment, education, support, and prevention of HIV and Hepatitis C (ANHN 2019). The 4 Rs are in the forefront of the practiced health research conducted in collaboration with ANHN. As Morning Star Lodge has engaged in collaborative research projects with ANHN over the past several years, the two organizations have created a mutually beneficial and culturally relevant relationship with one another. Morning Star Lodge conducts community-based research on the discretion and direction of ANHN while ANHN collaborates with MSL by providing space and resources for the organization to conduct ceremonies together. This relationship was built by building capacity and adhering to First Nations ceremonial protocol.

As a nonprofit charitable organization, All Nations Hope Network has been in existence since 1995 in the province of Saskatchewan, Canada. The Indigenous network strives to provide support for Indigenous Peoples to ensure they are meaningfully involved in the decision-making process relating to health services (ANHN 2019). Involvement in research, outreach, networking, and training have been at the forefront of seeking solutions for what exists within the Indigenous population. As an Indigenous network, they know the importance of hunting and gathering and that this is of vital importance to *pimâtisiwin* ᐱᒫᑎᓯᐃᐧᐣ (life). Over the many years they have hunted for many ways of knowing such as: the languages, the ceremonies, the songs, the sacred items, the medicines, and the teachings. They then have gathered the Knowledge Keepers, the ceremonialists, the linguists, the medicine people, the teachers, and the many *kihtehayah* ᑭᐦᑌᐦᐊᔭᐠ (Elders). As Indigenous Peoples of the land, they know the solutions are rooted deeply in Indigenous ways of knowing. Practicing *wicehtowin* ᐃᐧᒉᐦᑐᐃᐧᐣ (The act of having a partnership; unity) has always been at the foundation of the work accomplished through the Network. Many partnerships have been established within the institutions, systems, and agencies in Saskatchewan, in Canada and internationally. They know the importance of opening all the gifts given to humanity from the Creator, gifts from all the nations that have come to this land now called Canada.

Morning Star Lodge has undertaken a number of successful research projects with the FHQTC and ANHN. These projects are related to very specific and pressing health issues and health disparities that are faced by the Indigenous communities of the southern Saskatchewan regions, including HIV/AIDS, dementia, an aging population, and water governance.

The Canadian Institutes of Health Research funded project *Digging Deep: Examining the Root Causes of HIV/AIDS Among Aboriginal Women,* which began in 2014 and focused on the life experiences of Indigenous Women who tested positive for HIV/AIDS and/or Hepatitis C. The goal was to develop evidence-based community solutions that were culturally safe. A specific focus was placed on the strengths and assets of the Indigenous women who took part in the study. Approximately 150 Indigenous women participated in this study; the ground-breaking data served to develop toolkits to assist Indigenous women in accessing care, help sustain evidence for policy changes, and gave way to a spin-off project entitled *Kotawe (start a fire): Igniting cultural responsiveness through community-determined intervention research.*

A research project funded by AGEWELL: *Indigenous Technology Needs Exploration* includes working with older adults in the FHQTC to determine their technology needs through an Indigenous Technology Needs Assessment. From this work a journal article titled *Defining Technology User Needs of Indigenous Older Adults Requiring Dementia Care* was developed in collaboration with the CRAC (Starblanket et al. 2019a). AGEWELL further supported a symposium called "Knowing Your Health" through a grant *Introducing New Technology to Monitor the Health Data of Older Adults with Multi-Morbidities Related to Dementia in Indigenous Communities*; participants were surveyed on their health status in order to create baseline data. This baseline data will be used in upcoming research where three new technologies will be introduced to monitor health status. This research also led to another collaboration with the Centre for Aging and Brain Health Innovation (CABHI). The CABHI funded research project, *Testing Locally Developed Language Apps to Reduce Caregiver Stress and Promote Aging in Place Related to Dementia in Indigenous Populations*, was entered into with the direction of CRAC who were the ones to set the research agenda for the project to evaluate the impact of technology accessibility on Indigenous individuals living with symptoms of dementia and caregiver stress using Indigenous language apps created by the FHQTC: Cree, Nakota, Lakota, Dakota, and Saulteaux.

Morning Star Lodge is involved in national and international research consortiums. One such consortium is the Canadian Consortium on Neurodegeneration in Aging (CCNA). The CCNA Team 18 looks to find ways to address dementia needs of Indigenous Peoples across Canada. Through this consortium, MSL has helped in developing a culturally safe cognitive assessment tool, translated in the Nakota language. The Canadian Indigenous Cognitive Assessment (CICA) tool is based on the Kimberley Indigenous Cognitive Assessment (KICA) tool. KICA was developed in response to the need of a cognitive screening tool for older Indigenous adults living in rural and remote areas. While the tool was originally developed in Kimberly, Australia, it has since been adapted to fit the Canadian setting. CICA is currently being validated and translated to traditional Indigenous languages to ensure that the tool is culturally relevant. This project demonstrates the need for the development of further culturally relevant and appropriate healthcare measures to ensure all have pertinent assessment tools to achieve holistic health. The CICA sheds light on the importance of connecting with community members in order to accurately translate the tool to the local, Indigenous languages, and it emphasizes the need for further tools to be developed in this field.

Another research project undertaken by Morning Star Lodge and directed by the community is the 2018 project funded by the Water Economics, Policy, and Governance Network (WEPGN) titled *Indigenous Women and Water Governance in Canada: An Examination of Traditional Roles and Current Policy*. This project followed Indigenous Research protocols to examine the perspectives of Indigenous women in the FHQTC region as to how their traditional roles affect the water governance and water resource management practices in their communities. Water is important for Indigenous Peoples as identified during this project; one of the first relationships that is formed is with water in the mother's womb. Several barriers were identified in the water governance and management for First Nations communities, including accessing clean and safe water and financial support for sanitizing and managing water systems. Because of these barriers, people living on reserves are either reluctant or simply cannot use water for simple household tasks such as cooking, bathing, and cleaning. Water must be brought in from urban centers in bottles to be used at home. Again, this project was directed solely at the discretion of the community members of the FHQTC communities. The design of the project and the intended result is about self-determination; being able to access clean water is a right that most non-Indigenous Canadians have but that Indigenous Peoples must still fight for.

As part of its mandate, Morning Star Lodge seeks to develop and innovate Indigenous research methods and methodologies (Starblanket et al. 2019b). An Indigenous-based qualitative data analysis called Nanâtawihowin Âcimowina Kika-môsahkinikêhk Papiskîci-itascikêwin Astâcikowina (NAKPA), from the Cree meaning "Medicine/Healing Stories Picked, Sorted, Stored," was adapted from an earlier method developed by Métis physician Dr. Judith Bartlett called the Collective Consensual Data Analytic Procedure (CCDAP). NAKPA calls for phenomenology data in the form of one-on-one interviews and sharing circles to be broken down into singular thought quotes from participants. Using a panel comprising of community members, experts, research team members, and community partners, each quote is divided up on a wall in manner of likeness. Once each quote has been placed in a column, the panel comes to a consensus on the theme of each column. By analyzing the data in such a manner, the community is actively involved, which aligns itself with the worldviews of Indigenous research. A panel also eliminates the bias that any one person may add to qualitative data analysis (Starblanket et al. 2019b). In order to better help the communities, the NAKPA data analysis was published in an International open access journal (Nanâtawihowin Âcimowina Kika-Môsahkinikêhk Papiskîci-Itascikêwin Astâcikowina [Medicine/Healing Stories Picked, Sorted, Stored]: Adapting the Collective Consensual Data Analytic Procedure (CCDAP) as an Indigenous Research Method, International Journal of Qualitative Methods, December 23, 2019), and a toolkit is being developed to train community members to build capacity in qualitative data analysis.

The body map method has been an important educational and research tool in which MSL has been able to innovatively decolonize the method to ensure its relevancy for Indigenous communities. In the book *How to learn the Alexander Technique*, Bill Conable, a teacher of the Alexander Technique and the originator of the term body mapping, states:

"We all seem to have in the mind maps of our bodies and their workings. They include size, shape, and mechanics. These maps are what we use to interpret our kinesthetic and visceral sensations... we also guide our movement by them." (Gehman 2018).

The term body mapping has been used in the context of occupational health and safety for almost 50 years as a model of participatory research and awareness-raising to identify three elements: a life-size body map, testimony (brief story narrated in the first person), and the key to describe each visual element of the map. Body mapping began with the "Memory Box" project dealing with HIV/AIDS in South Africa by the organization Art2Be. Body mapping originated in South Africa as an art-therapy method for women living with HIV/AIDS in 2002 (Gastaldo et al. 2012). The method evolved from the Memory Box Project designed by Jonathan Morgan, a clinical psychologist from the University of Cape Town, South Africa. The Memory Box Project was a therapeutic way for women with HIV/AIDS to record their stories and provide a keepsake for their loved ones in a handmade memory box. Most importantly, it is essential to acknowledge how important one's personal story is to a family structure and community. It is these stories that provide information about an individual, family, and community identity. Body mapping can be used as a therapy and research tool, an advocacy tool, a starting point to begin an intergenerational dialogue, and as a way of recording a person's story (Gastaldo et al. 2012). It can identify the history of an individual, the family, and community structure, the present and the future, by recognizing how social determinants affect the lifestyles of that person, community, or a nation and can provide evidence to build bridges.

The process of body mapping involves "drawing (or having drawn) one's body outline onto a large surface and using colors, pictures, symbols, and words to represent experiences lived through the body (Gastaldo et al. 2012). Some of the exercises in creating body maps include body tracing, drawing where participants come from, drawing hopes for the future, and painting support systems. Body maps have participants first outline their bodies to create highly personal self-portraits. The process includes drawing, painting, visualization exercises, group discussions, sharing, and reflection. This workshop is one way to visualize factors that make up the events of one's life. A social determinant of health framework expands the idea of health from the individual to a bigger picture. Body Mapping is a tool that is art-based and focuses on the body as a way to represent experiences lived through the body. After participating in the ceremony with the Morning Star Lodge leading Elder Betty McKenna, the Body Map methodology was renamed *Kina Keko Kawasasek* (Everything is in a circle). The *Kina Keko Kawasasek* methodology is the model of acknowledging the key concepts of Indigenous Centeredness, Indigenous Consciousness, Indigenous Capacity for Total Responsiveness, Multi-Faculty Responsiveness (Spirit, Heart, Mind, Body), Responsiveness and Connectedness to the Collective Whole, Responsiveness and Connectedness to the Total Environment, Indigenous Value-based Seeing, Relating, Knowing & Doing (Dumont 2005). According to Elder Jim Dumont's report, First Nations Regional Longitudinal Health Survey: RHS Cultural Framework he states:

> Indigenous intelligence is the wise and conscientious embodiment of exemplary knowledge and the use of this knowledge in a good, beneficial and meaningful way. Within whatever world view one is operating intelligence has to do with more than the acquisition of knowledge and the mental manipulation of thoughts and ideas; intelligence has to do with activating knowledge into something useable within a system that is charged with meaning. (Dumont 2005)

It is with this understanding that it allows researchers to move forward in the research process. When addressing the Indigenized Body Map method, one needs to consider Elder Dumont's key concepts to addressing Indigenous Knowledges is to consider Indigenous Centeredness, Indigenous Consciousness, Indigenous Capacity for Total Responsiveness, Multi-Faculty Responsiveness (Spirit, Heart, Mind, Body), Responsiveness and Connectedness to the Collective Whole, Responsiveness and Connectedness to the Total Environment, Indigenous Value-based Seeing, Relating, Knowing and Doing (Dumont 2005). It is these concepts that help guide the *Kina Keko Kawasasek* (Body Map) research by excising the total capacity of body, mind, heart, and experience.

Morning Star Lodge has been an integral part of advancing Indigenous led, community-based health research in Southern Saskatchewan, and indeed, throughout Canada and the rest of the world. By empowering communities to take the lead and direct the research that is taking place, meaningful strides have been done in closing the gap in health disparities that are felt within these communities. Furthermore, by building capacity, MSL is effectively able to work with the member First Nations communities to lower rates of unemployment, illness, and education and in doing so, Morning Star Lodge demonstrates that Indigenous research is self-determination.

Conclusion

Indigenous Peoples have been undertaking research and passing down Knowledges for millennia. However, colonial systems have oppressed Indigenous Peoples, causing intergenerational trauma (Aguiar and Halseth 2015). To compound this, Western institutions conducted years of unethical research on Indigenous Peoples (Ferreira and Gendron 2011), leaving them without any of the benefits of that research and causing further trauma. Indigenous Knowledges are exploited by researchers who have used it to their own advantage resulting in almost no benefit to Indigenous Peoples (Ball and Janyst 2008). All this has left Indigenous Peoples with mistrust of research and research protocols, and a strong reluctance to engage with institutions conducting research (Ball and Janyst 2008). All the while, a gap in health disparities between Indigenous and non-Indigenous Peoples has grown larger despite leaps in health research and technology (Reading and Wien 2009).

However, in recent years, Indigenous Peoples have taken back control of their Knowledges and ways of knowing (Auger et al. 2016). Through actions such as the TRC and the UNDRIP, Indigenous Peoples have been empowered to rediscover their Traditional Ways, and this includes conducting research that is relevant to their communities. Indigenous research has changed over the past few decades, and it continues to change based on the needs of communities. Indigenous researchers are learning from the communities they serve and changing the way research is undertaken, disseminated, and mobilized. Indigenous communities have taken control of their own research, utilizing Indigenous researchers trained in contemporary research methods, but more importantly who are ingrained in Indigenous research methodologies and practices.

The discourse, however, must move beyond deficit research into the intersectoral impact between historic trauma and colonial policies. Indigenous Peoples are far too familiar with deficit research. Researchers must acknowledge deficit-focused biases when engaging in research while actively participating in decolonizing research methodologies through incorporating culturally safe research practices, which undertake a strength-based approach. In order to examine the root causes of ill health among Indigenous Peoples it must be recognized that society is steeped in colonization, which is not ending with historical trauma, it is ongoing. Canada still has the *Indian Act* policies that impact Indigenous Peoples across Canada. Colonial policies are still impacting Indigenous Peoples (Richmond and Cook 2016), many Indigenous Peoples are trying to heal and address the historical trauma, and yet the ongoing trauma continues to be a reality for them (Aguiar and Halseth 2015).

By leading and directing the research that their communities need, Indigenous Peoples have started closing the gap in health disparities that have appeared in the last few centuries. Several community-based Indigenous health research groups have emerged throughout the country and, indeed, the world, to lead this research in a respectful, reciprocal, responsible, and relevant manner. Utilizing Indigenous worldviews, ceremony, Elders and Knowledge Keepers to guide the research, the data that is collected is impacting the health of the community in a positive way. Research in the fields of HIV/AIDS, dementia, and aging at home has helped Indigenous Peoples find solutions to emerging health issues in a manner that is culturally safe and fit their worldviews. Indeed, reclaiming Indigenous research and Knowledges as self-determination has helped to bridge the gap in the social and health disparities that Indigenous Peoples face.

References

Aguiar W, Halseth R (2015) Aboriginal peoples and historic trauma: the process of intergenerational transmission. National Collaborating Centre for Aboriginal Health, Prince George

All Nations Hope Network (2019) About us. Retrieved from https://allnationshope.ca/pages/programs

Anaya J (2014) Report of the special Rapporteur on the rights of indigenous peoples. Human Rights Council. July 4, 2014. p. 6. http://unsr.jamesanaya.org/docs/countries/2014-report-canada-a-hrc-27-52-add-2-en.pdf

Archives Canada (2016) Report of the Royal Commission on aboriginal peoples. Retrieved from: https://www.bac-lac.gc.ca/eng/discover/aboriginal-heritage/royal-commission-aboriginal-peoples/Pages/final-report.aspx

Assembly of First Nations (2019) Education policy area – first nations. Retrieved from: https://www.afn.ca/policy-sectors/education/

Auger M, Howell T, Gomes T (2016) Moving toward holistic wellness, empowerment and self-determination for indigenous peoples in Canada: can traditional Indigenous health care practices increase ownership over health and health care decisions? Can J Public Health 107(4–5):e393–e398

Ball J, Janyst P (2008) Enacting research ethics in partnerships with indigenous communities in Canada: "Do it in a good way". J Empir Res Hum Res Ethics 3(2):33–51

Bourassa C (2011) Métis health: the invisible problem. J Charlton Publishing, Kanata

Bourassa C (2018) Knowledge, ceremony, and an Indigenous approach to research, T. C. Innovation, Interviewer

Bourassa C, McKay-McNabb K, Hampton M (2004) Racism, sexism and colonialism: the impact on the health of aboriginal women in Canada. Can Woman Stud 24(1):23–29

Bourassa C, Blind M, Dietrich D, Oleson E (2015) Understanding the intergenerational effects of colonization: aboriginal women with neurological conditions–their reality and resilience. Int J Indigenous Health 10(2):3–20. http://journals.uvic.ca/index.php/ijih/issue/view/770

Bourassa C, Boehme G, Dieter J, Tickell J (2017a) Innovative pathways to dementia care in First Nations communities: our strong partnership with File Hills Qu'Appelle Tribal Council. Alzheimers Dement 13(7):P1560. https://doi.org/10.1016/j.jalz.2017.07.708

Bourassa C, McKenna EB, Juschka D (eds) (2017b) Listening to the beat of our drum: indigenous parenting in contemporary society. Demeter Press, Bradford

Boyer BB, Dillard D, Woodahl EL, Whitener R, Thummel KE, Burke W (2011) Ethical issues in developing pharmacogenetic research partnerships in American indigenous communities. Clin Pharmacol Ther 89(3):343–345

Brown A, Davy C, Harfield S, McArthur A, Munn Z (2016) Access to primary health care services for indigenous peoples: a framework synthesis. Int J Equity Health 15(1):163. https://doi.org/10.1186/s12939-016-0450-5

Bruchac MM (2014) Indigenous knowledge and traditional knowledge. In: Smith C (ed) Encyclopedia of global archaeology. Springer Science and Business Media, New York, pp 3814–3824, chapter 10

Buchanan BR, Miller FG, Wallerstein NB (2007) Ethical issues in community based participatory research: balancing rigorous research with community participation. Prog Community Health Partnersh 1(2):153–116

Cauchie L (2017) National Collaborating Centre for Aboriginal Health Home Publications. Publication Search. Retrieved from: https://www.nccah-ccnsa.ca/495/Housing_as_a_social_determinant_for_First_Nations,_Inuit,_and_Métis_health.nccah?id=20

Center for Suicide Prevention (2013) Indigenous suicide prevention. Retrieved from: https://www.suicideinfo.ca/resource/indigenous-suicide-prevention/

Centers for Disease Control and Prevention (2018) Disparities. Retrieved from: https://www.cdc.gov/healthyyouth/disparities/index.htm

Dieter J, McKim L, Tickell J, Bourassa C, Lavallee J, Boehme G (2018) The path of creating co-researchers in the File Hills Qu'Appelle Tribal Council. Int Indigenous Policy J 9(4). https://doi.org/10.18584/iipj.2018.9.4.1

Dumont J (2005) First nations Regional Longitudinal Health Survey (RHS) cultural framework. Retrieved from: https://fnigc.ca/wp-content/uploads/2020/09/rhs_cultural_framework.pdf

Ermine W (2007) The ethical space of engagement. Ind Law J 6(1). Retrieved from: https://jps.library.utoronto.ca/index.php/ilj/article/view/27669/20400

Ferreira MP, Gendron F (2011) Community-based participatory research with traditional and indigenous communities of the Americas: historical context and future directions. Int J Crit Pedagogy 3(3):153–168

First Nations in BC Knowledge Network (2016) Measuring wellness: an indicator development guide for first nations. Retrieved from: https://fnbc.info/resource/measuring-wellness-indicator-development-guide-first-nations

First Nations Information Governance Centre (2019) OCAP®. Retrieved from https://fnigc.ca/ocap

Flicker S, O'Campo P, Monchalin R, Thisle J, Worthington C, Masching R, Guta A, Pooyak S, Whitebird W, Thomas C (2015) Research done in "a good way": the importance of indigenous elder involvement in HIV community-based research. Am J Public Health 105:1149–1154

Foulks EF (1989) Misalliances in the Barrow alcohol study. Am Indian Alsk Native Ment Health Res 2(3):7–17

Frideres J (2016) First nations people in Canada. Oxford University Press, Don Mills

Gastaldo D, Magalhães L, Carrasco C, Davy C (2012) Body-map storytelling as research: methodological considerations for telling the stories of undocumented workers through body mapping. Retrieved from http://www.migrationhealth.ca/undocumented-workers-ontario/body-mapping

Gehman S (2018) Body mapping. Retrieved from: www.alexandertechnique.com/articles/bodymap/

Government of Canada (2008) Interagency advisory panel on research ethics. Retrieved from: https://ethics.gc.ca/eng/home.html

Government of Canada (2010) Highlights from the report of the royal commission on aboriginal peoples. Retrieved from: https://www.rcaanc-cirnac.gc.ca/eng/1100100014597/1572547985018

Government of Canada (2017) Sixties Scoop agreement in principle. Retrieved from: https://www.canada.ca/en/indigenous-northern-affairs/news/2017/10/sixties_scoop_agreementinprinciple.html

Jewell E, Mosby I (2019) Calls to action accountability: a status update on reconciliation. Yellow Head Institute. Retrieved from https://yellowheadinstitute.org/2019/12/17/calls-to-action-accountability-a-status-update-on-reconciliation/

Joseph B (2018) Blog- 8 things you need to know about on-reserve housing issues, February 12. Retrieved from https://www.ictinc.ca/blog/8-things-you-need-to-know-about-on-reserve-housing-issues

Kirkness VJ, Barnhardt R (2001) First Nations and higher education: the four Rs – Respect, relevance, reciprocity, responsibility. In: Hayoe R, Pan J (eds) Knowledge across cultures: a contribution to dialogue among civilizations. Comparative Education Research Centre, The University of Hong Kong, Hong Kong, pp 1–18

Kovach M (2009) Indigenous methodologies: characteristics, conversations, and contexts. University of Toronto Press, Toronto

Kubik W, Bourassa C, Hampton M (2009) Stolen sisters and second class citizens: the legacy of colonization in Canada. Humanist Sociol 33(1, 2):18–34

LeVeaux D, Christopher S (2009) Contextualizing CBPR: key principles of CBPR meet the indigenous research context. Pimatisiwin 7(1):1–16

Lewis JP, Boyd K (2012) Forward steps and missteps: what we've learned through the process of conducting CBPR research in rural Alaska. J Indigenous Res 2(1):3

Long D, Dickason O (2016) Visions of the heart, 4th edn. Oxford University Press, Don Mills

MacDonald N, Stanwick R, Lynk A (2014) Canada's shameful history of nutrition research on residential school children: the need for strong medical ethics in aboriginal health research. Paediatr Child Health 19(2):64

McGregor D, Restoule JP, Johnston R (eds) (2018) Indigenous research: theories, practices, and relationships. Canadian Scholars' Press, Toronto

Minkler M, Wallerstein N (2003) Community based participatory research in health. Jossey-Bass, San Francisco

National Collaborating Centre for Aboriginal Health (2017) Housing as a Social Determinant of First Nations, Inuit and Métis Health. National Collaborating Centre for Aboriginal Health, Prince George

National Wellness Institute (n.d.) The six dimensions of wellness. Retrieved from: https://nationalwellness.org/resources/six-dimensions-of-wellness/

Nolan M, Palmer K, Tepper J (2017) Indigenous health services often hampered by legislative confusion. Retrieved from: https://healthydebate.ca/2017/09/topic/indigenous-health

Petrasek MacDonald J, Ward W, Halseth R (2018) Alzheimer's disease and related dementias in indigenous populations in Canada: prevalence and risk factors. National Collaborating Centre for Aboriginal Health, Prince George

Reading CL, Wien F (2009) Health inequalities and the social determinants of Aboriginal peoples' health. National Collaborating Centre for Aboriginal Health, Prince George, pp 1–47

Richmond CA, Cook C (2016) Creating conditions for Canadian aboriginal health equality: the promise of healthy public policy. Public Health Rev 3(2). https://doi.org/10.1186/s40985-016-0016-5

Ross R (2017) Indigenous healing: exploring traditional paths. Langara College, Vancouver

Starblanket D, O'Connell M, Gould B, Jardine M, Bourassa C, Ferguson M (2019a) Defining technology user needs of Indigenous older adults requiring dementia care. Gerontechnology 18(3):142–155

Starblanket D, Lefebvre S, Legare M, Billan J, Akan N, Goodpipe E, Bourassa C (2019b) Nanâtawihowin Âcimowina Kika-Môsahkinikêhk Papiskîci-Itascikêwin Astâcikowina [Medicine/Healing Stories Picked, Sorted, Stored]: adapting the Collective Consensual Data Analytic Procedure (CCDAP) as an indigenous research method. Int J Qual Methods 18:1–10

Statistics Canada (2015) Health at a glance. Retrieved from https://www150.statcan.gc.ca/n1/pub/82-624-x/2013001/article/11763-eng.htm

Statistics Canada (2018) First nations people, Métis and Inuit in Canada: diverse and growing populations. Retrieved from https://www150.statcan.gc.ca/n1/pub/89-659-x/89-659-x2018001-eng.htm

Truth and Reconciliation Committee of Canada (2015) Honouring the truth, reconciling for the future: summary of the final report of the Truth and Reconciliation Commission of Canada. Winnipeg, Canada

United Nations Educational, Scientific, and Cultural Organization (2018) What does indigenous knowledge mean? A compilation of attributes, April 6. Retrieved from https://www.ictinc.ca/blog/what-does-indigenous-knowledge-mean. 15 Dec 2018

Wîcihitowin (2019) Willie Ermine. Retrieved from: https://wicihitowin.ca/speaker/willie-ermine/

Wilson S (2001) What is an indigenous research methodology? Can J Nativ Educ 25(2):177

Wilson S (2008) Research is ceremony: Indigenous research methods. Fernwood Publishing, Black Point

Working Better Together (2015) Willie J. Ermine. A conference on indigenous research ethics. Retrieved from https://indigenousresearchethics2015.wordpress.com/program/speaker-bios/willie-j-ermine/

World Health Organization (2010) The determinants of health, December 1. Retrieved from: https://www.who.int/hia/evidence/doh/en/

World Health Organization (2019) Constitution. Retrieved from: https://www.who.int/about/who-we-are/constitution

The Limits of Public Health Approaches and Discourses of Masculinities in Violence Against Women Prevention

Bob Pease

Contents

Introduction	1832
Interrogating the Ecological Model of Violence Prevention	1834
Epidemiology and Risk Factors	1838
Gender as a Social Determinant or Risk Factor	1839
Science and the Evidence Base	1840
Using Traditional Masculinity in Public Health to Engage Men	1842
Problematizing "Toxic" Masculinity in Public Health	1843
Promoting "Healthy" Masculinity in Public Health	1844
Moving Beyond Masculinity in Public Health	1846
Conclusion	1847
References	1848

Abstract

This chapter critically interrogates the public health framing of violence against women and the social ecological model of violence prevention embedded within it. It outlines how the public health–informed ecological model has assumed primacy as the main conceptual framework for understanding violence against women in the West. It is argued that the social ecological approach lacks a coherent theoretical framework, and that consequently it is unable to make meaningful connections between the various levels of analysis of violence against women. The limitations of other components of the public health approach to violence prevention are identified, including epidemiology and risk factor analysis, social determinants of health, evidence-based policy and practice, and toxic and healthy masculinities discourses.

B. Pease (✉)
Institute for Social Change, University of Tasmania, Hobart, TAS, Australia

School of Humanities and Social Sciences, Deakin University, Geelong, VIC, Australia
e-mail: bob.pease@utas.edu.au; b.pease@deakin.edu.au

© The Author(s), under exclusive licence to Springer Nature Singapore Pte Ltd. 2022
D. McCallum (ed.), *The Palgrave Handbook of the History of Human Sciences*,
https://doi.org/10.1007/978-981-16-7255-2_31

Keywords

Public health · Violence prevention · Ecological model · Toxic masculinity · Healthy masculinity · Gender · Critique of masculinity discourse · Patriarchy

Introduction

Public health approaches to violence prevention have now become the dominant paradigm in the prevention of violence against women in the West. State-based anti-violence policies and programs are now largely framed by a wider public health promotion agenda (Messner et al. 2015; Flood 2018a). While the public health frame has allowed violence against women prevention to reach larger audiences, it has also limited and depoliticized anti-violence work.

Of course, public health is not a homogenous discipline. There are critical public health perspectives (Bunton and Willis 2004; Ratelo et al. 2010; Legge 2018; Mykhalovskiy et al. 2018 that emphasize the political and economic implications of structural inequality for health. There are also feminist engagements with public health which bring a gendered lens and a gender inequality framework to the health issues facing women (Hammarstrom 1999; Rogers 2006; Stewart et al. 2010; Our Watch 2015). However, as Mykhalovskiy et al. (2018) note, critical and feminist approaches *within* public health are subjected to institutional pressures and "trade-offs" that subordinate their critical social science insights to public health epistemologies and frameworks.

The public health model of prevention is grounded in prevention science which is primarily concerned with the control and prevention of disease. Primary prevention aims to prevent a disease before it occurs, and secondary prevention aims to reduce the impact of a disease while tertiary prevention aims to help people manage health problems and illness (Institute for Work and Health 2015). While, historically, public health prevention models have focused on the prevention of infectious diseases, more recently they have been concerned with promoting health and, within that wider public health framework, they have started to address violence against women.

The premise of the public health approach is that primary prevention approaches in other health-related fields such as HIV transmission and smoking can be applied to violence prevention (Casey and Lindhorst 2009**).** The primary prevention model may be appropriate for preventing infectious diseases because it aims to address a virus as a single agent of transmission of the disease (Woody 2006). It may also be appropriate in addressing unhealthy behaviors such as smoking, substance abuse, and physical inactivity. However, violence is not an illness or a disease and it is not something that goes into remission or something that someone has. Nor, this author argues, is it useful to frame it as an unhealthy behavior as in "toxic masculinity" (see discussion later in this chapter). When the public health frame is applied to violence against women, it removes reference to disease (although the language often remains, for example, in "the epidemic of violence against women"), but it locates violence in a health behaviors discourse.

Many public health approaches have moved beyond the biomedical model and embrace a social determinants of health approach. This approach acknowledges gender and income inequality but as "social gradients" in relation to social status. The constitution of the "social" in the social determinants approach relies more upon epidemiology and the quantification of risk factors than it does on critical sociological theory (Schofield 2015). Furthermore, men's violence against women is not comparable to these other public health problems and consequently primary prevention models in addressing these public health issues are not transferrable to tackling the underlying structural and discursive causes of men's violence against women.

The prevailing violence prevention approaches within public health argue that a Western science-based model is the most effective strategy to address violence against women (Storer et al. 2016). The concept of prevention provides reassurance to the public that a particular problem can be resolved. It is premised upon the notion of being able to understand cause and effect and prediction through science. It is an empiricist project at heart that involves social engineering and scientific expertise and is usually concerned with cost efficiency in the context of limited resources and neoliberal restructuring of social services.

Within this perspective, social problems are assessed, diagnosed, and predicted by the application of scientific knowledge (Stepney 2014). The public health approach of gathering data, identifying groups to target, developing intervention methods, and measuring effectiveness is consistent with managerial policymaking and has significantly shaped how violence against women is understood by government.

Primary prevention strategies within public health have historically focused mostly on behavior change interventions. More recently they have become concerned with "unlearning" violent behaviors and challenging the norms which legitimate violence. The focus is on eliminating the culture of violence. When structural gender relations are mentioned, they are framed primarily as rigid gender roles, gender norms, and unequal access to resources and support systems (VicHealth 2007; Our Watch 2015).

Change the Story (Our Watch 2015), Australia's primary prevention of violence against women framework is informed by the "evidence base" and techniques of public health and the social-ecological model of individual behavior. Primary prevention focuses on settings such as workplaces, universities, sports, religious institutions, media, and community services. However, as Castellino (2010) identifies, a focus on changing attitudes and behavior in settings does not address the structural causes of violence against women.

The failure of the primary prevention approach to violence against women is due in part to the institutional forces and ideologically based powerful interest groups that are threatened by the prevention initiatives. As eliminating men's violence against women requires the dismantling of men's privilege and the undoing of patriarchy, it threatens men's interests in maintaining unequal gender relations. There appears to be no political will to challenge these vested interests. Often the scope of prevention is limited to social policies that ignore the broader political and economic context in which the problem (in this case men's violence against women) is situated. The ideologies of governments and government instrumentalities shape the parameters of how prevention is understood and what strategies can be enacted (Gough 2013).

Interrogating the Ecological Model of Violence Prevention

Since the 1990s, primary prevention in public health has been conceptualized within an ecological model focused upon risk factors associated with multiple layers of an individual's social environment. In 2007, the Victorian Health Promotion Foundation in Australia claimed that the ecological model of violence "reflects the consensus of international opinion around the world" (VicHealth 2007: 29).

The ecological approach has historically been the dominant approach to the prevention of disease and the health promotion in public health. It focuses on environmental and individual determinants of behavior, drawing upon Bronfenbrenner's (1977) ecological framework of human development. Utilizing a systems metaphor, it emphasizes micro-, meso-, exo-, and macrosystem influences (Richards et al. 2011).

The ecological metaphor is adapted from the discipline of biological ecology and explores the relationship of individuals to the social environment. As it utilizes a systems approach, it focuses on the principles of adaptation and interdependence (Richards et al. 2011). Within the context of biology, ecology refers to the dependence of living creatures on their environment.

The premise of the ecological model is that the relationship between individuals and their societal context can be understood by reference to the principles informing biological ecology. Bronfenbrenner's (1977) ecology of human development purports to be a scientific study of the relationship between a human organism and environment in which it survives. In Brofenbrenner's view (1977:518), "[e]nvironmental structures and the processes taking place within and between them must be viewed as interdependent and must be analysed in systems terms."

Stanger (2011:168) defines an ecosystem of being "composed of the scientific observation of interactions, organisms and environments." He notes that ecosystem is used as a metaphor in various public discourses to describe interrelationships between humans and their cultural and political context. However, this is problematical because the ecosystem concepts of equilibrium, diversity, and resilience fail to capture the dynamics between men and women under patriarchy. Stanger argues that political and economic discourses are not comparable to ecosystems in terms of their complexity. Similarly, Stojanovic et al. (2016) argue that society cannot be adequately conceptualized as a set of social-ecological systems because they are unable to address power and privilege. They depoliticize the social and accept existing social relations as "natural."

Even biological scientists and ecologists have become critical of using the ecosystemsconcept for studying the ecological environment. Some are concerned that the ecosystem concept is a machine analogy derived from systems analysis within engineering. The analogy is challenged because it is at odds with the fact that ecological systems do not always achieve equilibrium as is the case with mechanical systems. The ecosystem is not an objective scientific observation about nature. Rather it is a paradigm for developing a particular perspective on nature. While it focuses on some properties of nature, it ignores others (O'Neal 2001).

Heise (1998) suggests that it is possible to incorporate feminist and social science insights about family violence into the ecological model. She argues that many feminists are reluctant to acknowledge causes other than patriarchy for explaining men's violence against women and attributes this in part to the resistance to feminist analysis within mainstream discourses of violence. However, she also believes that the feminist emphasis on male dominance and patriarchy is unable to explain why many men are not violent toward women, even though they are subject to the same cultural messages that socialize men into male superiority and privilege. While she acknowledges that patriarchy is important to theorize men's violence against women, in her view, it does not fully explain why particular men become violent toward women.

Heise (2006) argues that violence against women results from the interaction of a multitude of factors at different levels of the social ecology. It is seen to be the result of an interplay between personal, situational, and sociocultural factors. Individual microlevel factors include having an absent or rejecting father, alcohol abuse, witnessing family violence as a child, or being abused as a child. Exosystem factors include unemployment, low socioeconomic status, and delinquent peer groups. Macrosystem factors include masculinity, rigid gender roles, male entitlement, and cultural condoning of violence.

Sliwka and Macdonald (2005), who are less feminist sympathetic than Heise, argue also that the feminist approach to violence against women does not explore multiple layers of explanation and they frame "the feminist theory" (as if there is only one feminist theory) as being a linear-type model that excludes class and race, for example. For them, the ecological model is able to encompass a wider and more diverse range of factors and acknowledges the complexity of the causes of violence against women.

The ecological framework is referred to as a "multi-level theoretical formulation." In arguing that no single factor can adequately explain violence, the World Health Organization (WHO) Report extrapolates that no single theoretical framework can analyze the multiple causal factors at different levels that impact on violence against women. Wall (2013) also argues that an ecological model enables one to consider theoretically the diversity of influences on violence against women.

The implication of these views is that the ecological model is not a theory but a neutral framework that can be used to relate different theories to empirical data. The premise is that the ecological model is an unbiased framework that is able to accommodate a diversity of theoretical perspectives (Ratelo et al. 2010). The claimed scientific neutrality that underpins the ecological model fails to recognize how critical theoretical frames are marginalized.

Quadara and Wall (2012) maintain that the ecological model enables conceptualizing the interactive nature of the multiple causal factors in relation to violence against women. However, risk factors are static representations of determinants of violence without any analysis of how they intersect. Furthermore, when Quadara and Wall list factors at the individual level (including alcohol and drug use, antisocial tendencies, childhood history of sexual abuse, and witnessing family violence), the interpersonal level (including workplaces, schools, and neighborhoods), and the societal level (including government policies and laws and societal norms and cultural belief systems), they do not ground these in the context of overall theoretical understanding. Without an overarching theoretical framework, it is not possible to make coherent connections between the various levels.

An eclectic so-called integrated framework which draws upon approaches from many different disciplines and theoretical approaches is problematic because it attempts to integrate constructs from discourses of knowledge that are incommensurable with other discourses. This means that because it is not possible to construct an integrated intervention plan, some discourses are likely to be prioritized over others. This is often most evident when structural issues that are difficult to address are marginalized in intervention programs, where more attention is focused on behavioral and attitudinal change interventions which can be more easily measured and addressed.

While radical feminist approaches emphasize the political and social structures that reproduce gender inequality and men's violence against women, they are still concerned with other levels of explanation. However, gender-neutral theories that emphasize personality disorders, psychopathology, attachment disorders, substance abuse, depression, child abuse, and history of witnessing violent behavior are in tension, if not in contradiction, with a comprehensive feminist analysis. It is not that these issues are necessarily irrelevant; rather, their theorizing needs to be gendered and framed with a nuanced understanding of patriarchy.

The premises underpinning feminist analyses at the societal level in the ecological model are at odds with the premises underpinning gender-neutral psychological approaches to personality factors at the individual level. A feminist analysis has relevance at all levels in terms of explaining patriarchal structural forces, patriarchal ideologies, men's peer support for patriarchy, men's sexist practices with women, and men's internalized dominance. However, nonfeminist psychological and systems approaches that are gender neutral suggest alternative interventions to address violence against women. This means that the ecological model does not construct a theoretically integrated conceptual framework (Pease 2014). While adherents of the ecological model talk about the complexity of the causes of violence against women, they rarely identify the main causes. Instead, they talk about determinants of violence and risk factors. While gender inequality is sometimes presented as a social determinant of violence, when it comes to prevention approaches, proponents of the ecological approach tend to focus in the individual level of intervention (Our Watch 2015). For example, Quadara and Wall (2012) state that primary prevention is predominantly concerned with changing behavior.

Some writers refer to what they call a "feminist ecological model" (Meyer and Post 2006; Flood 2008; Powell 2014). Flood (2008) argues that feminist and ecological models are compatible because the ecological framework is able to address determinants of violence against women at multiple levels. Powell (2014) similarly makes the case for what she calls an integrated feminist ecological framework. However, one of the limitations of the ecological approach is that men's dominance is regarded as only a contributing variable in relation to violence against women rather than as the central organizing framework. It is the difference between regarding gender as one variable in a model versus developing a critical gender-centered theoretical framework that elucidates how violence against women is played out in a gendered social context with other social divisions such as class, race, sexuality, age, nationality, and religion (Hunnicutt 2009).

Combining conflicting theoretical explanations under one model is incoherent. As noted earlier, some of the theoretical explanations drawn upon in the ecological framework are at odds with a feminist analysis rather than complementing it. If it is recognized that the multiple dimensions of patriarchy are the primary cause of men's violence against women, then not all theories are compatible with this analysis and consequently, not all theories can be integrated into a feminist framework. It is important to incorporate interpersonal and psychic dimensions alongside structural analyses. However, these different emphases need to be integrated into a theoretically coherent framework. When gender or gender dominance is just a variable or a risk factor amidst many others, it is likely to become sidelined.

Contemporary feminist analyses recognize the importance of interrogating the intersections between gender and other social divisions. While early feminist approaches understandably foregrounded gender inequality in their analysis of violence against women, increasingly there has been a widespread recognition of the value of gender's intersection with other sources of oppression such as class, race, national origin, sexuality, and disability. This intersectionality framework is substantially different from considering gender as just one among many risk factors.

An intersectional feminist approach differs from the ecological approach advocated by Heise (1998), the World Health Organization (Kruger et al. 2002), VicHealth (2007), and Our Watch (2015) because the feminist approach is the lens through which other theories are held together rather than as simply being one component, as in the ecological model. Consequently, feminist analysis is enriched by other perspectives rather than marginalized and deradicalized, as happens in the ecological framework. Intersectional feminism should be the basis for deciding on the inclusion or otherwise of different theories. In this way, other perspectives are utilized when they complement and enrich a feminist understanding. Not all theoretical approaches are consistent with a feminist analysis and ungendered factors need to be reconsidered through a gendered lens.

It may be possible to do feminist work within an ecological framework and that such a framework may allow discussion of patriarchy and male dominance in ways that may be more acceptable to government. However, there is nothing specifically feminist about the model per se and it has lent itself equally to anti-feminist commentators and researchers (see, for example, Dutton 1994; Dutton and Nicholls 2005; Sliwka and Macdonald 2005). Also, as argued previously, at the level of epistemology, ecological systems models are at odds with structural, discursive, and interactional feminist understandings.

It is concerning that Heise (1998), who espouses a feminist approach, and Dutton (1994, 2006), who is anti-feminist, make the same argument that the ecological model is able to move beyond the limitations of feminist theories of patriarchy because they allegedly cannot explain why only some men commit violence against women. Such arguments do not do justice to nuanced and complex theories of patriarchy that acknowledge multiple and intersecting structures and discursive and interactional dimensions of men's domination. There are, of course, multiple feminist frameworks and many of them acknowledge the intersections of gender with other social divisions. In the context of a backlash against feminism, any association of feminism with linear and monocausal analysis is likely to further marginalize feminist theories.

Heise (1998) acknowledges that the ecological framework is a heuristic tool. The ecological model is a paradigm for organizing ideas and it is premised upon various assumptions about the social world that limit our thinking and our analysis. Ecological frameworks do not identify the links between psychological and sociological dimensions across the various levels. They simply provide a model for specifying a large range of factors that are seen to influence violence toward women. Theoretical and conceptual frameworks are necessary to understand how the various levels intersect and what is needed to address these levels.

The ecological framework is also presented in the form of a logic model. Logic models involve diagrams that purport to explain causal relationships that are represented in the form of a flow chart. They are premised upon the idea that one can analyze how certain factors influence other factors. However, they are unable to grasp the complexity of causal relationships in the material world and they restrict thinking about solutions (Lee 2017).

Epidemiology and Risk Factors

While public health models of violence prevention sometimes identify social determinants of violence (Our Watch 2015), in practice, most prevention programs address individual risk factors (Broom 2016). The framing of multiple dimensions across the individual, family, community, and societal levels in an ecological framework is the integration of multiple risk and protective factors (Fulu and Heise 2015). While "macro-social processes" are sometimes mentioned as risk factors, these risk factors are decontextualized from the wider social relations in which they are embedded.

The systematic evidence review Fulu and Heise (2015) conducted using the ecological model emphasised quantitative research and measurable dimensions of violence in assessing the various risk factors influencing violence against women. Such risk factors included in the review were genetic endowment, personality profile, developmental history, communication style, and the dynamics of relationships. Genetic and biological factors sit alongside social categories of analysis such as class, gender, and race. The focus on psychological and biological factors obscures the political, economic, and social causes of inequality. Broader social formations like capitalism, neoliberalism, or patriarchy do not appear in this analysis.

Within public health, violence against women in general and sexual violence against women in particular are conceptualized as a matter of risk of future perpetration or future victimization of violence (Carmody 2009). Chung et al. (2006) note that there has been a lot of attention devoted to the development of risk assessment procedures for measuring the risk of sexual violence recidivism. There has been an increase in the language of risk and "the risk society" since the 1990s. Cowburn (2010) identifies an enormous amount of psychological literature that deals with risk management and risk assessment of sex offenders.

Risk assessment frameworks profess to be able to predict the level of risk of either being a victim or perpetrator of violence. In this view, violence against women is framed as a consequence of various risk factors at multiple levels of the society, from the community to the individual. Men are framed in this discourse not only as gendered but also as an amalgamation of risk factors stemming from childhood experiences. Considerable attention is given to the influence of childhood experiences in shaping attitudes toward women (Boyd 2009). In this view, men's violence is understood as arising from the personal history of the perpetrator rather than from the wider context of gender inequality. Gender privilege and hegemonic masculinity disappear and individual men's experiences of victimization come to the fore. Furthermore, with the increased focus on risk assessment of convicted offenders, the larger unconvicted population of men are not addressed at all. Risk models do not address why men as group commit violence on women as a group (Bacchi 1999).

Carmody (2009) has appropriately asked whether a risk focus is the most effective way of addressing violence against women. Notwithstanding the claims of evidence-based policy and practice, there is no "evidence" that risk factor frameworks are effective when applied to violence prevention. Webb (2006) argues that the notion of risk is used to legitimate neoliberal policies of governance. There are links between risk-based models of violence and neoliberal agendas of individual responsibility for crime prevention (Hoyle 2007). Interventions addressing violence against women are increasingly framed within a neoliberal policy framework. Neoliberalism constructs discourses of risk for its own purposes. Policy interventions are aimed at lessening risk rather than meeting needs (Culpitt 1999).

Understanding causes is more than understanding correlation between risk factors and violence. To understand the causes of violence to women, we need to develop theoretical explanations of how so-called risk and protective factors shape outcomes (Hawkins 2006). The ecological model does not address how various risk factors occur. How is it that political, economic, and social forces reproduce the environment in which risk factors are manifested? Models of multiple causation in the web of numerous and interconnected risk and protective factors fail to identify the theoretical underpinnings of these models.

Gender as a Social Determinant or Risk Factor

One problem with a social determinants approach to public health is that the economic, political, and social processes which shape the social determinants are not identified (Hankivosky and Christoffersen 2008). The social determinants of health model have been critiqued for failing to explicate how various forms of power are reproduced within political institutions (Navarro 2009; Schofield 2015).

The social determinants of health approach are not informed by a critical sociological analysis of how the "social" creates social and health inequalities. Rather, drawing upon epidemiology as a supposedly scientific approach, the focus is on statistical correlations between social factors and health outcomes. What is not understood is how inequalities of resources and power produce social inequalities (Schofield 2015) or how power and privilege embedded in social structures are reinforced.

In mainstream public health, gender is treated as a variable in statistical analysis rather than as an explanatory theory (Fulu and Heise 2015). This decontextualizes and depoliticizes gender which is often conflated with sex as a social determinant or risk factor (Inhorn and Whittle 2001; Thurston and Vissandjee 2014). The use of gender in public health approaches to violence prevention is primarily to do with the gendered differences between men and women. Power disparities between men and women are often acknowledged as a macro-social factor. However, there is no recognition of the wider form of male dominance and male supremacy within which these gender inequalities are situated.

The public health approach to gender sits within what Connell (2012) refers to as "categoricalism," a view of gender as dichotomous categories of male and female and man and woman. This categorical approach to gender in public health framings is unable to understand gender as a process of enactment and consequently cannot conceptualize how gender inequalities are created and sustained. Numerous commentators have observed that public health frameworks do not have sufficiently nuanced understanding of sex and gender to connect the macrolevel of social structure to the microlevel of individual practices (Hankivosky and Christoffersen (2008); Austerberry 2011). The key theoretical developments in feminist theory and critical masculinity studies have generally not been adopted in public health approaches to violence against women.

Science and the Evidence Base

Evidence-based approaches to health issues are now central in public health. There is an increasing emphasis in public health promotion on building the evidence base, whereby the aim is to develop a scientific foundation for prevention (Broom 2016).

Consequently, the evidence base is an important part of public health framings of violence against women. This is premised upon an empiricist approach to knowledge that emphasizes experimental knowledge and randomized controlled trials. In the UK, this approach is evident in the "What works to Prevent Violence" program which assumes that violence can be prevented through the technical application of scientific knowledge. Fulu et al., for example, state that evidence demonstrates that violence is caused by the interaction of multiple factors at different levels of the "social ecology," as if the social ecology was an actual social formation that can be scientifically observed rather than a socially constructed metaphor, as discussed earlier. After reviewing a multitude of "risk factors," including those that focus on genetic endowment, personality profile, and developmental history of individual perpetrators, readers are told the quality of the evidence gathered is best if there is at least one randomized controlled trial.

In this frame, violence against women is understood as a phenomenon that is stable and able to be objectively measured. As noted earlier, however, violence is not a disease or a virus that can be measured. Putting aside critiques of evidence-based medicine (Gupta 2003; Griffiths 2005) from which evidence-based policy and practices are derived, what may work in relation to physical diseases will not necessarily work in relation to social phenomena such as violence against women.

Population health is located within epidemiology and so-called objective cause and effect relations (Raphael and Bryant 2002). With a focus on management and actuarial measurement, risk assessment policies rely upon rational models of policy development that purport to be able to objectively define and measure the level of risk among a given population (Webb 2006). Risk is portrayed as an objective entity that can be measured and calculated to determine probability (Cowbuurn 2010a, b). Hall (2004:5) frames this as the belief that "information will save you." If only we have enough information, then we can eliminate the problem. Beck (2007) argues that the prediction of risk cannot be subject to scientific assessment. Evidence-based policy frameworks do not provide the best basis to produce knowledge that is concerned with social justice and social change (Pease 2009).

Some feminists have endeavored to develop a feminist epidemiology that has embraced a form of feminist empiricism (Inhorn and Whittle 2001), which Harding (1986) noted over 30 years ago was an attempt to acknowledge gender within a positivist paradigm without challenging the epistemological basis of that paradigm. So while public health approaches may well be integrated with some forms of feminism, they are at odds with forms of feminism which challenge the epistemology of positivism.

The dominant perspective on public health approaches to violence prevention is that science is a neutral, nonideological mechanism for investigating social phenomena. Can positivist research paradigms, within which most public health promotion models are situated, address the complexity of the phenomena of violence against women? Evidence-based practice relies heavily upon experimental forms of knowledge which are located within a positivist scientific paradigm that emphasizes experimental research designs. It gives less attention to interactive and local knowledge derived from lived experience and critical knowledge that is concerned with the influence of social structures and power relations (Raphael and Bryant 2002).

Concepts of evidence and objectivity can be used for emancipatory projects and to demonstrate the prevalence of a social problem. However, the "what works?" mantra is premised upon a particular set of assumptions about the nature of knowledge. Such a project assumes the superiority of experimental knowledge above that of interactive knowledge from professionals and critical knowledge derived from sociopolitical analysis (Clegg 2005). Furthermore, when there is a "strong evidence base" for a particular intervention, it is neglected if it contradicts dominant ideas and challenges vested interests.

Systematic reviews are a tool of evidence-based practice. The process of systematic reviews privileges a narrow form of empiricist evidence and excludes or marginalizes other forms of knowledge (Clegg 2005). They are premised upon the assumption that knowledge can be extracted outside of context or social relations. They are a product of the "audit culture" which is located within scientific positivism (MacLure 2004). Systematic reviews offer no analysis or interpretation of the "evidence" and they ignore other forms of knowledge (MacLure 2004). Cornish (2015) demonstrates that systematic literature reviews decontextualize empirical reports. Such reviews restrict the definition of what constitutes evidence, ignore context and complexity, and aspire to scientific objectivity. Thus, the mantra of "what works" fails to consider the social relations and political context of the local site on which the study was developed.

Using Traditional Masculinity in Public Health to Engage Men

Some public health interventions aimed at engaging men emphasize strength-based approaches and what are seen to be positive aspects of traditional gender roles for men. Englar-Carlson and Kiselica (2013) outline what they assert are ten traditionally oriented strengths of men: male relational style, male ways of caring, generative fatherhood, male self-reliance, worker-provider tradition of men, male courage, daring and risk taking, group orientation of boys and men, humanitarian service of fraternal organizations, men's use of humor, and male heroism.

Fleming et al. (2014) have identified what they call "inadvertent harm in public health interventions," arising from the use of constructions of traditional masculinity to engage men in addressing health issues. Using the notion of "man up," which signifies strong will and courage, to appeal to men, many public health interventions end up reinforcing dominant forms of masculinity which are the source of men's ill-health. Fleming et al. (2014) note that the same "man up" language has been used in public health and elsewhere to challenge the violence and sexually aggressive behavior of men.

As it is noted in the 2019 publication by Our Watch (the Australian violence prevention organization), strategies that encourage men to "man up" reinforce rather than transform traditional gender roles. By asking men to draw upon traditional notions of toughness and aggression to challenge men's violence, they are being asked to exhibit the same behaviors that are at the heart of men's violence. While acknowledging these problems, Our Watch nevertheless supports such strategies because they are seen as successful in engaging men and they argue that they can be an important first step in getting men to acknowledge the problem of men's violence.

While strategies of using traditional masculinity to engage men are coming under increasing criticism in public health, alternative forms of masculinity are frequently promoted to encourage men and boys to adopt healthier and more positive expressions of masculinity (Kiselica and Englar-Carlson 2010; Sutton 2016; Abebe et al. 2018; Flood 2018b, c; Roberts et al. 2019). Here, the project is to transform what is called "toxic masculinity" into more inclusive and gender-equitable masculinities.

Problematizing "Toxic" Masculinity in Public Health

The concept of "toxic masculinity" defines a particular type of masculinity that is seen to encourage homophobia, misogyny, domination of women, and sexual violence. It involves physical conflict as a way of resolving disputes and validates an aggressive expression of masculinity (Watts 2019). Proponents of the term emphasize that it is only a particular form of masculinity that is problematic and that there are many diverse expressions of masculinity that are healthy for men. Such proponents want to assure men that there is nothing inherently wrong about being male, and that some forms of male socialization are inherently useful (Flood 2018b).

Flood (2018c) identifies what he sees as the benefits of using the term "toxic masculinity": an emphasis on the social construction of masculinity; highlighting that toxic masculinity is just one form of masculinity; implying that there are other healthier forms of masculinity; popularizing feminist critiques of gendered power; and usefulness in engaging men and boys in violence prevention.

It is not widely known that the term "toxic masculinity" was first coined by mythopoetic men's groups to signify a form of "warrior" masculinity which was at odds with an essential masculine essence which was claimed to be founded on care and compassion alongside strength (Bliss 1995). However, it is noted that the mythopoetic men's movement promoted a form of benevolent patriarchy with men as head of the household and economic provider (de Boise 2019). Such groups sought to rescue the concept of masculinity from feminist critiques (Salter 2019).

The concept of toxic is a biological metaphor in that it relates to the impact of toxicants on living beings. It implies that toxic masculinity is a disease that men have caught and are contaminated by rather than men choosing this behavior because of the benefits they receive (Waling 2019). There is often little acknowledgment by the proponents of "toxic masculinity," or the similar construct of the "man box," a rigid set of beliefs that pressure men to behave in a "manly way" (The Men's Project and Flood 2018), of the benefits and rewards for men in adhering to dominant forms of masculinity and a gender unequal society.

The American Psychological Association (APA Boys and Men Guidelines Group 2018), in their guidelines for working with men, argues that this form of masculinity is also unhealthy for men. Focusing on the unhealthy aspects of toxic masculinity for men may shift the focus off the consequences of men's behavior toward women. If we encourage men to rethink masculinity primarily from the point of view of the benefits they may attain, it is questionable whether men will change those behaviors and systems, which negatively impact on women, but provide benefits to them (Laporte 2019).

Furthermore, most men who do a cost-benefit analysis may recognize that aspects of dominant masculinity and gender inequality are not healthy for them but they tend to conclude that the benefits outweigh the costs (Derry 2019). Framing men as victims of toxic masculinity denies their agency in reproducing particular forms of masculinity. If we believe that men are "trapped" and pressured into toxic masculinity and the man box, rather than choosing to adhere to dominant conceptions of masculinity because of the privileges and rewards they receive, we will not be able to devise appropriate strategies to disengage men from their unearned privileges.

What toxic masculinity does is that it transforms structural problems into interpersonal behaviors and acts that are separated from their social context (de Boise 2019). As Laporte (2019) suggests, the concepts of toxic and healthy masculinity obscure the power relationships between men and women and disconnect men's behaviors from wider systems of privilege and oppression. He emphasizes the importance of situating masculinity within patriarchy, white supremacy, colonialism, and capitalism.

Masculinity is presented as the cause of men's behavior and unequal gender relations rather than a consequence of gendered social relations (Waling 2019). Men's attitudes and behavior are not free floating but are rather a product of patriarchy. While we continue to live in a patriarchal society, masculinity will always be an expression of dominance, just as femininity will always be an expression of subordination (Jensen 2019).

Thus, it is important to locate what is called "toxic masculinity" in patriarchy and consider whether the changes asked of men either challenge or support patriarchy (Jensen 2019). Flood (2018c) acknowledges these dangers in the term "toxic masculinity." Notwithstanding these concerns, Flood nevertheless continues to advocate the value of the toxic and healthy masculinity discourse (Flood 2018b; Flood 2019). Healthy masculinity is promoted as an important first step in encouraging men to acknowledge their emotions and their vulnerability.

While proponents of the term "toxic masculinity" argue that there are healthy forms of masculinity that men can express, many men experience the term as condemning of all men and seek to defend traditional forms of masculinity. Antifeminist critics of toxic masculinity argue that it is a critique of what it is to be male (Watts 2019). It is ironic that conservative critics suggest that the concept of toxic masculinity implicates all men, when proponents of the concept are at pains to emphasize that there are forms of masculinity which are not toxic. Thus, the reassurance strategy to mitigate male defensiveness has not lessened the backlash by men.

Promoting "Healthy" Masculinity in Public Health

A recent Our Watch (2019) discussion paper on engaging men in the prevention of violence against women emphasized the importance of promoting "masculinities" that are respectful and positive. These concepts of healthy and positive masculinities are increasingly being used in violence prevention work to engage men to challenge what are seen as problematical conceptions of masculinity (Flood 2019; Our Watch 2019; Roberts et al. 2019). The sought-after qualities of healthy masculinity, such as emotional vulnerability, acknowledgment of fear, expressions of joy and kindness, and respect for girls and women (Roberts et al. 2019), are admirable. The issue is why do such qualities need to be gendered or rebranded as masculine?

It is interesting that qualities such as empathy, compassion, caring, and emotional vulnerability, which are usually associated with women and with femininity, are reframed as "healthy masculinity." Why are these qualities deemed to be masculine rather than qualities of an ethical human being? Such a reframing discourages men to engage with femininity and ends up reproducing the gender hierarchy that privileges masculinity above femininity. It thus reinforces the gender binary, devalues what is seen as feminine, and appropriates human traits as masculinity specific (Waling 2019).

To the extent that masculinity is a series of traits and behaviors, it does not solely belong to men. It is also enacted by women, trans, nonbinary, and gender nonconforming people as well (Laporte 2019). Also, if they are human strengths, then associating them with men and masculinity reinforces gender differences and promotes the notion of an essentialist masculinity (Addis et al. 2010). Although it is sometimes acknowledged in discussions of healthy masculinity that masculinity is not only linked to male bodies (Roberts et al. 2019), in reality the concept of masculinity tends to reinforce the link between the male sex category and gendered performance and practice. To the extent that this is so, it is hard to see how any conception of masculinity can avoid this reinforcement of essential gender differences.

Qualities like courage, justice, love, empathy, and caring are important qualities to aspire to but why should these qualities be seen to be associated solely with male biology and/or masculine gendered norms? (Almassi 2015). This is not to suggest that men should deny their gendered positioning as men or fail to acknowledge the privileges associated with that positioning. In encouraging men to adopt an ungendered moral identity, men cannot escape their embodied and socially located position in gender hierarchies. However, this does not require men to cling to notions of masculinity. The biological and physical differences between men and women can be acknowledged without reverting to claims of male and female ways of feeling, thinking, and acting under the guise of masculinity and femininity (Jensen 2019).

Our Watch (2019) notes the concerns about potential reinforcement of traditional gender roles but argues against abandoning the concept of masculinity in engaging men in violence prevention. They draw upon O'Neil (2010) to argue against relinquishing the notion of positive masculinity because of its claimed usefulness in encouraging men to expand their definition of what a man can be. Flood (2018b) also acknowledges the dangers of using the notion of healthy masculinity because it encourages men's investment in a gendered identity as men and because it implies that certain qualities are only available to men. However, while he accepts that it is ultimately important to break down gender hierarchies and boundaries and lessen the connection between ethical selfhood and being perceived as masculine, he argues that it is a necessary stage in engaging men because it enables men to feel connected to other men.

To what extent, however, does the concept of "healthy masculinity" involve men in interrogating their power and privilege under patriarchy? Discussions of healthy masculinity are focused on the traits of individuals rather than the structures of oppression. Will men be more likely to address the harms that traditional masculinity does to them and neglect the damage and oppression they perpetrate against women? Also, when the focus is on the harms men experience from gender inequality and traditional masculinity, we need to consider how these harms may be mitigated by the benefits men receive from patriarchy. Because men gain benefits from gender inequality and traditional masculinity, will they be prepared to willingly give up these benefits if they recognize the harms they cause? Explaining men's abusive and violent behavior as resulting from suppression of emotions does not address men's complicity in the reproduction of patriarchy. It is not clear how men attaining healthier masculinities will lead them to relinquish their power and privilege or encourage them to challenge the system that privileges them.

The question must thus be asked as to whether the construct of masculinity has been useful in engaging men effectively in challenging men's violence and undermining gender inequality. While it is clear that we need to articulate alternative and more positive visions of how men can behave in the world, is promoting healthy masculinities the way forward?

Addis et al. (2010) argue that using masculinity to frame violent and abusive behaviors of men has reproduced gender inequality and has had a negative impact on both women and men. While many proponents of healthy and positive versions of masculinity do not advocate an essentialist view of gender, it is nevertheless the case that nonprofessional and "common sense" understandings of masculinity are based on the assumption that masculinity is a product of an inherent male nature. Consequently, discussions about masculinity tend to reinforce essentialist views about gender and reproduce the gender binary.

Encouraging men to pursue healthy masculinity implies that alternative forms of masculinity are the only way that men can address violent and abusive behaviors. There is no consideration of men pursuing feminine or androgynous expressions of gender and no attempt to break down the gender binary (Waling 2019) or challenge patriarchal structures which are at the heart of men's violence. If the aim is to move beyond gender binaries, we need to consider strategies of change for men that do not rely upon frameworks of masculinity. Alternative strategies would encourage men to either embrace these qualities as "feminine" or to widen their behaviors under the framework of what it means to be human.

Moving Beyond Masculinity in Public Health

When a number of critical masculinity scholars were invited to comment on what they thought constituted "healthy masculinity" (Okun 2014), many of them distanced themselves from the concept. Kaufman (2014) argued that there was no such thing as healthy masculinity. While he noted that there are healthier and less healthy forms of masculinity, he argued that any definition of masculinity limited men to behaviors supposedly associated with being male. Jensen (2014) argued that all traits claimed to be associated with healthy masculinity were not distinctive of men but were rather human traits unrelated to biological differences between male and female bodies. Katz (2014) sees healthy masculinity as an irreconcilable concept similar to notions of healthy whiteness. Johnson (2014) similarly describes healthy masculinity as oxymoronic because it privileges gender over our humanness and reinforces gender hierarchy. Furthermore, it encourages us to emphasize individual personality over patriarchal privilege and oppression.

By focusing on particular aspects of masculinity that are designated as problematic or toxic, we do not address the concept of masculinity as a whole (Laporte 2019). Stoltenberg (cited in Cooper 2018) does not use the term toxic masculinity for the very reason that it implies that there are positive forms of manhood that men can embrace. Thus, some who are critical of the term argue that all forms of masculinity *are* toxic (Stoltenberg 2013; Cooper 2018). This does not represent a condemnation of men per se or imply that there is something inherently bad about being male. This is to confuse the socially constructed notion of masculinity with men as sexually differentiated beings. Nor does it suggest that there are no alternative forms of selfhood for men. One can be critical of masculinity and still acknowledge alternative subjectivities for men to inhabit.

Conclusion

The public health approach to violence prevention, and the social-ecological model that underpins it, have enabled implicit feminist perspectives to be enacted in government and community agendas and have widened the opportunities for the involvement of men in anti-violence work. However, the dominance of this framework to the exclusion of other theoretical perspectives has limited, and in some cases subverted, progressive analyses and feminist activist work on violence prevention. The focus on discourses of "toxic" and "healthy" masculinities in engaging men in violence prevention in public health has reinforced gender binaries and essentialist understandings about gender.

A clearer conceptual understanding of violence against women outside of the public health framing is required. Any conceptual framework that aims to address men's violence against women needs to move beyond the limitations of the public health model. Such a framework must emphasize the centrality of gender as relational, enacted, institutionalized, and intersected with other social divisions; be theoretically integrated and internally coherent at multiple levels of intervention; specify the main causes of men's violence against women; and emphasize the primacy of structural factors alongside patriarchal ideology, men's peer support for violence against women, the exercise of coercive control in family life, and the patriarchal psyche of individual men. (For an alternative analysis of violence against women outside of the public health framework, see Pease, B. (2019). *Facing patriarchy: From a violent gender order to a culture of peace.* London: Zed Books.)

Acknowledgments Parts of this chapter have been adapted with permission from the publication, Pease, B. (2019). *Facing patriarchy: From a violent gender order to a culture of peace.* London: Zed Books.

References

Abebe K, Jones K, Culyba A, Felix N, Anderson H, Torres I, Zelazny S, Bamwine P, Boateng A, Cirba B, Detchon A, Devine D, Feinstein Z, Macak J, Massof M, Miller-Walfish S, Morrow S, Mulbah P, Miller E (2018) Engendering healthy masculinities to prevent sexual violence: rationale for and design of the Manhood 2.0 trial. Contemp Clin Trials 71:18–32

Addis M, Mansfield A, Syzdek M (2010) Is 'masculinity' a problem? Framing the effects of gendered learning in men. Psychol Men Mascul 11(2):77–90

Almassi B (2015) Feminist reclamations of normative masculinity: on democratic manhood, feminist masculinity and allyship practices. Fem Philos Quart 1(2):1–22

American Psychological Association, Boys and Men Guidelines Group (2018) APA Guidelines for psychological practice with boys and men. Retrieved from http//www.apa.org/about/policy/psychological-practice-boys-men-guidelines.pdf

Austerberry H (2011) Review of the Palgrave handbook of gender and healthcare by E Kuhlmann and E Annandale. Crit Public Health 22(1):109–110

Bacchi C (1999) Women, policy and politics: the construction of policy problems. Sage, London

Beck U (2007) World at risk. Polity Press, Cambridge

Bliss S (1995) Mythopoetic men's movements. In: Kimmel M (ed) The politics of manhood: Profeminist men respond to the mythopoetic men's movement (and the mythopoetic leaders answer). Temple University Press, Philadelphia

Boyd C (2009) Thinking about risk: preventing violence against women in Victoria. Unpublished paper

Bronfenbrenner U (1977) Toward an experimental ecology of human development. Am Psychol, July:513–531

Broom D (2016) Hazardous good intentions? Unintended consequences of the project of Prevention. Health Sociol Rev 17(2):129–140

Bunton R, Willis J (2004) 25 years of critical public health. Crit Public Health 14(2):79–80

Carmody M (2009) Conceptualising the prevention of sexual assault and the role of education. Australian Centre for the Study of Sexual Assault, ACSSA Issues, no 10

Casey E, Lindhorst T (2009) Towards a multilevel ecological approach to the primary prevention of sexual assault. Trauma Violence Abuse 10:91–114

Castellino T (2010) A feminist audit of men's engagement in the elimination of violence. Paper presented at the University of Melbourne, November

Chung D, O'Leary P, Hand T (2006) Sexual violence offenders: prevention and intervention approaches. Issues (5), June, Australian Centre for the Study of Sexual Assault

Clegg S (2005) Evidence-based practice in educational research: a critical realist critique of systematic reviews. Br J Sociol Educ 26(3):415–428

Connell R (2012) Gender, health and theory: conceptualising the issue in local and world perspectives. Soc Sci Med 74(11):1675–1683

Cooper W (2018) All masculinity is toxic. www.vice.com/em_us/article/zmk3ej/all-masculinity-is-toxic

Cornish F (2015) Evidence synthesis in international development: a critique of systematic review and a pragmatist alternative. Anthropol Med 22(3):263–277

Cowburn M (2010) Invisible men: social reactions to male sexual coercion: bringing men and masculinities into community safety and public policy. Crit Soc Policy 30 (2):225–244

Culpitt I (1999) Social policy and risk. Sage, London

de Boise S (2019) Editorial: is masculinity toxic? NORMA Int J Mascul Stud 14(3):147–151

Derry C (2019) Ending violence against women: male privilege 'toxic masculinity' and the goodies in the 'man box'. xyonline.net

Dutton D (1994) Patriarchy and wife assault: the ecological fallacy. Violence Vict 9(2):167–182

Dutton D (2006) Rethinking domestic violence. UBC Press, Vancouver

Dutton D, Nicholls T (2005) The gender paradigm in domestic violence research and theory: part 1 – the conflict of theory and data. Aggress Violent Behav 10:680–714

Englar-Carlson M, Kiselica M (2013) Affirming the strengths in men: a positive masculinity approach to assisting to male clients. J Couns Dev 91:399–409

Fleming P, Gruskin S, Rojo F, Dworkin S (2014) "Real men don't": constructions of masculinity and inadvertent harm in public health interventions. Am J Public Health 104(6):1029–1112

Flood M (2008) A response to Bob Pease's' engaging men in men's violence prevention. Paper presented at the Forum: Men's Role in Preventing Men's Violence Against Women. Melbourne, 20 November

Flood M (2018a) Engaging men and boys in violence prevention. Palgrave, New York

Flood M (2018b) Men and the man box: a commentary. In: The Men's Project and Flood M (ed) The man box: a study on being a young man in Australia. Jesuit Social Services, Melbourne, pp 46–54

Flood M (2018c) Toxic masculinity: a primer and commentary. xyonline.net

Flood M (2019) Men, gender and healthy masculinity, Wyndham City and Hobson's Bay City Council, Laverton, October 11

Fulu E, Heise L (2015) What works to prevent violence against women and girls? Evidence reviews: paper 1: state of the field of research on violence against women and girls, What Works to Prevent Violence Program

Gough I (2013) Understanding prevention policy: a theoretical approach. LSE Research Online, January, London School of Economics and Political Science

Griffiths P (2005) Evidence-based practice: a deconstruction critique. Int J Nurs Stud 42(3):355–368

Gupta M (2003) A critical appraisal of evidence-based medicine: some ethical considerations. J Eval Clin Pract 9(2):111–122

Hall R (2004) "It can happen to you": rape prevention in the age of risk management. Hypatia 19(3):1–19

Hammarstrom A (1999) Why feminism in public health? Scand J Public Health 27:241–244

Hankivosky O, Christoffersen A (2008) Intersectionality and the determinants of health: a Canadian perspective. Crit Public Health 18(3):271–283

Harding S (1986) The science question in feminism. Cornell University Press, New York

Hawkins J (2006) Science, social work, prevention: finding the intersections. Soc Work Res 30(3):137–152

Heise L (1998) Violence against women: an integrated ecological framework. Violence Against Women 4(3):262–290

Heise L (2006) Determinants of intimate partner violence: exploring variation in individual and population level risk, Department of Infectious Diseases Epidemiology, London School of Hygiene and Tropical Medicine

Hoyle C (2007) 'Will she be safe'? A crucial analysis of risk assessment in domestic violence cases. Child Youth Serv Rev 30:323–337

Hunnicutt G (2009) Varieties of patriarchy and violence against women: resurrecting 'patriarchy' as a theoretical tool. Violence Against Women 15(5):553–573

Inhorn M, Whittle L (2001) Feminism meets the 'new' epidemiologies: toward an appraisal of anti-feminist biases in epidemiological research on women's health. Soc Sci Med 53:553–567

Institute for Work and Health (2015) What researchers mean by primary, secondary and tertiary prevention. At Work, Issue 80, Spring

Jensen R (2014) Men are human first. In: Okun R (ed) Voice male: the untold story of the profeminist men's movement. Interlink Books, Northhampton

Jensen R (2019) It's not about 'toxic masculinity' or 'healthy masculinity', it's about masculinity under patriarchy. Fem Curr feminist current.com

Johnson A (2014) Healthy masculinity is oxymoronic. In: Okun R (ed) Voice male: the untold story of the profeminist men's movement. Interlink Books, Northhampton

Katz J (2014) Irreconcilable concepts. In: Okun R (ed) Voice male: the untold story of the profeminist men's movement. Interlink Books, Northhampton

Kaufman M (2014) Any gender is a drag. In: Okun R (ed) Voice male: the untold story of the profeminist men's movement. Interlink Books, Northhampton

Kiselica M, Englar-Carlson M (2010) Identifying, affirming and building upon male strengths: the positive psychology/positive masculinity model of psychotherapy with boys and men. Psychother Theory Res Pract Train 47(3):276–287

Kruger E, Dahlberg L, Mercy J, Zwi A, Lozano R (2002) World health report on health and violence. World Health Organisation, Geneva

Laporte J (2019) Disrupting the toxic vs healthy masculinity discourse: an autoethnographic study. Master of Arts, University of Massachusetts, Lowell

Lee P (2017) What's wrong with logic models? LCSA: occasional paper 1, Local Community Services Association

Legge D (2018) Capitalism, imperialism and class: essential foundations for a critical public health. Crit Public Health. https://doi.org/10.1080/09581596.2018.1478067

MacLure M (2004) Clarity bordering on stupidity: where's the quality in systematic review? Paper presented to the British Educational Research Association Annual Conference, Manchester, September

Messner M, Greenberg M, Peretz T (2015) Some men: feminist allies and the movement to end violence against women. Oxford University Press, Oxford

Meyer E, Post L (2006) Alone at night: a feminist ecological model of community violence. Fem Criminol 1(3):207–227

Mykhalovskiy E, Frohlich K, Poland B, Di Ruggiero E, Rock M, Comer L (2018) Critical social science with public health: Agonism, critique and engagement. Crit Public Health:1–12. https://doi.org/10.1080/0958196.2018.1474174

Navarro V (2009) What we mean by social determinants of health. Int J Health Serv 39(3):423–441

O'Neal R (2001) Is it time to bury the ecosystem concept? Ecology 82(12):3275–3284

O'Neil J (2010) Is criticism of generic masculinity, essentialism, and positive-healthy masculinity a problem for the psychology of men? Psychol Men Mascul 11(2):98–106

Okun R (ed) (2014) Voice male: the untold story of the profeminist men's movement. Interlink Books, Northhampton

Our Watch (2015) Change the story: a shared framework for the primary prevention of violence against women and their children in Australia. Our Watch, VicHealth, ANROWs Our Watch, Melbourne

Our Watch (2019) Men in focus: unpacking masculinities and engaging men in prevention of violence against women. Our Watch, Melbourne

Pease B (2009) From evidence-based practice to critical knowledge in post-positivist social work. In: Allan J, Briskman L, Pease B (eds) Critical social work: theories and practices for a socially just world, 2nd edn. Allen and Unwin, Sydney

Pease B (2014) Theorising men's violence prevention policies: limitations and possibilities of interventions in a patriarchal state. In: Henry N, Powell A (eds) Preventing sexual violence: interdisciplinary approaches to overcoming a rape culture. Palgrave Macmillan, Buckingham

Powell A (2014) Shifting upstream: bystander action against sexism and discrimination against women. In: Henry N, Powell A (eds) Preventing sexual violence: interdisciplinary approaches to overcoming a rape culture. Palgrave Macmillan, Houndmills

Quadara A, Wall L (2012) What is effective primary prevention in sexual assault? Translating the evidence for action, ACSSA Wrap no 11. Australian Centre for the Study of Sexual Assault, Australian Institute of Family Studies, Melbourne

Raphael D, Bryant T (2002) The limitations of population health as a model for new public health. Health Promot Int 17(2):189–199

Ratelo K, Suffla S, Lazarus S, van Niekerk A (2010) Towards the development of a responsive, social science-informed critical public health framework on male interpersonal violence. Soc Change 40(4):414–438

Richards L, Gauvin L, Raine K (2011) Ecological models revisited. Annu Rev Public Health 32:307–326

Roberts S, Bartlett T, Stewart R (2019) Healthier masculinities scoping review. A report prepared for VicHealth. Monash University, VicHealth

Rogers W (2006) Feminism and public health ethics. J Med Ethics 32(6):351–354
Salter M (2019) The problem with a fight against toxic masculinity. The Atlantic, February 27. www.theatlantic.com
Schofield T (2015) A sociological approach to health determinants. Cambridge University Press, Cambridge
Sliwka G, Macdonald J (2005) Pathways to couple violence: an ecological approach. Men's Health Information and Resource Centre, University of Western Sydney
Stanger N (2011) Moving 'eco' back into social-ecological models: a proposal to reorient ecological literacy into human development models and school systems. Hum Ecol Forum 18(2):167–173
Stepney P (2014) Prevention in social work: the final frontier? Crit Rad Soc Work 2(3):305–320
Stewart A, Lewis K, Neophytou, K, Bolton, T (2010) Does combining health promotion and feminist frameworks equal better health outcomes for women? Paper presented at the 6th Australian Women's Health Conference, Melbourne
Stojanovic T, McNae H, Tett P, Potts T, Reis J, Smith H, Dillingham I (2016) The 'social' aspect of social-ecological systems: a critique of analytical frameworks and findings from a multisite study of coastal sustainability. Ecol Soc 21(3):1–16
Stoltenberg J (2013) Talking about 'healthy masculinity' is like talking about 'healthy cancer'. Fem Curr, August 9. www.feministcurrent.com
Storer H, Casey E, Carlson J, Edleson J, Tolman R (2016) Primary prevention is? A global perspective on how organisations, engaging men in preventing gender-based violence conceptualize and operationalize their work. Violence Against Women 22(2):249–268
Sutton H (2016) Eradicate sexual assault by fostering healthy masculinities. Campus Secur Rep 13(5):3–5
The Men's Project and Flood M (2018) The man box: a study on being a young man in Australia. The Men's Project, Jesuit Social Services, Melbourne
Thurston W, Vissandjee B (2014) An ecological model for understanding culture as a determinant of women's health. Crit Public Health 15(3):229–242
VicHealth (2007) Preventing violence before it occurs: a framework and background paper to guide the primary prevention of violence against women. Victorian Government, Carlton
Waling A (2019) Problematising 'toxic' and 'healthy' masculinity for addressing gender inequalities. Aust Fem Stud 34(101):362375
Wall L (2013) Issues in evaluation of complex social change programs for sexual assault prevention. ACSSA Issues, No 14, May
Watts G (2019) Stop scolding men for being toxic. *The Conversation*, April 29. theconversation.com
Webb S (2006) Social work and risk society. Palgrave Macmillan, Houndmills, Basingstoke
Woody J (2006) Prevention: making a shadow component a real goal in social work. Adv Soc Work 7(2):44–46

The Madness of Women: Myth and Experience

73

Jane M. Ussher

Contents

Introduction	1854
The Madness of Woman: A Problem Peculiar to Her Sex?	1855
Raging Hormones and Reproductive Debilitation	1857
Resisting Biological Determinism: Rejecting Raging Biomedical Explanations for Women's Madness	1859
Gender-Based Diagnosis of Depression: Pathologizing Femininity	1864
Treatment or Torture? Controlling Deviant or Difficult Women	1866
Conclusion	1869
References	1870

Abstract

Women are more likely than men to be diagnosed as "mad," from eighteenth and nineteenth century diagnoses of "hysteria," to twentieth and twenty-first diagnoses of "neurotic" and mood disorders. Women are also more likely to receive psychiatric "treatment," including nineteenth and twentieth century psychiatric hospitalization, accompanied by restraint, electroconvulsive therapy (ECT) or psycho-surgery, as well as psychological therapy or psychotropic medication today. This chapter will expose the myths of women's madness, by exploring the genealogy of current psychiatric discourse as it applies to women, revealing the flaws and misogyny inherent within the "regimes of truth" promulgated by psy-professionals – the regulation of women deemed deviant or dysfunctional. This is not to deny the materiality of the prolonged misery or distress reported by many women. Acknowledging the origins of this distress in the context of women's lives means it can be conceptualized as a reasonable response, rather than a reflection of internal pathology. Some women are more at risk of distress and pathologization than others, including women of color, poor women, lesbian,

---

J. M. Ussher (✉)
Translational Health Research Institute, Western Sydney University, Penrith, NSW, Australia
e-mail: j.ussher@westernsydney.edu.au

© The Author(s), under exclusive licence to Springer Nature Singapore Pte Ltd. 2022
D. McCallum (ed.), *The Palgrave Handbook of the History of Human Sciences*,
https://doi.org/10.1007/978-981-16-7255-2_34

bisexual, queer, and transgender women, and those who have experienced trauma and abuse. A feminist intersectional analysis, that acknowledges the intersection of identities, and the material, discursive, and intrapsychic aspects of women's distress, allows us to address women's concerns and challenge their diagnosis as mad.

Keywords

Women's madness · Gender and mental health · Misogyny · Material-discursive-intrapsychic · Intersectionality

Introduction

In this chapter, the discursive construction of women's madness at the beginning of the twenty-first century will be examined, as well as the genealogy of current Western psy-profession, through turning a lens on the practices in the policies and prejudices of the past. This analysis draws on the work of the French philosopher Michel Foucault, who argued that power is irrevocably connected to knowledge, which in turn has a regulatory function – for example, through categorizing what is normal and abnormal; what is "sane" and what is "mad." Foucault argued that discursive representations and practices "systematically form the objects of which they speak" (Foucault 1972), actively constructing realities in specific ways, and normalizing regulation of society and individuals. In this framework, discourses produce identities, subject positions, and what are described as "institutional sites" from which a person can speak or be addressed. Subjectivity – our sense of self – is not conceptualized as coming from within but is deemed to be constructed within discourse – constituted by language and material practice. Therefore, representations of madness encapsulated within the "bible of psychiatry," the *Diagnostic and Statistical Manual of the American Psychiatric Association (DSM)* (American Psychiatric Association 2013) not only define what it means to be psychologically "disordered" but also function to construct the subject position of "mad woman," legitimating the right of psy-professionals (Rose 1996) – psychiatrists and psychologists – to diagnose and treat her condition, and defining the "truths" that are accepted as explanations for her disordered mind.

This chapter will explore competing explanations for women's greater propensity to be diagnosed and treated for madness, the "regimes of knowledge" (Rose 1996) circulated by the psy-professions that define what can be known, and what is offered as "treatment." This includes biomedical theories that position a woman's madness in the reproductive body (her raging hormones) and sociocultural theories that focus on a woman's social environment (such as poverty, discrimination, inequality, or abuse). Feminist social constructionist theories that dismiss the very notion of madness itself will also be explored, including the argument that psychiatric diagnosis and treatment serves to pathologize and regulate femininity; by defining what is mad we define what it is to be sane. Or more specifically, we define the boundaries of femininity for the "good woman" – the feminine ideal of a cisgender, heterosexual, White woman, who is self-controlled in every way. These arguments will provide an explanation for why women are significantly more likely to be deemed mad than men – more likely to be diagnosed as having specific "disorders" which are reified as "mental illness" through inclusion in the DSM (Ussher 2011).

The Madness of Woman: A Problem Peculiar to Her Sex?

Across the history of Western medicine and psychiatry, women's greater propensity to madness has been attributed to the reproductive body – with women's reproductive organs deemed to be "pre-eminent" in all aspects of her psyche and physical well-being. Thus medical officer John Haslam explained women's preponderance in the English Bethlem asylum during the nineteenth century due to the "natural processes which women undergo," claiming that "insanity" is "often connected with menstruation" and "parturition" (Haslam 1809). Similarly, George Man Burrows concluded that "many circumstances in the physical and moral condition of women, from the epoch of puberty to the critical period, would lead us to conclude that more women than men become insane in every country and every place." These "circumstances" included "women being exposed to more natural causes of physical excitation and irritation than men," as well as "menstruation, parturition and all its consequences" (Burrows 1828). Menopause was described as "a critical and dangerous time for women," a time when "the nervous system is so unhinged that the management of the mental and moral fibres often taxes the ingenuity of the medical confident" (Tilt 1882). These pronouncements follow in the Hippocratic tradition of locating pathological behavior in the womb, characterized by the disorder "hysteria," which derives from the Greek term for the womb υςερα (ustera) (Bronfen 1998).

While Hippocrates may have first coined the term "hysteria," it was Plato who memorably attributed women's disorders to the "wandering womb," in *Timaeus*, published in 91 CE. Plato saw the womb as "an animal which longs to generate children," claiming that it could be found "straying about in the body and cutting off the passages of breath... (where) it impedes respiration and brings the sufferer into the extremist anguish" (Borossa 2001).

While hysteria was first described by the ancient Greeks, in the seventeenth century, it emerged as one of the most prevalent diseases within Western medicine. Thomas Sydenham commented that "the frequency of hysteria is no less remarkable than the multiformity of the shapes which it puts on. Few of the maladies of miserable mortality are not imitated by it" (Sydenham 1679). While the causes of this ill-defined disorder were extended beyond the womb in the late eighteenth century to include the nervous system, which allowed men to be diagnosed of hysteria, it was always considered to be "woman's disease." Indeed, Thomas Laycock in 1840 described hysteria as a woman's "natural state," in contrast to being considered a "morbid state" in a man, and in 1903, Otto Weininger claimed that "hysteria is the organic crisis of the organic mendacity of woman" (Bronfen 1998).

Nineteenth-century physicians were critical of this feminine "temperament," describing hysterical women as difficult, impressionable, narcissistic, suggestible, labile, and egocentric (Smith-Rosenberg 1986). The typical hysteric was represented as an idle, affluent, self-indulgent, and deceitful woman, "craving for sympathy" (Smith-Rosenberg 1986), whose "unnatural" desire for privacy and independence was described as "personally and morally repulsive, idle, intractable, and manipulative" (Showalter 1987). Some went as far as to describe such women as "evil," with the physician Silas Weir Mitchell, declaring that "a hysterical girl is a vampire who sucks the blood of the healthy people around her" (Smith-Rosenberg 1986). In contrast, women who suffered from neurasthenia, a "nervous disorder" that shared many of symptoms with hysteria, described as an "ill defined set of symptoms – a form of nervous exhaustion" (Busfield 1996), were described as having a "refined and unselfish nature," and as being "just the kind of woman one likes to meet" (Showalter 1987). Politeness and compliance on the part of the patient was clearly of the essence, as is arguably still the case today, when "difficult" women are diagnosed with borderline personality disorder (Becker 2000).

During the Restoration and up until the eighteenth century, even healthy women were considered to be simply a "walking womb" (Bronfen 1998). However, at the end of the eighteenth century, the link between the womb and madness became established as scientific "fact," which, as Michel Foucault notes, resulted in "the entire female body" being "riddled" by "a perpetual possibility of hysteria" (Foucault 1967). This served to legitimate medical management of all errant women, described by Foucault as a "process whereby the feminine body was analysed – qualified and disqualified – as being thoroughly saturated with sexuality; whereby it was integrated into medical practices, by reasons of a pathology intrinsic to it" (Foucault 1978). Women's sexuality was often blamed for hysteria (and by implication, for women's madness), reflecting the nineteenth century view that female desire was morally dangerous. Historically, some women have been more likely to be sexualized, and thereby more likely to be deemed hysterical, deviant, and deficient, than others. Women of color have been positioned as subhuman, likened to animals with uncontrollable sexual appetites, since the time of the slave era (Hill Collins 2002). Nineteenth-century psychiatry pathologized and sexualized the "female invert," a woman we would today identify as lesbian, bisexual, or queer, and deemed her a neurotic degenerate, an aberration, and a nymphomaniac (Groneman 1994). Hysteria among poor or lower class women was attributed to the base sexuality which was believed to characterize their class, with prostitutes assumed to be at particular risk (Smith-Rosenberg 1986).

Hysteria has been described as a "veritable joker in the taxonomic pack" (Porter 1993) or the "wastebasket of medicine" (Bronfen 1998) largely because of its multifarious and nebulous nature. Hysteria has been described as a "mimetic disorder" because it mimics culturally permissible expressions of distress – hysterical limps, palsies, and paralyses were common symptoms of illness in the nineteenth century, but are uncommon today, as they no longer stand as a "sickness stylistics for expressing inner pain" (Porter 1993). As Edward Shorter argues, the symptoms that are deemed to be legitimate signs of illness, including madness, the "symptom pool," are specific to a particular culture and historical point in time. By categorizing certain symptoms "illegitimate," culture encourages people not to report them in order to avoid being seen as "undeserving" and not having a "real" medical problem. There is, as Shorter declares, "great pressure on the unconscious mind to produce only legitimate symptoms" (Shorter 1992). The same could be said of madness or mental illness women are diagnosed with today.

Notions of the wandering womb may seem laughable today, and hysteria is all but a defunct diagnostic category. However, the location of women's madness in the sexual or reproductive body is not a historical anachronism. The Hippocratic tradition is maintained by the biomedical monopoly over health and illness, that results in bio-psychiatry having jurisdiction over women's madness, and raging hormones being positioned as to blame (Ussher 2006).

Raging Hormones and Reproductive Debilitation

These practices of pathologizing and regulating femininity continue into the twenty-first century. Today the "legitimate" symptoms of madness are defined within the *Diagnostic and Statistical Manual of the American Psychiatric Association* (DSM) (American Psychiatric Association 2013). New diagnoses are added within each edition, and others, such as hysteria or homosexuality, are removed (Metcalfe and Caplan 2004). The necessary symptoms for diagnosis are circulated within popular consciousness through interactions with the psy-professions, drug company advertising, the media, and "self-help" diagnostic websites. Is it surprising that so many women self-diagnose with psychiatric disorders, and then seek professional confirmation of their pathological state? Their distress is as real as the women who were diagnosed as hysterics in the nineteenth century. However, the psychiatric disorders with which women are now diagnosed are no less mimetic than hysteria was then – women signal their psychic pain, their deep distress, through culturally sanctioned "symptoms," which allows distress to be considered to be "real." Or women are told by others that they have a problem, and are then effectively positioned within the realm of psychiatric diagnosis and treatment, with all the regulation and potential subjugation that follows.

Many contemporary feminist critics focus on the diagnostic category anorexia nervosa as the successor to hysteria, describing it as the modern manifestation of a "female malady" (Malson 1998; Showalter 1987). This is a persuasive argument: hysteria and anorexia nervosa are diagnoses predominantly given to women; the "typical patient" is often described as willful or immature in both contexts; and each has been conceptualized as a feminine "disorder," tied to femininity (Malson 1998). Anorexia nervosa is not, however, the only female malady to afflict women today; nor is it the only diagnostic category to be contested or located in cultural constructions of femininity. Depression is presently the diagnostic category widely accepted as a "woman's problem," described as a "menace to mood and national productivity" (Gardner 2003) and the most common diagnosis applied to women's distress (Ussher 2010).

If we look to epidemiological research, the evidence that women are more likely to experience "depressive illness" would appear to be indisputable. Firstly, researchers have examined lifetime occurrence in nonclinical contexts, by recruiting a representative sample of the population and asking whether they have ever experienced depression. Across various studies it has been reported that women outnumber men at a rate of 2:1 – 4:1 (Parker and Brotchie 2010; Van de Velde et al. 2010). Secondly, studies examining incidence rates in the previous 1–12 months report that women are between 1.3 and 3.8 times more likely than men to have experienced depression and anxiety (Bebbington 1996). Some women are at higher risk of diagnosis of depression than others, including women of color living in predominantly White societies (Hargrove et al. 2020), lesbian, bisexual, and queer women (Pitts et al. 2006), transgender and nonbinary women (Hyde et al. 2014), and poor women (Fisher et al. 2012). The intersection of identities increases risk for women who are Black and lesbian, bisexual, or queer (Calabrese et al. 2015), or are poor, Black, transgender, and queer (Ferlatte et al. 2019; Williams et al. 2017).

Diagnosis of depression is primarily the reason why women significantly outnumber men in first admission rates for psychiatric treatment, and in register studies where incidence of madness is calculated by contact with services (Ussher 2011). The fact that women are twice as likely as men to be prescribed psychotropic medication, in particular serotonin reuptake inhibitors (SSRIs) (Currie 2005), and twice as likely to be given ECT, is also largely attributable to diagnoses of depression.

Bio-psychiatric theories of depression focus on either genetic determinants or on neurochemical "imbalance," as the psychiatrists Klein and Wender (1993) declare: "the majority of cases of depression and manic-depressive illness appear to be genetically transmitted and chemically produced. Stated differently, the disorders seem to be hereditary and what is inherited is a tendency towards abnormal chemical functioning...in the brain" (Fee 2000). In this vein, it has been argued that genetic differences between women and men are the explanation for women's higher rates of depression. However, a meta-analysis of four community and two twin studies, containing 20,000 participants, did not find any consistent gender difference in heritability (Sullivan et al. 2000), leading to the conclusion that "the relative importance of genetic effects in major depression is the same in women and men" (Kuehner 2003).

An alternative strand of bio-psychiatric theorizing can be traced to the pronouncements of Plato, or to the nineteenth century prelates who proselytized about hysteria. Today, the focus is on reproductive hormones, exemplified by the comments of Mary Seaman (1997), in a review of hormonal causes of "psychopathology," that "evolutionary imperatives" have given "female hormones" a "special neuro-protective role but also a stress mediating role" which acts to shield women from psychosis, but make them "more vulnerable than men to depression and anxiety."

Justification for this view is located in the fact that gender differences in diagnoses of depression emerge at puberty and appear to be reduced postmenopause. Prior to adolescence, boys outnumber girls over the whole gamut of mental health diagnoses, including conduct disorders, language and speech disorders, autism and Asperger's disorder, attention deficit hyperactivity disorder, as well as enuresis and encopresis (bedwetting and soiling) (Cosgrove and Riddle 2004). In preadolescence, boys also outnumber girls in community studies of depressive symptoms and in rates of diagnosed depressive disorders. Postadolescence, however, this gender imbalance is reversed. This suggests that adolescence may be an important turning point in women's mental health.

In the minds of many experts, it is the biological event of puberty that is deemed to be the key factor, which is reminiscent of the advice given to mothers in the late nineteenth century, when adolescence was described as "naturally a time of restlessness and nerve irritability" for girls. A "period of storm and stress" and of "brooding, depression and morbid introspection" (Porter 1855). This view would initially appear to be supported by epidemiological research: A US National Comorbidity Survey reported that the increased rate of female depression occurred at age 10–14 (Kessler et al. 1993), whereas other studies are more precise, specifying age 13 (Peterson et al. 1991), or achievement of pubertal status, rather than simple chronological age (Angold et al. 1998). And while girls' propensity to report depression appears to increase at mid-puberty, boys are less likely to report depression postpuberty, which has led to the conclusion that "the transition to mid-puberty appeared transiently to protect boys from depression" (Angold et al. 1998).

At the same time, there is inconsistent evidence as to the continuity of gender differences in reporting of depression across the adult life span. On the one hand, it has been concluded that there is evidence for a female preponderance of depression in older age groups (Kuehner 2003). However, two large-scale population studies conducted in the USA (Kessler et al. 1993) and the UK (Bebbington et al. 1998) in the 1990s reported that gender differences in depression do not persist after 55 years of age, "due an absolute fall in female prevalence" (Bebbington et al. 1998). This may appear to confirm the essentialist view that women's madness is tied to the reproductive body, to the physiological changes that accompany fecundity, supporting the conclusion that "the female prevalence in depression is linked to women's reproductive years" (Cyranowski et al. 2000). However, the evidence for this viewpoint is more equivocal than this confident statement would suggest, as is outlined below.

Resisting Biological Determinism: Rejecting Raging Biomedical Explanations for Women's Madness

There have always been dissenters to the view that women's madness is located in the body. The seventeenth-century physician Thomas Willis stated that while diseases of "unknown nature and origins," including madness, were blamed on the "bad influence of the uterus," this organ is "for the most part, not responsible at all" (Foucault 1967). His contemporary, Thomas Sydnenham, was more specific, stating that hysteria was the result of social conditions that enslaved women (Rousseau 1993). In the nineteenth century, the British psychiatrist Henry Maudsley adopted a similar view, claiming that hysteria resulted from women having fewer outlets for their nervous energy: "The range of activities of women is so limited, and their available paths in life so few, compared with those men have in the present social arrangements, that they have not, like men, vicarious outlets for feelings in a variety of healthy aims and pursuits" (Maudsley 1879). The French psychiatrist Pinel agreed that hysteria was the product of a restrictive and rigid bourgeois family life; however, he believed that "neurosis" in women was also encouraged by "lascivious reading" (Bronfen 1998). Even Silas Wier Mitchell, who described hysterical women as "vampires," acknowledged that this disorder could be caused by "the daily fret and wearisomeness of their lives which... lack... distinct occupations and aims" (Mitchell 1885).

Thus the "symptoms" experienced by women diagnosed as hysterics in could be framed as reasonable responses to an untenable social situation – young middle-class women secluded from the world and courted prior to marriage, then expected to embark on the monotony of house-keeping, child rearing, and self-sacrifice afterwards, often in the context of an unhappy relationship (Smith-Rosenberg 1986). Or the urban working class women living a life of unrelenting drudgery, being paid at subsistence levels, with no support from family – two-thirds of the hysterics in the French asylum *Salpetriere* were from this group (Showalter 1997). Developing symptoms that required complete bed rest or hospitalization could be seen a form of resistance – serving to force others to take on domestic drudgery, which gave the "hysteric" some semblance of control (Smith-Rosenberg 1986). In this vein, feminists have celebrated hysteria as "a woman's response to a system in which she is expected to remain silent, a system in which her subjectivity is denied, kept invisible" (Herndl 1988). Jane Gallop described the hysteric as a "proto-feminist" because of her "calling into question constraining sexual identities" (1983), and Helene Cixous argued that hysteria was the "nuclear example of women's power to protest" (Cixous and Clement 1987).

The notion that modern manifestations of women's madness have an embodied etiology is equally questionable. For many biomedical researchers today, the key to understanding women's depression lies in understanding adolescent onset – which has led to investigation of the relationship between pubertal hormones and depressed mood, in the belief that the "turning on" of the endocrine system in girls as they emerge from pre- to postpuberty might explain increases in depression at this age. However, there is little evidence to support this hypothesis. For example, while some evidence for a relationship between depressed mood and testosterone (Paikoff et al. 1991) or estrogen levels (Susman et al. 1987) has been reported, neither study found a relationship between hormones and depression in adolescent girls. In a study of 4500 young women aged 9–13 conducted by Angold and colleagues, where it was reported that the onset of puberty predicted the emergence of the gender difference in depression more accurately than chronological age, an adrenarche explanation, tied to increased adrenal androgens, was ruled out, as "those changes occur in later childhood" (Angold et al. 1998). This should not be surprising, for a simple relationship between hormones and behavior can be easily contested, and indeed is contested by many who work in this field. As Carol Worthman argues: "hormones do not directly cause specific biological or behavioral effects. Rather, hormonal action is mediated through an array of other factors. These include: circulating binding proteins, metabolic enzymes, cellular receptors, nuclear binding sites, competing molecules, and presence of cofactors" (Worthman 1995). Equally, there is evidence from the field of psychoneuroimmunology that a reciprocal relationship exists between psychosocial and physiological events, producing an interdependency between social processes and health (Uchino et al. 1996) – rather than the body being the cause of problems.

The evidence for an association between hormones and distress experienced during adult stages of the reproductive life cycle – premenstrually, after childbirth, and during the menopause – is also equivocal. When "Premenstrual Tension" (PMT) first appeared in the medical literature in 1931, it was attributed to the "female sex hormone" estrogen (Frank 1931). In the intervening years, many competing biomedical explanations have been put forward for premenstrual syndrome (PMS) and premenstrual dysphoric disorder (PMDD), the successors to PMT, including gonadal steroids and gonadotrophins; neurovegetative signs (sleep, appetite changes); neuroendocrine factors; serotonin and other neurotransmitters; β-endorphin; and other potential substrates (including prostaglandins, vitamins, electrolytes, and CO_2) (Nevatte et al. 2013). However, there is no evidence for a consistent association between women's hormones and mood (Nolen-Hoeksema 1990) and women report higher rates of depression than men regardless of whether they also report premenstrual distress, or PMS (Popay et al. 1993). For example, in one study which reported premenstrual exacerbation of depressive symptoms in women who were diagnosed with PMS (Mira et al. 1995), the blunting of the growth hormone response found across the cycle was not found in non-PMS sufferers, which was interpreted as suggesting that "depression associated with premenstrual hormone changes is a consequence of a propensity to depression rather than offering an explanation of high rates of depression in women" (Bebbington 1996). Thus women who report depression may experience an exacerbation of symptoms premenstrually, but this does not mean that hormonal changes across the menstrual cycle cause the distress in the first place, or that hormones can explain women's higher overall reporting of depression.

Equally, while many women experience premenstrual *change*, the labelling of such change as a pathological condition PMS (or PMDD), and the experience of premenstrual *distress*, is not inevitable. The cultural context within which a woman lives (Chrisler and Caplan 2002) and the ways in which she negotiates and copes with premenstrual change (Ussher and Perz 2013) will influence whether this change is accepted as a normal part of her experience, or positioned as "symptoms" that should be eradicated because they leave her feeling "out of control" (Ussher and Perz 2017). At the same time, there is strong evidence that premenstrual distress is more strongly associated with women's social and relationship context than with her hormonal status (Ussher and Perz 2013), with over-responsibility, relationship dissatisfaction, and communication problems exacerbating distress, while social support and positive communication facilitate tolerance of premenstrual change outside of a pathological framework (Ussher 2008; Ussher et al. 2007). However, women continue to blame and hate their bodies when they experience negative moods during the premenstrual phase of the cycle, a process of self-objectification that suggests internalization of cultural discourse about PMS, a modern incarnation of the wandering womb (Ussher and Perz 2020).

A similar argument can be made for postpartum depression (Ussher 2004). Current psychiatric orthodoxy tells us that the risk factors for prolonged misery in the postpartum period are the same as at any other time in life (Whiffen 1992), with the added strains of early motherhood serving as a stressor, activating a preexisting risk for depression. Indeed, the majority of psychiatrists now concur that hormonal or obstetric factors are not associated with nonpsychotic postnatal depression (Albright 1993), and research suggests that there is no difference between women with and without children in the initial onset of depression (Nazroo et al. 1998). Postpartum blues may be linked to hormonal changes, although the mechanisms of this are not clear, but there is no consistent relationship between the blues and ongoing diagnosis of depression (Bebbington 1996).

One of the most notable factors associated with reporting of postpartum depression is a previous history of depression (Fisher et al. 2012); women who have previously been depressed have also been found to be slower to recover postpartum, and more likely to relapse, than those who have not been depressed previously (Bell et al. 1994). Other factors associated with prolonged misery at this time include stressful life events, maternity blues, poverty, infant temperament, and unwanted pregnancy (Fisher et al. 2012; O'Hara and Swain 1996). As is the case with PMS, the social and relational context of women's lives is also important, with reporting of postpartum depression associated with life stress (Boyce and Hickey 2005) and the quality of the couple relationship (Mauthner 1998). Idealized cultural constructions of motherhood, internalized by women, are also influential. For example, in a study of 40 women interviewed about motherhood and postpartum depression, Natasha Mauthner found that the common denominator between the 18 women who reported depression was high and unrealistic expectations of motherhood (Mauthner 2010). While women knew that there was no such thing as the perfect mother, they each had firm expectations of how they should "cope" and found it difficult to accept any failure in meeting these impossibly high standards. These findings were confirmed by a study of 71 new mothers, where reporting of depressive symptoms was higher in women whose experiences of motherhood were more negative than their expectations (Harwood et al. 2007). At the same time, in countries that provide strong support for new mothers, lower rates of depression are reported postnatally (Miller 2002), whereas in countries where women have few reproductive rights, postpartum depression is higher (Patel et al. 2002). When you look at the reality of women's experiences of early motherhood, this is not surprising. Indeed, as Wendy Hollway argues, "is it any surprise that women's identities go through a life-changing transition or that so many new mothers become depressed" (Hollway 2006), when women are faced with the non-negotiation and incessant demands of this dependent life?

What of the "psychological turmoil" experienced by the supposedly "wretchedly depressed women" (Studd 1997) at midlife? Researchers and clinicians confidently proclaim that these symptoms are caused by changing hormonal levels, which affect "hypothalamic function," or "neuropeptides and neurotransmitters," or cause "neuroendocrine dysfunction of the limbic system," or affect the "synchrony or coherence between components of the circadian system" (Deeks 2003). However, as outlined above, research that has examined depression over the life span suggests that women are less likely to be diagnosed with depression at midlife and beyond than in their younger years. Equally, systematic reviews fail to find an association between major depression and menopause (Yonkers et al. 2000) suggesting that the inevitability of menopausal (or perimenopausal) depression is a myth. For example, in a longitudinal study of 2565 women aged 45–55 living in Massachusetts, the majority of women who entered menopause did not become depressed; the women who *did* report depression at midlife were more likely to have been depressed earlier in life (Avis et al. 1994). Similarly, in a study of 2000 Australian women aged 45–55, the majority reported that most of the time they felt clear-headed (72%), good natured (71%), useful (68%), satisfied (61%), confident (58%), loving (55%), and optimistic (51%) (Dennerstein 1996). This is illustrated by the findings of a study conducted by Janette Perz and myself (Perz and Ussher 2008). All of the women we interviewed positioned midlife as a time of positivity, feeling comfortable with themselves and looking forward to the next phase of life, describing themselves as more confident in expressing their opinions and beliefs, regardless of the reactions of others, as a result of being "stronger" or "wiser," through having experienced life.

Social and relational context, and women's negotiation of midlife change, appear to be the factors associated with distress (or well-being) at midlife – rather than hormonal changes in the menopausal body (Ussher et al. 2019). Thus, in a study of 469 Manitoba women, family shifts and stresses were reported to be the strongest predictor of depression at midlife (Kaufert et al. 1992). Other studies have found that married women report less depression than unmarried women at midlife, although marital satisfaction was a greater predictor of emotional well-being than marital status per se (Robinson Kurpius et al. 2001). This suggests that relational factors are associated with menopausal depression, as is the case with depression at any stage in the life cycle, meaning that the notion of the menopausal body *causing* turmoil and depression is nothing more than a fiction. The cultural construction of women's aging is also important. For example, in a study which compared depression at midlife in North American and Japanese women, Avis et al. (1993) reported that there were much lower rates reported in the Japanese group, reflecting the different cultural meaning of menopause and aging in Japan. However, at the same time, in a study of midlife women living in New York, the majority of women said that they felt very happy (McQuaide 1998), with the factors which predicted well-being including higher income, having a close group of friends, good health, high self-esteem, goals for the future, and positive feelings about appearance. Indeed recent Western cultural representations of feeling "fabulous at 40" (or 50) (Maccaro 2007) challenge the age-old negative stereotypes of the decrepit menopausal woman in the West.

When the reproductive body is positioned as to blame for women's madness, this reinforces has been previously described as "the myth of the monstrous feminine"

(Ussher 2006), wherein fecundity marks women as "other" – as fickle, fearful, and potentially dangerous. This pathologizes woman for the very aspect of her being which most characterizes her difference from man, at the same time as the complex roots of her despair are denied.

Gender-Based Diagnosis of Depression: Pathologizing Femininity

So how can we understand women's higher rates of diagnosis of depression? Feminist critics have convincingly argued that such diagnosis is a gendered construct. Much attention has been given to research conducted by Broverman and colleagues in the late 1960s, where it was argued that women who conform to the feminine role, and paradoxically, also those who reject it, were likely to receive a psychiatric diagnosis (Broverman et al. 1970). At the same time, definitions of mental health were found to coincide with definitions of masculinity, whereas femininity was seen as psychologically unhealthy. The decade after this research produced a flurry of studies that confirmed these findings (Sherman 1980), reporting that psychiatrists positioned women who deviate from gender role stereotypes as the most disturbed.

At the same time, it has been reported that medical practitioners over diagnose depression in women, as a diagnosis is given even when women do not meet the criteria for depression (Potts et al. 1991). Conversely, when women and men *do* meet the criteria for depression, it has been reported that men are less likely than women to receive a diagnosis (Potts et al. 1991), suggesting that underdiagnosis operates for men. In one study, the overdiagnosis of women only operated with male psychiatrists (Loring and Powell 1988), suggesting that the gender (or perhaps the prejudices) of the physician may influence their judgment – in line with previous reports that clinicians' personal identities and demographic characteristics influence how they relate to clients (Poland and Caplan 2004). Indeed it has been suggested that gender role stereotypes used by clinicians lead to women being seen "as intrinsically more maladjusted" (Potts et al. 1991) – health professionals expect women to be mad (or "depressed"), so are more likely to look for it, and to see it even if it is not there. It has also been claimed that gender bias exists in the standardized questionnaires which measure depression, as many categorize experiences that are normative for women, or part of the feminine role (such as crying, sadness, or loss of interest in sex) as "symptoms" (Salokangas et al. 2002). Thus, instruments, that are often used in large-scale epidemiological surveys, may simply be overestimating depression in women (Salokangas et al. 2002), and thus distorting conclusions about gender differences in psycho-pathology. Conversely, it has been suggested that standardized measures of depression may *under*diagnose depression in particular cultural groups, such as South Asian women, as a result of being culturally or linguistically insensitive to the meanings of distress in a non-Western context (Nazroo 1997).

Women in specific demographic groups – in particular, working class women (Caplan and Cosgrove 2004), Black women (Loring and Powell 1988), older women (Siegal 2004), and lesbian or queer women (Metcalfe and Caplan 2004) – are at the highest risk of this (mis)diagnosis. For example, the landmark study reported that lower income clients were more likely to receive a diagnosis of severe mental illness, and were also more likely to receive electroconvulsive therapy (ECT), drugs, lobotomies or custodial care, when compared to more wealthy clients (Hollingshead and Redlich 1958). Equally, therapists have been found to rate depressed African-American clients more negatively than depressed Anglo-Americans (Jenkins-Hall and Sacco 1991), with White therapists rating African-America clients as more psychologically impaired than African-American therapists (Jones 1982). In the UK, the lower rates of depression in South Asian women were explained by one group of health professionals through the adoption of a range of "orientalist" stereotypes which acted to pathologize South Asian culture – positioning it as "other," patriarchal and repressive – compared to an idealized liberated West (Burr 2002). This can result in South Asian women's mental health needs going unrecognized and untreated (Williams et al. 2006), as well as adding to the high level of stigma associated with discussing psychological distress that already exists in this cultural group (Hussain and Cochrane 2004). Misdiagnosis of disadvantaged groups is both part of and compounds minority stress – social discrimination that leads to distress and influences mental health outcomes (Meyer 2003). Older women are also more likely to experience both over- and underdiagnosis (Siegal 2004). The tendency to pathologize everything about older people can lead to misery being labelled as mental illness, when it is not (Halleck 1971), while invisibility can lead to lack of recognition of symptoms and the withholding of necessary support services (Ginter 1995). While homosexuality was officially been removed from the *DSM*, many clinicians still view it as a pathology, and will pathologize lesbian, queer, and trans women who seek help for difficulties with relationships or work, seeing their sexuality as an issue of concern (Charter et al. 2020; Metcalfe and Caplan 2004). From the perspective of intersectionality theory (Crenshaw 1989; Davis 2008), where it is recognized that women have multiple cultural and social identities, being a member of multiple marginalized groups may exacerbate vulnerability to misdiagnosis. For example, identity development differences between White and African American or Latina lesbians (Parks et al. 2004) may have implications for mental health and well-being, and it has been reported that older women who are poor face the double discrimination of class and age (Siegal 2004).

A further, related, explanation for women's higher rates of diagnosis as mad is that men and women differ in their presentation of distress, meaning that women are more likely to come under the scrutinizing gaze of mental health professionals. This suggests that gender differences in depression are an artifact – the result of women being more likely than men to report either mild symptoms of depression (Newman 1984), or symptoms that last a few days (Craig and Van Notta 1979). Conversely, it has been suggested that men are more likely to forget their depressive symptoms than women, or to underplay the severity of past episodes (Wilhelm and Parker 1994), and to deny depression, seeing it as self-indulgent and unproductive. Men who experience depressed mood have been reported to be less likely than women to express their feelings openly, which can make it more difficult for clinicians to detect that there is a problem (Batty 2006). However, at the same time, belief in the veracity of these accounts can serve as a justification for gender bias in clinicians' judgments, and for the absence of self-reflection, as responsibility for overdiagnosis in women (as well as underdiagnosis in men) is placed firmly at the feet of the "patient."

Treatment or Torture? Controlling Deviant or Difficult Women

Gender-based diagnosis is not the only concern of feminist critics. Since psychiatrists took over from the lay asylum keepers in the mid-nineteenth century, women have been hospitalized, had their freedom curtailed, and have been subjected to myriad "treatments," because they are deemed mad. First-hand accounts of those incarcerated, invariably against their will, tell of women brutally removed from normal life, to be subjected to myriad deprivations and interventions which would drive anyone insane: being fed a diet of plain gruel, and forced to eat even when there is no appetite; being kept in isolation for days at a time; being assaulted by attendants, strapped into manacles, a boxed crib, or a straight jacket; force-fed drugs; subjected to the cold pack or hydra water treatment, involving submersion in icy cold water for hours at a time, held down with straps and layers of wet blankets (Geller and Harris 1994).

While all of the "treatments" described above were also offered to men (or rather, men were also subjected to them), there were also a range of invasive interventions practiced solely on women, focusing on the sexual or reproductive body. These included injections of ice water into the rectum; placement of ice in the vagina; leeching of the labia and cervix (Showalter 1987); removal of ovaries to calm raging hormones; enforced weight gain to keep the ovaries from slipping and causing discomfort; electrical charges applied to the uterus; hot water injections in the vagina; and clitoral cauterization (Geller and Harris 1994). For puerperal mania, the treatment recommended was "to shave and apply cold to the head, administer tartar-emetic, purge, and blister" (Barker 1883). Within the asylum, women patients who were "violent, mischievous, dirty (and used) bad language" were put in solitary confinement (Showalter 1987), a punishment not meted out to similarly behaved men. Noisy women were also kept silent through the use of a "scold's bridle" (Showalter 1987), a metal helmet framing the head with a metal "bit" placed in the mouth.

When we look to the "symptoms" which provoked these pronouncements and treatments, we can see how the very definition of madness functioned to control and arguably punish women for both enacting an exaggerated form of femininity, or for being "unacceptable" (Geller and Harris 1994) – contravening the ideals of femininity circulating at that particular time. Agnes E. was committed to the Auckland asylum in the late nineteenth century for having "used horrible language, having been up to that time a decent and even religious woman, fully self-respecting," in the words of her case notes (Stubbs and Tolmie 2005, p. 69). Phebe Davis spent 3 years in the New York Asylum, from 1850 to 1853, diagnosed as "insane" because she had the temerity to disagree with other people and make her views known.

One feminized treatment, made infamous to successive generations through its exposure in Charlotte Perkins Gilman's auto-biographical novel "*The Yellow Wallpaper*," was Silas Weir Mitchell's bed rest cure. This involved a woman (and it was always a woman) being confined to her bed in a darkened room for between 6 weeks and 2 months, forbidden from any mental or physical activity, including talking, reading, sewing, writing, or even sitting up, with a nurse undertaking feeding and bedpan cleaning. Some women were given also electrical massages to stimulate their limbs (Mitchell 1877). Gilman wrote how this treatment drove her mad: "I would crawl into remote closets and under beds – to hide from the grinding pressure of that profound distress" (Gilman 1935). She finally escaped. Other women were not so fortunate. Susannah E. was committed three times between 1869 and 1870 for attempting to leave her husband, and Helen C. was committed following a complaint of marital ill treatment, including being called a "bloody whore" and being ordered out of her home, after asking her husband for a drink of water (Labrum 2005). Wives were clearly expected to be compliant, respectful, and satisfied – as were mothers. Mary O. was committed to the Auckland asylum for refusing "to see her children in the room" and asking "to take them away from her," and Jessie N. was hospitalized in the same place for "being far more satisfied to be absent from [her children] than a same mother would be" (Labrum 2005). As Bronwyn Labrum writes in her analysis of the Auckland asylum cases, mothers were supposed to be "self-less and giving, servers not takers, not self-absorbed as another 'neglectful' mother was described" (Labrum 2005, p. 73). All of these women were incarcerated at the request of fathers or husbands.

Some would argue little has changed. The physical restraints of the Victorian asylum have been replaced by drug treatments and ECT. Today, women committed to secure hospitals are infantilized, subjected to ECT at three times the rate of men, as well as being more likely to receive forced ECT and medication (Williams 1996), and report high rates of emotional and physical abuse by staff – punishment for not "toeing the line" (Lloyd 2005). Bonnie Burstow has described ECT as "state sponsored violence against women" (Burstow 2006) and Carol Warren as literally a "shocking experience" (Warren 1988). As the benefits of ECT are highly questionable – there is evidence that it is no more effective than placebo in alleviating either depression or suicide risk (Black and Winoker 1989) – there is little justification for forcibly subjecting women to this form of treatment.

Within bio-psychiatry today, depression is considered to be a disorder of the brain, requiring "anti-depressant" medication. "Serotonergic abnormalities" are the most widely accepted explanation today (Butler and Meegan 2008), which we are told can be "corrected" by SSRIs (Zoloft 2009), the most common being *Prozac* – fluoxetine hydrochloride, or *Zoloft* – sertraline hydrochloride.

While efficacy of treatment is deemed proof of a depletion of serotonin, there is no consistent evidence of specific imbalances in serotonin in individuals reporting depression (Moncrieff 2009), and no evidence that making serotonin nerves more active, the aim of SSRIs, can help people overcome emotional problems (Breggin and Breggin 1991). Scientists themselves readily admit they are unsure how antidepressants work (Gardner 2003), with even those who are advocates of SSRI's, concluding that "there are complex interactions between the various neurotransmitters that are not yet fully understood" (Butler and Meegan 2008). Equally, the efficacy of placebos in comparison to SSRIs is so strong – the response to medication is duplicated in 80% of placebo groups – the pharmacological rationale for the treatment is undermined (Moncrieff and Kirsch 2005). At the same time, the fact that the original U.S. Food and Drug Administration (FDA) testing of SSRIs was conducted on small groups of men with diagnoses of major depression (Kramer 1993), yet the major market for these drugs today are women with minor or "shadow" depression (Gardner 2003) is also problematic. There is evidence for sex differences in response to SSRIs and other antidepressants (Bigos et al. 2009). And while the majority of the clinical trials for the various brands of SSRIs lasted only 6 weeks (Marcia 2004), many people are prescribed SSRIs on a long-term basis, with disregard for the unknown consequences of blocking serotonin long term (Currie 2005).

SSRI's have also been associated with serious side effects, including suicide, aggression, harm to relationships, dystonia (muscle spasm), sexual dysfunction, akathisia (inner agitation), chronic dyskinesis (abnormal muscle movements), gastrointestinal and dermatological problems, and "out of character" behavior (Liebert and Gavey 2008). The rates of side effects in populations taking SSRIs are startlingly high. For example, sexual dysfunction is estimated to affect 30–70% of SSRI users (Gregorian et al. 2002); agitation, dizziness, headaches, or sleep problems 10–32% (VanderKooy et al. 2002); and neurological problems 22% (Spigset 1999). The most serious side effect, suicide, denied for many years by the drug companies, is now accepted as double the risk of older antidepressants or nontreatment (Currie 2005), and three times the risk for low-risk depression treated in primary care contexts (Healy 2003) – the bulk of the prescription market. As women form the majority of those prescribed SSRIs (Currie 2005), and also experience a higher rate of the most serious side effects (Spigset 1999), this is clearly a gendered issue.

There is also a very high rate of "spontaneous remission" with depression (along with all so-called psychiatric disorders), meaning that the majority of people experience alleviation of "symptoms" in a relatively short period without any outside intervention at all (Andrews 2001b; Kendler et al. 1997). Diet and exercise can also be as effective as SSRI's in treating the "symptoms" that are described as depression (Dunn et al. 2005), as can psychological therapy, including cognitive behavior therapy or interpersonal therapy (Hales and Hales 1995; Williams 1992), as well as narrative therapy (Brown and Augusta-Scott 2007), or psycho-analytic psychotherapy (McWilliams 2004). These therapeutic approaches do have their own limitations, not least of which is the implicit acceptance of "depression" as a disorder (in contrast to or feminist therapy (Brown 2010; Ussher et al. 2002), but they do have the advantage of not being a pharmacological cure. However, those promoting noninvasive interventions (or the absence of intervention altogether) face the combined marketing might of BigPharma – formidable competition. Pharmaceutical companies are among the most profitable in the world – driven by the economic imperative to keep profits high through retaining and continuously expanding their market (Currie 2005). Psychotropic medication plays a key role in these profits, with the top five SSRIs earning between $1 billion and $3 billion each (Medawar and Hardon 2004), despite the drugs being almost identical (Currie 2005), amounting to total profits of over $10 billion per year (Nestler et al. 2002). And it is not just BigPharma marketers who promote a biomedical solution for misery and madness. In a study which examined the rhetorical strategies adopted by a range of professionals who worked in fields associated with depression, and who were supportive of SSRIs, Leibert and Gavey found that risk was minimized, or seen to be easily managed, benefits were believed to out-way risks, and the existence of side effects was questioned (Liebert and Gavey 2009). For example, increased incidences of suicidal thoughts were dismissed by a registered pharmacist, because "there wasn't increased risk of suicide."

As Leibert and Gavey (2009) conclude, these rhetorical strategies enable knowledge of adverse effects to be "contained within discourses that did not interrupt participants' ongoing support for the drugs" (p. 1890). The use of these trivializing, justifying, or pathologizing discourses clearly functions to protect those who advocate SSRIs from having to take any responsibility for addressing the complex issue of serious side effects, while further obscuring discussion of these issues in public life. It also means that these drugs continue to be prescribed, as the problem is established as more problematic than the cure.

Conclusion

All of these arguments support a social constructionist analysis of women's madness, wherein psychiatric diagnosis is seen as a historically and culturally specific process which functions to position those deemed to have "disorders" such as depression, or premenstrual dysphoric disorder as outsiders, as mad, serving to legitimate medical intervention and control. The very legitimacy of individual diagnostic categories is questioned, as can the objectivity and neutrality of those making diagnoses and offering treatment – in particular the bio-psychiatrists with their pecuniary connections to BigPharma, the industry that most profits from women's madness.

However, we need to understand why women experience distress, even if we are critical of diagnostic categories such as depression, or PMDD. Feminists who dismiss medicalization are also left with the dilemma that at an individual level, psychiatric diagnosis can serve to validate to women that there is a "real" problem, isolating prolonged misery from "the character of the sufferer" (LaFrance 2007). For some women, adopting a biomedical model to explain their experiences as "depression" means that they aren't simply "crazy" or "malingering," as depression is positioned as something they cannot avoid, rather than it being a personal failing.

In arguing that "madness" exists entirely at a discursive level, we may also implicitly deny the influence of biology or genetics, or appear to relegate the body to a passive subsidiary role, which has meaning or interpretation imposed upon it, or is seen as separate from the psychosocial domain. We cannot deny the material and intrapsychic concomitants of the experiences that are constructed as disorders such as "depression": the exhaustion, insomnia, problems in concentration, appetite disturbance, or the feelings of heaviness, agitation, despair, and desperation. Other material aspects of women's lives may also be negated in a discursive analysis: including the influence of social class, age, power, socioeconomic status, ethnicity, sexual and gender diversity, relationship status and social support, or a history of violence and abuse (Ussher 2011).

A material-discursive-intrapsychic (MDI) analysis allows us to acknowledge the "real" of women's psychological and somatic distress, whether this distress is either mild or severe, yet to conceptualize it as a complex phenomenon which is only discursively positioned as "madness" or as a diagnostic categories such as "depression" within a specific historical and cultural context (Ussher 2010). Equally, we can acknowledge that an individual woman, living in a particular place at a particular point in time, with a particular set of personal circumstances, beliefs, and coping strategies, may come to experience psychological distress, label it as depression, and then seek treatment, because of the complex interaction of these different factors within her life. However, none of these material, discursive, or intrapsychic levels of analysis is privileged above the other.

This material-discursive-intrapsychic (MDI) framework allows us to examine the ways in which the materiality of a woman's embodiment and life context, the discursive construction of madness and of gendered roles, and the intrapsychic negotiation that women, and their families, engage in, contributes to the lived experience of distress, or a woman being positioned as mad. It also allows us to acknowledge the intersection of identities, and the multiple layers of marginalization experienced by many women, such as those who are Black, queer, trans, and/or poor.

While critiquing medicalization of madness, it is important to acknowledge that psychotropic medication, particularly when used alongside therapy, may be beneficial for alleviating some cases of "extreme mental turmoil" (Moncrieff 2009, p. 308), when the distress has not been reduced by time, the greatest cure of all – "spontaneous remission" (Andrews 2001a). However, it is not necessary or appropriate for the "problems in everyday living" (Currie 2005, p. 19) that are positioned as "depression" or "anxiety" in pharmaceutical advertising (Metzl 2003). Medical treatments should also never stand alone as solutions for women's misery – in a clinical setting they need to be part of a multidisciplinary approach that involves "consumers/survivors and recovering persons in a meaningful way" (Morrow 2008), where women have *choice* in the treatments they receive, as well as a choice to have no treatment at all. And *any* individual treatment always needs to be part of a broader project that addresses the social, relational, and political causes of women's suffering, where the individual woman is not the sole focus of intervention, with the problem blamed on her body.

References

Albright A (1993) Postpartum depression: an overview. J Couns Dev 71(3):316–320
American Psychiatric Association (2013) Diagnostic and statistical manual of mental disorders, edition V. American Psychiatric Association, Washington, DC

Andrews G (2001a) Placebo response in depression: bane of research, boon to therapy. Br J Psychiatry 178(3):192–194

Andrews G (2001b) Should depression be managed as a chronic disease? BMJ 322(7283):419–421

Angold A, Costello EJ, Worthman CM (1998) Puberty and depression: the roles of age, pubertal status and pubertal timing. Psychol Med 28:51–61

Avis N, Kaufert PA, Lock M, McKinlay SM, Vass K (1993) The evolution of menopause symptoms. Int Pract Res 7(1):17–32

Avis N, Brambilla D, McKinlay SM, Vass K (1994) A longitudinal analysis of the association between menopause and depression. Results from the Massachutes women's health study. Ann Epidemiol 4(3):214–220

Barker F (1883) The puerperal diseases: clinical lectures delivered at Bellevue Hospital. Appleton, New York

Batty Z (2006) Masculinity and depression: Men's subjective experience of depression, coping and preferences for therapy and gender role conflict. PhD, University of Western Sydney

Bebbington PE (1996) The origins of sex differences in depressive disorder: bridging the gap. Int Rev Psychiatry 8(4):295–332

Bebbington PE, Dunn G, Jenkins R, Lewis G, Brugha T, Farrell M, Meltzer H (1998) The influence of age and sex on the prevalence of depression conditions: report from the National Survey of psychiatric morbidity. Psychol Med 28(1):9–19

Becker D (2000) When she was bad: borderline personality disorder in a posttraumatic age. Am J Orthopsychiatry 70(4):422–432

Bell AJ, Land NM, Milne S, Hassanyeh F (1994) Long-term outcome of post-partum psychiatric illness requiring admission. J Affect Disord 31(1):67–70

Bigos KL, Pollock BG, Stankevich BA, Bies RR (2009) Sex differences in the pharmacokinetics and pharmacodynamics of antidepressants: an updated review. Gend Med 6(4):522–543

Black D, Winoker G (1989) Does treatment influence mortality of depressives? Ann Clin Psychiatry 1:165–173

Borossa J (2001) Hysteria. Cambridge Icon Book

Boyce P, Hickey A (2005) Psychosocial risk factors to major depression after childbirth. Soc Psychiatry Psychiatr Epidemiol 40:605–612

Breggin PB, Breggin GR (1991) Toxic psychiatry: why therapy, empathy and love must replace the drugs, electroshock and biochemical theories of the "new psychiatry". St Martins Press, New York

Bronfen E (1998) The knotted subject: hysteria and its discontents. Princeton University Press, Princeton

Broverman K, Broverman D, Clarkson F, Rosenkrantz P, Vogel S (1970) Sex role steretypes and clinical judgements of mental health. J Consul Clin Psychol 34(1):1–7

Brown C, Augusta-Scott T (2007) Narrative therapy: making meaning, making lives. Sage Publications, Thousand Oaks

Brown LS (2010) Feminist therapy. American Psychological Association, Washington, DC

Burr J (2002) Cultural stereotypes of women from South Asian communities: mental health care professionals' explanations for patterns of suicide and depression. Soc Sci Med 55(5):835–845

Burrows GM (1828) Commentaries on the causes, forms symptoms and treatment, moral and medical, of insanity. Underwood, London

Burstow B (2006) Understanding and ending ECT. A feminist imperative. Can Women's Stud 25(1, 2):115–122

Busfield J (1996) Men, women and madness: understanding gender and mental disorder. New York University Press, New York

Butler S, Meegan M (2008) Recent developments in the design of anti-depressive therapies: targeting the serotonin transporter. Curr Med Chem 15(17):1737–1761

Calabrese SK, Meyer IH, Overstreet NM, Haile R, Hansen NB (2015) Exploring discrimination and mental health disparities faced by Black sexual minority women using a minority stress framework. Psychol Women Q 39(3):287–304

Caplan PJ, Cosgrove L (2004) Is this really necessary? In: Caplan PJ, Cosgrove L (eds) Bias in psychiatric diagnosis. Jason Aronson, Inc, Northvale, pp xiv–xxxiii

Charter R, Ussher JM, Perz J, Robinson K (2020) The transgender and gender diverse parent: negotiating mental health issues. Int J Transgender Health, in press

Chrisler JC, Caplan P (2002) The strange case of Dr. Jekyll and Ms Hyde: How PMS became a cultural phenomenon and a psychiatric disorder. Ann Rev Sex Res 13:274–306

Cixous H, Clement C (1987) The newly born women, trans Betsy Wing. University of Minnesota Press, Minneapolis

Cosgrove L, Riddle B (2004) Gender bias in sex distribution of mental disorders in the DSM-IV-TR. In: Caplan PJ, Cosgrove L (eds) Bias in psychiatric diagnosis. Jason Aronson, New York, pp 127–140

Craig TJ, Van Notta PA (1979) Influence of two demographic charachteristics on two measures of depressive symptoms. Arch Gen Psychiatry 36:149–154

Crenshaw K (1989) Demarginalizing the intersection of race and sex: a Black feminist critique of antidiscrimination doctrine, feminist theory and antiracist politics. Univ Chic Leg Forum 140: 139–167

Currie J (2005) The marketization of depression: the prescribing of SSRI antidepressants to women. Women Health Protect. Retrieved July 15, 2009

Cyranowski JM, Frank E, Young E, Shear MK (2000) Adolescent onset of the gender difference in lifetime rates of major depression. Arch Gen Psychiatry 57:21–27

Davis K (2008) Intersectionality as buzzword: a sociology of science perspective on what makes a feminist theory successful. Fem Theory 9(1):67–85

Deeks AA (2003) Psychological aspects of menopause management. Best Pract Res Clin Endocrinol Metabol 17(1):17–31

Dennerstein L (1996) Well-being, symptoms and the menopausal transition. Maturitas 23:147–157

Dunn A, Trivedi M, Kampers J, Clark C (2005) Exercise treatment for depression efficacy and dose response. Am J Prev Med 28(11):1–8

Fee D (2000) The project of pathology: reflexivity and depression in Elizabeth Wurtzel's Prozac Nation. In: Fee D (ed) Pathology and the postmodern: mental illness as discourse and experience. Sage, London, pp 74–99

Ferlatte O, Salway T, Rice SM, Oliffe JL, Knight R, Ogrodniczuk JS (2019) Inequities in depression within a population of sexual and gender minorities. J Ment Health

Fisher J, de Mello MC, Patel V, Rahman A, Tran T, Holton S, Holmesf W (2012) Prevalence and determinants of common perinatal mental disorders in women in low-and lower-middle-income countries: a systematic review. Bull World Health Organ 90(2):139–149

Foucault M (1967) Madness and civilisation: a history of insanity in the age of reason. Tavistock, London

Foucault M (1972) The archeology of knowledge and the discourse on language. Pantheon Books, New York

Foucault M (1978) The history of sexuality: an introduction. Penguin, London

Frank R (1931) The hormonal causes of premenstrual tension. Arch Neurol Psychiatr 26:1053–1057

Gallop J (1983) Nurse Freud: class struggle in the family. Miami University. Cited by Showalter 1993, p 288

Gardner P (2003) Distorted packaging: marketing depression as illness, drugs as cure. J Med Hum 24(1/2):105–130

Geller JL, Harris M (1994) Women of the asylum. Voices from behind the walls 1840–1945. Anchor Books, New York

Gilman CP (1935) The living of Charlotte Perkins Gilman. Arno Press, New York

Ginter GG (1995) Differential diagnosis in older adults: dementia, depression and delirium. J Couns Dev 73:346–351

Gregorian R, Golden K, Bahce A (2002) Antidepressant induced sexual dysfunction. Ann Phaarmacother 36:1577–1589

Groneman C (1994) Nymphomania: the historical construction of female sexuality. Signs J Women Cult Soc 19(2):337–367

Hales DR, Hales RE (1995) Caring for the mind: the comprehensive guide to mental health. Bantam Books, New York

Halleck SL (1971) The politics of therapy. Science House, New York

Hargrove TW, Halpern CT, Gaydosh L, Hussey JM, Whitsel EA, Dole N et al (2020) Race/ethnicity, gender, and trajectories of depressive symptoms across early- and mid-life among the add health cohort. J Racial Ethn Health Disparities 7(4):619–629

Harwood K, McLean N, Durkin K (2007) First time mothers' expectations of parenthood: what happens when optimistic expectations are not matched by later experiences? Dev Psychol 43(1):1–12

Haslam J (1809) Observations on madness and melancholy, London

Healy D (2003) Lines of evidence on the risk of suicide with selective serotonin reuptake inhibitors. Psychother Psychosom 72:71–76

Herndl DP (1988) The writing cure. NWSA J 53

Hill Collins P (2002) Black feminist thought: knowledge, consciousness, and the politics of empowerment. https://doi.org/10.4324/9780203900055

Hollingshead AB, Redlich FC (1958) Social class and mental illness: a community study. John Wiley and Sons, New York

Hollway W (2006) The capacity to care: gender and ethical subjectivity. Routledge, London

Hussain F, Cochrane R (2004) Depression in South Asian women living in the UK: a review of the literature with implications for service provision. Transcult Psychiatry 41(2):253–270

Hyde Z, Doherty M, Tilley P, McCaul K, Rooney R, Jancey J (2014) The first Australian National Trans Mental Health Study: summary of results. School of Public Health, Curtin University, Perth

Jenkins-Hall K, Sacco WP (1991) Effect of client race and depression on evaluations by white therapists. J Soc Clin Psychol 38:322–333

Jones EE (1982) Psychotherapists' impressions of treatment outcome as a function of race. J Clin Psychol 38:722–731

Kaufert PA, Gilbert P, Tate R (1992) The Manitoba project: a re-examination of the link between menopause and depression. Maturitas 14:143–155

Kendler KS, Walters EE, Kessler RC (1997) The prediction of length of major depressive episodes: results from an epidemiological survey of female twins. Psychol Med 27:107–117

Kessler RC, McGonagle KA, Swartz M, Blazer DG, Nelson CB (1993) Sex and depression in the National Comorbidity Survey I: lifetime prevalence, chronicity and recurrence. J Affect Disord 29:85–96

Klein DF, Wender PH (1993) Understanding depression. Oxford University Press.

Kramer P (1993) Listening to prozac: a psychiatrist explores antidepressant drugs and the remaking of the self. Viking Penguin, New York

Kuehner C (2003) Gender differences in unipolar depression: an update of epidemiological findings and possible explanations. Acta Psychiatr Scand 108:163–174

Labrum B (2005) The boundaries of femininity: madness and gender in New Zealand 1870–1910. In: Menzies R, Chunn DE, Chan W (eds) Women, madness and the law: a feminist reader. Glasshouse Press, London, pp 60–77

LaFrance MN (2007) A bitter pill. A discursive analysis of women's medicalized accounts of depression. J Health Psychol 12(1):127–140

Liebert R, Gavey N (2008) "I didn't just cross a line I tripped over an edge". Experiences of serious adverse side effects with selective serotonin reuptake inhibitor use. N Z J Psychiatry 37(1):38–48

Liebert R, Gavey N (2009) "There are always two sides to these things": managing the dilemma of serious side effects from SSRIs. Soc Sci Med 68:1882–1891

Lloyd A (2005) The treatment of women in secure hospitals. In: Menzies R, Chunn DE, Chan W (eds) Women, madness and the law: a feminist reader. Glasshouse Press, London, pp 227–244

Loring M, Powell B (1988) Gender, race and DSM-III: a study of the objectivity of psychiatric diagnostic behavior. J Health Soc Behav 29(1):1–22

Maccaro J (2007) Fabulous at 50. Siloam, Lake Mary
Malson H (1998) The thin woman: feminism, post-structuralism and the social psychology of anorexia nervosa. Routledge, London
Marcia A (2004) The truth about the drug companies: how they decive us and what to do about it. Random House, New York
Maudsley H (1879) The pathology of mind. Macmillan, London
Mauthner N (1998) 'It's a woman's cry for help': a relational perspective on post-natal depression. Fem Psychol 8(3):325–355
Mauthner N (2010) "I wasn't being true to myself". Women's narratives of postpartum depression. In: Jack DC, Ali A (eds) The depression epidemic: international perspectives on women's self-silencing and psychological distress. Oxford University Press, Oxford, pp 459–484
McQuaide S (1998) Women at midlife. Soc Work 43(1):21–31
McWilliams N (2004) Psychoanalytic psychotherapy: a practitioner's guide. Guilford Press, New York
Medawar D, Hardon A (2004) Medicines out of control? Antidepressants and the conspiracy of goodwill. Aksant Academic Publishers, Netherlands
Metcalfe WR, Caplan PJ (2004) Seeking "normal" sexuality on a complex matrix. In: Caplan PJ, Cosgrove L (eds) Bias in psychiatric diagnosis. Jason Aronson, New York, pp 121–126
Metzl JM (2003) Prozac on the couch: prescribing gender in the era of wonder drugs. Duke University Press, Durham
Meyer IH (2003) Prejudice, social stress, and mental health in lesbian, gay, and bisexual populations: conceptual issues and research evidence. Psychol Bull 129:674–697
Miller LJ (2002) Postpartum depression. JAMA 287:762–765
Mira M, Abraham S, McNeil D, Vizzard J (1995) The inter-relationship of premenstrual symptoms. Psychol Med 25(5):947–955
Mitchell SW (1877) Massage. J Nerv Ment Dis 4:636–638
Mitchell SW (1885) Lectures on diseases of the nervous system especially in women. J.A. Churchill, London
Moncrieff J (2009) Deconstructing psychiatric treatment. In: Reynolds J, Muston R, Heller T, Leach J, McCormick M, Wallcraft J, Walsh M (eds) Mental health still matters. Palgrave Macmillan, London, pp 301–309
Moncrieff J, Kirsch I (2005) Efficacy of anti-depressants in adults. Br Med J 331:155–157
Morrow M (2008) Women, violence and mental illness: an evolving feminist critique. In: Patton C, Loshny H (eds) Global science/women's health. Cambria Press, New York, pp 147–162
Nazroo JY (1997) Ethnicity and mental health. Policy Studies Institute, London
Nazroo JY, Edwards AC, Brown GW (1998) Gender differences in the prevalence of depression: artefact, alternative disorders, biology or roles? Sociol Health Illness 20(3):312–330
Nestler EJ, Barrot M, DiLeone RJ, Eisch AJ, Gold SJ, Monteggia LM (2002) Neurobiology of depression. Neuron 34(1):13–25
Nevatte T, O'Brien P, Bäckström T, Brown C, Dennerstein L, Endicott J et al (2013) ISPMD consensus on the management of premenstrual disorders. Arch Womens Ment Health 16(4): 279–291
Newman JP (1984) Sex differences in symptoms of depression: clinical disorder or normal distress? J Health Soc Behav 25:136–159
Nolen-Hoeksema S (1990) Sex differences in depression. Stanford University Press, Stanford
O'Hara MW, Swain AM (1996) Rates and risk of post-partum depression: a meta-analysis. Int Rev Psychiatry 8:37–54
Paikoff RL, Brooks-Gunn J, Worren MP (1991) Effects of girls' hormonal status on depressive and aggressive symptoms over the course of one year. J Youth Adolesc 20:1912–1915
Parker G, Brotchie H (2010) Gender differences in depression. Int Rev Psychiatry 22(5):429–436
Parks CL, Hughes TL, Matthews AK (2004) Race/ethnicity and sexual orientation: intersecting identities. Cult Divers Ethn Minor Psychol 10(3):241–254
Patel V, Rodrigues M, DeSouza N (2002) Gender, poverty, and postnatal depression: a study of mothers in Goa, India. Am J Psychiatr 159(1):43–47

Perz J, Ussher JM (2008) The horror of this living decay: Women's negotiation and resistance of medical discourses around menopause and midlife. Women's Stud Int Forum 31:293–299

Peterson AC, Sarigiani PA, Kennedy RE (1991) Adolescent depression: why more girls? J Youth Adolesc 20:247–271

Pitts M, Smith A, Mitchell A, Patel S (2006) Private lives: a report on the health and wellbeing of GLBTI Australians. Australian Research Centre in Sex, Health and Society, La Trobe, Melbourne

Poland J, Caplan PJ (2004) The deep structure of bias in psychiatric diagnosis. In: Caplan PJ, Cosgrove L (eds) Bias in psychiatric diagnosis. Jason Aronson, Inc, Northvale, pp 9–24

Popay J, Bartley M, Owen C (1993) Gender inequalities in health: social position, affective disorders and minor psychiatric morbidity. Soc Sci Med 36(1):21–32

Porter D (1855) Book of men, women and babies. De Witt and Davenport, New York

Porter R (1993) The body and the mind, the doctor and the patient: negotiating hysteria. In: Gilman SL, King H, Porter R, Rousseau GS, Showalter E (eds) Hysteria beyond Freud. University of California Press, Berkeley, pp 225–285

Potts MK, Burnam MA, Wells KB (1991) Gender differences in depression detection: a comparison of clinician diagnosis and standardized assessment. Psychol Assess J Consul Clin Psychol 3(4):609–615

Robinson Kurpius SE, Foley Nicpon M, Maresh SE (2001) Mood, marriage, and menopause. J Couns Psychol 48(1):77–84

Rose NS (1996) Inventing our selves: psychology, power, and personhood. Cambridge University Press, New York

Rousseau GS (1993) A strange pathology. hysteria in the early modern world 1500–1800. In: Gilman SL, King H, Porter R, Rousseau GS, Showalter E (eds) Hysteria beyond Freud. University of California Press, Berkeley, pp 91–221

Salokangas RKR, Vaahtera K, Pacriev S, Sohlman B, Lehtinen V (2002) Gender differences in depressive symptoms. an artefact caused by measurement instruments. J Affect Disord 68:215–220

Seaman MV (1997) Psychopathology in women and men: focus on female hormones. Am J Psychiatry 154(12):1641–1647

Sherman JA (1980) Therapist attitudes and sex role stereotyping. In: Brodsky AM, Hare-Mustin RT (eds) Women and psycho-therapy. Guilford, New York, pp 35–66

Shorter E (1992) From paralysis to fatigue. The Free Press, New York

Showalter E (1987) The female malady: women, madness and English culture 1830–1940. Virago, London

Showalter E (1997) Hystories: hysterical epidemics and modern culture. Columbia University Press, New York

Siegal RJ (2004) Ageism in psychiatric diagnosis. In: Caplan PJ, Cosgrove L (eds) Bias in psychiatric diagnosis. Jason Aronson, Inc., Northvale, pp 89–97

Smith-Rosenberg C (1986) Disorderly conduct: visions of gender in Victorian America. Oxford University Press, Oxford

Spigset O (1999) Adverse reactions of selective serotonin reuptake inhibitors – response from a spontaneous reporting system. Drug Saf 20:277–287

Stubbs J, Tolmie J (2005) Defending battered women on charges of homicide: the strucural and systematic versus the personal and particular. In: Menzies R, Chunn DE, Chan W (eds) Women, madness and the law: a feminist reader. Glasshouse Press, London, pp 191–209

Studd J (1997) Depression and the menopause. Br Med J 314:977

Sullivan PF, Neale MC, Kendler KS (2000) Genetic epidemiology of major depression: review and meta-analysis. Am J Psychiatry 157(10):1552–1562

Susman EJ, Nottelmann ED, Inoff-Germain G, Dorn L, Chrousos GP (1987) Hormonal influences on aspects of psychological development during adolescence. J Adolesc Health Care 8:492–504

Sydenham T (1679) Epistolary dissertation to Dr. Cole. The works of Thomas Sydenham, vol 2. The Sydenham Society, London. (1843)

Tilt EJ (1882) The change of life in health and disease. A clinical treatise on the diseases of the ganglionic nervous system incidental to women at the decline of life. Bermingham, New York

Uchino BN, Cacioppo JT, Kiecolt-Glaser JK (1996) The relationship between social support and physiological processes: a review with emphasis on underlying mechanisms and implications for health. Psychol Bull 119:499–531

Ussher JM (2004) Postnatal depression: a critical feminist perspective. In: Stewart M (ed) Pregnancy, birth and maternity care – a feminist perspective. Books for Midwives, London, pp 105–120

Ussher JM (2006) Managing the monstrous feminine: regulating the reproductive body. Routledge, London

Ussher JM (2008) Challenging the positioning of premenstrual change as PMS: the impact of a psychological intervention on women's self-policing. Qual Res Psychol 5(1):33–44

Ussher JM (2010) Are we medicalizing women's misery? A critical review of women's higher rates of reported depression. Fem Psychol 20(1):9–35

Ussher JM (2011) The madness of women: myth and experience. Routledge, London

Ussher JM, Perz J (2013) PMS as a gendered illness linked to the construction and relational experience of hetero-femininity. Sex Roles 68(1–2):132–150

Ussher JM, Perz J (2017) Evaluation of the relative efficacy of a couple cognitive-behaviour therapy (CBT) for premenstrual disorders (PMDs), in comparison to one-to-one CBT and a wait list control: a randomized controlled trial. PLoS One 12(4):e0175068

Ussher JM, Perz J (2020) "I feel fat and ugly and hate myself": self-objectification through negative constructions of premenstrual embodiment. Femin Psychol, online ahead of print (0)

Ussher JM, Hunter MS, Cariss M (2002) A woman-centred psychological intervention for premenstrual symptoms, drawing on cognitive-behavioural and narrative therapy. Clin Psychol Psychother 9:319–331

Ussher JM, Perz J, Mooney-Somers J (2007) The experience and positioning of affect in the context of intersubjectivity: the case of premenstrual syndrome. J Crit Psychol 21:145–165

Ussher JM, Hawkey AJ, Perz J (2019) 'Age of despair', or 'when life starts': migrant and refugee women negotiate constructions of menopause. Culture Health and Sexuality 21(7):741–756. https://doi.org/10.1080/13691058.13692018.11514069. Epub 13692018 Oct 13691053

Van de Velde S, Bracke P, Levecque K (2010) Gender differences in depression in 23 European countries. Cross-national variation in the gender gap in depression. Soc Sci Med 71(2):305–313

VanderKooy JD, Kennedy S, Bagby R (2002) Antidepressant side effects in depression patients treated in a naturalistic setting: a study of bupropien, moclobemide paroxetine, sertraline and venlafaxine. Western Canada J Psychiatry 47(2):174–180

Warren C (1988) Electroconvulsive therapy, the self and family relations. Res Sociol Health Care 7: 283–300

Whiffen VE (1992) Is postpartum depression a distinct diagnosis? Clin Psychol Rev 12(5):485–508

Wilhelm K, Parker G (1994) Sex differences in lifetime depression rates: fact or artifact? Psychol Med 24(1):97–11

Williams CC, Curling D, Steele LS, Gibson MF, Daley A, Green DC, Ross LE (2017) Depression and discrimination in the lives of women, transgender and gender liminal people in Ontario, Canada. Health Soc Care Commun 25(3):1139–1150

Williams J (1996) Social inequalities and mental health: developing services and developing knowledge. J Community Appl Soc Psychol 6(5):311–316

Williams JMG (1992) The psychological treatment of depression. Routledge, London

Williams PE, Turpin G, Hardy G (2006) Clinical psychology service provision and ethnic diversity within the UK: a review of the literature. Clin Psychol Psychother 13:324–338

Worthman CM (1995) Hormones, sex and gender. Annu Rev Anthropol 24:593–616

Yonkers KA, Bradshaw KD, Halbrieich U (2000) Oestrogens, progestins and mood. In: Steiner M, Yonkers KA, Eriksson E (eds) Mood disorders in women. Martin Dunitz, London, pp 207–232

Zoloft (2009). http://www.zoloft.com/common_questions.asp. Retrieved 18 Sep 2009

74. Feminine Hunger: A Brief History of Women's Food Restriction Practices in the West

Natalie Jovanovski

Contents

Introduction	1878
Food Restriction as a Harmful Gender Norm	1879
The WHM and the Physical and Psychological Harms of Food Restriction	1880
Anorexia Mirabilis: Hunger as a Sign of Women's Piety	1881
"The Weapon of Self-Hurt": Hunger as a Political Mouthpiece	1883
Anorexia Nervosa: Hunger as a Psychological Condition	1885
From Weight-Loss Dieting to "Wellness": Hunger as a Responsible Lifestyle Choice	1887
Visibilizing Women's Intentional Hunger: A Feminist Research Agenda for the WHM	1889
Recognize Women's Intentional Hunger as a Gendered Phenomenon Requiring Gender-Transformative Health Promotion	1890
Shift One's Focus from a Weight-Centric to Weight-Neutral Paradigm	1891
Reframe Individualized Narratives of Self-Care into Acts of Feminist Resistance	1891
Get Organized Through Feminist Consciousness-Raising	1892
Conclusion	1893
References	1893

Abstract

Intentional hunger, or the experience of voluntarily restricting one's food intake, has long been considered a women's issue. From tales of female fasting saints in thirteenth century Europe to today's Instagram celebrities sharing clean eating tips, the gendered connotations of food restriction and intentional hunger have been a consistent theme throughout Western history. While some sociologist and feminist writers have argued that the meanings ascribed to female food restriction practices should be located within their historical contexts and thus cannot be neatly compared, very few writers have situated intentional hunger within a broader feminist framework, citing patriarchy and the role of gender norms in relation to their diverse impact on women's eating behaviors and, subsequently,

N. Jovanovski (✉)
University of Melbourne, Melbourne, VIC, Australia
e-mail: jovanovskin@unimelb.edu.au

© The Author(s), under exclusive licence to Springer Nature Singapore Pte Ltd. 2022
D. McCallum (ed.), *The Palgrave Handbook of the History of Human Sciences*,
https://doi.org/10.1007/978-981-16-7255-2_29

their health. Indeed, despite largely affecting women, the women's health movement (WHM) has not yet played an active role in challenging these gendered norms in relation to diet culture. This chapter presents a feminist sociological analysis on women's intentional hunger using historical examples from a Western context. In doing so, the chapter shows that gender norms have played a central role in women's harmful and restrictive eating practices, and that contemporary challenges to "diet culture" must focus on challenging these gender norms in public health and health promotion materials.

Keywords

Intentional hunger · Weight-loss dieting · Diet culture · Gender norms · Patriarchy · Women's Health Movement (WHM) · Fasting · Hunger strike · Anorexia nervosa · Clean eating

Introduction

In historical accounts and contemporary representations of food and eating, tales of the hungry woman abound. From dark, satirical images of hunger striking suffragists imprisoned and force-fed over their right to vote, to millions of everyday social media posts showcasing women's "clean eating" tips, one feature that has remained historically consistent is the normalized discourse of women's intentional hunger (Nicholas 2008). Intentional hunger refers to the practice of voluntary food restriction – which includes the deliberate exclusion of some types of food over others – and differs from hunger experienced as a result of poverty or famine (e.g., Van Esterik 1999). Some sociologist and feminist writers have argued that the meanings ascribed to female food restriction practices should be located within their historical contexts and thus cannot be neatly compared (Brumberg 1989; Gooldin 2003; Walker Bynum 1985). However, very few writers have situated intentional hunger within a broader feminist framework, citing patriarchy and the role of gender norms in relation to their diverse impact on women's eating behaviors and, subsequently, their health, throughout various phases of Western history. This chapter presents a feminist sociological argument that food restriction – and by extension, women's intentional hunger – is a long-standing symbol of women's oppression under patriarchy; a topic that has received some attention in the feminist literature on eating psychopathology (e.g., Bordo 2004; Morgan 1977; Orbach 2005), but remained largely invisible as a feminist issue in women's health research and the women's health movement (WHM) in general. Given that food restriction practices have been associated with a host of adverse physical and psychological health outcomes in women (Omasu et al. 2019; Nordmo et al. 2019; Schilling 2018; Turner 2019), the chapter will demonstrate that the choice to engage in these behaviors is a harmful yet historically normalized act; one that reflects broader narratives about women's place in society and their right to nourishment.

Based on a summary of the existing literature, the chapter calls for the contemporary WHM to make visible women's food restriction practices and their intentional hunger. It will present a condensed history of self-induced food restriction practices, looking at the spiritual, political, and pathological dimensions of women's hunger in the West. By examining the meanings ascribed to these different forms of gendered food restriction practices, the chapter emphasizes that contemporary manifestations of self-induced food restriction, often in the form of "clean eating," are predominantly gendered issues; arguing that the WHM must play an important role in visibilizing women's intentional hunger and developing a language of resistance that encourages future generations of women to embrace food and eating.

Food Restriction as a Harmful Gender Norm

Food is considered a feminized object, but women have not been socialized to feel entitled to food. As has been argued previously (Jovanovski 2017, 2018), women's relationships with food are inherently gendered, and norms relating to female food practices (or "food femininities") tend to reinforce problematic narratives about women's right to eat and be nourished. For example, in popular cultural discourses, when women are depicted in relation to food, they are often shown policing their bodies, or preparing and consuming food for the approval of others (Jovanovski 2017, 2018), a feature that has managed to prevail across Western historical contexts through the reinforcement of gender norms. According to Weber et al. (2019, p. 2455), "gender norms are the spoken and unspoken rules of societies about the acceptable behaviors of girls and boys, women and men – how they should act, look, and even think or feel." Gender norms – either masculine or feminine – are socially constructed categories that create a sex-based hierarchy, with males socialized to be more dominant and females socialized to be submissive.

Gender norms and stereotypes are part of a larger sociocultural backdrop that, according to feminist theorists, are governed by a system of power relations called patriarchy (Bartky 1990; Weedon 1997). According to Weedon (1997, pp. 1–2), patriarchy refers to the "power relations in which women's interests are subordinated to the interests of men." Gender norms – such as those involved in shaping women's relationships with food – are thus used in patriarchal societies to control, suppress, and subordinate women (Jovanovski 2018). Feminine gender norms can manifest in diverse ways for women. For example, one feminine gender norm may involve cooking selflessly for others and feeding oneself last, while another may involve policing one's food intake to control one's weight. While seemingly disparate, these norms fall under the umbrella of "femininity" and are considered by feminist researchers to be forms of female control under patriarchy (Jovanovski 2017).

Women's intentional hunger is an important way to demonstrate the many ways that gender norms serve to construct women as passive, other-oriented and self-harming subjects. As sociologist Sigal Gooldin (2003, p. 31) has argued, the meanings ascribed to self-induced hunger "cannot be reduced to any single motivational factor," such as religion or psychopathology. Rather, as argued in this chapter, intentional hunger in women exists under the broad framework of patriarchy, and manifests in seemingly diverse gendered ways. This chapter will show that women's food restriction practices throughout history – and their experiences of intentional hunger – have been reinforced by gender norms (or food femininities) that tacitly influence how women should act in relation to food. These norms have played an important role in shaping women's health behaviors today and must be identified, foregrounded, and challenged by the WHM in a contemporary health context.

The WHM and the Physical and Psychological Harms of Food Restriction

Women's health researchers and activists – referred to broadly in this chapter as the "women's health movement" – are well-placed to address intentional hunger as a harmful gendered phenomenon, as they have played a significant role in identifying and challenging gendered norms and institutions that affect women's health for over five decades (e.g., Boston Women's Health Book Collective 1973). While key areas of struggle that women have addressed (and continue to address) include the fight for reproductive rights and the prevention of men's violence against women, women's food restriction practices have not featured as prominently in public health and health promotion initiatives. This is despite the work of radical thinkers in the WHM throughout the 1960s and 1970s, who explicitly addressed the harms of "sex-role stereotyping" (Gray Jamieson 2012, p. 27) and the importance of nourishing one's body with food and enjoyable movement (Boston Women's Health Book Collective 1973). Given that rigidly prescribed gender norms have been associated with adverse health outcomes (Weber et al. 2019), especially in relation to weight-loss dieting (Jovanovski 2017; Nagala et al. 2020), it is important for those within the WHM to challenge intentional hunger and to frame it as a harmful gendered phenomenon.

Food restriction for the purposes of weight loss (i.e., weight-loss dieting) is one of the most common reasons why women experience intentional hunger in the twenty-first century. Research shows that dieting behaviors have contributed to a host of preventable physical and psychological health problems in girls and women, such as bone density loss, a compromised immune system, and the development of eating disorders (Bombak et al. 2019; Omasu et al. 2019; Nordmo et al. 2019; Schilling 2018; Turner 2019). Women are at particular risk of these health problems because they are more likely to engage in food restriction practices than men. Research shows that women engage in food restriction regardless of where they are positioned on the body mass index (Sares-Jaske et al. 2019), whereas males are more likely to engage in food restriction practices once they reach a BMI of 25 and above. While not all examples of intentional hunger involve fasting for the purposes of weight loss, the physical risks associated with prologued and/or episodic phases of food restriction can cause long-lasting harm to one's body. As Turner (2019) explains, the human body's physical response to food restriction is the same, regardless of whether one is participating in a fad diet or in the grips of famine: it decreases its metabolic rate and increases the production of hormones responsible for hunger. Brief, chronic episodes of food restriction can lead to more prologued versions of food restriction (such as that experienced by women with eating disorders) or, paradoxically, to binge eating episodes (Bombak et al. 2019). A feminist research perspective looking at dieting (and intentional hunger) is ideal for critically unpacking the gender norms that underlie women's food restriction practices, because it "challenges the structural and social power inequalities within patriarchal societies that produce inequalities and disadvantage women" (Davies et al. 2019, p. 601). Understanding the different types of voluntary food restriction practices experienced by women throughout different phases of Western history helps us trace the way it has been gendered, and how it has continued to manifest in the twenty-first century.

Anorexia Mirabilis: Hunger as a Sign of Women's Piety

One of the earliest forms of intentional hunger recorded in women, documented from as early as the thirteenth century in Europe, involved fasting for spiritual and ascetic reasons (Russell 2005; Walker Bynum 1985). The stereotype of the self-sacrificial fasting woman served as a powerful gendered template for well over 400 years (Russell 2005), and was used by women to quell sexual urges, express Godly devotion, and even earn a living in some instances. In Judeo-Christian traditions, female saints were most likely to be characterized by their experiences of intentional hunger (Gooldin 2003; Walker Bynum 1985), engaging in food restriction practices to achieve self-discipline and to tame sexual (or humanly) desires (Ellmann 1993; Walker Bynum 1985). Episodes of fasting that lasted for long periods of time were understood to be miracles of God, and were referred to as "anorexia mirabilis," or the miraculous cessation of hunger (Gooldin 2003). The "economy of sacrifice" (Ellmann 1993, p. 13) involved in fasting served to position women's contributions to spiritual life in self-restrictive terms, and reinforced existing gender norms relating to women's selflessness and supposed other-oriented nature. Russell (2005) describes historical accounts of medieval fasting in her book *Hunger: An Unnatural History*. She explains that the gendered qualities embedded in saintly fasting could be read as a form of "literally fe[eding] from God." The "deepen[ing] [of one's] role as the Bride of Christ," she explains, is reflected by "becom[ing] a channel through which [one] could serve others... multiply[ing] crumbs into loaves, exud[ing] oil from [one's] breasts, and cur[ing] disease with [one's] saliva" (Russell 2005, p. 46). The archetype of the hungry woman, miraculously feeding others through her piety, served as a persuasive socializing force for women for hundreds of years thereafter. As Walker Bynum (1985, p. 4) explains, much of the writing on women's spirituality during medieval times discussed female miracles in relation to food motifs, and in doing so, alluded to self-sacrificial acts of fasting as ways of providing "a literary and psychological unity to the woman's way of seeing the world." The intentional hunger of female saints was thus seen as a normative and widely accepted lens through which women learned to understand the role of their appetites and desires, and as a way of unifying women as a class.

By the early nineteenth century, tales of anorexia mirabilis – and other forms of intentional hunger – proliferated throughout Europe, capturing the imaginations of both believers and nonbelievers alike (Nicholas 2008). As is emphasized in this chapter, these acts of self-starvation served to strengthen existing gender norms around food, eating, and the body as a social object, regardless of whether they were being performed by women or men. Masculine depictions of intentional hunger were often active and/or emphasized the suffering of being hungry. Living skeletons, for example, who were mostly male, reinforced their normative gendered status through their famous public performances, and were emboldened by their emaciated bodies. Far from being passive, the "Skeleton Dude" of Ohio, for example, actively "pranced around the stage" wearing a monocle and tight-fitting suit, flirting with women in the audience and telling stories about his life (Gooldin 2003, p. 42). The male "hunger artist," too, who sat locked in a cage at carnivals, turned his hunger into a public spectacle. Instead of normalizing his nonnormative body, however, the vision of his emaciated frame served to reinforce the suffering involved in being hungry and turned what was otherwise a lonely, physical experience into a social one. As argued here, these masculine acts of intentional hunger served to further reinforce men's social prowess and bravery, rather than stifling it; the living skeleton remained an active and vibrant part of the social world despite his hunger, and the hunger artist emphasized the full force of his hunger through his visible suffering. In contrast to these depictions of masculine intentional hunger, fasting women and girls were almost always house-bound (often laying in bed) and depicted as appetite-free, which serve as a reinforcement of women's passive, domestic, and selfless status in society. Ann Moore, also known as the fasting woman of Tutbury, was one such example. Moore, who lived in poverty in a small Midlands village in England, claimed to have gone without food for 5 years due to a lack of desire for nourishment, and regularly welcomed townspeople into her home to observe her lack of appetite as she sat still in a chair or rested in her bed (Gooldin 2003). As Gooldin (2003, p. 29) explains, Moore's actions fall "within the limitations of a patriarchal social structure," where "food, and its rejection, was an available channel for women to express their religious convictions." By gaining recognition for publicly denying their hunger, women expressed the extent to which their socialization was predicated on the naturalization of lack and restriction. This socialization process was bolstered by both religious and medical figures, often simultaneously.

Ann Moore's performance of intentional hunger, and the performances of many other fasting women during this time, were hailed as both miraculous by religious scholars, and as an illness or act of fraud by medical scientists. As Gooldin (2003, p. 33) explains, these perspectives served as "two paradigmatic frames of explanation," with scientific explanations gradually replacing religious views that naturalized women's self-induced hunger. This paradigmatic shift in understanding women's food restriction led to significant problems for some fasting women, especially when scientific explanations pointed to fraudulent behavior. As Nicholas

(2008) explains, there is evidence to suggest that many women, such as Moore, were burdened by poor financial circumstances, and sought fame and money for their performance of intentional hunger by playing on existing archetypes of sacrificial and pious femininity. As hunger due to poverty was viewed as an unpalatable reality and seldom reported in newspapers in the early nineteenth century, many women adopted the performance of intentional hunger for their family's survival. The medical-scientific paradigm, which sought to measure fasting women's urine, sweat, and pulse for signs of fraudulence, subsequently affected many women's chances of being recognized as miraculous fasting women. As Nicholas (2008, n.p.) explains:

> Surveillance of the miraculous fasters, often women in materially insecure positions who found economic success through the spectacle, was undertaken in the name of a medical science whose explanatory discursive power was rising as religious explanations waned.

Indeed, rather than explaining women's intentional hunger as a miracle of God – and rejecting gendered norms around women's self-sacrificial nature – those who relied on the scientific paradigm to explain women's food restrictive practices either discredited women as frauds, or used the emerging discourses of the Enlightenment to medicalize women's passivity and self-sacrifice in relation to food and their bodies. While problematic for their reinforcement of harmful gender norms, women's performances of intentional hunger were often enacted out of desperation, a theme that continued to resonate all throughout the nineteenth century.

"The Weapon of Self-Hurt": Hunger as a Political Mouthpiece

While the practice of voluntary food restriction was already etched into the cultural imaginary in the late 1800s as a sign of women's self-sacrifice and devotion to God, it was also being weaponized by women for political gain. During the "first-wave" of feminism, the militant branch of the women's suffrage movement, made famous by activist Emmeline Pankhurst and the Women's Social and Political Union (WSPU), used "hunger striking" as a way of advancing women's rights (Schlossberg 2012). Hunger striking refers to the practice of fasting for political purposes, and "forcing [ones] opponent to grant certain demands... without any serious effort to convert [them] or achieve a 'change of heart'" (Sharp 1973, p. 2; as cited in Scanlan 2008). First enacted by prisoners in tsarist Russia as a method of political agitation – a method that was taught to the suffragists by many dissenters during their emigration to England in the late 1800s – hunger striking was both popularized and propagandized by the suffragists to draw public attention to the dehumanization of women in otherwise democratic societies. According to Ziarek (2007), hunger striking in the women's suffrage movement was made particularly famous by Marion Wallace Dunlop, a militant suffragist and prominent member of the WSPU. Dunlop, who was sentenced to 1 month in prison for graffitiing the English Bill of Rights on the walls of Parliament, was arrested and denied status as a political prisoner. As an act of resistance, she stopped eating and was only released after 91 h "because prison officials... were afraid she would become a martyr for suffragettes" (Ziarek 2007, p. 99). Dunlop's version of political fasting was aggressive, sacrificing her own well-being to arouse anxiety in her opponents. It was also widely adopted. Since Dunlop's famous political fast, the practice of hunger striking has been used globally by women and other marginalized groups to raise awareness about their oppression and to demand tangible change (Ambruster-Sandoval 2017; Scanlan 2008; Ziarek 2007).

As physically harmful as prolonged and episodic food restriction can be, intentional hunger for the purposes of political gain plays a part in the rich history of the early feminist movement. The image of the defiant, hunger striking suffragist – force-fed by prison guards – has become emblematic of women's struggle for rights and equality, and has been romanticized as a sign of women's bravery in the face of adversity. This is despite the fact that hunger striking was often accompanied by pain and self-sacrifice; features that have been criticized by feminist writers for being gendered acts of self-harm (Starhawk 2002). As suffragist Lady Constance Lytton describes, hunger striking was indeed a "weapon of self-hurt" (as cited in Schlossberg 2012, p. 89), but to the suffragists, it was an acceptable form of self-sacrifice because it "served as a metaphor for the struggle for women's political equality itself" (Schlossberg 2012, p. 89). As Ziarek (2007) explains, these acts "exposed... in public the hidden irrational violence of the sovereign state against women's bodies" (Ziarek 2007, p. 100) and thus served to heighten women's voices in the public domain. The more women fasted for women's rights, the more violent men's actions toward them became, manifesting in force-feedings (by male prison guards) and satirical cartoons produced by male journalists that tacitly eroticized women's plight (Schlossberg 2012). Violent responses to women's intentional hunger eventually caught the attention of politicians and the wider public, and subsequently helped women gain some of the rights they were fighting for.

While acts of political fasting may have been effective at publicly raising awareness about women's plight, they were also undeniably gendered acts; acts of self-harm that the public were more accustomed to seeing in women rather than men. Ambruster-Sandoval (2017) views the distinction between public reactions to male and female fasting – of women's "hysterical" fasting and men's "brave" fasting – as a patriarchal dismissal of women's bravery. Comparing well-known male hunger strikers with the suffragettes, Ambruster-Sandoval (2017, pp. 21–22) states that: "Sands, Chavez, Gandhi, [and] even Jesus... What do these male bodies communicate – strength, valour, bravery, or courage perhaps? The gendered distinction between male and female bodies is significant because the latter are often seen as hysterical, irrational, insane, and anorexic, thus they do not typically generate much third-party support." Ambruster-Sandoval's (2017) assertion that the hungry female body elicits less praise than the hungry male body is an important and insightful observation. However, it also ignores the existence of other, more passive forms of voluntary food restriction that women were engaging in during this time, and the symbolic consequences these gendered acts had on women's political potency. While struggles to obtain rights in a patriarchal society serve as a strong symbol and powerful reminder of the importance of women's equality, one reason that women have not been as valorized for their political efforts is precisely because of the normalization and cultural acceptance of their physical passivity and gendered acts of self-sacrifice– in this case, through their denial of food. It is argued in this chapter that the long history of women's intentional hunger that precedes the political agitation of English suffragists weakens the political potency of their message; a message that led to ridicule and pity rather than calls of bravery. To borrow Audre Lorde's (1984) famous quote, hunger striking for women's rights can be understood as a form of attempting to use the master's tools to dismantle the master's house. Indeed, according to Schlossberg (2012), hunger striking gained the attention of male politicians precisely because it played on infantilizing and sexist notions of the "delicate, middle-class female body" (Schlossberg 2012, p. 89) and its various sensibilities, a stereotype of womanhood that had medicalized and confined women all throughout history. While women attempted to communicate their suffering through hunger striking, and received widespread attention for their plight, their "self-lacerating form of protest" (Ellmann 1993, p. 2) was loaded with pre-existing gendered connotations of passivity, self-sacrifice and "hysteria"; stereotypes of womanhood that were being pathologized at the same time that women protested for their rights.

Anorexia Nervosa: Hunger as a Psychological Condition

While the suffragists grew hungry for equality, and fasting girls sacrificed their nourishment for fame, money, and the holy spirit, some women were being pathologized for their intentional hunger. In the second half of the nineteenth century, physicians Ernst-Charles Lasegue and Sir William Gull observed that some of their female patients engaged in what they considered to be the mysterious practice of voluntary food restriction, exhibiting symptoms of "severe weight loss [and] amenorrhea" that could not be explained by "underlying organic pathology" (Vandereycken 2002, p. 152). In one of his famous public lectures in 1874, Gull coined the term "anorexia nervosa" to describe his patients' experiences and separated the condition from other maladies involving self-starvation (Bemporad 1996), including hysteria (Dittmar and Bates 1987). Although some preliminary theories suggested physiological causes for women's behaviors (i.e., Simmonds' theory of pituitary dysfunction in women presenting with anorexic symptoms), early explanations about the causes of women's self-starvation and intentional hunger were limited.

Despite these limitations in knowledge, physicians such as Gull and Lasegue continued to look for medical reasons to explain women's behavior, couching their theories in medical and scientific language that framed women as innately irrational and overemotional, and therefore, more likely to engage in food restriction than men (Hepworth 1999). Further interest in women's "prolonged inedia" (Bemporad 1996, p. 218) was slow for several decades until psychoanalytic theories started to emerge in the 1940s, attributing anorexia nervosa in women to a "phantasy," resulting from an unconscious fear of oral impregnation (Till 2011). According to Till (2011), early psychoanalysts argued that women came to associate food with sexuality, and that they refused to eat as a way of rejecting their emerging sexuality. These theories, however, focused predominantly on what was happening "inside" the minds of women, rather than on the external forces that both defined womanhood and shaped women's behavior.

It was not until the feminist movement of the 1970s, 1980s, and 1990s, however, that criticisms emerged about the medicalization and psychopathologization of women's intentional hunger (Hepworth 1999; Orbach 2005). These criticisms – from seminal writers such as Susie Orbach, Sandra Lee Bartky, and Susan Bordo – drew explicit links between women's restrictive eating practices (and scientific/medical explanations of women's intentional hunger) and their gendered socialization under patriarchy, paving the way for a resurgent feminist psychology movement (Bartky 1990; Hepworth 1999; Orbach 2005). Key themes identified in this movement were the sociocultural and gendered factors that informed women's self-starvation practices, and criticisms of the way that women's behaviors were explained in early medical and psychoanalytic texts. In *The Social Construction of Anorexia Nervosa*, Hepworth (1999, p. 3) highlights some of the sexist undercurrents of Enlightenment-era medical philosophies, and criticizes the early theorists of anorexia (such as Gull and Lasegue) for reinforcing the idea that women's "irrationality" could be explained through medical evaluation. Hepworth (1999) argues that "notions of anorexia nervosa accorded with dominant ideas of science, medicine and women," discourses that coalesced in the late nineteenth century to define men as rational and women as irrational and overemotive beings. She identifies anorexia nervosa as a symbol of this ideology, one where ideas of femininity were used to naturalize understandings of women's behavior and therefore, to subordinate women using the discourse of scientific progress. She argues that the search for organic causes of women's hunger failed to interrogate this existing social and historical backdrop, which ultimately served to "maintain... women's subordination" (p. 27).

Feminist psychological researchers sought to interrogate this social and historical backdrop by understanding women's psychopathologized experiences of intentional hunger through a politicized lens. In *Hunger Strike*, feminist psychotherapist Susie Orbach (2005, p. xx) explains that women's fear of food, and their experience of intentional hunger, can be understood as consequences of patriarchal socialization. She explains that:

> The woman... cannot digest the food she so desires. She cannot absorb it. She cannot let it nourish her. Her needs must be rejected. But before she can wait for rejection by the other, she disclaims the need herself so that her feelings of powerlessness are mitigated through a felt sense of agency.

As Orbach (2005) eloquently explains, women attempt to gain control over their bodies through practices of starvation because they lack control in other aspects of their lives. The rejection of her appetite – both physical and symbolic – is thus a direct result of a woman's powerless position in a male-dominated society.

Other feminist writers during this time interrogated patriarchy and gender norms by focusing on the association between women's restrictive food practices and the thin ideal (Bartky 1990; Bordo 2004; Orbach 2005). According to Bartky (1990), women come to internalize men's sexual objectification of their bodies and, through this process, internalize a critical monologue about their appearance. More often than not, this internalized criticism refers to the weight, shape, and size of one's body; a phenomenon known as internalized weight stigma (Meadows and Higgs 2019). Bolstered by multibillion-dollar diet and beauty industries (Bordo 2004), which capitalize on women's bodily insecurities and anxieties about food, women come to learn what the acceptable female body looks like, and alter their eating habits to ameliorate their internal distress. While the feminist psychological argument explaining women's experiences of intentional hunger was persuasive, and critically interrogated the role of patriarchy and gender norms in instantiating and reproducing sexist beliefs about women's bodies and their right to food, one criticism it has received is in relation to its focus on individual women's experiences and intrapsychic distress, at the expense of discussions looking at changing the sociocultural and structural factors that reinforce intentional hunger in the first place (Jovanovski 2017). Nevertheless, the feminist psychological analysis of "anorexia nervosa" as a form of intentional hunger has been one of the most powerful and persuasive critiques of patriarchy in relation to women's food restriction practices that currently exist.

From Weight-Loss Dieting to "Wellness": Hunger as a Responsible Lifestyle Choice

While body weight is not the only reason behind women's restrictive eating practices, it defined many women's relationships with food in the twentieth and twenty-first centuries. Fashion and beauty trends dictating a slim physique were popularized throughout much of the 1900s in Western countries, and criticized persuasively by many feminist writers as a sign of women's physical and psychological subordination under patriarchy (Bordo 2004; Orbach 2005). While the idealized female body shape has changed over the years, from the thin "flapper" look of the 1920s to the current accentuated hourglass body shape made famous by the Kardashians and Instagram fitness models, the mandate of "thinness," and the sexual objectification of the female body, has remained. The "fat" female body has continued to be stigmatized, and been the subject of both moralizing and medicalizing discourses, which some writers have argued has contributed to a culture of "lipoliteracy," or the process of superficially judging a person's health on the basis of their weight, shape, and size (Murray 2008). This cultural message has, in turn, contributed to the normalization of restrictive eating practices through the narrative of weight-loss dieting (Schilling 2018). Despite the continuing cultural discourse of weight stigma that pervades contemporary society, narratives that are largely directed at women, messages of weight-loss dieting involving calorie-counting and compulsive scale-checking have increasingly grown out of favor (Jovanovski 2017). Instead, these messages have gradually been replaced by the more palatable and health-oriented language of "wellness," proliferating throughout social media communities that are typically frequented by women (Hanganu-Bresch 2019). Based on these new "public pedagogies" (Camacho-Minano et al. 2019), a woman's ability to empower herself to achieve "good health," which often coincides with losing weight through restrictive food practices, is important (Dubriwny 2012; Jovanovski 2017; Roy 2008). As argued in this chapter, this contemporary narrative has reinforced new problematic norms of intentional hunger in women.

The rebranding of the diet industry – from one focused on calorie-counting to one focused on wellness – can be situated under the broader framework of healthism (Hanganu-Bresch 2019; Jimenez-Loaisa et al. 2019). First proposed by Crawford in the early 1980s, the term healthism refers to the "preoccupation with personal health as a primary – often the primary – focus for the definition and achievement of well-being; a goal which is to be attained primarily through the modification of lifestyles" (1980, p. 368). According to Arguedas (2020, n.p.), this preoccupation with health is considered harmful because of its reduction of well-being "to a single dimension. . . requiring adherence to various prescriptive and proscriptive norms." Indeed, healthism can be perceived as moralizing because of its central focus on individual interventions involving lifestyle choices and behavioral changes; ultimately conforming to a neoliberal ideology (Dieterle 2020). As Roy (2008) explains in her exploratory critical analysis on health writings in women's magazines, these discourses are also criticized for the gendered messages they promote. In her discussion of health articles from three English and Canadian women's magazines, she found that female readers are simultaneously positioned as being responsible for the health and well-being of others, and expected to be responsible for their own health in order to protect others. Gender norms associated with the attainment of "wellness" thus tend to principally frame women as carers; carers that focus on their own well-being only in relation to others.

Increasingly, however, discourses of healthism have taken on a markedly postfeminist ethic (Camacho-Minano et al. 2019). According to Camacho-Minano et al. (2019, p. 661), postfeminist biopedagogies – or problematic, gender-conforming discourses that "compel girls to work on their bodies and their minds toward a constant improvement of the self" – are disseminated via social media platforms and framed as empowering choices for women; choices that ultimately "reproduce the normative feminine body in restricted and disempowering ways" (p. 661). Through the gendered and neoliberal lens of healthism, new forms of restrictive eating practices have been reinforced in women. In the last decade, "clean eating," for example, has taken over the world of Instagram, and spearheaded a movement of often female social media influencers who "appear empowered and in control, representing the pinnacle of morality, righteousness, and success" (Irvine 2016, n. p.). Defined as the process of increasingly cutting out foods from one's diet that are considered "unhealthy" (e.g., contain salt, sugar, fat, and/or artificial preservatives), clean eating messages have been specifically directed at women via social media platforms (Camacho-Minano et al. 2019) and have been subsumed within a medicalised narrative (Allen et al. 2018), despite containing much of the same moralizing discourses that once defined the weight-loss diet industry (Hanganu-Bresch 2019). While little is known about clean eating as it is a relatively new social phenomenon, some have drawn a link between its key features and the intentional hunger experienced by women diagnosed with eating disorders. As Hanganu-Bresch (2019, p. 1) explains, "whereas the 'classic' eating disorders such as anorexia and bulimia have established historical precedents and have been thoroughly examined by critical feminist theory and sociologists of medicine, a pathological obsession with the healthfulness of food has not been previously recorded."

Increasingly, there have been signs to suggest that clean eating discourses contain the same self-restricting gendered narratives that have come to define "eating disorders" for over five decades. As Irvine (2016, n.p.) explains, Instagram fitness messages that reinforce clean eating often rely on body-policing slogans such as "eat clean, get lean" to "encourage women to make the right choices." From this perspective, many women are:

> Caught in a balancing act trying to manage conflicting ideals of femininity. Often holding themselves accountable to dominant images of 'fit' and 'healthy' women that are disproportionately white, slender, middle-class, and (hetero)sexually desirable, while simultaneously being encouraged to be a 'strong', 'empowered', and active consumer. (Irvine 2016, n.p.)

By focusing on the way that corporations, health promotion campaigns, and individuals (especially those with high profile social media platforms) reinforce women's intentional hunger under the guise of "wellness" in the twenty-first century, those involved in the WHM should work to counteract these discourses with resistant narratives using both feminist messaging critical of gender norms and narratives of health that promote nonjudgmental eating and exercise practices.

Visibilizing Women's Intentional Hunger: A Feminist Research Agenda for the WHM

Looking back at women's food restriction practices throughout Western history, it becomes clear that intentional hunger can be characterized by a number of gendered factors that ultimately reinforce regressive stereotypes of what it means to be a woman in relation to food. Understanding women's experiences of intentional hunger from a feminist sociological perspective focused on gender norms – norms that are driven by a patriarchal system of power – is a significant and important contribution to the field of women's health. Indeed, how we understand women's intentional hunger shapes how we tackle it in the twenty-first century. For much of the time, hunger is merely thought of as a physiological state. As Sharman Russell (2005, p. 26) explains, "to be hungry is to be uncomfortable, and most of us experience hunger in the same way we experience pain, as a signal to do something." However, as this chapter has attempted to demonstrate, hunger is more than just a visceral or embodied experience; it is also a social one (Boyce 2012; Russell 2005).

As Ellmann (1993, p. 3) explains, hunger is, "a form of speech, and speech is necessarily a dialogue whose meanings do not end with the intention of the speaker but depend upon the understanding of the interlocutor." Women's experiences of intentional hunger thus depend on the audiences witnessing their hunger or, indeed, whether anybody is witnessing it at all. As explained earlier in the chapter, the WHM have not yet played a dominant role in identifying intentional hunger as a gendered phenomenon, or tackling it as part of their broader agenda. In this part of the chapter, contemporary methods are suggested for WHM researchers and activists to visibilize and challenge narratives of women's intentional hunger using feminist theory. Specifically, a research agenda is offered for those working in the field of women's health regarding how to address intentional hunger in future health-related initiatives.

Recognize Women's Intentional Hunger as a Gendered Phenomenon Requiring Gender-Transformative Health Promotion

When women's health researchers and advocates talk about women's food restriction practices – and by extension, their intentional hunger – it is important that they identify these experiences as harmful gendered acts. As explained in this chapter, norms and stereotypes that normalize food restriction practices in women, such as those that paint women unquestionably as body-policing subjects (i.e., weight-loss diet messages) or other-oriented carers (i.e., food is for others), contribute to women's broader social and health inequalities. Rather than seeing these gendered norms and stereotypes as separate, fragmented messages, it is useful to consider them as part of an overarching framework of patriarchal socialization: one that controls and limits women's relationships with food and their bodies.

In order to dismantle these cultural messages of restrictive eating, women's health researchers and advocates need to explicitly foreground harmful gender norms, and avoid reinforcing them in their health promotion content. This will require a radical transformation (or even abolition) of gender roles, norms, and relations. Researchers Ann Pederson, Lorraine Greaves, and Nancy Poole advocate for a gender-transformative model of health promotion, one that encourages women's health researchers and advocates to create health content that both problematizes and avoids gendered norms and stereotypes (Pederson et al. 2014). Under the gender-transformative model of health promotion, health promotion experts design interventions in "women-centred [ways], embrac[ing] harm reduction principles, build[ing] on women's strengths and be[ing] explicitly equity-oriented" (p. 144). While the gender-transformative model of health promotion has not yet been used to address restrictive eating practices in women, it has been used extensively to address smoking and substance use in women, and acknowledges intersecting oppressions that women face. This approach would be a beneficial way of both challenging gender norms around women's intentional hunger, resisting the promotion of sexually objectifying and weight-stigmatizing portrayals of women's bodies (Jovanovski 2017, 2018; Murray 2008), and developing new, strength-based messages of health and well-being for women in relation to food and their bodies.

Shift One's Focus from a Weight-Centric to Weight-Neutral Paradigm

The main driver of women's food restriction practices and intentional hunger in the twenty-first century is weight-loss, which is increasingly being couched in discourses of "wellness" (Bombak et al. 2019; Schilling 2018; Turner 2019). The health content produced, and advice dispensed, by those working in the WHM must be responsible and recognize the high recidivism rates of weight-loss dieting and the physical and psychological harms of engaging in prolonged dieting behaviors. This means shifting the focus of health messages from a "weight-centric" to a "weight-neutral" approach (Tylka et al. 2014). According to Tylka et al. (2014, p. 2), the weight-normative (or "weight-centric") approach "rests on the assumption that weight and disease are related in a linear fashion... [and that] personal responsibility for 'healthy lifestyle choices' and the maintenance of 'healthy weights'" are important. Some researchers have argued that, due to the focus on weight loss as a health goal, there is some risk in promoting weight-stigmatizing advice (Schilling 2018). This is a particular problem in the context of current norms of intentional hunger, which reinforce "lifestyle choices" and narratives of "wellness" to female audiences and implicitly promote weight-focused eating behaviors. Approaches to health that do not focus on weight as a marker of well-being are increasingly being recommended by researchers (Bombak et al. 2019; Schilling 2018).

The weight-neutral (or "weight-inclusive") approach, which is seen as a useful alternative to the weight-normative approach, "rests on the assumption that everybody is capable of achieving health and well-being independent of weight [and] given access to nonstigmatizing health care" (Tylka et al. 2014, p. 6). Rather than focusing on the reduction of weight to achieve health, the weight-neutral approach focuses on "eating nutritious food when hungry, ceasing to eat when full, and engag[ing] in pleasurable (and thus more sustainable) exercise" (p. 6). For this reason, it is considered a conservative approach, reducing the risk of promoting harmful weight-loss dieting messages and body dissatisfaction in women (Bombak et al. 2019). Addressing women's restrictive relationships with food – and the cultural normalization of their intentional hunger – will thus require a paradigm shift to a weight-neutral approach by public health experts, women's health researchers, and advocates.

Reframe Individualized Narratives of Self-Care into Acts of Feminist Resistance

While adopting a weight-neutral paradigm encourages self-care behaviors through intuitive eating and enjoyable exercise, fostering messages of self-care from a feminist, rather than individualist (or health-oriented), perspective are also essential.

As discussed throughout the chapter, understanding women's intentional hunger throughout different periods of Western history requires an awareness of the role that patriarchy has in reinforcing the normalization of women's intentional hunger, and in distancing women from their bodies and their own pleasures. Rather than understanding intentional hunger as an individual problem, it is important for WHM researchers and activists to discuss it as a sociocultural one that spans hundreds of years of Western civilization. Self-care, from this perspective, is more than simply asking women to engage in healthy behaviors as responsible individuals, but rather, requires fostering an awareness that caring for oneself is a feminist act of rebellion; one that ultimately strengthens female solidarity, builds a community of care, and dismantles patriarchy. As radical feminist writer Audre Lorde (1988, p. 132) once described in relation to her feminist ethics of self-care: "caring for myself is not self-indulgence, it is self-preservation, and that is an act of political warfare." Conceptualizing self-care – through the consumption of food and the nourishing of the body – from a feminist perspective is one way. As Naomi Wolf (1990, p. 187) famously declared in *The Beauty Myth*, "dieting is the most potent political sedative in women's history; a quietly made population is a tractable one." Strengthening women's political potential as a class and further building on existing communities of resistance – through the promotion of self-care around food and eating – is an important goal for the WHM in the twenty-first century.

Get Organized Through Feminist Consciousness-Raising

As outlined in this chapter, feminist acts of resistance were sometimes marred by gender norms that valorized food restriction in women (e.g., hunger striking). In an effort to counteract messages of intentional hunger from a gender-transformative and feminist perspective, WHM scholars and activists must contribute to producing and encouraging physical spaces where women can share their experiences of diet culture, which will in turn, help them identify and dismantle the patriarchal industries and structures responsible. The WHM must play an active part in subverting Ellmann's (1993, p. 2) famous quote about eating disorders – "women get ill instead of getting organised" – helping women to get organized instead of getting ill.

The Western feminist movement of the 1960s and 1970s gained momentum due to the grassroots organizing of women in hundreds of small groups, called consciousness-raising groups. Women's liberation movement activist and founder of New York Radical Women, Pamela Allen (1970), describes consciousness-raising groups as physical spaces where women could be free to interrogate all gendered aspects of their lives with other women. Based on her own feminist consciousness-raising experiences, she outlines four group processes that help women "become autonomous in thought and behaviour" (p. 272). These processes include: Opening Up, Listening, Analyzing, and Abstracting. The first two steps of the consciousness-raising process, opening up and listening, involve sharing one's own, and listening to other women's, experiences of misogyny. This process enables women to understand that their oppression is a shared experience, and in the context of dismantling diet culture, allows women to have a free space to discuss their history of food restriction as a shared female history. Once women have spoken about their concerns and found commonalities with other women, they are required to engage in a process of analysis and abstraction, where they identify the root causes of their oppression and devise strategies to challenge (or overthrow) their oppressors. In the context of challenging the normalization of women's restrictive eating practices, those participating in consciousness-raising would identify a series of oppressive structures, institutions and systems, and devise tangible strategies to dismantle them. An important note to add here is that feminists such as Allen (1970) emphasize that consciousness-raising groups should not function as group therapy sessions, but rather, focus on politicizing one's pain and finding ways to challenge one's oppressors. The creation of face-to-face consciousness-raising sessions, outside of the diet-infused culture of social media, may prove to be beneficial in pushing back against the normalization of women's hunger.

Conclusion

This chapter has called for the visibility of women's food restriction practices and their intentional hunger in the WHM. Examples of women's practices of food restriction throughout Western history from a feminist sociological perspective were discussed, highlighting the ways in which patriarchy has operated through the reinforcement of gender norms and stereotypes that paint food restriction in women as a sign of virtue, political strength, and health. The feminist research agenda presented in this chapter, which looks at adopting a weight-neutral, and feminist movement-building approach, aims to start a conversation between WHM activists and scholars that challenges the cultural normalization of women's hunger. The ultimate goal for creating these agenda items is for the WHM to be a responsible, listening audience to women's intentional hunger, giving meaning to their suffering and finding ways – in collaboration with women collectively – to ameliorate it.

Acknowledgments This work was funded by the author's Australian Research Council (ARC) Discovery Early Career Researcher Award (DE200100357).

References

Allen P (1970) Free space: a perspective on the small group in women's liberation. Times Change Press, New York

Allen M, Dickinson KM, Prichard I (2018) The dirt on clean eating: a cross sectional analysis of dietary intake, restrained eating and opinions about clean eating among women. Nutrients 10: 1266–1276

Ambruster-Sandoval R (2017) Starving for justice: hunger strikes, spectacular speech, and the struggle for dignity. The University of Arizona Press, Tucson

Arguedas AAR (2020) "Can naughty be healthy?": healthism and its discontents in news coverage of orthorexia nervosa. Soc Sci Med 246:112784

Bartky SL (1990) Femininity and domination: studies in the phenomenology of oppression. Routledge/Taylor & Francis Group, New York

Bemporad JR (1996) Self-starvation through the ages: reflections on the pre-history of anorexia nervosa. Int J Eat Disord 19(3):217–237

Bombak A, Monaghan LF, Rich E (2019) Dietary approaches to weight-loss, Health At Every Size® and beyond: rethinking the war on obesity. Soc Theory Health 17:89–108

Bordo S (2004) Unbearable weight: feminism, Western culture, and the body. University of California Press, Berkeley

Boston Women's Health Book Collective (1973) Our bodies, ourselves: a book by and for women. Simon and Schuster, New York

Boyce C (2012) Representing the "hungry forties" in image and verse: the politics of hunger in early-Victorian illustrated periodicals. Vic Lit Cult 40:421–449

Brumberg JJ (1989) Fasting girl: the history of anorexia nervosa. Plume, New York

Camacho-Minano MJ, MacIsaac S, Rich E (2019) Postfeminist biopedagogies of Instagram: young women learning about bodies, health and fitness. Sport Educ Soc 24(6):651–664

Crawford R (1980) Healthism an the medicalisation of everyday life. Int J Health Serv 10:365–388

Davies SE, Harman S, Manjoo R, Tanyag M, Wenham C (2019) Why it must be a feminist global health agenda. Lancet 393:601–603

Dieterle JM (2020) Shifting the focus: food choice, paternalism, and state regulation. Food Ethics 5(2):1–16

Dittmar H, Bates B (1987) Humanistic approaches to the understanding and treatment of anorexia nervosa. J Adolesc 10:57–69

Dubriwny TN (2012) The vulnerable empowered woman: feminism, postfeminism, and women's health. Rutgers University Press, New Brunswick

Ellmann M (1993) The hunger artists: starving, writing and imprisonment. Virago Press, London

Gooldin S (2003) Fasting women, living skeletons and hunger artists: spectacles of body and miracles at the turn of a century. Body Soc 9(2):27–53

Gray Jamieson G (2012) Reaching for health: the Australian women's health movement and public policy. ANU E Press, Acton

Hanganu-Bresch C (2019) Orthorexia: eating right in the context of healthism. Med Humanit 46:1–12

Hepworth J (1999) The social construction of anorexia nervosa. Sage Publications, London

Irvine B (2016) On clean eating. Retrieved from: https://feministacademiccollective.com/2016/07/18/on-clean-eating/

Jimenez-Loaisa A, Carrillo VJB, Cutra DG, Jennings G (2019) Healthism and the experiences of social, healthcare and self-stigma of women with higher-weight. Soc Theory Health. https://doi.org/10.1057/s41285-019-00118-9

Jovanovski N (2017) Digesting femininities: the feminist politics of contemporary food culture. Palgrave Macmillan, Cham

Jovanovski N (2018) Slutburgers and sexual subjects: the re-sexualisation of women in fast-food advertising and culinary culture. In: Harrison K, Ogden C (eds) Pornographies: critical positions. University of Chester Press, Chester

Lorde A (1984) Sister outrider: essays and speeches. Crossing Press, Berkeley

Lorde A (1988) A burst of light: essays by Audre Lorde. Firebrand Books, Ithaca

Meadows A, Higgs S (2019) Internalised weight stigma moderates the impact of a stigmatising prime on eating in the absence of hunger in higher – but not lower-weight individuals. Front Psychol. https://doi.org/10.3389/fpsyg.2019.01022

Morgan HG (1977) Fasting girls and our attitudes to them. Br Med J 2(6103):1652–1655

Murray S (2008) The "fat" female body. Palgrave Macmillan, New York

Nagala JM, Domingye BW, Darmstadt GL, Weber AM, Meausoone V, Cislaghi B, Shakya HB (2020) Gender norms and weight control behaviours in US adolescents: a prospective cohort study (1994–2002). J Adolesc Health 66(1):S34–S41

Nicholas J (2008) Hunger politics: towards seeing voluntary self-starvation as an act of resistance. Thirdspace 8(1):n.p.

Nordmo M, Danielsen YS, Nordmo M (2019) The challenge of keeping it off, a descriptive systematic review of high-quality, follow-up studies of obesity treatments. Obes Rev 21:1–15

Omasu F, Aishima K, Nasu M, Hisatsugu Y, Fuchigami K (2019) Discussion on the relationship between dieting and bone density among female college students and the health guidance. Open J Prev Med 9:11–19

Orbach S (2005) Hunger strike: the anorectic's struggle as a metaphor for our age. Karnac Books, London

Pederson A, Greaves L, Poole N (2014) Gender-transformative health promotion for women: a framework for action. Health Promot Int 30(1):140–150

Roy SC (2008) 'Taking charge of your health': discourses of responsibility in English-Canadian women's magazines. Sociol Health Illn 30(3):463–477

Russell SA (2005) Hunger: an unnatural history. Basic Books, New York

Sares-Jaske L, Knekt P, Mannisto S, Lindfors O, Heliovaara M (2019) Self-report dieters: who are they? Nutrients 11:1789–1810

Scanlan SJ (2008) Women and nonviolent protest: the hunger strike and maternal versus feminist collective action frames. Annual Meeting of the American Sociological Association, Boston

Schilling LP (2018) Disorder in disguise: recognising the need for change when common diet trends cause harm. ACSMs Health Fit J 22(5):34–39

Schlossberg L (2012) Consuming images: women, hunger, and the vote. In: Heller T, Moran P (eds) Scenes of the apple: food and the female body in nineteenth- and twentieth-century women's writing. State University of New York Press, Albany

Starhawk (2002) Webs of power: notes from the global uprising. New Society, Gabriola Island

Till C (2011) The quantification of gender: anorexia nervosa and femininity. Health Sociol Rev 20(4):437–449

Turner P (2019) The no need to diet book: become a diet rebel and make friends with food. Head of Zeus, London

Tylka TL, Annunziato RA, Burgard D, Danielsdottir S, Shuman E, Davis C, Calogero RM (2014) The weight-inclusive versus weight-normative approach to health: evaluating the evidence for prioritising well-being over weight loss. J Obes. https://doi.org/10.1155/2014/983495

Van Esterik P (1999) Right to food; right to feed; right to be fed. The intersection of women's rights and the right to food. Agric Hum Values 16:225–232

Vandereycken W (2002) History of anorexia and bulimia nervosa. In: Fairburn CG, Brownell KD (eds) Eating disorders and obesity, second edition: a comprehensive handbook. The Guilford Press, New York

Walker Bynum C (1985) Fast, feast and flesh: the religious significance of food to medieval women. Representations 11:1–25

Weber AM, Cislaghi B, Meausoone V, Abdalla S, Mejla-Guevara I, Loftus P, Hallgren E, Seff I, Stark L, Victora CG, Buffarini R, Barros AJD, Domingue BW, Bhushan D, Gupta R, Nagata JM, Shakya HB, Richter LM, Norris SA, Ngo TD, Chae S, Haberland N, McCarthy K, Cullen MR, Darmstadt GL (2019) Gender norms and health: insights from global survey data. Lancet 393:2455–2468

Weedon C (1997) Feminist practice and poststructuralist theory, 2nd edn. Blackwell Publishers, Oxford, UK

Wolf N (1990) The beauty myth: how images of beauty are used against women. Vintage, London

Ziarek EP (2007) Bare life on strikes: notes on the biopolitics of race and gender. S Atl Q 107(1):89–105

Systems of Prostitution and Pornography: Harm, Health, and Gendered Inequalities

75

Meagan Tyler and Maddy Coy

Contents

Introduction	1898
Harms of Prostitution	1899
Violence	1900
Psychological Harms	1902
The Interconnectedness of Pornography and Prostitution	1904
Pornography as Connected to Other Forms of Prostitution	1904
Pornography as a Form of Prostitution	1906
Experiences of Abuse in Pornography Production	1908
Pornography and Harms of Distribution	1909
Understanding Systems of Prostitution	1911
Conclusion	1913
References	1913

Abstract

This chapter brings together a wide range of literature addressing issues of harm, health, and gendered inequalities with regard to prostitution and pornography (as a form of prostitution), from a variety of localities across the Global North and Global South. After beginning with an overview and summary of psychological and physical harm, violence, and trauma experienced by women in different forms of prostitution, pornography is considered. The issue of pornography is viewed through the lens of production and research on the harms of prostitution, rather than the more common focus on secondary harms created through consumption. The importance of such an approach is emphasized in the final section,

---

M. Tyler (✉)
RMIT, Melbourne, VIC, Australia
e-mail: Meagan.Tyler@rmit.edu.au

M. Coy
Center for Gender, Sexualities and Women's Studies Research, University of Florida, Gainesville, FL, USA
e-mail: m.coy@ufl.edu

© The Author(s), under exclusive licence to Springer Nature Singapore Pte Ltd. 2022
D. McCallum (ed.), *The Palgrave Handbook of the History of Human Sciences*,
https://doi.org/10.1007/978-981-16-7255-2_30

explaining how conceptualizing *systems of prostitution* and engaging with structural analysis can promote more holistic ways of understanding issues of health and wellbeing for women and girls in conditions of sexual exploitation.

Keywords

Prostitution · Pornography · Violence against women · Gendered violence · Sex work

Introduction

There is extensive evidence and a survivor-led movement demonstrating the ways in which the sex industry harms women's health and wellbeing. This chapter brings together literature on harms around violence and psychological trauma in prostitution, with theorizing around understanding pornography as interlinked to prostitution and, indeed, as a form of prostitution itself. This is important in recognizing social and individual harms across these elements of the sex industry. While harms associated with the consumption of pornography are significant, harms involved with those directly involved in production also need to be considered. Within the broader literature addressed here, the terms 'prostitution' and 'sex work' are used by different authors. In our own work, particularly in respect to sex trade survivors, we consciously and compassionately eschew the term "sex work" (for some further discussion of the associated language issues, specifically with regard to harm, please see: Gupta 2018; Smiley 2019). The term "prostitution" is therefore used throughout this chapter as the default and "sex work" is used only where a direct quote is employed. The chapter concludes with a consideration of systems of prostitution, setting out the need for a structural analysis of the inequalities of power that underpin prostitution and pornography: patriarchy, white supremacy, colonialism, and capitalism. By analyzing the harmful health consequences of prostitution and pornography as outcomes of structural inequalities, this chapter proposes a need to think in more holistic ways than a public health approach (see Pease in this section) and imagine deeper and more far-reaching responses to globalized sexual exploitation of women and girls.

It is documented internationally that the vast majority of people in prostitution are women and girls, with an over-representation of marginalized groups within this, including: Indigenous women, women of color, migrant women, women from ethnic minorities, women in poverty, women subject to domestic violence, homeless women and drug using women (Butler 2016; Cobbina and Oselin 2011; Farley 2020; Farley et al. 2003, 2005; Gupta 2018; Kempadoo 2001; Macy and Graham 2012; Monroe 2005; NWAC 2014; O'Connor and Yonkova 2018; Raphael 2004; Scully 2001; Stark and Hodgson 2004). Almost all sex buyers are male, although only a minority of men buy sex (IBISWorld 2015; Monto 2004; Shively et al. 2012).

It is important to note that systems of prostitution intersect and overlap with practices of domestic and transnational trafficking for sexual exploitation. What can be defined as trafficking reflects different law and policy regimes, and empirical research often does not use legally accurate definitions (Madden Dempsey 2017). In women's experiences, boundaries between prostitution and trafficking are often not clear cut (O'Connor 2019). While policy positions on how best to address the sex industry are deeply divided, there is consensus that women in the prostitution system (phrased as 'selling sex' in some literature) should not face criminal sanction and should have access to specialist support services and economic alternatives, as needed. It is also important to note that the research and examples discussed here are from a range of localities. As Dewey et al. (2018) observe, there is substantial diversity, in some ways, within overarching systems of prostitution around the world; from differences in legal and policy approaches across jurisdictions, to differences of experience between and within varying forms of prostitution. However, it is also important, as Dewey et al. go on to say, not to focus on difference and specificity of context at the expense of understanding patterns and similarities. It is therefore important not to

[O]veremphasise the uniqueness of each research site in ways that isolate and compartmentalise forms of knowledge by preventing meaningful comparisons. Such focused intensity risks falling into the kind of paralytic academic solipsism that prohibits the ready translation of empirical research into clear and concise policy recommendations. (Dewey et al. 2018: 7)

As discussed in the sections below, multinational research spanning countries of the Global North and South shows high levels of harm, violence and trauma. This suggests that the core ontology of being bought and sold for sex access is shared by women across geopolitical contexts, while the most negative impacts are experienced by the most socially marginalized women and are compounded by issues such as criminalization, migration regimes, and poor healthcare provision.

Harms of Prostitution

Prior to the 1980s, academic literature focused primarily on the "social harms" of prostitution, that is, potential harms that prostitution might bring to the broader community in the form of threats to the social order or public health (Farley and Kelly 2000). During the 1980s, especially in relevant medical and health literature, there was a shift to emphasize the spread of sexually transmitted infections (STIs) in the context of prostitution (Farley and Kelly 2000; Puri et al. 2017), an area where there continues to be significant contemporary research (e.g., Adriaenssens and Hendrickx 2012; Quast and Gonzalez 2017; Steen 2019; Swahn et al. 2016; Verscheijden et al. 2015). In line with the prevalence of perspectives that positioned prostitution as a "victimless crime" (c.f. Matthews 2015), during this time, only a very limited number of studies assessed the harms of prostitution from the perspective of survivors and those in systems of prostitution or considered the gendered dimensions (cf. Hoigard and Finstad 1986; Silbert and Pines 1984). However, in the last two decades, studies have begun to more thoroughly document the harms experienced by prostituted women and girls across a range of localities (e.g., Choi et al. 2009; Coy 2012, 2013; Deer 2012; Ditmore 2014; Farley 2003; Gorkoff and Runner 2003; Guha 2018; Karandikar and Prospero 2010; Kramer 2004; Nixon et al. 2002; Okal et al. 2011; Raphael 2004; Raphael and Shapiro 2002; Tutty and Nixon 2003). A number of powerful first-person accounts from survivors have also been published and a growing survivor-led movement is making visible the dehumanization and violation of prostitution (e.g., Grootboom 2018; Moran 2013; Norma and Tankard-Reist 2016; Sahu et al. 2017). Some of the most striking elements that emerge from this body of literature are the increased likelihood of experiencing a variety of forms of physical and sexual violence, and psychological harms, in particular, post-traumatic stress.

Violence

There can be little doubt about the high prevalence of physical and sexual violence in prostitution across a range of contexts. A systematic review of the available literature on experiences of violence in prostitution, concludes that there is a "high burden of violence against sex workers globally" (Deering et al. 2014: e42), including physical and sexual violence. Studies on the increased likelihood that women in prostitution will experience violence imply that abuse at the hands of pimps and men who purchase sexual access to women is common. As Choi et al. (2009) note:

> A growing literature has repeatedly revealed that physical, sexual, and emotional abuse are common in prostitution, both by pimps and by customers and that the dynamics of control and abuse in prostitution can be similar to that seen in other coercive relationships, such as intimate partner violence. (Choi et al. 2009: 945)

However, women in prostitution can also be increased targets of violence from intimate partners and police (Deering et al. 2014; Dasgupta 2020; Karandikar and Prospero 2010; Kinnell 2013; Nixon et al. 2002; Raphael and Shapiro 2002). Physical and sexual violence and sexual harassment by police, targeted at Black women, women of color and trans women, is documented by research from the USA (Ritchie 2017).These forms of harm have been commonly overlooked in non-feminist literature on prostitution, where the primary risk has often still been seen to be transmission of STIs (Farley and Kelly 2000), although more recent health research has begun to explore the connections between violence and STI transmission (e.g., Decker et al. 2010; Parcesepe et al. 2015). As Nixon et al. (2002: 1036) point out in their ground-breaking study of experiences of sexual violence in prostitution, while STIs are certainly a serious concern, it is not the transmission of disease that was foremost in their participants' accounts: "Violence and abuse were dominant themes in the women's narratives."

That the women in the study by Nixon et al. (2002) consistently spoke of violence, even when they were not directly asked about it, is indicative of the high rates of physical and sexual assault noted in other studies on the experiences of women in prostitution. In early studies, such as Silbert and Pines (1984), 70% of women in their study of street prostitution in the USA, reported they had been raped and approximately two-thirds had been physically assaulted. Parriotts' (1994) study of prostituted women, also in the USA, similarly found high levels of violence: 85% of participants reported having been raped while in prostitution, 90% reported being physically assaulted, and 50% of the overall sample reported being beaten once a month or more. Such findings indicate extremely high rates of physical violence, considerably higher than those within the general population (Parriott 1994). More recently, and in more varied jurisdictions, there have been findings consistently showing extreme levels of violence: in Korea, 70% of women in prostitution report they have experienced at least one "unexpected and uncontrollable sexual, physical attack" (Choi et al. 2009); in Vancouver (Canada) 90% of women in prostitution report being physically assaulted while in prostitution and 78% report being raped (Farley et al. 2005); in Andhra Pradesh (India), a large study – of more than 1000 women in prostitution – found that 77% reported being raped *in the last six months* (George et al. 2011); 97% of people in prostitution surveyed in Phnom Penh (Cambodia) report having being raped *at least once in the previous year* (Jenkins et al. 2006); and in Kenya – a study with 81 women in prostitution, across multiple locations – found that "almost all participants reported experiencing physical or sexual violence in their working lives" (Okal et al. 2011). Another study of 207 women and girls who had been trafficked primarily for sexual exploitation in Europe found that almost all (95%) had been subject to physical or sexual violence, with 71% experiencing both, and 58% having been injured (Zimmerman et al. 2006). The impacts for women's health were wide-ranging, including physical, sexual, and reproductive consequences and psychological harms.

These findings, from various localities, are corroborated by the most comprehensive study on prostitution and violence available, conducted by Farley and her research team (2003). They surveyed 854 women, girls, men, and transgender people (the majority of those surveyed were women), across nine countries (Canada, Columbia, Germany, Mexico, South Africa, Thailand, Turkey, the USA, and Zambia) about their experiences in prostitution. The study reported that 73% of respondents had been physically assaulted and more than half of the overall sample (57%) had been raped in prostitution. To give some further context of frequency and recurrence, approximately a third of the overall sample reported having been raped more than five times while in prostitution (Farley et al. 2003: 43).

It is sometimes argued by critics of this work, most prominently Weitzer (2000, 2005, 2007), that studies on violence against women in prostitution focus too heavily on street prostitution, which Weitzer claims is linked more to violent crime than to "indoor" forms of prostitution, such as brothel and escort prostitution. However, claims that legalized forms of indoor prostitution are inherently safer for those in prostitution have been undermined by other research based in legalized contexts. This work demonstrates that, even in contexts where prostitution / sex work is legal and regulated, it still carries many of the risks associated with illegal street prostitution, such as STIs and violence at the hands of men who purchase sexual access to women (Bindel 2017; Graham 2014; Raymond 2013; Sullivan 2004; Tyler and

Jovanovski 2018) which casts doubt on the possibility of ameliorating these harms through legalizing prostitution or shifting it to indoor environments (Farley 2004). Weitzer (2005) has also criticized a number of feminist researchers, including Farley, for focusing on street prostitution but extrapolating their findings to cover other forms of prostitution. It should be noted that much of Farley's work does, in fact, include women (and girls, men, and transgender people) from several forms of prostitution. Farley et al. (2003: 49) found, for example, that experiences of rape while in prostitution did not substantially differ across various forms of prostitution, including brothel, street, and strip-club prostitution. The authors note that: "Prostitution is multi-traumatic whether its location is in clubs, brothels, hotels / motels / john's homes (also called escort or high class call girl [sic] prostitution) motor vehicles or the streets" (Farley et al. 2003: 60). Monica O'Connor (2019: 46) notes that the silence in literature on "sex work" around "the harm of unwanted sexual intrusion and the evidence on the harmful consequences for women's health of multiple unwanted sexual acts is remarkable." As the following sections illustrate, these harms are documented in settings and contexts around the world.

Finally, the axiomatically most serious health consequence of prostitution is the strikingly disproportionate mortality rate. Studies indicate that the femicide rate for women in prostitution, perpetrated primarily by sex buyers, is significantly higher than for the general population (Brewer et al. 2006; Potterat et al. 2004; Salfati et al. 2008; Ward et al. 1999). Recent literature suggests that, even in legalized contexts – such as Germany – homicide and attempted homicide remain a significant and disproportionate danger to those in the sex trade (Schon and Hoheide 2021). Women in prostitution are also "the single most targeted group of women in serial homicide" (Sorochinski and Salfati 2019: 1791) and Brewer et al. (2006: 1101), writing in the North American context, contend: "Prostitute women [sic] have the highest homicide victimization rate of any set of women ever studied." Managing fear of lethal violence and hypervigilance contributes to the kinds of psychological harm and trauma, outlined in the following section.

Psychological Harms

The claim that all forms of prostitution can be multi-traumatic for prostituted persons is given added weight by studies on psychological harm. As noted by Tschoeke et al. (2019: 242): "worldwide sex work is frequently associated with serious physical and mental health problems." Given the high rates of violence and abuse experienced by prostituted women, it is perhaps not surprising that significant rates of psychological distress have also been found. A number of studies on the experiences of trauma and dissociation for women in prostitution show that it is often a coping mechanism to psychologically survive experiences in prostitution and/or a direct result of physical and sexual violence (see O'Connor 2019 for an overview). A key theme from these studies is that the ontology of being bought for sex – particularly with respect to touch, smell, and bodily invasion – requires women to develop dissociative mechanisms to manage emotional distress ranging from fear, shame, disgust, and numbness (O'Connor 2019; see also Moran 2013). Farley and colleagues (2005: 244), in their research with 100 women in prostitution in Canada, note: "The physical and emotional violence of prostitution leads to somatic dissociation." It has been noted that further research in this area is needed (Tschoeke et al. 2019), particularly around the intersecting nature of prostitution-related trauma and inequalities based on poverty, racial injustice, colonization, childhood sexual abuse, and other factors (Alschech et al. 2020).

The large, multinational study by Farley et al. (2003) is, again, one of the most comprehensive to date in this area. In that study, 68% of participants met the Diagnostic and Statistical Manual of Mental Disorders (DSM) criteria for a diagnosis of Post-Traumatic Stress Disorder (PTSD), a similar rate to that seen among war veterans who have faced active combat (Farley et al. 2003: 37, 47). The rates of PTSD were found to be similar across various types of prostitution, both legal and illegal, including street prostitution, brothel prostitution, and stripping (Farley et al. 2003: 48–49). Significant rates of PTSD have also been found in a Canadian study of 100 women in prostitution Vancouver (Farley et al. 2005) and an Australian study of 72 women in street prostitution in Sydney (Roxburgh et al. 2006). In the Canadian study, where a substantial over-representation of first nations women was noted, almost three quarters of those interviewed met the DSM criteria for PTSD (Farley et al. 2005). In the Australian study, all but one of the 72 women interviewed reported experiencing trauma and around half of the participants met the DSM criteria for PTSD (Roxburgh et al. 2006). The study of 207 women and girls who had been trafficked found that over half (56%) of women reported signs of PTSD when they reached support services, with extremely high levels of depression and anxiety (Zimmerman et al. 2006).

It has been suggested that such high rates of dissociation and PTSD could be explained by factors other than the experience of prostitution, including childhood sexual abuse. However, Waltman (2012) has pointed out that, in other studies – such as an investigation into the experiences of formerly prostituted persons in Korea (Choi et al. 2009) – prostitution has been found to be strongly related to PTSD symptoms, even when controlling for experiences of childhood sexual abuse (Waltman 2012: 9). Furthermore, such issues around childhood sexual abuse are additionally complicated by the age of entry into prostitution for many women and girls. Juvenile entry into systems of prostitution – sometimes referred to as child commercial sexual exploitation – is fairly routinely documented (Nixon et al. 2002; see also O'Connor and Yonkova 2018). Early research claiming that the average age of entry into prostitution was 13–14 years old (e.g., Silbert and Pines 1984; Weisberg 1985) has been challenged (Weitzer 2007). However, a summary of later work from Marianne Hester and Nicole Westmarland (2004: 55), bringing together findings from various jurisdictions, maintains that "most of the women and men involved in prostitution probably entered prostitution as young people, under the age of 18 or 21". While other studies put forward that at least a significant minority of people in systems of prostitution were first prostituted as children (Bindel et al. 2013; Clarke et al. 2012). Again, it is often those who are most marginalized who are at risk, and underage entry into prostitution has been associated with childhood abuse and inequalities around education, racism, and poverty (Clarke et al. 2012; Kramer and Berg 2003; O'Connor and Breslin 2020).

Psychological and physical harms are compounded by systemic racism in policing and criminal justice responses. Black women, women of color, and Indigenous women are more likely than white women to be criminalized in the USA, including being more likely to be arrested for carrying condoms (Butler 2016; Ritchie 2017). Many migrant women in the sex industry lack access to basic rights of healthcare and economic support and are at risk of being detained and/or deported (O'Connor and Yonkova 2018). As discussed in more depth in the final section, prostitution, pornography, and violence against women must be analyzed in terms of intersecting inequalities of patriarchy, white supremacy, migration regimes, and legacies of colonization (Butler 2016; Deer 2015; MacKinnon 2011; Smiley 2016).

The Interconnectedness of Pornography and Prostitution

It is important to note that the high levels of violence and trauma, recorded in traditionally recognized forms of prostitution, are generally not considered in public debates about pornography. A picture of endemic violence, rape, and PTSD is not one that fits well with the glamorized version of "porn-chic" that has been prominent in popular media in much of the West in decent decades (Coy et al. 2011; Tyler 2011). Indeed, pornography and prostitution are often, both popularly and legally, conceived as separate entities (Spector 2006a; Whisnant 2004). With pornography, when harms are discussed, the focus has largely been in terms of consumption; harms through the mainstreaming of violent content (e.g., Dines 2010) as well as harms to women through consumption (e.g., use of pornography in intimate partner violence, see Tarzia and Tyler 2020), and broader potential harms to women as a class through the eroticizing of women's subordination and violence against women (e.g., Bart 1985; Itzin 1992). It is also important, however, to consider pornography through the prism of production and as part of the global sex industry, including harms in those in the pornography industry (Tyler 2015). As Karen Boyle (2000) has noted, debates about the negative effects of pornography consumption (otherwise referred to as "effects" research) have largely side-lined discussions about the harms of pornography production. In contrast, this section emphasizes the way in which there are crossovers between pornography and other forms of prostitution before a discussion of harms that can be understood as specific to pornography, in terms of production. These elements are important in better recognizing harms regarding pornography as a form of prostitution.

Pornography as Connected to Other Forms of Prostitution

Since the early 2000s, there has been significant documentation – both within and outside of the academy – of the expansion and normalization of the pornography industry and pornographic content (Boyle 2018; Dines 2010; McNair 2002; Paul

2005; Rich 2001; Tyler and Quek 2016; Williams 2004). Through processes that have become known as the "mainstreaming of pornography" or "pornification" (Tyler and Quek 2016) in the Global North, there was (and is) a burgeoning amount of pornographic content created, particularly in the United States (US). As the pornification trend gained momentum at the turn of the century (before the rise of internet pornography) the US pornography industry was reportedly creating at least 10,000 titles a year (Williams 2004). While the exact worth of the pornography industry remains difficult to ascertain with any great precision, there was widespread agreement around this time that in the US, it grossed more than the Hollywood film and popular music industries combined (McNair 2002; Rich 2001). Since then, new methods of distribution, coupled with new technology, have facilitated increasing consumption and changing patterns of consumption, for example, streaming on smartphones (Lim et al. 2017). Even as this is published, rapid advances in technology will further enable the buying and selling of sexual images and acts (e.g., through online platforms such as OnlyFans), further blurring the differences across and between different types of prostitution.

It is important to understand that, prior to these significant technological changes, connections across various forms of prostitution have been evident. Traditionally recognized forms of prostitution – such as brothel and street prostitution – are often linked, not only to each other, but to other forms of prostitution such as pornography, stripping, "camming" (or other forms of computer-mediated, or increasingly "platform-based" prostitution), and mail-order bride services. As Stark (2006) contends, both legal and illegal strands of the prostitution industry – including pornography – intertwine:

> Women and girls in prostitution rings are often used simultaneously in multiple systems of prostitution. Prostitution ring pimps use women and girls in mainstream venues such as strip clubs as well as underground prostitution venues where attendance is restricted. For instance, there are women who travel the mainstream strip circuit and they are simultaneously used as sex slaves in pornography shoots carried out by prostitution ring pimps. Other women are prostituted in a brothel during the day and used in pornography during the evenings. (Stark 2006: 46)

The movement of women from one form of prostitution to another is also mentioned in other research investigating women's experiences in prostitution (Farley 2020; Farley and Kelly 2000; Raphael and Shapiro 2002; Raymond 1995) and the pornography industry (O'Neal 2016). A study of more than 200 women in prostitution in Chicago, for instance, found that 54% of women who began in street prostitution moved on to other forms of prostitution including escort services, "exotic dancing," and pornography (Raphael and Shapiro 2002: 25). Furthermore, the multinational study from Farley et al. (2003), mentioned above, found that almost half of all those surveyed (49%) reported having had pornography made of them while in other forms of prostitution (Farley et al. 2003: 46) and that 47% of those interviewed reported that they were "upset by an attempt to make them do what had been seen in pornography" (Farley et al. 2003: 46). In addition to women moving from traditionally recognized forms of prostitution to pornography, according to some of those within the industry, it is also common for women performing in pornography to work simultaneously in other forms of prostitution (Reed 2006; Simpson 2005). Sometimes, to supplement income from pornography, even well-known "porn stars," have been known to work in strip clubs and brothels (Berg 2016; Simpson 2005). In terms of lived experience, it can therefore be difficult to maintain a practical distinction between pornography and traditionally recognized forms of prostitution (Farley 2020). However, there is limited recent empirical research exploring these connections and the role of technological developments in further blurring the distinctions (cf. Berg 2016).

The integration of pornography with other forms of traditionally recognized prostitution is also evidenced through the use of pornography for "grooming" women and girls in prostitution (Giobbe 1990; Raymond 1995; MacKinnon 2006; Stark and Hodgson 2004). The use of pornography is a strategy often used by pimps (Stark and Hodgson 2004) to show women what will be expected of them in prostitution. In research conducted by the prostitution survivors' group WHISPER (Women Harmed in Systems of Prostitution Engaged in Revolt), 30% of women interviewed reported that pimps used pornography as a tool to instruct them (Giobbe 1990). Catharine MacKinnon (2006) also notes that the use of pornography as "seasoning" was mentioned by a number of prostituted women during the hearings into the anti-pornography ordinances in the US. She mentions one woman in particular who "told how pornography was used to train and season young girls in prostitution and how men would bring photographs of women in pornography being abused and say, in effect, "I want you to do this," and demand that the acts being inflicted on the women in the materials be specifically duplicated (MacKinnon 2006: 251). Such testimony also suggests that men who purchase sexual access to women in prostitution attempt to use pornography as a kind of instruction manual. A recent study of prostitution in Ireland, drawing on a wide range of sources including data from support services, case studies, interviews with practitioners, and analysis of the Escort Ireland website, reports that "[a]nother common fantasy that buyers demand is the 'porn star experience' (PSE), where women are required to perform the behaviours and sex acts typically seen in pornography" (O'Connor and Breslin 2020: 93). There are a variety of ways, therefore, that pornography and other forms of prostitution can be conceived of as mutually reinforcing (Tyler 2015).

Pornography as a Form of Prostitution

Spector (2006a: 9) states that in mainstream approaches to prostitution – both those that are pro-prostitution and those that favor restrictions – the focus remains on the individual woman in prostitution, while in approaches to pornography there is an observable shift away from this individual-centered approach, and instead the "social value of expressive liberty" is emphasized. In arguing for the social value of protecting pornography as expression, the focus shifts from production and the "rights" of the individuals involved in production, to the "rights" of consumers (Spector 2006b: 430). As Spector shows, this is a serious inconsistency which creates flawed analyses of pornography in which the "individual worker" fades from view completely, "as if no pornography were live-actor pornography at all"'(2006b: 435).

However, when considering pornography from the position of production – that is understanding that real people performed real acts in order for it to be produced – it becomes difficult to see how theorizing on pornography has become so abstracted from prostitution, particularly given the increasingly amorphous use of "sex work" in the wider academic and popular literature (Tyler 2015). Whisnant addresses this issue directly in "Confronting Pornography: Some conceptual basics" and argues that "[p]ornography is the documentation of prostitution" (Whisnant 2004: 19). By way of example, Whisnant provides the following hypothetical scenario:

> Suppose Fred is making money by selling Gertrude's sex act to Harvey and reaping part or all of the proceeds. In short, Fred is a pimp. It then occurs to him that with this new technological innovation called the camera (or video camera, or webcam, etc.) he could sell Gertrude's sex act not just once, to Harvey, but many thousands of times to many thousands of different men...The structure, logic, and purpose of Fred's activity have not changed. He is still a pimp. He has simply become more savvy and enterprising...The basic elements of Gertrude's experience, similarly, have not changed: she is still exchanging sex acts for money. The only member of our original trio now having a significantly different experience is Harvey, who now has his sexual experience 'with' (at, on) Gertrude at some technological remove. He may like it this way or he may not, but keep in mind that he is getting the goods at a much lower price, with greater anonymity, and with the added benefit of not having to see himself as a john. (Whisnant 2004: 20)

Whisnant also counteracts the claim that pornography is qualitatively different from prostitution because men are paid for sex in pornography, too: "So essentially, a male prostitute has entered the scene and is now participating alongside the female prostitute. But what of it? The basic structure of pimp, prostitute, and customer remains intact" (2004: 20). Therefore, it is arguable, that the only significant change between the original form of prostitution and the finished product of pornography is the experience of the consumer. Considering pornography from the production side makes it look a lot more like prostitution than something which should be considered as entirely separate (Tyler 2015).

A number of writers and theorists, particularly those employing radical feminist analyses, have highlighted the similarities between pornography and prostitution, and have, moreover, argued that pornography *is* prostitution (Dines and Jensen 2006; Dines et al. 1998; Farley et al. 2003; MacKinnon 2006; Russell 1993; Whisnant 2004). Dines and Jensen (2006), arguing along the same lines as Whisnant (2004), state that "[w]hile pornography has never been treated as prostitution by the law, it's fundamentally the same exchange. The fact that sex is mediated through a magazine or movie doesn't change that, nor does that fact that women sometimes use pornography. The fundamentals remain: Men pay to use women for sexual pleasure" (Dines and Jensen 2006, n.p.). Farley puts it more simply: "Pornography is a specific form of prostitution, in which prostitution occurs and is documented" (Farley 2003: xiv). Russell (1993) maintains the focus on what happens to the women in prostitution: "Does it really make sense that an act of prostitution in front of a camera is more acceptable than the same act performed in private...These women are not simulating sex. They are literally being fucked..." (Russell 1993: 18).

It is worth noting that the understanding that pornography *is* prostitution is also common to a number of works written by those from within the sex industry (e.g., Almodovar 2006; Lords 2004; Reed 2006). For instance, Almodovar (2006: 151), a self-described "retired prostitute" and "sex worker rights advocate" argues that a camera is the only difference between pornography and prostitution, even referring to pornography as "prostitution on camera". Furthermore, Almodovar (2006: 158) notes that while she disagrees with many aspects of a radical feminist understanding of prostitution, "I do agree with these [radical] feminists that pornography and prostitution are one and the same." That such differing perspectives converge on this particular issue is powerful evidence in support of the value of understanding pornography, and the harms of pornography, through the prism of systems of prostitution. In doing this, it forces a consideration of the above harms as potentially concomitant and provides a way to think about similar issues of violence and trauma that women in pornography might also experience.

Experiences of Abuse in Pornography Production

There are very limited studies on the experiences of harm and trauma in the pornography industry (Waltman 2012; Tyler 2015; cf. Choi et al. 2009). Given the comparison and connections in the sections above, however, there are a variety of reasons to understand that certain harms – such as exposure to physical and sexual violence as well as severe psychological stress – are spread across forms of prostitution, including pornography production. As Waltman (2012: 8) posits: "testimonial evidence on violence, coercion and trauma during pornography production ... mirror both quantitative and qualitative data on these subjects in the lives of prostituted women around the world." Furthermore, there are examples from first person accounts and industry reporting that indicate that experiences of harm, particularly around violence and sexual violence, are not uncommon.

Increasing numbers of women, particularly those based in the US pornography industry, have spoken about their experiences of abuse and assault while on set. Sometimes this has been in the form of memoir, as is the case with Jenna Jameson (2004; for commentary see: Dines 2010; Tyler 2011) – one of the most famous "porn stars" of the early 2000s – and, at others, in pornography industry publications (e.g., Ross 2000, 2004) and mainstream media publications (e.g., Amis 2001; Gira Grant 2015; Snow 2017). In such accounts, women have publicly provided descriptions of rape by male pornography performers (at times encouraged or intended by directors for the purposes of producing content), choking by directors, deception negating consent in terms of sex acts to be performed, physical harms, and psychological trauma. There is a consistent theme that emerges of a culture within the pornography industry where these kinds of harms are normalized and/or frequently accepted; an issue brought to greater public attention in the recent reporting of multiple sexual assault and rape charges against Ron Jeremy, arguably the most prominent figure in US pornography production over several decades (Jacobs 2020). As Boyle (2000: 193) concludes from such evidence: "the testimonies of women who have been harmed in the production and consumption of pornography demonstrate a strong link between pornography and violence."

The brutality of the treatment described in these public accounts is significantly at odds with the glamourized version of pornography prominent in popular culture. It has, however, been acknowledged in industry publications, such as *Adult Video News* (*AVN*), particularly since the rise of so-called gonzo pornography, which saw a significant increase in more "extreme" and violent content in the 2000s (Dines 2010; Jensen 2007). The following two accounts, for example, were published in *AVN*:

> He was allowed to hit hard, choke hard and to pull hair hard. There were times when he slapped me so hard that he left a mark. He choked me while lifting me off the ground. He shoved my face into the cement floor. He threw me over his shoulder, cutting off my air. Choking me was one of the worst things...He was hurting the girls. He was given the green light to hurt the women for the effect of the video. (Starr quoted in Ross 2000, n.p.)
>
> This is porn but they dig so deep into the person's head - this is supposed to be degrading...They want you to cry. They want you to really, really cry. They want you to have an emotional breakdown right there. They want to see it all and then they want to fuck you while you're crying. They will literally beat you up in the process...I got punched in the ribs. I am completely bruised up.... (Hunter quoted in Ross 2004, n.p.)

In research documenting the experiences of women in prostitution, such incidents tend to be recognized as part of broader patterns of abuse, but in the context of pornography they are often cast as "isolated incidents," seemingly unconnected to the functioning of the pornography industry itself (Waltman 2012). However, these examples from pornography production do need to be seen in the context of broader patterns of documented abuse that women experience in other parts of the sex industry (Waltman 2012) and can be seen to mirror women's experiences found in studies of violence against women in prostitution as detailed above. It is possible, therefore, for such experiences to be understood as part of a continuum of violence against women (Kelly 1988), perpetrated through different forms of prostitution.

Pornography and Harms of Distribution

Positioning pornography as a form of prostitution, and considering harms through the prism of production, also raises a consideration that subsequent distribution of pornography can cause additional kinds of harm to those who have been used in its making. Russell (1993: 18–19) has claimed, for example, that "because this abuse is photographed for public consumption, and because the women being photographed invariably have no control over what happens to the pictures, movies, or videos for the rest of their lives, the harm suffered by prostitutes [sic] who are used to make pornography is often significantly more severe." Moreover, drawing on studies regarding the psychological harms experienced by women in systems of prostitution,

Waltman (2012: 9) asserts that pornography, "as a specific branch of prostitution, is particularly vicious and cruel to women." To conceptualize pornography as a form of prostitution, and one that may carry additional types of harm for the women involved in its making, is not to suggest that there are forms of prostitution that are not harmful. It is also not intended to open an argument about the relative merits or of one form of prostitution over the other. Rather it is important to address claims that pornography as a form of "indoor prostitution" is either harmless, or clearly less harmful than other forms of prostitution (e.g., Johnson 2002; McElroy 1995; Weitzer 2005).

Again, there are insights to be gained from those who have made public accounts of their time in the pornography industry. For instance, Jameson (2004) – mentioned above – in her autobiography, remarks on the stress and fear associated with being recognized. As she describes:

> Most girls get their first experience in gonzo films – in which they're taken to a crappy studio apartment in Mission Hills and penetrated in every hole possible by some abusive asshole who thinks her name is Bitch. And these girls...go home afterward and pledge never to do it again because it was such a terrible experience. But, unfortunately, they can't take that experience back, so they live the rest of their days in fear that their relatives, their co-workers, or their children will find out, which they inevitably do. (Jameson as quoted in Dines 2010: 39)

There are points within the autobiography where Jameson recounts the experience of pornography production as unpleasant and at times as abusive, but here she also emphasizes that there is an additional threat, or potential harm, because a record of that experience has been created. Furthermore, the record has been created specifically for the purpose of sale and distribution and cannot be erased after someone has chosen to exit the industry. These conditions, coupled with high rates of pornography consumption, suggest the fear of being recognized or "found out" is well-founded. There are similarities with the ongoing fears and harms documented in more recent research dealing with elements of image-based sexual abuse (IBSA), also sometimes referred to as "revenge pornography" (e.g., McGlynn et al. 2021).

Unlike the recent literature on IBSA, however, questions of harms associated with pornography production and distribution are not especially new. Writing from a radical feminist perspective, for example, MacKinnon was challenging readers to think about the harms of pornography, from the perspective of women used in production, back in the early 1990s:

> You hear the camera clicking or whirring as you are being hurt, keeping in time to the rhythm of your pain. You always know that the pictures are out there somewhere, sold or traded or shown around or just kept in a drawer. In them, what was done to you is immortal. He has them; someone, anyone, has seen you there, that way. This is unbearable. What he felt as he watched you as he used you is always being done again and lived again, and felt again though the pictures.... (MacKinnon 1993: 4)

But overlapping with the contemporary concerns about IBSA, the potential for the ongoing nature of harm is highlighted in MacKinnon's analysis. In this case, she emphasizes that women – even after exiting the sex industry – bear the burden of knowing that the record of their time in the industry can still be viewed by others.

In commenting on the analysis put forward by MacKinnon, Tanya Horeck (2004: 83) notes "how technologies of vision not only replay and repeat women's original trauma, but produce a new dimension of pain". And as Tyler (2015: 121) draws out, this "new dimension of pain" can also involve the distress of knowing, not only that the pornographic images are still circulating, but also knowing that they may be used in the prostitution of others. It is not uncommon for pornography to be used by pimps as part of a process of "grooming" or "training" girls and women for prostitution (Giobbe 1990; Raymond 1995; MacKinnon 2006), as noted earlier, and the use of pornography can be a strategy used by pimps and johns (Stark and Hodgson 2004) to show women what will be expected of them. Reflecting on this cycle of abuse, one sex trade survivor is quoted by Christine Stark and Carol Hodgson, explaining how these connections weigh on her:

> The man who prostituted me showed me pictures of what he was going to do to me and he would "practice" on me what was happening in the picture. That's how I learned what to do for the trick. The hard thing is, I know the pornography he made of me is being used to hurt others. (quoted in Stark and Hodgson 2004: 21)

Circumstances such as this might also contribute to the distress or multi-traumatic experience of pornography production for some survivors. Indeed, the previously cited study from Farley et al. (2003) regarding psychological distress and prostitution found that women involved in pornography production "had significantly more severe symptoms of PTSD than did women who did not have pornography made of them" (Farley 2007: 146).

Therefore, for a range of reasons outlined in the above sub-sections, a clear distinction between prostitution and pornography as contexts for, and forms of, sexual exploitation, is difficult to maintain both conceptually and experientially. Pornography can be understood as a system of prostitution and the cross-cutting evidence on harms requires analyses that connect the impacts of such violation on women as a class. One way of bringing these aspects together, and developing relevant structural analyses, is through thinking about systems of prostitution.

Understanding Systems of Prostitution

The weight of evidence across different studies indicates that violence is integral to prostitution, with similar experiences in the production of pornography, and that the extensive psychological harms and trauma experienced by women and girls reflect the dehumanization of being bought and sold for sex. A radical feminist approach to theorizing from women's lived experience of prostitution builds from these insights to contend that prostitution exists on what Liz Kelly (1988) has conceptualized as the "continuum of violence against women," with a wide range of impacts for women's health and wellbeing. Radical feminist analyses of prostitution tend to be centered around the issue of harm, both the harm done directly to women in systems of prostitution and the harm done to women as a class through the existence of systems of prostitution (Barry 1995). These analyses often emphasizes the violence that many women in prostitution suffer at the hands of pimps and johns, but the harm caused through the prostitution acts themselves is also a focal point (Coy et al. 2019; Barry 1995; Farley et al. 2003; Jeffreys 1997; Raymond 2013). That is to say, acts of prostitution are seen as harmful in and of themselves, even if there is no additional physical violence. The sex required in systems of prostitution is seen as objectifying and dehumanizing (Barry 1995; Tyler 2012), and as a violation of women's human rights (Farley et al. 2003; Jeffreys 1997; Raymond 1995). Once prostitution is understood as violence, analytic attention turns to the ways in which both prostitution and pornography, as with all violence against women, are "systemic" and "systematic," "in the context of poverty, imperialism, colonialism and racism" (MacKinnon 2006: 29; see also: Gupta 2017). This recognition underpins Stark and Hodgson (2004) conceptualization of "systems of prostitution."

A radical feminist analysis considers prostitution within the structural context of patriarchy, and it is the structural inequalities between men and women under patriarchy that are considered to make prostitution possible (Dworkin 1993; Miriam 2005). It is crucial to understand how systems of prostitution also rely on racial and economic inequalities and that these intersect with the inequalities women as a class experience vis-à-vis men as a class (e.g., Butler 2016; Carter and Giobbe 1999; Nelson 1993): what bell hooks (2013) terms "imperialist white supremacist capitalist patriarchy." Racism and sexism in pornography, for example, interact in pornography in ways that reflect colonial dynamics of white supremacy and "patriarchal capitalism," where profit is made from exploitation and commodification of Black women's bodies (Benard 2016; Butler 2016). Writing with reference to trafficking of Native women in the US, Sarah Deer (2015) notes that prostitution has "colonial roots" in that "the legacy of relocation, chronic poverty and historical trauma significantly reduces the opportunities available to Native women and make them vulnerable to prostitution and sex trafficking" (p. 75). While Cheryl Nelson Butler (2016: 103) draws on critical race feminism to analyze prostitution and trafficking in the US, demonstrating how "race, gender, age and class biases systematically push people of colour into prostitution and close shut the escape paths." Framing engagement in pornography and prostitution in terms of women's agency decontextualizes both systems and the unequal historical and contemporary power dynamics that they are rooted in. While individual women may experience a sense of agency from being involved in systems of prostitution, focusing on individual choice obscures the contexts such as patriarchal capitalism and white supremacy (Barry 1995; Benard 2016). Akeia A.F. Benard and Sarah Deer both also highlight the role of white men (and women) as colonizers who institutionalized the sexual exploitation of women of color. At a transnational level, these power dynamics are evident in sex tourism, resting on the sexualization of young women of color by privileged Western men and the economic inequalities of globalized patriarchy (Rosario Sanchez 2015). Samhita Mukhopadhyay (2008: 157) summarizes these dynamics by referring to the prostitution system as the "entertain white men" industry.

Conclusion

What this means for our understanding of violence against women and health in systems of prostitution is multi-faceted. As Butler (2016) identifies, lack of access to health care is itself an aspect of class oppression and poverty that shapes how women are economically coerced into prostitution. The over-representation of minoritized women, women of color, and migrant women in systems of prostitution further demonstrates that the most socially marginalized women and girls are most likely to be exploited in systems of prostitution and are subject to the most negative consequences (Nelson 1993; Butler 2016; Gupta 2017). In the short-term, health services and specialist support services should be delivered in ways that are accessible to women irrespective of citizenship status and take into account that women's experiences and needs differ according to social location (Zimmerman et al. 2006). The extensive evidence base on how systems of prostitution harm women's health and wellbeing demonstrates the need for more far-reaching responses. Sexual exploitation undermines women's bodily integrity and rights to live free from violence, core elements of sexual health and human rights (Benard 2016). Psychological trauma limits women's capacity to achieve potential and seriously diminishes quality of life. Prevention of such harms is therefore imperative. Prevention must be grounded in addressing the underlying inequalities that fuel systems of prostitution and, furthermore, must acknowledge the centrality of male sexual entitlement and demand for sexual access to women's bodies through all forms of prostitution, including pornography.

References

Adriaenssens S, Hendrickx J (2012) Sex, price and preferences: accounting for unsafe sexual practices in prostitution markets. Sociol Health Illn 34(5):665–680

Almodovar NJ (2006) Porn stars, radical feminists, cops, and outlaw whores: the battle between feminist theory and reality, free speech and free spirits. In: Spector J (ed) Prostitution and pornography: philosophical debate about the sex industry. Stanford University Press, Stanford, pp 149–175

Alschech J, Regehr C, Logie C, Seto M (2020) Contributors to posttraumatic stress symptoms in women sex workers. Am J Orthopsychiatry 90(5):567–577

Amis M (2001) A rough trade. The Guardian (London). 17th March. Available from: https://www.theguardian.com/books/2001/mar/17/society.martinamis1

Barry K (1995) The prostitution of sexuality. New York University Press, New York

Bart P (1985) Pornography: Institutionalising woman-hating and eroticising dominance and submission for fun and profit. Justice Q 2(2):283–292

Benard AAF (2016) Colonizing black female bodies within patriarchal capitalism: feminist and human rights perspectives. Sex Media Soc 2(4):1–11

Berg H (2016) 'A scene is just a marketing tool': alternative income streams in porn's gig economy. Porn Stud 3(2):160–174

Bindel J (2017) The pimping of prostitution. Palgrave Macmillan, London

Bindel J, Breslin R, Brown L (2013) Capital exploits: a study of prostitution and trafficking in London. Eaves, London

Boyle K (2000) The pornography debates: beyond cause and effect. Women's Stud Int Forum 23(2): 187–195

Boyle K (2018) The implications of Pornification: pornography, the mainstream and false equivalences. In: Lombard N (ed) The Routledge handbook of gender and violence. Routledge, London, pp 85–96

Brewer DD, Dudek J, Potterat J, Muth S, Roberts J, Woodhouse D (2006) Extent, trends, and perpetrators of prostitution-related homicide in the United States. J Forensic Sci 51(5):1101–1108

Butler CN (2016) A critical race feminist perspective on prostitution and sex trafficking in America. Yale J Law Fem 27(1):95–139

Carter V, Giobbe E (1999) Duet: prostitution, racism and feminist discourse. Hastings Women's Law J 37(1):37–57

Choi H, Klein C, Shin M, Lee H (2009) Posttraumatic stress disorder (PTSD) and disorders of extreme stress (DESNOS) symptoms following prostitution and childhood abuse. Violence Against Women 15(8):933–951

Clarke R, Clarke E, Roe-Sepowitz, Fey R (2012) Age at entry into prostitution: relationship to drug use, race, suicide, education level, childhood abuse, and family experiences. J Human Behav Soc Environ 22(3):270–289

Cobbina J, Oselin S (2011) It's not only for the money: an analysis of adolescent versus adult entry into street prostitution. Sociol Inq 81(3):310–332

Coy M (2012) 'I am a person too': Women's accounts and images about body and self in prostitution. In: Coy M (ed) Prostitution, harm and gender inequality. Ashgate, London

Coy M (2013) Invaded spaces and feeling dirty: women's narratives of violation in prostitution and sexual violence. In: Horvath M, Brown J (eds) Rape: challenging contemporary thinking. Willan Publishing, Devon, pp 184–206

Coy M, Wakeling J, Garner M (2011) Selling sex sells: representations of prostitution and the sex industry in sexualised popular culture as symbolic violence. Women's Stud Int Forum 34(5): 441–448

Coy M, Smiley C, Tyler M (2019) Challenging the 'prostitution problem': dissenting voices, sex buyers and the myth of neutrality. Arch Sex Behav 48:1931–1935

Dasgupta S (2020) Violence in commercial sex work: a case study on the impact of violence among commercial female sex workers in India and strategies to combat violence. Violence Against Women, online first. https://doi.org/10.1177/1077801220969881

Decker M, McCauley H, Phuengsamran D, Janyam S, Seage G, Silverman J (2010) Violence victimisation, sexual risk and sexually transmitted infection symptoms among female sex workers in Thailand. Sex Transm Infect 86:236–240

Deer S (2012) Garden of truth. Federal Lawyer 59(3):44–48

Deer S (2015) The beginning and the end of rape: confronting sexual violence in native America. University of Minnesota Press, Minneapolis

Deering K, Amin A, Shoveller J, Nesbitt A, Garcia-Moreno C, Duff P, Argento E, Shannon K (2014) A systematic review of the correlates of violence against sex workers. Am J Public Health 2014(5):e43–e54

Dewey S, Crowhurst I, Izugbara C (2018) Introduction. In: Dewey S, Crowhurst I, Izugbara C (eds) Routledge international handbook of sex industry research. Routledge, New York, pp 1–10

Dines G (2010) Pornland: how porn has hijacked our sexuality. Beacon Press

Dines G, Jensen R (2006) So what do you give our society's most influential pimp? The Houston Chronicle. 9th April

Dines G, Jensen R, Russo A (1998) Pornography: the production and consumption of inequality. Routledge, New York

Ditmore M (2014) 'Caught between the Tiger and the crocodile': Cambodian sex workers' experiences of structural and physical violence. Stud Gend Sex 15(1):22–31

Dworkin A (1993) Against the male flood: censorship, pornography and equality. In Letters from a war zone. Lawrence Hill Books, New York, pp 253–275

Farley M (ed) (2003) Prostitution, trafficking, and traumatic stress. Harworth Press, New York

Farley M (2004) 'Bad for the body, bad for the heart': prostitution harms women even if legalised or decriminalised. Violence Against Women 10(10):1087–1125

Farley M (2007) 'Renting an organ for ten minutes': what tricks tell us about prostitution, pornography and trafficking. In: Guinn D (ed) Pornography: driving the demand for international sex trafficking. Los Angeles, Captive Daughters Media

Farley M (2020) Prostitution, the sex trade, and the COVID-19 pandemic. Logos: J Mod Soc Culture 19(1): n.p. Available from: http://logosjournal.com/2020/prostitution-the-sex-trade-and-the-covid-19-pandemic/

Farley M, Kelly V (2000) Prostitution: a critical review of the medical and social sciences literature. Women Crim Just 11(1):29–64

Farley M, Cotton A, Lynne J, Zumbeck S, Spiwak F, Reyes M, Alvarez D, Sezgin U (2003) Prostitution and trafficking in nine countries: an update on violence and posttraumatic stress disorder. In: Farley M (ed) Prostitution, trafficking, and traumatic stress. Harworth Press, New York, pp 33–75

Farley M, Lynne J, Cotton A (2005) Prostitution in Vancouver: violence and the colonization of First Nations women. Transcult Psychiatry v42(2):242–271

George A, Sabarwal S, Martin P (2011) Violence in contract work among female sex workers in Andhra Pradesh, India. J Infect Dis 204(5):s1235–s1240

Giobbe E (1990) Confronting the liberal lies about prostitution. In: Leidholdt D, Raymond J (eds) The sexual liberals and the attack on feminism. Pergamon Press, New York, pp 67–83

Gira Grant M (2015) How Stoya took on James Deen and broke the porn industry's silence. The Guardian (London), 5th December. Available from: https://www.theguardian.com/culture/2015/dec/04/how-stoya-took-on-james-deen-and-broke-the-porn-industrys-silence

Gorkoff K, Runner J (eds) (2003) Being heard: the experiences of young women in prostitution. Fernwood Publishing, Manitoba

Graham E (2014) More than condoms and sandwiches: a feminist investigation of the contradictory promises of harm reduction approaches to prostitution. PhD thesis, The University of British Columbia, Canada

Grootboom G (2018) Exit! A prostitution survivor voice from South Africa. ANTYAJAA: Indian J Women Soc Change 2(2):202–205

Guha M (2018) Disrupting the 'life-cycle' of violence in social relations: recommendations for anti-trafficking interventions from an analysis of pathways out of sex work for women in Eastern India. Gend Dev 26(1):53–69

Gupta R (2017) Understanding and undoing the legacies of sexual violence in India, USA and the world. ANTYAJAA: Indian J Women Soc Change 2(1):1–3

Gupta R (2018) The politics of language: why sex is not work. ANTYAJAA: Indian J Women Soc Change 2(2):222–231

Hester M, Westmarland N (2004) Tackling street prostitution: towards an holistic approach. Report prepared for the UK Home Office. Home Office Research Studies, London

Hoigard C, Finstad L (1986) Backstreets: prostitution, money, and love. Pennsylvania State University Press, University Park

hooks b (2013) Writing beyond race: living theory and practice. Routledge, New York

Horeck T (2004) Public rape: representations of violation in fiction and film. Routledge, London

IBISWorld (2015) Brothel keeping and sex worker services in Australia: market research report. IBISWorld, Melbourne

Itzin C (ed) (1992) Pornography: women, violence and civil liberties. Oxford University Press, New York

Jacobs J (2020) Ron Jeremy is charged with sexually assaulting 13 more women. The New York Times, 31st August. Available from: https://www.nytimes.com/2020/08/31/movies/ron-jeremy-sexual-assault-charges.html

Jameson J (2004) How to make love like a porn star: a cautionary tale. Regan Books, New York

Jeffreys S (1997) The idea of prostitution. Spinifex, Melbourne

Jenkins C, Cambodian Prostitutes Union, Women's Network for Unity, Sainsbury C (2006) Violence and exposure to HIV amongst sex Workers in Phnom Penh, Cambodia. USAID and Policy Project, Washington, DC

Jensen R (2007) Getting off: pornography and the end of masculinity. South End Press, Cambridge, MA

Johnson M (ed) (2002) Jane sexes it up: true confessions of feminist desire. Four Walls Eight Windows Publishing, New York

Karandikar S, Prospero M (2010) From client to pimp: male violence against female sex workers. J Interpers Violence 25(2):257–273

Kelly L (1988) Surviving sexual violence. Polity Press, Cambridge

Kempadoo K (2001) Women of color in the global sex trade: transnational feminist perspectives. Meridians 1(2):28–51

Kinnell H (2013) Violence and sex work in Britain. Routledge, London

Kramer L (2004) Emotional experiences of performing prostitution. J Trauma Prac 2(3–4):186–197

Kramer A, Berg E (2003) A survival analysis of timing of entry into prostitution: the differential impact of race, educational level, and childhood risk factors. Sociol Inq 73(4):511–528

Lords T (2004) Underneath it all. Harper Collins, New York

MacKinnon C (2006) Are women human? And other international dialogues. Harvard University Press, Cambridge, MA

MacKinnon C (1993) Only words. Harvard University Press, Cambridge, MA

MacKinnon C (2011) Trafficking, prostitution and inequality. Harv Civil Rights-Civil Liberties Law Review 46(2):271–309

Macy R, Graham L (2012) Identifying domestic and international sex-trafficking victims during human service provision. Trauma Violence Abuse 13(2):59–76

Madden Dempsey M (2017) What counts as trafficking for sexual exploitation? How legal methods can improve empirical research. J Hum Trafficking 3(1):61–80

Matthews R (2015) Female prostitution and victimization: a realist analysis. Int Rev Victimology 21(1):85–100

McElroy W (1995) XXX: a woman's right to pornography. St Martin's Press, New York

McGlynn C, Johnson K, Rackley E, Henry N, Flynn A, Powell A (2021) 'It's torture for the soul': the harms of image-based sexual abuse. Soc Leg Stud 30(4):541–562

McNair B (2002) Striptease culture: sex, media and the democratization of desire. Routledge, London

Miriam K (2005) Stopping the traffic in women: power, agency and abolition in feminist debates over sex-trafficking. J Soc Philos v36(1):1–17

Monroe J (2005) Women in street prostitution: the result of poverty and the brunt of inequality. J Poverty 9(3):69–88

Monto M (2004) Female prostitution, customers, and violence. Violence Against Women 10(1):160–188

Moran R (2013) Paid for: my journey through prostitution. Spinifex, Melbourne

Mukhopadhyay S (2008) Trial by media: black female lasciviousness and the question of consent. In: Friedman J, Valenti J (eds) Yes means yes: visions of female sexual power and a world without rape. Seal Press, Berkeley, pp 151–161

Nelson V (1993) Prostitution: where racism and sexism intersect. Mich J Gend Law 1(1):81–89

Nixon K, Tutty L, Downe P, Gorkoff K, Ursel J (2002) 'The everyday occurrence': violence in the lives of girls exploited through prostitution. Violence Against Women 8(9):1016–1043

Norma C, Tankard-Reist M (eds) (2016) Prostitution narratives: stories of survival in the sex trade. Melbourne, Spinifex

NWAC – Native Women's Association of Canada (2014) Sexual exploitation and trafficking of aboriginal women and girls: final report. Native Women's Association of Canada, Akwesasne

O'Connor M (2019) The sex economy. Agenda Publishing, Newcastle

O'Connor M, Breslin R (2020) Shifting the burden of criminality: an analysis of the Irish sex trade in the context of prostitution law reform. UCD, Dublin

O'Connor M, Yonkova N (2018) Gender, trafficking for sexual exploitation, and prostitution. In: Black L, Dunne P (eds) Law and gender in modern Ireland: critique and reform. Hart Publishing, London

O'Neal E (2016) Zooming in on the money shot: an exploratory quantitative analysis of pornographic film actors. PhD Thesis, University of Central Florida, United States

Okal J, Chersich M, Tsui S, Sutherland E, Temmerman M, Luchters S (2011) Sexual and physical violence against female sex workers in Kenya: a qualitative enquiry. AIDS Care 23(5):612–618

Parcesepe A, Toivgoob A, Changc M, Riedeld M, Carlsone C, DiBennardof R, Witte S (2015) Physical and sexual violence, childhood sexual abuse and HIV/STI risk behaviour among alcohol-using women engaged in sex work in Mongolia. Glob Public Health 10(1):88–102

Parriott R (1994) Health experiences of twin cities women used in prostitution: survey findings and recommendations. WHISPER, Minneapolis

Paul P (2005) Pornified: how pornography is transforming our lives, our relationships, and our families. Times Books, New York

Potterat J, Brewer D, Muth S, Rothenberg R, Woodhouse D, Muth J (2004) Mortality in a long-term open cohort of prostitute women. Am J Epidemiol 159:778–785

Puri N, Ngyuyen P, Goldenberg S (2017) Burden and correlates of mental health diagnoses among sex workers in an Urban Setting. BMC Women's Health 17(133):1–9

Quast T, Gonzalez F (2017) Sex work regulation and sexually transmitted infections in Tijuana, Mexico. Health Econ 26:656–670

Raphael J (2004) Listening to Olivia: violence, poverty and prostitution. Northeastern University Press, Boston

Raphael J, Shapiro D (2002) Sisters speak out: the lives and needs of prostituted women in Chicago. Center for Impact Research, Chicago

Raymond J (1995) Report to the special rapporteur on violence against women. Coalition against Trafficking in Women (CATW). Available from: http://www.iswface.org/coalitionagainsttraffick.html

Raymond J (2013) Not a choice, not a job: exposing the myths about prostitution and the global sex trade. Spinifex, Melbourne

Reed T (2006) Private versus public art: where prostitution ends and pornography begins. In: Spector J (ed) Prostitution and pornography: philosophical debate about the sex industry. Stanford University Press, Stanford, pp 249–258

Rich F (2001) Naked capitalists. The New York Times, 20th May

Ritchie A (2017) Invisible no more: policing violence against black women and women of colour. Beacon Press, Boston

Rosario Sanchez R (2015) Sex tourism is sexualized imperialism. Feminist Current, 25th November. Available from: https://www.feministcurrent.com/2015/11/25/sex-tourism-sexualized-imperialism/

Ross G (2000) Rough sex pulled in the wake of controversy. Adult Video News. Online 3 March. Available from: https://avn.com/business/articles/video/rough-sex-pulled-in-the-wake-of-controversy-33851.html

Ross G (2004) Who beat up Nicki hunter? Adult FYI, 9th November. Available from: https://web.archive.org/web/20041110191354/http://www.adultfyi.com/read.aspx?ID=6831

Roxburgh A, Degenhardt L, Copeland J (2006) Posttraumatic stress disorder among female street-based sex workers in the greater Sydney area, Australia. BMC Psychiatry 6(24):1–12

Russell D (1993) Introduction. In: Russell D (ed) Making violence sexy: feminist views on pornography. Teachers College Press, New York, pp 1–22

Sahu A, Mondol R, Khatoon F, Chettry N, Khatoon N (2017) The insider voice about prostitution. ANTYAJAA: Indian J Women Soc Change 2(1):81–90

Salfati C, James A, Ferguson L (2008) Prostitute homicides: a descriptive study. J Interpers Violence 23(4):505–543

Schon M, Hoheide A (2021) Murders in the German sex trade: 1920–2017. Dignity J Sex Exploitation 6(1):1–20

Scully E (2001) Pre-Cold War traffic in sexual labour and its foes: some contemporary lessons. In: Kyle D, Koslowski R (eds) Global human smuggling: comparative perspectives. Johns Hopkins Press, Baltimore, pp 75–106
Shively W, Kliorys M Wheeler K, Hunt D (2012) National overview of prostitution and sex trafficking demand reduction efforts: final report. Available from: http://www.ncjrs.gov/pdffiles1/nij/grants/238796.pdf
Silbert M, Pines A (1984) Victimization of street prostitutes. Victimology 7(1):122–133
Simpson N (2005) The money shot: the business of porn. Critical Sense 13:11–38
Smiley C (2016) A long road behind us, a long road ahead: towards an indigenous feminist national inquiry. Can J Women Law 28(2):308–313
Smiley C (2019) Why sex work doesn't work. Available from: https://web.archive.org/web/20200803111946/https://www.cherrysmiley.com/post/whysexworkdoesntwork
Snow A (2017) A famous Porn Star claims she was raped on set. Will she receive justice? The Daily Beast, 13th April. Available from: https://www.thedailybeast.com/a-famous-porn-star-claims-she-was-raped-on-set-will-she-receive-justice
Sorochinski M, Salfati CG (2019) Sex worker homicide series: profiling the crime scene. Int J Offender Ther Comp Criminol 63(9):1776–1793
Spector J (2006a) Introduction. In: Spector J (ed) Prostitution and pornography: philosophical debate about the sex industry. Stanford University Press, Stanford, pp 1–17
Spector J (2006b) Obscene division: feminist liberal assessments of prostitution versus feminist liberal defences of pornography. In: Spector J (ed) Prostitution and pornography: philosophical debate about the sex industry. Stanford University Press, Stanford, pp 419–445
Stark C (2006) Stripping as a system of prostitution. In: Spector J (ed) Prostitution and pornography: philosophical debate about the sex industry. Stanford University Press, Stanford, pp 40–50
Stark C, Hodgson C (2004) Sister oppressions: a comparison of wife battering and prostitution. J Trauma Pract 2(3–4):16–32
Steen R (2019) The condom paradox in Southern Africa: how to explain high HIV prevalence among sex workers reporting high rates of condom use? Report for UNAIDS. Available from: https://hivpreventioncoalition.unaids.org/resource/the-condom-paradox-in-southern-africa-how-to-explain-high-hiv-prevalence-among-sex-workers-reporting-high-rates-of-condom-use-and-what-to-do-about-it/
Sullivan M (2004) Can prostitution be safe?: applying occupational health and safety codes to Australia's legalised brothel prostitution. In: Stark C, Whisnant R (eds) Not for sale: feminists resisting prostitution and pornography. Melbourne, Spinifex, pp 252–268
Swahn M, Culbreth R, Salazar L, Kasirye R, Seeley J (2016) Prevalence of HIV and associated risks of sex work among youth in the slums of Kampala. AIDS Res Treat 2016:1–8
Tarzia L, Tyler M (2020) Recognizing connections between intimate partner sexual violence and pornography. Violence Against Women, online first. https://doi.org/10.1177/1077801220971352
Tschoeke S, Borbe R, Steinert T, Bichescu-Burian D (2019) A systematic review of dissociation in female sex workers. J Trauma Dissociation 20(2):242–257
Tutty L, Nixon K (2003) 'Selling sex? It's really like selling your soul': vulnerability to the experience of exploitation through child prostitution. In: Gorkoff K, Runner J (eds) Being heard: the experiences of young women in prostitution. Fernwood Publishing, Manitoba, pp 29–45
Tyler M (2011) Selling sex short: the pornographic and sexological construction of women's sexuality in the West. Cambridge Scholars Publishing, Newcastle
Tyler M (2012) Theorising harm through the sex of prostitution. In: Coy M (ed) Prostitution, harm and gender inequality: theory, research and policy. Ashgate, Farnham, pp 87–103
Tyler M (2015) Harms of production: theorising pornography as a form of prostitution. Women's Stud Int Forum 48(1):114–123
Tyler M, Jovanovski N (2018) The limits of ethical consumption in the sex industry: an analysis of online brothel reviews. Women's Stud Int Forum 66(1):9–17

Tyler M, Quek K (2016) Conceptualizing pornographication: ways forward for feminist analysis. Sex Media Soc 2(2):1–14

Verscheijden M, Woestenberg P, Gotz H, van Veen M, Koedijk F, van Benthem B (2015) Sexually transmitted infections among female sex workers tested at STI clinics in the Netherlands, 2006–2013. Emerg Themes Epidemiol 12(12):1–11

Waltman M (2012) The ideological obstacle: charging pornographers for sexual exploitation. Paper presented as the Midwest Political Science Association. Chicago, 12–15th April. Available from: https://papers.ssrn.com/sol3/papers.cfm?abstract_id=2050290

Ward H, Day S, Weber J (1999) Risky business: health and safety in the sex industry over a nine-year period. Sex Transm Infect 75(5):340–343

Weisberg D (1985) Children of the night: a study of adolescent prostitution. Lexington Books, Cambridge MA

Weitzer R (2000) Why we need more research on sex work. In: Weitzer R (ed) Sex for sale: prostitution, pornography and the sex industry. Routledge, New York, pp 1–17

Weitzer R (2005) Rehashing tired claims about prostitution. Violence Against Women 11(7): 971–977

Weitzer R (2007) Prostitution as a form of work. Sociol Compass 1(1):143–155

Whisnant R (2004) Confronting pornography: some conceptual basics. In: Stark C, Whisnant R (eds) Not for sale: feminists resisting prostitution and pornography. Melbourne, Spinifex, pp 15–27

Williams L (2004) Proliferating pornographies on/scene: an introduction. In: Williams L (ed) Porn studies. Duke University, Durham, pp 1–26

Zimmerman C, Hossain M, Yun K, Roche B, Morison L, Watts C (2006) Stolen smiles: the physical & psychological health consequences of women and adolescents trafficked in Europe. London School of Hygiene and Tropical Medicine, London

Index

A
Abbe, Ernst, 283
Abercrombie, Michael, 281
Aborigine, 1068
Abortion, 1792, 1794, 1799–1801
    rights, 1797
Abstract verbal reasoning, 1154
Academic dependency, 896
Academic freedom, 835
Academic historians, 236
Academic library, 880–882
Academic progress, 1461
Accommodating, 684
Accumulation of capital, 1490, 1493, 1497
Accuracy of representation, 247
Achievement, 1130
Acquired psychic nexus, 114
Activism, 1796–1798
Activist turn, 796–798
Actor-network theory (ANT), 840
Adaptive capacity, 488
A discourse on inequality, 462
Adler, A., 1132
Administration, 1730, 1731, 1734–1749, 1751
    See also Colonialism
Administrative law, 1276
Adolescence, 1038
Adorno, Theodor W., 165
Aerosol pollution, 485
Aesthetic dimension of science, 1177
Affective, 115
Africa in their natures, 1233–1234
African Americans, 1308, 1322
African knowledges, 899
Afro Asian Solidarity conference, 901
Agency, 333, 343, 1054, 1751
    vs. capabilities, 1596
    goals, 1605
Age of feudalism, 1245

The Age of Fitness, 173
Aggression, 986
Aging, 1404, 1405
AIST test, *see* Attitude-Interest Analysis
    Test (AIAT)
Akiba, Takashi, 1734, 1740
Akinsola, Akiwowo, 895
ALAS, 909
Alatas, Farid, 896
Alatas, Hussein, 896
Aldrovandi, Ulissee, 262
Algorithmic animation, 286–287
Alienation, 1326
Alienism, 1278–1282
Alienists, 1257, 1261, 1263, 1264
All Dogs are Blue, 1234, 1235
Altay region, 1152
Alternative truths, 1225
Alzheimer's disease, 1406, 1415, 1416
Amateur writer, 1731, 1732, 1738, 1744, 1749, 1750
Amauta, 962–964, 966
American anthropologist, 1730, 1735, 1745
American anthropology, 1619, 1747
American exceptionalism, 900
American gender history, 757
American Indians, diabolical representation, 1629, 1630
Americanist anthropology, 1670
American Law Institute (ALI), 1298
American Museum of Natural History (AMNH), 308, 1659
American Psychiatric Association, 1232, 1328
American Psychological Association, 1843
American social psychology, 1024
American sociology, 568, 848, 1009
Amin, Samir, 904
Anachronism, 1680
Analytic possibility, 387

Anatomical images, 231, 234–238, 243, 245–248, 250–252, 254
Anatomical knowledge, 246
Anatomical representations, 230, 231, 234, 235, 237–239, 246, 247, 249, 251–254
Anatomical sex, 435
Anatomy, 230
Ancient entanglements
　aging, 1405
　mental infirmity, 1405
Ando, Kiichirō, 1734
Androgyny, 441
Anglophone sociology, 563
Animal(s), 230
　bodies, 231, 235, 243, 245
　spirits, 1576
Animation, 281, 286–287
Animism, 508
Annales School, 604
Anorexia mirabilis, 1881–1883
Anorexia nervosa, 1857, 1885–1887
Anthropocene, 463, 485, 488, 490, 495
Anthropological findings, 88
Anthropological knowledge, 1615
Anthropologische Gesellschaft Wien, 1664
Anthropologists of religion, 190
Anthropology, 81, 87–89, 91, 94, 99, 102–104, 484, 485, 563, 1006, 1652, 1732–1734, 1736, 1740–1744, 1747, 1749
　damaged planet, 497–499
　human adaptation, 486, 487
　natural resource extraction, 488–490
　non-human agency, 494–496
　persistent colonial orders, 497, 498
　temporalities, 494
　toxicity, 491–493
Anthropometric measures, 1150
Anthroposophy, 68
Anti-austerity protests, 474
Anti-Dreyfusards, 154
Anti-foundationalism, 907
Anti-government protest, 474
Anti-narrative, 88
Anti-nepotism, 1050
　rules, 1036, 1044, 1050
Anti-psychiatry, 1420, 1422, 1426–1430, 1439, 1445–1448
Anti-semitism, 241
Anti-social reactions, 1243
Aoyagi, Machiko, 1748
Apartheid, 1319
Apes, 243
Applied science, 1183
A Question of Power, 1236
Arab Spring, 474

Arbitrage, 1554
Archaeology, 81, 87–89, 96, 102
Archaic love, 1103
Archives, 1749
Argonauts of the western pacific, 465
Aristotle, 51, 458, 467
Arnaud, S., 100, 101
Art, 246, 251, 253, 254, 1231–1233
　and anatomy, 253–254
Artificial intelligence (AI), 50, 91
Artisanal small-scale mining (ASM), 489
Artistic skills, 246
Artists, 246
Aryan race, 300
Aspiration, 1130
Assessment, 427, 431, 436, 441
The Association for Women in Psychology, 1041
Associationism, 613
Astell, M., 1123
Asuwada, 897, 898
Asylum-based care, 1358, 1371
Asylums, 1309, 1335, 1343
Atheism, 36
Attitude-Interest Analysis Test (AIAT), 432, 446
Aurier, A., 1231
Ausdruckspsychologie, 1200
Austerity, 461, 473
Austin, J.L., 45
Austrian methodology, 1555–1557
Austrian school of economics, 1542
　birth of, 1543–1544
　debate on economic planning, 1544–1546
　entrepreneurship, 1553–1555
　institutions re-examined, 1550–1553
　methodology, 1555–1557
　theory of competition, 1547–1548
Authority, 151, 154, 1042
　and power, asylum, 1313–1314
Autobiographical identity, 369
Autobiography, 90, 93, 100, 365–370
Autonomous self, 412
Autonomous social sciences, 896
Autonomy, 134, 154
Azam, E., 100, 101

B

Bachelard, G., 840
Bacon, Francis, 45, 1042
Baer, Karl Ernst von, 267–269
Bain, A., 1127
Bakhtin, M., 95
Balanced growth, 1586
Bandung conference, 895
Banking trade, 1480

Barbarization, 1629, 1632
Barrett, L.F., 340
Barthes, R., 93, 95
Bartlett, F., 91
Basaglia, Franco, 1421, 1422, 1427, 1430, 1439, 1443–1445
Bastian, A., 87, 88, 1655
Baxter, R., 1120
Bazerman, C., 85, 97, 98, 101
Beatty, J., 80, 95
Behavioral economics, 1459, 1590
Behavioral symmetry, 1546
Behavior Farm, 1050, 1051
Behaviorism, 86
Bekhterev, V.M., 1243
Belau National Museum, 1748
Belief, 145–148, 150, 151, 153, 155, 156
Belonging, 728, 730, 731, 736
Bem, Sandra, 441, 444
Benedict, Ruth, 1758, 1762–1764
Benevolence, 128
Benjamin, Harry, 432, 434, 439
Bergson, H., 90
Berlin, 1658
Berliner Gesellschaft für Anthropologie, Ethnologie und Urgeschichte, 1653
Bibliometric databases, 879
Biehl, J., 1228
Big Pharma, 1366, 1370, 1372, 1375
Big push, 1586
Bildung, 215
Binary, 429, 433, 439, 443–447
Binaryness, 425, 440–449
Binet-Simon, 1149
Binet's test, 1094
Biographical self, 365
Biography, 996
Biological evolution, 300
Biologically orientated theory, 1103
Biological model, 1265
Biological psychiatry, 1366
Biology, 6, 10, 433, 434, 438, 439
Biomedical discourse, 1242
Biomedical research, 1369
Bio-psychiatric theories of depression, 1858
Bio semiotically informed work, 527
Birth of modernity, 357
Bisexuality, 439, 444
Black Athena, 904
Black liberation movement, 1325
Black Lives Matter, 1321
Blackness, 1234
Boas, Franz, 1655, 1760–1762
Boasian cultural anthropologists, 1758, 1759
Boccacio's *Decameron*, 94

Book-length studies, 352
Boring, E.G., 1040, 1052, 1055
Borstal, 1345
Boston Women's Health Book Collective, 1787
Botany, 1047
Boundary infrastructure, 921
Boundary object, 920, 1009
Boundary organization, 921
Boundary-work, 919
    audience boundary-work, 920
    expansion, 920
    expulsion (or monopolization), 920
    protection of autonomy, 920
Bounded rationality, 1458
Bourdieu, P., 336, 664–666, 845, 853
Bowker, G., 921
Boyle, R., 922
Braat, M., 86
Braine, L., 1054
Brain functions, 979
Braslow, Joel, 1363
Brass penis, 1037, 1057
British anthropology, 1740, 1742
British colonialism, 910
British Psychoanalytical Society, 1350
British Psychological Society, 1342
British social psychology, 1026
Broca, Paul, 243
Bromberg, W., 1325
Brown vs Board of Education, 1133
Bruner, J., 91, 92
Budapest psychoanalysts' views on child development, 1098–1105
Buffon, Georges, 265
Bukovsky, V., 1247
Burckhardt, Jacob, 208, 212, 221
Bureaucratic apparatus, 1576
Bureaucratic domination, 588
Bureau of Ethnology, 1654
Burma, 469
Burnham, John, 1373
Buryats, 1155
Business principles, 1578
Butler, J., 331, 1225

C
California Task Force to Promote Self-Esteem, 1118
Cameralism, 842
Cameron, Hector Charles, 1345
Camper, Petrus, 240
Canadian Consortium on Neurodegeneration in Aging (CCNA), 1823
Canguilhem, G., 840

Canonical authors, 22
Canonization, 868, 875
Capabilities, 1596
Capability approach (CA), 1587, 1588
  development economics, 1595
  economic growth, 1587
  feminist scholarship, 1601
  gender inequalities, 1601–1603
  GNP, 1587
  human development, 1598
  human development framework, 1588
  intra-household inequalities, 1603
  poverty, 1587
  theoretical basis, 1593
Capital accumulation, 1491
Capital credit, 1579
Capitalism, 968, 1575
Capitalist, 1569
Capitalistic system, 1597
Captive mind, 896
Caracciolo, M., 96
Care of the self, 354
Carlyle, T., 1130
Carrard, P., 80, 82, 83, 99, 104
Carrier, M., 103
Carroy, J., 98
Cartesian dualism, 1008
Cartesian philosophy, 1007
Cartoons, 293, 314, 315
Case, 100
Case histories, 96, 99, 101, 102, 104
Castration, 430, 431
Catallaxy, 1552
Categoricalism, 1840
Categories, 985, 986, 991, 993
Category of intelligence, 989
Catharine MacKinnon, 1906, 1911
Cauldwell, David O., 432
Caveman, 308, 314, 315
Celebratory approach, 1044–1046
Central Europe, 1098
Central European anthropology, 1615, 1616
Centrality, 878
Centre-periphery divide, 883
Centres and peripheries, 876–878
CEPAL, 909
Ceremony, 1817, 1821, 1825
Certified Professional Midwives (CPMs), 1797
Chaney, S., 1227
Character, 1125
Characterology, 1208
Character types, 996
Charcot, J.-M., 1042

Charisma, 1571
Charming stubbornness, 1051
Charon, R., 89
Chemical industrial food production, 484
Cheyne, A., 92
Chief merit, 1564
Child and Adolescent Mental Health Services (CAMHS), 1352
Child-centered pedology, 1093
Child development, 986, 989, 1093
  Budapest psychoanalysts' view on, 1098–1105
  normative approach to, 1112
  observation of, 1094–1095
Child Guidance, 1346, 1347
Childhood, 1332, 1334, 1343, 1348
  as interiority, 362–364
Childism, 1334
Child-oriented marriage, 761
Child psychiatry, 1332–1335, 1348, 1351
  correspondence, 1338
  identifying, 1334
  institutional records, 1335, 1336
  interpreting primary materials, 1334
  legislation, 1336
  national health service, 1346–1348, 1350
  patient case files, 1335
  personal papers, 1338
  practitioners' accounts, 1336, 1337
  print and media, 1339
  public policy, 1336
  tangled roots, 1340–1342, 1344–1346
  voice, 1339, 1340
Child psychology in Hungary
  after World War 2, 1105–1107
  birth of, 1089–1092
  Budapest school of psychoanalysis, 1098–1105
  change of regime, 1110
  observation, child development, 1094–1095
  pedology and psychometry, 1092–1094
  reappearance of, 1107
  between world wars, 1095–1098
Child-rearing, 1097
Children, 1334, 1338
Children's Hospitals, 1344
Child welfare, 1333
Chimerical investors, 1480
Chinese Sociology, 826, 896
Chlorpromazine, 1365–1368, 1372
Chomsky, Noam, 35
Choulant, Ludwig, 235, 236, 246
Christian marriage system, 758

Christian theology, 352
Chromosome, 439, 449
Chronofetishism, 589
Chuvash children, 1158
Cinematography, 277
Circulation of knowledge, 866
Cisgender, 443
Cisgenderism, 445
Civilizational history, 897
Civilization, 565, 1741, 1751
Civilized races, 297
Civil rights, 238
Clandestine political violence, 631–636
Clandestinity, 632
CLAPCS, 909
Clark, K., 1134
Clark, M., 1134
Class, 231, 237, 251, 252, 790, 791
"Classical" experiments, 996
Classical political economy, 1474
  foreign trade, 1494–1496
  fundamental law of distribution, 1487–1489
  government and the state, taxation and public debt, 1496–1498
  grand orders of men, 1479–1480
  information asymmetries and banking trade, 1480–1481
  law of population, 1493–1494
  method and content of, 1475
  money and currency, 1481–1483
  natural vs. market price, 1485–1486
  necessaries vs. luxuries, 1484–1485
  quantities of labor embodied, 1486–1487
  scarce natural resources, 1489–1490
  surplus approach, 1483–1484
  technical and oragnizational change, 1490–1493
Classical sociology, 926
Classical theory, 947, 948, 950, 955
Classification, 426, 431, 437, 443, 449
  of knowledge, 9
Clean eating, 1888–1889
Cleansing force, 395
Climate change, 463
Clinical material, 1309
Clinical trials, 1369
Coercive rehabilitation, 1242–1243
Cognitive, 115
  authority, 1053, 1054
  geographical circuits, 894
  narratology, 96
  psychology, 82, 1009
Cognitively–oriented social psychology, 1007

Cohn, D., 81, 101, 102
Coiter, Volcher, 262
Cold War, 470, 1006, 1007, 1021, 1212–1214
Coleborne, C., 1226, 1227
Collective agency, 1604
Collective capabilities, 1605
Collective effervescence, 145, 148, 149, 151
Collective memory
  and collected memories, 786
  to collective forgetting, 788–789
  media and longue *durée*, 793–794
  and nation states, 790–792
  and power, 792–793
Collective representations, 783
Collectivist societies, 465
Collectivities, 1604
Collyer's study of Australian Sociology, 826
Colonial capitalism, 905, 968
Colonial difference, 907
Colonial ethnography, 1750
Colonialism, 394, 899, 1318, 1391, 1397, 1749, 1790
Coloniality, 904
  of knowledge, 1067, 1070, 1072–1077
  modernity, 906
Colonial psychiatry, 1390, 1392, 1395–1397
Colonial situation, 1074
Colonies, 1069
Colonization, 1007
  of knowledge, 1073
Columbian Museum, 1660
Combe, G., 1126
Committee on Transnational Social Psychology, 1021
Commodification of once common natural resources, 471
Commodities, 1594
Communication studies, 254
Communicative memory, 780
Communist countries, 792
Communities of memory, 791
Community-based NGOs, 1247
Community-based research (CBR), 1819, 1822, 1826
Community-based services, 1244
Community Research Advisory Committee (CRAC), 1821
Comparative Law in Germany, 1740
Comparative research, 1147
Comprehensive social science, 851
Compulsory schooling, 1092, 1094
Comte, A., 86
Confession, 353–355

Confocal microscopy, 282
Connell, R., 903, 925
Consciousness raising (CR), 1793, 1798
Conservative free market economists, 1600
Consternation, 1239
Constitutional inadequacy, 1224
Constraining, 679
Constructivism, 52
Contact zone, 1751
Contemporary feminist analyses, 1837
Contemporary social scientific theories, 186–187
Contemporary social theorists, 952
Context, 13, 18, 990
Context-bound and discontinuous model, 946
Contextual analysis, 843, 855
Contextual history, 1045
Contextualization, 123
Contextual understanding, 111, 990
Continuum, 438, 444
Contra postcolonialism, 902
Conventional rule, 1592
Convention on the Rights of Persons with Disabilities (CRPD), 1226
Conviviality, 516–518
Cooper, David, 1421, 1422, 1426–1430, 1439, 1443, 1445, 1447
Cooperation, 1071
Coordinated Investigation of Micronesian Anthropology (CIMA), 1735
  See also American anthropologist
Co-production, 922
Core gender identity, 432, 434, 439, 440, 449
Coronavirus (Covid-19), 143, 144, 150
Corporal punishment, 1342, 1352
Corporate capitalism, 1210
Correct action, 134
Correctional facilities, 1335
Cosmopolitanism, 30, 42
  circle of moral concern, anti-sentimental expansion of, 48–50
  machines, 50–54
Cosmopolitan memory, 795
Counseling center, 1109
Counter-discourse, 1225
Counter-knowledge, 1225
Covid-19-pandemic, 750, 751
Craig, C., 81
Craniometry, 243
Credit, 1578, 1579
Crichton-Miller, Hugh, 1348
Criminal codes, 1275
Criminality, 1341

Criminal law reform, 1274
Criminology, 1282–1285, 1293, 1296
"Crisis" thinking, 981
Critical psychologists, 996
Critical psychology, 992, 993, 1046
Critical social theory, 1075
Critical theory, 160, 164, 165
Critical thinking, 983
Critique, 5, 11, 13, 15, 23, 161, 162, 168, 169, 171, 172, 174–176, 981, 992, 996
  genealogy, 162–163
  history, 163–165
  of masculinity discourse, 1844, 1847
  problematizing, 165–169
Cro-Magnons, 295, 299, 315, 316
Cross-cultural analysis, 467
Cross-cultural approaches, 1364
Cross-cultural psychology, 1146, 1160
Cross-dressing, 425, 429, 430
Crude evolutionary models, 463
Cryptonormativism, 174
Crystallization, 60
Cuban Revolution, 965
Culler, J., 94
Cultural anthropology, 254
Cultural colonialism, 1645
Cultural critique, 468
Cultural forms, 237
Cultural history, 236
Cultural history of Hungary, 1644
Cultural-intellectual movement, 1614
Cultural memory, 780
Cultural paranoia, 1324–1326
Cultural primitivism, 1151
Cultural psychology, 981
Cultural relativism, 468
Cultural shift, 357
Cultural superego, 944
Cultural turn, 811
  in economic anthropology, 468–473
Cultural values, 897
Cultures of corruption, 464
Curves, 294, 297, 298
Customary law, 1734, 1738, 1740
Cybernetics, 1213

D

d'Alton, Joseph Wilhelm Eduard, 273
Damasio, A., 340
Danto, A., 83
Danziger, K., 1123
Darwin, 1660

Das Kapital, 595
Daston, L., 248
Dawson, D.W., 1347
The Death of Nature, 1042
The Death of the Family, 1428
Debt, 458, 459, 461, 474–476, 478
Decentralized elites, 600
Decolonialists, 906
Decoloniality, 903, 906, 907, 910
Decolonial knowledge, 894
Decolonial scientometrics, 881
Decolonial studies, 779
Decolonial theory, 1076
Decolonization, 394, 397, 570, 776, 780, 789, 870, 1070, 1365
Decolonizing psychology, 1076
Deconstruction, 13
Deconstructive perspectives, 910
Defensive realists, 590
Degeneration theory, 1282, 1283, 1289
Deinstitutionalization, 1317
De-link, 1077
Delmotte, Florence, 577
Dementia, 1264, 1404–1406, 1411, 1412
　praecox, 1238
Demonization, 1629, 1632
Denmark, 470
Dependency, 908, 967
　analysis, 963, 968, 971
Depression
　bio-psychiatric theories of, 1858
　gender based diagnosis, 1864–1866
　gender differences in, 1859
　menopausal, 1863
　North American *vs.* Japanese women, 1863
　post-partum, 1862
　in South Asian women, 1865
Deprivation discourse, 1324
The Descent of Man, 294
Descriptive psychology, 126
Desjarlais, R., 1227
Despair and dread, 364
Despotism, 602
Determinant judgments, 121
Development, 260–262, 265, 267–269, 271–278, 281–283, 286, 287
　and growth, 1498
　psychology, 1197
Deviance, 1307
Deviation, 144, 150, 152
The Division of labor in society, 465
Diabolical representations, China, 1629, 1631
Diagnosis, 427, 428, 432, 434, 438

Diagrams, 295, 297, 298, 300, 304
Dialectical reasoning, 390
Dialectic of Enlightenment, 165
Dias, Jorge, 1681, 1682, 1686–1691, 1694
Dichotomy, 383
die Geisteswissenschaften, 6, 14, 15, 21
Diet culture, 1892
Differential psychology, 1160
Diffusionism, 1691
Digital archaeology, 573
Digital embryo, 283, 286
Digital humanities, 573
Dillon, Michael, 432
Dillon, S., 81
Dilthey, Wilhelm, 110
　approach to history, 116
　formative social ethics, 127–129
　hermeneutical objectivity, 117–118
　historical reason as historical judgment, 123–124
　human sciences, understanding in, 111–115
　philosophical thinking, 111
　theoretical cognition and reflective knowledge, 115
　theory of worldviews, 124–126
　three kinds of objectifications, 118–121
Dioramas, 305, 310, 316
Disciplinary amnesia, 846
Disciplinary boundaries, 926–927
　boundaries of sociology, 919
Disciplinary doxa, 847
Disciplinary history, 931
Disciplinary repression, 848
Discipline, 5, 9, 808–810, 813–816, 818–822, 824–829
　boundaries, 9
　critique, 164
　formation, 9
　knowledge, 12
Disciplining of social psychology, 1006
Discrimination, 755, 760, 768
Discursive cognition, 115
Dis-embedded economy, 474, 475
Dis-embedding, 470
Dismembering, 62–66
Disorders of Sex Development (DSD), 439
Dispositif, 169
Dissection, 233
Distal determinants, 1812
Diverse economies, 477
Diversity, 912
Division of labor, 1479, 1496

Division of Social and Transcultural Psychiatry Research, 1364
DNA, 301, 302, 304, 313
Doctrine of progress, 837
Döllinger, Ignaz, 272
Domain, 122
Domestic citizenship, 1246
Donne, J., 1306
Douglas, Mary, 187, 197–203
Doyle, D., 1323
Drag effect, 738
Dramatization, 380
Drawings, 294, 295
3D-reconstructions, 309, 317
Dreyfus, Hubert, 53
Dreyfus Affair, 154
Dreyfusards, 154
du Bois, W.E.B., 903
Durkheim, Emile, 135–148, 150–157, 187–189, 192, 194, 196, 197, 199, 202, 203, 465, 659, 850, 926
Dynamic nominalism, 334
Dynamic stochastic general equilibrium (DSGE) model, 1534
Dynamic system, 1566

E

Ecological model
  ecosystem, 1834
  exo system factors, 1835
  feminist ecological model, 1835, 1836
  gender-neutral theories, 1836
  integrated framework, 1836
  intersectional feminism, 1837
  male dominance and patriarchy, 1835
  micro level factors, 1835
  multi-level theoretical formulation, 1835
  patriarchy, 1837
Economic(s), 13, 102
  crisis, 458–478
  development, 1568, 1587
  dualism, 1586
  equilibrium, 1566
  exchange, 1590
  insecurity, 1589
  planning, 1544–1546
  responsibilities, 1597
  systems, 1591
  theory, 1590
Economic anthropology, 458, 459
  cultural turn in, 468–473
  formalist and substantivist debates, 466–468
  history and key approaches of, 464–468
  individualist *vs.* systemic approaches, 464–466
  neoliberal economic policy, 473–475
  social and moral philosophy, influence of, 462–463
Economic Commission for Latin America (ECLA), 967
Economicism, 967
Economism, 43
Economy, 9
  vs. society, 1597
Ecosystem, 1834
Edinburgh school, 1179
Editorialism, 963
Education, 230, 232, 1097
Effeminacy, 428, 429, 435, 442, 444
Effervescence, 145, 148, 149, 151, 156
Ego documents, 359
ELACP, 909
ELAS, 909
Elbe, Lili, 431, 432
Electro-convulsive therapy (ECT), 1267, 1316, 1867
Electro-shock therapies, 1362
Elementary understanding, 117
Elias, N., 565, 568, 727, 737, 740
Eliasian approach, 733–740
Elias's theory of civilizing processes, 601
Elite, 1057
Ellis, Havelock, 426, 444
Embeddedness, 468
Embodied nature, 987
Embryology, 261
  algorithmic animation, 286–287
  chronological visualization, 262
  cinematography, 282
  embryos, in silico, 283–286
  epigenetic iconography, 267–272
  filmic methods, 281
  images, development in, 262–267
  living organisms and dead material, 277–281
  optical slicing, 282–283
  series of sections, 274–277
Emergency Committee in Psychology (ECP), 1049
Emotions, 367, 734–735, 743
Empire(s), 594, 600, 780, 1733
Empire of Japan, 1730–1732, 1735, 1737, 1739, 1744, 1747, 1750
Empirical (scientific) statements, 13
Enclosure, 1072

Endocrinology, 437, 444
Endogenous knowledge, 898
End Segregation in America, 1234
English aristocracy, 611
English literature, 900
English lunacy certificate, 1226
Enlightenment, 17, 165, 353, 356
  rationality, 1177
Enterprise, 1569
Enthüllen, 847
Entitlements, 1595
Entrepreneur, 1569, 1574
Entrepreneurship, 1553–1555, 1572
Entrepreneurs vs. capitalists, 1577
Entseelung, 1581
Environmental crisis, 484, 485
Eoliths, 295
Eon, Chevalier d, 426
Epicurean, 30, 35, 44–46, 48, 49
Epigenesis, 265, 266
Epigenetic iconography, 267–272
Epileptoid genius, 1231
Epimeleia heautou, 378
Epistemic authority, 1174
Epistemic trust, 1170
Epistemic values, 1178
Epistemologies, 1077
Epistemology, 230, 235, 236, 248–250
Epochs, 120
Eponychium, 521
Equilibrium, 1565
Equity, 474, 477
Esteem needs, 1136, 1137
Ethical standards, 1370
Ethical tenets, 377
Ethics, 140
Ethnic groups, 1146, 1706
Ethnic Research Institute, 1734, 1740, 1751
Ethnic self-consciousness, 1716
Ethnographic analysis, 1180
Ethnographic capital, 666
Ethnographic representation, 1731, 1749, 1751
Ethnography, 476, 1333, 1614–1618, 1705, 1731, 1732, 1735, 1737, 1738, 1742, 1743, 1748, 1750
Ethnology, 1614, 1615, 1617, 1681, 1682, 1686, 1733, 1740, 1742
Ethnopsychiatry, 1363, 1364, 1382
Ethnos, 1701, 1704, 1705, 1709–1711, 1717, 1733
Eugenic, 37, 41, 48
Eurocentric bias, 1614
Eurocentrism, 894, 904

Europe, 458–460, 466, 467, 472, 477, 478
European civilization, 762
European human sciences, 1613
European identity, 741–744
European imperialism, 241
European social psychology, 1025
  post-World War II development of, 1020–1025
European social theory, 946
Euroscepticism, 743
Evans, M.S., 920
Everyday psychology, 1066
Evidence-based approaches to health
  critique of, 1841
  systematic evidence reviews, 1842
Evidence-based medicine, 1365
  and randomization, 1368–1371
Evolution, 292–297, 299–302, 304, 309, 310, 312–314, 316
Evolutionary biology, 300
Evolutionary history, 308, 312
Evolutionary psychology, 1041, 1197
Evolutionism, 1616
Exceptionalism, 900
Exhibitions, 308, 310–313
Existential anxiety, 364
The Experimentalists, 1053–1055
Expertise, 1275, 1281, 1283, 1286–1288, 1296–1298, 1300
Explanation, 111, 113, 119, 125, 126, 129
Expressions, 118
Expressions of lived experience, 119
Extractivism, 485, 490
Extramural lives, 1245–1248
Extraversion, 823, 899
Extreme transvestite, 431

F

Factors enabling, 679
Factual narratology, 82, 94, 97, 99, 102
Fairy-tale, 88
Family, 790
Fanon, F., 1326
Fassin, D., 1224
Fasting, 1881–1884
Feedback effect, 334
Felida, 101
Fellow-feeling, 127
Female, 428–430, 432, 435, 437, 439, 441, 444, 445, 447
The Female Malady, 1429
The Feminine Mystique, 1043

Femininity, 424, 425, 432, 433, 437, 438, 440–441, 444–446, 448, 449, 1122
Feminism, 1045, 1055, 1792
  contemporary feminist analyses, 1837
  feminist ecological model, 1836
  feminist empiricism, 1841
  gender-neutral psychological approaches, 1836
  intersectional feminism, 1837
  violence against women, 1835, 1836
Feminist, 426, 428, 440, 441, 444
  ecological model, 1836
  economics, 1599–1601
  social constructionist theories, 1854
Fertilization, 279
Feudalism, 597, 605–607
Fiction, 81, 85, 93, 97–99, 101, 104, 368
"Fictitious" commodities, 474
Field, 122
Figurational Sociology, 682
Film, 279–282, 284, 286, 315–317
Financial incentives, 1590
Financial risk, 1579
Fin-de-siècle social theories, 941
Finnish culture, 361
First Nations, 1068, 1069, 1811, 1818
First World War, 1674, 1675
Fishing quota system, 471
Flap anatomies, 249
FLASCO, 909
Flechsig, P.E., 1229
Fleck, L., 840
Flood, DH., 100
Floto, Hartwig, 208
  career, 219–220
  confessional and political fault lines, 216–217
  priority of teaching, 211–213
  professionalization, 214–216
  Ranke's historical exercises, 210–211
  social and educational background, 209–210
  source editing projects, 220–221
  teachers of the nation, 213–214
  virtues and vices, 217–219
Flourishing of knowledge, 851–852
Fludernik, M., 81, 96, 97, 99, 102
Fluidity, 425, 441, 444, 445, 447
Fluorescence microscopy, 284
Foetal personhood, 412
Fold/folding, 267, 268
Folklore, 1732, 1734, 1737
Folk psychology, 982, 983
Food restriction, 1879–1880
Foreign trade, 1494–1496

Forensic psychiatry
  alienism, 1278–1282
  clinical assessment tools, 1293–1294
  degeneration theory and Lombrosian criminology, 1282–1285
  feeble-minded children and incorrigible youth, 1290–1292
  judicial gatekeeping, 1297–1298
  medical jurisprudence and penal reform, 1275–1278
  preventive detention, 1294–1298
  prophylactic assessments, 1289–1294
  psychiatric expertise, 1285–1289
  psychopaths, 1292–1293
Forgetting, 788–789
Formalist, 467
Formalist–substantivist debate, 468
Formalized disciplinary discourse, 851
Formative social ethics, 127–129
Fossil hydrocarbons, 484
Foucault, Michel, 161, 166–170, 172–178, 237, 328, 662–664, 840, 1310, 1421, 1427, 1428, 1439–1443, 1446
Foucault and Political Reason, 173
Foundations, 135–139, 141, 151, 152, 155, 156
France, A., 88
Frazer, J.G., 187, 191–196, 199
Fredline, C., 1047
Free competition, 1484
Free market, 466, 470, 475, 476
Free will, 986, 988
French historical school of epistemology, 840
French path of Enlightenment thinking, 586
French positivism, 836
French spiritualism, 138
Freud, Anna, 1350
Freud, Sigmund, 184, 187, 189, 190, 192, 196, 199, 202, 203, 1037, 1042, 1043
Freudian, 19
  psychoanalysis, 1372
  psychology, 577
Fujimura, J., 921
Function, 182–184, 186–188, 190–192, 195, 196, 199, 202, 203
Functionalism, 50, 1731, 1742–1744, 1749
Functioning, 1593, 1595
Fürst, J., 1247

G
Galison, P., 248, 922
Gallie, W., 83
Ganzheitspsychologie, 1200
Gate-keeper, 1236
Gay, 425–426, 428

Geertz, Clifford, 187, 197–203, 477
Geisteswissenschaften, 15, 981, 1197
Gender, 230, 237–239, 252, 297, 306, 308, 315, 316, 332, 424–429, 432–442, 444–446, 448, 449, 577, 790, 990, 1786, 1787, 1789, 1790
  affirmative surgery, 424
  binary, 1845, 1846
  categoricalism, 1840
  equality policy, 767
  essentialist, 1039–1041, 1048
  gender binary, 1845, 1846
  identity, 424, 425, 432, 436, 438, 439, 448
  inequality, 237, 1603
  norms, 1879–1880, 1887, 1892
  relations, 577, 751, 755, 756, 758
  sociology, 759
  stereotypes, 238, 1038–1040, 1042
  studies, 757, 758
Gender Identity Project, 434, 440
Gender-related assessments, 1603
Genealogical approach, 991
Genealogy, 22, 161–163, 165–173, 176–178, 376
Genealogy as Critique, 162
Genealogy of Morals, 165
Genealogy of the present, 161
General belief systems, 377
General equilibrium, 1517, 1518, 1522
  model, 1530
Generalization, 182, 190
General paralysis of insanity (GPI), 1360
General paralysis of the insane, 1265
General paresis, 1265
General Practitioners (GPs), 1352
General psychology, 982
Generations, 790
Genetic ancestry testing, 415
Genetic memory, 34
Genetic psychology, 1038
Genetics, 1352
Genette, G., 94, 95, 97, 99, 101
Genocide, 641–646
Genographic Project, 302, 304
Geography, 1333
Geology, 305
Geriatric psychiatry, 1404
  ancient entanglements, 1405
  dementia, 1405–1408, 1410, 1411
  imperative, 1408–1410
  senility, 1414
German and Austro-Hungarian empires, 1659
German criminology, 1297
German ethnographer, 1731
German ethnography, 1742

German historical scholarship, 208–223
German historicism, 843
German mandarins, 1210
German model, 1265
German sociology, 847
German tax law, 761
Germ layers, 268, 272, 274
Gerontology, 1413, 1414
G-I/R, 432, 434, 438, 448
Giacometti, Michel, 1687
Gibbons, M., 922
Gibson, E.J., 1048–1051, 1053, 1055
Gibson, J.J., 1048, 1049, 1055
Gieryn, T.F., 919
The Gift, 466
Gilman, S., 1225
Global financial crisis (GFC), 458–462, 469, 473, 475, 476, 478
Global historical sociology, 564
Globalization, 899, 1652
Global mental health, 1381, 1393, 1394
Global North, 900
Global South, 874, 882, 884, 885
Global women's health movement, 1799, 1800
Go-betweens, 1659
Goffman, Erving, 326, 1307, 1421, 1446
  Asylums, 1434–1436, 1438, 1443
  nature/reality of mental illness, 1435
  total institution, 1435, 1437
Gogh, V.V., 1230–1231
Gold standard, 461
Goldstein, J., 330
Goldthorpe, John, 569
Gonads, 433, 436–438, 443, 449
Good, 134, 143
Goodenough, F., 1045
Good science, 1170, 1180
Gossen, H.H., 1508–1512
Gould, S.J., 84
Governmentality, 172, 384
The Government Mental Hospital in Thiruvananthapuram in Kerala, 1244
Gradual, 433, 444, 447
Grand narratives, 86, 87
Graumann, Carl, 1007
Gray's Anatomy, 234
Great chain of being, 1201
Great confinement, 1314
Great Depression, 461, 463
Great Man, 1044–1046
Great power chauvinism, 1156
The Great transformation, 466
The Great Wall of China and Tartar riders, 1638, 1639
Greek city-states, 596

Greek psychology, 985, 986, 998
Green fluorescent protein (GFP), 283
Greenland Inuit, 1632, 1634
The Green New Deal, 463
Greimas, A., 94
Griesemer, J.R., 920
Griffiths, D., 98
Gross national product (GNP), 1586
Group identity, 1132
Guided capitalism, 1575
Guston, D.H., 921
Guthrie, Leonard, 1344

H
Habeas corpus, 1310
Habermas, Jürgen, 165
Habitat, 122
Habitus, 735–736, 846
Hacking, I., 333–336
Hajek, KM., 88, 94–97, 100, 102
Hale Bishop, S., 1323
Hall, G.S., 1038, 1321
Haller, Albrecht von, 265
Hans Spemann, 281
Haraway, Donna, 245
Haraway's effect, 519
Harmonious inequality, 682, 683
Harris's cultural materialism, 509
Hartley, David, 35
Hartmann, R., 87
Hartsock's standpoint theory, 817
Harvey, William, 262
Hayek's own analysis, 1545
Healing roles, 1248
Health, 388, 1786–1790
  disparities, 1809–1813
Healthy masculinity, 1844–1847
Hegel, Georg Wilhelm Friedrich, 566
Hegelian philosophy, 16
Hegel-Marxism, 160
Hegemonic psychological science, 1069–1071, 1075
Hegemony, 683
Heger, Franz, 1662
Hempel, CG., 83, 91
Hereditary taint, 1266
Heredity, 1344, 1352
Herman, D., 80, 91–94, 96
Hermaphrodite, 433, 436, 439, 443
Hermaphroditims, 433–435
Hermeneutic circle, 118
Hermeneutics, 10, 116, 117, 120, 125, 126, 130
The Hermeneutics of the Subject, 378
Heroes, 787

Herold, Johann Moritz David, 271, 272
Hess, D., 920
Heterogeneity, 161
Heteronomy, 34
Heteronormative, 436, 441
Heterosexual, 430, 437, 438, 444, 448, 1039, 1041
Hevern, VW., 91
Heymans, Gerard, 441, 445
Hierarchical gender relations, 755
Hierarchical organization, 1048
Hierarchies, 1046, 1047, 1056, 1058
Higher understanding, 117
Hijikata, Hisakatsu, 1732, 1738, 1740–1749, 1751
Hijikata's illustration, 1746
Hilbert, David, 42
Hilde Mangold, 281
Hilferding, Rudolf, 1564, 1566, 1568, 1570, 1574, 1575, 1577–1581
Hilferding's Hypothesis, 1571
Hindu/Sanskrit texts, 897
Hippocratic tradition, 1857
Hirano, Yoshitarō, 1739
Hirschfeld, Magnus, 425, 426, 428–432, 434–436, 438–445, 447
His, Wilhelm, 276, 277
Historians of science, 980
Historical constructionism, 343
Historical exercises, 208
Historical knowledge, 10
Historical psychology, 987, 993
Historical sociology, emotions
  authoritarianism, 720
  civilizing process, 708, 709
  definition, 700, 704, 705
  drivenness, 701, 719
  Freud, foundations, 709–711
  Freud's work, 711, 712
  historical psychology, 706, 707
  history, 702, 703
  homo clausus, 716, 717
  Norbert Elias, 704, 705
  psychoanalysis, 701, 719
  social relations, 712–714
  sociology, 700
  supergo, 717, 718
  valencies, 715
Historical sociology, 562, 564, 565, 573, 576, 578, 579, 695, 754, 811–812, 821
  beginnings of, 566–567
  brief history of, 565–573
  German, 589
  goals, 591
  insulation and marginalisation, 567–570

Marxist, 604
meanings, 584
revival and revision, 570–573
themes of, 574–578
Weber work, 597
Historical studies, 208, 209, 214, 217, 220–222
Historical understanding, 17, 996
Historic images, 235
Historicism, 16, 17
Historicist, 11, 17
Historicist writing, 16
Historicization, 1228
Historiography, 84, 85, 87, 95–99, 101, 102, 104, 208–223, 235–236, 238, 239, 379
History, 9, 12, 16, 81, 83, 84, 103, 104, 1461–1465
  of anthropology, 1623, 1629, 1632, 1721
  discipline of, 811
  of economics, 1512
  of emotions, 338
  of ethnography, 1622
  of Ideas, 811–815, 817, 820–822
  of mental illness, 1384–1387
  of philosophy, 837
  of psychology, 1196–1199
  of science, 23, 404, 407, 416, 836, 840, 848, 980, 990
  of social psychology, 1006, 1010, 1011
  of sociological theory, 936
History of economic thought
  and academic domains, 1454–1457
  divergencies and pluralism, 1457–1461
  problems and failures, 1466–1467
History of Sexuality, 378
History of social science, a guide to scientific flourishing, 851–852
History of sociology, 835, 844, 918, 919, 924
  attempt to locate, 866–873
  centers and peripheries, 876–878
  European modernity, 873–875
  in geopolitical power structures, 875
  history of, 842–846
  as sociological reflexivity, 853–855
  and the status quo-counter hegemonic potentials, 884–886
History of the present, 22–24, 161, 162, 168, 169, 172–178, 991
History of the second degree, 782
History of the self, 356, 357, 367
Hitler, A., 1233
Hobbes, T., 922
Hodiecentrism, 612
Holistic understanding, 111
Holocaust, 776
Holy selfhood, 360

Homo clausus, 716
Homo economicus, 463, 470
Homo economicus model, 1589, 1597
Homosexual, 327, 425–431, 436, 437, 439, 441, 443, 446, 1041
Homosexual transvestie, 427, 428, 430, 431
Homo sylvestris, 1636, 1637
Horkheimer, Max, 164
Horney, K., 1042
Hountondji, Paulin, 898
Household budget, 475
Household management, 458, 467
Housing, 473, 475
  market, 473, 475
Howes, E.P., 1049, 1052
Human adaptation, 486, 487
Human affairs, 1019
Human anatomical diversity, 248
Human anatomy and physiology, 245
"The Human" as a subject, 5
Human behavior, 1360, 1592
Human body, 243
Human-centered development frameworks, 1606
Human development
  definition, 1598
  feminist economics, 1599
  HDI, 1599
  HDR, 1599
Human Development Index (HDI), 1599
Human Development Report (HDR), 1598
Human economy, 476, 477
Human evolution, 293, 295, 297–299, 301, 302, 310, 312–314
Human experimentation, 1358, 1359, 1361, 1363, 1365, 1366, 1368, 1370, 1371, 1374, 1375
Human Genome Diversity Project, 302, 304
Human interiority, 357–361
Humanism, 15
Humanist, 994
  culture, 59
Humanistic evaluation, 10
Humanitarian narratives, 359
Humanity(ies), 6, 12, 30–32, 35, 37–39, 41, 42, 46–50, 52, 54, 988, 998
Human kinds, 985, 1198
Human mortality, 232
Human motivation, 1566
Human nature, 11
Human-nonhuman relations, 517
Human origins, 292, 295, 312, 313, 316, 317
Human phylogeny, 298
Human population genetics, 302, 304, 317
Human races, 294, 295, 299–301, 312

Human rights, 668
Human sciences, 4–22, 24, 25, 64, 111–115, 384, 1007, 1241–1243
Human self-understanding, 6, 11, 25, 979
Human sentiments, 1589
Human societies, 1627, 1628, 1641, 1644
Human vitality, 395
Hume, D., 1124
Hungary
    Jesuits, Lutherans and Calvinists as agents of world ethnography, 1634–1640
    socio-cultural strategies, 1629–1633
    world ethnography, 1624–1629
Hunger striking, 1883–1885, 1892
Hunt, L., 330, 340
Hunter, KM., 99
Hunter, William, 238
Hurwitz, B., 100
Hydrocarbons, 488
Hypnotism, 88, 90, 98, 988
Hysteria, 100, 101, 1042, 1264, 1855, 1856

I

Iberian colonization, 909
Iceland, 458–478
    banking collapse, 470
    virtualism and neoliberal economic transformation in, 470–473
Iconography, 246
Idealism of freedom, 124
Ideals, 134, 138, 145, 149, 150, 153–156
Ideational comprehension, 122
Identification processes, 727, 732, 743
Identity
    collective, 730, 735
    components, 731
    constitutive tension of, 731
    definition, 729
    European, 741–744
    habitus and filo pastry of, 735–736
    involvement and detachment, 734–735
    polysemous concept, 730
Ideological moment, 61
Ideology, 13, 1153
Idiocy, 1340, 1342
Image-based sexual abuse (IBSA), 1910
Images, 230–232, 234–238, 240–243, 245–254
Imagination, 1177
Imaginative completion, 114
Imanishi, Kinji, 1739
Immanent purposiveness, 116

Imperial difference, 907
Imperialism of economics, 1455
Inaugural period, 62
Income distribution, 1492
Incommensurability, 1186
Indian historiography, 901
Indian mental hospitals, 1243
Indians, 1069, 1076, 1321
Indigenism, 963, 969–971
Indigenization, 1066
Indigenous knowledge, 894
Indigenous patients, 1320
Indigenous Peoples
    Canada's relationship with, 1816
    health disparities, 1809–1813
    Indigenous research and Indigenous knowledges, 1820–1826
    knowledges, 1813
    Sixties Scoop, 1808
Indigenous psychologies (IP), 979, 984, 993
    anthropological and psychological perspectives, 1069
    histories of, 1066–1069
    ideas and practices for creating, 1070
    modernity/coloniality and psychological knowledge, 1072–1077
Indigenous research, 1816
    history in Canada, 1813–1816
    as self-determination, 1820–1826
Indigenous research methodologies (IRM), 1788, 1807, 1809, 1816
Indigenous social sciences, 896
Individual and collective identity, 727
Individualization processes, 756
Individualism, 155
Individualistic representational theory, 1007
Individual transferable quotas (ITQs), 471
Induction, 1661
Industrialization, 245
Industrialized societies, 466
Inequality, 462, 752, 756, 768, 1587
Infant cognitive development, 1095
Inferiority, 1132
Influence strategies, 1073
Information asymmetries, 1480
Inkblot test, 410
Inner child, 352
Inner life, 366, 367
Inner lives
    confession, 353–355
    contextual reading, 351
    emotional religious experience, 350

interiority, 350, 353–355
  knowledge-intensive societies, 350
  people's lives, 350
  secular orientation, 352
  secular period, 351
Innovative, 685
  manager, 1574
Insanity, 1335, 1340
Insanity Defense Reform Act, 1298
Insideness, 354, 356, 362
Institute of Psychiatry, 1347
Institute of Psychoanalysis, 1347
Institute of the Pacific, 1739, 1751
Institutional clinical experiments, 1267
Institutional economics, 1459
Institutional histories, 815–816
Institutionalization, 811, 816, 820–821
Institutionalized lives, 1243–1245
Institutional kind, 1571
Institutional reform, 1279, 1295
Institutions, 1543, 1547, 1553
  re-examination, 1550–1553
Instrumental system, 1576
Instrumentation, 995
Insulin coma therapy, 1316
Intellectual communities, 1048, 1053–1058
Intellectual disability, 1333
Intellectual histories, 815–816
Intellectual progeny, 52
Intellectual tradition, 1701
Intelligence, 989, 995, 1146
  tests, 408–410
Intentional hunger, 1882
  description, 1878
  masculine, 1882
  performance of, 1883
  psychopathologised experiences of, 1886
Inter-, or trans-disciplinary, 8
Inter-cultural psychology, 985
Interdisciplinary, 8, 979, 986, 1772
  initiatives, 1758–1760
  and intradisciplinary collaboration, 1186–1187
  social psychology, 1019–1020
Interiority, 351
  childhood, 362–364
  and confession, 353–355
  human, 357–361
  process of reason, 356–357
  and reflexivity, 364–366
Internal colonialism, 903
Internalism-externalism, 814, 821, 823, 825

International boundary organization, 921
International Committee for the Sociology of Sport (ICSS), 680
International Congress of Americanists (ICA), 1665, 1669
Internationalist strain of Americanist anthropology, 1676
International Monetary Fund (IMF), 460
International School for American Archeology and Ethnology (ISAAE), Mexico, 1673
International social psychology, 1024
International Social Science Research Council (ISSRC), 1021
International Sociological Association (ISA), 680
International Union of Scientific Psychology, 1199
Intersectional feminism, 1837
Intersectional feminist approach, 1837
Intersectionality, 1865
Intersectional realities, 1045
Intersex, 425, 431–441, 444, 448, 449
Intraversion, 823
Inuit, 1817
Invented traditions, 791
Inventor, 1569
Involuntary unemployment, 1529
Invulnerability, 364
IQ debate, 1108
Iranian Revolution, 396
Ireland, William W., 1343
Irradiant, 685
Isaacs, Susan, 1350
Islamic psychology, 1068
Islamic subjectivity, 396
Italian criminology, 1283
I-Thou relation, 126

J
Jahn, M., 91, 96
Jajdelska, E., 85, 102
Jakobson, R., 94
James, W., 1129, 1318
Japan, 1730–1737, 1740, 1741, 1744, 1747, 1748
Japanese administration, 1731, 1737, 1738, 1746, 1748, 1751
Japanese Empire, 1730, 1740, 1746, 1748, 1752
  See also Empire of Japan
Japanese ethnography, 1618

Japanese immigrant, 1737–1740, 1742, 1751
Jasanoff, S., 922
Jesuit missionaries, 1614, 1635
Jevons, W.S., 1512–1516
Jordanova, Ludmilla, 238
Judgmental contexts, 122
Judicial gatekeeping, 1297–1298
Jung, C.G., 187, 192, 196

K
Kanner, Leo, 1336
Kant, Immanuel, 30, 51, 121–125, 127–129, 135, 136, 138–142, 154
    anthropology, 31–33
    ethics, 50
    Stoics and Epicureans, 42–48
Kantianism, 136, 1588
Kantian philosophy, 143
Kardiner, A., 1134
Kazak children, 1155
Keijō Imperial University, 1734
Keynes, J.M., 1526–1530
Keynesianism, 1530–1535
Keynesian policy, 1529
Keynes's theory, 1548
Kina Keko Kawasasek methodology, 1825
King, M.L., 100, 1325
King, Truby, 1333
Kin group, 1738, 1742, 1743, 1745, 1748
Kinship, 1738, 1740
    analysis, 526
Kirzner's framework, 1554
Klein, Melanie, 1350
Kleinman, A., 1225
Klepper, M., 92
Knight, Charles, 308
Knowledge boundaries, 918
    five knowledge boundary concepts, 919
Knowledge-making, 1226
Knowledge production, 248
Know thyself, 250
Knox, Robert, 241
Komintern, 964, 965, 969
Königliches Museum für Völkerkunde, 1666
Koror, 1737, 1742, 1744, 1748, 1750
    See also Palau
Krafft-Ebing, Richard von, 426, 427
Krämer, Augustin, 1731, 1737
Kristeva, J., 95
Kritische Historie, 167
Kubary, John, 1737
Kuhn, T., 840

Kula ring, 465, 466, 468
Kulturkampf, 216

L
Labor, 386
    theory of value, 1486
Lachmannian approach, 1557
Ladd-Franklin, C., 1057
Laing, R.D., 1421–1426, 1429, 1430, 1439, 1443
Lamarck, Jean-Baptiste, 36, 38
Lamont, M., 918
Land Registration Record, 1743
Langdon Down, John, 1342, 1343
Language evolution, 300
Language/literature, 901
The Language of Madness, 1429
Language of the human sciences, 1119
Larsson, A., 930
Late Modernity, 364–366
Latina Health Initiative, 1796
Latin America, 962, 963, 965, 966, 968, 970, 971
Latin American Council for the Social Sciences, 967
Latin American Regional perspective, 905, 907–909
Latin American scholarship, 908
Latin American sociology, 569, 965, 966, 968, 972
Laura Spelman Rockefeller Memorial (LSRM), 1758
Lavater's approach, 405
Lawmaking, 656
Leadership, 1576
Leao, R.S., 1235
Learning disability, 1336, 1345, 1347
Lederer, Emil, 1564, 1566, 1568, 1579, 1580
Lederer static equilibrium, 1566
The Legacy of asylums, 1316–1318
Legal history, 652, 654, 661, 664, 671
Legal pluralism, 655
Lehman Brothers, 460
Leisure, 678, 680, 695
Lepenies, W., 925
Lesbian, gay, bisexual, trans, and intersex (LGBTI), 425–427, 431
Lesbian, 425–426, 428
László, J., 91
Lévi-Strauss, Claude, 94, 197, 199–203, 466
Lévy-Bruhl, Lucien, 194–195
Lewin, K., 1132

Leys, R., 341
LGBTQ rights, 1798
Liberal theory, 1590
Libido development, 438, 1103
Lieux de mémoire, 791
Life-manifestations, 118
Light-sheet fluorescence microscopy (LSFM), 282
Linguistics, 90, 91, 94, 102, 103
Linnaean equation, 32
Literacy, 1153
Lived experience, 114
Lobotomy, 1316, 1363, 1375
Locke, John, 35, 1123
Logical deduction, 1573
Logic models, 1838
Lombrosian criminology, 1284
Lombroso, C., 1232
Long period method, 1498
Looping, 19, 427, 429–431, 449, 995, 1036, 1198
  effect, 333–336, 852
Lorén Foundation, 925
Lowenfeld, Margaret, 1348
Luhmann, N., 666–669
Luhrmann, T., 1228
Lunacy reform, 1256–1259
Lunatic asylums, 1230, 1243, 1258, 1259
  rise of, 1254–1256
Lunatics, 1263
Luria, A.R., 1147
  psychological expedition to Uzbekistan, 1151–1152
  theory of cultural development, 1147
Lutheran scholars, 1635
Lyotard, 86

M

Machine, 245
MacIntosh, M., 326
Macrocosm, 243
Macroeconomics, 1525–1535
  linkages, 1588
Madness, 1332–1335, 1342, 1351
Madness of women
  depression diagnosis, 1864–1866
  raging hormones and reproductive debilitation, 1857–1859
  rejecting raging bio-medical explanations for, 1859–1864
  reproductive body, 1855–1857
  treatment, 1866–1869

Mainstream, 690
  development economics, 1606
  economics, 1466
Maladjustment, 1347, 1349
Malcolm X, 1325
Male, 428–430, 432, 435–439, 441, 442, 444, 445, 447
Malinowski, Bronisław, 465, 468, 1731, 1734, 1740, 1742, 1743
Malleable Self, 380
Malpighi, Marcello, 262
Man box, 1843
Mania, 1264
  with melancholia, 1264
Manifest destiny, 900
Mann's analysis, 600
Manual on Psychiatry for Differently Thinking People, 1247
Man up' language, 1842
Manuscripts, 387
Maori, 1320
Map, 302, 304
Marginal gains: Monetary transactions in Atlantic Africa, 478
Marginalism, 1503
Marginalist economic theory, 1475–1477
Marginality as exclusion, 1308
Mariana Islands, 1737, 1738
Mariátegui, José Carlos, 962–966, 968–971
Mariyan, 1750
Market economy, 466
Market space (*agora*), 468
Marriage, 751–753, 759, 761–765
Marrow, J., 1228
Marshall, Mac, 1747
Marx, 462, 463
Marxian, 18
Marxism, 6, 8, 379, 385–387, 394, 396, 397, 682, 962, 963, 965–971, 1156, 1706, 1707
Marxist(s), 10, 13, 18, 20, 993
  approach, 592, 595
  historical sociology, 604
Marx's theory, 658
Masculine honor culture, 1057
Masculinity, 424, 425, 428, 432, 433, 437, 438, 440, 441, 443–446, 448, 449, 1037, 1039, 1041
  healthy masculinity, 1844–1846
  toxic masculinity, 1843–1844
  traditional masculinity, 1842
Maslow, A., 1136
Mason, Otis T., 1663

Mass vaccinations, 41
Material(s), 253
　conditions, 1046, 1048
　exchange, 1666, 1670
　gains, 1590
　resources, 1037, 1048, 1051, 1058
Material-discursive-intrapsychic (MDI) analysis, 1870
Materialism, 389
Mathematical game-theory, 609
Matsumura, Akira, 1733, 1736
Matsuoka, Shizuo, 1737
Maudsley Hospital, 1347
Mauss, Marcel, 466, 467, 476, 477
Maxwell, James Clerk, 34
Mead, Margaret, 1758, 1765–1770
Meaning contexts, 117
Measure, 426, 432, 438, 440, 445–448
Mechanical solidarity, 465
Media, 793–794
Medial context, 120
Medical certification, 1262
Medical education, 232, 252
Medicalization, 1112
　of madness, 1306
Medical jurisprudence, 1276, 1283, 1285
Medical photography, 238
Medical research, 1359
Medical science, 246
Medical socialization, 254
Medicine, 9, 82, 87, 89, 97–99, 102, 104, 237, 810, 813, 817
Medieval theories, 586
Mediterranean civilization, 589
Medium of objective spirit, 117
Melancholia, 1263
Melanesia, 1735, 1738, 1739
Memory activism, 796–798
Memory studies, 777
　activist turn, 796–798
　and historical sociology, 778–782
　second wave of, 789–794
　transnational turn, 794–796
Mendes, G.N., 1322
Menger, C., 1519–1520
Menger's methodology, 1544
Mennell, Stephen, 566
Menopause, 1855
Mental deficiency, 1151
Mental development, 1149
Mental disorder, 1406
Mental functions, 996
Mental health, 1594, 1789

Mental health in segregated societies
　racial and colonial psychiatry, 1319–1320
　US cases, 1320–1324
Mental hospital as a small society, 1308
Mental hospitals, 1335, 1347, 1351
Mental hygiene, 1348, 1349
Mental hygiene law, 1245
Mental illness, 1232, 1242, 1246, 1306
Mental infirmity, 1404, 1405
Mental institutions, 1244
Mentalité, 987
Mental patient lives, 1241
Mental suffering, 1234–1240
Mental test, 1039, 1148
Merton, R.K., 838
Mesmerism, 1042
Metamorphosis, 262, 272
Metapsychology, 391
Method, 809–812, 815, 820, 822, 827–829
Methodenstreit, 1543–1544
Methodological individualism, 813, 1008, 1476, 1571, 1604
Methodological nationalism, 658, 795
Methodology(ies), 810, 814–815, 818–824, 826–829, 905
Métis, 1817
MeToo, 1328
#MeToo-Campaign, 750, 765
Metzl, J., 1227
Meunier, R., 85, 99
M-F testing, 426, 446
Microcosm, 243
Microeconomics, 1520–1525
Micronesia, 1731–1733, 1735–1744, 1746–1751
　Japanese administration, 1747
Micronesian, 1731, 1732, 1736–1738, 1740–1752
　ethnographies, 1619
Microscopic observation, 233
Microscopy, 234, 284
Microtome, 275
Mignolo, Walter, 906
Miles, Catherine Cox, 441, 445, 1040
Milieux de mémoire, 791
Mills, C.W., 1328
Milton, J., 1121
Ministry of Health and Welfare, 1246
Mink, L., 83
Minnesota Multiphasic Personality Inventory (MMPI), 1293
Minorities, 1147, 1150, 1151, 1156
Minority groups, 1133

Minzoku, 1733
   See also Ethnos
Minzokugaku, 1733, 1742
   See also Folklore; Ethonology
Miscegenation, 908
Mitochondrial or African Eve, 313
Mnemohistory, 782
Mnemonic solidarity, 796
Modekngei, 1748
Modeling, 276, 277, 284, 286
Models, 230–232, 234, 235, 238, 243, 247, 248, 250–253
Models of madness, 1261–1264
Modern context, 365
Modern economics, 1456, 1591
Modern human races, 295, 299
Modern imaging techniques, 253
Modernism, 367
Modernist attitude, 175
Modernity, 897, 978, 987, 1067, 1076–1077
Modernization, 563, 791, 1244
   theory, 564, 588, 589, 752, 854
Modern mind, 366–370
Modern moral culture, 352–353
Modern science, 237
Modern social psychology, 1024
Modern unbelief, 353
Mol, A., 337
Molecular anthropology, 301
Molnár, V., 923
Monetary policy, 1533
Money, John, 425, 432, 434, 438, 439
Monopolistic firms, 1568
Monopoly mechanism, 594, 609
Monsters, 245
Moral authority, 148
Moral community, 147
Moral disposition, 1589
Moral economy, 469, 475, 1597, 1598
Moral fact, 143
Moral foundation, 134
Moral ideals, 145, 149, 150, 153, 155, 156
Moral individualism, 154, 155
Moral insanity, 1264
Morality, 251, 463, 1598
Morally ordered society, 1127
Moral Management, 1342
Moral order, 1121, 1127
Moral philosophy, 135, 136
   in France, 138–139
Moral rules, 139, 141, 143, 144, 149, 150, 156
Moral sciences, 7, 14
Moral treatment, 1257, 1278, 1284, 1289, 1358

Moral values, 143, 144, 154
Morawski, J., 85
Morgan, MS., 87, 102, 103
Morning Star Lodge (MSL), 1820
Mortality, 254
Mortgage, 474
Mother-infant relationship, 1103
Motive, 1136, 1138
Movement for Socialism (MAS), 971
mtDNA, 302, 304
Mukerji, D.P., 895
Müller, Friedrich Max, 184–186
Multi-level theoretical formulation, 1835
Multiple contextualization, 1700
Multispecies ethnography, 516, 520
   conviviality, 516–518
   description, 508
   laboratories, 523–527
   and multiple politics, 512–516
   rights-oriented, 520
Murphy, B.E., 1318
Muscovite empire, 1641
Museum, 293, 295, 297, 308–310, 312, 313, 776
   tours, 1666
Museum of Ethnography and Anthropology (MAE), 1703
Mutualism, 1008
Myrdal, A., 928
Myrdal, G., 928
Mysticism, 195
Mythopoetic men's groups, 1843
Myths, 1071, 1072

N
Nakagawa, Zennosuke, 1738, 1740–1744, 1751
Nakajima, Atsushi, 1741, 1744, 1750
   Mariyan, 1750
Nanâtawihowin Âcimowina Kika-môsahkinikêhk Papiskîci-itascikêwin Astâ cikowina (NAKPA), 1824
Nan'yō, 1739
   See also Nan'yō Guntō
Nan'yōchō, 1732, 1735, 1737–1746, 1748, 1751
Nan'yō Guntō, 1731, 1735, 1738, 1744–1746, 1749, 1751
   See also Micronesia
Nan'yō Guntō Bunka Kyōkai, 1732
   See also South Sea Islands Cultural Society
Narrative, 80–84, 86–100, 102–104, 983
   medicine, 89, 90, 93, 100
   templates, 785

Narrativity, 86, 89, 96, 102
Narratology, 82, 85, 90, 93–95, 98–102, 104
Nation, 1701, 1707, 1709–1711, 1715, 1717, 1718, 1721, 1722
National Black Women's Health Project, 1794
National capitalism, 460
National Council of Women Psychologists, 1050
National Health Service (NHS), 1332, 1426
National histories of sociology
  in China, 824
  history and historical sociology, 811–812
  history of ideas and sociology of ideas, 812–815
  intellectual histories, institutional histories and sociology, 815–816
  in prestigious universities, 825
  sociology of knowledge, 816–818
National identity, 215, 1682, 1683, 1685, 1688, 1691–1694
Nationalism, 216, 729, 733, 740, 744, 899, 1156, 1617, 1716, 1720, 1722
National Museum of Ethnology, 1741
National Socialism, 1211
Nation-building anthropology, 1682, 1683
Nation states, 778, 779, 790–792
Nativization, 1156
Naturalism, 124, 125
Natural kinds, 985
Natural narratology, 96
Natural resource(s), 470
  extraction, 488, 490
Natural science, 111, 810–811, 1658
Naturhistorisches Hofmuseum (NHM), 1661
Naturhistorisches Museum, 1667
Naval administration, 1735–1737
Navy's occupation, 1733, 1736
Neanderthal, 295, 298–300, 302, 306–308, 310, 315, 316
Negative freedom, 1596
Negro, 1322
Neoclassical economics, 463, 467, 476, 1565
  Cournot, A.A., 1506–1508
  description, 1503–1505
  development of, 1520–1525
  economic crisis as challenge for, 1525–1526
  economic equilibrium, 1523–1524
  founding fathers, 1512–1520
  Gossen, H.H., 1508–1512
  Jevons, W.S., 1512–1516
  Keynes, J.M., 1526–1530
  Keynesianism and monetarism, 1530–1535
  Menger, C., 1519–1520

  predecessors, 1505–1512
  Walras, L., 1516–1519
Neoclassical synthesis, 1531
Neo-colonization, 1007
Neo-evolutionism, 599, 601, 603
Neofunctionalist theory, 614
Neo-Kantian approach, 594
Neo-Kantianism, 42
Neoliberal economic development, 1606
Neoliberal economic policy, 473–475
Neoliberal economic theory, 1589
Neoliberalism, 460, 461, 472, 1787
  discourse, 472
  force, 472
  geographic and social determinants of, 476
  ideology, 472
Neoliberalization, 470
Neoliberal policy framework, 1839
Nepotism rules, 1050, 1051, 1053, 1057
Nerves, 1264
Neugebauer, Franz-Ludwig von, 435
Neurasthenia, 1264, 1856
Neurath, Otto, 245
Neurobiological self, 413
Neuro-disciplines, 993
Neurofeedback, 414
Neurohistory, 339–342
Neurology, 100
Neuropsychologists, 983
Neuropsychology, 993
Neuroscience, 14, 90, 96
New consciousness, 358
The New Deal, 463
New Economic Policy, 1161
New Guinea, 1739
The New Left, 1328
New York City, 1657
Nietzsche, Friedrich, 161
Noble savages, 462
Noguchi, Masaaki, 1738, 1744–1746, 1749, 1751
Nominalist, 979
Non-aligned movement, 899
Non-ethical character, 1591
Non-settler colonialism, 894
Non-sites of memory, 788
Nordic Sociological Association, 930
Normal, 152, 245
  human development, 245
Normality, 152, 230
Normal-pathological, 153
Normative individualism, 1605
Normative theories, 585

Northern Baykal region, 1152
Northern theory, 875
Nostalgia, 798
Nowotny, H., 922

O
Object, 425–433, 438, 445
Objective idealism, 124
Objective self-fashioning, 414
Objectivity, 84, 87, 88, 103, 248, 425, 428, 440, 1040, 1045
Occidentalism, 903
Occidental rationalism, 597
Occupy Movement, 474
Oedipus complex, 1102
Official law, 654–656
Oka, Masao, 1733
Oligopoly, 1524
Olmos, P., 84
On Crimes and Punishments, 1276
One sex system, 433
On the Genealogy of Morality, 376
On the Process of Civilization, 565
Ontological-epistemic, 894
Ontological security, 364
Ontologies, 1077
Ontology-making practices, 986
The Open Society and its Enemies, 568
Operationalism, 1200
Oracular authority, 1588
Oral History, 1337
Oral traditions, 1072
Organic contingency, 390
Organic solidarity, 465
Organization of knowledge, 4
Organized violence, 575, 626, 642, 645
    clandestine political violence, 631–636
    genocide, 641–646
    revolutions, 636–640
    war, 627–631
Oriental archive, 902
Oriental despotism, 595
Orientalism, 895, 901, 1645
Orientalists, 899
Orientational framework, 112
Otis, L., 98
Our Bodies Ourselves (OBOS), 1794, 1796, 1797, 1799–1801
Our Bodies Ourselves Today (OBOS Today), 1797
Our Watch, 1837, 1842, 1844, 1845
Overcrowding, 1316
Overempower, 1075
Over-specialization, 992
Ownership, control, access, and possession (OCAP®), 1815, 1817
Oyrotsky people, 1153

P
Pacific Northwest Coast, 1658
Pacific War, 1731, 1737, 1739, 1746
Paddington Green Children's Hospital, 1335
Palau, 1731, 1737, 1738, 1740–1748, 1750
    traditional meeting houses (*bai*), 1741
Paleoanthropology, 295, 300, 304, 317
Paleontology, 305
Pander, Christian Heinrich, 267–269
Panspermism, 262
Papermaking, 234
Paresis, 1265
Paris Medical Faculty, 252
Parliamentarization, 688
Parsonian systems theory, 567
Parsons, Talcott, 567
Parsons's theoretical framework, 661
Partial equilibrium analysis, 1521
Paskins, M., 85
Passing, 430
Pathography, 1230
Pathological attitude of life, 1228–1230
Pathological modifiability, 383
Pathology, 152, 153, 388
Patriarchal capitalism, 1912
Patriarchal masculinity, 1056
Patriarchy, 1046, 1835–1838, 1841, 1843–1846, 1879, 1887, 1892
Patrimonialism, 598, 607, 609
Patriotism, 734
Pedagogy and psychology, 1091, 1109
Pedology, 1093
Pedroso, Consiglieri, 1683, 1684, 1689, 1693
Penal reform, 1276–1277, 1284
Penis gavel, 1057
Perestroika, 1720, 1721
Performativity, 331
Persona, 217
Personality and culture, 1758, 1760
    cultural patterning, 1762–1765
    founding figure, 1760–1762
    interpreter, 1771, 1772, 1778, 1779
    temperament, 1765–1771
Personality psychology, 1201–1203
Perspective, 14, 21, 22, 981, 992, 993
Perspective-taking, 1187

Perspectivist, 22
Peruvian Communist Party, 965
Phenomenology, 19
Philosophical anthropology, 15, 1205
Philosophical education, 392
Philosophical introspection, 63
Philosophy, 90, 92, 98, 134–136, 138–140, 980
 of history, 83, 86
Photography, 234, 238, 243, 250, 294, 295, 310, 1736
Phrenology, 242, 406, 989, 1125
Phylogenetic tree, 293, 299–301, 313, 317
Physical anthropology, 293, 294, 297, 316, 1617, 1700, 1705
Physical health, 1789
Physiognomy, 243, 404–406, 1042
Physiology, 87, 100
Pierce, C.M., 1325
Pinel, P., 1309
Pioneering, 685
Pithecanthropus, 297, 299, 306, 310
Placebo, 1368–1370, 1375
Places of knowledge production, 875, 878, 880
Plasticity, 381
Pléh, C., 90–92
Pluralists, 797
Pluriverse, 1077
Poetic metamorphosis, 114
Pohnpei, 1737–1739, 1742, 1743, 1745
Point of impression, 121
Polanyi, Karl, 466, 467, 469, 470, 476, 477
Police authorization, 430
Political activism, 1798
Political agendas, 237
Political democracy, 596
Political divisions, 67
Political economy, 1565
Political fasting, 1884
Political power, 237
Political traditions, 64
Political violence, 395, 575
Political *vs.* methodological individualism, 1571
Politics, 1231–1233
Pomata, G., 100
Popular forms of psychology, 988
Popular psychology, 979, 983
Population growth, 1493
Pornography, 1056
 experiences of abuse in, 1908–1909
 as a form of prostitution, 1906–1908
 and harms of distribution, 1909–1911
 MacKinnon's analysis of, 1906, 1910

Porn star experience (PSE), 1906
Porter, R., 1311
Portuguese anthropology, 1616
 anthropological theory and national identity, 1690–1694
 history of, 1681–1688
 theoretical influences, 1688–1690
Portuguese culture, 1616
Portuguese ethnography, 1615, 1616, 1685
 anthropological theory and national identity, 1690–1694
Portuguese folk culture, 1687
Positive freedom, 1596
Positivism, 87, 103, 836, 848, 849
Positivist, 24, 25
Positivistic standards, knowledge, 1700
Post-colonial, 818–822, 826
 knowledge, 894
 situation, 1074
 studies, 254, 779
Postcolonialism, 900–903, 906
Posthumanity, 86
Postmemory, 790
Postmodernism, 19
Postnational model, 742
Post-Pacific War era, 1730, 1746, 1747, 1749
Postpartum depression, 1793, 1862
Poststructuralist, 380, 910
Post-traumatic stress disorder (PTSD), 1903
Post-war period, 236
Potlach, 465
Poverty, 37, 237, 1587
Power, 791–793, 1037, 1039, 1045–1048
 knowledge, 18, 901, 906, 912
Practical innovations, 1569
Practice theory, 336
Pragmatic specialization, 851
Pragmatism, 839
Precariousness, 1241
Precarity, 1225, 1241
Prediction, 1174
Preferences, 1592
Preformation, 261, 265
Prehistoric anthropology, 1705
Prehistoric archeology, 317
Prehistoric Body Theater, 316
Prehistory, 292, 293, 306, 309, 312, 314
Premenstrual syndrome, 1861
Premenstrual tension (PMT), 1861
Pre-modernity, 762
Pre-modern societies, 1404, 1405

Prenatal bonding, 412
Pre-scientific, 115
Presentism, 161, 836, 990
Pre-social scientific theories, 182
Preventive detention, 1294–1298
Prewar era, 1730, 1733, 1747, 1749
Prices, 1485–1486, 1489, 1490, 1493
Price-specie flow mechanism, 1495
Primary adaptive strategies, 488
Primitive culture, 1741
Principle of comparative advantage, 1495
Printing, 234
Prinzhorn, H., 1232
Privatization of natural resources, 471
Problematization, 169
 modes of problematization, 169
Procedural justice, 1276
Process-based approach to markets, 1547
Process theory, 752
Productive systems, 116
Professional identity, 252, 253
Professionalization, 14, 17, 214–216, 231, 236, 1047, 1259–1261
Professional medical education, 230
Profitability-economic linkages, 1588
Profit-making medical care system, 1794
Profits, 1484, 1485, 1488, 1497
Progress, 295, 297, 299, 300, 302, 307–309, 311–313
Projective test, 410–411
Proletaroids, 1205
Proliferating expressions, 67
Propp, V., 94
Proprietorial sense, 1229
Prostitution
 harms of, 1899–1904
 pornography as a form of, 1906–1908
 psychological harms, 1902–1904
 systems of, 1911–1912
 violence, 1900–1902
Protestant interpretations, 351
Protest psychosis, 1324–1326
Proust, M., 95
Providence, 16
Prussian-like approach, 1093
Pseudo-hermaphroditism, 433–436, 439, 443
Pseudo-literality, 392
Psychiatric care, 1242
Psychiatric epidemiology, 1364
Psychiatric expertise, 1285–1289
Psychiatric medications, 1365
Psychiatric social workers (PSWs), 1347

Psychiatry, 98, 100, 1265–1268, 1308–1310, 1358, 1366, 1367, 1410
 and biology, 1359–1365
 clinical trials in, 1371–1372
 experience and expertise, 1372–1374
Psychic decomposition, 1229
Psychic habitus, 608
Psychic nexus, 114
Psychoanalysis, 82, 93, 98, 100–102, 391, 394, 988, 1047, 1098, 1361
Psychoanalytic psychotherapy, 393, 1868
Psychoanalytic spirituality, 390
Psychoanalytic theory, 846
Psychoanalytic therapy, 393
Psychoanalytic traditions, 386
Psychodynamic approaches, 1267
Psycho-historical approach, 601
Psychological character of psychologists, 996
Psychological expedition, 1148
Psychological humanities, 1188
Psychological medicine, 389
Psychological Round Table (PRT), 1037, 1055–1057
Psychological sex, 425, 426, 432, 433, 435, 438, 441, 449
Psychological social psychology (PSP), 1013–1016, 1018
Psychological society, 991
Psychology, 81, 82, 85–87, 90–92, 94, 96, 98, 100, 102, 104, 111, 126, 1130, 1132, 1133, 1136, 1138, 1461, 1466
 cultural, 981
 definition, 979
 diverse roots, 984
 history of, 979, 980
 of personality, 1207
 social, 981
 territories, 978
Psychopathologization, 1323
Psychopathology, 391, 1231–1233, 1241, 1858
Psychopathy, 1292, 1293
Psychopharmacology, 1366, 1372
Psycho-politics, 1224
Psycho-sexual role, 438
Psychosis, 1234–1240
Psychosurgery, 1315
Psychotherapy, 93, 1135, 1316
Psychotic breakdown, 1229
Psychotic intensity, 1230, 1236
Psychotropic drug, 1370
Psychotropic medication, 1868
Public debt, 1496–1498

Public health, 230, 245, 251, 1369, 1788, 1789
  critical public health perspectives, 1832
  epidemiology, 1838–1839
  evidence-based approaches, 1840–1842
  gender as social determinant, 1839–1840
  healthy masculinity, 1844–1846
  primary prevention, 1832–1834, 1836
  public health model of violence prevention, 1832
  risk factors, 1838–1839
  social determinants of health approach, 1833
  toxic masculinity, 1843–1844
  traditional masculinity, 1842
Public kind, 1571
Public psychiatry, 1225
Public school status-rivalry, 690
Puckett, K., 85, 94, 95
Puerperal melancholia, 1263
Pure economics, 1457
Putnam, Frederick Ward, 1659
Putnam, Hilary, 50

Q
Qualitative methods, 1183
Quantitative research, 569
Questionnaire, 88, 428–429, 431–433, 440, 441, 445, 448
Quijano, Anibal, 895

R
Race, 230, 239–243, 252, 294, 297, 299, 300, 302, 306, 310, 990, 1308
  science, 1343
Racial and colonial psychiatry, 1319–1320
Racial biology, 1321
Racial difference, 239, 241
Racial hierarchy, 239
Racial liberalism, 1233
Rackstrow's Museum, 234
Radcliffe-Browne, A.R., 465
Radical Enlightenment, 353
Radical feminist analysis, 1912
Radical transformation, 396
Radioactive fallout, 484
Randomized controlled trial (RCT), 1368, 1369, 1373, 1374
Ranke, Leopold von, 208–212, 214, 216, 218, 222, 223
Ranke, 87

Rationality, 382, 463
Rationalizing religion, 356
Readers, 90, 91, 95, 96, 99
Realism, 11
Reasoning madness, 382
Recapitulation, 1038
  theory, 1341
Reciprocity, 477, 1818
Reckwitz, A., 337
Reconstructions, 292, 307
Recovery, 1044
Reductionist claims, 21
Refining ethnos theory, 1713–1716, 1719, 1720
Reflection, 19, 981, 983, 992–994
Reflective awareness, 11
Reflective judgments, 121–122
Reflective knowledge, 115, 130
Reflexive, 19, 992, 994
  awareness, 112
Reflexivity, 19, 20, 69–71, 324, 342, 344, 364–366, 853–855, 993, 995, 997, 1036, 1182
Regimentation, 1313
Regions of memory, 795
Reich, R., 1247
Reimer, David, 439, 448
Relativism, 11
Relativity, 11
Religion, 135, 138, 142, 145–148, 153, 155, 182, 1734, 1748, 1750 1752
  anthropologists of religion, 190
  contemporary social scientific theories, 186–187
  Douglas, Mary, 197–199
  Frazer, J.G., 191–194
  Geertz, Clifford, 197–199
  Lévi-Strauss, Claude, 200–203
  Lévy-Bruhl, Lucien, 194–195
  origin and function, 183
  religionist argument, 187–190
  religious studies, 184–186
  social scientific theories, 186
  truth, 183–184
  Turner, Victor, 197–199
  Tylor, E.B., 190–192
Religious life, 146
Religious melancholy, 359
Religious morality, 145, 153
Religious studies, 184–186
Renaissance, 231, 248, 753
Renouvier, Charles, 136, 138, 151, 152
Rents, 1483, 1490, 1492, 1493

Representation(s), 230, 231, 233–239, 245–254
  history of anatomical representations in, 231–235
  of indigenous people, 1629–1633
Representational realism, 237
Repression, 1104
Reproductive labor, 1048, 1051–1053
Resilience, 1809
Respect, 1818
Responsibility, 1818, 1819
Restoration, 390
Restraint, 1352
Revenge pornography, 1910
Revolution, 386, 636–640
Revolutionary *praxis*, 388
Rhythm, 274
Ribeiro, A., 89
Ricoeur, P., 83, 92
Ries, Julius, 277–279
Right action, 134
Rights, 1136, 1137
Rigoli, J., 100
Risk factors
  neoliberalism, 1839
  risk assessment framework, 1839
  sexual violence against women, 1838
Rites, 156
Ritual, 145, 148, 155
  action, 148
RMPA, 1340
Rockefeller philanthrop, 1763, 1773
Rogers, C., 1135
Romantic confessions, 368
Romanticism, 1008
Rorschach test, 410
Rose, Nikolas, 161
Rousseau, Jean-Jacques, 40–42, 462, 463
Roux, Wilhelm, 277
Royal mechanism, 605, 609, 610
Royal Medico-Psychological Association, 1335, 1338
Rudenko, Sergei, 1705, 1706
Rules, 137, 141–145, 147, 149, 150
Rupture of continuity, 359
Russia, 1702, 1703
Russian anthropology, 1706, 1721
Ryan, J., 81, 92, 99
Ryan, M-L., 80, 96, 102

S
Sacred, 145–150, 155, 157
Said, Edward, 900
Samoyeds, 1642, 1643
Sanction, 143
Sanskrit concepts, 897
Sapir, Edward, 1758, 1771–1778
Satawal, 1741, 1742, 1745, 1746, 1748
Scala naturae, 1202
Schaffer, S., 922
Scheidt, CE., 102
Schiebinger, Londa, 237, 238
Schizophrenia, 1228, 1234
Schleiermacher, Friedrich, 110, 111, 116, 123, 124
Scholarly societies, 1653
School education, 1090
Schooling, 1153
Schools Medical Service, 1333
Schoolteachers, 1090
Schreber, D.P., 1228–1230, 1236
Schumpeter
  capitalism, 1575
  capitalist, 1569, 1572
  credit, 1578, 1579
  cyclical fluctuations, 1570
  economic development, 1567
  economic evolution, 1567
  enterprise, 1569
  entrepreneur, 1569, 1572–1574
  equilibrium, 1565
  external factors, 1567
  growth, 1568
  history, 1564, 1565
  hypothesis, 1571
  individual behavior, 1571
  innovation, 1568–1570
  invention *vs.* innovation, 1570
  inventor, 1569
  methodological individualism, 1571
  neoclassical economics, 1565
  stationary state, 1565
  technology, 1570
  utility, 1567
Science, 5, 11, 12
  aesthetic dimension of, 1177
  categories of, 1178–1180
  as communal practice, 1180
  enlightenment rationality, 1177
  humaines, 7, 66–69
  of man, 7
  progress, 1175
  psychological, 1170, 1173, 1181–1188
  and religion, 214
  sociales, 66–69
  of soul, 986

Science (*cont.*)
   understanding, 1176
   and values, 1171–1172
   values impacting, 1180–1181
Science and technology studies (STS), 18, 841
Science de l'homme, 7
   dismembering, 62–66
   evolutions, 58
   history, 58–60
   political crucible, 60–62
   reflexivity, 69–71
Scientific decay, 851
Scientific disciplines, 1653
Scientific fields, 841
Scientific method, 1172
Scientific observation, 1342
Scientific progress, 1460
Scientific psychology, 979
Scientific societies, 1662
Scott, P., 922
Scull, Andrew, 1360
Sculptures, 310
Sealey, A., 101
Second World War, 1006, 1009, 1019
Sectioning/section, 275, 277
Secularization, 356, 1309
Secular morality, 153–156
Seedbed-societies, 593
Segerstedt, T.T., 930
Segregation, 1134
Seler, Eduard, 1657
Self, 92, 96, 102, 104
Self-awareness, 436
Self-creating vitality, 388
Self-criticism, 1158
Self-determination, 1820–1826
Self-difference, 380
Self-esteem, 1119
   agreeableness, social order and, 1124–1127
   early uses of, 1120–1124
   formulaic approach to, 1129–1131
   inferiority, damage and discrimination, 1131–1135
   and mental philosophers, 1127–1129
   tests of, 1137–1138
   therapy and self-realization, 1135–1137
Self-explanation, 428
Self-help, 988
Selfhood, 380
Self-identification, 732
Self-identified victims of psychiatry, 1247
Self-identity, 376, 380, 986
Self-narratives, 89

Self-possession, 34
Self-reflection, 115
Self-reflexivity, 835
Self reinforced, 360
Self-sameness, 380, 381
Self-sufficient household economy, 1567
Self-transformation, 396
Semiotics, 94
Senile dementia, 1264
Senility, 1414
Sen's capability approach, 1465
Sense of what is right or just, 127
Sensibility, 357–361, 363
Sentimentalism, 358
Separation between rational (philosophical) statements, 13
Series, 265, 267–269, 272, 274–277, 283, 287
Serotonin reuptake inhibitors (SSRIs), 1858, 1867, 1868
Settler colonialism, 894
Seven Essays to Interpret the Peruvian Reality, 963
Sewell, William, 1006
Sex change, 431, 432, 434
Sex differences, 1039, 1040
Sex hormones, 437, 444
Sexism, 755, 758, 768, 1043, 1055, 1058
Sex morphology, 435
Sex of the self, 424, 425, 427, 429, 433–436, 440, 441, 445, 449
Sexologist, 425, 440
Sexology, 424–427
Sex roles, 1786, 1787
Sexual difference, 238
Sexual dysfunction, 1868
Sexual exploitation, 1789
Sexual harassment, 1036, 1054
Sexual intermediates, 429, 434, 441–443
Sexual inversion, 426–429
Sexuality, 237, 755, 1793, 1794, 1797–1801
Sexual politics, 1429
Sexual violence, 1838, 1843
Sexual Visions, 238
Sex work, 1901, 1907
Shapin, S., 922
Sherif, C.W., 1047
Shimizu, Akitoshi, 1732, 1737, 1739, 1743
Shimla conference, 896
Shinn, M., 1045
Shirokogoroff, 1707, 1708
Shklovsky, V., 94
Shock therapy, 1362
Sigerist, Henry, 246

Simmel, G., 926
Simon, F., 1325
Simon, H., 1315
Simon, Z.B., 86
SiMView, 283
Singularity, 60
Sin of pride, 1120
Sissay, L., 1239–1240
Sites of memory, 791
Situational improprieties, 1435, 1438
Skull, 241, 243
Slavery, 1321
Smail, D.L., 339
Smith, Adam, 463
Smith, R., 80, 81, 84, 86, 87, 91
Smithsonian Institution, 1653, 1666
Snezhnevskii, A., 1247
Social adjustment, 1313
Social and physical environment, 361
Social art, 61
Social boundaries, 918
Social capital, 1465
Social change, 572, 1796
Social class, 1159, 1789
  and therapy, 1314–1316
Social construction, 334
Social control, 1312
Social dangers, 1241–1242
Social defense thought, 1245
Social Democratic Worker's Party, 929
Social determinants of health
  critique of, 1840
  gender, 1839–1840
Social differentiation, 594
Social ecology, 1840
Social emancipation, 757
Social engineering, 927, 1760, 1762, 1763, 1765, 1771, 1772, 1776
Social epistemology, 18
Social ethics, 127–129
Social-ethnic organism, 1719
Social field, 841, 845
Social frameworks of memory, 783–786
Social history
  gender research, 756–760
  marriage, 761–765
  sociogenetic and psychogenetic approach, 752–754
  work and professional life, 765–768
Social inequalities, 758
Socialism, 1576
Sociality, 382
Socialization, 41, 1577

Social justice, 477, 1796
Social life, 1600
Socially abandoned, 1228
Social organization, 1308, 1312, 1323
Social policy, 245
Social psychiatry, 1365
Social psychology, 91, 981, 987, 1006–1010, 1017–1019, 1026
  cognitive, 1010
  definition, 1007
  dual disciplinary heritage, 1011–1017
  European social psychology, post-World War II development of, 1020–1025
  historical and rhetorical analysis, 1020
  interdepartmental doctoral program, 1020
  interdisciplinary graduate training programmes, 1019
  post-war, 1009
Social question, 925
Social reform movement, 925
Social relation, 1591
Social science(s), 61, 834, 877, 1008, 1786, 1787, 1790
  academic domains in, 1456
  as an orchestra, 1466
Social Science Research Council (SSRC), 1021, 1022, 1758, 1759
Social scientific approaches, 1788
Social scientification, 1458
Social scientific theories, 184, 186
Social status, 1789
Social studies of science, 992
Social subsystems, 758
Social theory, 1786
Social transformations, 1570, 1573
Social vaccine, 1118
Social welfare, 469
Social work, 1336
Societal practice, 173
Society for the Psychological Study of Social Issues (SPSSI), 1013
Socio-cultural system, 116
Socio-cultural theory, 1011
Socio-economic evolution, 1573
Sociological jurisprudence, 661
Sociological theory, 1008
Sociological traditions, 967, 968
Sociologies, 680
Sociologism, 1307
Sociology, 51, 93, 98, 99, 102–104, 562–564, 571, 578, 810, 877, 1454
  absences, 1071
  American, 822, 823, 828

Sociology (*cont.*)
   American social sciences, 944
   Argentinian, 825
   Australian, 821, 826
   Brazilian, 822
   Caribbean, 824
   Chinese, 822, 826
   classical theorists, 946
   cultural, 826
   decentering west, 952–957
   discipline of, 810
   economics, history, psychology and, 1461
   and economics, 1455
   engagement with knowledge, 825–827
   epistemological framework, 937
   Eurocentrism, 948, 952, 953, 955, 957
   French, 813
   historical, 811–812
   historical accounts, 936
   historical contextualization, 946
   historicism, 945
   histories of, 808–812, 815–816, 819, 821–824, 827–828
   humanistic tradition, 941
   ideas, 812–815, 817, 820, 825, 827
   industrial society, 942
   of institutions, 823–825
   intellectual histories, institutional histories and, 815–816
   intellectual history, 944
   intellectuals, 815–817, 827
   Italian, 820, 822
   knowledge, 813, 814, 816–818, 825–827
   medical, 809–810
   modernity, 944, 946, 947, 950, 953
   national histories of, 809, 811, 820, 822–829
   natural histories of, 819
   political and intellectual context, 943
   quantitative techniques, 944
   religion, 145
   rethinking canon, 947–951
   Russian, 819, 824
   science, 849, 940
   scientism, 940, 943, 952
   social science, 835
   social thought, 937–939
   sociological tradition, 940, 941
   sociology, 835
   SOSA, 938–940, 945
   South African, 822
   Swedish, 824
   systematic theory and method, 940
   theoretical analysis, 942
   theoretical approaches, 943
   theoretical schools and civilizational sources, 957
   theoretical synthesis, 946
   theory, 936, 938
   Urban, 810
   utility function, 1459
Sociology of law, 652
   Bourdieu, P., 664–666
   Foucault's approach, 662–664
   future perspectives, 669–671
   historical, 654, 655
   historicity of early, 658–660
   Luhmann, N., 666–669
   pre-history of, 656–658
Sociology of morality, 135
   Durkheim's social and intellectual background, 136–137
   moral philosophy, in France, 138–139
Sociology of scientific knowledge (SSK), 840
Socio-political problem, 1241
Soemmering, Samuel Thomas, 238
Solidarity, 127
Somatic therapies, 1360, 1363
Sombart, Werner, 1564, 1575, 1576, 1581
Soricelli, R.L., 100
Source criticism, 215
Source editing projects, 220–221
Souriau, P., 90
Southern Expansion Doctrine, 1739, 1744, 1751
South Sea Islands Cultural Society, 1738, 1741, 1744, 1746, 1751
South Seas Colonization Corporation, 1739
South Seas Court of Justice, 1738, 1745
South Seas Development Company, 1738, 1739
South Seas Government Exhibition Room, 1745
South-South axis, 900
Souvenirs d'enfance et de jeunesse, 65
Sovereign individual, 376
Soviet ethnography, 1709, 1713, 1716, 1723
Soviet psychology, 993
Soviet Union
   cross-cultural research, 1151–1156
   pedological research and mental testing, 1148–1151
Spain, 473
Specialities, 808–811, 813, 816, 819–821, 827
Specialization, 8, 58, 67, 68, 71
Specialized knowledge, 10
Species-life, 387

Specimen exchanges, 1654
Spectacle, 254
Speculative finance, 471
Spencer, H., 925, 1128
Spengler, Oswald, 566
Spiritual constitution, 1572
Spirituality
 Augustinian freedom, 378
 disenchanted modernity, 378
 dominant practices, 377
 human sciences, 384
 industrial populations, 377
 Nietzsche's tacit hypothesis, 376
 pathology and health, 388
 psychoanalytic, 390
 self-punishment, 376
 targeted periodization, 376
 transformative potentials, 379
Spontaneous orders, 1551–1553
Spontaneous remission, 1868, 1870
Sport
 diminishing contrasts, 686–688, 690, 691
 figurational sociology, 682
 habits of good sociology, 691–694
 harmonious inequality, 682–686
 historical sociology, 679
 increasing varieties, 687–689
 mapping the field, 680, 681
 non-utilitarian, 678
 organized, 678
 playful, 678
Sportization, 611, 688
Sports sciences, 694
Stability, 471, 472, 477
Stalinist dictatorship, 1106
Stalin's theory of nation, 1711, 1712
Standardized packages, 921
Standpoint theory, 817
Stanley Coopersmith, 1138
Star, S.L., 920
State formation, 584
 development paths, European states, 603–619
 European thinking, 585–591
 formation of early states and civilizations, 591–603
State-making, 629
Static representations, 247
Static system, 1566
Stationary flow, 1566
Statistic, 432, 436, 445, 446
Statistical methods, 995
Statistical state, 41

21st century, 88, 89, 97, 98, 103
Steffen, G., 924
Steinach, Eugen, 430, 435, 436
Steinmetz, George, 568
Stigma, 1314
Stirling County Study, 1364
Stockholm School of Economics, 927
Stoics, 34, 35, 43–45, 48, 49
Stoller, Robert, 425, 432, 434, 435, 439, 449
Storyboard, 1748
Stratification theory
 and characterology, 1200
 and search for holism, 1208–1211
 in wartime, 1211–1212
Streptomycin, 1369
Structural adjustment, 460, 461
Structural articulation, 114
Structural functionalism, 567, 584, 588, 1787
Structuralist, 17, 18
 legacy, 968
Structure of Social Action (SOSA), 938
Stukenbrock, A., 102
Style, 121
Subaltern studies, 901
Subject, 425–434, 436, 439, 440, 446, 449
Subjectification, 172
Subjectivation, 385
Subjectivity, 367, 429
Subsistence, 469, 475
Substantivist, 467
Sudo, Ken'ichi, 1732, 1747
Südsee-Expedition, 1735
 See also German ethnography
Sugiura, Ken'ichi, 1738, 1740–1747, 1749, 1751
Suicide, 431, 438, 439
Sully, James, 1129, 1341
Superego, 717
Survival value, 1128
Sweden, 924
Swedish model, 927
Swedish Sociological Association, 930
Sybel, Heinrich von, 211, 213, 215, 216, 221, 222
Symbolical analogies, 122
Symbolic anthropology, 202
Symbolic boundaries, 918
Symbolic capital, 846
Symbolic/linguistic capacity, 382
Sympathy, 49, 1125
Systematic investigations, 377
Systematic reviews, 1842
System of the disciplines, 164

Systems thinking, 981
Szasz, Thomas, 1421, 1422, 1430, 1433, 1434, 1439, 1446
　philosophy of freedom, 1431
　The Manufacture of Madness, 1432–1433
　The Myth of Mental Illness, 1431–1432

T
Tables, 294, 297, 304, 312
Taihoku Imperial University, 1734
Taiwan's indigenous peoples, 1734
Tajfel, H., 1047
Tanner, A., 1040, 1045
Tarde, Gabriel, 1564, 1566, 1569, 1570, 1572, 1573, 1580
Tartaria Magna, 1641
Tarulli, D., 92
Tatar children, 1155
Tavistock and Portman NHS Foundation Trust, 1352
Tavistock Clinic, 1348, 1352
Teaching, 211–213
Teather, A., 88
Technical change, 1491, 1492
Technical development, 1568
Technical superiority, 1568
Technological innovation, 250, 1575
Technological unemployment, 1492
Technologies of the self, 328–330
　advantages, 403–404
　brain scans, 412–414
　definition, 402
　IQ tests, 408–410
　physiognomy and phrenology, 404–408
　projective test, 410–411
　ultrasound scans, 411–412
Telos-driven, 352
Temperament, 1765–1768
Temperamental kind, 1571
Tempocentrism, 589
Temporalities, 782
Temporalization, 274
Terman, Lewis M., 441, 445, 446, 448, 1041
Terman-Miles test, 441, 447, 448
Territory, 122
Terrorism, 632
Tests of self-esteem, 1137–1138
Textbooks, 313
Textual analysis, 843, 855
18th century, 98, 101
19th century, 81, 85–88, 98, 100
20th century, 83, 86, 87, 92, 99

Theoretical cognition, 115
Theoretical foundation, 1587
Theory and systems, 980
Theory of Communicative Action, 165
Theory of cultural development, 1148
Theory of ethnos, 1700, 1701, 1707
Theory of germ layers, 272
Theory of meaning, 17
Theory of preformation, 261
The person, 10
The Politics of Experience and The Bird of Paradise, 1424
The Poverty of Historicism, 568
The Protestant Ethic and the Spirit of Capitalism, 567
The public understanding of science, 25
Therapy, 102
The science of Man, 987
The sciences, 12
The sciences of man, 15
Thesis on the Problem of Race, 969
The social contract, 462
The Structure of Social Action, 563
The theory of psychology, 982
The wealth of nations, 463
Thomas, D.S., 928
Thomas, W.I., 928
Three waves of historical sociology, 564
Time, 781
Time-consciousness, 359
Titchener, E.B., 1040, 1047, 1054
Todorov, T., 93, 95
Tokyo Imperial University, 1733, 1736, 1738, 1740–1742
Torii, Ryūzō, 1733
Total institutions, 1316
Totemism, 508–510
Toxicity, 491, 492
Toxic masculinity, 1832, 1842–1844, 1847
Trading zone, 921, 922
Traditional chief, 1738, 1745
Traditional development economics, 1594
Traditional knowledge, 1806, 1808, 1812, 1815
Traditional masculinity, 1842
Trans, 425–426, 434
Transaction sphere, 922
Transcultural psychiatry, 1363, 1380, 1390
　and colonial mind, 1390–1398
　and histories, 1381–1384
　universality of mental illness, 1384–1387
Transformation, 377, 378, 380, 393, 396
　process, 755
Transformative turn, 379

Transgender, 424, 443
Trans-human activity, 997
Transitional justice, 668
Translated, 1741, 1747–1751
Transnational approach, 1615
Transnational memory space, 795
Transnational networks, 1653
Transnational science, 1653
Transnational turn, 794–796
Transpositional understanding, 118
Transsexual, 430–432, 439, 440
Transvestism, 426–428, 430, 432, 439, 440, 449
Transvestite, 425, 426, 430, 431, 435, 443
Travelling memory, 780
Trobriand Islands, 465, 466
True gender self, 425, 449
True sex, 438, 449
Trustified capitalism, 1578
Trust Territory of the Pacific Archives, 1749
Truth, 183–184
  games, 384
Truth and Reconciliation Commission (TRC), 1319
Tsafendas, D., 1237–1238
Tsuboi, Shōgorō, 1732, 1733, 1736
Ttrajectory, 688
Tungus, 1153
Turner, Victor, 187, 197–203
Two sex system, 433
Tylor, E.B., 184, 187, 190–192, 194–196, 199, 202
Typicality, 123

U

Ukrainian people, 1704
Ulrichs, Karl Heinrich, 428
Unidimensionality, 447
Union of Soviet Socialist Republics (USSR), 1705, 1709, 1710, 1712, 1717–1719, 1722
United Nations Convention on the Rights of Persons with Disabilities (CRPD), 1301
Unity of the human and social sciences, 58, 63, 66, 71
Universal history, 16, 17
Universality, 1174
Universal psychological need, 1137
University-based academic psychiatry, 1245
Untold Lives, 1044
Uppsala School of Sociology, 930
USA, 460
US cases, 1320–1324
US housing market, 460
US National Museum (USNM), 1667
U.S. women's liberation movement, 1798
Utilitarianism, 135, 136, 138, 141, 154
Utilitarians, 138, 141
Utility function, 1592
Utsushikawa, Nenozō, 1734
Uzbek children, 1155

V

Valencies, 715
Values, 141, 143–147, 149–154, 156, 992, 1482, 1483
  categories of, 1170, 1178–1180
  definitions and distinctions, 1172–1174
  denial of value-free science, 1177
  epistemic authority, 1174
  impacting science, 1180–1181
  meaning of, 1170
  methods as value system, 1184–1186
  mystery and imagination, 1177
  predictive power, 1174
  problem-solving, 1175
  progress, 1175
  and psychological science, 1181–1184
  roles, 1169
  and science, 1171–1172
  statements, 1169
  truth and beauty, 1176–1177
  understanding, 1176
Vasconcelos, Leite de, 1681, 1683, 1684, 1687, 1689
Veblen, Thorstein, 1564, 1567, 1570, 1573, 1578, 1580
Vergeistung, 1575, 1581
Vesalius, Andreas, 231
Vicissitudes, 1224
Vienna, 245, 1662
Vietnam, 469
Vila, A., 98
Violence, 394, 751
  against women, 1901, 1904, 1909, 1912
  prevention, 1832–1841, 1843–1845, 1847
Virchow, R., 87
Virtual anthropology, 304
Virtualism, 459, 470, 471, 473, 476
Virtues, 218, 219
Virtuous and honorable conduct, 1124
Visual epistemology, 248
Visualizations, human origins and evolution, 292

Visualizations, human origins and
     evolution, 292 (*cont.*)
  cartoons and films, 314–316
  drawings, photographs, and graphs,
     293–299
  expositions and museums, 311–314
  life-scene and full-body reconstructions,
     305–311
  trees and maps, 299–304
Virtualization of the economy, 471
Visual stereotypes, 1633
Vitruvian Man, 32
Volitional, 115
Völkerpsychologie, 987, 1196
Voluntaristic theory of action, 592
von Humboldt, Alexander, 1663
Vulnerability, 488
Vygotsky, L.S., 1146, 1158
  criticism on theory of cultural
     develpment, 1159
  national minorities, 1150
  pedologiya, 1152
  theory of cultural development, 1154

W
Wagner, P., 924
Waitz, Georg, 211, 213, 216–218, 220
Wallerstein's whole account, 614
Walras, L., 1516–1519
Walrasian theory of general equilibrium, 1521
War, 627–631
Warriors, 797
Washburn, M.F., 1055
Washington, D.C., 1655
Washington Consensus, 461
Waste disposal, 484
Wawrzyniak, Joanna, 577
Wax models, 234, 238, 243, 247
Wax plate modeling, 276
Weapon of self-hurt, 1884
Weber, Max, 564, 566, 850, 926, 1564, 1571
Weberian approach, 603, 607
Weber's model of patrimonial domination, 609
Weber's theory, 660
Wechsler Adult Intelligence Scale (WAIS),
     1293
Weight-loss dieting, 1880, 1887–1889
Weight-neutral approach, 1891
Weimar Republic, 1209
Weisstein, N., 1037, 1049, 1056
Welfare economics, 1522–1523, 1592, 1593
Welfare policies, 237

Wertham, F., 1323
Western European model, 612
Western feminist movement, 1892
Western medicine, 230, 239
Western modernity, 564
Western psychiatry, 1365
Western psychology, 1070
White, H., 84, 85, 88, 90, 95
White supremacy, 1046, 1789
White working-class women, 757
Wilhelm Roux, 281
Will, 1109
Winnicott, Donald Woods, 1339
Wirklichkeitswissenschaft, 592
Wise, MN., 83, 87, 92
Wissenschaft, 167
Wissenssoziologie, 842
Witnessing, 1234–1240
Witness Seminars, 1337
Wittfogel's hydraulic approach, 599
Wittrock, B., 927
Wolff, Caspar Friedrich, 265–267
Woman as problem, 1042–1043
Womanless psychology, 1038–1042
Women and Health Documentation Center
     Network, 1797
Women and Madness, 1043
Women of color, 1794
Women's and Gender Studies, 751
Women's health activists, 1795
Women's Health Information Library, 1797
Women's health movement (WHM),
     1798–1801, 1880–1881
  feminist research agenda for, 1889–1893
  and physical and psychological harms of
     food restriction, 1880–1881
Women's studies, 757
Women's work, 1048
Woods, A., 80, 89
Woolley, H.T., 1039, 1058
Work and professional life, 765–768
Working-class health activists, 1794
Work therapy, 1315
World ethnography, 1614
World Health Organization (WHO), 1365, 1835
World history, 459, 465, 476, 477
World-science, 874
World's Columbian Exhibition (WCE), 1670
World's Fairs, 1654
World-system approach, 614, 904
Worldviews, 115, 119, 124–126
World War II, 1199
Worthiness of persons, 463

Wright, R., 1323
Writing, 80, 82, 89, 95, 100, 103, 104
   model, 1795

Y
Yanagita, Kunio, 1737, 1742
Yanaihara, Tadao, 1740, 1747
Yawata, Ichirō, 1740, 1741

Yerkes, R., 1049, 1053
Yoruba poetry, 898
Yurt, 1154

Z
Zashiki-rou, 1245
Zimmerman, A., 87, 88
Zola, E., 85